Recursive Macroeconomic Theory
Third edition

To our parents, Zabrina, and Carolyn

Recursive Macroeconomic Theory
Third edition

Lars Ljungqvist

Stockholm School of Economics

Thomas J. Sargent

New York University

and

Hoover Institution

The MIT Press

Cambridge, Massachusetts

London, England

© 2012 Massachusetts Institute of Technology

All rights reserved. No part of this book may be reproduced in any form by any electronic or mechanical means (including photocopying, recording, or information storage and retrieval) without permission in writing from the publisher.

Printed and bound in the United States of America.

Library of Congress Cataloging-in-Publication Data

Ljungqvist, Lars.
  Recursive macroeconomic theory / Lars Ljungqvist, Thomas J. Sargent. – 3rd ed.
    p.   cm.
  Includes bibliographical references and index.
  ISBN 978-0-262-01874-6 (hbk. : alk. paper)
    1. Macroeconomics. 2. Recursive functions. 3. Statics and dynamics
(Social sciences)
I. Sargent, Thomas J. II. Title.
HB172.5.L59 2012
339.01'51135–dc23

                                                                    2012016608

10 9 8 7 6 5 4 3 2

Contents

Part I: The imperialism of recursive methods

1. Overview

1.1. Warning. 1.2. A common ancestor. 1.3. The savings problem. 1.3.1. Linear quadratic permanent income theory. 1.3.2. Precautionary saving. 1.3.3. Complete markets, insurance, and the distribution of wealth. 1.3.4. Bewley models. 1.3.5. History dependence in standard consumption models. 1.3.6. Growth theory. 1.3.7. Limiting results from dynamic optimal taxation. 1.3.8. Asset pricing. 1.3.9. Multiple assets. 1.4. Recursive methods. 1.4.1. Methodology: dynamic programming issues a challenge. 1.4.2. Dynamic programming challenged. 1.4.3. Imperialistic response of dynamic programming. 1.4.4. History dependence and "dynamic programming squared". 1.4.5. Dynamic principal-agent problems. 1.4.6. More applications.

Part II: Tools

2. Time Series 29

3. Dynamic Programming 103

Part III: Competitive equilibria and applications

12. Recursive Competitive Equilibria   455

13. Asset Pricing Theory   481

14. Asset Pricing Empirics   515

15. Economic Growth   583

16. Optimal Taxation with Commitment   613

Part IV: The savings problem and Bewley models

17. Self-Insurance

18. Incomplete Markets Models

Part V: Recursive contracts

19. Dynamic Stackelberg Problems 775

20. Insurance Versus Incentives 797

21. Equilibrium without Commitment   859

22. Optimal Unemployment Insurance   913

23. Credible Government Policies, I   937

24. Credible Government Policies, II   985

25. Two Topics in International Trade 1005

Part VI: Classical monetary and labor economics

26. Fiscal-Monetary Theories of Inflation 1045

Part VII: Technical appendices

A. Functional Analysis   1257

B. Linear projections and hidden Markov models   1269

Acknowledgments

We wrote this book during the 1990s, 2000s, and early 2010s while teaching graduate courses in macro and monetary economics. We owe a substantial debt to the students in these classes for learning with us. We would especially like to thank Marco Bassetto, Victor Chernozhukov, Riccardo Colacito, Mariacristina De Nardi, William Dupor, William Fuchs, George Hall, Sagiri Kitao, Hanno Lustig, Monika Piazzesi, Navin Kartik, Martin Schneider, Yongseok Shin, Christopher Sleet, Stijn Van Nieuwerburgh, Laura Veldkamp, Neng Wang, Chao Wei, Mark Wright, and Sevin Yeltekin for commenting on drafts of earlier editions when they were graduate students. In prefaces to earlier editions, we forecast that they would soon be prominent economists, and we are happy that has come true. We also thank Isaac Bailey, Anmol Bhandari Saki Bigio, Jaroslav Borovicka, David Evans, Chrisopher Huckfeldt, Anna Orlik, Ignacio Presno, Cecilia Parlatore Siritto, and Håkòn Travoll for helpful comments on earlier drafts of this edition. Each of these people made substantial suggestions for improving this book. We expect much from members of this group, as we did from an earlier group of students that Sargent (1987b) thanked.

We received useful criticisms from Roberto Chang, Gary Hansen, Jonathan Heathcote, Berthold Herrendorf, Mark Huggett, Charles Jones, Narayana Kocherlakota, Dirk Krueger, Per Krusell, Francesco Lippi, Rodolfo Manuelli, Beatrix Paal, Adina Popescu, Jonathan Thomas, Nicola Tosini, and Jesus Fernandez-Villaverde.

Rodolfo Manuelli and Pierre Olivier Weill kindly allowed us to reproduce some of their exercises. We indicate the exercises that they donated. Some of the exercises in chapters 6, 9, and 27 are versions of ones in Sargent (1987b). François Velde provided substantial help with the TeX and Unix macros that produced this book. Maria Bharwada helped typeset the first edition.

For providing good environments to work on this book, Ljungqvist thanks the Stockholm School of Economics and Sargent thanks the Hoover Institution and the departments of economics at the University of Chicago, Stanford University, Princeton University, and New York University.

Preface to the third edition

Recursive Methods

Much of this book is about how to use recursive methods to study macro-economics. Recursive methods are very important in the analysis of dynamic systems in economics and other sciences. They originated after World War II in diverse literatures promoted by Wald (sequential analysis), Bellman (dynamic programming), and Kalman (Kalman filtering).

Dynamics

Dynamics studies sequences of vectors of random variables indexed by time, called *time series*. Time series are immense objects, with as many components as the number of variables times the number of time periods. A dynamic economic model characterizes and interprets the mutual covariation of all of these components in terms of the purposes and opportunities of economic agents. Agents *choose* components of the time series in light of their opinions about other components.

Recursive methods break a dynamic problem into pieces by forming a sequence of problems, each one being a constrained choice between utility today and utility tomorrow. The idea is to find a way to describe the position of the system now, where it might be tomorrow, and how agents care now about where it is tomorrow. Thus, recursive methods study dynamics indirectly by characterizing a pair of *functions*: a transition function mapping the *state* today into the state tomorrow, and another function mapping the state into the other endogenous variables of the model. The *state* is a vector of variables that characterizes the system's current position. Time series are generated from these objects by iterating the transition law.

Recursive approach

Recursive methods constitute a powerful approach to dynamic economics due to their described focus on a tradeoff between the current period's utility and a continuation value for utility in all future periods. As mentioned, the simplification arises from dealing with the evolution of state variables that capture the consequences of today's actions and events for all future periods, and in the case of uncertainty, for all possible realizations in those future periods. Not only is this a powerful approach to characterizing and solving complicated problems, but it also helps us to develop intuition, conceptualize, and think about dynamic economics. Students often find that half of the job in understanding how a complex economic model works is done once they understand what the set of state variables is. Thereafter, the students are soon on their way to formulating optimization problems and transition equations. Only experience from solving practical problems fully conveys the power of the recursive approach. This book provides many applications.

Still another reason for learning about the recursive approach is the increased importance of numerical simulations in macroeconomics, and most computational algorithms rely on recursive methods. When such numerical simulations are called for in this book, we give some suggestions for how to proceed but without saying too much on numerical methods.[1]

Philosophy

This book mixes tools and sample applications. Our philosophy is to present the tools with enough technical sophistication for our applications, but little more. We aim to give readers a taste of the power of the methods and to direct them to sources where they can learn more.

Macroeconomic dynamics is now an immense field with diverse applications. We do not pretend to survey the field, only to sample it. We intend our sample to equip the reader to approach much of the field with confidence. Fortunately for us, good books cover parts of the field that we neglect, for example, Aghion and Howitt (1998), Benassy (2011), Barro and Sala-i-Martin (1995), Blanchard and Fischer (1989), Cooley (1995), DeJong and Dave (2011), Farmer

---

[1] Judd (1998) and Miranda and Fackler (2002) provide good treatments of numerical methods in economics.

(1993), Azariadis (1993), Gali (2008), Majumdar (2009), Christensen and Kiefer (2009), Canova (2007), Romer (1996), Altug and Labadie (1994), Shimer (2010), Stachurski (2009), Walsh (1998), Cooper (1999), Adda and Cooper (2003), Pissarides (1990), and Woodford (2000). Stokey, Lucas, and Prescott (1989) and Bertsekas (1976) remain standard references for recursive methods in macroeconomics. Technical Appendix A in this book revises material from chapter 2 of Sargent (1987b).

Changes in the third edition

This edition contains three new chapters and substantial revisions of many parts of other chapters from earlier editions. New to this edition are chapter 14 on asset pricing empirics, chapter 24 on credible government policy in more sophisticated settings, and chapter 29 on foundations of the aggregate labor supply function. The new chapters and revisions cover exciting new topics that widen and deepen the message that recursive methods are pervasive and powerful.

New chapters

Chapter 14 extends and applies the asset pricing theory developed in chapter 8 on complete markets and chapter 18 on incomplete markets to devise empirical strategies for specifying and testing asset pricing models. Building on work by Darrell Duffie, Lars Peter Hansen, and their co-authors, chapter 14 discusses ways of characterizing asset pricing puzzles associated with the preference specifications and market structures commonly used in other parts of macroeconomics.[2] It then describes alterations of those structures that hold promise for resolving some of those puzzles.

Chapter 24 extends the chapter 23 analysis of credible government policies to a more sophisticated class of dynamic economies. While the economic environments of chapter 23 are purposefully cast in ways to make *all* interesting dynamics be associated with reputation, chapter 24 incorporates other sources

---

[2] By a 'puzzle', we mean a model's prediction that is not borne out by data. A puzzle is always relative to a prevailing model.

of dynamics associated, for example, with interactions between the demand and supply of money.

Chapter 29 describes new macroeconomic models of aggregate labor supply. Until recently, there was a sharp division between macroeconomists working in the real business cycle tradition and microeconomists studying data on individual workers' labor supply decisions and outcomes. On the one hand, real business cycle researchers had appealed to the Hansen (1985)-Rogerson (1988) employment lotteries model to justify a high aggregate labor supply elasticity. On the other hand, labor microeconomists had expressed skepticism about the empirical plausibility of those lotteries and the supplementary complete consumption insurance markets that the Hansen-Rogerson theory uses to obtain that high labor supply elasticity. The microeconomists typically based their much smaller labor supply elasticities by working within some version of an incomplete markets life cycle model. Chapter 29 describes how prominent real business cycle theorists have recently abandoned the employment lotteries model in favor of a life cycle model. The adoption of a common theoretical framework means that the quest for the aggregate labor supply elasticity is now in terms of the forces and institutions that shape individuals' career length choices.

Revisions of other chapters

We have added significant amounts of material to a number of chapters, including chapters 2, 6, 7, 8, 11, 13, 16, 17, 18, 19, 23, 26, and 28. Chapter 2 has a better treatment of laws of large numbers, a prettier derivation of the Kalman filter, and an extended economic example (a permanent income model of consumption) that illustrates some of the time series techniques introduced in the chapter. It also contains a new appendix describing the powerful Markov chain Monte Carlo method that has opened wide new avenues in empirical implementation of dynamic economic models. Chapter 6 on search models has been improved throughout. New material on a model of Burdett and Judd (1983) deepens the treatment of the basic McCall search model. Chapter 7 reaps more returns from the recursive competitive equilibrium concept in the context of partial equilibrium models in the tradition of Sherwin Rosen and his co-authors. The chapter shows how Rosen schooling models fit within the recursive competitive equilibrium framework. Chapter 8 has been revised and improved to say more about how to find a recursive structure within an Arrow-Debreu pure exchange economy. Chapter 11 has been extensively revised. It now extracts many more implications from the basic model, including links to the term structure of interest rate, explanations of how sequence and recursive formulations imply things about movements in government debt. It also includes a section on a two country model that illustrates the transmission abroad of effects of changes in tax policies in one country. A companion to this chapter entitled 'Practicing Dynare' contains dynare code for computing all of the quantitative examples in the chapter. Chapter 13 on asset pricing theory has been revised in a way designed to prepare the way for our new chapter 14 on asset pricing empirics. Chapter 16 on optimal taxation has been revised in several ways. For example, it includes additional exercises that bring out aspects of the theory of optimal taxation. We also use a general equilibrium model with money to attempt to shed light on contentious aspects of the fiscal theory of the price level. Chapter 17 extends our discussion of the supermartingale convergence theorem and how it implies precautionary saving to settings with endogenous labor supply. Chapter 18 on incomplete markets has been revised in parts. Chapter 19 on Stackelberg problems in linear quadratic models has been extended and improved. Chapter 23 on credible public policy has been set up in ways that lead naturally to our new followup chapter 24 on credible economic policy in models with other sources of dynamics. Chapter 26 has more to say about the fiscal theory of the price level including game theoretic arguments by Bassetto (2002). To elaborate on how congestion externalities are such a driving force in matching models, chapter 28 includes an analysis of age dynamics in an overlapping generations model with a single matching function.

Ideas

Beyond emphasizing recursive methods, the economics of this book revolves around several main ideas.

1. The competitive equilibrium model of a dynamic stochastic economy: This model contains complete markets, meaning that all commodities at different dates that are contingent on alternative random events can be traded in a market with a centralized clearing arrangement. In one version of the model, all trades occur at the beginning of time. In another, trading in one-period claims occurs sequentially. The model is a foundation for asset-pricing theory, growth theory, real business cycle theory, and normative public finance. There is no room for fiat money in the standard competitive equilibrium model, so we shall have to alter the model to let fiat money in.

2. A class of incomplete markets models with heterogeneous agents: The models arbitrarily restrict the types of assets that can be traded, thereby possibly igniting a precautionary motive for agents to hold those assets. Such models have been used to study the distribution of wealth and the evolution of an individual or family's wealth over time. One model in this class lets money in.

3. Several models of fiat money: We add a shopping time specification to a competitive equilibrium model to get a simple vehicle for explaining ten doctrines of monetary economics. These doctrines depend on the government's intertemporal budget constraint and the demand for fiat money, aspects that transcend many models. We also use Samuelson's overlapping generations model, Bewley's incomplete markets model, and Townsend's turnpike model to perform a variety of policy experiments.

4. Restrictions on government policy implied by the arithmetic of budget sets: Most of the ten monetary doctrines reflect properties of the government's budget constraint. Other important doctrines do too. These doctrines, known as Modigliani-Miller and Ricardian equivalence theorems, have a common structure. They embody an equivalence class of government policies that produce the same allocations. We display the structure of such theorems with an eye to finding the features whose absence causes them to fail, letting particular policies matter.

5. Ramsey taxation problem: What is the optimal tax structure when only distorting taxes are available? The primal approach to taxation recasts this question as a problem in which the choice variables are allocations rather than tax rates. Permissible allocations are those that satisfy resource constraints and implementability constraints, where the latter are budget constraints in which the consumer and firm first-order conditions are used to substitute out for prices and tax rates. We study labor and capital taxation, and examine the optimality of the inflation tax prescribed by the Friedman rule.

6. Social insurance with private information and enforcement problems: We use the recursive contracts approach to study a variety of problems in which a benevolent social insurer must balance providing insurance against providing proper incentives. Applications include the provision of unemployment insurance and the design of loan contracts when the lender has an imperfect capacity to monitor the borrower.

7. Time consistency and reputational models of macroeconomics: We study how reputation can substitute for a government's ability to commit to a policy. The theory describes multiple systems of expectations about its behavior to which a government wants to conform. The theory has many applications, including implementing optimal taxation policies and making monetary policy in the presence of a temptation to inflate offered by a Phillips curve.

8. Search theory: Search theory makes some assumptions opposite to ones in the complete markets competitive equilibrium model. It imagines that there is no centralized place where exchanges can be made, or that there are not standardized commodities. Buyers and/or sellers have to devote effort to search for commodities or work opportunities, which arrive randomly. We describe the basic McCall search model and various applications. We also describe some equilibrium versions of the McCall model and compare them with search models of another type that postulates the existence of a matching function. A matching function takes job seekers and vacancies as inputs, and maps them into a number of successful matches.

Theory and evidence

Though this book aims to give the reader the tools to read about applications, we spend little time on empirical applications. However, the empirical failures of one model have been a main force prompting development of another model. Thus, the perceived empirical failures of the standard complete markets general equilibrium model stimulated the development of the incomplete markets and recursive contracts models. For example, the complete markets model forms a standard benchmark model or point of departure for theories and empirical work on consumption and asset pricing. The complete markets model has these empirical problems: (1) there is too much correlation between individual income and consumption growth in micro data (e.g., Cochrane, 1991 and Attanasio and Davis, 1995); (2) the equity premium is larger in the data than is implied by a representative agent asset-pricing model with reasonable risk-aversion parameter (e.g., Mehra and Prescott, 1985); and (3) the risk-free interest rate is too low relative to the observed aggregate rate of consumption growth (Weil, 1989). While there have been numerous attempts to explain these puzzles by altering the preferences in the standard complete markets model, there has also been work that abandons the complete markets assumption and replaces it with some version of either exogenously or endogenously incomplete markets. The Bewley models of chapters 17 and 18 are examples of exogenously incomplete markets. By ruling out complete markets, this model structure helps with empirical problems 1 and 3 above (e.g., see Huggett, 1993), but not much with problem 2. In chapter 20, we study some models that can be thought of as having endogenously incomplete markets. They can also explain puzzle 1 mentioned earlier in this paragraph; at this time it is not really known how far they take us toward solving problem 2, though Alvarez and Jermann (1999) report promise.

Micro foundations

This book is about micro foundations for macroeconomics. Browning, Hansen, and Heckman (1999) identify two possible justifications for putting microfoundations underneath macroeconomic models. The first is aesthetic and preempirical: models with micro foundations are by construction coherent and explicit. And because they contain descriptions of agents' purposes, they allow us to analyze policy interventions using standard methods of welfare economics. Lucas (1987) gives a distinct second reason: a model with micro foundations broadens the sources of empirical evidence that can be used to assign numerical values to the model's parameters. Lucas endorses Kydland and Prescott's (1982) procedure of borrowing parameter values from micro studies. Browning, Hansen, and Heckman (1999) describe some challenges to Lucas's recommendation for an empirical strategy. Most seriously, they point out that in many contexts the specifications underlying the microeconomic studies cited by a calibrator conflict with those of the macroeconomic model being "calibrated." It is typically not obvious how to transfer parameters from one data set and model specification to another data set, especially if the theoretical and econometric specification differs.

Although we take seriously the doubts about Lucas's justification for microeconomic foundations that Browning, Hansen and Heckman raise, we remain strongly attached to micro foundations. For us, it remains enough to appeal to the first justification mentioned, the coherence provided by micro foundations and the virtues that come from having the ability to "see the agents" in the artificial economy. We see Browning, Hansen, and Heckman as raising many legitimate questions about empirical strategies for implementing macro models with micro foundations. We don't think that the clock will soon be turned back to a time when macroeconomics was done without micro foundations.

Road map

An economic agent is a pair of objects: a utility function (to be maximized) and a set of available choices. Chapter 2 has no economic agents, while chapters 3 through 6 and chapter 17 each contain a single agent. The remaining chapters all have multiple agents, together with an equilibrium concept rendering their choices coherent.

Chapter 2 describes two basic models of a time series: a Markov chain and a linear first-order difference equation. In different ways, these models use the algebra of first-order difference equations to form tractable models of time series. Each model has its own notion of the state of a system. These time series models define essential objects in terms of which the choice problems of later chapters are formed and their solutions are represented.

Chapters 3, 4, and 5 introduce aspects of dynamic programming, including numerical dynamic programming. Chapter 3 describes the basic functional equation of dynamic programming, the Bellman equation, and several of its properties. Chapter 4 describes some numerical ways for solving dynamic programs, based on Markov chains. Chapter 5 describes linear quadratic dynamic programming and some uses and extensions of it, including how to use it to approximate solutions of problems that are not linear quadratic. This chapter also describes the Kalman filter, a useful recursive estimation technique that is mathematically equivalent to the linear quadratic dynamic programming problem.[3] Chapter 6 describes a classic two-action dynamic programming problem, the McCall search model, as well as Jovanovic's extension of it, a good exercise in using the Kalman filter.

While single agents appear in chapters 3 through 6, systems with multiple agents, whose environments and choices must be reconciled through markets, appear for the first time in chapters 7 and 8. Chapter 7 uses linear quadratic dynamic programming to introduce two important and related equilibrium concepts: rational expectations equilibrium and Markov perfect equilibrium. Each of these equilibrium concepts can be viewed as a fixed point in a space of beliefs about what other agents intend to do; and each is formulated using recursive methods. Chapter 8 introduces two notions of competitive equilibrium in dynamic stochastic pure exchange economies, then applies them to pricing various consumption streams.

---

[3] The equivalence is through duality, in the sense of mathematical programming.

Chapter 9 first introduces the overlapping generations model as a version of the general competitive model with a peculiar preference pattern. It then goes on to use a sequential formulation of equilibria to display how the overlapping generations model can be used to study issues in monetary and fiscal economics, including Social Security.

Chapter 10 compares an important aspect of an overlapping generations model with an infinitely lived agent model with a particular kind of incomplete market structure. This chapter is thus our first encounter with an incomplete markets model. The chapter analyzes the Ricardian equivalence theorem in two distinct but isomorphic settings: one a model with infinitely lived agents who face borrowing constraints, another with overlapping generations of two-period-lived agents with a bequest motive. We describe situations in which the timing of taxes does or does not matter, and explain how binding borrowing constraints in the infinite-lived model correspond to nonoperational bequest motives in the overlapping generations model.

Chapter 11 studies fiscal policy within a nonstochastic growth model with distorting taxes. This chapter studies how foresight about future policies and transient responses to past ones contribute to current outcomes. In particular, this chapter describes 'feedforward' and 'feedback' components of expressions for equilibrium outcomes. Chapter 12 describes the recursive competitive equilibrium concept and applies it within the context of the stochastic growth model.

Chapter 13 studies asset pricing and a host of practical doctrines associated with asset pricing, including Ricardian equivalence again and Modigliani-Miller theorems for private and government finance. Chapter 14 studies empirical strategies for implementing asset pricing models. Chapter 15 is about economic growth. It describes the basic growth model, and analyzes the key features of the specification of the technology that allows the model to exhibit balanced growth.

Chapter 16 studies competitive equilibria distorted by taxes and our first mechanism design problems, namely, ones that seek to find the optimal temporal pattern of distorting taxes. In a nonstochastic economy, the most startling finding is that the optimal tax rate on capital is zero in the long run.

Chapter 17 is about self-insurance. We study a single agent whose limited menu of assets gives him an incentive to self-insure by accumulating assets. We study a special case of what has sometimes been called the "savings problem," and analyze in detail the motive for self-insurance and the surprising implications it has for the agent's ultimate consumption and asset holdings. The type of agent studied in this chapter will be a component of the incomplete markets models to be studied in chapter 18.

Chapter 18 studies incomplete markets economies with heterogeneous agents and imperfect markets for sharing risks. The models of market incompleteness in this chapter come from simply ruling out markets in many assets, without motivating the absence of those asset markets from the physical structure of the economy. We must wait until chapter 20 for a study of some of the reasons that such markets may not exist.

The next chapters describe various manifestations of recursive contracts. Chapter 19 describes how linear quadratic dynamic programming can sometimes be used to compute recursive contracts. Chapter 20 describes models in the mechanism design tradition, work that starts to provide a foundation for incomplete assets markets, and that recovers specifications bearing an incomplete resemblance to the models of chapter 18. Chapter 20 is about the optimal provision of social insurance in the presence of information and enforcement problems. Relative to earlier chapters, chapter 20 escalates the sophistication with which recursive methods are applied, by utilizing promised values as state variables. Chapter 21 extends the analysis to a general equilibrium setting and draws out some implications for asset prices, among other things. Chapter 22 uses recursive contracts to design optimal unemployment insurance and worker compensation schemes.

Chapters 23 and 24 apply some of the same ideas to problems in "reputational macroeconomics," using promised values to formulate the notion of credibility. We study how a reputational mechanism can make policies sustainable even when the government lacks the commitment technology that was assumed to exist in the policy analysis of chapter 16. This reputational approach is used in chapter 26 to assess whether or not the Friedman rule is a sustainable policy. Chapter 25 describes a model of gradualism in trade policy that has some features in common with the first model of chapter 20.

Chapter 26 switches gears by adding money to a very simple competitive equilibrium model, in a most superficial way; the excuse for that superficial device is that it permits us to present and unify ten more or less well-known monetary doctrines. Chapter 27 presents a less superficial model of money, the turnpike model of Townsend, which is basically a special nonstochastic version of one of the models of chapter 18. The specialization allows us to focus on a variety of monetary doctrines.

Chapter 28 describes multiple agent models of search and matching. Except for a section on money in a search model, the focus is on labor markets as a central application of these theories. To bring out the economic forces at work in different frameworks, we examine the general equilibrium effects of layoff taxes. Chapter 29 compares forces in an employment lotteries model with forces in a time-averaging model of aggregate labor supply.

Two appendixes collect various technical results on functional analysis and linear projections and hidden Markov models.

Alternative uses of the book

We have used parts of this book to teach both first- and second-year graduate courses in macroeconomics and monetary economics at the University of Chicago, Stanford University, New York University, Princeton University, and the Stockholm School of Economics. Here are some alternative plans for courses:

1. A one-semester first-year course: chapters 2–6, 8, 9, 10, and either chapter 13, 15, or 16.

2. A second-semester first-year course: add chapters 8, 12, 13, 14, 15, 16, parts of 17 and 18, and all of 20.

3. A first course in monetary economics: chapters 9, 23, 24, 25, 26, 27, and the last section of 28.

4. A second-year macroeconomics course: select from chapters 13–29.

5. A self-contained course about recursive contracts: chapters 19–25.

As an example, Sargent used the following structure for a one-quarter first-year course at the University of Chicago: for the first and last weeks of the quarter, students were asked to read the monograph by Lucas (1987). Students were "prohibited" from reading the monograph in the intervening weeks. During the middle eight weeks of the quarter, students read material from chapters 6 (about search theory); chapter 8 (about complete markets); chapters 9, 26, and 27 (about models of money); and a little bit of chapters 20, 21, and 22 (on social insurance with incentive constraints). The substantive theme of the course was the issues set out in a nontechnical way by Lucas (1987). However, to understand Lucas's arguments, it helps to know the tools and models studied in the middle weeks of the course. Those weeks also exposed students to a range of alternative models that could be used to measure Lucas's arguments against some of the criticisms made, for example, by Manuelli and Sargent (1988).

Another one-quarter course would assign Lucas's (1992) article on efficiency and distribution in the first and last weeks. In the intervening weeks of the course, assign chapters 17, 18, and 20.

As another example, Ljungqvist used the following material in a four-week segment on employment/unemployment in first-year macroeconomics at the Stockholm School of Economics. Labor market issues command a strong interest especially in Europe. Those issues help motivate studying the tools in chapters 6 and 28 (about search and matching models), and parts of 22 (on the optimal provision of unemployment compensation). On one level, both chapters 6 and 28 focus on labor markets as a central application of the theories presented, but on another level, the skills and understanding acquired in these chapters transcend the specific topic of labor market dynamics. For example, the thorough practice on formulating and solving dynamic programming problems in chapter 6 is generally useful to any student of economics, and the models of chapter 28 are an entry-pass to other heterogeneous-agent models like those in chapter 18. Further, an excellent way to motivate the study of recursive contracts in chapter 22 is to ask how unemployment compensation should optimally be provided in the presence of incentive problems.

As a final example, Sargent used versions of the material in 6, 11, and 14 to teach undergraduate classes at Princeton and NYU.

Matlab programs

Various exercises and examples use Matlab programs. These programs are referred to in a special index at the end of the book. They can be downloaded from < https://files.nyu.edu/ts43/public/books.html > .

Notation

We use the symbol ■ to denote the conclusion of a proof. The editors of this book requested that where possible, brackets and braces be used in place of multiple parentheses to denote composite functions. Thus, the reader will often encounter $f[u(c)]$ to express the composite function $f \circ u$.

Brief history of the notion of the state

This book reflects progress economists have made in refining the notion of state so that more and more problems can be formulated recursively. The art in applying recursive methods is to find a convenient definition of the state. It is often not obvious what the state is, or even whether a finite-dimensional state *exists* (e.g., maybe the entire infinite history of the system is needed to characterize its current position). Extending the range of problems susceptible to recursive methods has been one of the major accomplishments of macroeconomic theory since 1970. In diverse contexts, this enterprise has been about discovering a convenient state and constructing a first-order difference equation to describe its motion. In models equivalent to single-agent control problems, state variables are either capital stocks or information variables that help predict the future.[4] In single-agent models of optimization in the presence of measurement errors, the true state vector is latent or "hidden" from the optimizer and the economist, and needs to be estimated. Here *beliefs* come to serve as the patent state. For example, in a Gaussian setting, the mathematical expectation and covariance matrix of the latent state vector, conditioned on the available history of observations, serves as the state. In authoring his celebrated filter, Kalman

---

[4] Any available variables that *Granger cause* variables impinging on the optimizer's objective function or constraints enter the state as information variables. See C.W.J. Granger (1969).

(1960) showed how an estimator of the hidden state could be constructed recursively by means of a difference equation that uses the current observables to update the estimator of last period's hidden state.[5] Muth (1960); Lucas (1972), Kareken, Muench, and Wallace (1973); Jovanovic (1979); and Jovanovic and Nyarko (1996) all used versions of the Kalman filter to study systems in which agents make decisions with imperfect observations about the state.

For a while, it seemed that some very important problems in macroeconomics could not be formulated recursively. Kydland and Prescott (1977) argued that it would be difficult to apply recursive methods to macroeconomic policy design problems, including two examples about taxation and a Phillips curve. As Kydland and Prescott formulated them, the problems were not recursive: the fact that the public's forecasts of the government's future decisions influence the public's current decisions made the government's problem simultaneous, not sequential. But soon Kydland and Prescott (1980) and Hansen, Epple, and Roberds (1985) proposed a recursive formulation of such problems by expanding the state of the economy to include a Lagrange multiplier or *costate* variable associated with the government's budget constraint. The costate variable acts as the marginal cost of keeping a promise made earlier by the government. Marcet and Marimon (1999) extended and formalized a recursive version of such problems.

A significant breakthrough in the application of recursive methods was achieved by several researchers including Spear and Srivastava (1987); Thomas and Worrall (1988); and Abreu, Pearce, and Stacchetti (1990). They discovered a state variable for recursively formulating an infinitely repeated moral hazard problem. That problem requires the principal to track a history of outcomes and to use it to construct statistics for drawing inferences about the agent's actions. Problems involving self-enforcement of contracts and a government's reputation share this feature. A *continuation value* promised by the principal to the agent can summarize the history. Making the promised valued a state

---

[5] In competitive multiple-agent models in the presence of measurement errors, the dimension of the hidden state threatens to explode because beliefs about beliefs about ... naturally appear, a problem studied by Townsend (1983). This threat has been overcome through thoughtful and economical definitions of the state. For example, one way is to give up on seeking a purely "autoregressive" recursive structure and to include a moving average piece in the descriptor of beliefs. See Sargent (1991). Townsend's equilibria have the property that prices fully reveal the private information of diversely informed agents.

variable allows a recursive solution in terms of a function mapping the inherited promised value and random variables realized today into an action or allocation today and a promised value for tomorrow. The sequential nature of the solution allows us to recover history-dependent strategies just as we use a stochastic difference equation to find a "moving average" representation.[6]

It is now standard to use a continuation value as a state variable in models of credibility and dynamic incentives. We shall study several such models in this book, including ones for optimal unemployment insurance and for designing loan contracts that must overcome information and enforcement problems.

---

[6] Related ideas are used by Shavell and Weiss (1979); Abreu, Pearce, and Stacchetti (1986, 1990) in repeated games; and Green (1987) and Phelan and Townsend (1991) in dynamic mechanism design. Andrew Atkeson (1991) extended these ideas to study loans made by borrowers who cannot tell whether they are making consumption loans or investment loans.

Part I
The imperialism of recursive methods

Chapter 1
Overview

1.1. Warning

This chapter provides a nontechnical summary of some themes of this book. We debated whether to put this chapter first or last. A way to use this chapter is to read it twice, once before reading anything else in the book, then again after having mastered the techniques presented in the rest of the book. That second time, this chapter will be easy and enjoyable reading, and it will remind you of connections that transcend a variety of apparently disparate topics. But on first reading, this chapter will be difficult, partly because the discussion is mainly literary and therefore incomplete. Measure what you have learned by comparing your understandings after those first and second readings. Or just skip this chapter and read it after the others.

1.2. A common ancestor

Clues in our mitochondrial DNA tell biologists that we humans share a common ancestor called Eve who lived 200,000 years ago. All of macroeconomics too seems to have descended from a common source, Irving Fisher's and Milton Friedman's consumption Euler equation, the cornerstone of the permanent income theory of consumption. Modern macroeconomics records the fruit and frustration of a long love-hate affair with the permanent income mechanism. As a way of summarizing some important themes in our book, we briefly chronicle some of the high and low points of this long affair.

1.3. The savings problem

A consumer wants to maximize

$$E_0 \sum_{t=0}^{\infty} \beta^t u(c_t) \tag{1.3.1}$$

where $\beta \in (0,1)$, u is a twice continuously differentiable, increasing, strictly concave utility function, and E_0 denotes a mathematical expectation conditioned on time 0 information. The consumer faces a sequence of budget constraints[1]

$$A_{t+1} = R_{t+1}(A_t + y_t - c_t) \tag{1.3.2}$$

for $t \geq 0$, where $A_{t+1} \geq \underline{A}$ is the consumer's holdings of an asset at the beginning of period $t+1$, $\underline{A}$ is a lower bound on asset holdings, y_t is a random endowment sequence, c_t is consumption of a single good, and R_{t+1} is the gross rate of return on the asset between t and $t+1$. In the general version of the problem, both R_{t+1} and y_t can be random, though special cases of the problem restrict R_{t+1} further. A first-order necessary condition for this problem is

$$\beta E_t R_{t+1} \frac{u'(c_{t+1})}{u'(c_t)} \leq 1, \quad = \text{ if } A_{t+1} > \underline{A}. \tag{1.3.3}$$

This Euler inequality recurs as either the cornerstone or the straw man in many theories contained in this book.

Different modeling choices put (1.3.3) to work in different ways. One can restrict u, β, the return process R_{t+1}, the lower bound on assets $\underline{A}$, the income process y_t, and the consumption process c_t in various ways. By making alternative choices about restrictions to impose on subsets of these objects, macroeconomists have constructed theories about consumption, asset prices, and the distribution of wealth. Alternative versions of equation (1.3.3) also underlie Chamley's (1986) and Judd's (1985b) striking results about eventually not taxing capital.

---

[1] We use a different notation in chapter 17: A_t here conforms to $-b_t$ in chapter 17.

1.3.1. Linear quadratic permanent income theory

To obtain a version of the permanent income theory of Friedman (1955) and Hall (1978), set $R_{t+1} = R$, impose $R = \beta^{-1}$, assume the quadratic utility function $u(c_t) = -(c_t - \gamma)^2$, and allow consumption c_t to be negative. We also allow $\{y_t\}$ to be an arbitrary stationary process, and dispense with the lower bound $\underline{A}$. The Euler inequality (1.3.3) then implies that consumption is a martingale:

$$E_t c_{t+1} = c_t. \tag{1.3.4}$$

Subject to a boundary condition that[2] $E_0 \sum_{t=0}^{\infty} \beta^t A_t^2 < \infty$, equation (1.3.4) and the budget constraints (1.3.2) can be solved to yield

$$c_t = \left[\frac{r}{1+r} \right] \left[E_t \sum_{j=0}^{\infty} \left(\frac{1}{1+r} \right)^j y_{t+j} + A_t \right] \tag{1.3.5}$$

where $1 + r = R$. Equation (1.3.5) expresses consumption as a fixed marginal propensity to consume $\frac{r}{1+r}$ that is applied to the sum of human wealth – namely $\left[E_t \sum_{j=0}^{\infty} \left(\frac{1}{1+r} \right)^j y_{t+j} \right]$ – and financial wealth, $A)t$. This equation has the following notable features: (1) consumption is smoothed on average across time: current consumption depends only on the expected present value of nonfinancial income; (2) feature (1) opens the way to Ricardian equivalence: redistributions of lump-sum taxes over time that leave the expected present value of nonfinancial income unaltered do not affect consumption; (3) there is certainty equivalence: increases in the conditional variances of future incomes about their forecast values do not affect consumption (though they do diminish the consumer's utility); (4) a by-product of certainty equivalence is that the marginal propensities to consume out of financial and nonfinancial wealth are equal.

This theory continues to be a workhorse in much good applied work (see Ligon (1998) and Blundell and Preston (1999) for recent creative applications). Chapter 5 describes conditions under which certainty equivalence prevails, while chapters 2 and 5 also describe the structure of the cross-equation restrictions

---

[2] The motivation for using this boundary condition instead of a lower bound $\underline{A}$ on asset holdings is that there is no "natural" lower bound on asset holdings when consumption is permitted to be negative. Chapters 8 and 18 discuss what are called "natural borrowing limits," the lowest possible appropriate values of $\underline{A}$ in the case that c is nonnegative.

that the hypothesis of rational expectations imposes and that empirical studies
heavily exploit.

1.3.2. Precautionary saving

A literature on "the savings problem" or "precautionary saving" investigates
the consequences of altering the assumption in the linear quadratic permanent
income theory that u is quadratic, an assumption that makes the marginal utility of consumption become negative for large enough c. Rather than assuming
that u is quadratic, the literature on the savings problem assumes that u is
increasing and strictly concave. This assumption keeps the marginal utility of
consumption above zero. We retain other features of the linear quadratic model
($\beta R = 1$, $\{y_t\}$ is a stationary process), but now impose a borrowing limit
$A_t \geq \underline{a}$.

   With these assumptions, something amazing occurs: Euler inequality (1.3.3)
implies that the marginal utility of consumption is a *nonnegative* supermartingale.[3]   That gives the model the striking implication that $c_t \to_{as} +\infty$ and
$A_t \to_{as} +\infty$, where $\to_{as}$ means almost sure convergence. Consumption and
wealth will fluctuate randomly in response to income fluctuations, but so long
as randomness in income continues, they will drift upward over time without
bound. If randomness eventually expires in the tail of the income process, then
both consumption and income converge. But even small perpetual random
fluctuations in income are enough to cause both consumption and assets to diverge to $+\infty$. This response of the optimal consumption plan to randomness
is required by the Euler equation (1.3.3) and is called precautionary savings.
By keeping the marginal utility of consumption positive, precautionary savings
models arrest the certainty equivalence that prevails in the linear quadratic permanent income model. Chapter 17 studies the savings problem in depth and
struggles to understand the workings of the powerful martingale convergence

---

   [3] See chapter 17. The situation is simplest in the case that the y_t process is i.i.d. so
that the value function can be expressed as a function of level $y_t + A_t$ alone: $V(A + y)$.
Applying the Benveniste-Scheinkman formula from chapter 3 shows that $V'(A + y) = u'(c)$,
which implies that when $\beta R = 1$, (1.3.3) becomes $E_t V'(A_{t+1} + y_{t+1}) \leq V'(A_t + y_t)$, which
states that the derivative of the value function is a nonnegative supermartingale. That in turn
implies that A almost surely diverges to $+\infty$.

theorem. The supermartingale convergence theorem also plays an important role in the model insurance with private information in chapter 20.

1.3.3. *Complete markets, insurance, and the distribution of wealth*

To build a model of the distribution of wealth, we consider a setting with many consumers. To start, imagine a large number of *ex ante* identical consumers with preferences (1.3.1) who are allowed to share their income risk by trading one-period contingent claims. For simplicity, assume that the saving possibility represented by the budget constraint (1.3.2) is no longer available[4] but that it is replaced by access to an extensive set of insurance markets. Assume that household i has an income process $y_t^i = g_i(s_t)$ where s_t is a state vector governed by a Markov process with transition density $\pi(s'|s)$, where s and s' are elements of a common state space $\mathbf{S}$. (See chapters 2 and 8 for material about Markov chains and their uses in equilibrium models.) Each period every household can trade one-period state-contingent claims to consumption next period. Let $Q(s'|s)$ be the price of one unit of consumption next period in state s' when the state this period is s. When household i has the opportunity to trade such state-contingent securities, its first-order conditions for maximizing (1.3.1) are

$$Q\left(s_{t+1}|s_t\right) = \beta \frac{u'\left(c_{t+1}^i\left(s_{t+1}\right)\right)}{u'\left(c_t^i\left(s_t\right)\right)} \pi\left(s_{t+1}|s_t\right). \qquad (1.3.6)$$

Notice that $\int_{s_{t+1}} Q(s_{t+1}|s_t) ds_{t+1}$ is the price of a risk-free claim on consumption one period ahead: it is thus the reciprocal of the gross risk-free interest rate R. Therefore, if we sum both sides of (1.3.6) over s_{t+1}, we obtain our standard consumption Euler condition (1.3.3) at equality.[5] Thus, the complete markets equation (1.3.6) is consistent with our complete markets Euler equation (1.3.3), but (1.3.6) imposes more. We will exploit this fact extensively in chapter 16.

In a widely studied special case, there is no aggregate risk, so that $\int_i y_t^i d i = \int_i g_i(s_t) d i =$ constant. In that case, it can be shown that the competitive equilibrium state-contingent prices become

$$Q\left(s_{t+1}|s_t\right) = \beta \pi\left(s_{t+1}|s_t\right). \qquad (1.3.7)$$

---

[4] It can be shown that even if it were available, people would not want to use it.

[5] That the asset is risk-free becomes manifested in R_{t+1} being a function of s_t, so that it is known at t.

This, in turn, implies that the risk-free gross rate of return R is β^{-1}.[6] If we substitute (1.3.7) into (1.3.6), we discover that $c_{t+1}^i(s_{t+1}) = c_t^i(s_t)$ for all (s_{t+1}, s_t). Thus, the consumption of consumer i is constant across time and across states of nature s, so that in equilibrium, all idiosyncratic risk is insured away. Higher present-value-of-endowment consumers will have permanently higher consumption than lower present-value-of-endowment consumers, so that there is a non-degenerate cross-section distribution of wealth and consumption. In this model, the cross-section distributions of wealth and consumption replicate themselves over time, and furthermore each individual forever occupies the same position in that distribution.

A model that has the cross-section distribution of wealth and consumption being time invariant is not a bad approximation to the data. But there is ample evidence that individual households' positions *within* the distribution of wealth move over time.[7] Several models described in this book alter consumers' trading opportunities in ways designed to frustrate risk sharing enough to cause individuals' position in the distribution of wealth to change with luck and enterprise. One class that emphasizes luck is the set of incomplete markets models started by Truman Bewley. It eliminates the household's access to almost all markets and returns it to the environment of the precautionary savings model.

1.3.4. Bewley models

At first glance, the precautionary savings model with $\beta R = 1$ seems like a bad starting point for building a theory that aspires to explain a situation in which cross-section distributions of consumption and wealth are constant over time even as individuals experience random fluctuations within that distribution. A panel of households described by the precautionary savings model with $\beta R = 1$ would have cross-section distributions of wealth and consumption that march upward and never settle down. What have come to be called Bewley models are

---

[6] This follows because the price of a risk-free claim to consumption tomorrow at date t in state s_t is $\sum_{s_{t+1}} Q(s_{t+1}|s_t) = \beta \sum_{s_{t+1}} \pi(s_{t+1}|s_t) = \beta$.

[7] See Díaz-Giménez, Quadrini and Ríos-Rull (1997); Krueger and Perri (2004, 2006); Rodriguez, Díaz-Giménez, Quadrini and Ríos-Rull (2002); and Davies and Shorrocks (2000).

constructed by lowering the interest rate R to allow those cross-section distributions to settle down.[8] Bewley models are arranged so that the cross section distributions of consumption, wealth, and income are constant over time and so that the asymptotic stationary distributions of consumption, wealth, and income for an individual consumer across time equal the corresponding cross-section distributions across people. A Bewley model can thus be thought of as starting with a continuum of consumers operating according to the precautionary savings model with $\beta R = 1$ and its diverging individual asset process. We then lower the interest rate enough to make assets converge to a distribution whose cross-section average clears a market for a risk-free asset. Different versions of Bewley models are distinguished by what the risk-free asset is. In some versions it is a consumption loan from one consumer to another; in others it is fiat money; in others it can be either consumption loans or fiat money; and in yet others it is claims on physical capital. Chapter 18 studies these alternative interpretations of the risk-free asset.

As a function of a constant gross interest rate R, Figure 1.3.1 plots the time series average of asset holdings for an individual consumer. At $R = \beta^{-1}$, the time series mean of the individual's assets diverges, so that $Ea(R)$ is infinite. For $R < \beta^{-1}$, the mean exists. We require that a continuum of *ex ante* identical but *ex post* different consumers share the same time series average $Ea(R)$ and also that the distribution of a over time for a given agent equals the distribution of A_{t+1} at a point in time across agents. If the asset in question is a pure consumption loan, we require as an equilibrium condition that $Ea(R) = 0$, so that borrowing equals lending. If the asset is fiat money, then we require that $Ea(R) = \frac{M}{p}$, where M is a fixed stock of fiat money and p is the price level.

Thus, a Bewley model lowers the interest rate R enough to offset the precautionary savings force that with $\beta R = 1$ propels assets upward in the savings problem. Precautionary saving remains an important force in Bewley models:

---

[8] It is worth thinking about the sources of the following differences. In the complete markets model sketched in subsection 1.3.3, an *equilibrium* risk-free gross interest rate R satisfies $R\beta = 1$ and each consumer completely smooths consumption across both states and time, so that the distribution of consumption trivially converges. The precautionary savings model of section 1.3.2 *assumes* that $R\beta = 1$ and derives the outcome that each consumer's consumption and financial wealth both diverge toward $+\infty$. Why can $\beta R = 1$ be compatible with non-exploding individual consumption and wealth levels in the complete markets model of subsection 1.3.3, but not in the precautionary savings model of subsection 1.3.2?

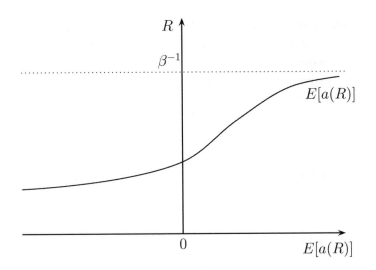

Figure 1.3.1: Mean of time series average of household consumption as function of risk-free gross interest rate R.

an increase in the volatility of income generally pushes the $Ea(R)$ curve to the right, driving the equilibrium R downward.

1.3.5. History dependence in standard consumption models

Individuals' positions in the wealth distribution are frozen in the complete markets model, but not in the Bewley model, reflecting the absence or presence, respectively, of *history dependence* in equilibrium allocation rules for consumption. The preceding version of the complete markets model erases history dependence, while the savings problem model and the Bewley model do not.

History dependence is present in these models in an easy to handle recursive way because the household's asset level completely encodes the history of endowment realizations that it has experienced. We want a way of representing history dependence more generally in contexts where a stock of assets does not suffice to summarize history. History dependence can be troublesome because without a convenient low-dimensional state variable to encode history, it requires that there be a separate decision rule for each date that expresses the time t decision as a function of the history at time t, an object with a number of arguments that grows exponentially with t. As analysts, we have a strong incentive to find a low-dimensional state variable. Fortunately, economists have made tremendous strides in handling history dependence with recursive methods that summarize a history with a single number and that permit compact time-invariant expressions for decision rules. We shall discuss history dependence later in this chapter and will encounter many such examples in chapters 19 through 25.

1.3.6. Growth theory

Equation (1.3.3) is also a key ingredient of growth theory (see chapters 11 and 15). In the one-sector growth model, a representative household solves a version of the savings problem in which the single asset is interpreted as a claim on the return from a physical capital stock K that enters a constant returns to scale production function $F(K, L)$, where L is labor input. When returns to capital are tax free, the theory equates the gross rate of return R_{t+1} to the gross marginal product of capital net of depreciation, namely, $F_{k,t+1} + (1 - \delta)$, where $F_k(k, t+1)$ is the marginal product of capital and δ is a depreciation rate. Suppose that we add leisure to the utility function, so that we replace $u(c)$ with the more general one-period utility function $U(c, \ell)$, where ℓ is the household's leisure. Then the appropriate version of the consumption Euler condition (1.3.3) at equality becomes

$$U_c(t) = \beta U_c(t + 1) \left[F_k(t + 1) + (1 - \delta)\right]. \tag{1.3.8}$$

The constant returns to scale property implies that $F_k(K, N) = f'(k)$, where $k = K/N$ and $F(K, N) = Nf(K/N)$. If there exists a steady state in which k and c are constant over time, then equation (1.3.8) implies that it must satisfy

$$\rho + \delta = f'(k) \tag{1.3.9}$$

where $\beta^{-1} \equiv (1 + \rho)$. The value of k that solves this equation is called the "augmented Golden rule" steady-state level of the capital-labor ratio. This celebrated equation shows how technology (in the form of f and δ) and time preference (in the form of β) are the determinants of the steady-state level of capital when income from capital is not taxed. However, if income from capital is taxed at the flat rate marginal rate $\tau_{k,t+1}$, then the Euler equation (1.3.8) becomes modified

$$U_c(t) = \beta U_c(t + 1) \left[F_k(t + 1)(1 - \tau_{k,t+1}) + (1 - \delta)\right]. \tag{1.3.10}$$

If the flat rate tax on capital is constant and if a steady-state k exists, it must satisfy

$$\rho + \delta = (1 - \tau_k) f'(k).\tag{1.3.11}$$

This equation shows how taxing capital diminishes the steady-state capital labor ratio. See chapter 11 for an extensive analysis of the one-sector growth model when the government levies time-varying flat rate taxes on consumption, capital, and labor, as well as offering an investment tax credit.

1.3.7. Limiting results from dynamic optimal taxation

Equations (1.3.9) and (1.3.11) are central to the dynamic theory of optimal taxes. Chamley (1986) and Judd (1985b) forced the government to finance an exogenous stream of government purchases, gave it the capacity to levy time-varying flat rate taxes on labor and capital at different rates, formulated an optimal taxation problem (a so-called Ramsey problem), and studied the possible limiting behavior of the optimal taxes. Two Euler equations play a decisive role in determining the limiting tax rate on capital in a nonstochastic economy: the household's Euler equation (1.3.10), and a similar consumption Euler equation for the Ramsey planner that takes the form

$$W_c(t) = \beta W_c(t+1) \left[F_k(t+1) + (1-\delta) \right]\tag{1.3.12}$$

where

$$W(c_t, \ell_t) = U(c_t, \ell_t) + \Phi \left[U_c(t) c_t - U_\ell(t)(1-\ell_t) \right]\tag{1.3.13}$$

and where Φ is a Lagrange multiplier on the government's intertemporal budget constraint. As Jones, Manuelli, and Rossi (1997) emphasize, if the function $W(c, \ell)$ is simply viewed as a peculiar utility function, then what is called the primal version of the Ramsey problem can be viewed as an ordinary optimal growth problem with period utility function W instead of U.[9]

In a Ramsey allocation, taxes must be such that *both* (1.3.8) and (1.3.12) always hold, among other equations. Judd and Chamley note the following

---

[9] Notice that so long as $\Phi > 0$ (which occurs whenever taxes are necessary), the objective in the primal version of the Ramsey problem disagrees with the preferences of the household over (c, ℓ) allocations. This conflict is the source of a time-inconsistency problem in the Ramsey problem with capital.

implication of the two Euler equations (1.3.8) and (1.3.12). If the government expenditure sequence converges and if a steady state exists in which $c_t, \ell_t, k_t, \tau_{kt}$ all converge, then it must be true that (1.3.9) holds *in addition* to (1.3.11). But both of these conditions can prevail only if $\tau_k = 0$. Thus, the steady-state properties of two versions of our consumption Euler equation (1.3.3) underlie Chamley and Judd's remarkable result that asymptotically it is optimal not to tax capital.

In stochastic versions of dynamic optimal taxation problems, we shall glean additional insights from (1.3.3) as embedded in the asset-pricing equations (1.3.16) and (1.3.18). In optimal taxation problems, the government has the ability to manipulate asset prices through its influence on the equilibrium consumption allocation that contributes to the stochastic discount factor $m_{t+1,t}$ defined in equation (1.3.16) below. The Ramsey government seeks a way wisely to use its power to revalue its existing debt by altering state-history prices. To appreciate what the Ramsey government is doing, it helps to know the theory of asset pricing.

1.3.8. Asset pricing

The dynamic asset pricing theory of Breeden (1979) and Lucas (1978) also starts with (1.3.3), but alters what is fixed and what is free. The Breeden-Lucas theory is silent about the endowment process $\{y_t\}$ and sweeps it into the background. It fixes a function u and a discount factor β, and takes a consumption process $\{c_t\}$ as given. In particular, assume that $c_t = g(X_t)$, where X_t is a Markov process with transition c.d.f. $F(X'|X)$. Given these inputs, the theory is assigned the task of restricting the rate of return on an asset, defined by Lucas as a claim on the consumption endowment:

$$R_{t+1} = \frac{p_{t+1} + c_{t+1}}{p_t}$$

where p_t is the price of the asset. The Euler inequality (1.3.3) becomes

$$E_t \beta \frac{u'(c_{t+1})}{u'(c_t)} \left(\frac{p_{t+1} + c_{t+1}}{p_t} \right) = 1. \qquad (1.3.14)$$

This equation can be solved for a pricing function $p_t = p(X_t)$. In particular, if we substitute $p(X_t)$ into (1.3.14), we get Lucas's functional equation for $p(X)$.

1.3.9. Multiple assets

If the consumer has access to several assets, a version of (1.3.3) holds for each asset:

$$E_t \beta \frac{u'(c_{t+1})}{u'(c_t)} R_{j,t+1} = 1 \qquad (1.3.15)$$

where $R_{j,t+1}$ is the gross rate of return on asset j. Given a utility function u, a discount factor β, and the hypothesis of rational expectations (which allows the researcher to use empirical projections as counterparts of the theoretical projections E_t), equations (1.3.15) put extensive restrictions across the moments of a vector time series for $[c_t, R_{1,t+1}, \ldots, R_{J,t+1}]$. A key finding of the literature (e.g., Hansen and Singleton, 1983) is that for u's with plausible curvature,[10] consumption is too smooth for $\{c_t, R_{j,t+1}\}$ to satisfy equation (1.3.15), where c_t is measured as aggregate consumption.

Lars Hansen and others have elegantly organized this evidence as follows. Define the stochastic discount factor

$$m_{t+1,t} = \beta \frac{u'(c_{t+1})}{u'(c_t)} \qquad (1.3.16)$$

and write (1.3.15) as

$$E_t m_{t+1,t} R_{j,t+1} = 1. \qquad (1.3.17)$$

Represent the gross rate of return as

$$R_{j,t+1} = \frac{o_{t+1}}{q_t}$$

where o_{t+1} is a one-period payout on the asset and q_t is the price of the asset at time t. Then (1.3.17) can be expressed as

$$q_t = E_t m_{t+1,t} o_{t+1}. \qquad (1.3.18)$$

The structure of (1.3.18) justifies calling $m_{t+1,t}$ a stochastic discount factor: to determine the price of an asset, multiply the random payoff for each state by the discount factor for that state, then add over states by taking a conditional expectation. Applying the definition of a conditional covariance and a Cauchy-Schwartz inequality to this equation implies

$$\frac{q_t}{E_t m_{t+1,t}} \geq E_t o_{t+1} - \frac{\sigma_t(m_{t+1,t})}{E_t m_{t+1,t}} \sigma_t(o_{t+1}) \qquad (1.3.19)$$

---

[10] Chapter 14 describes Pratt's (1964) mental experiment for deducing plausible curvature.

where $\sigma_t(y_{t+1})$ denotes the conditional standard deviation of y_{t+1}. Setting $o_{t+1} = 1$ in (1.3.18) shows that $E_t m_{t+1,t}$ must be the time t price of a risk-free one-period security. Inequality (1.3.19) bounds the ratio of the price of a risky security q_t to the price of a risk-free security $E_t m_{t+1,1}$ by the right side, which equals the expected payout on that risky asset *minus* its conditional standard deviation $\sigma_t(o_{t+1})$ *times* a "market price of risk" $\sigma_t(m_{t+1,t})/E_t m_{t+1,t}$. By using data only on payouts o_{t+1} and prices q_t, inequality (1.3.19) has been used to estimate the market price of risk without restricting how $m_{t+1,t}$ relates to consumption. If we take these atheoretical estimates of $\sigma_t(m_{t+1,t})/E_t m_{t+1,t}$ and compare them with the theoretical values of $\sigma_t(m_{t+1,t})/E_t m_{t+1,t}$ that we get with a plausible curvature for u, and by imposing $\hat{m}_{t+1,t} = \beta \frac{u'(c_{t+1})}{u'(c_t)}$ for aggregate consumption, we find that the theoretical $\hat{m}$ has far too little volatility to account for the atheoretical estimates of the conditional coefficient of variation of $m_{t+1,t}$. As we discuss extensively in chapter 14, this outcome reflects the fact that aggregate consumption is too smooth to account for atheoretical estimates of the market price of risk.

There have been two broad types of response to the empirical challenge. The first retains (1.3.17) but abandons (1.3.16) and instead adopts a statistical model for $m_{t+1,t}$. Even without the link that equation (1.3.16) provides to consumption, equation (1.3.17) imposes restrictions across asset returns and $m_{t+1,t}$ that can be used to identify the $m_{t+1,t}$ process. Equation (1.3.17) contains no-arbitrage conditions that restrict the joint behavior of returns. This has been a fruitful approach in the affine term structure literature (see Backus and Zin (1993), Piazzesi (2000), and Ang and Piazzesi (2003)).[11]

Another approach has been to disaggregate and to write the household-i version of (1.3.3):

$$\beta E_t R_{t+1} \frac{u'(c_{i,t+1})}{u'(c_{it})} \leq 1, \quad = \quad \text{if } A_{i,t+1} > \underline{A}_i. \tag{1.3.20}$$

If at time t, a subset of households are on the corner, (1.3.20) will hold with equality only for another subset of households. Households in the second set price assets.[12]

---

[11] Affine term structure models generalize earlier models that implemented rational expectations versions of the expectations theory of the term structure of interest rates. See Campbell and Shiller (1991), Hansen and Sargent (1991), and Sargent (1979).

[12] David Runkle (1991) and Gregory Mankiw and Steven Zeldes (1991) checked (1.3.20) for subsets of agents.

Chapter 21 describes a model of Harald Zhang (1997) and Alvarez and Jermann (2000, 2001). The model introduces participation (collateral) constraints and shocks in a way that makes a changing subset of agents i satisfy (1.3.20). Zhang and Alvarez and Jermann formulate these models by adding participation constraints to the recursive formulation of the consumption problem based on (1.4.7). Next we briefly describe the structure of these models and their attitude toward our theme equation, the consumption Euler equation (1.3.3). The idea of Zhang and Alvarez and Jermann was to meet the empirical asset pricing challenges by disrupting (1.3.3). As we shall see, that requires eliminating some of the assets that some of the households can trade. These advanced models exploit a convenient method for representing and manipulating history dependence.

1.4. Recursive methods

The pervasiveness of the consumption Euler inequality will be a significant substantive theme of this book. We now turn to a methodological theme, the imperialism of the recursive method called dynamic programming.

The notion that underlies dynamic programming is a finite-dimensional object called the *state* that, from the point of view of current and future payoffs, completely summarizes the current situation of a decision maker. If an optimum problem has a low-dimensional state vector, immense simplifications follow. A recurring theme of modern macroeconomics and of this book is that finding an appropriate state vector is an art.

To illustrate the idea of the state in a simple setting, return to the savings problem and assume that the consumer's endowment process is a time-invariant function of a state s_t that follows a Markov process with time-invariant one-period transition density $\pi(s'|s)$ and initial density $\pi_0(s)$, so that $y_t = y(s_t)$. To begin, recall the description (1.3.5) of consumption that prevails in the special linear quadratic version of the savings problem. Under our present assumption that y_t is a time-invariant function of the Markov state, (1.3.5) and the household's budget constraint imply the following representation of the household's decision rule:

$$c_t = f(A_t, s_t) \qquad (1.4.1a)$$

$$A_{t+1} = g\left(A_t, s_t\right).\tag{1.4.1b}$$

Equation $(1.4.1a)$ represents consumption as a time-invariant function of a state vector (A_t, s_t). The Markov component s_t appears in $(1.4.1a)$ because it contains all of the information that is useful in forecasting future endowments (for the linear quadratic model, $(1.3.5)$ reveals the household's incentive to forecast future incomes), and the asset level A_t summarizes the individual's current financial wealth. The s component is assumed to be exogenous to the household's decisions and has a stochastic motion governed by $\pi(s'|s)$. But the future path of A is chosen by the household and is described by $(1.4.1b)$. The system formed by $(1.4.1)$ and the Markov transition density $\pi(s'|s)$ is said to be *recursive* because it expresses a current decision c_t as a function of the state and tells how to update the state. By iterating $(1.4.1b)$, notice that A_{t+1} can be expressed as a function of the history $[s_t, s_{t-1}, \ldots, s_0]$ and A_0. The endogenous state variable financial wealth thus encodes all payoff-relevant aspects of the history of the exogenous component of the state s_t.

Define the value function $V(A_0, s_0)$ as the optimum value of the savings problem starting from initial state (A_0, s_0). The value function V satisfies the following functional equation, known as a Bellman equation:

$$V\left(A, s\right) = \max_{c, A'}\left\{u\left(c\right) + \beta E\left[V\left(A', s'\right)|s\right]\right\}\tag{1.4.2}$$

where the maximization is subject to $A' = R(A + y - c)$ and $y = y(s)$. Associated with a solution $V(A, s)$ of the Bellman equation is the pair of policy functions

$$c = f\left(A, s\right)\tag{1.4.3a}$$

$$A' = g\left(A, s\right)\tag{1.4.3b}$$

from $(1.4.1)$. The *ex ante* value (i.e., the value of $(1.3.1)$ before s_0 is drawn) of the savings problem is then

$$v\left(A_0\right) = \sum_s V\left(A_0, s\right)\pi_0\left(s\right).\tag{1.4.4}$$

We shall make ample use of the *ex ante* value function.

1.4.1. Methodology: dynamic programming issues a challenge

Dynamic programming is now recognized as a powerful method for studying private agents' decisions and also the decisions of a government that wants to design an optimal policy in the face of constraints imposed on it by private agents' best responses to that government policy. But it has taken a long time for the power of dynamic programming to be realized for government policy design problems.

Dynamic programming had been applied since the late 1950s to design government decision rules to control an economy whose transition laws included rules that described the decisions of private agents. In 1976 Robert E. Lucas, Jr., published his now famous critique of dynamic-programming-based econometric policy evaluation procedures. The heart of Lucas's critique was the implication for government policy evaluation of a basic property that pertains to any optimal decision rule for private agents with a form (1.4.3) that attains a Bellman equation like (1.4.2). The property is that the optimal decision rules (f, g) depend on the transition density $\pi(s'|s)$ for the exogenous component of the state s. As a consequence, any widely understood government policy that alters the law of motion for a state variable like s that appears in private agents' decision rules should alter those private decision rules. (In the applications that Lucas had in mind, the s in private agents' decision problems included variables useful for predicting tax rates, the money supply, and the aggregate price level.) Therefore, Lucas asserted that econometric policy evaluation procedures that assumed that private agents' decision rules are fixed in the face of alterations in government policy are flawed.[13] Most econometric policy evaluation procedures at the time were vulnerable to Lucas's criticism. To construct valid policy evaluation procedures, Lucas advocated building new models that would attribute rational expectations to decision makers.[14] Lucas's discussant Robert Gordon implied that after that ambitious task had been accomplished, we could then use dynamic programming to compute optimal policies, i.e., to solve Ramsey problems.

---

[13] They were flawed because they assumed "no response" when they should have assumed "best response" of private agents' decision rules to government decision rules.

[14] That is, he wanted private decision rules to solve dynamic programming problems with the correct transition density π for s.

1.4.2. Dynamic programming challenged

But Edward C. Prescott's 1977 paper entitled "Should Control Theory Be Used for Economic Stabilization?" asserted that Gordon was too optimistic. Prescott claimed that in his 1977 JPE paper with Kydland he had proved that it was "logically impossible" to use dynamic programming to find optimal government policies in settings where private traders face genuinely dynamic problems. Prescott said that dynamic programming was inapplicable to government policy design problems because the structure of the best response of *current private decisions* to *future* government policies prevents the government policy design problem from being recursive (a manifestation of the time inconsistency of optimal government plans). The optimal government plan would therefore require a government commitment technology, and the government policy must take the form of a sequence of history-dependent decision rules that could not be expressed as a function of natural state variables.

1.4.3. Imperialistic response of dynamic programming

Much of the subsequent history of macroeconomics belies Prescott's claim of "logical impossibility." More and more problems that smart people like Prescott in 1977 thought could not be attacked with dynamic programming *can* now be solved with dynamic programming. Prescott didn't put it this way, but now we would: in 1977 we lacked a way to handle history dependence within a dynamic programming framework. Finding a recursive way to handle history dependence is a major achievement of the past 25 years and an important methodological theme of this book that opens the way to a variety of important applications.

We shall encounter important traces of the fascinating history of this topic in various chapters. Important contributors to the task of overcoming Prescott's challenge seemed to work in isolation from one another, being unaware of the complementary approaches being followed elsewhere. Important contributors included Shavell and Weiss (1979); Kydland and Prescott (1980); Miller and Salmon (1985); Pearlman; Currie and Levine (1985); Pearlman (1992), and Hansen, Epple, and Roberds (1985). These researchers achieved truly independent discoveries of the same important idea.

As we discuss in detail in chapter 19, one important approach amounted to putting a government costate vector on the costate equations of the private decision makers, then proceeding as usual to use optimal control for the government's problem. (A costate equation is a version of an Euler equation.) Solved forward, the costate equation depicts the dependence of private decisions on forecasts of future government policies that Prescott was worried about. The key idea in this approach was to formulate the government's problem by taking the costate equations of the private sector as additional *constraints* on the government's problem. These amount to promise-keeping constraints (they are cast in terms of derivatives of value functions, not value functions themselves, because costate vectors are gradients of value functions). After adding the costate equations of the private sector (the "followers") to the transition law of the government (the "leader"), one could then solve the government's problem by using dynamic programming as usual. One simply writes down a Bellman equation for the government planner taking the private sector costate variables as pseudo-state variables. Then it is almost business as usual (Gordon was correct!). We say "almost" because after the Bellman equation is solved, there is one more step: to pick the initial value of the private sector's costate. To maximize the government's criterion, this initial condition should be set to zero because initially there are no promises to keep. The government's optimal decision is a function of the natural state variable *and* the costate variables. The date t costate variables encode history and record the "cost" to the government at t of confirming the private sector's prior expectations about the government's time t decisions, expectations that were embedded in the private sector's decisions before t. The solution is time inconsistent (the government would always like to reinitialize the time t multiplier to zero and thereby discard past promises, but that is ruled out by the assumption that the government is committed to follow the optimal plan). See chapter 19 for many technical details, computer programs, and an application.

1.4.4. History dependence and "dynamic programming squared"

Rather than pursue the "costate on the costate" approach further, we now turn to a closely related approach that we illustrate in a dynamic contract design problem. While superficially different from the government policy design problem, the contract problem has many features in common with it. What is again needed is a recursive way to encode history dependence. Rather than use costate variables, we move up a derivative and work with promised values. This leads to value functions appearing inside value functions or "dynamic programming squared."

Define the history s^t of the Markov state by $s^t = [s_t, s_{t-1}, \ldots, s_0]$ and let $\pi_t(s^t)$ be the density over histories induced by π, π_0. Define a consumption allocation rule as a sequence of functions, the time component of which maps s^t into a choice of time t consumption, $c_t = \sigma_t(s^t)$, for $t \geq 0$. Let $c = \{\sigma_t(s^t)\}_{t=0}^\infty$. Define the (*ex ante*) value associated with an allocation rule as

$$v\left(c\right) = \sum_{t=0}^{\infty} \sum_{s^t} \beta^t u\left(\sigma_t\left(s^t\right)\right) \pi_t\left(s^t\right). \tag{1.4.5}$$

For each possible realization of the period zero state $\bar{s}_0$, there is a *continuation history* $s^t|_{\bar{s}_0}$. The observation that a continuation history is itself a complete history is our first hint that a recursive formulation is possible.[15] For each possible realization of the first period s_0, a consumption allocation rule implies a one-period *continuation consumption rule* $c|_{\bar{s}_0}$. A continuation consumption rule is itself a consumption rule that maps histories into time series of consumption. The one-period continuation history treats the time $t+1$ component of the original history evaluated at $\bar{s}_0$ as the time t component of the continuation history. The period t consumption of the one-period continuation consumption allocation conforms to the time $t+1$ component of original consumption allocation evaluated at $\bar{s}_0$. The time and state separability of (1.4.5) then allow us to represent $v(c)$ recursively as

$$v\left(c\right) = \sum_{s_0} \left[u\left(c_0\left(s_0\right)\right) + \beta v\left(c|_{s_0}\right)\right] \pi_0\left(s_0\right), \tag{1.4.6}$$

where $v(c|_{s_0})$ is the value of the continuation allocation. We call $v(c|_{s_0})$ the continuation value. In a special case that successive components of s_t are i.i.d.

---

[15] See chapters 8 and 23 for discussions of the recursive structure of histories.

and have a discrete distribution, we can write $(1.4.6)$ as

$$v = \sum_s \left[u\left(c_s\right) + \beta w_s \right] \Pi_s, \qquad (1.4.7)$$

where $\Pi_s = \mathrm{Prob}(y_t = \bar{y}_s)$ and $[\bar{y}_1 < \bar{y}_2 \cdots < \bar{y}_S]$ is a grid on which the endowment resides, c_s is consumption in state s given v, and w_s is the continuation value in state s given v. Here we use v in $(1.4.7)$ to denote what was $v(c)$ in $(1.4.6)$ and w_s to denote what was $v(c|s)$ in $(1.4.6)$.

So far this has all been for an arbitrary consumption plan. Evidently, the *ex ante* value v attained by an *optimal* consumption program must satisfy

$$v = \max_{\{c_s, w_s\}_{s=1}^S} \sum_s \left[u\left(c_s\right) + \beta w_s \right] \Pi_s \qquad (1.4.8)$$

where the maximization is subject to constraints that summarize the individual's opportunities to trade current state-contingent consumption c_s against future state-contingent continuation values w_s. In these problems, the value of v is an outcome that depends, in the savings problem for example, on the household's initial level of assets. In fact, for the savings problem with i.i.d. endowment shocks, the outcome is that v is a monotone function of A. This monotonicity allows the following remarkable representation. After solving for the optimal plan, use the monotone transformation to let v replace A as a state variable and represent the optimal decision rule in the form

$$c_s = f\left(v, s\right) \qquad (1.4.9a)$$
$$w_s = g\left(v, s\right). \qquad (1.4.9b)$$

The promised value v (a forward-looking variable if there ever was one) is also the variable that functions as an index of history in $(1.4.9)$. Equation $(1.4.9b)$ reminds us that v is a "backward looking" variable that registers the cumulative impact of past states s_t. The definition of v as a promised value, for example in $(1.4.8)$, tells us that v is also a forward-looking variable that encodes expectations (promises) about future consumption.

1.4.5. *Dynamic principal-agent problems*

The right side of (1.4.8) tells the terms on which the household is willing to trade current utility for continuation utility. Models that confront enforcement and information problems use the trade-off identified by (1.4.8) to design intertemporal consumption plans that optimally balance risk sharing and intertemporal consumption smoothing against the need to offer correct incentives. Next we turn to such models.

We remove the household from the market and hand it over to a planner or principal who offers the household a contract that the planner designs to deliver an *ex ante* promised value v subject to enforcement or information constraints.[16] Now v becomes a state variable that occurs in the *planner's* value function. We assume that the only way the household can transfer his endowment over time is to deal with the planner. The saving or borrowing technology (1.3.2) is no longer available to the agent, though it might be to the planner. We continue to consider the i.i.d. case mentioned above. Let $P(v)$ be the *ex ante* optimal value of the planner's problem. The presence of a value function (for the agents) as an argument of the value function of the principal causes us sometimes to speak of "dynamic programming squared." The planner "earns" $y_t - c_t$ from the agent at time t by commandeering the agent's endowment but returning consumption c_t. The value function $P(v)$ for a planner who must deliver promised value v satisfies

$$P(v) = \max_{\{c_s, w_s\}_{s=1}^S} [\overline{y}_s - c_s + \beta P(w_s)] \, \Pi_s, \qquad (1.4.10)$$

where the maximization is subject to the promise-keeping constraint (1.4.7) and some other constraints that depend on details of the problem, as we indicate shortly. The other constraints are context-specific incentive-compatibility constraints and describe the best response of the agent to the arrangement offered by the principal. Condition (1.4.7) is a *promise-keeping* constraint. The planner is constrained to provide a vector of $\{c_s, w_s\}_{s=1}^S$ that delivers the value v.

We briefly describe two types of contract design problems and the constraints that confront the planner because of the opportunities that the environment grants the agent.

To model the problem of enforcement without an information problem, assume that while the planner can observe y_t each period, the household always

---

[16] Here we are sticking close to two models of Thomas and Worrall (1988, 1990).

has the option of consuming its endowment y_t and receiving an *ex ante* continuation value v_{aut} with which to enter the next period, where v_{aut} is the *ex ante* value the consumer receives by always consuming his endowment. The consumer's freedom to walk away induces the planner to structure the insurance contract so that it is never in the household's interest to defect from the contract (the contract must be "self-enforcing"). A self-enforcing contract requires that the following participation constraints be satisfied:

$$u\left(c_s\right) + \beta w_s \geq u\left(\overline{y}_s\right) + \beta v_{\text{aut}} \quad \forall s. \tag{1.4.11}$$

A self-enforcing contract provides imperfect insurance when occasionally some of these participation constraints are binding. When they are binding, the planner sacrifices consumption smoothing in the interest of providing incentives for the contract to be self-enforcing.

An alternative specification eliminates the enforcement problem by assuming that once the household enters the contract, it does not have the option to walk away. A planner wants to supply insurance to the household in the most efficient way, but now the planner cannot observe the household's endowment. The planner must trust the household to report its endowment. It is assumed that the household will truthfully report its endowment only if it wants to. This leads the planner to add to the promise-keeping constraint (1.4.7) the following truth-telling constraints:

$$u\left(c_s\right) + \beta w_s \geq u\left(\overline{y}_s - \overline{y}_\tau + c_\tau\right) + \beta w_\tau \quad \forall\left(s, \tau\right), \tag{1.4.12}$$

where constraint (1.4.12) pertains to a situation when the household's true endowment is $\overline{y}_s$ but the household considers to falsely report that the endowment instead is $\overline{y}_\tau$. The left and right sides of (1.4.12) are the utility of telling the truth and lying, respectively. If the household (falsely) reports $\overline{y}_\tau$, the planner awards the household a net transfer $c_\tau - \overline{y}_\tau$ and a continuation value w_τ. If (1.4.12) holds for all τ, the household will always choose to report the true state s.

As we shall see in chapters 20 and 21, the planner elicits truthful reporting by manipulating how continuation values vary with the reported state. Households that report a low income today might receive a transfer today, but they suffer an adverse consequence by getting a diminished continuation value starting tomorrow. The planner structures this menu of choices so that only low-endowment households, those that badly want a transfer today, are willing to accept the diminished continuation value that is the consequence of reporting that low income today.

At this point, a supermartingale convergence theorem raises its ugly head again. But this time it propels consumption and continuation utility *downward*. The super martingale result leads to what some people have termed the "immiseration" property of models in which dynamic contracts are used to deliver incentives to reveal information.

To enhance our appreciation for the immiseration result, we now touch on another aspect of macroeconomic's love-hate affair with the Euler inequality (1.3.3). In both of the incentive models just described, one with an enforcement problem, the other with an information problem, it is important that the household not have access to a good risk-free investment technology like that represented in the constraint (1.3.2) that makes (1.3.3) the appropriate first-order condition in the savings problem. Indeed, especially in the model with limited information, the planner makes ample use of his ability to reallocate consumption intertemporally in ways that can violate (1.3.2) in order to elicit accurate information from the household. In chapter 20, we shall follow Cole and Kocherlakota (2001) by allowing the household to save (but not to *dissave*) a risk-free asset that bears fixed gross interest rate $R = \beta^{-1}$. The Euler inequality comes back into play and alters the character of the insurance arrangement so that outcomes resemble ones that occur in a Bewley model, provided that the debt limit in the Bewley model is chosen appropriately.

1.4.6. More applications

We shall study many more applications of dynamic programming and dynamic programming squared, including models of search in labor markets, reputation and credible public policy, gradualism in trade policy, unemployment insurance, and monetary economies. It is time to get to work seriously studying the mathematical and economic tools that we need to approach these exciting topics. Let us begin.

Part II
Tools

Chapter 2
Time Series

2.1. Two workhorses

This chapter describes two tractable models of time series: Markov chains and first-order stochastic linear difference equations. These models are organizing devices that put restrictions on a sequence of random vectors. They are useful because they describe a time series with parsimony. In later chapters, we shall make two uses each of Markov chains and stochastic linear difference equations: (1) to represent the exogenous information flows impinging on an agent or an economy, and (2) to represent an optimum or equilibrium outcome of agents' decision making. The Markov chain and the first-order stochastic linear difference both use a sharp notion of a state vector. A state vector summarizes the information about the current position of a system that is relevant for determining its future. The Markov chain and the stochastic linear difference equation will be useful tools for studying dynamic optimization problems.

2.2. Markov chains

A stochastic process is a sequence of random vectors. For us, the sequence will be ordered by a time index, taken to be the integers in this book. So we study discrete time models. We study a discrete-state stochastic process with the following property:

MARKOV PROPERTY: A stochastic process $\{x_t\}$ is said to have the *Markov property* if for all $k \geq 1$ and all t,

$$\text{Prob}\left(x_{t+1}|x_t, x_{t-1}, \ldots, x_{t-k}\right) = \text{Prob}\left(x_{t+1}|x_t\right).$$

We assume the Markov property and characterize the process by a *Markov chain*. A time-invariant Markov chain is defined by a triple of objects, namely, an n-dimensional state space consisting of vectors $e_i, i = 1, \ldots, n$, where e_i is an $n \times 1$ unit vector whose ith entry is 1 and all other entries are zero; an $n \times n$ *transition matrix* P, which records the probabilities of moving from one value of the state to another in one period; and an $(n \times 1)$ vector π_0 whose ith element is the probability of being in state i at time 0: $\pi_{0i} = \text{Prob}(x_0 = e_i)$. The elements of matrix P are

$$P_{ij} = \text{Prob}\left(x_{t+1} = e_j | x_t = e_i\right).$$

For these interpretations to be valid, the matrix P and the vector π_0 must satisfy the following assumption:

ASSUMPTION M:

a. For $i = 1, \ldots, n$, the matrix P satisfies

$$\sum_{j=1}^{n} P_{ij} = 1. \qquad (2.2.1)$$

b. The vector π_0 satisfies

$$\sum_{i=1}^{n} \pi_{0i} = 1.$$

A matrix P that satisfies property $(2.2.1)$ is called a *stochastic matrix*. A stochastic matrix defines the probabilities of moving from one value of the state to another in one period. The probability of moving from one value of the state to another in *two* periods is determined by P^2 because

$$\text{Prob}\left(x_{t+2} = e_j | x_t = e_i\right)$$

$$= \sum_{h=1}^{n} \text{Prob}\left(x_{t+2} = e_j | x_{t+1} = e_h\right) \text{Prob}\left(x_{t+1} = e_h | x_t = e_i\right)$$

$$= \sum_{h=1}^{n} P_{ih} P_{hj} = P_{ij}^{(2)},$$

where $P_{ij}^{(2)}$ is the i, j element of P^2. Let $P_{i,j}^{(k)}$ denote the i, j element of P^k. By iterating on the preceding equation, we discover that

$$\text{Prob}\left(x_{t+k} = e_j | x_t = e_i\right) = P_{ij}^{(k)}.$$

The unconditional probability distributions of x_t are determined by

$$\pi_1' = \text{Prob}\,(x_1) = \pi_0'P$$
$$\pi_2' = \text{Prob}\,(x_2) = \pi_0'P^2$$
$$\vdots$$
$$\pi_k' = \text{Prob}\,(x_k) = \pi_0'P^k,$$

where $\pi_t' = \text{Prob}(x_t)$ is the $(1 \times n)$ vector whose ith element is $\text{Prob}(x_t = e_i)$.

2.2.1. Stationary distributions

Unconditional probability distributions evolve according to

$$\pi_{t+1}' = \pi_t'P. \tag{2.2.2}$$

An unconditional distribution is called *stationary* or *invariant* if it satisfies

$$\pi_{t+1} = \pi_t,$$

that is, if the unconditional distribution remains unaltered with the passage of time. From the law of motion (2.2.2) for unconditional distributions, a stationary distribution must satisfy

$$\pi' = \pi'P \tag{2.2.3}$$

or

$$\pi'\,(I - P) = 0.$$

Transposing both sides of this equation gives

$$(I - P')\,\pi = 0, \tag{2.2.4}$$

which determines π as an eigenvector (normalized to satisfy $\sum_{i=1}^n \pi_i = 1$) associated with a unit eigenvalue of P'. We say that P, π is a *stationary Markov chain* if the initial distribution π is such that (2.2.3) holds.

The fact that P is a stochastic matrix (i.e., it has nonnegative elements and satisfies $\sum_j P_{ij} = 1$ for all i) guarantees that P has at least one unit eigenvalue, and that there is at least one eigenvector π that satisfies equation

(2.2.4). This stationary distribution may not be unique because P can have a repeated unit eigenvalue.

Example 1. A Markov chain

$$P = \begin{bmatrix} 1 & 0 & 0 \\ .2 & .5 & .3 \\ 0 & 0 & 1 \end{bmatrix}$$

has two unit eigenvalues with associated stationary distributions $\pi' = \begin{bmatrix} 1 & 0 & 0 \end{bmatrix}$ and $\pi' = \begin{bmatrix} 0 & 0 & 1 \end{bmatrix}$. Here states 1 and 3 are both *absorbing* states. Furthermore, any initial distribution that puts zero probability on state 2 is a stationary distribution. See exercises *2.10* and *2.11*.

Example 2. A Markov chain

$$P = \begin{bmatrix} .7 & .3 & 0 \\ 0 & .5 & .5 \\ 0 & .9 & .1 \end{bmatrix}$$

has one unit eigenvalue with associated stationary distribution $\pi' = \begin{bmatrix} 0 & .6429 & .3571 \end{bmatrix}$. Here states 2 and 3 form an *absorbing subset* of the state space.

2.2.2. Asymptotic stationarity

We often ask the following question about a Markov process: for an arbitrary initial distribution π_0, do the unconditional distributions π_t approach a stationary distribution

$$\lim_{t \to \infty} \pi_t = \pi_\infty,$$

where π_∞ solves equation (2.2.4)? If the answer is yes, then does the limit distribution π_∞ depend on the initial distribution π_0? If the limit π_∞ is independent of the initial distribution π_0, we say that the process is *asymptotically stationary with a unique invariant distribution*. We call a solution π_∞ a *stationary distribution* or an *invariant distribution* of P.

We state these concepts formally in the following definition:

Definition 2.2.1. Let π_∞ be a unique vector that satisfies $(I - P')\pi_\infty = 0$. If for all initial distributions π_0 it is true that $P^{t'}\pi_0$ converges to the same

π_∞, we say that the Markov chain is asymptotically stationary with a unique invariant distribution.

The following theorems can be used to show that a Markov chain is asymptotically stationary.

Theorem 2.2.1. *Let P be a stochastic matrix with $P_{ij} > 0 \ \forall (i, j)$. Then P has a unique stationary distribution, and the process is asymptotically stationary.*

Theorem 2.2.2. *Let P be a stochastic matrix for which $P_{ij}^n > 0 \ \forall (i, j)$ for some value of $n \geq 1$. Then P has a unique stationary distribution, and the process is asymptotically stationary.*

The conditions of Theorem 2.2.1 (and Theorem 2.2.2) state that from any state there is a positive probability of moving to any other state in one (or n) steps. Please note that some of the examples below will violate the conditions of Theorem 2.2.2 for any n.

2.2.3. Forecasting the state

The minimum mean squared error forecast of the state next period is the conditional mathematical expectation:

$$E\left[x_{t+1} | x_t = e_i\right] = \begin{bmatrix} P_{i1} \\ P_{i2} \\ \vdots \\ P_{in} \end{bmatrix} = P' e_i = P'_{i, \cdot}. \tag{2.2.5}$$

where $P'_{i, \cdot}$ denotes the transpose of the ith row of the matrix P. In section B.2 of this book's appendix B, we use this equation to motivate the following first-order stochastic difference equation for the state:

$$x_{t+1} = P' x_t + v_{t+1} \tag{2.2.6}$$

where v_{t+1} is a random disturbance that evidently satisfies $E[v_{t+1} | x_t] = 0$.

Now let $\overline{y}$ be an $n \times 1$ vector of real numbers and define $y_t = \overline{y}' x_t$, so that $y_t = \overline{y}_i$ if $x_t = e_i$. Evidently, we can write

$$y_{t+1} = \overline{y}' P' x_t + \overline{y}' v_{t+1}. \tag{2.2.7}$$

The pair of equations (2.2.6), (2.2.7) becomes a simple example of a hidden Markov model when the observation y_t is too coarse to reveal the state. See section B.2 of technical appendix B for a discussion of such models.

2.2.4. Forecasting functions of the state

From the conditional and unconditional probability distributions that we have listed, it follows that the unconditional expectations of y_t for $t \geq 0$ are determined by $E y_t = (\pi_0' P^t) \overline{y}$. Conditional expectations are determined by

$$E\left(y_{t+1} | x_t = e_i\right) = \sum_j P_{ij} \overline{y}_j = (P\overline{y})_i \tag{2.2.8}$$

$$E\left(y_{t+2} | x_t = e_i\right) = \sum_k P_{ik}^{(2)} \overline{y}_k = \left(P^2 \overline{y}\right)_i \tag{2.2.9}$$

and so on, where $P_{ik}^{(2)}$ denotes the (i,k) element of P^2 and $(\cdot)_i$ denotes the ith row of the matrix $(\cdot)$. An equivalent formula from (2.2.6), (2.2.7) is $E[y_{t+1} | x_t] = \overline{y}' P' x_t = x_t' P \overline{y}$, which equals $(P\overline{y})_i$ when $x_t = e_i$. Notice that

$$E\left[E\left(y_{t+2} | x_{t+1} = e_j\right) | x_t = e_i\right] = \sum_j P_{ij} \sum_k P_{jk} \overline{y}_k$$

$$= \sum_k \left(\sum_j P_{ij} P_{jk}\right) \overline{y}_k = \sum_k P_{ik}^{(2)} \overline{y}_k = E\left(y_{t+2} | x_t = e_i\right).$$

Connecting the first and last terms in this string of equalities yields $E[E(y_{t+2}|x_{t+1})|x_t] = E[y_{t+2}|x_t]$. This is an example of the "law of iterated expectations." The law of iterated expectations states that for any random variable z and two information sets J, I with $J \subset I$, $E[E(z|I)|J] = E(z|J)$. As another example of the law of iterated expectations, notice that

$$E y_1 = \sum_j \pi_{1,j} \overline{y}_j = \pi_1' \overline{y} = (\pi_0' P) \overline{y} = \pi_0' (P\overline{y})$$

and that

$$E\left[E\left(y_1 | x_0 = e_i\right)\right] = \sum_i \pi_{0,i} \sum_j P_{ij} \overline{y}_j = \sum_j \left(\sum_i \pi_{0,i} P_{ij}\right) \overline{y}_j = \pi_1' \overline{y} = E y_1.$$

2.2.5. Forecasting functions

There are powerful formulas for forecasting functions of a Markov state. Again, let $\overline{y}$ be an $n \times 1$ vector and consider the random variable $y_t = \overline{y}'x_t$. Then

$$E\left[y_{t+k}|x_t = e_i\right] = \left(P^k \overline{y}\right)_i$$

where $\left(P^k \overline{y}\right)_i$ denotes the ith row of $P^k \overline{y}$. Stacking all n rows together, we express this as

$$E\left[y_{t+k}|x_t\right] = P^k \overline{y}. \tag{2.2.10}$$

We also have

$$\sum_{k=0}^{\infty} \beta^k E\left[y_{t+k}|x_t = \overline{e}_i\right] = \left[(I - \beta P)^{-1} \overline{y}\right]_i,$$

where $\beta \in (0,1)$ guarantees existence of $(I - \beta P)^{-1} = (I + \beta P + \beta^2 P^2 + \cdots)$.

2.2.6. Enough one-step-ahead forecasts determine P

One-step-ahead forecasts of a sufficiently rich set of random variables characterize a Markov chain. In particular, one-step-ahead conditional expectations of n independent functions (i.e., n linearly independent vectors $h_1, \ldots, h_n$) uniquely determine the transition matrix P. Thus, let $E[h_{k,t+1}|x_t = e_i] = (Ph_k)_i$. We can collect the conditional expectations of h_k for all initial states i in an $n \times 1$ vector $E[h_{k,t+1}|x_t] = Ph_k$. We can then collect conditional expectations for the n independent vectors $h_1, \ldots, h_n$ as $Ph = J$ where $h = [\, h_1 \quad h_2 \quad \ldots \quad h_n \,]$ and J is the $n \times n$ matrix consisting of all conditional expectations of all n vectors $h_1, \ldots, h_n$. If we know h and J, we can determine P from $P = Jh^{-1}$.

2.2.7. Invariant functions and ergodicity

Let P, π be a stationary n-state Markov chain with the state space $X = [e_i, i = 1, \ldots, n]$. An $n \times 1$ vector $\bar{y}$ defines a random variable $y_t = \bar{y}' x_t$. Let $E[y_\infty | x_0]$ be the expectation of y_s for s very large, conditional on the initial state. The following is a useful precursor to a law of large numbers:

Theorem 2.2.3. *Let $\bar{y}$ define a random variable as a function of an underlying state x, where x is governed by a stationary Markov chain (P, π). Then*

$$\frac{1}{T} \sum_{t=1}^{T} y_t \to E[y_\infty | x_0] \tag{2.2.11}$$

with probability 1.

To illustrate Theorem 2.2.3, consider the following example:

Example: Consider the Markov chain $P = \begin{bmatrix} 1 & 0 \\ 0 & 1 \end{bmatrix}$, $\pi_0 = \begin{bmatrix} p \\ (1-p) \end{bmatrix}$ for $p \in (0, 1)$. Consider the random variable $y_t = \bar{y}' x_t$ where $\bar{y} = \begin{bmatrix} 10 \\ 0 \end{bmatrix}$. The chain has two possible sample paths, $y_t = 10, t \geq 0$, which occurs with probability p and $y_t = 0, t \geq 0$, which occurs with probability $1 - p$. Thus, $\frac{1}{T} \sum_{t=1}^{T} y_t \to 10$ with probability p and $\frac{1}{T} \sum_{t=1}^{T} y_t \to 0$ with probability $(1-p)$.

The outcomes in this example indicate why we might want something more than (2.2.11). In particular, we would like to be free to replace $E[y_\infty | x_0]$ with the constant unconditional mean $E[y_t] = E[y_0]$ associated with the stationary distribution π. To get this outcome, we must strengthen what we assume about P by using the following concepts.

Suppose that (P, π) is a stationary Markov chain. Imagine repeatedly drawing x_0 from π and then generating $x_t, t \geq 1$ by successively drawing from transition densities given by the matrix P. We use

Definition 2.2.2. A random variable $y_t = \bar{y}' x_t$ is said to be *invariant* if $y_t = y_0, t \geq 0$, for all realizations of $x_t, t \geq 0$ that occur with positive probability under (P, π).

Thus, a random variable y_t is invariant (or "an invariant function of the state") if it remains constant at y_0 while the underlying state x_t moves through the state space X. Notice how the definition leaves open the possibility that y_0

itself might differ across sample paths indexed by different draws of the initial condition x_0 from the initial (and stationary) density π.

The stationary Markov chain (P, π) induces a joint density $f(x_{t+1}, x_t)$ over (x_{t+1}, x_t) that is independent of calendar time t; P, π and the definition $y_t = \overline{y}' x_t$ also induce a joint density $f_y(y_{t+1}, y_t)$ that is independent of calendar time. In what follows, we compute mathematical expectations with respect to the joint density $f_y(y_{t+1}, y_t)$.

For a finite-state Markov chain, the following theorem gives a convenient way to characterize invariant functions of the state.

Theorem 2.2.4. *Let (P, π) be a stationary Markov chain. If*

$$E\left[y_{t+1}|x_t\right] = y_t \tag{2.2.12}$$

then the random variable $y_t = \overline{y}' x_t$ is invariant.

Proof. By using the law of iterated expectations, notice that

$$
\begin{aligned}
E\left(y_{t+1} - y_t\right)^2 &= E\left[E\left(y_{t+1}^2 - 2y_{t+1}y_t + y_t^2\right)|x_t\right] \\
&= E\left[Ey_{t+1}^2|x_t - 2E\left(y_{t+1}|x_t\right)y_t + Ey_t^2|x_t\right] \\
&= Ey_{t+1}^2 - 2Ey_t^2 + Ey_t^2 \\
&= 0
\end{aligned}
$$

where the middle term on the right side of the second line uses that $E[y_t|x_t] = y_t$, the middle term on the right side of the third line uses the hypothesis (2.2.12), and the third line uses the hypothesis that π is a stationary distribution. In a finite Markov chain, if $E(y_{t+1} - y_t)^2 = 0$, then $y_{t+1} = y_t$ for all y_{t+1}, y_t that occur with positive probability under the stationary distribution. ∎

As we shall have reason to study in chapters 17 and 18, *any* (not necessarily stationary) stochastic process y_t that satisfies (2.2.12) is said to be a *martingale*. Theorem 2.2.4 tells us that a martingale that is a function of a finite-state stationary Markov state x_t must be constant over time. This result is a special case of the martingale convergence theorem that underlies some remarkable results about savings to be studied in chapter 17.[1]

---

[1] Theorem 2.2.4 tells us that a stationary martingale process has so little freedom to move that it has to be constant forever, not just eventually, as asserted by the martingale convergence theorem.

Equation (2.2.12) can be expressed as $P\overline{y} = \overline{y}$ or

$$(P - I)\overline{y} = 0, \tag{2.2.13}$$

which states that an invariant function of the state is a (right) eigenvector of P associated with a unit eigenvalue. Thus, associated with unit eigenvalues of P are (1) left eigenvectors that are stationary distributions of the chain (recall equation (2.2.4)), and (2) right eigenvectors that are invariant functions of the chain (from equation (2.2.13)).

Definition 2.2.3. Let (P, π) be a stationary Markov chain. The chain is said to be *ergodic* if the only invariant functions $\overline{y}$ are constant with probability 1 under the stationary unconditional probability distribution π, i.e., $\overline{y}_i = \overline{y}_j$ for all i, j with $\pi_i > 0, \pi_j > 0$.

REMARK: Let $\tilde{\pi}^{(1)}, \tilde{\pi}^{(2)}, \ldots, \tilde{\pi}^{(m)}$ be m distinct 'basis' stationary distributions for an n state Markov chain with transition matrix P. Each $\tilde{\pi}^{(k)}$ is an $(n \times 1)$ left eigenvector of P associated with a distinct unit eigenvalue. Each $\pi^{(j)}$ is scaled to be a probability vector (i.e., its components are nonnegative and sum to unity). The set S of *all* stationary distributions is convex. An element $\pi_b \in S$ can be represented as

$$\pi_b = b_1 \tilde{\pi}^{(1)} + b_2 \tilde{\pi}^{(2)} + \cdots + b_m \tilde{\pi}^{(m)},$$

where $b_j \geq 0, \sum_j b_j = 1$ is a probability vector.

REMARK: A stationary density π_b for which the pair (P, π_b) is an ergodic Markov chain is an extreme point of the convex set S, meaning that it can be represented as $\pi_b = \tilde{\pi}^{(j)}$ for one of the 'basis' stationary densities.

A law of large numbers for Markov chains is:

Theorem 2.2.5. *Let $\overline{y}$ define a random variable on a stationary and ergodic Markov chain (P, π). Then*

$$\frac{1}{T} \sum_{t=1}^{T} y_t \to E[y_0] \tag{2.2.14}$$

with probability 1.

This theorem tells us that the time series average converges to the population mean of the stationary distribution.

Three examples illustrate these concepts.

Example 1. A chain with transition matrix $P = \begin{bmatrix} 0 & 1 \\ 1 & 0 \end{bmatrix}$ has a unique stationary distribution $\pi = [.5 \quad .5]'$ and the invariant functions are $[\alpha \quad \alpha]'$ for any scalar α. Therefore, the process is ergodic and Theorem 2.2.5 applies.

Example 2. A chain with transition matrix $P = \begin{bmatrix} 1 & 0 \\ 0 & 1 \end{bmatrix}$ has a continuum of stationary distributions $\gamma \begin{bmatrix} 1 \\ 0 \end{bmatrix} + (1 - \gamma) \begin{bmatrix} 0 \\ 1 \end{bmatrix}$ for any $\gamma \in [0, 1]$ and invariant functions $\begin{bmatrix} 0 \\ \alpha_1 \end{bmatrix}$ and $\begin{bmatrix} \alpha_2 \\ 0 \end{bmatrix}$ for any scalars α_1, α_2. Therefore, the process is not ergodic when $\gamma \in (0, 1)$, for note that neither invariant function is constant across states that receive positive probability according to a stationary distribution associated with $\gamma \in (0, 1)$. Therefore, the conclusion $(2.2.14)$ of Theorem 2.2.5 does not hold for an initial stationary distribution associated with $\gamma \in (0, 1)$, although the weaker result Theorem 2.2.3 does hold. When $\gamma \in (0, 1)$, nature chooses state $i = 1$ or $i = 2$ with probabilities $\gamma, 1 - \gamma$, respectively, at time 0. Thereafter, the chain remains stuck in the realized time 0 state. Its failure ever to visit the unrealized state prevents the sample average from converging to the population mean of an arbitrary function $\bar{y}$ of the state. Notice that conclusion $(2.2.14)$ of Theorem 2.2.5 does hold for the stationary distributions associated with $\gamma = 0$ and $\gamma = 1$.

Example 3. A chain with transition matrix $P = \begin{bmatrix} .8 & .2 & 0 \\ .1 & .9 & 0 \\ 0 & 0 & 1 \end{bmatrix}$ has a continuum of stationary distributions $\gamma \begin{bmatrix} \frac{1}{3} & \frac{2}{3} & 0 \end{bmatrix}' + (1 - \gamma) \begin{bmatrix} 0 & 0 & 1 \end{bmatrix}'$ for $\gamma \in [0, 1]$ and invariant functions $\alpha_1 \begin{bmatrix} 1 & 1 & 0 \end{bmatrix}'$ and $\alpha_2 \begin{bmatrix} 0 & 0 & 1 \end{bmatrix}'$ for any scalars α_1, α_2. The conclusion $(2.2.14)$ of Theorem 2.2.5 does not hold for the stationary distributions associated with $\gamma \in (0, 1)$, but Theorem 2.2.3 does hold. But again, conclusion $(2.2.14)$ does hold for the stationary distributions associated with $\gamma = 0$ and $\gamma = 1$.

2.2.8. *Simulating a Markov chain*

It is easy to simulate a Markov chain using a random number generator. The Matlab program `markov.m` does the job. We'll use this program in some later chapters.[2]

2.2.9. *The likelihood function*

Let P be an $n \times n$ stochastic matrix with states $1, 2, \ldots, n$. Let π_0 be an $n \times 1$ vector with nonnegative elements summing to 1, with $\pi_{0,i}$ being the probability that the state is i at time 0. Let i_t index the state at time t. The Markov property implies that the probability of drawing the path $(x_0, x_1, \ldots, x_{T-1}, x_T) = (\overline{e}_{i_0}, \overline{e}_{i_1}, \ldots, \overline{e}_{i_{T-1}}, \overline{e}_{i_T})$ is

$$
\begin{aligned}
L &\equiv \mathrm{Prob}\left(\overline{x}_{i_T}, \overline{x}_{i_{T-1}}, \ldots, \overline{x}_{i_1}, \overline{x}_{i_0}\right) \\
&= P_{i_{T-1}, i_T} P_{i_{T-2}, i_{T-1}} \cdots P_{i_0, i_1} \pi_{0, i_0}.
\end{aligned}
\tag{2.2.15}
$$

The probability L is called the *likelihood*. It is a function of both the sample realization $x_0, \ldots, x_T$ and the parameters of the stochastic matrix P. For a sample $x_0, x_1, \ldots, x_T$, let n_{ij} be the number of times that there occurs a one-period transition from state i to state j. Then the likelihood function can be written

$$
L = \pi_{0, i_0} \prod_i \prod_j P_{i,j}^{n_{ij}},
$$

a *multinomial* distribution.

Formula $(2.2.15)$ has two uses. A first, which we shall encounter often, is to describe the probability of alternative histories of a Markov chain. In chapter 8, we shall use this formula to study prices and allocations in competitive equilibria.

A second use is for estimating the parameters of a model whose solution is a Markov chain. Maximum likelihood estimation for free parameters θ of a Markov process works as follows. Let the transition matrix P and the initial distribution π_0 be functions $P(\theta), \pi_0(\theta)$ of a vector of free parameters θ. Given a sample $\{x_t\}_{t=0}^T$, regard the likelihood function as a function of the parameters θ. As the estimator of θ, choose the value that maximizes the likelihood function L.

---

[2] An index in the back of the book lists Matlab programs.

2.3. Continuous-state Markov chain

In chapter 8, we shall use a somewhat different notation to express the same ideas. This alternative notation can accommodate either discrete- or continuous-state Markov chains. We shall let S denote the state space with typical element $s \in S$. Let state transitions be described by the cumulative distribution function $\Pi(s'|s) = \text{Prob}(s_{t+1} \leq s'|s_t = s)$ and let the initial state s_0 be described by the cumulative distribution function $\Pi_o(s) = \text{Prob}(s_0 \leq s)$. The *transition density* is $\pi(s'|s) = \frac{d}{ds'}\Pi(s'|s)$ and the initial density is $\pi_0(s) = \frac{d}{ds}\Pi_0(s)$. For all $s \in S, \pi(s'|s) \geq 0$ and $\int_{s'} \pi(s'|s)ds' = 1$; also $\int_s \pi_0(s)ds = 1$.[3] Corresponding to (2.2.15), the likelihood function or density over the history $s^t = [s_t, s_{t-1}, \ldots, s_0]$ is

$$\pi\left(s^t\right) = \pi\left(s_t|s_{t-1}\right) \cdots \pi\left(s_1|s_0\right) \pi_0\left(s_0\right). \qquad (2.3.1)$$

For $t \geq 1$, the time t unconditional distributions evolve according to

$$\pi_t\left(s_t\right) = \int_{s_{t-1}} \pi\left(s_t|s_{t-1}\right) \pi_{t-1}\left(s_{t-1}\right) d\, s_{t-1}.$$

A stationary or *invariant* distribution satisfies

$$\pi_\infty\left(s'\right) = \int_s \pi\left(s'|s\right) \pi_\infty\left(s\right) d\, s,$$

which is the counterpart to (2.2.3).

DEFINITION: A Markov chain $\left(\pi(s'|s), \pi_0(s)\right)$ is said to be *stationary* if π_0 satisfies

$$\pi_0\left(s'\right) = \int_s \pi\left(s'|s\right) \pi_0\left(s\right) d\, s.$$

DEFINITION: Paralleling our discussion of finite-state Markov chains, we can say that the function $\phi(s)$ is *invariant* if

$$\int \phi\left(s'\right) \pi\left(s'|s\right) ds' = \phi\left(s\right).$$

A stationary continuous-state Markov process is said to be *ergodic* if the only invariant functions $\phi(s')$ are constant with probability 1 under the stationary distribution π_∞.

---

[3] Thus, when S is discrete, $\pi(s_j|s_i)$ corresponds to $P_{i,j}$ in our earlier notation.

A law of large numbers for Markov processes states:

Theorem 2.3.1. *Let $y(s)$ be a random variable, a measurable function of s, and let $\big(\pi(s'|s), \pi_0(s)\big)$ be a stationary and ergodic continuous-state Markov process. Assume that $E|y| < +\infty$. Then*

$$\frac{1}{T} \sum_{t=1}^{T} y_t \to Ey = \int y(s)\, \pi_0(s)\, ds$$

with probability 1 with respect to the distribution π_0.

2.4. Stochastic linear difference equations

The first-order linear vector stochastic difference equation is a useful example of a continuous-state Markov process. Here we use $x_t \in I\!R^n$ rather than s_t to denote the time t state and specify that the initial distribution $\pi_0(x_0)$ is Gaussian with mean μ_0 and covariance matrix Σ_0, and that the transition density $\pi(x'|x)$ is Gaussian with mean $A_o x$ and covariance CC'.[4] This specification pins down the joint distribution of the stochastic process $\{x_t\}_{t=0}^{\infty}$ via formula (2.3.1). The joint distribution determines all moments of the process.

  This specification can be represented in terms of the first-order stochastic linear difference equation

$$x_{t+1} = A_o x_t + C w_{t+1} \tag{2.4.1}$$

for $t = 0, 1, \ldots$, where x_t is an $n \times 1$ state vector, x_0 is a random initial condition drawn from a probability distribution with mean $Ex_0 = \mu_0$ and covariance matrix $E(x_0 - \mu_0)(x_0 - \mu_0)' = \Sigma_0$, A_o is an $n \times n$ matrix, C is an $n \times m$ matrix, and w_{t+1} is an $m \times 1$ vector satisfying the following:

ASSUMPTION A1: w_{t+1} is an i.i.d. process satisfying $w_{t+1} \sim \mathcal{N}(0, I)$.

---

[4] An $n \times 1$ vector z that is multivariate normal has the density function

$$f(z) = (2\pi)^{-.5n} |\Sigma|^{-.5} \exp\left(-.5(z-\mu)' \Sigma^{-1} (z-\mu)\right)$$

where $\mu = Ez$ and $\Sigma = E(z-\mu)(z-\mu)'$.

We can weaken the Gaussian assumption A1. To focus only on first and second moments of the x process, it is sufficient to make the weaker assumption:

ASSUMPTION A2: w_{t+1} is an $m \times 1$ random vector satisfying:

$$Ew_{t+1}|J_t = 0 \qquad (2.4.2a)$$

$$Ew_{t+1}w'_{t+1}|J_t = I, \qquad (2.4.2b)$$

where $J_t = [\, w_t \ \cdots \ w_1 \ x_0 \,]$ is the information set at t, and $E[\,\cdot\,|J_t]$ denotes the conditional expectation. We impose no distributional assumptions beyond $(2.4.2)$. A sequence $\{w_{t+1}\}$ satisfying equation $(2.4.2a)$ is said to be a martingale difference sequence adapted to J_t. A sequence $\{z_{t+1}\}$ that satisfies $E[z_{t+1}|J_t] = z_t$ is said to be a martingale adapted to J_t.

An even weaker assumption is

ASSUMPTION A3: w_{t+1} is a process satisfying

$$Ew_{t+1} = 0$$

for all t and

$$Ew_t w'_{t-j} = \begin{cases} I, & \text{if } j = 0; \\ 0, & \text{if } j \neq 0. \end{cases}$$

A process satisfying assumption A3 is said to be a vector "white noise." [5]

Assumption A1 or A2 implies assumption A3 but not vice versa. Assumption A1 implies assumption A2 but not vice versa. Assumption A3 is sufficient to justify the formulas that we report below for second moments. We shall often append an observation equation $y_t = Gx_t$ to equation $(2.4.1)$ and deal with the augmented system

$$x_{t+1} = A_o x_t + C w_{t+1} \qquad (2.4.3a)$$

$$y_t = Gx_t. \qquad (2.4.3b)$$

Here y_t is a vector of variables observed at t, which may include only some linear combinations of x_t. The system $(2.4.3)$ is often called a linear *state-space system*.

---

[5] Note that $(2.4.2a)$ by itself allows the distribution of w_{t+1} conditional on J_t to be heteroskedastic.

Example 1. Scalar second-order autoregression: Assume that z_t and w_t are scalar processes and that

$$z_{t+1} = \alpha + \rho_1 z_t + \rho_2 z_{t-1} + w_{t+1}.$$

Represent this relationship as the system

$$\begin{bmatrix} z_{t+1} \\ z_t \\ 1 \end{bmatrix} = \begin{bmatrix} \rho_1 & \rho_2 & \alpha \\ 1 & 0 & 0 \\ 0 & 0 & 1 \end{bmatrix} \begin{bmatrix} z_t \\ z_{t-1} \\ 1 \end{bmatrix} + \begin{bmatrix} 1 \\ 0 \\ 0 \end{bmatrix} w_{t+1}$$

$$z_t = \begin{bmatrix} 1 & 0 & 0 \end{bmatrix} \begin{bmatrix} z_t \\ z_{t-1} \\ 1 \end{bmatrix}$$

which has form (2.4.3).

Example 2. First-order scalar mixed moving average and autoregression: Let

$$z_{t+1} = \rho z_t + w_{t+1} + \gamma w_t.$$

Express this relationship as

$$\begin{bmatrix} z_{t+1} \\ w_{t+1} \end{bmatrix} = \begin{bmatrix} \rho & \gamma \\ 0 & 0 \end{bmatrix} \begin{bmatrix} z_t \\ w_t \end{bmatrix} + \begin{bmatrix} 1 \\ 1 \end{bmatrix} w_{t+1}$$

$$z_t = \begin{bmatrix} 1 & 0 \end{bmatrix} \begin{bmatrix} z_t \\ w_t \end{bmatrix}.$$

Example 3. Vector autoregression: Let z_t be an $n \times 1$ vector of random variables. We define a vector autoregression by a stochastic difference equation

$$z_{t+1} = \sum_{j=1}^{4} A_j z_{t+1-j} + C_y w_{t+1}, \tag{2.4.4}$$

where w_{t+1} is an $n \times 1$ martingale difference sequence satisfying equation (2.4.2) with $x_0' = \begin{bmatrix} z_0 & z_{-1} & z_{-2} & z_{-3} \end{bmatrix}$ and A_j is an $n \times n$ matrix for each j. We can map equation (2.4.4) into equation (2.4.1) as follows:

$$\begin{bmatrix} z_{t+1} \\ z_t \\ z_{t-1} \\ z_{t-2} \end{bmatrix} = \begin{bmatrix} A_1 & A_2 & A_3 & A_4 \\ I & 0 & 0 & 0 \\ 0 & I & 0 & 0 \\ 0 & 0 & I & 0 \end{bmatrix} \begin{bmatrix} z_t \\ z_{t-1} \\ z_{t-2} \\ z_{t-3} \end{bmatrix} + \begin{bmatrix} C_y \\ 0 \\ 0 \\ 0 \end{bmatrix} w_{t+1}. \tag{2.4.5}$$

Define A_o as the state transition matrix in equation (2.4.5). Assume that A_o has all of its eigenvalues bounded in modulus below unity. Then equation (2.4.4) can be initialized so that z_t is *covariance stationary*, a term we now define.

2.4.1. First and second moments

We can use equation $(2.4.1)$ to deduce the first and second moments of the sequence of random vectors $\{x_t\}_{t=0}^{\infty}$. A sequence of random vectors is called a *stochastic process*.

Definition 2.4.1. A stochastic process $\{x_t\}$ is said to be *covariance stationary* if it satisfies the following two properties: (a) the mean is independent of time, $Ex_t = Ex_0$ for all t, and (b) the sequence of autocovariance matrices $E(x_{t+j} - Ex_{t+j})(x_t - Ex_t)'$ depends on the separation between dates $j = 0, \pm 1, \pm 2, \ldots$, but not on t.

We use

Definition 2.4.2. A square real valued matrix A_o is said to be *stable* if all of its eigenvalues modulus are strictly less than unity.

We shall often find it useful to assume that $(2.4.3)$ takes the special form

$$\begin{bmatrix} x_{1,t+1} \\ x_{2,t+1} \end{bmatrix} = \begin{bmatrix} 1 & 0 \\ 0 & \tilde{A} \end{bmatrix} \begin{bmatrix} x_{1,t} \\ x_{2t} \end{bmatrix} + \begin{bmatrix} 0 \\ \tilde{C} \end{bmatrix} w_{t+1} \qquad (2.4.6)$$

where $\tilde{A}$ is a stable matrix. That $\tilde{A}$ is a stable matrix implies that the only solution of $(\tilde{A} - I)\mu_2 = 0$ is $\mu_2 = 0$ (i.e., 1 is *not* an eigenvalue of $\tilde{A}$). It follows that the matrix $A_o = \begin{bmatrix} 1 & 0 \\ 0 & \tilde{A} \end{bmatrix}$ on the right side of $(2.4.6)$ has one eigenvector associated with a single unit eigenvalue: $(A_o - I) \begin{bmatrix} \mu_1 \\ \mu_2 \end{bmatrix} = 0$ implies μ_1 is an arbitrary scalar and $\mu_2 = 0$. The first equation of $(2.4.6)$ implies that $x_{1,t+1} = x_{1,0}$ for all $t \geq 0$. Picking the initial condition $x_{1,0}$ pins down a particular eigenvector $\begin{bmatrix} x_{1,0} \\ 0 \end{bmatrix}$ of A_o. As we shall see soon, this eigenvector is our candidate for the unconditional mean of x that makes the process covariance stationary.

We will make an assumption that guarantees that there exists an initial condition $(Ex_0, E(x - Ex_0)(x - Ex_0)')$ that makes the x_t process covariance stationary. Either of the following conditions works:

CONDITION A1: All of the eigenvalues of A_o in $(2.4.3)$ are strictly less than 1 in modulus.

CONDITION A2: The state-space representation takes the special form (2.4.6) and all of the eigenvalues of $\tilde{A}$ are strictly less than 1 in modulus.

To discover the first and second moments of the x_t process, we regard the initial condition x_0 as being drawn from a distribution with mean $\mu_0 = Ex_0$ and covariance $\Sigma_0 = E(x - Ex_0)(x - Ex_0)'$. We shall deduce starting values for the mean and covariance that make the process covariance stationary, though our formulas are also useful for describing what happens when we start from other initial conditions that generate transient behavior that stops the process from being covariance stationary.

Taking mathematical expectations on both sides of equation (2.4.1) gives

$$\mu_{t+1} = A_o \mu_t \tag{2.4.7}$$

where $\mu_t = Ex_t$. We will assume that all of the eigenvalues of A_o are strictly less than unity in modulus, except possibly for one that is affiliated with the constant terms in the various equations. Then x_t possesses a stationary mean defined to satisfy $\mu_{t+1} = \mu_t$, which from equation (2.4.7) evidently satisfies

$$(I - A_o)\,\mu = 0, \tag{2.4.8}$$

which characterizes the mean μ as an eigenvector associated with the single unit eigenvalue of A_o. Notice that

$$x_{t+1} - \mu_{t+1} = A_o\,(x_t - \mu_t) + Cw_{t+1}. \tag{2.4.9}$$

Also, the fact that the remaining eigenvalues of A_o are less than unity in modulus implies that starting from any μ_0, $\mu_t \to \mu$.[6]

From equation (2.4.9), we can compute that the law of motion of the unconditional covariance matrices $\Sigma_t \equiv E(x_t - \mu)(x_t - \mu)'$. Thus,

$$E\,(x_{t+1} - \mu)\,(x_{t+1} - \mu)' = A_o E\,(x_t - \mu)\,(x_t - \mu)'\,A_o' + CC'$$

---

[6] To understand this, assume that the eigenvalues of A_o are distinct, and use the representation $A_o = P\Lambda P^{-1}$ where Λ is a diagonal matrix of the eigenvalues of A_o, arranged in descending order of magnitude, and P is a matrix composed of the corresponding eigenvectors. Then equation (2.4.7) can be represented as $\mu_{t+1}^* = \Lambda \mu_t^*$, where $\mu_t^* \equiv P^{-1}\mu_t$, which implies that $\mu_t^* = \Lambda^t \mu_0^*$. When all eigenvalues but the first are less than unity, Λ^t converges to a matrix of zeros except for the $(1, 1)$ element, and μ_t^* converges to a vector of zeros except for the first element, which stays at $\mu_{0,1}^*$, its initial value, which we are free to set equal to 1, to capture the constant. Then $\mu_t = P\mu_t^*$ converges to $P_1\mu_{0,1}^* = P_1$, where P_1 is the eigenvector corresponding to the unit eigenvalue.

or

$$\Sigma_{t+1} = A_o \Sigma_t A_o' + CC'.$$

A fixed point of this matrix difference equation evidently satisfies

$$\Sigma_\infty = A_o \Sigma_\infty A_o' + CC'.$$

We shall use the notation

$$C_x(0) = \Sigma_\infty$$

for the fixed point, which, if it exists, is the covariance matrix $E(x_t - \mu)(x_t - \mu)'$ under the stationary distribution of x.

Thus, to compute $C_x(0)$, we must solve

$$C_x(0) = A_o C_x(0) A_o' + CC', \tag{2.4.10}$$

where $C_x(0) \equiv E(x_t - \mu)(x_t - \mu)'$. Equation (2.4.10) is a *discrete Lyapunov* equation in the $n \times n$ matrix $C_x(0)$. It can be solved with the Matlab program `doublej.m`.

By virtue of (2.4.1) and (2.4.7), note that

$$(x_{t+j} - \mu_{t+j}) = A_o^j(x_t - \mu_t) + Cw_{t+j} + \cdots + A_o^{j-1}Cw_{t+1}.$$

Postmultiplying both sides by $(x_t - \mu_t)'$ and taking expectations shows that the autocovariance sequence satisfies

$$C_x(j) \equiv E(x_{t+j} - \mu)(x_t - \mu)' = A_o^j C_x(0). \tag{2.4.11}$$

The autocovariance matrix sequence $\{C_x(j)\}_{j=-\infty}^{\infty}$ is also called the *autocovariogram*. Once (2.4.10), is solved, the remaining second moments $C_x(j)$ can be deduced from equation (2.4.11).[7]

Suppose that $y_t = Gx_t$. Then $\mu_{yt} = Ey_t = G\mu_t$ and

$$E(y_{t+j} - \mu_{yt+j})(y_t - \mu_{yt})' = GC_x(j)G', \tag{2.4.12}$$

for $j = 0, 1, \ldots$. Equations (2.4.12) show that the autocovariogram for a stochastic process governed by a stochastic linear difference equation obeys the nonstochastic version of that difference equation.

---

[7] Notice that $C_x(-j) = C_x(j)'$.

2.4.2. Summary of moment formulas

Object	Formula
unconditional mean	$\mu_{t+1} = A_o \mu_t$
unconditional covariance	$\Sigma_{t+1} = A_o \Sigma_t A_o' + CC'$
$E[x_t \| x_0]$	$A_o^t x_0$
$E(x_t - E_0 x_t)(x_t - E_0 x_t)'$	$\sum_{h=0}^{t-1} A_o^h CC'(A_o^h)'$
stationary mean	$(I - A_o)\mu = 0$
stationary variance	$C_x(0) = A_o C_x(0) A_o' + CC'$
stationary autocovariance	$C_x(j) = A_o^j C_x(0)$

The accompanying table summarizes some formulas for various conditional and unconditional first and second moments of the state x_t governed by our linear stochastic state space system A_o, C, G. In section 2.5, we select some moments and use them to form population linear regressions.

2.4.3. Impulse response function

Suppose that the eigenvalues of A_o not associated with the constant are bounded above in modulus by unity. Using the lag operator L defined by $Lx_{t+1} \equiv x_t$, express equation (2.4.1) as

$$(I - A_o L)\, x_{t+1} = C w_{t+1}. \tag{2.4.13}$$

Iterate equation (2.4.1) forward from $t = 0$ to get

$$x_t = A_o^t x_0 + \sum_{j=0}^{t-1} A_o^j C w_{t-j} \tag{2.4.14}$$

Evidently,

$$y_t = G A_o^t x_0 + G \sum_{j=0}^{t-1} A_o^j C w_{t-j} \tag{2.4.15}$$

and $E y_t | x_0 = G A_o^t x_0$. Equations (2.4.14) and (2.4.15) are examples of a moving average representation. Viewed as a function of lag j, $h_j = A_o^j C$ or $\tilde{h}_j = G A_o^j C$ is called the *impulse response function*. The moving average representation and the associated impulse response function show how x_{t+j} or y_{t+j}

is affected by lagged values of the shocks, the w_{t+1}'s. Thus, the contribution of a shock w_{t-j} to x_t is $A_o^j C$.[8]

Equation ($2.4.15$) implies that the t-step ahead conditional covariance matrices are given by

$$E\left(y_t - Ey_t|x_0\right)\left(y_t - Ey_t|x_0\right)' = G\left[\sum_{h=0}^{t-1} A_o^h CC' A_o^{h'}\right] G'. \qquad (2.4.16)$$

2.4.4. Prediction and discounting

From equation ($2.4.1$) we can compute the useful prediction formulas

$$E_t x_{t+j} = A_o^j x_t \qquad (2.4.17)$$

for $j \geq 1$, where $E_t(\cdot)$ denotes the mathematical expectation conditioned on $x^t = (x_t, x_{t-1}, \ldots, x_0)$. Let $y_t = Gx_t$, and suppose that we want to compute $E_t \sum_{j=0}^{\infty} \beta^j y_{t+j}$. Evidently,

$$E_t \sum_{j=0}^{\infty} \beta^j y_{t+j} = G\left(I - \beta A_o\right)^{-1} x_t, \qquad (2.4.18)$$

provided that the eigenvalues of βA_o are less than unity in modulus. Equation ($2.4.18$) tells us how to compute an expected discounted sum, where the discount factor β is constant.

---

[8] The Matlab programs `dimpulse.m` and `impulse.m` compute impulse response functions.

2.4.5. *Geometric sums of quadratic forms*

In some applications, we want to calculate

$$\alpha_t = E_t \sum_{j=0}^{\infty} \beta^j x'_{t+j} Y x_{t+j}$$

where x_t obeys the stochastic difference equation $(2.4.1)$ and Y is an $n \times n$ matrix. To get a formula for α_t, we use a guess-and-verify method. We guess that α_t can be written in the form

$$\alpha_t = x'_t \nu x_t + \sigma, \tag{2.4.19}$$

where ν is an $(n \times n)$ matrix and σ is a scalar. The definition of α_t and the guess $(2.4.19)$ imply[9]

$$\begin{aligned}
\alpha_t &= x'_t Y x_t + \beta E_t \left(x'_{t+1} \nu x_{t+1} + \sigma \right) \\
&= x'_t Y x_t + \beta E_t \left[\left(A_o x_t + C w_{t+1} \right)' \nu \left(A_o x_t + C w_{t+1} \right) + \sigma \right] \\
&= x'_t \left(Y + \beta A'_o \nu A_o \right) x_t + \beta \, \text{trace} \left(\nu C C' \right) + \beta \sigma.
\end{aligned}$$

It follows that ν and σ satisfy

$$\begin{aligned}
\nu &= Y + \beta A'_o \nu A_o \\
\sigma &= \beta \sigma + \beta \, \text{trace} \, \nu C C'.
\end{aligned} \tag{2.4.20}$$

The first equation of $(2.4.20)$ is a *discrete Lyapunov equation* in the square matrix ν and can be solved by using one of several algorithms.[10] After ν has been computed, the second equation can be solved for the scalar σ.

We mention two important applications of formulas $(2.4.19)$ and $(2.4.20)$.

---

[9] Here we are repeatedly using the fact that for two conformable matrices A, B, $\text{trace} AB = \text{trace} BA$ to conclude that $E(w'_{t+1} C' \nu C w_{t+1}) = E \text{trace}(\nu C w_{t+1} w'_{t+1} C') = \text{trace}(\nu C E w_{t+1} w'_{t+1} C') = \text{trace}(\nu C C')$.

[10] The Matlab control toolkit has a program called `dlyap.m` that works when all of the eigenvalues of A_O are strictly less than unity; the program called `doublej.m` works even when there is a unit eigenvalue associated with the constant.

2.4.5.1. Asset pricing

Let y_t be governed by the state-space system $(2.4.3)$. In addition, assume that there is a scalar random process z_t given by

$$z_t = Hx_t.$$

Regard the process y_t as a payout or dividend from an asset, and regard $\beta^t z_t$ as a stochastic discount factor. The price of a perpetual claim on the stream of payouts is

$$\alpha_t = E_t \sum_{j=0}^{\infty} \left(\beta^j z_{t+j} \right) y_{t+j}. \tag{2.4.21}$$

To compute α_t, we simply set $Y = H'G$ in $(2.4.19)$ and $(2.4.20)$. In this application, the term σ functions as a risk premium; it is zero when $C = 0$.

2.4.5.2. Evaluation of dynamic criterion

Let a state x_t be governed by

$$x_{t+1} = Ax_t + Bu_t + Cw_{t+1} \tag{2.4.22}$$

where u_t is a control vector that is set by a decision maker according to a fixed rule

$$u_t = -F_0 x_t. \tag{2.4.23}$$

Substituting $(2.4.23)$ into $(2.4.22)$ gives $(2.4.1)$ where $A_o = A - BF_0$. We want to compute the *value function*

$$v\left(x_0\right) = -E_0 \sum_{t=0}^{\infty} \beta^t \left[x_t'Rx_t + u_t'Qu_t\right]$$

for fixed positive definite matrices R and Q, fixed decision rule F_0 in $(2.4.23)$, $A_o = A - BF_0$, and arbitrary initial condition x_0. Formulas $(2.4.19)$ and $(2.4.20)$ apply with $Y = R + F_0'QF_0$ and $A_o = A - BF_0$. Express the solution as

$$v\left(x_0\right) = -x_0'Px_0 - \sigma \tag{2.4.24}$$

where by applying formulas $(2.4.19)$ and $(2.4.20)$, P satisfies formula $(2.4.29)$ below.

Now consider the following one-period problem. Suppose that we must use decision rule F_0 from time 1 onward, so that the value at time 1 on starting from state x_1 is

$$v(x_1) = -x_1' P_0 x_1 - \sigma. \qquad (2.4.25)$$

Taking $u_t = -F_0 x_t$ as given for $t \geq 1$, what is the best choice of u_0? This leads to the optimum problem:

$$\max_{u_0} -\{x_0' R x_0 + u_0' Q u_0 + \beta E (A x_0 + B u_0 + C w_1)' P_0 (A x_0 + B u_0 + C w_1) + \beta \sigma\}. \qquad (2.4.26)$$

The first-order conditions for this problem can be rearranged to attain

$$u_0 = -F_1 x_0 \qquad (2.4.27)$$

where

$$F_1 = \beta (Q + \beta B' P B)^{-1} B' P A. \qquad (2.4.28)$$

For convenience, we state the formula for P_0:

$$P_0 = R + F_0' Q F_0 + \beta (A - B F_0)' P (A - B F_0). \qquad (2.4.29)$$

Given F_0, formula $(2.4.29)$ determines the matrix P in the value function that describes the expected discounted value of the sum of payoffs from sticking forever with this decision rule. Given P, formula $(2.4.28)$ gives the best decision rule $u_0 = -F_1 x_0$ if at $t = 0$ you are permitted only a one-period deviation from the rule $u_t = -F_0 x_t$ that has to be used for $t \geq 1$. If $F_1 \neq F_0$, we say that the decision maker would accept the opportunity to deviate from F_0 for one period.

It is tempting to iterate on $(2.4.28)$ and $(2.4.29)$ as follows to seek a decision rule from which a decision maker would not want to deviate for one period: (1) given an F_0, find P; (2) reset F equal to the F_1 found in step 1, then to substitute it for F_0 in $(2.4.29)$ to compute a new P, call it P_1; (3) return to step 1 and iterate to convergence. This leads to the two equations

$$P_j = R + F_j' Q F_j + \beta (A - B F_j)' P_j (A - B F_j)$$
$$F_{j+1} = \beta (Q + \beta B' P_j B)^{-1} B' P_j A \qquad (2.4.30)$$

which are to be initialized from an arbitrary F_0 that ensures that $\sqrt{\beta}(A - B F_0)$ is a stable matrix. After this process has converged, one cannot find a value-increasing one-period deviation from the limiting decision rule $u_t = -F_\infty x_t$.[11]

---

[11] It turns out that if you don't want to deviate for one period, then you would never want to deviate, so that the limiting rule is optimal.

As we shall see in chapter 4, this is an excellent algorithm for solving a dynamic programming problem. It is an example of the Howard improvement algorithm. In chapter 5, we describe an alternative algorithm that iterates on the following equations

$$P_{j+1} = R + F_j'QF_j + \beta \left(A - BF_j\right)' P_j \left(A - BF_j\right)$$
$$F_j = \beta \left(Q + \beta B'P_j B\right)^{-1} B'P_j A, \tag{2.4.31}$$

that is to be initialized from an arbitrary positive semi-definite matrix P_0.[12]

2.5. Population regression

This section explains the notion of a population regression equation. Suppose that we have a state-space system (2.4.3) with initial conditions that make it covariance stationary. We can use the preceding formulas to compute the second moments of any pair of random variables. These moments let us compute a linear regression. Thus, let X be a $p \times 1$ vector of random variables somehow selected from the stochastic process $\{y_t\}$ governed by the system (2.4.3). For example, let $p = 2m$, where y_t is an $m \times 1$ vector, and take $X = \begin{bmatrix} y_t \\ y_{t-1} \end{bmatrix}$ for any $t \geq 1$. Let Y be any scalar random variable selected from the $m \times 1$ stochastic process $\{y_t\}$. For example, take $Y = y_{t+1,1}$ for the same t used to define X, where $y_{t+1,1}$ is the first component of y_{t+1}.

We consider the following least-squares approximation problem: find a $1 \times p$ vector of real numbers β that attain

$$\min_{\beta} E \left(Y - \beta X\right)^2. \tag{2.5.1}$$

Here βX is being used to estimate Y, and we want the value of β that minimizes the expected squared error. The first-order necessary condition for minimizing $E(Y - \beta X)^2$ with respect to β is

$$E \left(Y - \beta X\right) X' = 0, \tag{2.5.2}$$

which can be rearranged as[13]

$$\beta = \left(EYX'\right) \left[E \left(XX'\right)\right]^{-1}. \tag{2.5.3}$$

---

[12] $P_0 = 0$ is a popular choice.

[13] That $EX'X$ is nonnegative definite implies that the second-order conditions for a minimum of condition (2.5.1) are satisfied.

By using the formulas (2.4.8), (2.4.10), (2.4.11), and (2.4.12), we can compute EXX' and EYX' for whatever selection of X and Y we choose. The condition (2.5.2) is called the least-squares normal equation. It states that the projection error $Y - \beta X$ is orthogonal to X. Therefore, we can represent Y as

$$Y = \beta X + \epsilon \qquad (2.5.4)$$

where $E\epsilon X' = 0$. Equation (2.5.4) is called a population regression equation, and βX is called the least-squares projection of Y on X or the least-squares regression of Y on X. The vector β is called the population least-squares regression vector. The law of large numbers for continuous-state Markov processes, Theorem 2.3.1, states conditions that guarantee that sample moments converge to population moments, that is, $\frac{1}{S} \sum_{s=1}^{S} X_s X_s' \to EXX'$ and $\frac{1}{S} \sum_{s=1}^{S} Y_s X_s' \to EYX'$. Under those conditions, sample least-squares estimates converge to β.

There are as many such regressions as there are ways of selecting Y, X. We have shown how a model (e.g., a triple A_o, C, G, together with an initial distribution for x_0) restricts a regression. Going backward, that is, telling what a given regression tells about a model, is more difficult. Many regressions tell little about the model, and what little they have to say can be difficult to decode. As we indicate in sections 2.6 and 2.7.1, the likelihood function completely describes what a given data set says about a model in a way that is straightforward to decode.

2.5.1. Multiple regressors

Now let Y be an $n \times 1$ vector of random variables and think of regression solving the least squares problem for each of them to attain a representation

$$Y = \beta X + \epsilon \qquad (2.5.5)$$

where β is now $n \times p$ and ϵ is now an $n \times 1$ vector of least squares residuals. The population regression coefficients are again given by

$$\beta = E\left(YX'\right)\left[E\left(XX'\right)\right]^{-1}. \qquad (2.5.6)$$

We will use this formula repeatedly in section 2.7 to derive the Kalman filter.

2.6. Estimation of model parameters

We have shown how to map the matrices A_o, C into all of the second moments of the stationary distribution of the stochastic process $\{x_t\}$. Linear economic models typically give A_o, C as functions of a set of deeper parameters θ. We shall give examples of some such models in chapters 4 and 5. Those theories and the formulas of this chapter give us a mapping from θ to these theoretical moments of the $\{x_t\}$ process. That mapping is an important ingredient of econometric methods designed to estimate a wide class of linear rational expectations models (see Hansen and Sargent, 1980, 1981). Briefly, these methods use the following procedures to match theory to data. To simplify, we shall assume that at time t, observations are available on the entire state x_t. As discussed in section 2.7.1, the details are more complicated if only a subset of the state vector or a noisy signal of the state is observed, though the basic principles remain the same.

Given a sample of observations for $\{x_t\}_{t=0}^T \equiv x_t, t = 0, \ldots, T$, the likelihood function is defined as the joint probability distribution $f(x_T, x_{T-1}, \ldots, x_0)$. The likelihood function can be *factored* using

$$f(x_T, \ldots, x_0) = f(x_T | x_{T-1}, \ldots, x_0) f(x_{T-1} | x_{T-2}, \ldots, x_0) \cdots$$
$$f(x_1 | x_0) f(x_0), \tag{2.6.1}$$

where in each case f denotes an appropriate probability distribution. For system (2.4.1), $f(x_{t+1} | x_t, \ldots, x_0) = f(x_{t+1} | x_t)$, which follows from the Markov property possessed by equation (2.4.1). Then the likelihood function has the recursive form

$$f(x_T, \ldots, x_0) = f(x_T | x_{T-1}) f(x_{T-1} | x_{T-2}) \cdots f(x_1 | x_0) f(x_0). \tag{2.6.2}$$

If we assume that the w_t's are Gaussian, then the conditional distribution $f(x_{t+1} | x_t)$ is Gaussian with mean $A_o x_t$ and covariance matrix CC'. Thus, under the Gaussian distribution, the log of the conditional density of the n dimensional vector x_{t+1} becomes

$$\log f(x_{t+1} | x_t) = -.5n \log(2\pi) - .5 \log \det(CC')$$
$$- .5(x_{t+1} - A_o x_t)' (CC')^{-1} (x_{t+1} - A_o x_t) \tag{2.6.3}$$

Given an assumption about the distribution of the initial condition x_0, equations (2.6.2) and (2.6.3) can be used to form the likelihood function of a sample of observations on $\{x_t\}_{t=0}^T$. One computes maximum likelihood estimates by using a hill-climbing algorithm to maximize the likelihood function with respect to the free parameters that determine A_o, C.[14]

When the state x_t is not observed, we need to go beyond the likelihood function for $\{x_t\}$. One approach uses filtering methods to build up the likelihood function for the subset of observed variables.[15] In section 2.7, we derive the Kalman filter as an application of the population regression formulas of section 2.5. Then in section 2.7.1, we use the Kalman filter as a device that tells us how to find state variables that allow us recursively to form a likelihood function for observations of variables that are not themselves Markov.

2.7. The Kalman filter

As a fruitful application of the population regression formula $(2.5.6)$, we derive the celebrated Kalman filter for the state space system for $t \geq 0$:[16]

$$x_{t+1} = A_o x_t + C w_{t+1} \tag{2.7.1}$$

$$y_t = G x_t + v_t \tag{2.7.2}$$

where x_t is an $n \times 1$ state vector and y_t is an $m \times 1$ vector of signals on the hidden state; w_{t+1} is a $p \times 1$ vector iid sequence of normal random variables with mean 0 and identity covariance matrix, and v_t is another iid vector sequence of normal random variables with mean zero and covariance matrix R. We assume that w_{t+1} and v_s are orthogonal (i.e., $E w_{t+1} v_s' = 0$) for all $t+1$ and s greater than or equal to 0. We assume that

$$x_0 \sim \mathcal{N}\left(\hat{x}_0, \Sigma_0\right). \tag{2.7.3}$$

This specification implies that

$$y_0 \sim \mathcal{N}\left(G\hat{x}_0, G\Sigma_0 G' + R\right). \tag{2.7.4}$$

---

[14] For example, putting those free parameters into a vector θ, think of A_o, C as being the matrix functions $A_o(\theta), C(\theta)$.

[15] See Hamilton (1994), Canova (2007), DeJong and Dave (2011), and section 2.7.1 below.

[16] In exercise 2.22, we ask you to derive the Kalman filter for a state space system that uses a different timing convention and that allows the state and measurement noises to be correlated.

The decision maker is assumed to observe $y_t, \ldots, y_0$ but not $x_t, \ldots, x_0$ at time t. He knows the structure (2.7.1)-(2.7.2) and the first and second moments implied by this structure. We want to find recursive formulas for the population regressions $\hat{x}_t = E[x_t|y_{t-1}, \ldots, y_0]$ and the covariance matrices $\Sigma_t = E(x_t - \hat{x}_t)(x_t - \hat{x}_t)'$.

We use the insight that the new information in y_0 relative to what is already known $(\hat{x}_0)$ is $a_0 \equiv y_0 - G\hat{x}_0$ (and more generally, the new information at t relative to what can be inferred from the past is $a_t = y_t - G\hat{x}_t$). The decision maker regresses what he doesn't know on what he does. Thus, first apply (2.5.6) to compute the population regression

$$x_0 - \hat{x}_0 = L_0 (y_0 - G\hat{x}_0) + \eta \qquad (2.7.5)$$

where η is a matrix of least squares residuals. The least squares orthogonality conditions are

$$E (x_0 - \hat{x}_0) (y_0 - G\hat{x}_0)' = L_0 E (y_0 - G\hat{x}_0) (y_0 - G\hat{x}_0)'.$$

Evaluating the moment matrices and solving for L_0 gives the formula

$$L_0 = \Sigma_0 G' (G\Sigma_0 G' + R)^{-1}. \qquad (2.7.6)$$

Define $\hat{x}_1 = E[x_1|y_0]$. Equation (2.7.1) implies that $E[x_1|\hat{x}_0] = A_o\hat{x}_0$ and that

$$x_1 = A_o\hat{x}_0 + A_o (x_0 - \hat{x}_0) + Cw_1. \qquad (2.7.7)$$

Furthermore, applying (2.7.5) shows that $Ex_1|y_0 = A_o\hat{x}_0 + A_oL_0(y_0 - G\hat{x}_0)$, which we express as

$$\hat{x}_1 = A_o\hat{x}_0 + K_0 (y_0 - G\hat{x}_0), \qquad (2.7.8)$$

where

$$K_0 = A_o\Sigma_0 G' (G\Sigma_0 G' + R)^{-1}. \qquad (2.7.9)$$

Subtract (2.7.8) from (2.7.7) to get

$$x_1 - \hat{x}_1 = A_o (x_0 - \hat{x}_0) + Cw_1 - K_0 (y_0 - G\hat{x}_0). \qquad (2.7.10)$$

Use this equation to compute the following recursion for $E(x_1 - \hat{x}_1)(x_1 - \hat{x}_1)' = \Sigma_1$:

$$\Sigma_1 = (A_o - K_0G) \Sigma_0 (A_o - K_0G)' + (CC' + K_0RK_0'). \qquad (2.7.11)$$

Thus, we have that the distribution of $x_1|y_0 \sim \mathcal{N}(\hat{x}_1, \Sigma_1)$.

Iterating the above argument gives the recursion:[17]

$$a_t = y_t - G\hat{x}_t \tag{2.7.12a}$$

$$K_t = A_o\Sigma_t G' \left(G\Sigma_t G' + R\right)^{-1} \tag{2.7.12b}$$

$$\hat{x}_{t+1} = A_o\hat{x}_t + K_t a_t \tag{2.7.12c}$$

$$\Sigma_{t+1} = CC' + K_t R K_t' + \left(A_o - K_t G\right)\Sigma_t \left(A_o - K_t G\right)'. \tag{2.7.12d}$$

System $(2.7.12)$ is the celebrated Kalman filter, and K_t is called the Kalman gain. Equation $(2.7.12d)$ is known as a matrix Ricatti difference equation.

The process $a_t = y_t - E[y_t|y_{t-1}, \ldots, y_0]$ is called the 'innovation' process in y. It is the part of y_t that cannot be predicted from past values of y. Note that $Ea_t a_t' = (G\Sigma_t G' + R)$, the moment matrix whose inverse appears on the right side of the least squares regression formula $(2.7.12b)$. A direct calculation that uses the formulas $a_t = G(x_t - \hat{x}_t) + v_t$ and $a_{t-1} = G(x_{t-1} - \hat{x}_{t-1}) + v_{t-1}$ to compute expected values of products shows that $Ea_t a_{t-1}' = 0$, and more generally that $E[a_t|a_{t-1}, \ldots, a_0] = 0$. An alternative argument based on first principles proceeds as follows. Let $H(y^t)$ denote the linear space of all linear combinations of y^t. Note that $a_{t+1} = y_{t+1} - Ey_{t+1}|y^t$; that $a_t \in H(y^t)$; that by virtue of being a least-squares error, $a_{t+1} \perp H(y^t)$; and that therefore $a_{t+1} \perp a_t$, and more generally, $a_{t+1} \perp a^t$. Thus, $\{a_t\}$ is a 'white noise' process of innovations to the $\{y_t\}$ process.

Sometimes $(2.7.12)$ is called a 'whitening filter' that takes a $\{y_t\}$ process of signals as an input and produces a process $\{a_t\}$ of innovations as an output. The linear space $H(a^t)$ is evidently an orthogonal basis for the linear space $H(y^t)$.

In what will seem to be superficially very different contexts, we shall encounter equations that will remind us of $(2.7.12b)$, $(2.7.12d)$. See chapter 5, page 140.

---

[17] Substituting for K_t from $(2.7.12b)$ allows us to rewrite $(2.7.12d)$ as

$$\Sigma_{t+1} = A_o\Sigma_t A_o' + CC' - A_o\Sigma_t G' \left(G\Sigma_t G' + R\right)^{-1} G\Sigma_t A_o',$$

a formula that we shall be reminded of when we study dynamic programming for problems with linear constraints and quadratic return functions.

2.7.1. Estimation again

The innovations representation that emerges from the Kalman filter is

$$\hat{x}_{t+1} = A_o\hat{x}_t + K_t a_t \qquad (2.7.13a)$$

$$y_t = G\hat{x}_t + a_t \qquad (2.7.13b)$$

where for $t \geq 1$, $\hat{x}_t = E[x_t|y^{t-1}]$ and $E a_t a_t' = G\Sigma_t G' + R \equiv \Omega_t$. Evidently, for $t \geq 1$, $E[y_t|y^{t-1}] = G\hat{x}_t$ and the distribution of y_t conditional on y^{t-1} is $\mathcal{N}(G\hat{x}_t, \Omega_t)$. The objects $G\hat{x}_t, \Omega_t$ emerging from the Kalman filter are thus sufficient statistics for the distribution of y_t conditioned on y^{t-1} for $t \geq 1$. The sufficient conditions and also the innovation $a_t = y_t - G\hat{x}_t$ can be calculated recursively from (2.7.12). The unconditional distribution of y_0 is evidently $\mathcal{N}(G\hat{x}_0, \Omega_0)$.

As a counterpart to (2.6.2), we can factor the likelihood function for a sample $(y_T, y_{T-1}, \ldots, y_0)$ as

$$f(y_T, \ldots, y_0) = f\left(y_T|y^{T-1}\right) f\left(y_{T-1}|y^{T-2}\right) \cdots f(y_1|y_0) f(y_0). \qquad (2.7.14)$$

The log of the conditional density of the $m \times 1$ vector y_t is

$$\log f\left(y_t|y^{t-1}\right) = -.5m\log(2\pi) - .5\log\det(\Omega_t) - .5a_t'\Omega_t^{-1}a_t. \qquad (2.7.15)$$

We can use (2.7.15) and (2.7.12) to evaluate the likelihood function (2.7.14) recursively for a given set of parameter values θ that underlie the matrices A_o, G, C, R. Such calculations are at the heart of efficient strategies for computing maximum-likelihood and Bayesian estimators of free parameters.[18]

The likelihood function is also an essential object for a Bayesian statistician.[19] It completely summarizes how the data influence the posterior via the following application of Bayes' law. Where θ is our parameter vector, y_0^T our data record, and $\tilde{p}(\theta)$ a probability density that summarizes our prior 'views' or 'information' about θ *before* seeing y_0^T, our views about θ *after* seeing y_0^T is

---

[18] See Hansen (1982); Eichenbaum (1991); Christiano and Eichenbaum (1992); Burnside, Eichenbaum, and Rebelo (1993); and Burnside and Eichenbaum (1996a, 1996b) for alternative estimation strategies.

[19] See Canova (2007), Christensen and Kiefer (2009), and DeJong and Dave (2011) for extensive descriptions of how Bayesian and maximum likelihood methods can be applied to macroeconomic and other dynamic models.

summarized by a *posterior probability* $\tilde{p}(\theta|y_0^T)$ that is constructed from Bayes's law via

$$\tilde{p}\left(\theta|y_0^T\right) = \frac{f\left(y_0^T|\theta\right)\tilde{p}\left(\theta\right)}{\int f\left(y_0^T|\theta\right)\tilde{p}\left(\theta\right)d\theta}$$

where the denominator is the marginal joint density $f(y_0^T)$ of y_0^T.

In appendix B, we describe a simulation algorithm for approximating a Bayesian posterior. The algorithm constructs a Markov chain whose invariant distribution *is* the posterior, then iterates the Markov chain to convergence.

2.8. Vector autoregressions and the Kalman filter

2.8.1. Conditioning on the semi-infinite past of y

Under an interesting set of conditions summarized, for example, by Anderson, Hansen, McGrattan, and Sargent (1996), iterations on $(2.7.12b)$, $(2.7.12d)$ converge to time-invariant K, Σ for any positive semi-definite initial covariance matrix Σ_0. A time-invariant matrix $\Sigma_t = \Sigma$ that solves $(2.7.12d)$ is the covariance matrix of x_t around $Ex_t|\{y_{-\infty}^{t-1}\}$, where $\{y_{-\infty}^{t-1}\}$ denotes the semi-infinite history of y_s for all dates on or before $t-1$.[20]

---

[20] The Matlab program `kfilter.m` implements the time-invariant Kalman filter, allowing for correlation between the w_{t+1} and v_t Also see exercise 2.22.

2.8.2. A time-invariant VAR

Suppose that the fixed point of $(2.7.12d)$ just described exists. If we initiate $(2.7.12d)$ from this fixed point Σ, then the innovations representation becomes time invariant:

$$\hat{x}_{t+1} = A_o\hat{x}_t + Ka_t \tag{2.8.1a}$$

$$y_t = G\hat{x}_t + a_t \tag{2.8.1b}$$

where $Ea_ta_t' = G\Sigma G' + R$. Use $(2.8.1)$ to express $\hat{x}_{t+1} = (A - KG)\hat{x}_t + Ky_t$. If we assume that the eigenvalues of $A - KG$ are bounded in modulus below unity,[21] we can solve the preceding equation to get

$$\hat{x}_{t+1} = \sum_{j=0}^{\infty} (A - KG)^j \, Ky_{t-j}. \tag{2.8.2}$$

Then solving $(2.8.1b)$ for y_t gives the *vector autoregression*

$$y_t = G\sum_{j=0}^{\infty} (A - KG)^j \, Ky_{t-j-1} + a_t, \tag{2.8.3}$$

where by construction

$$E\left[a_ty_{t-j-1}'\right] = 0 \quad \forall j \geq 0. \tag{2.8.4}$$

The orthogonality conditions $(2.8.4)$ identify $(2.8.3)$ as a vector autoregression.

---

[21] Anderson, Hansen, McGrattan, and Sargent (1996) show assumptions that guarantee that the eigenvalue of $A - KG$ are bounded in modulus below unity.

2.8.3. *Interpreting VARs*

Equilibria of economic models (or linear or log-linear approximations to them –
see chapter 11 and the examples in section 2.11 of this chapter and appendix B
of chapter 14 – typically take the form of the state space system $(2.7.1),(2.7.2)$.
This hidden Markov model disturbs the evolution of the state x_t by the $p \times 1$
shock vector w_{t+1} and it perturbs the $m \times 1$ vector of observed variables y_t by
the $m \times 1$ vector of measurement errors. Thus, $p + m$ shocks impinge on y_t.
An economic theory typically makes w_{t+1}, v_t be directly interpretable as shocks
that impinge on preferences, technologies, endowments, information sets, and
measurements. The state space system $(2.7.1),(2.7.2)$ is a representation of the
stochastic process y_t in terms of these interpretable shocks. But the typical
situation is that these shocks can not be recovered directly from the y_ts, even
when we know the matrices A_o, G, C, R.

   The innovations representation $(2.8.1a)$, $(2.8.1b)$ represents the *same* stochastic process y_t in terms of an $m \times 1$ vector of shocks a_t that would be recovered by running an infinite-order (population) vector autoregression for y_t. Because of its role in constructing the mapping from the original representation
$(2.7.1),(2.7.2)$ to the one associated with the vector autoregression $(2.8.3)$, the
Kalman filter is a very useful tool for interpreting vector autoregressions.

2.9. Applications of the Kalman filter

2.9.1. *Muth's reverse engineering exercise*

Phillip Cagan (1956) and Milton Friedman (1956) posited that when people
wanted to form expectations of future values of a scalar y_t, they would use the
following "adaptive expectations" scheme:

$$y^*_{t+1} = K \sum_{j=0}^{\infty} (1 - K)^j \, y_{t-j} \qquad (2.9.1a)$$

or

$$y^*_{t+1} = (1 - K) \, y^*_t + K y_t, \qquad (2.9.1b)$$

where y_{t+1}^* is people's expectation.[22] Friedman used this scheme to describe people's forecasts of future income. Cagan used it to model their forecasts of inflation during hyperinflations. Cagan and Friedman did not assert that the scheme is an optimal one, and so did not fully defend it. Muth (1960) wanted to understand the circumstances under which this forecasting scheme would be optimal. Therefore, he sought a stochastic process for y_t such that equation (2.9.1) would be optimal. In effect, he posed and solved an "inverse optimal prediction" problem of the form "You give me the forecasting scheme; I have to find the stochastic process that makes the scheme optimal." Muth solved the problem using classical (nonrecursive) methods. The Kalman filter was first described in print in the same year as Muth's solution of this problem (Kalman, 1960). The Kalman filter lets us solve Muth's problem quickly.

Muth studied the model

$$x_{t+1} = x_t + w_{t+1} \qquad (2.9.2a)$$

$$y_t = x_t + v_t, \qquad (2.9.2b)$$

where y_t, x_t are scalar random processes, and w_{t+1}, v_t are mutually independent i.i.d. Gaussian random processes with means of zero and variances $Ew_{t+1}^2 = Q, Ev_t^2 = R$, and $Ev_s w_{t+1} = 0$ for all t, s. The initial condition is that x_0 is Gaussian with mean $\hat{x}_0$ and variance Σ_0. Muth sought formulas for $\hat{x}_{t+1} = E[x_{t+1}|y^t]$, where $y^t = [y_t, \ldots, y_0]$.

For this problem, $A = 1, CC' = Q, G = 1$, making the Kalman filtering equations become

$$K_t = \frac{\Sigma_t}{\Sigma_t + R} \qquad (2.9.3a)$$

$$\Sigma_{t+1} = \Sigma_t + Q - \frac{\Sigma_t^2}{\Sigma_t + R}. \qquad (2.9.3b)$$

The second equation can be rewritten

$$\Sigma_{t+1} = \frac{\Sigma_t (R + Q) + QR}{\Sigma_t + R}. \qquad (2.9.4)$$

For $Q = R = 1$, Figure 2.9.1 plots the function $f(\Sigma) = \frac{\Sigma(R+Q)+QR}{\Sigma+R}$ appearing on the right side of equation (2.9.4) for values $\Sigma \geq 0$ against the 45-degree line.

---

[22] See Hamilton (1994) and Kim and Nelson (1999) for diverse applications of the Kalman filter. Appendix B (see Technical Appendixes) briefly describes a discrete-state nonlinear filtering problem.

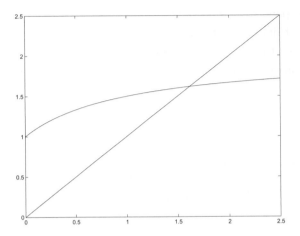

Figure 2.9.1: Graph of $f(\Sigma) = \frac{\Sigma(R+Q)+QR}{\Sigma+R}$, $Q = R = 1$,
against the 45-degree line. Iterations on the Riccati equation
for Σ_t converge to the fixed point.

Note that $f(0) = Q$. This graph identifies the fixed point of iterations on $f(\Sigma)$
as the intersection of $f(\cdot)$ and the 45-degree line. That the slope of $f(\cdot)$ is less
than unity at the intersection assures us that the iterations on f will converge
as $t \to +\infty$ starting from any $\Sigma_0 \geq 0$.

Muth studied the solution of this problem as $t \to \infty$. Evidently, $\Sigma_t \to$
$\Sigma_\infty \equiv \Sigma$ is the fixed point of a graph like Figure 2.9.1. Then $K_t \to K$ and the
formula for $\hat{x}_{t+1}$ becomes

$$\hat{x}_{t+1} = (1 - K)\,\hat{x}_t + K y_t \tag{2.9.5}$$

where $K = \frac{\Sigma}{\Sigma+R} \in (0, 1)$. This is a version of Cagan's adaptive expectations
formula. It can be shown that $K \in [0, 1]$ is an increasing function of $\frac{Q}{R}$. Thus,
K is the fraction of the innovation a_t that should be regarded as 'permanent'
and $1 - K$ is the fraction that is purely transitory. Iterating backward on
equation (2.9.5) gives $\hat{x}_{t+1} = K \sum_{j=0}^{t}(1 - K)^j y_{t-j} + (1 - K)^{t+1}\hat{x}_0$, which is
a version of Cagan and Friedman's geometric distributed lag formula. Using
equations (2.9.2), we find that $E[y_{t+j}|y^t] = E[x_{t+j}|y^t] = \hat{x}_{t+1}$ for all $j \geq 1$.
This result in conjunction with equation (2.9.5) establishes that the adaptive
expectation formula (2.9.5) gives the optimal forecast of y_{t+j} for all horizons

$j \geq 1$. This finding is remarkable because for most processes, the optimal forecast will depend on the horizon. That there is a single optimal forecast for all horizons justifies the term *permanent income* that Milton Friedman (1955) chose to describe the forecast of income.

The dependence of the forecast on horizon can be studied using the formulas

$$E\left[x_{t+j}|y^{t-1}\right] = A^j \hat{x}_t \qquad (2.9.6a)$$

$$E\left[y_{t+j}|y^{t-1}\right] = GA^j \hat{x}_t \qquad (2.9.6b)$$

In the case of Muth's example,

$$E\left[y_{t+j}|y^{t-1}\right] = \hat{y}_t = \hat{x}_t \quad \forall j \geq 0.$$

2.9.2. Jovanovic's application

In chapter 6, we will describe a version of Jovanovic's (1979) matching model, at the core of which is a "signal-extraction" problem that simplifies Muth's problem. Let x_t, y_t be scalars with $A = 1, C = 0, G = 1, R > 0$. Let x_0 be Gaussian with mean μ and variance Σ_0. Interpret x_t (which is evidently constant with this specification) as the hidden value of θ, a "match parameter". Let y^t denote the history of y_s from $s = 0$ to $s = t$. Define $m_t \equiv \hat{x}_{t+1} \equiv E[\theta|y^t]$ and $\Sigma_{t+1} = E(\theta - m_t)^2$. Then the Kalman filter becomes

$$m_t = (1 - K_t)\, m_{t-1} + K_t y_t \qquad (2.9.7a)$$

$$K_t = \frac{\Sigma_t}{\Sigma_t + R} \qquad (2.9.7b)$$

$$\Sigma_{t+1} = \frac{\Sigma_t R}{\Sigma_t + R}. \qquad (2.9.7c)$$

The recursions are to be initiated from (m_{-1}, Σ_0), a pair that embodies all "prior" knowledge about the position of the system. It is easy to see from Figure 2.9.1 that when $Q = 0$, $\Sigma = 0$ is the limit point of iterations on equation (2.9.7c) starting from any $\Sigma_0 \geq 0$. Thus, the value of the match parameter is eventually learned.

It is instructive to write equation $(2.9.7c)$ as

$$\frac{1}{\Sigma_{t+1}} = \frac{1}{\Sigma_t} + \frac{1}{R}. \tag{2.9.8}$$

The reciprocal of the variance is often called the *precision* of the estimate. According to equation $(2.9.8)$ the precision increases without bound as t grows, and $\Sigma_{t+1} \to 0$.[23]

We can represent the Kalman filter in the form

$$m_{t+1} = m_t + K_{t+1}a_{t+1}$$

which implies that

$$E\left(m_{t+1} - m_t\right)^2 = K_{t+1}^2 \sigma_{a,t+1}^2$$

where $a_{t+1} = y_{t+1} - m_t$ and the variance of a_t is equal to $\sigma_{a,t+1}^2 = (\Sigma_{t+1} + R)$ from equation $(5.6.5)$. This implies

$$E\left(m_{t+1} - m_t\right)^2 = \frac{\Sigma_{t+1}^2}{\Sigma_{t+1} + R}.$$

For the purposes of our discrete-time counterpart of the Jovanovic model in chapter 6, it will be convenient to represent the motion of m_{t+1} by means of the equation

$$m_{t+1} = m_t + g_{t+1}u_{t+1}$$

where $g_{t+1} \equiv \left(\frac{\Sigma_{t+1}^2}{\Sigma_{t+1}+R}\right)^{.5}$ and u_{t+1} is a standardized i.i.d. normalized and standardized with mean zero and variance 1 constructed to obey $g_{t+1}u_{t+1} \equiv K_{t+1}a_{t+1}$.

---

[23] As a further special case, consider when there is zero precision initially $(\Sigma_0 = +\infty)$. Then solving the difference equation $(2.9.8)$ gives $\frac{1}{\Sigma_t} = t/R$. Substituting this into equations $(2.9.7)$ gives $K_t = (t + 1)^{-1}$, so that the Kalman filter becomes $m_0 = y_0$ and $m_t = [1 - (t + 1)^{-1}]m_{t-1} + (t + 1)^{-1}y_t$, which implies that $m_t = (t + 1)^{-1}\sum_{s=0}^{t} y_t$, the sample mean, and $\Sigma_t = R/t$.

2.10. The spectrum

For a covariance stationary stochastic process, all second moments can be encoded in a complex-valued matrix called the *spectral density* matrix. The autocovariance sequence for the process determines the spectral density. Conversely, the spectral density can be used to determine the autocovariance sequence.

Under the assumption that A_o is a stable matrix,[24] the state x_t converges to a unique covariance stationary probability distribution as t approaches infinity. The spectral density matrix of this covariance stationary distribution $S_x(\omega)$ is defined to be the Fourier transform of the covariogram of x_t:

$$S_x(\omega) \equiv \sum_{\tau=-\infty}^{\infty} C_x(\tau) e^{-i\omega\tau}. \tag{2.10.1}$$

For the system (2.4.1), the spectral density of the stationary distribution is given by the formula

$$S_x(\omega) = \left[I - A_o e^{-i\omega}\right]^{-1} CC' \left[I - A'_o e^{+i\omega}\right]^{-1}, \quad \forall \omega \in [-\pi, \pi]. \tag{2.10.2}$$

The spectral density summarizes all covariances. They can be recovered from $S_x(\omega)$ by the Fourier inversion formula[25]

$$C_x(\tau) = (1/2\pi) \int_{-\pi}^{\pi} S_x(\omega) e^{+i\omega\tau} d\omega.$$

Setting $\tau = 0$ in the inversion formula gives

$$C_x(0) = (1/2\pi) \int_{-\pi}^{\pi} S_x(\omega) d\omega,$$

which shows that the spectral density decomposes covariance across frequencies.[26] A formula used in the process of generalized method of moments (GMM) estimation emerges by setting $\omega = 0$ in equation (2.10.1), which gives

$$S_x(0) \equiv \sum_{\tau=-\infty}^{\infty} C_x(\tau).$$

---

[24] It is sufficient that the only eigenvalue of A_O not strictly less than unity in modulus is that associated with the constant, which implies that A_O and C fit together in a way that validates (2.10.2).

[25] Spectral densities for continuous-time systems are discussed by Kwakernaak and Sivan (1972). For an elementary discussion of discrete-time systems, see Sargent (1987a). Also see Sargent (1987a, chap. 11) for definitions of the spectral density function and methods of evaluating this integral.

[26] More interestingly, the spectral density achieves a decomposition of covariance into components that are orthogonal across frequencies.

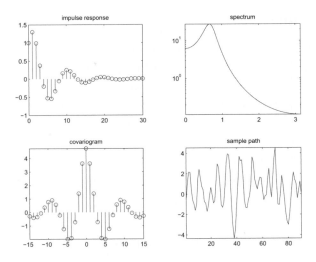

Figure 2.10.1: Impulse response, spectrum, covariogram, and sample path of process $(1 - 1.3L + .7L^2)y_t = w_t$.

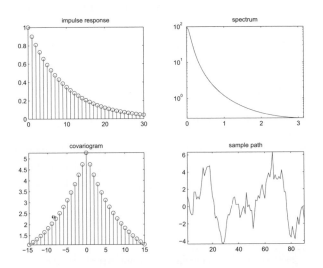

Figure 2.10.2: Impulse response, spectrum, covariogram, and sample path of process $(1 - .9L)y_t = w_t$.

2.10.1. Examples

To give some practice in reading spectral densities, we used the Matlab program `bigshow2.m` to generate Figures 2.10.2, 2.10.3, 2.10.1, and 2.10.4 The program

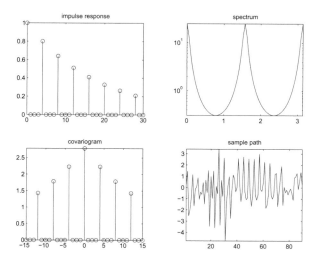

Figure 2.10.3: Impulse response, spectrum, covariogram, and sample path of process $(1 - .8L^4)y_t = w_t$.

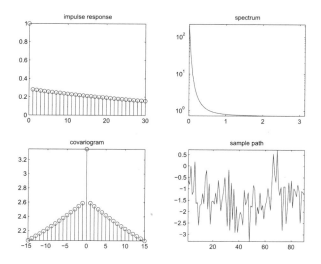

Figure 2.10.4: Impulse response, spectrum, covariogram, and sample path of process $(1 - .98L)y_t = (1 - .7L)w_t$.

takes as an input a univariate process of the form

$$a(L)y_t = b(L)w_t,$$

where w_t is a univariate martingale difference sequence with unit variance, where $a(L) = 1 - a_2 L - a_3 L^2 - \cdots - a_n L^{n-1}$ and $b(L) = b_1 + b_2 L + \cdots + b_n L^{n-1}$, and where we require that $a(z) = 0$ imply that $|z| > 1$. The program computes and displays a realization of the process, the impulse response function from w to y, and the spectrum of y. By using this program, a reader can teach himself to read spectra and impulse response functions. Figure 2.10.2 is for the pure autoregressive process with $a(L) = 1 - .9L, b = 1$. The spectrum sweeps downward in what C.W.J. Granger (1966) called the "typical spectral shape" for an economic time series. Figure 2.10.3 sets $a = 1 - .8L^4, b = 1$. This is a process with a strong seasonal component. That the spectrum peaks at π and $\pi/2$ is a telltale sign of a strong seasonal component. Figure 2.10.1 sets $a = 1 - 1.3L + .7L^2, b = 1$. This is a process that has a spectral peak in the interior of $(0, \pi)$ and cycles in its covariogram.[27] Figure 2.10.4 sets a $= 1 - .98L, b = 1 - .7L$. This is a version of a process studied by Muth (1960). After the first lag, the impulse response declines as $.99^j$, where j is the lag length.

2.11.   Example: the LQ permanent income model

To illustrate several of the key ideas of this chapter, this section describes the linear quadratic savings problem whose solution is a rational expectations version of the permanent income model of Friedman (1956) and Hall (1978). We use this model as a vehicle for illustrating impulse response functions, alternative notions of the state, the idea of cointegration, and an invariant subspace method.

The LQ permanent income model is a modification (and not quite a special case, for reasons that will be apparent later) of the following "savings problem" to be studied in chapter 17. A consumer has preferences over consumption streams that are ordered by the utility functional

$$E_0 \sum_{t=0}^{\infty} \beta^t u (c_t) \tag{2.11.1}$$

where E_t is the mathematical expectation conditioned on the consumer's time t information, c_t is time t consumption, $u(c)$ is a strictly concave one-period

---

[27] See Sargent (1987a) for a more extended discussion.

utility function, and $\beta \in (0, 1)$ is a discount factor. The consumer maximizes
$(2.11.1)$ by choosing a consumption, borrowing plan $\{c_t, b_{t+1}\}_{t=0}^{\infty}$ subject to the
sequence of budget constraints

$$c_t + b_t = R^{-1}b_{t+1} + y_t \tag{2.11.2}$$

where y_t is an exogenous stationary endowment process, R is a constant gross
risk-free interest rate, b_t is one-period risk-free debt maturing at t, and b_0 is a
given initial condition. We shall assume that $R^{-1} = \beta$. For example, we might
assume that the endowment process has the state-space representation

$$z_{t+1} = A_{22}z_t + C_2 w_{t+1} \tag{2.11.3a}$$
$$y_t = U_y z_t \tag{2.11.3b}$$

where w_{t+1} is an i.i.d. process with mean zero and identity contemporaneous
covariance matrix, A_{22} is a stable matrix, its eigenvalues being strictly below
unity in modulus, and U_y is a selection vector that identifies y with a particular linear combination of the z_t. We impose the following condition on the
consumption, borrowing plan:

$$E_0 \sum_{t=0}^{\infty} \beta^t b_t^2 < +\infty. \tag{2.11.4}$$

This condition suffices to rule out Ponzi schemes. The *state* vector confronting
the household at t is $\begin{bmatrix} b_t & z_t \end{bmatrix}'$, where b_t is its one-period debt falling due at the
beginning of period t and z_t contains all variables useful for forecasting its future
endowment. We impose this condition to rule out an always-borrow scheme that
would allow the household to enjoy bliss consumption forever. The rationale for
imposing this condition is to make the solution resemble the solution of problems
to be studied in chapter 17 that impose nonnegativity on the consumption path.
First-order conditions for maximizing $(2.11.1)$ subject to $(2.11.2)$ are[28]

$$E_t u'(c_{t+1}) = u'(c_t), \quad \forall t \geq 0. \tag{2.11.5}$$

For the rest of this section we assume the quadratic utility function $u(c_t) = -.5(c_t - \gamma)^2$, where γ is a bliss level of consumption. Then $(2.11.5)$ implies[29]

$$E_t c_{t+1} = c_t. \tag{2.11.6}$$

---

[28] We shall study how to derive this first-order condition in detail in later chapters.
[29] A linear marginal utility is essential for deriving $(2.11.6)$ from $(2.11.5)$. Suppose instead
that we had imposed the following more standard assumptions on the utility function: $u'(c) >$

Along with the quadratic utility specification, we allow consumption c_t to be negative.[30]

   To deduce the optimal decision rule, we have to solve the system of difference equations formed by (2.11.2) and (2.11.6) subject to the boundary condition (2.11.4). To accomplish this, solve (2.11.2) forward and impose $\lim_{T \to +\infty} \beta^T b_{T+1} = 0$ to get

$$b_t = \sum_{j=0}^{\infty} \beta^j \left(y_{t+j} - c_{t+j} \right). \tag{2.11.7}$$

Imposing $\lim_{T \to +\infty} \beta^T b_{T+1} = 0$ suffices to impose (2.11.4) on the debt path. Take conditional expectations on both sides of (2.11.7) and use (2.11.6) and the law of iterated expectations to deduce

$$b_t = \sum_{j=0}^{\infty} \beta^j E_t y_{t+j} - \frac{1}{1 - \beta} c_t \tag{2.11.8}$$

or

$$c_t = (1 - \beta) \left[\sum_{j=0}^{\infty} \beta^j E_t y_{t+j} - b_t \right]. \tag{2.11.9}$$

If we define the net rate of interest r by $\beta = \frac{1}{1+r}$, we can also express this equation as

$$c_t = \frac{r}{1 + r} \left[\sum_{j=0}^{\infty} \beta^j E_t y_{t+j} - b_t \right]. \tag{2.11.10}$$

Equation (2.11.9) or (2.11.10) expresses consumption as equaling economic *income*, namely, a constant marginal propensity to consume or interest factor $\frac{r}{1+r}$ times the sum of nonfinancial wealth $\sum_{j=0}^{\infty} \beta^j E_t y_{t+j}$ and financial wealth $-b_t$. Notice that (2.11.9) or (2.11.10) represents c_t as a function of the *state* $[b_t, z_t]$

$0, u''(c) < 0, u'''(c) > 0$ and required that $c \geq 0$. The Euler equation remains (2.11.5). But the fact that $u''' < 0$ implies via Jensen's inequality that $E_t u'(c_{t+1}) > u'(E_t c_{t+1})$. This inequality together with (2.11.5) implies that $E_t c_{t+1} > c_t$ (consumption is said to be a 'submartingale'), so that consumption stochastically diverges to $+\infty$. The consumer's savings also diverge to $+\infty$. Chapter 17 discusses this 'precautionary savings' divergence result in depth.

   [30] That c_t can be negative explains why we impose condition (2.11.4) instead of an upper bound on the level of borrowing, such as the natural borrowing limit of chapters 8, 17, and 18.

confronting the household, where from $(2.11.3)$ z_t contains the information useful for forecasting the endowment process.

2.11.1. Another representation

Pulling together our preceding results, we can regard z_t, b_t as the time t state, where z_t is an *exogenous* component of the state and b_t is an *endogenous* component of the state vector. The system can be represented as

$$z_{t+1} = A_{22}z_t + C_2 w_{t+1}$$
$$b_{t+1} = b_t + U_y \left[(I - \beta A_{22})^{-1} (A_{22} - I) \right] z_t$$
$$y_t = U_y z_t$$
$$c_t = (1 - \beta) \left[U_y (I - \beta A_{22})^{-1} z_t - b_t \right].$$

Another way to understand the solution is to show that *after* the optimal decision rule has been obtained, there is a point of view that allows us to regard the state as being c_t together with z_t and to regard b_t as an outcome. Following Hall (1978), this is a sharp way to summarize the implication of the LQ permanent income theory. We now proceed to transform the state vector in this way.

To represent the solution for b_t, substitute $(2.11.9)$ into $(2.11.2)$ and after rearranging obtain

$$b_{t+1} = b_t + \left(\beta^{-1} - 1 \right) \sum_{j=0}^{\infty} \beta^j E_t y_{t+j} - \beta^{-1} y_t. \qquad (2.11.11)$$

Next, shift $(2.11.9)$ forward one period and eliminate b_{t+1} by using $(2.11.2)$ to obtain

$$c_{t+1} = (1 - \beta) \sum_{j=0}^{\infty} E_{t+1} \beta^j y_{t+j+1} - (1 - \beta) \left[\beta^{-1} (c_t + b_t - y_t) \right].$$

If we add and subtract $\beta^{-1}(1 - \beta) \sum_{j=0}^{\infty} \beta^j E_t y_{t+j}$ from the right side of the preceding equation and rearrange, we obtain

$$c_{t+1} - c_t = (1 - \beta) \sum_{j=0}^{\infty} \beta^j \left(E_{t+1} y_{t+j+1} - E_t y_{t+j+1} \right). \qquad (2.11.12)$$

The right side is the time $t+1$ innovation to the expected present value of the endowment process y. It is useful to express this innovation in terms of a moving average representation for income y_t. Suppose that the endowment process has the moving average representation[31]

$$y_{t+1} = d(L) w_{t+1} \tag{2.11.13}$$

where w_{t+1} is an i.i.d. vector process with $Ew_{t+1} = 0$ and contemporaneous covariance matrix $Ew_{t+1}w'_{t+1} = I$, $d(L) = \sum_{j=0}^{\infty} d_j L^j$, where L is the lag operator, and the household has an information set[32] $w^t = [w_t, w_{t-1}, \ldots,]$ at time t. Then notice that

$$y_{t+j} - E_t y_{t+j} = d_0 w_{t+j} + d_1 w_{t+j-1} + \cdots + d_{j-1} w_{t+1}.$$

It follows that

$$E_{t+1} y_{t+j} - E_t y_{t+j} = d_{j-1} w_{t+1}. \tag{2.11.14}$$

Using $(2.11.14)$ in $(2.11.12)$ gives

$$c_{t+1} - c_t = (1 - \beta) d(\beta) w_{t+1}. \tag{2.11.15}$$

The object $d(\beta)$ is the present value of the moving average coefficients in the representation for the endowment process y_t.

After all of this work, we can represent the optimal decision rule for c_t, b_{t+1} in the form of the two equations $(2.11.12)$ and $(2.11.8)$, which we repeat here for convenience:

$$c_{t+1} = c_t + (1 - \beta) \sum_{j=0}^{\infty} \beta^j \left(E_{t+1} y_{t+j+1} - E_t y_{t+j+1} \right) \tag{2.11.16}$$

$$b_t = \sum_{j=0}^{\infty} \beta^j E_t y_{t+j} - \frac{1}{1-\beta} c_t. \tag{2.11.17}$$

Equation $(2.11.17)$ asserts that the household's debt due at t equals the expected present value of of its endowment minus the expected present value of its

---

[31] Representation $(2.11.3)$ implies that $d(L) = U_y(I - A_{22}L)^{-1}C_2$.

[32] A moving average representation for a process y_t is said to be *fundamental* if the linear space spanned by y^t is equal to the linear space spanned by w^t. A time-invariant innovations representation, attained via the Kalman filter as in section 2.7, is by construction fundamental.

consumption stream. A high debt thus indicates a large expected present value of 'surpluses' $y_t - c_t$.

Recalling the form of the endowment process $(2.11.3)$, we can compute

$$E_t \sum_{j=0}^{\infty} \beta^j z_{t+j} = (I - \beta A_{22})^{-1} z_t$$

$$E_{t+1} \sum_{j=0}^{\infty} \beta^j z_{t+j+1} = (I - \beta A_{22})^{-1} z_{t+1}$$

$$E_t \sum_{j=0}^{\infty} \beta^j z_{t+j+1} = (I - \beta A_{22})^{-1} A_{22} z_t.$$

Substituting these formulas into $(2.11.16)$ and $(2.11.17)$ and using $(2.11.3a)$ gives the following representation for the consumer's optimum decision rule:[33]

$$c_{t+1} = c_t + (1 - \beta) U_y (I - \beta A_{22})^{-1} C_2 w_{t+1} \qquad (2.11.18a)$$

$$b_t = U_y (I - \beta A_{22})^{-1} z_t - \frac{1}{1 - \beta} c_t \qquad (2.11.18b)$$

$$y_t = U_y z_t \qquad (2.11.18c)$$

$$z_{t+1} = A_{22} z_t + C_2 w_{t+1} \qquad (2.11.18d)$$

Representation $(2.11.18)$ reveals several things about the optimal decision rule. (1) The *state* consists of the endogenous part c_t and the exogenous part z_t. These contain all of the relevant information for forecasting future c, y, b. Notice that financial assets b_t have disappeared as a component of the state because they are properly encoded in c_t. (2) According to $(2.11.18)$, consumption is a random walk with innovation $(1 - \beta)d(\beta)w_{t+1}$ as implied also by $(2.11.15)$. This outcome confirms that the Euler equation $(2.11.6)$ is built into the solution. That consumption is a random walk of course implies that it does not possess an asymptotic stationary distribution, at least so long as z_t exhibits perpetual random fluctuations, as it will generally under $(2.11.3)$.[34] This feature is inherited partly from the assumption that $\beta R = 1$. (3) The impulse

---

[33] See section B of chapter 8 for a reinterpretation of precisely these outcomes in terms of a competitive equilibrium of a model with a complete set of markets in history- and date-contingent claims to consumption.

[34] The failure of consumption to converge will occur again in chapter 17 when we drop quadratic utility and assume that consumption must be nonnegative.

response function of c_t is a box: for all $j \geq 1$, the response of c_{t+j} to an increase in the innovation w_{t+1} is $(1 - \beta)d(\beta) = (1 - \beta)U_y(I - \beta A_{22})^{-1}C_2$. (4) Solution $(2.11.18)$ reveals that the joint process c_t, b_t possesses the property that Granger and Engle (1987) called *cointegration*. In particular, *both* c_t and b_t are non-stationary because they have unit roots (see representation $(2.11.11)$ for b_t), but there is a linear combination of c_t, b_t that *is* stationary provided that z_t is stationary. From $(2.11.17)$, the linear combination is $(1 - \beta)b_t + c_t$. Accordingly, Granger and Engle would call $[(1 - \beta)\quad 1]$ a cointegrating vector that, when applied to the nonstationary vector process $[b_t \quad c_t]'$, yields a process that is asymptotically stationary. Equation $(2.11.8)$ can be arranged to take the form

$$(1 - \beta) b_t + c_t = (1 - \beta) E_t \sum_{j=0}^{\infty} \beta^j y_{t+j}, \qquad (2.11.19)$$

which asserts that the 'cointegrating residual' on the left side equals the conditional expectation of the geometric sum of future incomes on the right.[35]

2.11.2. Debt dynamics

If we subtract equation $(2.11.18b)$ evaluated at time t from equation $(2.11.18b)$ evaluated at time $t + 1$ we obtain

$$b_{t+1} - b_t = U_y (I - \beta A_{22})^{-1} (z_{t+1} - z_t) - \frac{1}{1 - \beta} (c_{t+1} - c_t).$$

Substituting $z_{t+1} - z_t = (A_{22} - I)z_t + C_2 w_{t+1}$ and equation $(2.11.18a)$ into the above equation and rearranging gives

$$b_{t+1} - b_t = U_y (I - \beta A_{22})^{-1} (A_{22} - I) z_t. \qquad (2.11.20)$$

---

[35] See Campbell and Shiller (1988) and Lettau and Ludvigson (2001, 2004) for interesting applications of related ideas.

2.11.3. Two classic examples

We illustrate formulas $(2.11.18)$ with the following two examples. In both examples, the endowment follows the process $y_t = z_{1t} + z_{2t}$ where

$$
\begin{bmatrix} z_{1t+1} \\ z_{2t+1} \end{bmatrix} = \begin{bmatrix} 1 & 0 \\ 0 & 0 \end{bmatrix} \begin{bmatrix} z_{1t} \\ z_{2t} \end{bmatrix} + \begin{bmatrix} \sigma_1 & 0 \\ 0 & \sigma_2 \end{bmatrix} \begin{bmatrix} w_{1t+1} \\ w_{2t+1} \end{bmatrix}
$$

where w_{t+1} is an i.i.d. 2×1 process distributed as $\mathcal{N}(0, I)$. Here z_{1t} is a permanent component of y_t while z_{2t} is a purely transitory component.

Example 1. Assume that the consumer observes the state z_t at time t. This implies that the consumer can construct w_{t+1} from observations of z_{t+1} and z_t. Application of formulas $(2.11.18)$ implies that

$$
c_{t+1} - c_t = \sigma_1 w_{1t+1} + (1 - \beta) \sigma_2 w_{2t+1}. \tag{2.11.21}
$$

Since $1 - \beta = \frac{r}{1+r}$ where $R = (1+r)$, formula $(2.11.21)$ shows how an increment $\sigma_1 w_{1t+1}$ to the permanent component of income z_{1t+1} leads to a permanent one-for-one increase in consumption and no increase in savings $-b_{t+1}$; but how the purely transitory component of income $\sigma_2 w_{2t+1}$ leads to a permanent increment in consumption by a fraction $(1 - \beta)$ of transitory income, while the remaining fraction β is saved, leading to a permanent increment in $-b$. Application of formula $(2.11.20)$ to this example shows that

$$
b_{t+1} - b_t = -z_{2t} = -\sigma_2 w_{2t}, \tag{2.11.22}
$$

which confirms that none of $\sigma_1 w_{1t}$ is saved, while all of $\sigma_2 w_{2t}$ is saved.

Example 2. Assume that the consumer observes y_t, and its history up to t, but not z_t at time t. Under this assumption, it is appropriate to use an *innovation representation* to form A_{22}, C_2, U_y in formulas $(2.11.18)$. In particular, using our results from section 2.9.1, the pertinent state space representation for y_t is

$$
\begin{bmatrix} y_{t+1} \\ a_{t+1} \end{bmatrix} = \begin{bmatrix} 1 & -(1-K) \\ 0 & 0 \end{bmatrix} \begin{bmatrix} y_t \\ a_t \end{bmatrix} + \begin{bmatrix} 1 \\ 1 \end{bmatrix} a_{t+1}
$$

$$
y_t = \begin{bmatrix} 1 & 0 \end{bmatrix} \begin{bmatrix} y_t \\ a_t \end{bmatrix}
$$

where K is the Kalman gain and $a_t = y_t - E[y_t | y^{t-1}]$. From subsection 2.9.1, we know that $K \in [0, 1]$ and that K increases as $\frac{\sigma_1^2}{\sigma_2^2}$ increases, i.e., as the ratio of the variance of the permanent shock to the variance of the transitory shock to income increases. Applying formulas $(2.11.18)$ implies

$$c_{t+1} - c_t = [1 - \beta(1 - K)]a_{t+1} \qquad (2.11.23)$$

where the endowment process can now be represented in terms of the univariate innovation to y_t as

$$y_{t+1} - y_t = a_{t+1} - (1 - K)a_t. \qquad (2.11.24)$$

Equation $(2.11.24)$ indicates that the consumer regards a fraction K of an innovation a_{t+1} to y_{t+1} as *permanent* and a fraction $1-K$ as purely transitory. He permanently increases his consumption by the full amount of his estimate of the permanent part of a_{t+1}, but by only $(1 - \beta)$ times his estimate of the purely transitory part of a_{t+1}. Therefore, in total he permanently increments his consumption by a fraction $K + (1 - \beta)(1 - K) = 1 - \beta(1 - K)$ of a_{t+1} and saves the remaining fraction $\beta(1 - K)$ of a_{t+1}. According to equation $(2.11.24)$, the first difference of income is a first-order moving average, while $(2.11.23)$ asserts that the first difference of consumption is i.i.d. Application of formula $(2.11.20)$ to this example shows that

$$b_{t+1} - b_t = (K - 1)a_t, \qquad (2.11.25)$$

which indicates how the fraction K of the innovation to y_t that is regarded as permanent influences the fraction of the innovation that is saved.

2.11.4. Spreading consumption cross section

Starting from an arbitrary initial distribution for c_0 and say the asymptotic stationary distribution for z_0, if we were to apply formulas $(2.4.11)$ and $(2.4.12)$ to the state space system $(2.11.18)$, the common unit root affecting c_t, b_t would cause the time t variance of c_t to grow linearly with t. If we think of the initial distribution as describing the joint distribution of c_0, b_0 for a cross section of ex ante identical households 'born at time 0, then these formulas would describe the evolution of the cross-section for b_t, c_t as the population of households ages. The distribution would spread out.[36]

---

[36] See Deaton and Paxton (1994) and Storesletten, Telmer, and Yaron (2004) for evidence that cross section distributions of consumption spread out with age.

2.11.5. Invariant subspace approach

We can glean additional insights about the structure of the optimal decision rule by solving the decision problem in a mechanical but quite revealing way that easily generalizes to a host of problems, as we shall see later in chapter 5. We can represent the system consisting of the Euler equation (2.11.6), the budget constraint (2.11.2), and the description of the endowment process (2.11.3) as

$$
\begin{bmatrix} \beta & 0 & 0 \\ 0 & I & 0 \\ 0 & 0 & 1 \end{bmatrix} \begin{bmatrix} b_{t+1} \\ z_{t+1} \\ c_{t+1} \end{bmatrix} = \begin{bmatrix} 1 & -U_y & 1 \\ 0 & A_{22} & 0 \\ 0 & 0 & 1 \end{bmatrix} \begin{bmatrix} b_t \\ z_t \\ c_t \end{bmatrix} + \begin{bmatrix} 0 \\ C_2 \\ C_1 \end{bmatrix} w_{t+1} \qquad (2.11.26)
$$

where C_1 is an undetermined coefficient. Premultiply both sides by the inverse of the matrix on the left and write

$$
\begin{bmatrix} b_{t+1} \\ z_{t+1} \\ c_{t+1} \end{bmatrix} = \tilde{A} \begin{bmatrix} b_t \\ z_t \\ c_t \end{bmatrix} + \tilde{C} w_{t+1}. \qquad (2.11.27)
$$

We want to find solutions of (2.11.27) that satisfy the no-explosion condition (2.11.4). We can do this by using machinery to be introduced in chapter 5. The key idea is to discover what part of the vector $[\, b_t \quad z_t \quad c_t \,]'$ is truly a *state* from the view of the decision maker, being inherited from the past, and what part is a *costate* or *jump* variable that can adjust at t. For our problem b_t, z_t are truly components of the state, but c_t is free to adjust. The theory determines c_t at t as a function of the true state variables $[b_t, z_t]$. A powerful approach to determining this function is the following so-called invariant subspace method of chapter 5. Obtain the eigenvector decomposition of $\tilde{A}$:

$$
\tilde{A} = V \Lambda V^{-1}
$$

where Λ is a diagonal matrix consisting of the eigenvalues of $\tilde{A}$ and V is a matrix of the associated eigenvectors. Let $V^{-1} \equiv \begin{bmatrix} V^{11} & V^{12} \\ V^{21} & V^{22} \end{bmatrix}$. Then applying formula (5.5.11) of chapter 5 implies that if (2.11.4) is to hold, the jump variable c_t must satisfy

$$
c_t = - \left(V^{22} \right)^{-1} V^{21} \begin{bmatrix} b_t \\ z_t \end{bmatrix}. \qquad (2.11.28)
$$

Formula (2.11.28) gives the unique value of c_t that ensures that (2.11.4) is satisfied, or in other words, that the state remains in the "stabilizing subspace."

Notice that the variables on the right side of (2.11.28) conform with those called for by (2.11.10): $-b_t$ is there as a measure of financial wealth, and z_t is there because it includes all variables that are useful for forecasting the future endowments that appear in (2.11.10).

2.12.　Concluding remarks

In addition to giving us tools for thinking about time series, the Markov chain and the stochastic linear difference equation have each introduced us to the notion of the state vector as a description of the present position of a system.[37] Subsequent chapters use both Markov chains and stochastic linear difference equations. In the next chapter we study decision problems in which the goal is optimally to manage the evolution of a state vector that can be partially controlled.

A.　Linear difference equations

2.A.1. *A first-order difference equation*

This section describes the solution of a linear first-order scalar difference equation. First, let $|\lambda| < 1$, and let $\{u_t\}_{t=-\infty}^{\infty}$ be a bounded sequence of scalar real numbers. Let L be the lag operator defined by $Lx_t \equiv x_{t-1}$ and let L^{-1} be the forward shift operator defined by $L^{-1}x_t \equiv x_{t+1}$. Then

$$(1 - \lambda L)\, y_t = u_t, \forall t \tag{2.A.1}$$

has the solution

$$y_t = (1 - \lambda L)^{-1} u_t + k\lambda^t \tag{2.A.2}$$

---

[37]　See Quah (1990) and Blundell and Preston (1998) for applications of some of the tools of this chapter and of chapter 5 to studying some puzzles associated with a permanent income model.

for any real number k. You can verify this fact by applying $(1 - \lambda L)$ to both sides of equation $(2.A.2)$ and noting that $(1 - \lambda L)\lambda^t = 0$. To pin down k we need one condition imposed from outside (e.g., an initial or terminal condition) on the path of y.

Now let $|\lambda| > 1$. Rewrite equation $(2.A.1)$ as

$$y_{t-1} = \lambda^{-1} y_t - \lambda^{-1} u_t, \forall t \tag{2.A.3}$$

or

$$\left(1 - \lambda^{-1} L^{-1}\right) y_t = -\lambda^{-1} u_{t+1}. \tag{2.A.4}$$

A solution is

$$y_t = -\lambda^{-1} \left(\frac{1}{1 - \lambda^{-1} L^{-1}} \right) u_{t+1} + k\lambda^t \tag{2.A.5}$$

for any k. To verify that this is a solution, check the consequences of operating on both sides of equation $(2.A.5)$ by $(1 - \lambda L)$ and compare to $(2.A.1)$.

Solution $(2.A.2)$ exists for $|\lambda| < 1$ because the distributed lag in u converges. Solution $(2.A.5)$ exists when $|\lambda| > 1$ because the distributed lead in u converges. When $|\lambda| > 1$, the distributed lag in u in $(2.A.2)$ may diverge, so that a solution of this form does not exist. The distributed lead in u in $(2.A.5)$ need not converge when $|\lambda| < 1$.

2.A.2. A second-order difference equation

Now consider the second order difference equation

$$(1 - \lambda_1 L)(1 - \lambda_2 L) y_{t+1} = u_t \tag{2.A.6}$$

where $\{u_t\}$ is a bounded sequence, y_0 is an initial condition, $|\lambda_1| < 1$ and $|\lambda_2| > 1$. We seek a bounded sequence $\{y_t\}_{t=0}^{\infty}$ that satisfies $(2.A.6)$. Using insights from the previous subsection, operate on both sides of $(2.A.6)$ by the forward inverse of $(1 - \lambda_2 L)$ to rewrite equation $(2.A.6)$ as

$$(1 - \lambda_1 L) y_{t+1} = -\frac{\lambda_2^{-1}}{1 - \lambda_2^{-1} L^{-1}} u_{t+1}$$

or

$$y_{t+1} = \lambda_1 y_t - \lambda_2^{-1} \sum_{j=0}^{\infty} \lambda_2^{-j} u_{t+j+1}. \tag{2.A.7}$$

Thus, we obtained equation $(2.A.7)$ by "solving stable roots (in this case λ_1 backward, and unstable roots (in this case λ_2) forward". Equation $(2.A.7)$ has a form that we shall encounter often. $\lambda_1 y_t$ is called the 'feedback part' and $-\frac{\lambda_2^{-1}}{1-\lambda_2^{-1}L^{-1}}u_{t+1}$ is called the "feed-forward part' of the solution. We have already encountered solutions of this form. Thus, notice that equation $(2.11.20)$ from subsection 2.11.2 is almost of this form, 'almost' because in equation $(2.11.20)$, $\lambda_1 = 1$. In section 5.5 of chapter 5 we return to these ideas in a more general setting.

B. MCMC approximation of Bayesian posterior

The last twenty years witnessed impressive advances in numerical methods for computing Bayesian and maximum likelihood estimators. In this appendix, we briefly describe a Markov Chain Monte Carlo method that constructs a Bayesian posterior distribution by forming a Markov chain whose invariant distribution equals that posterior distribution.

In the Bayesian method, the following objects are in play:

1. A sample of data y_0^T.

2. A vector $\theta \in \Theta$ of free parameters describing the preferences, technology, and information sets of an economic model.

3. A prior probability distribution $\tilde{p}(\theta)$ over the parameters.

4. A mapping from θ to a state-space representation of an equilibrium of a economic dynamic model. We present examples of this mapping in sections 2.11 and appendix B of chapter 14 and chapters 5 and 7.

5. As described in section 2.7, a mapping from a state space representation of an equilibrium of an economic model to an innovations representation via the Kalman filter and thereby to a recursive representation of a Gaussian log likelihood function

$$\log L\left(\theta|y_0^T\right) = -.5\left(T+1\right)k\log\left(2\pi\right) - .5\sum_{t=0}^{T}\log|\Omega_t| - .5\sum_{t=0}^{T}a_t'\Omega_t^{-1}a_t,$$

where $\Omega_t = Ea_t a_t'$.

6. A posterior probability

$$\tilde{p}\left(\theta|y_0^T\right) = \frac{p\left(y_0^T|\theta\right)\tilde{p}\left(\theta\right)}{\int p\left(y_0^T|\theta\right)\tilde{p}\left(\theta\right)d\theta}$$

where the denominator is the marginal density of y_0^T.

Our goal is to compute the posterior $\tilde{p}(\theta|y_0^T)$. The Markov Chain Monte Carlo (MCMC) method constructs a Markov chain on a state space Θ for which $\theta \in \Theta$ and such that

a. The chain is easy to sample from.

b. The chain has a unique invariant distribution $\pi(\theta)$.

c. The invariant distribution equals the posterior: $\pi(\theta) = \tilde{p}(\theta|y_0^T)$.

Two key ingredients of the *Metropolis-Hastings algorithm* are

i. The *target* density $\tilde{p}(\theta|y_0^T)$.

ii. A *proposal* or *jumping* density $q(z|\theta; y_0^T)$.

The proposal density should be a good guess at $\tilde{p}(\theta|y_0^T)$. For our applications, a standard choice of a proposal density comes from adjusting the asymptotic distribution associated with the maximum likelihood estimator, $\theta \sim \mathcal{N}(\hat{\theta}_{ML}, \Sigma_\theta)$ where $\Sigma_\theta = V^{-1}$ and $V = \frac{\partial^2 \log L(\theta|y_0^T)}{\partial\theta\partial\theta'}\Big|_{\theta_{ML}}$. A common choice of a proposal density is:

$$q\left(\theta^*|\theta_j, y_0^T\right) = \mathcal{N}\left(\theta_j, c\Sigma_\theta\right) \tag{2.B.1}$$

where c is a scale parameter. Define the *kernel* $\kappa(\theta|y_0^T)$ by

$$\log\kappa\left(\theta|y_0^T\right) = \log L\left(\theta|y_0^T\right) + \log\tilde{p}\left(\theta\right).$$

Note that

$$\tilde{p}\left(\theta|y_0^T\right) \propto \kappa\left(\theta|y_0^T\right)$$

where the factor of proportionality is the integrating constant $\int p(y_0^T|\theta)\tilde{p}(\theta)d\theta$.

The Metropolis-Hastings algorithm defines a Markov chain on Θ by these steps:[38]

---

[38] See Robert and Casella (2004, ch. 7) for discussions of this algorithm and conditions for convergence.

1. Draw θ_0, $j = 0$.

2. For $j \geq 0$, draw θ^* from $q(\theta^*|\theta_j, y_0^T)$; θ^* is a "candidate" for the next draw of θ_j.

3. Randomly decide whether to accept this candidate by first computing the probability of acceptance

$$r = \frac{\tilde{p}\left(\theta^*|y_0^T\right)}{\tilde{p}\left(\theta_j|y_0^T\right)} = \frac{\kappa\left(\theta^*|y_0^T\right)}{\kappa\left(\theta_j|y_0^T\right)}.$$

(Note that in this step, we only have to compute the kernels, not the integrating constant $\int p(y_0^T|\theta)\tilde{p}(\theta)d\theta$.) Then set

$$\theta_{j+1} = \begin{cases} \theta^* & \text{with probability } \min\left(r, 1\right); \\ \theta_j & \text{otherwise.} \end{cases}$$

This algorithm defines the transition density of a Markov chain mapping θ_j into θ_{j+1}. Let the transition density be $\text{Prob}(\theta_{j+1} = \theta^*|\theta_j = \theta) = \Pi(\theta, \theta^*)$. Then we have the following

PROPOSITION: The invariant distribution of the chain is the posterior:

$$\tilde{p}\left(\theta^*|y_0^T\right) = \int \Pi\left(\theta, \theta^*\right)\tilde{p}\left(\theta|y_0^T\right)d\theta.$$

Two practical concerns associated with the Metropolis-Hastings algorithm are, first, whether the chain converges, and, second, the rate of convergence. The literature on MCMC has developed practical diagnostics for checking convergence, and it is important to use these thoughtfully. The rate of convergence is influenced by the acceptance rate, which can be influenced by choice of the scale parameter c when the proposal density is chosen as recommended above. Common piece chooses c to give an acceptance rate between .2 and .4.

Dynare computes maximum likelihood and Bayesian estimates using the above algorithm. See Barillas, Bhandari, Bigio, Colacito, Juillard, Kitao, Matthes, Sargent, and Shin (2012) for some examples.[39]

---

[39] See <http://www.dynare.org>.

Exercises

Exercise 2.1 Consider the Markov chain $(P, \pi_0) = \left(\begin{bmatrix} .9 & .1 \\ .3 & .7 \end{bmatrix}, \begin{bmatrix} .5 \\ .5 \end{bmatrix} \right)$, and a random variable $y_t = \bar{y} x_t$ where $\bar{y} = \begin{bmatrix} 1 \\ 5 \end{bmatrix}$. Compute the likelihood of the following three histories for y_t for $t = 0, 1, \ldots, 4$:

a. $1, 5, 1, 5, 1$.

b. $1, 1, 1, 1, 1$.

c. $5, 5, 5, 5, 5$.

Exercise 2.2 Consider a two-state Markov chain. Consider a random variable $y_t = \bar{y} x_t$ where $\bar{y} = \begin{bmatrix} 1 \\ 5 \end{bmatrix}$. It is known that $E(y_{t+1}|x_t) = \begin{bmatrix} 1.8 \\ 3.4 \end{bmatrix}$ and that $E(y_{t+1}^2|x_t) = \begin{bmatrix} 5.8 \\ 15.4 \end{bmatrix}$. Find a transition matrix consistent with these conditional expectations. Is this transition matrix unique (i.e., can you find another one that is consistent with these conditional expectations)?

Exercise 2.3 Consumption is governed by an n-state Markov chain P, π_0 where P is a stochastic matrix and π_0 is an initial probability distribution. Consumption takes one of the values in the $n \times 1$ vector $\bar{c}$. A consumer ranks stochastic processes of consumption $t = 0, 1 \ldots$ according to

$$E \sum_{t=0}^{\infty} \beta^t u(c_t)$$

where E is the mathematical expectation and $u(c) = \frac{c^{1-\gamma}}{1-\gamma}$ for some parameter $\gamma \geq 1$. Let $u_i = u(\bar{c}_i)$. Let $v_i = E[\sum_{t=0}^{\infty} \beta^t u(c_t)|x_0 = \bar{e}_i]$ and $V = Ev$, where $\beta \in (0, 1)$ is a discount factor.

a. Let u and v be the $n \times 1$ vectors whose ith components are u_i and v_i, respectively. Verify the following formulas for v and V: $v = (I - \beta P)^{-1} u$, and $V = \sum_i \pi_{0,i} v_i$.

b. Consider the following two Markov processes:

Process 1: $\pi_0 = \begin{bmatrix} .5 \\ .5 \end{bmatrix}$, $P = \begin{bmatrix} 1 & 0 \\ 0 & 1 \end{bmatrix}$.

Process 2: $\pi_0 = \begin{bmatrix} .5 \\ .5 \end{bmatrix}$, $P = \begin{bmatrix} .5 & .5 \\ .5 & .5 \end{bmatrix}$.

For both Markov processes, $\bar{c} = \begin{bmatrix} 1 \\ 5 \end{bmatrix}$.

Assume that $\gamma = 2.5, \beta = .95$. Compute the unconditional discounted expected utility V for each of these processes. Which of the two processes does the consumer prefer? Redo the calculations for $\gamma = 4$. Now which process does the consumer prefer?

c. An econometrician observes a sample of 10 observations of consumption rates for our consumer. He knows that one of the two preceding Markov processes generates the data, but he does not know which one. He assigns equal "prior probability" to the two chains. Suppose that the 10 successive observations on consumption are as follows: $1, 1, 1, 1, 1, 1, 1, 1, 1, 1$. Compute the likelihood of this sample under process 1 and under process 2. Denote the likelihood function $\text{Prob}(\text{data}|\text{Model}_i), i = 1, 2$.

d. Suppose that the econometrician uses Bayes' law to revise his initial probability estimates for the two models, where in this context Bayes' law states:

$$\text{Prob}(M_i)\,|\text{data} = \frac{(\text{Prob}(\text{data})|M_i) \cdot \text{Prob}(M_i)}{\sum_j \text{Prob}(\text{data})|M_j \cdot \text{Prob}(M_j)}$$

where M_i denotes model i. The denominator of this expression is the unconditional probability of the data. After observing the data sample, what probabilities does the econometrician place on the two possible models?

e. Repeat the calculation in part d, but now assume that the data sample is $1, 5, 5, 1, 5, 5, 1, 5, 1, 5$.

Exercise 2.4   Consider the univariate stochastic process

$$y_{t+1} = \alpha + \sum_{j=1}^{4} \rho_j y_{t+1-j} + c w_{t+1}$$

where w_{t+1} is a scalar martingale difference sequence adapted to $J_t = [w_t, \ldots, w_1, y_0, y_{-1}, y_{-2}, y_{-3}]$, $\alpha = \mu(1 - \sum_j \rho_j)$ and the ρ_j's are such that the matrix

$$A = \begin{bmatrix} \rho_1 & \rho_2 & \rho_3 & \rho_4 & \alpha \\ 1 & 0 & 0 & 0 & 0 \\ 0 & 1 & 0 & 0 & 0 \\ 0 & 0 & 1 & 0 & 0 \\ 0 & 0 & 0 & 0 & 1 \end{bmatrix}$$

has all of its eigenvalues in modulus bounded below unity.

a. Show how to map this process into a first-order linear stochastic difference equation.

b. For each of the following examples, if possible, assume that the initial conditions are such that y_t is covariance stationary. For each case, state the appropriate initial conditions. Then compute the covariance stationary mean and variance of y_t assuming the following parameter sets of parameter values:

i. $\rho = [1.2 \quad -.3 \quad 0 \quad 0]$, $\mu = 10, c = 1$.

ii. $\rho = [1.2 \quad -.3 \quad 0 \quad 0]$, $\mu = 10, c = 2$.

iii. $\rho = [.9 \quad 0 \quad 0 \quad 0]$, $\mu = 5, c = 1$.

iv. $\rho = [.2 \quad 0 \quad 0 \quad .5]$, $\mu = 5, c = 1$.

v. $\rho = [.8 \quad .3 \quad 0 \quad 0]$, $\mu = 5, c = 1$.

Hint 1: The Matlab program `doublej.m`, in particular, the command `X=doublej(A,C*C')` computes the solution of the matrix equation $AXA' + CC' = X$. This program can be downloaded from $<$https://files.nyu.edu/ts43/public/books.html$>$.

Hint 2: The mean vector is the eigenvector of A associated with a unit eigenvalue, scaled so that the mean of unity in the state vector is unity.

c. For each case in part b, compute the h_j's in $E_t y_{t+5} = \gamma_0 + \sum_{j=0}^{3} h_j y_{t-j}$.

d. For each case in part b, compute the $\tilde{h}_j$'s in $E_t \sum_{k=0}^{\infty} .95^k y_{t+k} = \sum_{j=0}^{3} \tilde{h}_j y_{t-j}$.

e. For each case in part b, compute the autocovariance $E(y_t - \mu_y)(y_{t-k} - \mu_y)$ for the three values $k = 1, 5, 10$.

Exercise 2.5 A consumer's rate of consumption follows the stochastic process

(1)
$$c_{t+1} = \alpha_c + \sum_{j=1}^{2} \rho_j c_{t-j+1} + \sum_{j=1}^{2} \delta_j z_{t+1-j} + \psi_1 w_{1,t+1}$$

$$z_{t+1} = \sum_{j=1}^{2} \gamma_j c_{t-j+1} + \sum_{j=1}^{2} \phi_j z_{t-j+1} + \psi_2 w_{2,t+1}$$

where w_{t+1} is a 2×1 martingale difference sequence, adapted to $J_t = \begin{bmatrix} w_t & \ldots w_1 & c_0 & c_{-1} & z_0 & z_{-1} \end{bmatrix}$, with contemporaneous covariance matrix $E w_{t+1} w'_{t+1} | J_t = I$, and the coefficients $\rho_j, \delta_j, \gamma_j, \phi_j$ are such that the matrix

$$A = \begin{bmatrix} \rho_1 & \rho_2 & \delta_1 & \delta_2 & \alpha_c \\ 1 & 0 & 0 & 0 & 0 \\ \gamma_1 & \gamma_2 & \phi_1 & \phi_2 & 0 \\ 0 & 0 & 1 & 0 & 0 \\ 0 & 0 & 0 & 0 & 1 \end{bmatrix}$$

has eigenvalues bounded strictly below unity in modulus.

The consumer evaluates consumption streams according to

(2) $$V_0 = E_0 \sum_{t=0}^{\infty} .95^t u(c_t),$$

where the one-period utility function is

(3) $$u(c_t) = -.5(c_t - 60)^2.$$

a. Find a formula for V_0 in terms of the parameters of the one-period utility function (3) and the stochastic process for consumption.

b. Compute V_0 for the following two sets of parameter values:

i. $\rho = \begin{bmatrix} .8 & -.3 \end{bmatrix}, \alpha_c = 1, \delta = \begin{bmatrix} .2 & 0 \end{bmatrix}, \gamma = \begin{bmatrix} 0 & 0 \end{bmatrix}, \phi = \begin{bmatrix} .7 & -.2 \end{bmatrix}, \psi_1 = \psi_2 = 1$.

ii. Same as for part i except now $\psi_1 = 2, \psi_2 = 1$.

Hint: Remember `doublej.m`.

Exercise 2.6 Consider the stochastic process $\{c_t, z_t\}$ defined by equations (1) in exercise 2.5. Assume the parameter values described in part b, item i. If possible, assume the initial conditions are such that $\{c_t, z_t\}$ is covariance stationary.

a. Compute the initial mean and covariance matrix that make the process covariance stationary.

b. For the initial conditions in part a, compute numerical values of the following population linear regression:

$$c_{t+2} = \alpha_0 + \alpha_1 z_t + \alpha_2 z_{t-4} + w_t$$

where $Ew_t [1 \quad z_t \quad z_{t-4}] = [0 \quad 0 \quad 0]$.

Exercise 2.7 Get the Matlab programs `bigshow2.m` and `freq.m` from < https://files.nyu.edu/ts43/public/books.html >. Use `bigshow2` to compute and display a simulation of length 80, an impulse response function, and a spectrum for each of the following scalar stochastic processes y_t. In each of the following, w_t is a scalar martingale difference sequence adapted to its own history and the initial values of lagged y's.

a. $y_t = w_t$.

b. $y_t = (1 + .5L)w_t$.

c. $y_t = (1 + .5L + .4L^2)w_t$.

d. $(1 - .999L)y_t = (1 - .4L)w_t$.

e. $(1 - .8L)y_t = (1 + .5L + .4L^2)w_t$.

f. $(1 + .8L)y_t = w_t$.

g. $y_t = (1 - .6L)w_t$.

Study the output and look for patterns. When you are done, you will be well on your way to knowing how to read spectral densities.

Exercise 2.8 This exercise deals with Cagan's money demand under rational expectations. A version of Cagan's (1956) demand function for money is

$$(1) \qquad\qquad m_t - p_t = -\alpha\,(p_{t+1} - p_t)\,, \alpha > 0,\ t \geq 0,$$

where m_t is the log of the nominal money supply and p_t is the price level at t. Equation (1) states that the demand for real balances varies inversely with the expected rate of inflation, $(p_{t+1} - p_t)$. There is no uncertainty, so the expected inflation rate equals the actual one. The money supply obeys the difference equation

$$(2) \qquad\qquad (1 - L)\,(1 - \rho L)\, m_t^s = 0$$

subject to initial condition for m_{-1}^s, m_{-2}^s. In equilibrium,

$$(3) \qquad\qquad m_t \equiv m_t^s \quad \forall t \geq 0$$

(i.e., the demand for money equals the supply). For now assume that

(4) $$|\rho\alpha/(1+\alpha)| < 1.$$

An *equilibrium* is a $\{p_t\}_{t=0}^\infty$ that satisfies equations (1), (2), and (3) for all t.

a. Find an expression for an equilibrium p_t of the form

(5) $$p_t = \sum_{j=0}^{n} w_j m_{t-j} + f_t.$$

Please tell how to get formulas for the w_j for all j and the f_t for all t.

b. How many equilibria are there?

c. Is there an equilibrium with $f_t = 0$ for all t?

d. Briefly tell where, if anywhere, condition (4) plays a role in your answer to part a.

e. For the parameter values $\alpha = 1, \rho = 1$, compute and display all the equilibria.

Exercise 2.9   The $n \times 1$ state vector of an economy is governed by the linear stochastic difference equation

(1) $$x_{t+1} = Ax_t + C_t w_{t+1}$$

where C_t is a possibly time-varying matrix (known at t) and w_{t+1} is an $m \times 1$ martingale difference sequence adapted to its own history with $E w_{t+1} w'_{t+1} | J_t = I$, where $J_t = \begin{bmatrix} w_t & \cdots & w_1 & x_0 \end{bmatrix}$. A scalar one-period payoff p_{t+1} is given by

(2) $$p_{t+1} = Px_{t+1}$$

The stochastic discount factor for this economy is a scalar m_{t+1} that obeys

(3) $$m_{t+1} = \frac{Mx_{t+1}}{Mx_t}.$$

Finally, the price at time t of the one-period payoff is given by $q_t = f_t(x_t)$, where f_t is some possibly time-varying function of the state. That m_{t+1} is a stochastic discount factor means that

(4) $$E(m_{t+1}p_{t+1}|J_t) = q_t.$$

a. Compute $f_t(x_t)$, describing in detail how it depends on A and C_t.

b. Suppose that an econometrician has a time series data set $X_t = [z_t \quad m_{t+1} \quad p_{t+1} \quad q_t]$, for $t = 1, \ldots, T$, where z_t is a strict subset of the variables in the state x_t. Assume that investors in the economy see x_t even though the econometrician sees only a subset z_t of x_t. Briefly describe a way to use these data to test implication (4). (Possibly but perhaps not useful hint: recall the law of iterated expectations.)

Exercise 2.10 Let P be a transition matrix for a Markov chain. Suppose that P' has two distinct eigenvectors π_1, π_2 corresponding to unit eigenvalues of P'. Scale π_1 and π_2 so that they are vectors of probabilities (i.e., elements are nonnegative and sum to unity). Prove for any $\alpha \in [0,1]$ that $\alpha\pi_1 + (1-\alpha)\pi_2$ is an invariant distribution of P.

Exercise 2.11 Consider a Markov chain with transition matrix

$$P = \begin{bmatrix} 1 & 0 & 0 \\ .2 & .5 & .3 \\ 0 & 0 & 1 \end{bmatrix}$$

with initial distribution $\pi_0 = [\pi_{1,0} \quad \pi_{2,0} \quad \pi_{3,0}]'$. Let $\pi_t = [\pi_{1t} \quad \pi_{2t} \quad \pi_{3t}]'$ be the distribution over states at time t. Prove that for $t > 0$

$$\pi_{1t} = \pi_{1,0} + .2 \left(\frac{1 - .5^t}{1 - .5} \right) \pi_{2,0}$$

$$\pi_{2t} = .5^t \pi_{2,0}$$

$$\pi_{3t} = \pi_{3,0} + .3 \left(\frac{1 - .5^t}{1 - .5} \right) \pi_{2,0}.$$

Exercise 2.12 Let P be a transition matrix for a Markov chain. For $t = 1, 2, \ldots$, prove that the jth column of $(P')^t$ is the distribution across states at t when the initial distribution is $\pi_{j,0} = 1, \pi_{i,0} = 0 \forall i \neq j$.

Exercise 2.13 A household has preferences over consumption processes $\{c_t\}_{t=0}^\infty$ that are ordered by

$$-.5 \sum_{t=0}^\infty \beta^t \left[(c_t - 30)^2 + .000001 b_t^2 \right] \tag{1}$$

where $\beta = .95$. The household chooses a consumption, borrowing plan to maximize (1) subject to the sequence of budget constraints

$$c_t + b_t = \beta b_{t+1} + y_t$$

for $t \geq 0$, where b_0 is an initial condition, β^{-1} is the one-period gross risk-free interest rate, b_t is the household's one-period debt that is due in period t, and y_t is its labor income, which obeys the second-order autoregressive process

$$\left(1 - \rho_1 L - \rho_2 L^2\right) y_{t+1} = (1 - \rho_1 - \rho_2)\, 5 + .05 w_{t+1}$$

where $\rho_1 = 1.3$, $\rho_2 = -.4$.

a. Define the *state* of the household at t as $x_t = \begin{bmatrix} 1 & b_t & y_t & y_{t-1} \end{bmatrix}'$ and the *control* as $u_t = (c_t - 30)$. Then express the transition law facing the household in the form $(2.4.22)$. Compute the eigenvalues of A. Compute the zeros of the characteristic polynomial $(1 - \rho_1 z - \rho_2 z^2)$ and compare them with the eigenvalues of A. (*Hint:* To compute the zeros in Matlab, set $a = \begin{bmatrix} .4 & -1.3 & 1 \end{bmatrix}$ and call `roots(a)`. The zeros of $(1 - \rho_1 z - \rho_2 z^2)$ equal the *reciprocals* of the eigenvalues of the associated A.)

b. Write a Matlab program that uses the Howard improvement algorithm $(2.4.30)$ to compute the household's optimal decision rule for $u_t = c_t - 30$. Tell how many iterations it takes for this to converge (also tell your convergence criterion).

c. Use the household's optimal decision rule to compute the law of motion for x_t under the optimal decision rule in the form

$$x_{t+1} = (A - BF^*)\, x_t + C w_{t+1},$$

where $u_t = -F^* x_t$ is the optimal decision rule. Using Matlab, compute the impulse response function of $\begin{bmatrix} c_t & b_t \end{bmatrix}'$ to w_{t+1}. Compare these with the theoretical expressions $(2.11.18)$.

Exercise 2.14   Consider a Markov chain with transition matrix

$$P = \begin{bmatrix} .5 & .5 & 0 & 0 \\ .1 & .9 & 0 & 0 \\ 0 & 0 & .9 & .1 \\ 0 & 0 & 0 & 1 \end{bmatrix}$$

with state space $X = \{e_i, i = 1, \ldots, 4\}$ where e_i is the ith unit vector. A random variable y_t is a function $y_t = \begin{bmatrix} 1 & 2 & 3 & 4 \end{bmatrix} x_t$ of the underlying state.

a. Find all stationary distributions of the Markov chain.

b. Can you find a stationary distribution for which the Markov chain ergodic?

c. Compute all possible limiting values of the sample mean $\frac{1}{T} \sum_{t=0}^{T-1} y_t$ as $T \to \infty$.

Exercise 2.15 Suppose that a scalar is related to a scalar white noise w_t with variance 1 by $y_t = h(L)w_t$ where $h(L) = \sum_{j=0}^{\infty} L^j h_j$ and $\sum_{j=0}^{\infty} h_j^2 < +\infty$. Then a special case of formula (2.10.2) coupled with the observer equation $y_t = G x_t$ implies that the spectrum of y is given by

$$S_y(\omega) = h(\exp(-i\omega)) \, h(\exp(i\omega)) = |h(\exp(-i\omega))|^2$$

where $h(\exp(-i\omega)) = \sum_{j=0}^{\infty} h_j \exp(-i\omega j)$.

In a famous paper, Slutsky investigated the consequences of applying the following filter to white noise: $h(L) = (1 + L)^n (1 - L)^m$ (i.e., the convolution of n two-period moving averages with m difference operators). Compute and plot the spectrum of y for $\omega \in [-\pi, \pi]$ for the following choices of m, n:

a. $m = 10, n = 10$.

b. $m = 10, n = 40$.

c. $m = 40, n = 10$.

d. $m = 120, n = 30$.

e. Comment on these results.

Hint: Notice that $h(\exp(-i\omega)) = (1 + \exp(-i\omega))^n (1 - \exp(-i\omega))^m$.

Exercise 2.16 Consider an n-state Markov chain with state space $X = \{e_i, i = 1, \ldots, n\}$ where e_i is the ith unit vector. Consider the indicator variable $I_{it} = e_i x_t$ which equals 1 if $x_t = e_i$ and 0 otherwise. Suppose that the chain has a unique stationary distribution and that it is ergodic. Let π be the stationary distribution.

a. Verify that $E I_{it} = \pi_i$.

b. Prove that

$$\frac{1}{T}\sum_{t=0}^{T-1} I_{it} = \pi_i$$

as $T \to \infty$ with probability one with respect to the stationary distribution π.

Exercise 2.17   **Lake model**

A worker can be in one of two states, state 1 (unemployed) or state 2 (employed). At the beginning of each period, a previously unemployed worker has probability $\lambda = \int_{\bar{w}}^{B} dF(w)$ of becoming employed. Here $\bar{w}$ is his reservation wage and $F(w)$ is the c.d.f. of a wage offer distribution. We assume that $F(0) = 0, F(B) = 1$. At the beginning of each period an unemployed worker draws one and only one wage offer from F. Successive draws from F are i.i.d. The worker's decision rule is to accept the job if $w \geq \bar{w}$, and otherwise to reject it and remain unemployed one more period. Assume that $\overline{w}$ is such that $\lambda \in (0,1)$. At the beginning of each period, a previously employed worker is fired with probability $\delta \in (0,1)$. Newly fired workers must remain unemployed for one period before drawing a new wage offer.

a.   Let the state space be $X = \{e_i, i = 1, 2\}$ where e_i is the ith unit vector. Describe the Markov chain on X that is induced by the description above. Compute all stationary distributions of the chain. Under what stationary distributions, if any, is the chain ergodic?

b. Suppose that $\lambda = .05, \delta = .25$. Compute a stationary distribution. Compute the fraction of his life that an infinitely lived worker would spend unemployed.

c. Drawing the initial state from the stationary distribution, compute the joint distribution $g_{ij} = \text{Prob}(x_t = e_i, x_{t-1} = e_j)$ for $i = 1, 2, j = 1, 2$.

d. Define an indicator function by letting $I_{ij,t} = 1$ if $x_t = e_i, x_{t-1} = e_j$ at time t, and 0 otherwise. Compute

$$\lim_{T\to\infty} \frac{1}{T}\sum_{t=1}^{T} I_{ij,t}$$

for all four i, j combinations.

e. Building on your results in part d, construct method of moments estimators of λ and δ. Assuming that you know the wage offer distribution F, construct a method of moments estimator of the reservation wage $\bar{w}$.

f. Compute maximum likelihood estimators of λ and δ.

g. Compare the estimators you derived in parts e and f.

h. *Extra credit.* Compute the asymptotic covariance matrix of the maximum likelihood estimators of λ and δ.

Exercise 2.18 **Random walk**

A Markov chain has state space $X = \{e_i, i = 1, \ldots, 4\}$ where e_i is the unit vector and transition matrix

$$
P = \begin{bmatrix}
1 & 0 & 0 & 0 \\
.5 & 0 & .5 & 0 \\
0 & .5 & 0 & .5 \\
0 & 0 & 0 & 1
\end{bmatrix}.
$$

A random variable $y_t = \bar{y} x_t$ is defined by $\bar{y} = \begin{bmatrix} 1 & 2 & 3 & 4 \end{bmatrix}$.

a. Find all stationary distributions of this Markov chain.

b. Under what stationary distributions, if any, is this chain ergodic? Compute invariant functions of P.

c. Compute $E[y_{t+1}|x_t]$ for $x_t = e_i, i = 1, \ldots, 4$.

d. Compare your answer to part (c) with (2.2.12). Is $y_t = \bar{y}' x_t$ invariant? If not, what hypothesis of Theorem 2.2.4 is violated?

e. The stochastic process $y_t = \bar{y}' x_t$ is evidently a bounded martingale. Verify that y_t converges almost surely to a constant. To what constant(s) does it converge?

Exercise 2.19 **IQ**

An infinitely lived person's 'true intelligence' $\theta \sim \mathcal{N}(100, 10)$. For each date $t \geq 0$, the person takes a 'test' with the outcome being a univariate random variable $y_t = \theta + v_t$, where v_t is an iid process with distribution $\mathcal{N}(0, 100)$. The person's initial IQ is $IQ_0 = 100$ and at date $t \geq 1$ before the date t test is taken it is $IQ_t = E\theta|y^{t-1}$, where y^{t-1} is the history of test scores from date 0 until date $t - 1$.

a. Give a recursive formula for IQ_t and for $E(IQ_t - \theta)^2$.

b. Use Matlab to simulate 10 draws of θ and associated paths of y_t, IQ_t for $t = 0, \ldots, 50$.

c. Prove that $\lim_{t \to \infty} E(\mathrm{IQ}_t - \theta)^2 = 0$.

Exercise 2.20   **Random walk**

A scalar process x_t follows the process

$$x_{t+1} = x_t + w_{t+1}$$

where w is an iid $\mathcal{N}(0, 1)$ scalar process and $x_0 \sim \mathcal{N}(\hat{x}_0, \Sigma_0)$. Each period, an observer receives two signals in the form of a 2×1 vector y_t that obeys

$$y_t = \begin{bmatrix} 1 \\ 1 \end{bmatrix} x_t + v_t$$

where the 2×1 process v_t is iid with distribution $v_t \sim \mathcal{N}(0, R)$ where $R = \begin{bmatrix} 1 & 0 \\ 0 & 1 \end{bmatrix}$.

a. Suppose that $\Sigma_0 = 1.36602540378444$. For $t \geq 0$, find formulas for $E[x_t | y^{t-1}]$, where y^{t-1} is the history of y_s for s from 0 to $t - 1$.

b. Verify numerically that the matrix $A - KG$ in formula (2.8.3) is stable.

c. Find an infinite-order vector autoregression for y_t.

Exercise 2.21   **Impulse response for VAR**

Find the impulse response function for the state space representation (2.8.1) associated with a vector autoregression.

Exercise 2.22   **Kalman filter with cross-products**

Consider the state space system

$$x_{t+1} = A_o x_t + C w_{t+1}$$
$$y_{t+1} = G x_t + D w_{t+1}$$

where x_t is an $n \times 1$ state vector w_{t+1} is an $m \times 1$ iid process with distribution $\mathcal{N}(0, I)$, y_t is an $m \times 1$ vector of observed variables, and $x_0 \sim \mathcal{N}(\hat{x}_0, \Sigma_0)$. For $t \geq 1$, $\hat{x}_t = E[x_t | y^t]$ where $y^t = [y_t, \ldots, y_1]$ and $\Sigma_t = E(x_t - \hat{x}_t)(x_t - \hat{x}_t)'$.

a. Show how to select w_{t+1}, C, and D so that Cw_{t+1} and Dw_{t+1} are mutually uncorrelated processes. Also give an example in which Cw_{t+1} and Dw_{t+1} are correlated.

b. Construct a recursive representation for $\hat{x}_t$ of the form:

$$\hat{x}_{t+1} = A_o \hat{x}_t + K_t a_{t+1}$$
$$y_{t+1} = G\hat{x}_t + a_{t+1}$$

where $a_{t+1} = y_{t+1} - E[y_{t+1}|y^t]$ for $t \geq 0$ and verify that

$$K_t = (CD' + A\Sigma_t G')(DD' + G\Sigma_t G')^{-1}$$
$$\Sigma_{t+1} = (A - K_t G)\Sigma_t (A - K_t G)' + (C - K_t D)(C - K_t D)'$$

and $Ea_{t+1}a'_{t+1} = G\Sigma_t G' + DD'$. *Hint:* apply the population regression formula.

Exercise 2.23   **A monopolist, learning, and ergodicity**

A monopolist produces a quantity Q_t of a single good in every period $t \geq 0$ at zero cost. At the beginning of each period $t \geq 0$, before output price p_t is observed, the monopolist sets quantity Q_t to maximize

$$(1) \qquad\qquad E_{t-1} p_t Q_t$$

where p_t satisfies the linear inverse demand curve

$$(2) \qquad\qquad p_t = a - bQ_t + \sigma_p \epsilon_t$$

where $b > 0$ is a constant known to the firm, ϵ_t is an i.i.d. scalar with distribution $\epsilon_t \sim \mathcal{N}(0,1)$, and the constant in the inverse demand curve a is a scalar random variable unknown to the firm and whose unconditional distribution is $a \sim \mathcal{N}(\mu_a, \sigma_a^2)$, where $\mu_a > 0$ is large relative to $\sigma_a > 0$. Assume that the random variable a is independent of ϵ_t for all t. Before the firm chooses Q_0, it knows the unconditional distribution of a, but not the realized value of a. For each $t \geq 0$, the firm wants to estimate a because it wants to make a good decision about output Q_t. At the end of each period t, when it must set Q_{t+1}, the firm observes p_t and also of course knows the value of Q_t that it had set. In (1), for $t \geq 1$, $E_{t-1}(\cdot)$ denotes the mathematical expectation conditional on the history of signals $p_s, q_s, s = 0, \ldots, t-1$; for $t = 0$, $E_{-1}(\cdot)$ denotes the expectation conditioned on no previous observations of p_t, Q_t.

a. What is the optimal setting for Q_0? For each date $t \geq 0$, determine the firm's optimal setting for Q_t as a function of the information p^{t-1}, Q^{t-1} that the firm has when it sets Q_t.

b. Under the firm's optimal policy, is the pair (p_t, Q_t) Markov?

c. 'Finding the state is an art.' Find a recursive representation of the firm's optimal policy for setting Q_t for $t \geq 0$. Interpret the state variables that you propose.

d. Under the firm's optimal rule for setting Q_t, does the random variable $E_{t-1}p_t$ converge to a constant as $t \to +\infty$? If so, prove that it does and find the limiting value. If not, tell why it does not converge.

e. Now suppose that instead of maximizing (1) each period, there is a single infinitely lived monopolist who once and for all before time 0 chooses a plan for an entire sequence $\{Q_t\}_{t=0}^{\infty}$, where the Q_t component has to be a measurable function of (p^{t-1}, q^{t-1}), and where the monopolist's objective is to maximize

$$(3) \qquad\qquad E_{-1} \sum_{t=0}^{\infty} \beta^t p_t Q_t$$

where $\beta \in (0, 1)$ and E_{-1} denotes the mathematical expectation conditioned on the null history. Get as far as you can in deriving the monopolist's optimal sequence of decision rules.

Exercise 2.24 **Stationarity**

A pair of scalar stochastic processes (z_t, y_t) evolves according to the state system for $t \geq 0$:

$$z_{t+1} = .9z_t + w_{t+1}$$

$$y_t = z_t + v_t$$

where w_{t+1} and v_t are mutually uncorrelated scalar Gaussian random variables with means of 0 and variances of 1. Furthermore, $Ew_{t+1}v_s = 0$ for all t, s pairs. In addition, $z_0 \sim \mathcal{N}(\hat{z}_0, \Sigma_0)$.

a. Is $\{z_t\}$ Markov? Explain.

b. Is $\{y_t\}$ Markov? Explain.

c. Define what it would mean for the scalar process $\{z_t\}$ to be *covariance stationary*.

d. Find values of $(\hat{z}_0, \Sigma_0)$ that make the process for $\{z_t\}$ covariance stationary.

e. Assume that y_t is observable, but that z_t is not. Define what it would mean for the scalar process y_t to be *covariance stationary*.

f. Describe in as much detail as you can how to represent the distribution of y_t conditional on the infinite history y^{t-1} in the form $y_t \sim \mathcal{N}(E[y_t|y^{t-1}], \Omega_t)$.

Exercise 2.25 **Consumption**

a. Please use formulas (2.11.18) to verify formulas (2.11.21) and (2.11.23)-(2.11.24) of subsection 2.11.3.

b. Please use formulas (2.9.3) to compute the decision rules in formulas (2.11.21) and (2.11.23) for the following parameter values: $\beta = .95, \sigma_1 = \sigma_2 = 1$.

c. Please use formulas (2.9.3) to compute the decision rules in formulas (2.11.21) and (2.11.23) for the following parameter values: $\beta = .95, \sigma_1 = 2, \sigma_2 = 1$.

d. Please use formula (2.11.20) to confirm formulas (2.11.22) and (2.11.25).

Exercise 2.26 **Math and verbal IQ's**

An infinitely lived person's 'true intelligence' θ has two components, math ability θ_1 and verbal ability θ_2, where $\theta \sim \mathcal{N}\left(\begin{bmatrix} 100 \\ 100 \end{bmatrix}, \begin{bmatrix} 10 & 0 \\ 0 & 10 \end{bmatrix} \right)$. For each date $t \geq 0$, the person takes a single 'test' with the outcome being a univariate random variable $y_t = G_t\theta + v_t$, where v_t is an iid process with distribution $\mathcal{N}(0, 50)$ and $G_t = [.9 \quad .1]$ for $t = 0, 2, 4, \ldots$ and $G_t = [.01 \quad .99]$ for $t = 1, 3, 5, \ldots$. Here the person takes a math test at t even and a verbal test at t odd (but you have to know how to read English to survive the math test, and you have to know how to tell time in order to plan your time allocation well for the verbal test). The person's initial IQ vector is $IQ_0 = \begin{bmatrix} 100 \\ 100 \end{bmatrix}$ and at date $t \geq 1$ before the date t test is taken it is $IQ_t = E\theta|y^{t-1}$, where y^{t-1} is the history of test scores from date 0 until date $t - 1$.

a. Give a recursive formula for IQ_t and for $E(IQ_t - \theta)(IQ_t - \theta)'$.

b. Use Matlab to simulate 10 draws of θ and associated paths of y_t, IQ_t for $t = 0, \ldots, 50$.

c. Show computationally or analytically that $\lim_{t \to +\infty} E(\mathrm{IQ}_t - \theta)(\mathrm{IQ}_t - \theta)' = \begin{bmatrix} 0 & 0 \\ 0 & 0 \end{bmatrix}$.

Exercise 2.27 **Permanent income model again**

Each of two consumers named $i = 1, 2$ has preferences over consumption streams that are ordered by the utility functional

(1)
$$E_0 \sum_{t=0}^{\infty} \beta^t u\left(c_t^i\right)$$

where E_t is the mathematical expectation conditioned on the consumer's time t information, c_t^i is time t consumption of consumer i at time t, $u(c)$ is a strictly concave one-period utility function, and $\beta \in (0, 1)$ is a discount factor. The consumer maximizes (1) by choosing a consumption, borrowing plan $\{c_t^i, b_{t+1}^i\}_{t=0}^{\infty}$ subject to the sequence of budget constraints

$$c_t^i + b_t^i = R^{-1} b_{t+1}^i + y_t^i$$

where y_t is an exogenous stationary endowment process, R is a constant gross risk-free interest rate, b_t^i is one-period risk-free debt maturing at t, and $b_0^i = 0$ is a given initial condition. Assume that $R^{-1} = \beta$. We impose the following condition on the consumption, borrowing plan of consumer i:

$$E_0 \sum_{t=0}^{\infty} \beta^t \left(b_t^i\right)^2 < +\infty.$$

Assume the quadratic utility function $u(c_t) = -.5(c_t - \gamma)^2$, where $\gamma > 0$ is a bliss level of consumption. Negative consumption rates are allowed.

Let $s_t \in \{0, 1\}$ be an i.i.d. process with $\mathrm{Prob}(s_t = 1) = \mathrm{Prob}(s_1 = 0) = .5$. The endowment process of consumer 1 is $y_t^1 = 1 - .5 s_t$ and the endowment process of person 2 is $y_t^2 = .5 + .5 s_t$. Thus, the two consumers' endowment processes are perfectly negatively correlated i.i.d. processes with means of .75.

a. Find optimal decision rules for consumption for both consumers. Prove that the consumers' optimal decisions imply the following laws of motion for b_t^1, b_t^2:

$$b_{t+1}^1 (s_t = 0) = b_t^1 - .25$$
$$b_{t+1}^1 (s_t = 1) = b_t^1 + .25$$
$$b_{t+1}^2 (s_t = 0) = b_t^2 + .25$$
$$b_{t+1}^2 (s_t = 1) = b_t^2 - .25$$

b. Show that for each consumer, c_t^i, b_t^i are co-integrated.

c. Verify that b_{t+1}^i is risk-free in the sense that conditional on information available at time t, it is independent of news arriving at time $t+1$.

d. Verify that with the initial conditions $b_0^1 = b_0^2 = 0$, the following two equalities obtain:

$$b_t^1 + b_t^2 = 0 \quad \forall t \geq 1$$
$$c_t^1 + c_t^2 = 1.5 \quad \forall t \geq 1$$

Use these conditions to interpret the decision rules that you have computed as describing a closed pure consumption loans economy in which consumers 1 and 2 borrow and lend with each other and in which the risk-free asset is a one-period IOU from one of the consumers to the other.

e. Define the 'stochastic discount factor of consumer i' as $m_{t+1}^i = \frac{\beta u'(c_{t+1}^i)}{u'(c_t^i)}$. Show that the stochastic discount factors of consumer 1 and 2 are

$$
m_{t+1}^1 =
\begin{cases}
\beta + .25\frac{\beta(1-\beta)}{(\gamma - c_t^1)}, & \text{if } s_{t+1} = 0; \\
\beta - .25\frac{\beta(1-\beta)}{(\gamma - c_t^1)}, & \text{if } s_{t+1} = 1; .
\end{cases}
$$

$$
m_{t+1}^2 =
\begin{cases}
\beta - .25\frac{\beta(1-\beta)}{(\gamma - c_t^2)}, & \text{if } s_{t+1} = 0; \\
\beta + .25\frac{\beta(1-\beta)}{(\gamma - c_t^2)}, & \text{if } s_{t+1} = 1; .
\end{cases}
$$

Are the stochastic discount factors of the two consumers equal?

f. Verify that $E_t m_{t+1}^1 = E_t m_{t+1}^2 = \beta$.

Exercise 2.28 **Invertibility**

A univariate stochastic process y_t has a first-order moving average representation

$$
(1) \qquad\qquad y_t = \epsilon_t - 2\epsilon_{t-1}
$$

where $\{\epsilon_t\}$ is an i.i.d. process distributed $\mathcal{N}(0,1)$.

a. Argue that ϵ_t cannot be expressed as as linear combination of $y_{t-j}, j \geq 0$ where the sum of the squares of the weights is finite. This means that ϵ_t is not in the space spanned by square summable linear combinations of the infinite history y^t.

b. Write equation (1) as a state space system, indicating the matrices A, C, G.

c. Using the matlab program `kfilter.m` to compute an innovations representation for $\{y_t\}$. Verify that the innovations representation for y_t can be represented as

(2) $$y_t = a_t - .5a_{t-1}$$

where $a_t = y_t - E[y_t|y^{t-1}]$ is a serially uncorrelated process. Compute the variance of a_t. Is it larger or smaller than the variance of ϵ_t?

d. Find an autoregressive representation for y_t of the form

(3) $$y_t = \sum_{j=1}^{\infty} A_j y_{t-j} + a_t$$

where $E a_t y_{t-j} = 0$ for $j \geq 1$. (*Hint:* either use formula (2) or else remember formula (2.8.3).)

e. Is y_t Markov? Is $\begin{bmatrix} y_t & y_{t-1} \end{bmatrix}'$ Markov? Is $\begin{bmatrix} y_t & y_{t-1} & \cdots & y_{t-10} \end{bmatrix}'$ Markov?

f. Extra credit. Verify that ϵ_t *can* be expressed as a square summable linear combination of $y_{t+j}, j \geq 1$.

Chapter 3
Dynamic Programming

This chapter introduces basic ideas and methods of dynamic programming.[1]
It sets out the basic elements of a recursive optimization problem, describes a
key functional equation called the Bellman equation, presents three methods for
solving the Bellman equation, and gives the Benveniste-Scheinkman formula for
the derivative of the optimal value function. Let's dive in.

3.1. Sequential problems

Let $\beta \in (0,1)$ be a discount factor. We want to choose an infinite sequence of
"controls" $\{u_t\}_{t=0}^{\infty}$ to maximize

$$\sum_{t=0}^{\infty} \beta^t r\left(x_t, u_t\right), \qquad (3.1.1)$$

subject to $x_{t+1} = g(x_t, u_t)$, with $x_0 \in I\!\!R^n$ given. We assume that $r(x_t, u_t)$
is a concave function and that the set $\{(x_{t+1}, x_t) : x_{t+1} \leq g(x_t, u_t), u_t \in I\!\!R^k\}$
is convex and compact. Dynamic programming seeks a time-invariant *policy
function* h mapping the *state* x_t into the control u_t, such that the sequence
$\{u_s\}_{s=0}^{\infty}$ generated by iterating the two functions

$$\begin{aligned} u_t &= h\left(x_t\right) \\ x_{t+1} &= g\left(x_t, u_t\right), \end{aligned} \qquad (3.1.2)$$

starting from initial condition x_0 at $t = 0$, solves the original problem. A
solution in the form of equations $(3.1.2)$ is said to be *recursive*. To find the
policy function h we need to know another function $V(x)$ that expresses the
optimal value of the original problem, starting from an arbitrary initial condition
$x \in X$. This is called the *value function*. In particular, define

$$V\left(x_0\right) = \max_{\{u_s\}_{s=0}^{\infty}} \sum_{t=0}^{\infty} \beta^t r\left(x_t, u_t\right), \qquad (3.1.3)$$

---

[1] This chapter aims to the reader to start using the methods quickly. We hope to promote
demand for further and more rigorous study of the subject. In particular see Bertsekas (1976),
Bertsekas and Shreve (1978), Stokey and Lucas (with Prescott) (1989), Bellman (1957), and
Chow (1981). This chapter covers much of the same material as Sargent (1987b, chapter 1).

where again the maximization is subject to $x_{t+1} = g(x_t, u_t)$, with x_0 given. Of course, we cannot possibly expect to know $V(x_0)$ until after we have solved the problem, but let's proceed on faith. If we knew $V(x_0)$, then the policy function h could be computed by solving for each $x \in X$ the problem

$$\max_u \{r(x, u) + \beta V(\tilde{x})\}, \qquad (3.1.4)$$

where the maximization is subject to $\tilde{x} = g(x, u)$ with x given, and $\tilde{x}$ denotes the state next period. Thus, we have exchanged the original problem of finding an infinite *sequence* of controls that maximizes expression (3.1.1) for the problem of finding the optimal value function $V(x)$ and a function h that solves the continuum of maximum problems (3.1.4)—one maximum problem for each value of x. This exchange doesn't look like progress, but we shall see that it often is.

Our task has become jointly to solve for $V(x), h(x)$, which are linked by the *Bellman equation*

$$V(x) = \max_u \{r(x, u) + \beta V[g(x, u)]\}. \qquad (3.1.5)$$

The maximizer of the right side of equation (3.1.5) is a *policy function* $h(x)$ that satisfies

$$V(x) = r[x, h(x)] + \beta V\{g[x, h(x)]\}. \qquad (3.1.6)$$

Equation (3.1.5) or (3.1.6) is a *functional equation* to be solved for the pair of unknown functions $V(x), h(x)$.

Methods for solving the Bellman equation are based on mathematical structures that vary in their details depending on the precise nature of the functions r and g.[2] All of these structures contain versions of the following four findings. Under various particular assumptions about r and g, it turns out that

---

[2] There are alternative sets of conditions that make the maximization (3.1.4) well behaved. One set of conditions is as follows: (1) r is concave and bounded, and (2) the constraint set generated by g is convex and compact, that is, the set of $\{(x_{t+1}, x_t) : x_{t+1} \leq g(x_t, u_t)\}$ for admissible u_t is convex and compact. See Stokey, Lucas, and Prescott (1989) and Bertsekas (1976) for further details of convergence results. See Benveniste and Scheinkman (1979) and Stokey, Lucas, and Prescott (1989) for the results on differentiability of the value function. In Appendix A (see Technical Appendixes), we describe the mathematics for one standard set of assumptions about (r, g). In chapter 5, we describe it for another set of assumptions about (r, g).

1. The functional equation $(3.1.5)$ has a unique strictly concave solution.

2. This solution is approached in the limit as $j \to \infty$ by iterations on

$$V_{j+1}(x) = \max_{u}\{r(x, u) + \beta V_j(\tilde{x})\},$$

subject to $\tilde{x} = g(x, u), x$ given, starting from any bounded and continuous initial V_0.

3. There is a unique and time-invariant optimal policy of the form $u_t = h(x_t)$, where h is chosen to maximize the right side of $(3.1.5)$.

4. Off corners, the limiting value function V is differentiable.

Since the value function is differentiable, the first-order necessary condition for problem $(3.1.4)$ becomes[3]

$$r_2(x, u) + \beta V'\{g(x, u)\} g_2(x, u) = 0. \qquad (3.1.7)$$

If we also assume that the policy function $h(x)$ is differentiable, differentiation of expression $(3.1.6)$ yields[4]

$$V'(x) = r_1[x, h(x)] + r_2[x, h(x)] h'(x)$$
$$+ \beta V'\{g[x, h(x)]\} \Big\{g_1[x, h(x)] + g_2[x, h(x)] h'(x)\Big\}. \qquad (3.1.8)$$

When the states and controls can be defined in such a way that only u appears in the transition equation, i.e., $\tilde{x} = g(u)$: the derivative of the value function becomes, after substituting expression $(3.1.7)$ with $u = h(x)$ into $(3.1.8)$,

$$V'(x) = r_1[x, h(x)]. \qquad (3.1.9)$$

This is a version of a formula of Benveniste and Scheinkman (1979).

At this point, we describe three broad computational strategies that apply in various contexts.

---

[3] Here and below, subscript 1 denotes the vector of derivatives with respect to the x components and subscript 2 denotes the derivatives with respect to the u components.

[4] Benveniste and Scheinkman (1979) proved differentiability of $V(x)$ under broad conditions that do not require that $h(x)$ be differentiable. For conditions under which $h(x)$ is differentiable, see Santos (1991,1993).

3.1.1. Three computational methods

There are three main types of computational methods for solving dynamic programs. All aim to solve the functional equation (3.1.4).

Value function iteration.   The first method proceeds by constructing a sequence of value functions and associated policy functions. The sequence is created by iterating on the following equation, starting from $V_0 = 0$, and continuing until V_j has converged:

$$V_{j+1}(x) = \max_u \{r(x, u) + \beta V_j(\tilde{x})\}, \qquad (3.1.10)$$

subject to $\tilde{x} = g(x, u), x$ given.[5] This method is called *value function iteration* or *iterating on the Bellman equation.*

Guess and verify.   A second method involves guessing and verifying a solution V to equation (3.1.5). This method relies on the uniqueness of the solution to the equation, but because it relies on luck in making a good guess, it is not generally available.

Howard's improvement algorithm. A third method, known as *policy function iteration* or *Howard's improvement algorithm*, consists of the following steps:

1. Pick a feasible policy, $u = h_0(x)$, and compute the value associated with operating forever with that policy:

$$V_{h_j}(x) = \sum_{t=0}^{\infty} \beta^t r[x_t, h_j(x_t)],$$

   where $x_{t+1} = g[x_t, h_j(x_t)]$, with $j = 0$.

2. Generate a new policy $u = h_{j+1}(x)$ that solves the two-period problem

$$\max_u \{r(x, u) + \beta V_{h_j}[g(x, u)]\},$$

   for each x.

---

[5] See Appendix A on functional analysis (see Technical Appendixes) for what it means for a sequence of functions to converge. A proof of the uniform convergence of iterations on equation (3.1.10) is contained in that appendix.

3. Iterate over j to convergence on steps 1 and 2.

In Appendix A (see Technical Appendixes), we describe some conditions under which the policy improvement algorithm converges to the solution of the Bellman equation. The policy improvement algorithm often converges faster than does value function iteration (e.g., see exercise *3.1* at the end of this chapter).[6] The policy improvement algorithm is also a building block for methods used to study government policy in chapter 23.

Each of our three methods for solving dynamic programming problems has its uses. Each is easier said than done, because it is typically impossible analytically to compute even *one* iteration on equation (3.1.10). This fact thrusts us into the domain of computational methods for approximating solutions: pencil and paper are insufficient. Chapter 4 describes computational methods that can applied to problems that cannot be solved by hand. Here we shall describe the first of two special types of problems for which analytical solutions *can* be obtained. It involves Cobb-Douglas constraints and logarithmic preferences. Later, in chapter 5, we shall describe a specification with linear constraints and quadratic preferences. For that special case, many analytic results are available. These two classes have been important in economics as sources of examples and as inspirations for approximations.

3.1.2. Cobb-Douglas transition, logarithmic preferences

Brock and Mirman (1972) used the following optimal growth example.[7] A planner chooses sequences $\{c_t, k_{t+1}\}_{t=0}^{\infty}$ to maximize

$$\sum_{t=0}^{\infty} \beta^t \ln\left(c_t\right)$$

subject to a given value for k_0 and a transition law

$$k_{t+1} + c_t = A k_t^{\alpha}, \tag{3.1.11}$$

where $A > 0, \alpha \in (0,1), \beta \in (0,1)$.

---

[6] The speed of the policy improvement algorithm comes from its implementing Newton's method, which converges quadratically while iteration on the Bellman equation converges at a linear rate. See chapter 4 and Appendix A (see Technical Appendixes).

[7] See also Levhari and Srinivasan (1969).

This problem can be solved "by hand," using any of our three methods. We begin with iteration on the Bellman equation. Start with $v_0(k) = 0$, and solve the one-period problem: choose c to maximize $\ln(c)$ subject to $c + \tilde{k} = Ak^\alpha$. The solution is evidently to set $c = Ak^\alpha, \tilde{k} = 0$, which produces an optimized value $v_1(k) = \ln A + \alpha \ln k$. At the second step, we find $c = \frac{1}{1+\beta\alpha} Ak^\alpha, \tilde{k} = \frac{\beta\alpha}{1+\beta\alpha} Ak^\alpha, v_2(k) = \ln \frac{A}{1+\alpha\beta} + \beta \ln A + \alpha\beta \ln \frac{\alpha\beta A}{1+\alpha\beta} + \alpha(1+\alpha\beta) \ln k$. Continuing, and using the algebra of geometric series, gives the limiting policy functions $c = (1-\beta\alpha)Ak^\alpha, \tilde{k} = \beta\alpha Ak^\alpha$, and the value function $v(k) = (1-\beta)^{-1}\{\ln[A(1-\beta\alpha)] + \frac{\beta\alpha}{1-\beta\alpha} \ln(A\beta\alpha)\} + \frac{\alpha}{1-\beta\alpha} \ln k$.

Here is how the guess-and-verify method applies to this problem. Since we already know the answer, we'll guess a function of the correct form, but leave its coefficients undetermined.[8]   Thus, we make the guess

$$v(k) = E + F \ln k, \tag{3.1.12}$$

where E and F are undetermined constants. The left and right sides of equation $(3.1.12)$ must agree for all values of k. For this guess, the first-order necessary condition for the maximum problem on the right side of equation $(3.1.10)$ implies the following formula for the optimal policy $\tilde{k} = h(k)$, where $\tilde{k}$ is next period's value and k is this period's value of the capital stock:

$$\tilde{k} = \frac{\beta F}{1 + \beta F} Ak^\alpha. \tag{3.1.13}$$

Substitute equation $(3.1.13)$ into the Bellman equation and equate the result to the right side of equation $(3.1.12)$. Solving the resulting equation for E and F gives $F = \alpha/(1 - \alpha\beta)$ and $E = (1 - \beta)^{-1}[\ln A(1 - \alpha\beta) + \frac{\beta\alpha}{1-\alpha\beta} \ln A\beta\alpha]$. It follows that

$$\tilde{k} = \beta\alpha Ak^\alpha. \tag{3.1.14}$$

Note that the term $F = \alpha/(1 - \alpha\beta)$ can be interpreted as a geometric sum $\alpha[1 + \alpha\beta + (\alpha\beta)^2 + \ldots]$.

Equation $(3.1.14)$ shows that the optimal policy is to have capital move according to the difference equation $k_{t+1} = A\beta\alpha k_t^\alpha$, or $\ln k_{t+1} = \ln A\beta\alpha + \alpha \ln k_t$. That α is less than 1 implies that k_t converges as t approaches infinity for any positive initial value k_0. The stationary point is given by the solution of $k_\infty = A\beta\alpha k_\infty^\alpha$, or $k_\infty^{\alpha-1} = (A\beta\alpha)^{-1}$.

---

[8]   This is called the *method of undetermined coefficients*.

3.1.3. Euler equations

In many problems, there is no unique way of defining states and controls, and several alternative definitions lead to the same solution of the problem. When the states and controls can be defined in such a way that only u appears in the transition equation, i.e., $\tilde{x} = g(u)$: the first-order condition for the problem on the right side of the Bellman equation (expression (3.1.7)) in conjunction with the Benveniste-Scheinkman formula (expression (3.1.9)) implies

$$r_2(x_t, u_t) + \beta\, r_1(x_{t+1}, u_{t+1})\, g'(u_t) = 0, \qquad x_{t+1} = g\,(u_t)\,.$$

The first equation is called an *Euler equation*. Under circumstances in which the second equation can be inverted to yield u_t as a function of x_{t+1}, using the second equation to eliminate u_t from the first equation produces a second-order difference equation in x_t, since eliminating u_{t+1} brings in x_{t+2}.

3.1.4. A sample Euler equation

As an example of an Euler equation, consider the Ramsey problem of choosing $\{c_t, k_{t+1}\}_{t=0}^{\infty}$ to maximize $\sum_{t=0}^{\infty} \beta^t u(c_t)$ subject to $c_t + k_{t+1} = f(k_t)$, where k_0 is given and the one-period utility function satisfies $u'(c) > 0, u''(c) < 0, \lim_{c_t \searrow 0} u'(c_t) = \infty$, and where $f'(k) > 0, f''(k) < 0$. Let the state be k and the control be $\tilde{k}$, where $\tilde{k}$ denotes next period's value of k. Substitute $c = f(k) - \tilde{k}$ into the utility function and express the Bellman equation as

$$v\,(k) = \max_{\tilde{k}} \left\{ u\left[f\,(k) - \tilde{k}\right] + \beta v\left(\tilde{k}\right)\right\}. \tag{3.1.15}$$

Application of the Benveniste-Scheinkman formula gives

$$v'\,(k) = u'\left[f\,(k) - \tilde{k}\right] f'\,(k)\,. \tag{3.1.16}$$

Notice that the first-order condition for the maximum problem on the right side of equation (3.1.15) is $-u'[f(k) - \tilde{k}] + \beta v'(\tilde{k}) = 0$, which, using equation (3.1.16), gives

$$u'\left[f\,(k) - \tilde{k}\right] = \beta u'\left[f\left(\tilde{k}\right) - \hat{k}\right] f'\left(\tilde{k}\right), \tag{3.1.17}$$

where $\hat{k}$ denotes the two-period-ahead value of k. Equation (3.1.17) can be expressed as

$$1 = \beta \frac{u'(c_{t+1})}{u'(c_t)} f'(k_{t+1}),$$

an Euler equation that is exploited extensively in the theories of finance, growth, and real business cycles.

3.2. Stochastic control problems

We now consider a modification of problem (3.1.1) to permit uncertainty. Essentially, we add some well-placed shocks to the previous nonstochastic problem. So long as the shocks are either independently and identically distributed or Markov, straightforward modifications of the method for handling the nonstochastic problem will work.

Thus, we modify the transition equation and consider the problem of maximizing

$$E_0 \sum_{t=0}^{\infty} \beta^t r(x_t, u_t), \qquad 0 < \beta < 1, \tag{3.2.1}$$

subject to

$$x_{t+1} = g(x_t, u_t, \epsilon_{t+1}), \tag{3.2.2}$$

with x_0 known and given at $t = 0$, where ϵ_t is a sequence of independently and identically distributed random variables with cumulative probability distribution function $\text{prob}\{\epsilon_t \leq e\} = F(e)$ for all t; $E_t(y)$ denotes the mathematical expectation of a random variable y, given information known at t. At time t, x_t is assumed to be known, but $x_{t+j}, j \geq 1$ is not known at t. That is, ϵ_{t+1} is realized at $(t + 1)$, after u_t has been chosen at t. In problem (3.2.1)–(3.2.2), uncertainty is injected by assuming that x_t follows a random difference equation.

Problem (3.2.1)–(3.2.2) continues to have a recursive structure, stemming jointly from the additive separability of the objective function (3.2.1) in pairs (x_t, u_t) and from the difference equation characterization of the transition law (3.2.2). In particular, controls dated t affect returns $r(x_s, u_s)$ for $s \geq t$ but not earlier. This feature implies that dynamic programming methods remain appropriate.

The problem is to maximize expression (3.2.1) subject to equation (3.2.2) by choice of a "policy" or "contingency plan" $u_t = h(x_t)$. The Bellman equation (3.1.5) becomes

$$V(x) = \max_u \{r(x, u) + \beta E\left[V\left[g(x, u, \epsilon)\right] | x\right]\},\qquad (3.2.3)$$

where $E\{V[g(x, u, \epsilon)]|x\} = \int V[g(x, u, \epsilon)]dF(\epsilon)$ and where $V(x)$ is the optimal value of the problem starting from x at $t = 0$. The solution $V(x)$ of equation (3.2.3) can be computed by iterating on

$$V_{j+1}(x) = \max_u \{r(x, u) + \beta E\left[V_j\left[g(x, u, \epsilon)\right] | x\right]\},\qquad (3.2.4)$$

starting from any bounded continuous initial V_0. Under various particular regularity conditions, there obtain versions of the same four properties listed earlier. [9]

The first-order necessary condition for the problem on the right side of equation (3.2.3) is

$$r_2(x, u) + \beta E\left\{V'[g(x, u, \epsilon)]\ g_2(x, u, \epsilon)\ \Big| x\right\} = 0,$$

which we obtained simply by differentiating the right side of equation (3.2.3), passing the differentiation operation under the E (an integration) operator. Off corners, the value function satisfies

$$V'(x) = r_1[x, h(x)] + r_2[x, h(x)]\ h'(x)$$
$$+ \beta E\left\{V'\{g[x, h(x), \epsilon]\}\ \{g_1[x, h(x), \epsilon] + g_2[x, h(x), \epsilon]\ h'(x)\} \Big| x\right\}.$$

When the states and controls can be defined in such a way that x does not appear in the transition equation, the formula for $V'(x)$ becomes

$$V'(x) = r_1[x, h(x)].$$

Substituting this formula into the first-order necessary condition for the problem gives the stochastic Euler equation

$$r_2(x, u) + \beta E\left[r_1(\tilde{x}, \tilde{u})\ g_2(x, u, \epsilon)\ \Big| x\right] = 0,$$

where tildes over x and u denote next-period values.

---

[9] See Stokey and Lucas (with Prescott) (1989), or the framework presented in Appendix A (see Technical Appendixes).

3.3. Concluding remarks

This chapter has put forward basic tools and findings: the Bellman equation and several approaches to solving it; the Euler equation; and the Benveniste-Scheinkman formula. To appreciate and believe in the power of these tools requires more words and more practice than we have yet supplied. In the next several chapters, we put the basic tools to work in different contexts with particular specification of return and transition equations designed to render the Bellman equation susceptible to further analysis and computation.

Exercise

Exercise 3.1   **Howard's policy iteration algorithm**

Consider the Brock-Mirman problem: to maximize

$$E_0 \sum_{t=0}^{\infty} \beta^t \ln c_t,$$

subject to $c_t + k_{t+1} \le A k_t^\alpha \theta_t$, k_0 given, $A > 0$, $1 > \alpha > 0$, where $\{\theta_t\}$ is an i.i.d. sequence with $\ln \theta_t$ distributed according to a normal distribution with mean zero and variance σ^2.

Consider the following algorithm. Guess at a policy of the form $k_{t+1} = h_0(A k_t^\alpha \theta_t)$ for any constant $h_0 \in (0, 1)$. Then form

$$J_0(k_0, \theta_0) = E_0 \sum_{t=0}^{\infty} \beta^t \ln \left(A k_t^\alpha \theta_t - h_0 A k_t^\alpha \theta_t \right).$$

Next choose a new policy h_1 by maximizing

$$\ln \left(A k^\alpha \theta - k' \right) + \beta E J_0(k', \theta'),$$

where $k' = h_1 A k^\alpha \theta$. Then form

$$J_1(k_0, \theta_0) = E_0 \sum_{t=0}^{\infty} \beta^t \ln \left(A k_t^\alpha \theta_t - h_1 A k_t^\alpha \theta_t \right).$$

Continue iterating on this scheme until successive h_j have converged.

Show that, for the present example, this algorithm converges to the optimal policy function in one step.

Chapter 4
Practical Dynamic Programming

4.1. The curse of dimensionality

We often encounter problems where it is impossible to attain closed forms for iterating on the Bellman equation. Then we have to adopt numerical approximations. This chapter describes two popular methods for obtaining numerical approximations. The first method replaces the original problem with another problem that forces the state vector to live on a finite and discrete grid of points, then applies discrete-state dynamic programming to this problem. The "curse of dimensionality" impels us to keep the number of points in the discrete state space small. The second approach uses polynomials to approximate the value function. Judd (1998) is a comprehensive reference about numerical analysis of dynamic economic models and contains many insights about ways to compute dynamic models.

4.2. Discrete-state dynamic programming

We introduce the method of discretization of the state space in the context of a particular discrete-state version of an optimal savings problem. An infinitely lived household likes to consume one good that it can acquire by spending labor income or accumulated savings. The household has an endowment of labor at time t, s_t, that evolves according to an m-state Markov chain with transition matrix $\mathcal{P}$ and state space $[\bar{s}_1, \bar{s}_2, \ldots, \bar{s}_m]$. If the realization of the process at t is $\bar{s}_i$, then at time t the household receives labor income of amount $w\bar{s}_i$. The wage w is fixed over time. We shall sometimes assume that m is 2, and that s_t takes on value 0 in an unemployed state and 1 in an employed state. In this case, w has the interpretation of being the wage of employed workers.

The household can choose to hold a single asset in discrete amounts $a_t \in \mathcal{A}$ where $\mathcal{A}$ is a grid $[a_1 < a_2 < \cdots < a_n]$. How the model builder chooses the end points of the grid $\mathcal{A}$ is important, as we describe in detail in chapter 18 on incomplete market models. The asset bears a gross rate of return r that is fixed over time.

The household's maximum problem, for given values of (w, r) and given initial values (a_0, s_0), is to choose a policy for $\{a_{t+1}\}_{t=0}^{\infty}$ to maximize

$$E \sum_{t=0}^{\infty} \beta^t u(c_t), \qquad (4.2.1)$$

subject to

$$c_t + a_{t+1} = (r+1) a_t + w s_t$$
$$c_t \geq 0 \qquad (4.2.2)$$
$$a_{t+1} \in \mathcal{A}$$

where $\beta \in (0, 1)$ is a discount factor and r is fixed rate of return on the assets. We assume that $\beta(1 + r) < 1$. Here $u(c)$ is a strictly increasing, concave one-period utility function. Associated with this problem is the Bellman equation

$$v(a, s) = \max_{a' \in \mathcal{A}} \left\{ u\left[(r+1) a + ws - a'\right] + \beta E v(a', s') \,|s \right\},$$

where a is next period's value of asset holdings, and s' is next period's value of the shock; here $v(a, s)$ is the optimal value of the objective function, starting from asset, employment state (a, s). We seek a value function $v(a, s)$ that satisfies equation (18.2.3) and an associated policy function $a' = g(a, s)$ mapping this period's (a, s) pair into an optimal choice of assets to carry into next period. Let assets live on the grid $\mathcal{A} = [a_1, a_2, \ldots, a_n]$. Then we can express the Bellman equation as

$$v(a_i, \bar{s}_j) = \max_{a_h \in \mathcal{A}} \left\{ u\left[(r+1) a_i + w\bar{s}_j - a_h\right] + \beta \sum_{l=1}^{m} \mathcal{P}_{jl} v(a_h, \bar{s}_l) \right\}, \qquad (4.2.3)$$

for each $i \in [1, \ldots, n]$ and each $j \in [1, \ldots, m]$.

4.3. Bookkeeping

For a discrete state space of small size, it is easy to solve the Bellman equation numerically by manipulating matrices. Here is how to write a computer program to iterate on the Bellman equation in the context of the preceding model of asset accumulation.[1] Let there be n states $[a_1, a_2, \ldots, a_n]$ for assets and two states $[s_1, s_2]$ for employment status. For $j = 1, 2$, define $n \times 1$ vectors $v_j, j = 1, 2$, whose ith rows are determined by $v_j(i) = v(a_i, s_j), i = 1, \ldots, n$. Let $\mathbf{1}$ be the $n \times 1$ vector consisting entirely of ones. For $j = 1, 2$, define two $n \times n$ matrices R_j whose (i, h) elements are

$$R_j(i, h) = u[(r + 1)a_i + ws_j - a_h], \quad i = 1, \ldots, n, h = 1, \ldots, n.$$

Define an operator $T([v_1, v_2])$ that maps a pair of $n \times 1$ vectors $[v_1, v_2]$ into a pair of $n \times 1$ vectors $[tv_1, tv_2]$:[2]

$$tv_j(i) = \max_h \left\{ R_j(i, h) + \beta \mathcal{P}_{j1} v_1(h) + \beta \mathcal{P}_{j2} v_2(h) \right\}$$

for $j = 1, 2$, or

$$\begin{aligned}
tv_1 &= \max\{R_1 + \beta \mathcal{P}_{11} \mathbf{1} v_1' + \beta \mathcal{P}_{12} \mathbf{1} v_2'\} \\
tv_2 &= \max\{R_2 + \beta \mathcal{P}_{21} \mathbf{1} v_1' + \beta \mathcal{P}_{22} \mathbf{1} v_2'\}.
\end{aligned} \tag{4.3.1}$$

Here it is understood that the "max" operator applied to an $(n \times m)$ matrix M returns an $(n \times 1)$ vector whose ith element is the maximum of the ith row of the matrix M. These two equations can be written compactly as

$$\begin{bmatrix} tv_1 \\ tv_2 \end{bmatrix} = \max \left\{ \begin{bmatrix} R_1 \\ R_2 \end{bmatrix} + \beta \left(\mathcal{P} \otimes \mathbf{1} \right) \begin{bmatrix} v_1' \\ v_2' \end{bmatrix} \right\}, \tag{4.3.2}$$

where $\otimes$ is the Kronecker product.[3]

---

[1] Matlab versions of the program have been written by Gary Hansen, Selahattin İmrohoroğlu, George Hall, and Chao Wei.

[2] Programming languages like Gauss and Matlab execute maximum operations over vectors very efficiently. For example, for an $n \times m$ matrix A, the Matlab command `[r,index] =max(A)` returns the two $(1 \times m)$ row vectors `r,index`, where $r_j = \max_i A(i, j)$ and index_j is the row i that attains $\max_i A(i, j)$ for column j [i.e., $\text{index}_j = \text{argmax}_i A(i, j)$]. This command performs m maximizations simultaneously.

[3] If A is an m-by-n matrix and B is a p-by-q matrix, then the Kronecker product $A \otimes B$ is the mp-by-nq block matrix $A \otimes B = \begin{bmatrix} a_{11}B & \cdots & a_{1n}B \\ \vdots & \ddots & \vdots \\ a_{m1}B & \cdots & a_{mn}B \end{bmatrix}.$

The Bellman equation $[v_1 v_2] = T([v_1, v_2])$ can be solved by iterating to convergence on $[v_1, v_2]_{m+1} = T([v_1, v_2]_m)$.

4.4. Application of Howard improvement algorithm

Often computation speed is important. Exercise 3.1 showed that the policy improvement algorithm can be much faster than iterating on the Bellman equation. It is also easy to implement the Howard improvement algorithm in the present setting. At time t, the system resides in one of N predetermined positions, denoted x_i for $i = 1, 2, \ldots, N$. There exists a predetermined set $\mathcal{M}$ of $(N \times N)$ stochastic matrices P that are the objects of choice. Here $P_{ij} = \text{Prob}\,[x_{t+1} = x_j \mid x_t = x_i]$, $i = 1, \ldots, N$; $j = 1, \ldots, N$.

The matrices P satisfy $P_{ij} \geq 0$, $\sum_{j=1}^{N} P_{ij} = 1$, and additional restrictions dictated by the problem at hand that determine the set $\mathcal{M}$. The one-period return function is represented as c_P, a vector of length N, and is a function of P. The ith entry of c_P denotes the one-period return when the state of the system is x_i and the transition matrix is P. The Bellman equation is

$$v_P(x_i) = \max_{P \in \mathcal{M}} \left\{ c_P(x_i) + \beta \sum_{j=1}^{N} P_{ij}\, v_P(x_j) \right\}$$

or

$$v_P = \max_{P \in \mathcal{M}} \left\{ c_P + \beta P v_P \right\}. \tag{4.4.1}$$

We can express this as

$$v_P = T v_P \, ,$$

where T is the operator defined by the right side of (4.4.1). Following Putterman and Brumelle (1979) and Putterman and Shin (1978), define the operator

$$B = T - I,$$

so that

$$Bv = \max_{P \in \mathcal{M}} \left\{ c_P + \beta P v \right\} - v.$$

In terms of the operator B, the Bellman equation is

$$Bv = 0. \tag{4.4.2}$$

The policy improvement algorithm consists of iterations on the following two steps.

1. For fixed P_n, solve

$$(I - \beta P_n)\, v_{P_n} = c_{P_n} \tag{4.4.3}$$

for v_{P_n}.

2. Find P_{n+1} such that

$$c_{P_{n+1}} + (\beta P_{n+1} - I)\, v_{P_n} = B v_{P_n} \tag{4.4.4}$$

Step 1 is accomplished by setting

$$v_{P_n} = (I - \beta P_n)^{-1} c_{P_n}. \tag{4.4.5}$$

Step 2 amounts to finding a policy function (i.e., a stochastic matrix $P_{n+1} \in \mathcal{M}$) that solves a two-period problem with v_{P_n} as the terminal value function.

Following Putterman and Brumelle, the policy improvement algorithm can be interpreted as a version of Newton's method for finding the zero of $Bv = v$. Using equation $(4.4.3)$ for $n+1$ to eliminate $c_{P_{n+1}}$ from equation $(4.4.4)$ gives

$$(I - \beta P_{n+1})\, v_{P_{n+1}} + (\beta P_{n+1} - I)\, v_{P_n} = B v_{P_n}$$

which implies

$$v_{P_{n+1}} = v_{P_n} + (I - \beta P_{n+1})^{-1} B v_{P_n}. \tag{4.4.6}$$

From equation $(4.4.4)$, $(\beta P_{n+1} - I)$ can be regarded as the gradient of $B v_{P_n}$, which supports the interpretation of equation $(4.4.6)$ as implementing Newton's method.[4]

---

[4] Newton's method for finding the solution of $G(z) = 0$ is to iterate on $z_{n+1} = z_n - G'(z_n)^{-1} G(z_n)$.

4.5. Numerical implementation

We shall illustrate Howard's policy improvement algorithm by applying it to our savings example. Consider a feasible policy function $a' = g(k, s)$. For each j, define the $n \times n$ matrices J_j by

$$J_j\left(a, a'\right) = \begin{cases} 1 & \text{if } g\left(a, s_j\right) = a' \\ 0 & \text{otherwise .} \end{cases}$$

Here $j = 1, 2, \ldots, m$ where m is the number of possible values for s_t, and $J_j(a, a')$ is the element of J_j with rows corresponding to initial assets a and columns to terminal assets a'. For a given policy function $a' = g(a, s)$ define the $n \times 1$ vectors r_j with rows corresponding to

$$r_j\left(a\right) = u\left[(r + 1)\, a + w s_j - g\left(a, s_j\right)\right], \tag{4.5.1}$$

for $j = 1, \ldots, m$.

   Suppose the policy function $a' = g(a, s)$ is used forever. Let the value associated with using $g(a, s)$ forever be represented by the m $(n \times 1)$ vectors $[v_1, \ldots, v_m]$, where $v_j(a_i)$ is the value starting from state (a_i, s_j). Suppose that $m = 2$. The vectors $[v_1, v_2]$ obey

$$\begin{bmatrix} v_1 \\ v_2 \end{bmatrix} = \begin{bmatrix} r_1 \\ r_2 \end{bmatrix} + \begin{bmatrix} \beta \mathcal{P}_{11} J_1 & \beta \mathcal{P}_{12} J_1 \\ \beta \mathcal{P}_{21} J_2 & \beta \mathcal{P}_{22} J_2 \end{bmatrix} \begin{bmatrix} v_1 \\ v_2 \end{bmatrix}.$$

Then

$$\begin{bmatrix} v_1 \\ v_2 \end{bmatrix} = \left[I - \beta \begin{pmatrix} \mathcal{P}_{11} J_1 & \mathcal{P}_{12} J_1 \\ \mathcal{P}_{21} J_2 & \mathcal{P}_{22} J_2 \end{pmatrix} \right]^{-1} \begin{bmatrix} r_1 \\ r_2 \end{bmatrix}. \tag{4.5.2}$$

Here is how to implement the Howard policy improvement algorithm.

   Step 1. For an initial feasible policy function $g_\tau(a, j)$ for $\tau = 1$, form the r_j matrices using equation $(4.5.1)$, then use equation $(4.5.2)$ to evaluate the vectors of values $[v_1^\tau, v_2^\tau]$ implied by using that policy forever.

   Step 2. Use $[v_1^\tau, v_2^\tau]$ as the terminal value vectors in equation $(4.3.2)$, and perform one step on the Bellman equation to find a new policy function $g_{\tau+1}(a, s)$ for $\tau + 1 = 2$. Use this policy function, increment τ by 1, and repeat step 1.

   Step 3. Iterate to convergence on steps 1 and 2.

4.5.1. Modified policy iteration

Researchers have had success using the following modification of policy iteration: for $k \geq 2$, iterate k times on the Bellman equation. Take the resulting policy function and use equation (4.5.2) to produce a new candidate value function. Then starting from this terminal value function, perform another k iterations on the Bellman equation. Continue in this fashion until the decision rule converges.

4.6. Sample Bellman equations

This section presents some examples. The first two examples involve no optimization, just computing discounted expected utility. Appendix A of chapter 6 describes some related examples based on search theory.

4.6.1. Example 1: calculating expected utility

Suppose that the one-period utility function is the constant relative risk aversion form $u(c) = c^{1-\gamma}/(1-\gamma)$. Suppose that $c_{t+1} = \lambda_{t+1} c_t$ and that $\{\lambda_t\}$ is an n-state Markov process with transition matrix $P_{ij} = \text{Prob}(\lambda_{t+1} = \bar{\lambda}_j | \lambda_t = \bar{\lambda}_i)$. Suppose that we want to evaluate discounted expected utility

$$V(c_0, \lambda_0) = E_0 \sum_{t=0}^{\infty} \beta^t u(c_t), \qquad (4.6.1)$$

where $\beta \in (0, 1)$. We can express this equation recursively:

$$V(c_t, \lambda_t) = u(c_t) + \beta E_t V(c_{t+1}, \lambda_{t+1}) \qquad (4.6.2)$$

We use a guess-and-verify technique to solve equation (4.6.2) for $V(c_t, \lambda_t)$. Guess that $V(c_t, \lambda_t) = u(c_t) w(\lambda_t)$ for some function $w(\lambda_t)$. Substitute the guess into equation (4.6.2), divide both sides by $u(c_t)$, and rearrange to get

$$w(\lambda_t) = 1 + \beta E_t \left(\frac{c_{t+1}}{c_t} \right)^{1-\gamma} w(\lambda_{t+1})$$

or

$$w_i = 1 + \beta \sum_j P_{ij} (\lambda_j)^{1-\gamma} w_j. \qquad (4.6.3)$$

Equation (4.6.3) is a system of linear equations in $w_i, i = 1, \ldots, n$ whose solution can be expressed as

$$w = \left[1 - \beta P \ \text{diag} \left(\lambda_1^{1-\gamma}, \ldots, \lambda_n^{1-\gamma}\right)\right]^{-1} \mathbf{1}$$

where $\mathbf{1}$ is an $n \times 1$ vector of ones.

4.6.2. Example 2: risk-sensitive preferences

Suppose we modify the preferences of the previous example to be of the recursive form

$$V(c_t, \lambda_t) = u(c_t) + \beta \mathcal{R}_t V(c_{t+1}, \lambda_{t+1}), \tag{4.6.4}$$

where

$$\mathcal{R}_t(V) = \left(\frac{2}{\sigma}\right) \log E_t \left[\exp\left(\frac{\sigma V_{t+1}}{2}\right)\right] \tag{4.6.5}$$

is an operator used by Jacobson (1973), Whittle (1990), and Hansen and Sargent (1995) to induce a preference for robustness to model misspecification.[5] Here $\sigma \leq 0$; when $\sigma < 0$, it represents a concern for model misspecification, or an extra sensitivity to risk.

We leave it to the reader to propose a method for computing an approximation to a value function that solves the functional equation (4.6.4). (Hint: the method used in example 1 will not apply directly because the homogeneity property exploited there fails to prevail now.)

---

[5] Also see Epstein and Zin (1989) and Weil (1989) for a version of the $\mathcal{R}_t$ operator.

4.6.3. Example 3: costs of business cycles

Robert E. Lucas, Jr., (1987) proposed that the cost of business cycles be measured in terms of a proportional upward shift in the consumption process that would be required to make a representative consumer indifferent between its random consumption allocation and a nonrandom consumption allocation with the same mean. This measure of business cycles is the fraction Ω that satisfies

$$E_0 \sum_{t=0}^{\infty} \beta^t u\left[(1+\Omega)\, c_t\right] = \sum_{t=0}^{\infty} \beta^t u\left[E_0\left(c_t\right)\right]. \tag{4.6.6}$$

Suppose that the utility function and the consumption process are as in example 1. Then for given Ω, the calculations in example 1 can be used to calculate the left side of equation (4.6.6). In particular, the left side just equals $u[(1 + \Omega)c_0]w(\lambda)$, where $w(\lambda)$ is calculated from equation (4.6.3). To calculate the right side, we have to evaluate

$$E_0 c_t = c_0 \sum_{\lambda_t,\ldots,\lambda_1} \lambda_t \lambda_{t-1} \cdots \lambda_1 \pi\left(\lambda_t | \lambda_{t-1}\right) \pi\left(\lambda_{t-1} | \lambda_{t-2}\right) \cdots \pi\left(\lambda_1 | \lambda_0\right), \tag{4.6.7}$$

where the summation is over all possible *paths* of growth rates between 0 and t. In the case of i.i.d. λ_t, this expression simplifies to

$$E_0 c_t = c_0 \left(E\lambda\right)^t, \tag{4.6.8}$$

where $E\lambda_t$ is the unconditional mean of λ. Under equation (4.6.8), the right side of equation (4.6.6) is easy to evaluate.

Given γ, π, a procedure for constructing the cost of cycles—more precisely, the costs of deviations from mean trend—to the representative consumer is first to compute the right side of equation (4.6.6). Then we solve the following equation for Ω:

$$u\left[(1+\Omega)\, c_0\right] w\left(\lambda_0\right) = \sum_{t=0}^{\infty} \beta^t u\left[E_0\left(c_t\right)\right].$$

Using a closely related but somewhat different stochastic specification, Lucas (1987) calculated Ω. He assumed that the endowment is a geometric trend with growth rate μ plus an i.i.d. shock with mean zero and variance σ_z^2. Starting from a base $\mu = \mu_0$, he found μ, σ_z pairs to which the household is indifferent, assuming various values of γ that he judged to be within a reasonable range.[6] Lucas found that for reasonable values of γ, it takes a very small adjustment in the trend rate of growth μ to compensate for even a substantial increase in the "cyclical noise" σ_z, which meant to him that the costs of business cycle fluctuations are small.

Subsequent researchers have studied how other preference specifications would affect the calculated costs. Tallarini (1996, 2000) used a version of the preferences described in example 2 and found larger costs of business cycles when parameters are calibrated to match data on asset prices. Hansen, Sargent, and Tallarini (1999) and Alvarez and Jermann (1999) considered local measures of the cost of business cycles and provided ways to link them to the equity premium puzzle, to be studied in chapter 14.

4.7. Polynomial approximations

Judd (1998) describes a method for iterating on the Bellman equation using a polynomial to approximate the value function and a numerical optimizer to perform the optimization at each iteration. We describe this method in the context of the Bellman equation for a particular problem that we shall encounter later.

In chapter 20, we shall study Hopenhayn and Nicolini's (1997) model of optimal unemployment insurance. A planner wants to provide incentives to an unemployed worker to search for a new job while also partially insuring the worker against bad luck in the search process. The planner seeks to deliver discounted expected utility V to an unemployed worker at minimum cost while providing proper incentives to search for work. Hopenhayn and Nicolini show that the minimum cost $C(V)$ satisfies the Bellman equation

$$C\left(V\right) = \min_{V^u}\left\{c + \beta\left[1 - p\left(a\right)\right]C\left(V^u\right)\right\} \tag{4.7.1}$$

where c, a are given by

$$c = u^{-1}\left[\max\left(0, V + a - \beta\{p\left(a\right)V^e + \left[1 - p\left(a\right)\right]V^u\}\right)\right]. \tag{4.7.2}$$

---

[6] See chapter 14 for a discussion of reasonable values of γ. See Table 1 of Manuelli and Sargent (1988) for a correction to Lucas's calculations.

and

$$a = \max \left\{ 0, \frac{\log \left[r\beta \left(V^e - V^u \right) \right]}{r} \right\}. \tag{4.7.3}$$

Here V is a discounted present value that an insurer has promised to an unemployed worker, V_u is a value for next period that the insurer promises the worker if he remains unemployed, $1 - p(a)$ is the probability of remaining unemployed if the worker exerts search effort a, and c is the worker's consumption level. Hopenhayn and Nicolini assume that $p(a) = 1 - \exp(ra)$, $r > 0$.

4.7.1. Recommended computational strategy

To approximate the solution of the Bellman equation (4.7.1), we apply a computational procedure described by Judd (1996, 1998). The method uses a polynomial to approximate the ith iterate $C_i(V)$ of $C(V)$. This polynomial is stored on the computer in terms of $n + 1$ coefficients. Then at each iteration, the Bellman equation is to be solved at a small number $m \geq n + 1$ values of V. This procedure gives values of the ith iterate of the value function $C_i(V)$ at those particular V's. Then we interpolate (or "connect the dots") to fill in the continuous function $C_i(V)$. Substituting this approximation $C_i(V)$ for $C(V)$ in equation (4.7.1), we pass the minimum problem on the right side of equation (4.7.1) to a numerical minimizer. Programming languages like Matlab and Gauss have easy-to-use algorithms for minimizing continuous functions of several variables. We solve one such numerical problem minimization for each node value for V. Doing so yields optimized value $C_{i+1}(V)$ at those node points. We then interpolate to build up $C_{i+1}(V)$. We iterate on this scheme to convergence. Before summarizing the algorithm, we provide a brief description of Chebyshev polynomials.

4.7.2. Chebyshev polynomials

Where n is a nonnegative integer and $x \in \mathbb{R}$, the nth Chebyshev polynomial, is

$$T_n(x) = \cos\left(n \cos^{-1} x\right). \qquad (4.7.4)$$

Given coefficients $c_j, j = 0, \ldots, n$, the nth-order Chebyshev polynomial approximator is

$$C_n(x) = c_0 + \sum_{j=1}^{n} c_j T_j(x). \qquad (4.7.5)$$

We are given a real-valued function f of a single variable $x \in [-1, 1]$. For computational purposes, we want to form an approximator to f of the form (4.7.5). Note that we can store this approximator simply as the $n + 1$ coefficients $c_j, j = 0, \ldots, n$. To form the approximator, we evaluate $f(x)$ at $n + 1$ carefully chosen points, then use a least-squares formula to form the c_j's in equation (4.7.5). Thus, to interpolate a function of a single variable x with domain $x \in [-1, 1]$, Judd (1996, 1998) recommends evaluating the function at the $m \geq n + 1$ points $x_k, k = 1, \ldots, m$, where

$$x_k = \cos\left(\frac{2k-1}{2m}\pi\right), k = 1, \ldots, m. \qquad (4.7.6)$$

Here x_k is the zero of the kth Chebyshev polynomial on $[-1, 1]$. Given the $m \geq n + 1$ values of $f(x_k)$ for $k = 1, \ldots, m$, choose the least-squares values of c_j

$$c_j = \frac{\sum_{k=1}^{m} f(x_k) T_j(x_k)}{\sum_{k=1}^{m} T_j(x_k)^2}, \ j = 0, \ldots, n \qquad (4.7.7)$$

4.7.3. Algorithm: summary

In summary, applied to the Hopenhayn-Nicolini model, the numerical procedure consists of the following steps:

1. Choose upper and lower bounds for V^u, so that V and V^u will be understood to reside in the interval $[\underline{V}^u, \overline{V}^u]$. In particular, set $\overline{V}^u = V^e - \frac{1}{\beta p'(0)}$, the bound required to assure positive search effort, computed in chapter 20. Set $\underline{V}^u = V_{rmaut}$.

2. Choose a degree n for the approximator, a Chebyshev polynomial, and a number $m \geq n + 1$ of nodes or grid points.

3. Generate the m zeros of the Chebyshev polynomial on the set $[1, -1]$, given by (4.7.6).

4. By a change of scale, transform the z_i's to corresponding points V_ℓ^u in $[\underline{V}^u, \overline{V}^u]$.

5. Choose initial values of the $n + 1$ coefficients in the Chebyshev polynomial, for example, $c_j = 0, \ldots, n$. Use these coefficients to define the function $C_i(V^u)$ for iteration number $i = 0$.

6. Compute the function $\tilde{C}_i(V) \equiv c + \beta[1 - p(a)]C_i(V^u)$, where c, a are determined as functions of (V, V^u) from equations (4.7.2) and (4.7.3). This computation builds in the functional forms and parameters of $u(c)$ and $p(a)$, as well as β.

7. For each point V_ℓ^u, use a numerical minimization program to find $C_{i+1}(V_\ell^u) = \min_{V^u} \tilde{C}_i(V_u)$.

8. Using these m values of $C_{j+1}(V_\ell^u)$, compute new values of the coefficients in the Chebyshev polynomials by using "least squares" [formula (4.7.7)]. Return to step 5 and iterate to convergence.

4.7.4. Shape-preserving splines

Judd (1998) points out that because they do not preserve concavity, using Chebyshev polynomials to approximate value functions can cause problems. He recommends the Schumaker quadratic shape-preserving spline. It ensures that the objective in the maximization step of iterating on a Bellman equation will be concave and differentiable (Judd, 1998, p. 441). Using Schumaker splines avoids the type of internodal oscillations associated with other polynomial approximation methods. The exact interpolation procedure is described in Judd (1998, p. 233). A relatively small number of nodes usually is sufficient. Judd and Solnick (1994) find that this approach outperforms linear interpolation and discrete-state approximation methods in a deterministic optimal growth problem.[7]

4.8. Concluding remarks

This chapter has described two of three standard methods for approximating solutions of dynamic programs numerically: discretizing the state space and using polynomials to approximate the value function. The next chapter describes the third method: making the problem have a quadratic return function and linear transition law. A benefit of making the restrictive linear-quadratic assumptions is that they make solving a dynamic program easy by exploiting the ease with which stochastic linear difference equations can be manipulated.

---

[7] The Matlab program `schumaker.m` (written by Leonardo Rezende of the University of Illinois) can be used to compute the spline. Use the Matlab command `ppval` to evaluate the spline.

Chapter 5
Linear Quadratic Dynamic Programming

5.1. Introduction

This chapter describes the class of dynamic programming problems in which the return function is quadratic and the transition function is linear. This specification leads to the widely used optimal linear regulator problem, for which the Bellman equation can be solved quickly using linear algebra. We consider the special case in which the return function and transition function are both time invariant, though the mathematics is almost identical when they are permitted to be deterministic functions of time.

Linear quadratic dynamic programming has two uses for us. A first is to study optimum and equilibrium problems arising for linear rational expectations models. Here the dynamic decision problems naturally take the form of an optimal linear regulator. A second is to use a linear quadratic dynamic program to approximate one that is not linear quadratic.

Later in the chapter, we tell how the Kalman filtering problem from chapter 2 relates to the linear-quadratic dynamic programming problem. Suitably reinterpreted, formulas that solve the optimal linear regulator are the Kalman filter.

5.2. The optimal linear regulator problem

The undiscounted optimal linear regulator problem is to maximize over choice of $\{u_t\}_{t=0}^{\infty}$ the criterion

$$-\sum_{t=0}^{\infty}\{x_t' R x_t + u_t' Q u_t\}, \tag{5.2.1}$$

subject to $x_{t+1} = A x_t + B u_t$, x_0 given. Here x_t is an $(n \times 1)$ vector of state variables, u_t is a $(k \times 1)$ vector of controls, R is a positive semidefinite symmetric matrix, Q is a positive definite symmetric matrix, A is an $(n \times n)$ matrix, and B is an $(n \times k)$ matrix. We guess that the value function is quadratic, $V(x) = -x' P x$, where P is a positive semidefinite symmetric matrix.

Using the transition law to eliminate next period's state, the Bellman equation becomes

$$-x' P x = \max_{u}\{-x' R x - u' Q u - (A x + B u)' P (A x + B u)\}. \tag{5.2.2}$$

The first-order necessary condition for the maximum problem on the right side of equation (5.2.2) is [1]

$$(Q + B' P B) u = -B' P A x, \tag{5.2.3}$$

which implies the feedback rule for u:

$$u = -(Q + B' P B)^{-1} B' P A x \tag{5.2.4}$$

or $u = -F x$, where

$$F = (Q + B' P B)^{-1} B' P A. \tag{5.2.5}$$

Substituting the optimizer (5.2.4) into the right side of equation (5.2.2) and rearranging gives

$$P = R + A' P A - A' P B (Q + B' P B)^{-1} B' P A. \tag{5.2.6}$$

Equation (5.2.6) is called the *algebraic matrix Riccati* equation. It expresses the matrix P as an implicit function of the matrices R, Q, A, B. Solving this equation for P requires a computer whenever P is larger than a 2×2 matrix.

---

[1] We use the following rules for differentiating quadratic and bilinear matrix forms: $\frac{\partial x' A x}{\partial x} = (A + A') x$; $\frac{\partial y' B z}{\partial y} = B z$, $\frac{\partial y' B z}{\partial z} = B' y$.

In exercise *5.1*, you are asked to derive the Riccati equation for the case where the return function is modified to

$$- (x_t' R x_t + u_t' Q u_t + 2 u_t' H x_t) .$$

5.2.1. Value function iteration

Under particular conditions to be discussed in the section on stability, equation (5.2.6) has a unique positive semidefinite solution that is approached in the limit as $j \to \infty$ by iterations on the matrix Riccati difference equation[2]

$$P_{j+1} = R + A' P_j A - A' P_j B \left(Q + B' P_j B \right)^{-1} B' P_j A, \qquad (5.2.7a)$$

starting from $P_0 = 0$. The policy function associated with P_j is

$$F_{j+1} = \left(Q + B' P_j B \right)^{-1} B' P_j A. \qquad (5.2.7b)$$

Equation (5.2.7) is derived much like equation (5.2.6) except that one starts from the iterative version of the Bellman equation rather than from the asymptotic version.

5.2.2. Discounted linear regulator problem

The discounted optimal linear regulator problem is to maximize

$$- \sum_{t=0}^{\infty} \beta^t \{ x_t' R x_t + u_t' Q u_t \}, \qquad 0 < \beta < 1, \qquad (5.2.8)$$

subject to $x_{t+1} = A x_t + B u_t, x_0$ given. This problem leads to the following matrix Riccati difference equation modified for discounting:

$$P_{j+1} = R + \beta A' P_j A - \beta^2 A' P_j B \left(Q + \beta B' P_j B \right)^{-1} B' P_j A. \qquad (5.2.9)$$

The algebraic matrix Riccati equation is modified correspondingly. The value function for the infinite horizon problem is $V(x_0) = -x_0' P x_0$, where P is the

---

[2] If the eigenvalues of A are bounded in modulus below unity, this result obtains, but much weaker conditions suffice. See Bertsekas (1976, chap. 4) and Sargent (1980).

limiting value of P_j resulting from iterations on equation (5.2.9) starting from $P_0 = 0$. The optimal policy is $u_t = -Fx_t$, where $F = \beta(Q + \beta B'PB)^{-1}B'PA$.

The Matlab program `olrp.m` solves the discounted optimal linear regulator problem. Matlab has a variety of other programs that solve both discrete- and continuous-time versions of undiscounted optimal linear regulator problems.

5.2.3. Policy improvement algorithm

The policy improvement algorithm can be applied to solve the discounted optimal linear regulator problem. We discussed aspects of this algorithm earlier in section 2.4.5.2. Starting from an initial F_0 for which the eigenvalues of $A - BF_0$ are less than $1/\sqrt{\beta}$ in modulus, the algorithm iterates on the two equations

$$P_j = R + F_j'QF_j + \beta \left(A - BF_j\right)' P_j \left(A - BF_j\right) \tag{5.2.10}$$

$$F_{j+1} = \beta \left(Q + \beta B'P_j B\right)^{-1} B'P_j A. \tag{5.2.11}$$

The first equation pins down the matrix for the quadratic form in the value function associated with using a fixed rule F_j forever. The second equation gives the matrix for the optimal first-period decision rule for a two-period problem with second-period value function $-x^{*'}P_j x^*$ where x^* is the second-period state. The first equation is an example of a *discrete Lyapunov* or *Sylvester* equation, which is to be solved for the matrix P_j that determines the value $-x_t'P_j x_t$ that is associated with following policy F_j forever. The solution of this equation can be represented in the form

$$P_j = \sum_{k=0}^{\infty} \beta^k \left(A - BF_j\right)'^k \left(R + F_j'QF_j\right) \left(A - BF_j\right)^k.$$

If the eigenvalues of the matrix $A - BF_j$ are bounded in modulus by $1/\sqrt{\beta}$, then a solution of this equation exists. There are several methods available for solving this equation.[3] The Matlab program `policyi.m` solves the undiscounted optimal linear regulator problem using policy iteration. This algorithm is typically much faster than the algorithm that iterates on the matrix Riccati equation. Later we shall present a third method for solving for P that rests on the link between P and shadow prices for the state vector.

---

[3] The Matlab programs `dlyap.m` and `doublej.m` solve discrete Lyapunov equations. See Anderson, Hansen, McGrattan, and Sargent (1996).

5.3.   The stochastic optimal linear regulator problem

The stochastic discounted linear optimal regulator problem is to choose a decision rule for u_t to maximize

$$-E_0 \sum_{t=0}^{\infty} \beta^t \{x_t' R x_t + u_t' Q u_t\}, \qquad 0 < \beta < 1, \tag{5.3.1}$$

subject to x_0 given, and the law of motion

$$x_{t+1} = A x_t + B u_t + C \epsilon_{t+1}, \qquad t \geq 0, \tag{5.3.2}$$

where ϵ_{t+1} is an $(n \times 1)$ vector of random variables that is independently and identically distributed according to the normal distribution with mean vector zero and covariance matrix

$$E \epsilon_t \epsilon_t' = I. \tag{5.3.3}$$

(See Kwakernaak and Sivan, 1972, for an extensive study of the continuous-time version of this problem; also see Chow, 1981.)

The value function for this problem is

$$v(x) = -x' P x - d, \tag{5.3.4}$$

where P is the unique positive semidefinite solution of the discounted algebraic matrix Riccati equation corresponding to equation (5.2.9). As before, it is the limit of iterations on equation (5.2.9) starting from $P_0 = 0$. The scalar d is given by

$$d = \beta (1 - \beta)^{-1} \operatorname{trace} (PCC)'. \tag{5.3.5}$$

Furthermore, the optimal policy continues to be given by $u_t = -F x_t$, where

$$F = \beta (Q + \beta B' P' B)^{-1} B' P A. \tag{5.3.6}$$

A notable feature of this solution is:

CERTAINTY EQUIVALENCE PRINCIPLE: The decision rule (5.3.6) that solves the stochastic optimal linear regulator problem is identical with the decision rule for the corresponding nonstochastic linear optimal regulator problem.

PROOF: Substitute guess (5.3.4) into the Bellman equation to obtain

$$v(x) = \max_u \left\{ -x' R x - u' Q u - \beta E \left[(Ax + Bu + C\epsilon)' P (Ax + Bu + C\epsilon) \right] - \beta d \right\},$$

where ϵ is the realization of ϵ_{t+1} when $x_t = x$ and where $E\epsilon|x = 0$. The preceding equation implies

$$
\begin{aligned}
v(x) = \max_u \{ &-x'Rx - u'Qu - \beta E \{ x'A'PAx + x'A'PBu \\
&+ x'A'PC\epsilon + u'B'PAx + u'B'PBu + u'B'PC\epsilon \\
&+ \epsilon'C'PAx + \epsilon'C'PBu + \epsilon'C'PC\epsilon \} - \beta d \}.
\end{aligned}
$$

Evaluating the expectations inside the braces and using $E\epsilon|x = 0$ gives

$$
\begin{aligned}
v(x) = \max_u - \{ &x'Rx + u'Qu + \beta x'A'PAx + \beta 2x'A'PBu \\
&+ \beta u'B'PBu + \beta E\epsilon'C'PC\epsilon \} - \beta d.
\end{aligned}
$$

The first-order condition for u is

$$
(Q + \beta B'PB)u = -\beta B'PAx,
$$

which implies equation (5.3.6). Using $E\epsilon'C'PC\epsilon = \text{trace}(PCC)'$, substituting equation (5.3.6) into the preceding expression for $v(x)$, and using equation (5.3.4) gives

$$
P = R + \beta A'PA - \beta^2 A'PB(Q + \beta B'PB)^{-1}B'PA,
$$

and

$$
d = \beta(1 - \beta)^{-1}\text{trace}(PCC'). \quad \blacksquare
$$

5.3.1. Discussion of certainty equivalence

The remarkable thing is that, although through d the objective function (5.3.3) depends on CC', the optimal decision rule $u_t = -Fx_t$ is independent of CC'. This is the message of equation (5.3.6) and the discounted algebraic Riccati equation for P, which are identical with the formulas derived earlier under certainty. In other words, the optimal decision rule $u_t = h(x_t)$ is independent of the problem's noise statistics.[4] The certainty equivalence principle is

---

[4] Therefore, in linear quadratic versions of the optimum savings problem, there are no precautionary savings. Compare outcomes from section 2.11 of chapter 2 and chapters 17 and 18.

a special property of the optimal linear regulator problem and comes from the quadratic objective function, the linear transition equation, and the property $E(\epsilon_{t+1}|x_t) = 0$. Certainty equivalence does not characterize stochastic control problems generally.

5.4. Shadow prices in the linear regulator

For several purposes,[5] it is helpful to interpret the gradient $-2Px_t$ of the value function $-x_t'Px_t$ as a shadow price or Lagrange multiplier. Thus, associate with the Bellman equation the Lagrangian

$$-x_t'Px_t = V(x_t) = \min_{\mu_{t+1}} \max_{u_t, x_{t+1}} -\left\{ x_t'Rx_t + u_t'Qu_t + x_{t+1}'Px_{t+1} \right. $$
$$\left. + 2\mu_{t+1}'\left[Ax_t + Bu_t - x_{t+1}\right] \right\},$$

where $2\mu_{t+1}$ is a vector of Lagrange multipliers. The first-order necessary conditions for an optimum with respect to u_t and x_{t+1} are

$$2Qu_t + 2B'\mu_{t+1} = 0$$
$$2Px_{t+1} - 2\mu_{t+1} = 0. \tag{5.4.1}$$

Using the transition law and rearranging gives the usual formula for the optimal decision rule, namely, $u_t = -(Q + B'PB)^{-1}B'PAx_t$. Notice that by (5.4.1), the shadow price vector satisfies $\mu_{t+1} = Px_{t+1}$.

In section 5.5, we shall describe a computational strategy that solves for P by directly finding the optimal multiplier process $\{\mu_t\}$ and representing it as $\mu_t = Px_t$. This strategy exploits the *stability* properties of optimal solutions of the linear regulator problem, which we now briefly take up.

---

[5] In a planning problem in a linear quadratic economy, the gradient of the value function has information from which competitive equilibrium prices can be coaxed. See Hansen and Sargent (2000).

5.4.1. Stability

Upon substituting the optimal control $u_t = -Fx_t$ into the law of motion $x_{t+1} = Ax_t + Bu_t$, we obtain the optimal "closed-loop system" $x_{t+1} = (A - BF)x_t$. This difference equation governs the evolution of x_t under the optimal control. The system is said to be *stable* if $\lim_{t\to\infty} x_t = 0$ starting from any initial $x_0 \in R^n$. Assume that the eigenvalues of $(A - BF)$ are distinct, and use the eigenvalue decomposition $(A - BF) = D\Lambda D^{-1}$ where the columns of D are the eigenvectors of $(A - BF)$ and Λ is a diagonal matrix of eigenvalues of $(A - BF)$. Write the "closed-loop" equation as $x_{t+1} = D\Lambda D^{-1}x_t$. The solution of this difference equation for $t > 0$ is readily verified by repeated substitution to be $x_t = D\Lambda^t D^{-1}x_0$. Evidently, the system is stable for all $x_0 \in R^n$ if and only if the eigenvalues of $(A - BF)$ are all strictly less than unity in absolute value. When this condition is met, $(A - BF)$ is said to be a "stable matrix." [6]

A vast literature is devoted to characterizing the conditions on A, B, R, and Q that imply that F is such that the optimal closed-loop system matrix $(A - BF)$ is stable. These conditions are surveyed by Anderson, Hansen, McGrattan, and Sargent (1996) and can be briefly described here for the undiscounted case $\beta = 1$. Roughly speaking, the conditions on A, B, R, and Q are as follows: First, A and B must be such that it is *possible* to pick a control law $u_t = -Fx_t$ that drives x_t to zero eventually, starting from any $x_0 \in R^n$ ["the pair (A, B) must be stabilizable"]. Second, the matrix R must be such that it is *desirable* to drive x_t to zero as $t \to \infty$.

It would take us too far afield to go deeply into this body of theory, but we can give a flavor of the results by considering the following special assumptions and their implications. Similar results can obtain under weaker conditions relevant for economic problems. [7]

ASSUMPTION A.1: The matrix R is positive definite.

There immediately follows:

PROPOSITION 1: Under assumption A.1, if a solution to the undiscounted regulator exists, it satisfies $\lim_{t\to\infty} x_t = 0$.

---

[6] It is possible to amend the statements about stability in this section to permit $A - BF$ to have a single unit eigenvalue associated with a constant in the state vector. See chapter 2 for examples.

[7] See Kwakernaak and Sivan (1972) and Anderson, Hansen, McGrattan, and Sargent (1996) for much weaker conditions.

PROOF: If $x_t \not\to 0$, then $\sum_{t=0}^{\infty} x_t' R x_t \to \infty$. ∎

ASSUMPTION A.2: The matrix R is positive semidefinite.

Under assumption A.2, R is similar to a triangular matrix R^*:

$$R = T' \begin{pmatrix} R_{11}^* & 0 \\ 0 & 0 \end{pmatrix} T$$

where R_{11}^* is positive definite and T is nonsingular. Notice that $x_t' R x_t = x_{1t}^* R_{11}^* x_{1t}^*$ where $x_t^* = T x_t = \begin{pmatrix} T_1 \\ T_2 \end{pmatrix} x_t = \begin{pmatrix} x_{1t}^* \\ x_{2t}^* \end{pmatrix}$. Let $x_{1t}^* \equiv T_1 x_t$. These calculations support:

PROPOSITION 2: Suppose that a solution to the optimal linear regulator exists under assumption A.2. Then $\lim_{t \to \infty} x_{1t}^* = 0$.

The following definition is used in control theory:

DEFINITION: The pair (A, B) is said to be *stabilizable* if there exists a matrix F for which $(A - BF)$ is a stable matrix.

The following indicates the flavor of a variety of stability theorems from control theory:[8],[9]

THEOREM: If (A, B) is stabilizable and R is positive definite, then under the optimal rule F, $(A - BF)$ is a stable matrix.

In the next section, we assume that A, B, Q, R satisfy conditions sufficient to invoke such a stability proposition, and we use that assumption to justify a solution method that solves the undiscounted linear regulator by searching among the many solutions of the *Euler equations* for a stable solution.

---

[8] These conditions are discussed under the subjects of controllability, stabilizability, reconstructability, and detectability in the literature on linear optimal control. (For continuous-time linear system, these concepts are described by Kwakernaak and Sivan, 1972; for discrete-time systems, see Sargent, 1980.) These conditions subsume and generalize the transversality conditions used in the discrete-time calculus of variations (see Sargent, 1987a). That is, the case when $(A - BF)$ is stable corresponds to the situation in which it is optimal to solve "stable roots backward and unstable roots forward." See Sargent (1987a, chap. 9). Hansen and Sargent (1981) describe the relationship between Euler equation methods and dynamic programming for a class of linear optimal control systems. Also see Chow (1981).

[9] The conditions under which $(A - BF)$ is stable are also the conditions under which x_t converges to a unique stationary distribution in the stochastic version of the linear regulator problem.

5.5.   A Lagrangian formulation

This section describes a Lagrangian formulation of the optimal linear regulator.[10]   Besides being useful computationally, this formulation carries insights
about the connections between stability and optimality and also opens the way
to constructing solutions of dynamic systems not coming directly from an intertemporal optimization problem.[11]

For the undiscounted optimal linear regulator problem, form the Lagrangian

$$
\mathcal{L} = -\sum_{t=0}^{\infty} \left\{ x_t' R x_t + u_t' Q u_t \right.
$$

$$
\left. + 2\mu_{t+1}' \left[A x_t + B u_t - x_{t+1} \right] \right\}.
$$

First-order conditions for maximization with respect to $\{u_t, x_{t+1}\}$ are

$$
2Q u_t + 2B' \mu_{t+1} = 0
$$
$$
\mu_t = R x_t + A' \mu_{t+1} \ , \ t \geq 0. \tag{5.5.1}
$$

The Lagrange multiplier vector μ_{t+1} is often called the *costate* vector. Recall
from the second equation of (5.4.1) that $\mu_{t+1} = P x_{t+1}$ where P is the matrix
that solves the algebraic Riccati equation. Thus, μ_{t+1} is the gradient of the
value function. Solve the first equation for u_t in terms of μ_{t+1}; substitute into
the law of motion $x_{t+1} = A x_t + B u_t$; arrange the resulting equation and the
second equation of (5.5.1) into the form

$$
L \begin{pmatrix} x_{t+1} \\ \mu_{t+1} \end{pmatrix} = N \begin{pmatrix} x_t \\ \mu_t \end{pmatrix} \ , \ t \geq 0,
$$

where

$$
L = \begin{pmatrix} I & BQ^{-1}B' \\ 0 & A' \end{pmatrix}, N = \begin{pmatrix} A & 0 \\ -R & I \end{pmatrix}.
$$

When L is of full rank (i.e., when A is of full rank), we can write this system
as

$$
\begin{pmatrix} x_{t+1} \\ \mu_{t+1} \end{pmatrix} = M \begin{pmatrix} x_t \\ \mu_t \end{pmatrix} \tag{5.5.2}
$$

---

[10]   Such formulations are recommended by Chow (1997) and Anderson, Hansen, McGrattan,
and Sargent (1996).

[11]   Blanchard and Kahn (1980); Whiteman (1983); Hansen, Epple, and Roberds (1985); and
Anderson, Hansen, McGrattan and Sargent (1996) use and extend such methods.

where

$$M \equiv L^{-1}N = \begin{pmatrix} A + BQ^{-1}B'A'^{-1}R & -BQ^{-1}B'A'^{-1} \\ -A'^{-1}R & A'^{-1} \end{pmatrix}. \tag{5.5.3}$$

We seek to solve the difference equation system (5.5.2) for a sequence $\{x_t\}_{t=0}^{\infty}$ that satisfies the initial condition for x_0 and a terminal condition $\lim_{t\to+\infty} x_t = 0$ that expresses our wish for a *stable* solution. We inherit our wish for stability of the $\{x_t\}$ sequence in this sense for a desire to maximize $-\sum_{t=0}^{\infty}[x_t'Rx_t + u_t'Qu_t]$, which requires that $x_t'Rx_t$ converge to zero.

To proceed, we study properties of the $(2n \times 2n)$ matrix M. It is helpful to introduce a $(2n \times 2n)$ matrix

$$J = \begin{pmatrix} 0 & -I_n \\ I_n & 0 \end{pmatrix}.$$

The rank of J is $2n$.

DEFINITION: A matrix M is called *symplectic* if

$$MJM' = J. \tag{5.5.4}$$

It can be verified directly that M in equation (5.5.3) is symplectic.

It follows from equation (5.5.4) and from the fact $J^{-1} = J' = -J$ that for any symplectic matrix M,

$$M' = J^{-1}M^{-1}J. \tag{5.5.5}$$

Equation (5.5.5) states that M' is related to the inverse of M by a similarity transformation. For square matrices, recall that (a) similar matrices share eigenvalues; (b) the eigenvalues of the inverse of a matrix are the inverses of the eigenvalues of the matrix; and (c) a matrix and its transpose have the same eigenvalues. It then follows from equation (5.5.5) that the eigenvalues of M occur in reciprocal pairs: if λ is an eigenvalue of M, so is λ^{-1}.

Write equation (5.5.2) as

$$y_{t+1} = My_t \tag{5.5.6}$$

where $y_t = \begin{pmatrix} x_t \\ \mu_t \end{pmatrix}$. Consider the following triangularization of M

$$V^{-1}MV = \begin{pmatrix} W_{11} & W_{12} \\ 0 & W_{22} \end{pmatrix}$$

where each block on the right side is $(n \times n)$, where V is nonsingular, and where W_{22} has all its eigenvalues exceeding 1 in modulus and W_{11} has all of its eigenvalues less than 1 in modulus. The *Schur decomposition* and the *eigenvalue decomposition* are two such decompositions.[12] Write equation (5.5.6) as

$$y_{t+1} = VWV^{-1}y_t. \tag{5.5.7}$$

The solution of equation (5.5.7) for arbitrary initial condition y_0 is evidently

$$y_t = V \begin{bmatrix} W_{11}^t & W_{12,t} \\ 0 & W_{22}^t \end{bmatrix} V^{-1}y_0 \tag{5.5.8}$$

where $W_{12,t} = W_{12}$ for $t = 1$ and for $t \geq 2$ obeys the recursion

$$W_{12,t} = W_{11}^{t-1}W_{12,t-1} + W_{12,t-1}W_{22}^{t-1}$$

and where W_{ii}^t is W_{ii} raised to the tth power.

Write equation (5.5.8) as

$$\begin{pmatrix} y_{1t}^* \\ y_{2t}^* \end{pmatrix} = \begin{bmatrix} W_{11}^t & W_{12,t} \\ 0 & W_{22}^t \end{bmatrix} \begin{pmatrix} y_{10}^* \\ y_{20}^* \end{pmatrix}$$

where $y_t^* = V^{-1}y_t$, and in particular where

$$y_{2t}^* = V^{21}x_t + V^{22}\mu_t, \tag{5.5.9}$$

and where V^{ij} denotes the (i, j) piece of the partitioned V^{-1} matrix.

Because W_{22} is an unstable matrix, unless $y_{20}^* = 0$, y_t^* will diverge. Let V^{ij} denote the (i, j) piece of the partitioned V^{-1} matrix. To attain stability, we must impose $y_{20}^* = 0$, which from equation (5.5.9) implies

$$V^{21}x_0 + V^{22}\mu_0 = 0$$

---

[12] Evan Anderson's Matlab program `schurg.m` attains a convenient Schur decomposition and is very useful for solving linear models with distortions. See McGrattan (1994) for examples of distorted economies whose equilibria can be computed using a Schur decomposition.

or

$$\mu_0 = -\left(V^{22}\right)^{-1} V^{21} x_0.$$

This equation replicates itself over time in the sense that it implies

$$\mu_t = -\left(V^{22}\right)^{-1} V^{21} x_t. \tag{5.5.10}$$

But notice that because $\left(V^{21}\ V^{22}\right)$ is the second row block of the inverse of V,

$$\left(V^{21}\ V^{22}\right)\ \begin{pmatrix} V_{11} \\ V_{21} \end{pmatrix} = 0$$

which implies

$$V^{21} V_{11} + V^{22} V_{21} = 0.$$

Therefore,

$$-\left(V^{22}\right)^{-1} V^{21} = V_{21} V_{11}^{-1}.$$

So we can write

$$\mu_0 = V_{21} V_{11}^{-1} x_0 \tag{5.5.11}$$

and

$$\mu_t = V_{21} V_{11}^{-1} x_t.$$

However, we know from equations $(5.4.1)$ that $\mu_t = P x_t$, where P occurs in the matrix that solves the Riccati equation $(5.2.6)$. Thus, the preceding argument establishes that

$$P = V_{21} V_{11}^{-1}. \tag{5.5.12}$$

This formula provides us with an alternative, and typically computationally very efficient, way of computing the matrix P.

This same method can be applied to compute the solution of any system of the form $(5.5.2)$ if a solution exists, even if the eigenvalues of M fail to occur in reciprocal pairs. The method will typically work so long as the eigenvalues of M split half inside and half outside the unit circle.[13] Systems in which the eigenvalues (properly adjusted for discounting) fail to occur in reciprocal pairs arise when the system being solved is an equilibrium of a model in which there are distortions that prevent there being any optimum problem that the equilibrium solves. See Woodford (1999) for an application of such methods to solve

---

[13] See Whiteman (1983); Blanchard and Kahn (1980); and Anderson, Hansen, McGrattan, and Sargent (1996) for applications and developments of these methods.

for linear approximations of equilibria of a monetary model with distortions. See chapter 11 for some applications to an economy with distorting taxes.

5.6. The Kalman filter again

Suitably reinterpreted, the same recursion (5.2.7) that solves the optimal linear regulator also determines the celebrated *Kalman filter* that we derived in section 2.7 of chapter 2. Recall that the Kalman filter is a recursive algorithm for computing the mathematical expectation $E[x_t|y_{t-1}, \ldots, y_0]$ of a hidden state vector x_t, conditional on observing a history $y_t, \ldots, y_0$ of a vector of noisy signals on the hidden state. The Kalman filter can be used to formulate or simplify a variety of signal-extraction and prediction problems in economics.

   We briefly remind the reader that the setting for the Kalman filter is the following linear state-space system.[14] Given $x_0 \sim \mathcal{N}(\hat{x}_0, \Sigma_0)$, let

$$x_{t+1} = Ax_t + Cw_{t+1} \tag{5.6.1a}$$

$$y_t = Gx_t + v_t \tag{5.6.1b}$$

where x_t is an $(n \times 1)$ state vector, w_t is an i.i.d. sequence Gaussian vector with $Ew_t w_t' = I$, and v_t is an i.i.d. Gaussian vector orthogonal to w_s for all t, s with $Ev_t v_t' = R$; and A, C, and G are matrices conformable to the vectors they multiply. Assume that the initial condition x_0 is unobserved but is known to have a Gaussian distribution with mean $\hat{x}_0$ and covariance matrix Σ_0. At time t, the history of observations $y^t \equiv [y_t, \ldots, y_0]$ is available to estimate the location of x_t and the location of x_{t+1}. The Kalman filter is a recursive algorithm for computing $\hat{x}_{t+1} = E[x_{t+1}|y^t]$. The algorithm is

$$\hat{x}_{t+1} = (A - K_t G)\,\hat{x}_t + K_t y_t \tag{5.6.2}$$

where

$$K_t = A\Sigma_t G'\,(G\Sigma_t G' + R)^{-1} \tag{5.6.3a}$$

$$\Sigma_{t+1} = A\Sigma_t A' + CC' - A\Sigma_t G'\,(G\Sigma_t G' + R)^{-1} G\Sigma_t A. \tag{5.6.3b}$$

---

[14] We derived the Kalman filter as a recursive application of population regression in chapter 2, page 56.

Here $\Sigma_t = E(x_t - \hat{x}_t)(x_t - \hat{x}_t)'$, and K_t is called the *Kalman gain*. Sometimes the Kalman filter is written in terms of the "innovation representation"

$$\hat{x}_{t+1} = A\hat{x}_t + K_t a_t \tag{5.6.4a}$$

$$y_t = G\hat{x}_t + a_t \tag{5.6.4b}$$

where $a_t \equiv y_t - G\hat{x}_t \equiv y_t - E[y_t|y^{t-1}]$. The random vector a_t is called the *innovation* in y_t, being the part of y_t that cannot be forecast linearly from its own past. Subtracting equation $(5.6.4b)$ from $(5.6.1b)$ gives $a_t = G(x_t - \hat{x}_t) + v_t$; multiplying each side by its own transpose and taking expectations gives the following formula for the innovation covariance matrix:

$$Ea_t a_t' = G\Sigma_t G' + R. \tag{5.6.5}$$

Equations $(5.6.3)$ display extensive similarities to equations $(5.2.7)$, the recursions for the optimal linear regulator. Indeed, the mathematical structures are identical when viewed properly. Note that equation $(5.6.3b)$ is a Riccati equation. With the judicious use of matrix transposition and reversal of time, the two systems of equations $(5.6.3)$ and $(5.2.7)$ can be made to match.[15] See chapter 2, especially section 2.9, for some applications of the Kalman filter.[16]

---

[15] See Hansen and Sargent (ch. 4, 2008) for an account of how the LQ dynamic programming problem and the Kalman filter are connected through duality. That chapter formulates the Kalman filtering problem in terms of a Lagrangian, then judiciously transforms the first-order conditions into an associated optimal linear regulator.

[16] The Matlab program `kfilter.m` computes the Kalman filter. Matlab has several programs that compute the Kalman filter for discrete time and continuous time models.

5.7. Concluding remarks

In exchange for their restrictions, the linear quadratic dynamic optimization problems of this chapter acquire tractability. The Bellman equation leads to Riccati difference equations that are so easy to solve numerically that the curse of dimensionality loses most of its force. It is easy to solve linear quadratic control or filtering with many state variables. That it is difficult to solve those problems otherwise is why linear quadratic approximations are widely used. We describe those approximations in Appendix B to this chapter.

In chapter 7, we go beyond the single-agent optimization problems of this chapter to study systems with multiple agents who simultaneously solve linear quadratic dynamic programming problems, with the decision rules of some agents influencing transition laws of variables appearing in other agents' decision problems. We introduce two related equilibrium concepts to reconcile different agents' decisions.

A. Matrix formulas

Let (z, x, a) each be $n \times 1$ vectors, $A, C, D,$ and V each be $(n \times n)$ matrices, B an $(m \times n)$ matrix, and y an $(m \times 1)$ vector. Then $\frac{\partial a'x}{\partial x} = a$, $\frac{\partial x'Ax}{\partial x} = (A + A')x$, $\frac{\partial^2 (x'Ax)}{\partial x \partial x'} = (A + A')$, $\frac{\partial x'Ax}{\partial A} = xx'$, $\frac{\partial y'Bz}{\partial y} = Bz$, $\frac{\partial y'Bz}{\partial z} = B'y$, $\frac{\partial y'Bz}{\partial B} = yz'$.

The equation

$$A'VA + C = V$$

to be solved for V is called a *discrete Lyapunov equation*, and its generalization

$$A'VD + C = V$$

is called the discrete *Sylvester equation*. The discrete Sylvester equation has a unique solution if and only if the eigenvalues $\{\lambda_i\}$ of A and $\{\delta_j\}$ of D satisfy the condition $\lambda_i \delta_j \neq 1 \ \forall \ i, \ j$.

B. Linear quadratic approximations

This appendix describes an important use of the optimal linear regulator: to approximate the solution of more complicated dynamic programs.[17] Optimal linear regulator problems are often used to approximate problems of the following form: maximize over $\{u_t\}_{t=0}^{\infty}$

$$E_0 \sum_{t=0}^{\infty} \beta^t r\left(z_t\right) \tag{5.B.1}$$

$$x_{t+1} = Ax_t + Bu_t + Cw_{t+1} \tag{5.B.2}$$

where $\{w_{t+1}\}$ is a vector of i.i.d. random disturbances with mean zero and finite variance, and $r(z_t)$ is a concave and twice continuously differentiable function of $z_t \equiv \begin{pmatrix} x_t \\ u_t \end{pmatrix}$. All nonlinearities in the original problem are absorbed into the composite function $r(z_t)$.

5.B.1. An example: the stochastic growth model

Take a parametric version of Brock and Mirman's stochastic growth model, whose social planner chooses a policy for $\{c_t, a_{t+1}\}_{t=0}^{\infty}$ to maximize

$$E_0 \sum_{t=0}^{\infty} \beta^t \ln c_t$$

where

$$c_t + i_t = Aa_t^{\alpha}\theta_t$$

$$a_{t+1} = (1 - \delta)\, a_t + i_t$$

$$\ln \theta_{t+1} = \rho \ln \theta_t + w_{t+1}$$

where $\{w_{t+1}\}$ is an i.i.d. stochastic process with mean zero and finite variance, θ_t is a technology shock, and $\tilde{\theta}_t \equiv \ln \theta_t$. To get this problem into the form

---

[17] Kydland and Prescott (1982) used such a method, and so do many of their followers in the real business cycle literature. See King, Plosser, and Rebelo (1988) for related methods of real business cycle models.

(5.B.1)–(5.B.2), take $x_t = \begin{pmatrix} a_t \\ \tilde{\theta}_t \end{pmatrix}$, $u_t = i_t$, and $r(z_t) = \ln(Aa_t^\alpha \exp \tilde{\theta}_t - i_t)$, and we write the laws of motion as

$$\begin{pmatrix} 1 \\ a_{t+1} \\ \tilde{\theta}_{t+1} \end{pmatrix} = \begin{pmatrix} 1 & 0 & 0 \\ 0 & (1-\delta) & 0 \\ 0 & 0 & \rho \end{pmatrix} \begin{pmatrix} 1 \\ a_t \\ \tilde{\theta}_t \end{pmatrix} + \begin{pmatrix} 0 \\ 1 \\ 0 \end{pmatrix} i_t + \begin{pmatrix} 0 \\ 0 \\ 1 \end{pmatrix} w_{t+1}$$

where it is convenient to add the constant 1 as the first component of the state vector.

5.B.2. *Kydland and Prescott's method*

We want to replace $r(z_t)$ by a quadratic $z_t' M z_t$. We choose a point $\bar{z}$ and approximate with the first two terms of a Taylor series:[18]

$$\hat{r}(z) = r(\bar{z}) + (z - \bar{z})' \frac{\partial r}{\partial z} \qquad (5.B.3)$$
$$+ \frac{1}{2} (z - \bar{z})' \frac{\partial^2 r}{\partial z \partial z'} (z - \bar{z}).$$

If the state x_t is $n \times 1$ and the control u_t is $k \times 1$, then the vector z_t is $(n+k) \times 1$. Let e be the $(n+k) \times 1$ vector with 0's everywhere except for a 1 in the row corresponding to the location of the constant unity in the state vector, so that $1 \equiv e' z_t$ for all t.

Repeatedly using $z'e = e'z = 1$, we can express equation (5.B.3) as

$$\hat{r}(z) = z' M z,$$

where

$$M = e \left[r(\bar{z}) - \left(\frac{\partial r}{\partial z} \right)' \bar{z} + \frac{1}{2} \bar{z}' \frac{\partial^2 r}{\partial z \partial z'} \bar{z} \right] e'$$
$$+ \frac{1}{2} \left(\frac{\partial r}{\partial z} e' - e\bar{z}' \frac{\partial^2 r}{\partial z \partial z'} - \frac{\partial^2 r}{\partial z \partial z'} \bar{z} e' + e \frac{\partial r}{\partial z}' \right)$$
$$+ \frac{1}{2} \left(\frac{\partial^2 r}{\partial z \partial z'} \right)$$

---

[18] This setup is taken from McGrattan (1994) and Anderson, Hansen, McGrattan, and Sargent (1996).

where the partial derivatives are evaluated at $\bar{z}$. Partition M, so that

$$z'Mz \equiv \begin{pmatrix} x \\ u \end{pmatrix}' \begin{pmatrix} M_{11} & M_{12} \\ M_{21} & M_{22} \end{pmatrix} \begin{pmatrix} x \\ u \end{pmatrix}$$
$$= \begin{pmatrix} x \\ u \end{pmatrix}' \begin{pmatrix} R & W \\ W' & Q \end{pmatrix} \begin{pmatrix} x \\ u \end{pmatrix}.$$

5.B.3. Determination of $\bar{z}$

Usually, the point $\bar{z}$ is chosen as the (optimal) stationary state of the *nonstochastic* version of the original nonlinear model:

$$\sum_{t=0}^{\infty} \beta^t r\,(z_t)$$
$$x_{t+1} = Ax_t + Bu_t.$$

This stationary point is obtained in these steps:

1. Find the Euler equations.
2. Substitute $z_{t+1} = z_t \equiv \bar{z}$ into the Euler equations and transition laws, and solve the resulting system of nonlinear equations for $\bar{z}$. This purpose can be accomplished, for example, by using the nonlinear equation solver `fsolve.m` in Matlab.

5.B.4. *Log linear approximation*

For some problems, Christiano (1990) has advocated a quadratic approximation in logarithms. We illustrate his idea with the stochastic growth example. Define

$$\tilde{a}_t = \log a_t \ , \ \tilde{\theta}_t = \log \theta_t.$$

Christiano's strategy is to take $\tilde{a}_t, \tilde{\theta}_t$ as the components of the state and write the law of motion as

$$\begin{pmatrix} 1 \\ \tilde{a}_{t+1} \\ \tilde{\theta}_{t+1} \end{pmatrix} = \begin{pmatrix} 1 & 0 & 0 \\ 0 & 0 & 0 \\ 0 & 0 & \rho \end{pmatrix} \begin{pmatrix} 1 \\ \tilde{a}_t \\ \tilde{\theta}_t \end{pmatrix}$$
$$+ \begin{pmatrix} 0 \\ 1 \\ 0 \end{pmatrix} u_t \ + \begin{pmatrix} 0 \\ 0 \\ 1 \end{pmatrix} w_{t+1}$$

where the control u_t is $\tilde{a}_{t+1}$.

Express consumption as

$$c_t = A \left(\exp \tilde{a}_t \right)^\alpha \left(\exp \tilde{\theta}_t \right) + (1 - \delta) \exp \tilde{a}_t - \exp \tilde{a}_{t+1}.$$

Substitute this expression into $\ln c_t \equiv r(z_t)$, and proceed as before to obtain the second-order Taylor series approximation about $\bar{z}$.

5.B.5. *Trend removal*

It is conventional in the real business cycle literature to specify the law of motion for the technology shock θ_t by

$$\tilde{\theta}_t = \log \left(\frac{\theta_t}{\gamma^t} \right), \ \gamma > 1$$

$$\tilde{\theta}_{t+1} = \rho \tilde{\theta}_t + w_{t+1}, \qquad |\rho| < 1.$$

This inspires us to write the law of motion for capital as

$$\gamma \frac{a_{t+1}}{\gamma^{t+1}} = (1 - \delta) \frac{a_t}{\gamma^t} + \frac{i_t}{\gamma^t}$$

or

$$\gamma \exp \tilde{a}_{t+1} = (1 - \delta) \exp \tilde{a}_t + \exp \left(\tilde{i}_t \right)$$

where $\tilde{a}_t \equiv \log\left(\frac{a_t}{\gamma^t}\right), \tilde{i}_t = \log\left(\frac{i_t}{\gamma_t}\right)$. By studying the Euler equations for a model with a growing technology shock ($\gamma > 1$), we can show that there exists a steady state for $\tilde{a}_t$, but not for a_t. Researchers often construct linear quadratic approximations around the nonstochastic steady state of $\tilde{a}$.

Exercises

Exercise 5.1 Consider the modified version of the optimal linear regulator problem where the objective is to maximize

$$-\sum_{t=0}^{\infty} \beta^t \left\{ x_t' R x_t + u_t' Q u_t + 2u_t' H x_t \right\}$$

subject to the law of motion:

$$x_{t+1} = Ax_t + Bu_t.$$

Here x_t is an $n \times 1$ state vector, u_t is a $k \times 1$ vector of controls, and x_0 is a given initial condition. The matrices R, Q are positive definite and symmetric. The maximization is with respect to sequences $\{u_t, x_t\}_{t=0}^{\infty}$.

a. Show that the optimal policy has the form

$$u_t = - \left(Q + \beta B' P B \right)^{-1} \left(\beta B' P A + H \right) x_t,$$

where P solves the algebraic matrix Riccati equation

$$P = R + \beta A' P A - \left(\beta A' P B + H' \right) \left(Q + \beta B' P B \right)^{-1} \left(\beta B' P A + H \right). \quad (1)$$

b. Write a Matlab program to solve equation (1) by iterating on P starting from P being a matrix of zeros.

Exercise 5.2 Verify that equations (5.2.10) and (5.2.11) implement the policy improvement algorithm for the discounted linear regulator problem.

Exercise 5.3   A household chooses $\{c_t, a_{t+1}\}_{t=0}^{\infty}$ to maximize

$$-\sum_{t=0}^{\infty} \beta^t \left\{ (c_t - b)^2 + \gamma i_t^2 \right\}$$

subject to

$$c_t + i_t = ra_t + y_t$$
$$a_{t+1} = a_t + i_t$$
$$y_{t+1} = \rho_1 y_t + \rho_2 y_{t-1}.$$

Here c_t, i_t, a_t, y_t are the household's consumption, investment, asset holdings, and exogenous labor income at t; while $b > 0, \gamma > 0, r > 0, \beta \in (0, 1)$, and ρ_1, ρ_2 are parameters, and a_0, y_0, y_{-1} are initial conditions. Assume that ρ_1, ρ_2 are such that $(1 - \rho_1 z - \rho_2 z^2) = 0$ implies $|z| > 1$.

a. Map this problem into an optimal linear regulator problem.

b. For parameter values $[\beta, (1 + r), b, \gamma, \rho_1, \rho_2] = (.95, .95^{-1}, 30, 1, 1.2, -.3)$, compute the household's optimal policy function using your Matlab program from exercise *5.1*.

Exercise 5.4   Modify exercise *5.3* by assuming that the household seeks to maximize

$$-\sum_{t=0}^{\infty} \beta^t \left\{ (s_t - b)^2 + \gamma i_t^2 \right\}$$

Here s_t measures consumption services that are produced by durables or habits according to

$$s_t = \lambda h_t + \pi c_t$$
$$h_{t+1} = \delta h_t + \theta c_t$$

where h_t is the stock of the durable good or habit, $(\lambda, \pi, \delta, \theta)$ are parameters, and h_0 is an initial condition.

a. Map this problem into a linear regulator problem.

b. For the same parameter values as in exercise *5.3* and $(\lambda, \pi, \delta, \theta) = (1, .05, .95, 1)$, compute the optimal policy for the household.

c. For the same parameter values as in exercise *5.3* and $(\lambda, \pi, \delta, \theta) = (-1, 1, .95, 1)$, compute the optimal policy.

d. Interpret the parameter settings in part b as capturing a model of durable consumption goods, and the settings in part c as giving a model of habit persistence.

Exercise 5.5   A household's labor income follows the stochastic process

$$y_{t+1} = \rho_1 y_t + \rho_2 y_{t-1} + w_{t+1} + \gamma w_t,$$

where w_{t+1} is a Gaussian martingale difference sequence with unit variance. Calculate

$$E \sum_{j=0}^{\infty} \beta^j \left[y_{t+j} | y^t, w^t \right], \tag{1}$$

where y^t, w^t denotes the history of y, w up to t.

a. Write a Matlab program to compute expression (1).

b.   Use your program to evaluate expression (1) for the parameter values $(\beta, \rho_1, \rho_2, \gamma) = (.95, 1.2, -.4, .5)$.

Exercise 5.6   **Finding the state is an art**

For $t \geq 0$, the endowment for a one-good economy d_t is governed by the second order stochastic difference equation

$$d_{t+1} = \rho_0 + \rho_1 d_t + \rho_2 d_{t-1} + \sigma_d \epsilon_{t+1}$$

where ϵ_{t+1} is an i.i.d. process and $\epsilon_{t+1} \sim \mathcal{N}(0, 1)$, ρ_0, ρ_1, and ρ_2 are scalars, and d_0, d_1 are given initial conditions. A *stochastic discount factor* is given by $s_t = \beta^t (b_0 - b_1 d_t)$, where b_0 is a positive scalar, $b_1 \geq 0$, and $\beta \in (0, 1)$. The value of the endowment at time 0 is defined to be

$$(1) \qquad\qquad\qquad v_0 = E_0 \sum_{t=0}^{\infty} s_t d_t$$

and E_0 is the mathematical expectation operator conditioned on d_0, d_{-1}.

a.   Assume that v_0 in equation (1) is finite. Carefully describe a recursive algorithm for computing v_0.

b. Describe conditions on β, ρ_1, ρ_2 that are sufficient to make v_0 finite.

Exercise 5.7   **Dynamic Laffer curves**

The demand for currency in a small country is described by

$$(1) \qquad\qquad M_t/p_t = \gamma_1 - \gamma_2 p_{t+1}/p_t,$$

where $\gamma_1 > \gamma_2 > 0$, M_t is the stock of currency held by the public at the end of period t, and p_t is the price level at time t. There is no randomness in the country, so that there is perfect foresight. Equation (1) is a Cagan-like demand function for currency, expressing real balances as an inverse function of the expected gross rate of inflation.

Speaking of Cagan, the government is running a permanent real deficit of g per period, measured in goods, all of which it finances by currency creation. The government's budget constraint at t is

$$(2) \qquad\qquad (M_t - M_{t-1})/p_t = g,$$

where the left side is the real value of the new currency printed at time t. The economy starts at time $t = 0$, with the initial level of nominal currency stock $M_{-1} = 100$ being given.

For this model, define an *equilibrium* as a pair of *positive* sequences $\{p_t > 0, M_t > 0\}_{t=0}^{\infty}$ that satisfy equations (1) and (2) (portfolio balance and the government budget constraint, respectively) for $t \geq 0$, and the initial condition assigned for M_{-1}.

a. Let $\gamma_1 = 100, \gamma_2 = 50, g = .05$. Write a computer program to compute equilibria for this economy. Describe your approach and display the program.

b. Argue that there exists a continuum of equilibria. Find the *lowest* value of the initial price level p_0 for which there exists an equilibrium. (*Hint 1:* Notice the positivity condition that is part of the definition of equilibrium. *Hint 2:* Try using the general approach to solving difference equations described in section 5.5.)

c. Show that for all of these equilibria except the one that is associated with the minimal p_0 that you calculated in part b, the gross inflation rate and the gross money creation rate both eventually converge to the *same* value. Compute this value.

d. Show that there is a unique equilibrium with a lower inflation rate than the one that you computed in part b. Compute this inflation rate.

e. Increase the level of g to .075. Compare the (eventual or asymptotic) inflation rate that you computed in part b and the inflation rate that you computed in part c. Are your results consistent with the view that "larger permanent deficits cause larger inflation rates"?

f. Discuss your results from the standpoint of the Laffer curve.

Hint: A Matlab program `dlqrmon.m` performs the calculations. It is available from the web site for the book.

Exercise 5.8   A government faces an exogenous stream of government expenditures $\{g_t\}$ that it must finance. Total government expenditures at t consist of two components:

$$(1) \qquad\qquad g_t = g_{Tt} + g_{Pt}$$

where g_{Tt} is transitory expenditures and g_{Pt} is permanent expenditures. At the beginning of period t, the government observes the history up to t of both g_{Tt} and g_{Pt}. Further, it knows the stochastic laws of motion of both, namely,

$$(2) \qquad \begin{aligned} g_{Pt+1} &= g_{Pt} + c_1 \epsilon_{1,t+1} \\ g_{Tt+1} &= (1-\rho)\,\mu_T + \rho g_{Tt} + c_2 \epsilon_{2t+1} \end{aligned}$$

where $\epsilon_{t+1} = \begin{bmatrix} \epsilon_{1t+1} \\ \epsilon_{2t+1} \end{bmatrix}$ is an i.i.d. Gaussian vector process with mean zero and identity covariance matrix. The government finances its budget with a distorting taxes. If it collects T_t total revenues at t, it bears a dead weight loss of $W(T_t)$ where $W(T) = w_1 T_t + .5 w_2 T_t^2$, where $w_1, w_2 > 0$. The government's loss functional is

$$(3) \qquad\qquad E \sum_{t=0}^{\infty} \beta^t W(T_t), \quad \beta \in (0,1).$$

The government can purchase or issue one-period risk-free loans at a constant price q. Therefore, it faces a sequence of budget constraints

$$(4) \qquad\qquad g_t + q b_{t+1} = T_t + b_t,$$

where q^{-1} is the gross rate of return on one-period risk-free government loans. Assume that $b_0 = 0$. The government also faces the terminal value condition

$$\lim_{t \to +\infty} \beta^t W'(T_t) b_{t+1} = 0,$$

which prevents it from running a Ponzi scheme. The government wants to design a tax collection strategy expressing T_t as a function of the history of g_{Tt}, g_{Pt}, b_t that *minimizes* (3) subject to (1), (2), and (4).

a. Formulate the government's problem as a dynamic programming problem. Please carefully define the state and control for this problem. Write the Bellman equation in as much detail as you can. Tell a computational strategy for solving the Bellman equation. Tell the forms of the optimal value function and the optimal decision rule.

b. Using objects that you computed in part **a**, please state the form of the law of motion for the joint process of $g_{Tt}, g_{Pt}, T_t, b_{t+1}$ under the optimal government policy.

Some background: Assume now that the optimal tax rule that you computed above has been in place for a very long time. A macroeconomist who is studying the economy observes time series on g_t, T_t, but *not* on b_t or the breakdown of g_t into its components g_{Tt}, g_{Pt}. The macroeconomist has a very long time series for $[g_t, T_t]$ and proceeds to compute a *vector autoregression* for this vector.

c. Define a population vector autoregression for the $[g_t, T_t]$ process. (Feel free to assume that lag lengths are infinite if this simplifies your answer.)

d. Please tell precisely how the vector autoregression for $[g_t, T_t]$ depends on the parameters $[\rho, \beta, \mu, q, w_1, w_2, c_1, c_2]$ that determine the joint $[g_t, T_t]$ process according to the economic theory you used in part a.

e. Now suppose that in addition to his observations on $[T_t, g_t]$, the economist gets an error-ridden time series on government debt b_t:

$$\tilde{b}_t = b_t + c_3 w_{3t+1}$$

where w_{3t+1} is an i.i.d. scalar Gaussian process with mean zero and unit variance that is orthogonal to w_{is+1} for $i = 1, 2$ for all s and t. Please tell how the vector autoregression for $[g_t, T_t, \tilde{b}_t]$ is related to the parameters $[\rho, \beta, \mu, q, w_1, w_2, c_1, c_2, c_3]$.

Is there any way to use the vector autoregression to make inferences about those parameters?

Exercise 5.9

A planner chooses a contingency plan for $\{c_t, k_{t+1}\}_{t=0}^{\infty}$ to maximize

$$-.5E_0 \sum_{t=0}^{\infty} \beta^t \left[(c_t - b_t)^2 + ei_t^2\right]$$

subject to the technology

$$c_t + i_t = \gamma k_t + d_t$$
$$k_{t+1} = (1 - \delta) k_t + i_t,$$

the laws of motion for the exogenous shock processes

$$b_{t+1} = \mu_b (1 - \rho_b) + \rho_b b_t + \sigma_b \epsilon_{b,t+1}$$
$$d_{t+1} = \mu_d (1 - \rho_d) + \rho_d d_t + \sigma_d \epsilon_{d,t+1},$$

and given initial conditions k_0, b_0, d_0. Here k_t is physical capital, c_t is consumption, b_t is a scalar stochastic process for bliss consumption, and d_t is an exogenous endowment process, $\beta \in (0,1)$, $e > 0$, $\delta \in (0,1)$, $\rho_b \in (0,1)$, $\rho_d \in (0,1)$, and the adjustment cost parameter $e > 0$. Also, $\begin{bmatrix} \epsilon_{b,t+1} \\ \epsilon_{d,t+1} \end{bmatrix}$ is an i.i.d. process that is distributed $\sim \mathcal{N}(0, I)$. We assume that $\beta\gamma(1 - \delta) = 1$, a condition that Hall and Friedman imposed to form permanent income models of consumption. For convenience, group all parameters into the vector

$$\theta = \begin{bmatrix} \beta & \delta & \gamma & e & \mu_b & \mu_d & \rho_b & \rho_d & \sigma_b & \sigma_d \end{bmatrix}.$$

Part I. Assume that the planner knows all parameters of the model. At time t, the planner observes the history of d_s, b_s, k_s for $s \leq t$.

a. Formulate the planning problem as a discounted dynamic programming problem.

b. Use the Bellman equation for the planning problem to describe the effects on the decision rule for c_t and k_{t+1} of an increase in σ_b. Tell the effects of an increase in σ_d.

c. Describe an algorithm to solve the Bellman equation.

Part II. An econometrician observes a time series $\{c_t, i_t\}_{t=0}^T$ for the economy described in part I. (This economy is either a socialist economy with a benevolent planner or a competitive economy with complete markets.) The econometrician does not observe b_t, d_t, k_t for any t but believes that

$$
\begin{bmatrix} k_0 \\ b_0 \\ d_0 \end{bmatrix} \sim \mathcal{N}\left(\mu_0, \Sigma_0\right).
$$

The econometrician knows the value of β but not the remaining parameters in θ.

a. Describe as completely as you can how the econometrician can form maximum likelihood estimates of the remaining parameters in θ given his sample $\{c_t, i_t\}_{t=0}^T$. If possible, find a recursive representation of the likelihood function.

b. Suppose that the econometrician has a Bayesian prior distribution over the unknown parameters in θ. Please describe an algorithm for constructing the Bayesian posterior distribution for these parameters.

Exercise 5.10

A consumer values consumption, asset streams $\{c_t, k_{t+1}\}_{t=0}^\infty$ according to

$$
(1) \qquad\qquad -.5E_0 \sum_{t=0}^\infty \beta^t \left(c_t - b\right)^2
$$

where $\beta \in (0,1)$ and

$$
k_{t+1} = R\left(k_t + y_t - c_t\right)
$$
$$
y_{t+1} = \mu_y\left(1 - \rho_1 - \rho_2\right) + \rho_1 y_t + \rho_2 y_{t-1} + \sigma_y \epsilon_{t+1}
$$
$$
c_t = \alpha y_t + \left(R - 1\right) k_t, \quad \alpha \in (0,1)
$$

and k_0, y_0, y_{-1} are given initial conditions, and ϵ_{t+1} is an i.i.d. shock with $\epsilon_{t+1} \sim \mathcal{N}(0,1)$.

a. Tell how to compute the value of the objective function (1) under the prescribed decision rule for c_t. In particular, write a Bellman equation and get as far as you can in solving it.

b. Tell how to use the Howard policy improvement algorithm to get a better decision rule.

Exercise 5.11   **Firm level adjustment costs**

A competitive firms sells output y_t at price p_t and chooses a production plan to maximize

$$(1) \qquad\qquad\qquad \sum_{t=0}^{\infty} \beta^t R_t$$

where

$$(2) \qquad\qquad R_t = p_t y_t - .5d \left(y_{t+1} - y_t\right)^2$$

subject to y_0 being a given initial condition. Here $\beta \in (0,1)$ is a discount factor, and $d > 0$ measures a cost of adjusting the rate of output. The firm is a price taker. The price p_t lies on the demand curve

$$(3) \qquad\qquad\qquad p_t = A_0 - A_1 Y_t$$

where $A_0 > 0, A_1 > 0$ and Y_t is the market-wide level of output, being the sum of output of n identical firms. The firm believes that market-wide output follows the law of motion

$$(4) \qquad\qquad\qquad Y_{t+1} = H_0 + H_1 Y_t$$

where Y_0 is a known initial condition. The firm observes Y_t and y_t at time t when it chooses y_{t+1}.

a. Formulate a Bellman equation for the firm.

b. For parameter values $\beta = .95, d = 2, A_0 = 100, A_1 = 1, H_0 = 200, H_1 = .8$, compute the firm's optimal value function and optimal decision rule.

Exercise 5.12   **Firm level adjustment cost, II**

A competitive firms sells output y_t at price p_t and chooses a production plan to maximize

$$(1) \qquad\qquad\qquad E_0 \sum_{t=0}^{\infty} \beta^t R_t$$

where E_0 denotes a mathematical expectation conditional on time 0 information,

$$(2) \qquad\qquad R_t = p_t y_t - .5d \left(y_{t+1} - y_t \right)^2$$

subject to y_0 being a given initial condition. Here $\beta \in (0,1)$ is a discount factor, and $d > 0$ measures a cost of adjusting the rate of output. The firm is a price taker. The price p_t lies on the demand curve

$$(3) \qquad\qquad p_t = A_0 - A_1 Y_t + u_t$$

where $A_0 > 0, A_1 > 0$ and Y_t is the market-wide level of output, being the sum of output of n identical firms. In (3), u_t is a demand shock that follows the first-order autoregressive process

$$(4) \qquad\qquad u_{t+1} = \rho u_t + \sigma_u \epsilon_{t+1}$$

where ϵ_{t+1} is an i.i.d. scalar process with $\epsilon_{t+1} \sim \mathcal{N}(0,1)$ and $|\rho| < 1$. The firm believes that market-wide output follows the law of motion

$$(5) \qquad\qquad Y_{t+1} = H_0 + H_1 Y_t + H_2 u_t$$

where Y_0 is a known initial condition. The firm observes P_t, Y_t, and y_t at time t when it chooses y_{t+1}.

a. Formulate a Bellman equation for the firm.

b. For parameter values $\beta = .95, d = 2, A_0 = 100, A_1 = 1, H_0 = 200, H_1 = .8, H_2 = 2, \rho = .9, \sigma_u = .05$, compute the firm's optimal value function and optimal decision rule.

Exercise 5.13   **Permanent income model**

A household chooses a process $\{c_t, a_{t+1}\}_{t=0}^{\infty}$ to maximize

$$E_0 \sum_{t=0}^{\infty} \beta^t \{ -.5 \left(c_t - b \right)^2 - .5\epsilon a_t^2 \}, \quad \beta \in (0,1)$$

subject to

$$a_{t+1} + c_t = R a_t + y_t$$
$$y_{t+1} = (1 - \rho_1 - \rho_2) + \rho_1 y_t + \rho_2 y_{t-1} + \sigma_y \epsilon_{t+1}$$

where c_t is consumption, $b > 0$ is a bliss level of consumption, a_t is financial assets at the beginning of t, $R = \beta^{-1}$ is the gross rate of return on assets held from t to $t+1$, and ϵ_{t+1} is an i.i.d. scalar process with $\epsilon_{t+}] \sim \mathcal{N}(0,1)$. The household faces known initial conditions a_0, y_0, y_{-1}.

a. Write a Bellman equation for the household's problem.

b. Compute the household's value function and optimal decision rule for the following parameter values: $b = 1000, \beta = .95, R = \beta^{-1}, \rho_1 = 1.2, \rho_2 = -.4, \sigma_y = .05, \epsilon = .000001$.

c. Compute the eigenvalues of $A - BF$.

d. Compute the household's value function and optimal decision rule for the following parameter values: $b = 1000, \beta = .95, R = \beta^{-1}, \rho_1 = 1.2, \rho_2 = -.4, \sigma_y = .05, \epsilon = 0$. Compare what you obtain with your answers in part **b**.

Exercise 5.14 **Permanent income model again**

A household chooses a process $\{c_t, a_{t+1}\}_{t=0}^{\infty}$ to maximize

$$E_0 \sum_{t=0}^{\infty} \beta^t \{-.5 \left(c_t - b\right)^2 - .5\epsilon a_t^2\}, \quad \beta \in (0,1)$$

subject to

$$a_{t+1} + c_t = R a_t + y_t$$

$$y_{t+1} = \left(1 - \rho_1 - \rho_2\right) + \rho_1 y_t + \rho_2 y_{t-3} + \sigma_y \epsilon_{t+1}$$

where c_t is consumption, $b > 0$ is a bliss level of consumption, a_t is financial assets at the beginning of t, $R = \beta^{-1}$ is the gross rate of return on assets held from t to $t+1$, and ϵ_{t+1} is an i.i.d. scalar process with $\epsilon_{t+1} \sim \mathcal{N}(0,1)$. The household faces known initial conditions $a_0, y_0, y_{-1}, y_{-2}, y_{-3}$.

a. Write a Bellman equation for the household's problem.

b. Compute the household's value function and optimal decision rule for the following parameter values: $b = 1000, \beta = .95, R = \beta^{-1}, \rho_1 = .55, \rho_2 = .3, \sigma_y = .05, \epsilon = .000001$.

c. Compute the eigenvalues of $A - BF$.

d. Compute the household's value function and optimal decision rule for the following parameter values: $b = 1000, \beta = .95, R = \beta^{-1}, \rho_1 = .55, \rho_2 = .3, \sigma_y = .05, \epsilon = 0$. Compare what you obtain with your answers in part **b**.

Chapter 6
Search, Matching, and Unemployment

6.1. Introduction

This chapter applies dynamic programming to a choice between only two actions, to accept or reject a take-it-or-leave-it job offer. An unemployed worker faces a probability distribution of wage offers or job characteristics, from which a limited number of offers are drawn each period. Given his perception of the probability distribution of offers, the worker must devise a strategy for deciding when to accept an offer.

The theory of search is a tool for studying unemployment. Search theory puts unemployed workers in a setting where sometimes they choose to reject available offers and to remain unemployed now because they prefer to wait for better offers later. We use the theory to study how workers respond to variations in the rate of unemployment compensation, the perceived riskiness of wage distributions, the probability of being fired, the quality of information about jobs, and the frequency with which the wage distribution can be sampled.

This chapter provides an introduction to the techniques used in the search literature and a sampling of search models. The chapter studies ideas introduced in two important papers by McCall (1970) and Jovanovic (1979a). These papers differ in the search technologies with which they confront an unemployed worker.[1] We also study a related model of occupational choice by Neal (1999).

We hope to convey some of the excitement that Robert E. Lucas, Jr. (1987, p.57) expressed when he wrote this about the McCall search model: "Questioning a McCall worker is like having a conversation with an out-of-work friend: 'Maybe you are setting your sights too high' or 'Why did you quit your old job before you had a new one lined up?' This is real social science: an attempt to

---

[1] Stigler's (1961) important early paper studied a search technology different from both McCall's and Jovanovic's. In Stigler's model, an unemployed worker has to choose in advance a number n of offers to draw, from which he takes the highest wage offer. Stigler's formulation of the search problem was not sequential.

model, to *understand*, human behavior by visualizing the situations people find themselves in, the options they face and the pros and cons as they themselves see them." The modifications of the basic McCall model by Jovanovic, Neal, and in the various sections and exercises of this chapter all come from visualizing aspects of the situations in which workers find themselves.

6.2. Preliminaries

This section describes elementary properties of probability distributions that are used extensively in search theory.

6.2.1. Nonnegative random variables

We begin with some properties of nonnegative random variables that possess first moments. Consider a random variable p with a cumulative probability distribution function $F(P)$ defined by $\text{Prob}\{p \leq P\} = F(P)$. We assume that $F(0) = 0$, that is, that p is nonnegative. We assume that F, a nondecreasing function, is continuous from the right. We also assume that there is an upper bound $B < \infty$ such that $F(B) = 1$, so that p is bounded with probability 1.

The mean of p, Ep, is defined by

$$Ep = \int_0^B p \, dF(p). \tag{6.2.1}$$

Let $u = 1 - F(p)$ and $v = p$ and use the integration-by-parts formula $\int_a^b u \, dv = uv \Big|_a^b - \int_a^b v \, du$, to verify that

$$\int_0^B [1 - F(p)] \, dp = \int_0^B p \, dF(p).$$

Thus, we have the following formula for the mean of a nonnegative random variable:

$$Ep = \int_0^B [1 - F(p)] \, dp = B - \int_0^B F(p) \, dp. \tag{6.2.2}$$

Now consider two independent random variables p_1 and p_2 drawn from the distribution F. Consider the event $\{(p_1 < p) \cap (p_2 < p)\}$, which by the independence assumption has probability $F(p)^2$. The event $\{(p_1 < p) \cap (p_2 < p)\}$ is equivalent to the event $\{\max(p_1, p_2) < p\}$, where "max" denotes the maximum. Therefore, if we use formula $(6.2.2)$, the random variable $\max(p_1, p_2)$ has mean

$$E \max (p_1, p_2) = B - \int_0^B F(p)^2 \, dp. \qquad (6.2.3)$$

Similarly, if $p_1, p_2, \ldots, p_n$ are n independent random variables drawn from F, we have $\operatorname{Prob}\{\max(p_1, p_2, \ldots, p_n) < p\} = F(p)^n$ and

$$M_n \equiv E \max (p_1, p_2, \ldots, p_n) = B - \int_0^B F(p)^n \, dp, \qquad (6.2.4)$$

where M_n is defined as the expected value of the maximum of $p_1, \ldots, p_n$.

6.2.2. Mean-preserving spreads

Rothschild and Stiglitz introduced the idea of a mean-preserving spread as a convenient way to characterize the riskiness of two distributions with the same mean. Consider a class of distributions with the same mean. We index this class by a parameter r belonging to some set R. For the rth distribution we denote $\operatorname{Prob}\{p \leq P\} = F(P, r)$ and assume that $F(P, r)$ is differentiable with respect to r for all $P \in [0, B]$. We assume that there is a single finite B such that $F(B, r) = 1$ for all r in R and that $F(0, r) = 0$ for all r in R, so that we are considering a class of distributions R for nonnegative, bounded random variables.

From equation $(6.2.2)$, we have

$$Ep = B - \int_0^B F(p, r) \, dp. \qquad (6.2.5)$$

Therefore, two distributions with the same value of $\int_0^B F(\theta, r) d\theta$ have identical means. We write this as the identical means condition:

(i) $$\int_0^B [F(\theta, r_1) - F(\theta, r_2)] \, d\theta = 0.$$

Two distributions r_1, r_2 are said to satisfy the *single-crossing property* if there exists a $\hat{\theta}$ with $0 < \hat{\theta} < B$ such that

(ii) $$F(\theta, r_2) - F(\theta, r_1) \leq 0 \, (\geq 0) \qquad \text{when} \quad \theta \geq (\leq) \hat{\theta}.$$

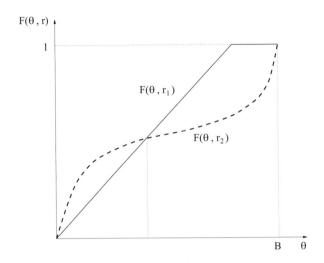

Figure 6.2.1: Two distributions, r_1 and r_2, that satisfy the single-crossing property.

Figure 6.2.1 illustrates the single-crossing property. If two distributions r_1 and r_2 satisfy properties (i) and (ii), we can regard distribution r_2 as having been obtained from r_1 by a process that shifts probability toward the tails of the distribution while keeping the mean constant.

Properties (i) and (ii) imply the following property:

(iii)
$$\int_0^y \left[F\left(\theta, r_2\right) - F\left(\theta, r_1\right) \right] d\theta \geq 0, \qquad 0 \leq y \leq B.$$

Rothschild and Stiglitz regard properties (i) and (iii) as defining the concept of a "mean-preserving spread." In particular, a distribution indexed by r_2 is said to have been obtained from a distribution indexed by r_1 by a mean-preserving spread if the two distributions satisfy (i) and (iii). [2]

---

[2] Rothschild and Stiglitz (1970, 1971) use properties (i) and (iii) to characterize mean-preserving spreads rather than (i) and (ii) because (i) and (ii) fail to possess transitivity. That is, if $F(\theta, r_2)$ is obtained from $F(\theta, r_1)$ via a mean-preserving spread in the sense that the term has in (i) and (ii), and $F(\theta, r_3)$ is obtained from $F(\theta, r_2)$ via a mean-preserving spread in the sense of (i) and (ii), it does not follow that $F(\theta, r_3)$ satisfies the single-crossing property (ii) vis-à-vis distribution $F(\theta, r_1)$. A definition based on (i) and (iii), however, does provide a transitive ordering, which is a desirable feature for a definition designed to order distributions according to their riskiness.

For infinitesimal changes in r, Diamond and Stiglitz use the differential versions of properties (i) and (iii) to rank distributions with the same mean in order of riskiness. An increment in r is said to represent a mean-preserving increase in risk if

(iv)
$$\int_0^B F_r(\theta, r)\, d\theta = 0$$

(v)
$$\int_0^y F_r(\theta, r)\, d\theta \geq 0, \qquad 0 \leq y \leq B\,,$$

where $F_r(\theta, r) = \partial F(\theta, r)/\partial r$.

6.3. McCall's model of intertemporal job search

We now consider an unemployed worker who is searching for a job under the following circumstances: Each period the worker draws one offer w from the same wage distribution $F(W) = \text{Prob}\{w \leq W\}$, with $F(0) = 0$, $F(B) = 1$ for $B < \infty$. The worker has the option of rejecting the offer, in which case he or she receives c this period in unemployment compensation and waits until next period to draw another offer from F; alternatively, the worker can accept the offer to work at w, in which case he or she receives a wage of w per period forever. Neither quitting nor firing is permitted.

Let y_t be the worker's income in period t. We have $y_t = c$ if the worker is unemployed and $y_t = w$ if the worker has accepted an offer to work at wage w. The unemployed worker devises a strategy to maximize the mathematical expectation of $\sum_{t=0}^\infty \beta^t y_t$ where $0 < \beta < 1$ is a discount factor.

Let $v(w)$ be the expected value of $\sum_{t=0}^\infty \beta^t y_t$ for a worker who has offer w in hand, who is deciding whether to accept or to reject it, and who behaves optimally. We assume no recall. The value function $v(w)$ satisfies the Bellman equation

$$v(w) = \max_{\text{accept,reject}} \left\{ \frac{w}{1 - \beta}, c + \beta \int_0^B v(w')\, dF(w') \right\}, \qquad (6.3.1)$$

where the maximization is over the two actions: (1) *accept* the wage offer w and work forever at wage w, or (2) *reject* the offer, receive c this period, and draw a new offer w' from distribution F next period. Figure 6.3.1 graphs the functional equation (6.3.1) and reveals that its solution is of the form

$$v\left(w\right) = \begin{cases} \dfrac{\overline{w}}{1-\beta} = c + \beta \displaystyle\int_{0}^{B} v\left(w'\right) dF\left(w'\right) & \text{if} \quad w \le \overline{w} \\[2ex] \dfrac{w}{1-\beta} & \text{if} \quad w \ge \overline{w}. \end{cases} \qquad (6.3.2)$$

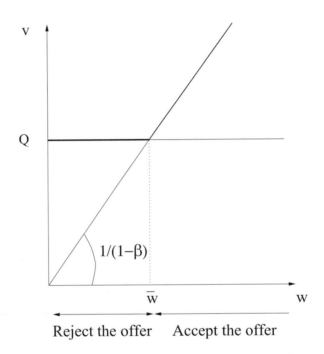

Reject the offer   Accept the offer

Figure 6.3.1: The function $v(w) = \max\{w/(1-\beta), c + \beta \int_0^B v(w')dF(w')\}$. The reservation wage $\overline{w} = (1-\beta)[c + \beta \int_0^B v(w')dF(w')]$.

Using equation (6.3.2), we can convert the functional equation (6.3.1) in the value function $v(w)$ into an ordinary equation in the reservation wage $\overline{w}$. Evaluating $v(\overline{w})$ and using equation (6.3.2), we have

$$\frac{\overline{w}}{1-\beta} = c + \beta \int_0^{\overline{w}} \frac{\overline{w}}{1-\beta} dF\left(w'\right) + \beta \int_{\overline{w}}^B \frac{w'}{1-\beta} dF\left(w'\right)$$

or

$$\frac{\overline{w}}{1-\beta}\int_0^{\overline{w}}dF\left(w'\right)+\frac{\overline{w}}{1-\beta}\int_{\overline{w}}^B dF\left(w'\right)$$

$$=c+\beta\int_0^{\overline{w}}\frac{\overline{w}}{1-\beta}dF\left(w'\right)+\beta\int_{\overline{w}}^B\frac{w'}{1-\beta}dF\left(w'\right)$$

or

$$\overline{w}\int_0^{\overline{w}}dF\left(w'\right)-c=\frac{1}{1-\beta}\int_{\overline{w}}^B\left(\beta w'-\overline{w}\right)dF\left(w'\right).$$

Adding $\overline{w}\int_{\overline{w}}^B dF(w')$ to both sides gives

$$\left(\overline{w}-c\right)=\frac{\beta}{1-\beta}\int_{\overline{w}}^B\left(w'-\overline{w}\right)dF\left(w'\right). \tag{6.3.3}$$

Equation (6.3.3) is often used to characterize the reservation wage $\overline{w}$. The left side is the cost of searching one more time when an offer $\overline{w}$ is in hand. The right side is the expected benefit of searching one more time in terms of the expected present value associated with drawing $w' > \overline{w}$. Equation (6.3.3) instructs the agent to set $\overline{w}$ so that the cost of searching one more time equals the benefit.

6.3.1. Characterizing reservation wage

Let us define the function on the right side of equation (6.3.3) as

$$h\left(w\right)=\frac{\beta}{1-\beta}\int_w^B\left(w'-w\right)dF\left(w'\right). \tag{6.3.4}$$

Notice that $h(0)=Ew\beta/(1-\beta)$, that $h(B)=0$, and that $h(w)$ is differentiable, with derivative given by[3]

$$h'\left(w\right)=-\frac{\beta}{1-\beta}\left[1-F\left(w\right)\right]<0.$$

---

[3] To compute $h'(w)$, we apply Leibniz's rule to equation (6.3.4). Let $\phi(t)=\int_{\alpha(t)}^{\beta(t)}f(x,t)dx$ for $t\in[c,d]$. Assume that f and f_t are continuous and that α,β are differentiable on $[c,d]$. Then Leibniz's rule asserts that $\phi(t)$ is differentiable on $[c,d]$ and

$$\phi'\left(t\right)=f\left[\beta\left(t\right),t\right]\beta'\left(t\right)-f\left[\alpha\left(t\right),t\right]\alpha'\left(t\right)+\int_{\alpha(t)}^{\beta(t)}f_t\left(x,t\right)dx.$$

To apply this formula to the equation in the text, let w play the role of t.

We also have

$$h''(w) = \frac{\beta}{1-\beta}F'(w) > 0,$$

so that $h(w)$ is convex to the origin. Figure 6.3.2 graphs $h(w)$ against $(w-c)$ and indicates how $\overline{w}$ is determined. From Figure 6.3.2 it is apparent that an increase in unemployment compensation c leads to an increase in $\overline{w}$.

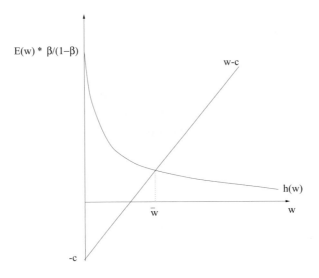

Figure 6.3.2: The reservation wage, $\overline{w}$, that satisfies $\overline{w}-c = [\beta/(1-\beta)]\int_{\overline{w}}^{B}(w'-\overline{w})dF(w') \equiv h(\overline{w})$.

To get another useful characterization of $\overline{w}$, we express equation (6.3.3) as

$$\overline{w} - c = \frac{\beta}{1-\beta}\int_{\overline{w}}^{B}(w'-\overline{w})\,dF(w') + \frac{\beta}{1-\beta}\int_{0}^{\overline{w}}(w'-\overline{w})\,dF(w')$$

$$-\frac{\beta}{1-\beta}\int_{0}^{\overline{w}}(w'-\overline{w})\,dF(w')$$

$$=\frac{\beta}{1-\beta}Ew - \frac{\beta}{1-\beta}\overline{w} - \frac{\beta}{1-\beta}\int_{0}^{\overline{w}}(w'-\overline{w})\,dF(w')$$

or

$$\overline{w} - (1-\beta)\,c = \beta Ew - \beta\int_{0}^{\overline{w}}(w'-\overline{w})\,dF(w').$$

Applying integration by parts to the last integral on the right side and rearranging, we have

$$\overline{w} - c = \beta \left(Ew - c\right) + \beta \int_0^{\overline{w}} F\left(w'\right) dw'. \tag{6.3.5}$$

At this point it is useful to define the function

$$g\left(s\right) = \int_0^s F\left(p\right) dp. \tag{6.3.6}$$

This function has the characteristics that $g(0) = 0$, $g(s) \geq 0$, $g'(s) = F(s) > 0$, and $g''(s) = F'(s) > 0$ for $s > 0$. Then equation $(6.3.5)$ can be represented as $\overline{w} - c = \beta(Ew - c) + \beta g(\overline{w})$. Figure 6.3.3 uses equation $(6.3.5)$ to determine $\overline{w}$.

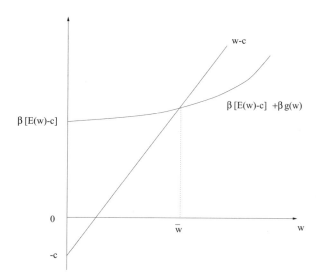

Figure 6.3.3: The reservation wage, $\overline{w}$, that satisfies $\overline{w} - c = \beta(Ew - c) + \beta \int_0^{\overline{w}} F(w')dw' \equiv \beta(Ew - c) + \beta g(\overline{w})$.

6.3.2. Effects of mean-preserving spreads

Figure 6.3.3 can be used to establish two propositions about $\overline{w}$. First, given F, $\overline{w}$ increases when the rate of unemployment compensation c increases. Second, given c, a mean-preserving increase in risk causes $\overline{w}$ to increase. This second proposition follows directly from Figure 6.3.3 and the characterization (iii) or (v) of a mean-preserving increase in risk. From the definition of g in equation (6.3.6) and the characterization (iii) or (v), a mean-preserving spread causes an upward shift in $\beta(Ew - c) + \beta g(w)$.

Since either an increase in unemployment compensation or a mean-preserving increase in risk raises the reservation wage, it follows from the expression for the value function in equation (6.3.2) that unemployed workers are also better off in those situations. It is obvious that an increase in unemployment compensation raises the welfare of unemployed workers but it might seem surprising that a mean-preserving increase in risk does too. Intuition for this latter finding can be gleaned from the result in option pricing theory that the value of an option is an increasing function of the variance in the price of the underlying asset. This is so because the option holder chooses to accept payoffs only from the right tail of the distribution. In our context, the unemployed worker has the option to accept a job and the asset value of a job offering wage rate w is equal to $w/(1 - \beta)$. Under a mean-preserving increase in risk, the higher incidence of very good wage offers increases the value of searching for a job while the higher incidence of very bad wage offers is less detrimental because the option to work will not be exercised at such low wages.

6.3.3. Allowing quits

Thus far, we have supposed that the worker cannot quit. It happens that had we given the worker the option to quit and search again, after being unemployed one period, he would never exercise that option. To see this point, recall that the reservation wage $\overline{w}$ in (6.3.2) satisfies

$$v(\overline{w}) = \frac{\overline{w}}{1 - \beta} = c + \beta \int v(w')\, dF(w') . \tag{6.3.7}$$

Suppose the agent has in hand an offer to work at wage w. Assuming that the agent behaves optimally after any rejection of a wage w, we can compute the lifetime utility associated with three mutually exclusive alternative ways of responding to that offer:

A1.   Accept the wage and keep the job forever:

$$\frac{w}{1-\beta}.$$

A2.   Accept the wage but quit after t periods:

$$\frac{w - \beta^t w}{1 - \beta} + \beta^t \left(c + \beta \int v\left(w'\right) dF\left(w'\right) \right) = \frac{w}{1-\beta} - \beta^t \frac{w - \overline{w}}{1 - \beta}.$$

A3.   Reject the wage:

$$c + \beta \int v\left(w'\right) dF\left(w'\right) = \frac{\overline{w}}{1-\beta}.$$

We conclude that if $w < \overline{w}$,

$$A1 \prec A2 \prec A3,$$

and if $w > \overline{w}$,

$$A1 \succ A2 \succ A3.$$

The three alternatives yield the same lifetime utility when $w = \overline{w}$.

6.3.4. Waiting times

It is straightforward to derive the probability distribution of the waiting time until a job offer is accepted. Let N be the random variable "length of time until a successful offer is encountered," with the understanding that $N = 1$ if the first job offer is accepted. Let $\lambda = \int_0^{\overline{w}} dF(w')$ be the probability that a job offer is rejected. Then we have $\text{Prob}\{N = 1\} = (1 - \lambda)$. The event that $N = 2$ is the event that the first draw is less than $\overline{w}$, which occurs with probability λ, and that the second draw is greater than $\overline{w}$, which occurs with probability $(1 - \lambda)$. By virtue of the independence of successive draws, we have $\text{Prob}\{N = 2\} = (1 - \lambda)\lambda$. More generally, $\text{Prob}\{N = j\} = (1 - \lambda)\lambda^{j-1}$, so the waiting time is geometrically distributed. The mean waiting time $\bar{N}$ is given by

$$\bar{N} = \sum_{j=1}^{\infty} j \cdot \text{Prob}\{N = j\} = \sum_{j=1}^{\infty} j\left(1 - \lambda\right)\lambda^{j-1} = (1 - \lambda)\sum_{j=1}^{\infty}\sum_{k=1}^{j}\lambda^{j-1}$$

$$= (1 - \lambda)\sum_{k=0}^{\infty}\sum_{j=1}^{\infty}\lambda^{j-1+k} = (1 - \lambda)\sum_{k=0}^{\infty}\lambda^k\left(1 - \lambda\right)^{-1} = (1 - \lambda)^{-1}.$$

That is, the mean waiting time to a successful job offer equals the reciprocal of the probability of an accepted offer on a single trial.[4]

As an illustration of the power of using a recursive approach, we can also compute the mean waiting time $\bar{N}$ as follows. First, given that our search environment is stationary and therefore is associated with a constant reservation wage and a constant probability of escaping unemployment, it follows that the "remaining" mean waiting time for all unemployed workers is equal to $\bar{N}$ in any given period. That is, all unemployed workers face a remaining mean waiting time of $\bar{N}$ regardless of how long of an unemployment spell they have suffered so far. Second, the mean waiting time $\bar{N}$ must then be equal to the weighted sum of two possible outcomes: either the worker accepts a job next period, with probability $(1 - \lambda)$; or she remains unemployed in the next period, with probability λ. In the first case, the worker will have ended her unemployment after one last period of unemployment while in the second case, the worker will have suffered one period of unemployment *and* will face a remaining mean waiting time of $\bar{N}$ periods. Hence, the mean waiting time must satisfy the following recursive formula:

$$\bar{N} = (1 - \lambda) \cdot 1 + \lambda \cdot \left(1 + \bar{N}\right) \qquad \Longrightarrow \qquad \bar{N} = (1 - \lambda)^{-1}.$$

We invite the reader to prove that, given F, the mean waiting time increases with increases in the rate of unemployment compensation, c.

---

[4] An alternative way of deriving the mean waiting time is to use the algebra of z transforms; we say that $h(z) = \sum_{j=0}^{\infty} h_j z^j$ and note that $h'(z) = \sum_{j=1}^{\infty} j h_j z^{j-1}$ and $h'(1) = \sum_{j=1}^{\infty} j h_j$. (For an introduction to z transforms, see Gabel and Roberts, 1973.) The z transform of the sequence $(1 - \lambda)\lambda^{j-1}$ is given by $\sum_{j=1}^{\infty} (1 - \lambda)\lambda^{j-1} z^j = (1 - \lambda)z/(1 - \lambda z)$. Evaluating $h'(z)$ at $z = 1$ gives, after some simplification, $h'(1) = 1/(1 - \lambda)$. Therefore, we have that the mean waiting time is given by $(1 - \lambda) \sum_{j=1}^{\infty} j\lambda^{j-1} = 1/(1 - \lambda)$.

6.3.5. Firing

We now consider a modification of the job search model in which each period after the first period on the job the worker faces probability α of being fired, where $1 > \alpha > 0$. The probability α of being fired next period is assumed to be independent of tenure. A previously unemployed worker samples wage offers from a time-invariant and known probability distribution F. Unemployed workers receive unemployment compensation in the amount c. The worker receives a time-invariant wage w on a job until she is fired. A worker who is fired becomes unemployed for one period before drawing a new wage. Only previously employed workers are fired. A previously employed worker who is fired at the beginning of a period cannot draw a new wage offer that period but must be unemployed for one period.

We let $\hat{v}(w)$ be the expected present value of income of a previously unemployed worker who has offer w in hand and who behaves optimally. If she rejects the offer, she receives c in unemployment compensation this period and next period draws a new offer w', whose value to her now is $\beta \int \hat{v}(w')dF(w')$. If she rejects the offer, $\hat{v}(w) = c + \beta \int \hat{v}(w')dF(w')$. If she accepts the offer, she receives w this period; next period with probability $1 - \alpha$, she is not fired and therefore what she receives is worth $\beta\hat{v}(w)$ today; with probability α, she is fired next period, which has the consequence that after one period of unemployment she draws a new wage, an outcome that today is worth $\beta[c + \beta \int \hat{v}(w')dF(w')]$. Therefore, if she accepts the offer, $\hat{v}(w) = w + \beta(1 - \alpha)\hat{v}(w) + \beta\alpha[c + \beta \int \hat{v}(w')dF(w')]$. Thus, the Bellman equation becomes [5]

$$\hat{v}(w) = \max_{\text{accept,reject}} \left\{ w + \beta(1 - \alpha)\hat{v}(w) + \beta\alpha[c + \beta E\hat{v}], \ c + \beta E\hat{v} \right\},$$

where $E\hat{v} = \int \hat{v}(w')dF(w')$. Here the appearance of $\hat{v}(w)$ on the right side recognizes that if the worker had accepted wage offer w last period with expected discounted present value $\hat{v}(w)$, the stationarity of the problem (i.e., the fact that

---

[5] If a worker who is fired at the beginning of a period were to have the opportunity to draw a new offer that same period, then the Bellman equation would instead be

$$\tilde{v}(w) = \max_{\text{accept,reject}} \left\{ w + \beta(1 - \alpha)\tilde{v}(w) + \beta\alpha \int \tilde{v}(w')dF(w'), c + \beta \int \tilde{v}(w')dF(w') \right\}.$$

F, α, c are all fixed) makes $\hat{v}(w)$ also be the continuation value associated with retaining this job next period. This equation has a solution of the form[6]

$$\hat{v}\left(w\right) = \begin{cases} \dfrac{w + \beta\alpha\left[c + \beta E\hat{v}\right]}{1 - \beta\left(1 - \alpha\right)}, & \text{if } w \geq \overline{w} \\[2ex] c + \beta E\hat{v}, & w \leq \overline{w} \end{cases}$$

where $\overline{w}$ solves

$$\frac{\overline{w} + \beta\alpha\left[c + \beta E\hat{v}\right]}{1 - \beta\left(1 - \alpha\right)} = c + \beta E\hat{v},$$

which can be rearranged as

$$\frac{\overline{w}}{1 - \beta} = c + \beta \int \hat{v}\left(w'\right) dF\left(w'\right). \tag{6.3.8}$$

We can compare the reservation wage in (6.3.8) to the reservation wage in expression (6.3.7) when there was no risk of being fired. The two expressions look identical but the reservation wages differ because the value functions differ. In particular, $\hat{v}(w)$ is strictly less than $v(w)$. This is an immediate implication of our argument that it cannot be optimal to quit if you have accepted a wage strictly greater than the reservation wage in the situation without possible firings (see section 6.3.3). So even though workers who face no possible firings can mimic outcomes in situations where they would facing possible firings by occasionally "firing themselves" by quitting into unemployment, they choose not to do so because that would lower their expected present value of income. Since the employed workers in the situation where they face possible firings are worse off than employed workers in the situation without possible firings, it follows that $\hat{v}(w)$ lies strictly below $v(w)$ over the whole domain because, even at wages that are rejected, the value function partly reflects a stream of future outcomes whose expectation is less favorable in the situation in which workers face a chance of being fired.

Since the value function $\hat{v}(w)$ with firings lies strictly below the value function $v(w)$ without firings, it follows from (6.3.8) and (6.3.7) that the reservation wage $\overline{w}$ is strictly lower with firings. There is less of a reason to hold out for high-paying jobs when a job is expected to last for a shorter period of time.

---

[6] That it takes this form can be established by guessing that $\hat{v}(w)$ is nondecreasing in w. This guess implies the equation in the text for $\hat{v}(w)$, which is nondecreasing in w. This argument verifies that $\hat{v}(w)$ is nondecreasing, given the uniqueness of the solution of the Bellman equation.

That is, unemployed workers optimally invest less in search when the payoffs associated with wage offers have gone down because of the probability of being fired.

6.4. A lake model

Consider an economy consisting of a continuum of *ex ante* identical workers living in the environment described in the previous section. These workers move recurrently between unemployment and employment. The mean duration of each spell of employment is α^{-1} and the mean duration of unemployment is $[1 - F(\overline{w})]^{-1}$. The average unemployment rate U_t across the continuum of workers obeys the difference equation

$$U_{t+1} = \alpha \left(1 - U_t\right) + F\left(\overline{w}\right) U_t,$$

where α is the hazard rate of escaping employment and $[1 - F(\overline{w})]$ is the hazard rate of escaping unemployment. Solving this difference equation for a stationary solution, i.e., imposing $U_{t+1} = U_t = U$, gives

$$U = \frac{\alpha}{\alpha + 1 - F\left(\overline{w}\right)} \qquad \Longrightarrow \qquad U = \frac{\dfrac{1}{1 - F\left(\overline{w}\right)}}{\dfrac{1}{1 - F\left(\overline{w}\right)} + \dfrac{1}{\alpha}}. \tag{6.4.1}$$

Equation (6.4.1) expresses the stationary unemployment rate in terms of the ratio of the average duration of unemployment to the sum of average durations of unemployment and employment. The unemployment rate, being an average across workers at each moment, thus reflects the average outcomes experienced by workers *across time*. This way of linking economy-wide averages at a point in time with the time-series average for a representative worker is our first encounter with a class of models sometimes referred to as Bewley models, which we shall study in depth in chapter 18.

This model of unemployment is sometimes called a lake model and can be depicted as in Figure 6.4.1, with two lakes denoted U and $1 - U$ representing volumes of unemployment and employment, and streams of rate α from the $1 - U$ lake to the U lake and of rate $1 - F(\overline{w})$ from the U lake to the $1 - U$ lake. Equation (6.4.1) allows us to study the determinants of the unemployment rate in terms of the hazard rate of becoming unemployed α and the hazard rate of escaping unemployment $1 - F(\overline{w})$.

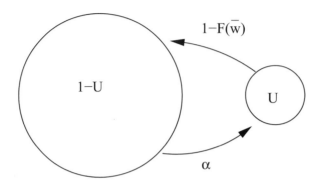

Figure 6.4.1: Lake model with flows of rate α from employment state $1 - U$ to unemployment state U and of rate $[1 - F(\overline{w})]$ from U to $1 - U$.

6.5. A model of career choice

This section describes a model of occupational choice that Derek Neal (1999) used to study the employment histories of recent high school graduates. Neal wanted to explain why young men switch jobs *and* careers often early in their work histories, then later focus their searches for jobs within a single career, and finally settle down in a particular job. Neal's model can be regarded as a simplified version of Brian McCall's (1991) model.

A worker chooses career-job (θ, ϵ) pairs subject to the following conditions: There is no unemployment. The worker's earnings at time t equal $\theta_t + \epsilon_t$, where θ_t is a component specific to a *career* and ϵ_t is a component specific to a particular *job*. The worker maximizes $E \sum_{t=0}^{\infty} \beta^t (\theta_t + \epsilon_t)$. A *career* is a draw of θ from c.d.f. F; a *job* is a draw of ϵ from c.d.f. G. Successive draws are independent, and $G(0) = F(0) = 0$, $G(B_\epsilon) = F(B_\theta) = 1$. The worker can draw a new career only if he also draws a new job. However, the worker is free to retain his existing career (θ), and to draw a new job (ϵ'). The worker decides at the beginning of a period whether to stay in a career-job pair inherited from the past, stay in the inherited career but draw a new job, or draw a new career-job pair. There is no opportunity to recall past jobs or careers.

Let $v(\theta, \epsilon)$ be the optimal value of the problem at the beginning of a period for a worker currently having inherited career-job pair (θ, ϵ) and who is about to decide whether to draw a new career and or job. The Bellman equation is

$$v(\theta, \epsilon) = \max \left\{ \theta + \epsilon + \beta v(\theta, \epsilon), \ \theta + \int [\epsilon' + \beta v(\theta, \epsilon')] \, dG(\epsilon'), \right.$$
$$\left. \int \int [\theta' + \epsilon' + \beta v(\theta', \epsilon')] \, dF(\theta') \, dG(\epsilon') \right\}. \qquad (6.5.1)$$

The maximization is over the three possible actions: (1) retain the present job-career pair; (2) retain the present career but draw a new job; and (3) draw both a new job and a new career. We might nickname these three alternatives 'stay put', 'new job', 'new life'. The value function is increasing in both θ and ϵ.

Figures 6.5.1 and 6.5.2 display the optimal value function and the optimal decision rule for Neal's model where F and G are each distributed according to discrete uniform distributions on $[0, 5]$ with 50 evenly distributed discrete values for each of θ and ϵ and $\beta = .95$. We computed the value function by iterating to convergence on the Bellman equation. The optimal policy is characterized by three regions in the (θ, ϵ) space. For high enough values of $\epsilon + \theta$, the worker stays put. For high θ but low ϵ, the worker retains his career but searches for a better job. For low values of $\theta + \epsilon$, the worker finds a new career and a new job. In figures 6.5.1 and 6.5.2, the decision to *retain* both job and career occurs in the high θ, high ϵ region of the state space; the decision to retain career θ but search for a new job ϵ occurs in the high θ and low ϵ region of the state space; and the decision to 'get a new life' by drawing both a new θ and a new ϵ occurs in the low θ, low ϵ region.[7]

When the career-job pair (θ, ϵ) is such that the worker chooses to stay put, the value function in (6.5.1) attains the value $(\theta + \epsilon)/(1 - \beta)$. Of course, this happens when the decision to stay put weakly dominates the other two actions, which occurs when

$$\frac{\theta + \epsilon}{1 - \beta} \geq \max \{C(\theta), Q\}, \qquad (6.5.2)$$

where Q is the value of drawing both a new job and a new career,

$$Q \equiv \int \int [\theta' + \epsilon' + \beta v(\theta', \epsilon')] \, dF(\theta') \, dG(\epsilon'),$$

---

[7] The computations were performed by the Matlab program `neal2.m`.

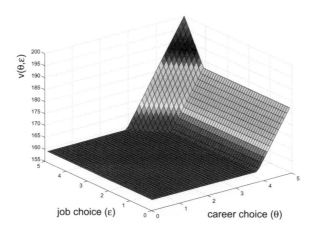

Figure 6.5.1: Optimal value function for Neal's model with $\beta = .95$. The value function is flat in the reject (θ, ϵ) region; increasing in θ only in the keep-career-but-draw-new-job region; and increasing in both θ and ϵ in the stay-put region.

and $C(\theta)$ is the value of drawing a new job but keeping θ:

$$C(\theta) = \theta + \int \left[\epsilon' + \beta v(\theta, \epsilon') \right] dG(\epsilon').$$

For a given career θ, a job $\bar{\epsilon}(\theta)$ makes equation (6.5.2) hold with equality. Evidently, $\bar{\epsilon}(\theta)$ solves

$$\bar{\epsilon}(\theta) = \max \left[(1 - \beta) C(\theta) - \theta, (1 - \beta) Q - \theta \right].$$

The decision to stay put is optimal for any career-job pair (θ, ϵ) that satisfies $\epsilon \geq \bar{\epsilon}(\theta)$. When this condition is not satisfied, the worker will draw either a new career-job pair (θ', ϵ') or only a new job ϵ'. Retaining the current career θ is optimal when

$$C(\theta) \geq Q. \tag{6.5.3}$$

We can solve (6.5.3) for the critical career value $\bar{\theta}$ satisfying

$$C(\bar{\theta}) = Q. \tag{6.5.4}$$

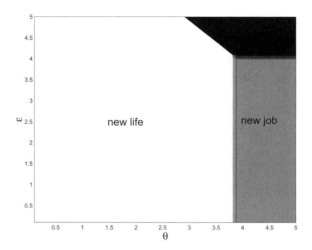

Figure 6.5.2: Optimal decision rule for Neal's model. For (θ, ϵ)'s within the white area, the worker changes both jobs and careers. In the grey area, the worker retains his career but draws a new job. The worker accepts (θ, ϵ) in the black area.

Thus, independently of ϵ, the worker will never abandon any career $\theta \geq \bar{\theta}$. The decision rule for accepting the current career can thus be expressed as follows: accept the current career θ if $\theta \geq \bar{\theta}$ or if the current career-job pair (θ, ϵ) satisfies $\epsilon \geq \bar{\epsilon}(\theta)$.

We can say more about the cutoff value $\bar{\epsilon}(\theta)$ in the retain-θ region $\theta \geq \bar{\theta}$. When $\theta \geq \bar{\theta}$, because we know that the worker will keep θ forever, it follows that

$$C(\theta) = \frac{\theta}{1 - \beta} + \int J(\epsilon') \, dG(\epsilon'),$$

where $J(\epsilon)$ is the optimal value of $\sum_{t=0}^{\infty} \beta^t \epsilon_t$ for a worker who has just drawn ϵ, who has already decided to keep his career θ, and who is deciding whether to try a new job next period. The Bellman equation for J is

$$J(\epsilon) = \max \left\{ \frac{\epsilon}{1 - \beta}, \epsilon + \beta \int J(\epsilon') \, dG(\epsilon') \right\}. \qquad (6.5.5)$$

This resembles the Bellman equation for the optimal value function for the basic McCall model, with a slight modification. The optimal policy is of the reservation-job form: keep the job ϵ for $\epsilon \geq \bar{\epsilon}$, otherwise try a new job next period. The absence of θ from (6.5.5) implies that in the range $\theta \geq \bar{\theta}$, $\bar{\epsilon}$ is independent of θ.

These results explain some features of the value function plotted in Figure 6.5.1 At the boundary separating the "new life" and "new job" regions of the (θ, ϵ) plane, equation (6.5.4) is satisfied. At the boundary separating the "new job" and "stay put" regions, $\frac{\theta+\epsilon}{1-\beta} = C(\theta) = \frac{\theta}{1-\beta} + \int J(\epsilon')dG(\epsilon')$. Finally, between the "new life" and "stay put" regions, $\frac{\theta+\epsilon}{1-\beta} = Q$, which defines a diagonal line in the (θ, ϵ) plane (see Figure 6.5.2). The value function is the constant value Q in the "get a new life" region (i.e., the region in which the optimal decision is to draw a new (θ, ϵ) pair). Equation (6.5.3) helps us understand why there is a set of high θ's in Figure 6.5.2 for which $v(\theta, \epsilon)$ rises with θ but is flat with respect to ϵ.

Probably the most interesting feature of the model is that it is possible to draw a (θ, ϵ) pair such that the value of keeping the career (θ) and drawing a new job match (ϵ') exceeds both the value of stopping search, and the value of starting again to search from the beginning by drawing a new (θ', ϵ') pair. This outcome occurs when a large θ is drawn with a small ϵ. In this case, it can occur that $\theta \geq \bar{\theta}$ and $\epsilon < \bar{\epsilon}(\theta)$.

Viewed as a normative model for young workers, Neal's model tells them: don't shop for a firm until you have found a career you like. As a positive model, it predicts that workers will not switch careers after they have settled on one. Neal presents data indicating that while this stark prediction does not hold up perfectly, it is a good first approximation. He suggests that extending the model to include learning, along the lines of Jovanovic's model to be described in section 6.8, could help explain the later career switches that his model misses. [8]

---

[8]   Neal's model can be used to deduce waiting times to the event $(\theta \geq \bar{\theta}) \cup (\epsilon \geq \bar{\epsilon}(\theta))$. The first event within the union is choosing a career that is never abandoned. The second event is choosing a permanent job. Neal used the model to approximate and interpret observed career and job switches of young workers.

6.6. Offer distribution unknown

Consider the following modification of the McCall search model. An unemployed worker wants to maximize the expected present value of $\sum_{t=0}^{\infty} \beta^t y_t$ where y_t equals wage w when employed and c when unemployed. Each period the worker receives one offer to work forever at a wage w drawn from one of two cumulative distribution functions F and G, where $F(0) = G(0) = 0$ and $F(B) = G(B) = 1$ for $B > 0$. Nature draws from the same distribution, either F or G, at all dates and the worker knows this, but he or she does not know whether it is F of G. At time 0 *before* drawing a wage offer, the worker attaches probability $\pi_{-1} \in (0, 1)$ to the distribution being F. We assume that the distributions have densities f and g, respectively, and that they have common support. Before drawing a wage at time 0, the worker thus believes that the density of w_0 is $h(w_0; \pi_{-1}) = \pi_{-1} f(w_0) + (1 - \pi_{-1}) g(w_0)$. After drawing w_0, the worker uses Bayes' law to deduce that the posterior probability that the density is $f(w)$ is[9]

$$\pi_0 = \frac{\pi_{-1} f(w_0)}{\pi_{-1} f(w_0) + (1 - \pi_{-1}) g(w_0)}.$$

More generally, after observing w_t for the tth draw, the worker believes that the probability that w_{t+1} is to be drawn from distribution F is

$$\pi_t = \frac{\pi_{t-1} f(w_t) / g(w_t)}{\pi_{t-1} f(w_t) / g(w_t) + (1 - \pi_{t-1})} \tag{6.6.1}$$

and that the density of w_{t+1} is

$$h(w_{t+1}; \pi_t) = \pi_t f(w_{t+1}) + (1 - \pi_t) g(w_{t+1}). \tag{6.6.2}$$

Notice that

$$E(\pi_t | \pi_{t-1}) = \int \left[\frac{\pi_{t-1} f(w)}{\pi_{t-1} f(w) + (1 - \pi_{t-1}) g(w)} \right] \left[\pi_{t-1} f(w) + (1 - \pi_{t-1}) g(w) \right] dw$$

$$= \pi_{t-1} \int f(w) dw$$

$$= \pi_{t-1},$$

---

[9] The worker's initial beliefs induce a joint probability distribution over a potentially infinite sequence of draws $w_0, w_1, \ldots$. Bayes' law is simply an application of the laws of probability to compute the conditional distribution of the tth draw w_t conditional on $[w_0, \ldots, w_{t-1}]$. Since we assume from the start that the decision maker *knows* the joint distribution and the laws of probability, one respectable view is that Bayes' law is less a 'theory of learning' than a statement about the consequences of information inflows for a decision maker who thinks he knows the truth (i.e., a joint probability distribution) from the beginning.

so that the process π_t is a *martingale* bounded by 0 and 1. (In the first line in the above string of equalities, the term in the first set of brackets is just π_t as a function of w_t, while the term in the second set of brackets is the density of w_t conditional on π_{t-1}.) Notice that here we are computing $E(\pi_t | \pi_{t-1})$ under the subjective density described in the second term in brackets. It follows from the martingale convergence theorem (see appendix A of chapter 17) that π_t converges almost surely to a random variable in $[0, 1]$. Practically, this means if the probability attached to all sample paths $\{\pi_t\}_{t=0}^{\infty}$ that converge is unity. However, in general different sample paths converge to different limiting values. The limit points of $\{\pi_t\}_{t=0}^{\infty}$ as $t \to +\infty$ thus constitute a random variable with what is in general a non-trivial distribution.

Let $v(w_t, \pi_t)$ be the optimal value of the problem for a previously unemployed worker who has just drawn w and updated π according to (6.6.1). The Bellman equation is

$$v\left(w, \pi_t\right) = \max_{\text{accept,reject}} \left\{ \frac{w}{1 - \beta}, c + \beta \int v\left(w', \pi_{t+1}\left(w'\right)\right) h\left(w'; \pi_t\right) dw' \right\} \quad (6.6.3)$$

subject to (6.6.1) and (6.6.2). The state vector is the worker's current draw w and his post-draw estimate of the probability that the distribution is f. The second term on the right side of (6.6.3) integrates the value function evaluated at next period's state vector with respect to the worker's subjective distribution $h(w'; \pi_t)$ of next period's draw w'. The value function for next period recognizes that π_{t+1} will be updated in a way that depends on w' via Bayes' law as captured by equation (6.6.1). Evidently, the optimal policy is to set a reservation wage $\bar{w}(\pi_t)$ that depends on π_t.

As an example, we have computed the optimal policy by backward induction assuming that f is a uniform distribution on $[0, 2]$ while g is a beta distribution with parameters $(3, 1.2)$.[10] We set unemployment compensation $c = .6$ and the discount factor $\beta = .95$.[11] The two densities are plotted in figure 6.6.1, which shows that the g density provides better prospects for the worker than does the uniform f density. It stands to reason that the worker's reservation wage falls as the posterior probability π that he places on density f rises, as figure 6.6.2 confirms.

---

[10] The beta distribution for w is characterized by a density $g(w; \alpha, \gamma) \propto w^{\alpha-1}(1-w)^{(\gamma-1)}$, where the factor of proportionality is chosen to make the density integrate to 1.

[11] The matlab programs `search_learn_francisco_3.m` and `search_learn_beta_2.m` perform these calculations.

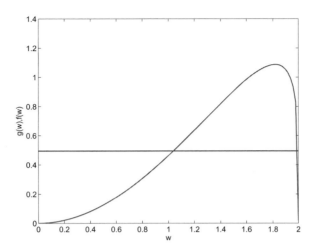

Figure 6.6.1: Two densities for wages, a uniform $f(w)$ and a $g(w)$ corresponding to a beta distribution with parameters 3, 1.2.

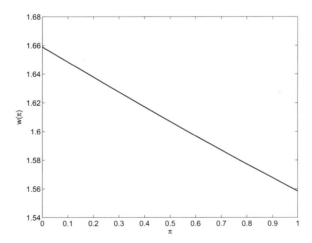

Figure 6.6.2: The reservation as a function of the posterior probability π that the worker thinks that the wage is drawn from the uniform density f.

Figure 6.6.3 shows empirical cumulative distribution functions for durations of unemployment and π at time of job acceptance under two alternative assumptions about whether the uniform distribution F or the beta distribution G permanently governs the wage. We constructed these by simulating the model 10,000 times at the parameter values just given, starting from a common initial condition for beliefs $\pi_{-1} = .5$ and assuming that, unbeknownst to the worker, either the uniform density $f(w)$ or the beta density $g(w)$ truly governs successive wage draws. Only when π_t approaches 1 will workers have learned that nature is drawing from f and not g. Evidently, most workers accept jobs long before a law of large numbers has enough time to teach them for sure which of the two densities from which nature draws wage offers. Thus, workers usually choose not to collect enough observations for them to learn for sure which distribution governs wage offers. In both panels, the lower line shows the cumulative distribution function when nature draws from F and the lower panel shows the c.d.f. when nature draws from G. [12]

A comparison of the CDF's when nature draws from F and G, respectively, is revealing. When G prevails, the cumulative distribution functions in the top panel reveal that workers typically accept jobs earlier than when F prevails. This captures what the interrogator of an unemployed McCall worker in the passage of Lucas cited in the introduction might have had in mind when he said 'Maybe you are setting your sights too high'. The bottom panel reveals that when nature permanently draws from G, employed workers put a higher probability on their having actually sampled from G than from F, while the reverse is true when nature draws permanently from F.

---

[12] It is a useful exercise to use recall formula (6.2.2) for the mean of a nonnegative random variable and then glance at the CDFs in the bottom panel to approximate the mean π_t at time of job acceptance.

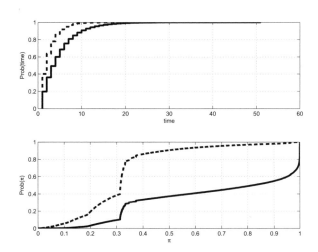

Figure 6.6.3: Top panel: CDF of duration of unemployment; bottom panel: CDF of π at time worker accepts wage and leaves unemployment. In each panel, the lower filled line is the CDF when nature permanently draws from the uniform density f while the dotted line is the CDF when nature permanently draws from the beta density g.

6.7. An equilibrium price distribution

The McCall search model confronts a worker with a given distribution of wages. In this section, we ask why firms might conceivably choose to confront an *ex ante* homogenous collection of workers with a nontrivial distribution of wages. Knowing that the workers have a reservation wage policy, why would a firm ever offer a worker *more* than the reservation wage? That question challenges us to think about whether it is possible to conceive of a coherent setting in which it would be optimal for a collection of profit maximizing firms somehow to make decisions that generate a distribution of wages.

In this section, we take up this question, but for historical reasons investigate it in the context of a sequential search model in which buyers seek the lowest price.[13] Buyers can draw additional offers from a known distribution at a fixed cost c for each additional *batch* of n independent draws from a known

---

[13] See Burdett and Mortensen (1998) for a parallel analysis of the analogous issues in a model of job search.

price distribution. Both within and across batches, successive draws are independent. The buyer's optimal strategy is to set a reservation price and to continue drawing until the first time a price less than the reservation price has been offered. Let $\tilde{p}$ be the reservation price.

Rothschild (1973) posed the following challenge for a model in which there is a large number of identical buyers each of whom has reservation price $\tilde{p}$. If all sellers know the reservation price $\tilde{p}$, why would any of them offer a price less than $\tilde{p}$? This cogent question points to a force for the price distribution to collapse, an outcome that would destroy the motive for search behavior on the part of buyers. Thus, the challenge is to construct an *equilibrium* version of a search model in which it is in firms' interest to generate the non-trivial price distribution that sustains buyers' search activities.

Burdett and Judd (1983) met this challenge by creating an environment in which *ex ante* identical buyers *ex post* receive differing numbers of price offers that are drawn from a common distribution set by firms. They construct an equilibrium in which a continuum of profit maximizing sellers are content to generate this distribution of prices. Sellers set their prices to maximize expected profit per customer. But sellers don't know the number of other offers that a prospective customer has received. Heterogeneity in the number of offers received by buyers together with seller's ignorance of the number and nature of *other* offers received by a *particular* customer creates a tradeoff between profit per customer and volume that makes possible a non-degenerate equilibrium price distribution. Firms that post higher prices are lower-volume sellers. Firms that post lower prices are higher-volume sellers. There exists an equilibrium distribution of prices in which all types of firms expect to earn the same profit per potential customer.

6.7.1. A Burdett-Judd setup

A continuum of buyers purchases a single good from one among a continuum of firms. Each firm contacts a fixed measure ν of potential buyers. The firms produce a homogeneous good at zero marginal cost. Each firm takes the c.d.f. of prices charged by other firms as given and chooses a price. The firm wants to maximize its expected profits per consumer. A firm's expected profit per consumer equals its price times the probability that its price is the minimum among the set of acceptable offers received by the buyer. The distribution of prices set by other firms impinges on a firm's expected profits because it affects the probability that its offer will be accepted by a buyer.

6.7.2. Consumer problem with noisy search

A consumer wants to purchase a good for a minimum price. Firms make offers that buyers can view as being drawn from a distribution of nonnegative prices with cumulative distribution function $G(P) = \text{Prob}(p \leq P)$ with $G(\underline{p}) = 0, G(B) = 1$. Assume that G is continuously differentiable and so has an associated probability density. A buyer's search activity is divided into batches. Within each batch the buyer receives a random number of offers drawn from the same distribution G. Burdett and Judd call this structure 'noisy search'. In particular, at a cost of $c > 0$ per search round, with probability $q \in (0,1)$ a buyer receives one offer drawn from G and with probability $1 - q$ receives two offers. Thus, a 'round' consists of a 'compound lottery' first of a random number of draws, then that number of i.i.d. random draws price offers from the c.d.f. G. A buyer can recall offers within a round but not across rounds. Evidently, $\text{Prob}\{\min(p_1, p_2) \geq p\} = (1 - G(p))^2$ and $\text{Prob}\{\min(p_1, p_2) \leq p\} = 1 - (1 - G(p))^2$. Then *ex ante* the c.d.f. of low prices drawn in a single round is

$$H(p) = qG(p) + (1 - q)\left(1 - (1 - G(p))^2\right). \tag{6.7.1}$$

Let $v(p)$ be the expected price including future search costs of a consumer who has already paid c, has offer p in hand, and is about to decide whether to accept or reject the offer. The Bellman equation is

$$v(p) = \min_{\text{accept,reject}}\left\{p, c + \int_{\underline{p}}^{B} v(p')\, dH(p')\right\}. \tag{6.7.2}$$

The reservation price $\tilde{p}$ satisfies $v(\tilde{p}) = \tilde{p} = c + \int_{\underline{p}}^{\tilde{p}} p' dH(p')$, which implies [14]

$$c = \int_{\underline{p}}^{\tilde{p}} H\left(p\right) dp. \tag{6.7.3}$$

Combining equation (6.7.3) with the formula $Ep = \int_{\underline{p}}^{\tilde{p}} (1 - H(p)) dp$ for the mean of a nonnegative random variable implies that the reservation price $\tilde{p}$ satisfies

$$\tilde{p} = c + Ep,$$

which states that the reservation price equals the cost of one additional round of search plus the mean price drawn from one more round of noisy search. The challenge is to construct an equilibrium price distribution G, and thus an implied distribution H, in which most firms choose to post prices less than the buyer's reservation price $\tilde{p}$.

6.7.3. Firms

For simplicity and to focus our attention entirely on the search problem, we assume that the good costs firms nothing to produce. In setting its price, we assume that a firm seeks to maximize expected profit per customer. A firm makes an offer to a customer without knowing whether this is the only offer available to the customer or whether the customer, having drawn two offers, possibly has a lower offer in hand. The firm begins by computing the fraction of its customers who will have received one offer and the fraction of its customers who will have received only one offer. Let there be a large number ν of total potential buyers per batch, consisting of νq persons each of whom receives one offer and $\nu(1 - q)$ people each of whom receives two offers. The total number of offers is evidently $\nu(1q + 2(1 - q)) = \nu(2 - q)$. Evidently, the fraction of all offers that is received by customers who have received one offer is $\frac{\nu q}{\nu(2-q)} = \frac{q}{2-q}$. This calculation induces a typical firm to believe that the fraction of its customers

---

[14] The Bellman equation implies $\tilde{p} = c + \int_{\tilde{p}}^{B} v(p') dH(p)$, which can be rearranged to become $\int_{\underline{p}}^{\tilde{p}} (\tilde{p} - p) dH(p) = c$. Let $u = \tilde{p} - p$ and $dv = dH(p)$ and apply the integration by parts formula $\int u\, dv = uv - \int v\, du$ to the previous equality to get $\int_{\underline{p}}^{\tilde{p}} H(p) dp = c$.

who receive one offer is

$$\hat{q} = \frac{q}{2-q} \qquad (6.7.4)$$

and the fraction who receive two offers is $1 - \hat{q} = \frac{2(1-q)}{2-q}$. The firm regards $\hat{q}$ as its estimate of the probability that a given customer has received only its offer, while it thinks that a fraction $1 - \hat{q}$ of its customers has also received a competing offer from another firm.

There is a continuum of firms each of which takes as given a price offer distribution of other firms with c.d.f. $G(p)$, where $G(\underline{p}) = 0, G(\tilde{p}) = 1$. We have assume that G is differentiable.[15] This distribution satisfies the outcome that in equilibrium no firm makes an offer exceeding the buyer's reservation price $\tilde{p}$. Let $Q(p)$ be the probability that a consumer will accept an offer p, where $\underline{p} \le p \le \tilde{p}$. Evidently, a consumer who receives *one* offer $p < \tilde{p}$ will accept it with probability 1. But only a fraction $1 - G(p)$ of consumer who receive *two* offers will accept an offer $p < \tilde{p}$. Why? because $1 - G(p)$ is the fraction of consumers whose *other* offer exceeds p; so a fraction $G(p)$ of two-offer customers who receive offer p will reject it because they have received an offer lower than p. Therefore, the overall probability that a randomly encountered consumer will accept an offer $p \in [\underline{p}, \tilde{p}]$ is

$$Q(p) = \hat{q} + (1 - \hat{q})(1 - G(p)). \qquad (6.7.5)$$

---

[15] Burdett and Judd (1983, p. 959, lemma 1) show that an equilibrium G is differentiable when $q \in (0,1)$ and $\tilde{p} > 0$. Their argument goes as follows. Suppose to the contrary that there is a positive probability attached to a single price $p' \in (0, \tilde{p})$. Consider a firm that contemplates charging p'. When $q < 1$, the firm knows that there is a positive probability that a prospective consumer has received another offer also of p'. If the firm lowers its offer infinitesimally, it can expect to steal that customer and thereby increase its expected profits. Therefore, a decision to charge p' can't maximize expected profits for a typical firm. We have been led to a contradiction by assuming that G has a discontinuity at p'.

6.7.4. Equilibrium

The objects in play are a reservation price $\tilde{p}$ and a value function $v(p)$ for a typical buyer; and a c.d.f. $G(p)$ of prices that is the outcome of the independent price-setting decisions of individual firms and that is taken as given by all buyers and sellers.

DEFINITION: An equilibrium is a c.d.f. of price offers $G(p)$ on domain $[\underline{p}, \tilde{p}]$, a c.d.f. of per-batch price offers to consumers $H(p)$, and a reservation price $\tilde{p}$ such that (i) the c.d.f. of offers to buyers $H(p)$ satisfies (6.7.1); (ii) $\tilde{p}$ is an optimal reservation price for buyers that satisfies $c = \int_{\underline{p}}^{\tilde{p}} dH(p)$; and (iii) firms are indifferent with respect to charging any $p \in [\underline{p}, \tilde{p}]$; therefore, firms choose p by randomizing using $G(p)$.

We confirm an equilibrium by using a guess-and-verify method. Make the following guess for an equilibrium c.d.f. $G(p)$.[16] First, set

$$\underline{p} = \hat{q}\tilde{p} \tag{6.7.6}$$

and then set

$$G\left(p\right) = \begin{cases} 0 & \text{if } p \leq \underline{p} \\ 1 - \frac{\tilde{p}-p}{p}\frac{\hat{q}}{1-\hat{q}} & \text{if } p \in \left[\underline{p}, \tilde{p}\right] \\ 1 & \text{if } p > \tilde{p} . \end{cases} \tag{6.7.7}$$

Under this guess, $Q(p)$ becomes

$$Q\left(p\right) = \frac{\tilde{p}\hat{q}}{p} \quad \forall p \in \left[\underline{p}, \tilde{p}\right] .$$

Therefore, the expected profit per customer for a firm that sets price $p \in [\underline{p}, \tilde{p}]$ is

$$pQ\left(p\right) = \tilde{p}\hat{q}, \tag{6.7.8}$$

which is evidently independent of the firm's choice of offer p in the interval $[\underline{p}, \tilde{p}]$. The firm is indifferent about the price it offers on this interval. In particular, notice that The right side of equality (6.7.8) is the product of the fraction of a firm's buyers receiving one offer, $\hat{q}$, times the reservation price $\tilde{p}$. This is the expected profit per customer of a firm that charges the reservation price. The

---

[16] We can make sure that the buyer's search problem is consistent with this guess by setting c to confirm (6.7.3).

left side of equality $(6.7.8)$ is the product of the price p times probability $Q(p)$ that a buyer will accept price p, which as we have noted equals the expected profit per customer for a firm that sets price p.

We assume that firms randomize over choices of p in such a way that $G(p)$ given by $(6.7.7)$ emerges as the c.d.f. for prices.

6.7.5. Special cases

The Burdett-Judd model isolates forces for the price distribution to collapse and countervailing forces that can sustain a nontrivial price distribution.

1. Consider the special case in which $q = 1$ (and therefore $\hat{q} = 1$). Here, $\underline{p} = \tilde{p}$. The formula $(6.7.7)$ shows that the distribution of prices collapses. This case exhibits the Rothschild challenge with which we began.
2. Next, consider the opposite special case in which $q = 0$ (and therefore $\hat{q} = 0$). Here, $\underline{p} = 0$ and the c.d.f. $G(p) = 1 \forall p \in [\underline{p}, \tilde{p}]$. Bertrand competition drives all prices down to the marginal cost of production, which we have assumed to be zero. This case exhibits another force for the price distribution to collapse, again in the spirit of Rothschild's challenge.
3. Finally, consider the general case in which $q \in (0, 1)$ and therefore $\hat{q} \in (0, 1)$). When q is strictly in the interior of $[\underline{p}, \tilde{p}]$, we can sustain a nontrivial distribution of prices. Firms are indifferent between being high volume, low price sellers and high price, low volume sellers. The equilibrium price distribution $G(p)$ renders a firm's expected profits per prospective customer $pQ(p)$ independent of p.

6.8.   Jovanovic's matching model

Another interesting effort to confront Rothschild's questions about the source of the equilibrium wage (or price) distribution comes from matching models, in which the main idea is to reinterpret w not as a wage but instead, more broadly, as a parameter characterizing the entire quality of a match occurring between a pair of agents. The variable w is regarded as a summary measure of the productivities or utilities jointly generated by the activities of the match. We can consider pairs consisting of a firm and a worker, a man and a woman, a house and an owner, or a person and a hobby. The idea is to analyze the way in which matches form and maybe also dissolve by viewing both parties to the match as being drawn from populations that are statistically homogeneous to an outside observer, even though the match is idiosyncratic from the perspective of the parties to the match.

Jovanovic (1979a) has used a model of this kind supplemented by a hypothesis that both sides of the match behave optimally but only gradually learn about the quality of the match. Jovanovic was motivated by a desire to explain three features of labor market data: (1) on average, wages rise with tenure on the job, (2) quits are negatively correlated with tenure (that is, a quit has a higher probability of occurring earlier in tenure than later), and (3) the probability of a subsequent quit is negatively correlated with the current wage rate. Jovanovic's insight was that each of these empirical regularities could be interpreted as reflecting the operation of a matching process with gradual learning about match quality. We consider a simplified version of Jovanovic's model of matching. (Prescott and Townsend, 1980, describe a discrete-time version of Jovanovic's model, which has been simplified here.) A market has two sides that could be variously interpreted as consisting of firms and workers, or men and women, or owners and renters, or lakes and fishermen. Following Jovanovic, we shall adopt the firm-worker interpretation here. An unmatched worker and a firm form a pair and jointly draw a random match parameter θ from a probability distribution with cumulative distribution function $\text{Prob}\{\theta \leq s\} = F(s)$. Here the match parameter reflects the marginal productivity of the worker in the match. In the first period, before the worker decides whether to work at this match or to wait and to draw a new match next period from the same distribution F, the worker and the firm both observe only $y = \theta + u$, where u is a random noise that is uncorrelated with θ. Thus, in the first period, the worker-firm pair receives only a noisy observation on θ. This situation corresponds to that when both sides of the market form only an error-ridden impression of the quality of the match at first. On the basis of this noisy observation, the firm, which is imagined to operate competitively under constant returns to scale, offers to pay the worker the conditional expectation of θ, given $(\theta + u)$, for the first period, with the understanding that in subsequent periods it will pay the worker the expected value of θ, depending on whatever additional information both sides of the match receive.[17] Given this policy of the firm, the worker decides whether to accept the match and to work this period for $E[\theta|(\theta+u)]$ or to refuse the offer and draw a new match parameter θ' and noisy observation on it, $(\theta' + u')$, next period. If the worker decides to accept the offer in the first period, then in the second period both the firm and the worker are assumed to observe the true value of θ. This situation corresponds to that in which both sides learn about each other and about the quality of the match. In the second period the firm offers to pay the worker θ then and forever more. The worker next decides whether to accept this offer or to quit, be unemployed this period, and draw a new match parameter and a noisy observation on it next period.

We can conveniently think of this process as having three stages. Stage 1 is the "predraw" stage, in which a previously unemployed worker has yet to draw the one match parameter and the noisy observation on it that he is entitled to draw after being unemployed the previous period. We let Q denote the expected present value of wages, before drawing, of a worker who was unemployed last period and who behaves optimally. The second stage of the process occurs after the worker has drawn a match parameter θ, has received the noisy observation of $(\theta + u)$ on it, and has received the firm's wage offer of $E[\theta|(\theta+u)]$ for this period. At this stage, the worker decides whether to accept this wage for this period and the prospect of receiving θ in all subsequent periods. The third stage occurs in the next period, when the worker and firm discover the true value of θ and the worker must decide whether to work at θ this period and in all subsequent periods that he remains at this job (match).

---

[17] Jovanovic assumed firms to be risk neutral and to maximize the expected present value of profits. They compete for workers by offering wage contracts. In a long-run equilibrium the payments practices of each firm would be well understood, and this fact would support the described implicit contract as a competitive equilibrium.

We now add some more specific assumptions about the probability distribution of θ and u. We assume that θ and u are independently distributed random variables. Both are normally distributed, θ being normal with mean μ and variance σ_0^2, and u being normal with mean 0 and variance σ_u^2. Thus, we write

$$\theta \sim N\left(\mu, \sigma_0^2\right), \qquad u \sim N\left(0, \sigma_u^2\right) . \tag{6.8.1}$$

In the first period, after drawing a θ, the worker and firm both observe the noise-ridden version of θ, $y = \theta + u$. Both worker and firm are interested in making inferences about θ, given the observation $(\theta + u)$. They are assumed to use Bayes' law and to calculate the posterior probability distribution of θ, that is, the probability distribution of θ conditional on $(\theta + u)$. The probability distribution of θ, given $\theta + u = y$, is known to be normal, with mean m_0 and variance σ_1^2. Using the Kalman filtering formula in chapter 2, we have[18]

$$m_0 = E\left(\theta | y\right) = E\left(\theta\right) + \frac{\operatorname{cov}\left(\theta, y\right)}{\operatorname{var}\left(y\right)}\left[y - E\left(y\right)\right]$$

$$= \mu + \frac{\sigma_0^2}{\sigma_0^2 + \sigma_u^2}\left(y - \mu\right) \equiv \mu + K_0\left(y - \mu\right), \tag{6.8.2}$$

$$\sigma_1^2 = E\left[\left(\theta - m_0\right)^2 | y\right] = \frac{\sigma_0^2}{\sigma_0^2 + \sigma_u^2}\sigma_u^2 = K_0\sigma_u^2 .$$

After drawing θ and observing $y = \theta + u$ the first period, the firm is assumed to offer the worker a wage of $m_0 = E[\theta | (\theta + u)]$ the first period and a promise to pay θ for the second period and thereafter. The worker has the choice of accepting or rejecting the offer.

From equation (6.8.2) and the property that the random variable $y - \mu = \theta + u - \mu$ is normal, with mean zero and variance $(\sigma_0^2 + \sigma_u^2)$, it follows that m_0 is itself normally distributed, with mean μ and variance $\sigma_0^4 / (\sigma_0^2 + \sigma_u^2) = K_0\sigma_0^2$:

$$m_0 \sim N\left(\mu, K_0\sigma_0^2\right) . \tag{6.8.3}$$

Note that $K_0\sigma_0^2 < \sigma_0^2$, so that m_0 has the same mean but a smaller variance than θ.

---

[18] In the special case in which random variables are jointly normally distributed, linear least-squares projections equal conditional expectations.

6.8.1. Recursive formulation and solution

The worker seeks to maximize the expected present value of wages. We now proceed to solve the worker's problem by working backward. At stage 3, the worker knows θ and is confronted by the firm with an offer to work this period and forever more at a wage of θ. We let $J(\theta)$ be the expected present value of wages of a worker at stage 3 who has a known match θ in hand and who behaves optimally. The worker who accepts the match this period receives θ this period and faces the same choice at the same θ next period. (The worker can quit next period, though it will turn out that the worker who does not quit this period never will.) Therefore, if the worker accepts the match, the value of match θ is given by $\theta + \beta J(\theta)$, where β is the discount factor. The worker who rejects the match must be unemployed this period and must draw a new match next period. The expected present value of wages of a worker who was unemployed last period and who behaves optimally is Q. Therefore, the Bellman equation is $J(\theta) = \max\{\theta + \beta J(\theta), \beta Q\}$. This equation is graphed in Figure 6.8.1 and evidently has the solution

$$
J\left(\theta\right) = \begin{cases} \theta + \beta J\left(\theta\right) = \frac{\theta}{1-\beta} & \text{for } \theta \geq \overline{\theta} \\ \beta Q & \text{for } \theta \leq \overline{\theta}. \end{cases} \tag{6.8.4}
$$

The optimal policy is a reservation wage policy: accept offers $\theta \geq \overline{\theta}$, and reject offers $\theta \leq \overline{\theta}$, where θ satisfies

$$
\frac{\overline{\theta}}{1-\beta} = \beta Q. \tag{6.8.5}
$$

We now turn to the worker's decision in stage 2, given the decision rule in stage 3. In stage 2, the worker is confronted with a current wage offer $m_0 = E[\theta|(\theta + u)]$ and a conditional probability distribution function that we write as $\text{Prob}\{\theta \leq s|\theta + u\} = F(s|m_0, \sigma_1^2)$. (Because the distribution is normal, it can be characterized by the two parameters m_0, σ_1^2.) We let $V(m_0)$ be the expected present value of wages of a worker at the second stage who has offer m_0 in hand and who behaves optimally. The worker who rejects the offer is unemployed this period and draws a new match parameter next period. The expected present value of this option is βQ. The worker who accepts the offer receives a wage of m_0 this period and a probability distribution of wages of $F(\theta'|m_0, \sigma_1^2)$ for next period. The expected present value of this option is $m_0 + \beta \int J(\theta') dF(\theta'|m_0, \sigma_1^2)$.

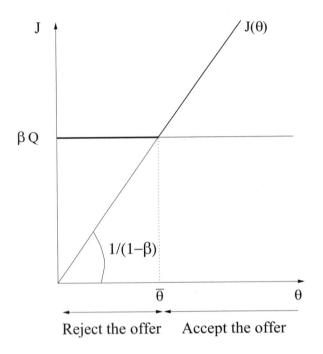

Figure 6.8.1: The function $J(\theta) = \max\{\theta + \beta J(\theta), \beta Q\}$.
The reservation wage in stage 3, $\bar{\theta}$, satisfies $\bar{\theta}/(1-\beta) = \beta Q$.

The Bellman equation for the second stage therefore becomes

$$V\left(m_{0}\right) = \max\left\{m_{0} + \beta \int J\left(\theta'\right) dF\left(\theta'|m_{0}, \sigma_{1}^{2}\right), \beta Q\right\}. \qquad (6.8.6)$$

Note that both m_0 and $\beta \int J(\theta')dF(\theta'|m_0, \sigma_1^2)$ are increasing in m_0, whereas βQ is a constant. For this reason a reservation wage policy will be an optimal one. The functional equation evidently has the solution

$$V\left(m_{0}\right) = \begin{cases} m_{0} + \beta \int J\left(\theta'\right) dF\left(\theta'|m_{0}, \sigma_{1}^{2}\right) & \text{for } m_{0} \geq \overline{m}_{0} \\ \beta Q & \text{for } m_{0} \leq \overline{m}_{0}. \end{cases} \qquad (6.8.7)$$

If we use equation (6.8.7), an implicit equation for the reservation wage $\overline{m}_0$ is then

$$V\left(\overline{m}_{0}\right) = \overline{m}_{0} + \beta \int J\left(\theta'\right) dF\left(\theta'|\overline{m}_{0}, \sigma_{1}^{2}\right) = \beta Q. \qquad (6.8.8)$$

Using equations (6.8.8) and (6.8.4), we shall show that $\overline{m}_0 < \overline{\theta}$, so that the worker becomes choosier over time with the firm. This force makes wages rise with tenure.

Using equations (6.8.4) and (6.8.5) repeatedly in equation (6.8.8), we obtain

$$\overline{m}_0 + \beta \frac{\overline{\theta}}{1-\beta} \int_{-\infty}^{\overline{\theta}} dF\left(\theta'|\overline{m}_0, \sigma_1^2\right) + \frac{\beta}{1-\beta} \int_{\overline{\theta}}^{\infty} \theta' dF\left(\theta'|\overline{m}_0, \sigma_1^2\right)$$

$$= \frac{\overline{\theta}}{1-\beta} = \frac{\overline{\theta}}{1-\beta} \int_{-\infty}^{\overline{\theta}} dF\left(\theta'|\overline{m}_0, \sigma_1^2\right)$$

$$+ \frac{\overline{\theta}}{1-\beta} \int_{\overline{\theta}}^{\infty} dF\left(\theta'|\overline{m}_0, \sigma_1^2\right).$$

Rearranging this equation, we get

$$\overline{\theta} \int_{-\infty}^{\overline{\theta}} dF\left(\theta'|\overline{m}_0, \sigma_1^2\right) - \overline{m}_0 = \frac{1}{1-\beta} \int_{\overline{\theta}}^{\infty} \left(\beta\theta' - \overline{\theta}\right) dF\left(\theta'|\overline{m}_0, \sigma_1^2\right). \quad (6.8.9)$$

Now note the identity

$$\overline{\theta} = \int_{-\infty}^{\overline{\theta}} \overline{\theta} dF\left(\theta'|\overline{m}_0, \sigma_1^2\right) + \left(\frac{1}{1-\beta} - \frac{\beta}{1-\beta}\right) \int_{\overline{\theta}}^{\infty} \overline{\theta} dF\left(\theta'|\overline{m}_0, \sigma_1^2\right). \quad (6.8.10)$$

Adding equation (6.8.10) to (6.8.9) gives

$$\overline{\theta} - \overline{m}_0 = \frac{\beta}{1-\beta} \int_{\overline{\theta}}^{\infty} \left(\theta' - \overline{\theta}\right) dF\left(\theta'|\overline{m}_0, \sigma_1^2\right). \quad (6.8.11)$$

The right side of equation (6.8.11) is positive. The left side is therefore also positive, so that we have established that

$$\overline{\theta} > \overline{m}_0. \quad (6.8.12)$$

Equation (6.8.11) resembles equation (6.3.3) and has a related interpretation. Given $\overline{\theta}$ and $\overline{m}_0$, the right side is the expected benefit of a match $\overline{m}_0$, namely, the expected present value of the match in the event that the match parameter eventually turns out to exceed the reservation match $\overline{\theta}$ so that the match endures. The left side is the one-period cost of temporarily staying in a match paying less than the eventual reservation match value $\bar{\theta}$: having remained unemployed for a period in order to have the privilege of drawing the match parameter θ, the worker has made an investment to acquire this opportunity and must make a similar investment to acquire a new one. Having only the noisy observation of $(\theta + u)$ on θ, the worker is willing to stay in matches m_0 with $\overline{m}_0 < m_0 < \bar{\theta}$ because it is worthwhile to speculate that the match is really better than it seems now and will seem next period.

Now turning briefly to stage 1, we have defined Q as the predraw expected present value of wages of a worker who was unemployed last period and who is about to draw a match parameter and a noisy observation on it. Evidently, Q is given by

$$Q = \int V(m_0)\, dG\left(m_0 | \mu, K_0 \sigma_0^2\right) \tag{6.8.13}$$

where $G(m_0 | \mu, K_0 \sigma_0^2)$ is the normal distribution with mean μ and variance $K_0 \sigma_0^2$, which, as we saw before, is the distribution of m_0.

Collecting some of the equations, we see that the worker's optimal policy is determined by

$$J(\theta) = \begin{cases} \theta + \beta J(\theta) = \frac{\theta}{1-\beta} & \text{for } \theta \geq \bar{\theta} \\ \beta Q & \text{for } \theta \leq \bar{\theta} \end{cases} \tag{6.8.14}$$

$$V(m_0) = \begin{cases} m_0 + \beta \int J(\theta')\, dF\left(\theta' | m_0, \sigma_1^2\right) & \text{for } m_0 \geq \overline{m}_0 \\ \beta Q & \text{for } m_0 \leq \overline{m}_0 \end{cases} \tag{6.8.15}$$

$$\bar{\theta} - \overline{m}_0 = \frac{\beta}{1-\beta} \int_{\bar{\theta}}^{\infty} (\theta' - \bar{\theta})\, dF\left(\theta' | \overline{m}_0, \sigma_1^2\right) \tag{6.8.16}$$

$$Q = \int V(m_0)\, dG\left(m_0 | \mu, K_0 \sigma_0^2\right). \tag{6.8.17}$$

To analyze formally the existence and uniqueness of a solution to these equations, one would proceed as follows. Use equations (6.8.14), (6.8.15), and (6.8.16) to write a single functional equation in V,

$$V(m_0) = \max\left\{ m_0 + \beta \int \max\left[\frac{\theta}{1-\beta}, \right. \right.$$

$$\left. \beta \int V(m_1')\, dG\left(m_1' | \mu, K_0 \sigma_0^2\right) \right] dF(\theta | m_0, \sigma_1^2),$$

$$\left. \beta \int V(m_1')\, dG\left(m_1' | \mu, K_0 \sigma_0^2\right) \right\}.$$

The expression on the right defines an operator, T, mapping continuous functions V into continuous functions TV. This functional equation can be expressed $V = TV$. The operator T can be directly verified to satisfy the following two properties: (1) it is monotone, that is, $v(m) \geq z(m)$ for all m implies $(Tv)(m) \geq (Tz)(m)$ for all m; (2) for all positive constants c, $T(v+c) \leq Tv + \beta c$. These are Blackwell's sufficient conditions for the functional equation $Tv = v$ to have a unique continuous solution. See Appendix A on functional analysis (see Technical Appendixes).

6.8.2. Endogenous statistics

We now proceed to calculate probabilities and expectations of some interesting events and variables. The probability that a previously unemployed worker accepts an offer is given by

$$\text{Prob}\{m_0 \geq \overline{m}_0\} = \int_{\overline{m}_0}^{\infty} dG\left(m_0 | \mu, K_0 \sigma_0^2\right).$$

The probability that a previously unemployed worker accepts an offer and then quits the second period is given by

$$\text{Prob}\{\left(\theta \leq \overline{\theta}\right) \cap \left(m_0 \geq \overline{m}_0\right)\} = \int_{\overline{m}_0}^{\infty} \int_{-\infty}^{\overline{\theta}} dF\left(\theta | m_0, \sigma_1^2\right) dG\left(m_0 | \mu, K_0 \sigma_0^2\right).$$

The probability that a previously unemployed worker accepts an offer the first period and also elects not to quit the second period is given by

$$\text{Prob}\{\left(\theta \geq \overline{\theta}\right) \cap \left(m_0 \geq \overline{m}\right)\} = \int_{\overline{m}_0}^{\infty} \int_{\overline{\theta}}^{\infty} dF\left(\theta | m_0, \sigma_1^2\right) dG\left(m_0 | \mu, K_0 \sigma_0^2\right).$$

The mean wage of those employed the first period is given by

$$\overline{w}_1 = \frac{\displaystyle\int_{\overline{m}_0}^{\infty} m_0 \, dG\left(m_0 | \mu, K_0 \sigma_0^2\right)}{\displaystyle\int_{\overline{m}_0}^{\infty} dG\left(m_0 | \mu, K_0 \sigma_0^2\right)}, \tag{6.8.18}$$

whereas the mean wage of those workers who are in the second period of tenure is given by

$$\overline{w}_2 = \frac{\displaystyle\int_{\overline{m}_0}^{\infty} \int_{\overline{\theta}}^{\infty} \theta \, dF\left(\theta | m_0, \sigma_1^2\right) dG\left(m_0 | \mu, K_0 \sigma_0^2\right)}{\displaystyle\int_{\overline{m}_0}^{\infty} \int_{\overline{\theta}}^{\infty} dF\left(\theta | m_0, \sigma_1^2\right) dG\left(m_0 | \mu, K_0 \sigma_0^2\right)}. \tag{6.8.19}$$

We shall now prove that $\overline{w}_2 > \overline{w}_1$, so that wages rise with tenure. After substituting $m_0 \equiv \int \theta dF(\theta|m_0, \sigma_1^2)$ into equation (6.8.18),

$$
\overline{w}_1 = \frac{\displaystyle\int_{\overline{m}_0}^{\infty} \int_{-\infty}^{\infty} \theta \, dF\left(\theta|m_0, \sigma_1^2\right) dG\left(m_0|\mu, K_0\sigma_0^2\right)}{\displaystyle\int_{\overline{m}_0}^{\infty} dG\left(m_0|\mu, K_0\sigma_0^2\right)}
$$

$$
= \frac{1}{\displaystyle\int_{\overline{m}_0}^{\infty} dG\left(m_0|\mu, K_0\sigma_0^2\right)} \left\{ \int_{\overline{m}_0}^{\infty} \int_{-\infty}^{\overline{\theta}} \theta \, dF\left(\theta|m_0, \sigma_1^2\right) dG\left(m_0|\mu, K_0\sigma_0^2\right) \right.
$$

$$
\left. + \, \overline{w}_2 \int_{\overline{m}_0}^{\infty} \int_{\overline{\theta}}^{\infty} dF\left(\theta|m_0, \sigma_1^2\right) dG\left(m_0|\mu, K_0\sigma_0^2\right) \right\}
$$

$$
< \frac{\displaystyle\int_{\overline{m}_0}^{\infty} \left\{ \overline{\theta} \, F\left(\overline{\theta}|m_0, \sigma_1^2\right) + \overline{w}_2 \left[1 - F\left(\overline{\theta}|m_0, \sigma_1^2\right)\right] \right\} dG\left(m_0|\mu, K_0\sigma_0^2\right)}{\displaystyle\int_{\overline{m}_0}^{\infty} dG\left(m_0|\mu, K_0\sigma_0^2\right)}
$$

$$
< \overline{w}_2.
$$

It is quite intuitive that the mean wage of those workers who are in the second period of tenure must exceed the mean wage of all employed in the first period. The former group is a subset of the latter group where workers with low productivities, $\theta < \overline{\theta}$, have left. Since the mean wages are equal to the true average productivity in each group, it follows that $\overline{w}_2 > \overline{w}_1$.

The model thus implies that "wages rise with tenure," both in the sense that mean wages rise with tenure and in the sense that $\overline{\theta} > \overline{m}_0$, which asserts that the lower bound on second-period wages exceeds the lower bound on first-period wages. That wages rise with tenure was observation 1 that Jovanovic sought to explain.

Jovanovic's model also explains observation 2, that quits are negatively correlated with tenure. The model implies that quits occur between the first and second periods of tenure. Having decided to stay for two periods, the worker never quits.

The model also accounts for observation 3, namely, that the probability of a subsequent quit is negatively correlated with the current wage rate. The probability of a subsequent quit is given by

$$
\text{Prob}\{\theta' < \overline{\theta}|m_0\} = F\left(\overline{\theta}|m_0, \sigma_1^2\right),
$$

which is evidently negatively correlated with m_0, the first-period wage. Thus, the model explains each observation that Jovanovic sought to interpret. In the version of the model that we have studied, a worker eventually becomes permanently matched with probability 1. If we were studying a population of such workers of fixed size, all workers would eventually be absorbed into the state of being permanently matched. To provide a mechanism for replenishing the stock of unmatched workers, one could combine Jovanovic's model with the "firing" model in section 6.3.5. By letting matches θ "go bad" with probability λ each period, one could presumably modify Jovanovic's model to get the implication that, with a fixed population of workers, a fraction would remain unmatched each period because of the dissolution of previously acceptable matches.

6.9. A longer horizon version of Jovanovic's model

Here we consider a $T+1$ period version of Jovanovic's model, in which learning about the quality of the match continues for T periods before the quality of the match is revealed by "nature." (Jovanovic assumed that $T = \infty$.) We use the recursive projection technique (the Kalman filter) of chapter 2 to handle the firm's and worker's sequential learning. The prediction of the true match quality can then easily be updated with each additional noisy observation.

A firm-worker pair jointly draws a match parameter θ at the start of the match, which we call the beginning of period 0. The value θ is revealed to the pair only at the beginning of the $(T + 1)$th period of the match. After θ is drawn but before the match is consummated, the firm-worker pair observes $y_0 = \theta + u_0$, where u_0 is random noise. At the beginning of each period of the match, the worker-firm pair draws another noisy observation $y_t = \theta + u_t$ on the match parameter θ. The worker then decides whether or not to continue the match for the additional period. Let $y^t = \{y_0, \dots, y_t\}$ be the firm's and worker's information set at time t. We assume that θ and u_t are independently distributed random variables with $\theta \sim \mathcal{N}(\mu, \Sigma_0)$ and $u_t \sim \mathcal{N}(0, \sigma_u^2)$. For $t \geq 0$ define $m_t = E[\theta|y^t]$ and $m_{-1} = \mu$. The conditional means m_t and variances $E(\theta - m_t)^2 = \Sigma_{t+1}$ can be computed with the Kalman filter via the formulas from chapter 2:

$$m_t = (1 - K_t) m_{t-1} + K_t y_t \qquad (6.9.1a)$$

$$K_t = \frac{\Sigma_t}{\Sigma_t + R} \tag{6.9.1b}$$

$$\Sigma_{t+1} = \frac{\Sigma_t R}{\Sigma_t + R}, \tag{6.9.1c}$$

where $R = \sigma_u^2$ and Σ_0 is the unconditional variance of θ. The recursions are to be initiated from $m_{-1} = \mu$, and given Σ_0.

Using the formulas from chapter 2, we have that conditional on y^t, $m_{t+1} \sim \mathcal{N}(m_t, K_{t+1}\Sigma_{t+1})$ and $\theta \sim \mathcal{N}(m_t, \Sigma_{t+1})$ where Σ_0 is the unconditional variance of θ.

6.9.1. The Bellman equations

For $t \geq 0$, let $v_t(m_t)$ be the value of the worker's problem at the beginning of period t for a worker who optimally estimates that the match value is m_t after having observed y^t. At the start of period $T + 1$, we suppose that the value of the match is revealed without error. Thus, at time T, $\theta \sim \mathcal{N}(m_T, \Sigma_{T+1})$. The firm-worker pair estimates θ by m_t for $t = 0, \ldots, T$, and by θ for $t \geq T + 1$. Then the following functional equations characterize the solution of the problem:

$$v_{T+1}(\theta) = \max\left\{\frac{\theta}{1-\beta}, \beta Q\right\}, \tag{6.9.2}$$

$$v_T(m) = \max\left\{m + \beta \int v_{T+1}(\theta)\, dF(\theta \mid m, \Sigma_{T+1}), \beta Q\right\}, \tag{6.9.3}$$

$$v_t(m) = \max\left\{m + \beta \int v_{t+1}(m')\, dF(m' \mid m, K_{t+1}\Sigma_{t+1}), \beta Q\right\},$$
$$t = 0, \ldots, T-1, \tag{6.9.4}$$

$$Q = \int v_0(m)\, dF(m \mid \mu, K_0\Sigma_0), \tag{6.9.5}$$

with K_t and Σ_t from the Kalman filter. Starting from v_{T+1} and reasoning backward, it is evident that the worker's optimal policy is to set reservation wages $\overline{m}_t, t = 0, \ldots, T$ that satisfy

$$\overline{m}_{T+1} = \overline{\theta} = \beta(1-\beta)Q,$$
$$\overline{m}_T + \beta \int v_{T+1}(\theta)\, dF(\theta \mid \overline{m}_T, \Sigma_{T+1}) = \beta Q, \tag{6.9.6}$$
$$\overline{m}_t + \beta \int v_{t+1}(m')\, dF(m' \mid \overline{m}_t, K_{t+1}\Sigma_{t+1}) = \beta Q, \quad t = 1, \ldots, T-1.$$

To compute a solution to the worker's problem, we can define a mapping from Q into itself, with the property that a fixed point of the mapping is the optimal value of Q. Here is an algorithm:

a. Guess a value of Q, say Q^i with $i = 1$.

b. Given Q^i, compute sequentially the value functions in equations (6.9.2) through (6.9.4). Let the solutions be denoted $v^i_{T+1}(\theta)$ and $v^i_t(m)$ for $t = 0, \ldots, T$.

c. Given $v^i_1(m)$, evaluate equation (6.9.5) and call the solution $\tilde{Q}^i$.

d. For a fixed "relaxation parameter" $g \in (0,1)$, compute a new guess of Q from

$$Q^{i+1} = gQ^i + (1-g)\,\tilde{Q}^i\,.$$

e. Iterate on this scheme to convergence.

We now turn to the case where the true θ is never revealed by nature, that is, $T = \infty$. Note that $(\Sigma_{t+1})^{-1} = (\sigma_u^2)^{-1} + (\Sigma_t)^{-1}$, so $\Sigma_{t+1} < \Sigma_t$ and $\Sigma_{t+1} \to 0$ as $t \to \infty$. In other words, the accuracy of the prediction of θ becomes arbitrarily good as the information set y^t becomes large. Consequently, the firm and worker eventually learn the true θ, and the value function "at infinity" becomes

$$v_\infty(\theta) = \max\left\{\frac{\theta}{1-\beta}, \beta Q\right\},$$

and the Bellman equation for any finite tenure t is given by equation (6.9.4), and Q in equation (6.9.5) is the value of an unemployed worker. The optimal policy is a reservation wage $\overline{m}_t$, one for each tenure t. In fact, in the absence of a final date $T + 1$ when θ is revealed by nature, the solution is actually a time-invariant policy function $\overline{m}(\sigma_t^2)$ with an acceptance and a rejection region in the space of (m, σ^2).

To compute a numerical solution when $T = \infty$, we would still have to rely on the procedure that we have outlined based on the assumption of some finite date when the true θ is revealed, say in period $\hat{T} + 1$. The idea is to choose a sufficiently large $\hat{T}$ so that the conditional variance of θ at time $\hat{T}$, $\sigma_{\hat{T}}^2$, is close to zero. We then examine the approximation that $\sigma_{\hat{T}+1}^2$ is equal to zero. That is, equations (6.9.2) and (6.9.3) are used to truncate an otherwise infinite series of value functions.

6.10. Concluding remarks

The situations analyzed in this chapter are ones in which a currently unemployed worker rationally chooses to refuse an offer to work, preferring to remain unemployed today in exchange for better prospects tomorrow. The worker is voluntarily unemployed in one sense, having chosen to reject the current draw from the distribution of offers. In this model, the activity of unemployment is an investment incurred to improve the situation faced in the future. A theory in which unemployment is voluntary permits an analysis of the forces impinging on the choice to remain unemployed. Thus we can study the response of the worker's decision rule to changes in the distribution of offers, the rate of unemployment compensation, the number of offers per period, and so on.

Chapter 22 studies the optimal design of unemployment compensation. That issue is a trivial one in the present chapter with risk-neutral agents and no externalities. Here the government should avoid any policy that affects the workers' decision rules since it would harm efficiency, and the first-best way of pursuing distributional goals is through lump-sum transfers. In contrast, chapter 22 assumes risk-averse agents and incomplete insurance markets, which together with information asymmetries, make for an intricate contract design problem in the provision of unemployment insurance.

Chapter 28 presents various equilibrium models of search and matching. We study workers searching for jobs in an island model, workers and firms forming matches in a model with a "matching function," and how a medium of exchange can overcome the problem of "double coincidence of wants" in a search model of money.

A. More numerical dynamic programming

This appendix describes two more examples using the numerical methods of chapter 4.

6.A.1. Example 4: search

An unemployed worker wants to maximize $E_0 \sum_{t=0}^{\infty} \beta^t y_t$ where $y_t = w$ if the worker is employed at wage w, $y_t = 0$ if the worker is unemployed, and $\beta \in (0,1)$. Each period an unemployed worker draws a positive wage from a discrete-state Markov chain with transition matrix P. Thus, wage offers evolve according to a Markov process with transition probabilities given by

$$P(i,j) = \text{Prob}\left(w_{t+1} = \tilde{w}_j | w_t = \tilde{w}_i\right).$$

Once he accepts an offer, the worker works forever at the accepted wage. There is no firing or quitting. Let v be an $(n \times 1)$ vector of values v_i representing the optimal value of the problem for a worker who has offer $w_i, i = 1, \ldots, n$ in hand and who behaves optimally. The Bellman equation is

$$v_i = \max_{\text{accept,reject}} \left\{ \frac{w_i}{1-\beta}, \beta \sum_{j=1}^{n} P_{ij} v_j \right\}$$

or

$$v = \max\{\tilde{w}/(1-\beta), \beta P v\}.$$

Here $\tilde{w}$ is an $(n \times 1)$ vector of possible wage values. This matrix equation can be solved using the numerical procedures described earlier. The optimal policy depends on the structure of the Markov chain P. Under restrictions on P making w positively serially correlated, the optimal policy has the following reservation wage form: there is a $\overline{w}$ such that the worker should accept an offer w if $w \geq \overline{w}$.

6.A.2. Example 5: a Jovanovic model

Here is a simplified version of the search model of Jovanovic (1979a). A newly unemployed worker draws a job offer from a distribution given by $\mu_i = \text{Prob}(w_1 = \tilde{w}_i)$, where w_1 is the first-period wage. Let μ be the $(n \times 1)$ vector with ith component μ_i. After an offer is drawn, subsequent wages associated with the job evolve according to a Markov chain with time-varying transition matrices

$$P_t(i,j) = \text{Prob}(w_{t+1} = \tilde{w}_j | w_t = \tilde{w}_i),$$

for $t = 1, \ldots, T$. We assume that for times $t > T$, the transition matrices $P_t = I$, so that after T a job's wage does not change anymore with the passage of time. We specify the P_t matrices to capture the idea that the worker-firm pair is learning more about the quality of the match with the passage of time. For example, we might set

$$P_t = \begin{bmatrix} 1 - q^t & q^t & 0 & 0 & \cdots & 0 & 0 \\ q^t & 1 - 2q^t & q^t & 0 & \cdots & 0 & 0 \\ 0 & q^t & 1 - 2q^t & q^t & \cdots & 0 & 0 \\ \vdots & \vdots & \vdots & \vdots & \vdots & \vdots & \vdots \\ 0 & 0 & 0 & 0 & \cdots & 1 - 2q^t & q^t \\ 0 & 0 & 0 & 0 & \cdots & q^t & 1 - q^t \end{bmatrix},$$

where $q \in (0,1)$. In the following numerical examples, we use a slightly more general form of transition matrix in which (except at endpoints of the distribution),

$$\text{Prob}(w_{t+1} = \tilde{w}_{k \pm m} | w_t = \tilde{w}_k) = P_t(k, k \pm m) = q^t$$
$$P_t(k,k) = 1 - 2q^t. \tag{6.A.1}$$

Here $m \geq 1$ is a parameter that indexes the spread of the distribution.

At the beginning of each period, a previously matched worker is exposed with probability $\lambda \in (0,1)$ to the event that the match dissolves. We then have a set of Bellman equations

$$v_t = \max\{\tilde{w} + \beta(1-\lambda)P_t v_{t+1} + \beta\lambda Q, \beta Q + \bar{c}\}, \tag{6.A.2a}$$

for $t = 1, \ldots, T$, and

$$v_{T+1} = \max\{\tilde{w} + \beta(1-\lambda)v_{T+1} + \beta\lambda Q, \beta Q + \bar{c}\}, \tag{6.A.2b}$$

$$Q = \mu' v_1 \otimes \mathbf{1}$$

$$\bar{c} = c \otimes \mathbf{1}$$

where $\otimes$ is the Kronecker product, and $\mathbf{1}$ is an $(n \times 1)$ vector of ones. These equations can be solved by using calculations of the kind described previously. The optimal policy is to set a sequence of reservation wages $\{\overline{w}_j\}_{j=1}^T$.

Wage distributions

We can use recursions to compute probability distributions of wages at tenures $1, 2, \ldots, n$. Let the reservation wage for tenure j be $\overline{w}_j \equiv \tilde{w}_{\rho(j)}$, where $\rho(j)$ is the index associated with the cutoff wage. For $i \geq \rho(1)$, define

$$\delta_1(i) = \text{Prob}\{w_1 = \tilde{w}_i \mid w_1 \geq \overline{w}_1\} = \frac{\mu_i}{\sum_{h=\rho(1)}^n \mu_h}.$$

Then

$$\gamma_2(j) = \text{Prob}\{w_2 = \tilde{w}_j \mid w_1 \geq \overline{w}_1\} = \sum_{i=\rho(1)}^n P_1(i,j)\,\delta_1(i).$$

For $i \geq \rho(2)$, define

$$\delta_2(i) = \text{Prob}\{w_2 = \tilde{w}_i \mid w_2 \geq \overline{w}_2 \cap w_1 \geq \overline{w}_1\}$$

or

$$\delta_2(i) = \frac{\gamma_2(i)}{\sum_{h=\rho(2)}^n \gamma_2(h)}.$$

Then

$$\gamma_3(j) = \text{Prob}\{w_3 = \tilde{w}_j \mid w_2 \geq \overline{w}_2 \cap w_1 \geq \overline{w}_1\} = \sum_{i=\rho(2)}^n P_2(i,j)\,\delta_2(i).$$

Next, for $i \geq \rho(3)$, define $\delta_3(i) = \text{Prob}\{w_3 = \tilde{w}_i \mid (w_3 \geq \overline{w}_3) \cap (w_2 \geq \overline{w}_2) \cap (w_1 \geq \overline{w}_1)\}$. Then

$$\delta_3(i) = \frac{\gamma_3(i)}{\sum_{h=\rho(3)}^n \gamma_3(h)}.$$

Continuing in this way, we can define the wage distributions $\delta_1(i), \delta_2(i), \delta_3(i), \ldots$. The mean wage at tenure k is given by

$$\sum_{i \geq \rho(k)} \tilde{w}_i \delta_k(i).$$

Separation probabilities

The probability of rejecting a first period offer is $Q(1) = \sum_{h < \rho(1)} \mu_h$. The probability of separating at the beginning of period $j \geq 2$ is $Q(j) = \sum_{h < \rho(j)} \gamma_j(h)$.

Numerical examples

Figures 6.A.1, 6.A.2, and 6.A.3 report some numerical results for three versions of this model. For all versions, we set $\beta = .95, c = 0, q = .5$, and $T + 1 = 21$. For all three examples, we used a wage grid with 60 equispaced points on the interval $[0, 10]$.

For the initial distribution μ we used the uniform distribution. We used a sequence of transition matrices of the form $(6.A.1)$, with a "gap" parameter of m. For the first example, we set $m = 6$ and $\lambda = 0$, while the second sets $m = 10$ and $\lambda = 0$ and third sets $m = 10$ and $\lambda = .1$.

Figure 6.A.1 shows the reservation wage falls as m increases from 6 to 10, and that it falls further when the probability of being fired λ rises from zero to .1. Figure 6.A.2 shows the same pattern for average wages. Figure 6.A.3 displays quit probabilities for the first two models. They fall with tenure, with shapes and heights that depend to some degree on m, λ.

Exercises

Exercise 6.1 **Being unemployed with a chance of an offer**

An unemployed worker samples wage offers on the following terms: each period, with probability ϕ, $1 > \phi > 0$, she receives no offer (we may regard this as a wage offer of zero forever). With probability $(1 - \phi)$ she receives an offer to work for w forever, where w is drawn from a cumulative distribution function $F(w)$. Assume that $F(0) = 0, F(B) = 1$ for some $B > 0$. Successive draws across periods are independently and identically distributed. The worker chooses a strategy to maximize

$$E \sum_{t=0}^{\infty} \beta^t y_t, \qquad \text{where} \quad 0 < \beta < 1,$$

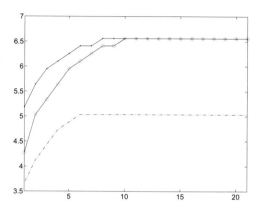

Figure 6.A.1: Reservation wages as a function of tenure for model with three different parameter settings $[m = 6, \lambda = 0]$ (the dots), $[m = 10, \lambda = 0]$ (the line with circles), and $[m = 10, \lambda = .1]$ (the dashed line).

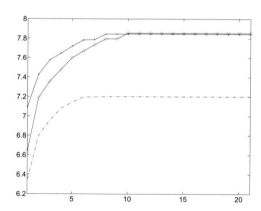

Figure 6.A.2: Mean wages as a function of tenure for model with three different parameter settings $[m = 6, \lambda = 0]$ (the dots), $[m = 10, \lambda = 0]$ (the line with circles), and $[m = 10, \lambda = .1]$ (the dashed line).

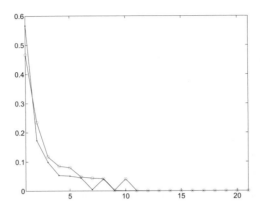

Figure 6.A.3: Quit probabilities as a function of tenure for Jovanovic model with $[m = 6, \lambda = 0]$ (line with dots) and $[m = 10, \lambda = .1]$ (the line with circles).

$y_t = w$ if the worker is employed, and $y_t = c$ if the worker is unemployed. Here c is unemployment compensation, and w is the wage at which the worker is employed. Assume that, having once accepted a job offer at wage w, the worker stays in the job forever.

Let $v(w)$ be the expected value of $\sum_{t=0}^{\infty} \beta^t y_t$ for an unemployed worker who has offer w in hand and who behaves optimally. Write the Bellman equation for the worker's problem.

Exercise 6.2   **Two offers per period**

Consider an unemployed worker who each period can draw *two* independently and identically distributed wage offers from the cumulative probability distribution function $F(w)$. The worker will work forever at the same wage after having once accepted an offer. In the event of unemployment during a period, the worker receives unemployment compensation c. The worker derives a decision rule to maximize $E \sum_{t=0}^{\infty} \beta^t y_t$, where $y_t = w$ or $y_t = c$, depending on whether she is employed or unemployed. Let $v(w)$ be the value of $E \sum_{t=0}^{\infty} \beta^t y_t$ for a currently unemployed worker who has best offer w in hand.

a. Formulate the Bellman equation for the worker's problem.

b. Prove that the worker's reservation wage is *higher* than it would be had the worker faced the same c and been drawing only *one* offer from the same distribution $F(w)$ each period.

Exercise 6.3 **A random number of offers per period**

An unemployed worker is confronted with a random number, n, of job offers each period. With probability π_n, the worker receives n offers in a given period, where $\pi_n \geq 0$ for $n \geq 1$, and $\sum_{n=1}^{N} \pi_n = 1$ for $N < +\infty$. Each offer is drawn independently from the same distribution $F(w)$. Assume that the number of offers n is independently distributed across time. The worker works forever at wage w after having accepted a job and receives unemployment compensation of c during each period of unemployment. He chooses a strategy to maximize $E \sum_{t=0}^{\infty} \beta^t y_t$ where $y_t = c$ if he is unemployed, $y_t = w$ if he is employed.

Let $v(w)$ be the value of the objective function of an unemployed worker who has best offer w in hand and who proceeds optimally. Formulate the Bellman equation for this worker.

Exercise 6.4 **Cyclical fluctuations in number of job offers**

Modify exercise *6.3* as follows: Let the number of job offers n follow a Markov process, with

$$\text{Prob}\{\text{Number of offers next period} = m | \text{Number of offers this period} = n\}$$
$$= \pi_{mn}, \qquad m = 1, \ldots, N, \quad n = 1, \ldots, N$$
$$\sum_{m=1}^{N} \pi_{mn} = 1 \qquad \text{for} \quad n = 1, \ldots, N.$$

Here $[\pi_{mn}]$ is a "stochastic matrix" generating a Markov chain. Keep all other features of the problem as in exercise *6.3*. The worker gets n offers per period, where n is now generated by a Markov chain so that the number of offers is possibly correlated over time.

a. Let $v(w, n)$ be the value of $E \sum_{t=0}^{\infty} \beta^t y_t$ for an unemployed worker who has received n offers this period, the best of which is w. Formulate the Bellman equation for the worker's problem.

b. Show that the optimal policy is to set a reservation wage $\overline{w}(n)$ that depends on the number of offers received this period.

Exercise 6.5 **Choosing the number of offers**

An unemployed worker must choose the number of offers n to solicit. At a cost of $k(n)$ the worker receives n offers this period. Here $k(n+1) > k(n)$ for $n \geq 1$. The number of offers n must be chosen in advance at the beginning of the period and cannot be revised during the period. The worker wants to maximize $E \sum_{t=0}^{\infty} \beta^t y_t$. Here y_t consists of w each period she is employed but not searching, $[w - k(n)]$ the first period she is employed but searches for n offers, and $[c - k(n)]$ each period she is unemployed but solicits and rejects n offers. The offers are each independently drawn from $F(w)$. The worker who accepts an offer works forever at wage w.

Let Q be the value of the problem for an unemployed worker who has not yet chosen the number of offers to solicit. Formulate the Bellman equation for this worker.

Exercise 6.6 **Mortensen externality**

Two parties to a match (say, worker and firm) jointly draw a match parameter θ from a c.d.f. $F(\theta)$. Once matched, they stay matched forever, each one deriving a benefit of θ per period from the match. Each unmatched pair of agents can influence the number of offers received in a period in the following way. The worker receives n offers per period, where $n = f(c_1 + c_2)$ and c_1 represents the resources the worker devotes to searching and c_2 represents the resources the typical firm devotes to searching. Symmetrically, the representative firm receives n offers per period where $n = f(c_1 + c_2)$. (We shall define the situation so that firms and workers have the same reservation θ so that there is never unrequited love.) Both c_1 and c_2 must be chosen at the beginning of the period, prior to searching during the period. Firms and workers have the same preferences, given by the expected present value of the match parameter θ, net of search costs. The discount factor β is the same for worker and firm.

a. Consider a Nash equilibrium in which party i chooses c_i, taking c_j, $j \neq i$, as given. Let Q_i be the value for an unmatched agent of type i before the level of c_i has been chosen. Formulate the Bellman equation for agents of types 1 and 2.

b. Consider the social planning problem of choosing c_1 and c_2 sequentially so as to maximize the criterion of λ times the utility of agent 1 plus $(1 - \lambda)$ times the utility of agent 2, $0 < \lambda < 1$. Let $Q(\lambda)$ be the value for this problem for two unmatched agents before c_1 and c_2 have been chosen. Formulate the Bellman equation for this problem.

c. Comparing the results in a and b, argue that, in the Nash equilibrium, the optimal amount of resources has not been devoted to search.

Exercise 6.7 **Variable labor supply**

An unemployed worker receives each period a wage offer w drawn from the distribution $F(w)$. The worker has to choose whether to accept the job— and therefore to work forever—or to search for another offer and collect c in unemployment compensation. The worker who decides to accept the job must choose the number of hours to work in each period. The worker chooses a strategy to maximize

$$E \sum_{t=0}^{\infty} \beta^t u\left(y_t, l_t\right), \qquad \text{where} \quad 0 < \beta < 1,$$

and $y_t = c$ if the worker is unemployed, and $y_t = w(1 - l_t)$ if the worker is employed and works $(1 - l_t)$ hours; l_t is leisure with $0 \le l_t \le 1$.

Analyze the worker's problem. Argue that the optimal strategy has the reservation wage property. Show that the number of hours worked is the same in every period.

Exercise 6.8 **Wage growth rate and the reservation wage**

An unemployed worker receives each period an offer to work for wage w_t forever, where $w_t = w$ in the first period and $w_t = \phi^t w$ after t periods on the job. Assume $\phi > 1$, that is, wages increase with tenure. The initial wage offer is drawn from a distribution $F(w)$ that is constant over time (entry-level wages are stationary); successive drawings across periods are independently and identically distributed.

The worker's objective function is to maximize

$$E \sum_{t=0}^{\infty} \beta^t y_t, \qquad \text{where} \quad 0 < \beta < 1,$$

and $y_t = w_t$ if the worker is employed and $y_t = c$ if the worker is unemployed, where c is unemployment compensation. Let $v(w)$ be the optimal value of the objective function for an unemployed worker who has offer w in hand. Write the Bellman equation for this problem. Argue that, if two economies differ only in the growth rate of wages of employed workers, say $\phi_1 > \phi_2$, the economy with the higher growth rate has the smaller reservation wage. *Note:* Assume that $\phi_i \beta < 1$, $i = 1, 2$.

Exercise 6.9   **Search with a finite horizon**

Consider a worker who lives two periods. In each period the worker, if unemployed, receives an offer of lifetime work at wage w, where w is drawn from a distribution F. Wage offers are identically and independently distributed over time. The worker's objective is to maximize $E\{y_1 + \beta y_2\}$, where $y_t = w$ if the worker is employed and is equal to c—unemployment compensation—if the worker is not employed.

Analyze the worker's optimal decision rule. In particular, establish that the optimal strategy is to choose a reservation wage in each period and to accept any offer with a wage at least as high as the reservation wage and to reject offers below that level. Show that the reservation wage decreases over time.

Exercise 6.10   **Finite horizon and mean-preserving spread**

Consider a worker who draws every period a job offer to work forever at wage w. Successive offers are independently and identically distributed drawings from a distribution $F_i(w)$, $i = 1, 2$. Assume that F_1 has been obtained from F_2 by a mean-preserving spread. The worker's objective is to maximize

$$E \sum_{t=0}^{T} \beta^t y_t, \qquad 0 < \beta < 1,$$

where $y_t = w$ if the worker has accepted employment at wage w and is zero otherwise. Assume that both distributions, F_1 and F_2, share a common upper bound, B.

a. Show that the reservation wages of workers drawing from F_1 and F_2 coincide at $t = T$ and $t = T - 1$.

b. Argue that for $t \leq T - 2$ the reservation wage of the workers that sample wage offers from the distribution F_1 is higher than the reservation wage of the workers that sample from F_2.

c. Now introduce unemployment compensation: the worker who is unemployed collects c dollars. Prove that the result in part a no longer holds; that is, the reservation wage of the workers that sample from F_1 is higher than the one corresponding to workers that sample from F_2 for $t = T - 1$.

Exercise 6.11 **Pissarides' analysis of taxation and variable search intensity**

An unemployed worker receives each period a zero offer (or no offer) with probability $[1 - \pi(e)]$. With probability $\pi(e)$ the worker draws an offer w from the distribution F. Here e stands for effort—a measure of search intensity—and $\pi(e)$ is increasing in e. A worker who accepts a job offer can be fired with probability α, $0 < \alpha < 1$. The worker chooses a strategy, that is, whether to accept an offer or not and how much effort to put into search when unemployed, to maximize

$$E \sum_{t=0}^{\infty} \beta^t y_t, \qquad 0 < \beta < 1,$$

where $y_t = w$ if the worker is employed with wage w and $y_t = 1 - e + z$ if the worker spends e units of leisure searching and does not accept a job. Here z is unemployment compensation. For the worker who searched and accepted a job, $y_t = w - e - T(w)$; that is, in the first period the wage is net of search costs. Throughout, $T(w)$ is the amount paid in taxes when the worker is employed. We assume that $w - T(w)$ is increasing in w. Assume that $w - T(w) = 0$ for $w = 0$, that if $e = 0$, then $\pi(e) = 0$—that is, the worker gets no offers—and that $\pi'(e) > 0$, $\pi''(e) < 0$.

a. Analyze the worker's problem. Establish that the optimal strategy is to choose a reservation wage. Display the condition that describes the optimal choice of e, and show that the reservation wage is independent of e.

b. Assume that $T(w) = t(w - a)$ where $0 < t < 1$ and $a > 0$. Show that an increase in a decreases the reservation wage and increases the level of effort, increasing the probability of accepting employment.

c. Show under what conditions a change in t has the opposite effect.

Exercise 6.12 **Search and financial income**

An unemployed worker receives every period an offer to work forever at wage w, where w is drawn from the distribution $F(w)$. Offers are independently and identically distributed. Every agent has another source of income, which we denote ϵ_t, that may be regarded as financial income. In every period all agents get a realization of ϵ_t, which is independently and identically distributed over time, with distribution function $G(\epsilon)$. We also assume that w_t and ϵ_t are independent. The objective of a worker is to maximize

$$E \sum_{t=0}^{\infty} \beta^t y_t, \qquad 0 < \beta < 1,$$

where $y_t = w + \phi \epsilon_t$ if the worker has accepted a job that pays w, and $y_t = c + \epsilon_t$ if the worker remains unemployed. We assume that $0 < \phi < 1$ to reflect the fact that an employed worker has less time to collect financial income. Assume $1 > \text{Prob}\{w \geq c + (1 - \phi)\epsilon\} > 0$.

Analyze the worker's problem. Write down the Bellman equation, and show that the reservation wage increases with the level of financial income.

Exercise 6.13   **Search and asset accumulation**

A previously unemployed worker receives an offer to work forever at wage w, but only if he chooses to do so, where w is drawn from the distribution $F(w)$. Previously employed workers receive no offers to work. But a previously employed worker is free to quit in any period, receive unemployment compensation that period, and so become a previously unemployed worker in the following period. Wage offers are identically and independently distributed over time. The worker maximizes

$$E \sum_{t=0}^{\infty} \beta^t \left(u\left(c_t\right) + v\left(l_t\right) \right), \qquad 0 < \beta < 1,$$

where c_t is consumption and l_t is leisure. Assume that $u(c)$ is strictly increasing, twice continuously differentiable, bounded, and strictly concave, while $v(l)$ is strictly increasing, twice continuously differentiable, and strictly concave; that $c_t \geq 0$; and that $l_t \in \{0, 1\}$, so that the person can either work full time (here $l_t = 0$) or not at all (here $l_t = 1$). A gross return on assets a_t held between t and $t+1$ is R_{t+1} and is i.i.d. with c.d.f. $H(R)$. The budget constraint is given by

$$a_{t+1} \leq R_{t+1} \left(a_t + w_t - c_t\right)$$

if the worker has a job that pays w_t. The random gross return R_{t+1} is observed at the beginning of period $t + 1$ before the worker chooses n_{t+1}, c_{t+1}. If the worker is unemployed, the budget constraint is $a_{t+1} \leq R_{t+1}(a_t + z - c_t)$ and

$l_t = 1$. Here z is unemployment compensation. It is assumed that a_t, the worker's asset position, cannot be negative. This is a no-borrowing assumption. Write a Bellman equation for this problem.

Exercise 6.14 **Temporary unemployment compensation**

Each period an unemployed worker draws one, and only one, offer to work forever at wage w. Wages are i.i.d. draws from the c.d.f. F, where $F(0) = 0$ and $F(B) = 1$. The worker seeks to maximize $E \sum_{t=0}^{\infty} \beta^t y_t$, where y_t is the sum of the worker's wage and unemployment compensation, if any. The worker is entitled to unemployment compensation in the amount $\gamma > 0$ only during the *first* period that she is unemployed. After one period on unemployment compensation, the worker receives none.

a. Write the Bellman equations for this problem. Prove that the worker's optimal policy is a time-varying reservation wage strategy.

b. Show how the worker's reservation wage varies with the duration of unemployment.

c. Show how the worker's "hazard of leaving unemployment" (i.e., the probability of accepting a job offer) varies with the duration of unemployment.

Now assume that the worker is also entitled to unemployment compensation if she quits a job. As before, the worker receives unemployment compensation in the amount of γ during the first period of an unemployment spell, and zero during the remaining part of an unemployment spell. (To qualify again for unemployment compensation, the worker must find a job and work for at least one period.)

The timing of events is as follows. At the very beginning of a period, a worker who was employed in the previous period must decide whether or not to quit. The decision is irreversible; that is, a quitter cannot return to an old job. If the worker quits, she draws a new wage offer as described previously, and if she accepts the offer she immediately starts earning that wage without suffering any period of unemployment.

d. Write the Bellman equations for this problem. *Hint*: At the very beginning of a period, let $v^e(w)$ denote the value of a worker who was employed in the previous period with wage w (before any wage draw in the current period). Let $v_1^u(w')$ be the value of an unemployed worker who has drawn wage offer

w' and who is entitled to unemployment compensation, if she rejects the offer. Similarly, let $v_+^u(w')$ be the value of an unemployed worker who has drawn wage offer w' but who is not eligible for unemployment compensation.

e. Characterize the three reservation wages, $\overline{w}^e$, $\overline{w}_1^u$, and $\overline{w}_+^u$, associated with the value functions in part d. How are they related to γ? (*Hint*: Two of the reservation wages are straightforward to characterize, while the remaining one depends on the actual parameterization of the model.)

Exercise 6.15 **Seasons, I**

An unemployed worker seeks to maximize $E \sum_{t=0}^{\infty} \beta^t y_t$, where $\beta \in (0,1)$, y_t is her income at time t, and E is the mathematical expectation operator. The person's income consists of one of two parts: unemployment compensation of c that she receives each period she remains unemployed, or a fixed wage w that the worker receives if employed. Once employed, the worker is employed forever with no chance of being fired. Every odd period (i.e., $t = 1, 3, 5, \ldots$) the worker receives one offer to work forever at a wage drawn from the c.d.f. $F(W) = \text{Prob}(w \leq W)$. Assume that $F(0) = 0$ and $F(B) = 1$ for some $B > 0$. Successive draws from F are independent. Every even period (i.e., $t = 0, 2, 4, \ldots$), the unemployed worker receives two offers to work forever at a wage drawn from F. Each of the two offers is drawn independently from F.

a. Formulate the Bellman equations for the unemployed person's problem.

b. Describe the form of the worker's optimal policy.

Exercise 6.16 **Seasons, II**

Consider the following problem confronting an unemployed worker. The worker wants to maximize

$$E_0 \sum_0^{\infty} \beta^t y_t, \quad \beta \in (0,1),$$

where $y_t = w_t$ in periods in which the worker is employed and $y_t = c$ in periods in which the worker is unemployed, where w_t is a wage rate and c is a constant level of unemployment compensation. At the start of each period, an unemployed worker receives one and only one offer to work at a wage w drawn from a c.d.f. $F(W)$, where $F(0) = 0, F(B) = 1$ for some $B > 0$. Successive draws from F are identically and independently distributed. There is no recall of past offers. Only unemployed workers receive wage offers. The wage is fixed as long as the worker remains in the job. The only way a worker can leave a job is if she is fired. At the *beginning* of each odd period ($t = 1, 3, \ldots$), a previously employed worker faces the probability of $\pi \in (0, 1)$ of being fired. If a worker is fired, she immediately receives a new draw of an offer to work at wage w. At each even period ($t = 0, 2, \ldots$), there is no chance of being fired.

a. Formulate a Bellman equation for the worker's problem.

b. Describe the form of the worker's optimal policy.

Exercise 6.17 **Gittins indexes for beginners**

At the end of each period, [19] a worker can switch between two jobs, A and B, to begin the following period at a wage that will be drawn at the beginning of next period from a wage distribution specific to job A or B, and to the worker's history of past wage draws from jobs of either type A or type B. The worker must decide to stay or leave a job at the end of a period after his wage for this period on his current job has been received, but before knowing what his wage would be next period in either job. The wage at either job is described by a job-specific n-state Markov chain. Each period the worker works at either job A or job B. At the end of the period, before observing next period's wage on either job, he chooses which job to go to next period. We use lowercase letters ($i, j = 1, \ldots, n$) to denote states for job A, and uppercase letters ($I, J = 1, \ldots n$) for job B. There is no option of being unemployed.

Let $w_a(i)$ be the wage on job A when state i occurs and $w_b(I)$ be the wage on job B when state I occurs. Let $A = [A_{ij}]$ be the matrix of one-step transition probabilities between the states on job A, and let $B = [B_{ij}]$ be the matrix for job B. If the worker leaves a job and later decides to return to it, he draws the wage for his first new period on the job from the conditional distribution determined by his last wage working at that job.

The worker's objective is to maximize the expected discounted value of his lifetime earnings, $E_0 \sum_{t=0}^{\infty} \beta^t y_t$, where $\beta \in (0, 1)$ is the discount factor, and where y_t is his wage from whichever job he is working at in period t.

a. Consider a worker who has worked at both jobs before. Suppose that $w_a(i)$ was the last wage the worker receives on job A and $w_b(I)$ the last wage on job B. Write the Bellman equation for the worker.

---

[19] See Gittins (1989) for more general versions of this problem.

b. Suppose that the worker is just entering the labor force. The first time he works at job A, the probability distribution for his initial wage is $\pi_a = (\pi_{a1}, \ldots, \pi_{an})$. Similarly, the probability distribution for his initial wage on job B is $\pi_b = (\pi_{b1}, \ldots, \pi_{bn})$ Formulate the decision problem for a new worker, who must decide which job to take initially. *Hint*: Let $v_a(i)$ be the expected discounted present value of lifetime earnings for a worker who was last in state i on job A and has never worked on job B; define $v_b(I)$ symmetrically.

Exercise 6.18   **Jovanovic (1979b)**

An employed worker in the tth period of tenure on the current job receives a wage $w_t = x_t(1 - \phi_t - s_t)$ where x_t is job-specific human capital, $\phi_t \in (0, 1)$ is the fraction of time that the worker spends investing in job-specific human capital, and $s_t \in (0, 1)$ is the fraction of time that the worker spends searching for a new job offer. If the worker devotes s_t to searching at t, then with probability $\pi(s_t) \in (0, 1)$ at the beginning of $t + 1$ the worker receives a new job offer to begin working at new job-specific capital level μ' drawn from the c.d.f. $F(\cdot)$. That is, searching for a new job offer promises the prospect of instantaneously reinitializing job-specific human capital at μ'. Assume that $\pi'(s) > 0, \pi''(s) < 0$. While on a given job, job-specific human capital evolves according to

$$x_{t+1} = G(x_t, \phi_t) = g(x_t \phi_t) - \delta x_t,$$

where $g'(\cdot) > 0, g''(\cdot) < 0$, $\delta \in (0, 1)$ is a depreciation rate, and $x_0 = \mu$ where t is tenure on the job, and μ is the value of the "match" parameter drawn at the start of the current job. The worker is risk neutral and seeks to maximize $E_0 \sum_{\tau=0}^{\infty} \beta^\tau y_\tau$, where y_τ is his wage in period τ.

a. Formulate the worker's Bellman equation.

b. Describe the worker's decision rule for deciding whether to accept an offer μ' at the beginning of next period.

c. Assume that $g(x\phi) = A(x\phi)^\alpha$ for $A > 0, \alpha \in (0, 1)$. Assume that $\pi(s) = s^{.5}$. Assume that F is a discrete n-valued distribution with probabilities f_i; for example, let $f_i = n^{-1}$. Write a Matlab program to solve the Bellman equation. Compute the optimal policies for ϕ, s and display them.

Exercise 6.19 **Value function iteration and policy improvement algorithm**, donated by Pierre-Olivier Weill

The goal of this exercise is to study, in the context of a specific problem, two methods for solving dynamic programs: value function iteration and Howard's policy improvement. Consider McCall's model of intertemporal job search. An unemployed worker draws one offer from a c.d.f. F, with $F(0) = 0$ and $F(B) = 1$, $B < \infty$. If the worker rejects the offer, she receives unemployment compensation c and can draw a new wage offer next period. If she accepts the offer, she works forever at wage w. The objective of the worker is to maximize the expected discounted value of her earnings. Her discount factor is $0 < \beta < 1$.

a. Write the Bellman equation. Show that the optimal policy is of the reservation wage form. Write an equation for the reservation wage w^*.

b. Consider the value function iteration method. Show that at each iteration, the optimal policy is of the reservation wage form. Let w_n be the reservation wage at iteration n. Derive a recursion for w_n. Show that w_n converges to w^* at rate β.

c. Consider Howard's policy improvement algorithm. Show that at each iteration, the optimal policy is of the reservation wage form. Let w_n be the reservation wage at iteration n. Derive a recursion for w_n. Show that the rate of convergence of w_n towards w^* is locally quadratic. Specifically use a Taylor expansion to show that, for w_n close enough to w^*, there is a constant K such that $w_{n+1} - w^* \cong K(w_n - w^*)^2$.

Exercise 6.20

Different types of unemployed workers are identical, except that they sample from different wage distributions. Each period an unemployed worker of type α draws a single new offer to work forever at a wage w from a cumulative distribution function F_α that satisfies $F_\alpha(w) = 0$ for $w < 0$, $F_\alpha(0) = \alpha$, $F_\alpha(B) = 1$, where $B > 0$ and F_α is a right continuous function mapping $[0, B]$ into $[0, 1]$. The c.d.f. of a type α worker is given by

$$
F_\alpha(w) = \begin{cases} \alpha & \text{for } 0 \le w \le \alpha B \text{ ;} \\ w/B & \text{for } \alpha B < w < B - \alpha B \text{ ;} \\ 1 - \alpha & \text{for } B - \alpha B \le w < B; \\ 1 & \text{for } w = B \end{cases}
$$

where $\alpha \in [0, .5)$. An unemployed α worker seeks to maximize the expected value of $\sum_{t=0}^{\infty} \beta^t y_t$, where $\beta \in (0, 1)$ and $y_t = w$ if the worker is employed and $y_t = c$ if he or she is unemployed, where $0 < c < B$ is a constant level of unemployment compensation. By choosing a strategy for accepting or rejecting job offers, the worker affects the distribution with respect to which the expected value of $\sum_{t=0}^{\infty} \beta^t y_t$ is calculated. The worker cannot recall past offers. If a previously unemployed worker accepts an offer to work at wage w this period, he must work forever at that wage (there is neither quitting nor firing nor searching for a new job while employed).

a. Formulate a Bellman equation for a type α worker. Prove that the worker's optimal strategy is to set a time-invariant reservation wage.

b. Consider two types of workers, $\alpha = 0$ and $\alpha = .3$. Can you tell which type of worker has a higher reservation wage?

c. Which type of worker would you expect to find a job more quickly?

Exercise 6.21   **Searching for the lowest price**

A buyer wants to purchase an item at the lowest price, net of total search costs. At a cost of $c > 0$ per draw, the buyer can draw an offer to buy the item at a price p that is drawn from the c.d.f. $F(P) = \text{Prob}(p \leq P)$ where P is a non-decreasing, right-continuous function with $F(\underline{B}) = 0, F(\overline{B}) = 1$, where $0 < \underline{B} < \overline{B} < +\infty$. All search occurs within one period.

a. Find the buyer's optimal strategy.

b. Find an expression for the expected value of the purchase price net of all search costs.

Exercise 6.22   **Quits**

Each period an unemployed worker draws one offer to work at a nonnegative wage w, where w is governed by a c.d.f F that satisfies $F(0) = 0$ and $F(B) = 1$ for some $B > 0$. The worker seeks to maximize the expected value of $\sum_{t=0}^{\infty} \beta^t y_t$ where $y_t = w$ if the worker is employed and c if the worker is unemployed. At the beginning of each period a worker employed at wage w the previous period is exposed to a probability of $\alpha \in (0, 1)$ of having his job reclassified, which means that he will be given a new wage w' drawn from F. A reclassified worker has the option of working at wage w' until reclassified again, or quitting, receiving unemployment compensation of c this period, and drawing a new wage offer the next period.

a. Formulate the Bellman equation for an unemployed worker.

b. Describe the decision rule for a previously *unemployed* worker.

c. Describe the decision rule for quitting or staying for a previously *employed* worker.

d. Describe how to compute the probability that a previously employed worker will quit.

Exercise 6.23 **A career ladder**

Each period a previously unemployed worker draws one offer to work at a non-negative wage w, where w is governed by a c.d.f F that satisfies $F(0) = 0$ and $F(B) = 1$ for some $B > 0$. The worker seeks to maximize the expected value of $\sum_{t=0}^{\infty} \beta^t y_t$ where $\beta \in (0, 1)$ and $y_t = w$ if the worker is employed and c if the worker is unemployed. At the beginning of each period a worker employed at wage w the previous period is exposed to a probability of $\alpha \in (0, 1)$ of getting a promotion, which means that he will be given a new wage γw where $\gamma > 1$. This new wage will prevail until a next promotion.

a. Formulate a Bellman equation for a previously employed worker.

b. Formulate a Bellman equation for a previously unemployed worker.

c. Describe the decision rule for an unemployed worker.

d. Describe the decision rule for a previously employed worker.

Exercise 6.24 **Human capital**

A previously unemployed worker draws one offer to work at a wage wh, where h is his level of human capital and w is drawn from a c.d.f. F where $F(0) = 0, F(B) = 1$ for $B > 0$. The worker retains w, but not h, so long as he remains in his current job (or employment spell). The worker knows his current level of h before he draws w. Wage draws are independent over time. When employed, the worker's human capital h evolves according to a discrete state Markov chain on the space $[\bar{h}_1, \bar{h}_2, \ldots, \bar{h}_n]$ with transition density H^e where $H^e(i, j) = \text{Prob}[h_{t+1} = \bar{h}_j | h_t = \bar{h}_i]$. When unemployed, the worker's human capital h evolves according to a discrete state Markov chain on the same space $[\bar{h}_1, \bar{h}_2, \ldots, \bar{h}_n]$ with transition density H^u where $H^u(i,j) = \text{Prob}[h_{t+1} = \bar{h}_j | h_t = \bar{h}_i]$. The two transition matrices H^e and H^u are such that an employed worker's human capital grows probabilistically (meaning that it is more likely that next period's human capital will be higher than this period's) and an unemployed worker's human capital decays probabilistically (meaning that it is more likely that next period's human capital will be lower than this period's). An unemployed worker receives unemployment compensation of c per period. The worker wants to maximize the expected value of $\sum_{t=0}^{\infty} \beta^t y_t$ where $y_t = w h_t$ when employed, and c when unemployed. At the beginning of each period, employed workers receive their new human capital realization from the Markov chain H^e. Then they are free to quit, meaning that they surrender their previous w, retain their newly realized level of human capital but immediately become unemployed, and can immediately draw a new w from F. They can accept that new draw immediately or else choose to be unemployed for at least one period while waiting for new opportunities to draw one w offer per period from F.

a. Obtain a Bellman equation or Bellman equations for the worker's problem.

b. Describe qualitatively the worker's optimal decision rule. Do you think employed workers might ever decide to quit?

c. Describe an algorithm to solve the Bellman equation or equations.

Exercise 6.25   **Markov wages**

Each period, a previously unemployed worker draws one offer to work forever at wage w. The worker wants to maximize $E \sum_{t=0}^{\infty} \beta^t y_t$, where $\beta \in (0,1)$ and $y_t = c > 0$ if the worker is unemployed, and $y_t = w$ if the worker is employed. Quitting is not allowed and once hired the worker cannot be fired. Successive draws of the wage are from a Markov chain with transition probabilities arranged in the $n \times n$ transition matrix P with (i,j) element $P_{ij} = \text{Prob}(w_{t+1} = \overline{w}_j | w_t = \overline{w}_i)$ where $\overline{w}_1 < \overline{w}_2 < \cdots < \overline{w}_n$.

a. Construct a Bellman equation for the worker.

b. Can you prove that the worker's optimal strategy is to set a reservation wage?

c. Assume that $\beta = .95$, $c = 1$, $\begin{bmatrix} \overline{w}_1 & \overline{w}_2 & \cdots & \overline{w}_n \end{bmatrix} = \begin{bmatrix} 1 & 2 & 3 & 4 & 5 \end{bmatrix}$ and

$$P = \begin{bmatrix} .8 & .2 & 0 & 0 & 0 \\ .18 & .8 & .02 & 0 & 0 \\ .25 & .25 & 0 & .25 & .25 \\ 0 & 0 & .02 & .8 & .18 \\ 0 & 0 & 0 & .2 & .8 \end{bmatrix}.$$

Please write a Matlab or R or C++ program to solve the Bellman equation. Show the optimal policy function and the value function.

d. Assume that all parameters are the same as in part **c** except for β, which now equals .99. Please find the optimal policy function and the optimal value function.

e. Please discuss whether, why, and how your answers to parts **c** and **d** differ.

Exercise 6.26 **Neal model with unemployment**

Consider the following modification of the Neal (1999) model. A worker chooses career-job (θ, ϵ) pairs subject to the following conditions. If employed, the worker's earnings at time t equal $\theta_t + \epsilon_t$, where θ_t is a component specific to a *career* and ϵ_t is a component specific to a particular *job*. If unemployed, the worker receives unemployment compensation equal to c. The worker maximizes $E \sum_{t=0}^{\infty} \beta^t y_t$ where $y_t = (\theta_t + \epsilon_t)$ if the worker is employed and $y_t = c$ if the worker is unemployed. A *career* is a draw of θ from c.d.f. F; a *job* is a draw of ϵ from c.d.f. G. Successive draws are independent, and $G(0) = F(0) = 0$, $G(B_\epsilon) = F(B_\theta) = 1$. The worker can draw a new career only if he also draws a new job. However, the worker is free to retain his existing career (θ), and to draw a new job (ϵ') next period. The worker decides at the beginning of a period whether to stay in a career-job pair inherited from the past, stay in the inherited career but draw a new job *for next period*, or draw a new career-job pair (θ', ϵ') *for next period*. If the worker decides to draw either a new θ' or a new ϵ' for next period, he or she must become unemployed this period.

a. Let $v(\theta, \epsilon)$ be the optimal value of the problem at the beginning of a period for a worker currently having inherited career-job pair (θ, ϵ) and who is about to decide whether to decide whether to become unemployed in order to draw a new career and or job next period. Formulate a Bellman equation.

b. Characterize the worker's optimal policy.

Part III
Competitive equilibria and applications

Chapter 7
Recursive (Partial) Equilibrium

7.1. An equilibrium concept

This chapter formulates competitive and oligopolistic equilibria in some dynamic settings. Up to now, we have studied single-agent problems where components of the state vector not under the control of the agent were taken as given. In this chapter, we describe multiple-agent settings in which components of the state vector that one agent takes as exogenous are determined by the decisions of other agents. We study partial equilibrium models of a kind applied in microeconomics.[1] We describe two closely related equilibrium concepts for such models: a rational expectations or recursive competitive equilibrium, and a Markov perfect equilibrium. The first equilibrium concept jointly restricts a Bellman equation and a transition law that is taken as given in that Bellman equation. The second equilibrium concept leads to pairs (in the duopoly case) or sets (in the oligopoly case) of Bellman equations and transition equations that are to be solved by simultaneous backward induction.

Though the equilibrium concepts introduced in this chapter transcend linear quadratic setups, we choose to present them in the context of linear quadratic examples because this renders the Bellman equations tractable.

---

[1] For example, see Rosen and Topel (1988) and Rosen, Murphy, and Scheinkman (1994)

7.2. Example: adjustment costs

This section describes a model of a competitive market with producers who face adjustment costs.[2] In the course of the exposition, we introduce and exploit a version of the 'big K, little k' trick that is widely used in macroeconomics and applied economic dynamics.[3] The model consists of n identical firms whose profit function makes them want to forecast the aggregate output decisions of other firms just like them in order to choose their own output. We assume that n is a large number so that the output of any single firm has a negligible effect on aggregate output and, hence, firms are justified in treating their forecast of aggregate output as unaffected by their own output decisions. Thus, one of n competitive firms sells output y_t and chooses a production plan to maximize

$$\sum_{t=0}^{\infty} \beta^t R_t \tag{7.2.1}$$

where

$$R_t = p_t y_t - .5d\left(y_{t+1} - y_t\right)^2 \tag{7.2.2}$$

subject to y_0 being a given initial condition. Here $\beta \in (0,1)$ is a discount factor, and $d > 0$ measures a cost of adjusting the rate of output. The firm is a price taker. The price p_t lies on the inverse demand curve

$$p_t = A_0 - A_1 Y_t \tag{7.2.3}$$

where $A_0 > 0, A_1 > 0$ and Y_t is the market-wide level of output, being the sum of output of n identical firms. The firm believes that market-wide output follows the law of motion

$$Y_{t+1} = H_0 + H_1 Y_t \equiv H\left(Y_t\right), \tag{7.2.4}$$

where Y_0 is a known initial condition. The belief parameters H_0, H_1 are equilibrium objects, but for now we proceed on faith and take them as given. The firm observes Y_t and y_t at time t when it chooses y_{t+1}. The adjustment cost $d(y_{t+1} - y_t)^2$ gives the firm the incentive to forecast the market price, but since

---

[2] The model is a version of one analyzed by Lucas and Prescott (1971) and Sargent (1987a). The recursive competitive equilibrium concept was used by Lucas and Prescott (1971) and described further by Prescott and Mehra (1980).

[3] Also see section 12.8 of chapter 12.

the market price is a function of market output Y_t via the demand equation (7.2.3), this in turn motivates the firm to forecast future values of Y. To state the firm's optimization problem completely requires that we specify laws of motion for all state variables, including ones like Y that it cares about but does not control. For this reason, the perceived law of motion (7.2.4) for Y is among the constraints that the firm faces.

Substituting equation (7.2.3) into equation (7.2.2) gives

$$R_t = (A_0 - A_1 Y_t) y_t - .5d (y_{t+1} - y_t)^2 .$$

The firm's incentive to forecast the market price translates into an incentive to forecast the level of market output Y. We can write the Bellman equation for the firm as

$$v (y, Y) = \max_{y'} \left\{ A_0 y - A_1 y Y - .5d (y' - y)^2 + \beta v (y', Y') \right\} \qquad (7.2.5)$$

where the maximization is subject to the perceived law of motion $Y' = H(Y)$. Here $'$ denotes next period's value of a variable. The Euler equation for the firm's problem is

$$-d (y' - y) + \beta v_y (y', Y') = 0. \qquad (7.2.6)$$

Noting that for this problem the control is y' and applying the Benveniste-Scheinkman formula from chapter 3 gives

$$v_y (y, Y) = A_0 - A_1 Y + d (y' - y) .$$

Substituting this equation into equation (7.2.6) gives

$$-d (y_{t+1} - y_t) + \beta [A_0 - A_1 Y_{t+1} + d (y_{t+2} - y_{t+1})] = 0. \qquad (7.2.7)$$

In the process of solving its Bellman equation, the firm sets an output path that satisfies equation (7.2.7), taking equation (7.2.4) as given, subject to the initial conditions (y_0, Y_0) as well as an extra terminal condition. The terminal condition is

$$\lim_{t \to \infty} \beta^t y_t v_y (y_t, Y_t) = 0. \qquad (7.2.8)$$

This is called the transversality condition and acts as a first-order necessary condition "at infinity." The firm's decision rule solves the difference equation (7.2.7) subject to the given initial condition y_0 and the terminal condition

(7.2.8). Solving Bellman equation (7.2.5) by backward induction automatically incorporates both equations (7.2.7) and (7.2.8).

The firm's optimal policy function is

$$y_{t+1} = h\left(y_t, Y_t\right). \tag{7.2.9}$$

Then with n identical firms, setting $Y_t = ny_t$ makes the actual law of motion for output for the market

$$Y_{t+1} = nh\left(Y_t/n, Y_t\right). \tag{7.2.10}$$

Thus, when firms believe that the law of motion for market-wide output is equation (7.2.4), their optimizing behavior makes the actual law of motion equation (7.2.10).

For this model, we adopt the following definition:

DEFINITION: A recursive competitive equilibrium[4] of the model with adjustment costs is a value function $v(y, Y)$, an optimal policy function $h(y, Y)$, and a law of motion $H(Y)$ such that

a. Given H, $v(y, Y)$ satisfies the firm's Bellman equation and $h(y, Y)$ is the optimal policy function.

b. The law of motion H satisfies $H(Y) = nh(Y/n, Y)$.

A recursive competitive equilibrium equates the actual and perceived laws of motion (7.2.4) and (7.2.10). The firm's optimum problem induces a mapping $\mathcal{M}$ from a perceived law of motion for output H to an actual law of motion $\mathcal{M}(H)$. The mapping is summarized in equation (7.2.10). The H component of a rational expectations equilibrium is a fixed point of the operator $\mathcal{M}$.

This is a special case of a recursive competitive equilibrium, to be defined more generally in section 7.3. How might we find an equilibrium? The mapping $\mathcal{M}$ is not a contraction and there is no guarantee that direct iterations on $\mathcal{M}$ will converge.[5] In fact, in many contexts, including the present one, there exist admissible parameter values for which divergence of iterations on $\mathcal{M}$ prevails.

---

[4] This is also often called a rational expectations equilibrium.

[5] A literature that studies whether models populated with agents who learn can converge to rational expectations equilibria features iterations on a modification of the mapping $\mathcal{M}$ that can be approximated as $\gamma\mathcal{M} + (1-\gamma)I$ where I is the identity operator and $\gamma \in (0,1)$ is a relaxation parameter. See Marcet and Sargent (1989) and Evans and Honkapohja (2001) for

The next subsection shows another method that works when the equilibrium solves an associated planning problem. For convenience, we'll assume from now on that the number of firms n is one, while retaining the assumption of price-taking behavior.

7.2.1. A planning problem

Our approach to computing an equilibrium is to seek to match the Euler equations of the market problem with those for a planning problem that can be posed as a single-agent dynamic programming problem. The optimal quantities from the planning problem are then the recursive competitive equilibrium quantities, and the equilibrium price is a shadow price in the planning problem.

For convenience we set $n = 1$. To construct a planning problem, we first compute the sum S_t of consumer and producer surplus at time t, defined as

$$S_t = S\left(Y_t, Y_{t+1}\right) = \int_0^{Y_t} \left(A_0 - A_1 x\right) dx - .5d\left(Y_{t+1} - Y_t\right)^2. \qquad (7.2.11)$$

The first term is the area under the demand curve. The planning problem is to choose a production plan to maximize

$$\sum_{t=0}^{\infty} \beta^t S\left(Y_t, Y_{t+1}\right) \qquad (7.2.12)$$

subject to an initial condition Y_0. The Bellman equation for the planning problem is

$$V(Y) = \max_{Y'} \left\{ A_0 Y - \frac{A_1}{2} Y^2 - .5d\left(Y' - Y\right)^2 + \beta V\left(Y'\right) \right\}. \qquad (7.2.13)$$

The Euler equation is

$$-d\left(Y' - Y\right) + \beta V'\left(Y'\right) = 0. \qquad (7.2.14)$$

---

statements and applications of this approach to establish conditions under which collections of adaptive agents who use least squares learning converge to a rational expectations equilibrium. The Marcet-Sargent-Evans-Honkapohja approach provides foundations for a method that Krusell and Smith (1998) use to approximation a rational expectations equilibrium of an incomplete-markets economy. See chapter 18.

Applying the Benveniste-Scheinkman formula gives

$$V'(Y) = A_0 - A_1 Y + d(Y' - Y). \tag{7.2.15}$$

Substituting this into equation (7.2.14) and rearranging gives

$$\beta A_0 + dY_t - [\beta A_1 + d(1 + \beta)] Y_{t+1} + d\beta Y_{t+2} = 0. \tag{7.2.16}$$

Return to equation (7.2.7) and set $y_t = Y_t$ for all t. (Remember that we have set $n = 1$. When $n \neq 1$ we have to adjust pieces of the argument for n.) Notice that with $y_t = Y_t$, equations (7.2.16) and (7.2.7) are identical. The Euler equation for the planning problem matches the second-order difference equation that we derived by first finding the Euler equation of the representative firm and substituting into it the expression $Y_t = ny_t$ that "makes the representative firm representative". Thus, if it is appropriate to apply the same terminal conditions for these two difference equations, which it is, then we have verified that a solution of the planning problem also is an equilibrium. Setting $y_t = Y_t$ in equation (7.2.7) amounts to dropping equation (7.2.4) and instead solving for the coefficients H_0, H_1 that make $y_t = Y_t$ true and that jointly solve equations (7.2.4) and (7.2.7).

It follows that for this example we can compute an equilibrium by forming the optimal linear regulator problem corresponding to the Bellman equation (7.2.13). The optimal policy function for this problem is the law of motion $Y' = H(Y)$ that a firm faces within a rational expectations equilibrium.[6]

---

[6] Lucas and Prescott (1971) used the method of this section. The method exploits the connection between equilibrium and Pareto optimality expressed in the fundamental theorems of welfare economics. See Mas-Colell, Whinston, and Green (1995).

7.3. Recursive competitive equilibrium

The equilibrium concept of the previous section is widely used. Following Prescott and Mehra (1980), it is useful to define the equilibrium concept more generally as a *recursive competitive equilibrium*. Let x be a vector of state variables under the control of a representative agent and let X be the vector of those same variables chosen by "the market." Let Z be a vector of other state variables chosen by "nature," that is, determined outside the model. The representative agent's problem is characterized by the Bellman equation

$$v\left(x, X, Z\right) = \max_{u}\{R\left(x, X, Z, u\right) + \beta v\left(x', X', Z'\right)\} \tag{7.3.1}$$

where $'$ denotes next period's value, and where the maximization is subject to the restrictions:

$$x' = g\left(x, X, Z, u\right) \tag{7.3.2}$$

$$X' = G\left(X, Z\right) \tag{7.3.3}$$

$$Z' = \zeta\left(Z\right). \tag{7.3.4}$$

Here g describes the impact of the representative agent's controls u on his state x'; G and ζ describe his beliefs about the evolution of the aggregate state. The solution of the representative agent's problem is a decision rule

$$u = h\left(x, X, Z\right). \tag{7.3.5}$$

To make the representative agent representative, we impose $X = x$, but only "after" we have solved the agent's decision problem. Substituting equation (7.3.5) and $X = x_t$ into equation (7.3.2) gives the *actual* law of motion

$$X' = G_A\left(X, Z\right), \tag{7.3.6}$$

where $G_A(X, Z) \equiv g[X, X, Z, h(X, X, Z)]$. We are now ready to propose a definition:

DEFINITION: A *recursive competitive equilibrium* is a policy function h, an actual aggregate law of motion G_A, and a perceived aggregate law G such that (a) given G, h solves the representative agent's optimization problem; and (b) h implies that $G_A = G$.

This equilibrium concept is also sometimes called a *rational expectations equilibrium*. The equilibrium concept makes G an outcome. The functions giving the representative agent's expectations about the aggregate state variables contribute no free parameters and are *outcomes* of the analysis. There are no free parameters that characterize expectations.[7]

7.4. Equilibrium human capital accumulation

As an example of a recursive competitive equilibrium, we formulate what we regard as a schooling model of the type used by Sherwin Rosen. A household chooses an amount of labor to send to a school that takes four periods to produce an educated worker. Time is a principal input into the schooling technology.

7.4.1. Planning problem

A planner chooses a contingency plan for new entrants n_t to maximize

$$E_0 \sum_{t=0}^{\infty} \beta^t \left\{ f_0 + (f_1 + \theta_t) N_t - \frac{f_2}{2} N_t^2 - \frac{d}{2} n_t^2 \right\}$$

subject to the laws of motion

$$\begin{aligned} \theta_{t+1} &= \rho \theta_t + \sigma_\theta \epsilon_{t+1} \\ N_{t+1} &= \delta N_t + n_{t-3}, \end{aligned} \tag{7.4.1}$$

where N_t is the stock of educated labor at time t, n_t is the number of new entrants into school at time t, $\delta \in (0,1)$ is one minus a depreciation rate, θ_t is a technology shock, and ϵ_{t+1} is an i.i.d. random process distributed as $\mathcal{N}(0,1)$. The planner confronts initial conditions $\theta_0, N_0, n_{-1}, n_{-2}, n_{-3}$. Notice how (7.4.1) incorporates a four period time to build stocks of labor. The planner's problem can be formulated as a stochastic discounted optimal linear regulator problem, i.e., a linear-quadratic dynamic programming problem of the type studied in chapter 5. We ask the reader to verify that it suffices to take $X_t = \begin{bmatrix} \theta_t \\ N_{t+3} \end{bmatrix}$ as the state for the planner's problem. A solution of the planner's

---

[7] This is the sense in which rational expectations models make expectations disappear.

problem is then a law of motion $X_{t+1} = (A - BF)X_t + C\epsilon_{t+1}$ and a decision rule $n_t = -FX_t$.

For the purpose of defining a recursive competitive equilibrium, it is useful to note that it is also possible to define the state for the planner's problem more profligately as $\tilde{X}_t = [\theta \quad N_t \quad n_{t-1} \quad n_{t-2} \quad n_{t-3}]'$ with associated decision rule $n_t = -\tilde{F}\tilde{X}_t$ and law of motion

$$\tilde{X}_{t+1} = \left(\tilde{A} - \tilde{B}\tilde{F}\right)\tilde{X}_t + \tilde{C}\epsilon_{t+1}. \tag{7.4.2}$$

We can use this representation to express a shadow wage $\tilde{w}_t = f_1 - f_2 N_t + \theta_t$ as $\tilde{w}_t = S_w \tilde{X}_t$.

7.4.2. Decentralization

A firm and a representative household are price takers in a recursive competitive equilibrium. The firm faces a competitive wage process $\{w_t\}_{t=0}^{\infty}$ as a price taker and chooses a contingency plan for $\{N_t\}_{t=0}^{\infty}$ to maximize

$$E_0 \sum_{t=0}^{\infty} \beta^t \left\{ f_0 + (f_1 + \theta_t) N_t - \frac{f_2}{2} N_t^2 - w_t N_t \right\}.$$

The first-order condition for the firm's problem is

$$w_t = f_1 - f_2 N_t + \theta_t, \tag{7.4.3}$$

which we can regard as an inverse demand function for the stock of labor.

A representative household chooses a contingency plan for $\{n_t, N_{t+4}\}_{t=0}^{\infty}$ to maximize

$$E_0 \sum_{t=0}^{\infty} \beta^t \left\{ w_t N_t - \frac{d}{2} n_t^2 \right\} \tag{7.4.4}$$

subject to (7.4.1) and initial conditions in the form of given values for N_t for $t = 0, 1, 2, 3$. To deduce first order conditions for this problem, it is helpful first to notice that (7.4.1) implies that for $j \geq 4$,

$$N_{t+j} = \delta^{j-3} N_{t+1} + \delta^{j-4} n_t + \delta^{j-3} n_{t+1} + \ldots \delta n_{t+j-5} + n_{t+j-4}, \tag{7.4.5}$$

so that

$$\frac{\partial \sum_{j=0}^{\infty} \beta^j w_{t+j} N_{t+j}}{\partial n_t} = \beta^4 \sum_{j=0}^{\infty} (\beta\delta)^j w_{t+j+4}.$$

It follows that the first-order conditions for maximizing $(7.4.4)$ subject to $(7.4.1)$ are

$$n_t = d^{-1} E_t \beta^4 \sum_{j=0}^{\infty} (\beta\delta)^j \, w_{t+j+4}, \quad t \geq 0 \qquad (7.4.6)$$

We can regard $(7.4.6)$ as a supply curve for a flow of new entrants into the schooling technology. It expresses the supply of new entrants into school n_t as a linear function of the expected present value of wages.

A rational expectations equilibrium is a stochastic process $\{w_t, N_t, n_t\}$ such that (a) given the w_t process, N_t, n_t solves the household's problem, and (b) given the w_t process, N_t solves the firms' problem. Evidently, a rational expectations equilibrium can also be characterized as a $\{w_t, N_t, n_t\}$ process that equates demand for labor (equation $(7.4.3)$) to supply of labor (equations $(7.4.5)$ and $(7.4.6)$).

To formulate the firm's and household's problems within a recursive competitive equilibrium, we can guess that the shadow wage $\tilde{w}_t$ mentioned above equals the competitive equilibrium wage. We can then confront the household with an exogenous wage governed by the stochastic process for w_t governed by the state space representation

$$\tilde{X}_{t+1} = \left(\tilde{A} - \tilde{B}\tilde{F} \right) \tilde{X}_t + \tilde{C}\epsilon_{t+1}$$
$$w_t = S_w \tilde{X}_t.$$

7.5. Equilibrium occupational choice

As another example of a recursive competitive equilibrium, we formulate a modification of a Rosen schooling model designed to focus on occupational choice.[8] Like the model in the previous section, this one focuses on the cost of acquiring human capital via a time-to-build technology. Investment times now differ across occupations.

Output of a single good is produced via the following production function:

$$Y_t = f_0 + f_1 \begin{bmatrix} U_t \\ S_t \end{bmatrix} - \begin{bmatrix} U_t \\ S_t \end{bmatrix}' f_2 \begin{bmatrix} U_t \\ S_t \end{bmatrix} \qquad (7.5.1)$$

---

[8] For applications see Siow (1984) and Ryoo and Rosen (2004).

where U_t is a stock of skilled labor and S_t is a stock of unskilled labor, and f_2 is a positive semi-definite matrix parameterizing whether skilled and unskilled labor are complements or substitutes in production. Stocks of the two types of labor evolve according to the laws of motion

$$U_{t+1} = \delta_U U_t + n_{Ut}$$
$$S_{t+1} = \delta_S S_t + n_{St-2}$$

$$(7.5.2)$$

where flows into the two types of skills are restricted by

$$n_{Ut} + n_{St} = n_t, \qquad (7.5.3)$$

where n_t is an exogenous flow of new entrants into the labor market governed by the stochastic process

$$n_{t+1} = \mu_n (1 - \rho) + \rho n_t + \sigma_n \epsilon_{t+1} \qquad (7.5.4)$$

where ϵ_{t+1} is an i.i.d. scalar stochastic process with time $t+1$ component distributed as $\mathcal{N}(0, 1)$. Equations $(7.5.2)$, $(7.5.3)$, $(7.5.4)$ express a time-to-build or schooling technology for converting new entrants n_t into increments in stocks of unskilled labor (this takes one period of waiting) and of skilled labor (this takes three periods of waiting). Stocks of skilled and unskilled labors depreciate, say through death or retirement, at the rates $(1 - \delta_S), (1 - \delta_U)$, respectively, where $\delta_S \in (0, 1)$ and $\delta_U \in (0, 1)$. In addition, we assume that there is an output cost of $\frac{e}{2} n_{st}^2$ associated with allocating new workers (or 'students') to the skilled worker pool.

7.5.1. A planning problem

Let's start with a planning problem, then construct a competitive equilibrium. Given initial conditions $(U_0, S_0, n_{S,-1}, n_{S,-2}, n_0)$, a planner chooses n_{St}, n_{Ut} to maximize

$$E_0 \sum_{t=0}^{\infty} \beta^t \left\{ f_0 + f_1 \begin{bmatrix} U_t \\ S_t \end{bmatrix} - .5 \begin{bmatrix} U_t \\ S_t \end{bmatrix}' f_2 \begin{bmatrix} U_t \\ S_t \end{bmatrix} - \frac{e}{2} n_{St}^2 \right\} \tag{7.5.5}$$

subject to (7.5.2), (7.5.3), (7.5.4). This is a stochastic discounted optimal linear regulator problem. Define the state as $X_t = \begin{bmatrix} U_t & S_t & 1 & n_{S,t-1} & n_{S,t-2} & n_t \end{bmatrix}$ and the control as n_{St}. An optimal decision rule has the form $n_{St} = -F X_t$ and the law of motion of the state under the optimal decision is

$$X_{t+1} = (A - BF) X_t + C \epsilon_{t+1}. \tag{7.5.6}$$

Define shadow wages

$$\begin{bmatrix} \tilde{w}_{Ut} \\ \tilde{w}_{St} \end{bmatrix} = f_1 - f_2 \begin{bmatrix} U_t \\ S_t \end{bmatrix} \equiv \begin{bmatrix} S_U \\ S_S \end{bmatrix} X_t, \tag{7.5.7}$$

where S_U and S_S are the appropriate selector vectors. The expected present value of entering school to become an unskilled worker is evidently

$$E_t \beta \sum_{j=1}^{\infty} (\beta \delta_U)^{j-1} \tilde{w}_{U,t+j} = \beta S_U \left(I - (A - BF) \beta \delta_U \right)^{-1} (A - BF) X_t$$

and the expected present value of entering school at t to become a skilled worker is

$$E_t \beta^3 \sum_{j=3}^{\infty} (\beta \delta_S)^{j-3} \tilde{w}_{S,t+j} = \beta^3 S_S \left(I - (A - BF) \beta \delta_S \right)^{-1} (A - BF)^3 X_t.$$

7.5.2. Decentralization

We can decentralize the planning problem by finding a recursive competitive equilibrium whose allocation matches that associated with the planning problem. A competitive firm hires stocks of skilled and unskilled workers at competitive wages w_{St}, w_{Ut} each period. Taking those wages as given, it chooses S_t, U_t to maximize

$$E_0 \sum_{t=0}^{\infty} \beta^t \left\{ f_0 + f_1 \begin{bmatrix} U_t \\ S_t \end{bmatrix} - .5 \begin{bmatrix} U_t \\ S_t \end{bmatrix}' f_2 \begin{bmatrix} U_t \\ S_t \end{bmatrix} - w_{Ut} U_t - w_{St} S_t \right\}. \tag{7.5.8}$$

Notice that the absence of intertemporal linkages in this problem makes it break into a sequence of static problems. The firm doesn't have to know the law of motion for wages. The firm equates the marginal products of each type of labor to that type's wage.

A representative family faces wages $\{w_{St}, w_{Ut}\}$ as a price taker and chooses contingency plans for $\{n_{St}, U_{t+1}, S_{t+1}\}_{t=0}^{\infty}$ to maximize

$$E_0 \sum_{t=0}^{\infty} \beta^t \left\{ w_{Ut} U_t + w_{st} S_t - \frac{e}{2} n_{St}^2 \right\} \tag{7.5.9}$$

subject to the perceived law of motion for w_{Ut}, w_{St}

$$\begin{bmatrix} w_{Ut} \\ w_{St} \end{bmatrix} = U_w \tilde{X}_t$$
$$\tilde{X}_{t+1} = \tilde{A} \tilde{X}_t + \tilde{C} \epsilon_{t+1} \tag{7.5.10}$$

and (7.5.2) and (7.5.3). According to (7.5.9), the family allocates n_t between n_{Ut} and n_{St} to maximize the expected present value of earnings from both types of labor, minus the present value of 'adjustment costs' $\frac{e}{2} n_{St}^2$. The state vector confronting the representative family is $[U_t \quad S_t \quad \tilde{X}_t]$ where $\tilde{X}_t$ has dimension comparable to X_t; $\tilde{A} X + C \varepsilon$ is a perceived law of motion for $\tilde{X}$. In a recursive competitive equilibrium, it will turn out that $\tilde{A} = A - BF$, where $A - BF$ is the optimal law of motion obtained from the planning problem.

In the spirit of Siow (1984) and Sherwin Rosen, it is interesting to focus on the special case in which $e = 0$. Here the competitive equilibrium features the outcome that

$$\beta E_t \sum_{j=1}^{\infty} (\beta \delta_U)^j w_{U,t+j} = E_t \beta^3 \sum_{j=3}^{\infty} (\beta \delta_S)^{j-3} w_{S,t+j}. \tag{7.5.11}$$

This condition says that the family allocates new entrants to equate the present values of earnings across occupations, a calculation that takes into account that it takes longer to train for some occupations than for others. The laws of motion of competitive equilibrium quantities adjust to equalize the present values of wages in the two occupations.

7.6. Markov perfect equilibrium

It is instructive to consider a dynamic model of duopoly. A market has two firms. Each firm recognizes that its output decision will affect the aggregate output and therefore influence the market price. Thus, we drop the assumption of price-taking behavior.[9] The one-period return function of firm i is

$$R_{it} = p_t y_{it} - .5d \left(y_{it+1} - y_{it} \right)^2. \tag{7.6.1}$$

There is a demand curve

$$p_t = A_0 - A_1 \left(y_{1t} + y_{2t} \right). \tag{7.6.2}$$

Substituting the demand curve into equation (7.6.1) lets us express the return as

$$R_{it} = A_0 y_{it} - A_1 y_{it}^2 - A_1 y_{it} y_{-i,t} - .5d \left(y_{it+1} - y_{it} \right)^2, \tag{7.6.3}$$

where $y_{-i,t}$ denotes the output of the firm other than i. Firm i chooses a decision rule that sets y_{it+1} as a function of $(y_{it}, y_{-i,t})$ and that maximizes

$$\sum_{t=0}^{\infty} \beta^t R_{it}.$$

Temporarily assume that the maximizing decision rule is $y_{it+1} = f_i(y_{it}, y_{-i,t})$. Given the function f_{-i}, the Bellman equation of firm i is

$$v_i \left(y_{it}, y_{-i,t} \right) = \max_{y_{it+1}} \left\{ R_{it} + \beta v_i \left(y_{it+1}, y_{-i,t+1} \right) \right\}, \tag{7.6.4}$$

---

[9] One consequence of departing from the price-taking framework is that the market outcome will no longer maximize welfare, measured as the sum of consumer and producer surplus. See exercise *7.4* for the case of a monopoly.

where the maximization is subject to the perceived decision rule of the other firm

$$y_{-i,t+1} = f_{-i}\left(y_{-i,t}, y_{it}\right). \tag{7.6.5}$$

Note the cross-reference between the two problems for $i = 1, 2$.

We now advance the following definition:

DEFINITION: A Markov perfect equilibrium is a pair of value functions v_i and a pair of policy functions f_i for $i = 1, 2$ such that

a. Given f_{-i}, v_i satisfies the Bellman equation (7.6.4).

b. The policy function f_i attains the right side of the Bellman equation (7.6.4).

The adjective Markov denotes that the equilibrium decision rules depend on the current values of the state variables y_{it} only, not other parts of their histories. Perfect means 'complete', i.e., that the equilibrium is constructed by backward induction and therefore builds in optimizing behavior for each firm for all possible future states, including many that will not be realized when we iterate forward on the pair of equilibrium strategies f_i.

7.6.1. Computation

If it exists, a Markov perfect equilibrium can be computed by iterating to convergence on the pair of Bellman equations (7.6.4). In particular, let v_i^j, f_i^j be the value function and policy function for firm i at the jth iteration. Then imagine constructing the iterates

$$v_i^{j+1}\left(y_{it}, y_{-i,t}\right) = \max_{y_{i,t+1}} \left\{ R_{it} + \beta v_i^j\left(y_{it+1}, y_{-i,t+1}\right) \right\}, \tag{7.6.6}$$

where the maximization is subject to

$$y_{-i,t+1} = f_{-i}^j\left(y_{-i,t}, y_{it}\right). \tag{7.6.7}$$

In general, these iterations are difficult.[10] In the next section, we describe how the calculations simplify for the case in which the return function is quadratic and the transition laws are linear.

---

[10] See Levhari and Mirman (1980) for how a Markov perfect equilibrium can be computed conveniently with logarithmic returns and Cobb-Douglas transition laws. Levhari and Mirman construct a model of fish and fishers.

7.7. Linear Markov perfect equilibria

In this section, we show how the optimal linear regulator can be used to solve a model like that in the previous section. That model should be considered to be an example of a dynamic game. A dynamic game consists of these objects: (a) a list of players; (b) a list of dates and actions available to each player at each date; and (c) payoffs for each player expressed as functions of the actions taken by all players.

The optimal linear regulator is a good tool for formulating and solving dynamic games. The standard equilibrium concept—subgame perfection—in these games requires that each player's strategy be computed by backward induction. This leads to an interrelated pair of Bellman equations. In linear quadratic dynamic games, these "stacked Bellman equations" become "stacked Riccati equations" with a tractable mathematical structure.

We now consider the following two-player, linear quadratic *dynamic game*. An $(n \times 1)$ state vector x_t evolves according to a transition equation

$$x_{t+1} = A_t x_t + B_{1t} u_{1t} + B_{2t} u_{2t} \tag{7.7.1}$$

where u_{jt} is a $(k_j \times 1)$ vector of controls of player j. We start with a finite horizon formulation, where t_0 is the initial date and t_1 is the terminal date for the common horizon of the two players. Player 1 maximizes

$$-\sum_{t=t_0}^{t_1-1} \left(x_t^T R_1 x_t + u_{1t}^T Q_1 u_{1t} + u_{2t}^T S_1 u_{2t} \right) \tag{7.7.2}$$

where R_1 and S_1 are positive semidefinite and Q_1 is positive definite. Player 2 maximizes

$$-\sum_{t=t_0}^{t_1-1} \left(x_t^T R_2 x_t + u_{2t}^T Q_2 u_{2t} + u_{1t}^T S_2 u_{1t} \right) \tag{7.7.3}$$

where R_2 and S_2 are positive semidefinite and Q_2 is positive definite.

We formulate a Markov perfect equilibrium as follows. Player j employs linear decision rules

$$u_{jt} = -F_{jt} x_t, \quad t = t_0, \ldots, t_1 - 1$$

where F_{jt} is a $(k_j \times n)$ matrix. Assume that player i knows $\{F_{-i,t}; t = t_0, \ldots, t_1 - 1\}$. Then player 1's problem is to maximize expression (7.7.2) subject to the known law of motion (7.7.1) *and* the known control law $u_{2t} = -F_{2t} x_t$

of player 2. Symmetrically, player 2's problem is to maximize expression $(7.7.3)$ subject to equation $(7.7.1)$ and $u_{1t} = -F_{1t}x_t$. A Markov perfect equilibrium is a pair of sequences $\{F_{1t}, F_{2t}; t = t_0, t_0 + 1, \ldots, t_1 - 1\}$ such that $\{F_{1t}\}$ solves player 1's problem, given $\{F_{2t}\}$, and $\{F_{2t}\}$ solves player 2's problem, given $\{F_{1t}\}$. We have restricted each player's strategy to depend only on x_t, and not on the *history* $h_t = \{(x_s, u_{1s}, u_{2s}), s = t_0, \ldots, t\}$. This restriction on strategy spaces accounts for the adjective "Markov" in the phrase "Markov perfect equilibrium."

Player 1's problem is to maximize

$$-\sum_{t=t_0}^{t_1-1} \left\{ x_t^T \left(R_1 + F_{2t}^T S_1 F_{2t}\right) x_t + u_{1t}^T Q_1 u_{1t} \right\}$$

subject to

$$x_{t+1} = (A_t - B_{2t}F_{2t}) x_t + B_{1t}u_{1t}.$$

This is an optimal linear regulator problem, and it can be solved by working backward. Evidently, player 2's problem is also an optimal linear regulator problem.

The solution of player 1's problem is given by

$$F_{1t} = \left(B_{1t}^T P_{1t+1} B_{1t} + Q_1\right)^{-1} B_{1t}^T P_{1t+1} \left(A_t - B_{2t} F_{2t}\right) \qquad (7.7.4)$$

$$t = t_0, t_0 + 1, \ldots, t_1 - 1$$

where P_{1t} is the solution of the following matrix Riccati difference equation with terminal condition $P_{1t_1} = 0$:

$$P_{1t} = (A_t - B_{2t}F_{2t})^T P_{1t+1} (A_t - B_{2t}F_{2t}) + \left(R_1 + F_{2t}^T S_1 F_{2t}\right)$$

$$- (A_t - B_{2t}F_{2t})^T P_{1t+1} B_{1t} \left(B_{1t}^T P_{1t+1} B_{1t} + Q_1\right)^{-1} B_{1t}^T P_{1t+1} (A_t - B_{2t}F_{2t}).$$

$$(7.7.5)$$

The solution of player 2's problem is

$$F_{2t} = \left(B_{2t}^T P_{2t+1} B_{2t} + Q_2\right)^{-1} B_{2t}^T P_{2t+1} \left(A_t - B_{1t} F_{1t}\right) \qquad (7.7.6)$$

where P_{2t} solves the following matrix Riccati difference equation, with terminal condition $P_{2t_1} = 0$:

$$P_{2t} = (A_t - B_{1t}F_{1t})^T P_{2t+1} (A_t - B_{1t}F_{1t}) + \left(R_2 + F_{1t}^T S_2 F_{1t}\right)$$

$$- (A_t - B_{1t}F_{1t})^T P_{2t+1} B_{2t} \qquad\qquad\qquad\qquad (7.7.7)$$

$$\left(B_{2t}^T P_{2t+1} B_{2t} + Q_2\right)^{-1} B_{2t}^T P_{2t+1} \left(A_t - B_{1t}F_{1t}\right).$$

The equilibrium sequences $\{F_{1t}, F_{2t}; t = t_0, t_0 + 1, \ldots, t_1 - 1\}$ can be calculated from the pair of coupled Riccati difference equations (7.7.5) and (7.7.7). In particular, we use equations (7.7.4), (7.7.5), (7.7.6), and (7.7.7) to "work backward" from time $t_1 - 1$. Notice that given P_{1t+1} and P_{2t+1}, equations (7.7.4) and (7.7.6) are a system of $(k_2 \times n) + (k_1 \times n)$ *linear* equations in the $(k_2 \times n) + (k_1 \times n)$ unknowns in the matrices F_{1t} and F_{2t}.

Notice how j's control law F_{jt} is a function of $\{F_{is}, s \geq t, i \neq j\}$. Thus, agent i's choice of $\{F_{it}; t = t_0, \ldots, t_1 - 1\}$ influences agent j's choice of control laws. However, in the Markov perfect equilibrium of this game, each agent is assumed to ignore the influence that his choice exerts on the other agent's choice.[11]

We often want to compute the solutions of such games for infinite horizons, in the hope that the decision rules F_{it} settle down to be time invariant as $t_1 \to +\infty$. In practice, we usually fix t_1 and compute the equilibrium of an infinite horizon game by driving $t_0 \to -\infty$. Judd followed that procedure in the following example.

7.7.1. An example

This section describes the Markov perfect equilibrium of an infinite horizon linear quadratic game proposed by Kenneth Judd (1990). The equilibrium is computed by iterating to convergence on the pair of Riccati equations defined by the choice problems of two firms. Each firm solves a linear quadratic optimization problem, taking as given and known the sequence of linear decision rules used by the other player. The firms set prices and quantities of two goods interrelated through their demand curves. There is no uncertainty. Relevant variables are defined as follows:

I_{it} = inventories of firm i at beginning of t.

q_{it} = production of firm i during period t.

p_{it} = price charged by firm i during period t.

S_{it} = sales made by firm i during period t.

E_{it} = costs of production of firm i during period t.

---

[11] In an equilibrium of a *Stackelberg* or *dominant player* game, the timing of moves is so altered relative to the present game that one of the agents called the *leader* takes into account the influence that his choices exert on the other agent's choices. See chapter 19.

C_{it} = costs of carrying inventories for firm i during t.

The firms' cost functions are

$$C_{it} = c_{i1} + c_{i2}I_{it} + .5c_{i3}I_{it}^2$$
$$E_{it} = e_{i1} + e_{i2}q_{it} + .5e_{i3}q_{it}^2$$

where e_{ij}, c_{ij} are positive scalars.

Inventories obey the laws of motion

$$I_{i,t+1} = (1 - \delta)\,I_{it} + q_{it} - S_{it}$$

Demand is governed by the linear schedule

$$S_t = dp_{it} + B$$

where $S_t = [\,S_{1t} \quad S_{2t}\,]'$, d is a (2×2) negative definite matrix, and B is a vector of constants. Firm i maximizes the undiscounted sum

$$\lim_{T \to \infty} \frac{1}{T} \sum_{t=0}^{T} (p_{it}S_{it} - E_{it} - C_{it})$$

by choosing a decision rule for price and quantity of the form

$$u_{it} = -F_i x_t$$

where $u_{it} = [\,p_{it} \quad q_{it}\,]'$, and the state is $x_t = [\,I_{1t} \quad I_{2t}\,]$.

In the web site for the book, we supply a Matlab program `nnash.m` that computes a Markov perfect equilibrium of the linear quadratic dynamic game in which player i maximizes

$$-\sum_{t=0}^{\infty}\{x_t'r_ix_t + 2x_t'w_iu_{it} + u_{it}'q_iu_{it} + u_{jt}'s_iu_{jt} + 2u_{jt}'m_iu_{it}\}$$

subject to the law of motion

$$x_{t+1} = ax_t + b_1u_{1t} + b_2u_{2t}$$

and a control law $u_{jt} = -f_j x_t$ for the other player; here a is $n \times n$; b_1 is $n \times k_1$; b_2 is $n \times k_2$; r_1 is $n \times n$; r_2 is $n \times n$; q_1 is $k_1 \times k_1$; q_2 is $k_2 \times k_2$; s_1 is $k_2 \times k_2$; s_2 is $k_1 \times k_1$; w_1 is $n \times k_1$; w_2 is $n \times k_2$; m_1 is $k_2 \times k_1$; and m_2 is $k_1 \times k_2$. The equilibrium of Judd's model can be computed by filling in the matrices appropriately. A Matlab tutorial `judd.m` uses `nnash.m` to compute the equilibrium.

7.8. Concluding remarks

This chapter has introduced two equilibrium concepts and illustrated how dynamic programming algorithms are embedded in each. For the linear models we have used as illustrations, the dynamic programs become optimal linear regulators, making it tractable to compute equilibria even for large state spaces. We chose to define these equilibria concepts in partial equilibrium settings that are more natural for microeconomic applications than for macroeconomic ones. In the next chapter, we use the recursive equilibrium concept to analyze a general equilibrium in an endowment economy. That setting serves as a natural starting point for addressing various macroeconomic issues.

Exercises

These problems aim to teach about (1) mapping problems into recursive forms, (2) different equilibrium concepts, and (3) using Matlab. Computer programs are available from the web site for the book.[12]

Exercise 7.1 **A competitive firm**

A competitive firm seeks to maximize

$$\sum_{t=0}^{\infty} \beta^t R_t \tag{1}$$

where $\beta \in (0,1)$, and time t revenue R_t is

$$R_t = p_t y_t - .5d\left(y_{t+1} - y_t\right)^2, \quad d > 0, \tag{2}$$

where p_t is the price of output, and y_t is the time t output of the firm. Here $.5d(y_{t+1} - y_t)^2$ measures the firm's cost of adjusting its rate of output. The firm starts with a given initial level of output y_0. The price lies on the market demand curve

$$p_t = A_0 - A_1 Y_t, A_0, A_1 > 0 \tag{3}$$

---

[12] The web site is $<$https://files.nyu.edu/ts43/public/books.html$>$.

where Y_t is the market level of output, which the firm takes as exogenous, and which the firm believes follows the law of motion

$$Y_{t+1} = H_0 + H_1 Y_t, \tag{4}$$

with Y_0 as a fixed initial condition.

a. Formulate the Bellman equation for the firm's problem.

b. Formulate the firm's problem as a discounted optimal linear regulator problem, being careful to describe all of the objects needed. What is the *state* for the firm's problem?

c. Use the Matlab program `olrp.m` to solve the firm's problem for the following parameter values: $A_0 = 100, A_1 = .05, \beta = .95, d = 10, H_0 = 95.5$, and $H_1 = .95$. Express the solution of the firm's problem in the form

$$y_{t+1} = h_0 + h_1 y_t + h_2 Y_t, \tag{5}$$

giving values for the h_j's.

d. If there were n identical competitive firms all behaving according to equation (5), what would equation (5) imply for the *actual* law of motion (4) for the market supply Y?

e. Formulate the Euler equation for the firm's problem.

Exercise 7.2 **Rational expectations**

Now assume that the firm in problem 1 is "representative." We implement this idea by setting $n = 1$. In equilibrium, we will require that $y_t = Y_t$, but we don't want to impose this condition at the stage that the firm is optimizing (because we want to retain competitive behavior). Define a rational expectations equilibrium to be a pair of numbers H_0, H_1 such that if the representative firm solves the problem ascribed to it in problem 1, then the firm's optimal behavior given by equation (5) implies that $y_t = Y_t \ \forall \ t \geq 0$.

a. Use the program that you wrote for exercise *7.1* to determine which if any of the following pairs (H_0, H_1) is a rational expectations equilibrium: (i) (94.0888, .9211); (ii) (93.22, .9433), and (iii) (95.08187459215024, .95245906270392)?

b. Describe an iterative algorithm that uses the program that you wrote for exercise *7.1* to compute a rational expectations equilibrium. (You are not being asked actually to use the algorithm you are suggesting.)

Exercise 7.3   **Maximizing welfare**

A planner seeks to maximize the welfare criterion

$$\sum_{t=0}^{\infty} \beta^t S_t, \tag{1}$$

where S_t is "consumer surplus plus producer surplus" defined to be

$$S_t = S\left(Y_t, Y_{t+1}\right) = \int_0^{Y_t} \left(A_0 - A_1 x\right) dx - .5d\left(Y_{t+1} - Y_t\right)^2.$$

a. Formulate the planner's Bellman equation.

b. Formulate the planner's problem as an optimal linear regulator, and, for the same parameter values in exercise 7.1, solve it using the Matlab program `olrp.m`. Represent the solution in the form $Y_{t+1} = s_0 + s_1 Y_t$.

c. Compare your answer in part b with your answer to part a of exercise 7.2.

Exercise 7.4   **Monopoly**

A monopolist faces the industry demand curve (3) and chooses Y_t to maximize $\sum_{t=0}^{\infty} \beta^t R_t$ where $R_t = p_t Y_t - .5d(Y_{t+1} - Y_t)^2$ and where Y_0 is given.

a. Formulate the firm's Bellman equation.

b. For the parameter values listed in exercise 7.1, formulate and solve the firm's problem using `olrp.m`.

c. Compare your answer in part b with the answer you obtained to part b of exercise 7.3.

Exercise 7.5   **Duopoly**

An industry consists of two firms that jointly face the industry-wide inverse demand curve $p_t = A_0 - A_1 Y_t$, where now $Y_t = y_{1t} + y_{2t}$. Firm $i = 1, 2$ maximizes

$$\sum_{t=0}^{\infty} \beta^t R_{it} \tag{1}$$

where $R_{it} = p_t y_{it} - .5d(y_{i,t+1} - y_{it})^2$.

a. Define a Markov perfect equilibrium for this industry.

b. Formulate the Bellman equation for each firm.

c. Use the Matlab program `nash.m` to compute an equilibrium, assuming the parameter values listed in exercise *7.1*.

Exercise 7.6 **Self-control**

This is a model of a human who has time inconsistent preferences, of a type proposed by Phelps and Pollak (1968) and used by Laibson (1994).[13] The human lives from $t = 0, \ldots, T$. Think of the human as actually consisting of $T+1$ personalities, one for each period. Each personality is a distinct agent (i.e., a distinct utility function and constraint set). Personality T has preferences ordered by $u(c_T)$ and personality $t < T$ has preferences that are ordered by

$$u\left(c_t\right) + \delta \sum_{j=1}^{T-t} \beta^j u\left(c_{t+j}\right),$$

where $u(\cdot)$ is a twice continuously differentiable, increasing, and strictly concave function of consumption of a single good; $\beta \in (0, 1)$, and $\delta \in (0, 1]$. When $\delta < 1$, preferences of the sequence of personalities are time inconsistent (that is, not recursive). At each t, let there be a savings technology described by

$$k_{t+1} + c_t \leq f\left(k_t\right),$$

where f is a production function with $f' > 0, f'' \leq 0$.

a. Define a Markov perfect equilibrium for the $T + 1$ personalities.

b. Argue that the Markov perfect equilibrium can be computed by iterating on the following functional equations:

$$V_{j+1}\left(k\right) = \max_c \left\{ u\left(c\right) + \beta\delta W_j\left(k'\right)\right\}$$
$$W_{j+1}\left(k\right) = u\left[c_{j+1}\left(k\right)\right] + \beta W_j\left[f\left(k\right) - c_{j+1}\left(k\right)\right]$$

where $c_{j+1}(k)$ is the maximizer of the right side of the first equation above for $j + 1$, starting from $W_0(k) = u[f(k)]$. Here $W_j(k)$ is the value of $u(c_{T-j}) + \beta u(c_{T-j+1}) + \ldots + \beta^{T-j}u(c_T)$, taking the decision rules $c_h(k)$ as given for $h = 0, 1, \ldots, j$.

---

[13] See Gul and Pesendorfer (2000) for a single-agent recursive representation of preferences exhibiting temptation and self-control.

c. State the optimization problem of the time 0 person who is given the power to dictate the choices of all subsequent persons. Write the Bellman equations for this problem. The time 0 person is said to have a commitment technology for "self-control" in this problem.

Exercise 7.7   **Equilibrium search**

An economy consists of a continuum of *ex ante* identical workers each of whom is either employed or unemployed. A worker wants to maximize the expected value of $\sum_{t=0}^{\infty} \beta^t y_t$ where $\beta \in (0, 1)$ and

$$y_t = \begin{cases} w & \text{if employed} \\ c(U) & \text{if unemployed .} \end{cases}$$

Each period, an unemployed worker draws one and only one offer to work (until fired) at a wage w drawn from a c.d.f. $F(W) = \text{Prob}(w \leq W)$ where $F(0) = 0, F(B) = 1$ for $B > 0$. Successive draws from F are i.i.d. If a worker accepts a job, he receives w this period and enters the beginning of next period as 'employed'. At the beginning of each period, each such previously employed worker is exposed to a probability of $\lambda \in (0, 1)$ of being fired; with probability $1 - \lambda$ he is not fired and again receives the previously drawn w as a wage. If fired, the worker becomes newly unemployed and has the same opportunity as all other unemployed workers, i.e., he draws an offer w from c.d.f. F. If an unemployed worker rejects that offer, he receives unemployment compensation $c(U) = c\left[\frac{1}{1+\exp(-6U)} - .5\right]$ and enters next period unemployed. Here U is the aggregate unemployment rate at the beginning of the period. The unemployment rate tomorrow U^* is related to the unemployment rate U today by the law of motion

$$U^* = \lambda (1 - U) + (1 - \phi(U)) U,$$

where $\phi(U)$ is the fraction of unemployed workers who accept a wage offer this period.

a. Write a Bellman equation for an unemployed worker.

b. Describe the form of an unemployed worker's optimal decision rule.

c. Describe how $\phi(U)$ is implied by a typical worker's optimal decision rule.

d. Define a recursive competitive equilibrium for this environment.

Chapter 8
Equilibrium with Complete Markets

8.1. Time 0 versus sequential trading

This chapter describes competitive equilibria of a pure exchange infinite horizon economy with stochastic endowments. These are useful for studying risk sharing, asset pricing, and consumption. We describe two systems of markets: an *Arrow-Debreu* structure with complete markets in dated contingent claims all traded at time 0, and a sequential-trading structure with complete one-period *Arrow securities*. These two entail different assets and timings of trades, but have identical consumption allocations. Both are referred to as complete markets economies. They allow more comprehensive sharing of risks than do the incomplete markets economies to be studied in chapters 17 and 18, or the economies with imperfect enforcement or imperfect information, studied in chapters 20 and 21.

8.2. The physical setting: preferences and endowments

In each period $t \geq 0$, there is a realization of a stochastic event $s_t \in S$. Let the history of events up and until time t be denoted $s^t = [s_0, s_1, \ldots, s_t]$. The unconditional probability of observing a particular sequence of events s^t is given by a probability measure $\pi_t(s^t)$. For $t > \tau$, we write the probability of observing s^t conditional on the realization of s^τ as $\pi_t(s^t|s^\tau)$. In this chapter, we shall assume that trading occurs after observing s_0, which we capture by setting $\pi_0(s_0) = 1$ for the initially given value of s_0.[1]

   In section 8.9 we shall follow much of the literatures in macroeconomics and econometrics and assume that $\pi_t(s^t)$ is induced by a Markov process. We wait to impose that special assumption until section 8.9 because some important findings do not require making that assumption.

---

[1] Most of our formulas carry over to the case where trading occurs before s_0 has been realized; just postulate a nondegenerate probability distribution $\pi_0(s_0)$ over the initial state.

There are I agents named $i = 1, \ldots, I$. Agent i owns a stochastic endowment of one good $y_t^i(s^t)$ that depends on the history s^t. The history s^t is publicly observable. Household i purchases a history-dependent consumption plan $c^i = \{c_t^i(s^t)\}_{t=0}^{\infty}$ and orders these consumption streams by [2]

$$U\left(c^i\right) = \sum_{t=0}^{\infty} \sum_{s^t} \beta^t u\left[c_t^i\left(s^t\right)\right] \pi_t\left(s^t\right), \tag{8.2.1}$$

where $0 < \beta < 1$. The right side is equal to $E_0 \sum_{t=0}^{\infty} \beta^t u(c_t^i)$, where E_0 is the mathematical expectation operator, conditioned on s_0. Here $u(c)$ is an increasing, twice continuously differentiable, strictly concave function of consumption $c \geq 0$ of one good. The utility function satisfies the Inada condition [3]

$$\lim_{c \downarrow 0} u'\left(c\right) = +\infty.$$

Notice that in assuming (8.2.1), we are imposing identical preference orderings across all individuals i that can be represented in terms of discounted expected utility with common β, common utility function $u(\cdot)$, and common probability distributions $\pi_t(s^t)$. As we proceed through this chapter, watch for results that would evaporate if we were instead to allow $\beta, u(\cdot)$, or $\pi_t(s^t)$ to depend on i.

A *feasible allocation* satisfies

$$\sum_i c_t^i\left(s^t\right) \leq \sum_i y_t^i\left(s^t\right) \tag{8.2.2}$$

for all t and for all s^t.

---

[2] Exercises 8.13 - 8.17 consider examples in which we replace (8.2.1) with

$$U^i\left(c^i\right) = \sum_{t=0}^{\infty} \sum_{s^t} \beta^t u\left[c_t^i\left(s^t\right)\right] \pi_t^i\left(s^t\right),$$

where $\pi^i(s^t)$ is a personal probability distribution specific to agent i. Blume and Easley (2006) study such settings, focusing particularly on which agents' beliefs ultimately influence the tails of allocations and prices. Throughout most of this chapter, we adopt the assumption, routinely employed in much of macroeconomics, that all agents share probabilities.

[3] One role of this Inada condition is to make the consumption of each agent strictly positive in every date-history pair. A related role is to deliver a state-by-state borrowing limit to impose in economies with sequential trading of Arrow securities.

8.3. Alternative trading arrangements

For a two-event stochastic process $s_t \in S = \{0, 1\}$, the trees in Figures 8.3.1 and 8.3.2 give two portraits of how histories s^t unfold. From the perspective of time 0 given $s_0 = 0$, Figure 8.3.1 portrays all prospective histories possible up to time 3. Figure 8.3.2 portrays a *particular* history that it is known the economy has indeed followed up to time 2, together with the two possible one-period continuations into period 3 that can occur after that history.

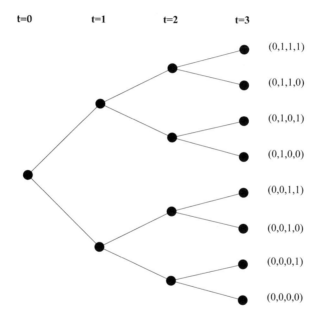

Figure 8.3.1: The Arrow-Debreu commodity space for a two-state Markov chain. At time 0, there are trades in time $t = 3$ goods for each of the eight nodes that signify histories that can possibly be reached starting from the node at time 0.

In this chapter we shall study two distinct trading arrangements that correspond, respectively, to the two views of the economy in Figures 8.3.1 and 8.3.2. One is what we shall call the Arrow-Debreu structure. Here markets meet at time 0 to trade claims to consumption at all times $t > 0$ and that are contingent on all possible histories up to t, s^t. In that economy, at time 0 and

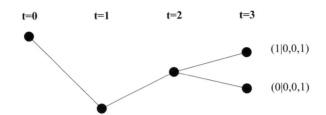

Figure 8.3.2: The commodity space with Arrow securities. At date $t = 2$, there are trades in time 3 goods for only those time $t = 3$ nodes that can be reached from the realized time $t = 2$ history $(0, 0, 1)$.

for all $t \geq 1$, households trade claims on the time t consumption good *at all nodes s^t*. After time 0, no further trades occur. The other economy has *sequential* trading of only one-period-ahead state-contingent claims. Here trades of one-period ahead state-contingent claims occur at each date $t \geq 0$. Trades for history s^{t+1}–contingent date $t + 1$ goods occur only at the *particular* date t history s^t that has been reached at t, as in Figure 8.3.2. It turns out that these two trading arrangements support identical equilibrium allocations. Those allocations share the notable property of being functions only of the *aggregate* endowment realization $\sum_{i=1}^{I} y_t^i(s^t)$ and time-invariant parameters describing the initial distribution of wealth.

8.3.1. History dependence

Before trading, the situation of household i at time t depends on the history s^t. A natural measure of household i's luck in life is $\{y_0^i(s_0), y_1^i(s^1), \ldots, y_t^i(s^t)\}$, which evidently in general depends on the history s^t. A question that will occupy us in this chapter and in chapters 18 and 20 is whether, after trading, the household's consumption allocation at time t is also history dependent. Remarkably, in the complete markets models of this chapter, the consumption allocation at time t depends only on the aggregate endowment realization at time t and some time-invariant parameters that describe the time 0 *initial* distribution of wealth. The market incompleteness of chapter 18 and the information and enforcement frictions of chapter 20 will break that result and put history dependence into equilibrium allocations.

8.4. Pareto problem

As a benchmark against which to measure allocations attained by a market economy, we seek efficient allocations. An allocation is said to be efficient if it is Pareto optimal: it has the property that any reallocation that makes one household strictly better off also makes one or more other households worse off. We can find efficient allocations by posing a Pareto problem for a fictitious social planner. The planner attaches nonnegative *Pareto weights* $\lambda_i, i = 1, \ldots, I$ to the consumers' utilities and chooses allocations $c^i, i = 1, \ldots, I$ to maximize

$$W = \sum_{i=1}^{I} \lambda_i U\left(c^i\right) \tag{8.4.1}$$

subject to (8.2.2). We call an allocation *efficient* if it solves this problem for some set of nonnegative λ_i's. Let $\theta_t(s^t)$ be a nonnegative Lagrange multiplier on the feasibility constraint (8.2.2) for time t and history s^t, and form the Lagrangian

$$L = \sum_{t=0}^{\infty} \sum_{s^t} \left\{ \sum_{i=1}^{I} \lambda_i \beta^t u\left(c_t^i\left(s^t\right)\right) \pi_t\left(s^t\right) + \theta_t\left(s^t\right) \sum_{i=1}^{I} \left[y_t^i\left(s^t\right) - c_t^i\left(s^t\right)\right] \right\}.$$

The first-order condition for maximizing L with respect to $c_t^i(s^t)$ is

$$\beta^t u'\left(c_t^i\left(s^t\right)\right) \pi_t\left(s^t\right) = \lambda_i^{-1} \theta_t\left(s^t\right) \tag{8.4.2}$$

for each i, t, s^t. Taking the ratio of (8.4.2) for consumers i and 1, respectively, gives

$$\frac{u'\left(c_t^i\left(s^t\right)\right)}{u'\left(c_t^1\left(s^t\right)\right)} = \frac{\lambda_1}{\lambda_i}$$

which implies

$$c_t^i\left(s^t\right) = u'^{-1}\left(\lambda_i^{-1}\lambda_1 u'\left(c_t^1\left(s^t\right)\right)\right). \tag{8.4.3}$$

Substituting (8.4.3) into feasibility condition (8.2.2) at equality gives

$$\sum_i u'^{-1}\left(\lambda_i^{-1}\lambda_1 u'\left(c_t^1\left(s^t\right)\right)\right) = \sum_i y_t^i\left(s^t\right). \tag{8.4.4}$$

Equation (8.4.4) is one equation in the one unknown $c_t^1(s^t)$. The right side of (8.4.4) is the realized aggregate endowment, so the left side is a function only of the aggregate endowment. Thus, given $\{\lambda_i\}_{i=1}^I$, $c_t^1(s^t)$ depends only on the current realization of the aggregate endowment and not separately either on the date t or on the specific history s^t leading up to that aggregate endowment or the cross-section distribution of individual endowments realized at t. Equation (8.4.3) then implies that for all i, $c_t^i(s^t)$ depends only on the aggregate endowment realization. We thus have:

PROPOSITION 1: An efficient allocation is a function of the realized aggregate endowment and does not depend separately on either the specific history s^t leading up to that aggregate endowment or on the cross-section distribution of individual endowments realized at t: $c_t^i(s^t) = c_\tau^i(\tilde{s}^\tau)$ for s^t and $\tilde{s}^\tau$ such that $\sum_j y_t^j(s^t) = \sum_j y_\tau^j(\tilde{s}^\tau)$.

To compute the optimal allocation, first solve (8.4.4) for $c_t^1(s^t)$, then solve (8.4.3) for $c_t^i(s^t)$. Note from (8.4.3) that only the ratios of the Pareto weights matter, so that we are free to normalize the weights, e.g., to impose $\sum_i \lambda_i = 1$.

8.4.1. Time invariance of Pareto weights

Through equations (8.4.3) and (8.4.4), the allocation $c_t^i(s^t)$ assigned to consumer i depends in a time-invariant way on the aggregate endowment $\sum_j y_t^j(s^t)$. Consumer i's share of the aggregate endowment varies directly with his Pareto weight λ_i. In chapter 20, we shall see that the constancy through time of the Pareto weights $\{\lambda_j\}_{j=1}^I$ is a telltale sign that there are no enforcement- or information-related incentive problems in this economy. In chapter 20, when we inject those imperfections into the environment, the time invariance of the Pareto weights evaporates.

8.5. Time 0 trading: Arrow-Debreu securities

We now describe how an optimal allocation can be attained by a competitive equilibrium with the Arrow-Debreu timing. Households trade dated history-contingent claims to consumption. There is a complete set of securities. Trades occur at time 0, after s_0 has been realized. At $t = 0$, households can exchange claims on time t consumption, contingent on history s^t at price $q_t^0(s^t)$, measured in some unit of account. The superscript 0 refers to the date at which trades occur, while the subscript t refers to the date that deliveries are to be made. The household's budget constraint is

$$\sum_{t=0}^{\infty} \sum_{s^t} q_t^0\left(s^t\right) c_t^i\left(s^t\right) \leq \sum_{t=0}^{\infty} \sum_{s^t} q_t^0\left(s^t\right) y_t^i\left(s^t\right). \tag{8.5.1}$$

The household's problem is to choose c^i to maximize expression (8.2.1) subject to inequality (8.5.1).

Underlying the *single* budget constraint (8.5.1) is the fact that multilateral trades are possible through a clearing operation that keeps track of net claims.[4] All trades occur at time 0. After time 0, trades that were agreed to at time 0 are executed, but no more trades occur.

Attach a Lagrange multiplier μ_i to each household's budget constraint (8.5.1). We obtain the first-order conditions for the household's problem:

$$\frac{\partial U\left(c^i\right)}{\partial c_t^i\left(s^t\right)} = \mu_i q_t^0\left(s^t\right), \tag{8.5.2}$$

for all i, t, s^t. The left side is the derivative of total utility with respect to the time t, history s^t component of consumption. Each household has its own Lagrange multiplier μ_i that is independent of time. With specification (8.2.1) of the utility functional, we have

$$\frac{\partial U\left(c^i\right)}{\partial c_t^i\left(s^t\right)} = \beta^t u'\left[c_t^i\left(s^t\right)\right] \pi_t\left(s^t\right). \tag{8.5.3}$$

This expression implies that equation (8.5.2) can be written

$$\beta^t u'\left[c_t^i\left(s^t\right)\right] \pi_t\left(s^t\right) = \mu_i q_t^0\left(s^t\right). \tag{8.5.4}$$

---

[4] In the language of modern payments systems, this is a system with net settlements, not gross settlements, of trades.

We use the following definitions:

DEFINITIONS: A *price system* is a sequence of functions $\{q_t^0(s^t)\}_{t=0}^\infty$. An *allocation* is a list of sequences of functions $c^i = \{c_t^i(s^t)\}_{t=0}^\infty$, one for each i.

DEFINITION: A *competitive equilibrium* is a feasible allocation and a price system such that, given the price system, the allocation solves each household's problem.

Notice that equation (8.5.4) implies

$$\frac{u'\left[c_t^i\left(s^t\right)\right]}{u'\left[c_t^j\left(s^t\right)\right]} = \frac{\mu_i}{\mu_j} \tag{8.5.5}$$

for all pairs (i,j). Thus, ratios of marginal utilities between pairs of agents are constant across all histories and dates.

An equilibrium allocation solves equations (8.2.2), (8.5.1), and (8.5.5). Note that equation (8.5.5) implies that

$$c_t^i\left(s^t\right) = u'^{-1}\left\{u'\left[c_t^1\left(s^t\right)\right]\frac{\mu_i}{\mu_1}\right\}. \tag{8.5.6}$$

Substituting this into equation (8.2.2) at equality gives

$$\sum_i u'^{-1}\left\{u'\left[c_t^1\left(s^t\right)\right]\frac{\mu_i}{\mu_1}\right\} = \sum_i y_t^i\left(s^t\right). \tag{8.5.7}$$

The right side of equation (8.5.7) is the current realization of the aggregate endowment. Therefore, the left side, and so $c_t^1(s^t)$, must also depend only on the current aggregate endowment, as well as on the ratio $\frac{\mu_i}{\mu_1}$. It follows from equation (8.5.6) that the equilibrium allocation $c_t^i(s^t)$ for each i depends only on the economy's aggregate endowment as well as on $\frac{\mu_i}{\mu_1}$. We summarize this analysis in the following proposition:

PROPOSITION 2: The competitive equilibrium allocation is a function of the realized aggregate endowment and does not depend on time t or the specific history or on the cross section distribution of endowments: $c_t^i(s^t) = c_\tau^i(\tilde{s}^\tau)$ for all histories s^t and $\tilde{s}^\tau$ such that $\sum_j y_t^j(s^t) = \sum_j y_\tau^j(\tilde{s}^\tau)$.

8.5.1. Equilibrium pricing function

Suppose that c^i, $i = 1, \ldots, I$ is an equilibrium allocation. Then the marginal condition (8.5.2) or (8.5.4) can be regarded as determining the price system $q_t^0(s^t)$ as a function of the equilibrium allocation assigned to household i, for any i. But to exploit this fact in computation, we need a way first to compute an equilibrium allocation without simultaneously computing prices. As we shall see soon, solving the planning problem provides a convenient way to do that.

Because the units of the price system are arbitrary, one of the prices can be normalized at any positive value. We shall set $q_0^0(s_0) = 1$, putting the price system in units of time 0 goods. This choice implies that $\mu_i = u'[c_0^i(s_0)]$ for all i.

8.5.2. Optimality of equilibrium allocation

A competitive equilibrium allocation is a particular Pareto optimal allocation, one that sets the Pareto weights $\lambda_i = \mu_i^{-1}$. These weights are unique up to multiplication by a positive scalar. Furthermore, at a competitive equilibrium allocation, the *shadow prices* $\theta_t(s^t)$ for the associated planning problem equal the prices $q_t^0(s^t)$ for goods to be delivered at date t contingent on history s^t associated with the Arrow-Debreu competitive equilibrium. That allocations for the planning problem and the competitive equilibrium are identical reflects the two fundamental theorems of welfare economics (see Mas-Colell, Whinston, and Green (1995)). The first welfare theorem states that a competitive equilibrium allocation is efficient. The second welfare theorem states that any efficient allocation can be supported by a competitive equilibrium with an appropriate initial distribution of wealth.

8.5.3. Interpretation of trading arrangement

In the competitive equilibrium, all trades occur at $t = 0$ in one market. Deliveries occur after $t = 0$, but no more trades. A vast clearing or credit system operates at $t = 0$. It ensures that condition (8.5.1) holds for each household i. A symptom of the once-and-for-all and net-clearing trading arrangement is that each household faces one budget constraint that accounts for trades across all dates and histories.

In section 8.8, we describe another trading arrangement with more trading dates but fewer securities at each date.

8.5.4. Equilibrium computation

To compute an equilibrium, we have somehow to determine ratios of the Lagrange multipliers, μ_i/μ_1, $i = 1, \ldots, I$, that appear in equations (8.5.6) and (8.5.7). The following *Negishi algorithm* accomplishes this.[5]

1. Fix a positive value for one μ_i, say μ_1, throughout the algorithm. Guess some positive values for the remaining μ_i's. Then solve equations (8.5.6) and (8.5.7) for a candidate consumption allocation $c^i, i = 1, \ldots, I$.

2. Use (8.5.4) for any household i to solve for the price system $q_t^0(s^t)$.

3. For $i = 1, \ldots, I$, check the budget constraint (8.5.1). For those i's for which the cost of consumption exceeds the value of their endowment, raise μ_i, while for those i's for which the reverse inequality holds, lower μ_i.

4. Iterate to convergence on steps 1-3.

Multiplying all of the μ_i's by a positive scalar simply changes the units of the price system. That is why we are free to normalize as we have in step 1.

In general, the equilibrium price system and distribution of wealth are mutually determined. Along with the equilibrium allocation, they solve a vast system of simultaneous equations. The Negishi algorithm provides one way to solve those equations. In applications, it can be complicated to implement. Therefore, in order to simplify things, most of the examples and exercises in this chapter specialize preferences in a way that eliminates the dependence of equilibrium prices on the distribution of wealth.

---

[5] See Negishi (1960).

8.6. Simpler computational algorithm

The preference specification in the following example enables us to avoid iterating on Pareto weights as in the Negishi algorithm.

8.6.1. Example 1: risk sharing

Suppose that the one-period utility function is of the constant relative risk-aversion (CRRA) form

$$u\left(c\right) = \left(1 - \gamma\right)^{-1} c^{1-\gamma}, \ \gamma > 0.$$

Then equation (8.5.5) implies

$$\left[c_t^i\left(s^t\right)\right]^{-\gamma} = \left[c_t^j\left(s^t\right)\right]^{-\gamma} \frac{\mu_i}{\mu_j}$$

or

$$c_t^i\left(s^t\right) = c_t^j\left(s^t\right) \left(\frac{\mu_i}{\mu_j}\right)^{-\frac{1}{\gamma}}. \tag{8.6.1}$$

Equation (8.6.1) states that time t elements of consumption allocations to distinct agents are constant fractions of one another. With a power utility function, it says that individual consumption is perfectly correlated with the aggregate endowment or aggregate consumption.[6]

  The fractions of the aggregate endowment assigned to each individual are independent of the realization of s^t. Thus, there is extensive cross-history and cross-time consumption sharing. The constant-fractions-of-consumption characterization comes from two aspects of the theory: (1) complete markets and (2) a homothetic one-period utility function.

---

[6] Equation (8.6.1) implies that conditional on the history s^t, time t consumption $c_t^i(s^t)$ is independent of the household's individual endowment at t, s^t, $y_t^i(s^t)$. Mace (1991), Cochrane (1991), and Townsend (1994) have tested and rejected versions of this conditional independence hypothesis. In chapter 20, we study how particular impediments to trade explain these rejections.

8.6.2. Implications for equilibrium computation

Equation $(8.6.1)$ and the pricing formula $(8.5.4)$ imply that an equilibrium price vector satisfies

$$q_t^0\left(s^t\right) = \mu_i^{-1}\alpha_i^{-\gamma}\beta^t\left(\bar{y}_t\left(s^t\right)\right)^{-\gamma}\pi_t\left(s^t\right),\tag{8.6.2}$$

where $c_t^i(s_t) = \alpha_i\bar{y}_t(s^t)$, $\bar{y}_t(s^t) = \sum_i y_t^i(s^t)$, and α_i is consumer i's fixed consumption share of the aggregate endowment. We are free to normalize the price system by setting $\mu_i\alpha_i^{-\gamma}$ for one consumer to an arbitrary positive number.

The homothetic CRRA preference specification that leads to equation $(8.6.2)$ allows us to compute an equilibrium using the following steps:

1. Use $(8.6.2)$ to compute an equilibrium price system.

2. Use this price system and consumer i's budget constraint to compute

$$\alpha_i = \frac{\sum_{t=0}^{\infty}\sum_{s^t}q_t^0\left(s^t\right)y_t^i\left(s^t\right)}{\sum_{t=0}^{\infty}\sum_{s^t}q_t^0\left(s^t\right)\bar{y}_t\left(s^t\right)}.$$

Thus, consumer i's fixed consumption share α_i equals its share of aggregate wealth evaluated at the competitive equilibrium price vector.

8.6.3. Example 2: no aggregate uncertainty

In this example, the endowment structure is sufficiently simple that we can compute an equilibrium without assuming a homothetic one-period utility function. Let the stochastic event s_t take values on the unit interval $[0,1]$. There are two households, with $y_t^1(s^t) = s_t$ and $y_t^2(s^t) = 1 - s_t$. Note that the aggregate endowment is constant, $\sum_i y_t^i(s^t) = 1$. Then equation $(8.5.7)$ implies that $c_t^1(s^t)$ is constant over time and across histories, and equation $(8.5.6)$ implies that $c_t^2(s^t)$ is also constant. Thus, the equilibrium allocation satisfies $c_t^i(s^t) = \bar{c}^i$ for all t and s^t, for $i = 1,2$. Then from equation $(8.5.4)$,

$$q_t^0\left(s^t\right) = \beta^t\pi_t\left(s^t\right)\frac{u'\left(\bar{c}^i\right)}{\mu_i},\tag{8.6.3}$$

for all t and s^t, for $i = 1,2$. Household i's budget constraint implies

$$\frac{u'\left(\bar{c}^i\right)}{\mu_i}\sum_{t=0}^{\infty}\sum_{s^t}\beta^t\pi_t\left(s^t\right)\left[\bar{c}^i - y_t^i\left(s^t\right)\right] = 0.$$

Solving this equation for $\bar{c}^i$ gives

$$\bar{c}^i = (1 - \beta) \sum_{t=0}^{\infty} \sum_{s^t} \beta^t \pi_t \left(s^t\right) y_t^i \left(s^t\right). \tag{8.6.4}$$

Summing equation (8.6.4) verifies that $\bar{c}^1 + \bar{c}^2 = 1$.[7]

8.6.4. Example 3: periodic endowment processes

Consider the special case of the previous example in which s_t is deterministic and alternates between the values 1 and 0; $s_0 = 1$, $s_t = 0$ for t odd, and $s_t = 1$ for t even. Thus, the endowment processes are perfectly predictable sequences $(1, 0, 1, \ldots)$ for the first agent and $(0, 1, 0, \ldots)$ for the second agent. Let $\tilde{s}^t$ be the history of $(1, 0, 1, \ldots)$ up to t. Evidently, $\pi_t(\tilde{s}^t) = 1$, and the probability assigned to all other histories up to t is zero. The equilibrium price system is then

$$q_t^0 \left(s^t\right) = \begin{cases} \beta^t, & \text{if } s^t = \tilde{s}^t; \\ 0, & \text{otherwise}; \end{cases}$$

when using the time 0 good as numeraire, $q_0^0(\tilde{s}_0) = 1$. From equation (8.6.4), we have

$$\bar{c}^1 = (1 - \beta) \sum_{j=0}^{\infty} \beta^{2j} = \frac{1}{1 + \beta}, \tag{8.6.5a}$$

$$\bar{c}^2 = (1 - \beta) \beta \sum_{j=0}^{\infty} \beta^{2j} = \frac{\beta}{1 + \beta}. \tag{8.6.5b}$$

---

[7] If we let $\beta^{-1} = 1 + r$, where r is interpreted as the risk-free rate of interest, then note that (8.6.4) can be expressed as

$$\bar{c}^i = \left(\frac{r}{1+r}\right) E_0 \sum_{t=0}^{\infty} (1 + r)^{-t} y_t^i \left(s^t\right).$$

Hence, equation (8.6.4) is a version of Friedman's permanent income model, which asserts that a household with zero financial assets consumes the annuity value of its human wealth defined as the expected discounted value of its labor income (which for present purposes we take to be $y_t^i(s^t)$). In the present example, the household completely smooths its consumption across time and histories, something that the household in Friedman's model typically cannot do. See chapter 17.

Consumer 1 consumes more every period because he is richer by virtue of receiving his endowment earlier.

8.6.5. *Example 4*

In this example, we assume that the one-period utility function is $\frac{c^{1-\gamma}}{1-\gamma}$. There are two consumers named $i = 1, 2$. Their endowments are $y_t^1 = y_t^2 = .5$ for $t = 0, 1$ and $y_t^1 = s_t$ and $y_t^2 = 1 - s_t$ for $t \geq 2$. The state space $s_t = \{0, 1\}$ and s_t is governed by a Markov chain with probability $\pi(s_0 = 1) = 1$ for the initial state and time-varying transition probabilities $\pi_1(s_1 = 1|s_0 = 1) = 1, \pi_2(s_2 = 1|s_1 = 1) = \pi_2(s_2 = 0|s_1 = 1) = .5, \pi_t(s_t = 1|s_{t-1} = 1) = 1, \pi_t(s_t = 0|s_{t-1} = 0) = 1$ for $t > 2$. This specification implies that $\pi_t(1, 1, \ldots, 1, 1, 1) = .5$ and $\pi_t(0, 0, \ldots, 0, 1, 1) = .5$ for all $t > 2$.

We can apply the method of subsection 8.6.2 to compute an equilibrium. The aggregate endowment is $\bar{y}_t(s^t) = 1$ for all t and all s^t. Therefore, an equilibrium price vector is $q_1^0(1, 1) = \beta, q_2^0(0, 1, 1) = q_2^0(1, 1, 1) = .5\beta^2$ and $q_t^0(1, 1, \ldots, 1, 1) = q_t^0(0, 0, \ldots, 1, 1) = .5\beta^t$ for $t > 2$. Use these prices to compute the value of agent i's endowment: $\sum_t \sum_{s^t} q_t^0(s^t) y_t^i(s^t) = \sum_t \beta^t .5[.5 + .5 + 0 + \ldots + 0] + \sum_t \beta^t .5[.5 + .5 + 1 + \ldots + 1] = 2 \sum_t \beta^t .5[.5 + .5 + \ldots + .5] = .5 \sum_t \beta^t = \frac{.5}{1-\beta}$. Consumer i's budget constraint is satisfied when he consumes a constant consumption of .5 each period in each state: $c_t^i(s^t) = .5$ for all t for all s^t.

In subsection 8.9.4, we shall use the equilibrium allocation from the Arrow-Debreu economy in this example to synthesize an equilibrium in an economy with sequential trading.

8.7. Primer on asset pricing

Many asset-pricing models assume complete markets and price an asset by breaking it into a sequence of history-contingent claims, evaluating each component of that sequence with the relevant "state price deflator" $q_t^0(s^t)$, then adding up those values. The asset is *redundant*, in the sense that it offers a bundle of history-contingent dated claims, each component of which has already been priced by the market. While we shall devote chapters 13 and 14 entirely to asset-pricing theories, it is useful to give some pricing formulas at this point because they help illustrate the complete market competitive structure.

8.7.1. Pricing redundant assets

Let $\{d_t(s^t)\}_{t=0}^\infty$ be a stream of claims on time t, history s^t consumption, where $d_t(s^t)$ is a measurable function of s^t. The price of an asset entitling the owner to this stream must be

$$p_0^0(s_0) = \sum_{t=0}^\infty \sum_{s^t} q_t^0\left(s^t\right) d_t\left(s^t\right). \tag{8.7.1}$$

If this equation did not hold, someone could make unbounded profits by synthesizing this asset through purchases or sales of history-contingent dated commodities and then either buying or selling the asset. We shall elaborate this arbitrage argument below and later in chapter 13 on asset pricing.

8.7.2. Riskless consol

As an example, consider the price of a *riskless consol*, that is, an asset offering to pay one unit of consumption for sure each period. Then $d_t(s^t) = 1$ for all t and s^t, and the price of this asset is

$$\sum_{t=0}^\infty \sum_{s^t} q_t^0\left(s^t\right). \tag{8.7.2}$$

8.7.3. Riskless strips

As another example, consider a sequence of *strips* of payoffs on the riskless consol. The time t strip is just the payoff process $d_\tau = 1$ if $\tau = t \geq 0$, and 0 otherwise. Thus, the owner of the strip is entitled to the time t coupon only. The value of the time t strip at time 0 is evidently

$$\sum_{s^t} q_t^0\left(s^t\right).$$

Compare this to the price of the consol (8.7.2). We can think of the t-period riskless strip as a t-period zero-coupon bond. See appendix B of chapter 14 for an account of a closely related model of yields on such bonds.

8.7.4. Tail assets

Return to the stream of dividends $\{d_t(s^t)\}_{t \geq 0}$ generated by the asset priced in equation (8.7.1). For $\tau \geq 1$, suppose that we strip off the first $\tau - 1$ periods of the dividend and want the time 0 value of the remaining dividend stream $\{d_t(s^t)\}_{t \geq \tau}$. Specifically, we seek the value of this asset for a particular possible realization of s^τ. Let $p_\tau^0(s^\tau)$ be the time 0 price of an asset that entitles the owner to dividend stream $\{d_t(s^t)\}_{t \geq \tau}$ if history s^τ is realized,

$$p_\tau^0\left(s^\tau\right) = \sum_{t \geq \tau} \sum_{s^t \mid s^\tau} q_t^0\left(s^t\right) d_t\left(s^t\right), \tag{8.7.3}$$

where the summation over $s^t \mid s^\tau$ means that we sum over all possible subsequent histories $\tilde{s}^t$ such that $\tilde{s}^\tau = s^\tau$. When the units of the price are time 0, state s_0 goods, the normalization is $q_0^0(s_0) = 1$. To convert the price into units of time τ, history s^τ consumption goods, divide by $q_\tau^0(s^\tau)$ to get

$$p_\tau^\tau\left(s^\tau\right) \equiv \frac{p_\tau^0\left(s^\tau\right)}{q_\tau^0\left(s^\tau\right)} = \sum_{t \geq \tau} \sum_{s^t \mid s^\tau} \frac{q_t^0\left(s^t\right)}{q_\tau^0\left(s^\tau\right)} d_t\left(s^t\right). \tag{8.7.4}$$

Notice that[8]

$$
\begin{aligned}
q_t^\tau\left(s^t\right) &\equiv \frac{q_t^0\left(s^t\right)}{q_\tau^0\left(s^\tau\right)} = \frac{\beta^t u'\left[c_t^i\left(s^t\right)\right] \pi_t\left(s^t\right)}{\beta^\tau u'\left[c_\tau^i\left(s^\tau\right)\right] \pi_\tau\left(s^\tau\right)} \\
&= \beta^{t-\tau} \frac{u'\left[c_t^i\left(s^t\right)\right]}{u'\left[c_\tau^i\left(s^\tau\right)\right]} \pi_t\left(s^t \mid s^\tau\right).
\end{aligned}
\tag{8.7.5}
$$

---

[8] Because the marginal conditions hold for all consumers, this condition holds for all i.

Here $q_t^\tau(s^t)$ is the price of one unit of consumption delivered at time t, history s^t in terms of the date τ, history s^τ consumption good; $\pi_t(s^t|s^\tau)$ is the probability of history s^t conditional on history s^τ at date τ. Thus, the price at time τ, history s^τ for the "tail asset" is

$$p_\tau^\tau(s^\tau) = \sum_{t \geq \tau} \sum_{s^t|s^\tau} q_t^\tau(s^t) d_t(s^t). \qquad (8.7.6)$$

When we want to create a time series of, say, equity prices, we use the "tail asset" pricing formula (8.7.6). An equity purchased at time τ entitles the owner to the dividends from time τ forward. Our formula (8.7.6) expresses the asset price in terms of prices with time τ, history s^τ good as numeraire.

8.7.5. One-period returns

The one-period version of equation (8.7.5) is

$$q_{\tau+1}^\tau(s^{\tau+1}) = \beta \frac{u'[c_{\tau+1}^i(s^{\tau+1})]}{u'[c_\tau^i(s^\tau)]} \pi_{\tau+1}(s^{\tau+1}|s^\tau).$$

The right side is the one-period *pricing kernel* at time τ. If we want to find the price at time τ at history s^τ of a claim to a random payoff $\omega(s_{\tau+1})$, we use

$$p_\tau^\tau(s^\tau) = \sum_{s_{\tau+1}} q_{\tau+1}^\tau(s^{\tau+1}) \omega(s_{\tau+1})$$

or

$$p_\tau^\tau(s^\tau) = E_\tau \left[\beta \frac{u'(c_{\tau+1})}{u'(c_\tau)} \omega(s_{\tau+1}) \right], \qquad (8.7.7)$$

where E_τ is the conditional expectation operator. We have deleted the i superscripts on consumption, with the understanding that equation (8.7.7) is true for any consumer i; we have also suppressed the dependence of c_τ on s^τ, which is implicit.

Let $R_{\tau+1} \equiv \omega(s_{\tau+1})/p_\tau^\tau(s^\tau)$ be the one-period gross *return* on the asset. Then for any asset, equation (8.7.7) implies

$$1 = E_\tau \left[\beta \frac{u'(c_{\tau+1})}{u'(c_\tau)} R_{\tau+1} \right] \equiv E_\tau [m_{\tau+1} R_{\tau+1}]. \qquad (8.7.8)$$

The term $m_{\tau+1} \equiv \beta u'(c_{\tau+1})/u'(c_\tau)$ functions as a *stochastic discount factor*. Like $R_{\tau+1}$, it is a random variable measurable with respect to $s_{\tau+1}$, given s^τ. Equation (8.7.8) is a restriction on the conditional moments of returns and m_{t+1}. Applying the law of iterated expectations to equation (8.7.8) gives the unconditional moments restriction

$$1 = E\left[\beta \frac{u'(c_{\tau+1})}{u'(c_\tau)} R_{\tau+1}\right] \equiv E\left[m_{\tau+1} R_{\tau+1}\right]. \qquad (8.7.9)$$

In chapters 13 and 14 we shall many more instances of this equation.

In the next section, we display another market structure in which the one-period pricing kernel $q_{t+1}^t(s^{t+1})$ also plays a decisive role. This structure uses the celebrated one-period "Arrow securities," the sequential trading of which substitutes perfectly for the comprehensive trading of long horizon claims at time 0.

8.8. Sequential trading: Arrow securities

This section describes an alternative market structure that preserves both the equilibrium allocation and the key one-period asset-pricing formula (8.7.7).

8.8.1. Arrow securities

We build on an insight of Arrow (1964) that one-period securities are enough to implement complete markets, provided that new one-period markets are re-opened for trading each period and provided that time t, history s^t wealth is properly assigned to each agent. Thus, at each date $t \geq 0$, but only at the history s^t actually realized, trades occur in a set of claims to one-period-ahead state-contingent consumption. We describe a competitive equilibrium of this sequential-trading economy. With a full array of these one-period-ahead claims, the sequential-trading arrangement attains the same allocation as the competitive equilibrium that we described earlier.

8.8.2. *Financial wealth as an endogenous state variable*

A key step in constructing a sequential-trading arrangement is to identify a variable to serve as the state in a value function for the household at date t. We find this state by taking an equilibrium allocation and price system for the (Arrow-Debreu) time 0 trading structure and applying a guess-and-verify method. We begin by asking the following question. In the competitive equilibrium where all trading takes place at time 0, what is the implied continuation wealth of household i at time t after history s^t, but before adding in its time t, history s^t endowment $y_t^i(s^t)$? To answer this question, in period t, conditional on history s^t, we sum up the value of the household's purchased claims to current and future goods net of its outstanding liabilities. Since history s^t has been realized, we discard all claims and liabilities contingent on time t histories $\tilde{s}^t \neq s^t$ that were not realized. Household i's net claim to delivery of goods in a future period $\tau \geq t$ contingent on history $\tilde{s}^\tau$ whose time t partial history $\tilde{s}^t = s^t$ is $[c_\tau^i(\tilde{s}^\tau) - y_\tau^i(\tilde{s}^\tau)]$. Thus, the household's financial wealth, or the value of all its current and future net claims, expressed in terms of the date t, history s^t consumption good is

$$\Upsilon_t^i\left(s^t\right) = \sum_{\tau=t}^{\infty} \sum_{s^\tau \mid s^t} q_\tau^t\left(s^\tau\right) \left[c_\tau^i\left(s^\tau\right) - y_\tau^i\left(s^\tau\right)\right]. \tag{8.8.1}$$

Notice that feasibility constraint $(8.2.2)$ at equality implies that

$$\sum_{i=1}^{I} \Upsilon_t^i\left(s^t\right) = 0, \qquad \forall t, s^t.$$

8.8.3. Financial and non-financial wealth

Define $\Upsilon_t^i(s^t)$ as *financial wealth* and $\sum_{\tau=t}^{\infty} \sum_{s^\tau | s^t} q_\tau^t(s^\tau) y_\tau^i(s^\tau)$ as *non-financial wealth*.[9] In terms of these concepts, (8.8.1) implies

$$\Upsilon_t^i\left(s^t\right) + \sum_{\tau=t}^{\infty} \sum_{s^\tau | s^t} q_\tau^t\left(s^\tau\right) y_\tau^i\left(s^\tau\right) = \sum_{\tau=t}^{\infty} \sum_{s^\tau | s^t} q_\tau^t\left(s^\tau\right) c_\tau^i\left(s^\tau\right), \qquad (8.8.2)$$

which states that at each time and each history, the sum of financial and non-financial wealth equals the present value of current and future consumption claims. At time 0, we have set $\Upsilon_t^i(s^0) = 0$ for all i. At $t > 0$, financial wealth $\Upsilon_t^i(s^t)$ typically differs from zero for individual i, but it sums to zero across i.

8.8.4. Reopening markets

Formula (8.7.5) takes the form of a pricing function for a complete markets economy with date- and history-contingent commodities whose markets can be regarded as having been reopened at date τ, history s^τ, starting from wealth levels implied by the tails of each household's endowment and consumption streams for a complete markets economy that originally convened at $t = 0$. We leave it as an exercise to the reader to prove the following proposition.

PROPOSITION 3: Start from the distribution of time t, history s^t wealth that is implicit in a time 0 Arrow-Debreu equilibrium. If markets are reopened at date t after history s^t, no trades occur. That is, given the price system (8.7.5), all households choose to continue the tails of their original consumption plans.

---

[9] In some applications, financial wealth is also called 'non-human wealth' and non-financial wealth is called 'human wealth'.

8.8.5. Debt limits

In moving from the Arrow-Debreu economy to one with sequential trading, we propose to match the time t, history s^t wealth of the household in the sequential economy with the equilibrium tail wealth $\Upsilon_t^i(s^t)$ from the Arrow-Debreu economy computed in equation (8.8.2). But first we have to say something about debt limits, a feature that was only implicit in the time 0 budget constraint (8.5.1) in the Arrow-Debreu economy. In moving to the sequential formulation, we restrict asset trades to prevent Ponzi schemes. We want the weakest possible restrictions. We synthesize restrictions that work by starting from the equilibrium allocation of the Arrow-Debreu economy (with time 0 markets), and find some state-by-state debt limits that support the equilibrium allocation that emerged from the Arrow-Debreu economy under a sequential trading arrangement. Often we'll refer to these weakest possible debt limits as the "natural debt limits." These limits come from the common sense requirement that it has to be *feasible* for the consumer to repay his state contingent debt in every possible state. Together with our assumption that $c_t^i(s^t)$ must be nonnegative, that feasibility requirement leads to the natural debt limits.

Let $q_\tau^t(s^\tau)$ be the Arrow-Debreu price, denominated in units of the date t, history s^t consumption good. Consider the value of the tail of agent i's endowment sequence at time t in history s^t:

$$A_t^i\left(s^t\right) = \sum_{\tau=t}^{\infty} \sum_{s^\tau | s^t} q_\tau^t\left(s^\tau\right) y_\tau^i\left(s^\tau\right). \qquad (8.8.3)$$

We call $A_t^i(s^t)$ the *natural debt limit* at time t and history s^t. It is the maximal value that agent i can repay starting from that period, assuming that his consumption is zero always. With sequential trading, we shall require that household i at time $t-1$ and history s^{t-1} cannot promise to pay more than $A_t^i(s^t)$ conditional on the realization of s_t tomorrow, because it will not be feasible to repay more. Household i at time $t-1$ faces one such borrowing constraint for each possible realization of s_t tomorrow.

8.8.6. Sequential trading

There is a sequence of markets in one-period-ahead state-contingent claims. At each date $t \geq 0$, households trade claims to date $t + 1$ consumption, whose payment is contingent on the realization of s_{t+1}. Let $\tilde{a}^i_t(s^t)$ denote the claims to time t consumption, other than its time t endowment $y^i_t(s^t)$, that household i brings into time t in history s^t. Suppose that $\tilde{Q}_t(s_{t+1}|s^t)$ is a *pricing kernel* to be interpreted as follows: $\tilde{Q}_t(s_{t+1}|s^t)$ is the price of one unit of time $t + 1$ consumption, contingent on the realization s_{t+1} at $t + 1$, when the history at t is s^t. The household faces a sequence of budget constraints for $t \geq 0$, where the time t, history s^t budget constraint is

$$\tilde{c}^i_t \left(s^t \right) + \sum_{s_{t+1}} \tilde{a}^i_{t+1} \left(s_{t+1}, s^t \right) \tilde{Q}_t \left(s_{t+1}|s^t \right) \leq y^i_t \left(s^t \right) + \tilde{a}^i_t \left(s^t \right). \tag{8.8.4}$$

At time t, a household chooses $\tilde{c}^i_t(s^t)$ and $\{\tilde{a}^i_{t+1}(s_{t+1}, s^t)\}$, where $\{\tilde{a}^i_{t+1}(s_{t+1}, s^t)\}$ is a vector of claims on time $t + 1$ consumption, there being one element of the vector for each value of the time $t + 1$ realization of s_{t+1}. To rule out Ponzi schemes, we impose the state-by-state borrowing constraints

$$-\tilde{a}^i_{t+1} \left(s^{t+1} \right) \leq A^i_{t+1} \left(s^{t+1} \right), \tag{8.8.5}$$

where $A^i_{t+1}(s^{t+1})$ is computed in equation (8.8.3).

Let $\eta^i_t(s^t)$ and $\nu^i_t(s^t; s_{t+1})$ be nonnegative Lagrange multipliers on the budget constraint (8.8.4) and the borrowing constraint (8.8.5), respectively, for time t and history s^t. Form the Lagrangian

$$L^i = \sum_{t=0}^{\infty} \sum_{s^t} \Big\{ \beta^t u(\tilde{c}^i_t(s^t)) \pi_t(s^t)$$

$$+ \eta^i_t(s^t) \Big[y^i_t(s^t) + \tilde{a}^i_t(s^t) - \tilde{c}^i_t(s^t) - \sum_{s_{t+1}} \tilde{a}^i_{t+1}(s_{t+1}, s^t) \tilde{Q}_t(s_{t+1}|s^t) \Big]$$

$$+ \sum_{s_{t+1}} \nu^i_t(s^t; s_{t+1}) \Big[A^i_{t+1}(s^{t+1}) + \tilde{a}^i_{t+1}(s^{t+1}) \Big] \Big\},$$

for a given initial wealth $\tilde{a}^i_0(s_0)$. The first-order conditions for maximizing L^i with respect to $\tilde{c}^i_t(s^t)$ and $\{\tilde{a}^i_{t+1}(s_{t+1}, s^t)\}_{s_{t+1}}$ are

$$\beta^t u'(\tilde{c}^i_t(s^t)) \pi_t(s^t) - \eta^i_t(s^t) = 0, \tag{8.8.6a}$$

$$-\eta^i_t(s^t) \tilde{Q}_t(s_{t+1}|s^t) + \nu^i_t(s^t; s_{t+1}) + \eta^i_{t+1}(s_{t+1}, s^t) = 0, \tag{8.8.6b}$$

for all s_{t+1}, t, s^t. In the optimal solution to this problem, the natural debt limit (8.8.5) will not be binding, and hence the Lagrange multipliers $\nu_t^i(s^t; s_{t+1})$ all equal zero for the following reason: if there were any history s^{t+1} leading to a binding natural debt limit, the household would from then on have to set consumption equal to zero in order to honor its debt. Because the household's utility function satisfies the Inada condition $\lim_{c\downarrow 0} u'(c) = +\infty$, that would mean that all future marginal utilities would be infinite. Thus, it would be easy to find alternative affordable allocations that yield higher expected utility by postponing earlier consumption to periods after such a binding constraint.

After setting $\nu_t^i(s^t; s_{t+1}) = 0$ in equation (8.8.6b), the first-order conditions imply the following restrictions on the optimally chosen consumption allocation,

$$\tilde{Q}_t(s_{t+1}|s^t) = \beta \frac{u'(\tilde{c}_{t+1}^i(s^{t+1}))}{u'(\tilde{c}_t^i(s^t))} \pi_t(s^{t+1}|s^t), \tag{8.8.7}$$

for all s_{t+1}, t, s^t.

DEFINITION: A *distribution of wealth* is a vector $\vec{\tilde{a}}_t(s^t) = \{\tilde{a}_t^i(s^t)\}_{i=1}^I$ satisfying $\sum_i \tilde{a}_t^i(s^t) = 0$.

DEFINITION: A *competitive equilibrium with sequential trading of one-period Arrow securities* is an initial distribution of wealth $\vec{\tilde{a}}_0(s_0)$, a collection of borrowing limits $\{A_t^i(s^t)\}$ satisfying (8.8.3) for all i, for all t, and for all s^t, a feasible allocation $\{\tilde{c}^i\}_{i=1}^I$, and pricing kernels $\tilde{Q}_t(s_{t+1}|s^t)$ such that
(a) for all i, given $\tilde{a}_0^i(s_0)$, the borrowing limits $\{A_t^i(s^t)$, and the pricing kernels, the consumption allocation $\tilde{c}^i$ solves the household's problem for all i;
(b) for all realizations of $\{s^t\}_{t=0}^\infty$, the households' consumption allocations and implied portfolios $\{\tilde{c}_t^i(s^t), \{\tilde{a}_{t+1}^i(s_{t+1}, s^t)\}_{s_{t+1}}\}_i$ satisfy $\sum_i \tilde{c}_t^i(s^t) = \sum_i y_t^i(s^t)$ and $\sum_i \tilde{a}_{t+1}^i(s_{t+1}, s^t) = 0$ for all s_{t+1}.

This definition leaves open the initial distribution of wealth. We'll say more about the initial distribution of wealth soon.

8.8.7. Equivalence of allocations

By making an appropriate guess about the form of the pricing kernels, it is easy to show that a competitive equilibrium allocation of the complete markets model with time 0 trading is also an allocation for a competitive equilibrium with sequential trading of one-period Arrow securities, one with a particular initial distribution of wealth. Thus, take $q_t^0(s^t)$ as given from the Arrow-Debreu equilibrium and suppose that the pricing kernel $\tilde{Q}_t(s_{t+1}|s^t)$ makes the following recursion true:

$$q_{t+1}^0(s^{t+1}) = \tilde{Q}_t(s_{t+1}|s^t)q_t^0(s^t),$$

or

$$\tilde{Q}_t(s_{t+1}|s^t) = q_{t+1}^t(s^{t+1}), \tag{8.8.8}$$

where recall that $q_{t+1}^t(s^{t+1}) = \frac{q_{t+1}^0(s^{t+1})}{q_t^0(s^t)}$.

Let $\{c_t^i(s^t)\}$ be a competitive equilibrium allocation in the Arrow-Debreu economy. If equation (8.8.8) is satisfied, that allocation is also a sequential-trading competitive equilibrium allocation. To show this fact, take the household's first-order conditions (8.5.4) for the Arrow-Debreu economy from two successive periods and divide one by the other to get

$$\frac{\beta u'[c_{t+1}^i(s^{t+1})]\pi(s^{t+1}|s^t)}{u'[c_t^i(s^t)]} = \frac{q_{t+1}^0(s^{t+1})}{q_t^0(s^t)} = \tilde{Q}_t(s_{t+1}|s^t). \tag{8.8.9}$$

If the pricing kernel satisfies equation (8.8.8), this equation is equivalent with the first-order condition (8.8.7) for the sequential-trading competitive equilibrium economy. It remains for us to choose the initial wealth of the sequential-trading equilibrium so that the sequential-trading competitive equilibrium duplicates the Arrow-Debreu competitive equilibrium allocation.

We conjecture that the initial wealth vector $\vec{a}_0(s_0)$ of the sequential-trading economy should be chosen to be the zero vector. This is a natural conjecture, because it means that each household must rely on its own endowment stream to finance consumption, in the same way that households are constrained to finance their history-contingent purchases for the infinite future at time 0 in the Arrow-Debreu economy. To prove that the conjecture is correct, we must show that the zero initial wealth vector enables household i to finance $\{c_t^i(s^t)\}$ and leaves no room to increase consumption in any period after any history.

The proof proceeds by guessing that, at time $t \geq 0$ and history s^t, household i chooses a portfolio given by $\tilde{a}^i_{t+1}(s_{t+1}, s^t) = \Upsilon^i_{t+1}(s^{t+1})$ for all s_{t+1}. The value of this portfolio expressed in terms of the date t, history s^t consumption good is

$$\sum_{s_{t+1}} \tilde{a}^i_{t+1}(s_{t+1}, s^t) \tilde{Q}_t(s_{t+1}|s^t) = \sum_{s^{t+1}|s^t} \Upsilon^i_{t+1}(s^{t+1}) q^t_{t+1}(s^{t+1})$$

$$= \sum_{\tau=t+1}^{\infty} \sum_{s^\tau|s^t} q^t_\tau(s^\tau) \left[c^i_\tau(s^\tau) - y^i_\tau(s^\tau) \right], \quad (8.8.10)$$

where we have invoked expressions $(8.8.2)$ and $(8.8.8)$.[10] To demonstrate that household i can afford this portfolio strategy, we now use budget constraint $(8.8.4)$ to compute the implied consumption plan $\{\tilde{c}^i_\tau(s^\tau)\}$. First, in the initial period $t = 0$ with $\tilde{a}^i_0(s_0) = 0$, the substitution of equation $(8.8.10)$ into budget constraint $(8.8.4)$ at equality yields

$$\tilde{c}^i_0(s_0) + \sum_{t=1}^{\infty} \sum_{s^t} q^0_t(s^t) \left[c^i_t(s^t) - y^i_t(s^t) \right] = y^i_0(s_0) + 0 \,.$$

This expression together with budget constraint $(8.5.1)$ at equality imply $\tilde{c}^i_0(s_0) = c^i_0(s_0)$. In other words, the proposed portfolio is affordable in period 0 and the associated consumption plan is the same as in the competitive equilibrium of the Arrow-Debreu economy. In all consecutive future periods $t > 0$ and histories s^t, we replace $\tilde{a}^i_t(s^t)$ in constraint $(8.8.4)$ by $\Upsilon^i_t(s^t)$, and after noticing that the value of the asset portfolio in $(8.8.10)$ can be written as

$$\sum_{s_{t+1}} \tilde{a}^i_{t+1}(s_{t+1}, s^t) \tilde{Q}_t(s_{t+1}|s^t) = \Upsilon^i_t(s^t) - \left[c^i_t(s^t) - y^i_t(s^t) \right], \quad (8.8.11)$$

it follows immediately from $(8.8.4)$ that $\tilde{c}^i_t(s^t) = c^i_t(s^t)$ for all periods and histories.

---

[10] We have also used the following identities,

$$q^{t+1}_\tau(s^\tau) q^t_{t+1}(s^{t+1}) = \frac{q^0_\tau(s^\tau)}{q^0_{t+1}(s^{t+1})} \frac{q^0_{t+1}(s^{t+1})}{q^0_t(s^t)} = q^t_\tau(s^\tau) \text{ for } \tau > t.$$

We have shown that the proposed portfolio strategy attains the same consumption plan as in the competitive equilibrium of the Arrow-Debreu economy, but what precludes household i from further increasing current consumption by reducing some component of the asset portfolio? The answer lies in the debt limit restrictions to which the household must adhere. In particular, if the household wants to ensure that consumption plan $\{c_\tau^i(s^\tau)\}$ can be attained starting next period in all possible future states, the household should subtract the value of this commitment to future consumption from the natural debt limit in (8.8.3). Thus, the household is facing a state-by-state borrowing constraint that is more restrictive than restriction (8.8.5): for any s^{t+1},

$$-\tilde{a}_{t+1}^i(s^{t+1}) \leq A_{t+1}^i(s^{t+1}) - \sum_{\tau=t+1}^{\infty} \sum_{s^\tau | s^{t+1}} q_\tau^{t+1}(s^\tau) c_\tau^i(s^\tau) = -\Upsilon_{t+1}^i(s^{t+1}),$$

or

$$\tilde{a}_{t+1}^i(s^{t+1}) \geq \Upsilon_{t+1}^i(s^{t+1}).$$

Hence, household i does not want to increase consumption at time t by reducing next period's wealth below $\Upsilon_{t+1}^i(s^{t+1})$ because that would jeopardize attaining the preferred consumption plan that satisfies first-order conditions (8.8.7) for all future periods and histories.

8.9. Recursive competitive equilibrium

We have established that equilibrium allocations are the same in the Arrow-Debreu economy with complete markets in dated contingent claims all traded at time 0 and in a sequential-trading economy with a complete set of one-period Arrow securities. This finding holds for arbitrary individual endowment processes $\{y_t^i(s^t)\}_i$ that are measurable functions of the history of events s^t, which in turn are governed by some arbitrary probability measure $\pi_t(s^t)$. At this level of generality, the pricing kernels $\tilde{Q}_t(s_{t+1}|s^t)$ and the wealth distributions $\vec{a}_t(s^t)$ in the sequential-trading economy both depend on the history s^t, so both are time-varying functions of all past events $\{s_\tau\}_{\tau=0}^t$. This can make it difficult to formulate an economic model that can be used to confront empirical observations. We want a framework in which economic outcomes are functions of a limited number of "state variables" that summarize the effects of past events and current information. This leads us to make the following specialization of the exogenous forcing processes that facilitates a recursive formulation of the sequential-trading equilibrium.

8.9.1. Endowments governed by a Markov process

Let $\pi(s'|s)$ be a Markov chain with given initial distribution $\pi_0(s)$ and state space $s \in S$. That is, $\text{Prob}(s_{t+1} = s'|s_t = s) = \pi(s'|s)$ and $\text{Prob}(s_0 = s) = \pi_0(s)$. As we saw in chapter 2, the chain induces a sequence of probability measures $\pi_t(s^t)$ on histories s^t via the recursions

$$\pi_t(s^t) = \pi(s_t|s_{t-1})\pi(s_{t-1}|s_{t-2})\ldots\pi(s_1|s_0)\pi_0(s_0). \tag{8.9.1}$$

In this chapter, we have assumed that trading occurs after s_0 has been observed, which we capture by setting $\pi_0(s_0) = 1$ for the initially given value of s_0.

Because of the Markov property, the conditional probability $\pi_t(s^t|s^\tau)$ for $t > \tau$ depends only on the state s_τ at time τ and does not depend on the history before τ,

$$\pi_t(s^t|s^\tau) = \pi(s_t|s_{t-1})\pi(s_{t-1}|s_{t-2})\ldots\pi(s_{\tau+1}|s_\tau). \tag{8.9.2}$$

Next, we assume that households' endowments in period t are time invariant measurable functions of s_t, $y_t^i(s^t) = y^i(s_t)$ for each i. Of course, all of our previous results continue to hold, but the Markov assumption for s_t imparts further structure to the equilibrium.

8.9.2. *Equilibrium outcomes inherit the Markov property*

Proposition 2 asserted a particular kind of history independence of the equilibrium allocation that prevails under any stochastic process for the endowments. In particular, each individual's consumption is a function only of the current realization of the aggregate endowment and does not depend on the specific history leading to that outcome.[11] Now, under our present assumption that $y_t^i(s^t) = y^i(s_t)$ for each i, it follows immediately that

$$c_t^i(s^t) = \bar{c}^i(s_t). \tag{8.9.3}$$

Substituting (8.9.2) and (8.9.3) into (8.8.7) shows that the pricing kernel in the sequential-trading equilibrium is a function only of the current state,

$$\tilde{Q}_t(s_{t+1}|s^t) = \beta \frac{u'(\bar{c}^i(s_{t+1}))}{u'(\bar{c}^i(s_t))} \pi(s_{t+1}|s_t) \equiv Q(s_{t+1}|s_t). \tag{8.9.4}$$

After similar substitutions with respect to equation (8.7.5), we can also establish history independence of the relative prices in the Arrow-Debreu economy:

PROPOSITION 4: If time t endowments are a function of a Markov state s_t, the Arrow-Debreu equilibrium price of date-$t \geq 0$, history s^t consumption goods expressed in terms of date τ ($0 \leq \tau \leq t$), history s^τ consumption goods is not history dependent: $q_t^\tau(s^t) = q_k^j(\tilde{s}^k)$ for $j, k \geq 0$ such that $t - \tau = k - j$ and $[s_\tau, s_{\tau+1}, \ldots, s_t] = [\tilde{s}_j, \tilde{s}_{j+1}, \ldots, \tilde{s}_k]$.

Using this proposition, we can verify that both the natural debt limits (8.8.3) and households' wealth levels (8.8.2) exhibit history independence,

$$A_t^i(s^t) = \bar{A}^i(s_t), \tag{8.9.5}$$

$$\Upsilon_t^i(s^t) = \bar{\Upsilon}^i(s_t). \tag{8.9.6}$$

The finding concerning wealth levels (8.9.6) conveys a useful insight into how the sequential-trading competitive equilibrium attains the first-best outcome in which no idiosyncratic risk is borne by individual households. In particular, each household enters every period with a wealth level that is independent of past realizations of his endowment. That is, his past trades have fully insured

---

[11] Of course, the equilibrium allocation also depends on the distribution of $\{y_t^i(s^t)\}$ *processes* across agents i, as reflected in the relative values of the Lagrange multipliers μ_i.

him against the idiosyncratic outcomes of his endowment. And from that very same insurance motive, the household now enters the present period with a wealth level that is a function of the current state s_t. It is a state-contingent wealth level that was chosen by the household in the previous period $t - 1$, and this wealth will be just sufficient to continue a trading strategy previously designed to insure against future idiosyncratic risks. The optimal holding of wealth is a function of s_t alone because the current state s_t determines the current endowment and the current pricing kernel and contains all information relevant for predicting future realizations of the household's endowment process as well as future prices. It can be shown that a household especially wants higher wealth levels for those states next period that either make his next period endowment low or more generally signal poor future prospects for its endowment into the more distant future. Of course, individuals' desires are tempered by differences in the economy's aggregate endowment across states (as reflected in equilibrium asset prices). Aggregate shocks cannot be diversified away but must be borne somehow by all of the households. The pricing kernel $Q(s_t|s_{t-1})$ and the assumed clearing of all markets set into action an "invisible hand" that coordinates households' transactions at time $t - 1$ in such a way that only aggregate risk and no idiosyncratic risk is borne by the households.

8.9.3. Recursive formulation of optimization and equilibrium

The fact that the pricing kernel $Q(s'|s)$ and the endowment $y^i(s)$ are functions of a Markov process s motivates us to seek a recursive formulation of the household's optimization problem. Household i's state at time t is its wealth a_t^i and the current realization s_t. We seek a pair of optimal policy functions $h^i(a, s)$, $g^i(a, s, s')$ such that the household's optimal decisions are

$$c_t^i = h^i(a_t^i, s_t), \tag{8.9.7a}$$

$$a_{t+1}^i(s_{t+1}) = g^i(a_t^i, s_t, s_{t+1}). \tag{8.9.7b}$$

Let $v^i(a, s)$ be the optimal value of household i's problem starting from state (a, s); $v^i(a, s)$ is the maximum expected discounted utility household that household i with current wealth a can attain in state s. The Bellman equation for the household's problem is

$$v^i(a, s) = \max_{c, \hat{a}(s')} \left\{ u(c) + \beta \sum_{s'} v^i[\hat{a}(s'), s'] \pi(s'|s) \right\} \qquad (8.9.8)$$

where the maximization is subject to the following version of constraint (8.8.4):

$$c + \sum_{s'} \hat{a}(s') Q(s'|s) \leq y^i(s) + a \qquad (8.9.9)$$

and also

$$c \geq 0, \qquad (8.9.10a)$$

$$-\hat{a}(s') \leq \bar{A}^i(s'), \qquad \forall s'. \qquad (8.9.10b)$$

Let the optimum decision rules be

$$c = h^i(a, s), \qquad (8.9.11a)$$

$$\hat{a}(s') = g^i(a, s, s'). \qquad (8.9.11b)$$

Note that the solution of the Bellman equation implicitly depends on $Q(\cdot|\cdot)$ because it appears in the constraint (8.9.9). In particular, use the first-order conditions for the problem on the right of equation (8.9.8) and the Benveniste-Scheinkman formula and rearrange to get

$$Q(s_{t+1}|s_t) = \frac{\beta u'(c_{t+1}^i) \pi(s_{t+1}|s_t)}{u'(c_t^i)}, \qquad (8.9.12)$$

where it is understood that $c_t^i = h^i(a_t^i, s_t)$ and $c_{t+1}^i = h^i(a_{t+1}^i(s_{t+1}), s_{t+1}) = h^i(g^i(a_t^i, s_t, s_{t+1}), s_{t+1})$.

DEFINITION: A *recursive competitive equilibrium* is an initial distribution of wealth $\vec{a}_0$, a set of borrowing limits $\{\bar{A}^i(s)\}_{i=1}^I$, a pricing kernel $Q(s'|s)$, sets of value functions $\{v^i(a, s)\}_{i=1}^I$, and decision rules $\{h^i(a, s), g^i(a, s, s')\}_{i=1}^I$ such that

(a) The state-by-state borrowing constraints satisfy the recursion

$$\bar{A}^i(s) = y^i(s) + \sum_{s'} Q(s'|s) \bar{A}^i(s'|s). \qquad (8.9.13)$$

(b) For all i, given a_0^i, $\bar{A}^i(s)$, and the pricing kernel, the value functions and decision rules solve the household's problem;

(c) For all realizations of $\{s_t\}_{t=0}^{\infty}$, the consumption and asset portfolios $\{\{c_t^i, \{\hat{a}_{t+1}^i(s')\}_{s'}\}_i\}_t$ implied by the decision rules satisfy $\sum_i c_t^i = \sum_i y^i(s_t)$ and $\sum_i \hat{a}_{t+1}^i(s') = 0$ for all t and s'.

We shall use the recursive competitive equilibrium concept extensively in our discussion of asset pricing in chapter 13.

8.9.4. *Computing an equilibrium with sequential trading of Arrow-securities*

We use example 4 from subsection 8.6.5 to illustrate the following algorithm for computing an equilibrium in an economy with sequential trading of a complete set of Arrow securities:

1. Compute an equilibrium of the Arrow-Debreu economy with time 0 trading.

2. Set the equilibrium allocation for the sequential trading economy to the equilibrium allocation from Arrow-Debreu time 0 trading economy.

3. Compute equilibrium prices from formula (8.9.12) for a Markov economy or the corresponding formula (8.8.9) for a non-Markov economy.

4. Compute debt limits from (8.9.13).

5. Compute portfolios of one-period Arrow securities by first computing implied time t, history s^t wealth $\Upsilon_t^i(s^t)$ from (8.8.2) evaluated at the Arrow-Debreu equilibrium prices, then set $a_t^i(s_t) = \Upsilon_t^i(s^t)$.

Applying this procedure to example 4 from section 8.6.5 gives us the price system $Q_0(s_1 = 1|s_0 = 1) = \beta, Q_0(s_1 = 0|s_0 = 1) = 0, Q_1(s_2 = 1|s_1 = 1) = .5\beta, Q_1(s_2 = 0|s_1 = 0) = .5\beta$ and $Q_t(s_{t+1} = 1|s_t = 1) = Q_t(s_{t+1} = 0|s_t = 0) = \beta$ for $t \geq 2$. Also, $\Upsilon_t^i(s^t) = 0$ for $i = 1, 2$ and $t = 0, 1$. For $t \geq 2$, $\Upsilon_t^1(s_t = 1) = \sum_{\tau \geq t} \beta^{\tau-t}[.5 - 1] = \frac{-.5}{1-\beta}$ and $\Upsilon_t^2(s_t = 1) = \sum_{\tau \geq t} \beta^{\tau-t}[.5 - 0] = \frac{.5}{1-\beta}$. Therefore, in period 1, the first consumer trades Arrow securities in amounts $a_2^1(s_2 = 1) = \frac{-.5}{1-\beta}, a_2^1(s_2 = 0) = \frac{.5}{1-\beta}$, while the second consumer trades Arrow securities in amounts $a_2^1(s_2 = 1) = \frac{.5}{1-\beta}, a_2^1(s_2 = 0) = \frac{-.5}{1-\beta}$ After period 2, the consumers perpetually roll over their debts or assets of either $\frac{.5}{1-\beta}$ or $\frac{-.5}{1-\beta}$.

8.10. *j*-step pricing kernel

We are sometimes interested in the price at time t of a claim to one unit of consumption at date $\tau > t$ contingent on the time τ state being s_τ, *regardless* of the particular history by which s_τ is reached at τ. We let $Q_j(s'|s)$ denote the j-step pricing kernel to be interpreted as follows: $Q_j(s'|s)$ gives the price of one unit of consumption j periods ahead, contingent on the state in that future period being s', given that the current state is s. For example, $j = 1$ corresponds to the one-step pricing kernel $Q(s'|s)$.

With markets in all possible j-step-ahead contingent claims, the counterpart to constraint (8.8.4), the household's budget constraint at time t, is

$$c_t^i + \sum_{j=1}^{\infty} \sum_{s_{t+j}} Q_j(s_{t+j}|s_t) z_{t,j}^i(s_{t+j}) \leq y^i(s_t) + a_t^i. \tag{8.10.1}$$

Here $z_{t,j}^i(s_{t+j})$ is household i's holdings at the end of period t of contingent claims that pay one unit of the consumption good j periods ahead at date $t+j$, contingent on the state at date $t+j$ being s_{t+j}. The household's wealth in the next period depends on the chosen asset portfolio and the realization of s_{t+1},

$$a_{t+1}^i(s_{t+1}) = z_{t,1}^i(s_{t+1}) + \sum_{j=2}^{\infty} \sum_{s_{t+j}} Q_{j-1}(s_{t+j}|s_{t+1}) z_{t,j}^i(s_{t+j}).$$

The realization of s_{t+1} determines which element of the vector of one-period-ahead claims $\{z_{t,1}^i(s_{t+1})\}$ pays off at time $t+1$, and also the capital gains and losses inflicted on the holdings of longer horizon claims implied by equilibrium prices $Q_j(s_{t+j+1}|s_{t+1})$.

With respect to $z_{t,j}^i(s_{t+j})$ for $j > 1$, use the first-order condition for the problem on the right of (8.9.8) and the Benveniste-Scheinkman formula and rearrange to get

$$Q_j(s_{t+j}|s_t) = \sum_{s_{t+1}} \frac{\beta u'[c_{t+1}^i(s_{t+1})] \pi(s_{t+1}|s_t)}{u'(c_t^i)} Q_{j-1}(s_{t+j}|s_{t+1}). \tag{8.10.2}$$

This expression, evaluated at the competitive equilibrium consumption allocation, characterizes two adjacent pricing kernels.[12]   Together with first-order

---

[12] According to expression (8.9.3), the equilibrium consumption allocation is not history dependent, so that $(c_t^i, \{c_{t+1}^i(s_{t+1})\}_{s_{t+1}}) = (\bar{c}^i(s_t), \{\bar{c}^i(s_{t+1})\}_{s_{t+1}})$. Because marginal conditions hold for all households, the characterization of pricing kernels in (8.10.2) holds for any i.

condition (8.9.12), formula (8.10.2) implies that the kernels $Q_j, j = 2, 3, \ldots,$ can be computed recursively:

$$Q_j(s_{t+j}|s_t) = \sum_{s_{t+1}} Q_1(s_{t+1}|s_t)Q_{j-1}(s_{t+j}|s_{t+1}). \qquad (8.10.3)$$

8.10.1. Arbitrage-free pricing

It is useful briefly to describe how arbitrage free pricing theory deduces restrictions on asset prices by manipulating budget sets with redundant assets. We now present an arbitrage argument as an alternative way of deriving restriction (8.10.3) that was established above by using households' first-order conditions evaluated at the equilibrium consumption allocation. In addition to j-step-ahead contingent claims, we illustrate the arbitrage pricing theory by augmenting the trading opportunities in our Arrow securities economy by letting the consumer also trade an ex-dividend Lucas tree. Because markets are already complete, these additional assets are redundant. They have to be priced in a way that leaves the budget set unaltered.[13]

Assume that at time t, in addition to purchasing a quantity $z_{t,j}(s_{t+j})$ of j-step-ahead claims paying one unit of consumption at time $t+j$ if the state takes value s_{t+j} at time $t+j$, the consumer also purchases N_t units of a stock or Lucas tree. Let the ex-dividend price of the tree at time t be $p(s_t)$. Next period, the tree pays a dividend $d(s_{t+1})$ depending on the state s_{t+1}. Ownership of the N_t units of the tree at the beginning of $t+1$ entitles the consumer to a claim on $N_t[p(s_{t+1}) + d(s_{t+1})]$ units of time $t+1$ consumption.[14] As before, let a_t be the wealth of the consumer, apart from his endowment, $y(s_t)$. In this setting, the augmented version of constraint (8.10.1), the consumer's budget constraint, is

$$c_t + \sum_{j=1}^{\infty}\sum_{s_{t+j}} Q_j(s_{t+j}|s_t)z_{t,j}(s_{t+j}) + p(s_t)N_t \le a_t + y(s_t) \qquad (8.10.4a)$$

---

[13] That the additional assets are redundant follows from the fact that trading Arrow securities is sufficient to complete markets.

[14] We calculate the price of this asset using a different method in chapter 13.

and

$$a_{t+1}(s_{t+1}) = z_{t,1}(s_{t+1}) + [p(s_{t+1}) + d(s_{t+1})] N_t$$

$$+ \sum_{j=2}^{\infty} \sum_{s_{t+j}} Q_{j-1}(s_{t+j}|s_{t+1}) z_{t,j}(s_{t+j}). \qquad (8.10.4b)$$

Multiply equation $(8.10.4b)$ by $Q_1(s_{t+1}|s_t)$, sum over s_{t+1}, solve for $\sum_{s_{t+1}} Q_1(s_{t+1}|s_t) z_{t,1}(s_t)$, and substitute this expression in $(8.10.4a)$ to get

$$c_t + \left\{ p(s_t) - \sum_{s_{t+1}} Q_1(s_{t+1}|s_t)[p(s_{t+1}) + d(s_{t+1})] \right\} N_t$$

$$+ \sum_{j=2}^{\infty} \sum_{s_{t+j}} \left\{ Q_j(s_{t+j}|s_t) - \sum_{s_{t+1}} Q_{j-1}(s_{t+j}|s_{t+1}) Q_1(s_{t+1}|s_t) \right\} z_{t,j}(s_{t+j})$$

$$+ \sum_{s_{t+1}} Q_1(s_{t+1}|s_t) a_{t+1}(s_{t+1}) \leq a_t + y(s_t). \qquad (8.10.5)$$

If the two terms in braces are not zero, the consumer can attain unbounded consumption and future wealth by purchasing or selling either the stock (if the first term in braces is not zero) or a state-contingent claim (if any of the terms in the second set of braces is not zero). Therefore, so long as the utility function has no satiation point, in any equilibrium, the terms in the braces must be zero. Thus, we have the arbitrage pricing formulas

$$p(s_t) = \sum_{s_{t+1}} Q_1(s_{t+1}|s_t)[p(s_{t+1}) + d(s_{t+1})], \qquad (8.10.6a)$$

$$Q_j(s_{t+j}|s_t) = \sum_{s_{t+1}} Q_{j-1}(s_{t+j}|s_{t+1}) Q_1(s_{t+1}|s_t). \qquad (8.10.6b)$$

These are called *arbitrage pricing formulas* because if they were violated, there would exist an *arbitrage*. An arbitrage is defined as a risk-free transaction that earns positive profits.

8.11. Recursive version of Pareto problem

At the very outset of this chapter, we characterized Pareto optimal allocations. This section considers how to formulate a Pareto problem recursively, which will give a preview of things to come in chapters 20 and 23. For this purpose, we consider a special case of the earlier section 8.6.3 example 2 of an economy with a constant aggregate endowment and two types of household with $y_t^1 = s_t, y_t^2 = 1 - s_t$. We now assume that the s_t process is i.i.d., so that $\pi_t(s^t) = \pi(s_t)\pi(s_{t-1})\cdots\pi(s_0)$. Also, let's assume that s_t has a discrete distribution so that $s_t \in [\bar{s}_1, \ldots, \bar{s}_S]$ with probabilities $\Pi_i = \text{Prob}(s_t = \bar{s}_i)$ where $\bar{s}_{i+1} > \bar{s}_i$ and $\bar{s}_1 \geq 0$ and $\bar{s}_S \leq 1$.

In our recursive formulation, each period a planner delivers a pair of previously promised discounted utility streams by assigning a state-contingent consumption allocation today and a pair of state-contingent promised discounted utility streams starting tomorrow. Both the state-contingent consumption today and the promised discounted utility tomorrow are functions of the initial promised discounted utility levels.

Define v as the expected discounted utility of a type 1 person and $P(v)$ as the maximal expected discounted utility that can be offered to a type 2 person, given that a type 1 person is offered at least v. Each of these expected values is to be evaluated before the realization of the state at the initial date.

The Pareto problem is to choose stochastic processes $\{c_t^1(s^t), c_t^2(s^t)\}_{t=0}^{\infty}$ to maximize $P(v)$ subject to the utility constraint $\sum_{t=0}^{\infty}\sum_{s^t}\beta^t u(c_t^1(s^t))\pi_t(s^t) \geq v$ and $c_t^1(s^t) + c_t^2(s^t) = 1$. In terms of the competitive equilibrium allocation calculated for the section 8.6.3 example 2 economy above, let $\bar{c} = \bar{c}^1$ be the constant consumption allocated to a type 1 person and $1 - \bar{c} = \bar{c}^2$ be the constant consumption allocated to a type 2 person. Since we have shown that the competitive equilibrium allocation is a Pareto optimal allocation, we already know one point on the Pareto frontier $P(v)$. In particular, when a type 1 person is promised $v = u(\bar{c})/(1 - \beta)$, a type 2 person attains life-time utility $P(v) = u(1 - \bar{c})/(1 - \beta)$.

We can express the discounted values v and $P(v)$ recursively [15] as

$$v = \sum_{i=1}^{S} [u(c_i) + \beta w_i] \, \Pi_i$$

and

$$P(v) = \sum_{i=1}^{S} [u(1 - c_i) + \beta P(w_i)] \, \Pi_i,$$

where c_i is consumption of the type 1 person in state i, w_i is the continuation value assigned to the type 1 person in state i; and $1 - c_i$ and $P(w_i)$ are the consumption and the continuation value, respectively, assigned to a type 2 person in state i. Assume that the continuation values $w_i \in V$, where V is a set of admissible discounted values of utility. In this section, we assume that $V = [u(\epsilon)/(1 - \beta), \, u(1)/(1 - \beta)]$ where $\epsilon \in (0, 1)$ is an arbitrarily small number.

In effect, before the realization of the current state, a Pareto optimal allocation offers the type 1 person a state-contingent vector of consumption c_i in state i and a state-contingent vector of continuation values w_i in state i, with each w_i itself being a present value of one-period future utilities. In terms of the pair of values $(v, P(v))$, we can express the Pareto problem recursively as

$$P(v) = \max_{\{c_i, w_i\}_{i=1}^{S}} \sum_{i=1}^{S} [u(1 - c_i) + \beta P(w_i)] \Pi_i \qquad (8.11.1)$$

where the maximization is subject to

$$\sum_{i=1}^{S} [u(c_i) + \beta w_i] \Pi_i \geq v \qquad (8.11.2)$$

where $c_i \in [0, 1]$ and $w_i \in V$.

To solve the Pareto problem, form the Lagrangian

$$L = \sum_{i=1}^{S} \Pi_i [u(1 - c_i) + \beta P(w_i) + \theta(u(c_i) + \beta w_i)] - \theta v$$

---

[15] This is our first example of a 'dynamic program squared'. We call it that because the state variable v that appears in the Bellman equation for $P(v)$ itself satisfies another Bellman equation.

where θ is a Lagrange multiplier on constraint (8.11.2). First-order conditions with respect to c_i and w_i, respectively, are

$$-u'(1 - c_i) + \theta u'(c_i) = 0, \qquad (8.11.3a)$$

$$P'(w_i) + \theta = 0. \qquad (8.11.3b)$$

The envelope condition is $P'(v) = -\theta$. Thus, (8.11.3b) becomes $P'(w_i) = P'(v)$. But $P(v)$ happens to be strictly concave, so this equality implies $w_i = v$. Therefore, any solution of the Pareto problem leaves the continuation value w_i independent of the state i. Equation (8.11.3a) implies that

$$\frac{u'(1 - c_i)}{u'(c_i)} = -P'(v). \qquad (8.11.4)$$

Since the right side of (8.11.4) is independent of i, so is the left side, and therefore c is independent of i. And since v is constant over time (because $w_i = v$ for all i), it follows that c is constant over time.

Notice from (8.11.4) that $P'(v)$ serves as a relative Pareto weight on the type 1 person. The recursive formulation brings out that, because $P'(w_i) = P'(v)$, the relative Pareto weight remains constant over time and is independent of the realization of s_t. The planner imposes complete risk sharing.

In chapter 20, we shall encounter recursive formulations again. Impediments to risk sharing that occur in the form either of enforcement or of information constraints will impel the planner sometimes to make continuation values respond to the current realization of shocks to endowments or preferences.

8.12. Concluding remarks

The framework in this chapter serves much of macroeconomics either as foundation or straw man ("benchmark model" is a kinder phrase than "straw man"). It is the foundation of extensive literatures on asset pricing and risk sharing. We describe the literature on asset pricing in more detail in chapters 13 and 14. The model also serves as benchmark, or point of departure, for a variety of models designed to confront observations that seem inconsistent with complete markets. In particular, for models with exogenously imposed incomplete markets, see chapters 17 on precautionary saving and 18 on incomplete markets. For models with endogenous incomplete markets, see chapters 20 and 21 on enforcement and information problems. For models of money, see chapters 26 and 27. To take monetary theory as an example, complete markets models dispose of any need for money because they contain an efficient multilateral trading mechanism, with such extensive netting of claims that no medium of exchange is required to facilitate bilateral exchanges. Any modern model of money introduces frictions that impede complete markets. Some monetary models (e.g., the cash-in-advance model of Lucas, 1981) impose minimal impediments to complete markets, to preserve many of the asset-pricing implications of complete markets models while also allowing classical monetary doctrines like the quantity theory of money. The shopping time model of chapter 26 is constructed in a similar spirit. Other monetary models, such as the Townsend turnpike model of chapter 27 or the Kiyotaki-Wright search model of chapter 28, impose more extensive frictions on multilateral exchanges and leave the complete markets model farther behind. Before leaving the complete markets model, we'll put it to work in several of the following chapters.

A. Gaussian asset-pricing model

The theory of this chapter is readily adapted to a setting in which the state of
the economy evolves according to a continuous-state Markov process. We use
such a version in chapter 14. Here we give a taste of how such an adaptation can
be made by describing an economy in which the state follows a linear stochastic
difference equation driven by a Gaussian disturbance. If we supplement this
with the specification that preferences are quadratic, we get a setting in which
asset prices can be calculated swiftly.

Suppose that the state evolves according to the stochastic difference equation

$$s_{t+1} = As_t + Cw_{t+1} \qquad (8.A.1)$$

where A is a matrix whose eigenvalues are bounded from above in modulus by
$1/\sqrt{\beta}$ and w_{t+1} is a Gaussian martingale difference sequence adapted to the
history of s_t. Assume that $Ew_{t+1}w_{t+1} = I$. The conditional density of s_{t+1} is
Gaussian:

$$\pi(s_t|s_{t-1}) \sim \mathcal{N}(As_{t-1}, CC'). \qquad (8.A.2)$$

More precisely,

$$\pi(s_t|s_{t-1}) = K \exp\left\{-.5(s_t - As_{t-1})(CC')^{-1}(s_t - As_{t-1})\right\}, \qquad (8.A.3)$$

where $K = (2\pi)^{\frac{-k}{72}} \det(CC')^{\frac{-1}{72}}$ and s_t is $k \times 1$. We also assume that $\pi_0(s_0)$ is
Gaussian.[16]

If $\{c_t^i(s_t)\}_{t=0}^\infty$ is the equilibrium allocation to agent i, and the agent has
preferences represented by (8.2.1), the equilibrium pricing function satisfies

$$q_t^0(s^t) = \frac{\beta^t u'[c_t^i(s_t)]\pi(s^t)}{u'[c_0^i(s_0)]}. \qquad (8.A.4)$$

Once again, let $\{d_t(s_t)\}_{t=0}^\infty$ be a stream of claims to consumption. The
time 0 price of the asset with this dividend stream is

$$p_0 = \sum_{t=0}^\infty \int_{s^t} q_t^0(s^t)d_t(s_t)d\,s^t.$$

---

[16] If s_t is stationary, $\pi_0(s_0)$ can be specified to be the stationary distribution of the
process.

Substituting equation (8.A.4) into the preceding equation gives

$$p_0 = \sum_t \int_{s^t} \beta^t \frac{u'[c_t^i(s_t)]}{u'[c_0^i(s_0)]} d_t(s_t) \pi(s^t) ds^t$$

or

$$p_0 = E \sum_{t=0}^{\infty} \beta^t \frac{u'[c_t(s_t)]}{u'[c_0(s_0)]} d_t(s_t). \qquad (8.A.5)$$

This formula expresses the time 0 asset price as an inner product of a discounted marginal utility process and a dividend process.[17]

This formula becomes especially useful in the case that the one-period utility function $u(c)$ is quadratic, so that marginal utilities become linear, and the dividend process d_t is linear in s_t. In particular, assume that

$$u(c_t) = -.5(c_t - b)^2 \qquad (8.A.6)$$

$$d_t = S_d s_t, \qquad (8.A.7)$$

where $b > 0$ is a bliss level of consumption. Furthermore, assume that the equilibrium allocation to agent i is

$$c_t^i = S_{ci} s_t, \qquad (8.A.8)$$

where S_{ci} is a vector conformable to s_t.

The utility function (8.A.6) implies that $u'(c_t^i) = b - c_t^i = b - S_{ci} s_t$. Suppose that unity is one element of the state space for s_t, so that we can express $b = S_b s_t$. Then $b - c_t = S_f s_t$, where $S_f = S_b - S_{ci}$, and the asset-pricing formula becomes

$$p_0 = \frac{E_0 \sum_{t=0}^{\infty} \beta^t s_t' S_f' S_d s_t}{S_f s_0}. \qquad (8.A.9)$$

Thus, to price the asset, we have to evaluate the expectation of the sum of a discounted quadratic form in the state variable. This is easy to do by using results from chapter 2.

In chapter 2, we evaluated the conditional expectation of the geometric sum of the quadratic form

$$\alpha_0 = E_0 \sum_{t=0}^{\infty} \beta^t s_t' S_f' S_d s_t.$$

---

[17] For two scalar stochastic processes x, y, the inner product is defined as $< x, y > = E \sum_{t=0}^{\infty} \beta^t x_t y_t$.

We found that it could be written in the form

$$\alpha_0 = s_0' \mu s_0 + \sigma, \tag{8.A.10}$$

where μ is an $(n \times n)$ matrix and σ is a scalar that satisfy

$$\mu = S_f' S_d + \beta A' \mu A$$
$$\sigma = \beta \sigma + \beta \text{ trace } (\mu C C') \tag{8.A.11}$$

The first equation of $(8.A.11)$ is a *discrete Lyapunov equation* in the square matrix μ, and can be solved by using one of several algorithms.[18] After μ has been computed, the second equation can be solved for the scalar σ.

B. The permanent income model revisited

This appendix is a variation on the theme that 'many single agent models can be reinterpreted as general equilibrium models'.

8.B.1. Reinterpreting the single-agent model

In this appendix, we cast the single-agent linear quadratic permanent income model of section 2.11 of chapter 2 as a competitive equilibrium with time 0 trading of a complete set of history-contingent securities. We begin by reformulating the model in that chapter as a planning problem. The planner has utility functional

$$E_0 \sum_{t=0}^{\infty} \beta^t u(\bar{c}_t) \tag{8.B.1}$$

where E_t is the mathematical expectation conditioned on the consumer's time t information, $\bar{c}_t$ is time t consumption, $u(c) = -.5(\gamma - \bar{c}_t)^2$, and $\beta \in (0,1)$ is a discount factor. The planner maximizes $(8.B.1)$ by choosing a consumption, borrowing plan $\{\bar{c}_t, b_{t+1}\}_{t=0}^{\infty}$ subject to the sequence of budget constraints

$$\bar{c}_t + b_t = R^{-1} b_{t+1} + y_t \tag{8.B.2}$$

---

[18] The Matlab control toolkit has a program called `dlyap.m`; also see a program called `doublej.m`.

where y_t is an exogenous stationary endowment process, R is a constant gross risk-free interest rate, $-R^{-1}b_t \equiv \bar{k}_t$ is the stock of an asset that bears a risk free one-period gross return of R, and b_0 is a given initial condition. We assume that $R^{-1} = \beta$ and that the endowment process has the state-space representation

$$z_{t+1} = A_{22}z_t + C_2 w_{t+1} \qquad (8.B.3a)$$

$$y_t = U_y z_t \qquad (8.B.3b)$$

where w_{t+1} is an i.i.d. process with mean zero and identity contemporaneous covariance matrix, A_{22} is a stable matrix, its eigenvalues being strictly below unity in modulus, and U_y is a selection vector that identifies y with a particular linear combination of z_t. As shown in chapter 2, the solution of what we now interpret as a planning problem can be represented as the following versions of equations $(2.11.9)$ and $(2.11.20)$, respectively:

$$\bar{c}_t = (1-\beta)\left[U_y(I - \beta A_{22})^{-1}z_t - R\bar{k}_t\right] \qquad (8.B.4)$$

$$\bar{k}_{t+1} = \bar{k}_t + RU_y(I - \beta A_{22})^{-1}(A_{22} - I)z_t. \qquad (8.B.5)$$

We can represent the optimal consumption, capital accumulation path compactly as

$$\begin{bmatrix} \bar{k}_{t+1} \\ z_{t+1} \end{bmatrix} = A \begin{bmatrix} \bar{k}_t \\ z_t \end{bmatrix} + \begin{bmatrix} 0 \\ C_2 \end{bmatrix} w_{t+1} \qquad (8.B.6)$$

$$\bar{c}_t = S_c \begin{bmatrix} \bar{k}_t \\ z_t \end{bmatrix} \qquad (8.B.7)$$

where the matrices A, S_c can readily be constructed from the solutions and specifications just mentioned. In addition, it is useful to have at our disposal the marginal utility of consumption process $p_t^0 \equiv (\gamma - \bar{c}_t)$, which can be represented as

$$p_t^0(z^t) = S_p \begin{bmatrix} \bar{k}_t \\ z_t \end{bmatrix} \qquad (8.B.8)$$

and where S_p can be constructed easily from S_c. Solving equation $(8.B.5)$ recursively shows that k_{t+1} is a function $k_{t+1}(z^t; k_0)$ of history z^t. In equation $(8.B.8)$, $\bar{k}_t$ encodes the history dependence of $p_t^0(z^t)$.

Equations $(8.B.6)$, $(8.B.7)$, $(8.B.8)$ together with the equation $r_t^0 = \alpha$ to be explained below turn out to be representations of the equilibrium price system in the competitive equilibrium to which we turn next.

8.B.2. *Decentralization and scaled prices*

Let $q_t^0(z^t)$ the time 0 price of a unit of time t consumption at history z^t. Let $\pi_t(z^t)$ the probability density of the history z^t induced by the state-space representation $(8.B.3)$. Define the adjusted Arrow-Debreu price scaled by discounting and probabilities as

$$p_t^0(z^t) = \frac{q_t^0(z^t)}{\beta^t \pi_t(z^t)}. \tag{8.B.9}$$

We find it convenient to express a representative consumer's problem and a representative firm's problem in terms of these scaled Arrow-Debreu prices.

Evidently, the present value of consumption, for example, can be represented as

$$\sum_{t=0}^{\infty} \sum_{z^t} q_t^0(z^t) c_t(z^t) = \sum_{t=0}^{\infty} \sum_{z^t} \beta^t p_t^0(z^t) c_t(z^t) \pi_t(z^t)$$

$$= E_0 \sum_{t=0}^{\infty} \beta^t p_t(z^t) c_t(z^t).$$

Below, it will be convenient for us to represent present values as conditional expectations of discounted sums as is done in the second line.

We let $r_t^0(z^t)$ be the rental rate on capital, again scaled analogously to $(8.B.9)$. Both the consumer and the firm take these processes as given.

The consumer owns and operates the technology for accumulating capital. The consumer owns the endowment process $\{y_t\}_{t=0}^{\infty}$, which it sells to a firm that operates a production technology. The consumer rents capital to the firm. The firm uses the endowment and capital to produce output that it sells to the consumer at a competitive price. The consumer divides his time t purchases between consumption c_t and gross investment x_t.

8.B.2.1.   The consumer

Let $\{p_t^0(z^t), r_t^0(z^t)\}_{t=0}^{\infty}$ be a price system, each component of which takes the form of a 'scaled Arrow-Debreu price' (attained by dividing a time-0 Arrow-Debreu price by a discount factor times a probability, as in the previous subsection). The representative consumer's problem is to choose processes $\{c_t, k_{t+1}\}_{t=0}^{\infty}$ to maximize

$$-.5E_0 \sum_{t=0}^{\infty} \beta^t (\gamma - c_t)^2 \qquad (8.B.10)$$

subject to

$$E_0 \sum_{t=0}^{\infty} \beta^t p_t^0(z^t) o_t(z^t) = E_0 \sum_{t=0}^{\infty} \beta^t \left(p_t^0(z^t) y_t + r_t^0(z^t) k_t(z^t) \right) \qquad (8.B.11)$$

$$k_{t+1} = (1 - \delta)k_t + x_t \qquad (8.B.12)$$

$$o_t(z^t) = c_t(z^t) + x_t(z^t) \qquad (8.B.13)$$

where k_0 is a given initial condition. Here x_t is gross investment and k_t is physical capital owned by the household and rented to firms. The consumer purchases output $o_t = c_t + x_t$ from competitive firms. The consumer sells its endowment y_t and rents its capital k_t to firms at prices $p_t^0(z^t)$ and $r_t^0(z^t)$. Equation (8.B.12) is the law of motion for physical capital, where $\delta \in (0, 1)$ is a depreciation rate.

8.B.2.2.   The firm

A competitive representative firm chooses processes $\{k_t, c_t, x_t\}_{t=0}^{\infty}$ to maximize

$$E_0 \sum_{t=0}^{\infty} \beta^t \{ p_t^0(z^t) o_t(z^t) - p_t^0(z^t) y_t - r_t^0(z^t) k_t \} \qquad (8.B.14)$$

subject to the physical technology

$$o_t(z^t) = \alpha k_t + y_t(z_t), \qquad (8.B.15)$$

where $\alpha > 0$. Since the marginal product of capital is α, a good guess is that

$$r_t^0(z^t) = \alpha. \qquad (8.B.16)$$

8.B.3. Matching equilibrium and planning allocations

We impose the condition

$$\alpha + (1 - \delta) = R. \tag{8.B.17}$$

This makes the gross rates of return in investment identical in the planning and decentralized economies. In particular, if we substitute equation $(8.B.12)$ into equation $(8.B.15)$ and remember that $b_t \equiv Rk_t$, we obtain $(8.B.2)$.

It is straightforward to verify that the allocation $\{\bar{k}_{t+1}, \bar{c}_t\}_{t=0}^{\infty}$ that solves the planning problem is a competitive equilibrium allocation.

As in chapter 7, we have distinguished between the planning allocation $\{\bar{k}_{t+1}, \bar{c}_t\}_{t=0}^{\infty}$ that determines the equilibrium price functions defined in subsection 8.B.1 and the allocation chosen by the representative firm and the representative consumer who face those prices as price takers. This is yet another example of the 'big K, little k' device from chapter 7.

8.B.4. Interpretation

As we saw in section 2.11 of chapter 2 and also in representation $(8.B.4)$ $(8.B.5)$ here, what is now *equilibrium* consumption is a random walk. Why, despite his preference for a *smooth* consumption path, does the representative consumer accept fluctuations in his consumption? In the complete markets economy of this appendix, the consumer believes that it is possible for him completely to smooth consumption over time and across histories by purchasing and selling history contingent claims. But at the equilibrium prices facing him, the consumer prefers to tolerate fluctuations in consumption over time and across histories.

Exercises

Exercise 8.1 **Existence of representative consumer**

Suppose households 1 and 2 have one-period utility functions $u(c^1)$ and $w(c^2)$, respectively, where u and w are both increasing, strictly concave, twice differentiable functions of a scalar consumption rate. Consider the Pareto problem:

$$v_\theta(c) = \max_{\{c^1, c^2\}} \left[\theta u(c^1) + (1 - \theta) w(c^2) \right]$$

subject to the constraint $c^1 + c^2 = c$. Show that the solution of this problem has the form of a concave utility function $v_\theta(c)$, which depends on the Pareto weight θ. Show that $v'_\theta(c) = \theta u'(c^1) = (1 - \theta) w'(c^2)$.

The function $v_\theta(c)$ is the utility function of the *representative consumer*. Such a representative consumer always lurks within a complete markets competitive equilibrium even with heterogeneous preferences. At a competitive equilibrium, the marginal utilities of the representative agent and each and every agent are proportional.

Exercise 8.2 **Term structure of interest rates**

Consider an economy with a single consumer. There is one good in the economy, which arrives in the form of an exogenous endowment obeying [19]

$$y_{t+1} = \lambda_{t+1} y_t,$$

where y_t is the endowment at time t and $\{\lambda_{t+1}\}$ is governed by a two-state Markov chain with transition matrix

$$P = \begin{bmatrix} p_{11} & 1 - p_{11} \\ 1 - p_{22} & p_{22} \end{bmatrix},$$

and initial distribution $\pi_\lambda = [\pi_0 \quad 1 - \pi_0]$. The value of λ_t is given by $\bar{\lambda}_1 = .98$ in state 1 and $\bar{\lambda}_2 = 1.03$ in state 2. Assume that the history of y_s, λ_s up to t is observed at time t. The consumer has endowment process $\{y_t\}$ and has preferences over consumption streams that are ordered by

$$E_0 \sum_{t=0}^{\infty} \beta^t u(c_t)$$

---

[19] Such a specification was made by Mehra and Prescott (1985).

where $\beta \in (0, 1)$ and $u(c) = \frac{c^{1-\gamma}}{1-\gamma}$, where $\gamma \geq 1$.

a. Define a competitive equilibrium, being careful to name all of the objects of which it consists.

b. Tell how to compute a competitive equilibrium.

For the remainder of this problem, suppose that $p_{11} = .8$, $p_{22} = .85$, $\pi_0 = .5$, $\beta = .96$, and $\gamma = 2$. Suppose that the economy begins with $\lambda_0 = .98$ and $y_0 = 1$.

c. Compute the (unconditional) average growth rate of consumption, computed before having observed λ_0.

d. Compute the time 0 prices of three risk-free discount bonds, in particular, those promising to pay one unit of time j consumption for $j = 0, 1, 2$, respectively.

e. Compute the time 0 prices of three bonds, in particular, ones promising to pay one unit of time j consumption contingent on $\lambda_j = \bar{\lambda}_1$ for $j = 0, 1, 2$, respectively.

f. Compute the time 0 prices of three bonds, in particular, ones promising to pay one unit of time j consumption contingent on $\lambda_j = \bar{\lambda}_2$ for $j = 0, 1, 2$, respectively.

g. Compare the prices that you computed in parts d, e, and f.

Exercise 8.3 An economy consists of two infinitely lived consumers named $i = 1, 2$. There is one nonstorable consumption good. Consumer i consumes c_t^i at time t. Consumer i ranks consumption streams by

$$\sum_{t=0}^{\infty} \beta^t u(c_t^i),$$

where $\beta \in (0, 1)$ and $u(c)$ is increasing, strictly concave, and twice continuously differentiable. Consumer 1 is endowed with a stream of the consumption good $y_t^i = 1, 0, 0, 1, 0, 0, 1, \ldots$. Consumer 2 is endowed with a stream of the consumption good $0, 1, 1, 0, 1, 1, 0, \ldots$. Assume that there are complete markets with time 0 trading.

a. Define a competitive equilibrium.

b. Compute a competitive equilibrium.

c. Suppose that one of the consumers markets a derivative asset that promises to pay .05 units of consumption each period. What would the price of that asset be?

Exercise 8.4 Consider a pure endowment economy with a single representative consumer; $\{c_t, d_t\}_{t=0}^{\infty}$ are the consumption and endowment processes, respectively. Feasible allocations satisfy

$$c_t \leq d_t.$$

The endowment process is described by [20]

$$d_{t+1} = \lambda_{t+1} d_t.$$

The growth rate λ_{t+1} is described by a two-state Markov process with transition probabilities

$$P_{ij} = \text{Prob}(\lambda_{t+1} = \bar{\lambda}_j | \lambda_t = \bar{\lambda}_i).$$

Assume that

$$P = \begin{bmatrix} .8 & .2 \\ .1 & .9 \end{bmatrix},$$

and that

$$\bar{\lambda} = \begin{bmatrix} .97 \\ 1.03 \end{bmatrix}.$$

In addition, $\lambda_0 = .97$ and $d_0 = 1$ are both known at date 0. The consumer has preferences over consumption ordered by

$$E_0 \sum_{t=0}^{\infty} \beta^t \frac{c_t^{1-\gamma}}{1-\gamma},$$

where E_0 is the mathematical expectation operator, conditioned on information known at time 0, $\gamma = 2, \beta = .95$.

Part I

At time 0, after d_0 and λ_0 are known, there are complete markets in date- and history-contingent claims. The market prices are denominated in units of time 0 consumption goods.

---

[20] See Mehra and Prescott (1985).

a. Define a competitive equilibrium, being careful to specify all the objects composing an equilibrium.

b. Compute the equilibrium price of a claim to one unit of consumption at date 5, denominated in units of time 0 consumption, contingent on the following history of growth rates: $(\lambda_1, \lambda_2, \ldots, \lambda_5) = (.97, .97, 1.03, .97, 1.03)$. Please give a numerical answer.

c. Compute the equilibrium price of a claim to one unit of consumption at date 5, denominated in units of time 0 consumption, contingent on the following history of growth rates: $(\lambda_1, \lambda_2, \ldots, \lambda_5) = (1.03, 1.03, 1.03, 1.03, .97)$.

d. Give a formula for the price at time 0 of a claim on the entire endowment sequence.

e. Give a formula for the price at time 0 of a claim on consumption in period 5, contingent on the growth rate λ_5 being .97 (regardless of the intervening growth rates).

Part II

Now assume a different market structure. Assume that at each date $t \geq 0$ there is a complete set of one-period forward Arrow securities.

f. Define a (recursive) competitive equilibrium with Arrow securities, being careful to define all of the objects that compose such an equilibrium.

g. For the representative consumer in this economy, for each state compute the "natural debt limits" that constrain state-contingent borrowing.

h. Compute a competitive equilibrium with Arrow securities. In particular, compute both the pricing kernel and the allocation.

i. An entrepreneur enters this economy and proposes to issue a new security each period, namely, a risk-free two-period bond. Such a bond issued in period t promises to pay one unit of consumption at time $t+1$ for sure. Find the price of this new security in period t, contingent on λ_t.

Exercise 8.5

An economy consists of two consumers, named $i = 1, 2$. The economy exists in discrete time for periods $t \geq 0$. There is one good in the economy, which is not storable and arrives in the form of an endowment stream owned by each consumer. The endowments to consumers $i = 1, 2$ are

$$y_t^1 = s_t$$
$$y_t^2 = 1$$

where s_t is a random variable governed by a two-state Markov chain with values $s_t = \bar{s}_1 = 0$ or $s_t = \bar{s}_2 = 1$. The Markov chain has time invariant transition probabilities denoted by $\pi(s_{t+1} = s' | s_t = s) = \pi(s' | s)$, and the probability distribution over the initial state is $\pi_0(s)$. The *aggregate endowment* at t is $Y(s_t) = y_t^1 + y_t^2$.

Let c^i denote the stochastic process of consumption for agent i. Household i orders consumption streams according to

$$U(c^i) = \sum_{t=0}^{\infty} \sum_{s^t} \beta^t \ln[c_t^i(s^t)] \pi_t(s^t),$$

where $\pi_t(s^t)$ is the probability of the history $s^t = (s_0, s_1, \ldots, s_t)$.

a. Give a formula for $\pi_t(s^t)$.

b. Let $\theta \in (0, 1)$ be a Pareto weight on household 1. Consider the planning problem

$$\max_{c^1, c^2} \left\{ \theta \ln(c^1) + (1 - \theta) \ln(c^2) \right\}$$

where the maximization is subject to

$$c_t^1(s^t) + c_t^2(s^t) \le Y(s_t).$$

Solve the Pareto problem, taking θ as a parameter.

c. Define a *competitive equilibrium* with history-dependent Arrow-Debreu securities traded once and for all at time 0. Be careful to define all of the objects that compose a competitive equilibrium.

d. Compute the competitive equilibrium price system (i.e., find the prices of all of the Arrow-Debreu securities).

e. Tell the relationship between the solutions (indexed by θ) of the Pareto problem and the competitive equilibrium allocation. If you wish, refer to the two welfare theorems.

f. Briefly tell how you can compute the competitive equilibrium price system *before* you have figured out the competitive equilibrium allocation.

g. Now define a recursive competitive equilibrium with trading every period in one-period Arrow securities only. Describe all of the objects of which such an equilibrium is composed. (Please denominate the prices of one-period time $t+1$ state-contingent Arrow securities in units of time t consumption.) Define the "natural borrowing limits" for each consumer in each state. Tell how to compute these natural borrowing limits.

h. Tell how to compute the prices of one-period Arrow securities. How many prices are there (i.e., how many numbers do you have to compute)? Compute all of these prices in the special case that $\beta = .95$ and $\pi(s_j|s_i) = P_{ij}$ where
$$P = \begin{bmatrix} .8 & .2 \\ .3 & .7 \end{bmatrix}.$$

i. Within the one-period Arrow securities economy, a new asset is introduced. One of the households decides to market a one-period-ahead riskless claim to one unit of consumption (a one-period real bill). Compute the equilibrium prices of this security when $s_t = 0$ and when $s_t = 1$. Justify your formula for these prices in terms of first principles.

j. Within the one-period Arrow securities equilibrium, a new asset is introduced. One of the households decides to market a two-period-ahead riskless claim to one unit of consumption (a two-period real bill). Compute the equilibrium prices of this security when $s_t = 0$ and when $s_t = 1$.

k. Within the one-period Arrow securities equilibrium, a new asset is introduced. One of the households decides at time t to market five-period-ahead claims to consumption at $t+5$ contingent on the value of s_{t+5}. Compute the equilibrium prices of these securities when $s_t = 0$ and $s_t = 1$ and $s_{t+5} = 0$ and $s_{t+5} = 1$.

Exercise 8.6 **Optimal taxation**

The government of a small country must finance an exogenous stream of government purchases $\{g_t\}_{t=0}^\infty$. Assume that g_t is described by a discrete-state Markov chain with transition matrix P and initial distribution π_0. Let $\pi_t(g^t)$ denote the probability of the history $g^t = g_t, g_{t-1}, \ldots, g_0$, conditioned on g_0. The state of the economy is completely described by the history g^t. There are complete markets in date-history claims to goods. At time 0, after g_0 has been realized, the government can purchase or sell claims to time t goods contingent on the history g^t at a price $p_t^0(g^t) = \beta^t \pi_t(g^t)$, where $\beta \in (0,1)$. The date-state prices are exogenous to the small country. The government finances its expenditures by raising history-contingent tax revenues of $R_t = R_t(g^t)$ at time t. The present value of its expenditures must not exceed the present value of its revenues.

Raising revenues by taxation is distorting. The government confronts a dead weight loss function $W(R_t)$ that measures the distortion at time t. Assume that W is an increasing, twice differentiable, strictly convex function that satisfies $W(0) = 0, W'(0) = 0, W'(R) > 0$ for $R > 0$ and $W''(R) > 0$ for $R \geq 0$. The government devises a state-contingent taxation and borrowing plan to minimize

$$E_0 \sum_{t=0}^{\infty} \beta^t W(R_t), \tag{1}$$

where E_0 is the mathematical expectation conditioned on g_0.

Suppose that g_t takes two possible values, $\bar{g}_1 = .2$ (peace) and $\bar{g}_2 = 1$ (war) and that $P = \begin{bmatrix} .8 & .2 \\ .5 & .5 \end{bmatrix}$. Suppose that $g_0 = .2$. Finally, suppose that $W(R) = .5R^2$.

a. Please write out (1) long hand, i.e., write out an explicit expression for the mathematical expectation E_0 in terms of a summation over the appropriate probability distribution.

b. Compute the optimal tax and borrowing plan. In particular, give analytic expressions for $R_t = R_t(g^t)$ for all t and all g^t.

c. There is an equivalent market setting in which the government can buy and sell one-period Arrow securities each period. Find the price of one-period Arrow securities at time t, denominated in units of the time t good.

d. Let $B_t(g_t)$ be the one-period Arrow securities at t that the government issued for state g_t at time $t-1$. For $t > 0$, compute $B_t(g_t)$ for $g_t = \bar{g}_1$ and $g_t = \bar{g}_2$.

e. Use your answers to parts b and d to describe the government's optimal policy for taxing and borrowing.

Exercise 8.7   **A competitive equilibrium**

An endowment economy consists of two type of consumers. Consumers of type 1 order consumption streams of the one good according to

$$\sum_{t=0}^{\infty} \beta^t c_t^1$$

and consumers of type 2 order consumption streams according to

$$\sum_{t=0}^{\infty} \beta^t \ln(c_t^2)$$

where $c_t^i \geq 0$ is the consumption of a type i consumer and $\beta \in (0,1)$ is a common discount factor. The consumption good is tradable but nonstorable. There are equal numbers of the two types of consumer. The consumer of type 1 is endowed with the consumption sequence

$$y_t^1 = \mu > 0 \quad \forall t \geq 0$$

where $\mu > 0$. The consumer of type 2 is endowed with the consumption sequence

$$y_t^2 = \begin{cases} 0 & \text{if } t \geq 0 \text{ is even} \\ \alpha & \text{if } t \geq 0 \text{ is odd} \end{cases}$$

where $\alpha = \mu(1 + \beta^{-1})$.

a. Define a competitive equilibrium with time 0 trading. Be careful to include definitions of all of the objects of which a competitive equilibrium is composed.

b. Compute a competitive equilibrium allocation with time 0 trading.

c. Compute the time 0 wealths of the two types of consumers using the competitive equilibrium prices.

d. Define a competitive equilibrium with sequential trading of Arrow securities.

e. Compute a competitive equilibrium with sequential trading of Arrow securities.

Exercise 8.8   **Corners**

A pure endowment economy consists of two type of consumers. Consumers of type 1 order consumption streams of the one good according to

$$\sum_{t=0}^{\infty} \beta^t c_t^1$$

and consumers of type 2 order consumption streams according to

$$\sum_{t=0}^{\infty} \beta^t \ln(c_t^2)$$

where $c_t^i \geq 0$ is the consumption of a type i consumer and $\beta \in (0,1)$ is a common discount factor. Please note the nonnegativity constraint on consumption of each person (the force of this is that c_t^i is *consumption*, not *production*). The consumption good is tradable but nonstorable. There are equal numbers of the two types of consumer. The consumer of type 1 is endowed with the consumption sequence

$$y_t^1 = \mu > 0 \quad \forall t \geq 0$$

where $\mu > 0$. The consumer of type 2 is endowed with the consumption sequence

$$y_t^2 = \begin{cases} 0 & \text{if } t \geq 0 \text{ is even} \\ \alpha & \text{if } t \geq 0 \text{ is odd} \end{cases}$$

where

$$\alpha = \mu(1 + \beta^{-1}). \tag{1}$$

a. Define a competitive equilibrium with time 0 trading. Be careful to include definitions of all of the objects of which a competitive equilibrium is composed.

b. Compute a competitive equilibrium allocation with time 0 trading. Compute the equilibrium price system. Please also compute the sequence of one-period gross interest rates. Do they differ between odd and even periods?

c. Compute the time 0 wealths of the two types of consumers using the competitive equilibrium prices.

d. Now consider an economy identical to the preceding one except in one respect. The endowment of consumer 1 continues to be 1 each period, but we assume that the endowment of consumer 2 is larger (though it continues to be zero in every even period). In particular, we alter the assumption about endowments in condition (1) to the new condition

$$\alpha > \mu(1 + \beta^{-1}).$$

Compute the competitive equilibrium allocation and price system for this economy.

e. Compute the sequence of one-period interest rates implicit in the equilibrium price system that you computed in part d. Are interest rates higher or lower than those you computed in part b?

Exercise 8.9 **Equivalent martingale measure**

Let $\{d_t(s_t)\}_{t=0}^{\infty}$ be a stream of payouts. Suppose that there are complete markets. From (8.5.4) and (8.7.1), the price at time 0 of a claim on this stream of dividends is

$$a_0 = \sum_{t=0}^{\infty} \sum_{s^t} \beta^t \frac{u'(c_t^i(s^t))}{\mu_i} \pi_t(s^t) d_t(s_t).$$

Show that this a_0 can also be represented as

$$a_0 = \sum_t b_t \sum_{s^t} d_t(s_t) \tilde{\pi}_t(s^t) \tag{1}$$

$$= \tilde{E}_0 \sum_{t=0}^{\infty} b_t d_t(s_t)$$

where $\tilde{E}$ is the mathematical expectation with respect to the twisted measure $\tilde{\pi}_t(s^t)$ defined by

$$\tilde{\pi}_t(s^t) = b_t^{-1} \beta^t \frac{u'(c_t^i(s^t))}{\mu_i} \pi_t(s^t)$$

$$b_t = \sum_{s^t} \beta^t \frac{u'(c_t^i(s^t))}{\mu_i} \pi_t(s^t).$$

Prove that $\tilde{\pi}_t(s^t)$ is a probability measure. Interpret b_t itself as a price of particular asset. Note: $\tilde{\pi}_t(s^t)$ is called an *equivalent martingale measure*. See chapters 13 and 14.

Exercise 8.10 **Harrison-Kreps prices**

Show that the asset price in (1) of the previous exercise can also be represented
as

$$a_0 = \sum_{t=0}^{\infty} \sum_{s^t} \beta^t p_t^0(s^t) d_t(s^t) \pi_t(s^t)$$

$$= E_0 \sum_{t=0}^{\infty} \beta^t p_t^0 d_t$$

where $p_t^0(s^t) = q_t^0(s^t)/[\beta^t \pi_t(s^t)]$.

Exercise 8.11 **Early resolution of uncertainty**

An economy consists of two households named $i = 1, 2$. Each household evalu-
ates streams of a single consumption good according to $\sum_{t=0}^{\infty} \sum_{s^t} \beta^t u[c_t^i(s^t)] \pi_t(s^t)$.
Here $u(c)$ is an increasing, twice continuously differentiable, strictly concave
function of consumption c of one good. The utility function satisfies the In-
ada condition $\lim_{c \downarrow 0} u'(c) = +\infty$. A feasible allocation satisfies $\sum_i c_t^i(s^t) \leq$
$\sum_i y^i(s_t)$. The households' endowments of the one nonstorable good are both
functions of a state variable $s_t \in \mathbf{S} = \{0, 1, 2\}$; s_t is described by a time invariant Markov chain with initial distribution $\pi_0 = \begin{bmatrix} 0 & 1 & 0 \end{bmatrix}'$ and transition
density defined by the stochastic matrix

$$P = \begin{bmatrix} 1 & 0 & 0 \\ .5 & 0 & .5 \\ 0 & 0 & 1 \end{bmatrix}.$$

The endowments of the two households are

$$y_t^1 = s_t/2$$
$$y_t^2 = 1 - s_t/2.$$

a. Define a competitive equilibrium with Arrow securities.

b. Compute a competitive equilibrium with Arrow securities.

c. By hand, simulate the economy. In particular, for every possible realization
of the histories s^t, describe time series of c_t^1, c_t^2 and the wealth levels a_t^i of the
households. (Note: Usually this would be an impossible task by hand, but this
problem has been set up to make the task manageable.)

Exercise 8.12   donated by Pierre-Olivier Weill

An economy is populated by a continuum of infinitely lived consumers of types $j \in \{0, 1\}$, with a measure one of each. There is one nonstorable consumption good arriving in the form of an endowment stream owned by each consumer. Specifically, the endowments are

$$y_t^0(s_t) = (1 - s_t)\bar{y}^0$$
$$y_t^1(s_t) = s_t\bar{y}^1,$$

where s_t is a two-state time-invariant Markov chain valued in $\{0, 1\}$ and $\bar{y}^0 < \bar{y}^1$. The initial state is $s_0 = 1$. Transition probabilities are denoted $\pi(s'|s)$ for $(s, s') \in \{0, 1\}^2$, where $'$ denotes a next period value. The aggregate endowment is $y_t(s_t) \equiv (1 - s_t)\bar{y}^0 + s_t\bar{y}^1$. Thus, this economy fluctuates stochastically between recessions $y_t(0) = \bar{y}^0$ and booms $y_t(1) = \bar{y}^0$. In a recession, the aggregate endowment is owned by type 0 consumers, while in a boom it is owned by a type 1 consumers. A consumer orders consumption streams according to:

$$U(c^j) = \sum_{t=0}^{\infty} \sum_{s^t} \beta^t \pi(s^t|s_0) \frac{c_t^j(s^t)^{1-\gamma}}{1 - \gamma},$$

where $s^t = (s_t, s_{t-1}, \ldots, s_0)$ is the history of the state up to time t, $\beta \in (0, 1)$ is the discount factor, and $\gamma > 0$ is the coefficient of relative risk aversion.

a. Define a competitive equilibrium with time 0 trading. Compute the price system $\{q_t^0(s^t)\}_{t=0}^{\infty}$ and the equilibrium allocation $\{c^j(s^t)\}_{t=0}^{\infty}$, for $j \in \{0, 1\}$.

b. Find a utility function $\bar{U}(c) = E_0\left(\sum_{t=0}^{\infty} \beta^t u(c_t)\right)$ such that the price system $q_t^0(s^t)$ and the aggregate endowment $y_t(s_t)$ is an equilibrium allocation of the single-agent economy $\left(\bar{U}, \{y_t(s_t)\}_{t=0}^{\infty}\right)$. How does your answer depend on the initial distribution of endowments $y_t^j(s_t)$ among the two types $j \in \{0, 1\}$? How would you defend the representative agent assumption in this economy?

c. Describe the equilibrium allocation under the following three market structures: (i) at each node s^t, agents can trade only claims on their entire endowment streams; (ii) at each node s^t, there is a complete set of one-period ahead Arrow securities; and (iii) at each node s^t, agents can only trade two risk-free assets, namely, a one-period zero-coupon bond that pays one unit of consumption for sure at $t + 1$ and a two-period zero-coupon bond that pays one unit of the consumption good for sure at $t + 2$. How would you modify your answer in the absence of aggregate uncertainty?

d. Assume that $\pi(1|0) = 1$, $\pi(0|1) = 1$, and as before $s_0 = 1$. Compute the allocation in an equilibrium with time 0 trading. Does the type $j = 1$ agent always consume the largest share of the aggregate endowment? How does it depend on parameter values? Provide economic intuition for your results.

e. Assume that $\pi(1|0) = 1$ and $\pi(0|1) = 1$. Remember that $s_0 = 1$. Assume that at $t = 1$ agent $j = 0$ is given the option to default on her financial obligation. For example, in the time 0 trading economy, these obligations are deliveries of goods. Upon default, it is assumed that the agent is excluded from the market and has to consume her endowment forever. Will the agent ever exercise her option to default?

Exercise 8.13   **Diverse beliefs, I**

A pure endowment economy is populated by two consumers. Consumer i has preferences over history-contingent consumption sequences $\{c_t^i(s^t)\}$ that are ordered by

$$\sum_{t=0}^{\infty} \sum_{s^t} \beta^t u(c_t^i(s^t)) \pi_t^i(s^t),$$

where $u(c) = \ln(c)$ and where $\pi_t^i(s^t)$ is a density that consumer i assigns to history s^t. The state space is time invariant. In particular, $s_t \in S = \{0, .5, 1\}$ for all $t \geq 0$. Only two histories are possible for $t = 0, 1, 2, \ldots$:

$$\text{history } 1 : .5, 1, 1, 1, 1, \ldots$$
$$\text{history } 2 : .5, 0, 0, 0, 0, \ldots$$

Consumer 1 assigns probability $1/3$ to history 1 and probability $2/3$ to history 2, while consumer 2 assigns probability $2/3$ to history 1 and probability $1/3$ to history 2. Nature assigns equal probabilities to the two histories. The endowments of the two consumers are:

$$y_t^1 = s_t$$
$$y_t^2 = 1 - s_t.$$

a. Define a competitive equilibrium with sequential trading of a complete set of one-period Arrow securities.

b. Compute a competitive equilibrium with sequential trading of a complete set of one-period Arrow securities.

c. Is the equilibrium allocation Pareto optimal?

Exercise 8.14 **Diverse beliefs, II**

Consider the following I person pure endowment economy. There is a state variable $s_t \in S$ for all $t \geq 0$. Let s^t denote a history of s from 0 to t. The time t aggregate endowment is a function of the history, so $Y_t = Y_t(s^t)$. Agent i attaches a personal probability of $\pi_t^i(s^t)$ to history s^t. The history s^t is observed by all I people at time t. Assume that for all i, $\pi_t^i(s_t) > 0$ if and only if $\pi_t^1(s_t) > 0$ (so the consumers agree about which histories have positive probability). Consumer i ranks consumption plans $c_t^i(s^t)$ that are measurable functions of histories via the expected utility functional

$$
(1) \qquad \sum_{t=0}^{\infty} \sum_{s^t} \beta^t \ln(c_t^i(s^t)) \pi_t^i(s^t)
$$

The ownership structure of the economy is not yet determined.

A planner puts positive Pareto weights $\lambda_i > 0$ on consumers $i = 1, \ldots, I$ and solves a time 0 Pareto problem that respects each consumer's preferences as represented by (1).

a. Show how to solve for a Pareto optimal allocation. Display an expression for $c_t^i(s^t)$ as a function of $Y_t(s^t)$ and other pertinent variables.

b. Under what circumstances does the Pareto plan imply complete risk-sharing among the I consumers?

c. Under what circumstances does the Pareto plan imply an allocation that is not history dependent? By 'not history dependent', we mean that $Y_t(s^t) = Y_t(\tilde{s}^t)$ would imply the same allocation at time t?

d. For a given set of Pareto weights, find an associated equilibrium price vector and an initial distribution of wealth among the I consumers that makes the Pareto allocation be the allocation associated with a competitive equilibrium with time 0 trading of history-contingent claims on consumption.

e. Find a formula for the equilibrium price vector in terms of equilibrium quantities and the beliefs of consumers.

f. Suppose that $I = 2$. Show that as $\lambda_2/\lambda_1 \to +\infty$, the planner would distribute initial wealth in a way that makes consumer 2's beliefs more and more influential in determining equilibrium prices.

Exercise 8.15　　**Diverse beliefs, III**

An economy consists of two consumers named $i = 1, 2$. Each consumer evaluates streams of a single nonstorable consumption good according to

$$\sum_{t=0}^{\infty} \sum_{s^t} \beta^t \ln[c_t^i(s^t)] \pi_t^i(s^t).$$

Here $\pi_t^i(s^t)$ is consumer i's subjective probability over history s^t. A feasible allocation satisfies $\sum_i c_t^i(s^t) \le \sum_i y^i(s_t)$ for all $t \ge 0$ and for all s^t. The consumers' endowments of the one good are functions of a state variable $s_t \in \mathbf{S} = \{0, 1, 2\}$. In truth, s_t is described by a time invariant Markov chain with initial distribution $\pi_0 = \begin{bmatrix} 0 & 1 & 0 \end{bmatrix}'$ and transition density defined by the stochastic matrix

$$P = \begin{bmatrix} 1 & 0 & 0 \\ .5 & 0 & .5 \\ 0 & 0 & 1 \end{bmatrix}$$

where $P_{ij} = \text{Prob}[s_{t+1} = j - 1 | s_t = i - 1]$ for $i = 1, 2, 3$ and $j = 1, 2, 3$. The endowments of the two consumers are

$$y_t^1 = s_t/2$$
$$y_t^2 = 1 - s_t/2.$$

In part I, both consumers know the true probabilities over histories s^t (i.e., they know both π_0 and P). In part II, the two consumers have different subjective probabilities.

Part I:

Assume that both consumers know (π_0, P), so that $\pi_t^1(s^t) = \pi_t^2(s^t)$ for all $t \ge 0$ for all s^t.

a. Show how to deduce $\pi_t^i(s^t)$ from (π_0, P).

b. Define a competitive equilibrium with sequential trading of Arrow securities.

c. Compute a competitive equilibrium with sequential trading of Arrow securities.

d. By hand, simulate the economy. In particular, for every possible realization of the histories s^t, describe time series of c_t^1, c_t^2 and the wealth levels for the two consumers.

Part II:

Now assume that while consumer 1 knows (π_0, P), consumer 2 knows π_0 but thinks that P is

$$\hat{P} = \begin{bmatrix} 1 & 0 & 0 \\ .4 & 0 & .6 \\ 0 & 0 & 1 \end{bmatrix}.$$

e. Deduce $\pi_t^2(s^t)$ from $(\pi_0, \hat{P})$ for all $t \geq 0$ for all s^t.

f. Formulate and solve a Pareto problem for this economy.

g. Define an equilibrium with time 0 trading of a complete set of Arrow-Debreu history-contingent securities.

h. Compute an equilibrium with time 0 trading of a complete set of Arrow-Debreu history-contingent securities.

i. Compute an equilibrium with sequential trading of Arrow securities. For every possible realization of s^t for all $t \geq 0$, please describe time series of c_t^1, c_t^2 and the wealth levels for the two consumers.

Exercise 8.16   **Diverse beliefs, IV**

A pure exchange economy is populated by two consumers. Consumer i has preferences over history-contingent consumption sequences $\{c_t^i(s^t)\}$ that are ordered by

$$\sum_{t=0}^{\infty} \sum_{s^t} \beta^t u(c_t^i(s^t)) \pi_t^i(s^t),$$

where $u(c) = \ln(c)$, $\beta \in (0, 1)$, and $\pi_t^i(s^t)$ is a density that consumer i assigns to history s^t. The state space is time invariant. In particular, $s_t \in S = \{0, .5, 1\}$ for all $t \geq 0$. Only two histories are possible for $t = 0, 1, 2, \ldots$:

$$\text{history } 1 : .5, 1, 1, 1, 1, \ldots$$

$$\text{history } 2 : .5, 0, 0, 0, 0, \ldots$$

Consumer 1 assigns probability 1 to history 1 and probability 0 to history 2, while consumer 2 assigns probability 0 to history 1 and probability 1 to history

2. Nature assigns equal probabilities to the two histories. The endowments of the two consumers are:

$$y_t^1 = s_t$$
$$y_t^2 = 1 - s_t.$$

a. Formulate and solve a Pareto problem for this economy.

b. Define a competitive equilibrium with sequential trading of a complete set of one-period Arrow securities.

c. Does a competitive equilibrium with sequential trading of a complete set of one-period Arrow securities exist for this economy? If it does, compute it. If it does not, explain why.

Exercise 8.17   **Diverse beliefs, V**

A pure exchange economy is populated by two consumers. Consumer i has preferences over history-contingent consumption sequences $\{c_t^i(s^t)\}$ that are ordered by

$$\sum_{t=0}^{\infty} \sum_{s^t} \beta^t u(c_t^i(s^t)) \pi_t^i(s^t),$$

where $u(c) = \ln(c)$, $\beta \in (0, 1)$, and $\pi_t^i(s^t)$ is a density that consumer i assigns to history s^t. The state space is time invariant. In particular, $s_t \in S = \{0, .5, 1\}$ for all $t \geq 0$. Only two histories are possible for $t = 0, 1, 2, \ldots$:

$$\text{history } 1 : .5, 1, 1, 1, 1, \ldots$$
$$\text{history } 2 : .5, 0, 0, 0, 0, \ldots$$

Consumer 1 assigns probability 1 to history 1 and probability 0 to history 2, while consumer 2 assigns probability 0 to history 1 and probability 1 to history 2. Nature assigns equal probabilities to the two histories. The endowments of the two consumers are:

$$y_t^1 = 1 - s_t$$
$$y_t^2 = s_t.$$

a. Formulate and solve a Pareto problem for this economy.

b. Define a competitive equilibrium with sequential trading of a complete set of one-period Arrow securities.

c. Does a competitive equilibrium with sequential trading of a complete set of one-period Arrow securities exist for this economy? If it does, compute it. If it does not, explain why.

Exercise 8.18   **Risk-free bonds**

An economy consists of a single representative consumer who ranks streams of a single nonstorable consumption good according to $\sum_{t=0}^{\infty} \sum_{s^t} \beta^t \ln[c_t(s^t)] \pi_t(s^t)$. Here $\pi_t(s^t)$ is the subjective probability that the consumer attaches to a history s^t of a Markov state s_t, where $s_t \in \{1,2\}$. Assume that the subjective probability $\pi_t(s^t)$ equals the objective probability. Feasibility for this pure endowment economy is expressed by the condition $c_t \leq y_t$, where y_t is the endowment at time t. The endowment is exogenous and governed by

$$y_{t+1} = \lambda_{t+1}\lambda_t \cdots \lambda_1 y_0$$

for $t \geq 0$ where $y_0 > 0$. Here λ_t is a function of the Markov state s_t. Assume that $\lambda_t = 1$ when $s_t = 1$ and $\lambda_t = 1 + \zeta$ when $s_t = 2$, where $\zeta > 0$. States $s^t = [s_t, \ldots, s_0]$ are known at time t, but future states are not. The state s_t is described by a time invariant Markov chain with initial probability distribution $\pi_0 = [1 \quad 0]'$ and transition density defined by the stochastic matrix

$$P = \begin{bmatrix} P_{11} & P_{12} \\ P_{21} & P_{22} \end{bmatrix}$$

where $P_{ij} = \text{Prob}[s_{t+1} = j | s_t = i]$ for $i = 1, 2$ and $j = 1, 2$. Assume that $P_{ij} \geq 0$ for all pairs (i, j).

a. Show how to deduce $\pi_t(s^t)$ from (π_0, P).

b. Define a competitive equilibrium with sequential trading of Arrow securities.

c. Compute a competitive equilibrium with sequential trading of Arrow securities.

d. Let p_t^b be the time t price of a risk-free claim to one unit of consumption at time $t+1$. Define a competitive equilibrium with sequential trading of risk-free claims to consumption one period ahead.

e. Let $R_t = (p_t^b)^{-1}$ be the one-period risk-free gross interest rate. Give a formula for R_t and tell how it depends on the history s^t.

f. Suppose that $\beta = .95, \zeta = .02$ and $P = \begin{bmatrix} 1 & 0 \\ .5 & .5 \end{bmatrix}$. Please compute R_t when $s_t = 1$. Then compute R_t when $s_t = 2$.

g. What parts of your answers depend on assuming that the subjective probability $\pi_t(s^t)$ equals the objective probability?

Chapter 9
Overlapping Generations Models

This chapter describes the pure exchange overlapping generations model of Paul Samuelson (1958). We begin with an abstract presentation that treats the overlapping generations model as a special case of the chapter 8 general equilibrium model with complete markets and all trades occurring at time 0. A peculiar type of heterogeneity across agents distinguishes the model. Each individual cares about consumption only at two adjacent dates, and the set of individuals who care about consumption at a particular date includes some who care about consumption one period earlier and others who care about consumption one period later. We shall study how this special preference and demographic pattern affects some of the outcomes of the chapter 8 model.

While it helps to reveal the fundamental structure, allowing complete markets with time 0 trading in an overlapping generations model strains credulity. The formalism envisions that equilibrium price and quantity sequences are set at time 0, before the participants who are to execute the trades have been born. For that reason, most applied work with the overlapping generations model adopts a sequential-trading arrangement, like the sequential trade in Arrow securities described in chapter 8. The sequential-trading arrangement has all trades executed by agents living in the here and now. Nevertheless, equilibrium quantities and intertemporal prices are equivalent between these two trading arrangements. Therefore, analytical results found in one setting transfer to the other.

Later in the chapter, we use versions of the model with sequential trading to tell how the overlapping generations model provides a framework for thinking about equilibria with government debt and/or valued fiat currency, intergenerational transfers, and fiscal policy.

9.1. Endowments and preferences

Time is discrete, starts at $t = 1$, and lasts forever, so $t = 1, 2, \ldots$. There is an infinity of agents named $i = 0, 1, \ldots$. We can also regard i as agent i's period of birth. There is a single good at each date. The good is not storable. There is no uncertainty. Each agent has a strictly concave, twice continuously differentiable, one-period utility function $u(c)$, which is strictly increasing in consumption c of the one good. Agent i consumes a vector $c^i = \{c_t^i\}_{t=1}^{\infty}$ and has the special utility function

$$U^i(c^i) = u(c_i^i) + u(c_{i+1}^i), \quad i \geq 1, \qquad (9.1.1a)$$
$$U^0(c^0) = u(c_1^0). \qquad (9.1.1b)$$

Notice that agent i only wants goods dated i and $i + 1$. The interpretation of equations $(9.1.1)$ is that agent i lives during periods i and $i + 1$ and wants to consume only when he is alive.

Each household has an endowment sequence y^i satisfying $y_i^i \geq 0, y_{i+1}^i \geq 0, y_t^i = 0 \ \forall t \neq i$ or $i + 1$. Thus, households are endowed with goods only when they are alive.

9.2. Time 0 trading

We use the definition of competitive equilibrium from chapter 8. Thus, we temporarily suspend disbelief and proceed in the style of Debreu (1959) with time 0 trading. Specifically, we imagine that there is a "clearinghouse" at time 0 that posts prices and, at those prices, aggregates demands and supplies for goods in different periods. An equilibrium price vector makes markets for all periods $t \geq 2$ clear, but there may be excess supply in period 1; that is, the clearinghouse might end up with goods left over in period 1. Any such excess supply of goods in period 1 can be given to the initial old generation without any effects on the equilibrium price vector, since those old agents optimally consume all their wealth in period 1 and do not want to buy goods in future periods. The reason for our special treatment of period 1 will become clear as we proceed.

Thus, at date 0, there are complete markets in time t consumption goods with date 0 price q_t^0. A household's budget constraint is

$$\sum_{t=1}^{\infty} q_t^0 c_t^i \leq \sum_{t=1}^{\infty} q_t^0 y_t^i. \tag{9.2.1}$$

Letting μ^i be a Lagrange multiplier attached to consumer i's budget constraint, the consumer's first-order conditions are

$$\mu^i q_i^0 = u'(c_i^i), \tag{9.2.2a}$$

$$\mu^i q_{i+1}^0 = u'(c_{i+1}^i), \tag{9.2.2b}$$

$$c_t^i = 0 \text{ if } t \notin \{i, i+1\}. \tag{9.2.2c}$$

Evidently an allocation is feasible if for all $t \geq 1$,

$$c_t^t + c_t^{t-1} \leq y_t^t + y_t^{t-1}. \tag{9.2.3}$$

DEFINITION: An allocation is *stationary* if $c_{i+1}^i = c_o, c_i^i = c_y \ \forall i \geq 1$.

Here the subscript o denotes old and y denotes young. Note that we do not require that $c_1^0 = c_o$. We call an equilibrium with a stationary allocation a *stationary equilibrium*.

9.2.1. Example equilibria

Let $\epsilon \in (0, .5)$. The endowments are

$$y_i^i = 1 - \epsilon, \ \forall i \geq 1,$$

$$y_{i+1}^i = \epsilon, \ \forall i \geq 0, \tag{9.2.4}$$

$$y_t^i = 0 \text{ otherwise.}$$

This economy has many equilibria. We describe two stationary equilibria now, and later we shall describe some nonstationary equilibria. We can use a guess-and-verify method to confirm the following two equilibria.

1. Equilibrium H: a high-interest-rate equilibrium. Set $q_t^0 = 1 \ \forall t \geq 1$ and $c_i^i = c_{i+1}^i = .5$ for all $i \geq 1$ and $c_1^0 = \epsilon$. To verify that this is an equilibrium,

notice that each household's first-order conditions are satisfied and that the allocation is feasible. Extensive intergenerational trade occurs at time 0 at the equilibrium price vector q_t^0. Constraint (9.2.3) holds with equality for all $t \geq 2$ but with strict inequality for $t = 1$. Some of the $t = 1$ consumption good is left unconsumed.

2. Equilibrium L: a low-interest-rate equilibrium. Set $q_1^0 = 1$, $\frac{q_{t+1}^0}{q_t^0} = \frac{u'(\epsilon)}{u'(1-\epsilon)} = \alpha > 1$. Set $c_t^i = y_t^i$ for all i, t. This equilibrium is autarkic, with prices being set to eradicate all trade.

9.2.2. Relation to welfare theorems

As we shall explain in more detail later, equilibrium H Pareto dominates equilibrium L. In equilibrium H every generation after the initial old one is better off and no generation is worse off than in equilibrium L. The equilibrium H allocation is strange because some of the time 1 good is not consumed, leaving room to set up a giveaway program to the initial old that makes them better off and costs subsequent generations nothing. We shall see how the institution of either perpetual government debt or of fiat money can accomplish this purpose.[1]

Equilibrium L is a competitive equilibrium that evidently fails to satisfy one of the assumptions needed to deliver the first fundamental theorem of welfare economics, which identifies conditions under which a competitive equilibrium allocation is Pareto optimal.[2] The condition of the theorem that is violated by equilibrium L is the assumption that the value of the aggregate endowment at the equilibrium prices is finite.[3]

---

[1] See Karl Shell (1971) for an investigation that characterizes why some competitive equilibria in overlapping generations models fail to be Pareto optimal. Shell cites earlier studies that had sought reasons why the welfare theorems seem to fail in the overlapping generations structure.

[2] See Mas-Colell, Whinston, and Green (1995) and Debreu (1954).

[3] Note that if the horizon of the economy were finite, then the counterpart of equilibrium H would not exist and the allocation of the counterpart of equilibrium L would be Pareto optimal.

9.2.3. Nonstationary equilibria

Our example economy has more equilibria. To construct more equilibria, we summarize preferences and consumption decisions in terms of an offer curve. We describe a graphical apparatus proposed by David Gale (1973) and used to good advantage by William Brock (1990).

DEFINITION: The household's *offer curve* is the locus of (c_i^i, c_{i+1}^i) that solves

$$\max_{\{c_i^i, c_{i+1}^i\}} U(c^i)$$

subject to

$$c_i^i + \alpha_i c_{i+1}^i \leq y_i^i + \alpha_i y_{i+1}^i.$$

Here $\alpha_i \equiv \frac{q_{i+1}^0}{q_i^0}$, the reciprocal of the one-period gross rate of return from period i to $i+1$, is treated as a parameter.

Evidently, the offer curve solves the following pair of equations:

$$c_i^i + \alpha_i c_{i+1}^i = y_i^i + \alpha_i y_{i+1}^i \qquad (9.2.5a)$$

$$\frac{u'(c_{i+1}^i)}{u'(c_i^i)} = \alpha_i \qquad (9.2.5b)$$

for $\alpha_i > 0$. We denote the offer curve by

$$\psi(c_i^i, c_{i+1}^i) = 0.$$

The graphical construction of the offer curve is illustrated in Figure 9.2.1. We trace it out by varying α_i in the household's problem and reading tangency points between the household's indifference curve and the budget line. The resulting locus depends on the endowment vector and lies above the indifference curve through the endowment vector. By construction, the following property is also true: at the intersection between the offer curve and a straight line through the endowment point, the straight line is tangent to an indifference curve.[4]

---

[4] Given our assumptions on preferences and endowments, the conscientious reader will note that Figure 9.2.1 appears distorted because the offer curve really ought to intersect the feasibility line along the 45 degree line with $c_t^t = c_{t+1}^t$, i.e., at the allocation affiliated with equilibrium H above.

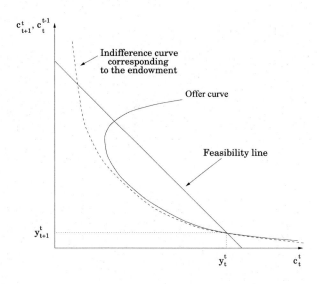

Figure 9.2.1: The offer curve and feasibility line.

Following Gale (1973), we can use the offer curve and a straight line depicting feasibility in the (c_i^i, c_i^{i-1}) plane to construct a machine for computing equilibrium allocations and prices. In particular, we can use the following pair of difference equations to solve for an equilibrium allocation. For $i \geq 1$, the equations are[5]

$$\psi(c_i^i, c_{i+1}^i) = 0, \qquad\qquad (9.2.6a)$$

$$c_i^i + c_i^{i-1} = y_i^i + y_i^{i-1}. \qquad\qquad (9.2.6b)$$

We take c_1^1 as an initial condition. After the allocation has been computed, the equilibrium price system can be computed from

$$q_i^0 = u'(c_i^i)$$

for all $i \geq 1$.

---

[5] By imposing equation (9.2.6b) with equality, we are implicitly possibly including a give-away program to the initial old.

9.2.4. Computing equilibria

Example 1:   Gale's equilibrium computation machine: A procedure for constructing an equilibrium is illustrated in Figure 9.2.2, which reproduces a version of a graph of David Gale (1973). Start with a proposed c_1^1, a time 1 allocation to the initial young. Then use the feasibility line to find the *maximal* feasible value for c_0^1, the time 1 allocation to the initial old. In the Arrow-Debreu equilibrium, the allocation to the initial old will be less than this maximal value, so that some of the time 1 good is thrown away. The reason for this is that the budget constraint of the initial old, $q_1^0(c_1^0 - y_1^0) \leq 0$, implies that $c_1^0 = y_1^0$.[6] The candidate time 1 allocation is thus feasible, but the time 1 young will choose c_1^1 only if the price α_1 is such that (c_2^1, c_1^1) lies on the offer curve. Therefore, we choose c_2^1 from the point on the offer curve that cuts a vertical line through c_1^1. Then we proceed to find c_2^2 from the intersection of a horizontal line through c_2^1 and the feasibility line. We continue recursively in this way, choosing c_i^i as the intersection of the feasibility line with a horizontal line through c_i^{i-1}, then choosing c_{i+1}^i as the intersection of a vertical line through c_i^i and the offer curve. We can construct a sequence of α_i's from the slope of a straight line through the endowment point and the sequence of (c_i^i, c_{i+1}^i) pairs that lie on the offer curve.

If the offer curve has the shape drawn in Figure 9.2.2, any c_1^1 between the upper and lower intersections of the offer curve and the feasibility line is an equilibrium setting of c_1^1. Each such c_1^1 is associated with a distinct allocation and α_i sequence, all but one of them converging to the *low*-interest-rate stationary equilibrium allocation and interest rate.

Example 2:   Endowment at $+\infty$: Take the preference and endowment structure of the previous example and modify only one feature. Change the endowment of the initial old to be $y_1^0 = \epsilon > 0$ *and* "$\delta = 1 - \epsilon > 0$ units of consumption at $t = +\infty$," by which we mean that we take

$$\sum_t q_t^0 y_t^0 = q_1^0 \epsilon + \delta \lim_{t \to \infty} q_t^0.$$

It is easy to verify that the only competitive equilibrium of the economy with this specification of endowments has $q_t^0 = 1 \ \forall t \geq 1$, and thus $\alpha_t = 1 \ \forall t \geq 1$. The

---

[6] Soon we shall discuss another market structure that avoids throwing away any of the initial endowment by augmenting the endowment of the initial old with a particular zero-dividend infinitely durable asset.

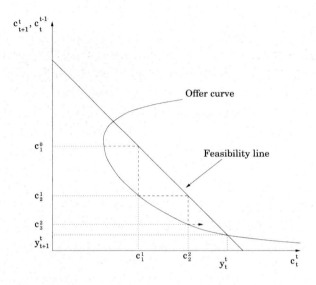

Figure 9.2.2: A nonstationary equilibrium allocation.

reason is that all the "low-interest-rate" equilibria that we computed in example 1 would assign an infinite value to the endowment of the initial old. Confronted with such prices, the initial old would demand unbounded consumption. That is not feasible. Therefore, such a price system cannot be an equilibrium.

Example 3:   A Lucas tree: Take the preference and endowment structure to be the same as example 1 and modify only one feature. Endow the initial old with a "Lucas tree," namely, a claim to a constant stream of $d > 0$ units of consumption for each $t \geq 1$.[7] Thus, the budget constraint of the initial old person now becomes

$$q_1^0 c_1^0 = d \sum_{t=1}^{\infty} q_t^0 + q_1^0 y_1^0.$$

The offer curve of each young agent remains as before, but now the feasibility line is

$$c_i^i + c_i^{i-1} = y_i^i + y_i^{i-1} + d$$

---

[7]   This is a version of an example of Brock (1990). The 'Lucas tree' refers to a colorful interpretation of a dividend stream as 'fruit' falling from a 'tree' in a pure exchange economy studied by Lucas (1978). See chapter 13.

for all $i \geq 1$. Note that young agents are endowed below the feasibility line. From Figure 9.2.3, it seems that there are two candidates for stationary equilibria, one with constant $\alpha < 1$, another with constant $\alpha > 1$. The one with $\alpha < 1$ is associated with the steeper budget line in Figure 9.2.3. However, the candidate stationary equilibrium with $\alpha > 1$ cannot be an equilibrium for a reason similar to that encountered in example 2. At the price system associated with an $\alpha > 1$, the wealth of the initial old would be unbounded, which would prompt them to consume an unbounded amount, which is not feasible. This argument rules out not only the stationary $\alpha > 1$ equilibrium but also all nonstationary candidate equilibria that converge to that constant α. Therefore, there is a unique equilibrium; it is stationary and has $\alpha < 1$.

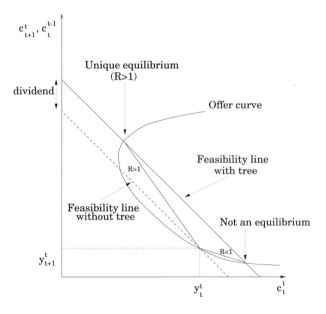

Figure 9.2.3: Unique equilibrium with a fixed-dividend asset.

If we interpret the gross rate of return on the tree as $\alpha^{-1} = \frac{p+d}{p}$, where $p = \sum_{t=1}^{\infty} q_t^0 d$, we can compute that $p = \frac{d}{R-1}$ where $R = \alpha^{-1}$. Here p is the price of the Lucas tree.

In terms of the logarithmic preference example 5 below, the difference equation (9.2.9) becomes modified to

$$\alpha_i = \frac{1 + 2d}{\epsilon} - \frac{\epsilon^{-1} - 1}{\alpha_{i-1}}. \tag{9.2.7}$$

Example 4:   Government expenditures: Take the preferences and endowments to be as in example 1 again, but now alter the feasibility condition to be

$$c_i^i + c_i^{i-1} + g = y_i^i + y_i^{i-1}$$

for all $i \geq 1$ where $g > 0$ is a positive level of government purchases. The "clearinghouse" is now looking for an equilibrium price vector such that this feasibility constraint is satisfied. We assume that government purchases do not give utility. The offer curve and the feasibility line look as in Figure 9.2.4. Notice that the endowment point (y_i^i, y_{i+1}^i) lies *outside* the relevant feasibility line. Formally, this graph looks like example 3, but with a "negative dividend d." Now there are two stationary equilibria with $\alpha > 1$, and a continuum of equilibria converging to the higher α equilibrium (the one with the lower slope α^{-1} of the associated budget line). Equilibria with $\alpha > 1$ cannot be ruled out by the argument in example 3 because no one's endowment sequence receives infinite value when $\alpha > 1$.

Later, we shall interpret this example as one in which a government finances a constant deficit either by money creation or by borrowing at a negative real net interest rate. We shall discuss this and other examples in a setting with sequential trading.

Example 5:   Log utility: Suppose that $u(c) = \ln c$ and that the endowment is described by equations (9.2.4). Then the offer curve is given by the recursive formulas $c_i^i = .5(1 - \epsilon + \alpha_i \epsilon), c_{i+1}^i = \alpha_i^{-1} c_i^i$. Let α_i be the gross rate of return facing the young at i. Feasibility at i and the offer curves then imply

$$\frac{1}{2\alpha_{i-1}}(1 - \epsilon + \alpha_{i-1}\epsilon) + .5(1 - \epsilon + \alpha_i\epsilon) = 1. \tag{9.2.8}$$

This implies the difference equation

$$\alpha_i = \epsilon^{-1} - \frac{\epsilon^{-1} - 1}{\alpha_{i-1}}. \tag{9.2.9}$$

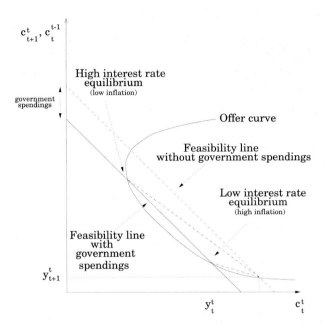

Figure 9.2.4: Equilibria with debt- or money-financed government deficit finance.

See Figure 9.2.2. An equilibrium α_i sequence must satisfy equation (9.2.8) and have $\alpha_i > 0$ for all i. Evidently, $\alpha_i = 1$ for all $i \geq 1$ is an equilibrium α sequence. So is any α_i sequence satisfying equation (9.2.8) and $\alpha_1 \geq 1$; $\alpha_1 < 1$ will not work because equation (9.2.8) implies that the tail of $\{\alpha_i\}$ is an unbounded negative sequence. The limiting value of α_i for any $\alpha_1 > 1$ is $\frac{1-\epsilon}{\epsilon} = u'(\epsilon)/u'(1-\epsilon)$, which is the interest factor associated with the stationary autarkic equilibrium. Notice that Figure 9.2.2 suggests that the stationary $\alpha_i = 1$ equilibrium is not stable, while the autarkic equilibrium is.

9.3. Sequential trading

We now alter the trading arrangement to bring them into line with standard presentations of the overlapping generations model. We abandon the time 0, complete markets trading arrangement and replace it with sequential trading in which a durable asset, either government debt or unbacked fiat money or claims on a Lucas tree, is passed from old to young. Some cross-generation transfers occur with voluntary exchanges, while others are engineered by government tax and transfer programs.

9.4. Money

In Samuelson's (1958) version of the model, trading occurs sequentially through a medium of exchange, an inconvertible (or "fiat") currency. In Samuelson's model, preferences and endowments are as described above, with one important additional component of the endowment. At date $t = 1$, old agents are endowed in the aggregate with $M > 0$ units of intrinsically worthless currency. No one has promised to redeem the currency for goods. The currency is not "backed" by any government promise to redeem it for goods. But as Samuelson showed, there exists a system of expectations that makes unbacked currency be valued. Currency will be valued today if people expect it to be valued tomorrow. Samuelson thus envisioned a situation in which currency is backed by expectations without promises.

For each date $t \geq 1$, young agents purchase m_t^i units of currency at a price of $1/p_t$ units of the time t consumption good. Here $p_t \geq 0$ is the time t price level. At each $t \geq 1$, each old agent exchanges his holdings of currency for the time t consumption good. The budget constraints of a young agent born in period $i \geq 1$ are

$$c_i^i + \frac{m_i^i}{p_i} \leq y_i^i, \tag{9.4.1}$$

$$c_{i+1}^i \leq \frac{m_i^i}{p_{i+1}} + y_{i+1}^i, \tag{9.4.2}$$

$$m_i^i \geq 0. \tag{9.4.3}$$

If $m_i^i \geq 0$, inequalities (9.4.1) and (9.4.2) imply

$$c_i^i + c_{i+1}^i \left(\frac{p_{i+1}}{p_i} \right) \leq y_i^i + y_{i+1}^i \left(\frac{p_{i+1}}{p_i} \right). \tag{9.4.4}$$

Provided that we set

$$\frac{p_{i+1}}{p_i} = \alpha_i = \frac{q_{i+1}^0}{q_i^0},$$

this budget set is identical with equation (9.2.1).

We use the following definitions:

DEFINITION: A nominal price sequence is a positive sequence $\{p_i\}_{i \geq 1}$.

DEFINITION: An equilibrium with valued fiat money is a feasible allocation and a nominal price sequence with $p_t < +\infty$ for all t such that given the price sequence, the allocation solves the household's problem for each $i \geq 1$.

The qualification that $p_t < +\infty$ for all t means that fiat money is valued. Sometimes we call an equilibrium with valued fiat money a 'monetary equilibrium'. If $\frac{1}{p_t} = +\infty$, we sometimes call it a 'nonmonetary equilibrium'.

9.4.1. Computing more equilibria with valued fiat currency

Summarize the household's optimal decisions with a saving function

$$y_i^i - c_i^i = s(\alpha_i; y_i^i, y_{i+1}^i). \tag{9.4.5}$$

Then the equilibrium conditions for the model are

$$\frac{M}{p_i} = s(\alpha_i; y_i^i, y_{i+1}^i) \tag{9.4.6a}$$

$$\alpha_i = \frac{p_{i+1}}{p_i}, \tag{9.4.6b}$$

where it is understood that $c_{i+1}^i = y_{i+1}^i + \frac{M}{p_{i+1}}$. Equation (9.4.6a) states that at time i the net of saving of generation i (the expression on the right side) equals the net dissaving of generation $i-1$ (the expression on the left side). To compute an equilibrium, we solve the difference equations (9.4.6) for $\{p_i\}_{i=1}^{\infty}$, then get the allocation from the household's budget constraints evaluated at equality at the equilibrium level of real balances. As an example, suppose that

$u(c) = \ln(c)$, and that $(y_i^i, y_{i+1}^i) = (w_1, w_2)$ with $w_1 > w_2$. The saving function is $s(\alpha_i) = .5(w_1 - \alpha_i w_2)$. Then equation $(9.4.6a)$ becomes

$$.5(w_1 - w_2 \frac{p_{t+1}}{p_t}) = \frac{M}{p_t}$$

or

$$p_t = 2M/w_1 + \left(\frac{w_2}{w_1}\right) p_{t+1}. \tag{9.4.7}$$

This is a difference equation whose solutions with a positive price level are

$$p_t = \frac{2M}{w_1(1 - \frac{w_2}{w_1})} + c \left(\frac{w_1}{w_2}\right)^t, \tag{9.4.8}$$

for any scalar $c > 0$.[8] The solution for $c = 0$ is the unique stationary solution. The solutions with $c > 0$ have uniformly higher price levels than the $c = 0$ solution, and have the value of currency approaching zero in the limit as $t \to +\infty$.

9.4.2. Equivalence of equilibria

We briefly look back at the equilibria with time 0 trading and note that the equilibrium allocations are the same under time 0 and sequential trading. Thus, the following proposition asserts that with an adjustment to the endowment and the consumption allocated to the initial old, a competitive equilibrium allocation with time 0 trading is an equilibrium allocation in the fiat money economy (with sequential trading).

PROPOSITION: Let $\bar{c}^i$ denote a competitive equilibrium allocation (with time 0 trading) and suppose that it satisfies $\bar{c}_1^1 < y_1^1$. Then there exists an equilibrium (with sequential trading) of the monetary economy with allocation that satisfies $c_i^i = \bar{c}_i^i, c_{i+1}^i = \bar{c}_{i+1}^i$ for $i \geq 1$.

PROOF: Take the competitive equilibrium allocation and price system and form $\alpha_i = q_{i+1}^0/q_i^0$. Set $m_i^i/p_i = y_i^i - \bar{c}_i^i$. Set $m_i^i = M$ for all $i \geq 1$, and determine p_1 from $\frac{M}{p_1} = y_1^1 - \bar{c}_1^1$. This last equation determines a positive initial price level p_1 provided that $y_1^1 - \bar{c}_1^1 > 0$. Determine subsequent price levels from $p_{i+1} = \alpha_i p_i$.

---

[8] See the appendix to chapter 2.

Determine the allocation to the initial old from $c_1^0 = y_1^0 + \frac{M}{p_1} = y_1^0 + (y_1^1 - \bar{c}_1^1)$.

∎

In the monetary equilibrium, time t real balances equal the per capita *saving* of the young and the per capita *dissaving* of the old. To be a monetary equilibrium, both quantities must be positive for all $t \geq 1$.

A converse of the proposition is true.

PROPOSITION: Let $\bar{c}^i$ be an equilibrium allocation for the fiat money economy. Then there is a competitive equilibrium with time 0 trading with the same allocation, provided that the endowment of the initial old is augmented with an appropriate transfer from the clearinghouse.

To verify this proposition, we have to construct the required transfer from the clearinghouse to the initial old. Evidently, it is $y_1^1 - \bar{c}_1^1$. We invite the reader to complete the proof.

9.5. Deficit finance

For the rest of this chapter, we shall assume sequential trading. With sequential trading of fiat currency, this section reinterprets one of our earlier examples with time 0 trading, the example with government spending.

Consider the following overlapping generations model: The population is constant. At each date $t \geq 1$, N identical young agents are endowed with $(y_t^t, y_{t+1}^t) = (w_1, w_2)$, where $w_1 > w_2 > 0$. A government levies lump-sum taxes of τ_1 on each young agent and τ_2 on each old agent alive at each $t \geq 1$. There are N old people at time 1 each of whom is endowed with w_2 units of the consumption good and $M_0 > 0$ units of inconvertible, perfectly durable fiat currency. The initial old have utility function c_1^0. The young have utility function $u(c_t^t) + u(c_{t+1}^t)$. For each date $t \geq 1$ the government augments the currency supply according to

$$M_t - M_{t-1} = p_t(g - \tau_1 - \tau_2), \tag{9.5.1}$$

where g is a constant stream of government expenditures per capita and $0 < p_t \leq +\infty$ is the price level. If $p_t = +\infty$, we intend that equation (9.5.1) be interpreted as

$$g = \tau_1 + \tau_2. \tag{9.5.2}$$

For each $t \geq 1$, each young person's behavior is summarized by

$$s_t = f(R_t; \tau_1, \tau_2) = \arg\max_{s \geq 0} \left[u(w_1 - \tau_1 - s) + u(w_2 - \tau_2 + R_t s) \right]. \qquad (9.5.3)$$

DEFINITION: An equilibrium with valued fiat currency is a pair of positive sequences $\{M_t, p_t\}$ such that (a) given the price level sequence, $M_t/p_t = f(R_t)$ (the dependence on τ_1, τ_2 being understood); (b) $R_t = p_t/p_{t+1}$; and (c) the government budget constraint $(9.5.1)$ is satisfied for all $t \geq 1$.

The condition $f(R_t) = M_t/p_t$ can be written as $f(R_t) = M_{t-1}/p_t + (M_t - M_{t-1})/p_t$. The left side is the saving of the young. The first term on the right side is the dissaving of the old (the real value of currency that they exchange for time t consumption). The second term on the right is the dissaving of the government (its deficit), which is the real value of the additional currency that the government prints at t and uses to purchase time t goods from the young.

To compute an equilibrium, define $d = g - \tau_1 - \tau_2$ and write equation $(9.5.1)$ as

$$\frac{M_t}{p_t} = \frac{M_{t-1}}{p_{t-1}} \frac{p_{t-1}}{p_t} + d$$

for $t \geq 2$ and

$$\frac{M_1}{p_1} = \frac{M_0}{p_1} + d$$

for $t = 1$. Substitute the equilibrium condition $M_t/p_t = f(R_t)$ into these equations to get

$$f(R_t) = f(R_{t-1})R_{t-1} + d \qquad (9.5.4a)$$

for $t \geq 2$ and

$$f(R_1) = \frac{M_0}{p_1} + d. \qquad (9.5.4b)$$

Given p_1, which determines an initial R_1 by means of equation $(9.5.4b)$, equations $(9.5.4)$ form an autonomous difference equation in R_t. With appropriate transformations of variables, this system can be solved using Figure 9.2.4.

9.5.1. Steady states and the Laffer curve

Let's seek a stationary solution of equations (9.5.4), a quest rendered reasonable by the fact that $f(R_t)$ is time invariant (because the endowment and the tax patterns as well as the government deficit d are time-invariant). Guess that $R_t = R$ for $t \geq 1$. Then equations (9.5.4) become

$$f(R)(1 - R) = d, \qquad (9.5.5a)$$

$$f(R) = \frac{M_0}{p_1} + d. \qquad (9.5.5b)$$

For example, suppose that $u(c) = \ln(c)$. Then $f(R) = \frac{w_1 - \tau_1}{2} - \frac{w_2 - \tau_2}{2R}$. We have graphed $f(R)(1 - R)$ against d in Figure 9.5.1. Notice that if there is one solution for equation (9.5.5a), then there are at least two.

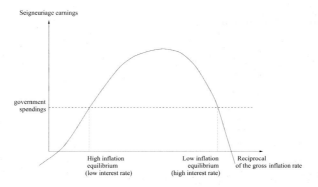

Figure 9.5.1: The Laffer curve in revenues from the inflation tax.

Here $(1 - R)$ can be interpreted as a tax rate on real balances, and $f(R)(1 - R)$ is a Laffer curve for the inflation tax rate. The high-return (low-tax) $R = \overline{R}$ is associated with the good Laffer curve stationary equilibrium, and the low-return (high-tax) $R = \underline{R}$ comes with the bad Laffer curve stationary equilibrium. Once R is determined, we can determine p_1 from equation (9.5.5b).

Figure 9.5.1 is isomorphic with Figure 9.2.4. The saving rate function $f(R)$ can be deduced from the offer curve. Thus, a version of Figure 9.2.4 can be used to solve the difference equation (9.5.4a) graphically. If we do so, we discover a continuum of nonstationary solutions of equation (9.5.4a), all but one of which have $R_t \to \underline{R}$ as $t \to \infty$. Thus, the bad Laffer curve equilibrium is stable.

The stability of the bad Laffer curve equilibrium arises under perfect foresight dynamics. Bruno and Fischer (1990) and Marcet and Sargent (1989) analyze how the system behaves under two different types of adaptive dynamics. They find that either under a crude form of adaptive expectations or under a least-squares learning scheme, R_t converges to $\overline{R}$. This finding is comforting because the comparative dynamics are more plausible at $\overline{R}$ (larger deficits bring higher inflation). Furthermore, Marimon and Sunder (1993) present experimental evidence pointing toward the selection made by the adaptive dynamics. Marcet and Nicolini (2003) build and calibrate an adaptive model of several Latin American hyperinflations that rests on this selection. Sargent, Williams, and Zha (2009) extend and estimate the model.

9.6. Equivalent setups

This section describes some alternative asset structures and trading arrangements that support the same equilibrium allocation. We take a model with a government deficit and show how it can be supported with sequential trading in government-indexed bonds, sequential trading in fiat currency, or time 0 trading of Arrow-Debreu dated securities.

9.6.1. The economy

Consider an overlapping generations economy with one agent born at each $t \geq 1$ and an initial old person at $t = 1$. Young agents born at date t have endowment pattern (y_t^t, y_{t+1}^t) and the utility function described earlier. The initial old person is endowed with $M_0 > 0$ units of unbacked currency and y_1^0 units of the consumption good. There is a stream of per-young-person government purchases $\{g_t\}$.

DEFINITION: An equilibrium with money-financed government deficits is a sequence $\{M_t, p_t\}_{t=1}^{\infty}$ with $0 < p_t < +\infty$ and $M_t > 0$ that satisfies (a) given $\{p_t\}$,

$$M_t = \arg\max_{\tilde{M} \geq 0} \left[u(y_t^t - \tilde{M}/p_t) + u(y_{t+1}^t + \tilde{M}/p_{t+1}) \right]; \qquad (9.6.1a)$$

and (b)

$$M_t - M_{t-1} = p_t g_t. \tag{9.6.1b}$$

Now consider a version of the same economy in which there is no currency but rather indexed government bonds. The demographics and endowments are identical with the preceding economy, but now each initial old person is endowed with B_1 units of a maturing bond, denominated in units of time 1 consumption good. In period t, the government sells new one-period bonds to the young to finance its purchases g_t of time t goods and to pay off the one-period debt falling due at time t. Let $R_t > 0$ be the gross real one-period rate of return on government debt between t and $t + 1$.

DEFINITION: An equilibrium with bond-financed government deficits is a sequence $\{B_{t+1}, R_t\}_{t=1}^{\infty}$ that satisfies (a) given $\{R_t\}$,

$$B_{t+1} = \arg\max_{\tilde{B}} [u(y_t^t - \tilde{B}/R_t) + u(y_{t+1}^t + \tilde{B})]; \tag{9.6.2a}$$

and (b)

$$B_{t+1}/R_t = B_t + g_t, \tag{9.6.2b}$$

with $B_1 \geq 0$ given.

These two types of equilibria are isomorphic in the following sense: Take an equilibrium of the economy with money-financed deficits and transform it into an equilibrium of the economy with bond-financed deficits as follows: set $B_t = M_{t-1}/p_t, R_t = p_t/p_{t+1}$. It can be verified directly that these settings of bonds and interest rates, together with the original consumption allocation, form an equilibrium of the economy with bond-financed deficits.

Each of these two types of equilibria is evidently also isomorphic to the following equilibrium formulated with time 0 markets:

DEFINITION: Let B_1^g represent claims to time 1 consumption owed by the government to the old at time 1. An equilibrium with time 0 trading is an initial level of government debt B_1^g, a price system $\{q_t^0\}_{t=1}^{\infty}$, and a sequence $\{s_t\}_{t=1}^{\infty}$ such that (a) given the price system,

$$s_t = \arg\max_{\tilde{s}} \left\{ u(y_t^t - \tilde{s}) + u\left[y_{t+1}^t + \left(\frac{q_t^0}{q_{t+1}^0} \right) \tilde{s} \right] \right\};$$

and (b)

$$q_1^0 B_1^g + \sum_{t=1}^{\infty} q_t^0 g_t = 0. \tag{9.6.3}$$

Condition b is the Arrow-Debreu version of the government budget constraint. Condition a is the optimality condition for the intertemporal consumption decision of the young of generation t.

The government budget constraint in condition b can be represented recursively as

$$q_{t+1}^0 B_{t+1}^g = q_t^0 B_t^g + q_t^0 g_t. \tag{9.6.4}$$

If we solve equation (9.6.4) forward and impose $\lim_{T \to \infty} q_{t+T}^0 B_{t+T}^g = 0$, we obtain the budget constraint (9.6.3) for $t = 1$. Condition (9.6.3) makes it evident that when $\sum_{t=1}^{\infty} q_t^0 g_t > 0$, $B_1^g < 0$, so that the government has negative net worth. This negative net worth corresponds to the unbacked claims that the market nevertheless values in the sequential-trading version of the model.

9.6.2. Growth

It is easy to extend these models to the case in which there is growth in the population. Let there be $N_t = nN_{t-1}$ identical young people at time t, with $n > 0$. For example, consider the economy with money-financed deficits. The total money supply is $N_t M_t$, and the government budget constraint is

$$N_t M_t - N_{t-1} M_{t-1} = N_t p_t g,$$

where g is per-young-person government purchases. Dividing both sides of the budget constraint by N_t and rearranging gives

$$\frac{M_t}{p_{t+1}} \frac{p_{t+1}}{p_t} = n^{-1} \frac{M_{t-1}}{p_t} + g. \tag{9.6.5}$$

This equation replaces equation (9.6.1b) in the definition of an equilibrium with money-financed deficits. (Note that in a steady state, $R = n$ is the high-interest-rate equilibrium.) Similarly, in the economy with bond-financed deficits, the government budget constraint would become

$$\frac{B_{t+1}}{R_t} = n^{-1} B_t + g_t.$$

It is also easy to modify things to permit the government to tax young and old people at t. In that case, with government bond finance the government budget constraint becomes

$$\frac{B_{t+1}}{R_t} = n^{-1}B_t + g_t - \tau_t^t - n^{-1}\tau_t^{t-1},$$

where τ_t^s is the time t tax on a person born in period s.

9.7. Optimality and the existence of monetary equilibria

Wallace (1980) discusses the connection between nonoptimality of the equilibrium without valued money and existence of monetary equilibria. Abstracting from his assumption of a storage technology, we study how the arguments apply to a pure endowment economy. The environment is as follows. At any date t, the population consists of N_t young agents and N_{t-1} old agents where $N_t = nN_{t-1}$ with $n > 0$. Each young person is endowed with $y_1 > 0$ goods, and an old person receives the endowment $y_2 > 0$. Preferences of a young agent at time t are given by the utility function $u(c_t^t, c_{t+1}^t)$, which is twice differentiable with indifference curves that are convex to the origin. The two goods in the utility function are normal goods, and

$$\theta(c_1, c_2) \equiv u_1(c_1, c_2)/u_2(c_1, c_2),$$

the marginal rate of substitution function, approaches infinity as c_2/c_1 approaches infinity and approaches zero as c_2/c_1 approaches zero. The welfare of the initial old agents at time 1 is strictly increasing in c_1^0, and each one of them is endowed with y_2 goods and $m_0^0 > 0$ units of fiat money. Thus, the beginning-of-period aggregate nominal money balances in the initial period 1 are $M_0 = N_0 m_0^0$.

For all $t \geq 1$, M_t, the post-transfer time t stock of fiat money obeys $M_t = zM_{t-1}$ with $z > 0$. The time t transfer (or tax), $(z-1)M_{t-1}$, is divided equally at time t among the N_{t-1} members of the current old generation. The transfers (or taxes) are fully anticipated and are viewed as lump-sum: they do not depend on consumption and saving behavior. The budget constraints of a young agent born in period t are

$$c_t^t + \frac{m_t^t}{p_t} \leq y_1, \tag{9.7.1}$$

$$c_{t+1}^t \leq y_2 + \frac{m_t^t}{p_{t+1}} + \frac{(z-1)}{N_t} \frac{M_t}{p_{t+1}}, \qquad (9.7.2)$$

$$m_t^t \geq 0, \qquad (9.7.3)$$

where $p_t > 0$ is the time t price level. In a nonmonetary equilibrium, the price level is infinite, so the real values of both money holdings and transfers are zero. Since all members in a generation are identical, the nonmonetary equilibrium is autarky with a marginal rate of substitution equal to

$$\theta_{\text{aut}} \equiv \frac{u_1(y_1, y_2)}{u_2(y_1, y_2)}.$$

We ask two questions about this economy. Under what circumstances does a monetary equilibrium exist? And, when it exists, under what circumstances does it improve matters?

Let $\hat{m}_t$ denote the equilibrium real money balances of a young agent at time t, $\hat{m}_t \equiv M_t/(N_t p_t)$. Substitution of equilibrium money holdings into budget constraints (9.7.1) and (9.7.2) at equality yield $c_t^t = y_1 - \hat{m}_t$ and $c_{t+1}^t = y_2 + n\hat{m}_{t+1}$. In a monetary equilibrium, $\hat{m}_t > 0$ for all t and the marginal rate of substitution $\theta(c_t^t, c_{t+1}^t)$ satisfies

$$\theta(y_1 - \hat{m}_t,\ y_2 + n\hat{m}_{t+1}) = \frac{p_t}{p_{t+1}} > \theta_{\text{aut}}, \quad \forall t \geq 1. \qquad (9.7.4)$$

The equality part of (9.7.4) is the first-order condition for money holdings of an agent born in period t evaluated at the equilibrium allocation. Since $c_t^t < y_1$ and $c_{t+1}^t > y_2$ in a monetary equilibrium, the inequality in (9.7.4) follows from the assumption that the two goods in the utility function are normal goods.

Another useful characterization of the equilibrium rate of return on money, p_t/p_{t+1}, can be obtained as follows. By the rule generating M_t and the equilibrium condition $M_t/p_t = N_t\hat{m}_t$, we have for all t,

$$\frac{p_t}{p_{t+1}} = \frac{M_{t+1}}{zM_t} \frac{p_t}{p_{t+1}} = \frac{N_{t+1}\hat{m}_{t+1}}{zN_t\hat{m}_t} = \frac{n}{z} \frac{\hat{m}_{t+1}}{\hat{m}_t}. \qquad (9.7.5)$$

We are now ready to address our first question, under what circumstances does a monetary equilibrium exist?

PROPOSITION: $\theta_{\text{aut}}z < n$ is necessary and sufficient for the existence of at least one monetary equilibrium.

PROOF: We first establish necessity. Suppose to the contrary that there is a monetary equilibrium and $\theta_{\text{aut}} z/n \geq 1$. Then, by the inequality part of (9.7.4) and expression (9.7.5), we have for all t,

$$\frac{\hat{m}_{t+1}}{\hat{m}_t} > \frac{z\theta_{\text{aut}}}{n} \geq 1. \tag{9.7.6}$$

If $z\theta_{\text{aut}}/n > 1$, one plus the net growth rate of $\hat{m}_t$ is bounded uniformly above one and, hence, the sequence $\{\hat{m}_t\}$ is unbounded, which is inconsistent with an equilibrium because real money balances per capita cannot exceed the endowment y_1 of a young agent. If $z\theta_{\text{aut}}/n = 1$, the strictly increasing sequence $\{\hat{m}_t\}$ in (9.7.6) might not be unbounded but converge to some constant $\hat{m}_\infty$. According to (9.7.4) and (9.7.5), the marginal rate of substitution will then converge to n/z, which by assumption is now equal to θ_{aut}, the marginal rate of substitution in autarky. Thus, real balances must be zero in the limit, which contradicts the existence of a strictly increasing sequence of positive real balances in (9.7.6).

To show sufficiency, we prove the existence of a unique equilibrium with constant per capita real money balances when $\theta_{\text{aut}} z < n$. Substitute our candidate equilibrium, $\hat{m}_t = \hat{m}_{t+1} \equiv \hat{m}$, into (9.7.4) and (9.7.5), which yields two equilibrium conditions,

$$\theta(y_1 - \hat{m},\ y_2 + n\hat{m}) = \frac{n}{z} > \theta_{\text{aut}}.$$

The inequality part is satisfied under the parameter restriction of the proposition, and we only have to show the existence of $\hat{m} \in [0, y_1]$ that satisfies the equality part. But the existence (and uniqueness) of such a $\hat{m}$ is trivial. Note that the marginal rate of substitution on the left side of the equality is equal to θ_{aut} when $\hat{m} = 0$. Next, our assumptions on preferences imply that the marginal rate of substitution is strictly increasing in $\hat{m}$, and approaches infinity when $\hat{m}$ approaches y_1. ∎

The stationary monetary equilibrium in the proof will be referred to as the $\hat{m}$ equilibrium. In general, there are other nonstationary monetary equilibria when the parameter condition of the proposition is satisfied. For example, in the case of logarithmic preferences and a constant population, recall the continuum of equilibria indexed by the scalar $c > 0$ in expression (9.4.8). But here we choose to focus solely on the stationary $\hat{m}$ equilibrium and its welfare implications. The $\hat{m}$ equilibrium will be compared to other feasible allocations using the Pareto criterion. Evidently, an allocation $C = \{c_1^0; (c_t^t, c_{t+1}^t), t \geq 1\}$ is feasible if

$$N_t c_t^t + N_{t-1} c_t^{t-1} \leq N_t y_1 + N_{t-1} y_2, \quad \forall t \geq 1,$$

or, equivalently,

$$n c_t^t + c_t^{t-1} \leq n y_1 + y_2, \quad \forall t \geq 1. \tag{9.7.7}$$

The definition of Pareto optimality is:

DEFINITION: A feasible allocation C is Pareto optimal if there is no other feasible allocation $\tilde{C}$ such that

$$\tilde{c}_1^0 \geq c_1^0,$$
$$u(\tilde{c}_t^t, \tilde{c}_{t+1}^t) \geq u(c_t^t, c_{t+1}^t), \quad \forall t \geq 1,$$

and at least one of these weak inequalities holds with strict inequality.

We first examine under what circumstances the nonmonetary equilibrium (autarky) is Pareto optimal.

PROPOSITION: $\theta_{\text{aut}} \geq n$ is necessary and sufficient for the optimality of the nonmonetary equilibrium (autarky).

PROOF: To establish sufficiency, suppose to the contrary that there exists another feasible allocation $\tilde{C}$ that is Pareto superior to autarky and $\theta_{\text{aut}} \geq n$. Without loss of generality, assume that the allocation $\tilde{C}$ satisfies (9.7.7) with equality. (Given an allocation that is Pareto superior to autarky but that does not satisfy (9.7.7), one can easily construct another allocation that is Pareto superior to the given allocation, and hence to autarky.) Let period t be the first period when this alternative allocation $\tilde{C}$ differs from the autarkic allocation. The requirement that the old generation in this period is not made worse off, $\tilde{c}_t^{t-1} \geq y_2$, implies that the first perturbation from the autarkic allocation must be $\tilde{c}_t^t < y_1$, with the subsequent implication that $\tilde{c}_{t+1}^t > y_2$. It follows that the consumption of young agents at time $t+1$ must also fall below y_1, and we define

$$\epsilon_{t+1} \equiv y_1 - \tilde{c}_{t+1}^{t+1} > 0. \tag{9.7.8}$$

Now, given $\tilde{c}_{t+1}^{t+1}$, we compute the smallest number c_{t+2}^{t+1} that satisfies

$$u(\tilde{c}_{t+1}^{t+1}, c_{t+2}^{t+1}) \geq u(y_1, y_2).$$

Let $\bar{c}_{t+2}^{t+1}$ be the solution to this problem. Since the allocation $\tilde{C}$ is Pareto superior to autarky, we have $\tilde{c}_{t+2}^{t+1} \geq \bar{c}_{t+2}^{t+1}$. Before using this inequality, though, we want to derive a convenient expression for $\bar{c}_{t+2}^{t+1}$.

Consider the indifference curve of $u(c_1, c_2)$ that yields a fixed utility equal to $u(y_1, y_2)$. In general, along an indifference curve, $c_2 = h(c_1)$, where $h' = -u_1/u_2 = -\theta$ and $h'' > 0$. Therefore, applying the intermediate value theorem to h, we have

$$h(c_1) = h(y_1) + (y_1 - c_1)[-h'(y_1) + f(y_1 - c_1)], \qquad (9.7.9)$$

where the function f is strictly increasing and $f(0) = 0$.

Now, since $(\tilde{c}_{t+1}^{t+1}, \bar{c}_{t+2}^{t+1})$ and (y_1, y_2) are on the same indifference curve, we can use $(9.7.8)$ and $(9.7.9)$ to write

$$\bar{c}_{t+2}^{t+1} = y_2 + \epsilon_{t+1}[\theta_{\text{aut}} + f(\epsilon_{t+1})],$$

and after invoking $\tilde{c}_{t+2}^{t+1} \geq \bar{c}_{t+2}^{t+1}$, we have

$$\tilde{c}_{t+2}^{t+1} - y_2 \geq \epsilon_{t+1}[\theta_{\text{aut}} + f(\epsilon_{t+1})]. \qquad (9.7.10)$$

Since $\tilde{C}$ satisfies $(9.7.7)$ at equality, we also have

$$\epsilon_{t+2} \equiv y_1 - \tilde{c}_{t+2}^{t+2} = \frac{\tilde{c}_{t+2}^{t+1} - y_2}{n}. \qquad (9.7.11)$$

Substitution of $(9.7.10)$ into $(9.7.11)$ yields

$$\epsilon_{t+2} \geq \epsilon_{t+1} \frac{\theta_{\text{aut}} + f(\epsilon_{t+1})}{n} \qquad (9.7.12)$$
$$> \epsilon_{t+1},$$

where the strict inequality follows from $\theta_{\text{aut}} \geq n$ and $f(\epsilon_{t+1}) > 0$ (implied by $\epsilon_{t+1} > 0$). Continuing these computations of successive values of ϵ_{t+k} yields

$$\epsilon_{t+k} \geq \epsilon_{t+1} \prod_{j=1}^{k-1} \frac{\theta_{\text{aut}} + f(\epsilon_{t+j})}{n} > \epsilon_{t+1} \left[\frac{\theta_{\text{aut}} + f(\epsilon_{t+1})}{n} \right]^{k-1}, \quad \text{for } k > 2,$$

where the strict inequality follows from the fact that $\{\epsilon_{t+j}\}$ is a strictly increasing sequence. Thus, the ϵ sequence is bounded below by a strictly increasing exponential and hence is unbounded. But such an unbounded sequence violates feasibility because ϵ cannot exceed the endowment y_1 of a young agent. It follows that we can rule out the existence of a Pareto superior allocation $\tilde{C}$, and conclude that $\theta_{\text{aut}} \geq n$ is a sufficient condition for the optimality of autarky.

To establish necessity, we prove the existence of an alternative feasible allocation $\hat{C}$ that is Pareto superior to autarky when $\theta_{\text{aut}} < n$. First, pick an $\epsilon > 0$ sufficiently small so that

$$\theta_{\text{aut}} + f(\epsilon) \leq n, \tag{9.7.13}$$

where f is defined implicitly by equation $(9.7.9)$. Second, set $\hat{c}_t^t = y_1 - \epsilon \equiv \hat{c}_1$, and

$$\hat{c}_{t+1}^t = y_2 + \epsilon[\theta_{\text{aut}} + f(\epsilon)] \equiv \hat{c}_2, \quad \forall t \geq 1. \tag{9.7.14}$$

That is, we have constructed a consumption bundle $(\hat{c}_1, \hat{c}_2)$ that lies on the same indifference curve as (y_1, y_2), and from $(9.7.13)$ and $(9.7.14)$, we have

$$\hat{c}_2 \leq y_2 + n\epsilon,$$

which ensures that the condition for feasibility $(9.7.7)$ is satisfied for $t \geq 2$. By setting $\hat{c}_1^0 = y_2 + n\epsilon$, feasibility is also satisfied in period 1 and the initial old generation is then strictly better off under the alternative allocation $\hat{C}$. ∎

With a constant nominal money supply, $z = 1$, the two propositions show that a monetary equilibrium exists if and only if the nonmonetary equilibrium is suboptimal. In that case, the following proposition establishes that the stationary $\hat{m}$ equilibrium is optimal.

PROPOSITION: Given $\theta_{\text{aut}} z < n$, then $z \leq 1$ is necessary and sufficient for the optimality of the stationary monetary equilibrium $\hat{m}$.

PROOF: The class of feasible stationary allocations with $(c_t^t, c_{t+1}^t) = (c_1, c_2)$ for all $t \geq 1$, is given by

$$c_1 + \frac{c_2}{n} \leq y_1 + \frac{y_2}{n}, \tag{9.7.15}$$

i.e., the condition for feasibility in $(9.7.7)$. It follows that the $\hat{m}$ equilibrium satisfies $(9.7.15)$ at equality, and we denote the associated consumption allocation of an agent born at time $t \geq 1$ by $(\hat{c}_1, \hat{c}_2)$. It is also the case that $(\hat{c}_1, \hat{c}_2)$ maximizes an agent's utility subject to budget constraints $(9.7.1)$ and $(9.7.2)$. The consolidation of these two constraints yields

$$c_1 + \frac{z}{n} c_2 \leq y_1 + \frac{z}{n} y_2 + \frac{z(z-1)}{n} \frac{M_t}{N_t} \frac{1}{p_{t+1}}, \tag{9.7.16}$$

where we have used the stationary rate or return in $(9.7.5)$, $p_t/p_{t+1} = n/z$. After also invoking $zM_t = M_{t+1}$, $n = N_{t+1}/N_t$, and the equilibrium condition $M_{t+1}/(p_{t+1}N_{t+1}) = \hat{m}$, expression $(9.7.16)$ simplifies to

$$c_1 + \frac{z}{n}c_2 \leq y_1 + \frac{z}{n}y_2 + (z-1)\hat{m}. \qquad (9.7.17)$$

To prove the statement about necessity, Figure 9.7.1 depicts the two curves $(9.7.15)$ and $(9.7.17)$ when condition $z \leq 1$ fails to hold, i.e., we assume that $z > 1$. The point that maximizes utility subject to $(9.7.15)$ is denoted $(\bar{c}_1, \bar{c}_2)$. Transitivity of preferences and the fact that the slope of budget line $(9.7.17)$ is flatter than that of $(9.7.15)$ imply that $(\hat{c}_1, \hat{c}_2)$ lies southeast of $(\bar{c}_1, \bar{c}_2)$. By revealed preference, then, $(\bar{c}_1, \bar{c}_2)$ is preferred to $(\hat{c}_1, \hat{c}_2)$ and all generations born in period $t \geq 1$ are better off under the allocation $\overline{C}$. The initial old generation can also be made better off under this alternative allocation since it is feasible to strictly increase their consumption,

$$\bar{c}_1^0 = y_2 + n(y_1 - \bar{c}_1^1) > y_2 + n(y_1 - \hat{c}_1^1) = \hat{c}_1^0.$$

Thus, we have established that $z \leq 1$ is necessary for the optimality of the stationary monetary equilibrium $\hat{m}$.

To prove sufficiency, note that $(9.7.4)$, $(9.7.5)$ and $z \leq 1$ imply that

$$\theta(\hat{c}_1, \hat{c}_2) = \frac{n}{z} \geq n.$$

We can then construct an argument that is analogous to the sufficiency part of the proof to the preceding proposition. ∎

As pointed out by Wallace (1980), the proposition implies no connection between the path of the price level in an $\hat{m}$ equilibrium and the optimality of that equilibrium. Thus, there may be an optimal monetary equilibrium with positive inflation, for example, if $\theta_{\mathrm{aut}} < n < z \leq 1$; and there may be a nonoptimal monetary equilibrium with a constant price level, for example, if $z = n > 1 > \theta_{\mathrm{aut}}$. What counts is the nominal quantity of fiat money. The proposition suggests that the quantity of money should not be increased. In particular, if $z \leq 1$, then an optimal $\hat{m}$ equilibrium exists whenever the nonmonetary equilibrium is nonoptimal.

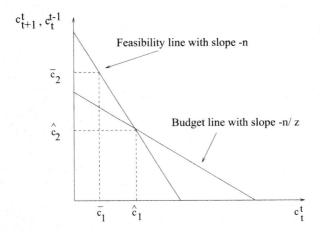

Figure 9.7.1: The feasibility line $(9.7.15)$ and the budget line $(9.7.17)$ when $z > 1$. The consumption allocation in the monetary equilibrium is $(\hat{c}_1, \hat{c}_2)$, and the point that maximizes utility subject to the feasibility line is denoted $(\bar{c}_1, \bar{c}_2)$.

9.7.1. Balasko-Shell criterion for optimality

For the case of constant population, Balasko and Shell (1980) have established a convenient general criterion for testing whether allocations are optimal.[9] Balasko and Shell permit diversity among agents in terms of endowments $[w_t^{th}, w_{t+1}^{th}]$ and utility functions $u^{th}(c_t^{th}, c_{t+1}^{th})$, where w_s^{th} is the time s endowment of an agent named h who is born at t and c_s^{th} is the time s consumption of agent named h born at t. Balasko and Shell assume fixed populations of types h over time. They impose several kinds of technical conditions that serve to rule out possible pathologies. The two main ones are these. First, they assume that indifference curves have neither flat parts nor kinks, and they also rule out indifference curves with flat parts or kinks as limits of sequences of indifference curves for given h as $t \to \infty$. Second, they assume that the aggregate endowments $\sum_h (w_t^{th} + w_t^{t-1,h})$ are uniformly bounded from above and that there exists an $\epsilon > 0$ such that $w_t^{sh} > \epsilon$ for all s, h, and for $t \in \{s, s+1\}$. They consider consumption allocations uniformly bounded away from the axes.

---

[9]   Balasko and Shell credit David Cass (1971) with having authored a version of their criterion.

With these conditions, Balasko and Shell consider the class of allocations in which all young agents at t share a common marginal rate of substitution $1 + r_t \equiv u_1^{th}(c_t^{th}, c_{t+1}^{th})/u_2^{th}(c_t^{th}, c_{t+1}^{th})$ and in which all of the endowments are consumed. Then Balasko and Shell show that an allocation is Pareto optimal if and only if

$$\sum_{t=1}^{\infty} \prod_{s=1}^{t} [1 + r_s] = +\infty, \qquad (9.7.18)$$

that is, if and only if the infinite sum of t-period gross interest rates, $\prod_{s=1}^{t}[1 + r_s]$, diverges.

The Balasko-Shell criterion for optimality succinctly summarizes the sense in which low-interest-rate economies are not optimal. We have already encountered repeated examples of the situation that, before an equilibrium with valued currency can exist, the equilibrium without valued currency must be a low-interest-rate economy in just the sense identified by Balasko and Shell's criterion, (9.7.18). Furthermore, by applying the Balasko-Shell criterion, (9.7.18), or generalizations of it that allow for a positive net growth rate of population n, it can be shown that, among equilibria with valued currency, only equilibria with high rates of return on currency are optimal.

9.8. Within-generation heterogeneity

This section describes an overlapping generations model having within-generation heterogeneity of endowments. We shall follow Sargent and Wallace (1982) and Smith (1988) and use this model as a vehicle for talking about some issues in monetary theory that require a setting in which government-issued currency coexists with and is a more-or-less good substitute for private IOUs.

We now assume that within each generation born at $t \geq 1$, there are J groups of agents. There is a constant number N_j of group j agents. Agents of group j are endowed with $w_1(j)$ when young and $w_2(j)$ when old. The saving function of a household of group j born at time t solves the time t version of problem (9.5.3). We denote this savings function $f(R_t, j)$. If we assume that all households of generation t have preferences $U^t(c^t) = \ln c_t^t + \ln c_{t+1}^t$, the saving function is

$$f(R_t, j) = .5 \left(w_1(j) - \frac{w_2(j)}{R_t} \right).$$

At $t = 1$, there are old people who are endowed in the aggregate with $H = H(0)$ units of an inconvertible currency.

For example, assume that $J = 2$, that $(w_1(1), w_2(1)) = (\alpha, 0), (w_1(2), w_2(2)) = (0, \beta)$, where $\alpha > 0, \beta > 0$. The type 1 people are lenders, while the type 2 are borrowers. For the case of log preference we have the savings functions $f(R, 1) = \alpha/2, f(R, 2) = -\beta/(2R)$.

9.8.1. Nonmonetary equilibrium

A nonmonetary equilibrium consists of sequences (R, s_j) of rates of return R and savings rates for $j = 1, \ldots, J$ and $t \geq 1$ that satisfy $(1) s_{tj} = f(R_t, j)$, and $(2) \sum_{j=1}^{J} N_j f(R_t, j) = 0$. Condition (1) builds in household optimization; condition (2) says that aggregate net savings equals zero (borrowing equals lending).

For the case in which the endowments, preferences, and group sizes are constant across time, the interest rate is determined at the intersection of the aggregate savings function with the R axis, depicted as R_1 in Figure 9.8.1. No intergenerational transfers occur in the nonmonetary equilibrium. The equilibrium consists of a sequence of separate two-period pure consumption loan economies of a type analyzed by Irving Fisher (1907).

9.8.2. Monetary equilibrium

In an equilibrium with valued fiat currency, at each date $t \geq 1$ the old receive goods from the young in exchange for the currency stock H. For any variable x, $\vec{x} = \{x_t\}_{t=1}^{\infty}$. An equilibrium with valued fiat money is a set of sequences $\vec{R}, \vec{p}, \vec{s}$ such that (1) $\vec{p}$ is a positive sequence, (2) $R_t = p_t/p_{t+1}$, (3) $s_{jt} = f(R_t, j)$, and (4) $\sum_{j=1}^{J} N_j f(R_t, j) = \frac{H}{p_t}$. Condition (1) states that currency is valued at all dates. Condition (2) states that currency and consumption loans are perfect substitutes. Condition (3) requires that saving decisions are optimal. Condition (4) equates the net saving of the young (the left side) to the net dissaving of the old (the right side). The old supply currency inelastically.

We can determine a stationary equilibrium graphically. A stationary equilibrium satisfies $p_t = p$ for all t, which implies $R = 1$ for all t. Thus, if it exists, a stationary equilibrium solves

$$\sum_{j=1}^{J} N_j f(1,j) = \frac{H}{p} \tag{9.8.1}$$

for a positive price level. (See Figure 9.8.1.) Evidently, a stationary monetary equilibrium exists if the net savings of the young are positive for $R = 1$.

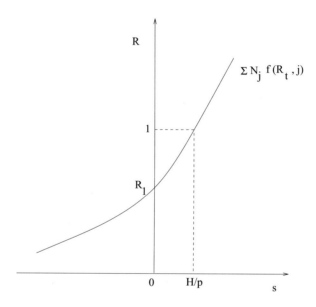

Figure 9.8.1: The intersection of the aggregate savings function with a horizontal line at $R = 1$ determines a stationary equilibrium value of the price level, if positive.

For the special case of logarithmic preferences and two classes of young people, the aggregate savings function of the young is time invariant and equal to

$$\sum_{j} f(R,j) = .5(N_1 \alpha - N_2 \frac{\beta}{R}).$$

Note that the equilibrium condition (9.8.1) can be written

$$.5 N_1 \alpha = .5 N_2 \frac{\beta}{R} + \frac{H}{p}.$$

The left side is the demand for savings or the demand for "currency" while the right side is the supply, consisting of privately issued IOU's (the first term) and government-issued currency. The right side is thus an abstract version of what is called M1, which is a sum of privately issued IOUs (demand deposits) and government-issued reserves and currency.

9.8.3. Nonstationary equilibria

Mathematically, the equilibrium conditions for the model with log preferences and two groups have the same structure as the model analyzed previously in equations (9.4.7) and (9.4.8), with simple reinterpretations of parameters. We leave it to the reader here and in an exercise to show that if there exists a stationary equilibrium with valued fiat currency, then there exists a continuum of equilibria with valued fiat currency, all but one of which have the real value of government currency approaching zero asymptotically. A linear difference equation like (9.4.7) supports this conclusion.

9.8.4. The real bills doctrine

In nineteenth-century Europe and the early days of the Federal Reserve system in the United States, central banks conducted open market operations not by purchasing government securities but by purchasing safe (risk-free) short-term private IOUs. We now analyze this old-fashioned type of open market operation. We allow the government to issue additional currency each period. It uses the proceeds exclusively to purchase private IOUs (make loans to private agents) in the amount L_t at time t. Such open market operations are subject to the sequence of restrictions

$$L_t = R_{t-1}L_{t-1} + \frac{H_t - H_{t-1}}{p_t} \qquad (9.8.2)$$

for $t \geq 1$ and $H_0 = H > 0$ given, $L_0 = 0$. Here L_t is the amount of the time t consumption good that the government lends to the private sector from period t to period $t+1$. Equation (9.8.2) states that the government finances these loans in two ways: first, by rolling over the proceeds $R_{t-1}L_{t-1}$ from the repayment of last period's loans, and second, by injecting new currency in the amount

$H_t - H_{t-1}$. With the government injecting new currency and purchasing loans in this way each period, the equilibrium condition in the loan market becomes

$$\sum_{j=1}^{J} N_j f(R_t, j) + L_t = \frac{H_{t-1}}{p_t} + \frac{H_t - H_{t-1}}{p_t} \qquad (9.8.3)$$

where the first term on the right is the real dissaving of the old at t (their real balances) and the second term is the real value of the new money printed by the monetary authority to finance purchases of private IOUs issued by the young at t. The left side is the net savings of the young plus the savings of the government.

Under several guises, the effects of open market operations like this have concerned monetary economists for centuries.[10] We state the following proposition:

IRRELEVANCE OF OPEN MARKET OPERATIONS: Open market operations are irrelevant: all positive sequences $\{L_t, H_t\}_{t=0}^{\infty}$ that satisfy the constraint (9.8.2) are associated with the same equilibrium allocation, interest rate, and price level sequences.

PROOF: Evidently, we can write the equilibrium condition (9.8.3) as

$$\sum_{j=1}^{J} N_j f(R_t, j) + L_t = \frac{H_t}{p_t}. \qquad (9.8.4)$$

For $t \geq 1$, iterating (9.8.2) once and using $R_{t-1} = \frac{p_{t-1}}{p_t}$ gives

$$L_t = R_{t-1} R_{t-2} L_{t-2} + \frac{H_t - H_{t-2}}{p_t}.$$

Iterating back to time 0 and using $L_0 = 0$ gives

$$L_t = \frac{H_t - H_0}{p_t}. \qquad (9.8.5)$$

Substituting (9.8.5) into (9.8.4) gives

$$\sum_{j=1}^{J} N_j f(R_t, j) = \frac{H_0}{p_t}. \qquad (9.8.6)$$

---

[10] One issue concerned the effects on the price level of allowing banks to issue private bank notes. Nothing in our model makes us take seriously that the notes H_t are issued by the government. We can also think of them as being issued by a private bank.

This is the same equilibrium condition in the economy with no open market operations, i.e., the economy with $L_t \equiv 0$ for all $t \geq 1$. Any price level and rate of return sequence that solves (9.8.6) also solves (9.8.3) for any L_t sequence satisfying (9.8.2). ∎

This proposition captures the spirit of Adam Smith's real bills doctrine, which states that if the government issues bank notes to purchase safe evidences of private indebtedness, it is not inflationary. Sargent and Wallace (1982) go on to analyze settings in which the money market is initially separated from the credit market by some legal restrictions that inhibit intermediation. Then open market operations are no longer irrelevant because they can partially undo the legal restrictions. Sargent and Wallace show how those legal restrictions can help stabilize the price level at a cost in terms of economic efficiency. Kahn and Roberds (1998) extend the Sargent and Wallace model to study issues about regulating electronic payments systems.

9.9. Gift-giving equilibrium

Michihiro Kandori (1992) and Lones Smith (1992) have used ideas from the literature on reputation (see chapter 23) to study whether there exist history-dependent sequences of gifts that support an optimal allocation. Their idea is to set up the economy as a game played with a sequence of players. We briefly describe a gift-giving game for an overlapping generations economy in which voluntary intergenerational gifts support an optimal allocation. Suppose that the consumption of an initial old person is

$$c_1^0 = y_1^0 + s_1$$

and the utility of each young agent is

$$u(y_i^i - s_i) + u(y_{i+1}^i + s_{i+1}), \quad i \geq 1 \tag{9.9.1}$$

where $s_i \geq 0$ is the gift from a young person at i to an old person at i. Suppose that the endowment pattern is $y_i^i = 1 - \epsilon, y_{i+1}^i = \epsilon$, where $\epsilon \in (0, .5)$.

Consider the following system of expectations, to which a young person chooses whether to conform:

$$s_i = \begin{cases} .5 - \epsilon & \text{if } v_i = \overline{v}; \\ 0 & \text{otherwise.} \end{cases} \tag{9.9.2a}$$

$$v_{i+1} = \begin{cases} \overline{v} & \text{if } v_i = \overline{v} \text{ and } s_i = .5 - \epsilon; \\ \underline{v} & \text{otherwise.} \end{cases} \qquad (9.9.2b)$$

Here we are free to take $\overline{v} = 2u(.5)$ and $\underline{v} = u(1 - \epsilon) + u(\epsilon)$. These are "promised utilities." We make them serve as "state variables" that summarize the history of intergenerational gift giving. To start, we need an initial value v_1. Equations (9.9.2) act as the transition laws that young agents face in choosing s_i in (9.9.1).

An initial condition v_1 and the rule (9.9.2) form a system of expectations that tells the young person of each generation what he is expected to give. His gift is immediately handed over to an old person. A system of expectations is called an *equilibrium* if for each $i \geq 1$, each young agent chooses to conform.

We can immediately compute two equilibrium systems of expectations. The first is the "autarky" equilibrium: give nothing yourself and expect all future generations to give nothing. To represent this equilibrium within equations (9.9.2), set $v_1 \neq \overline{v}$. It is easy to verify that each young person will confirm what is expected of him in this equilibrium. Given that future generations will not give, each young person chooses not to give.

For the second equilibrium, set $v_1 = \overline{v}$. Here each household chooses to give the expected amount, because failure to do so causes the next generation of young people not to give; whereas affirming the expectation to give passes that expectation along to the next generation, which affirms it in turn. Each of these equilibria is credible, in the sense of subgame perfection, to be studied extensively in chapter 23.

Narayana Kocherlakota (1998) has compared gift giving and monetary equilibria in a variety of environments and has used the comparison to provide a precise sense in which "money" substitutes for "memory".

9.10.   Concluding remarks

The overlapping generations model is a workhorse in analyses of public finance, welfare economics, and demographics. Diamond (1965) studied some fiscal policy issues within a version of the model with a neoclassical production. He showed that, depending on preference and productivity parameters, equilibria of the model can have too much capital, and that such capital overaccumulation can be corrected by having the government issue and perpetually roll over unbacked debt.[11] Auerbach and Kotlikoff (1987) formulated a long-lived overlapping generations model with capital, labor, production, and various kinds of taxes. They used the model to study a host of fiscal issues. Rios-Rull (1994a) used a calibrated overlapping generations growth model to examine the quantitative importance of market incompleteness for insuring against aggregate risk. See Attanasio (2000) for a review of theories and evidence about consumption within life-cycle models.

Several authors in a 1980 volume edited by John Kareken and Neil Wallace argued through example that the overlapping generations model is useful for analyzing a variety of issues in monetary economics. We refer to that volume, McCandless and Wallace (1992), Champ and Freeman (1994), Brock (1990), and Sargent (1987b) for a variety of applications of the overlapping generations model to issues in monetary economics.

Exercises

Exercise 9.1 At each date $t \geq 1$, an economy consists of overlapping generations of a constant number N of two-period-lived agents. Young agents born in t have preferences over consumption streams of a single good that are ordered by $u(c_t^t) + u(c_{t+1}^t)$, where $u(c) = c^{1-\gamma}/(1-\gamma)$, and where c_t^i is the consumption of an agent born at i in time t. It is understood that $\gamma > 0$, and that when $\gamma = 1$, $u(c) = \ln c$. Each young agent born at $t \geq 1$ has identical preferences and endowment pattern (w_1, w_2), where w_1 is the endowment when young and w_2 is the endowment when old. Assume $0 < w_2 < w_1$. In addition, there are some initial old agents at time 1 who are endowed with w_2 of the time 1

---

[11] Abel, Mankiw, Summers, and Zeckhauser (1989) propose an empirical test of whether there is capital overaccumulation in the U.S. economy, and conclude that there is not.

consumption good, and who order consumption streams by c_1^0. The initial old (i.e., the old at $t = 1$) are also endowed with M units of unbacked fiat currency. The stock of currency is constant over time.

a. Find the saving function of a young agent.

b. Define an equilibrium with valued fiat currency.

c. Define a stationary equilibrium with valued fiat currency.

d. Compute a stationary equilibrium with valued fiat currency.

e. Describe how many equilibria with valued fiat currency there are. (You are not being asked to compute them.)

f. Compute the limiting value as $t \to +\infty$ of the rate of return on currency in each of the nonstationary equilibria with valued fiat currency. Justify your calculations.

Exercise 9.2 Consider an economy with overlapping generations of a constant population of an even number N of two-period-lived agents. New young agents are born at each date $t \geq 1$. Half of the young agents are endowed with w_1 when young and 0 when old. The other half are endowed with 0 when young and w_2 when old. Assume $0 < w_2 < w_1$. Preferences of all young agents are as in problem 1, with $\gamma = 1$. Half of the N initial old are endowed with w_2 units of the consumption good and half are endowed with nothing. Each old person orders consumption streams by c_1^0. Each old person at $t = 1$ is endowed with M units of unbacked fiat currency. No other generation is endowed with fiat currency. The stock of fiat currency is fixed over time.

a. Find the saving function of each of the two types of young person for $t \geq 1$.

b. Define an equilibrium without valued fiat currency. Compute all such equilibria.

c. Define an equilibrium with valued fiat currency.

d. Compute all the (nonstochastic) equilibria with valued fiat currency.

e. Argue that there is a unique stationary equilibrium with valued fiat currency.

f. How are the various equilibria with valued fiat currency ranked by the Pareto criterion?

Exercise 9.3   Take the economy of exercise *9.1*, but make one change. Endow the initial old with a tree that yields a constant dividend of $d > 0$ units of the consumption good for each $t \geq 1$.

a. Compute all the equilibria with valued fiat currency.

b. Compute all the equilibria without valued fiat currency.

c. If you want, you can answer both parts of this question in the context of the following particular numerical example: $w_1 = 10, w_2 = 5, d = .000001$.

Exercise 9.4   Take the economy of exercise *9.1* and make the following two changes. First, assume that $\gamma = 1$. Second, assume that the number of young agents born at t is $N(t) = nN(t-1)$, where $N(0) > 0$ is given and $n \geq 1$. Everything else about the economy remains the same.

a. Compute an equilibrium without valued fiat money.

b. Compute a stationary equilibrium with valued fiat money.

Exercise 9.5   Consider an economy consisting of overlapping generations of two-period-lived consumers. At each date $t \geq 1$ there are born $N(t)$ identical young people each of whom is endowed with $w_1 > 0$ units of a single consumption good when young and $w_2 > 0$ units of the consumption good when old. Assume that $w_2 < w_1$. The consumption good is not storable. The population of young people is described by $N(t) = nN(t-1)$, where $n > 0$. Young people born at t rank utility streams according to $\ln(c_t^t) + \ln(c_{t+1}^t)$ where c_t^i is the consumption of the time t good of an agent born in i. In addition, there are $N(0)$ old people at time 1, each of whom is endowed with w_2 units of the time 1 consumption good. The old at $t = 1$ are also endowed with one unit of unbacked pieces of infinitely durable but intrinsically worthless pieces of paper called fiat money.

a. Define an equilibrium without valued fiat currency. Compute such an equilibrium.

b. Define an equilibrium with valued fiat currency.

c. Compute all equilibria with valued fiat currency.

d. Find the limiting rates of return on currency as $t \to +\infty$ in each of the equilibria that you found in part c. Compare them with the one-period interest rate in the equilibrium in part a.

e. Are the equilibria in part c ranked according to the Pareto criterion?

Exercise 9.6 **Exchange rate determinacy**

The world consists of two economies, named $i = 1, 2$, which except for their governments' policies are "copies" of one another. At each date $t \geq 1$, there is a single consumption good, which is storable, but only for rich people. Each economy consists of overlapping generations of two-period-lived agents. For each $t \geq 1$, in economy i, N poor people and N rich people are born. Let $c_t^h(s), y_t^h(s)$ be the time s (consumption, endowment) of a type h agent born at t. Poor agents are endowed with $[y_t^h(t), y_t^h(t+1)] = (\alpha, 0)$; rich agents are endowed with $[y_t^h(t), y_t^h(t+1)] = (\beta, 0)$, where $\beta >> \alpha$. In each country, there are $2N$ initial old who are endowed in the aggregate with $H_i(0)$ units of an unbacked currency and with $2N\epsilon$ units of the time 1 consumption good. For the rich people, storing k units of the time t consumption good produces Rk units of the time $t+1$ consumption good, where $R > 1$ is a fixed gross rate of return on storage. Rich people can earn the rate of return R either by storing goods or by lending to either government by means of indexed bonds. We assume that poor people are prevented from storing capital or holding indexed government debt by the sort of denomination and intermediation restrictions described by Sargent and Wallace (1982).

For each $t \geq 1$, all young agents order consumption streams according to $\ln c_t^h(t) + \ln c_t^h(t+1)$.

For $t \geq 1$, the government of country i finances a stream of purchases (to be thrown into the ocean) of $G_i(t)$ subject to the following budget constraint:

$$
(1) \qquad G_i(t) + RB_i(t-1) = B_i(t) + \frac{H_i(t) - H_i(t-1)}{p_i(t)} + T_i(t),
$$

where $B_i(0) = 0$; $p_i(t)$ is the price level in country i; $T_i(t)$ are lump-sum taxes levied by the government on the *rich* young people at time t; $H_i(t)$ is the stock of i's fiat currency at the end of period t; $B_i(t)$ is the stock of indexed government interest-bearing debt (held by the rich of either country). The government does not explicitly tax poor people, but might tax through an inflation tax. Each government levies a lump-sum tax of $T_i(t)/N$ on each young rich citizen of its own country.

Poor people in both countries are free to hold whichever currency they prefer. Rich people can hold debt of either government and can also store; storage and both government debts bear a constant gross rate of return R.

a. Define an *equilibrium* with valued fiat currencies (in both countries).

b. In a nonstochastic equilibrium, verify the following proposition: if an equilibrium exists in which both fiat currencies are valued, the exchange rate between the two currencies must be constant over time.

c. Suppose that government policy in each country is characterized by specified (exogenous) levels $G_i(t) = G_i, T_i(t) = T_i, \; B_i(t) = 0, \forall t \geq 1$. (The remaining elements of government policy adjust to satisfy the government budget constraints.) Assume that the exogenous components of policy have been set so that an equilibrium with two valued fiat currencies exists. Under this description of policy, show that the equilibrium exchange rate is indeterminate.

d. Suppose that government policy in each country is described as follows: $G_i(t) = G_i, T_i(t) = T_i, H_i(t+1) = H_i(1), B_i(t) = B_i(1) \; \forall t \geq 1$. Show that if there exists an equilibrium with two valued fiat currencies, the exchange rate is determinate.

e. Suppose that government policy in country 1 is specified in terms of exogenous levels of $s_1 = [H_1(t) - H_1(t-1)]/p_1(t) \; \forall t \geq 2$, and $G_1(t) = G_1 \; \forall t \geq 1$. For country 2, government policy consists of exogenous levels of $B_2(t) = B_2(1), G_2(t) = G_2 \forall t \geq 1$. Show that if there exists an equilibrium with two valued fiat currencies, then the exchange rate is determinate.

Exercise 9.7 **Credit controls**

Consider the following overlapping generations model. At each date $t \geq 1$ there appear N two-period-lived young people, said to be of generation t, who live and consume during periods t and $(t+1)$. At time $t = 1$ there exist N old people who are endowed with $H(0)$ units of paper "dollars," which they offer to supply inelastically to the young of generation 1 in exchange for goods. Let $p(t)$ be the price of the one good in the model, measured in dollars per time t good. For each $t \geq 1$, $N/2$ members of generation t are endowed with $y > 0$ units of the good at t and 0 units at $(t+1)$, whereas the remaining $N/2$ members of generation t are endowed with 0 units of the good at t and $y > 0$ units when they are old. All members of all generations have the same utility function:

$$u[c_t^h(t), c_t^h(t+1)] = \ln c_t^h(t) + \ln c_t^h(t+1),$$

where $c_t^h(s)$ is the consumption of agent h of generation t in period s. The old at $t = 1$ simply maximize $c_0^h(1)$. The consumption good is nonstorable. The currency supply is constant through time, so $H(t) = H(0)$, $t \geq 1$.

a. Define a competitive equilibrium without valued currency for this model. Who trades what with whom?

b. In the equilibrium without valued fiat currency, compute competitive equilibrium values of the gross return on consumption loans, the consumption allocation of the old at $t = 1$, and that of the "borrowers" and "lenders" for $t \geq 1$.

c. Define a competitive equilibrium with valued currency. Who trades what with whom?

d. Prove that for this economy there does not exist a competitive equilibrium with valued currency.

e. Now suppose that the government imposes the restriction that $l_t^h(t)[1 + r(t)] \geq -y/4$, where $l_t^h(t)[1 + r(t)]$ represents claims on $(t+1)$–period consumption purchased (if positive) or sold (if negative) by household h of generation t. This is a restriction on the amount of borrowing. For an equilibrium without valued currency, compute the consumption allocation and the gross rate of return on consumption loans.

f. In the setup of part e, show that there exists an equilibrium with valued currency in which the price level obeys the quantity theory equation $p(t) = qH(0)/N$. Find a formula for the undetermined coefficient q. Compute the consumption allocation and the equilibrium rate of return on consumption loans.

g. Are lenders better off in economy b or economy f? What about borrowers? What about the old of period 1 (generation 0)?

Exercise 9.8 **Inside money and real bills**

Consider the following overlapping generations model of two-period-lived people. At each date $t \geq 1$ there are born N_1 individuals of type 1 who are endowed with $y > 0$ units of the consumption good when they are young and zero units when they are old; there are also born N_2 individuals of type 2 who are endowed with zero units of the consumption good when they are young and $Y > 0$ units when they are old. The consumption good is nonstorable. At time $t = 1$, there are N old people, all of the same type, each endowed with zero units of the consumption good and H_0/N units of unbacked paper called "fiat currency." The populations of type 1 and 2 individuals, N_1 and N_2, remain constant for all $t \geq 1$. The young of each generation are identical in preferences and maximize the utility function $\ln c_t^h(t) + \ln c_t^h(t+1)$ where $c_t^h(s)$ is consumption in the sth period of a member h of generation t.

a. Consider the equilibrium without valued currency (that is, the equilibrium in which there is no trade between generations). Let $[1 + r(t)]$ be the gross rate of return on consumption loans. Find a formula for $[1 + r(t)]$ as a function of N_1, N_2, y, and Y.

b. Suppose that N_1, N_2, y, and Y are such that $[1+r(t)] > 1$ in the equilibrium without valued currency. Then prove that there can exist no quantity-theory-style equilibrium where fiat currency is valued and where the price level $p(t)$ obeys the quantity theory equation $p(t) = q \cdot H_0$, where q is a positive constant and $p(t)$ is measured in units of currency per unit good.

c. Suppose that N_1, N_2, y, and Y are such that in the nonvalued-currency equilibrium $1 + r(t) < 1$. Prove that there exists an equilibrium in which fiat currency is valued and that there obtains the quantity theory equation $p(t) = q \cdot H_0$, where q is a constant. Construct an argument to show that the equilibrium with valued currency is not Pareto superior to the nonvalued-currency equilibrium.

d. Suppose that N_1, N_2, y, and Y are such that, in the preceding nonvalued-currency economy, $[1 + r(t)] < 1$, there exists an equilibrium in which fiat currency is valued. Let $\bar{p}$ be the stationary equilibrium price level in that economy. Now consider an alternative economy, identical with the preceding one in all respects except for the following feature: a government each period purchases a constant amount L_g of consumption loans and pays for them by issuing debt on itself, called "inside money" M_I, in the amount $M_I(t) = L_g \cdot p(t)$. The government never retires the inside money, using the proceeds of the loans to finance new purchases of consumption loans in subsequent periods. The quantity of outside money, or currency, remains H_0, whereas the "total high-power money" is now $H_0 + M_I(t)$.

(i) Show that in this economy there exists a valued-currency equilibrium in which the price level is constant over time at $p(t) = \bar{p}$, or equivalently, with $\bar{p} = qH_0$ where q is defined in part c.

(ii) Explain why government purchases of private debt are not inflationary in this economy.

(iii) In many models, once-and-for-all government open-market operations in private debt normally affect real variables and/or price level. What accounts for the difference between those models and the one in this exercise?

Exercise 9.9 **Social security and the price level**

Consider an economy ("economy I") that consists of overlapping generations of two-period-lived people. At each date $t \geq 1$ there is born a constant number N of young people, who desire to consume both when they are young, at t, and when they are old, at $(t+1)$. Each young person has the utility function $\ln c_t(t) + \ln c_t(t+1)$, where $c_s(t)$ is time t consumption of an agent born at s. For all dates $t \geq 1$, young people are endowed with $y > 0$ units of a single nonstorable consumption good when they are young and zero units when they are old. In addition, at time $t = 1$ there are N old people endowed in the aggregate with H units of unbacked fiat currency. Let $p(t)$ be the nominal price level at t, denominated in dollars per time t good.

a. Define and compute an equilibrium with valued fiat currency for this economy. Argue that it exists and is unique. Now consider a second economy ("economy II") that is identical to economy I except that economy II possesses a social security system. In particular, at each date $t \geq 1$, the government taxes $\tau > 0$ units of the time t consumption good away from each young person and at the same time gives τ units of the time t consumption good to each old person then alive.

b. Does economy II possess an equilibrium with valued fiat currency? Describe the restrictions on the parameter τ, if any, that are needed to ensure the existence of such an equilibrium.

c. If an equilibrium with valued fiat currency exists, is it unique?

d. Consider the *stationary* equilibrium with valued fiat currency. Is it unique? Describe how the value of currency or price level would vary across economies with differences in the size of the social security system, as measured by τ.

Exercise 9.10 **Seignorage**

Consider an economy consisting of overlapping generations of two-period-lived agents. At each date $t \geq 1$, there are born N_1 "lenders" who are endowed with $\alpha > 0$ units of the single consumption good when they are young and zero units when they are old. At each date $t \geq 1$, there are also born N_2 "borrowers" who are endowed with zero units of the consumption good when they are young and $\beta > 0$ units when they are old. The good is nonstorable, and N_1 and N_2 are constant through time. The economy starts at time 1, at which time there are N old people who are in the aggregate endowed with $H(0)$ units of unbacked, intrinsically worthless pieces of paper called dollars. Assume that α, β, N_1, and N_2 are such that

$$\frac{N_2\beta}{N_1\alpha} < 1.$$

Assume that everyone has preferences

$$u[c_t^h(t), c_t^h(t+1)] = \ln c_t^h(t) + \ln c_t^h(t+1),$$

where $c_t^h(s)$ is consumption of time s good of agent h born at time t.

a. Compute the equilibrium interest rate on consumption loans in the equilibrium without valued currency.

b. Construct a *brief* argument to establish whether or not the equilibrium without valued currency is Pareto optimal.

The economy also contains a government that purchases and destroys G_t units of the good in period t, $t \geq 1$. The government finances its purchases entirely by currency creation. That is, at time t,

$$G_t = \frac{H(t) - H(t-1)}{p(t)},$$

where $[H(t) - H(t-1)]$ is the additional dollars printed by the government at t and $p(t)$ is the price level at t. The government is assumed to increase $H(t)$ according to

$$H(t) = zH(t-1), \qquad z \geq 1,$$

where z is a constant for all time $t \geq 1$.

At time t, old people who carried $H(t-1)$ dollars between $(t-1)$ and t offer these $H(t-1)$ dollars in exchange for time t goods. Also at t the government offers $H(t) - H(t-1)$ dollars for goods, so that $H(t)$ is the total supply of dollars at time t, to be carried over by the young into time $(t+1)$.

c. Assume that $1/z > N_2\beta/N_1\alpha$. Show that under this assumption there exists a continuum of equilibria with valued currency.

d. Display the unique stationary equilibrium with valued currency in the form of a "quantity theory" equation. Compute the equilibrium rate of return on currency and consumption loans.

e. Argue that if $1/z < N_2\beta/N_1\alpha$, then there exists no valued-currency equilibrium. Interpret this result. (*Hint:* Look at the rate of return on consumption loans in the equilibrium without valued currency.)

f. Find the value of z that *maximizes* the government's G_t in a stationary equilibrium. Compare this with the largest value of z that is compatible with the existence of a valued-currency equilibrium.

Exercise 9.11   **Unpleasant monetarist arithmetic**

Consider an economy in which the aggregate demand for government currency for $t \geq 1$ is given by $[M(t)p(t)]^d = g[R_1(t)]$, where $R_1(t)$ is the gross rate of return on currency between t and $(t+1)$, $M(t)$ is the stock of currency at t, and $p(t)$ is the value of currency in terms of goods at t (the reciprocal of the price level). The function $g(R)$ satisfies

(1) $g(R)(1 - R) = h(R) > 0$ for $R \in (\underline{R}, 1)$,

where $h(R) \leq 0$ for $R < \underline{R}, R \geq 1, \underline{R} > 0$ and where $h'(R) < 0$ for $R > R_m$, $h'(R) > 0$ for $R < R_m$ $h(R_m) > D$, where D is a positive number to be defined shortly. The government faces an infinitely elastic demand for its interest-bearing bonds at a constant-over-time gross rate of return $R_2 > 1$. The government finances a budget deficit D, defined as government purchases minus explicit taxes, that is constant over time. The government's budget constraint is

(2) $D = p(t)[M(t) - M(t-1)] + B(t) - B(t-1)R_2,$ $t \geq 1,$

subject to $B(0) = 0, M(0) > 0$. In equilibrium,

(3) $M(t)p(t) = g[R_1(t)].$

The government is free to choose paths of $M(t)$ and $B(t)$, subject to equations (2) and (3).

a. Prove that, for $B(t) = 0$, for all $t > 0$, there exist two stationary equilibria for this model.

b. Show that there exist values of $B > 0$, such that there exist stationary equilibria with $B(t) = B$, $M(t)p(t) = Mp$.

c. Prove a version of the following proposition: among stationary equilibria, the lower the value of B, the lower the stationary rate of inflation consistent with equilibrium. (You will have to make an assumption about Laffer curve effects to obtain such a proposition.)

This problem displays some of the ideas used by Sargent and Wallace (1981). They argue that, under assumptions like those leading to the proposition stated in part c, the "looser" money is today [that is, the higher $M(1)$ and the lower $B(1)$], the lower the stationary inflation rate.

Exercise 9.12   **Grandmont-Hall**

Consider a nonstochastic, one-good overlapping generations model consisting of two-period-lived young people born in each $t \geq 1$ and an initial group of old people at $t = 1$ who are endowed with $H(0) > 0$ units of unbacked currency at the beginning of period 1. The one good in the model is not storable. Let the aggregate first-period saving function of the young be time-invariant and be denoted $f[1 + r(t)]$ where $[1 + r(t)]$ is the gross rate of return on consumption loans between t and $(t + 1)$. The saving function is assumed to satisfy $f(0) = -\infty$, $f'(1 + r) > 0$, $f(1) > 0$.

Let the government pay interest on currency, starting in period 2 (to holders of currency between periods 1 and 2). The government pays interest on currency at a nominal rate of $[1 + r(t)]p(t + 1)/\bar{p}$, where $[1 + r(t)]$ is the real gross rate of return on consumption loans, $p(t)$ is the price level at t, and $\bar{p}$ is a target price level chosen to satisfy

$$\bar{p} = H(0)/f(1).$$

The government finances its interest payments by printing new money, so that the government's budget constraint is

$$H(t + 1) - H(t) = \left\{ [1 + r(t)]\frac{p(t + 1)}{\bar{p}} - 1 \right\} H(t), \qquad t \geq 1,$$

given $H(1) = H(0) > 0$. The gross rate of return on consumption loans in this economy is $1 + r(t)$. In equilibrium, $[1 + r(t)]$ must be at least as great as the real rate of return on currency

$$1 + r(t) \geq [1 + r(t)]p(t)/\bar{p} = [1 + r(t)]\frac{p(t + 1)}{\bar{p}} \frac{p(t)}{p(t + 1)}$$

with equality if currency is valued,

$$1 + r(t) = [1 + r(t)]p(t)/\bar{p}, \qquad 0 < p(t) < \infty.$$

The loan market-clearing condition in this economy is

$$f[1 + r(t)] = H(t)/p(t).$$

a. Define an equilibrium.

b. Prove that there exists a unique monetary equilibrium in this economy and compute it.

Exercise 9.13   **Bryant-Keynes-Wallace**

Consider an economy consisting of overlapping generations of two-period-lived agents. There is a constant population of N young agents born at each date $t \geq 1$. There is a single consumption good that is not storable. Each agent born in $t \geq 1$ is endowed with w_1 units of the consumption good when young and with w_2 units when old, where $0 < w_2 < w_1$. Each agent born at $t \geq 1$ has identical preferences $\ln c_t^h(t) + \ln c_t^h(t + 1)$, where $c_t^h(s)$ is time s consumption of agent h born at time t. In addition, at time 1, there are alive N old people who are endowed with $H(0)$ units of unbacked paper currency and who want to maximize their consumption of the time 1 good.

A government attempts to finance a constant level of government purchases $G(t) = G > 0$ for $t \geq 1$ by printing new base money. The government's budget constraint is

$$G = [H(t) - H(t-1)]/p(t),$$

where $p(t)$ is the price level at t, and $H(t)$ is the stock of currency carried over from t to $(t+1)$ by agents born in t. Let $g = G/N$ be government purchases per young person. Assume that purchases $G(t)$ yield no utility to private agents.

a. Define a stationary equilibrium with valued fiat currency.

b. Prove that, for g sufficiently small, there exists a stationary equilibrium with valued fiat currency.

c. Prove that, in general, if there exists one stationary equilibrium with valued fiat currency, with rate of return on currency $1 + r(t) = 1 + r_1$, then there exists at least one other stationary equilibrium with valued currency with $1 + r(t) = 1 + r_2 \neq 1 + r_1$.

d. Tell whether the equilibria described in parts b and c are Pareto optimal, among allocations among private agents of what is left after the government takes $G(t) = G$ each period. (A proof is not required here: an informal argument will suffice.)

Now let the government institute a forced saving program of the following form. At time 1, the government redeems the outstanding stock of currency $H(0)$, exchanging it for government bonds. For $t \geq 1$, the government offers each young consumer the option of saving at least F worth of time t goods in the form of bonds bearing a constant rate of return $(1 + r_2)$. A legal prohibition against private intermediation is instituted that prevents two or more private agents from sharing one of these bonds. The government's budget constraint for $t \geq 2$ is

$$G/N = B(t) - B(t-1)(1 + r_2),$$

where $B(t) \geq F$. Here $B(t)$ is the saving of a young agent at t. At time $t = 1$, the government's budget constraint is

$$G/N = B(1) - \frac{H(0)}{Np(1)},$$

where $p(1)$ is the price level at which the initial currency stock is redeemed at $t = 1$. The government sets F and r_2.

Consider stationary equilibria with $B(t) = B$ for $t \geq 1$ and r_2 and F constant.

e. Prove that if g is small enough for an equilibrium of the type described in part a to exist, then a stationary equilibrium with forced saving exists. (Either a graphical argument or an algebraic argument is sufficient.)

f. Given g, find the values of F and r_2 that maximize the utility of a representative young agent for $t \geq 1$.

g. Is the equilibrium allocation associated with the values of F and $(1 + r_2)$ found in part f optimal among those allocations that give $G(t) = G$ to the government for all $t \geq 1$? (Here an informal argument will suffice.)

Chapter 10
Ricardian Equivalence

10.1. Borrowing limits and Ricardian equivalence

This chapter studies whether the timing of taxes matters. Under some assumptions it does and under others it does not. The Ricardian doctrine describes assumptions under which the timing of lump taxes does not matter. In this chapter, we will study how the timing of taxes interacts with restrictions on the ability of households to borrow. We study the issue in two equivalent settings: (1) an infinite horizon economy with an infinitely lived representative agent; and (2) an infinite horizon economy with a sequence of one-period-lived agents, each of whom cares about its immediate descendant. We assume that the interest rate is exogenously given. For example, the economy might be a small open economy that faces a given interest rate determined in the international capital market. Chapters 11 amd 13 will describe general equilibrium analyses of the Ricardian doctrine where the interest rate is determined within the model.

The key findings of the chapter are that in the infinite horizon model, Ricardian equivalence holds under what we earlier called the natural borrowing limit, but not under more stringent ones. The natural borrowing limit lets households borrow up to the capitalized value of their endowment sequences. These results have limited counterparts in the overlapping generations model, since that model is equivalent to an infinite horizon model with a no-borrowing constraint.[1] In the overlapping generations model, a no-borrowing constraint translates into a requirement that bequests be nonnegative. Thus, in the overlapping generations model, the domain of the Ricardian proposition is restricted, at least relative to the infinite horizon model under the natural borrowing limit.

---

[1] This is one of the insights in the influential paper of Barro (1974) that reignited modern interest in Ricardian equivalence.

10.2. Infinitely lived agent economy

Each of N identical households orders a consumption stream by

$$\sum_{t=0}^{\infty} \beta^t u(c_t), \qquad (10.2.1)$$

where $\beta \in (0,1)$ and $u(\cdot)$ is a strictly increasing, strictly concave, twice-differentiable one-period utility function. We impose the Inada condition $\lim_{c\downarrow 0} u'(c) = +\infty$. This condition is important because we will be stressing the feature that $c \geq 0$. There is no uncertainty. The household can invest in a single risk-free asset bearing a fixed gross one-period rate of return $R > 1$, a loan either to foreigners or to the government. At time t, the household faces the budget constraint

$$c_t + R^{-1} b_{t+1} \leq y_t + b_t, \qquad (10.2.2)$$

where b_0 is given. Throughout this chapter, we assume that $R\beta = 1$. Here $\{y_t\}_{t=0}^{\infty}$ is a given nonstochastic nonnegative endowment sequence and $\sum_{t=0}^{\infty} \beta^t y_t < \infty$.

We investigate two alternative restrictions on asset holdings $\{b_t\}_{t=0}^{\infty}$. One is that $b_t \geq 0$ for all $t \geq 0$, which allows the household to lend but not borrow. The alternative is to permit the household to borrow, but only an amount that it is feasible to repay. To discover this amount, set $c_t = 0$ for all t in formula (10.2.2) and solve forward for b_t to get

$$\tilde{b}_t = -\sum_{j=0}^{\infty} R^{-j} y_{t+j}, \qquad (10.2.3)$$

where we have ruled out Ponzi schemes by imposing the transversality condition

$$\lim_{T\to\infty} R^{-T} b_{t+T} = 0. \qquad (10.2.4)$$

Following Aiyagari (1994), we call $\tilde{b}_t$ the *natural debt limit*.[2] Thus, our alternative restriction on assets is

$$b_t \geq \tilde{b}_t, \qquad (10.2.5)$$

which is evidently weaker than $b_t \geq 0$.[3]

---

[2] Even with $c_t = 0$, the consumer cannot repay more than $\tilde{b}_t$.

[3] We encountered a more general version of equation (10.2.5) in chapter 8 when we discussed Arrow securities.

10.2.1. Optimal consumption/savings decision when $b_{t+1} \geq 0$

Consider the household's problem of choosing $\{c_t, b_{t+1}\}_{t=0}^{\infty}$ to maximize expression (10.2.1) subject to a given initial condition for initial assets b_0, the budget constraint (10.2.2), and $b_{t+1} \geq 0$ for all t. The first-order conditions for this problem are

$$u'(c_t) \geq \beta R u'(c_{t+1}), \quad \forall t \geq 0; \tag{10.2.6a}$$

and

$$u'(c_t) > \beta R u'(c_{t+1}) \quad \text{implies} \quad b_{t+1} = 0. \tag{10.2.6b}$$

Because $\beta R = 1$, these conditions and the constraint (10.2.2) imply that $c_{t+1} = c_t$ when $b_{t+1} > 0$, but when the consumer is borrowing constrained, $b_{t+1} = 0$ and $y_t + b_t = c_t < c_{t+1}$. The optimal consumption plan evidently depends on the $\{y_t\}$ path, as the following examples illustrate.

Example 1: Assume $b_0 = 0$ and the endowment path $\{y_t\}_{t=0}^{\infty} = \{y_h, y_l, y_h, y_l, \ldots\}$, where $y_h > y_l > 0$. The present value of the household's endowment is

$$\sum_{t=0}^{\infty} \beta^t y_t = \sum_{t=0}^{\infty} \beta^{2t}(y_h + \beta y_l) = \frac{y_h + \beta y_l}{1 - \beta^2}.$$

The annuity value $\bar{c}$ that has the same present value as the endowment stream is given by

$$\frac{\bar{c}}{1 - \beta} = \frac{y_h + \beta y_l}{1 - \beta^2}, \quad \text{or} \quad \bar{c} = \frac{y_h + \beta y_l}{1 + \beta}.$$

The solution to the household's optimization problem is the constant consumption stream $c_t = \bar{c}$ for all $t \geq 0$, and using the budget constraint (10.2.2), we can back out the associated savings plan; $b_{t+1} = (y_h - y_l)/(1 + \beta)$ for even t, and $b_{t+1} = 0$ for odd t. The consumer is never borrowing constrained.[4]

Example 2: Assume $b_0 = 0$ and the endowment path $\{y_t\}_{t=0}^{\infty} = \{y_l, y_h, y_l, y_h, \ldots\}$, where $y_h > y_l > 0$. The optimal plan is $c_0 = y_l$ and $b_1 = 0$, and from period 1 onward, the solution is the same as in example 1. Hence, the consumer is borrowing constrained the first period.[5]

---

[4] Note $b_t = 0$ does not imply that the consumer is borrowing constrained. We say that he is borrowing constrained if the Lagrange multiplier on the constraint $b_t \geq 0$ is not zero.

[5] Examples 1 and 2 illustrate a general result stated in chapter 17. Given a borrowing constraint and a nonstochastic endowment stream, the impact of the borrowing constraint will not vanish until the household reaches the period with the highest annuity value of the remainder of the endowment stream.

Example 3:   Assume $b_0 = 0$ and $y_t = \lambda^t$ where $1 < \lambda < R$. Notice that $\lambda\beta < 1$. The solution with the borrowing constraint $b_t \geq 0$ is $c_t = \lambda^t, b_t = 0$ for all $t \geq 0$. The consumer is always borrowing constrained.

10.2.2. Optimal consumption/savings decision when $b_{t+1} \geq \tilde{b}_{t+1}$

Example 4:   Assume the same b_0 and same endowment sequence $y_t = \lambda^t$ as in example 3, but now impose only the natural borrowing constraint (10.2.5). The present value of the household's endowment is

$$\sum_{t=0}^{\infty} \beta^t \lambda^t = \frac{1}{1 - \lambda\beta}.$$

The household's budget constraint for each t is satisfied at a constant consumption level $\hat{c}$ satisfying

$$\frac{\hat{c}}{1 - \beta} = \frac{1}{1 - \lambda\beta}, \qquad \text{or} \quad \hat{c} = \frac{1 - \beta}{1 - \lambda\beta}.$$

Substituting this consumption rate into formula (10.2.2) and solving forward gives

$$b_t = \frac{1 - \lambda^t}{1 - \beta\lambda}. \tag{10.2.7}$$

The consumer issues more and more debt as time passes and uses his rising endowment to service it. The consumer's debt always satisfies the natural debt limit at t; in particular, $b_t > \tilde{b}_t = -\lambda^t/(1 - \beta\lambda)$.

Example 5:   Take the specification of example 3, but now impose $\lambda < 1$. Note that the solution (10.2.7) implies $b_t \geq 0$, so that the constant consumption path $c_t = \hat{c}$ in example 4 is now the solution even if the borrowing constraint $b_t \geq 0$ is imposed.

10.3. Government

Add a government to the model in a way that leaves the consumer's preferences over consumption plans continue to be ordered by (10.2.1). The government purchases a stream $\{g_t\}_{t=0}^{\infty}$ per household. This stream does not appear in the consumer's utility functional. The government levies a stream of lump-sum taxes $\{\tau_t\}_{t=0}^{\infty}$ on the household, subject to the sequence of budget constraints

$$B_t + g_t = \tau_t + R^{-1}B_{t+1}, \tag{10.3.1}$$

where B_t is one-period debt due at t, denominated in the time t consumption good, that the government owes the households or foreign investors. Notice that we allow the government to borrow, even though in some of the preceding specifications (e.g., examples 1, 2, and 3) we did not permit the household to borrow. (If $B_t < 0$, the government lends to households or to foreign investors.) Solving the government's budget constraint forward gives the intertemporal constraint

$$B_t = \sum_{j=0}^{\infty} R^{-j}(\tau_{t+j} - g_{t+j}) \tag{10.3.2}$$

for $t \geq 0$, where we have ruled out Ponzi schemes by imposing the transversality condition

$$\lim_{T \to \infty} R^{-T}B_{t+T} = 0.$$

10.3.1. Effect on household

We must now deduct τ_t from the household's endowment in equation (10.2.2),

$$c_t + R^{-1}b_{t+1} \leq y_t - \tau_t + b_t. \tag{10.3.3}$$

Solving this tax-adjusted budget constraint forward and invoking transversality condition (10.2.4) yield

$$b_t = \sum_{j=0}^{\infty} R^{-j}(c_{t+j} + \tau_{t+j} - y_{t+j}). \tag{10.3.4}$$

The natural debt limit is obtained by setting $c_t = 0$ for all t in (10.3.4),

$$\tilde{b}_t \geq \sum_{j=0}^{\infty} R^{-j}(\tau_{t+j} - y_{t+j}). \tag{10.3.5}$$

A comparison of equations ($10.2.3$) and ($10.3.5$) indicates how taxes affect $\tilde{b}_t$

We use the following definition:

DEFINITION: Given initial conditions (b_0, B_0), an *equilibrium* is a household plan $\{c_t, b_{t+1}\}_{t=0}^{\infty}$ and a government policy $\{g_t, \tau_t, B_{t+1}\}_{t=0}^{\infty}$ such that (a) the government policy satisfies the government budget constraint ($10.3.1$), and (b) given $\{\tau_t\}_{t=0}^{\infty}$, the household's plan is optimal.

We can now state a Ricardian proposition under the natural debt limit.

PROPOSITION 1: Suppose that the natural debt limit prevails. Given initial conditions (b_0, B_0), let $\{\bar{c}_t, \bar{b}_{t+1}\}_{t=0}^{\infty}$ and $\{\bar{g}_t, \bar{\tau}_t, \bar{B}_{t+1}\}_{t=0}^{\infty}$ be an equilibrium. Consider any other tax policy $\{\hat{\tau}_t\}_{t=0}^{\infty}$ satisfying

$$\sum_{t=0}^{\infty} R^{-t}\hat{\tau}_t = \sum_{t=0}^{\infty} R^{-t}\bar{\tau}_t. \tag{10.3.6}$$

Then $\{\bar{c}_t, \hat{b}_{t+1}\}_{t=0}^{\infty}$ and $\{\bar{g}_t, \hat{\tau}_t, \hat{B}_{t+1}\}_{t=0}^{\infty}$ is also an equilibrium where

$$\hat{b}_t = \sum_{j=0}^{\infty} R^{-j}(\bar{c}_{t+j} + \hat{\tau}_{t+j} - y_{t+j}) \tag{10.3.7}$$

and

$$\hat{B}_t = \sum_{j=0}^{\infty} R^{-j}(\hat{\tau}_{t+j} - \bar{g}_{t+j}). \tag{10.3.8}$$

PROOF: The first point of the proposition is that the same consumption plan $\{\bar{c}_t\}_{t=0}^{\infty}$, but adjusted borrowing plan $\{\hat{b}_{t+1}\}_{t=0}^{\infty}$, solve the household's optimum problem under the altered government tax scheme. Under the natural debt limit, the household in effect faces a single intertemporal budget constraint ($10.3.4$). At time 0, the household can be thought of as choosing an optimal consumption plan subject to the single constraint,

$$b_0 = \sum_{t=0}^{\infty} R^{-t}(c_t - y_t) + \sum_{t=0}^{\infty} R^{-t}\tau_t.$$

Thus, the household's budget set, and therefore its optimal plan, does not depend on the timing of taxes, only their present value. The altered tax plan leaves the household's intertemporal budget set unaltered and therefore doesn't affect its optimal consumption plan. Next, we construct the adjusted borrowing plan $\{\hat{b}_{t+1}\}_{t=0}^{\infty}$ by solving the budget constraint (10.3.3) forward to obtain (10.3.7).[6] The adjusted borrowing plan trivially satisfies the (adjusted) natural debt limit in every period, since the consumption plan $\{\bar{c}_t\}_{t=0}^{\infty}$ is a nonnegative sequence.

The second point of the proposition is that the altered government tax and borrowing plans continue to satisfy the government's budget constraint. In particular, we see that the government's budget set at time 0 does not depend on the timing of taxes, only their present value,

$$B_0 = \sum_{t=0}^{\infty} R^{-t}\tau_t - \sum_{t=0}^{\infty} R^{-t}g_t.$$

Thus, under the altered tax plan with an unchanged present value of taxes, the government can finance the same expenditure plan $\{\bar{g}_t\}_{t=0}^{\infty}$. The adjusted borrowing plan $\{\hat{B}_{t+1}\}_{t=0}^{\infty}$ is computed in a similar way as above to arrive at (10.3.8). ∎

---

[6] It is straightforward to verify that the adjusted borrowing plan $\{\hat{b}_{t+1}\}_{t=0}^{\infty}$ must satisfy the transversality condition (10.2.4). In any period $(k-1) \geq 0$, solving the budget constraint (10.3.3) backward yields

$$b_k = \sum_{j=1}^{k} R^j \left[y_{k-j} - \tau_{k-j} - c_{k-j} \right] + R^k b_0.$$

Evidently, the difference between $\bar{b}_k$ of the initial equilibrium and $\hat{b}_k$ is equal to

$$\bar{b}_k - \hat{b}_k = \sum_{j=1}^{k} R^j \left[\hat{\tau}_{k-j} - \bar{\tau}_{k-j} \right],$$

and after multiplying both sides by R^{1-k},

$$R^{1-k} \left(\bar{b}_k - \hat{b}_k \right) = R \sum_{t=0}^{k-1} R^{-t} [\hat{\tau}_t - \bar{\tau}_t].$$

The limit of the right side is zero when k goes to infinity due to condition (10.3.6), and hence, the fact that the equilibrium borrowing plan $\{\bar{b}_{t+1}\}_{t=0}^{\infty}$ satisfies transversality condition (10.2.4) implies that so must $\{\hat{b}_{t+1}\}_{t=0}^{\infty}$.

This proposition depends on imposing on the household the natural debt limit, which is weaker than the no-borrowing constraint. Under the no-borrowing constraint, we require that the asset choice b_{t+1} at time t satisfy budget constraint (10.3.3) and not fall below zero. That is, under the no-borrowing constraint, we have to check more than just a single intertemporal budget constraint for the household at time 0. Changes in the timing of taxes that obey equation (10.3.6) evidently alter the right side of equation (10.3.3) and can, for example, cause a previously binding borrowing constraint no longer to be binding, and vice versa. Binding borrowing constraints in either the initial $\{\bar{\tau}_t\}_{t=0}^{\infty}$ equilibrium or the new $\{\hat{\tau}_t\}_{t=0}^{\infty}$ equilibria eliminates a Ricardian proposition as general as Proposition 1. More restricted versions of the proposition evidently hold across restricted equivalence classes of taxes that do not alter when the borrowing constraints are binding across the two equilibria being compared.

PROPOSITION 2: Consider an initial equilibrium with consumption path $\{\bar{c}_t\}_{t=0}^{\infty}$ in which $b_{t+1} > 0$ for all $t \geq 0$. Let $\{\bar{\tau}_t\}_{t=0}^{\infty}$ be the tax rate in the initial equilibrium, and let $\{\hat{\tau}_t\}_{t=0}^{\infty}$ be any other tax-rate sequence for which

$$\hat{b}_t = \sum_{j=0}^{\infty} R^{-j}(\bar{c}_{t+j} + \hat{\tau}_{t+j} - y_{t+j}) \geq 0$$

for all $t \geq 0$. Then $\{\bar{c}_t\}_{t=0}^{\infty}$ is also an equilibrium allocation for the $\{\hat{\tau}_t\}_{t=0}^{\infty}$ tax sequence.

We leave the proof of this proposition to the reader.

10.4. Linked generations interpretation

Much of the preceding analysis with borrowing constraints applies to a setting with overlapping generations linked by a bequest motive. Assume that there is a sequence of one-period-lived agents. For each $t \geq 0$ there is a one-period-lived agent who values consumption and the utility of his direct descendant, a young person at time $t + 1$. Preferences of a young person at t are ordered by

$$u(c_t) + \beta V(b_{t+1}),$$

where $u(c)$ is the same utility function as in the previous section, $b_{t+1} \geq 0$ are bequests from the time t person to the time $t + 1$ person, and $V(b_{t+1})$ is the maximized utility function of a time $t+1$ agent. The maximized utility function is defined recursively by

$$V(b_t) = \max_{c_t, b_{t+1}} \{u(c_t) + \beta V(b_{t+1})\}_{t=0}^{\infty} \qquad (10.4.1)$$

where the maximization is subject to

$$c_t + R^{-1}b_{t+1} \leq y_t - \tau_t + b_t \qquad (10.4.2)$$

and $b_{t+1} \geq 0$. The constraint $b_{t+1} \geq 0$ requires that bequests cannot be negative. Notice that a person cares about his direct descendant, but not vice versa. We continue to assume that there is an infinitely lived government whose taxes and purchasing and borrowing strategies are as described in the previous section.

In consumption outcomes, this model is equivalent to the previous model with a no-borrowing constraint. Bequests here play the role of savings b_{t+1} in the previous model. A positive savings condition $b_{t+1} > 0$ in the previous version of the model becomes an "operational bequest motive" in the overlapping generations model.

It follows that we can obtain a restricted Ricardian equivalence proposition, qualified as in Proposition 2. The qualification is that the initial equilibrium must have an operational bequest motive for all $t \geq 0$, and that the new tax policy must not be so different from the initial one that it renders the bequest motive inoperative.

10.5. Concluding remarks

The arguments in this chapter were cast in a setting with an exogenous interest rate R and a capital market that is outside of the model. When we discussed potential failures of Ricardian equivalence due to households facing no-borrowing constraints, we were also implicitly contemplating changes in the government's outside asset position. For example, consider an altered tax plan $\{\hat{\tau}_t\}_{t=0}^{\infty}$ that satisfies (10.3.6) and shifts taxes away from the future toward the present. A large enough change will definitely ensure that the government is a lender in early periods. But since the households are not allowed to become indebted, the government must lend abroad and we can show that Ricardian equivalence breaks down.

The readers might be able to anticipate the nature of the general equilibrium proof of Ricardian equivalence in chapter 13. First, private consumption and government expenditures must then be consistent with the aggregate endowment in each period, $c_t + g_t = y_t$, which implies that an altered tax plan cannot affect the consumption allocation as long as government expenditures are kept the same. Second, interest rates are determined by intertemporal marginal rates of substitution evaluated at the equilibrium consumption allocation, as studied in chapter 8. Hence, an unchanged consumption allocation implies that interest rates are also unchanged. Third, at those very interest rates, it can be shown that households would like to choose asset positions that exactly offset any changes in the government's asset holdings implied by an altered tax plan. For example, in the case of the tax change contemplated in the preceding paragraph, the households would demand loans exactly equal to the rise in government lending generated by budget surpluses in early periods. The households would use those loans to meet the higher taxes and thereby finance an unchanged consumption plan.

The finding of Ricardian equivalence in the infinitely lived agent model is a useful starting point for identifying alternative assumptions under which the irrelevance result might fail to hold,[7] such as our imposition of borrowing constraints that are tighter than the "natural debt limit." Another deviation from the benchmark model is finitely lived agents, as analyzed by Diamond (1965) and Blanchard (1985). But as suggested by Barro (1974) and shown in this chapter, Ricardian equivalence will continue to hold if agents are altruistic toward

---

[7] Seater (1993) reviews the theory and empirical evidence on Ricardian equivalence.

their descendants and there is an operational bequest motive. Bernheim and Bagwell (1988) take this argument to its extreme and formulate a model where all agents become interconnected because of linkages across dynastic families. They show how those linkages can become extensive enough to render neutral all redistributive policies, including ones attained via distortionary taxes. But, in general, replacing lump-sum taxes by distortionary taxes is a sure-fire way to undo Ricardian equivalence (see, e.g., Barsky, Mankiw, and Zeldes, 1986). We will return to the question of the timing of distortionary taxes in chapter 16. Kimball and Mankiw (1989) describe how incomplete markets can make the timing of taxes interact with a precautionary savings motive in a way that disposes of Ricardian equivalence. We take up precautionary savings and incomplete markets in chapters 17 and 18. Finally, by allowing distorting taxes to be history dependent, Bassetto and Kocherlakota (2004) attain a Ricardian equivalence result for a variety of taxes.

Chapter 11
Fiscal Policies in a Growth Model

11.1. Introduction

This chapter studies effects of technology and fiscal shocks on equilibrium outcomes in a nonstochastic growth model. We use the model to state some classic doctrines about the effects of various types of taxes and also as a laboratory to exhibit numerical techniques for approximating equilibria and to display the structure of dynamic models in which decision makers have perfect foresight about future government decisions. Foresight imparts effects on prices and allocations that precede government actions that cause them.

Following Hall (1971), we augment a nonstochastic version of the standard growth model with a government that purchases a stream of goods and that finances itself with an array of distorting flat-rate taxes. We take government behavior as exogenous,[1] which means that for us a *government* is simply a list of sequences for government purchases $\{g_t\}_{t=0}^{\infty}$ and taxes $\{\tau_{ct}, \tau_{kt}, \tau_{nt}, \tau_{ht}\}_{t=0}^{\infty}$. Here $\tau_{ct}, \tau_{kt}, \tau_{nt}$ are, respectively, time-varying flat-rate rates on consumption, earnings from capital, and labor earnings; and τ_{ht} is a lump-sum tax (a "head tax" or "poll tax").

Distorting taxes prevent a competitive equilibrium allocation from solving a planning problem. Therefore, to compute an equilibrium allocation and price system, we solve a system of nonlinear difference equations consisting of the first-order conditions for decision makers and the other equilibrium conditions. We first use a method called shooting. It produces an accurate approximation. Less accurate but in some ways more revealing approximations can be found by following Hall (1971), who solved a linear approximation to the equilibrium conditions. We apply the lag operators described in appendix A of chapter 2 to find and represent the solution in a way that is especially helpful in revealing the dynamic effects of perfectly foreseen alterations in taxes and expenditures and

---

[1] In chapter 16, we take up a version of the model in which the government chooses taxes to maximize the utility of a representative consumer.

how current values of endogenous variables respond to paths of future exogenous variables.[2]

11.2.   Economy

11.2.1. Preferences, technology, information

There is no uncertainty, and decision makers have perfect foresight. A representative household has preferences over nonnegative streams of a single consumption good c_t and leisure $1 - n_t$ that are ordered by

$$\sum_{t=0}^{\infty} \beta^t U(c_t, 1 - n_t), \quad \beta \in (0, 1) \tag{11.2.1}$$

where U is strictly increasing in c_t and $1 - n_t$, twice continuously differentiable, and strictly concave. We require that $c_t \geq 0$ and $n_t \in [0, 1]$. We'll typically assume that $U(c, 1 - n) = u(c) + v(1 - n)$. Common alternative specifications in the real business cycle literature are $U(c, 1 - n) = \log c + \zeta \log(1 - n)$ and $U(c, 1-n) = \log c + \zeta(1-n)$.[3] We shall also focus on another frequently studied special case that has $v = 0$ so that $U(c, 1 - n) = u(c)$.

The technology is

$$g_t + c_t + x_t \leq F(k_t, n_t) \tag{11.2.2a}$$

$$k_{t+1} = (1 - \delta)k_t + x_t \tag{11.2.2b}$$

where $\delta \in (0, 1)$ is a depreciation rate, k_t is the stock of physical capital, x_t is gross investment, and $F(k, n)$ is a linearly homogeneous production function with positive and decreasing marginal products of capital and labor.[4]   It is

---

[2] See Sargent (1987a) for a more comprehensive account of lag operators. By using lag operators, we extend Hall's results to allow arbitrary fiscal policy paths.

[3] See Hansen (1985) for a comparison of these two specifications. Both of these specifications fulfill the necessary conditions for the existence of a balance growth path set forth by King, Plosser, and Rebelo (1988), which require that income and substitution effects cancel in an appropriate way.

[4] In section 11.11, we modify the production function to admit labor augmenting technical change, a form that respects the King, Plosser, and Rebelo (1988) necessary conditions for the existence of a balance growth path.

sometimes convenient to eliminate x_t from (11.2.2) and express the technology as

$$g_t + c_t + k_{t+1} \leq F(k_t, n_t) + (1 - \delta)k_t. \tag{11.2.3}$$

11.2.2. Components of a competitive equilibrium

There is a competitive equilibrium with all trades occurring at time 0. The household owns capital, makes investment decisions, and rents capital and labor to a representative production firm. The representative firm uses capital and labor to produce goods with the production function $F(k_t, n_t)$. A *price system* is a triple of sequences $\{q_t, \eta_t, w_t\}_{t=0}^{\infty}$, where q_t is the time 0 pretax price of one unit of investment or consumption at time t (x_t or c_t), η_t is the pretax price at time t that the household receives from the firm for renting capital at time t, and w_t is the pretax price at time t that the household receives for renting labor to the firm at time t. The prices w_t and η_t are expressed in terms of time t goods, while q_t is expressed in terms of the numeraire at time 0.

We extend the chapter 8 definition of a competitive equilibrium to include activities of a government. We say that a government expenditure and tax plan that satisfy a budget constraint is *budget feasible*. A set of competitive equilibria is indexed by alternative budget-feasible government policies.

The household faces the budget constraint:

$$\begin{aligned}
&\sum_{t=0}^{\infty} q_t \left\{ (1 + \tau_{ct})c_t + [k_{t+1} - (1 - \delta)k_t] \right\} \\
&\leq \sum_{t=0}^{\infty} q_t \left\{ \eta_t k_t - \tau_{kt}(\eta_t - \delta_t)k_t + (1 - \tau_{nt})w_t n_t - \tau_{ht} \right\}.
\end{aligned} \tag{11.2.4}$$

Here we have assumed that the government gives a depreciation allowance δk_t from the gross rentals on capital $\eta_t k_t$ and so collects taxes $\tau_{kt}(\eta_t - \delta)k_t$ on rentals from capital. The government faces the budget constraint

$$\sum_{t=0}^{\infty} q_t g_t \leq \sum_{t=0}^{\infty} q_t \left\{ \tau_{ct}c_t + \tau_{kt}(\eta_t - \delta)k_t + \tau_{nt}w_t n_t + \tau_{ht} \right\}. \tag{11.2.5}$$

There is a sense in which we have given the government access to too many kinds of taxes, because when lump-sum taxes are available, the government should not use any distorting taxes. We include all of these taxes because, like Hall (1971), we want a framework that is sufficiently general to allow us to analyze how the various taxes distort production and consumption decisions.

11.3. The term structure of interest rates

The price system $\{q_t\}_{t=0}^{\infty}$ evidently embeds within it a term structure of interest rates. It is convenient to represent q_t as

$$q_t = q_0 \frac{q_1}{q_0} \frac{q_2}{q_1} \cdots \frac{q_t}{q_{t-1}}$$
$$= q_0 m_{0,1} m_{1,2} \cdots m_{t-1,t}$$

where $m_{t,t+1} = \frac{q_{t+1}}{q_t}$. We can represent the one-period *discount factor* $m_{t,t+1}$ as

$$m_{t,t+1} = R_{t,t+1}^{-1} = \frac{1}{1 + r_{t,t+1}} \approx \exp(-r_{t,t+1}). \qquad (11.3.1)$$

Here $R_{t,t+1}$ is the gross one-period rate of interest between t and $t+1$ and $r_{t,t+1}$ is the net one-period rate of interest between t and $t+1$. Notice that q_t can also be expressed as

$$q_t = q_0 \exp(-r_{0,1}) \exp(-r_{1,2}) \cdots \exp(-r_{t-1,t})$$
$$= q_0 \exp\big(-(r_{0,1} + r_{1,2} + \cdots + r_{t-1,t})\big)$$
$$= q_0 \exp(-t r_{0,t})$$

where

$$r_{0,t} = t^{-1}(r_{0,1} + r_{1,2} + \cdots + r_{t-1,t}). \qquad (11.3.2)$$

Here $r_{0,t}$ is the net t-period rate of interest between 0 and t. Since q_t is the time 0 price of one unit of time t consumption, $r_{0,t}$ is said to be the yield to maturity on a 'zero coupon bond' that matures at time t. A zero coupon bond promises no coupons before the date of maturity and pays only the principal due at the date of maturity. Equation (11.3.2) expresses the expectations theory of the term structure of interest rates, according to which interest rates on t-period (long) loans are averages of rates on one period (short) loans expected to prevail over the horizon of the long loan. More generally, the s-period long rate at time t is

$$r_{t,t+s} = \frac{1}{s}(r_{t,t+1} + r_{t+1,t+2} + \cdots + r_{t+s-1,t+s}). \qquad (11.3.3)$$

A graph of $r_{t,t+s}$ against s for $s = 1, 2, \ldots, S$ is called the (real) yield curve at t.

An insight about the expectations theory of the term structure of interest rates can be gleaned from computing gross one-period holding period returns on zero coupon bonds of maturities $1, 2, \ldots$. Consider the gross return earned by someone who at time 0 purchases one unit of time t consumption for q_t units of the numeraire and then sells it at time 1. The person pays $\frac{q_t}{q_0}$ units of time 0 consumption goods to earn $\frac{q_t}{q_1}$ units of time 1 consumption goods. The gross rate of return from this trade measured in time 1 consumption goods per unit of time 0 consumption goods is $\frac{q_0}{q_1}$, which does not depend on the date t of the good bought at time 0 and then sold at time 1. Evidently, at time 0 the one-period return is *identical* for pure discount bonds of *all* maturities $t \geq 1$. More generally, at time t the one-period holding period gross return on zero coupon bonds of all maturities equals $\frac{q_t}{q_{t+1}}$.

A way to characterize the expectations theory of the term structure of interest rates is by the requirement that the price vector $\{q_t\}_{t=0}^{\infty}$ of zero coupon bonds must be such that one-period holding period yields are equated across zero coupon bonds of all maturities. Note also how the price system $\{q_t\}_{t=0}^{\infty}$ contains forecasts of one-period holding period yields on zero coupon bonds of all maturities at all dates $t \geq 0$.

In subsequent sections, we'll indicate how the growth model with taxes and government expenditures links the term structure of interest rates to aspects of government fiscal policy.

11.4. Digression: sequential version of government budget constraint

We have used the time 0 trading abstraction described in chapter 8. Sequential trading of one-period risk-free debt can also support the equilibrium allocations that we shall study in this chapter. It is especially useful explicitly to describe the sequence of one-period government debt that is implicit in the equilibrium tax policies here.

We presume that the government enters period 0 with no government debt.[5] Define total tax collections as $T_t = \tau_{ct}c_t + \tau_{kt}(\eta_t - \delta)k_t + w_t\tau_{nt}n_t + \tau_{ht}$ and express the government budget constraint as

$$\sum_{t=0}^{\infty} q_t(g_t - T_t) = 0. \tag{11.4.1}$$

This can be written as

$$g_0 - T_0 = \sum_{t=1}^{\infty} \frac{q_t}{q_0}(T_t - g_t),$$

which states that the government deficit $g_0 - T_0$ at time 0 equals the present value of future government surpluses. Here $B_0 \equiv \sum_{t=1}^{\infty} \frac{q_t}{q_0}(T_t - g_t)$ is the value of government debt issued at time 0, denominated in units of time 0 goods. We can use this definition of B_0 to deduce

$$B_0\frac{q_0}{q_1} = T_1 - g_1 + \sum_{t=2}^{\infty} \frac{q_t}{q_1}(T_t - g_t)$$

or, by recalling from the previous subsection that $R_{0,1} \equiv \frac{q_0}{q_1}$ denotes the gross one-period real interest rate between time 0 and time 1,

$$B_0 R_{0,1} = T_1 - g_1 + B_1$$

where now

$$B_1 \equiv \sum_{t=2}^{\infty} \frac{q_t}{q_1}(T_t - g_t)$$

is the value of government debt issued in period 1 in units of time 1 consumption. Iterating this construction forward gives us a sequence of period-by-period government budget constraints

$$g_t + R_{t-1,t}B_{t-1} = T_t + B_t \tag{11.4.2}$$

for $t \geq 1$, where $R_{t-1,t} = \frac{q_{t-1}}{q_t}$ and

$$B_t \equiv \sum_{s=t+1}^{\infty} \frac{q_s}{q_t}(T_s - g_s). \tag{11.4.3}$$

---

[5] Letting $B_{-1} = 0$ be the government debt owed at time -1 allows us to apply equation (11.4.2) to date $t = 0$ too.

The left side of equation (11.4.2) is time t government expenditures including interest and principal payments on its debt, while the right side is total revenues including those raised by issuing new one-period debt in the amount B_t.

Thus, embedded in a government policy that satisfies (11.2.5) is a sequence of one-period government debts satisfying (11.4.3). The value of government debt at t is the present value of government surpluses from date $t + 1$ onward. Equation (11.4.3) states that government *debts* at time t signal future *surpluses*.

Equation (11.4.2) can be represented as

$$B_t - B_{t-1} = g_t - T_t + r_{t-1,t} B_{t-1}. \qquad (11.4.4)$$

Here $g_t - T_t$ is what is commonly called either the *net-of-interest* government deficit or the *operational* government deficit or the *primary* government deficit, while $r_{t-1,t} B_{t-1}$ are net interest payments on the government debt and $g_t - T_t + r_{t-1,t} B_{t-1}$ is the gross-of-interest government deficit. Equation (11.4.4) asserts that the change in government debt equals the gross-of-interest government deficit.

The Arrow-Debreu budget constraint (11.4.1) automatically enforces a 'no-Ponzi scheme' condition on the path of government debt $\{B_t\}$. To see this, first recall that $\frac{q_s}{q_t} = R_{t,t+1}^{-1} \cdots R_{s-1,s}^{-1}$ and write (11.4.3) as

$$B_t = \sum_{s=t+1}^{T} \frac{q_s}{q_t} (T_s - g_s) + \sum_{s=T+1}^{\infty} \frac{q_s}{q_t} (T_s - g_s)$$

or

$$B_t \equiv \sum_{s=t+1}^{T} \frac{q_s}{q_t} (T_s - g_s) + \frac{q_T}{q_t} B_T$$

or

$$B_t \equiv \sum_{s=t+1}^{T} \frac{q_s}{q_t} (T_s - g_s) + R_{t,t+1}^{-1} \cdots R_{T-1,T}^{-1} B_T.$$

An argument like that in subsection 11.5.1 can be applied to show that in an equilibrium $\lim_{T \to +\infty} q_T B_{T+1} = 0$.

11.4.1. Irrelevance of maturity structure of government debt

At time t, the government issues a list of *bonds* that in the aggregate promise to pay a stream $\{\xi_s^t\}_{s=1}^\infty$ of goods at time $s > t$ satisfying

$$B_t = \sum_{s=t+1}^\infty \frac{q_s}{q_t} \xi_s^t. \qquad (11.4.5)$$

The only restriction that our model puts on the term structure of payments $\{\xi_s^t\}_{s=1}^\infty$ is that it must satisfy

$$\sum_{s=t+1}^\infty \frac{q_s}{q_t} \xi_s^t = \sum_{s=t+1}^\infty \frac{q_s}{q_t} (T_s - g_s) \equiv B_t \qquad (11.4.6)$$

The model of this chapter asserts that one payment stream $\{\xi_s^t\}_{s=t+1}^\infty$ that satisfies (11.4.6) is as good as any other. The model pins down the total value of the continuation government IOU stream $\{\xi_s^t\}_{s=t+1}^\infty$ at each t, but it leaves the maturity structure of payments, whether early or late, for example, undetermined.[6] Two polar examples of maturity structures of the government debt are:

1. All debt consists of *one-period pure discount bonds* that are rolled over every period:

$$\xi_s^t = \begin{cases} \bar{\xi}^t & \text{if } s = t+1 \\ 0 & \text{if } s \geq t+2 \end{cases}$$

where $\bar{\xi}^t$ satisfies $\frac{q_{t+1}}{q_t} \bar{\xi}^t = B_t$.

2. All debt consists of *consols* that in the aggregate promise to pay a constant total coupon $\hat{\xi}^t$ for $s \geq t+1$, where $\hat{\xi}^t$ satisfies

$$\hat{\xi}^t \sum_{s=t+1}^\infty \frac{q_s}{q_t} = B_t.$$

---

[6] For models that restrict the maturity structure of government debt by imposing more imperfections than we analyze in this chapter, see Lucas and Stokey (1983), Angeletos (2002), Buera and Nicolini (2004), and Shin (2007). Lucas and Stokey show how to set the maturity structure of debt payments in a way that induces an authority responsible for sequentially choosing flat rate taxes on labor to implement a Ramsey plan. Angeletos (2002), Buera and Nicolini (2004), and Shin (2007) use variations over time in the maturity structure of risk-free government debt to complete markets.

The sequence of period-by-period net returns on the government debt $\{r_{t,t+1}B_t\}_{t=0}^{\infty}$ is independent of the government's choice of sequences $\{\{\xi_s^t\}_{s=t+1}^{\infty}\}_{t=0}^{\infty}$.

11.5. Competitive equilibria with distorting taxes

A representative household chooses a sequence $\{c_t, n_t, k_{t+1}\}_{t=0}^{\infty}$ to maximize (11.2.1) subject to (11.2.4). A representative firm chooses $\{k_t, n_t\}_{t=0}^{\infty}$ to maximize $\sum_{t=0}^{\infty} q_t[F(k_t, n_t) - \eta_t k_t - w_t n_t]$.[7] A budget-feasible government policy is an expenditure plan $\{g_t\}_{t=0}^{\infty}$ and a tax plan that satisfy (11.2.5). A feasible allocation is a sequence $\{c_t, x_t, n_t, k_t\}_{t=0}^{\infty}$ that satisfies (11.2.3).

DEFINITION: A *competitive equilibrium with distorting taxes* is a budget-feasible government policy, a feasible allocation, and a price system such that, given the price system and the government policy, the allocation solves the household's problem and the firm's problem.

11.5.1. The household: no-arbitrage and asset-pricing formulas

A no-arbitrage argument implies a restriction on prices and tax rates across time from which there emerges a formula for the "user cost of capital" (see Hall and Jorgenson, 1967). Collect terms in similarly dated capital stocks and thereby rewrite the household's budget constraint (11.2.4) as

$$\sum_{t=0}^{\infty} q_t\big[(1 + \tau_{ct})c_t\big] \leq \sum_{t=0}^{\infty} q_t(1 - \tau_{nt})w_t n_t - \sum_{t=0}^{\infty} q_t \tau_{ht}$$

$$+ \sum_{t=1}^{\infty} \big[((1 - \tau_{kt})(\eta_t - \delta) + 1)q_t - q_{t-1}\big]k_t \qquad (11.5.1)$$

$$+ \big[(1 - \tau_{k0})(\eta_0 - \delta) + 1\big]q_0 k_0 - \lim_{T \to \infty} q_T k_{T+1}$$

The terms $\big[(1 - \tau_{k0})(\eta_0 - \delta) + 1\big]q_0 k_0$ and $-\lim_{T\to\infty} q_T k_{T+1}$ remain after creating the weighted sum in k_t's for $t \geq 1$.

---

[7] Note the contrast with the setup in chapter 12, which has two types of firms. Here we assign to the household the physical investment decisions made by the type II firms of chapter 12.

The household inherits a given k_0 that it takes as an initial condition, and it is free to choose any sequence $\{c_t, n_t, k_{t+1}\}_{t=0}^{\infty}$ that satisfies (11.5.1) where all prices and tax rates are taken as given. The objective of the household is to maximize lifetime utility (11.2.1), which is increasing in consumption $\{c_t\}_{t=0}^{\infty}$ and, for one of our preference specifications below, also increasing in leisure $\{1 - n_t\}_{t=0}^{\infty}$.

All else equal, the household would be happier with larger values on the right side of (11.5.1), preferably plus infinity, which would enable it to purchase unlimited amounts of consumption goods. Because resources are finite, we know that the right side of the household's budget constraint must be bounded in an equilibrium. This fact leads to an important restriction on the price and tax sequences. If the right side of the household's budget constraint is to be bounded, then the terms multiplying k_t for $t \geq 1$ must all equal zero because if any of them were strictly positive (negative) for some date t, the household could make the right side of (11.5.1) an arbitrarily large positive number by choosing an arbitrarily large positive (negative) value of k_t. On the one hand, if one such term were strictly positive for some date t, the household could purchase an arbitrarily large capital stock k_t assembled at time $t-1$ with a present-value cost of $q_{t-1}k_t$ and then sell the rental services and the undepreciated part of that capital stock to be delivered at time t, with a present-value income of $[(1 - \tau_{kt})(\eta_t - \delta) + 1]q_t k_t$. If such a transaction were to yield a strictly positive profit, it would offer the consumer a pure arbitrage opportunity and the right side of (11.5.1) would become unbounded. On the other hand, if there is one term multiplying k_t that is strictly negative for some date t, the household can make the right side of (11.5.1) arbitrarily large and positive by "short selling" capital by setting $k_t < 0$. The household could turn to purchasers of capital assembled at time $t-1$ and sell "synthetic" units of capital to them. Such a transaction need not involve any actual physical capital: the household could merely undertake trades that would give the other party to the transaction the same costs and incomes as those associated with purchasing capital assembled at time $t-1$. If such short sales of capital yield strictly positive profits, it would provide the consumer with a pure arbitrage opportunity and the right side of (11.5.1) would become unbounded. Therefore, the terms multiplying k_t must *equal* zero for all $t \geq 1$, so that

$$\frac{q_t}{q_{t+1}} = [(1 - \tau_{kt+1})(\eta_{t+1} - \delta) + 1] \tag{11.5.2}$$

for all $t \geq 0$. These are zero-profit or no-arbitrage conditions. We have derived these conditions by using only the weak property that $U(c, 1 - n)$ is increasing in consumption (i.e., that the household always prefers more to less).

It remains to be determined how the household sets the last term on the right side of (11.5.1), $-\lim_{T \to \infty} q_T k_{T+1}$. According to our preceding argument, the household would not purchase an amount of capital that would make this term strictly negative in the limit because that would reduce the right side of (11.5.1) and hence diminish the household's resources available for consumption. Instead, the household would like to make this term strictly positive and unbounded, so that the household could purchase unlimited amounts of consumption goods. But the market would stop the household from undertaking such a short sale in the limit, since no party would like to be on the other side of the transaction. This is obvious when considering a finite-horizon model where everyone would like to short sell capital in the very last period because there would then be no future period in which to fulfil the obligations of those short sales. Therefore, in our infinite-horizon model, as a condition of optimality, we impose the terminal condition that $-\lim_{T \to \infty} q_T k_{T+1} = 0$. Once we impose formula (11.5.5a) below linking q_t to U_{1t}, this terminal condition puts the following restriction on the equilibrium allocation:

$$-\lim_{T \to \infty} \beta^T \frac{U_{1T}}{(1 + \tau_{cT})} k_{T+1} = 0. \tag{11.5.3}$$

The household's initial capital stock k_0 is given. According to (11.5.1), its value is $[(1 - \tau_{k0})(\eta_0 - \delta) + 1]q_0 k_0$.

11.5.2. *User cost of capital formula*

The no-arbitrage conditions (11.5.2) can be rewritten as the following expression for the "user cost of capital" η_{t+1}:

$$\eta_{t+1} = \delta + \left(\frac{1}{1 - \tau_{kt+1}} \right) \left(\frac{q_t}{q_{t+1}} - 1 \right). \tag{11.5.4}$$

Recalling from (11.3.1) that $m_{t,t+1}^{-1} = R_{t,t+1} = (1 + r_{t,t+1}) = \frac{q_t}{q_{t+1}}$, equation (11.5.4) can be expressed as

$$\eta_{t+1} = \delta + \left(\frac{r_{t,t+1}}{1 - \tau_{kt+1}} \right).$$

The user cost of capital takes into account the rate of taxation of capital earnings, the capital gain or loss from t to $t+1$, and a depreciation cost.[8]

11.5.3. Household first-order conditions

So long as the no-arbitrage conditions (11.5.2) prevail, households are indifferent about how much capital they hold. Recalling that the one-period utility function is $U(c, 1-n)$, let $U_1 = \frac{\partial U}{\partial c}$ and $U_2 = \frac{\partial U}{\partial 1-n}$ so that $\frac{\partial U}{\partial n} = -U_2$. Then we have that the household's first-order conditions with respect to c_t, n_t are:

$$\beta^t U_{1t} = \mu q_t (1 + \tau_{ct}) \tag{11.5.5a}$$

$$\beta^t U_{2t} \leq \mu q_t w_t (1 - \tau_{nt}), \quad = \text{if } n_t < 1, \tag{11.5.5b}$$

where μ is a nonnegative Lagrange multiplier on the household's budget constraint (11.2.4). Multiplying the price system by a positive scalar simply rescales the multiplier μ, so we are free to choose a numeraire by setting μ to an arbitrary positive number.

11.5.4. A theory of the term structure of interest rates

Equation (11.5.5a) allows us to solve for q_t as a function of consumption

$$\mu q_t = \beta^t U_{1t}/(1 + \tau_{ct}) \tag{11.5.6a}$$

or in the special case that $U(c_t, 1-n_t) = u(c_t)$

$$\mu q_t = \beta^t u'(c_t)/(1 + \tau_{ct}). \tag{11.5.6b}$$

In conjunction with the observations made in subsection 11.3, these formulas link the term structure of interest rates to the paths of c_t, τ_{ct}. The government policy $\{g_t, \tau_{ct}, \tau_{nt}, \tau_{kt}, \tau_{ht}\}_{t=0}^\infty$ affects the term structure of interest rates directly via τ_{ct} and indirectly via its impact on the path for $\{c_t\}_{t=0}^\infty$.

---

[8] This is a discrete-time version of a continuous-time formula derived by Hall and Jorgenson (1967).

11.5.5. Firm

Zero-profit conditions for the representative firm impose additional restrictions on equilibrium prices and quantities. The present value of the firm's profits is

$$\sum_{t=0}^{\infty} q_t \left[F(k_t, n_t) - w_t n_t - \eta_t k_t \right].$$

Applying Euler's theorem on linearly homogeneous functions to $F(k, n)$, the firm's present value is:

$$\sum_{t=0}^{\infty} q_t \left[(F_{kt} - \eta_t) k_t + (F_{nt} - w_t) n_t \right].$$

No-arbitrage (or zero-profit) conditions are:

$$\begin{aligned} \eta_t &= F_{kt} \\ w_t &= F_{nt}. \end{aligned} \tag{11.5.7}$$

11.6. Computing equilibria

The definition of a competitive equilibrium and the concavity conditions that we have imposed on preferences imply that an equilibrium is a price system $\{q_t, \eta_t, w_t\}$, a budget feasible government policy $\{g_t, \tau_t\} \equiv \{g_t, \tau_{ct}, \tau_{nt}, \tau_{kt}, \tau_{ht}\}$, and an allocation $\{c_t, n_t, k_{t+1}\}$ that solve the system of nonlinear difference equations consisting of (11.2.3), (11.5.2), (11.5.5), and (11.5.7) subject to the initial condition that k_0 is given and the terminal condition (11.5.3). In this chapter, we shall simplify things by treating $\{g_t, \tau_t\} \equiv \{g_t, \tau_{ct}, \tau_{nt}, \tau_{kt}\}$ as exogenous and then use $\sum_{t=0}^{\infty} q_t \tau_{ht}$ as a slack variable that we choose to balance the government's budget. We now attack this system of difference equations.

11.6.1. *Inelastic labor supply*

We'll start with the following special case. (The general case is just a little more complicated, and we'll describe it below.) Set $U(c, 1 - n) = u(c)$, so that the household gets no utility from leisure, and set $n = 1$. We define $f(k) = F(k, 1)$ and express feasibility as

$$k_{t+1} = f(k_t) + (1 - \delta)k_t - g_t - c_t. \tag{11.6.1}$$

Notice that $F_k(k, 1) = f'(k)$ and $F_n(k, 1) = f(k) - f'(k)k$. Substitute $(11.5.5a)$, $(11.5.7)$, and $(11.6.1)$ into $(11.5.2)$ to get

$$
\begin{aligned}
&\frac{u'\big(f(k_t) + (1 - \delta)k_t - g_t - k_{t+1}\big)}{(1 + \tau_{ct})} \\
&- \beta \frac{u'(f(k_{t+1}) + (1 - \delta)k_{t+1} - g_{t+1} - k_{t+2})}{(1 + \tau_{ct+1})} \times \\
&[(1 - \tau_{kt+1})(f'(k_{t+1}) - \delta) + 1] = 0.
\end{aligned}
\tag{11.6.2}
$$

Given the government policy sequences, $(11.6.2)$ is a second-order difference equation in capital. We can also express $(11.6.2)$ as

$$u'(c_t) = \beta u'(c_{t+1}) \frac{(1 + \tau_{ct})}{(1 + \tau_{ct+1})} \left[(1 - \tau_{kt+1})(f'(k_{t+1}) - \delta) + 1\right]. \tag{11.6.3}$$

To compute an equilibrium, we must find a solution of the difference equation $(11.6.2)$ that satisfies two boundary conditions. As mentioned above, one boundary condition is supplied by the given level of k_0 and the other by $(11.5.3)$. To determine a particular terminal value k_∞, we restrict the path of government policy so that it converges, a way to impose $(11.5.3)$.

11.6.2. The equilibrium steady state

Tax rates and government expenditures serve as forcing functions for the difference equations (11.6.1) and (11.6.3). Let $z_t = \begin{bmatrix} g_t & \tau_{kt} & \tau_{ct} \end{bmatrix}'$ and write (11.6.2) as

$$H(k_t, k_{t+1}, k_{t+2}; z_t, z_{t+1}) = 0. \tag{11.6.4}$$

To allow convergence to a steady state, we assume government policies that are eventually constant, i.e., that satisfy

$$\lim_{t \to \infty} z_t = \overline{z}. \tag{11.6.5}$$

When we actually solve our models, we'll set a date T after which all components of the forcing sequences that comprise z_t are constant. A terminal steady-state capital stock $\overline{k}$ evidently solves

$$H(\overline{k}, \overline{k}, \overline{k}, \overline{z}, \overline{z}) = 0. \tag{11.6.6}$$

For our model, we can solve (11.6.6) by hand. In a steady state, (11.6.3) becomes

$$1 = \beta[(1 - \overline{\tau}_k)(f'(\overline{k}) - \delta) + 1].$$

Notice that an eventually constant consumption tax $\overline{\tau}_c$ does not distort $\overline{k}$ *vis-a-vis* its value in an economy without distorting taxes. Letting $\beta = \frac{1}{1+\rho}$, we can express the preceding equation as

$$\delta + \frac{\rho}{1 - \tau_k} = f'(\overline{k}). \tag{11.6.7}$$

When $\tau_k = 0$, equation (11.6.7) becomes $(\rho + \delta) = f'(\overline{k})$, which is a celebrated formula for the so-called "augmented Golden Rule" capital-labor ratio.

When the exogenous sequence $\{g_t\}_{t=0}^\infty$ converges, the steady state capital-labor ratio that solves $(\rho + \delta) = f'(\overline{k})$ is the asymptotic value of the capital-labor ratio that would be approached by a benevolent planner who chooses $\{c_t, k_{t+1}\}_{t=0}^\infty$ to maximize $\sum_{t=0}^\infty \beta^t u(c_t)$ subject to k_0 given and the sequence of constraints $c_t + k_{t+1} + g_t \le f(k_t) + (1 - \delta)k_t$.

11.6.3. *Computing the equilibrium path with the shooting*
algorithm

Having computed the terminal steady state, we are now in a position to apply the *shooting algorithm* to compute an equilibrium path that starts from an arbitrary initial condition k_0, assuming a possibly time-varying path of government policy.[9] The shooting algorithm solves the two-point boundary value problem by searching for an initial c_0 that makes the Euler equation (11.6.2) and the feasibility condition (11.2.3) imply that $k_S \approx \overline{k}$, where S is a finite but large time index meant to approximate infinity and $\overline{k}$ is the terminal steady value associated with the policy being analyzed. We let T be the value of t after which all components of z_t are constant. Here are the steps of the algorithm.[10]

1. Solve (11.6.4) for the terminal steady-state $\overline{k}$ that is associated with the permanent policy vector $\overline{z}$ (i.e., find the solution of (11.6.7)).

2. Select a large time index $S >> T$ and guess an initial consumption rate c_0. (A good guess comes from the linear approximation to be described in section 11.10.) Compute $u'(c_0)$ and solve (11.6.1) for k_1.

3. For $t = 0$, use (11.6.3) to solve for $u'(c_{t+1})$. Then invert u' and compute c_{t+1}. Use (11.6.1) to compute k_{t+2}.

4. Iterate on step 3 to compute candidate values $\hat{k}_t, t = 1, \ldots, S$.

5. Compute $\hat{k}_S - \overline{k}$.

6. If $\hat{k}_S > \overline{k}$, raise c_0 and compute a new $\hat{k}_t, t = 1, \ldots, S$.

7. If $\hat{k}_S < \overline{k}$, lower c_0.

8. In this way, search for a value of c_0 that makes $\hat{k}_S \approx \overline{k}$.

9. Compute $\sum_{t=0}^{\infty} q_t \tau_{ht}$ that satisfies the government budget constraint at equality.

---

[9] We recommend a suite of computer programs called **dynare**. We have used **dynare** to execute the numerical experiments described in this chapter. See Barillas, Bhandari, Bigio, Colacito, Juillard, Kitao, Matthes, Sargent, and Shin (2012) for dynare code that performs these and other calculations. See < http://www.dynare.org >.

[10] This algorithm proceeds in the spirit of the invariant-subspace method (implemented via a Schur decomposition) for solving the first-order conditions associated with the optimal linear regulator that we described in section 5.5 of chapter 5.

11.6.4. Other equilibrium quantities

After we solve (11.6.2) for an equilibrium $\{k_t\}$ sequence, we can recover other equilibrium quantities and prices from the following equations:

$$c_t = f(k_t) + (1 - \delta)k_t - k_{t+1} - g_t \tag{11.6.8a}$$

$$q_t = \beta^t u'(c_t)/(1 + \tau_{ct}) \tag{11.6.8b}$$

$$\eta_t = f'(k_t) \tag{11.6.8c}$$

$$w_t = f(k_t) - k_t f'(k_t) \tag{11.6.8d}$$

$$\bar{R}_{t+1} = \frac{(1 + \tau_{ct})}{(1 + \tau_{ct+1})} \left[(1 - \tau_{kt+1})(f'(k_{t+1}) - \delta) + 1 \right]$$

$$= \frac{(1 + \tau_{ct})}{(1 + \tau_{ct+1})} R_{t,t+1} \tag{11.6.8e}$$

$$R_{t,t+1}^{-1} = m_{t,t+1} = \beta \frac{u'(c_{t+1})}{u'(c_t)} \frac{(1 + \tau_{ct})}{(1 + \tau_{ct+1})} \tag{11.6.8f}$$

$$r_{t,t+1} \equiv R_{t,t+1} - 1 = (1 - \tau_{k,t+1})(f'(k_{t+1}) - \delta) \tag{11.6.8g}$$

It is convenient to express (11.6.3) as

$$u'(c_t) = \beta u'(c_{t+1}) \bar{R}_{t+1} \tag{11.6.8h}$$

or

$$\bar{R}_{t+1}^{-1} = \beta u'(c_{t+1}) / u'(c_t).$$

The left side of this equation is the rate which the market and the tax system allow the household to substitute consumption at t for consumption at $t+1$. The right side is the rate at which the household is willing to substitute consumption at t for consumption at $t+1$.

An equilibrium satisfies equations (11.6.8). In the case of constant relative risk aversion (CRRA) utility $u(c) = (1 - \gamma)^{-1} c^{1-\gamma}, \gamma \geq 1$, (11.6.8h) implies

$$\log\left(\frac{c_{t+1}}{c_t}\right) = \gamma^{-1} \log \beta + \gamma^{-1} \log \bar{R}_{t+1}, \tag{11.6.9}$$

which shows that the log of consumption growth varies directly with $\bar{R}_{t+1}$. Variations in distorting taxes have effects on consumption and investment that are intermediated through this equation, as several experiments below highlight.

11.6.5. Steady-state $\bar{R}$

Using (11.6.7) and formula (11.6.8e), we can determine that the steady state value of $\bar{R}_{t+1}$ is[11]

$$\bar{R}_{t+1} = (1 + \rho).\tag{11.6.10}$$

11.6.6. Lump-sum taxes available

If the government can impose lump-sum taxes, we can implement the shooting algorithm for a specified g, τ_k, τ_c, solve for equilibrium prices and quantities, and then find an associated value for $q \cdot \tau_h = \sum_{t=0}^{\infty} q_t \tau_{ht}$ that balances the government budget. This calculation treats the present value of lump-sum taxes as a residual that balances the government budget. In calculations presented later in this chapter, we shall assume that lump-sum taxes are available and so shall use this procedure.

11.6.7. No lump-sum taxes available

If lump-sum taxes are not available, then an additional step is required to compute an equilibrium. In particular, we have to ensure that taxes and expenditures are such that the government budget constraint (11.2.5) is satisfied at an equilibrium price system with $\tau_{ht} = 0$ for all $t \geq 0$. Braun (1994) and Mc-Grattan (1994b) accomplish this by employing an iterative algorithm that alters a particular distorting tax until (11.2.5) is satisfied. The idea is first to compute a candidate equilibrium for one arbitrary tax policy with possibly nonzero lump sum taxes, then to check whether the government budget constraint is satisfied. Usually we will find that lump sum taxes must be levied to balance the government budget in this candidate equilibrium. To find an equilibrium with zero lump sum taxes, we can proceed as follows. If the government budget would have have a deficit in present value without lump sum taxes (i.e., if the present value of lump sum taxes is positive in the candidate equilibrium), then either decrease some elements of the government expenditure sequence or increase some elements of the tax sequence and try again. Because there exist so

---

[11] To compute steady states, we assume that all tax rates and government expenditures are constant from some date T forward.

many equilibria, the class of tax and expenditure processes has to be restricted drastically to narrow the search for an equilibrium. [12]

11.7. A digression on back-solving

The shooting algorithm takes sequences for g_t and the various tax rates as given and finds paths of the allocation $\{c_t, k_{t+1}\}_{t=0}^{\infty}$ and the price system that solve the system of difference equations formed by (11.6.3) and (11.6.8). Thus, the shooting algorithm views government policy as exogenous and the price system and allocation as endogenous. Sims (1989) proposed another way to solve the growth model that exchanges the roles of some exogenous and endogenous variables. In particular, his *back-solving* approach takes a path $\{c_t\}_{t=0}^{\infty}$ as given, and then proceeds as follows.

Step 1: Given k_0 and sequences for the various tax rates, solve (11.6.3) for a sequence $\{k_{t+1}\}$.

Step 2: Given the sequences for $\{c_t, k_{t+1}\}$, solve the feasibility condition (11.6.8a) for a sequence of government expenditures $\{g_t\}_{t=0}^{\infty}$.

Step 3: Solve formulas (11.6.8b)–(11.6.8e) for an equilibrium price system.

The present model can be used to illustrate other applications of back-solving. For example, we could start with a given process for $\{q_t\}$, use (11.6.8b) to solve for $\{c_t\}$, and proceed as in steps 1 and 2 above to determine processes for $\{k_{t+1}\}$ and $\{g_t\}$, and then finally compute the remaining prices from the as yet unused equations in (11.6.8).

Sims recommended this method because it adopts a flexible or "symmetric" attitude toward exogenous and endogenous variables. Diaz-Giménez, Prescott, Fitzgerald, and Alvarez (1992), Sargent and Smith (1997), and Sargent and Velde (1999) have all used the method. We shall not use it in the remainder of this chapter, but it is a useful method to have in our toolkit. [13]

---

[12] See chapter 16 for theories about how to choose taxes in socially optimal ways.

[13] Constantinides and Duffie (1996) used back-solving to reverse engineer a cross-section of endowment processes that, with incomplete markets, would prompt households to consume their endowments at a given stochastic process of asset prices.

11.8. Effects of taxes on equilibrium allocations and prices

We use the model to analyze the effects of government expenditure and tax sequences. The household can alter his payments of a *distorting* by altering a decision. The household cannot alter his payments of a *nondistorting* tax. In the present model, τ_k, τ_c, τ_n are distorting taxes and the lump-sum tax τ_h is nondistorting. We can deduce the following outcomes from (11.6.8) and (11.6.7).

1. Lump-sum taxes and Ricardian equivalence. Suppose that the distorting taxes are all zero and that only lump-sum taxes are used to raise government revenues. Then the equilibrium allocation is identical with one that solves a version of a planning problem in which g_t is taken as an exogenous stream that is deducted from output. To verify this claim, notice that lump-sum taxes appear nowhere in formulas (11.6.8), and that these equations are identical with the first-order conditions and feasibility conditions for a planning problem. The timing of lump-sum taxes is irrelevant because only the present value of taxes $\sum_{t=0}^{\infty} q_t \tau_{ht}$ appears in the budget constraints of the government and the household.

2. When the labor supply is inelastic, constant τ_c and τ_n are not distorting. When the labor supply is inelastic, τ_n is not a distorting tax. A *constant* level of τ_c is not distorting.

3. Variations in τ_c over time are distorting. They affect the path of capital and consumption through equation (11.6.8g).

4. Capital taxation is distorting. Constant levels of the capital tax τ_k are distorting (see (11.6.8g) and (11.6.7)).

11.9. Transition experiments with inelastic labor supply

We continue to study the special case with $U(c, 1 - n) = u(c)$. Figures 11.9.1 through 11.9.5 apply the shooting algorithm to an economy with $u(c) = (1 - \gamma)^{-1}c^{1-\gamma}$, $f(k) = k^\alpha$ with parameter values $\alpha = .33, \delta = .2, \beta = .95$ and an initial constant level of g of $.2$. All of the experiments except one to be described in figure 11.9.2 set the critical utility curvature parameter $\gamma = 2$. We initially set all distorting taxes to zero and consider perturbations of them that we describe in the experiments below.

Figures 11.9.1 to 11.9.5 show responses to foreseen once-and-for-all increases in g, τ_c, and τ_k, that occur at time $T = 10$, where $t = 0$ is the initial time period. Prices induce effects that precede the policy changes that cause them. We start all of our experiments from an initial steady state that is appropriate for the pre-jump settings of all government policy variables. In each panel, a dashed line displays a value associated with the steady state at the initial constant values of the policy vector. A solid line depicts an equilibrium path under the new policy. It starts from the value that was associated with an initial steady state that prevailed before the policy change at $T = 10$ was announced. *Before* date $t = T = 10$, the response of each variable is entirely due to expectations about future policy changes. *After* date $t = 10$, the response of each variable represents a purely transient response to a new stationary level of the "forcing function" in the form of the exogenous policy variables. That is, before $t = T$, the forcing function is changing as date T approaches; after date T, the policy vector has attained its new permanent level, so that the only sources of dynamics are transient.

Discounted *future* values of fiscal variables impinge on current outcomes, where the discount rate in question is endogenous, while departures of the capital stock from its terminal steady-state value set in place a force for it to decay toward its steady state rate at a particular rate. These two forces, discounting of the future and transient decay back toward the terminal steady state, are evident in the experiments portrayed in Figures 11.9.1–11.9.5. In section 11.10.6, we express the decay rate as a function of the key curvature parameter γ in the one-period utility function $u(c) = (1 - \gamma)^{-1}c^{1-\gamma}$, and we note that the endogenous rate at which future fiscal variables are discounted is tightly linked to that decay rate.

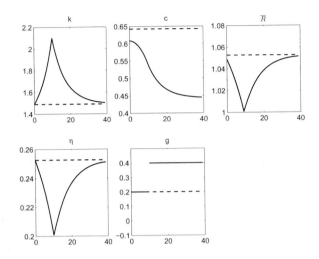

Figure 11.9.1: Response to foreseen once-and-for-all increase in g at $t = 10$. From left to right, top to bottom: $k, c, \bar{R}, \eta, g$. The dashed line is the original steady state.

Foreseen jump in g_t. Figure 11.9.1 shows the effects of a foreseen permanent increase in g at $t = T = 10$ that is financed by an increase in lump-sum taxes. Although the steady-state value of the capital stock is unaffected (this follows from the fact that g disappears from the steady state version of the Euler equation (11.6.2)), consumers make the capital stock vary over time. Consumers choose immediately to increase their saving in response to the adverse wealth effect that they suffer from the increase in lump-sum taxes that finances the permanently higher level of government expenditures. If the government consumes more, the household must consume less. The adverse wealth effect precedes the actual rise in government expenditures because consumers care about the present value of lump-sum taxes and are indifferent to their timing. Because the present value of lump-sum taxes jumps immediately, consumption also falls immediately in anticipation of the increase in government expenditures. This leads to a gradual build-up of capital in the dates between 0 and T, followed by a gradual fall after T. Variation over time in the capital stock helps smooth consumption over time, so that the main force at work is the consumption-smoothing motive featured in Milton Friedman's permanent income theory. The variation over time in $\bar{R}$ reconciles the consumer to a consumption path that is

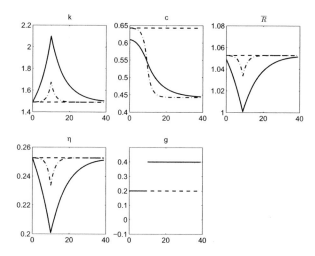

Figure 11.9.2: Response to foreseen once-and-for-all increase in g at $t = 10$. From left to right, top to bottom: $k, c, \bar{R}, \eta, g$. The dashed lines show the original steady state. The solid lines are for $\gamma = 2$, while the dashed-dotted lines are for $\gamma = .2$

not completely smooth. According to (11.6.9), the gradual increase and then the decrease in capital are inversely related to variations in the gross interest rate that reconcile the household to a consumption path that varies over time.

Figure 11.9.2 compares the responses to a foreseen increase in g at $t = 10$ for two economies, our original economy with $\gamma = 2$, shown in the solid line, and an otherwise identical economy with $\gamma = .2$, shown in the dashed-dotted line. The utility curvature parameter γ governs the household's willingness to substitute consumption across time. Lowering γ increases the household's willingness to substitute consumption across time. This shows up in the equilibrium outcomes in figure 11.9.2. For $\gamma = .2$, consumption is much less smooth than when $\gamma = 2$, and is closer to being a mirror image of the government expenditure path, staying high until government expenditures rise at $t = 10$. There are much smaller build ups and draw downs of capital, and this leads to smaller fluctuations in $\bar{R}$ and η. These two experiments reveal the dependence of the

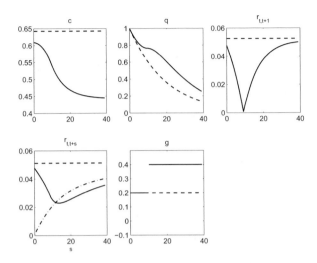

Figure 11.9.3: Response to foreseen once-and-for-all increase in g at $t = 10$. From left to right, top to bottom: $c, q, r_{t,t+1}$ and yield curves $r_{t,t+s}$ for $t = 0$ (solid line), $t = 10$ (dash-dotted line) and $t = 60$ (dashed line); term to maturity s is on the x axis for the yield curve, time t for the other panels.

strength of both the 'feedforward' anticipation effect and the 'feedback' transient effect that wears off initial conditions on the magnitude of γ. We discuss this more later in section 11.10.6 with the aid of equation (11.10.16).

For $\gamma = 2$ again, figure 11.9.3 describes the response of q_t and the term structure of interest rates to a foreseen increase in g_t at $t = 10$. The second panel on the top compares q_t for the initial steady state with q_t after the increase in g is foreseen at $t = 0$, while the third panel compares the implied short rate r_t computed via the section 11.3 formula $r_{t,t+1} = -\log(q_{t+1}/q_t) = -\log\left[\beta \frac{u'(c_{t+1})}{u'(c_t)} \frac{(1+\tau_{ct})}{(1+\tau_{c,t+1})}\right]$ and the fourth panel reports the term structure of interest rates $r_{t,t+s}$ computed via formula (11.3.3) for $t = 0, 10$ and $t = 60$ in three separate yield curves for those three dates. In this panel, the term to maturity s is on the x axis, while in the other panels, calendar time t is on the x axis. In this model, $q_t = \beta^t c_t^{-\gamma}$ and $r_{t,t+1} = -\log\beta\left(\frac{c_{t+1}}{c_t}\right)^{-\gamma}$, so the term structure of interest rates reflects the equilibrium path for $\{c_t\}_{t=0}^{\infty}$.

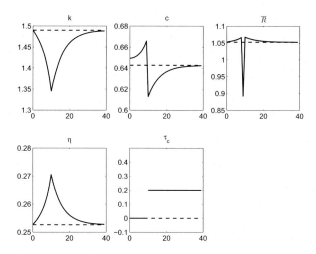

Figure 11.9.4: Response to foreseen once-and-for-all increase in τ_c at $t = 10$. From left to right, top to bottom: $k, c, \bar{R}, \eta, \tau_c$.

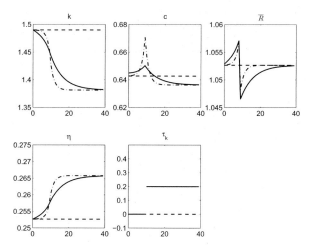

Figure 11.9.5: Response to foreseen increase in τ_k at $t = 10$. From left to right, top to bottom: $k, c, \bar{R}, \eta, \tau_k$. The solid lines depict equilibrium outcomes when $\gamma = 2$, the dashed-dotted lines when $\gamma = .2$.

At $t = 60$, the system has converged to the new steady state and the term structure of interest rates is flat. At $t = 10$, the term structure of interest rates is upward sloping because, as the top left panel showing consumption reveals, the *rate of growth* of consumption is expected to increase over time. At $t = 0$, the term structure of interest rate is 'U-shaped', declining until maturity 10, then increasing for longer maturities. This pattern reflects the pattern for consumption growth, which declines at an increasing rate until $t = 10$, then at a decreasing rate after that.

Foreseen jump in τ_c. Figure 11.9.4 portrays the response to a foreseen increase in the consumption tax. As we have remarked, with an inelastic labor supply, the Euler equation (11.6.2) and the other equilibrium conditions show that *constant* consumption taxes do not distort decisions, but that anticipated *changes* in them do. Indeed, (11.6.2) or (11.6.3) indicates that a foreseen increase in τ_{ct} (i.e., a decrease in $\frac{(1+\tau_{ct})}{(1+\tau_{ct+1})}$) operates like an *increase* in τ_{kt}. Notice that while all variables in Figure 11.9.4 eventually return to their initial steady-state values, the anticipated increase in τ_{ct} leads to an immediate jump in consumption at time 0, followed by a consumption binge that sends the capital stock downward until the date $t = T = 10$, at which τ_{ct} rises. The fall in capital causes $\bar{R}$ to rise over time, which via (11.6.9) requires the growth rate of consumption to rise until $t = T$. The jump in τ_c at $t = T = 10$ causes $\bar{R}$ to be depressed below 1, which via (11.6.9) accounts for the drastic fall in consumption at $t = 10$. From date $t = T$ onward, the effects of the *anticipated* distortion stemming from the fluctuation in τ_{ct} are over, and the economy is governed by the transient dynamic response associated with a capital stock that is now below the appropriate terminal steady-state capital stock. From date T onward, capital must rise. That requires austerity: consumption plummets at date $t = T = 10$. As the interest rate gradually falls, consumption grows at a diminishing rate along the path toward the terminal steady state.

Foreseen jump in τ_{kt}. For the two γ values 2 and .2, Figure 11.9.5 shows the response to a foreseen permanent jump in τ_{kt} at $t = T = 10$. Because the path of government expenditures is held fixed, the increase in τ_{kt} is accompanied by a reduction in the present value of lump-sum taxes that leaves the government budget balanced. The increase in τ_{kt} has effects that precede it. Capital starts declining immediately due to a rise in current consumption and a growing flow of consumption. The after-tax gross rate of return on capital starts rising at

$t = 0$ and increases until $t = 9$. It falls precipitously at $t = 10$ (see formula (11.6.8e) because of the foreseen jump in τ_k. Thereafter, $\bar{R}$ rises, as required by the transition dynamics that propel k_t toward its new lower steady state. Consumption is lower in the new steady state because the new lower steady-state capital stock produces less output. Consumption is smoother when $\gamma = 2$ than when $\gamma = .2$. Alterations in $\bar{R}$ accompany effects of the tax increase at $t = 10$ on consumption at earlier and later dates.

So far we have explored consequences of foreseen once-and-for-all changes in government policy. Next we describe some experiments in which there is a foreseen one-time change in a policy variable (a "pulse").

Foreseen one-time pulse in g_{10}. Figure 11.9.6 shows the effects of a foreseen one-time increase in g_t at date $t = 10$ that is financed entirely by alterations in lump sum taxes. Consumption drops immediately, then falls further over time in anticipation of the one-time surge in g. Capital is accumulated before $t = 10$. At $t = T = 10$, capital jumps downward because the government consumes it. The reduction in capital is accompanied by a jump in $\bar{R}$ above its steady-state value. The gross return $\bar{R}$ then falls toward its steady rate level and consumption rises at a diminishing rate toward its steady-state value. This experiment highlights what again looks like a version of a permanent income theory response to a foreseen decrease in the resources available for the public to spend (that is what the increase in g is about), with effects that are modified by the general equilibrium adjustments of the gross return $\bar{R}$.

11.10. Linear approximation

The present model is simple enough that it is very easy to apply the shooting algorithm. But for models with larger state spaces, it can be more difficult to apply the method. For those models, a frequently used procedure is to obtain a linear or log linear approximation around a steady state of the difference equation for capital, then to solve it to get an approximation of the dynamics in the vicinity of that steady state. The present model is a good laboratory for illustrating how to construct linear approximations. In addition to providing an easy way to approximate a solution, the method illuminates important features

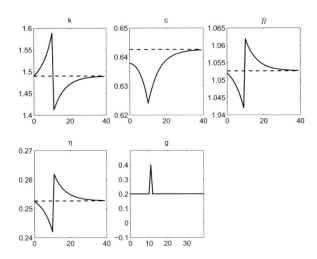

Figure 11.9.6: Response to foreseen one-time pulse increase in g at $t = 10$. From left to right, top to bottom: $k, c, \bar{R}, \eta, g$.

of the solution by partitioning it into two parts:[14] (1) a "feedback" part that portrays the transient response of the system to an initial condition k_0 that deviates from an asymptotic steady state, and (2) a "feedforward" part that shows the current effects of foreseen tax rates and expenditures.[15]

To obtain a linear approximation, perform the following steps:[16]

1. Set the government policy $z_t = \bar{z}$, a constant level. Solve $H(\bar{k}, \bar{k}, \bar{k}, \bar{z}, \bar{z}) = 0$ for a steady-state $\bar{k}$.

2. Obtain a first-order Taylor series approximation around $(\bar{k}, \bar{z})$:

$$\begin{aligned} H_{k_t} \left(k_t - \bar{k}\right) + H_{k_{t+1}} \left(k_{t+1} - \bar{k}\right) + H_{k_{t+2}} \left(k_{t+2} - \bar{k}\right) \\ + H_{z_t} \left(z_t - \bar{z}\right) + H_{z_{t+1}} \left(z_{t+1} - \bar{z}\right) = 0 \end{aligned} \tag{11.10.1}$$

3. Write the resulting system as

$$\phi_0 k_{t+2} + \phi_1 k_{t+1} + \phi_2 k_t = A_0 + A_1 z_t + A_2 z_{t+1} \tag{11.10.2}$$

---

[14] Hall (1971) employed linear approximations to exhibit some of this structure.

[15] Vector autoregressions embed the consequences of both backward-looking (transient) and forward-looking (foresight) responses to government policies.

[16] For an extensive treatment of lag operators and their uses, see Sargent (1987a).

or

$$\phi(L)\,k_{t+2} = A_0 + A_1 z_t + A_2 z_{t+1} \tag{11.10.3}$$

where L is the lag operator (also called the backward shift operator) defined by $Lx_t = x_{t-1}$. Factor the characteristic polynomial on the left as

$$\phi(L) = \phi_0 + \phi_1 L + \phi_2 L^2 = \phi_0 (1 - \lambda_1 L)(1 - \lambda_2 L). \tag{11.10.4}$$

For most of our problems, it will turn out that one of the λ_i's exceeds unity and that the other is less than unity. We shall therefore adopt the convention that $|\lambda_1| > 1$ and $|\lambda_2| < 1$. At this point, we ask the reader to accept that the values of λ_i split in this way. We discuss why they do so in section 11.10.2. Notice that equation (11.10.4) implies that $\phi_2 = \lambda_1 \lambda_2 \phi_0$. To obtain the factorization (11.10.4), we proceed as follows. Note that $(1 - \lambda_i L) = -\lambda_i \left(L - \frac{1}{\lambda_i} \right)$. Thus,

$$\phi(L) = \lambda_1 \lambda_2 \phi_0 \left(L - \frac{1}{\lambda_1} \right)\left(L - \frac{1}{\lambda_2} \right) = \phi_2 \left(L - \frac{1}{\lambda_1} \right)\left(L - \frac{1}{\lambda_2} \right) \tag{11.10.5}$$

because $\phi_2 = \lambda_1 \lambda_2 \phi_0$. Equation (11.10.5) identifies $\frac{1}{\lambda_1}, \frac{1}{\lambda_2}$ as the zeros of the polynomial $\phi(\zeta)$, i.e., $\lambda_i = \zeta_0^{-1}$ where $\phi(\zeta_0) = 0$.[17] We want to operate on both sides of (11.10.3) with the inverse of $(1 - \lambda_1 L)$, but that inverse is unstable backward (i.e., the power series $\sum_{j=0}^{\infty} \lambda_1^j L^j$ has coefficients that diverge in higher powers of L). Fortunately $(1 - \lambda_1 L)$ has a stable inverse in the *forward* direction, i.e., in terms of the forward shift operator L^{-1}.[18] In particular, notice that $(1 - \lambda_1 L) = -\lambda_1 L(1 - \lambda_1^{-1} L^{-1})$.[19] Using this result and $\phi_2 = \lambda_1 \lambda_2 \phi_0$, we can rewrite $\phi(L)$ as[20]

$$\phi(L) = -\frac{1}{\lambda_2} \phi_2 \left(1 - \lambda_1^{-1} L^{-1} \right)(1 - \lambda_2 L)\, L.$$

Represent equation (11.10.2) as

$$-\lambda_2^{-1} \phi_2 L \left(1 - \lambda_1^{-1} L^{-1} \right)(1 - \lambda_2 L)\, k_{t+2} = A_0 + A_1 z_t + A_2 z_{t+1}. \tag{11.10.6}$$

---

[17] The Matlab roots command `roots(phi)` finds zeros of polynomials, but you must arrange the polynomial as $\phi = [\phi_2 \quad \phi_1 \quad \phi_0]$.

[18] See appendix A of chapter 2.

[19] Notice that we can express $(1 - \lambda_1 L)^{-1} = -\lambda_1^{-1} L^{-1}(1 - \lambda^{-1} L^{-1})^{-1}$ as $-\lambda_1^{-1} L^{-1} \sum_{j=0}^{\infty} \lambda_1^{-j} L^{-j}$.

[20] Justifications for these steps are described at length in Sargent (1987a) and with rigor in Gabel and Roberts (1973).

Operate on both sides of (11.10.6) by $-(\phi_2/\lambda_2)^{-1}(1-\lambda_1^{-1}L^{-1})^{-1}$ to get the following representation:[21]

$$(1-\lambda_2 L)\,k_{t+1} = \frac{-\lambda_2\phi_2^{-1}}{1-\lambda_1^{-1}L^{-1}}\left[A_0 + A_1 z_t + A_2 z_{t+1}\right]. \qquad (11.10.7)$$

This concludes the procedure.

Equation (11.10.7) is our linear approximation to the equilibrium k_t sequence. It can be expressed as

$$k_{t+1} = \lambda_2 k_t - \lambda_2\phi_2^{-1}\sum_{j=0}^{\infty}(\lambda_1)^{-j}\left[A_0 + A_1 z_{t+j} + A_2 z_{t+j+1}\right]. \qquad (11.10.8)$$

We can summarize the process of obtaining this approximation as solving stable roots backward and unstable roots forward. Solving the unstable root forward imposes the terminal condition (11.5.3). This step corresponds to the step in the shooting algorithm that adjusts the initial investment rate to ensure that the capital stock eventually approaches the terminal steady-state capital stock.[22]

The term $\lambda_2 k_t$ is sometimes called the "feedback" part. The coefficient λ_2 measures the transient response rate, in particular, the rate at which capital returns to a steady state when it starts away from it. The remaining terms on the right side of (11.10.8) are sometimes called the "feedforward" parts. They depend on the infinite *future* of the exogenous z_t (which for us contain the components of government policy) and measure the effect on the current capital stock k_t of perfectly foreseen paths of fiscal policy. The decay parameter λ_1^{-1} measures the rate at which expectations of future fiscal policies are discounted in terms of their effects on current investment decisions. To a linear approximation, every rational expectations model has embedded within it both feedforward and feedback parts. The decay parameters λ_2 and λ_1^{-1} of the feedback and feedforward parts are determined by the roots of the characteristic polynomial. Equation (11.10.8) thus neatly exhibits the mixture of the pure foresight and the pure transient responses that are reflected in our examples in Figures 11.9.1 through 11.9.5. The feedback part captures the purely transient response and the feedforward part captures the perfect foresight component.

---

[21]　We have thus solved the stable root backward and the unstable root forward.

[22]　The invariant subspace methods described in chapter 5 are also all about solving stable roots backward and unstable roots forward.

11.10.1. Relationship between the λ_i's

It is a remarkable fact that if an equilibrium solves a planning problem, then the roots are linked by $\lambda_1 = \frac{1}{\beta \lambda_2}$, where $\beta \in (0, 1)$ is the planner's discount factor.[23] In this case, the feedforward decay rate $\lambda_1^{-1} = \beta \lambda_2$. Therefore, when the equilibrium allocation solves a planning problem, one of the λ_i's is less than $\frac{1}{\sqrt{\beta}}$ and the other exceeds $\frac{1}{\sqrt{\beta}}$ (this follows because $\lambda_1 \lambda_2 = \frac{1}{\beta}$).[24] From this it follows that one of the λ_i's, say λ_1, satisfies $\lambda_1 > \frac{1}{\sqrt{\beta}} > 1$ and the other λ_i, say λ_2 satisfies $\lambda_2 < \frac{1}{\sqrt{\beta}}$. Thus, for β close to 1, the condition $\lambda_1 \lambda_2 = \frac{1}{\beta}$ almost implies our earlier assumption that $\lambda_1 \lambda_2 = 1$, but not quite. Having $\lambda_2 < \frac{1}{\sqrt{\beta}}$ is sufficient to allow our linear approximation for k_t to satisfy $\sum_{t=0}^{\infty} \beta^t k_t^2 < +\infty$ for all z_t sequences that satisfy $\sum_{t=0}^{\infty} \beta^t z_t \cdot z_t < +\infty$.

A relationship between the feedforward and feedback decay rates appears evident in the experiments depicted in Figure 11.9.2. In particular, when the utility curvature parameter $\gamma = 2$, the rates at which future events are discounted in influencing outcomes before $t = 10$ and the rates of convergence back to steady state after $t = 10$ are both lower than when $\gamma = .2$.

11.10.2. Conditions for existence and uniqueness

For equilibrium allocations that do not solve planning problems, it ceases to be true that $\lambda_1 \lambda_2 = \frac{1}{\beta}$. In this case, the location of the zeros of the characteristic polynomial can be used to assess the existence and uniqueness of an equilibrium up to a linear approximation. If both λ_i's exceed $\frac{1}{\sqrt{\beta}}$, there exists no equilibrium allocation for which $\sum_{t=0}^{\infty} \beta^t k_t^2 < \infty$. If both λ_i's are less than $\frac{1}{\sqrt{\beta}}$, there exists a continuum of equilibria that satisfy that inequality. If the λ_i's split, with one exceeding and the other being less than $\frac{1}{\sqrt{\beta}}$, there exists a unique equilibrium.

---

[23] See Sargent (1987a, chap. XI) for a discussion.

[24] Notice that this means that the solution (11.10.8) remains valid for those divergent z_t processes, provided that they satisfy $\sum_{t=0}^{\infty} \beta^t z_{jt}^2 < +\infty$.

11.10.3. Once-and-for-all jumps

Next we specialize (11.10.7) to capture some examples of foreseen policy changes that we have studied above. Consider the special case treated by Hall (1971) in which the jth component of z_t follows the path

$$z_{jt} = \begin{cases} 0 & \text{if } t \leq T-1 \\ \overline{z}_j & \text{if } t \geq T \end{cases} \tag{11.10.9}$$

We define

$$v_t \equiv \sum_{i=0}^{\infty} \lambda_1^{-i} z_{t+i,j}$$

$$= \begin{cases} \dfrac{\left(\frac{1}{\lambda_1}\right)^{T-t} \overline{z}_j}{1-\left(\frac{1}{\lambda_1}\right)} & \text{if } t \leq T \\[4mm] \dfrac{1}{1-\left(\frac{1}{\lambda_1}\right)} \overline{z}_j & \text{if } t \geq T \end{cases} \tag{11.10.10}$$

$$h_t \equiv \sum_{i=0}^{\infty} \left(\frac{1}{\lambda_1}\right)^i z_{t+i+1,j}$$

$$= \begin{cases} \dfrac{\left(\frac{1}{\lambda_1}\right)^{T-(t+1)} \overline{z}_j}{1-\left(\frac{1}{\lambda_1}\right)} & \text{if } t \leq T-1 \\[4mm] \dfrac{1}{1-\left(\frac{1}{\lambda_1}\right)} \overline{z}_j & \text{if } t \geq T-1. \end{cases} \tag{11.10.11}$$

Using these formulas, let the vector z_t follow the path

$$z_t = \begin{cases} 0 & \text{if } t \leq T-1 \\ \overline{z} & \text{if } t \geq T \end{cases}$$

where $\overline{z}$ is a vector of constants. Then applying (11.10.10) and (11.10.11) to (11.10.7) gives the formulas

$$k_{t+1} = \begin{cases} \lambda_2 k_t - \dfrac{(\phi_0\lambda_1)^{-1} A_0}{1-\left(\frac{1}{\lambda_1}\right)} - \dfrac{(\phi_0\lambda_1)^{-1}\left(\frac{1}{\lambda_1}\right)^{T-t}}{1-\left(\frac{1}{\lambda_1}\right)} (A_1 + A_2\lambda_1)\overline{z} & \text{if } t \leq T-1 \\[4mm] \lambda_2 k_t - \dfrac{(\phi_0\lambda_1)^{-1}}{1-\left(\frac{1}{\lambda_1}\right)} [A_0 + (A_1 + A_2)\overline{z}] & \text{if } t \geq T. \end{cases}$$

11.10.4. Simplification of formulas

These formulas can be simultaneously generalized and simplified by using the following trick. Let z_t be governed by the state-space system

$$\bar{x}_{t+1} = A_x \bar{x}_t \tag{11.10.12a}$$

$$z_t = G_z \bar{x}_t, \tag{11.10.12b}$$

with initial condition $\bar{x}_0$ given. In chapter 2, we saw that many finite-dimensional linear time series models could be represented in this form, so that we are accommodating a large class of tax and expenditure processes. Then notice that

$$\left(\frac{A_1}{1 - \lambda_1^{-1} L^{-1}} \right) z_t = A_1 G_z \left(I - \lambda_1^{-1} A_x \right)^{-1} \bar{x}_t \tag{11.10.13a}$$

$$\left(\frac{A_2}{1 - \lambda_1^{-1} L^{-1}} \right) z_{t+1} = A_2 G_z \left(I - \lambda_1^{-1} A_x \right)^{-1} A_x \bar{x}_t \tag{11.10.13b}$$

Substituting these expressions into $(11.10.8)$ gives

$$k_{t+1} = \lambda_2 k_t - \lambda_2 \phi_2^{-1} \left[(1 - \lambda_1^{-1})^{-1} A_0 + A_1 G_z (I - \lambda_1^{-1} A_x)^{-1} \bar{x}_t \right.$$
$$\left. + A_2 G_z (I - \lambda_1^{-1} A_x)^{-1} A_x \bar{x}_t \right]. \tag{11.10.13c}$$

Taken together, system $(11.10.13)$ gives a complete description of the joint evolution of the exogenous state variables $\bar{x}_t$ driving z_t (our government policy variables) and the capital stock. System $(11.10.13)$ concisely displays the cross-equation restrictions that are the hallmark of rational expectations models: non-linear functions of the parameter occurring in G_z, A_x in the law of motion for the exogenous processes appear in the equilibrium representation $(11.10.13c)$ for the endogenous state variables.

We can easily use the state space system $(11.10.13)$ to capture the special case $(11.10.9)$. In particular, to portray $\bar{x}_{j,t+1} = \bar{x}_{j+1,t}$, set the $T \times T$ matrix A to be

$$A = \begin{bmatrix} 0_{T-1 \times 1} & I_{T-1 \times T-1} \\ 0_{1 \times T-1} & 1 \end{bmatrix} \tag{11.10.14}$$

and take the initial condition $\bar{x}_0 = \begin{bmatrix} 0 & 0 & \cdots & 0 & 1 \end{bmatrix}'$. To represent an element of z_t that jumps once and for all from 0 to $\bar{z}_j$ at $T = 0$, set the jth component of G_z equal to $G_{zj} = \begin{bmatrix} \bar{z}_j & 0 \cdots & 0 \end{bmatrix}$.

11.10.5. A one-time pulse

We can modify the transition matrix $(11.10.14)$ to model a one-time pulse in a component of z_t that occurs at and only at $t = T$. To do this, we simply set

$$A = \begin{bmatrix} 0_{T-1\times 1} & I_{T-1\times T-1} \\ 0_{1\times T-1} & 0 \end{bmatrix}. \qquad (11.10.15)$$

11.10.6. Convergence rates and anticipation rates

Equation $(11.10.8)$ shows that up to a linear approximation, the feedback co-efficient λ_2 equals the geometric rate at which the model returns to a steady state after a transient displacement away from a steady state. For our bench-mark values of our other parameters $\delta = .2, \beta = .95, \alpha = .33$ and all distorting taxes set to zero, we can compute that λ_2 is the following function of the utility curvature parameter γ that appears in $u(c) = (1 - \gamma)^{-1} c^{1-\gamma}$: [25]

$$\lambda_2 = \frac{\gamma}{a_1 \gamma^{-1} + a_2 + a_3 (\gamma^{-1} + a_4 \gamma^{-2} + a_5)^{\frac{1}{2}}} \qquad (11.10.16)$$

where $a_1 = .975, a_2 = .0329, a_3 = .0642, a_4 = .00063, a_5 = .0011$. Figure 11.10.1 plots this function. When $\gamma = 0$, the period utility function is linear and the household's willingness to substitute consumption over time is unlimited. In this case, $\lambda_2 = 0$, which means that in response to a perturbation of the capital stock away from a steady state, the return to a steady state is immediate. Furthermore, as mentioned above, because there are no distorting taxes in the initial steady state, we know that $\lambda_1 = \frac{1}{\beta\lambda_2}$, so that according to $(11.10.8)$, the feedforward response to future z's is a discounted sum that decays at rate $\beta\lambda_2$. Thus, when $\gamma = 0$, anticipations of *future* z's have no effect on current k. This is the other side of the coin of the immediate adjustment associated with the feedback part.

As the curvature parameter γ increases, λ_2 increases, more rapidly at first, more slowly later. As γ increases, the household values a smooth consumption path more and more highly. Higher values of γ impart to the equilibrium capital sequence both a more sluggish feedback response and a feedforward response that puts relatively more weight on prospective values of the z's in the more distant future.

---

[25] We used the Matlab symbolic toolkit to compute this expression.

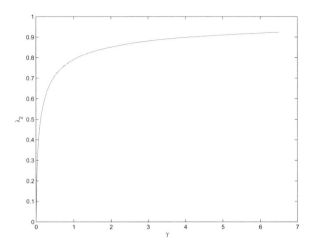

Figure 11.10.1: Feedback coefficient λ_2 as a function γ, evaluated at $\alpha = .33, \beta = .95, \delta = .2, g = .2$.

11.10.7. A remark about accuracy: Euler equation errors

It is important to estimate the accuracy of approximations. One simple diagnostic tool is to take a candidate solution for a sequence c_t, k_{t+1}, substitute them into the two Euler equations (11.12.1) and (11.12.2), and call the deviations between the left sides and the right sides the Euler equation errors.[26] An accurate method makes these errors small.[27]

---

[26] For more about this method, see Den Haan and Marcet (1994) and Judd (1998).

[27] Calculating Euler equation errors, but for a different purpose, goes back a long time. In chapter 2 of *The General Theory of Interest, Prices, and Money*, John Maynard Keynes noted that plugging in *data* (not a candidate *simulation*) into (11.12.2) gives big residuals. Keynes therefore assumed that (11.12.2) does not hold ("workers are off their labor supply curve").

11.11. Growth

It is straightforward to alter the model to allow for exogenous growth. We modify the production function to be

$$Y_t = F(K_t, A_t n_t) \tag{11.11.1}$$

where Y_t is aggregate output, N_t is total employment, A_t is labor-augmenting technical change, and $F(K, AN)$ is the same linearly homogeneous production function as before. We assume that A_t follows the process

$$A_{t+1} = \mu_{t+1} A_t \tag{11.11.2}$$

and will usually but not always assume that $\mu_{t+1} = \bar\mu > 1$. We exploit the linear homogeneity of $(11.11.1)$ to express the production function as

$$y_t = f(k_t) \tag{11.11.3}$$

where $f(k) = F(k, 1)$ and now $k_t = \frac{K_t}{n_t A_t}, y_t = \frac{Y_t}{n_t A_t}$. We say that k_t and y_t are measured per unit of "effective labor" $A_t n_t$. We also let $c_t = \frac{C_t}{A_t n_t}$ and $g_t = \frac{G_t}{A_t n_t}$ where C_t and G_t are total consumption and total government expenditures, respectively. We consider the special case in which labor is inelastically supplied. Then feasibility can be summarized by the following modified version of $(11.6.1)$:

$$k_{t+1} = \mu_{t+1}^{-1}[f(k_t) + (1 - \delta)k_t - g_t - c_t]. \tag{11.11.4}$$

Noting that per capita consumption is $c_t A_t$, we obtain the following counterpart to equation $(11.6.3)$:

$$\begin{aligned}
u'(c_t A_t) = &\beta u'(c_{t+1} A_{t+1})\frac{(1 + \tau_{ct})}{(1 + \tau_{ct+1})} \\
&[(1 - \tau_{kt+1})(f'(k_{t+1}) - \delta) + 1].
\end{aligned} \tag{11.11.5}$$

We assume the power utility function $u'(c) = c^{-\gamma}$, which makes the Euler equation become

$$(c_t A_t)^{-\gamma} = \beta(c_{t+1} A_{t+1})^{-\gamma} \bar R_{t+1},$$

where $\bar R_{t+1}$ continues to be defined by $(11.6.8e)$, except that now k_t is capital per effective unit of labor. The preceding equation can be represented as

$$\left(\frac{c_{t+1}}{c_t}\right)^{\gamma} = \beta \mu_{t+1}^{-\gamma} \bar R_{t+1}. \tag{11.11.6}$$

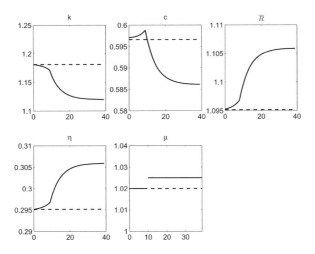

Figure 11.11.1: Response to foreseen once-and-for-all increase in rate of growth of productivity μ at $t = 10$. From left to right, top to bottom: $k, c, \bar{R}, \eta, \mu$, where now k, c are measured in units of effective unit of labor.

In a steady state, $c_{t+1} = c_t$. Then the steady-state version of the Euler equation (11.11.5) is

$$1 = \mu^{-\gamma}\beta[(1 - \tau_k)(f'(k) - \delta) + 1], \tag{11.11.7}$$

which can be solved for the steady-state capital stock. It is easy to compute that the steady-state level of capital per unit of effective labor satisfies

$$f'(k) = \delta + \left(\frac{(1 + \rho)\mu^{\gamma} - 1}{1 - \tau_k}\right) \tag{11.11.8}$$

and that

$$\bar{R} = (1 + \rho)\mu^{\gamma}. \tag{11.11.9}$$

Equation (11.11.9) immediately shows that *ceteris paribus*, a jump in the rate of technical change raises $\bar{R}$.

Next we apply the shooting algorithm to compute equilibria. We augment the vector of forcing variables z_t by including μ_t, so that it becomes $z_t = \begin{bmatrix} g_t & \tau_{kt} & \tau_{ct} & \mu_t \end{bmatrix}'$, where g_t is understood to be measured in effective units of labor, then proceed as before.

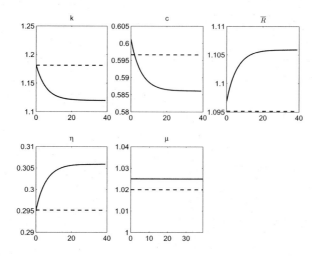

Figure 11.11.2: Response to increase in rate of growth of productivity μ at $t = 0$. From left to right, top to bottom: $k, c, \bar{R}, \eta, \mu$, where now k, c are measured in units of effective unit of labor.

Foreseen jump in productivity growth at $t = 10$. Figure 11.11.1 shows effects of a permanent increase from 1.02 to 1.025 in the productivity gross growth rate μ_t at $t = 10$. This figure and also Figure 11.11.2 now measure c and k in effective units of labor. The steady-state Euler equation (11.11.7) guides main features of the outcomes, and implies that a permanent increase in μ will lead to a decrease in the steady-state value of capital per unit of effective labor. Because capital is more efficient, even with less of it, consumption per capita can be raised, and that is what individuals care about. Consumption jumps immediately because people are wealthier. The increased productivity of capital spurred by the increase in μ leads to an increase in the gross return $\bar{R}$. Perfect foresight makes the effects of the increase in the growth of capital precede it.

Immediate (unforeseen) jump in productivity growth at $t = 1$. Figure 11.11.2 shows effects of an immediate jump in μ at $t = 0$. It is instructive to compare these with the effects of the foreseen increase in Figure 11.11.1. In Figure 11.11.2, the paths of all variables are entirely dominated by the feedback part of the solution, while before $t = 10$ those in Figure 11.11.1 have contributions from the feedforward part. The absence of feedforward effects makes the paths of all variables in Figure 11.11.2 smooth. Consumption per effective unit of labor jumps immediately then declines smoothly toward its steady state as the economy moves to a lower level of capital per unit of effective labor. The after-tax gross return $\bar{R}$ once again comoves with the consumption growth rate to verify the Euler equation (11.11.7).

11.12. Elastic labor supply

We now again shut down productivity growth by setting the gross productivity growth rate $\mu = 1$, but we allow a possibly nonzero labor supply elasticity by specifying $U(c, 1-n)$ to include a preference for leisure. Again, we let U_i be the partial derivative of U with respect to its ith argument. We assume an interior solution for $n \in (0,1)$. Now we have to carry along equilibrium conditions both for the intertemporal evolution of capital and for the labor-leisure choice. These are the two difference equations:

$$\frac{1}{(1 + \tau_{ct})} U_1\big(F(k_t, n_t) + (1 - \delta)k_t - g_t - k_{t+1}, 1 - n_t\big)$$
$$= \beta(1 + \tau_{ct+1})^{-1} U_1\big(F(k_{t+1}, n_{t+1}) + (1 - \delta)k_{t+1} - g_{t+1} - k_{t+2}, 1 - n_{t+1}\big)$$
$$\times [(1 - \tau_{kt+1})(F_k(k_{t+1}, n_{t+1}) - \delta) + 1]$$
$$\tag{11.12.1}$$

$$\frac{U_2\big(F(k_t, n_t) + (1 - \delta)k_t - g_t - k_{t+1}, 1 - n_t\big)}{U_1\big(F(k_t, n_t) + (1 - \delta)k_t - g_t - k_{t+1}, 1 - n_t\big)}$$
$$= \frac{(1 - \tau_{nt})}{(1 + \tau_{ct})} F_n(k_t, n_t).$$
$$\tag{11.12.2}$$

The linear approximation method applies equally well to this more general setting with just one additional step. We obtain a linear approximation to this dynamical system by proceeding as follows. First, find steady-state values $(\bar{k}, \bar{n})$ by solving the two steady-state versions of equations (11.12.1), (11.12.2). (Now $(\bar{k}, \bar{n})$ are steady-state values of capital per person and labor supplied per person, respectively.) Then take the following linear approximations to

(11.12.1), (11.12.2), respectively, around the steady state:

$$H_{k_t}(k_t - \overline{k}) + H_{k_{t+1}}(k_{t+1} - \overline{k}) + H_{n_{t+1}}(n_{t+1} - \overline{n}) + H_{k_{t+2}}(k_{t+2} - \overline{k})$$
$$+ H_{n_t}(n_t - \overline{n}) + H_{z_t}(z_t - \overline{z}) + H_{z_{t+1}}(z_{t+1} - \overline{z}) = 0$$

$$\text{(11.12.3)}$$

$$G_k(k_t - \overline{k}) + G_{n_t}(n_t - \overline{n}) + G_{k_{t+1}}(k_{t+1} - \overline{k}) + G_z(z_t - \overline{z}) = 0 \qquad \text{(11.12.4)}$$

Solve (11.12.4) for $(n_t - \overline{n})$ as functions of the remaining terms, substitute into (11.12.3) to get a version of equation (11.10.2), and proceed as before with a difference equation of the form (11.6.4).

11.12.1. Steady-state calculations

To compute a steady state for this version of the model, assume that government expenditures and all flat-rate taxes are constant over time. Steady-state versions of (11.12.1), (11.12.2) are

$$1 = \beta[(1 + (1 - \tau_k)(F_k(\overline{k}, \overline{n}) - \delta))] \qquad \text{(11.12.5)}$$

$$\frac{U_2(\overline{c}, 1 - \overline{n})}{U_1(\overline{c}, 1 - \overline{n})} = \frac{(1 - \tau_n)}{(1 + \tau_c)} F_n(\overline{k}, \overline{n}) \qquad \text{(11.12.6)}$$

and the steady state version of the feasibility condition (11.2.2) is

$$\overline{c} + \overline{g} + \delta \overline{k} = F(\overline{k}, \overline{n}). \qquad \text{(11.12.7)}$$

The linear homogeneity of $F(k, n)$ means that equation (11.12.5) by itself determines the steady-state capital-labor ratio $\frac{\overline{k}}{\overline{n}}$. In particular, where $\tilde{k} = \frac{k}{n}$, notice that $F(k, n) = nf(\tilde{k})$ and $F_k(k, n) = f'(\tilde{k})$. It is useful to use these facts to write (11.12.7) as

$$\frac{\overline{c} + \overline{g}}{\overline{n}} = f(\tilde{k}) - \delta \tilde{k}. \qquad \text{(11.12.8)}$$

Next, letting $\beta = \frac{1}{1+\rho}$, (11.12.5) can be expressed as

$$\delta + \frac{\rho}{(1 - \tau_k)} = f'(\tilde{k}), \qquad \text{(11.12.9)}$$

an equation that determines a steady-state capital-labor ratio $\tilde{k}$. An increase in $\frac{1}{(1-\tau_k)}$ decreases the capital-labor ratio, but the steady-state capital-labor ratio is independent of the steady state values of τ_c, τ_n. However, given the steady state value of the capital-labor ratio $\tilde{k}$, flat rate taxes on consumption and labor income influence the steady-state levels of consumption and labor via the steady state equations (11.12.6) and (11.12.7). Formula (11.12.6) reveals how both τ_c and τ_n distort the same labor-leisure margin.

If we define $\check{\tau}_c = \frac{\tau_n + \tau_c}{1 + \tau_c}$ and $\check{\tau}_k = \frac{\tau_k}{1 - \tau_k}$, then it follows that $\frac{(1-\tau_n)}{(1+\tau_c)} = 1 - \check{\tau}_c$ and $\frac{1}{(1-\tau_k)} = 1 + \check{\tau}_k$. The wedge $1 - \check{\tau}_c$ distorts the steady-state labor-leisure decision via (11.12.6) and the wedge $1 + \check{\tau}_k$ distorts the steady-state capital-labor ratio via (11.12.9).

11.12.2. Some experiments

To make things concrete, we use the following preference specification popularized by Hansen (1985) and Rogerson (1988):

$$U(c, 1 - n) = \ln c + B(1 - n) \tag{11.12.10}$$

where we set B substantially greater than 1 to assure an interior solution $n \in (0, 1)$ for labor supply. In particular, we set $B = 3$ in the experiments below. In terms of steady states, equation (11.12.6) becomes

$$B\bar{c} = \frac{(1 - \tau_n)}{(1 + \tau_c)} \left[f(\tilde{k}) - \tilde{k} f'(\tilde{k}) \right]. \tag{11.12.11}$$

It is useful to collect equations (11.12.9), (11.12.11), and (11.12.8), into the following system that recursively determines steady-state outcomes for $\tilde{k}, \bar{c}$, and $\bar{n}$ in the experiments to follow:

$$\delta + \frac{\rho}{(1 - \tau_k)} = f'(\tilde{k}) \tag{11.12.12}$$

$$B\bar{c} = \frac{(1 - \tau_n)}{(1 + \tau_c)} \left[f(\tilde{k}) - \tilde{k} f'(\tilde{k}) \right] \tag{11.12.13}$$

$$\bar{c} = \bar{n} \left(f(\tilde{k}) - \delta \tilde{k} \right) - \bar{g}. \tag{11.12.14}$$

Unforeseen jump in g. Figure 11.12.1 displays the consequences of an unforeseen and permanent jump in g at $t = 0$, financed entirely by adjustments in lump sum taxes. Equation $(11.12.12)$ determines $\tilde{k}$, which is unaltered. Equation $(11.12.13)$ then implies that $\bar{c}$ is unaltered. Equation $(11.12.14)$ determines $\bar{k}$ and $\bar{n}$, their ratio having been determined by $(11.12.12)$. The consequences of an unforeseen increase in g differ markedly from those analyzed above for the case in which the labor elasticity is zero. Then, the consequence was immediately and permanently to *lower* consumption per capita by the amount of the *increase* in government purchases per capita. Now the effect is to leave unaltered both steady state consumption per capita and the steady state capital/labor ratio. This is accomplished by raising the steady state levels of both capital and the labor supply. Thus, now the consequence of the increase in g is to 'grow the economy' enough eventually to leave consumption unaffected despite the increase in g.

These asymptotic outcomes immediately drop out of our steady state equations. The increase in g is accompanied by increases in k and n that leave the steady state capital/labor ratio unaltered, as required by equation $(11.12.9)$. Equation $(11.12.11)$ then dictates that steady-state consumption per capita also remain unaltered.

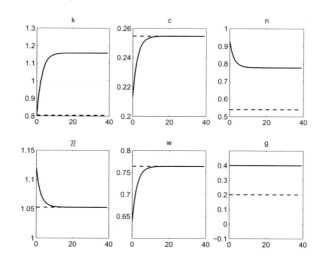

Figure 11.12.1: Elastic labor supply: response to unforeseen increase in g at $t = 0$. From left to right, top to bottom: $k, c, n, \bar{R}, w, g$. The dashed line is the original steady state.

Unforeseen jump in τ_n**.** Figure 11.12.2 shows outcomes from an unforeseen increase in the marginal tax rate on labor τ_n. Here the effect is to shrink the economy. As required by equation (11.12.9), the steady state capital labor ratio is unaltered. But equation (11.12.11) then requires that steady state consumption per capita must fall in response to the increase in τ_n. Both labor supplied n and capital fall in the new steady state.

Countervailing forces contributing to Prescott (2002) The preceding two experiments isolate forces that Prescott (2002) combines to reach his conclusion that Europe's economic activity has been depressed relative to the U.S. because its tax rates have been higher. Prescott's numerical calculations activate the forces that shrink the economy in our second experiment that increases τ_n while shutting down the force to grow the economy implied by a larger g. In particular, Prescott assumes that cross-country outcomes are generated by second experiment, with lump sum transfers being used to rebate the revenues raised from the larger labor tax rate τ_n that he estimates to prevail in Europe. If instead one assumes that higher taxes in Europe are used to pay for larger per capita government purchases, then forces to grow the economy identified in our first experiment are unleashed, making the adverse consequences for the level of economic activity of larger g, τ_n pairs in Europe become much smaller than Prescott calculated.

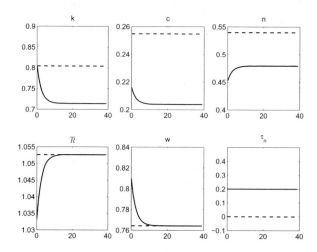

Figure 11.12.2: Elastic labor supply: response to unforeseen increase in τ_n at $t = 0$. From left to right, top to bottom: $k, c, n, \bar{R}, w, \tau_n$. The dashed line is the original steady state.

Foreseen jump in τ_n. Figure 11.12.3 describes consequences of a foreseen increase in τ_n that occurs at time $t = 10$. While the ultimate effects are identical with those described in the preceding experiment, transient outcomes differ. The immediate effect of the foreseen increase in τ_n is to spark a boom in employment and capital accumulation, while leaving consumption unaltered before time $t = 10$. People work more in response to the anticipation that rewards to working will decrease permanently at $t = 10$. Thus, the foreseen increase in τ_n sparks a temporary employment and investment boom.

To interpret what is going on here, we begin by noting that with preference specification (11.12.10), the following system of difference equations determines the dynamics of equilibrium allocations:

$$c_{t+1} = \beta \bar{R}_{t+1} c_t \tag{11.12.15a}$$

$$\bar{R}_{t+1} = \frac{1 + \tau_{ct}}{1 + \tau_{ct+1}} \left[1 + (1 - \tau_{kt+1})(f'(k_{t+1}/n_{t+1}) - \delta) \right] \tag{11.12.15b}$$

$$Bc_t = \frac{(1 - \tau_{nt})}{(1 + \tau_{ct})} F_n(k_t, n_t) \tag{11.12.15c}$$

$$k_{t+1} = F(k_t, n_t) + (1 - \delta)k_t - g_t - c_t \tag{11.12.15d}$$

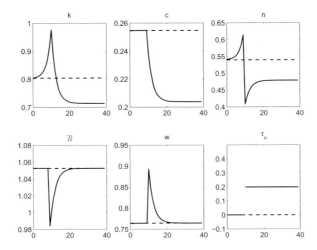

Figure 11.12.3: Elastic labor supply: response to foreseen increase in τ_n at $t = 10$. From left to right, top to bottom: $k, c, n, \bar{R}, w, \tau_n$. The dashed line is the original steady state.

These equations teach us that the foreseen increase in τ_n sparks a substantial rearrangement in how the household distributes its work over time. The effect of the permanent increase in τ_n at $t = 10$ is to reduce the after-tax wage from $t = 10$ onward, though initially the real wage falls by less than the decrease in $(1 - \tau_n)$ because of the increase in the capital labor ratio induced by the drastic fall in n at $t = 10$. Eventually, as the pre-tax real wage w returns to its initial value, the real wage falls by the entire amount of the decrease in $(1 - \tau_n)$. The decrease in the after-tax wage after $t = 10$ makes it relatively more attractive to work before $t = 10$. As a consequence, n_t rises above its initial steady state value before $t = 10$. The household uses the extra income to purchase enough capital to keep the capital-labor ratio and consumption equal to their respective initial steady state values for the first nine periods. This force increases n_t in the periods before $t = 10$. The effect of the build up of capital in the periods before $t = 0$ is to attenuate the decrease in the after tax wage that occurs at $t = 10$ because the equilibrium marginal product of labor has been raised higher than it would have been if capital had remained at its initial steady state value. From $t = 10$ onward, the capital stock is drawn down and the marginal product of labor falls, making the pre-tax real wage eventually return to its value in the initial steady state.

Mertens and Ravn (2011) use these to offer an interpretation of contractionary contributions that the Reagan tax cuts made to the U.S. recession of the early 1980s.

11.13.   A two-country model

This section describes a two country version of the basic model of this chapter. The model has a structure similar to ones used in the international real business cycle literature (e.g., Backus, Kehoe, and Kydland (1992)) and is in the spirit of an analysis of distorting taxes by Mendoza and Tesar (1998), though our presentation differs from theirs. We paste two countries together and allow them freely to trade goods, claims on future goods, but not labor. We shall have to be careful in how we specify taxation of earnings by non residents.

There are now two countries like the one in previous sections. Objects for the first country are denoted without asterisks, while those for the second country bear asterisks. There is international trade in goods, capital, and debt, but not in labor. We assume that leisure generates utility in neither country. Preferences over consumption streams in the two countries are ordered by $\sum_{t=0}^{\infty} \beta^t u(c_t)$ and $\sum_{t=0}^{\infty} \beta^t u(c_t^*)$, respectively, where $u(c) = \frac{c^{1-\gamma}}{1-\gamma}$ with $\gamma > 0$. Feasibility for the world economy is

$$(c_t + c_t^*) + (g_t + g_t^*) + (k_{t+1} - (1-\delta)k_t) + (k_{t+1}^* - (1-\delta)k_t^*) = f(k_t) + f(k_t^*) \quad (11.13.1)$$

where $f(k) = Ak^{\alpha}$ with $\alpha \in (0,1)$.

A consumer in country one can hold capital in either country, but pays taxes on rentals from foreign holdings of capital at the rate set by the foreign country. At time 0, residents in both countries can purchase consumption at date t at a common Arrow-Debreu price q_t. Let $\tilde{k}_t$ be capital in country 2 held by a representative consumer of country 1. Temporarily, in this paragraph only, let k_t denote the amount of domestic capital owned by the domestic consumer. (In all other paragraphs of our exposition of the two-country model, k_t denotes the amount of capital in country 1.) Let B_t^f be the amount of time t goods that the representative domestic consumer raises by issuing a one-period IOU to the representative foreign consumer; so $B_t^f > 0$ indicates that the domestic consumer is borrowing from abroad at t and $B_t^f < 0$ indicates that the domestic consumer is lending abroad at t. For $t \geq 1$ let $R_{t-1,t}$ be the gross return on a one-period loan from period $t - 1$ to period t. Define $R_{-1,0} \equiv 1$ and let the domestic consumer's initial debt to the foreign consumer be $R_{-1,0}B_{-1}^f$. We assume that returns on loans are not taxed by either country. The budget constraint of a country 1 consumer is

$$\sum_{t=0}^{\infty} q_t \left(c_t + (k_{t+1} - (1 - \delta)k_t) + (\tilde{k}_{t+1} - (1 - \delta)\tilde{k}_t) + R_{t-1,t}B_{t-1}^f \right)$$

$$\leq \sum_{t=0}^{\infty} q_t \left((\eta_t - \tau_{kt}(\eta_t - \delta))k_t + (\eta_t^* - \tau_{kt}^*(\eta_t^* - \delta))\tilde{k}_t + (1 - \tau_{nt})w_t n_t - \tau_{ht} + B_t^f \right).$$

$$(11.13.2)$$

A no-arbitrage condition for k_0 and $\tilde{k}_0$ is

$$(1 - \tau_{k0})\eta_0 + \delta\tau_{k0} = (1 - \tau_{k0}^*)\eta_0^* + \delta\tau_{k0}^*.$$

No-arbitrage conditions for k_t and $\tilde{k}_t$ for $t \geq 1$ imply

$$q_{t-1} = [(1 - \tau_{kt})(\eta_t - \delta) + 1] \, q_t$$
$$q_{t-1} = [(1 - \tau_{kt}^*)(\eta_t^* - \delta) + 1] \, q_t, \qquad (11.13.3)$$

which together imply that after-tax rental rates on capital are equalized across the two countries:

$$(1 - \tau_{kt}^*)(\eta_t^* - \delta) = (1 - \tau_{kt})(\eta_t - \delta). \qquad (11.13.4)$$

No arbitrage conditions for B_t^f for $t \geq 0$ are $q_t = q_{t+1}R_{t,t+1}$, which implies that

$$q_{t-1} = q_t R_{t-1,t} \qquad (11.13.5)$$

for $t \geq 1$.

Since domestic capital, foreign capital, and consumption loans bear the same rates of return by virtue of (11.13.4) and (11.13.5), portfolios are indeterminate. We are free to set holdings of foreign capital equal to zero in each country if we allow B_t^f to be nonzero. Adopting this way of resolving portfolio indeterminacy is convenient because it economizes on the number of initial conditions we have to specify. Therefore, we set holdings of foreign capital equal to zero in both countries but allow international lending. Then given an initial level B^f_{-1} of debt from the domestic country to the foreign country *, and where $R_{t-1,t} = \frac{q_{t-1}}{q_t}$, international debt dynamics satisfy

$$B^f_t = R_{t-1,t}B^f_{t-1} + c_t + (k_{t+1} - (1-\delta)k_t) + g_t - f(k_t) \tag{11.13.6}$$

and

$$c^*_t + (k^*_{t+1} - (1-\delta)k^*_t) + g^*_t - R_{t-1,t}B^f_{t-1} = f(k^*_t) - B^f_t. \tag{11.13.7}$$

Firms' first-order conditions in the two countries are:

$$\eta_t = f'(k_t), \quad w_t = f(k_t) - k_t f'(k_t)$$
$$\eta^*_t = f'(k^*_t), \quad w^*_t = f(k^*_t) - k^*_t f'(k^*_t). \tag{11.13.8}$$

International trade in goods establishes

$$\frac{q_t}{\beta^t} = \frac{u'(c_t)}{1 + \tau_{ct}} = \mu^* \frac{u'(c^*_t)}{1 + \tau^*_{ct}}, \tag{11.13.9}$$

where μ^* is a nonnegative number that is a function of the Lagrange multiplier on the budget constraint for a consumer in country * and where we have normalized the Lagrange multiplier on the budget constraint of the domestic country to set the corresponding μ for the domestic country to unity. Equilibrium requires that the following two national Euler equations be satisfied for $t \geq 0$:

$$u'(c_t) = \beta u'(c_{t+1}) \left[(1 - \tau_{kt+1})(f'(k_{t+1}) - \delta) + 1\right] \left[\frac{1 + \tau_{ct+1}}{1 + \tau_{ct}}\right] \tag{11.13.10}$$

$$u'(c^*_t) = \beta u'(c^*_{t+1}) \left[(1 - \tau^*_{kt+1})(f'(k^*_{t+1}) - \delta) + 1\right] \left[\frac{1 + \tau^*_{ct+1}}{1 + \tau^*_{ct}}\right] \tag{11.13.11}$$

Given that equation (11.13.9) holds for all $t \geq 0$, either equation (11.13.10) or equation (11.13.11) is redundant.

11.13.1. Initial conditions

As initial conditions, we take the pre-international-trade allocation of capital across countries $(\check{k}_0, \check{k}_0^*)$ and an initial level $B_{-1}^f = 0$ of international debt owed by the unstarred (domestic) country to the starred (foreign) country.

11.13.2. Equilibrium steady state values

The following two equations determine steady values for k and k^*.

$$f'(\overline{k}) = \delta + \frac{\rho}{1 - \tau_k} \tag{11.13.12}$$

$$f'(\overline{k}^*) = \delta + \frac{\rho}{1 - \tau_k^*} \tag{11.13.13}$$

Given the steady state capital-labor ratios $\overline{k}$ and $\overline{k}^*$, the following two equations determine steady state values of domestic and foreign consumption $\overline{c}$ and $\overline{c}^*$ as functions of a steady state value $\overline{B}^f$ of debt from the domestic country to country $*$:

$$(\overline{c} + \overline{c}^*) = f(\overline{k}) + f(\overline{k}^*) - \delta(\overline{k} + \overline{k}^*) - (\overline{g} + \overline{g}^*) \tag{11.13.14}$$

$$\overline{c} = f(\overline{k}) - \delta\overline{k} - \overline{g} - \rho\overline{B}^f \tag{11.13.15}$$

Equation (11.13.14) expresses feasibility at a steady state while (11.13.15) expresses trade balance, including interest payments, at a steady state.

11.13.3. Initial equilibrium values

Trade in physical capital and time 0 debt takes place before production and trade in other goods occurs at time 0. We shall always initialize international debt at zero: $B_{-1}^f = 0$, a condition that we use to express that international trade in capital begins at time 0. Given an initial total world-wide capital stock $\check{k}_0 + \check{k}_0^*$, initial values of k_0 and k_0^* satisfy

$$k_0 + k_0^* = \check{k}_0 + \check{k}_0^* \tag{11.13.16}$$

$$(1 - \tau_{k0})f'(k_0) + \delta\tau_{k0} = (1 - \tau_{k0}^*)f'(k_0^*) + \delta\tau_{k0}^*. \tag{11.13.17}$$

The price of a unit of capital in either country at time 0 is

$$p_{k0} = [(1 - \tau_{k0})f'(k_0) + (1 - \delta) + \delta\tau_{k0}]. \tag{11.13.18}$$

It follows that

$$B_{k0} = p_{k0}[k_0 - \check{k}_0], \tag{11.13.19}$$

which says that the domestic country finances imports of physical capital from abroad by borrowing from the foreign country *.

11.13.4. Shooting algorithm

To apply a shooting algorithm, we would search for pairs c_0, μ^* that yield a pair (k_0, k_0^*) and paths $\{c_t, c_t^*, k_t, k_t^*, B_t^f\}_{t=0}^T$ that solve equations (11.13.16), (11.13.17), (11.13.18), (11.13.19), (11.13.6), (11.13.9), and (11.13.18). The shooting algorithm 'aims' for $(\overline{k}, \overline{k}^*)$ that satisfy the steady-state equations (11.13.12), (11.13.13).

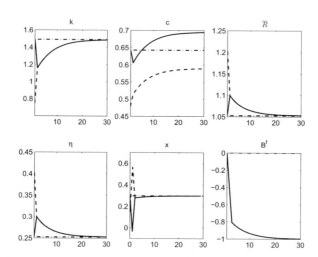

Figure 11.13.1: Response to unforeseen opening of trade at time 1. From left to right, top to bottom: $k, c, \overline{R}, \eta, x,$ and B^f. The solid line is the domestic country, the dashed line is the foreign country and the dashed dotted line is the original steady state.

11.13.5. Transition exercises

In the one-country exercises earlier in this chapter, announcements of new policies always occurred at time 0. In the two-country exercises to follow, we assume that announcements of new paths of tax rates and/or expenditures or trade regimes all occur at time 1. We do this to show some dramatic jumps in particular variables that occur at time 1 in response to announcements about changes that will occur at time 10 and later. Showing variables at times 0 and 1 helps display some of the outcomes on which we shall focus here. The production function is $f(k) = Ak^\alpha$. Parameter values are $\beta = .95, \gamma = 2, \delta = .2, \alpha = .33, A = 1$; g is initially $.2$ in both countries and all distorting taxes are initially 0 in both countries.

We describe outcomes from three exercises that illustrate two economic forces. The first force is consumers' desire to smooth consumption over time, expressed through households' consumption Euler equations. The second force is that equilibrium outcomes must offer no opportunities for arbitrage, expressed through equations that equate rates of returns on bonds and capital in both countries.

In the first two experiments, all taxes are lump sum in both countries. In the third experiment we activate a tax on capital in the domestic but not the foreign country. In all experiments, we allow lump sum taxes in both countries to adjust to satisfy government budget constraints in both countries.

11.13.5.1. Opening International Flows

In our first example, we study the transition dynamics for two countries when in period one newly produced output and stocks of capital, but not labor, suddenly become internationally mobile. The two economies are initially identical in all aspects except for one: we start the domestic economy at its autarkic steady state, while we start the foreign economy at an initial capital stock below its autarkic steady state. Because there are no distorting taxes on returns to physical capital, capital stocks in both economies converge to the same level.

In this experiment the domestic country is at its steady state capital stock while the poorer foreign country has a capital stock that is $.5$ less. This means that initially, before trade is opened at $t = 1$, the marginal product of capital in the foreign country exceeded the marginal product capital in the domestic country, that the foreign interest rate $R_{0,1}^*$ exceeded the domestic rate $R_{0,1}$, and that consequently the foreign consumption growth rate exceeded the domestic consumption growth rate. The disparity of interest rates before trade is opened is a force for physical capital to flow from the domestic country to the foreign country once when trade is opened at $t = 1$. Figure 11.13.2 presents the transitional dynamics. When countries become open to trade in goods and capital in period one, there occurs an immediate reallocation of capital from the capital-rich domestic country to the capital-poor foreign country. This transfer of capital has to take place because if it didn't, capital in different countries would yield different returns, providing consumers in both countries with arbitrage opportunities. Those cannot occur in equilibrium.

Before international trade had opened, rental rates on capital and interest rates differed across country because marginal products of capital differed and consumption growth rates differed. When trade opens at time 1 and capital is reallocated across countries to equalize returns, the interest rate in the domestic country jumps at time one. Because $\gamma = 2$, this means that consumption c in the domestic country must fall. The opposite is true for the foreign economy. Notice also that figure 11.13.2 shows an investment spike abroad while there is a large decline in investment in the domestic economy. This occurs because capital is reallocated from the domestic country to the foreign one. This transfer is feasible because investment in capital is reversible. The foreign country finances this import of physical capital by borrowing from the domestic country, so $-B^f$ increases. Foreign debt $-B^f$ continues to increase as both economies converge smoothly towards a steady-state with a positive level of $-\overline{B}^f$. Ultimately, these differences account for differences in steady-state consumption by $2\rho\overline{B}^f$.

Opening trade in goods and capital at time 1 benefits consumers in both economies. By opening up to capital flows, the foreign country achieves convergence to a steady-state consumption level at an accelerated rate. This steady-state consumption rate is lower than what it would be had the economy remained closed, but this reduction in long-run consumption is more than compensated by the rapid increase in consumption and output in the short-run. In contrast, domestic consumption falls in the short-run as trade allows domestic consumers to accumulate foreign assets that eventually support greater steady-state consumption.

This experiment shows the importance of studying transitional dynamics for welfare analysis. In this example, focusing only on steady-state consumption would lead to the false conclusion that opening markets are detrimental for poorer economies.

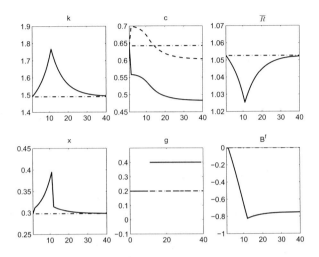

Figure 11.13.2: Response to increase in g at time 10 foreseen at time 1. From left to right, top to bottom: $k, c, \bar{R}, x, g, B^f$. The dashed-dotted line is the original steady state in the domestic country. The dashed line denotes the foreign country.

11.13.5.2. Foreseen Increase in g

Figure 11.13.2 presents transition dynamics after an increase in g in the domestic economy from .2 to .4 that is announced ten periods in advance. We start both economies from a steady-state with $B_0^f = 0$. When the new g path is announced at time 1, consumption smoothing motives induce domestic households to increase their savings in response to the adverse shock to domestic private wealth that is caused at time 1 by the foreseen increase in domestic government purchases g. Domestic households plan to use those savings to dampen the impact on consumption in periods after g will have increased ten periods ahead. Households save partly by accumulating more domestic capital in the short-run, their only source of assets in the closed economy version of this experiment. In an open economy, they have other ways to save, namely, by lending abroad.

The no-arbitrage conditions connect adjustments of both types of saving: the increase in savings by domestic households will reduce the equilibrium return on bonds and capital in the foreign economy to prevent arbitrage opportunities. Confronting the revised interest rate path that now begins with lower interest rates, foreign households increase their rates of consumption and investment in physical capital. These increases in foreign absorbtion are funded by increases in foreign consumers' external debt. After the announcement of the increase in g, the paths for consumption (and capital) in both countries follow the same patterns because no-arbitrage conditions equate the ratios of their marginal utilities of consumption. Both countries continue to accumulate capital until the increase in g occurs. After that, domestic households begin consuming some of their capital. Again by no-arbitrage conditions, when g actually increases both countries reduce their investment rates. The domestic economy, in turn, starts running current-account deficits partially to fund the increase in g. This means that foreign households begin repaying part of their external debt by reducing their capital stock. Although not plotted in figure 11.13.2, there is a sharp reduction in gross investment x in both countries when the increase in g occurs. After $t = 10$, all variable converge smoothly towards a new steady state where the domestic economy persists with positive asset holdings $-B^f$. Ultimately, this explains why the foreign country ends with lower steady state consumption than in the initial steady state. In the new steady state, minus the sum of the *decreases* of consumption rates across the two countries equals the *increase* in steady state government expenditures in the domestic country.[28]

The experiment teaches valuable economic lessons. First, it shows how the consequences of the foreseen increase in g will be distributed across time and households. Second, it tells how this distribution takes place: through time by accumulating or reducing the capital stock, and across households in different countries by running current-account deficits and surpluses.

---

[28] Despite the decrease in its steady state consumption, we have calculated that $\sum_{t=0}^{\infty} \beta^t u(c_t^*)$ is higher in the new equilibrium than in the old.

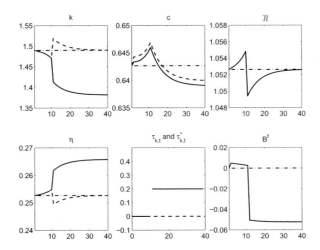

Figure 11.13.3: Response to once-and-for-all increase in τ_k at $t = 10$ foreseen at time $t = 1$. From left to right, top to bottom: $k, c, \bar{R}, \eta, \tau_k$ and τ_k^*, and B^f. Domestic country (solid line), foreign country (dotted line) and steady-state values (dot-line).

11.13.5.3. Foreseen increase in τ_k

We now explore the impact of an increase in capital taxation in the domestic economy 10 periods after its announcement at $t = 1$. Figure 11.13.3 shows equilibrium outcomes. When the increase in τ_k is announced, domestic households become aware that the domestic capital stock will eventually decline to increase gross returns to equalize after-tax returns across countries despite a higher domestic tax rate on returns from capital. Domestic households will reduce their capital stock by increasing their rate of consumption. The consequent higher equilibrium world interest rates then also induces foreign households to increase consumption. Prior to the increase in τ_k, the domestic country runs a current account deficit. When τ_k is eventually increased, capital is rapidly reallocated across borders to preclude arbitrage opportunities, leading to a lower interest rate on bonds. The fall in the return on bonds occurs because the capital returns tax τ_k in the domestic country will reduce the after-tax return on capital, and because the foreign economy has a higher capital stock. Foreign households fund this large purchase of capital with a sharp increase in external debt, to be interpreted as a current account deficit. After τ_k has increased, the economies smoothly converge to a new steady state that features lower consumption rates in both countries and where the differences in the capital stock equate after-tax returns. It is useful to note that steady-state consumption in the foreign economy is higher than in the domestic country despite its perpetually having positive liabilities. This occurs because foreign output is larger because the capital stock held abroad is also larger.

This example shows how, via the no-arbitrage conditions, both countries share the impact of the shock and how fluctuations in capital stocks smooth over time the adjustments in consumption in both countries.

11.14. Concluding remarks

In chapter 12 we shall describe a stochastic version of the basic growth model and alternative ways of representing its competitive equilibrium.[29] Stochastic and nonstochastic versions of the growth model are widely used throughout aggregative economics to study a range of policy questions. Brock and Mirman (1972), Kydland and Prescott (1982), and many others have used a stochastic version of the model to approximate features of the business cycle. In much of the earlier literature on real business cycle models, the phrase "features of the business cycle" has meant "particular moments of some aggregate time series that have been filtered in a particular way to remove trends." Lucas (1990) uses a nonstochastic model like the one in this chapter to prepare rough quantitative estimates of the eventual consequences of lowering taxes on capital and raising those on consumption or labor. Prescott (2002) uses a version of the model in this chapter with leisure in the utility function together with some illustrative (high) labor supply elasticities to construct the argument that in the last two decades, Europe's economic activity has been depressed relative to that of the United States because Europe has taxed labor more highly that the United States. Ingram, Kocherlakota, and Savin (1994) and Hall (1997) use actual data to construct the errors in the Euler equations associated with stochastic versions

---

[29] It will be of particular interest to learn how to achieve a recursive representation of an equilibrium by finding an appropriate formulation of a state vector in terms of which to cast an equilibrium. Because there are endogenous state variables in the growth model, we shall have to extend the method used in chapter 8.

of the basic growth model and interpret them not as computational errors, as in the procedure recommended in section 11.10.7, but as measures of additional shocks that have to be added to the basic model to make it fit the data. In the basic stochastic growth model described in chapter 12, the technology shock is the only shock, but it cannot by itself account for the discrepancies that emerge in fitting all of the model's Euler equations to the data. A message of Ingram, Kocherlakota, and Savin (1994) and Hall (1997) is that more shocks are required to account for the data. Wen (1998) and Otrok (2001) build growth models with more shocks and additional sources of dynamics, fit them to U.S. time series using likelihood function-based methods, and discuss the additional shocks and sources of data that are required to match the data. See Christiano, Eichenbaum, and Evans (2003) and Christiano, Motto, and Rostagno (2003) for papers that add a number of additional shocks and measure their importance. Greenwood, Hercowitz, and Krusell (1997) introduced what seems to be an important additional shock in the form of a technology shock that impinges directly on the relative price of investment goods. Jonas Fisher (2006) develops econometric evidence attesting to the importance of this shock in accounting for aggregate fluctuations. Davig, Leeper, and Walker (2012) use stochastic versions of the types of models discussed in this chapter to study issues of intertemporal fiscal balance.

Schmitt-Grohe and Uribe (2004b) and Kim and Kim (2003) warn that the linear and log linear approximations described in this chapter can be treacherous when they are used to compare the welfare under alternative policies of economies, like the ones described in this chapter, in which distortions prevent equilibrium allocations from being optimal ones. They describe ways of attaining locally more accurate welfare comparisons by constructing higher order approximations to decision rules and welfare functions.

A. Log linear approximations

Following Christiano (1990), a widespread practice is to obtain log linear rather than linear approximations. Here is how this would be done for the model of this chapter.

Let $\log k_t = \tilde{k}_t$ so that $k_t = \exp \tilde{k}_t$; similarly, let $\log g_t = \tilde{g}_t$. Represent z_t as $z_t = [\exp(\tilde{g}_t) \quad \tau_{kt} \quad \tau_{ct}]'$ (note that only g_t has been replaced by it's log here). Then proceed as follows to get a log linear approximation.

1. Compute the steady state as before. Set the government policy $z_t = \bar{z}$, a constant level. Solve $H(\exp(\tilde{k}_\infty), \exp(\tilde{k}_\infty), \exp(\tilde{k}_\infty), \bar{z}, \bar{z}) = 0$ for a steady state $\tilde{k}_\infty$. (Of course, this will give the same steady state for the original unlogged variables as we got earlier.)

2. Take first-order Taylor series approximation around $(\tilde{k}_\infty, \bar{z})$:

$$
\begin{aligned}
&H_{\tilde{k}_t}(\tilde{k}_t - \tilde{k}_\infty) + H_{\tilde{k}_{t+1}}(\tilde{k}_{t+1} - \tilde{k}_\infty) + H_{\tilde{k}_{t+2}}(\tilde{k}_{t+2} - \tilde{k}_\infty) \\
&+ H_{z_t}(z_t - \bar{z}) + H_{z_{t+1}}(z_{t+1} - \bar{z}) = 0.
\end{aligned}
\tag{11.A.1}
$$

(But please remember here that the first component of z_t is now $\tilde{g}_t$.)

3. Write the resulting system as

$$
\phi_0 \tilde{k}_{t+2} + \phi_1 \tilde{k}_{t+1} + \phi_2 \tilde{k}_t = A_0 + A_1 z_t + A_2 z_{t+1}
\tag{11.A.2}
$$

or

$$
\phi(L)\tilde{k}_{t+2} = A_0 + A_1 z_t + A_2 z_{t+1}
\tag{11.A.3}
$$

where L is the lag operator (also called the backward shift operator). Solve the linear difference equation (11.A.3) exactly as before, but for the sequence $\{\tilde{k}_{t+1}\}$.

4. Compute $k_t = \exp(\tilde{k}_t)$, and also remember to exponentiate $\tilde{g}_t$, then use equations (11.6.8) to compute the associated prices and quantities. Compute the Euler equation errors as before.

Exercises

Exercise 11.1   **Tax reform: I**

Consider the following economy populated by a government and a representative household. There is no uncertainty, and the economy and the representative household and government within it last forever. The government consumes a constant amount $g_t = g > 0, t \geq 0$. The government also sets sequences for two types of taxes, $\{\tau_{ct}, \tau_{ht}\}_{t=0}^{\infty}$. Here τ_{ct}, τ_{ht} are, respectively, a possibly time-varying flat-rate tax on consumption and a time-varying lump-sum or "head" tax. The preferences of the household are ordered by

$$\sum_{t=0}^{\infty} \beta^t u(c_t),$$

where $\beta \in (0, 1)$ and $u(\cdot)$ is strictly concave, increasing, and twice continuously differentiable. The feasibility condition in the economy is

$$g_t + c_t + k_{t+1} \leq f(k_t) + (1 - \delta)k_t$$

where k_t is the stock of capital owned by the household at the beginning of time t and $\delta \in (0, 1)$ is a depreciation rate. At time 0, there are complete markets for dated commodities. The household faces the budget constraint:

$$\sum_{t=0}^{\infty} \{q_t[(1 + \tau_{ct})c_t + k_{t+1} - (1 - \delta)k_t]\} \leq \sum_{t=0}^{\infty} q_t \{\eta_t k_t + w_t - \tau_{ht}\}$$

where we assume that the household inelastically supplies one unit of labor, and q_t is the price of date t consumption goods measured in the numeraire at time 0, η_t is the rental rate of date t capital measured in consumption goods at time t, and w_t is the wage rate of date t labor measured in consumption goods at time t. Capital is neither taxed nor subsidized.

A production firm rents labor and capital. The production function is $f(k)n$, where $f' > 0, f'' < 0$. The value of the firm is

$$\sum_{t=0}^{\infty} q_t \left[f(k_t)n_t - w_t n_t - \eta_t k_t \right],$$

where k_t is the firm's capital-labor ratio and n_t is the amount of labor it hires.

The government sets g_t exogenously and must set τ_{ct}, τ_{ht} to satisfy the budget constraint:

(1)
$$\sum_{t=0}^{\infty} q_t(\tau_{ct}c_t + \tau_{ht}) = \sum_{t=0}^{\infty} q_t g_t.$$

a. Define a competitive equilibrium.

b. Suppose that historically the government had unlimited access to lump-sum taxes and availed itself of them. Thus, for a long time the economy had $g_t = \bar{g} > 0, \tau_{ct} = 0$. Suppose that this situation had been expected to go on forever. Tell how to find the steady-state capital-labor ratio for this economy.

c. In the economy depicted in b, prove that the timing of lump-sum taxes is irrelevant.

d. Let $\bar{k}_0$ be the steady value of k_t that you found in part b. Let this be the initial value of capital at time $t = 0$ and consider the following experiment. Suddenly and unexpectedly, a court decision rules that lump-sum taxes are illegal and that starting at time $t = 0$, the government must finance expenditures using the consumption tax τ_{ct}. The value of g_t remains constant at $\bar{g}$. Policy advisor number 1 proposes the following tax policy: find a *constant* consumption tax that satisfies the budget constraint (1), and impose it from time 0 onward. Please compute the new steady-state value of k_t under this policy. Also, get as far as you can in analyzing the transition path from the old steady state to the new one.

e. Policy advisor number 2 proposes the following alternative policy. Instead of imposing the increase in τ_{ct} suddenly, he proposes to ease the pain by postponing the increase for 10 years. Thus, he/she proposes to set $\tau_{ct} = 0$ for $t = 0, \ldots, 9$, then to set $\tau_{ct} = \bar{\tau}_c$ for $t \geq 10$. Please compute the steady-state level of capital associated with this policy. Can you say anything about the transition path to the new steady-state k_t under this policy?

f. Which policy is better, the one recommended in d or the one in e?

Exercise 11.2   **Tax reform: II**

Consider the following economy populated by a government and a representative household. There is no uncertainty, and the economy and the representative household and government within it last forever. The government consumes a constant amount $g_t = g > 0, t \geq 0$. The government also sets sequences of two types of taxes, $\{\tau_{ct}, \tau_{kt}\}_{t=0}^{\infty}$. Here τ_{ct}, τ_{kt} are, respectively, a possibly time-varying flat-rate tax on consumption and a time-varying flat-rate tax on earnings from capital. The preferences of the household are ordered by

$$\sum_{t=0}^{\infty} \beta^t u(c_t),$$

where $\beta \in (0,1)$ and $u(\cdot)$ is strictly concave, increasing, and twice continuously differentiable. The feasibility condition in the economy is

$$g_t + c_t + k_{t+1} \leq f(k_t) + (1-\delta)k_t$$

where k_t is the stock of capital owned by the household at the beginning of time t and $\delta \in (0,1)$ is a depreciation rate. At time 0, there are complete markets for commodities at all dates. The household faces the budget constraint:

$$\sum_{t=0}^{\infty} \{q_t[(1+\tau_{ct})c_t + k_{t+1} - (1-\delta)k_t]\}$$

$$\leq \sum_{t=0}^{\infty} q_t \{\eta_t k_t - \tau_{kt}(\eta_t - \delta)k_t + w_t\}$$

where we assume that the household inelastically supplies one unit of labor, and q_t is the price of date t consumption goods in units of the numeraire at time 0, η_t is the rental rate of date t capital in units of time t goods, and w_t is the wage rate of date t labor in units of time t goods.

A production firm rents labor and capital. The value of the firm is

$$\sum_{t=0}^{\infty} q_t \left[f(k_t)n_t - w_t n_t - \eta_t k_t n_t \right],$$

where here k_t is the firm's capital-labor ratio and n_t is the amount of labor it hires.

The government sets $\{g_t\}$ exogenously and must set the sequences $\{\tau_{ct}, \tau_{kt}\}$ to satisfy the budget constraint:

(1) $$\sum_{t=0}^{\infty} q_t \left(\tau_{ct} c_t + \tau_{kt}(\eta_t - \delta)k_t \right) = \sum_{t=0}^{\infty} q_t g_t.$$

a. Define a competitive equilibrium.

b. Assume an initial situation in which from time $t \geq 0$ onward, the government finances a constant stream of expenditures $g_t = \bar{g}$ entirely by levying a constant tax rate τ_k on capital and a zero consumption tax. Tell how to find steady-state levels of capital, consumption, and the rate of return on capital.

c. Let $\bar{k}_0$ be the steady value of k_t that you found in part b. Let this be the initial value of capital at time $t = 0$ and consider the following experiment. Suddenly and unexpectedly, a new party comes into power that repeals the tax on capital, sets $\tau_k = 0$ forever, and finances the same constant level of $\bar{g}$ with a flat-rate tax on consumption. Tell what happens to the new steady-state values of capital, consumption, and the return on capital.

d. Someone recommends comparing the two alternative policies of (1) relying completely on the taxation of capital as in the initial equilibrium and (2) relying completely on the consumption tax, as in our second equilibrium, by comparing the discounted utilities of consumption in steady state, i.e., by comparing $\frac{1}{1-\beta} u(\bar{c})$ in the two equilibria, where $\bar{c}$ is the steady-state value of consumption. Is this a good way to measure the costs or gains of one policy *vis-a-vis* the other?

Exercise 11.3   **Anticipated productivity shift**

An infinitely lived representative household has preferences over a stream of consumption of a single good that are ordered by

$$\sum_{t=0}^{\infty} \beta^t u(c_t), \quad \beta \in (0, 1)$$

where u is a strictly concave, twice continuously differentiable, one-period utility function, β is a discount factor, and c_t is time t consumption. The technology is:

$$c_t + x_t \leq f(k_t)n_t$$

$$k_{t+1} = (1 - \delta)k_t + \psi_t x_t$$

where for $t \geq 1$

$$\psi_t = \begin{cases} 1 & \text{for } t < 4 \\ 2 & \text{for } t \geq 4. \end{cases}$$

Here $f(k_t)n_t$ is output, where $f > 0, f' > 0, f'' < 0$, k_t is capital per unit of labor input, and n_t is labor input. The household supplies one unit of labor inelastically. The initial capital stock k_0 is given and is owned by the representative household. In particular, assume that k_0 is at the optimal steady value for k presuming that ψ_t had been equal to 1 forever. There is no uncertainty. There is no government.

a. Formulate the planning problem for this economy in the space of sequences and form the pertinent Lagrangian. Find a formula for the optimal steady-state level of capital. How does a permanent increase in ψ affect the steady values of k, c, and x?

b. Formulate the planning problem for this economy recursively (i.e., compose a Bellman equation for the planner). Be careful to give a complete description of the state vector and its law of motion. ("Finding the state is an art.")

c. Formulate an (Arrow-Debreu) competitive equilibrium with time 0 trades, assuming the following decentralization. Let the household own the stocks of capital and labor and in each period let the household rent them to the firm. Let the household choose the investment rate each period. Define an appropriate price system and compute the first-order necessary conditions for the household and for the firm.

d. What is the connection between a solution of the planning problem and the competitive equilibrium in part c? Please link the prices in part c to corresponding objects in the planning problem.

e. Assume that k_0 is given by the steady-state value that corresponds to the assumption that ψ_t had been equal to 1 forever, and had been expected to remain equal to 1 forever. Qualitatively describe the evolution of the economy from time 0 on. Does the jump in ψ at $t = 4$ have any effects that precede it?

Exercise 11.4　**A capital levy**

A nonstochastic economy produces one good that can be allocated among consumption, c_t, government purchases, g_t, and gross investment, x_t. The economy-wide resource constraints are

$$c_t + g_t + x_t \leq f(k_t)$$
$$k_{t+1} = (1 - \delta)k_t + x_t,$$

where $\delta \in (0, 1)$ is a depreciation rate, k_t is the capital stock, and $f(k_t)$ gives production as a function of capital, where $f(k) = Ak^\alpha$ with $\alpha \in (0, 1)$. A single representative consumer owns the capital stock and one unit of labor. The consumer rents capital and labor to a competitive firm each period. The consumer ranks consumption plans according to

$$\sum_{t=0}^{\infty} \beta^t u(c_t)$$

where $u(c) = \frac{c^{1-\gamma}}{1-\gamma}$, with $\gamma \geq 1$. The household supplies one unit of labor inelastically each period.

The government has only one tax at its disposal, a one-time capital levy in which it confiscates part of the capital stock from the private sector. When the government imposes a capital levy, we assume that it sends the consumer a tax bill for a fraction ϕ of the beginning of period capital stock. Below, you will be asked to compare consequences of levying this tax either at the beginning of time $T = 0$ or at the beginning of time $T = 10$. The fraction can exceed 1 if that is necessary to finance the government budget. The government is allowed to impose no other taxes. In particular, it cannot impose a direct lump sum or 'head' tax .

a. Define a competitive equilibrium with time 0 trading.

b. Suppose that before time 0 the economy had been in a steady state in which g had always been zero and had been expected always to equal zero. Find a formula for the initial steady state capital stock in a competitive equilibrium with time zero trading. Let this value be $\overline{k}_0$.

c. At time 0, everyone suddenly wakes up to discover that from time 0 on, government expenditures will be $\overline{g} > 0$, where $\overline{g} + \delta \overline{k}_0 < f(\overline{k}_0)$, which implies that the new level of government expenditures would be feasible in the old steady state. Suppose that the government finances the new path of expenditures by a capital levy at time $T = 0$. The government imposes a capital levy by sending the household a bill for a fraction of the value of its capital at the time indicated. Find the new steady state value of the capital stock in a competitive equilibrium. Describe an algorithm to compute the fraction of the capital stock that the government must tax away at time 0 to finance its budget. Find the new steady state value of the capital stock in a competitive equilibrium. Describe the time paths of capital, consumption, and the interest rate from $t = 0$ to $t = +\infty$ in the new equilibrium and compare them with their counterparts in the initial $g_t \equiv 0$ equilibrium.

d. Assume the same new path of government expenditures indicated in part **c**, but now assume that the government imposes the one-time capital levy at time $T = 10$, and that this is foreseen at time 0. Find the new steady state value of the capital stock in a competitive equilibrium that is associated with this tax policy. Describe an algorithm to compute the fraction of the capital stock that the government must tax away at time $T = 10$ to finance its budget. Describe the time paths of capital, consumption, and the interest rate in this new equilibrium and compare them with their counterparts in part **b** and in the initial $g_t \equiv 0$ equilibrium.

e. Define a competitive equilibrium with sequential trading of one-period Arrow securities. Describe how to compute such an equilibrium. Describe the time path of the consumer's holdings of one-period securities in a competitive equilibrium with one period Arrow securities under the government tax policy assumed in part **d**. Describe the time path of government debt.

Exercise 11.5

A representative consumer has preferences ordered by

$$\sum_{t=0}^{\infty} \beta^t \log(c_t) \quad , \quad 0 < \beta < 1$$

where $\beta = \frac{1}{1+\rho}, \rho > 0$, and c_t is the consumption per worker. The technology is

$$y_t = f(k_t) = z k_t^{\alpha} \quad , \quad 0 < \alpha < 1 \quad , \quad z > 0$$

where y_t is the output per unit labor and k_t is capital per unit labor

$$y_t = c_t + x_t + g_t$$
$$k_{t+1} = (1 - \delta)k_t + x_t \quad , \quad 0 < \delta < 1$$

and x_t is gross investment per unit of labor and g_t is government expenditures per unit of labor. Assume a competitive equilibrium with a price system $\{q_t, r_t, w_t\}_{t=0}^{\infty}$ and a government policy $\{g_t, \tau_{ht}\}_{t=0}^{\infty}$

Assume that the government finances its expenditures by levying lump sum taxes. There are no distorting taxes. Assume that at time 0, the economy starts out with a capital per unit of labor k_0 that equals the steady state value appropriate for an economy in which g_t had been zero forever.

a. Find a formula for the steady state capital stock when $g_t = 0 \ \forall t$.

b. Compare the steady state capital labor ratio $\bar{k}$ in the competitive equilibrium with the capital labor ratio $\tilde{k}$ that maximizes steady state consumption per capita, i.e , $\tilde{k}$ solves

$$\tilde{c} = \max_k \big(f(k) - \delta k \big)$$

Is $\tilde{k}$ greater than or less that $\bar{k}$? If they differ, why ? Is $\tilde{c}$ greater or less than $\bar{c} = f(\bar{k}) - \delta \bar{k}$? Explain why.

c. Now assume that at time 0, g_t suddenly jumps to the value $g = \frac{1}{2}\bar{c}$ where $\bar{c}$ is the value of consumption per capita in the initial steady state in which g was zero forever. Starting from $k_0 = \bar{k}$ for the old $g = 0$ steady state, find the time paths of $\{c_t, k_{t+1}\}_{t=0}^{\infty}$ associated with the new path $g_t = g > 0 \ \forall t$ for government expenditures per capita. Also show the time path for $\bar{R}_{t+1} \equiv (1 - \delta) + f'(k_{t+1})$ Explain why the new paths are as they are.

Exercise 11.6   **Trade and growth**

Part I

Consider the problem of the planner in a small economy. When the economy is <u>closed to international trade</u>, the planner chooses $\{c_t, k_{t+1}\}_{t=0}^{\infty}$ to maximize

$$\sum_{t=0}^{\infty} \beta^t u(c_t),$$

where $0 < \beta < 1, \beta = \frac{1}{1+\rho}, \rho > 0$ subject to

$$c_t + k_{t+1} = f(k_t) + (1 - \delta)k_t, \quad \delta \in (0, 1)$$

where $u(c_t) = \frac{c_t^{1-\gamma}}{1-\gamma}$, $\gamma > 0$, $f(k_t) = zk_t^{\alpha}$, and $0 < \alpha < 1$.

Let $\bar{k}$ be the steady state value of k_t under the optimal plan.

a. Find a formula for $\bar{k}$.

b. Assume that $k_0 < \bar{k}$. Describe time paths for $\{c_t, k_{t+1}\}_{t=0}^{\infty}$ and $\bar{R}_{t+1} = (1 - \delta) + f'(k_{t+1})$.

c. What is the steady state value of $\bar{R}_{t+1}$?

d. Is $\bar{R}_{t+1}$ less or greater than its steady state value when $k_{t+1} < \bar{k}$?

Part II

Now assume that the economy is <u>open to international trade</u> in capital and financial assets. Assume that there is a fixed world gross rate of return $R = \beta^{-1}$ at which the planner can borrow or lend, what is often called a 'small open economy' assumption. The planner can use the proceeds of borrowing to purchase goods on the international market. These goods can be used to augment capital or to consume.

Let $\bar{k}_0$ be the level of initial capital $(\bar{k}_0 < \bar{k})$ just before the country opens up to trade just before time 0. Let $\bar{k}_0$ be the same initial capital per capita $\bar{k}_0 < \bar{k}$ studied in parts **a-d**.

At time $t = -1$, after $\bar{k}_0$ was set, trade opens up. At time $t = -1$, the planner can issue IOU's or bonds in amount B_{-1} and use the proceeds to purchase capital, thereby setting

$$k_0 = \bar{k}_0 + B_{-1}$$

where B_{-1} is denominated in time -1 consumption goods. The bonds are one period in duration and bear the constant world gross interest rate $R = \beta^{-1}$. For $t \geq 0$, the planner faces the constraints

$$c_t + k_{t+1} + RB_{t-1} = f(k_t) + (1 - \delta)k_t + B_t$$

Here RB_{t-1} is the interest and the principal on the bonds issued at $t - 1$ and B_t is the amount of one-period bonds issued at $t - 1$.

The planning problem in the small open economy is now to choose $\{c_t, k_{t+1}, B_t\}_{t=0}^{\infty}$ and B_{-1}, subject to $\bar{k}_0$ given. (Notice that k_0 is a choice variable and that $\bar{k}_0$ is an initial condition.) Please solve the planning problem in the small open economy and compare the solution to the solution in the closed economy. Is welfare $\sum_{t=0}^{\infty} \beta^t u(c_t)$ higher in the "open" or "closed" economy.

The next several problems assume the following environment. A representative consumer has preferences ordered by

$$\sum_{t=0}^{\infty} \beta^t u(c_t), \qquad 0 < \beta < 1, \qquad \beta = \frac{1}{1 + \rho}, \qquad \rho > 0$$

where c_t is the consumption per worker and where

$$u(c) = \begin{cases} \frac{c^{1-\gamma}}{1-\gamma} & \text{for } \gamma > 0 \text{ and } \gamma \neq 1 \\ \log(c) & \text{if } \gamma = 1. \end{cases}$$

The technology is $y_t = f(k_t) = zk_t^\alpha, 0 < \alpha < 1, z > 0$, where y_t is the output per unit labor and k_t is capital per unit labor and

$$y_t = c_t + x_t + g_t$$
$$k_{t+1} = (1 - \delta)k_t + x_t \quad, \quad 0 < \delta < 1$$

where x_t is gross investment per unit of labor and g_t is government expenditures per unit of labor. The government finances its expenditures by levying some combination of a flat rate tax τ_{ct} on the value of consumption goods purchased at t, a flat rate tax τ_{nt} on the value of labor earnings at t, a flat rate tax τ_{kt} on earnings from capital at t and a lump sum tax of τ_{ht} in time t consumption goods per worker at time t.

Let $\{q_t, r_t, w_t\}_{t=0}^\infty$ be a price system.

Exercise 11.7

Consider an economy in which $g_t = \bar{g} > 0 \ \forall \ t \geq 0$ and in which initially the government finances all expenditures by lump sum taxes.

a. Find a formula for the steady state capital labor ratio k_t for this economy. Find formulas for the steady state level of c_t and $\bar{R}_t = [(1 - \delta) + (1 - \tau_{kt+1})f'(k_{t+1})]$

b. Now suppose that starting from $k_0 = \bar{k}$, i.e., the steady state that you computed in part **a**, the government suddenly increases the tax on earnings from capital to a constant level $\tau_k > 0$. The government adjusts lump sum taxes to keep the government budget balances. Describe competitive equilibrium time paths for $c_t, k_{t+1}, \bar{R}_t$ and their relationship to corresponding values in the old steady state that you described in part **a**.

c. Describe how the <u>shapes</u> of the paths that you found in part **b** depend on the curvature parameter γ in the utility function $u(c) = \frac{c^{1-\gamma}}{1-\gamma}$. Higher values of γ imply higher curvature and more aversion to consumption path that fluctuate. Higher values of γ imply that the consumer values smooth consumption paths even more.

d. Starting from the steady state $\bar{k}$ that you computed in part a, now consider a situation in which the government announces at time 0 that starting in period 10 the tax on earnings from capital τ_k will rise permanently to $\overline{\tau_k} > 0$. The government adjusts its lump sum taxes to balance it's budget.

i) Find the new steady state values for $k_t, c_t, \bar{R}_t$.

ii) Describe the shapes of the transition paths from the initial steady states to the new one for $k_t, c_t, \bar{R}_t$.

iii) Describe how the shapes of the transition paths depend on the curvature parameter γ in the utility function u(c).

Hint : When γ is bigger, consumers more strongly prefer smoother consumption paths. Recall the forces behind formula (11.10.16) in section 11.10.6.

Exercise 11.8   **Trade and growth, version II**

Part I

Consider the problem of the planner in a small economy. When the economy is closed to trade, the planner chooses $\{c_t, k_{t+1}\}_{t=0}^{\infty}$ to maximize

$$\sum_{t=0}^{\infty} \beta^t u(c_t), \quad 0 < \beta < 1, \quad \beta = \frac{1}{1+\rho}, \rho > 0$$

subject to

$$c_t + k_t + 1 = f(k_t) + (1 - \delta)k_t, \delta \in (0, 1)$$

where

Let $\bar{k}$ be the steady state value of k_t under the optimal plan.

a. Find a formula for $\bar{k}$.

b. Assume that $k_0 > \bar{k}$. Describe time paths for $\{c_t, k_{t+1}\}_{t=0}^{\infty}$ and $\bar{R}_{t+1} = (1 - \delta) + f'(k_{t+1})$.

c. What is the steady state value of $\bar{R}_{t+1}$?

d. Is $\bar{R}_{t+1}$ less or greater than its steady state value when $k_{t+1} > \bar{k}$?

Part II

e. Now assume that the economy is open to international trade in capital and financial assets. Assume that there is a fixed world gross rate of return $R = \beta^{-1}$

at which the planner can borrow or <u>lend</u>. In particular, the planner is free to use the following plan. The planner can sell all of its capital $\bar{k}_0$ and simply consume the interest payments. Let $\bar{k}_0$ be the level of initial capital $(\bar{k}_0 > \bar{k})$ just before the country opens up to trade just before time 0. Let $\bar{k}_0$ be the initial capital per capita $\bar{k}_0 > \bar{k}$ studied in parts **a**- **d**. At time $t = -1$, after $\bar{k}_0$ was set, trade opens up. At time $t = -1$, the planner <u>sells</u> $\bar{k}_0$ in exchange for IOU's or bonds from the rest of the world in the amount $A_{-1} = \bar{k}_0$.

The bonds A_{-1} are one-period in duration and bear the constant world gross interest rate $R = \beta^{-1}$. After the sale of $\bar{k}_0 = A_{-1}$, the planner has <u>zero</u> capital and so shuts down the technology. Instead, the planner uses the asset market to smooth consumption. The planner chooses $\{A_{t+1}, c_t\}$ to maximize

$$\sum_{t=0}^{\infty} \beta^t u(c_t), \quad 0 < \beta < 1, \quad \beta = \frac{1}{1+\rho}, \quad \rho > 0$$

subject to $c_t + \beta A_{t+1} = A_t, A_0 = \bar{k}_0$. Please find the optimal path for consumption $\{c_t\}$.

f. Compare the path of $(c_t, k_{t+1}, \bar{R}_t)$ that you computed in parts **a-d** with "no trade" with the part **e** path "with trade" in which the government "shuts down the home technology" and lives entirely from the returns on foreign assets. Can you say which path the representative consumer will prefer?

g. Now return to the economy in part **e** with $\bar{k}_0 > \bar{k}$ from part **d**. Assume that the planner is free to borrow or lend capital at the fixed gross interest of $\beta^{-1} = 1+\rho$, as before. But now assume the planner chooses the <u>optimal</u> amount of $\bar{k}_0$ to sell off and so does not necessarily sell off the entire $\bar{k}_0$ and possibly continues to operate the technology.

   i) Find the solution of the planning problem.

   ii) Explain why it is optimal not to shut down the technology. **Hint:** Starting from having shut the technology down, think of putting a small amount ϵ of capital into the technology-this earns $z\epsilon^\alpha$ and costs $\rho\epsilon$ in terms of foregone interest. Because $0 < \alpha < 1$, $\rho\epsilon < z\epsilon^\alpha$ for small ϵ. Thus, the technology is <u>very productive</u> for small ϵ, so it is efficient to use it.

Exercise 11.9

Consider a consumer who wants to choose $\{c_t\}_{t=0}^{T}$ to maximize

$$\sum_{t=0}^{T} \beta^t c_t \quad , \quad 0 < \beta < 1$$

subject to the intertemporal budget constraint

$$\sum_{t=0}^{T} R^{-t}[c_t - y_t] = 0 \quad , \quad R > 1$$

where $c_t \geq 0$ and $y_t > 0$ for $t = 0, \cdots, T$. Here $R > 1$ is the gross interest rate $(1 + r)$, $r > 0$. Assume R is constant. Here $\{y_t\}_{t=0}^{T}$ is an exogenous sequence of "labor income".

a. Assume that $\beta R < 1$. Find the optimal path $\{c_t\}_{t=0}^{T}$.

b. Assume that $\beta R > 1$. Find the optimal path $\{c_t\}_{t=0}^{T}$.

c. Assume that $\beta R = 1$. Find the optimal path $\{c_t\}_{t=0}^{T}$.

Exercise 11.10   **Term Structure of Interest Rates**

This problem assumes the following environment. A representative consumer has preferences ordered by

$$\sum_{t=0}^{\infty} \beta^t u(c_t) \quad , \quad 0 < \beta < 1 \quad , \quad \beta = \frac{1}{1 + \rho} \quad , \quad \rho > 0$$

where c_t is the consumption per worker and where

$$u(c) = \begin{cases} \frac{c^{1-\gamma}}{1-\gamma} & \text{if } \gamma > 0 \text{ and } \gamma \neq 1 \\ \log(c) & \text{if } \gamma = 1 \end{cases}$$

The technology is

$$y_t = f(k_t) = z k_t^{\alpha} \quad , \quad 0 < \alpha < 1 \quad , \quad z > 0$$

where y_t is the output per unit labor and k_t is capital per unit labor

$$y_t = c_t + x_t + g_t$$
$$k_{t+1} = (1 - \delta)k_t + x_t \quad , \quad 0 < \delta < 1$$

where x_t is gross investment per unit of labor and g_t is government expenditures per unit of labor. The government finances its expenditures by levying some combination of a flat rate tax τ_{ct} on the value of consumption goods purchased at t, a flat rate tax τ_{nt} on the value of labor earnings at t, a flat rate tax τ_{kt} on earnings from capital at t and a lump sum tax of τ_{ht} in time t consumption goods per worker at time t. Define $\bar{R}_{t+1} = \frac{(1+\tau_{ct})}{(1+\tau_{ct+1})} \left[1 + (1 - \tau_{kt+1})(f'(k_{t+1} - \delta)) \right]$.

a. Recall that we can represent

$$q_t^0 = q_0^0 m_{0,1} m_{1,2} \cdots m_{t-1,t}$$

where $m_{t-1,t} = \frac{q_t^0}{q_{t-1}^0}$ and $m_{t-1,t} \equiv \exp(-r_{t-1,t}) \approx \frac{1}{1+r_{t-1,t}}$. Further, recall that the t period long yield satisfies $q_t^0 = \exp(-tr_{0,t})$ and $r_{0,t} = \frac{1}{t}[r_{0,1} + r_{1,2} + \cdots + r_{t-1,t}]$. Now suppose that at $t = 0$, $k_0 = \bar{k}$, where $\bar{k}$ is the steady state appropriate for an economy with constant $g_t = \bar{g} > 0$ and all expenditures are financed by lump sum taxes. Find q_t^0 for this economy.

b. Plot $r_{t-1,t}$ for this economy for $t = 1, 2, \cdots, 10$.

c. Plot $r_{0,t}$ for this economy for $t = 1, 2, \cdots, 10$ (this is what Bloomberg plots).

d. Now assume that at time 0, starting from $k_0 = \bar{k}$ for the steady state you computed in part a, the government unexpectedly and permanently raises the tax rate on income from capital $\tau_{kt} = \tau_k > 0$ to a positive rate.

    i) Plot r_{t-1t} for this economy for $t = 1, 2, \cdots, 10$. Explain how you got this outcome.

    ii) Plot $r_{0,t}$ for this economy for $t = 1, 2, \cdots, 10$. Explain how you got this outcome.

Exercise 11.11

This problem assumes the same economic environment as the previous exercise i.e, the "growth model" with fiscal policy. Suppose that you observe the path for consumption per capita in figure 11.1. Say what you can about the likely behavior over time of k_t, $\bar{R}_t = [1 + (1 - \tau_{kt})(f'(k_t) - \delta)]$, g_t and τ_{kt}. (You are free to make up any story that is consistent with the model.)

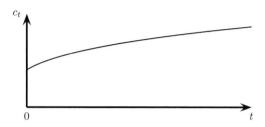

Figure 11.1: Consumption per capita.

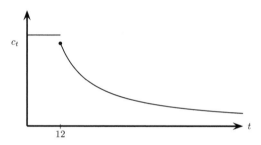

Figure 11.2: Consumption per capita.

Exercise 11.12

Assume the same economic environment as in the previous two problems. Assume that someone has observed the time path for c_t in figure 11.2:

a. Describe a consistent set of assumptions about the fiscal policy that explains this time path for c_t. In doing so, please distinguish carefully between changes in taxes and expenditures that are <u>foreseen</u> versus <u>unforeseen</u>.

b. Describe what is happening to k_t, $\bar{R}_t$ and g_t over time. Make whatever assumptions you must to get a complete but consistent story - here "consistent" means "consistent with the economic environment we have assumed".

Exercise 11.13

Consider the optimal growth model with a representative consumer with preferences

$$\sum_{t=0}^{\infty} \beta^t c_t, \ \ 0 < \beta < 1$$

with technology

$$c_t + k_{t+1} = f(k_t) + (1 - \delta)k_t, \quad \delta \in (0, 1)$$

$$c_t \geq 0, \quad k_0 > 0 \text{ given}$$

$$f' > 0, \quad f'' < 0, \quad \lim_{k \to 0} f'(k) = +\infty, \quad \lim_{k \to +\infty} f'(k) = 0$$

Let $\bar{k}$ be the steady state capital stock for the optimal planning problem.

a. For k_0 given, formulate and solve the optimal planning problem.

b. For $k_0 > \bar{k}$, describe the optimal time path of $\{c_t, k_{t+1}\}_{t=0}^{\infty}$.

c. For $k_0 < \bar{k}$, describe the optimal path of $\{c_t, k_{t+1}\}_{t=0}^{\infty}$.

d. Let the <u>saving rate</u> s_t be defined as the s_t that satisfies

$$k_{t+1} = s_t f(k_t) + (1 - \delta)k_t.$$

Here s_t in general varies along a $\{c_t, k_{t+1}\}$ sequence. Say what you can about how s_t varies as a function of k_t.

Exercise 11.14

A representative consumer has preferences over consumption streams ordered by $\sum_{t=0}^{\infty} \beta^t u(c)$, $\quad 0 < \beta < 1$, where $\beta \equiv \frac{1}{1+\rho}$, c_t is consumption per worker, $\gamma > 0$, and

$$u(c) = \begin{cases} \frac{c^{1-\gamma}}{1-\gamma} & \text{if } \gamma > 0 \text{ and } \gamma \neq 1 \\ \log(c) & \text{if } \gamma = 1. \end{cases}$$

The consumer supplies one unit of labor inelastically. The technology is

$$y_t = f(k_t) = z k_t^{\alpha}, \quad 0 < \alpha < 1$$

where y_t is output per worker, k_t is capital per worker, x_t is gross investment per worker, g_t = government expenditures per worker, and

$$y_t = c_t + x_t + g_t$$

$$k_{t+1} = (1 - \delta)k_t + x_t, \quad 0 < \delta < 1.$$

The government finances its expenditure stream $\{g_t\}$ by levying a stream of flat rate taxes $\{\tau_{ct}\}$ on the value of the consumption good purchased at t, a stream of flat rate taxes $\{\tau_{kt}\}$ on earnings from capital at t, and a stream of lump sum taxes $\{\tau_{ht}\}$. Let $\{q_t, q_t \eta_t, q_t w_t\}_{t=0}^{\infty}$ be a *price system*, where q_t is the price of time t consumption and investment goods, $q_t \eta_t$ is the price of renting capital at time t, and $q_t w_t$ is the price of renting labor at time t. All trades occur at time 0 and all prices are measured in units of the time 0 consumption good. The initial capital stock k_0 is given.

a. Define a competitive equilibrium with taxes and government purchases.

b. Assume that $g_t = 0$ for all $t \geq 0$ and that all taxes are also zero. Find the value $\bar{k}$ of the steady state capital per worker. Find a formula for the saving rate $\frac{x_t}{f(k_t)}$ at the steady state value of the capital stock.

c. Suppose that the initial capital stock $k_0 = .5\bar{k}$, so that the economy starts below its steady state level. Describe (i.e., draw graphs showing) the time paths of $\{c_t, k_{t+1}, \bar{R}_{t+1}\}_{t=0}^{\infty}$ where $\bar{R}_{t+1} \equiv [(1 - \delta) + f'(k_{t+1})]$.

d. Starting from the same initial k_0 as in part **c**, assume now that $g_t = \bar{g} = \phi f(k_0) > 0$ for all $t \geq 0$ where $\phi \in (0, 1 - \delta)$. Assume that the government finances its purchases by imposing lump sum taxes. Describe (i.e., draw graphs) showing the time paths of $\{c_t, k_{t+1}, \bar{R}_{t+1}\}_{t=0}^{\infty}$ and compare them to the outcomes that you obtained in part **c**. What outcomes differ? What outcomes, if any, are identical across the two economies? Please explain.

e. Starting from the same initial k_0 assumed in part **c**, assume now that $g_t = \bar{g} = \phi f(k_0) > 0$ for all $t \geq 0$ where $\phi \in (0, 1 - \delta)$. Assume that the government must now finance these purchases by imposing a time-invariant tax rate $\bar{\tau}_k$ on capital each period. The government *cannot* impose lump sum taxes or any other kind of taxes to balance its budget. Please describe how to find a competitive equilibrium.

Exercise 11.15

The structure of the economy is identical to that described in the previous exercise. Let $r_{0,t}$ be the yield to maturity on a t period bond at time 0, $t = 1, 2, \ldots,$. At time 0, Bloomberg reports the term structure of interest rates in figure 11.3. Please say what you can about the evolution of $\{c_t, k_{t+1}\}$ in this economy. Feel free to make any assumptions you need about fiscal policy $\{g_t, \tau_{kt}, \tau_{ct}, \tau_{ht}\}_{t=0}^{\infty}$ to make your answer coherent.

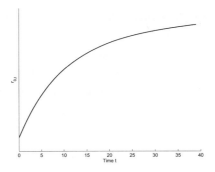

Figure 11.3: Yield to maturity $r_{0,t}$ at time 0 as a function of term to maturity t.

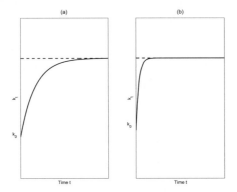

Figure 11.4: Capital stock as function of time in two economies with different values of γ.

Exercise 11.16

The structure of the economy is identical to that described in exercise 11.14. Assume that $\{g_t, \tau_{ct}, \tau_{kt}\}_{t=0}^{\infty}$ are all constant sequences (their values don't change over time). In this problem, we ask you to infer differences across two economies in which all aspects of the economy are identical *except* the parameter γ in the utility function.[30] In both economies, $\gamma > 0$. In one economy, $\gamma > 0$ is high

---

[30] It is possible that lump sum taxes differ across the two economies. Assume that lump sum taxes are adjusted to balance the government budget.

and in the other it is low. Among other identical features, the two economies have identical government policies and identical initial capital stocks.

a. Please look at figure 11.4. Please tell which outcome for $\{k_{t+1}\}_{t=0}^{\infty}$ describes the low γ economy, and which describes the high γ economy. Please explain your reasoning.

b. Please look at figure 11.5. Please tell which outcome for $\{f'(k_t)\}_{t=0}^{\infty}$ describes the low γ economy, and which describes the high γ economy. Please explain your reasoning.

c. Please plot time paths of consumption for the low γ and the high γ economies.

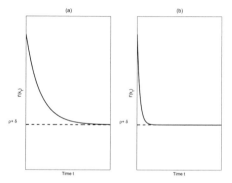

Figure 11.5: Marginal product of capital as function of time in two economies with different values of γ.

Exercise 11.17

The structure of the economy is identical to that described in exercise 11.16. Assume that $\{\tau_{ct}, \tau_{kt}\}_{t=0}^{\infty}$ are constant sequences (their values don't change over time) but that $\{g_t\}_{t=0}^{\infty}$ follows the path described in panel **c** of figure 11.6 – it takes a once and for all jump at time $t = 10$. Lump sum taxes adjust to balance the government budget. Panels **a** and **b** give consumption paths for two economies that are identical except in one respect. In one of the economies, the time 10 jump in g had been anticipated since time 0, while in the other, the jump in g that occurs at time 10 is completely unanticipated at time 10.

Please tell which panel corresponds to which view of the arrival of news about the path of g_t. Please say as much as you can about how $\{k_{t+1}\}_{t=0}^{\infty}$ and the interest rate behave in these two economies.

Exercise 11.18

A planner chooses sequences $\{c_t, k_{t+1}\}_{t=0}^{\infty}$ to maximize

$$\sum_{t=0}^{\infty} \beta^t u(c_t - \alpha c_{t-1}), \quad \alpha \in (-1, 1), \quad \beta \in (0, 1)$$

subject to

$$k_{t+1} + c_t = f(k_t) + (1 - \delta)k_t, \quad \delta \in (0, 1), \quad c_t \geq 0$$

where

$$u(x) = \begin{cases} \frac{x^{1-\gamma}}{1-\gamma} & \text{if } \gamma > 0 \text{ and } \gamma \neq 1 \\ \log(x) & \text{if } \gamma = 1 \end{cases}$$

and (k_0, c_{-1}) are given initial conditions. Here $u' > 0, u'' < 0$, and $\lim_{c \downarrow 0} u'(c) = 0$. If $\alpha > 0$, it indicates that the consumer has a 'habit'; if $\alpha < 0$, it indicates that the consumption good is somewhat durable.

a. Find a complete set of first-order necessary conditions for the planner's problem.

b. Define an optimal *steady state*.

c. Find optimal steady state values for $(k, f'(k), c)$.

d. For given initial conditions (k_0, c_{-1}), describe as completely as you can an algorithm for computing a path $\{c_t, k_{t+1}\}_{t=0}^{\infty}$ that solves the planning problem.

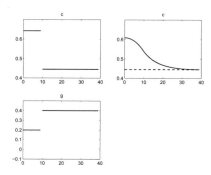

Figure 11.6: Panels **a** and **b**: consumption c_t as function of time in two economies. Panel **c**: government expenditures g_t as a function of time.

Chapter 12
Recursive Competitive Equilibria

12.1. Endogenous aggregate state variable

For pure endowment stochastic economies, chapter 8 described two types of competitive equilibria, one in the style of Arrow and Debreu with markets that convene at time 0 and trade a complete set of history-contingent securities, another with markets that meet each period and trade a complete set of one-period-ahead state-contingent securities called Arrow securities. Though their price systems and trading protocols differ, both types of equilibria support identical equilibrium allocations. Chapter 8 described how to transform the Arrow-Debreu price system into one for pricing Arrow securities. The key step in transforming an equilibrium with time 0 trading into one with sequential trading was to account for how individuals' wealth evolve as time passes in a time 0 trading economy. In a time 0 trading economy, individuals do not make any trades other than those executed in period 0, but the present value of those portfolios change as time passes and as uncertainty gets resolved. So in period t after some history s^t, we used the Arrow-Debreu prices to compute the value of an individual's purchased claims to current and future goods net of his outstanding liabilities. We could then show that these wealth levels (and the associated consumption choices) could also be attained in a sequential-trading economy where there are only markets in one-period Arrow securities that reopen in each period.

In chapter 8 we also demonstrated how to obtain a recursive formulation of the equilibrium with sequential trading. This required us to assume that individuals' endowments were governed by a Markov process. Under that assumption we could identify a state vector in terms of which the Arrow securities could be cast. This (aggregate) state vector then became a component of the state vector for each individual's problem. This transformation of price systems is easy in the pure exchange economies of chapter 8 because in equilibrium, the relevant state variable, wealth, is a function solely of the current realization of the exogenous Markov state variable. The transformation is more subtle in economies in which part of the aggregate state is endogenous in the sense that it emerges from the *history* of equilibrium interactions of agents' decisions. In this chapter, we use the basic stochastic growth model (sometimes also called the real business cycle model) as a laboratory for moving from an equilibrium with time 0 trading to a sequential equilibrium with trades of Arrow securities.[1] We also formulate a recursive competitive equilibrium with trading in Arrow securities by using a version of the "Big K, little k" trick that is often used in macroeconomics.

12.2. The stochastic growth model

Here we spell out the basic ingredients of the stochastic growth model: preferences, endowment, technology, and information. The environment is the same as in chapter 11 except that we now allow for a stochastic technology level. In each period $t \geq 0$, there is a realization of a stochastic event $s_t \in S$. Let the history of events up to time t be denoted $s^t = [s_t, s_{t-1}, \ldots, s_0]$. The unconditional probability of observing a particular sequence of events s^t is given by a probability measure $\pi_t(s^t)$. We write conditional probabilities as $\pi_\tau(s^\tau | s^t)$, which is the probability of observing s^τ conditional on the realization of s^t. In this chapter, we assume that the state s_0 in period 0 is nonstochastic, and hence $\pi_0(s_0) = 1$ for a particular $s_0 \in \mathbf{S}$. We use s^t as a commodity space in which goods are differentiated by histories.

A representative household has preferences over nonnegative streams of consumption $c_t(s^t)$ and leisure $\ell_t(s^t)$ that are ordered by

$$\sum_{t=0}^{\infty} \sum_{s^t} \beta^t u[c_t(s^t), \ell_t(s^t)] \pi_t(s^t) \tag{12.2.1}$$

where $\beta \in (0,1)$ and u is strictly increasing in its two arguments, twice continuously differentiable, strictly concave, and satisfies the Inada conditions

$$\lim_{c \to 0} u_c(c, \ell) = \lim_{\ell \to 0} u_\ell(c, \ell) = \infty.$$

---

[1] The stochastic growth model was formulated and fully analyzed by Brock and Mirman (1972). It is a workhorse for studying macroeconomic fluctuations. Kydland and Prescott (1982) used the framework to study quantitatively the importance of persistent technology shocks for business cycle fluctuations. Other researchers have used the stochastic growth model as a point of departure when exploring the implications of injecting various frictions into that otherwise frictionless environment.

In each period, the representative household is endowed with one unit of time that can be devoted to leisure $\ell_t(s^t)$ or labor $n_t(s^t)$:

$$1 = \ell_t(s^t) + n_t(s^t). \tag{12.2.2}$$

The only other endowment is a capital stock k_0 at the beginning of period 0.

The technology is

$$c_t(s^t) + x_t(s^t) \leq A_t(s^t) F(k_t(s^{t-1}), n_t(s^t)), \tag{12.2.3a}$$

$$k_{t+1}(s^t) = (1 - \delta)k_t(s^{t-1}) + x_t(s^t), \tag{12.2.3b}$$

where F is a twice continuously differentiable, constant-returns-to-scale production function with inputs capital $k_t(s^{t-1})$ and labor $n_t(s^t)$, and $A_t(s^t)$ is a stochastic process of Harrod-neutral technology shocks. Outputs are the consumption good $c_t(s^t)$ and the investment good $x_t(s^t)$. In (12.2.3b), the investment good augments a capital stock that is depreciating at the rate δ. Negative values of $x_t(s^t)$ are permissible, which means that the capital stock can be reconverted into the consumption good.

We assume that the production function satisfies standard assumptions of positive but diminishing marginal products,

$$F_i(k, n) > 0, \quad F_{ii}(k, n) < 0, \quad \text{for } i = k, n;$$

and the Inada conditions,

$$\lim_{k \to 0} F_k(k, n) = \lim_{n \to 0} F_n(k, n) = \infty,$$

$$\lim_{k \to \infty} F_k(k, n) = \lim_{n \to \infty} F_n(k, n) = 0.$$

Since the production function has constant returns to scale, we can define

$$F(k, n) \equiv n f(\hat{k}) \quad \text{where} \quad \hat{k} \equiv \frac{k}{n}. \tag{12.2.4}$$

Another property of a linearly homogeneous function $F(k, n)$ is that its first derivatives are homogeneous of degree 0, and thus the first derivatives are functions only of the ratio $\hat{k}$. In particular, we have

$$F_k(k, n) = \frac{\partial \, n f \, (k/n)}{\partial \, k} = f'(\hat{k}), \tag{12.2.5a}$$

$$F_n(k, n) = \frac{\partial \, n f \, (k/n)}{\partial \, n} = f(\hat{k}) - f'(\hat{k})\hat{k}. \tag{12.2.5b}$$

12.3. Lagrangian formulation of the planning problem

The planner chooses an allocation $\{c_t(s^t), \ell_t(s^t), x_t(s^t), n_t(s^t), k_{t+1}(s^t)\}_{t=0}^{\infty}$ to maximize (12.2.1) subject to (12.2.2) and (12.2.3), the initial capital stock k_0, and the stochastic process for the technology level $A_t(s^t)$. To solve this planning problem, we form the Lagrangian

$$
L = \sum_{t=0}^{\infty}\sum_{s^t}\beta^t \pi_t(s^t)\{u(c_t(s^t), 1 - n_t(s^t))
$$
$$
+ \mu_t(s^t)[A_t(s^t)F(k_t(s^{t-1}), n_t(s^t)) + (1 - \delta)k_t(s^{t-1}) - c_t(s^t) - k_{t+1}(s^t)]\}
$$

where $\mu_t(s^t)$ is a process of Lagrange multipliers on the technology constraint. First-order conditions with respect to $c_t(s^t)$, $n_t(s^t)$, and $k_{t+1}(s^t)$, respectively, are

$$
u_c\left(s^t\right) = \mu_t(s^t), \tag{12.3.1a}
$$
$$
u_\ell\left(s^t\right) = u_c\left(s^t\right)A_t(s^t)F_n\left(s^t\right), \tag{12.3.1b}
$$
$$
u_c\left(s^t\right)\pi_t(s^t) = \beta \sum_{s^{t+1}|s^t} u_c\left(s^{t+1}\right)\pi_{t+1}\left(s^{t+1}\right)
$$
$$
\left[A_{t+1}\left(s^{t+1}\right)F_k\left(s^{t+1}\right) + (1 - \delta)\right], \tag{12.3.1c}
$$

where the summation over $s^{t+1}|s^t$ means that we sum over all possible histories $\tilde{s}^{t+1}$ such that $\tilde{s}^t = s^t$.

12.4. Time 0 trading: Arrow-Debreu securities

In the style of Arrow and Debreu, we can support the allocation that solves the planning problem by a competitive equilibrium with time 0 trading of a complete set of date- and history-contingent securities. Trades occur among a representative household and two types of representative firms.[2]

We let $[q^0, w^0, r^0, p_{k0}]$ be a price system where p_{k0} is the price of a unit of the initial capital stock, and each of q^0, w^0, and r^0 is a stochastic process of

---

[2]   One can also support the allocation that solves the planning problem with a less decentralized setting with only the first of our two types of firms, and in which the decision for making physical investments is assigned to the household. We assign that decision to a second type of firm because we want to price more items, in particular, the capital stock.

prices for output and for renting labor and capital, respectively, and the time t component of each is indexed by the history s^t. A representative household purchases consumption goods from a type I firm and sells labor services to the type I firm that operates the production technology $(12.2.3a)$. The household owns the initial capital stock k_0 and at date 0 sells it to a type II firm. The type II firm operates the capital storage technology $(12.2.3b)$, purchases new investment goods x_t from a type I firm, and rents stocks of capital back to the type I firm.

We now describe the problems of the representative household and the two types of firms in the economy with time 0 trading.

12.4.1. Household

The household maximizes

$$\sum_t \sum_{s^t} \beta^t u \left[c_t(s^t), 1 - n_t(s^t) \right] \pi_t(s^t) \tag{12.4.1}$$

subject to

$$\sum_{t=0}^{\infty} \sum_{s^t} q_t^0(s^t) c_t(s^t) \leq \sum_{t=0}^{\infty} \sum_{s^t} w_t^0(s^t) n_t(s^t) + p_{k0} k_0. \tag{12.4.2}$$

First-order conditions with respect to $c_t(s^t)$ and $n_t(s^t)$, respectively, are

$$\beta^t u_c \left(s^t \right) \pi_t(s^t) = \eta q_t^0(s^t), \tag{12.4.3a}$$

$$\beta^t u_\ell \left(s^t \right) \pi_t(s^t) = \eta w_t^0(s^t), \tag{12.4.3b}$$

where $\eta > 0$ is a multiplier on the budget constraint $(12.4.2)$.

12.4.2. Firm of type I

The representative firm of type I operates the production technology $(12.2.3a)$ with capital and labor that it rents at market prices. For each period t and each realization of history s^t, the firm enters into state-contingent contracts at time 0 to rent capital $k_t^I(s^t)$ and labor services $n_t(s^t)$. The type I firm seeks to maximize

$$\sum_{t=0}^{\infty} \sum_{s^t} \left\{ q_t^0(s^t) \left[c_t(s^t) + x_t(s^t) \right] - r_t^0(s^t) k_t^I\left(s^t\right) - w_t^0(s^t) n_t(s^t) \right\} \qquad (12.4.4)$$

subject to

$$c_t(s^t) + x_t(s^t) \le A_t(s^t) F\left(k_t^I\left(s^t\right), n_t(s^t) \right). \qquad (12.4.5)$$

After substituting $(12.4.5)$ into $(12.4.4)$ and invoking $(12.2.4)$, the firm's objective function can be expressed alternatively as

$$\sum_{t=0}^{\infty} \sum_{s^t} n_t(s^t) \left\{ q_t^0(s^t) A_t(s^t) f\left(\hat{k}_t^I\left(s^t\right) \right) - r_t^0(s^t) \hat{k}_t^I\left(s^t\right) - w_t^0(s^t) \right\} \qquad (12.4.6)$$

and the maximization problem can then be decomposed into two parts. First, conditional on operating the production technology in period t and history s^t, the firm solves for the profit-maximizing capital-labor ratio, denoted $k_t^{I\star}(s^t)$. Second, given that capital-labor ratio $k_t^{I\star}(s^t)$, the firm determines the profit-maximizing level of its operation by solving for the optimal employment level, denoted $n_t^\star(s^t)$.

The firm finds the profit-maximizing capital-labor ratio by maximizing the expression in curly brackets in $(12.4.6)$. The first-order condition with respect to $\hat{k}_t^I(s^t)$ is

$$q_t^0(s^t) A_t(s^t) f'\left(\hat{k}_t^I\left(s^t\right) \right) - r_t^0(s^t) = 0. \qquad (12.4.7)$$

At the optimal capital-labor ratio $\hat{k}_t^{I\star}(s^t)$ that satisfies $(12.4.7)$, the firm evaluates the expression in curly brackets in $(12.4.6)$ in order to determine the optimal level of employment $n_t(s^t)$. In particular, $n_t(s^t)$ is optimally set equal to zero or infinity if the expression in curly brackets in $(12.4.6)$ is strictly negative or strictly positive, respectively. However, if the expression in curly brackets is zero in some period t and history s^t, the firm would be indifferent to the level of $n_t(s^t)$, since profits are then equal to zero for all levels of operation in that period and state. Here, we summarize the optimal employment decision by using equation $(12.4.7)$ to eliminate $r_t^0(s^t)$ in the expression in curly brackets in $(12.4.6)$;

$$\text{if } \left\{ q_t^0(s^t) A_t(s^t) \left[f\left(\hat{k}_t^{I\star}\left(s^t \right) \right) - f'\left(\hat{k}_t^{I\star}\left(s^t \right) \right) \hat{k}_t^{I\star}\left(s^t \right) \right] - w_t^0(s^t) \right\}$$

$$\begin{cases} < 0, & \text{then } n_t^\star\left(s^t \right) = 0; \\ = 0, & \text{then } n_t^\star\left(s^t \right) \text{ is indeterminate;} \\ > 0, & \text{then } n_t^\star\left(s^t \right) = \infty. \end{cases} \qquad (12.4.8)$$

In an equilibrium, both $k_t^I(s^t)$ and $n_t(s^t)$ are strictly positive and finite, so expressions $(12.4.7)$ and $(12.4.8)$ imply the following equilibrium prices:

$$q_t^0(s^t) A_t(s^t) F_k\left(s^t \right) = r_t^0(s^t) \qquad (12.4.9a)$$

$$q_t^0(s^t) A_t(s^t) F_n\left(s^t \right) = w_t^0(s^t) \qquad (12.4.9b)$$

where we have invoked $(12.2.5)$.

12.4.3. Firm of type II

The representative firm of type II operates technology $(12.2.3b)$ to transform output into capital. The type II firm purchases capital at time 0 from the household sector and thereafter invests in new capital, earning revenues by renting capital to the type I firm. It maximizes

$$-p_{k0} k_0^{II} + \sum_{t=0}^{\infty} \sum_{s^t} \left\{ r_t^0(s^t) k_t^{II}\left(s^{t-1} \right) - q_t^0(s^t) x_t(s^t) \right\} \qquad (12.4.10)$$

subject to

$$k_{t+1}^{II}\left(s^t \right) = (1-\delta) k_t^{II}\left(s^{t-1} \right) + x_t\left(s^t \right). \qquad (12.4.11)$$

Note that the firm's capital stock in period 0, k_0^{II}, is bought without any uncertainty about the rental price in that period while the investment in capital for a future period t, $k_t^{II}(s^{t-1})$, is conditioned on the realized history s^{t-1}. Thus, the type II firm manages the risk associated with technology constraint $(12.2.3b)$ that states that capital must be assembled one period prior to becoming an input for production. In contrast, the type I firm of the previous subsection can choose how much capital $k_t^I(s^t)$ to rent in period t conditioned on history s^t.

After substituting $(12.4.11)$ into $(12.4.10)$ and rearranging, the type II firm's objective function can be written as

$$k_0^{II}\left\{-p_{k0} + r_0^0\left(s_0\right) + q_0^0\left(s_0\right)\left(1-\delta\right)\right\} + \sum_{t=0}^{\infty}\sum_{s^t} k_{t+1}^{II}\left(s^t\right)$$

$$\cdot\left\{-q_t^0\left(s^t\right) + \sum_{s^{t+1}\mid s^t}\left[r_{t+1}^0\left(s^{t+1}\right) + q_{t+1}^0\left(s^{t+1}\right)\left(1-\delta\right)\right]\right\}, \qquad (12.4.12)$$

where the firm's profit is a linear function of investments in capital. The profit-maximizing level of the capital stock $k_{t+1}^{II}(s^t)$ in expression $(12.4.12)$ is equal to zero or infinity if the associated multiplicative term in curly brackets is strictly negative or strictly positive, respectively. However, for any expression in curly brackets in $(12.4.12)$ that is zero, the firm would be indifferent to the level of $k_{t+1}^{II}(s^t)$, since profits then equal zero for all levels of investment. In an equilibrium, k_0^{II} and $k_{t+1}^{II}(s^t)$ are strictly positive and finite, so each expression in curly brackets in $(12.4.12)$ must equal zero, and hence equilibrium prices must satisfy

$$p_{k0} = r_0^0\left(s_0\right) + q_0^0\left(s_0\right)\left(1-\delta\right), \qquad (12.4.13a)$$

$$q_t^0\left(s^t\right) = \sum_{s^{t+1}\mid s^t}\left[r_{t+1}^0\left(s^{t+1}\right) + q_{t+1}^0\left(s^{t+1}\right)\left(1-\delta\right)\right]. \qquad (12.4.13b)$$

12.4.4. Equilibrium prices and quantities

According to equilibrium conditions $(12.4.9)$, each input in the production technology is paid its marginal product, and hence profit maximization of the type I firm ensures an efficient allocation of labor services and capital. But nothing is said about the equilibrium quantities of labor and capital. Profit maximization of the type II firm imposes no-arbitrage restrictions $(12.4.13)$ across prices p_{k0} and $\{r_t^0(s^t), q_t^0(s^t)\}$. But nothing is said about the specific equilibrium value of an individual price. To solve for equilibrium prices and quantities, we turn to the representative household's first-order conditions $(12.4.3)$.

After substituting $(12.4.9b)$ into the household's first-order condition $(12.4.3b)$, we obtain

$$\beta^t u_\ell\left(s^t\right)\pi_t(s^t) = \eta q_t^0\left(s^t\right) A_t\left(s^t\right) F_n\left(s^t\right); \qquad (12.4.14a)$$

and then by substituting $(12.4.13b)$ and $(12.4.9a)$ into $(12.4.3a)$,

$$\beta^t u_c\left(s^t\right) \pi_t(s^t) = \eta \sum_{s^{t+1}|s^t} \left[r^0_{t+1}\left(s^{t+1}\right) + q^0_{t+1}\left(s^{t+1}\right)(1-\delta)\right]$$

$$= \eta \sum_{s^{t+1}|s^t} q^0_{t+1}\left(s^{t+1}\right)\left[A_{t+1}\left(s^{t+1}\right) F_k\left(s^{t+1}\right) + (1-\delta)\right]. \quad (12.4.14b)$$

Next, we use $q^0_t(s^t) = \beta^t u_c(s^t)\pi_t(s^t)/\eta$ as given by the household's first-order condition $(12.4.3a)$ and the corresponding expression for $q^0_{t+1}(s^{t+1})$ to substitute into $(12.4.14a)$ and $(12.4.14b)$, respectively. This step produces expressions identical to the planner's first-order conditions $(12.3.1b)$ and $(12.3.1c)$, respectively. In this way, we have verified that the allocation in the competitive equilibrium with time 0 trading is the same as the allocation that solves the planning problem.

Given the equivalence of allocations, it is standard to compute the competitive equilibrium allocation by solving the planning problem since the latter problem is a simpler one. We can compute equilibrium prices by substituting the allocation from the planning problem into the household's and firms' first-order conditions. All relative prices are then determined, and in order to pin down absolute prices, we would also have to pick a numeraire. Any such normalization of prices is tantamount to setting the multiplier η on the household's present value budget constraint equal to an arbitrary positive number. For example, if we set $\eta = 1$, we are measuring prices in units of marginal utility of the time 0 consumption good. Alternatively, we can set $q^0_0(s_0) = 1$ by setting $\eta = u_c(s_0)$. We can compute $q^0_t(s^t)$ from $(12.4.3a)$, $w^0_t(s^t)$ from $(12.4.3b)$, and $r^0_t(s^t)$ from $(12.4.9a)$. Finally, we can compute p_{k0} from $(12.4.13a)$ to get $p_{k0} = r^0_0(s_0) + q^0_0(s_0)(1-\delta)$.

12.4.5. Implied wealth dynamics

Even though trades are only executed at time 0 in the Arrow-Debreu market structure, we can study how the representative household's wealth evolves over time. For that purpose, after a given history s^t, we convert all prices, wages, and rental rates that are associated with current and future deliveries so that they are expressed in terms of time t, history s^t consumption goods, i.e., we change the numeraire:

$$q_\tau^t(s^\tau) \equiv \frac{q_\tau^0(s^\tau)}{q_t^0(s^t)} = \beta^{\tau-t}\frac{u_c(s^\tau)}{u_c(s^t)}\,\pi_\tau(s^\tau|s^t), \qquad (12.4.15a)$$

$$w_\tau^t(s^\tau) \equiv \frac{w_\tau^0(s^\tau)}{q_t^0(s^t)}, \qquad (12.4.15b)$$

$$r_\tau^t(s^\tau) \equiv \frac{r_\tau^0(s^\tau)}{q_t^0(s^t)}. \qquad (12.4.15c)$$

In chapter 8 we asked the question, what is the implied wealth of a household at time t after history s^t when excluding the endowment stream? Here we ask the same question except that now instead of endowments, we exclude the value of labor. For example, the household's net claim to delivery of goods in a future period $\tau \geq t$, contingent on history s^τ, is given by $[q_\tau^t(s^\tau)c_\tau(s^\tau) - w_\tau^t(s^\tau)n_\tau(s^\tau)]$, as expressed in terms of time t, history s^t consumption goods. Thus, the household's wealth, or the value of all its current and future net claims, expressed in terms of the date t, history s^t consumption good, is

$$\Upsilon_t(s^t) \equiv \sum_{\tau=t}^{\infty} \sum_{s^\tau|s^t} \left\{ q_\tau^t(s^\tau)c_\tau(s^\tau) - w_\tau^t(s^\tau)n_\tau(s^\tau) \right\}$$

$$= \sum_{\tau=t}^{\infty} \sum_{s^\tau|s^t} \left\{ q_\tau^t(s^\tau)\left[A_\tau(s^\tau)F(k_\tau(s^{\tau-1}), n_\tau(s^\tau)) \right. \right.$$

$$\left. \left. + (1-\delta)k_\tau(s^{\tau-1}) - k_{\tau+1}(s^\tau) \right] - w_\tau^t(s^\tau)n_\tau(s^\tau) \right\}$$

$$= \sum_{\tau=t}^{\infty} \sum_{s^\tau|s^t} \left\{ q_\tau^t(s^\tau)\left[A_\tau(s^\tau)\left(F_k(s^\tau)k_\tau(s^{\tau-1}) + F_n(s^\tau)n_\tau(s^\tau) \right) \right. \right.$$

$$\left. \left. + (1-\delta)k_\tau(s^{\tau-1}) - k_{\tau+1}(s^\tau) \right] - w_\tau^t(s^\tau)n_\tau(s^\tau) \right\}$$

$$= \sum_{\tau=t}^{\infty} \sum_{s^\tau|s^t} \left\{ r_\tau^t(s^\tau)k_\tau(s^{\tau-1}) + q_\tau^t(s^\tau)\left[(1-\delta)k_\tau(s^{\tau-1}) - k_{\tau+1}(s^\tau) \right] \right\}$$

$$= r_t^t(s^t)k_t(s^{t-1}) + q_t^t(s^t)(1-\delta)k_t(s^{t-1})$$

$$+ \sum_{\tau=t+1}^{\infty} \sum_{s^{\tau-1}|s^t} \left\{ \sum_{s^\tau|s^{\tau-1}} \left[r_\tau^t(s^\tau) + q_\tau^t(s^\tau)(1-\delta) \right] - q_{\tau-1}^t(s^{\tau-1}) \right\} k_\tau(s^{\tau-1})$$

$$= \left[r_t^t(s^t) + (1-\delta) \right] k_t(s^{t-1}), \tag{12.4.16}$$

where the first equality uses the equilibrium outcome that consumption is equal to the difference between production and investment in each period, the second equality follows from Euler's theorem on linearly homogeneous functions,[3] the third equality invokes equilibrium input prices in (12.4.9), the fourth equality is merely a rearrangement of terms, and the final, fifth equality acknowledges that $q_t^t(s^t) = 1$ and that each term in curly brackets is zero because of equilibrium price condition (12.4.13b).

12.5. Sequential trading: Arrow securities

As in chapter 8, we now demonstrate that sequential trading in one-period Arrow securities provides an alternative market structure that preserves the allocation from the time 0 trading equilibrium. In the production economy with sequential trading, we will also have to include markets for labor and capital services that reopen in each period.

We guess that at time t after history s^t, there exist a wage rate $\tilde{w}_t(s^t)$, a rental rate $\tilde{r}_t(s^t)$, and Arrow security prices $\tilde{Q}_t(s_{t+1}|s^t)$. The *pricing kernel* $\tilde{Q}_t(s_{t+1}|s^t)$ is to be interpreted as follows: $\tilde{Q}_t(s_{t+1}|s^t)$ gives the price of one unit of time $t+1$ consumption, contingent on the realization s_{t+1} at $t+1$, when the history at t is s^t.

---

[3] According to Euler's theorem on linearly homogeneous functions, our constant-returns-to-scale production function satisfies

$$F(k,n) = F_k(k,n)\,k + F_n(k,n)\,n.$$

12.5.1. Household

At each date $t \geq 0$ after history s^t, the representative household buys consumption goods $\tilde{c}_t(s^t)$, sells labor services $\tilde{n}_t(s^t)$, and trades claims to date $t+1$ consumption, whose payment is contingent on the realization of s_{t+1}. Let $\tilde{a}_t(s^t)$ denote the claims to time t consumption that the household brings into time t in history s^t. Thus, the household faces a sequence of budget constraints for $t \geq 0$, where the time t, history s^t budget constraint is

$$\tilde{c}_t(s^t) + \sum_{s_{t+1}} \tilde{a}_{t+1}(s_{t+1}, s^t)\tilde{Q}_t(s_{t+1}|s^t) \leq \tilde{w}_t(s^t)\tilde{n}_t(s^t) + \tilde{a}_t(s^t), \qquad (12.5.1)$$

where $\{\tilde{a}_{t+1}(s_{t+1}, s^t)\}$ is a vector of claims on time $t+1$ consumption, one element of the vector for each value of the time $t+1$ realization of s_{t+1}.

To rule out Ponzi schemes, we must impose borrowing constraints on the household's asset position. We could follow the approach of chapter 8 and compute state-contingent *natural debt limits,* where the counterpart to the earlier present value of the household's endowment stream would be the present value of the household's time endowment. Alternatively, we just impose that the household's indebtedness in any state next period, $-\tilde{a}_{t+1}(s_{t+1}, s^t)$, is bounded by some arbitrarily large constant. Such an arbitrary debt limit works well for the following reason. As long as the household is constrained so that it cannot run a true Ponzi scheme with an unbounded budget constraint, equilibrium forces will ensure that the representative household willingly holds the market portfolio. In the present setting, we can for example set that arbitrary debt limit equal to zero, as will become clear as we go along.

Let $\eta_t(s^t)$ and $\nu_t(s^t; s_{t+1})$ be the nonnegative Lagrange multipliers on the budget constraint (12.5.1) and the borrowing constraint with an arbitrary debt limit of zero, respectively, for time t and history s^t. The Lagrangian can then be formed as

$$\begin{aligned}
L = \sum_{t=0}^{\infty} \sum_{s^t} \Big\{ &\beta^t u(\tilde{c}_t(s^t), 1 - \tilde{n}_t(s^t))\,\pi_t(s^t) \\
&+ \eta_t(s^t)\Big[\tilde{w}_t(s^t)\tilde{n}_t(s^t) + \tilde{a}_t(s^t) - \tilde{c}_t(s^t) - \sum_{s_{t+1}} \tilde{a}_{t+1}(s_{t+1}, s^t)\tilde{Q}_t(s_{t+1}|s^t)\Big] \\
&+ \nu_t(s^t; s_{t+1})\tilde{a}_{t+1}(s^{t+1})\Big\},
\end{aligned}$$

for a given initial wealth level $\tilde{a}_0$. In an equilibrium, the representative household will choose interior solutions for $\{\tilde{c}_t(s^t), \tilde{n}_t(s^t)\}_{t=0}^{\infty}$ because of the assumed

Inada conditions. The Inada conditions on the utility function ensure that the household will set neither $\tilde{c}_t(s^t)$ nor $\ell_t(s^t)$ equal to zero, i.e., $\tilde{n}_t(s^t) < 1$. The Inada conditions on the production function guarantee that the household will always find it desirable to supply some labor, $\tilde{n}_t(s^t) > 0$. Given these interior solutions, the first-order conditions for maximizing L with respect to $\tilde{c}_t(s^t)$, $\tilde{n}_t(s^t)$ and $\{\tilde{a}_{t+1}(s_{t+1}, s^t)\}_{s_{t+1}}$ are

$$\beta^t u_c(\tilde{c}_t(s^t), 1 - \tilde{n}_t(s^t)) \pi_t(s^t) - \eta_t(s^t) = 0, \tag{12.5.2a}$$

$$-\beta^t u_\ell(\tilde{c}_t(s^t), 1 - \tilde{n}_t(s^t)) \pi_t(s^t) + \eta_t(s^t)\tilde{w}_t(s^t) = 0, \tag{12.5.2b}$$

$$-\eta_t(s^t)\tilde{Q}_t(s_{t+1}|s^t) + \nu_t(s^t; s_{t+1}) + \eta_{t+1}(s_{t+1}, s^t) = 0, \tag{12.5.2c}$$

for all s_{t+1}, t, s^t. Next, we proceed under the conjecture that the arbitrary debt limit of zero will not be binding, and hence the Lagrange multipliers $\nu_t(s^t; s_{t+1})$ are all equal to zero. After setting those multipliers equal to zero in equation (12.5.2c), the first-order conditions imply the following conditions for the optimal choices of consumption and labor:

$$\tilde{w}_t(s^t) = \frac{u_\ell(\tilde{c}_t(s^t), 1 - \tilde{n}_t(s^t))}{u_c(\tilde{c}_t(s^t), 1 - \tilde{n}_t(s^t))}, \tag{12.5.3a}$$

$$\tilde{Q}_t(s_{t+1}|s^t) = \beta \frac{u_c(\tilde{c}_{t+1}(s^{t+1}), 1 - \tilde{n}_{t+1}(s^{t+1}))}{u_c(\tilde{c}_t(s^t), 1 - \tilde{n}_t(s^t))} \pi_t(s^{t+1}|s^t), \tag{12.5.3b}$$

for all t, s^t, and s_{t+1}.

12.5.2. Firm of type I

At each date $t \geq 0$ after history s^t, a type I firm is a production firm that chooses a quadruple $\{\tilde{c}_t(s^t), \tilde{x}_t(s^t), \tilde{k}_t^I(s^t), \tilde{n}_t(s^t)\}$ to solve a static optimum problem:

$$\max\left\{\tilde{c}_t(s^t) + \tilde{x}_t(s^t) - \tilde{r}_t(s^t)\tilde{k}_t^I(s^t) - \tilde{w}_t(s^t)\tilde{n}_t(s^t)\right\} \tag{12.5.4}$$

subject to

$$\tilde{c}_t(s^t) + \tilde{x}_t(s^t) \leq A_t(s^t)F(\tilde{k}_t^I(s^t), \tilde{n}_t(s^t)). \tag{12.5.5}$$

The zero-profit conditions are

$$\tilde{r}_t(s^t) = A_t(s^t)F_k(s^t), \tag{12.5.6a}$$

$$\tilde{w}_t(s^t) = A_t(s^t)F_n(s^t). \tag{12.5.6b}$$

If conditions (12.5.6) are violated, the type I firm either makes infinite profits by hiring infinite capital and labor, or else it makes negative profits for any positive output level, and therefore shuts down. If conditions (12.5.6) are satisfied, the firm makes zero profits and its size is indeterminate. The firm of type I is willing to produce any quantities of $\tilde{c}_t(s^t)$ and $\tilde{x}_t(s^t)$ that the market demands, provided that conditions (12.5.6) are satisfied.

12.5.3. Firm of type II

A type II firm transforms output into capital, stores capital, and earns its revenues by renting capital to the type I firm. Because of the technological assumption that capital can be converted back into the consumption good, we can without loss of generality consider a two-period optimization problem where a type II firm decides how much capital $\tilde{k}^{II}_{t+1}(s^t)$ to store at the end of period t after history s^t in order to earn a stochastic rental revenue $\tilde{r}_{t+1}(s^{t+1})\,\tilde{k}^{II}_{t+1}(s^t)$ and liquidation value $(1-\delta)\,\tilde{k}^{II}_{t+1}(s^t)$ in the following period. The firm finances itself by issuing state-contingent debt to the households, so future income streams can be expressed in today's values by using prices $\tilde{Q}_t(s_{t+1}|s^t)$. Thus, at each date $t \geq 0$ after history s^t, a type II firm chooses $\tilde{k}^{II}_{t+1}(s^t)$ to solve the optimum problem

$$\max \tilde{k}^{II}_{t+1}(s^t)\Big\{-1 + \sum_{s_{t+1}} \tilde{Q}_t(s_{t+1}|s^t)\left[\tilde{r}_{t+1}(s^{t+1}) + (1-\delta)\right]\Big\}. \qquad (12.5.7)$$

The zero-profit condition is

$$1 = \sum_{s_{t+1}} \tilde{Q}_t(s_{t+1}|s^t)\left[\tilde{r}_{t+1}(s^{t+1}) + (1-\delta)\right]. \qquad (12.5.8)$$

The size of the type II firm is indeterminate. So long as condition (12.5.8) is satisfied, the firm breaks even at any level of $\tilde{k}^{II}_{t+1}(s^t)$. If condition (12.5.8) is not satisfied, either it can earn infinite profits by setting $\tilde{k}^{II}_{t+1}(s^t)$ to be arbitrarily large (when the right side exceeds the left), or it earns negative profits for any positive level of capital (when the right side falls short of the left), and so chooses to shut down.

12.5.4. *Equilibrium prices and quantities*

We leave it to the reader to follow the approach taken in chapter 8 to show the equivalence of allocations attained in the sequential equilibrium and the time 0 equilibrium, $\{\tilde{c}_t(s^t), \tilde{\ell}_t(s^t), \tilde{x}_t(s^t), \tilde{n}_t(s^t), \tilde{k}_{t+1}(s^t)\}_{t=0}^{\infty} = \{c_t(s^t), \ell_t(s^t), x_t(s^t), n_t(s^t), k_{t+1}(s^t)\}_{t=0}^{\infty}$. The trick is to guess that the prices in the sequential equilibrium satisfy

$$\tilde{Q}_t(s_{t+1}|s^t) = q_{t+1}^t(s^{t+1}), \tag{12.5.9a}$$

$$\tilde{w}_t(s^t) = w_t^t(s^t), \tag{12.5.9b}$$

$$\tilde{r}_t(s^t) = r_t^t(s^t). \tag{12.5.9c}$$

The other set of guesses is that the representative household chooses asset portfolios given by $\tilde{a}_{t+1}(s_{t+1}, s^t) = \Upsilon_{t+1}(s^{t+1})$ for all s_{t+1}. When showing that the household can afford these asset portfolios together with the prescribed quantities of consumption and leisure, we will find that the required initial wealth is equal to

$$\tilde{a}_0 = [r_0^0(s_0) + (1-\delta)]k_0 = p_{k0}k_0,$$

i.e., the household in the sequential equilibrium starts out at the beginning of period 0 owning the initial capital stock, which is then sold to a type II firm at the same competitive price as in the time 0 trading equilibrium.

12.5.5. *Financing a type II firm*

A type II firm finances purchases of $\tilde{k}_{t+1}^{II}(s^t)$ units of capital in period t after history s^t by issuing one-period state-contingent claims that promise to pay

$$\left[\tilde{r}_{t+1}(s^{t+1}) + (1-\delta)\right]\tilde{k}_{t+1}^{II}(s^t)$$

consumption goods tomorrow in state s_{t+1}. In units of today's time t consumption good, these payouts are worth

$$\sum_{s_{t+1}} \tilde{Q}_t(s_{t+1}|s^t)\left[\tilde{r}_{t+1}(s^{t+1}) + (1-\delta)\right]\tilde{k}_{t+1}^{II}(s^t)$$

(by virtue of (12.5.8)). The firm breaks even by issuing these claims. Thus, the firm of type II is entirely owned by its creditor, the household, and it earns zero profits.

Note that the economy's end-of-period wealth as embodied in $\tilde{k}^{II}_{t+1}(s^t)$ in period t after history s^t is willingly held by the representative household. This follows immediately from fact that the household's desired beginning-of-period wealth next period is given by $\tilde{a}_{t+1}(s^{t+1})$ and is equal to $\Upsilon_{t+1}(s^{t+1})$, as given by (12.4.16). Thus, the equilibrium prices entice the representative household to enter each future period with a strictly positive net asset level that is equal to the value of the type II firm. We have then confirmed the correctness of our earlier conjecture that the arbitrary debt limit of zero is not binding in the household's optimization problem.

12.6. Recursive formulation

Following the approach taken in chapter 8, we have established that the equilibrium allocations are the same in the Arrow-Debreu economy with complete markets at time 0 and in a sequential-trading economy with complete one-period Arrow securities. This finding holds for an arbitrary technology process $A_t(s^t)$, defined as a measurable function of the history of events s^t which in turn are governed by some arbitrary probability measure $\pi_t(s^t)$. At this level of generality, all prices $\{\tilde{Q}_t(s_{t+1}|s^t),\ \tilde{w}_t(s^t),\ \tilde{r}_t(s^t)\}$ and the capital stock $k_{t+1}(s^t)$ in the sequential-trading economy depend on the history s^t. That is, these objects are time-varying functions of all past events $\{s_\tau\}^t_{\tau=0}$.

In order to obtain a recursive formulation and solution to both the social planning problem and the sequential-trading equilibrium, we make the following specialization of the exogenous forcing process for the technology level.

12.6.1. *Technology is governed by a Markov process*

Let the stochastic event s_t be governed by a Markov process, $[s \in \mathbf{S}, \pi(s'|s),$ $\pi_0(s_0)]$. We keep our earlier assumption that the state s_0 in period 0 is non-stochastic and hence $\pi_0(s_0) = 1$ for a particular $s_0 \in \mathbf{S}$. The sequences of probability measures $\pi_t(s^t)$ on histories s^t are induced by the Markov process via the recursions

$$\pi_t(s^t) = \pi(s_t|s_{t-1})\pi(s_{t-1}|s_{t-2})\ldots\pi(s_1|s_0)\pi_0(s_0).$$

Next, we assume that the aggregate technology level $A_t(s^t)$ in period t is a time-invariant measurable function of its level in the last period and the current stochastic event s_t, i.e., $A_t(s^t) = A\left(A_{t-1}(s^{t-1}), s_t\right)$. For example, here we will proceed with the multiplicative version

$$A_t(s^t) = s_t A_{t-1}(s^{t-1}) = s_0 s_1 \cdots s_t A_{-1},$$

given the initial value A_{-1}.

12.6.2. *Aggregate state of the economy*

The specialization of the technology process enables us to adapt the recursive construction of chapter 8 to incorporate additional components of the state of the economy. Besides information about the current value of the stochastic event s, we need to know last period's technology level, denoted A, in order to determine current technology level, $s A$, and to forecast future technology levels. This additional element A in the aggregate state vector does not constitute any conceptual change from what we did in chapter 8. We are merely including one more state variable that is a direct mapping from exogenous stochastic events, and it does not depend on any endogenous outcomes.

But we also need to expand the aggregate state vector with an endogenous component of the state of the economy, namely, the beginning-of-period capital stock K. Given the new state vector $X \equiv [K \, A \, s]$, we are ready to explore recursive formulations of both the planning problem and the sequential-trading equilibrium. This state vector is a complete summary of the economy's current position. It is all that is needed for a planner to compute an optimal allocation and it is all that is needed for the "invisible hand" to call out prices and implement the first-best allocation as a competitive equilibrium.

We proceed as follows. First, we display the Bellman equation associated with a recursive formulation of the planning problem. Second, we use the same state vector X for the planner's problem as a state vector in which to cast the Arrow securities in a competitive economy with sequential trading. Then we define a competitive equilibrium and show how the prices for the sequential equilibrium are embedded in the decision rules and the value function of the planning problem.

12.7. Recursive formulation of the planning problem

We use capital letters C, N, K to denote objects in the planning problem that correspond to c, n, k, respectively, in the household's and firms' problems. We shall eventually equate them, but not until we have derived an appropriate formulation of the household's and firms' problems in a recursive competitive equilibrium. The Bellman equation for the planning problem is

$$v(K, A, s) = \max_{C,N,K'} \left\{ u(C, 1 - N) + \beta \sum_{s'} \pi(s'|s)v(K', A', s') \right\} \qquad (12.7.1)$$

subject to

$$K' + C \leq AsF(K, N) + (1 - \delta)K, \qquad (12.7.2a)$$

$$A' = As. \qquad (12.7.2b)$$

Using the definition of the state vector $X = [K\ A\ s]$, we denote the optimal policy functions as

$$C = \Omega^C(X), \qquad (12.7.3a)$$

$$N = \Omega^N(X), \qquad (12.7.3b)$$

$$K' = \Omega^K(X). \qquad (12.7.3c)$$

Equations $(12.7.2b)$, $(12.7.3c)$, and the Markov transition density $\pi(s'|s)$ induce a transition density $\Pi(X'|X)$ on the state X.

For convenience, define the functions

$$U_c(X) \equiv u_c(\Omega^C(X), 1 - \Omega^N(X)), \qquad (12.7.4a)$$

$$U_\ell(X) \equiv u_\ell(\Omega^C(X), 1 - \Omega^N(X)), \qquad (12.7.4b)$$

$$F_k(X) \equiv F_k(K, \Omega^N(X)), \qquad (12.7.4c)$$

$$F_n(X) \equiv F_n(K, \Omega^N(X)). \qquad (12.7.4d)$$

The first-order conditions for the planner's problem can be represented as[4]

$$U_\ell(X) = U_c(X) A s F_n(X), \tag{12.7.5a}$$

$$1 = \beta \sum_{X'} \Pi(X'|X) \frac{U_c(X')}{U_c(X)} [A's'F_K(X') + (1-\delta)]. \tag{12.7.5b}$$

12.8. Recursive formulation of sequential trading

We seek a competitive equilibrium with sequential trading of one-period-ahead state-contingent securities (i.e., Arrow securities). To do this, we must use a "Big K, little k" trick of the type used in chapter 7.

12.8.1. A "Big K, little k" trick

Relative to the setup described in chapter 8, we have augmented the time t state of the economy by both last period's technology level A_{t-1} and the current aggregate value of the endogenous state variable K_t. We assume that decision makers act as if their decisions do not affect current or future prices. In a sequential market setting, prices depend on the state, of which K_t is part. Of course, *in the aggregate*, decision makers choose the motion of K_t, so that we require a device that makes them ignore this fact when they solve their decision problems (we want them to behave as perfectly competitive price takers, not monopolists). This consideration induces us to carry along both "Big K" and "little k" in our computations. Big K is an endogenous state variable[5] that is useful for forecasting prices. Big K is a component of the state that agents regard as beyond their control when solving their optimum problems. Values of little k are chosen by firms and consumers. While we distinguish k and K when

---

[4] We are using the envelope condition

$$v_K(K, A, s) = U_c(X)[AsF_k(X) + (1-\delta)].$$

[5] More generally, Big K can be a vector of endogenous state variables that impinge on equilibrium prices.

posing the decision problems of the household and firms, to impose equilibrium we set $K = k$ *after* firms and consumers have optimized.

12.8.2. Price system

To decentralize the economy in terms of one-period Arrow securities, we need a description of the aggregate state in terms of which one-period state-contingent payoffs are defined. We proceed by guessing that the appropriate description of the state is the same vector X that constitutes the state for the planning problem. We temporarily forget about the optimal policy functions for the planning problem and focus on a decentralized economy with sequential trading and one-period prices that depend on X. We specify *price functions* $r(X), w(X), Q(X'|X)$, that represent, respectively, the rental price of capital, the wage rate for labor, and the price of a claim to one unit of consumption next period when next period's state is X' and this period's state is X. (Forgive us for recycling the notation for r and w from the previous sections on the formulation of history-dependent competitive equilibria with commodity space s^t.) The prices are all measured in units of this period's consumption good. We also take as given an arbitrary candidate for the law of motion for K:

$$K' = G(X). \qquad (12.8.1)$$

Equation (12.8.1) together with (12.7.2b) and a given subjective transition density $\hat{\pi}(s'|s)$ induce a subjective transition density $\hat{\Pi}(X'|X)$ for the state X. For now, G and $\hat{\pi}(s'|s)$ are arbitrary. We wait until later to impose other equilibrium conditions, including rational expectations in the form of some restrictions on G and $\hat{\pi}$.

12.8.3. Household problem

The perceived law of motion $(12.8.1)$ for K and the induced transition probabilities $\hat{\Pi}(X'|X)$ describe the beliefs of a representative household. The Bellman equation of the household is

$$J(a, X) = \max_{c,n,\bar{a}(X')} \left\{ u(c, 1-n) + \beta \sum_{X'} J(\bar{a}(X'), X') \hat{\Pi}(X'|X) \right\} \qquad (12.8.2)$$

subject to

$$c + \sum_{X'} Q(X'|X) \bar{a}(X') \leq w(X) n + a. \qquad (12.8.3)$$

Here a represents the wealth of the household denominated in units of current consumption goods and $\bar{a}(X')$ represents next period's wealth denominated in units of next period's consumption good. Denote the household's optimal policy functions as

$$c = \sigma^c(a, X), \qquad (12.8.4a)$$

$$n = \sigma^n(a, X), \qquad (12.8.4b)$$

$$\bar{a}(X') = \sigma^a(a, X; X'). \qquad (12.8.4c)$$

Let

$$\bar{u}_c(a, X) \equiv u_c(\sigma^c(a, X), 1 - \sigma^n(a, X)), \qquad (12.8.5a)$$

$$\bar{u}_\ell(a, X) \equiv u_\ell(\sigma^c(a, X), 1 - \sigma^n(a, X)). \qquad (12.8.5b)$$

Then we can represent the first-order conditions for the household's problem as

$$\bar{u}_\ell(a, X) = \bar{u}_c(a, X) w(X), \qquad (12.8.6a)$$

$$Q(X'|X) = \beta \frac{\bar{u}_c(\sigma^a(a, X; X'), X')}{\bar{u}_c(a, X)} \hat{\Pi}(X'|X). \qquad (12.8.6b)$$

12.8.4. *Firm of type I*

Recall from subsection 12.5.2 the static optimum problem of a type I firm in a sequential equilibrium. In the recursive formulation of that equilibrium, the optimum problem of a type I firm can be written as

$$\max_{c,x,k,n} \{c + x - r(X)k - w(X)n\} \tag{12.8.7}$$

subject to

$$c + x \leq AsF(k,n). \tag{12.8.8}$$

The zero-profit conditions are

$$r(X) = AsF_k(k,n), \tag{12.8.9a}$$
$$w(X) = AsF_n(k,n). \tag{12.8.9b}$$

12.8.5. *Firm of type II*

Recall from subsection 12.5.3 the optimum problem of a type II firm in a sequential equilibrium. In the recursive formulation of that equilibrium, the optimum problem of a type II firm can be written as

$$\max_{k'} \; k' \left\{ -1 + \sum_{X'} Q(X'|X)\left[r(X') + (1-\delta)\right] \right\}. \tag{12.8.10}$$

The zero-profit condition is

$$1 = \sum_{X'} Q(X'|X)\left[r(X') + (1-\delta)\right]. \tag{12.8.11}$$

12.9. Recursive competitive equilibrium

So far, we have taken the price functions $r(X)$, $w(X)$, $Q(X'|X)$ and the perceived law of motion (12.8.1) for K' and the associated induced state transition probability $\hat{\Pi}(X'|X)$ as given arbitrarily. We now impose equilibrium conditions on these objects and make them outcomes of the analysis.[6]

When solving their optimum problems, the household and firms take the endogenous state variable K as given. However, we want K to be determined by the equilibrium interactions of households and firms. Therefore, we impose $K = k$ *after* solving the optimum problems of the household and the two types of firms. Imposing equality afterward makes the household and the firms be price takers.

12.9.1. Equilibrium restrictions across decision rules

We shall soon define an equilibrium as a set of pricing functions, a perceived law of motion for the K', and an associated $\hat{\Pi}(X'|X)$ such that when the firms and the household take these as given, the household's and firms' decision rules *imply* the law of motion for K (12.8.1) after substituting $k = K$ and other market clearing conditions. We shall remove the arbitrary nature of both G and $\hat{\pi}$ and therefore also $\hat{\Pi}$ and thereby impose rational expectations.

We now proceed to find the restrictions that this notion of equilibrium imposes across agents' decision rules, the pricing functions, and the perceived law of motion (12.8.1). If the state-contingent debt issued by the type II firm is to match that demanded by the household, we must have

$$\overline{a}(X') = [r(X') + (1 - \delta)]K', \qquad (12.9.1a)$$

and consequently beginning-of-period assets in a household's budget constraint (12.8.3) have to satisfy

$$a = [r(X) + (1 - \delta)]K. \qquad (12.9.1b)$$

By substituting equations (12.9.1) into a household's budget constraint (12.8.3), we get

$$\sum_{X'} Q(X'|X)[r(X') + (1 - \delta)]K'$$

---

[6] An important function of the rational expectations hypothesis is to remove agents' expectations in the form of $\hat{\pi}$ and $\hat{\Pi}$ from the list of free parameters of the model.

$$= [r(X) + (1 - \delta)]K + w(X)n - c. \tag{12.9.2}$$

Next, by recalling equilibrium condition (12.8.11) and the fact that K' is a predetermined variable when entering next period, it follows that the left side of (12.9.2) is equal to K'. After also substituting equilibrium prices (12.8.9) into the right side of (12.9.2), we obtain

$$
\begin{aligned}
K' &= \left[A s F_k(k, n) + (1 - \delta) \right] K + A s F_n(k, n) n - c \\
&= A s F(K, \sigma^n(a, X)) + (1 - \delta) K - \sigma^c(a, X),
\end{aligned} \tag{12.9.3}
$$

where the second equality invokes Euler's theorem on linearly homogeneous functions and equilibrium conditions $K = k$, $N = n = \sigma^n(a, X)$ and $C = c = \sigma^c(a, X)$. To express the right side of equation (12.9.3) solely as a function of the current aggregate state $X = [K \ A \ s]$, we also impose equilibrium condition (12.9.1b)

$$
\begin{aligned}
K' &= A s F \left(K, \sigma^n([r(X) + (1 - \delta)]K, X) \right) \\
&\quad + (1 - \delta) K - \sigma^c([r(X) + (1 - \delta)]K, X).
\end{aligned} \tag{12.9.4}
$$

Given the arbitrary perceived law of motion (12.8.1) for K' that underlies the household's optimum problem, the right side of (12.9.4) is the *actual* law of motion for K' that is implied by the household's and firms' optimal decisions. In equilibrium, we want G in (12.8.1) not to be arbitrary but to be an *outcome*. We want to find an equilibrium perceived law of motion (12.8.1). By way of imposing rational expectations, we require that the perceived and actual laws of motion be identical. Equating the right sides of (12.9.4) and the perceived law of motion (12.8.1) gives

$$
\begin{aligned}
G(X) =& A s F \left(K, \sigma^n([r(X) + (1 - \delta)]K, X) \right) \\
&+ (1 - \delta) K - \sigma^c([r(X) + (1 - \delta)]K, X).
\end{aligned} \tag{12.9.5}
$$

Please remember that the right side of this equation is itself implicitly a function of G, so that (12.9.5) is to be regarded as instructing us to find a fixed point equation of a mapping from a perceived G and a price system to an actual G. This functional equation requires that the perceived law of motion for the capital stock $G(X)$ equals the actual law of motion for the capital stock that is determined jointly by the decisions of the household and the firms in a competitive equilibrium.

DEFINITION: A *recursive competitive equilibrium with Arrow securities* is a price system $r(X)$, $w(X)$, $Q(X'|X)$, a perceived law of motion $K' = G(X)$ and associated induced transition density $\hat{\Pi}(X'|X)$, a household value function $J(a, X)$, and decision rules $\sigma^c(a, X)$, $\sigma^n(a, x)$, $\sigma^a(a, X; X')$ such that:

a. Given $r(X)$, $w(X)$, $Q(X'|X)$, $\hat{\Pi}(X'|X)$, the functions $\sigma^c(a, X)$, $\sigma^n(a, X)$, $\sigma^a(a, X; X')$ and the value function $J(a, X)$ solve the household's optimum problem;

b. For all X, $\quad r(X) = AF_k\Big(K, \sigma^n([r(X) + (1 - \delta)]K, X)\Big)$, and

$$w(X) = AF_n\Big(K, \sigma^n([r(X) + (1 - \delta)]K, X)\Big);$$

c. $Q(X'|X)$ and $r(X)$ satisfy (12.8.11);

d. The functions $G(X)$, $r(X)$, $\sigma^c(a, X)$, $\sigma^n(a, X)$ satisfy (12.9.5);

e. $\hat{\pi} = \pi$.

Item a enforces optimization by the household, given the prices it faces and its expectations. Item b requires that the type I firm break even at every capital stock and at the labor supply chosen by the household. Item c requires that the type II firm break even. Market clearing is implicit when item d requires that the perceived and actual laws of motion of capital are equal. Item e and the equality of the perceived and actual G imply that $\hat{\Pi} = \Pi$. Thus, items d and e impose rational expectations.

12.9.2. *Using the planning problem*

Rather than directly attacking the fixed point problem (12.9.5) that is the heart of the equilibrium definition, we'll guess a candidate G as well as a price system, then describe how to verify that they form an equilibrium. As our candidate for G, we choose the decision rule (12.7.3c) for K' from the planning problem. As sources of candidates for the pricing functions, we again turn to the planning problem and choose:

$$r(X) = AF_k(X), \tag{12.9.6a}$$

$$w(X) = AF_n(X), \tag{12.9.6b}$$

$$Q(X'|X) = \beta\Pi(X'|X)\frac{U_c(X')}{U_c(X)}[A's'F_K(X') + (1 - \delta)]. \tag{12.9.6c}$$

In an equilibrium it will turn out that the household's decision rules for consumption and labor supply will match those chosen by the planner:[7]

$$\Omega^C(X) = \sigma^c([r(X) + (1 - \delta)]K, X), \qquad (12.9.7a)$$

$$\Omega^N(X) = \sigma^n([r(X) + (1 - \delta)]K, X). \qquad (12.9.7b)$$

The key to verifying these guesses is to show that the first-order conditions for both types of firms and the household are satisfied at these guesses. We leave the details to an exercise. Here we are exploiting some consequences of the welfare theorems, transported this time to a recursive setting with an endogenous aggregate state variable.

12.10. Concluding remarks

The notion of a recursive competitive equilibrium was introduced by Lucas and Prescott (1971) and Mehra and Prescott (1979). The application in this chapter is in the spirit of those papers but differs substantially in details. In particular, neither of those papers worked with Arrow securities, while the focus of this chapter has been to manage an endogenous state vector in terms of which it is appropriate to cast Arrow securities.

---

[7] The two functional equations (12.9.7) state restrictions that a recursive competitive equilibrium imposes across the household's decision rules σ and the planner's decision rules Ω.

Chapter 13
Asset Pricing Theory

13.1. Introduction

Chapter 8 showed how an equilibrium price system for an economy with a complete markets model could be used to determine the price of any redundant asset. That approach allowed us to price any asset whose payoff could be synthesized as a measurable function of the economy's state. We could use either the Arrow-Debreu time 0 prices or the prices of one-period Arrow securities to price redundant assets.

We shall use this complete markets approach again later in this chapter and in chapter 14. However, we begin with another frequently used approach, one that does not require the assumption that there are complete markets. This approach spells out fewer aspects of the economy and assumes fewer markets, but nevertheless derives testable intertemporal restrictions on prices and returns of different assets, and also across those prices and returns and consumption allocations. This approach uses only the Euler equations for a maximizing consumer, and supplies stringent restrictions without specifying a complete general equilibrium model. In fact, the approach imposes only a *subset* of the restrictions that would be imposed in a complete markets model. As we shall see in chapter 14, even these restrictions have proved difficult to reconcile with the data, the equity premium being a widely discussed example.

Asset-pricing ideas have had diverse ramifications in macroeconomics. In this chapter, we describe some of these ideas, including the important Modigliani-Miller theorem asserting the irrelevance of firms' asset structures. We describe a closely related kind of Ricardian equivalence theorem.[1]

---

[1] See Duffie (1996) for a comprehensive treatment of discrete- and continuous-time asset-pricing theories. See Campbell, Lo, and MacKinlay (1997) for a summary of recent work on empirical implementations.

13.2. Asset Euler equations

We now describe the optimization problem of a single agent who has the opportunity to trade two assets. Following Hansen and Singleton (1983), the household's optimization by itself imposes ample restrictions on the comovements of asset prices and the household's consumption. These restrictions remain true even if additional assets are made available to the agent, and so do not depend on specifying the market structure completely. Later we shall study a general equilibrium model with a large number of identical agents. Completing a general equilibrium model may impose additional restrictions, but will leave intact individual-specific versions of the ones to be derived here.

The agent has wealth $A_t > 0$ at time t and wants to use this wealth to maximize expected lifetime utility,

$$E_t \sum_{j=0}^{\infty} \beta^j u(c_{t+j}), \qquad 0 < \beta < 1, \tag{13.2.1}$$

where E_t denotes the mathematical expectation conditional on information known at time t, β is a subjective discount factor, and c_{t+j} is the agent's consumption in period $t + j$. The utility function $u(\cdot)$ is concave, strictly increasing, and twice continuously differentiable.

To finance future consumption, the agent can transfer wealth over time through bond and equity holdings. One-period bonds earn a risk-free real gross interest rate R_t, measured in units of time $t + 1$ consumption good per time t consumption good. Let L_t be gross payout on the agent's bond holdings between periods t and $t + 1$, payable in period $t + 1$ with a present value of $R_t^{-1} L_t$ at time t. The variable L_t is negative if the agent issues bonds and thereby borrows funds. The agent's holdings of equity shares between periods t and $t+1$ are denoted N_t, where a negative number indicates a short position in shares. We impose the borrowing constraints $L_t \geq -b_L$ and $N_t \geq -b_N$, where $b_L \geq 0$ and $b_N \geq 0$.[2] A share of equity entitles the owner to its stochastic dividend stream y_t. Let p_t be the share price in period t net of that period's dividend. The budget constraint becomes

$$c_t + R_t^{-1} L_t + p_t N_t \leq A_t, \tag{13.2.2}$$

---

[2] See chapters 8 and 18 for further discussions of natural and ad hoc borrowing constraints.

and next period's wealth is

$$A_{t+1} = L_t + (p_{t+1} + y_{t+1})N_t. \qquad (13.2.3)$$

The stochastic dividend is the only source of exogenous fundamental uncertainty, with properties to be specified as needed later. The agent's maximization problem is then a dynamic programming problem with the state at t being A_t and current and past y,[3] and the controls being L_t and N_t. At interior solutions, the Euler equations associated with controls L_t and N_t are

$$u'(c_t)R_t^{-1} = E_t\beta u'(c_{t+1}), \qquad (13.2.4)$$

$$u'(c_t)p_t = E_t\beta(y_{t+1} + p_{t+1})u'(c_{t+1}). \qquad (13.2.5)$$

These Euler equations give a number of insights into asset prices and consumption. Before turning to these, we first note that an optimal solution to the agent's maximization problem must also satisfy the following transversality conditions:[4]

$$\lim_{k\to\infty} E_t\beta^k u'(c_{t+k})R_{t+k}^{-1}L_{t+k} = 0, \qquad (13.2.6)$$

$$\lim_{k\to\infty} E_t\beta^k u'(c_{t+k})p_{t+k}N_{t+k} = 0. \qquad (13.2.7)$$

Heuristically, if any of the expressions in equations $(13.2.6)$ and $(13.2.7)$ were strictly positive, the agent would be overaccumulating assets so that a higher expected lifetime utility could be achieved by, for example, increasing consumption today. The counterpart to such nonoptimality in a finite horizon model would be that the agent dies with positive asset holdings. For reasons like those in a finite horizon model, the agent would be happy if the two conditions $(13.2.6)$ and $(13.2.7)$ could be violated on the negative side. But the market would stop the agent from financing consumption by accumulating the debts that would be associated with such violations of $(13.2.6)$ and $(13.2.7)$. No other agent would want to make those loans.

---

[3] Current and past y's enter as information variables. How many past y's appear in the Bellman equation depends on the stochastic process for y.

[4] For a discussion of transversality conditions, see Benveniste and Scheinkman (1982) and Brock (1982).

13.3. Martingale theories of consumption and stock prices

In this section, we briefly recall some early theories of asset prices and consumption, each of which is derived by making special assumptions about either R_t or $u'(c)$ in equations (13.2.4) and (13.2.5). These assumptions are too strong to be consistent with much empirical evidence, but they are instructive benchmarks.

First, suppose that the risk-free interest rate is constant over time, $R_t = R > 1$, for all t. Then equation (13.2.4) implies that

$$E_t u'(c_{t+1}) = (\beta R)^{-1} u'(c_t), \tag{13.3.1}$$

which is Robert Hall's (1978) result that the marginal utility of consumption follows a univariate linear first-order Markov process, so that no other variables in the information set help to predict (to Granger cause) $u'(c_{t+1})$, once lagged $u'(c_t)$ has been included.[5]

As an example, with the constant-relative-risk-aversion utility function $u(c_t) = (1 - \gamma)^{-1} c_t^{1-\gamma}$, equation (13.3.1) becomes

$$(\beta R)^{-1} = E_t \left(\frac{c_{t+1}}{c_t} \right)^{-\gamma}.$$

Using aggregate data, Hall tested implication (13.3.1) for the special case of quadratic utility by testing for the absence of Granger causality from other variables to c_t.

Efficient stock markets are sometimes construed to mean that the price of a stock ought to follow a martingale. Euler equation (13.2.5) shows that a number of simplifications must be made to get a martingale property for the stock price. We can transform the Euler equation

$$E_t \beta (y_{t+1} + p_{t+1}) \frac{u'(c_{t+1})}{u'(c_t)} = p_t$$

by noting that for any two random variables x, z, we have the formula $E_t xz = E_t x E_t z + \text{cov}_t(x, z)$, where $\text{cov}_t(x, z) \equiv E_t(x - E_t x)(z - E_t z)$. This formula

---

[5] See Granger (1969) for his definition of causality. A random process z_t is said *not* to cause a random process x_t if $E(x_{t+1} | x_t, x_{t-1}, \ldots, z_t, z_{t-1}, \ldots) = E(x_{t+1} | x_t, x_{t-1}, \ldots)$. The absence of Granger causality can be tested in several ways. A direct way is to compute the two regressions mentioned in the preceding definition and test for their equality. An alternative test was described by Sims (1972).

defines the conditional covariance $\operatorname{cov}_t(x, z)$. Applying this formula in the preceding equation gives

$$\beta E_t(y_{t+1} + p_{t+1}) E_t \frac{u'(c_{t+1})}{u'(c_t)} + \beta \operatorname{cov}_t \left[(y_{t+1} + p_{t+1}), \ \frac{u'(c_{t+1})}{u'(c_t)} \right] = p_t. \ (13.3.2)$$

To obtain a martingale theory of stock prices, it is necessary to assume, first, that $E_t u'(c_{t+1})/u'(c_t)$ is a constant, and second, that

$$\operatorname{cov}_t \left[(y_{t+1} + p_{t+1}), \ \frac{u'(c_{t+1})}{u'(c_t)} \right] = 0.$$

These conditions are obviously very restrictive and will only hold under very special circumstances. For example, a sufficient assumption is that agents are risk neutral, so that $u(c_t)$ is linear in c_t and $u'(c_t)$ becomes independent of c_t. In this case, equation (13.3.2) implies that

$$E_t \beta (y_{t+1} + p_{t+1}) = p_t. \tag{13.3.3}$$

Equation (13.3.3) states that, adjusted for dividends and discounting, the share price follows a first-order univariate Markov process and that no other variables Granger cause the share price. These implications have been tested extensively in the literature on efficient markets.[6]

We also note that the stochastic difference equation (13.3.3) has the class of solutions

$$p_t = E_t \sum_{j=1}^{\infty} \beta^j y_{t+j} + \xi_t \left(\frac{1}{\beta} \right)^t, \tag{13.3.4}$$

where ξ_t is any random process that obeys $E_t \xi_{t+1} = \xi_t$ (that is, ξ_t is a "martingale"). Equation (13.3.4) expresses the share price p_t as the sum of discounted expected future dividends and a "bubble term" unrelated to any fundamentals. In the general equilibrium model that we will describe later, this bubble term always equals zero.

---

[6] For a survey of this literature, see Fama (1976a). See Samuelson (1965) for the theory and Roll (1970) for an application to the term structure of interest rates.

13.4. Equivalent martingale measure

This section describes adjustments for risk and dividends that convert an asset price into a martingale. We return to the setting of chapter 8 and assume that the state s_t evolves according to a Markov chain with transition probabilities $\pi(s_{t+1}|s_t)$. Let an asset pay a stream of dividends $\{d(s_t)\}_{t\geq0}$. The cum-dividend[7] time t price of this asset, $a(s_t)$, can be expressed recursively as

$$a(s_t) = d(s_t) + \beta \sum_{s_{t+1}} \frac{u'[c^i_{t+1}(s_{t+1})]}{u'[c^i_t(s_t)]} a(s_{t+1}) \pi(s_{t+1}|s_t), \qquad (13.4.1)$$

where c^i_t is the consumption of agent i at date t in state s_t. This equation holds for every agent i. Equation $(13.4.1)$ can be written

$$a(s_t) = d(s_t) + R_t^{-1} \sum_{s_{t+1}} a(s_{t+1}) \tilde{\pi}(s_{t+1}|s_t) \qquad (13.4.2)$$

or

$$a(s_t) = d(s_t) + R_t^{-1} \tilde{E}_t a(s_{t+1}), \qquad (13.4.3)$$

where R_t^{-1} is the reciprocal of the one period gross risk-free interest rate

$$R_t^{-1} = R_t^{-1}(s_t) \equiv \beta \sum_{s_{t+1}} \frac{u'[c^i_{t+1}(s_{t+1})]}{u'[c^i_t(s_t)]} \pi(s_{t+1}|s_t) \qquad (13.4.4)$$

and $\tilde{E}$ in equation $(13.4.3)$ is the mathematical expectation with respect to the distorted transition density

$$\tilde{\pi}(s_{t+1}|s_t) = R_t \beta \frac{u'[c^i_{t+1}(s_{t+1})]}{u'[c^i_t(s_t)]} \pi(s_{t+1}|s_t). \qquad (13.4.5a)$$

The transformed transition probabilities are rendered probabilities—that is, made to sum to 1—through the multiplication by βR_t in equation $(13.4.5a)$. The transformed or "twisted" transition measure $\tilde{\pi}(s_{t+1}|s_t)$ can be used to define the twisted measure

$$\tilde{\pi}_t(s^t) = \tilde{\pi}(s_t|s_{t-1}) \dots \tilde{\pi}(s_1|s_0) \tilde{\pi}(s_0). \qquad (13.4.5b)$$

---

[7] Cum-dividend means that the person who owns the asset at the end of time t is entitled to the time t dividend. Ex-dividend means that the person who owns the asset at the end of the period does not receive the time t dividend.

For example,

$$\tilde{\pi}(s_{t+2}, s_{t+1}|s_t) = R_t(s_t)R_{t+1}(s_{t+1})\beta^2 \frac{u'[c^i_{t+2}(s_{t+2})]}{u'[c^i_t(s_t)]} \pi(s_{t+2}|s_{t+1})\pi(s_{t+1}|s_t).$$

The twisted measure $\tilde{\pi}_t(s^t)$ is called an *equivalent martingale measure*. We explain the meaning of the two adjectives. "Equivalent" means that $\tilde{\pi}$ assigns positive probability to any event that is assigned positive probability by π, and vice versa. The equivalence of π and $\tilde{\pi}$ is guaranteed by the assumption that $u'(c) > 0$ in $(13.4.5a)$.[8]

We now turn to the adjective "martingale."[9] To understand why this term is applied to $\tilde{\pi}_t(s^t)$, consider the particular case of an asset with dividend stream $d_T = d(s_T)$ and $d_t = 0$ for $t \neq T$. Using the arguments in chapter 8 or iterating on equation $(13.4.1)$, the cum-dividend price of this asset can be expressed as

$$a_T(s_T) = d(s_T), \tag{13.4.6a}$$

$$a_{T-1}(s_{T-1}) = R_{T-1}^{-1}\tilde{E}_{T-1}a_T(s_T) \tag{13.4.6b}$$

$$\vdots \qquad \vdots$$

$$a_t(s_t) = R_t^{-1}\tilde{E}_t R_{t+1}^{-1} R_{t+2}^{-1} \cdots R_{T-1}^{-1} a_T(s_t) \tag{13.4.6c}$$

where $\tilde{E}_t$ denotes the conditional expectation under the $\tilde{\pi}$ probability measure. Now fix $t < T$ and define the "deflated" or "interest-adjusted" asset price process

$$\tilde{a}_{t,t+j} = \frac{a_{t+j}}{R_t R_{t+1} \ldots R_{t+j-1}}, \tag{13.4.7}$$

---

[8] The existence of an equivalent martingale measure implies both the existence of a *positive* stochastic discount factor (see the discussion of Hansen and Jagannathan bounds later in this chapter), and the absence of arbitrage opportunities; see Kreps (1979) and Duffie (1996).

[9] Another insight is that the likelihood ratio $L(s^t) \equiv \frac{\tilde{\pi}_t(s^t)}{\pi_t(s^t)}$ is a martingale with respect to the measure $\pi_t(s^t)$. To verify this, notice that $L(s^t) = \frac{\tilde{\pi}(s_t|s_{t-1})}{\pi(s_t|s_{t-1})}L(s^{t-1})$. Therefore,

$$E[L(s^t)|s^{t-1}] = L(s^{t-1})E\left[\frac{\tilde{\pi}(s_t|s_{t-1})}{\pi(s_t|s_{t-1})}\Big|s^{t-1}\right] \quad \text{But}$$

$$E\left[\frac{\tilde{\pi}(s_t|s_{t-1})}{\pi(s_t|s_{t-1})}\Big|s^{t-1}\right] = \sum_{s_{t+1} \in S}\frac{\tilde{\pi}(s_t|s_{t-1})}{\pi(s_t|s_{t-1})}\pi(s_t|s_{t-1}) = 1.$$

Therefore, $E[L(s^t)|s^{t-1}] = L(s^{t-1})$, so the likelihood ratio is a martingale.

for $j = 1, \ldots, T - t$. It follows from equation $(13.4.6c)$ that

$$\tilde{E}_t \tilde{a}_{t,t+j} = a_t(s_t) \equiv \tilde{a}_{t,t}. \tag{13.4.8}$$

Equation $(13.4.8)$ asserts that relative to the twisted measure $\tilde{\pi}$, the interest-adjusted asset price is a martingale: using the twisted measure, the best prediction of the future interest-adjusted asset price is its current value.

Thus, when the equivalent martingale measure is used to price assets, we have so-called risk-neutral pricing. Notice that in equation $(13.4.2)$ the adjustment for risk is absorbed into the twisted transition measure. We can write equation $(13.4.8)$ as

$$\tilde{E}[a(s_{t+1})|s_t] = R_t[a(s_t) - d(s_t)], \tag{13.4.9}$$

where $\tilde{E}$ is the expectation operator for the twisted transition measure. Equation $(13.4.9)$ is another way of stating that, after adjusting for risk-free interest and dividends, the price of the asset is a *martingale* relative to the equivalent martingale measure.

Under the equivalent martingale measure, asset pricing reduces to calculating the conditional expectation of the stream of dividends that defines the asset. For example, consider a European call option written on the asset described earlier that is priced by equations $(13.4.6)$. The owner of the call option has the right but not the obligation to purchase the "asset" at time T at a price K. The owner of the call option will exercise this option only if $a_T \geq K$. The value at T of the option is therefore $Y_T = \max(0, a_T - K) \equiv (a_T - K)^+$. The price of the option at $t < T$ is then

$$Y_t = \tilde{E}_t \left[\frac{(a_T - K)^+}{R_t R_{t+1} \cdots R_{t+T-1}} \right]. \tag{13.4.10}$$

Black and Scholes (1973) used a particular continuous-time specification of $\tilde{\pi}$ that made it possible to solve equation $(13.4.10)$ analytically for a function Y_t. Their solution is known as the Black-Scholes formula for option pricing.

13.5. Equilibrium asset pricing

The preceding discussion of the Euler equations (13.2.4) and (13.2.5) leaves open how the economy generates, for example, the constant gross interest rate assumed in Hall's work. We now explore equilibrium asset pricing in a simple representative agent endowment economy, Lucas's asset-pricing model.[10] We imagine an economy consisting of a large number of identical agents with preferences as specified in expression (13.2.1). The only durable good in the economy is a set of identical "trees," one for each person in the economy. At the beginning of period t, each tree yields fruit or dividends in the amount y_t. The fruit is not storable, but the tree is perfectly durable. Each agent starts life at time zero with one tree.

The dividend y_t is assumed to be governed by a Markov process and the dividend is the sole state variable s_t of the economy, i.e., $s_t = y_t$. The time-invariant transition probability distribution function is given by $\text{Prob}\{s_{t+1} \leq s'|s_t = s\} = F(s', s)$.

All agents maximize expression (13.2.1) subject to the budget constraint (13.2.2)–(13.2.3) and transversality conditions (13.2.6)–(13.2.7). In an equilibrium, asset prices clear the markets. That is, the bond holdings of all agents sum to zero, and their total stock positions are equal to the aggregate number of shares. As a normalization, let there be one share per tree.

Due to the assumption that all agents are identical with respect to both preferences and endowments, we can work with a representative agent.[11] Lucas's model shares features with a variety of representative agent asset-pricing models (see Brock, 1982, and Altug, 1989, for example). These use versions of stochastic optimal growth models to generate allocations and price assets.

Such asset-pricing models can be constructed by the following steps:

1. Describe the preferences, technology, and endowments of a dynamic economy, then solve for the equilibrium intertemporal consumption allocation. Sometimes there is a particular planning problem whose solution equals the competitive allocation.

---

[10] See Lucas (1978). Also see the important early work by Stephen LeRoy (1971, 1973). Breeden (1979) was an early work on the consumption-based capital-asset-pricing model.

[11] In chapter 8, we showed that some heterogeneity is also consistent with the notion of a representative agent.

2. Set up a competitive market in some particular asset that represents a specific claim on future consumption goods. Permit agents to buy and sell at equilibrium asset prices subject to particular borrowing and short-sales constraints. Find an agent's Euler equation, analogous to equations (13.2.4) and (13.2.5), for this asset.

3. Equate the consumption that appears in the Euler equation derived in step 2 to the equilibrium consumption derived in step 1. This procedure will give the asset price at t as a function of the state of the economy at t.

In our endowment economy, a planner that treats all agents the same would like to maximize $E_0 \sum_{t=0}^{\infty} \beta^t u(c_t)$ subject to $c_t \leq y_t$. Evidently the solution is to set c_t equal to y_t. After substituting this consumption allocation into equations (13.2.4) and (13.2.5), we arrive at expressions for the risk-free interest rate and the share price:

$$u'(y_t)R_t^{-1} = E_t \beta u'(y_{t+1}), \qquad (13.5.1)$$

$$u'(y_t)p_t = E_t \beta(y_{t+1} + p_{t+1})u'(y_{t+1}). \qquad (13.5.2)$$

13.6. Stock prices without bubbles

Using recursions on equation (13.5.2) and the law of iterated expectations, which states that $E_t E_{t+1}(\cdot) = E_t(\cdot)$, we arrive at the following expression for the equilibrium share price:

$$u'(y_t)p_t = E_t \sum_{j=1}^{\infty} \beta^j u'(y_{t+j})y_{t+j} + E_t \lim_{k \to \infty} \beta^k u'(y_{t+k})p_{t+k}. \qquad (13.6.1)$$

Moreover, equilibrium share prices have to be consistent with market clearing; that is, agents must be willing to hold their endowments of trees forever. It follows immediately that the last term in equation (13.6.1) must be zero. Suppose to the contrary that the term is strictly positive. That is, the marginal utility gain of selling shares, $u'(y_t)p_t$, exceeds the marginal utility loss of holding the asset forever and consuming the future stream of dividends, $E_t \sum_{j=1}^{\infty} \beta^j u'(y_{t+j})y_{t+j}$. Thus, all agents would like to sell some of their shares and the price would be driven down. Analogously, if the last term in equation (13.6.1) were strictly negative, we would find that all agents would like to purchase more shares and the price would necessarily be driven up. We can therefore conclude that the equilibrium price must satisfy

$$p_t = E_t \sum_{j=1}^{\infty} \beta^j \frac{u'(y_{t+j})}{u'(y_t)} y_{t+j}, \qquad (13.6.2)$$

which is a generalization of equation (13.3.4) in which the share price is an expected discounted stream of dividends but with time-varying and stochastic discount rates.

Note that asset bubbles could also have been ruled out by directly referring to transversality condition (13.2.7) and market clearing. In an equilibrium, the representative agent holds the per capita outstanding number of shares. (We have assumed one tree per person and one share per tree.) After dividing transversality condition (13.2.7) by this constant time-invariant number of shares and replacing c_{t+k} by equilibrium consumption y_{t+k}, we arrive at the implication that the last term in equation (13.6.1) must vanish.[12]

Moreover, after invoking our assumption that the endowment follows a Markov process, it follows that the equilibrium price in equation (13.6.2) can be expressed as a function of the current state s_t,

$$p_t = p(s_t). \qquad (13.6.3)$$

---

[12] Brock (1982) and Tirole (1982) use the transversality condition when proving that asset bubbles cannot exist in economies with a constant number of infinitely lived agents. However, Tirole (1985) shows that asset bubbles can exist in equilibria of overlapping generations models that are dynamically inefficient, that is, when the growth rate of the economy exceeds the equilibrium rate of return. O'Connell and Zeldes (1988) derive the same result for a dynamically inefficient economy with a growing number of infinitely lived agents. Abel, Mankiw, Summers, and Zeckhauser (1989) provide international evidence suggesting that dynamic inefficiency is not a problem in practice.

13.7. Computing asset prices

We now turn to three examples in which it is easy to calculate an asset-pricing function by solving the expectational difference equation (13.5.2).

13.7.1. Example 1: logarithmic preferences

Take the special case of equation (13.6.2) that emerges when $u(c_t) = \ln c_t$. Then equation (13.6.2) becomes

$$p_t = \frac{\beta}{1 - \beta} y_t. \tag{13.7.1}$$

Equation (13.7.1) is our asset-pricing function. It maps the state of the economy at t, y_t, into the price of a Lucas tree at t.

13.7.2. Example 2: a finite-state version

Mehra and Prescott (1985) consider a discrete-state version of Lucas's one-kind-of-tree model. Let dividends assume the n possible distinct values $[\sigma_1, \sigma_2, \ldots, \sigma_n]$. Let dividends evolve through time according to a Markov chain, with

$$\text{prob}\{y_{t+1} = \sigma_l | y_t = \sigma_k\} = P_{kl} > 0.$$

The $(n \times n)$ matrix P with element P_{kl} is called a stochastic matrix. The matrix satisfies $\sum_{l=1}^{n} P_{kl} = 1$ for each k. Express equation (13.5.2) of Lucas's model as

$$p_t u'(y_t) = \beta E_t p_{t+1} u'(y_{t+1}) + \beta E_t y_{t+1} u'(y_{t+1}). \tag{13.7.2}$$

Express the price at t as a function of the state σ_k at t, $p_t = p(\sigma_k)$. Define $p_t u'(y_t) = p(\sigma_k) u'(\sigma_k) \equiv v_k$, $k = 1, \ldots, n$. Also define $\alpha_k = \beta E_t y_{t+1} u'(y_{t+1}) = \beta \sum_{l=1}^{n} \sigma_l u'(\sigma_l) P_{kl}$. Then equation (13.7.2) can be expressed as

$$p(\sigma_k) u'(\sigma_k) = \beta \sum_{l=1}^{n} p(\sigma_l) u'(\sigma_l) P_{kl} + \beta \sum_{l=1}^{n} \sigma_l u'(\sigma_l) P_{kl}$$

or

$$v_k = \alpha_k + \beta \sum_{l=1}^{n} P_{kl} v_l,$$

or in matrix terms, $v = \alpha + \beta P v$, where v and α are column vectors. The equation can be represented as $(I - \beta P)v = \alpha$. This equation has a unique solution given by [13]

$$v = (I - \beta P)^{-1}\alpha. \tag{13.7.3}$$

The price of the asset in state σ_k—call it p_k—can then be found from $p_k = v_k/[u'(\sigma_k)]$. Notice that equation (13.7.3) can be represented as

$$v = (I + \beta P + \beta^2 P^2 + \ldots)\alpha$$

or

$$p(\sigma_k) = p_k = \sum_l (I + \beta P + \beta^2 P^2 + \ldots)_{kl} \frac{\alpha_l}{u'(\sigma_k)},$$

where $(I + \beta P + \beta^2 P^2 + \ldots)_{kl}$ is the (k, l) element of the matrix $(I + \beta P + \beta^2 P^2 + \ldots)$. We ask the reader to interpret this formula in terms of a geometric sum of expected future variables.

13.7.3. Example 3: asset pricing with growth

Let's price a Lucas tree in a pure endowment economy with $c_t = y_t$ and $y_{t+1} = \lambda_{t+1} y_t$, where λ_t is Markov with transition matrix P. Let p_t be the ex-dividend price of the Lucas tree. Assume the CRRA utility $u(c) = c^{1-\gamma}/(1-\gamma)$. Evidently, the price of the Lucas tree satisfies

$$p_t = E_t \left[\beta \left(\frac{c_{t+1}}{c_t} \right)^{-\gamma} (p_{t+1} + y_{t+1}) \right].$$

Dividing both sides by y_t and rearranging gives

$$\frac{p_t}{y_t} = E_t \left[\beta(\lambda_{t+1})^{1-\gamma} \left(\frac{p_{t+1}}{y_{t+1}} + 1 \right) \right]$$

or

$$w_i = \beta \sum_j P_{ij} \lambda_j^{1-\gamma}(w_j + 1), \tag{13.7.4}$$

---

[13] Uniqueness follows from the fact that, because P is a nonnegative matrix with row sums all equaling unity, the eigenvalue of maximum modulus P has modulus unity. The maximum eigenvalue of βP then has modulus β. (This point follows from Frobenius's theorem.) The implication is that $(I - \beta P)^{-1}$ exists and that the expansion $I + \beta P + \beta^2 P^2 + \ldots$ converges and equals $(I - \beta P)^{-1}$.

where w_i represents the price-dividend ratio. Equation $(13.7.4)$ was used by Mehra and Prescott (1985) to compute equilibrium prices.

13.8. The term structure of interest rates

We will now explore the term structure of interest rates by pricing bonds with different maturities.[14] We continue to assume that the time t state of the economy is the current dividend on a Lucas tree $y_t = s_t$, which is Markov with transition $F(s', s)$. The risk-free real gross return between periods t and $t + j$ is denoted R_{jt}, measured in units of time $(t + j)$ consumption good per time t consumption good. Thus, R_{1t} replaces our earlier notation R_t for the one-period gross interest rate. At the beginning of t, the return R_{jt} is known with certainty and is risk free from the viewpoint of the agents. That is, at t, R_{jt}^{-1} is the price of a perfectly sure claim to one unit of consumption at time $t + j$. For simplicity, we only consider such zero-coupon bonds, and the extra subscript j on gross earnings L_{jt} now indicates the date of maturity. The subscript t still refers to the agent's decision to hold the asset between period t and $t + 1$.

As an example with one- and two-period safe bonds, the budget constraint and the law of motion for wealth in $(13.2.2)$ and $(13.2.3)$ are augmented as follows,

$$c_t + R_{1t}^{-1}L_{1t} + R_{2t}^{-1}L_{2t} + p_t N_t \le A_t, \tag{13.8.1}$$

$$A_{t+1} = L_{1t} + R_{1t+1}^{-1}L_{2t} + (p_{t+1} + y_{t+1})N_t. \tag{13.8.2}$$

Even though safe bonds represent sure claims to future consumption, these assets are subject to price risk prior to maturity. For example, two-period bonds from period t, L_{2t}, are traded at the price R_{1t+1}^{-1} in period $t+1$, as shown in wealth expression $(13.8.2)$. At time t, an agent who buys such assets and plans to sell them next period would be uncertain about the proceeds, since R_{1t+1}^{-1} is not known at time t. The price R_{1t+1}^{-1} follows from a simple arbitrage argument, since, in period $t+1$, these assets represent identical sure claims to time $(t+2)$ consumption goods as newly issued one-period bonds in period $t + 1$. The variable L_{jt} should therefore be understood as the agent's net holdings between

---

[14] Dynamic asset-pricing theories for the term structure of interest rates have been developed by Cox, Ingersoll, and Ross (1985a, 1985b) and by LeRoy (1982).

periods t and $t + 1$ of bonds that each pay one unit of consumption good at time $t + j$, without identifying when the bonds were initially issued.

Given wealth A_t and current dividend $y_t = s_t$, let $v(A_t, s_t)$ be the optimal value of maximizing expression (13.2.1) subject to equations (13.8.1) and (13.8.2), the asset-pricing function for trees $p_t = p(s_t)$, the stochastic process $F(s_{t+1}, s_t)$, and stochastic processes for R_{1t} and R_{2t}. The Bellman equation can be written as

$$v(A_t, s_t) = \max_{L_{1t}, L_{2t}, N_t} \left\{ u \left[A_t - R_{1t}^{-1} L_{1t} - R_{2t}^{-1} L_{2t} - p(s_t) N_t \right] \right.$$
$$\left. + \beta E_t v \left(L_{1t} + R_{1t+1}^{-1} L_{2t} + [p(s_{t+1}) + s_{t+1}] N_t, s_{t+1} \right) \right\},$$

where we have substituted for consumption c_t and wealth A_{t+1} from formulas (13.8.1) and (13.8.2), respectively. The first-order necessary conditions with respect to L_{1t} and L_{2t} are

$$u'(c_t) R_{1t}^{-1} = \beta E_t v_1 (A_{t+1}, s_{t+1}), \tag{13.8.3}$$

$$u'(c_t) R_{2t}^{-1} = \beta E_t \left[v_1 (A_{t+1}, s_{t+1}) R_{1t+1}^{-1} \right]. \tag{13.8.4}$$

After invoking Benveniste and Scheinkman's result and equilibrium allocation $c_t = y_t (= s_t)$, we arrive at the following equilibrium rates of return

$$R_{1t}^{-1} = \beta E_t \left[\frac{u'(s_{t+1})}{u'(s_t)} \right] \equiv R_1(s_t)^{-1}, \tag{13.8.5}$$

$$R_{2t}^{-1} = \beta E_t \left[\frac{u'(s_{t+1})}{u'(s_t)} R_{1t+1}^{-1} \right] = \beta^2 E_t \left[\frac{u'(s_{t+2})}{u'(s_t)} \right] \equiv R_2(s_t)^{-1}, \tag{13.8.6}$$

where the second equality in (13.8.6) is obtained by using (13.8.5) and the law of iterated expectations. Because of our Markov assumption, interest rates can be written as time-invariant functions of the economy's current state s_t. The general expression for the price at time t of a bond that yields one unit of the consumption good in period $t + j$ is

$$R_{jt}^{-1} = \beta^j E_t \left[\frac{u'(s_{t+j})}{u'(s_t)} \right]. \tag{13.8.7}$$

The term structure of interest rates is commonly defined as the collection of yields to maturity for bonds with different dates of maturity. In the case of zero-coupon bonds, the yield to maturity is simply

$$\tilde{R}_{jt} \equiv R_{jt}^{1/j} = \beta^{-1} \left\{ u'(s_t) \left[E_t u'(s_{t+j}) \right]^{-1} \right\}^{1/j}. \tag{13.8.8}$$

As an example, let us assume that dividends are independently and identically distributed over time. The yields to maturity for a j-period bond and a k-period bond are then related as follows:

$$\tilde{R}_{jt} = \tilde{R}_{kt} \left\{ u'(s_t) \left[Eu'(s) \right]^{-1} \right\}^{\frac{k-j}{kj}}.$$

The term structure of interest rates is therefore upward sloping whenever $u'(s_t)$ is less than $Eu'(s)$, that is, when consumption is relatively high today with a low marginal utility of consumption, and agents would like to save for the future. In an equilibrium, the short-term interest rate is therefore depressed if there is a diminishing marginal rate of physical transformation over time or, as in our model, there is no investment technology at all.

A classical theory of the term structure of interest rates is that long-term interest rates should be determined by expected future short-term interest rates. For example, the pure expectations theory hypothesizes that $R_{2t}^{-1} = R_{1t}^{-1} E_t R_{1t+1}^{-1}$. Let us examine if this relationship holds in our general equilibrium model. From equation $(13.8.6)$ and by using equation $(13.8.5)$, we obtain

$$
\begin{aligned}
R_{2t}^{-1} &= \beta E_t \left[\frac{u'(s_{t+1})}{u'(s_t)} \right] E_t R_{1t+1}^{-1} + \mathrm{cov}_t \left[\beta \frac{u'(s_{t+1})}{u'(s_t)}, \, R_{1t+1}^{-1} \right] \\
&= R_{1t}^{-1} E_t R_{1t+1}^{-1} + \mathrm{cov}_t \left[\beta \frac{u'(s_{t+1})}{u'(s_t)}, \, R_{1t+1}^{-1} \right],
\end{aligned}
\tag{13.8.9}
$$

which is a generalized version of the pure expectations theory, adjusted for the risk premium $\mathrm{cov}_t[\beta u'(s_{t+1})/u'(s_t), R_{1t+1}^{-1}]$. The formula implies that the pure expectations theory holds only in special cases. One special case occurs when utility is linear in consumption, so that $u'(s_{t+1})/u'(s_t) = 1$. In this case, R_{1t}, given by equation $(13.8.5)$, is a constant, equal to β^{-1}, and the covariance term is zero. A second special case occurs when there is no uncertainty, so that the covariance term is zero for that reason. Recall that the first special case of risk neutrality is the same condition that suffices to eradicate the risk premium appearing in equation $(13.3.2)$ and thereby sustain a martingale theory for a stock price.

13.9. State-contingent prices

Thus far, this chapter has taken a different approach to asset pricing than we took in chapter 8. Recall that in chapter 8 we described two alternative complete markets models, one with once-and-for-all trading at time 0 of date- and history-contingent claims, the other with sequential trading of a complete set of one-period Arrow securities. After these state-contingent prices had been computed, we were able to price any asset whose payoffs were linear combinations of the basic state-contingent commodities, just by taking a weighted sum. That approach would work easily for the Lucas tree economy, which by its simple structure with a representative agent can readily be cast as an economy with complete markets. The pricing formulas that we derived in chapter 8 apply to the Lucas tree economy, adjusting only for the way we have altered the specification of the Markov process describing the state of the economy.

Thus, in chapter 8, we gave formulas for a pricing kernel for j-step-ahead state-contingent claims. In the notation of that chapter, we called $Q_j(s_{t+j}|s_t)$ the price when the time t state is s_t of one unit of consumption in state s_{t+j}. In this chapter we have chosen to let the state be governed by a continuous-state Markov process. But we continue to use the notation $Q_j(s_j|s)$ to denote the j-step-ahead state-contingent price. We have the following version of the formula from chapter 8 for a j-period contingent claim

$$Q_j(s_j|s) = \beta^j \frac{u'(s_j)}{u'(s)} f^j(s_j, s), \tag{13.9.1}$$

where the j-step-ahead transition function obeys

$$f^j(s_j, s) = \int f(s_j, s_{j-1}) f^{j-1}(s_{j-1}, s) ds_{j-1}, \tag{13.9.2}$$

and

$$\text{prob}\{s_{t+j} \le s'|s_t = s\} = \int_{-\infty}^{s'} f^j(w, s) dw.$$

In subsequent sections, we use the state-contingent prices to give expositions of several important ideas, including the Modigliani-Miller theorem and a Ricardian theorem.

13.9.1. Insurance premium

We shall now use the contingent claims prices to construct a model of insurance. Let $q_\alpha(s)$ be the price in current consumption goods of a claim on one unit of consumption next period, contingent on the event that next period's dividends fall below α. We think of the asset being priced as "crop insurance," a claim to consumption when next period's crops fall short of α per tree.

From the preceding section, we have

$$q_\alpha(s) = \beta \int_0^\alpha \frac{u'(s')}{u'(s)} f(s', s) ds'. \tag{13.9.3}$$

Upon noting that

$$\int_0^\alpha u'(s') f(s', s) ds' = \text{prob}\{s_{t+1} \leq \alpha | s_t = s\} \, E\{u'(s_{t+1}) \,|\, s_{t+1} \leq \alpha, s_t = s\},$$

we can represent the preceding equation as

$$q_\alpha(s) = \frac{\beta}{u'(s)} \text{prob}\{s_{t+1} \leq \alpha | s_t = s\} \, E\{u'(s_{t+1}) \,|\, s_{t+1} \leq \alpha, s_t = s\}. \tag{13.9.4}$$

Notice that, in the special case of risk neutrality [$u'(s)$ is a constant], equation (13.9.4) collapses to

$$q_\alpha(s) = \beta \, \text{prob}\{s_{t+1} \leq \alpha | s_t = s\},$$

which is an intuitively plausible formula for the risk-neutral case. When $u'' < 0$ and $s_t \geq \alpha$, equation (13.9.4) implies that $q_\alpha(s) > \beta \text{prob}\{s_{t+1} \leq \alpha | s_t = s\}$ (because then $E\{u'(s_{t+1}) | s_{t+1} \leq \alpha, s_t = s\} > u'(s_t)$ for $s_t \geq \alpha$). In other words, when the representative consumer is risk averse ($u'' < 0$) and when $s_t \geq \alpha$, the price of crop insurance $q_\alpha(s)$ exceeds the "actuarially fair" price of $\beta \text{prob}\{s_{t+1} \leq \alpha | s_t = s\}$.

Another way to represent equation (13.9.3) that is perhaps more convenient for purposes of empirical testing is

$$1 = \frac{\beta}{u'(s_t)} E\left[u'(s_{t+1}) R_t(\alpha) \big| s_t\right] \tag{13.9.5}$$

where

$$R_t(\alpha) = \begin{cases} 0 & \text{if } s_{t+1} > \alpha \\ 1/q_\alpha(s_t) & \text{if } s_{t+1} \leq \alpha. \end{cases}$$

13.9.2. Man-made uncertainty

In addition to pricing assets with returns made risky by nature, we can use the model to price arbitrary man-made lotteries as demonstrated by Lucas (1982). Suppose that there is a market for one-period lottery tickets paying a stochastic prize ω in next period, and let $h(\omega, s', s)$ be a probability density for ω, conditioned on s' and s. The price of a lottery ticket in state s is denoted $q_L(s)$. To obtain an equilibrium expression for this price, we follow the steps in section 13.5, and include purchases of lottery tickets in the agent's budget constraint. (Quantities are negative if the agent is selling lottery tickets.) Then by reasoning similar to that leading to the arbitrage pricing formulas of chapter 8, we arrive at the lottery ticket price formula:

$$q_L(s) = \beta \int \int \frac{u'(s')}{u'(s)} \omega h(\omega, s', s) f(s', s) d\omega \, ds'. \tag{13.9.6}$$

Notice that if ω and s' are independent, the integrals of equation (13.9.6) can be factored and, recalling equation (13.8.5), we obtain

$$q_L(s) = \beta \int \frac{u'(s')}{u'(s)} f(s', s) \, ds' \cdot \int \omega h(\omega, s) d\omega = R_1(s)^{-1} E\{\omega | s\}. \tag{13.9.7}$$

Thus, the price of a lottery ticket is the price of a sure claim to one unit of consumption next period, times the expected payoff on a lottery ticket. There is no risk premium, since in a competitive market no one is in a position to impose risk on anyone else, and no premium need be charged for risks not borne.

13.9.3. The Modigliani-Miller theorem

The Modigliani and Miller theorem[15] describes circumstances under which the total value of a firm is independent of the firm's financial structure, that is, the particular evidences of debt and equity that it issues. Following Hirshleifer (1966) and Stiglitz (1969), the Modigliani-Miller theorem can be proved directly in a setting with complete state-contingent markets.

Suppose that an agent starts a firm at time t with a tree as its sole asset, and then immediately sells the firm to the public by issuing N number of shares

---

[15] See Modigliani and Miller (1958).

and B number of bonds as follows. Each bond promises to pay off r per period, and r is chosen so that rB is less than all possible realizations of future crops $y_{t+j}(s_{t+j})$, so that $y_{t+j} - rB$ is positive with probability one. After payments to bondholders, the owners of equity are entitled to the residual crop. Thus, the dividend of a share of equity is $(y_{t+j} - rB)/N$ in period $t+j$. Let p_t^B and p_t^N be the equilibrium prices of a bond and a share, respectively, which can be obtained by using the contingent claims prices:

$$p_t^B = \sum_{j=1}^{\infty} \int rQ_j(s_{t+j}|s_t)ds_{t+j}, \qquad (13.9.8)$$

$$p_t^N = \sum_{j=1}^{\infty} \int \frac{y_{t+j} - rB}{N} Q_j(s_{t+j}|s_t)ds_{t+j}. \qquad (13.9.9)$$

The total value of bonds and shares is then

$$p_t^B B + p_t^N N = \sum_{j=1}^{\infty} \int y_{t+j} Q_j(s_{t+j}|s_t)ds_{t+j}, \qquad (13.9.10)$$

which, by equations $(13.6.2)$ and $(13.9.1)$, is equal to the tree's initial value p_t. Equation $(13.9.10)$ exhibits the Modigliani-Miller proposition that the value of the firm, that is, the total value of the firm's bonds and equities, is independent of the number of bonds B outstanding. The total value of the firm is also independent of the coupon rate r.

The total value of the firm is independent of the financing scheme because the equilibrium prices of bonds and shares adjust to reflect the riskiness inherent in any mix of liabilities. To illustrate these equilibrium effects, let us assume that $u(c_t) = \ln c_t$ and y_{t+j} is i.i.d. over time so that $E_t(y_{t+j}) = E(y)$, and y_{t+j}^{-1} is also i.i.d. for all $j \geq 1$. With logarithmic preferences, we can define a stochastic discount factor as $m_{t+1} \equiv \beta \left(\frac{c_t}{c_{t+1}}\right)$ and express Euler equations like $(13.2.4)$ and $(13.2.5)$ in the unified form

$$Em_{t+1}R_{j,t+1} = 1 \qquad (13.9.11)$$

where $R_{j,t+1}$ is the one-period gross rate of return on asset j between t and $t+1$. It follows that with logarithmic preferences, the price of a tree p_t is given by equation $(13.7.1)$, and the other two asset prices are now

$$p_t^B = \sum_{j=1}^{\infty} E_t \left[r\beta^j \frac{u'(s_{t+j})}{u'(s_t)} \right] = \frac{\beta}{1-\beta} r E(y^{-1}) y_t, \qquad (13.9.12)$$

$$p_t^N = \sum_{j=1}^{\infty} E_t \left[\frac{y_{t+j} - rB}{N} \beta^j \frac{u'(s_{t+j})}{u'(s_t)} \right] = \frac{\beta}{1-\beta} \left[1 - rBE(y^{-1}) \right] \frac{y_t}{N}, \quad (13.9.13)$$

where we have used equations $(13.9.8)$, $(13.9.9)$, and $(13.9.1)$ and $y_t = s_t$. (The expression $[1 - rBE(y^{-1})]$ is positive because rB is less than the lowest possible realization of y.) As can be seen, the price of a share depends negatively on the number of bonds B and the coupon r, and also the number of shares N. We now turn to the expected rates of return on different assets, which should be related to their riskiness. First, notice that, with our special assumptions, the expected capital gains on issued bonds and shares are all equal to that of the underlying tree asset,

$$E_t \left[\frac{p_{t+1}^B}{p_t^B} \right] = E_t \left[\frac{p_{t+1}^N}{p_t^N} \right] = E_t \left[\frac{p_{t+1}}{p_t} \right] = E_t \left[\frac{y_{t+1}}{y_t} \right]. \quad (13.9.14)$$

It follows that any differences in expected total rates of return on assets must arise from the expected yields due to next period's dividends and coupons. Use equations $(13.7.1)$, $(13.9.12)$, and $(13.9.13)$ to get

$$\frac{r}{p_t^B} = \left\{ \left[1 - E_t(y_{t+1}) E_t(y_{t+1}^{-1}) \right] + E_t(y_{t+1}) E_t(y_{t+1}^{-1}) \right\} \frac{r}{p_t^B}$$

$$= \frac{1 - E(y)E(y^{-1})}{E(y^{-1})p_t} + \frac{E_t(y_{t+1})}{p_t} < E_t \left[\frac{y_{t+1}}{p_t} \right], \quad (13.9.15)$$

$$E_t \left[\frac{(y_{t+1} - rB)/N}{p_t^N} \right]$$

$$= \left\{ \left[1 - rBE(y^{-1}) \right] + rBE(y^{-1}) \right\} E_t \left[\frac{(y_{t+1} - rB)/N}{p_t^N} \right]$$

$$= \frac{E_t(y_{t+1} - rB)}{p_t} + \frac{rBE(y^{-1})E_t(y_{t+1} - rB)}{[1 - rBE(y^{-1})]p_t}$$

$$= \frac{E_t(y_{t+1})}{p_t} + \frac{rB[E(y^{-1})E(y) - 1]}{[1 - rBE(y^{-1})]p_t} > E_t \left[\frac{y_{t+1}}{p_t} \right], \quad (13.9.16)$$

where the two inequalities follow from Jensen's inequality, which states that $E(y^{-1}) > [E(y)]^{-1}$ for a nontrivial random variable y (i.e., one with a positive variance). Thus, from equations $(13.9.14)$–$(13.9.16)$, we can conclude that the firm's bonds (shares) earn a lower (higher) expected rate of return as compared to the underlying asset. Moreover, equation (13.9.16) shows that the expected rate of return on the shares is positively related to payments to bondholders rB. In other words, equity owners demand a higher expected return from a more leveraged firm because of the greater risk borne. Thus, despite the fact that Euler equation (13.9.11) holds for both the bond and equity, it is true that the expected return on equity exceeds the expected return on the risk-free bond.

13.10. Government debt

13.10.1. The Ricardian proposition

We now use a version of Lucas's tree model to describe the Ricardian proposition that tax financing and bond financing of a given stream of government expenditures are equivalent.[16] This proposition may be viewed as an application of the Modigliani-Miller theorem to government finance and obtains under circumstances in which the government is essentially like a firm in the constraints that it confronts with respect to its financing decisions.

We add to Lucas's model a government that spends current output according to a nonnegative stochastic process $\{g_t\}$ that satisfies $g_t < y_t$ for all t. The variable g_t denotes per capita government expenditures at t. For analytical convenience we assume that g_t is thrown away, giving no utility to private agents. The state $s_t = (y_t, g_t)$ of the economy is now a vector including the dividend y_t and government expenditures g_t. We assume that y_t and g_t are jointly described by a Markov process with transition density $f(s_{t+1}, s_t) = f(\{y_{t+1}, g_{t+1}\}, \{y_t, g_t\})$ where

$$\text{prob}\{y_{t+1} \leq y', g_{t+1} \leq g' | y_t = y, g_t = g\} = \int_0^{y'} \int_0^{g'} f(\{z, w\}, \{y, g\}) \, dw \, dz.$$

---

[16] An article by Robert Barro (1974) promoted strong interest in the Ricardian proposition. Barro described the proposition in a context distinct from the present one but closely related to it. Barro used an overlapping generations model but assumed altruistic agents who cared about their descendants. Restricting preferences to ensure an operative bequest motive, Barro described an overlapping generations structure that is equivalent to a model with an infinitely lived representative agent. See chapter 10 for more on Ricardian equivalence.

To emphasize that the dividend y_t and government expenditures g_t are solely functions of the current state s_t, we will use the notation $y_t = y(s_t)$ and $g_t = g(s_t)$.

The government finances its expenditures by issuing one-period debt that is permitted to be state contingent, and with a stream of lump-sum per capita taxes $\{\tau_t\}$, a stream that we assume is a stochastic process expressible at time t as a function of $s_t = (y_t, g_t)$ and any debt from last period. A general way of capturing that taxes and new issues of debt depend upon the current state s_t and the government's beginning-of-period debt, is to index both these government instruments by the history of all past states, $s^t = [s_0, s_1, \ldots, s_t]$. Hence, $\tau_t(s^t)$ is the lump-sum per capita tax in period t, given history s^t, and $b_t(s_{t+1}|s^t)$ is the amount of $(t+1)$ goods that the government promises at t to deliver, provided the economy is in state s_{t+1} at $(t+1)$, where this issue of debt is also indexed by the history s^t. In other words, we are adopting the "commodity space" s^t as we also did in chapter 8. For example, we let $c_t(s^t)$ denote the representative agent's consumption at time t, after history s^t.

We can here apply the three steps outlined earlier to construct equilibrium prices. Since taxation is lump sum without any distortionary effects, the competitive equilibrium consumption allocation still equals that of a planning problem where all agents are assigned the same Pareto weight. Thus, the social planning problem for our purposes is to maximize $E_0 \sum_{t=0}^{\infty} \beta^t u(c_t)$ subject to $c_t \leq y_t - g_t$, whose solution is $c_t = y_t - g_t$ which can alternatively be written as $c_t(s^t) = y(s_t) - g(s_t)$. Proceeding as we did in earlier sections, the equilibrium share price, interest rates, and state-contingent claims prices are described by

$$p(s_t) = E_t \sum_{j=1}^{\infty} \beta^j \frac{u'(y(s_{t+j}) - g(s_{t+j}))}{u'(y(s_t) - g(s_t))} y(s_{t+j}), \qquad (13.10.1)$$

$$R_j(s_t)^{-1} = \beta^j E_t \frac{u'(y(s_{t+j}) - g(s_{t+j}))}{u'(y(s_t) - g(s_t))}, \qquad (13.10.2)$$

$$Q_j(s_{t+j}|s_t) = \beta^j \frac{u'(y(s_{t+j}) - g(s_{t+j}))}{u'(y(s_t) - g(s_t))} f^j(s_{t+j}, s_t), \qquad (13.10.3)$$

where $f^j(s_{t+j}, s_t)$ is the j-step-ahead transition function that, for $j \geq 2$, obeys equation (13.9.2). It also useful to compute another set of state-contingent claims prices from chapter 8,

$$q_{t+j}^t(s^{t+j}) = Q_1(s_{t+j}|s_{t+j-1}) Q_1(s_{t+j-1}|s_{t+j-2}) \ldots Q_1(s_{t+1}|s_t)$$

$$= \beta^j \frac{u'(y(s_{t+j}) - g(s_{t+j}))}{u'(y(s_t) - g(s_t))} f(s_{t+j}, s_{t+j-1})$$
$$\cdot f(s_{t+j-1}, s_{t+j-2}) \cdots f(s_{t+1}, s_t). \qquad (13.10.4)$$

Here $q^t_{t+j}(s^{t+j})$ is the price of one unit of consumption delivered at time $t + j$, history s^{t+j}, in terms of date-t, history-s^t consumption good. Expression (13.10.4) can be derived from an arbitrage argument or an Euler equation evaluated at the equilibrium allocation. Notice that equilibrium prices (13.10.1)–(13.10.4) are independent of the government's tax and debt policy. Our next step in showing Ricardian equivalence is to demonstrate that the private agents' budget sets are also invariant to government financing decisions.

Turning first to the government's budget constraint, we have

$$g(s_t) = \tau_t(s^t) + \int Q_1(s_{t+1}|s_t) b_t(s_{t+1}|s^t) ds_{t+1} - b_{t-1}(s_t|s^{t-1}), \qquad (13.10.5)$$

where $b_t(s_{t+1}|s^t)$ is the amount of $(t + 1)$ goods that the government promises at t to deliver, provided the economy is in state s_{t+1} at $(t + 1)$, where this quantity is indexed by the history s^t at the time of issue. If the government decides to issue only one-period risk-free debt, for example, we have $b_t(s_{t+1}|s^t) = b_t(s^t)$ for all s_{t+1}, so that

$$\int Q_1(s_{t+1}|s_t) b_t(s^t) ds_{t+1} = b_t(s^t) \int Q_1(s_{t+1}|s_t) ds_{t+1} = b_t(s^t)/R_1(s_t).$$

Equation (13.10.5) then becomes

$$g(s_t) = \tau_t(s^t) + b_t(s^t)/R_1(s_t) - b_{t-1}(s^{t-1}). \qquad (13.10.6)$$

Equation (13.10.6) is a standard form of the government's budget constraint under conditions of certainty.

We can write the budget constraint (13.10.5) in the form

$$b_{t-1}(s_t|s^{t-1}) = \tau_t(s^t) - g(s_t) + \int Q_1(s_{t+1}|s_t) b_t(s_{t+1}|s^t) ds_{t+1}. \qquad (13.10.7)$$

Then we multiply the corresponding budget constraint in period $t + 1$ by $Q_1(s_{t+1}|s_t)$ and integrate over s_{t+1},

$$\int Q_1(s_{t+1}|s_t) b_t(s_{t+1}|s^t) ds_{t+1} = \int Q_1(s_{t+1}|s_t) \left[\tau_{t+1}(s^{t+1}) - g(s_{t+1}) \right] ds_{t+1}$$

$$+ \int \int Q_1(s_{t+1}|s_t) Q_1(s_{t+2}|s_{t+1}) b_{t+1}(s_{t+2}|s^{t+1}) ds_{t+2} ds_{t+1},$$

$$= \int q^t_{t+1}(s^{t+1}) \left[\tau_{t+1}(s^{t+1}) - g(s_{t+1}) \right] d(s^{t+1}|s^t)$$

$$+ \int q^t_{t+2}(s^{t+2}) b_{t+1}(s_{t+2}|s^{t+1}) d(s^{t+2}|s^t), \tag{13.10.8}$$

where we have introduced the following notation for taking multiple integrals,

$$\int x(s^{t+j}) d(s^{t+j}|s^t) \equiv \int \int \cdots \int x(s^{t+j}) ds_{t+j} \, ds_{t+j-1} \ldots ds_{t+1}.$$

Expression (13.10.8) can be substituted into budget constraint (13.10.7) by eliminating the bond term $\int Q_1(s_{t+1}|s_t) b_t(s_{t+1}|s^t) ds_{t+1}$. After repeated substitutions of consecutive budget constraints, we eventually arrive at the present value budget constraint [17]

$$b_{t-1}(s_t|s^{t-1}) = \tau_t(s^t) - g(s_t)$$

$$+ \sum_{j=1}^{\infty} \int q^t_{t+j}(s^{t+j}) \left[\tau_{t+j}(s^{t+j}) - g(s_{t+j}) \right] d(s^{t+j}|s^t)$$

$$= \tau_t(s^t) - g(s_t) - \sum_{j=1}^{\infty} \int Q_j(s_{t+j}|s_t) g(s_{t+j}) ds_{t+j}$$

$$+ \sum_{j=1}^{\infty} \int q^t_{t+j}(s^{t+j}) \tau_{t+j}(s^{t+j}) d(s^{t+j}|s^t) \tag{13.10.9}$$

as long as

$$\lim_{k \to \infty} \int q^t_{t+k+1}(s^{t+k+1}) b_{t+k}(s_{t+k+1}|s^{t+k}) d(s^{t+k+1}|s^t) = 0. \tag{13.10.10}$$

A strictly positive limit of equation (13.10.10) can be ruled out by using the transversality conditions for private agents' holdings of government bonds that we here denote $b^d_t(s_{t+1}|s^t)$. (The superscript d stands for demand and distinguishes the variable from government's supply of bonds.) Next, we simply assume away the case of a strictly negative limit of expression (13.10.10), since it would correspond to a rather uninteresting situation where the government

---

[17] The second equality follows from the expressions for j-step-ahead contingent-claim-pricing functions in (13.10.3) and (13.10.4), and exchanging orders of integration.

accumulates "paper claims" against the private sector by setting taxes higher
than needed for financial purposes. Thus, equation (13.10.9) states that the
value of government debt maturing at time t equals the present value of the
stream of government surpluses.

It is a key implication of the government's present value budget constraint
(13.10.9) that all government debt has to be backed by future primary surpluses
$[\tau_{t+j}(s^{t+j}) - g(s_{t+j})]$, i.e., government debt is the capitalized value of government net-of-interest surpluses. A government that starts out with a positive debt
must run a primary surplus for some state realization in some future period. It
is an implication of the fact that the economy is dynamically efficient.[18]

We now turn to a private agent's budget constraint at time t,

$$c_t(s^t) + \tau_t(s^t) + p(s_t)N_t(s^t) + \int Q_1(s_{t+1}|s_t)b_t^d(s_{t+1}|s^t)ds_{t+1}$$

$$\leq [p(s_t) + y(s_t)]\, N_{t-1}(s^{t-1}) + b_{t-1}^d(s_t|s^{t-1}). \qquad (13.10.11)$$

We multiply the corresponding budget constraint in period $t+1$ by $Q_1(s_{t+1}|s_t)$
and integrate over s_{t+1}. The resulting expression is substituted into equation
(13.10.11) by eliminating the purchases of government bonds in period t. The
two consolidated budget constraints become

$$c_t(s^t) + \tau_t(s^t) + \int \left[c_{t+1}(s^{t+1}) + \tau_{t+1}(s^{t+1})\right] Q_1(s_{t+1}|s_t)ds_{t+1}$$

$$+ \left\{p(s_t) - \int [p(s_{t+1}) + y(s_{t+1})]\, Q_1(s_{t+1}|s_t)ds_{t+1}\right\} N_t(s^t)$$

$$+ \int p(s_{t+1})N_{t+1}(s^{t+1})Q_1(s_{t+1}|s_t)ds_{t+1}$$

$$+ \int\int Q_1(s_{t+1}|s_t)Q_1(s_{t+2}|s_{t+1})b_{t+1}^d(s_{t+2}|s^{t+1})ds_{t+2}ds_{t+1}$$

$$\leq [p(s_t) + y(s_t)]\, N_{t-1}(s^{t-1}) + b_{t-1}^d(s_t|s^{t-1}), \qquad (13.10.12)$$

where the expression in braces is zero by an arbitrage argument. When continuing the consolidation of all future budget constraints, we eventually find

---

[18] In contrast, compare to our analysis in chapter 9 where we demonstrated that unbacked
government debt or fiat money can be valued by private agents when the economy is dynamically inefficient. These different findings are related to the question of whether or not there
can exist asset bubbles. See footnote 12.

that

$$c_t(s^t) + \tau_t(s^t) + \sum_{j=1}^{\infty} \int \left[c_{t+j}(s^{t+j}) + \tau_{t+j}(s^{t+j}) \right] q_{t+j}^t(s^{t+j}) d(s^{t+j}|s^t)$$

$$\leq [p(s_t) + y(s_t)] N_{t-1}(s^{t-1}) + b_{t-1}^d(s_t|s^{t-1}), \qquad (13.10.13)$$

where we have imposed limits equal to zero for the two terms involving $N_{t+k}(s^{t+k})$ and $b_{t+k}^d(s_{t+k+1}|s^{t+k})$ when k goes to infinity. The two terms vanish because of transversality conditions and the reasoning in the preceding paragraph. Thus, equation (13.10.13) states that the present value of the stream of consumption and taxes cannot exceed the agent's initial wealth at time t.

Finally, we substitute the government's present value budget constraint (13.10.9) into that of the representative agent (13.10.13) by eliminating the present value of taxes. Thereafter, we invoke equilibrium conditions $N_{t-1}(s^{t-1}) = 1$ and $b_{t-1}^d(s_t|s^{t-1}) = b_{t-1}(s_t|s^{t-1})$ and we use the equilibrium expressions for prices (13.10.1) and (13.10.3) to express $p(s_t)$ as the sum of all future dividends discounted by the j-step-ahead pricing kernel $Q_j(s_{t+j}|s_t)$. The result is

$$c_t(s^t) + \sum_{j=1}^{\infty} \int c_{t+j}(s^{t+j}) q_{t+j}^t(s^{t+j}) d(s^{t+j}|s^t)$$

$$\leq y(s_t) - g(s_t) + \sum_{j=1}^{\infty} \int [y(s_{t+j}) - g(s_{t+j})] Q_j(s_{t+j}|s_t) ds_{t+j}. \quad (13.10.14)$$

Given that equilibrium prices have been shown to be independent of the government's tax and debt policy, the implication of formula (13.10.14) is that the representative agent's budget set is also invariant to government financing decisions. Having no effects on prices and private agents' budget constraints, taxes and government debt do not affect private consumption decisions. [19]

We can summarize this discussion with the following proposition:

---

[19] We have indexed choice variables by the history s^t which is the commodity space for this economy. But it is instructive to verify that private agents will not choose history-dependent consumption when facing equilibrium prices (13.10.4). At time t after history s^t, an agent's first-order with respect to $c_{t+j}(s^{t+j})$ is given by

$$u'\left(c_t(s^t) \right) q_{t+j}^t(s^{t+j}) = \beta^j u'\left(c_{t+j}(s^{t+j}) \right) f(s_{t+j}, s_{t+j-1})$$
$$\cdot f(s_{t+j-1}, s_{t+j-2}) \dots f(s_{t+1}, s_t).$$

RICARDIAN PROPOSITION: Equilibrium consumption and prices depend only on the stochastic process for output y_t and government expenditure g_t. In particular, consumption and state-contingent prices are both independent of the stochastic process τ_t for taxes.

In this model, the choices of the time pattern of taxes and government bond issues have no effect on any "relevant" equilibrium price or quantity. The reason is that, as indicated by equations (13.10.5) and (13.10.9), larger deficits $(g_t - \tau_t)$, accompanied by larger values of government debt $b_t(s_{t+1})$, now signal future government surpluses. The agents in this model accumulate these government bond holdings and expect to use their proceeds to pay off the very future taxes whose prospects support the value of the bonds. Notice also that, given the stochastic process for (y_t, g_t), the way in which the government finances its deficits (or invests its surpluses) is irrelevant. Thus, it does not matter whether it borrows using short-term, long-term, safe, or risky instruments. This irrelevance of financing is an application of the Modigliani-Miller theorem. Equation (13.10.9) may be interpreted as stating that the present value of the government is independent of such financing decisions.

The next section elaborates on the significance that future government surpluses in equation (13.10.9) are discounted with contingent claims prices and not the risk-free interest rate, even though the government may choose to issue only safe debt. This distinction is made clear by using equations (13.10.4) and (13.10.2) to rewrite equation (13.10.9) as follows,

$$
b_{t-1}(s_t) = \tau_t - g_t + \sum_{j=1}^{\infty} E_t \left[\beta^j \frac{u'(y_{t+j} - g_{t+j})}{u'(y_t - g_t)} (\tau_{t+j} - g_{t+j}) \right]
$$

$$
= \tau_t - g_t + \sum_{j=1}^{\infty} \left\{ R_{jt}^{-1} E_t [\tau_{t+j} - g_{t+j}] \right.
$$

After dividing this expression by the corresponding first-order condition with respect to $c_{t+j}(\tilde{s}^{t+j})$ where $\tilde{s}^t = s^t$ and $\tilde{s}_{t+j} = s_{t+j}$, and invoking (13.10.4), we obtain

$$
1 = \frac{u'\left(c_{t+j}(s^{t+j})\right)}{u'\left(c_{t+j}(\tilde{s}^{t+j})\right)} \qquad \Longrightarrow \qquad c_{t+j}(s^{t+j}) = c_{t+j}(\tilde{s}^{t+j}).
$$

Hence, the agent finds it optimal to choose $c_{t+j}(s^{t+j}) = c_{t+j}(\tilde{s}^{t+j})$ whenever $s_{t+j} = \tilde{s}_{t+j}$, regardless of the history leading up to that state in period $t + j$.

$$+ \operatorname{cov}_t \left[\beta^j \frac{u'(y_{t+j} - g_{t+j})}{u'(y_t - g_t)}, \ \tau_{t+j} - g_{t+j} \right] \Bigg\} . \tag{13.10.15}$$

13.10.2. No Ponzi schemes

Bohn (1995) considers a nonstationary discrete-state-space version of Lucas's tree economy to demonstrate the importance of using a proper criterion when assessing long-run sustainability of fiscal policy, that is, determining whether the government's present-value budget constraint and the associated transversality condition are satisfied as in equations (13.10.9) and (13.10.10) of the earlier model. The present-value budget constraint says that any debt at time t must be repaid with future surpluses because the transversality condition rules out Ponzi schemes—financial trading strategies that involve rolling over an initial debt with interest forever.

At each date t, there is now a finite set of possible states of nature, and s^t is the history of all past realizations, including the current one. Let $\pi_{t+j}(s^{t+j}|s^t)$ be the probability of a history s^{t+j}, conditional on history s^t having been realized up until time t. The dividend of a tree in period t is denoted $y_t(s^t) > 0$, and can depend on the whole history of states of nature. The stochastic process is such that a private agent's expected utility remains bounded for any fixed fraction $c \in (0, 1]$ of the stream $y_t(s^t)$, implying

$$\lim_{j \to \infty} E_t \beta^j u'(c_{t+j}) c_{t+j} = 0 \tag{13.10.16}$$

for $c_t = c \cdot y_t(s^t)$.[20]

Bohn (1995) examines the following government policy. Government spending is a fixed fraction $(1 - c) = g_t/y_t$ of income. The government issues safe one-period debt so that the ratio of end-of-period debt to income is constant

---

[20] Expected lifetime utility is bounded if the sequence of "remainders" converges to zero,

$$0 = \lim_{k \to \infty} E_t \sum_{j=k}^{\infty} \beta^j u(c_{t+j}) \geq \lim_{k \to \infty} E_t \sum_{j=k}^{\infty} \beta^j \left\{ u'(c_{t+j}) c_{t+j} \right\} \geq 0,$$

where the first inequality is implied by concavity of $u(\cdot)$. We obtain equation (13.10.16) because $u'(c_{t+j})c_{t+j}$ is positive at all dates.

at some level $b = R_{1t}^{-1} b_t / y_t$, i.e., $b_t(s^t) = R_{1t}\, b\, y_t(s^t)$. Given any initial debt, taxes can then be computed from budget constraint (13.10.6). It is intuitively clear that this policy can be sustained forever, but let us formally show that the government's transversality condition holds in any period t, given history s^t,

$$\lim_{j \to \infty} \sum_{s^{t+j+1}|s^t} \tilde{q}_{t+j+1}^t \left(s^{t+j+1}\right) b_{t+j} \left(s^{t+j}\right) = 0, \qquad (13.10.17)$$

where the summation over $s^{t+j}|s^t$ means that we sum over all possible histories $\tilde{s}^{t+j}$ such that $\tilde{s}^t = s^t$, and $\tilde{q}_{t+j}^t(s^{t+j})$ is the price at t, given history s^t, of a unit of consumption good to be delivered in period $t + j$, contingent on the realization of history s^{t+j}. In an equilibrium, we have

$$\tilde{q}_{t+j}^t \left(s^{t+j}\right) = \beta^j \frac{u' \left[c \cdot y_{t+j} \left(s^{t+j}\right)\right]}{u' \left[c \cdot y_t \left(s^t\right)\right]} \pi_{t+j} \left(s^{t+j}|s^t\right). \qquad (13.10.18)$$

After substituting equation (13.10.18), the debt policy, and $c_t = c \cdot y_t$ into the left-hand side of equation (13.10.17),

$$\lim_{j \to \infty} E_t \left[\beta^{j+1} \frac{u' \left(c_{t+j+1}\right)}{u' \left(c_t\right)} R_{1,t+j}\, b \frac{c_{t+j}}{c}\right]$$

$$= \lim_{j \to \infty} E_t E_{t+j} \left[\beta^j \frac{u' \left(c_{t+j}\right)}{u' \left(c_t\right)} \beta \frac{u' \left(c_{t+j+1}\right)}{u' \left(c_{t+j}\right)} R_{1,t+j}\, b \frac{c_{t+j}}{c}\right]$$

$$= \frac{b}{c\, u' \left(c_t\right)} \lim_{j \to \infty} E_t \left[\beta^j u' \left(c_{t+j}\right) c_{t+j}\right] = 0.$$

The first of these equalities invokes the law of iterated expectations; the second equality uses the equilibrium expression for the one-period interest rate, which is still given by expression (13.10.2); and the final equality follows from (13.10.16). Thus, we have shown that the government's transversality condition and therefore its present-value budget constraint are satisfied.

Bohn (1995) cautions us that this conclusion of fiscal sustainability might erroneously be rejected if we instead use the risk-free interest rate to compute present values. To derive expressions for the safe interest rate, we assume that preferences are given by the constant relative risk-aversion utility function $u(c_t) = (c_t^{1-\gamma} - 1)/(1 - \gamma)$, and the dividend y_t grows at the rate $\tilde{y}_t = y_t/y_{t-1}$ which is i.i.d. with mean $E(\tilde{y})$. Thus, risk-free interest rates given by equation (13.10.2) become

$$R_{jt}^{-1} = E_t \left[\beta^j \left(\prod_{i=1}^{j} \tilde{y}_{t+i}\right)^{-\gamma}\right] = \prod_{i=1}^{j} E \left(\beta \tilde{y}^{-\gamma}\right) = R_1^{-j},$$

where R_1 is the time-invariant one-period risk-free interest rate. That is, the term structure of interest rates obeys the pure expectations theory, since interest rates are nonstochastic. (The analogue to expression $(13.8.9)$ for this economy would therefore be one where the covariance term is zero.)

For the sake of the argument, we now compute the expected value of future government debt discounted at the safe interest rate and take the limit

$$\lim_{j\to\infty} E_t\Big(\frac{b_{t+j}}{R_{j+1,t}}\Big) = \lim_{j\to\infty} E_t\Big(\frac{R_{1,t+j}\, by_{t+j}}{R_{j+1,t}}\Big)$$

$$= \lim_{j\to\infty} E_t\Big(\frac{R_1\, by_t\, \prod_{i=1}^{j} \tilde{y}_{t+i}}{R_1^{j+1}}\Big)$$

$$= by_t \lim_{j\to\infty}\Big[\frac{E(\tilde{y})}{R_1}\Big]^j = \begin{cases} 0, & \text{if } R_1 > E(\tilde{y}); \\ by_t, & \text{if } R_1 = E(\tilde{y}); \\ \infty, & \text{if } R_1 < E(\tilde{y}). \end{cases} \qquad (13.10.19)$$

The limit is infinity if the expected growth rate of dividends $E(\tilde{y})$ exceeds the risk-free rate R_1. The level of the safe interest rate depends on risk aversion and on the variance of dividend growth. This dependence is best illustrated with an example. Suppose there are two possible states of dividend growth that are equally likely to occur with a mean of 1 percent, $E(\tilde{y}) - 1 = .01$, and let the subjective discount factor be $\beta = .98$. Figure 13.10.1 depicts the equilibrium interest rate R_1 as a function of the standard deviation of dividend growth and the coefficient of relative risk aversion γ. For $\gamma = 0$, agents are risk neutral, so the interest rate is given by $\beta^{-1} \approx 1.02$ regardless of the amount of uncertainty. When making agents risk averse by increasing γ, there are two opposing effects on the equilibrium interest rate. On the one hand, higher risk aversion implies also that agents are less willing to substitute consumption over time. Therefore, there is an upward pressure on the interest rate to make agents accept an upward-sloping consumption profile. This fact completely explains the positive relationship between R_1 and γ when the standard deviation of growth is zero, that is, when deterministic growth is 1 percent. On the other hand, higher risk aversion in an uncertain environment means that agents attach a higher value to sure claims to future consumption, which tends to increase the bond price R_1^{-1}. As a result, Figure 13.10.1 shows how the risk-free interest R_1 falls below the expected gross growth rate of the economy when agents are sufficiently risk averse and the standard deviation of dividend growth is sufficiently large.[21]

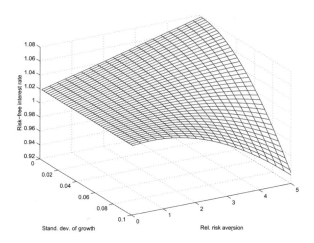

Figure 13.10.1: The risk-free interest rate R_1 as a function of the coefficient of relative risk aversion γ and the standard deviation of dividend growth. There are two states of dividend growth that are equally likely to occur with a mean of 1 percent, $E(\tilde{y}) - 1 = .01$, and the subjective discount factor is $\beta = .98$.

If $R_1 \leq E(\tilde{y})$ so that the expected value of future debt discounted at the safe interest rate does not converge to zero in equation (13.10.19), it follows that the expected sum of all future government surpluses discounted at the safe interest rate in equation (13.10.15) falls short of the initial debt. In fact, our example is then associated with negative expected surpluses at all future horizons,

$$E_t \left(\tau_{t+j} - g_{t+j} \right) = E_t \left(b_{t+j-1} - b_{t+j}/R_{1,t+j} \right) = E_t \left[\left(R_1 - \tilde{y}_{t+j} \right) b y_{t+j-1} \right]$$

---

[21] A risk-free interest rate less than the growth rate would indicate dynamic inefficiency in a deterministic steady state but not necessarily in a stochastic economy. Our model here of an infinitely lived representative agent is dynamically efficient. For discussions of dynamic inefficiency, see Diamond (1965) and Romer (1996, chap. 2).

$$= [R_1 - E(\tilde{y})] \, b \, [E(\tilde{y})]^{j-1} \, y_t \begin{cases} > 0, & \text{if } R_1 > E(\tilde{y}); \\ = 0, & \text{if } R_1 = E(\tilde{y}); \\ < 0, & \text{if } R_1 < E(\tilde{y}); \end{cases} \qquad (13.10.20)$$

where the first equality invokes budget constraint (13.10.6). Thus, for $R_1 \leq E(\tilde{y})$, the sum of covariance terms in equation (13.10.15) must be positive. The described debt policy also clearly has this implication where, for example, a low realization of $\tilde{y}_{t+j}$ implies a relatively high marginal utility of consumption and at the same time forces taxes up in order to maintain the targeted debt-income ratio in the face of a relatively low y_{t+j}.

As pointed out by Bohn (1995), this example illustrates the problem with empirical studies, such as Hamilton and Flavin (1986), Wilcox (1989), Hansen, Roberds and Sargent (1991), Gali (1991), and Roberds (1996), which rely on safe interest rates as discount factors when assessing the sustainability of fiscal policy. Such an approach would only be justified if future government surpluses were uncorrelated with future marginal utilities so that the covariance terms in equation (13.10.15) would vanish. This condition is trivially true in a non-stochastic economy or if agents are risk neutral; otherwise, it is difficult, in practice, to imagine a tax and spending policy that is uncorrelated with the difference between aggregate income and government spending that determines the marginal utility of consumption.

Chapter 14
Asset Pricing Empirics

14.1. Introduction

In chapter 13, we repeatedly encountered an object that in this chapter we shall call a stochastic discount factor m_{t+1}, namely

$$m_{t+1} = \beta \left(\frac{C_{t+1}}{C_t} \right)^{-\gamma}, \tag{14.1.1}$$

where β is a discount factor, γ is a coefficient of relative risk aversion, and C_t is the consumption of a representative consumer. The asset pricing theories in chapter 13 can be summarized in a nutshell as asserting that for any asset j traded by a representative consumer, its one period gross return $R_{j,t+1}$ must satisfy

$$E_t \left(m_{t+1} R_{j,t+1} \right) = 1. \tag{14.1.2}$$

Empirically, for the stochastic discount factor (14.1.1), restriction (14.1.2) fails to work well when applied to data on returns of stocks and risk-free bonds. Mehra and Prescott (1985) called this difficulty the 'equity premium puzzle.' As we explain in this chapter, a substantial part of the problem is that with aggregate U.S. data for C_t and 'reasonable' values for γ, $\left(\frac{C_{t+1}}{C_t} \right)^{-\gamma}$ is simply insufficiently volatile. This chapter first describes what is commonly meant by 'reasonable' values for γ. Then we describe the equity premium puzzle, other affiliated asset pricing puzzles, and some approaches to explaining them. Our major theme is how to modify the standard CRRA stochastic discount factor (14.1.1) in ways that can make (14.1.2) fit key features of the returns data better. We shall study some theories that increase the volatility of the stochastic discount factor by multiplying $\beta \left(\frac{C_{t+1}}{C_t} \right)^{-\gamma}$ with a volatile random object that reflects either aspects of the preferences of a representative consumer or heterogeneity in the distribution of consumption within a collection of consumers.

14.2. Interpretation of risk-aversion parameter

To understand why the measured equity premium is a puzzle, it is important to interpret γ in (14.1.1), a parameter that measures attitudes about gambles. Economists' prejudice that reasonable values of the coefficient of relative risk aversion must be below 3 comes from experiments that confront people with gambles drawn from well understood probability distributions.

The asset-pricing literature often uses the constant relative risk-aversion utility function

$$u\left(C\right) = \frac{C^{1-\gamma}}{1-\gamma}.$$

Note that

$$\gamma = \frac{-Cu''\left(C\right)}{u'\left(C\right)},$$

which is the individual's coefficient of relative risk aversion.

We want to interpret the parameter γ in terms of a preference for avoiding risk. Following Pratt (1964), consider offering two alternatives to a consumer who starts off with risk-free consumption level C: he can receive $C - \Delta_C$ with certainty or a lottery paying $C-y$ with probability .5 and $C+y$ with probability .5. For given values of y and C, we want to find the value of $\Delta_C = \Delta_C(y, C)$ that leaves the consumer indifferent between these two choices. That is, we want to find the function $\Delta_C(y, C)$ that solves

$$u\left[C - \Delta_C\left(y, C\right)\right] = .5u\left(C + y\right) + .5u\left(C - y\right). \tag{14.2.1}$$

For given values of C, y, we can solve the nonlinear equation (14.2.1) for Δ_C.

Alternatively, for small values of y, we can appeal to Pratt's local argument. Taking a Taylor series expansion of $u(C - \Delta_C)$ around the point c gives[1]

$$u\left(C - \Delta_C\right) = u\left(C\right) - \Delta_C u'\left(C\right) + O\left(\Delta_C^2\right). \tag{14.2.2}$$

Taking a Taylor series expansion of $u(C + \tilde{y})$ gives

$$u\left(C + \tilde{y}\right) = u\left(C\right) + \tilde{y}u'\left(C\right) + \frac{1}{2}\tilde{y}^2 u''\left(C\right) + O\left(\tilde{y}^3\right), \tag{14.2.3}$$

---

[1] Here $O(\cdot)$ means terms of order at most $(\cdot)$, while $o(\cdot)$ means terms of smaller order than $(\cdot)$.

$\gamma \setminus y$	10	100	1,000	5,000
2	.02	.2	20	500
5	.05	5	50	1,217
10	.1	1	100	2,212

Table 14.2.1: Risk premium $\Delta_C(y, C)$ for various values of y and γ

where $\tilde{y}$ is the random variable that takes value y with probability .5 and $-y$ with probability .5. Taking expectations on both sides gives

$$Eu\left(C + \tilde{y}\right) = u\left(C\right) + \frac{1}{2}y^2 u''\left(C\right) + o\left(y^2\right). \qquad (14.2.4)$$

Equating formulas (14.2.2) and (14.2.4) and ignoring the higher-order terms gives

$$\Delta_C\left(y, C\right) \approx \frac{1}{2}y^2 \left[\frac{-u''\left(C\right)}{u'\left(C\right)}\right].$$

For the constant relative risk-aversion utility function, we have

$$\Delta_C\left(y, C\right) \approx \frac{1}{2}y^2 \frac{\gamma}{C}.$$

This can be expressed as

$$\Delta_C/y \approx \frac{1}{2}\gamma\left(y/C\right). \qquad (14.2.5)$$

The left side is the percentage premium that the consumer is willing to pay to avoid a fair bet of size y; the right side is one-half γ times the ratio of the size of the bet y to his initial consumption level C.

Following Cochrane (1997), think of confronting someone with initial consumption of \$50,000 per year with a 50–50 chance of winning or losing y dollars. How much would the person be willing to pay to avoid that risk? For $C = 50,000$, we calculated Δ_C from equation (14.2.1) for values of $y = 10, 100, 1,000, 5,000$ (see Table 14.2.1). A common reaction to these premiums is that for values of γ even as high as 5, they are too big. This result is one important source of macroeconomists' prejudice that γ should not be much higher than 2 or 3.

14.3. The equity premium puzzle

Table 14.3.1 depicts empirical first and second moments of yields on relatively riskless bonds and risky equity in the U.S. data over the 90-year period 1889–1978. The average real yield on the Standard & Poor's 500 index was 7 percent, while the average yield on short-term debt was only 1 percent. The equity premium puzzle is that with aggregate consumption data, it takes an extraordinarily large value of the coefficient of relative risk aversion to generate such a large gap between the returns on equities and risk-free securities.[2]

	Mean	Variance-Covariance		
		$1 + r^s_{t+1}$	$1 + r^b_{t+1}$	c_{t+1}/c_t
$1 + r^s_{t+1}$	1.070	0.0274	0.00104	0.00219
$1 + r^b_{t+1}$	1.010		0.00308	−0.000193
c_{t+1}/c_t	1.018			0.00127

Table 14.3.1: Summary statistics for U.S. annual data, 1889–1978. The quantity $1 + r^s_{t+1}$ is the real return to stocks, $1 + r^b_{t+1}$ is the real return to relatively riskless bonds, and c_{t+1}/c_t is the growth rate of per capita real consumption of nondurables and services. Source: Kocherlakota (1996a, Table 1), who uses the same data as Mehra and Prescott (1985).

We choose to proceed in the fashion of Hansen and Singleton (1983) and to illuminate the equity premium puzzle by studying unconditional averages of Euler equations under the assumptions that returns are log normal. Let the real rates of return on stocks and bonds between periods t and $t + 1$ be denoted $1 + r^s_{t+1}$ and $1 + r^b_{t+1}$, respectively. In our Lucas tree model, these numbers would be given by $1 + r^s_{t+1} = (y_{t+1} + p_{t+1})/p_t$ and $1 + r^b_{t+1} = R_{1t}$. Concerning the real rate of return on bonds, we now use time subscript $t + 1$ to allow for uncertainty at time t about its realization. Since the numbers in Table 14.3.1

---

[2] For insightful reviews and lists of possible resolutions of the equity premium puzzle, see Aiyagari (1993), Kocherlakota (1996a), and Cochrane (1997).

are computed on the basis of nominal bonds, real bond yields are subject to inflation uncertainty. To allow for such uncertainty and to switch notation, we rewrite Euler equations (13.2.4) and (13.2.5) as

$$1 = \beta E_t \left[\left(1 + r_{t+1}^i \right) \frac{u'(c_{t+1})}{u'(c_t)} \right], \quad \text{for} \quad i = s, b. \tag{14.3.1}$$

We posit exogenous stochastic processes for both endowments (consumption) and rates of return,

$$\frac{C_{t+1}}{C_t} = \bar{c}_\triangle \exp \left\{ \varepsilon_{c,t+1} - \sigma_c^2/2 \right\}, \tag{14.3.2}$$

$$1 + r_{t+1}^i = \left(1 + \bar{r}^i \right) \exp \left\{ \varepsilon_{i,t+1} - \sigma_i^2/2 \right\}, \quad \text{for} \quad i = s, b, \tag{14.3.3}$$

where exp is the exponential function and $\{\varepsilon_{c,t+1}, \varepsilon_{s,t+1}, \varepsilon_{b,t+1}\}$ are jointly normally distributed with zero means and variances $\{\sigma_c^2, \sigma_s^2, \sigma_b^2\}$. Thus, the logarithm of consumption growth and the logarithms of rates of return are jointly normally distributed. When the logarithm of a random variable η is normally distributed with some mean μ and variance σ^2, the mean of η is $\exp(\mu + \sigma^2/2)$. Thus, the mean of consumption growth and the means of real yields on stocks and bonds are here equal to $\bar{c}_\triangle$, $1 + \bar{r}^s$, and $1 + \bar{r}^b$, respectively.

Assume the constant relative risk-aversion utility function $u(C_t) = (C_t^{1-\gamma} - 1)/(1 - \gamma)$. After substituting this utility function and the stochastic processes (14.3.2) and (14.3.3) into equation (14.3.1), we take unconditional expectations of equation (14.3.1). By the law of iterated expectations, we obtain

$$1 = \beta E \left[\left(1 + r_{t+1}^i \right) \left(\frac{c_{t+1}}{c_t} \right)^{-\gamma} \right],$$
$$= \beta \left(1 + \bar{r}^i \right) \bar{c}_\triangle^{-\gamma} E \left\{ \exp \left[\varepsilon_{i,t+1} - \sigma_i^2/2 - \gamma \left(\varepsilon_{c,t+1} - \sigma_c^2/2 \right) \right] \right\}$$
$$= \beta \left(1 + \bar{r}^i \right) \bar{c}_\triangle^{-\gamma} \exp \left[(1 + \gamma) \gamma \sigma_c^2/2 - \gamma \operatorname{cov}(\varepsilon_i, \varepsilon_c) \right],$$
$$\text{for} \quad i = s, b, \tag{14.3.4}$$

where the second equality follows from the expression in braces being log normally distributed. Taking logarithms of equation (14.3.4) yields

$$\log \left(1 + \bar{r}^i \right) = -\log(\beta) + \gamma \log(\bar{c}_\triangle) - (1 + \gamma) \gamma \sigma_c^2/2 + \gamma \operatorname{cov}(\varepsilon_i, \varepsilon_c),$$
$$\text{for} \quad i = s, b. \tag{14.3.5}$$

It is informative to interpret equation (14.3.5) for the risk-free interest rate in Bohn's model of section 13.10.2 under the auxiliary assumption of log normally distributed dividend growth, so that equilibrium consumption growth is given by equation (14.3.2). Since interest rates are time invariant, we have $\text{cov}(\varepsilon_b, \varepsilon_c) = 0$. In the case of risk-neutral agents ($\gamma = 0$), equation (14.3.5) has the familiar implication that the interest rate is equal to the inverse of the subjective discount factor β, regardless of any uncertainty. In the case of deterministic growth ($\sigma_c^2 = 0$), the second term of equation (14.3.5) says that the safe interest rate is positively related to the coefficient of relative risk aversion γ, as we also found in the example of Figure 13.10.1. Likewise, the downward pressure on the interest rate due to uncertainty in Figure 13.10.1 shows up as the third term of equation (14.3.5).[3] This downward pressure as σ_c^2 grows reflects the workings of a precautionary savings motive of the type to be discussed in chapter 17. At a given γ, a higher σ_c^2 induces people to want to save more. The risk-free rate must decline to prevent them from doing so.

We now turn to the equity premium by taking the difference between the expressions for the rates of return on stocks and bonds, as given by equation (14.3.5),

$$\log\left(1 + \bar{r}^s\right) - \log\left(1 + \bar{r}^b\right) = \gamma\left[\text{cov}\left(\varepsilon_s, \varepsilon_c\right) - \text{cov}\left(\varepsilon_b, \varepsilon_c\right)\right]. \tag{14.3.6}$$

Using the approximation $\log(1+r) \approx r$, and noting that the covariance between consumption growth and real yields on bonds in Table 14.3.1 is virtually zero, we can write the theory's interpretation of the historical equity premium as

$$\bar{r}^s - \bar{r}^b \approx \gamma \, \text{cov}\left(\varepsilon_s, \varepsilon_c\right). \tag{14.3.7}$$

After approximating $\text{cov}(\varepsilon_s, \varepsilon_c)$ with the covariance between consumption growth and real yields on stocks in Table 14.3.1, equation (14.3.7) states that an equity premium of 6 percent would require a γ of 27. Kocherlakota (1996a, p. 52) summarizes the prevailing view that "a vast majority of economists believe that values of $[\gamma]$ above ten (or, for that matter, above five) imply highly implausible behavior on the part of individuals." That statement is a reference to the argument of Pratt, described in the preceding section. This constitutes the equity

---

[3] Since the term involves the square of γ, the safe interest rate must eventually be a decreasing function of the coefficient of relative risk aversion when $\sigma_c^2 > 0$, but only at very high and therefore uninteresting values for the coefficient of relative risk aversion.

premium puzzle. Mehra and Prescott (1985) and Weil (1989) point out that an additional part of the puzzle relates to the low observed historical mean of the riskless rate of return. We describe this *risk-free rate puzzle* in section 14.6.[4]

Expression (14.3.7) indicates how excess returns compensate for risk. Assets that give low returns in bad consumption states (i.e., assets for which $\text{cov}(\varepsilon_s, \varepsilon_c) > 0$) are not useful for hedging consumption risk. Therefore, such assets have low prices, meaning that they are associated with high excess returns.

14.4. Market price of risk

Gallant, Hansen, and Tauchen (1990) and Hansen and Jagannathan (1991) interpret the equity premium puzzle in terms of the high "market price of risk" implied by time series data on asset returns. The market price of risk is defined in terms of asset prices and their one-period payoffs. Let q_t be the time t price of an asset bearing a one-period payoff p_{t+1}. A household's Euler equation for holdings of this asset can be represented as

$$q_t = E_t \left(m_{t+1} p_{t+1} \right) \tag{14.4.1}$$

where $m_{t+1} = \frac{\beta u'(C_{t+1})}{u'(C_t)}$ serves as a stochastic discount factor for discounting the stochastic payoff p_{t+1}. Using the definition of a conditional covariance, equation (14.4.1) can be written

$$q_t = E_t m_{t+1} E_t p_{t+1} + \text{cov}_t \left(m_{t+1}, p_{t+1} \right).$$

Applying the Cauchy-Schwarz inequality[5] to the covariance term in the preceding equation gives

$$\frac{q_t}{E_t m_{t+1}} \geq E_t p_{t+1} - \left(\frac{\sigma_t \left(m_{t+1} \right)}{E_t m_{t+1}} \right) \sigma_t \left(p_{t+1} \right), \tag{14.4.2}$$

---

[4] For $\beta < 0.99$, equation (14.3.5) for bonds with data from Table 14.3.1 produces a coefficient of relative risk aversion of at least 27. If we use the lower variance of the growth rate of U.S. consumption in post–World War II data, the implied γ exceeds 200, as noted by Aiyagari (1993).

[5] The Cauchy-Schwarz inequality is $\frac{|\text{cov}_t(m_{t+1}, p_{t+1})|}{\sigma_t(m_{t+1})\sigma_t(p_{t+1})} \leq 1$. To get equation (14.4.2) from the preceding equation, we use the $\text{cov}_t(m_{t+1}, p_{t+1}) \geq -\sigma_t(p_{t+1})\sigma_t(m_{t+1})$ branch of the Cauchy-Schwarz inequality.

where σ_t denotes a conditional standard deviation. As an example of (14.4.2), let the payoff p_{t+1} be a *return* R_{t+1} on an asset, so that $q_t = 1$. In this case, (14.4.2) implies

$$E_t R_{t+1} \leq R_{f,t+1} + \left(\frac{\sigma_t\left(m_{t+1}\right)}{E_t m_{t+1}}\right) \sigma_t\left(R_{t+1}\right), \qquad (14.4.3)$$

where $R_{f,t+1}^{-1} = E_t m_{t+1}$ is the reciprocal of the risk-free interest rate. Inequality (14.4.3) says that the return on any security is bounded by the sum of the risk-free rate $R_{f,t+1}$ and the market price of risk times the conditional standard deviation of the return. Assets (or portfolios of assets) whose returns attain the bound are said to be on the efficient mean-standard deviation frontier.

Gallant, Hansen, and Tauchen (1990) and Hansen and Jagannathan (1991) used asset prices and returns alone to estimate the market price of risk, without imposing the link to consumption data implied by any particular specification of a stochastic discount factor. Their version of the equity premium puzzle is that the market price of risk implied by the asset market data alone is much higher than can be reconciled with the aggregate consumption data, say, with a specification that $m_{t+1} = \beta \left(\frac{C_{t+1}}{C_t}\right)^{-\gamma}$. Aggregate consumption is not volatile enough to make the standard deviation of the object high enough for the reasonable values of γ that we have discussed.

In the next section, we describe how Hansen and Jagannathan coaxed evidence about the market price of risk from asset prices and one-period returns.

14.5. Hansen-Jagannathan bounds

The section 14.3 exposition of the equity premium puzzle based on the log normal specification of returns was highly parametric, being tied to particular specifications of preferences and the distribution of asset returns. Hansen and Jagannathan (1991) described a nonparametric way of summarizing the equity premium puzzle. Their work can be regarded as extending Robert Shiller's and Stephen LeRoy's earlier work on variance bounds to handle stochastic discount factors.[6] We present one of Hansen and Jagannathan's bounds.

Hansen and Jagannathan are interested in restricting asset prices in possibly more general settings than we have studied so far. We have described a theory

---

[6] See Hansen's (1982a) early call for such a generalization.

that prices assets by using a particular "stochastic discount factor," defined as $m_{t+1} = \beta \frac{u'(C_{t+1})}{u'(C_t)}$. The theory asserted that the price at t of an asset with one-period random payoff p_{t+1} is $E_t m_{t+1} p_{t+1}$. Hansen and Jagannathan were interested in more general models, in which the stochastic discount factor could assume other forms.

14.5.1. Law of one price implies that $EmR = 1$

This section briefly indicates how a very weak theoretical restriction on prices and returns is sufficient to imply that there exists a stochastic discount factor m that satisfies $EmR_j = 1$ for the return R_j on any asset j. In fact, when markets are incomplete there exist *many* different random variables m that satisfy $EmR_j = 1$. We have to say almost nothing about consumers' preferences to get this result, a 'law of one price' being enough.

Following Hansen and Jagannathan, let x_j be a random payoff on a security. Let there be J primitive securities, so $j = 1, \ldots, J$. Thus, let $x \in \mathbb{R}^J$ be a vector of random payoffs on the primitive securities. Assume that the $J \times J$ matrix Exx' exists and that so does its inverse $(Exx')^{-1}$. Also assume that a $J \times 1$ vector q of prices of the primitive securities is observed, where the jth component of q is the price of the jth component of the payoff vector x. Consider forming portfolios of the primitive securities, i.e., linear combinations of the primitive securities. How do prices of portfolios relate to the prices of the primitive securities from which they have been formed?

Let $c \in \mathbb{R}^J$ be a vector of portfolio weights. The random payoff on a portfolio with weights c is $c \cdot x$. Define the space of payoffs attainable from portfolios of the primitive securities:

$$P \equiv \left\{ p : p = c \cdot x \text{ for some } c \in \mathbb{R}^J \right\}.$$

We want to price pay outs on portfolios, that is, pay outs, in P. We seek a price functional ϕ mapping P into $\mathbb{R}$: $\phi : P \to \mathbb{R}$.

The observed price of the jth primitive security must satisfy $q_j = \phi(x_j)$, so the $J \times 1$ vector q of observed prices of primitive securities satisfies $q = \phi(x)$. The pricing functional ϕ values a portfolio with payoff $c \cdot x \in P$ at $\phi(c \cdot x)$. We can replicate the payoff of the portfolio $p = c \cdot x$ by purchasing primitive securities in amounts $c_1, \ldots, c_J$ and paying $c_1 q_1 + \cdots + c_J q_J = c \cdot q$. A *law of one price* asserts that these two ways of purchasing payoff $p \in P$ should have the same cost:

$$\phi(p) = \phi(c \cdot x) = c_1 \phi(x_1) + c_2 \phi(x_2) + \cdots + c_J \phi(x_J). \qquad (14.5.1)$$

Equation $(14.5.1)$ asserts that the pricing functional ϕ is linear on P.

An aspect of the law of one price is that $\phi(c \cdot x)$ depends on $c \cdot x$, not on c. If any other portfolio has return $c \cdot x$, it should also be priced at $\phi(c \cdot x)$. Thus, two portfolios with the same payoff have the same price:

$$\phi(c_1 \cdot x) = \phi(c_2 \cdot x) \text{ if } c_1 \cdot x = c_2 \cdot x.$$

If the x's are *returns*, then $q = \mathbf{1}$, the unit vector, and

$$\phi(c \cdot x) = c \cdot \mathbf{1}.$$

14.5.2. Inner product representation of the pricing kernel

If y is a scalar random variable, $E(yx)$ is the vector whose jth component is $E(yx_j)$. The cross-moments $E(yx)$ are called the inner product of x and y. According to the Riesz representation theorem, the linear functional ϕ can be represented as the inner product of the random payoff x with *some* scalar random variable y.[7] This random variable is called a stochastic discount factor. Thus, a *stochastic discount factor* is a scalar random variable y that makes the following equation true:

$$\phi(p) = E(yp) \ \forall p \in P. \qquad (14.5.2)$$

Equality $(14.5.2)$ implies that the vector of prices of the primitive securities, q, satisfies

$$q = E(yx). \qquad (14.5.3)$$

Because it implies that the pricing functional is linear, the law of one price implies that there exists a stochastic discount factor. In fact, there exist many stochastic discount factors. Hansen and Jagannathan sought to characterize the set of admissible stochastic discount factors.

---

[7] See appendix A for a statement and proof of the Riesz representation theorem.

Note
$$\text{cov}\,(y, p) = E\,(yp) - E\,(y)\,E\,(p)\,,$$

which implies that the price functional can be represented as

$$\phi\,(p) = E\,(y)\,E\,(p) + \text{cov}\,(y, p)\,.$$

This expresses the price of a portfolio as the expected payoff times the expected value of the stochastic discount factor plus the covariance between the payoff and the stochastic discount factor. Notice that the expected value of the stochastic discount factor is simply the price of a sure scalar payoff of unity:

$$\phi\,(1) = E\,(y)\,.$$

The linearity of the pricing functional leaves open the possibility that prices of some portfolios are negative. This would open up arbitrage opportunities. David Kreps (1979) showed that the principle that the price system should offer *no arbitrage* opportunities requires that the stochastic discount factor be strictly positive. For most of this section, we shall not impose the principle of no arbitrage, just the law of one price. Thus, we do not require stochastic discount factors to be positive.

14.5.3. Classes of stochastic discount factors

In previous sections we constructed structural models of the stochastic discount factor. In particular, for the stochastic discount factor, our theories typically implied that

$$y = m_{t+1} \equiv \frac{\beta u'\,(C_{t+1})}{u'\,(C_t)}, \tag{14.5.4}$$

the intertemporal marginal rate of substitution of consumption today for consumption tomorrow. For a particular utility function, this specification leads to a parametric form of the stochastic discount factor that depends on the random consumption of a particular consumer or set of consumers.

Hansen and Jagannathan want to approach the data with a *class* of stochastic discount factors. To begin, Hansen and Jagannathan note that one candidate for a stochastic discount factor is

$$y^* = x'\,(Exx')^{-1}\,q. \tag{14.5.5}$$

This can be verified directly, by substituting into equation (14.5.3) and verifying that $q = E(y^*x)$.

Besides equation (14.5.5), many other stochastic discount factors work, in the sense of pricing the random returns x correctly, that is, recovering q as their price. It can be verified directly that any other y that satisfies

$$y = y^* + e$$

is also a stochastic discount factor, where e is orthogonal to x.[8] Let $\mathcal{Y}$ be the space of all stochastic discount factors.

14.5.4. A Hansen-Jagannathan bound

Given data on q and the distribution of returns x, Hansen and Jagannathan wanted to infer properties of y while imposing no more structure than linearity of the pricing functional (the law of one price). Imposing only this, they constructed bounds on the first and second moments of stochastic discount factors y that are consistent with a given distribution of payoffs on a set of primitive securities. Here is how they constructed one of their bounds.

Let y be an unobserved stochastic discount factor. Though y is unobservable, we can represent it in terms of the population linear regression[9]

$$y = a + x'b + e \tag{14.5.6}$$

where e is orthogonal to x and

$$b = [\text{cov}\,(x, x)]^{-1}\,\text{cov}\,(x, y)$$
$$a = Ey - Ex'b.$$

Here $\text{cov}(x, x) = E(xx') - E(x)E(x)'$. We have data that allow us to estimate the second-moment matrix of x, but no data on y and therefore on $\text{cov}(x, y)$. But we do have data on q, the vector of security prices. So Hansen and Jagannathan proceeded indirectly to use the data on q, x to infer something about y. Notice that $q = E(yx)$ implies $\text{cov}(x, y) = q - E(y)E(x)$. Therefore,

$$b = [\text{cov}\,(x, x)]^{-1}\,[q - E\,(y)\,E\,(x)]. \tag{14.5.7}$$

---

[8] Let y_1 and y_2 be two stochastic discount factors, so that $Ey_1x = Ey_2x$, which implies that $E(y_1 - y_2)x = 0$. Thus, the difference between two stochastic discount factors is orthogonal to x, as asserted in the text.

[9] See chapter 2 for the definition and construction of a population linear regression.

Thus, *given* a guess about $E(y)$, asset returns and prices can be used to estimate b. Because the residuals in equation $(14.5.6)$ are orthogonal to x,

$$\text{var}(y) = \text{var}(x'b) + \text{var}(e).$$

Therefore

$$[\text{var}(x'b)]^{.5} \leq \sigma(y), \tag{14.5.8}$$

where $\sigma(y)$ denotes the standard deviation of the random variable y. This is the lower bound on the standard deviation of all[10] stochastic discount factors with prespecified mean $E(y)$. For various specifications, Hansen and Jagannathan used expressions $(14.5.7)$ and $(14.5.8)$ to compute the bound on $\sigma(y)$ as a function of $E(y)$, tracing out a frontier of admissible stochastic discount factors in terms of their means and standard deviations.

Here are two such specifications. First, recall that a (gross) return for an asset with price q and payoff x is defined as $z = x/q$. A return is risk free if z is constant. Then note that if there is an asset with risk-free return $z^{RF} \in x$, it follows that $E(yz^{RF}) = z^{RF}Ey = 1$, and therefore Ey is a known constant. Then there is only one point on the frontier that is of interest, the one with the known $E(y)$. If there is no risk-free asset, we can calculate a different bound for every specified value of $E(y)$.

Second, take a case where $E(y)$ is not known because there is no risk-free payout in the set of returns. Suppose, for example, that the data set consists of "excess returns." Let x^s be a return on a stock portfolio and x^b be a return on a risk-free bond. Let $z = x^s - x^b$ be the excess return. Then

$$E[yz] = 0.$$

Thus, for an excess return, $q = 0$, so formula $(14.5.7)$ becomes

$$b = - [\text{cov}(z, z)]^{-1} E(y) E(z).$$

Then[11]

$$\text{var}(z'b) = E(y)^2 E(z)' \left[\text{cov}(z, z)^{-1}\right] E(z).$$

---

[10] The stochastic discount factors are not necessarily positive. Hansen and Jagannathan (1991) derive another bound that imposes positivity.

[11] This formula follows from $\text{var}(z'b) = b'\text{cov}(z, z)b$.

Therefore, the Hansen-Jagannathan bound becomes

$$\sigma\left(y\right) \geq \left[E\left(z\right)' \operatorname{cov}\left(z, z\right)^{-1} E\left(z\right)\right]^{.5} E\left(y\right). \qquad (14.5.9)$$

In the special case of a scalar excess return, (14.5.9) becomes

$$\frac{\sigma\left(y\right)}{E\left(y\right)} \geq \frac{\left|E\left(z\right)\right|}{\sigma\left(z\right)}. \qquad (14.5.10)$$

The left side, the ratio of the standard deviation of the discount factor to its mean, is called the *market price of risk*. Thus, the bound (14.5.10) says that the market price of risk is at least $\frac{E(z)}{\sigma(z)}$. The ratio $\frac{E(z)}{\sigma(z)}$ thus determines a straight-line frontier in the $[E(y), \sigma(y)]$ plane above which the stochastic discount factor must reside.

The market price of risk $\frac{\sigma(y)}{E[y]}$ is the increase in the expected rate of return needed to compensate an investor for bearing a unit increase in the standard deviation of return along the efficient frontier.[12]

For a set of returns, $q = \mathbf{1}$ so that equation (14.5.7) becomes

$$b = \left[\operatorname{cov}\left(x, x\right)\right]^{-1} \left[\mathbf{1} - E\left(y\right) E\left(x\right)\right]. \qquad (14.5.11)$$

The bound is computed by solving equation (14.5.11) and

$$\sqrt{b' \operatorname{cov}\left(x, x\right) b} \leq \sigma\left(y\right). \qquad (14.5.12)$$

In more detail, we compute the bound for various values of $E(y)$ by using equation (14.5.11) to compute b, then using that b in expression (14.5.12) to compute the lower bound on $\sigma(y)$.

The bound (14.5.11) is a parabola, while formula (14.5.9) is a straight line in the $[E(y), \sigma(y)]$ plane. In the next section, we shall use quarterly data on two returns, the real return on a value-weighted NYSE stock return and the real return on U.S. Treasury bills over the period 1948-2005 in conjunction with inequality (14.5.11) to compute the Hansen-Jagannathan bound. We report the bound in figure 14.6.1, which contains other information about the predicted behavior of the stochastic discount factor of a consumer with time-separable CRRA preferences, to be explained in the next section.

---

[12] A Sharpe ratio measures the excess return relative to the standard deviation. The market price of risk is the maximal Sharpe ratio.

14.6. Failure of CRRA to attain HJ bounds

For time-separable CRRA preferences with discount factor β, the stochastic discount factor m_{t+1} is simply the marginal rate of substitution:

$$m_{t+1} = \beta \left(\frac{C_{t+1}}{C_t} \right)^{-\gamma} \tag{14.6.1}$$

where γ is the coefficient of relative risk aversion and C_t is consumption. Let $c_t = \log C_t$ and express (14.6.1) as

$$m_{t+1} = \beta \exp\left(-\gamma\left(c_{t+1} - c_t\right)\right). \tag{14.6.2}$$

For aggregate U.S. data on per capita consumption of nondurables and services, a good approximation to the data is the following model that makes the growth in the log of per capita consumption a random walk with drift:

$$c_t = \mu + c_{t-1} + \sigma_c \varepsilon_t, \quad \{\varepsilon_t\} \ i.i.d. \sim \mathcal{N}(0,1). \tag{14.6.3}$$

With this model for consumption growth, (14.6.2) becomes

$$m_{t+1} = \beta \exp\left(-\gamma\mu - \gamma\sigma_c \varepsilon_{t+1}\right), \tag{14.6.4}$$

and the log of the stochastic discount factor is

$$\log m_{t+1} = \log \beta - \gamma\mu - \gamma\sigma_c \varepsilon_{t+1}, \tag{14.6.5}$$

which is a normal random variable with mean $\log \beta - \gamma\mu$ and variance $\gamma^2 \sigma_c^2$. To compute the mean and standard deviation of the particular stochastic discount factor (14.6.4), we use:

PROPERTY: If $\log X \sim \mathcal{N}(\mu_x, \sigma_x^2)$, then $E(X) = \exp\left(\mu_x + \frac{1}{2}\sigma_x^2\right)$ and $\mathrm{std}(X) = E(X)\sqrt{(\exp(\sigma_x^2) - 1)}$. Here std denotes a standard deviation.

Applying this property, we find that the mean $E(m)$ and standard deviation $\sigma(m)$ are [13]

$$Em_{t+1} = E\left[m\right] = \beta \exp\left[-\gamma\mu + \frac{\sigma_\varepsilon^2 \gamma^2}{2}\right] \tag{14.6.6}$$

---

[13] Let log consumption growth be denoted x and have probability density $\phi(\cdot)$ with finite moments $\zeta_j = \int x^j \phi(x)dx$ for all orders $j \geq 1$. Then note that $Em \equiv \beta E \exp(-\gamma x)$, where $E \exp(-\gamma x)$ is a *moment generating function* with expansion $1 - \gamma\zeta_1 + \frac{\gamma^2}{2}\zeta_2 - \frac{\gamma^3}{3!}\zeta_3 + \cdots$. This equation asserts that the gross risk-free rate $E(m)$ depends on moments of the log consumption growth process of all orders. Stanley Zin (2002) named this the 'never a dull moment' fact and indicated how one could adjust higher moments of log consumption growth to fit asset pricing observations while fitting lower moments of a log consumption growth process.

and

$$\text{std}\,(m_{t+1}) \equiv \sigma\,(m) = E\,(m) \left\{\exp\left[\sigma_\varepsilon^2 \gamma^2\right] - 1\right\}^{\frac{1}{2}} \tag{14.6.7}$$

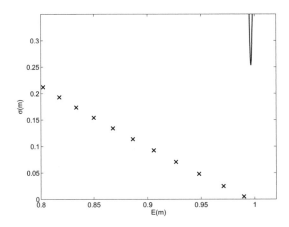

Figure 14.6.1: Solid line: Hansen-Jagannathan volatility bounds for quarterly returns on the value-weighted NYSE and Treasury Bill, 1948-2005. Crosses: Mean and standard deviation for intertemporal marginal rate of substitution for CRRA time separable preferences. The coefficient of relative risk aversion, γ takes on the values 1, 5, 10, 15, 20, 25, 30, 35, 40, 45, 50 and the discount factor β=0.995.

Figure 14.6.1 plots the Hansen and Jagannathan bound (the parabola in the upper right corner of the graph) constructed using quarterly data on two returns, the real return on a value-weighted NYSE stock return and the real return on U.S. Treasury bills over the period 1948-2005 in conjunction with inequality (14.5.11). The figure also reports the locus of $E(m)$ and $\sigma(m)$ implied by equations (14.6.6) and (14.6.7) traced out by different values of γ. The figure shows that while high values of γ deliver high $\sigma(m)$, high values of γ also push $E(m)$, the reciprocal of the risk-free rate down and away from the Hansen and Jagannathan bounds. The *equity premium puzzle* is the observation that it takes a very high value of γ to make $\sigma(m)$ high with CRRA preferences. The *risk-free rate puzzle* of Weil (1990) is that with the CRRA stochastic discount factor (14.6.4), setting γ higher pushes $E(m)$ downward and increasingly to the left of the Hansen-Jagannathan bounds.

The risk free rate problem can be understood by focusing on equations (14.6.6) and (14.6.7). The parameter γ expresses two distinct forces on the risk-free interest rate that come from two conceptually distinct consumer attitudes.

1. *Effects of γ on $\frac{\sigma(m)}{E(m)}$*   Equation (14.6.7) shows that increases in γ unambiguously increase the market price of risk $\frac{\sigma(m)}{E(m)}$. Here γ is playing its role of expressing the consumer's distaste for atemporal gambles. Higher values of γ indicate more hatred of risk and a higher price of risk.

2. *Effects of γ on $E(m)$*   Countervailing effects of increases in γ on $E(m)$ are visible in equation (14.6.6). Through the term $\exp(-\gamma\mu)$, γ expresses the representative consumer's distaste for deviations of consumption from a smooth path across time. The growth parameter μ induces deviations from intertemporal consumption smoothness, while γ multiplicatively affects the compensation in terms of a risk-free interest needed to compensate the consumer for accepting paths that are not smooth intertemporally, regardless of how risky they are. Here, γ is expressing views about intertemporal substitution of consumption today for consumption tomorrow. The parameter γ also affects $E(m)$ through the term $\exp\left(\frac{\sigma_\varepsilon^2\gamma^2}{2}\right)$, which reflects the consumer's dislike of risky consumption streams, a dislike that increases with γ and is compensated for by a *higher* $E(m)$ via a precautionary savings motive to be analyzed in chapter 17.

Empirically, the estimates of μ and σ_c in Table 14.6.1 are the same order of magnitude. Thus, μ is two orders of magnitude larger than σ_c^2, which makes increases in γ drive $E(m)$ *down* through its effect on the term $\exp(-\gamma\mu)$ much faster than it drives $E(m)$ *up* through its effect on $\exp\left(\frac{\sigma_\varepsilon^2\gamma^2}{2}\right)$. This is why increases in γ push $E(m)$ downward, at least for all but extraordinarily large γ's.[14]

In conclusion, it is the fact that the same parameter γ expresses two attitudes – atemporal risk aversion and intertemporal substitution aversion – that leads to Weil's risk-free rate puzzle as captured by our figure 14.6.1. In the next

---

[14] This observation underlies the insight of Kocherlakota (1990), who pointed out that by adjusting (β, γ) pairs suitably, it is possible to attain the Hansen-Jagannathan bounds for the random walk model of log consumption and CRRA time-separable preferences, thus explaining both the equity premium and the risk-free rate. Doing so requires a very high γ and $\beta > 1$.

Table 14.6.1: Estimates from quarterly U.S. data 1948:2-2005:4.

Parameter	Random Walk
μ	0.004952
σ_c	0.005050

section, we describe how Tallarini (2000) made progress by assigning γ only to the one job of describing risk aversion and assigning to a new parameter η the job of describing attitudes toward intertemporal substitution. By proceeding this way, Tallarini was able to find values of the risk aversion γ that pushed the $(E(m), \sigma(m))$ pair toward the Hansen and Jagannathan bounds.

14.7. Non-expected utility

To separate risk aversion from intertemporal substitution, Tallarini (2000) assumed preferences that can be described by a recursive non-expected utility value function iteration à la Kreps and Porteus (1978), Epstein and Zin (1989), and Weil (1990), namely,[15]

$$V_t = W\left(C_t, \mu\left(V_{t+1}\right)\right).$$

Here W is an aggregator function that maps today's consumption C and a function μ of tomorrow's random continuation value V_{t+1} into a value V_t today. Here $\mu\left(\cdot\right)$ is a 'certainty equivalent' function that maps a random variable V_{t+1} that is measurable with respect to next period's information into a random variable that is measurable with respect to this period's information:

$$\mu\left(V_{t+1}\right) = f^{-1}\left(E_t f\left(V_{t+1}\right)\right),$$

where f is a function that describes attitudes toward atemporal risk:

$$f\left(z\right) = \begin{cases} z^{1-\gamma} & \text{if } 0 < \gamma \neq 1 \\ \log z & \text{if } \gamma = 1, \end{cases} \tag{14.7.1}$$

---

[15] Obstfeld (1994) and Dolmas (1998) used recursive preferences to study costs of consumption fluctuations.

and γ is the coefficient of relative risk aversion.

Epstein and Zin (1991) used the CES aggregator

$$W\left(C,\mu\right) = \begin{cases} \left[(1-\beta)\,C^{1-\eta} + \beta\mu^{1-\eta}\right]^{\frac{1}{1-\eta}} & \text{if } 0 < \eta \neq 1 \\ C^{1-\beta}\mu^{\beta} & \text{if } \eta = 1, \end{cases} \tag{14.7.2}$$

where $\frac{1}{\eta}$ is the intertemporal elasticity of substitution. Setting $\gamma = \eta$, gives the special case of additive power utility with discount factor β.

Tallarini (2000) used a special case of this model. He set $\eta = 1$ in the aggregator function (14.7.2) and used the power function (14.7.1) for his certainty equivalent function. These choices led Tallarini to use the following recursion to define preferences over risky consumption streams:

$$V_t = C_t^{1-\beta}\left[\left(E_t\left(V_{t+1}^{1-\gamma}\right)\right)^{\frac{1}{1-\gamma}}\right]^{\beta}.$$

Taking logs gives

$$\log V_t = (1-\beta)\,c_t + \frac{\beta}{1-\gamma}\log E_t\left(V_{t+1}^{1-\gamma}\right)$$

or

$$\frac{\log V_t}{(1-\beta)} = c_t + \frac{\beta}{(1-\gamma)\,(1-\beta)}\log E_t\left(V_{t+1}^{1-\gamma}\right). \tag{14.7.3}$$

For our purposes, it is convenient to represent (14.7.3) in an alternative way. Define $U_t \equiv \log V_t/(1-\beta)$ and

$$\theta = \frac{-1}{(1-\beta)\,(1-\gamma)}. \tag{14.7.4}$$

Then (14.7.3) is equivalent with

$$U_t = c_t - \beta\theta\log E_t\left[\exp\left(\frac{-U_{t+1}}{\theta}\right)\right]. \tag{14.7.5}$$

For now, θ is just a convenient way of expressing a particular function of the interpretable parameters β and γ. But later in subsection 14.8.4, θ will be an interpretable parameter of independent interest.

Equation (14.7.5) is the risk-sensitive recursion of Hansen and Sargent (1995).[16] In the special case that $\gamma = 1$ (or $\theta = +\infty$), application of L'Hopital's

---

[16] Tallarini defined $\sigma = 2\,(1-\beta)\,(1-\gamma)$ in order to interpret his recursion in terms of the risk-sensitivity parameter σ of Hansen and Sargent (1995), who regarded negative values of σ as enhancing risk aversion.

rule verifies that recursion (14.7.5) becomes a discounted expected utility recursion

$$U_t = c_t + \beta E_t U_{t+1}.$$

To solve recursion (14.7.5), we use a guess and verify method. We continue to assume that the log consumption growth process is a random walk with drift (14.6.3) and seek to solve (14.7.5). Guess a linear value function

$$U_t = k_0 + k_1 c_t, \qquad (14.7.6)$$

where k_0, k_1 are scalar constants. Then solve the Bellman equation:

$$k_0 + k_1 c = c - \beta \theta \log E \exp \left(\frac{-(k_0 + k_1 [\mu + c + \sigma_c \varepsilon_{t+1}])}{\theta} \right) \qquad (14.7.7)$$

for k_0 and k_1. Direct calculations that again use the properties of log normal random variables show that $k_0 = \frac{\beta}{(1-\beta)^2} \left[\mu - \frac{\sigma_\varepsilon^2}{2\theta(1-\beta)} \right]$ and $k_1 = \frac{1}{1-\beta}$, so that

$$U_t = \frac{\beta}{(1-\beta)^2} \left[\mu - \frac{\sigma_\varepsilon^2}{2\theta(1-\beta)} \right] + \frac{1}{1-\beta} c_t. \qquad (14.7.8)$$

14.7.1. Another representation of the utility recursion

When log consumption follows the random walk with drift (14.6.3) or obeys a member of a class of models that makes the conditional distribution of c_{t+1} be Gaussian, another way to express recursion (14.7.5) is

$$U_t = c_t + \beta E_t U_{t+1} - \frac{\beta}{2\theta} \text{var}_t (U_{t+1}), \qquad (14.7.9)$$

where $\text{var}_t(U_{t+1})$ denotes the conditional variance of continuation utility U_{t+1}. Using (14.7.4) to eliminate θ in favor of γ, we can also express (14.7.9) as

$$U_t = c_t + \beta E_t U_{t+1} + \frac{\beta(1-\gamma)(1-\beta)}{2} \text{var}_t (U_{t+1}). \qquad (14.7.10)$$

When $\theta < +\infty$ or $\gamma > 1$, representation (14.7.9) generalizes the ordinary time separable expected utility recursion by making the consumer care not only

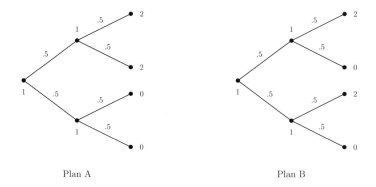

Plan A   Plan B

Figure 14.7.1: Plan A has early resolution of uncertainty. Plan B has late resolution of uncertainty.

about the conditional expectation of the continuation utility but also its conditional variance. According to (14.7.9), when $\theta < +\infty$ the consumer dislikes conditional variance in continuation utility.[17]

That the consumer dislikes conditional volatility of continuation utility when $\theta < +\infty$ means that he has preferences over both the timing of the resolution of uncertainty and the persistence risk. Figures 14.7.1 and 14.7.2 display consumption payoffs and transition probabilities (the fractions above the lines connecting nodes) for four plans. First, when $0 < \theta < +\infty$ the consumer prefers early resolution of risk (he prefers plan A to plan B in figure 14.7.1), while he is indifferent to it when $\theta = +\infty$.[18] Second, when $0 < \theta < +\infty$, the consumer dislikes *persistence* of risk (he prefers plan C to plan D in figure 14.7.2), while when $\theta = +\infty$ he is indifferent to it.[19]

---

[17] Equation (14.7.9) is a discrete time version of the stochastic differential utility model of Duffie and Epstein (1992).

[18] See Kreps and Porteus (1978).

[19] See Duffie and Epstein (1992).

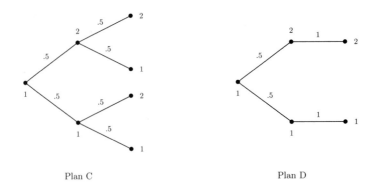

Plan C   Plan D

Figure 14.7.2: Plan C has i.i.d. risk. Plan D has persistent risk.

14.7.2. Stochastic discount factor

With preferences induced by the risk-sensitive recursion (14.7.5), calculating the intertemporal rate of substitution shows that the stochastic discount factor is

$$m_{t+1} = \left(\beta \frac{C_t}{C_{t+1}} \right) \left(\frac{\exp\left(-\theta^{-1}U_{t+1}\right)}{E_t\left[\exp\left(-\theta^{-1}U_{t+1}\right)\right]} \right), \qquad (14.7.11)$$

or

$$m_{t+1} = \left(\beta \frac{C_t}{C_{t+1}} \right) \left(\frac{\exp\left((1-\beta)(1-\gamma)U_{t+1}\right)}{E_t\left[\exp\left((1-\beta)(1-\gamma)U_{t+1}\right)\right]} \right). \qquad (14.7.12)$$

The term $g(\varepsilon_{t+1}) \equiv \left(\frac{\exp((1-\beta)(1-\gamma)U_{t+1})}{E_t[\exp((1-\beta)(1-\gamma)U_{t+1})]} \right)$ is a nonnegative random variable whose conditional expectation is unity by construction. This makes it interpretable as a *likelihood ratio*, i.e., the ratio of one probability density to another. Direct calculations show that the likelihood ratio g satisfies

$$g\left(\varepsilon_{t+1}\right) \equiv \left(\frac{\exp\left((1-\beta)(1-\gamma)U_{t+1}\right)}{E_t\left[\exp\left((1-\beta)(1-\gamma)U_{t+1}\right)\right]} \right) = \exp\left(w\varepsilon_{t+1} - \frac{1}{2}w^2 \right) \quad (14.7.13)$$

where w is given by

$$w = -\frac{\sigma_\varepsilon}{\theta(1-\beta)}. \qquad (14.7.14)$$

Using the definition of θ in (14.7.4), we can also express w as

$$w = \sigma_c(1-\gamma). \qquad (14.7.15)$$

Thus, for a log consumption process described by a random walk with drift $(14.6.3)$, the stochastic discount factor is

$$
\begin{aligned}
m_{t+1} &= \beta \exp\left(-\left(c_{t+1} - c_t\right)\right) \exp\left(w\varepsilon_{t+1} - \frac{1}{2}w^2\right) \\
&= \beta \exp\left(-\left(\mu + \sigma_c\varepsilon_{t+1} - \sigma_c\left(1 - \gamma\right)\varepsilon_{t+1} + \frac{1}{2}\sigma_c^2\left(1 - \gamma\right)^2\right)\right)
\end{aligned}
\tag{14.7.16}
$$

where we have used $(14.7.15)$ in moving from the first line to the second. Therefore,

$$
\log m_{t+1} = \log\beta - \mu - \gamma\sigma_c\varepsilon_{t+1} - \frac{1}{2}\sigma_c^2\left(1 - \gamma\right)^2, \tag{14.7.17}
$$

so that $\log m_{t+1} \sim \mathcal{N}(\log\beta - \mu - \frac{1}{2}\sigma_c^2(1 - \gamma)^2, \gamma^2\sigma_c^2)$.

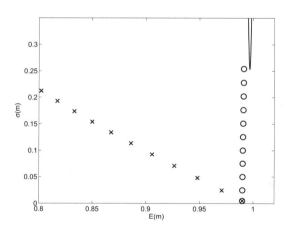

Figure 14.7.3: Solid line: Hansen-Jagannathan volatility bounds for quarterly returns on the value-weighted NYSE and Treasury bill, 1948–2005. Circles: Mean and standard deviation for intertemporal marginal rate of substitution generated by Epstein-Zin preferences with random walk consumption. Crosses: Mean and standard deviation for intertemporal marginal rate of substitution for CRRA time separable preferences. The coefficient of relative risk aversion γ takes on the values 1, 5, 10, 15, 20, 25, 30, 35, 40, 45, 50 and the discount factor $\beta = 0.995$.

Applying again the standard property of log normal random variables to formula $(14.7.16)$ for the stochastic discount factor gives

$$Em_{t+1} = \beta \exp\left[-\mu + \frac{\sigma_\varepsilon^2}{2}(2\gamma - 1)\right] \tag{14.7.18}$$

$$\frac{\sigma(m)}{E[m]} = \left\{\exp\left[\sigma_\varepsilon^2\gamma^2\right] - 1\right\}^{\frac{1}{2}}. \tag{14.7.19}$$

Please compare these two to the corresponding formulas $(14.6.6)$, $(14.6.7)$ for the time-separable CRRA specification. The salient difference is that γ no longer appears in the key term $\exp(-\mu)$ for $E(m)$ in $(14.7.18)$, while it does appear in the corresponding term in the CRRA formula $(14.6.6)$. Tallarini made γ disappear in this way by locking the inverse intertemporal rate of substitution parameter η at unity while allowing what is now a pure risk aversion parameter γ to vary. This arrests the force causing $E(m)$ in $(14.6.6)$ to fall as γ rises and allows Tallarini to avoid the risk-free rate puzzle and to approach the Hansen-Jagannathan bounds as he increases the risk aversion parameter γ.

Figure 14.7.3 is a version of Tallarini's (2000) key figure for our data on quarterly returns on the value-weighted NYSE and Treasury Bill, 1948-2005. It uses formulas $(14.6.6)$, $(14.6.7)$ to compute loci of $(E(m), \sigma(m))$ pairs for different values of the risk-aversion parameter γ. When it is compared to the corresponding figure 14.6.1 for time separable CRRA preferences, this figure registers a striking success for Tallarini. Notice how increasing γ pushes the volatility of the stochastic discount factor upward toward the Hansen-Jagannathan bounds while leaving $E(m)$ unaffected, thus avoiding the risk-free rate puzzle of Weil (1990). A value of the risk-aversion $\gamma = 50$ for the risk aversion coefficient almost succeeds in attaining the Hansen-Jagannathan bounds.[20]

---

[20] By adjusting the calibrated value of β upward, it would be possible to move all the circles in figure 14.7.3 to the right, thus moving them closer to the Hansen-Jagannathan bound.

14.8. Reinterpretation of the utility recursion

To succeed in approaching the Hansen-Jagannathan bounds required that Tallarini set the risk-aversion parameter γ to such a high value, namely 50, that it provoked Lucas (2003) not to regard Tallarini's findings as providing reliable evidence of high risk aversion:

> *No one has found risk aversion parameters of 50 or 100 in the diversification of individual portfolios, in the level of insurance deductibles, in the wage premiums associated with occupations with high earnings risk, or in the revenues raised by state-operated lotteries. It would be good to have the equity premium resolved, but I think we need to look beyond high estimates of risk aversion to do it.*
> — *Robert Lucas, Jr., "Macroeconomic Priorities," 2003*

On the basis of this reading of what he regarded to be more pertinent evidence about risk aversion, for the purpose of measuring the costs of aggregate fluctuations along lines to be described in section 14.9, Lucas (1987, 2003) preferred to use a value of γ of 1 or 2 rather than the value of γ of 50 that Tallarini said was needed to reconcile his model of preferences with both the consumption data and the asset returns data as summarized by the Hansen-Jagannathan bounds.[21]

14.8.1. Risk aversion or model misspecification aversion

To respond to Lucas's reluctance to use Tallarini's findings as a source of evidence about the representative consumer's attitude about consumption fluctuations, we now reinterpret γ as a parameter that expresses distress about model specification doubts rather than risk aversion. Fearing *risk* means disliking randomness with a *known* probability distribution. Fearing *uncertainty* or *model misspecification* means disliking *not knowing* the probability distribution itself. We will reinterpret the forward-looking term[22] $g(\varepsilon_{t+1}) \equiv \frac{\exp((1-\beta)(1-\gamma)U_{t+1})}{E_t[\exp((1-\beta)(1-\gamma)U_{t+1})]}$ that multiplies the ordinary logarithmic stochastic discount factor $\beta\frac{C_t}{C_{t+1}}$ in

---

[21] For another perspective on the evidence, see Barseghyan, Molinari, O'Donoghue, and Teitelbaum (2011), whose findings challenge the notion that purchases of insurance understand the probability distributions they are facing.

[22] The presence of the continuation value U_{t+1} is our reason for saying 'forward-looking'.

(14.7.12) as an adjustment that reflects a consumer's fears about model misspecification.

14.8.2. Recursive representation of probability distortions

It is convenient to use the following notation that is similar to some used in chapter 8. Let $s_t = c_t - c_{t-1}$ and $s^t = s_t, \ldots, s_1$. Denote a joint probability density over s^t as $F_t(s^t)$. Where $f_{t+1}(s_{t+1}|s^t)$ is a conditional density, the recursion

$$F_{t+1}\left(s^{t+1}\right) = f_{t+1}\left(s_{t+1}|s^t\right) F_t\left(s^t\right), \quad t \geq 1 \tag{14.8.1}$$

governs the evolution of the joint densities $F_t(s^t)$. For our random walk with drift model of consumption growth (14.6.3), $f_{t+1}(s_{t+1}|s^t) \sim \mathcal{N}(\mu, \sigma_c^2)$. Let $G_t(s^t)$ be the ratio of another joint density $\tilde{F}_t(s^t)$ to $F_t(s^t)$, so that

$$\tilde{F}_t\left(s^t\right) = G_t\left(s^t\right) F_t\left(s^t\right). \tag{14.8.2}$$

We can factor the likelihood ratio $G_{t+1}(s^{t+1})$ in a way analogous to the factorization of $F_{t+1}(s^{t+1})$ in (14.8.1):

$$G_{t+1}\left(s^{t+1}\right) = g_{t+1}\left(s_{t+1}|s^t\right) G_t\left(s^t\right), \quad t \geq 1. \tag{14.8.3}$$

Here $g_{t+1}(s_{t+1}|s^t)$ is a likelihood ratio of the conditional densities, namely,

$$g_{t+1}\left(s_{t+1}|s^t\right) = \frac{\tilde{f}_{t+1}\left(s_{t+1}|s^t\right)}{f_{t+1}\left(s_{t+1}|s^t\right)}.$$

Because $g_{t+1}(s_{t+1}|s^t)$ is a likelihood ratio, its expectation under $f_{t+1}(s_{t+1}|s^t)$ is unity:

$$\int g_{t+1}\left(s_{t+1}|s^t\right) f_{t+1}\left(s_{t+1}|s^t\right) ds_{t+1} = 1.$$

This in turn implies that under the $F_t(s^t)$ density, the likelihood ratio $G_t(s^t)$ is a martingale with respect to the filtration generated by s^t:

$$E\left[G_{t+1}\left(s^{t+1}\right)|s^t\right] = \left[\int g_{t+1}\left(s_{t+1}|s^t\right) f_{t+1}\left(s_{t+1}|s^t\right) ds_{t+1}\right] G_t\left(s^t\right)$$

$$= G_t\left(s^t\right).$$

14.8.3. Entropy

To measure the proximity of two conditional densities $\tilde{f}_{t+1}(s_{t+1}|s^t)$ and $f_{t+1}(s_{t+1}|s^t)$ we use the expected log likelihood ratio

$$
\begin{aligned}
\text{ent}_{t+1}\left(g\right) &\equiv \int \log\left(g_{t+1}\left(s_{t+1}|s^t\right)\right) \tilde{f}_{t+1}\left(s_{t+1}|s^t\right) ds_{t+1} \\
&= \int \log\left(g_{t+1}\left(s_{t+1}|s^t\right)\right) g_{t+1}\left(s_{t+1}|s^t\right) f_{t+1}\left(s_{t+1}|s^t\right) ds_{t+1},
\end{aligned}
\tag{14.8.4}
$$

where the mathematical expectation on the right side of the first line is with respect to the conditional density $\tilde{f}_{t+1}$ and the integration on the right side of the second line is with respect to the conditional density f_{t+1}. The object ent_{t+1} is called the *relative entropy* of conditional density $\tilde{f}_{t+1}$ with respect to conditional density f_{t+1}.

Entropy is a natural measure of the statistical proximity of the two densities because it determines the behavior of tests for statistically discriminating between two densities using samples of finite length. Entropy is nonnegative and equals zero if $g_{t+1} = 1$, so that $\tilde{f}_{t+1} = f_{t+1}$. If entropy is very small, it takes a very large number of observations to distinguish the two densities with high statistical confidence, while if entropy is large, it requires fewer observations.[23] We elaborate on this connection in section 14.8.7. But first we reinterpret recursion (14.7.5) as expressing a decision maker's fears about model misspecification.

14.8.4. Expressing ambiguity

Let ε be a random variable with normalized Gaussian density $\phi(\varepsilon) \sim \mathcal{N}(0,1)$. Let $\tilde{\phi}(\varepsilon) = g(\varepsilon)\phi(\varepsilon)$ be some other density for ε. Let $Z(\varepsilon)$ be a value function. Consider the minimization problem:

$$
\begin{aligned}
\mathbf{T}\left(Z\right) &= \min_{g(\varepsilon) \geq 0} \int Z\left(\varepsilon\right) g\left(\varepsilon\right) \phi\left(\varepsilon\right) d\varepsilon + \theta \, \text{ent}\left(g\left(\varepsilon\right)\right) \\
&= \min_{g(\varepsilon) \geq 0} \int \left[Z\left(\varepsilon\right) + \theta \log g\left(\varepsilon\right)\right] g\left(\varepsilon\right) \phi\left(\varepsilon\right) d\varepsilon
\end{aligned}
\tag{14.8.5}
$$

where $0 < \theta \leq +\infty$ and the minimization is subject to $\int g(\varepsilon)\phi(\varepsilon)d\varepsilon = 1$, so that g is a likelihood ratio. The minimization problem on the right side of (14.8.5)

---

[23] Anderson, Hansen, and Sargent (2003) develop the connection between entropy and statistical model discrimination.

has the following interpretation. *Ex ante*, a decision maker values random outcomes according to the value function $Z(\varepsilon)$. The decision maker's best guess about the probability density over ε is $\phi(\varepsilon)$. But he does not completely trust $\phi(\varepsilon)$, meaning that he thinks that outcomes might actually be drawn from some unknown distribution $g(\varepsilon)\phi(\varepsilon)$. He wants an *ex ante* valuation that will work 'well enough' for a set of densities in a proximity of $\phi(\varepsilon)$. He measures proximity by relative entropy. He attains such a 'robust' valuation, i.e., one suitable for a *set* of probability distributions, by solving the minimization problem on the right side of (14.8.5). The problem is to choose a probability distortion g that minimizes expected utility under the 'distorted' density $\tilde{\phi} = g\phi$ plus a positive penalty parameter θ times relative entropy. The purpose of the entropy penalty term (θ ent) is to constrain the probability distortions g to have small entropy. Smaller values of θ imply bigger probability distortions because the penalty on entropy is smaller in the minimization problem on the right side of the equations in (14.8.5). We regard the penalty parameter θ as an inverse index of the decision maker's distrust of his baseline probability model ϕ.

Performing the minimization on the right side of (14.8.5) gives the following minimizing value of $g(\varepsilon)$:

$$\hat{g}(\varepsilon) = \frac{\exp\left(-\theta^{-1}Z(\varepsilon)\right)}{E\exp\left(-\theta^{-1}Z(\varepsilon)\right)}. \tag{14.8.6}$$

Substituting the minimizer (14.8.6) into the right side of (14.8.5) gives the indirect utility function

$$\mathbf{T}(Z) = -\theta \log E \exp\left(-\theta^{-1}Z(\varepsilon)\right), \tag{14.8.7}$$

which we recognize to be the same risk-sensitivity operator that appears on the right side of recursion (14.7.5). We rely on the fact that (14.8.7) is the indirect utility function for the minimization problem on the right side of (14.8.5) to interpret the penalty parameter θ in $\mathbf{T}(Z)$ as expressing distrust of the distribution $\phi(\varepsilon)$. An application of l'Hopital's rule shows that when $\theta = +\infty$, $\mathbf{T}(Z) = \int Z(\varepsilon)\phi(\varepsilon)d\varepsilon$, which is expected utility under complete trust in the density ϕ.

14.8.5. Ambiguity averse preferences

We return to our representative consumer and use the following recursion to express preferences over consumption streams. But we now endow the consumer with distrust of the density $\phi(\varepsilon_{t+1})$ in (14.6.3). He expresses that distrust by using the following recursion to order consumption streams:

$$W_t = c_t + \beta \mathbf{T}_t \left(W_{t+1} \right),$$

where $\mathbf{T}_t$ is a conditional version of $\mathbf{T}$. This is equivalent with

$$W_t = c_t - \beta\theta \log E_t \left[\exp \left(\frac{-W_{t+1}}{\theta} \right) \right], \qquad (14.8.8)$$

which is identical with recursion (14.7.5), namely, the risk-sensitive recursion of Hansen and Sargent (1995) once again. When interpreted in terms of model ambiguity, (14.7.7) is said to describe 'multiplier preferences' that express model ambiguity through the penalty or 'multiplier' parameter θ.[24]

The identity of recursions (14.7.5) and (14.7.7) means that so far as choices among risky consumption plans indexed by (μ, σ_c) are concerned, the risk-sensitive representative consumer of Tallarini (2000) is observationally equivalent to a representative consumer who is concerned about model misspecification. But the *motivations* behind their choices differ and that would in principle allow us to distinguish them if we were able to confront the consumer with choices between gambles with known distributions and gambles with unknown distributions.[25]

Recall formula (14.7.4) that for risk-sensitive preferences defines θ in terms of the elementary parameters β and γ:

$$\theta = \frac{-1}{(1-\beta)(1-\gamma)}.$$

For Tallarini, γ is the fundamental parameter, and it is interpreted as describing the consumer's attitude toward atemporal risky choices under a *known* probability distribution. But under the probability ambiguity or 'robustness' interpretation, θ is an elementary parameter in its own right, one that measures the

---

[24] See Hansen and Sargent (2001) for the origin of the term 'multiplier preferences'.

[25] The classic experiments of Daniel Ellsberg (1961) have often been interpreted as indicating that people are averse to not knowing probability distributions of risks. Hansen and Sargent (2011) discuss various authors whose doubts about model specifications in macroeconomics have prompted them to decline to use expected utility.

consumer's concerns about not knowing the probability model that describes the consumption risk that he faces. The evidence cited in the above quote from Lucas (2003) and the introspective reasoning of Cochrane (1997) and Pratt (1964) that we described above on page 517 explain why many economists think that only small positive values of γ are plausible when it is interpreted as a risk-aversion parameter. Pratt's experiment confronts a decision maker with choices between gambles with *known* probability distributions.

How should we think about plausible values of γ, or rather, θ, when it is instead interpreted as encoding responses to gambles that involve *unknown* probability distributions? Hansen, Sargent, and Wang (2002) and Anderson, Hansen, and Sargent (2003) answer this question by recognizing the role of entropy in statistical tests for discriminating one probability distribution from another based on a sample of size T drawn from one or the other of the two distributions. They use the probability of making an error in discriminating between the two models as a way of disciplining the calibration of θ. That led them to argue that it is not appropriate to regard θ as a parameter that remains fixed when we vary the stochastic process for consumption under the consumer's approximating model. We take up this issue again in section 14.8.7.

14.8.6. Market price of model uncertainty

Recall the stochastic discount factor (14.7.11)

$$m_{t+1} = \left(\beta \frac{C_t}{C_{t+1}} \right) \left(\frac{\exp\left(-\theta^{-1} U_{t+1}\right)}{E_t \left[\exp\left(-\theta^{-1} U_{t+1}\right)\right]} \right),$$

or

$$m_{t+1} = \beta \exp\left(-\left(c_{t+1} - c_t\right)\right) g\left(\varepsilon_{t+1}\right), \tag{14.8.9}$$

where g is the likelihood ratio given by (14.7.13). One way to state the equity premium puzzle is that the data show that for U.S. data on per capita consumption, the conditional coefficient of variation of $\exp(-(c_{t+1} - c_t))$ is small while the conditional standard deviation of m_{t+1} revealed by asset market prices and returns is large. If the stochastic discount factor (14.8.9) is to explain the large observed equity premium, most of the job has to be done by volatility in $g(\varepsilon_{t+1})$.

Direct calculations show that the conditional standard deviation of the likelihood ratio $g(\varepsilon_{t+1})$ given by formula (14.7.13) is

$$\text{std}_t\left(g\right) = \left[\exp\left(w'w\right) - 1\right]^{\frac{1}{2}} \approx |w|, \tag{14.8.10}$$

where recall that w is given by (14.7.15) and that because $g(\varepsilon_{t+1})$ is a likelihood ratio, $E_t g = 1$. Therefore, the ratio of $\text{std}_t(g)$ to $E_t(g)$ equals $\text{std}_t(g)$. It can be verified directly that $|w_{t+1}|$ given by the above formulas comprises the lion's share of what Tallarini interpreted as the market price of risk given by formulas (14.6.6) and (14.6.7). This is because the first difference in the log of consumption has a small conditional coefficient of variation in our data, the heart of the equity premium puzzle. Thus, formula (14.8.10) is a good approximation to Tallarini's formula (14.7.19).[26]

However, Hansen, Sargent, and Tallarini (1999) and Hansen and Sargent (2008) advocated calling

$$\text{std}_t(g) \approx |w| = \frac{\sigma_c}{\theta(1 - \beta)}$$

the *market price of model uncertainty* and interpreted it as compensation that the representative consumer requires for bearing uncertainty about the probability distribution that governs $c_{t+1} - c_t$. Barillas, Hansen, and Sargent (2009) show that setting θ to capture what they regard as moderate and plausible amounts of model uncertainty goes a long way toward pushing what Tallarini would measure as the market price of uncertainty, but which they instead interpret as the market price of model uncertainty, toward the Hansen-Jagannathan bounds.

14.8.7. Measuring model uncertainty

Anderson, Hansen, and Sargent (2003) took the following approach to measuring plausible amounts of model uncertainty. The decision maker's baseline approximating model is the random walk with drift model (14.6.3). However, the decision maker doubts this model and surrounds it with a cloud of models characterized by likelihood ratios $g(\varepsilon)$. To get a robust valuation, he constructs a worst-case model, the model associated with the minimizing likelihood ratio $g(\varepsilon)$ in the appropriate version of (14.8.5). When his approximating model is (14.6.3), this worst-case model for log consumption growth is

$$c_{t+1} = c_t + (\mu + \sigma_c w) + \sigma_c \varepsilon_{t+1}, \tag{14.8.11}$$

---

[26] In particular, using $w = (1 - \gamma)\sigma_c$, direct computations show that $\text{std}(g) = [\exp((\gamma - 1)^2 \sigma_c^2) - 1]^{\frac{1}{2}}$ which for large γ approximates (14.7.19).

where ε_{t+1} is again distributed according to a Gaussian density with mean zero and unit variance. Equation (14.8.11) says that the mean of consumption growth is not μ but $\mu + \sigma_c w$, where w is again given by (14.7.14) or (14.7.15). Evidently, the approximating model is the $\gamma = 1$ version of (14.8.11). It can be verified that the ambiguity averse consumer has a stochastic discount factor with respect to the approximating model (14.6.3) that looks as if he believes (14.8.11) instead of (14.6.3). It is as if he evaluates utility according to the ordinary utility recursion $U_t = c_t + \beta \tilde{E}_t U_{t+1}$, where $\tilde{E}_t$ is the mathematical expectation taken with respect to the probability distribution generated by (14.8.11).

When it is to be interpreted as a measure of model uncertainty, rather than risk aversion, Anderson, Hansen, and Sargent recommend calibrating γ or θ by using an object called a 'detection error probability'. In the present context, this object answers the following question. Given a sample of size T drawn from either (14.6.3) (call it model A) or (14.8.11) (call it model B), what is the probability that a likelihood ratio test would incorrectly testify *either* that model A generated the data when in fact model B generated the data, *or* that model B generated the data when in fact model A did?[27] It is easy to compute detection error probabilities by simulating likelihood ratios for samples of size T and counting the frequency of such model discrimination mistakes.[28]

Evidently, when $\gamma = 1$ (which means $\theta = +\infty$), $w = 0$, so models A and B are identical, and therefore statistically indistinguishable. In this case, the detection error probability is .5, signifying that, via rounding error, the computer essentially flips a coin in deciding which model generated the data. A detection error probability of .5 thus means that it is impossible to distinguish the models from sample data. But as we increase γ above 1, i.e., drive the penalty parameter below $+\infty$, the detection error probability falls. The idea here is to guide our choice of γ or θ as follows. Set a detection error probability that reflects an amount of model specification uncertainty about which it seems plausible for the decision maker to be concerned, then in the context of the particular approximating model at hand (which for us is (14.6.3)), find the γ associated with that detection error probability.

A plausible value for the detection error probability is a matter of judgement. If the detection error probability is .5, it means that the two models are

---

[27] Anderson, Hansen, and Sargent (2003) describe the close links between entropy and such detection error probabilities.

[28] See Barillas, Hansen, and Sargent (2009).

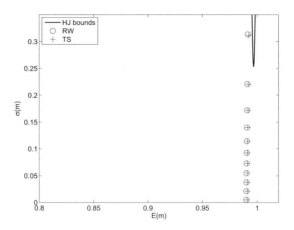

Figure 14.8.1: Reciprocal of risk free rate, market price of risk pairs for the random walk ($\circ$) model for values of detection error probabilities of 50, 45, 40, 35, 30, 25, 20, 15, 10, 5 and 1 percent.

statistically identical and can't be distinguished. A detection error probability of .25 means that there is a one in four chance of making the wrong decision about which model is generating the data. From our own experience fitting models to data, a person whose specification doubts include perturbed models with a detection error of .25 or .1 or even .05 could be said to have a plausible amount of model uncertainty.

Figure 14.8.1 redraws Tallarini's figure in terms of detection error probabilities for a sample size equal to the number of quarterly observations between 1948 and 2005 used to compute the Hansen and Jagannathan bounds. The figure again plots $(E(m), \sigma(m))$ pairs given by formulas $(14.7.18)\,(14.7.19)$ for γ's chosen to deliver the indicated detection error probabilities. The figure shows that moderate detection error probabilities of 10 or 15 percent take us more than half way to the Hansen and Jagannathan bounds, while 1 percent gets us there. The sense of these calculations is that moderate amounts of model uncertainty are able to substitute for huge amounts of risk aversion from the point of view of pushing the $(E(m), \sigma(m))$ toward the Hansen-Jagannathan bounds. In the next section, we revisit the quote from Lucas in light of this finding.

14.9. Costs of aggregate fluctuations

We now take up the important substantive issue that prompted Lucas to dismiss Tallarini's evidence about γ for the particular purpose then at hand for Lucas (1987, 2003). Lucas wanted to measure the gains to eliminating further unpredictable fluctuations in aggregate U.S. per capita consumption beyond reductions that had already been achieved by post World War II aggregate stabilization policies. His method was to find an upper bound on possible additional gains by computing the reduction in initial consumption that a representative consumer with time-separable preferences would be willing to accept in order to eliminate *all* unpredictable fluctuations that post WWII consumption around have exhibited. In this section, we describe Tallarini's version of Lucas's calculation and spotlight how γ influences the calculation.[29]

For the random walk with drift model of log consumption described by equation (14.6.3), the level of consumption $C_t = \exp(c_t)$ obeys $C_{t+1} = \exp(\mu + \sigma_c \varepsilon_{t+1})C_t$. A deterministic[30] process with same expected growth rate as the process $\{C_t\}$ is evidently

$$C_{t+1}^d = \exp\left(\mu + \frac{1}{2}\sigma_c^2\right) C_t^d \tag{14.9.1}$$

because $E \exp(\mu + \sigma_c \varepsilon_{t+1}) = \exp(\mu + \frac{1}{2}\sigma_c^2)$. Now using $\theta^{-1} = -(1-\gamma)(1-\beta)$ in our formula (14.7.8) allows us to express a risk-sensitive value function in terms of γ as

$$U_t = \frac{\beta}{(1-\beta)^2}\left[\mu + \frac{\sigma_\varepsilon^2(1-\gamma)}{2}\right] + \frac{1}{1-\beta}c_t, \tag{14.9.2}$$

Equating a time zero risk-sensitive value function for the deterministic process $\{C_t^d\}$ (on the left side) with a time zero risk-sensitive value function for the random process $\{C_t\}$ governed by the geometric random walk with drift (14.6.3) (on the right side) gives

$$\frac{\beta}{(1-\beta)^2}\left[\mu + \frac{\sigma_\varepsilon^2}{2}\right] + \frac{1}{1-\beta}c_o^d = \frac{\beta}{(1-\beta)^2}\left[\mu + \frac{\sigma_\varepsilon^2(1-\gamma)}{2}\right] + \frac{1}{1-\beta}c_0$$

where $c_0 = \log C_0$ and $c_0^d = \log C_0^d$. Solving for $c_0 - c_0^d$ gives

$$c_0 - c_0^d = \left(\frac{\beta}{1-\beta}\right)\frac{\gamma\sigma_c^2}{2}. \tag{14.9.3}$$

---

[29] For another perspective and the topic of this section, see Alvarez and Jermann (2004).

[30] A stochastic process is said to be *deterministic* if its future is perfectly predictable given its past and present.

The left side is the proportionate once-and-for-all reduction in initial consumption that a consumer would accept to trade a random process (14.6.3) for a perfectly predictable process with the same conditional mean growth rate as (14.6.3). The formula shows the utility costs of random fluctuations to be proportional to γ. The pertinent sources of evidence about the magnitude of γ and whether they can be interpreted as measures of attitudes toward risk, or maybe something else, are the issues under scrutiny in the above quote from Lucas. In showing that getting into the Hansen-Jagannathan bounds requires a value of γ of about 50 rather than Lucas's preferred number of 1 or 2, Tallarini's formula (14.9.3) brings a very large increase in estimates of the costs of aggregate consumption fluctuations. In dismissing the high value of γ that Tallarini found is necessary to attain the Hansen-Jagannathan bounds as measuring attitudes toward risks with known probabilities, Lucas was arguing for a lower estimate of the costs of aggregate risk than Tallarini had inferred.

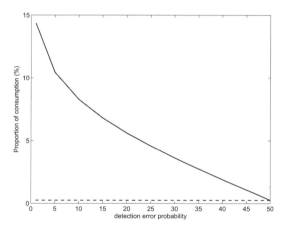

Figure 14.9.1: Proportions $c_0 - c_0^d$ of initial consumption that a representative consumer with model-uncertainty averse (multiplier) preferences would surrender not to confront risk (dotted line) and model uncertainty (solid line) for random-walk model of log consumption growth, plotted as a function of detection error probability.

In section 14.8.6 we argued that most of what Tallarini interpreted as the market price of *risk* should instead be interpreted as a market price of *model uncertainty*. The section 14.8.6 argument is one possible way of fulfilling Lucas's hope that "It would be good to have the equity premium resolved, but I think we need to look beyond high estimates of risk aversion to do it." And it supports Lucas's judgement that Tallarini's values of γ's calibrated to get into the Hansen-Jagannathan bounds are not suitable for mental experiments about risks with *known* probabilities of the kind that Lucas performed. Those high estimates of γ are relevant to *other* mental experiments about eliminating the consumer's concern about *model uncertainty*, but not about Lucas's experiment. Model uncertainty experiments are a subject of Barillas, Hansen, and Sargent (2009).

Continuing in this vein, figure 14.9.1 shows Barillas, Hansen, and Sargent's measures of the costs of removing random fluctuations in aggregate consumption per capita (the dotted line) as well as costs of removing model uncertainty (the solid line). The figure reports these costs as a function of the detection error probability described above. The costs of consumption risk drawn from a known distribution are small, as Lucas asserted, but for moderate values of detection error probabilities, the costs of model uncertainty are substantial.[31]

---

[31] See De Santis (2007) for a modification of the baseline around which the costs of aggregate fluctuations are measured. De Santis adopts a specification according to which a typical consumer's consumption process consists of an aggregate component and an uninsurable idiosyncratic component, modeled in the same fashion Constantinides and Duffie (1996) do in the model described in the next section. De Santis describes the welfare consequences of eliminating *aggregate* fluctuations, while leaving *idiosyncratic* fluctuations unaltered at their calibrated value. For a coefficient of relative risk aversion of 3, De Santis finds that the benefits of removing aggregate fluctuations are much larger when idiosyncratic fluctuations are not removed first. If one were to repeat De Santis's exercise for a coefficient of risk aversion of 1, the effect that he finds would disappear.

14.10. Reverse engineered consumption heterogeneity

In earlier sections, we explored how risk-sensitive preferences or a fear of model misspecification would increase the volatility of the stochastic discount factor by multiplying the ordinary stochastic discount factor $m_{t+1} = \beta \exp(-(c_{t+1} - c_t))$ with a random variable $g(\varepsilon_{t+1})$ that can be interpreted as a likelihood ratio. In this section, we describe how Constantinides and Duffie (1996) showed how to get such a volatility-increasing multiplicative adjustment in another way, namely, by introducing incomplete markets and stochastic volatility in the cross-sectional distribution of consumption growth.

Let $R_{j,t+1}, j = 1, \ldots, J$ be a list of returns on assets and let $m_{t+1} \geq 0$ be a stochastic discount for which

$$E_t m_{t+1} R_{j,t+1} = 1 \tag{14.10.1}$$

for $j = 1, \ldots, J$. As discussed in section 14.5, we know that such a discount factor exists on the weak assumptions that returns obey the law of one price and a no-arbitrage outcome. Constantinides and Duffie (1996)[32] reverse engineer stochastic consumption processes for a large collection of heterogeneous consumers $i \in I$ with the properties that

1. For each i, the personal stochastic discount factor m^i_{t+1} satisfies

$$E_t m^i_{t+1} R_{j,t+1} = 1, \quad j = 1, \ldots, J;$$

2. Where $\{C^i_t\}^\infty_{t=0}$ is consumer i's consumption process, each consumer's i's personal stochastic discount factor m^i_{t+1} is given by

$$m^i_{t+1} = \beta \left(\frac{C^i_{t+1}}{C^i_t} \right)^{-\gamma}; \tag{14.10.2}$$

   and

3. Depending on the evolution of the conditional volatility of the cross section of relative consumptions across agents, the γ that works in (14.10.2) can be much lower than the γ that would be estimated using a stochastic discount factor $m^a_{t+1} = \beta \left(\frac{C_{t+1}}{C_t} \right)^{-\gamma}$ based on *aggregate* consumption data $\{C_t\}$.

---

[32] Also see Mankiw (1986) and Attanasio and Weber (1993) for analyses that anticipate elements of the setup of this section.

Constantinides and Duffie's construction uses the properties of the log normal distribution in the following way.[33] Consider an economy with a large number (strictly speaking, a continuum) of consumers named i. Let C_t be the average across i of the individual consumption levels C_{it}. For a given vector return process $\{R_{j,t+1}, j = 1, \ldots, J\}$, take as given a 'successful' stochastic discount factor $\{m_{t+1}\}$ (i.e., one that satisfies (14.10.1) for $j = 1, \ldots, J$). (In section 14.11, we shall describe one widely used statistical specification of such an empirically successful stochastic discount factor.) For that empirically successful $\{m_{t+1}\}$ process, define the random y_{t+1} process[34]

$$y_{t+1} = \left[\frac{2}{\gamma(\gamma+1)} \right]^{\frac{1}{2}} \left[\log m_{t+1} - \log \beta + \gamma \log \left(\frac{C_{t+1}}{C_t} \right) \right]^{\frac{1}{2}}. \qquad (14.10.3)$$

For a given $\beta \in (0, 1)$ and a given $\gamma \geq 0$, define a candidate stochastic discount factor as the following function of given stochastic processes for $\{C_t, m_{t+1}\}$:

$$\tilde{m}_{t+1} = \beta \left(\frac{C_{t+1}}{C_t} \right)^{-\gamma} \exp \left(\frac{\gamma(\gamma+1)}{2} y_{t+1}^2 \right). \qquad (14.10.4)$$

Evidently, the right side of (14.10.4) can be said to condition on knowledge of y_{t+1}, which via (14.10.3) depends on m_{t+1}. By substituting (14.10.3) into (14.10.4), it can be verified that

$$\tilde{m}_{t+1} = m_{t+1}, \qquad (14.10.5)$$

so in equation (14.10.4) we have succeeded in generating a multiplicative adjustment $\exp\left(\frac{\gamma(\gamma+1)}{2} y_{t+1}^2 \right)$ to the standard CRRA stochastic discount factor that is capable of reconciling the returns data $R_{j,t+1}, j = 1, \ldots, J$ with the theoretical restrictions $E m_{t+1} R_{j,t+1} = 1$ for $j = 1, \ldots, J$.

But how can we interpret (14.10.4) and the associated multiplicative adjustment $\exp\left(\frac{\gamma(\gamma+1)}{2} y_{t+1}^2 \right)$? To answer that question, we proceed as follows. Let $\eta_{t+1}^i \sim \mathcal{N}(0, 1)$ and assume that the η_{t+1}^i processes are statistically independent across i. Using once again the properties of log normal random variables, note that

$$E \exp \left[-\gamma \left(\eta_{t+1}^i y_{t+1} - \frac{y_{t+1}^2}{2} \right) \Big| y_{t+1} \right] = \exp \left(\frac{\gamma(\gamma+1)}{2} y_{t+1}^2 \right). \qquad (14.10.6)$$

---

[33] This section can be viewed as an application of *back-solving*. See section 11.7 of chapter 11.

[34] Constantinides and Duffie (1996) assume that the $\{m_{t+1}, \frac{C_{t+1}}{C_t}\}$ processes satisfy $\log m_{t+1} - \log \beta + \gamma \log \left(\frac{C_{t+1}}{C_t} \right) \geq 0$.

The right side of (14.10.6) is the multiplicative adjustment to the ordinary CRRA discount factor with aggregate consumption that appears on the right side of (14.10.4). Therefore, by removing the conditioning on y_{t+1}^2, we can form a stochastic discount factor measurable with respect to *three* random processes $\{C_{t+1}, m_{t+1}, \eta_{t+1}^i\}$:

$$\tilde{m}_{t+1}^i = \beta \left(\frac{C_{t+1}}{C_t}\right)^{-\gamma} \exp\left[-\gamma\left(\eta_{t+1}^i y_{t+1} - \frac{y_{t+1}^2}{2}\right)\right]. \qquad (14.10.7)$$

Imagine doing this for each $i \in I$, taking note again of our assumption that the η_{t+1}^i processes are independent across the i's.[35] For each personal stochastic discount factor $\tilde{m}_{t+1}^i$, an application of the law of iterated expectations establishes that

$$E_t\left[\tilde{m}_{t+1}^i R_{j,t+1}\right] = 1, \quad j = 1, \ldots. J \qquad (14.10.8)$$

In this way, Constantinides and Duffie have reverse engineered a collection of personal stochastic discount factors $\tilde{m}_{t+1}^i$ that, for all i, succeed in reconciling the returns data to $E\tilde{m}_{t+1}^i R_{j,t+1} = 1$.

To interpret m_{t+1}^i for person i, define

$$\delta_{t+1}^i = \exp\left[\eta_{t+1}^i y_{t+1} - \frac{y_{t+1}^2}{2}\right]\delta_t^i. \qquad (14.10.9)$$

Evidently, $E\left[\delta_{t+1}^i | \delta_t^i, y_{t+1}\right] = \delta_t^i$, so that $\{\delta_t^i\}$ is a geometric random walk process. Also, an appeal to a law of large numbers for a cross section implies that the mean of δ_t^i across agents is 1 for all t.[36] Constantinides and Duffie posit that C_t^i is related to per capita consumption C_t by

$$C_t^i = \delta_t^i C_t, \qquad (14.10.10)$$

so that the nonnegative exponential random walk $\{\delta_t^i\}$ process governs consumer i's share of aggregate consumption. With this interpretation,

$$\frac{C_{t+1}^i}{C_t^i} = \exp\left[\eta_{t+1}^i y_{t+1} - \frac{y_{t+1}^2}{2}\right]\frac{C_{t+1}}{C_t}, \qquad (14.10.11)$$

---

[35] See Constantinides and Duffie (1993, p. 227) for important technical caveats about this independence assumption.

[36] Again, see Constantinides and Duffie (1993, p. 227) for qualifications about such an argument.

and y_{t+1}^2 is the conditional variance of the cross-section distribution of $\frac{C_{t+1}^i}{C_{t+1}} / \frac{C_t^i}{C_t} = \frac{\delta_{t+1}^i}{\delta_t^i}$. Substituting $(14.10.11)$ into $(14.10.7)$ shows that we can rewrite $\tilde{m}_{t+1}^i$ as

$$\tilde{m}_{t+1}^i = \beta \left(\frac{C_{t+1}^i}{C_t^i} \right)^{-\gamma}, \tag{14.10.12}$$

which is the ordinary discount factor for a consumer with discounted CRRA preferences and consumption process $\{C_t^i\}$.

To attain $(14.10.10)$ and $(14.10.12)$ as equilibrium outcomes requires a setting with the following features:

1. Markets are incomplete. In particular, consumers are allowed to trade only risk-free zero coupon bonds of various maturities together with our J securities, with security j have given return processes $\{R_{j,t+1}\}$.[37]

2. Each individual consumer's idiosyncratic risk is very persistent, as captured by the geometric random walk δ_t^i for household i's consumption share.

These two features are sufficient to describe an incomplete markets equilibrium with no trade in which $\{C_t^i\}$ is interpreted as an exogenous endowment process assigned to agent i.[38]

To illustrate key forces, Constantinides and Duffie consider the following model for cross-section consumption volatility:

$$y_{t+1}^2 = a + b \log \left(\frac{C_{t+1}}{C_t} \right) \tag{14.10.13}$$

Then it follows directly from $(14.10.11)$ that

$$E_t \left[R_{j,t+1} \hat{\beta} \left(\frac{C_{t+1}}{C_t} \right)^{-\hat{\gamma}} \right] = 1 \tag{14.10.14}$$

---

[37] We know from chapter 8 that if markets were complete, $(14.10.12)$ would be inconsistent with $(14.10.10)$ in equilibrium. Since $E_t m_{t+1}^i$ are equal across all i's are given at the assumed allocation $\{C_t^i\}$, it follows that the risk-free bonds are not traded.

[38] Heathcote, Storesletten, and Violante (2012) and De Santis (2007) put aspects of the Constantinides and Duffie specification of the consumption process to work in other contexts. By reinterpreting an individual consumption process in the Constantinides and Duffie as the outcome of an island in which there are complete insurance markets for heterogeneously endowed consumers, Heathcote, Storesletten, and Violante model partial consumption insurance: there is complete insurance within islands, incomplete insurance across islands. De Santis uses the Constantinides and Duffie consumption process to get a candidate for an alternative benchmark for measuring the costs of removing aggregate, but for De Santis, not idiosyncratic consumption fluctuations.

where

$$\log \hat{\beta} = \log \beta + \frac{\gamma(\gamma+1)}{2} a \qquad (14.10.15a)$$

$$\hat{\gamma} = \gamma - \frac{\gamma(\gamma+1)}{2} b. \qquad (14.10.15b)$$

Formula $(14.10.15b)$ implies that if $b < 0$, so that the cross-section dispersion of consumption increases during downturns in $\left(\frac{C_{t+1}}{C_t}\right)$, then $\hat{\gamma} > \gamma$.

Storesletten, Telmer, and Yaron (1998,2004) and Cogley (1999) pursued some of the ideas of Constantinides and Duffie by using evidence from the panel study of income dynamics (PSID) to estimate the persistence of endowment shocks and the volatility of consumption innovations.

14.11. Exponential affine stochastic discount factors

This section describes another class of reactions to the empirical problems encountered using the law of one price implication $E(m_{t+1}R_{j,t+1}) = 1$ coupled with the particular CRRA based stochastic discount factor $(14.6.4)$, which we repeat here for convenience

$$m_{t+1} = \beta \exp\left(-\gamma\mu - \gamma\sigma_c\varepsilon_{t+1}\right).$$

This model of the stochastic discount factor asserts that exposure to the random part of consumption growth, $\sigma_c\varepsilon_{t+1}$, is what causes risky securities to have higher rates of returns.

An influential asset pricing literature maintains $E(m_{t+1}R_{t+1}) = 1$ but divorces the stochastic discount factor m_{t+1} from consumption risk. Instead, the philosophy of this literature is no-holds-barred to specify a stochastic discount factor that (a) is analytically tractable (a test passed by $(14.6.4)$), and (b) can be calibrated to fit observed asset prices without provoking skeptical comments about nonsensical parameter values and interpretations of the risks that are priced (a test that critics like Lucas in the above quote say that $(14.6.4)$ fails to pass).

The alternative approach is to impose the law of one price via $E(m_{t+1}R_{t+1}) = 1$ (and often also the no-arbitrage principle via $m_{t+1} > 0$) but to abandon the link between the stochastic discount factor and a consumption growth process.

Instead, the idea is to posit a model-free process for the stochastic discount factor, and use the overidentifying restrictions from the household's Euler equations from a set of N returns $R_{it+1}, i = 1, \ldots, N$, to let the data indicate the relevant risks and the prices that the returns data attach to those risks.

The model has two main components. The first component is a vector autoregression that is designed to describe the underlying risks to be priced:

$$z_{t+1} = \mu + \phi z_t + C\varepsilon_{t+1} \tag{14.11.1}$$

$$r_t = \delta_0 + \delta_1' z_t, \tag{14.11.2}$$

where ϕ is a stable $m \times m$ matrix, C is an $m \times n$ matrix, $\varepsilon_{t+1} \sim \mathcal{N}(0, I)$ is an i.i.d. $m \times 1$ random vector, and z_t is an $m \times 1$ state vector. The second component is a stochastic discount factor whose log is affine (i.e., linear plus a constant) in the state vector z_t from the vector autoregresssion:

$$\Lambda_t = \Lambda_0 + \Lambda_z z_t \tag{14.11.3}$$

$$\log(m_{t+1}) = -r_t - \frac{1}{2}\Lambda_t'\Lambda_t - \Lambda_t'\varepsilon_{t+1}. \tag{14.11.4}$$

Here Λ_0 is $m \times 1$ and Λ_z is $m \times m$. Evidently,

$$E_t(m_{t+1}) = \exp\left(-(\delta_0 + \delta_1' z_t)\right) = \exp(-r_t) \tag{14.11.5}$$

$$\text{std}_t(m_{t+1}) = E_t(m_{t+1})\left(\exp(\Lambda_t'\Lambda_t) - 1\right)^{\frac{1}{2}}. \tag{14.11.6}$$

Equation (14.11.5) asserts that r_t is the yield on a risk-free one-period bond. That is why it is often called 'the short rate' in the literature on exponential affine models of the stochastic discount factor. The free parameters of the model are (μ, ϕ, C) for the vector autoregression, and $(\delta_0, \delta_1, \Lambda_0, \Lambda_z)$ pinning down the stochastic discount factor as a function of (z_t, ε_{t+1}). The 'loadings' or components of Λ_t that multiply corresponding components of the risks ε_{t+1} are sometimes called 'risk-prices', for reasons that we now explain.

14.11.1. *General application*

We can gather insights by applying the stochastic discount factor (14.11.4) to a risky one-period gross return $R_{j,t+1}$. Our standard pricing formula is

$$E_t \left(m_{t+1} R_{j,t+1} \right) = 1. \tag{14.11.7}$$

Let $R_{j,t+1}$ be described by

$$R_{j,t+1} = \exp\left(\nu_t \left(j \right) - \frac{1}{2} \alpha_t \left(j \right)' \alpha_t \left(j \right) + \alpha_t \left(j \right)' \varepsilon_{t+1} \right), \tag{14.11.8}$$

where $\nu_t(j)$ is a function of z_t that makes (14.11.7) become satisfied and

$$\alpha_t \left(j \right) = \alpha_0 \left(j \right) + \alpha_z \left(j \right) z_t, \tag{14.11.9}$$

where $\alpha_0(j)$ is an $n \times 1$ vector and $\alpha_z(j)$ is an $n \times n$ matrix. Evidently, (14.11.8) implies

$$E_t R_{j,t+1} = \exp \left(\nu_t \left(j \right) \right).$$

We want to find how Λ_t and $\alpha_t(j)$ affect the expected return $\nu_t(j)$. In (14.11.8), the components of the loading vector $\alpha_t(j)$ express the *exposure* of $R_{j,t+1}$ to the corresponding components of the vector of risks ε_{t+1}.

The formula for the mean of a log normal random variable implies that (14.11.7) becomes $\exp\left(\nu_t(j) - r_t - \alpha_t(j)'\Lambda_t \right) = 1$ or

$$\nu_t \left(j \right) = r_t + \alpha_t \left(j \right)' \Lambda_t. \tag{14.11.10}$$

According to (14.11.10), the net expected return $\nu_t(j)$ equals the net risk free return r_t plus the transposed vector Λ_t' of risk-prices times the vector $\alpha_t(j)$ of the risk-exposures of the gross return $R_{j,t+1}$. This outcome expresses how Λ_t is a vector of risk prices that tell how the expected return on an asset adjusts to exposure to the underlying risks ε_{t+1}.

14.11.2. *Term structure application*

A good representative of exponential affine models of stochastic discount factors is the model of the term structure of risk-free yields constructed and estimated by Ang and Piazzesi (2003). Let $p_t(n)$ be the price at time t of a risk-free claim to one unit of the consumption good at time $t + n$. The one-period gross *return* on holding such an $n + 1$ period pure discount bond from t to $t + 1$ is $R_{t+1} = \frac{p_{t+1}(n)}{p_t(n+1)}$. Evidently, in this case $E_t(m_{t+1}R_{t+1}) = 1$ implies[39]

$$p_t(n + 1) = E_t(m_{t+1}p_{t+1}(n)) \qquad (14.11.11)$$

and

$$p_t(1) = E_t(m_{t+1}) = \exp(-\delta_0 - \delta_1' z_t) = \exp(-r_t). \qquad (14.11.12)$$

From $(14.11.11)$ and $(14.11.12)$, it follows that

$$p_t(n) = \exp\left(\bar{A}_n + \bar{B}_n z_t\right), \qquad (14.11.13)$$

where $(\bar{A}_n, \bar{B}_n)$ can be computed recursively from

$$\bar{A}_{n+1} = \bar{A}_n + \bar{B}_n'(\mu - C\Lambda_0) + \frac{1}{2}\bar{B}_n'CC'\bar{B}_n - \delta_0 \qquad (14.11.14)$$

$$\bar{B}_{n+1}' = \bar{B}_n'(\phi - C\Lambda_z) - \delta_1', \qquad (14.11.15)$$

subject to the initial conditions $\bar{A}_1 = -\delta_0, \bar{B}_1 = -\delta_1$.

The *yield* to maturity on an n period pure discount bond is defined as

$$y_t(n) = -\frac{\log(p_t(n))}{n}, \qquad (14.11.16)$$

which is equivalent with $p_t(n) = \exp(-ny_t(n))$. Evidently,

$$y_t(n) = A_n + B_n' z_t, \qquad (14.11.17)$$

where $A_n = -\bar{A}_n/n, B_n = -\bar{B}_n/n$.

This model for yields fits within a class of linear Gaussian models that can be estimated by maximum likelihood using methods described in section 2.7.1 of chapter 2. See Ang and Piazzesi (2003) and Piazzesi (2005) for examples where

---

[39] See exercise 14.14.

some of the factors are interpreted in terms of a monetary policy authority's rule for setting a short rate.[40]

14.12. Concluding remarks

In this chapter, we have gone beyond chapter 8 in studying how, in the spirit of Hansen and Singleton (1983), consumer optimization alone puts restrictions on asset returns and consumption, without requiring complete markets or a fully articulated general equilibrium model. At various points in this chapter, especially in section 14.10, we have alluded to incomplete markets models. In chapters 18 and 20, we describe the ingredients of such models.

A. Riesz representation theorem

The version of the Riesz representation theorem used in this chapter is yet another ramification of population least squares regression.

Let X be a Hilbert space (i.e., a complete inner product space) of random variables with inner product $< x_1, x_2 >= E(x_1 x_2)$. Let $\phi : X \to I\!R$ be a continuous linear functional mapping X into $I\!R$, the real line. We say that ϕ is *linear* because (i) for $x_1 \in X, x_2 \in X$, $\phi(x_1 + x_2) = \phi(x_1) + \phi(x_2)$, and (ii) for $a \in I\!R, x \in X, \phi(ax) = a\phi(x)$. We want to prove

Theorem 14.A.1. *(Riesz representation) Let ϕ be a continuous linear functional $\phi : X \to I\!R$. There exists a unique element $y \in X$ such that*

$$\phi(x) = E(yx). \tag{14.A.1}$$

Proof. The null space $N \equiv N(\phi)$ is defined as:

$$N(\phi) = \big\{ x \in X : \phi(x) = 0 \big\}.$$

$N(\phi)$ is a closed linear subspace of X. If $N = X$, then evidently $\phi(x) = 0$ for all $x \in X$. If $X \neq N(\phi)$, then there exists a non-zero vector $x_1 \in N^{\perp}$, where

---

[40] Appendix B of this chapter uses methods of chapter 2 to bring out some of the implications of a simple affine term structure model of Backus and Zin (1994).Also see Chen and Scott (1993) and Dai and Singleton (2000), and Piazzesi and Schneider (2006).

$N^\perp$ is the orthogonal complement of N, i.e., the set of vectors $y \in X$ for which $< x, y >= 0$ for every $x \in N$. In fact, $N^\perp$ consists of scalar multiples of one vector x_1 that is a basis for $N^\perp$. To prove this, assume to the contrary that there are two linearly independent vector x_1 and x_2, both of which are elements of $N^\perp$. The linear independence of the vectors x_1 and x_2 implies that we can choose two nonzero real scalars a, b such that $\phi(ax_1 - bx_2) = a\phi(x_1) - b\phi(x_2) = 0$. But this implies that $ax_1 - bx_2$ belongs to both N and to $N^\perp$. This is possible only if $ax_1 - bx_2 = 0$, contradicting the premise that x_1 and x_2 are linearly independent. Thus, there is a unique linearly independent vector x_1 that is a basis for $N^\perp$.

We propose the following scaled version of the basis vector x_1 as our candidate for the vector y in representation (14.A.1):

$$y = \frac{\phi(x_1)}{< x_1, x_1 >} x_1. \tag{14.A.2}$$

Evidently, $y \in N^\perp$. By computing a population linear least squares regression of $x \in X$ on $y \in X$, we can represent x as the sum of the linear least squares projection of x_1 on y and an orthogonal residual:

$$x = ay + (x - ay), \tag{14.A.3}$$

where a is the scalar regression coefficient

$$a = \frac{< x, y >}{< y, y >}. \tag{14.A.4}$$

Both $ay \in N^\perp$ and $(x - ay) \in N$ are unique in representation (14.A.4). In (14.A.3), $ay \in N^\perp$ and $(x - ay) \in N$ because the least squares residual $x - ay$ is orthogonal to the regressor y. Therefore, applying ϕ to both sides of (14.A.3) gives

$$\phi(x) = a\phi(y)$$

by the linearity of ϕ and the fact that $\phi(x - ay) = 0$ because $(x - ay) \in N$. Direct computations show that $a = \frac{<x,y><x_1,x_1>}{\phi(x_1)^2}$ and from definition (14.A.2) that $\phi(y) = \frac{<x,y>}{<x_1,x_1>}$. Therefore,

$$\phi(x) = a\phi(y) = < x, y >. \tag{14.A.5}$$

REMARK: Suppose that $x \in M$ where M is a closed linear subspace of X. Then a simple corollary of Theorem 14.A.1 asserts that there exist multiple random variables $\tilde{y} \in X$ for which

$$\phi(x) = E(\tilde{y}x).$$

The random variable $\tilde{y}$ can be constructed as $\tilde{y} = y + \varepsilon$ where y is constructed as in Theorem 14.A.1 (except that now it is required to be in the linear subspace M) and ε is any random vector in the orthogonal complement of M (i.e., the space of random vectors in X that are orthogonal to the closed linear subspace M). This fact is pertinent for thinking about economies with incomplete markets.

B. A log normal bond pricing model

Following Backus and Zin (1994), we study the following log normal bond pricing model that is a special case of the affine term structure model of section 14.11. A one-period stochastic discount factor at t is m_{t+1} and an n-period stochastic discount factor at t is $m_{t+1}m_{t+2}\cdots m_{t+n}$.[41] The logarithm of the one-period stochastic discount factor follows the stochastic process

$$\log m_{t+1} = -\delta - e_z z_{t+1} \tag{14.B.1a}$$

$$z_{t+1} = A_z z_t + C_z w_{t+1} \tag{14.B.1b}$$

where w_{t+1} is an i.i.d. Gaussian random vector with $Ew_{t+1} = 0$, $Ew_{t+1}w'_{t+1} = I$, and A_z is an $m \times m$ matrix all of whose eigenvalues are bounded by unity in modulus. Soon we shall describe a particular process for the log of the nominal stochastic discount factor that Backus and Zin (1994) used to emulate the term structure of nominal interest rates in the United States during the post-World War II period.

This can be viewed as a special case of the section 14.11.2 model with the following settings mapping the general model into the special model:

$$(\mu, \phi, C, \Lambda'_0, \Lambda_z, \delta'_1) = (0, C_z, e_z C_z, 0, e_z A_z)$$

$$\delta_0 + \frac{1}{2}\Lambda'_0 \Lambda_0 = \delta.$$

---

[41] Some authors use the notation $m_{t+j,t}$ to denote a j-period stochastic discount factor at time t. The transformation between that notation and ours is $m_{t+1,t} = m_{t+1}, \ldots, m_{t+j,t} = m_{t+1}\cdots m_{t+j}$.

Then formulas $(14.11.14)$ and $(14.11.15)$ can be applied to compute the coefficients $\bar{A}_n, \bar{B}_n$ that appear in the price of an n period zero coupon bond of the form $(14.11.13)$, namely, $p_t(n) = \exp\left(\bar{A}_n + \bar{B}_n z_t\right)$, so that yields are of the form

$$y_t(n) = A_n + B_n' z_t$$

where $A_n = -\bar{A}_n/n, B_n = -\bar{B}_n/n$. Here we will proceed to obtain formulas of this form by working directly with representation $(14.B.1a)$-$(14.B.1b)$. We do this partly to give practice in applying some of the formulas from chapter 2.

Applying the properties of the log normal distribution to the conditional distribution of m_{t+1} induced by $(14.B.1)$ gives

$$\log E_t m_{t+1} = -\delta - e_z A_z z_t + \frac{e_z C_z C_z' e_z'}{2}. \qquad (14.B.2)$$

By iterating on $(14.B.1)$, we can obtain the following expression that is useful for characterizing the conditional distribution of $\log(m_{t+1} \cdots m_{t+n})$:

$$
\begin{aligned}
-\left(\log\left(m_{t+1}\right) + \cdots \log\left(m_{t+n}\right)\right) = {} & n\delta + e_z \left(A_z + A_z{}^2 + \cdots A_z{}^n\right) z_t \\
& + e_z C_z w_{t+n} + e_z \left[C_z + A_z C_z\right] w_{t+n-1} \\
& + \cdots + e_z \left[C_z + A_z C_z + \cdots + A_z{}^{n-1} C_z\right] w_{t+1} \\
& \hspace{8cm} (14.B.3)
\end{aligned}
$$

The distribution of $\log m_{t+1} + \cdots + \log m_{t+n}$ conditional on z_t is thus $\mathcal{N}(\mu_{nt}, \sigma_n^2)$, where [42]

$$\mu_{nt} = -\left[n\delta + e_z \left(A_z + \cdots A_z{}^n\right) z_t\right] \qquad (14.B.4a)$$

$$\sigma_1^2 = e_z C_z C_z' e_z' \qquad (14.B.4b)$$

$$\sigma_n^2 = \sigma_{n-1}^2 + e_z \left[I + \cdots + A_z{}^{n-1}\right] C_z C_z' \left[I + \cdots + A_z{}^{n-1}\right]' e_z' \qquad (14.B.4c)$$

where the recursion $(14.B.4c)$ holds for $n \geq 2$. Notice that the conditional means μ_{nt} vary over time but that the conditional covariances σ_n^2 are constant over time. Applying $(14.11.16)$ or $y_t(n) = -n^{-1} \log E_t[m_{t+1} \cdots m_{t+n}]$ and the formula for the log of the expectation of a log normally distributed random variable gives the following formula for bond yields:

$$y_t(n) = \left(\delta - \frac{\sigma_n^2}{2 \times n}\right) + n^{-1} e_z \left(A_z + \cdots + A_z{}^n\right) z_t. \qquad (14.B.5)$$

---

[42] For the purpose of programming these formulas, it is useful to note that $(I + A_z + \cdots + A_z{}^{n-1}) = (I - A_z)^{-1}(I - A_z{}^n)$.

The vector $y_t = \begin{bmatrix} y_{1t} & y_{2t} & \cdots & y_{nt} \end{bmatrix}'$ is called the term structure of nominal interest rates at time t. A specification known as the *expectations theory of the term structure* resembles but differs from $(14.B.5)$. The expectations theory asserts that n-period yields are averages of expected future values of one-period yields, which translates to

$$y_t(n) = \delta + n^{-1}e_z\left(A_z + \cdots + A_z{}^n\right)z_t \qquad (14.B.6)$$

because evidently the conditional expectation $E_t y_{1t+j} = \delta + e_z A_z^j z_t$. The expectations theory $(14.B.6)$ can be viewed as an approximation to the log normal yield model $(14.B.5)$ that neglects the contributions of the variance terms σ_n^2 to the constant terms.

Returning to the log normal bond pricing model, we evidently have the following compact state-space representation for the term structure of interest rates and its dependence on the law of motion for the stochastic discount factor:

$$X_{t+1} = A_o X_t + C w_{t+1} \qquad (14.B.7a)$$

$$Y_t \equiv \begin{bmatrix} y_t \\ \log(m_t) \end{bmatrix} = G X_t \qquad (14.B.7b)$$

where

$$X_t = \begin{bmatrix} 1 \\ z_t \end{bmatrix} \qquad A_o = \begin{bmatrix} 1 & 0 \\ 0 & A_z \end{bmatrix} \qquad C = \begin{bmatrix} 0 \\ C_z \end{bmatrix}$$

and

$$G = \begin{bmatrix} \delta - \frac{\sigma_1^2}{2} & e_z A_z \\ \delta - \frac{\sigma_2^2}{2 \times 2} & 2^{-1} e_z \left(A_z + A_z{}^2\right) \\ \vdots & \vdots \\ \delta - \frac{\sigma_n^2}{2 \times n} & n^{-1} e_z \left(A_z + \cdots + A_z{}^n\right) \\ -\delta & -e_z \end{bmatrix}.$$

14.B.1. Slope of yield curve depends on serial correlation of $\log m_{t+1}$

From $(14.B.7)$, it follows immediately that the unconditional mean of the term structure is

$$Ey_t' = [\,\delta - \frac{\sigma_1^2}{2} \quad \cdots \quad \delta - \frac{\sigma_n^2}{2 \times n}\,]',$$

so that the term structure on average rises with horizon only if σ_j^2/j falls as j increases. By interpreting our formulas for the σ_j^2's, it is possible to show that a term structure that on average *rises* with maturity implies that the log of the stochastic discount factor is *negatively* serially correlated. Thus, it can be verified from $(14.B.3)$ that the term σ_j^2 in $(14.B.4)$ and $(14.B.5)$ satisfies

$$\sigma_j^2 = \mathrm{var}_t\,(\log m_{t+1} + \cdots + \log m_{t+j})$$

where var_t denotes a variance conditioned on time t information z_t. Notice, for example, that

$$\mathrm{var}_t\,(\log m_{t+1} + \log m_{t+2}) = \mathrm{var}_t\,(\log m_{t+1}) + \mathrm{var}_t\,(\log m_{t+2})$$
$$+ 2\mathrm{cov}_t\,(\log m_{t+1}, \log m_{t+2}) \tag{14.B.8}$$

where cov_t is a conditional covariance. It can then be established that $\sigma_1^2 > \frac{\sigma_2^2}{2}$ can occur only if $\mathrm{cov}_t(\log m_{t+1}, \log m_{t+2}) < 0$. Thus, a yield curve that is upward sloping on average reveals that the log of the stochastic discount factor is negatively serially correlated. (See the spectrum of the log stochastic discount factor in Figure 14.B.5.)

14.B.2. Backus and Zin's stochastic discount factor

For a specification of A_z, C_z, δ for which the eigenvalues of A_z are all less than unity, we can use the formulas presented above to compute moments of the stationary distribution EY_t, as well as the autocovariance function $\mathrm{Cov}_Y(\tau)$ and the impulse response function given in $(2.4.14)$ or $(2.4.15)$. For the term structure of nominal U.S. interest rates over much of the post-World War II period, Backus and Zin (1994) provide us with an empirically plausible specification of A_z, C_z, e_z. In particular, they specify that $\log m_{t+1}$ is a stationary autoregressive moving average process

$$-\phi\,(L) \log m_{t+1} = \phi\,(1)\,\delta + \theta\,(L)\,\sigma w_{t+1}$$

where w_{t+1} is a scalar Gaussian white noise with $Ew_{t+1}^2 = 1$ and

$$\phi(L) = 1 - \phi_1 L - \phi_2 L^2 \qquad\qquad (14.B.9a)$$

$$\theta(L) = 1 + \theta_1 L + \theta_2 L^2 + \theta_3 L^3. \qquad\qquad (14.B.9b)$$

Backus and Zin specified parameter values that imply that all of the zeros of both $\phi(L)$ and $\theta(L)$ *exceed* unity in modulus,[43] a condition that ensures that the eigenvalues of A_o are all *less than* unity in modulus. Backus and Zin's specification can be captured by setting

$$z_t = \begin{bmatrix} \log m_t & \log m_{t-1} & w_t & w_{t-1} & w_{t-2} \end{bmatrix}$$

and

$$A_z = \begin{bmatrix} \phi_1 & \phi_2 & \theta_1\sigma & \theta_2\sigma & \theta_3\sigma \\ 1 & 0 & 0 & 0 & 0 \\ 0 & 0 & 0 & 0 & 0 \\ 0 & 0 & 1 & 0 & 0 \\ 0 & 0 & 0 & 1 & 0 \end{bmatrix}$$

and $C_z = \begin{bmatrix} \sigma & 0 & 1 & 0 & 0 \end{bmatrix}'$ where $\sigma > 0$ is the standard deviation of the innovation to $\log m_{t+1}$ and $e_z = \begin{bmatrix} 1 & 0 & 0 & 0 & 0 \end{bmatrix}$.

14.B.3. Reverse engineering a stochastic discount factor

Backus and Zin use time series data on y_t together with the restrictions implied by the log normal bond pricing model to deduce implications about the stochastic discount factor m_{t+1}. They call this procedure "reverse engineering the yield curve," but what they really do is use time series observations on the *yield curve* to reverse engineer a *stochastic discount factor*. They used the generalized method of moments to estimate (some people say "calibrate") the following values for monthly United States nominal interest rates on pure discount bonds: $\delta = .528, \sigma = 1.023, \theta(L) = 1 - 1.031448L + .073011L^2 + .000322L^3$, $\phi(L) = 1 - 1.031253L + .073191L^2$. Why do Backus and Zin carry along so many digits? To explain why, first notice that with these particular values $\frac{\theta(L)}{\phi(L)} \approx 1$, so that the log of the stochastic discount factor is well approximated by an i.i.d. process:

$$-\log m_{t+1} \approx \delta + \sigma w_{t+1}.$$

---

[43] A complex variable z_0 is said to be a zero of $\phi(z)$ if $\phi(z_0) = 0$.

This means that fluctuations in the log stochastic discount factor are difficult to predict. Backus and Zin argue convincingly that to match observed features that are summarized by estimated first and second moments of the nominal term structure y_t process and for yields on other risky assets for the United States after World War II, it is important that $\theta(L), \phi(L)$ have two properties: (a) first, $\theta(L) \approx \phi(L)$, so that the stochastic discount factor is a volatile variable whose fluctuations are difficult to predict variable; and (b) nevertheless that $\theta(L) \neq \phi(L)$, so that the stochastic discount factor has subtle predictable components. Feature (a) is needed to match observed prices of risky securities, as we shall discuss in chapter 14. In particular, observations on returns on risky securities can be used to calculate a so-called market price of risk that in theory should equal $\frac{\sigma_t(m_{t+1})}{E_t m_{t+1}}$, where σ_t denotes a conditional standard deviation and E_t a conditional mean, conditioned on time t information. Empirical estimates of the stochastic discount factor from the yield curve and other asset returns suggest a value of the market price of risk that is relatively large, in a sense that we explore in depth in chapter 14. A high volatility of m_{t+1} delivers a high market price of risk. Backus and Zin use feature (b) to match the shape of the yield curve over time. Backus and Zin's estimates of $\phi(L), \theta(L)$ imply term structure outcomes that display both features (a) and (b). For their values of $\theta(L), \phi(L), \sigma$, Figures 14.B.1–14.B.5 show various aspects of the theoretical yield curve. Figure 14.B.1 shows the theoretical value of the mean term structure of interest rates, which we have calculated by applying our chapter 2 formula for $\mu_Y = G\mu_X$ to (14.B.7). The theoretical value of the yield curve is on average upward sloping, as is true also in the data. For yields of durations $j = 1, 3, 6, 12, 24, 36, 48, 60, 120, 360$, where duration is measured in *months*, Figure 14.B.2 shows the impulse response of y_{jt} to a shock w_{t+1} in the log of the stochastic discount factor. We use formula (2.4.15) to compute this impulse response function. In Figure 14.B.2, bigger impulse response functions are associated with *shorter* horizons. The shape of the impulse response function for the short rate differs from the others: it is the only one with a humped shape. Figures 14.B.3 and 14.B.4 show the impulse response function of the log of the stochastic discount factor. Figure 14.B.3 confirms that $\log m_{t+1}$ is approximately i.i.d. (the impulse response occurs mostly at zero lag), but Figure 14.B.4 shows the impulse response coefficients for lags of 1 and greater and confirms that the stochastic discount factor is not quite i.i.d. Since the initial response is a large negative number, these small positive responses for positive lags impart *negative* serial correlation to the log stochastic discount factor. As noted above and as stressed by Backus and Zin (1992), negative serial correlation of the stochastic discount factor is needed to account for a yield curve that is upward sloping on average.

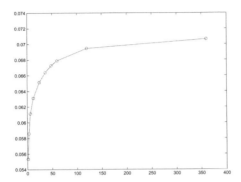

Figure 14.B.1: Mean term structure of interest rates with Backus-Zin stochastic discount factor (months on horizontal axis).

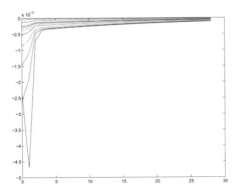

Figure 14.B.2: Impulse response of yields y_{nt} to innovation in stochastic discount factor. Bigger responses are for shorter maturity yields.

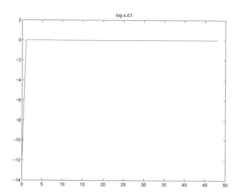

Figure 14.B.3: Impulse response of log of stochastic discount factor.

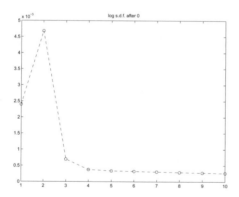

Figure 14.B.4: Impulse response of log stochastic discount factor from lag 1 on.

Figure 14.B.5 applies the Matlab program `bigshow2` to Backus and Zin's specified values of $(\sigma, \delta, \theta(L), \phi(L))$. The panel on the upper left is the impulse response again. The panel on the lower left shows the covariogram, which as expected is very close to that for an i.i.d. process. The spectrum of the log stochastic discount factor is not completely flat and so reveals that the log stochastic discount factor is serially correlated. (Remember that the spectrum for a serially uncorrelated process, a white noise, is perfectly flat.) That the

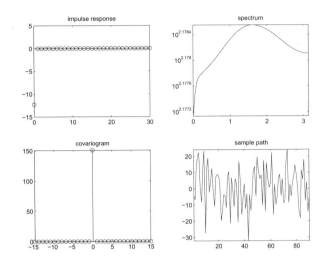

Figure 14.B.5: `bigshow2` for Backus and Zin's log stochastic discount factor.

spectrum is generally rising as frequency increases from $\omega = 0$ to $\omega = \pi$ indicates that the log stochastic discount factor is *negatively* serially correlated. But the negative serial correlation is subtle, so that the realization plotted in the panel on the lower right is difficult to distinguish from a white noise.

Exercises

Exercise 14.1   **Hansen-Jagannathan bounds**

Consider the following annual data for annual gross returns on U.S. stocks and U.S. Treasury bills from 1890 to 1979. These are the data used by Mehra and Prescott. The mean returns are $\mu = [\,1.07 \quad 1.02\,]$ and the covariance matrix of returns is $\begin{bmatrix} .0274 & .00104 \\ .00104 & .00308 \end{bmatrix}$.

a. For data on the excess return of stocks over bonds, compute Hansen and Jagannathan's bound on the stochastic discount factor y. Plot the bound for $E(y)$ on the interval $[.9, 1.02]$.

b. Using data on both returns, compute and plot the bound for $E(y)$ on the interval $[.9, 1.02]$. Plot this bound on the same figure as you used in part a.

c. At $<$https://files.nyu.edu/ts43/public/books.html$>$, there is a Matlab file epdata.m with Kydland and Prescott's time series. The series epdata(:,4) is the annual growth rate of aggregate consumption C_t/C_{t-1}. Assume that $\beta = .99$ and that $m_t = \beta u'(C_t)/u'(C_{t-1})$, where $u(\cdot)$ is the CRRA utility function. For the three values of $\gamma = 0, 5, 10$, compute the standard deviation and mean of m_t and plot them on the same figure as in part b. What do you infer from where the points lie?

Exercise 14.2 **The term structure and regime switching**, donated by Rodolfo Manuelli

Consider a pure exchange economy in which the stochastic process for per capita consumption is given by

$$C_{t+1} = C_t \exp\left[\alpha_0 - \alpha_1 s_t + \varepsilon_{t+1}\right],$$

where

(i) $\alpha_0 > 0$, $\alpha_1 > 0$, and $\alpha_0 - \alpha_1 > 0$.

(ii) ε_t is a sequence of i.i.d. random variables distributed $\mathcal{N}(\mu, \tau^2)$. Note: given this specification, it follows that $E[e^\varepsilon] = \exp[\mu + \tau^2/2]$.

(iii) s_t is a Markov process independent from ε_t that can take only two values, $\{0, 1\}$. The transition probability matrix is completely summarized by

$$\text{Prob}\left[s_{t+1} = 1 | s_t = 1\right] = \pi\left(1\right),$$
$$\text{Prob}\left[s_{t+1} = 0 | s_t = 0\right] = \pi\left(0\right).$$

(iv) The information set at time t, Ω_t, contains $\{C_{t-j}, s_{t-j}, \varepsilon_{t-j}; j \geq 0\}$.

There is a large number of identical individuals, each a representative agent, with the following utility function

$$U = E_0 \sum_{t=0}^{\infty} \beta^t u\left(C_t\right),$$

where $u(C) = C^{(1-\gamma)}/(1-\gamma)$. Assume that $\gamma > 0$ and $0 < \beta < 1$. As usual, $\gamma = 1$ corresponds to the log utility function.

a. Compute the "short-term" (one-period) risk-free interest rate.

b. Compute the "long-term" (two-period) risk-free interest rate measured in the same time units as the rate you computed in **a**. (That is, take the appropriate square root.)

c. Note that the log of the rate of growth of consumption is given by

$$\log(C_{t+1}) - \log(C_t) = \alpha_0 - \alpha_1 s_t + \varepsilon_{t+1}.$$

Thus, the conditional expectation of this growth rate is just $\alpha_0 - \alpha_1 s_t + \mu$. Note that when $s_t = 0$, growth is high, and when $s_t = 1$, growth is low. Thus, loosely speaking, we can identify $s_t = 0$ with the peak of the cycle (or good times) and $s_t = 1$ with the trough of the cycle (or bad times). Assume $\mu > 0$. Go as far as you can describing the implications of this model for the cyclical behavior of the term structure of interest rates.

d. Are short term rates pro- or countercyclical?

e. Are long rates pro- or countercyclical? If you cannot give a definite answer to this question, find conditions under which they are either pro- or countercyclical, and interpret your conditions in terms of the "permanence" (you get to define this) of the cycle.

Exercise 14.3 **Growth slowdowns and stock market crashes**, donated by Rodolfo Manuelli[44]

Consider a simple one-tree pure exchange economy. The only source of consumption is the fruit that grows on the tree. This fruit is called dividends by the tribe inhabiting this island. The stochastic process for dividend d_t is described as follows: If d_t is not equal to d_{t-1}, then $d_{t+1} = \gamma d_t$ with probability π, and $d_{t+1} = d_t$ with probability $(1 - \pi)$. If in any pair of periods j and $j+1$, $d_j = d_{j+1}$, then for all $t > j$, $d_t = d_j$. In words, if not stopped, the process grows at a gross rate ν in every period. However, once it stops growing for one period, it remains constant forever after. Let d_0 equal 1.

Preferences over stochastic processes for consumption are given by

$$U = E_0 \sum_{t=0}^{\infty} \beta^t u(C_t),$$

---

[44] See also Joseph Zeira (1999).

where $u(C) = C^{(1-\gamma)}/(1-\gamma)$. Assume that $\gamma > 0$, $0 < \beta < 1$, $\nu > 1$, and $\beta\nu^{(1-\gamma)} < 1$.

a. Define a competitive equilibrium in which shares to this tree are traded.

b. Display the equilibrium process for the price of shares in this tree p_t as a function of the history of dividends. Is the price process a Markov process in the sense that it depends just on the last period's dividends?

c. Let T be the first time in which $d_{T-1} = d_T = \gamma^{(T-1)}$. Is $p_{T-1} > p_T$? Show conditions under which this is true. What is the economic intuition for this result? What does it say about stock market declines or crashes?

d. If this model is correct, what does it say about the behavior of the aggregate value of the stock market in economies that switched from high to low growth (e.g., Japan)?

Exercise 14.4   **The term structure and consumption**, donated by Rodolfo Manuelli

Consider an economy populated by a large number of identical households. The (common) utility function is

$$\sum_{t=0}^{\infty} \beta^t u\,(C_t),$$

where $0 < \beta < 1$, and $u(x) = x^{(1-\gamma)}/(1-\gamma)$, for some $\gamma > 0$. (If $\gamma = 1$, the utility is logarithmic.) Each household owns one tree. Thus, the number of households and the number of trees coincide. The amount of consumption that grows in a tree satisfies

$$C_{t+1} = C^* C_t^{\varphi} \varepsilon_{t+1},$$

where $0 < \varphi < 1$, and ε_t is a sequence of i.i.d. log normal random variables with mean 1, and variance σ^2. Assume that, in addition to shares in trees, in this economy bonds of all maturities are traded.

a. Define a competitive equilibrium.

b. Go as far as you can calculating the term structure of interest rates, $\tilde{R}_{jt}$, for $j = 1, 2, \ldots$.

c. Economist A argues that economic theory predicts that the variance of the log of short-term interest rates (say, one-period) is always lower than the variance of long-term interest rates, because short rates are "riskier." Do you agree? Justify your answer.

d. Economist B claims that short-term interest rates, i.e., $j = 1$, are "more responsive" to the state of the economy, i.e., C_t, than are long-term interest rates, i.e., j large. Do you agree? Justify your answer.

e. Economist C claims that the Fed should lower interest rates because whenever interest rates are low, consumption is high. Do you agree? Justify your answer.

f. Economist D claims that in economies in which output (consumption in our case) is very persistent ($\varphi \approx 1$), changes in output (consumption) do not affect interest rates. Do you agree? Justify your answer and, if possible, provide economic intuition for your argument.

Exercise 14.5 **Ambiguity averse multiplier preferences**

Consider the recursion (14.7.5), namely,

$$U_t = c_t - \beta\theta \log E_t \left[\exp\left(\frac{-U_{t+1}}{\theta} \right) \right], \tag{1}$$

where $c_t = \log C_t, \beta \in (0,1)$, and $0 < \theta$. Let c_t follow the stochastic process

$$c_{t+1} = c_t + \mu + \sigma_\varepsilon \varepsilon_{t+1}, \tag{2}$$

where $\varepsilon_{t+1} \sim \mathcal{N}(0,1)$ is an i.i.d. random process.

a. Guess a value function of the form $U_t = k_0 + k_1 c_t$, where k_0 and k_1 are scalar constants. In detail, derive formulas for k_0, k_1 that verify the recursion (14.7.5) (or (1) above).

b. Using the formulas for k_0, k_1 that you have derived, verify that when c_t obeys (2), another way to express recursion (1) is

$$U_t = c_t + \beta E_t U_{t+1} - \frac{\beta}{2\theta} \mathrm{var}_t(U_{t+1}). \tag{3}$$

Use representation (3) to offer an interpretation of why the consumer prefers plan A to plan B in figure 14.7.1 and also why he prefers Plan C to Plan D in figure 14.7.2.

c. For recursion (1), verify in detail that

$$\frac{\partial U_t}{\partial C_t} = \frac{k_1}{C_t}$$

$$\frac{\partial U_t}{\partial C_{t+1}} = \beta \left(\frac{\exp\left(-\theta^{-1} U_{t+1}\right)}{E_t \exp\left(-\theta^{-1} U_{t+1}\right)} \right) \frac{k_1}{C_{t+1}}.$$

d. Explicitly verify that the likelihood ratio defined as g satisfies

$$g\left(\varepsilon_{t+1}\right) \equiv \left(\frac{\exp\left((1-\beta)\left(1-\gamma\right) U_{t+1}\right)}{E_t\left[\exp\left((1-\beta)\left(1-\gamma\right) U_{t+1}\right)\right]} \right) = \exp\left(w\varepsilon_{t+1} - \frac{1}{2} w^2 \right)$$

where w is given by

$$w = -\frac{\sigma_\varepsilon}{\theta\left(1-\beta\right)}.$$

Hint: Once again, you might want to use the formula for the mean of a log normal random variable.

Exercise 14.6   **Exponential affine sdf**

Consider an exponential affine model of a stochastic discount factor driven by the following vector autoregression for z_t:

$$z_{t+1} = \mu + \phi z_t + C\varepsilon_{t+1}$$

$$r_t = \delta_0 + \delta_1' z_t,$$

where ϕ is a stable $n \times n$ matrix, C is an $n \times n$ matrix, $\varepsilon_{t+1} \sim \mathcal{N}(0, I)$ is an i.i.d. $n \times 1$ random vector, and z_t is an $n \times 1$ state vector. The logarithm of the stochastic discount factor is affine (i.e., linear plus a constant) in the state vector z_t from the vector autoregresssion:

$$\Lambda_t = \Lambda_0 + \Lambda_z z_t$$

$$\log\left(m_{t+1}\right) = -r_t - \frac{1}{2}\Lambda_t'\Lambda_t - \Lambda_t'\varepsilon_{t+1}.$$

Let the gross risky return $R_{j,t+1}$ on an asset be

$$R_{j,t+1} = \exp\left(\nu_t\left(j\right) - \frac{1}{2}\alpha_t\left(j\right)' \alpha_t\left(j\right) + \alpha_t\left(j\right)' \varepsilon_{t+1} \right), \tag{1}$$

where $\nu_t(j)$ is a function of z_t that makes $E_t(m_{t+1}R_{j,t+1}) = 1$ be satisfied and $\alpha_t(j) = \alpha_0(j) + \alpha_z(j)z_t$, where $\alpha_0(j)$ is an $n \times 1$ vector and $\alpha_z(j)$ is an $n \times n$ matrix.

a. Show that (1) implies

$$E_t R_{j,t+1} = \exp(\nu_t(j)).$$

b. Prove that $E_t(m_{t+1}R_{j,t+1}) = 1$ implies that

$$\nu_t(j) = r_t + \alpha_t(j)' \Lambda_t. \tag{2}$$

c. Interpret formula (2).

d. Specify stochastic dynamics for aggregate consumption C_t that would make the 'ordinary' stochastic discount factor $m_{t+1} = \beta(\frac{C_{t+1}}{C_t})^{-\gamma}$ become (a special case of) an exponential affine model of a stochastic discount factor.

Exercise 14.7   **Value function for CRRA**

Consumption C_t follows the stochastic process

$$C_{t+1} = \exp(\mu + \sigma_\varepsilon \varepsilon_{t+1}) C_t$$

where $\varepsilon_{t+1} \sim \mathcal{N}(0,1)$ is an i.i.d. scalar random process, $\mu > 0$, and C_0 is an initial condition. The consumer ranks consumption streams according to

$$U_0 = E_0 \sum_{t=0}^{\infty} \beta^t \left(\frac{C_t^{(1-\gamma)}}{1 - \gamma} \right), \tag{1}$$

where $\beta \in (0,1)$, E_0 denotes the mathematical expectation conditional on C_0, and $\gamma \geq 0$.

a. Find a value function U_0 that satisfies (1), giving conditions on $\mu, \sigma_\varepsilon, \gamma, \beta$ that ensure that the right side of (1) exists.

b. Describe a recursion

$$U_t = \frac{C_t^{(1-\gamma)}}{1 - \gamma} + \beta E_t U_{t+1}$$

whose solution at time 0 satisfies (1).

Exercise 14.8   **Lucas tree economy**

A representative consumer in a Lucas tree economy has preferences over consumption streams ordered by

$$E_0 \sum_{t=0}^{\infty} \beta^t \log C_t. \tag{1}$$

There is one asset, a tree that yields dividends y_t at time t, which are governed by a two-state Markov chain P with states $s_t \in \{0, 1\}$ and

$$y_t = \begin{cases} y_L, & \text{if } s_t = 0; \\ y_H, & \text{if } s_t = 1, \end{cases}$$

where $y_H > y_L > 0$. The tree is the only source of goods in the economy. There is a single representative consumer, so that $C_t = y_t$. The transition matrix $P = \begin{bmatrix} \pi_L & 1 - \pi_L \\ 1 - \pi_H & \pi_H \end{bmatrix}$, where $\pi_L \in (0, 1)$ and $\pi_H \in (0, 1)$.

There is a market in trees. (There will be zero volume in equilibrium.) Let p_t be the (ex-dividend) price of a claim to the fruit of the tree from time $t + 1$ on. Ownership of the tree at the beginning of t entitles the owner of the tree to receive dividend y_t at t and then to sell the tree if he wants at price p_{t+1} after collecting the dividend at time $t + 1$.

a. Find a stochastic discount factor for this economy. Please tell how to compute the rate of return on one-period risk-free bonds in this economy.

b. Find an equilibrium pricing function mapping the state of the economy at t into the price of the tree p_t.

c. Describe the behavior of the one period gross return on the tree.

A zero-net worth outside entrepreneur comes into this economy and purchases all trees at time $\bar{t} > 0$. The outsider has no resources and finances his tree purchases by issuing (1) infinite duration risk-free bonds promising to pay $\eta \in (0, y_L)$ each period $t \geq \bar{t}+1$, and (2) equity that pays share-holders $y_t - \eta$ in period $t \geq \bar{t}+1$. The asset of the entrepreneur is the tree while his/her liabilities are the bonds and equities.

d. Please compute the equilibrium value of the risk-free bonds as a function of the promised coupon payment η and the state of the economy.

e. Please compute the equilibrium value of equity as a function of the bond coupon payment η and the state of the economy.

f. How do your answers in **d** and **e** compare with the value of the tree that you computed in part **b**?

Exercise 14.9   **Long-run risk, I**

Consider the following pure exchange economy with one representative consumer and an exogenous consumption process. Consumption $C_t = \exp(c_t)$ follows the stochastic process

$$c_{t+1} = \mu + z_t + c_t + \sigma_\varepsilon \varepsilon_{t+1}$$

$$z_{t+1} = \rho z_t + \sigma_z \varepsilon_{t+1},$$

where $\varepsilon_{t+1} \sim \mathcal{N}(0,1)$ is an i.i.d. scalar random process, $\mu > 0$, and c_0, z_0 are initial conditions, $\rho \in (0,1)$. By setting ρ close to but less than unity and σ_z to be small, z_t becomes a slowly moving component of the conditional mean of $E_t(c_{t+1} - c_t)$. The consumer ranks consumption streams according to

$$U_0 = E_0 \sum_{t=0}^{\infty} \beta^t \left(\frac{C_t^{(1-\gamma)}}{1-\gamma} \right), \tag{1}$$

where $\beta \in (0,1)$, E_0 denotes the mathematical expectation conditional on C_0, and $\gamma \geq 0$.

a. Find a formula for the consumer's stochastic discount factor m_{t+1}.

b. Compute $E_t(m_{t+1})$ and interpret it.

c. Compute $E_t(m_{t+1}m_{t+2})$ and interpret it.

d. Use your answers to parts **b** and **c** to tell how you would expect the term structure of interest rates to behave over time in this economy.

e. Find a value function U_0 that satisfies (1), giving conditions on $\mu, \sigma_\varepsilon, \sigma_z, \gamma, \beta$ that ensure that the right side of (1) exists.

Exercise 14.10   **Long-run risk, II**

Consider again the recursion (14.7.5), namely,

$$U_t = c_t - \beta\theta \log E_t \left[\exp \left(\frac{-U_{t+1}}{\theta} \right) \right], \tag{1}$$

where $c_t = \log C_t$, $\beta \in (0,1)$, and $0 < \theta$. Consumption $C_t = \exp(c_t)$ follows the stochastic process

$$c_{t+1} = \mu + z_t + c_t + \sigma_\varepsilon \varepsilon_{t+1}$$

$$z_{t+1} = \rho z_t + \sigma_z \varepsilon_{t+1},$$

where $\varepsilon_{t+1} \sim \mathcal{N}(0,1)$ is an i.i.d. scalar random process, $\mu > 0$, and c_0, z_0 are initial conditions, $\rho \in (0,1)$. By setting ρ close to but less than unity and σ_z to be small, z_t becomes a slowly moving component of the conditional mean of $E_t(c_{t+1} - c_t)$.

a. Guess a value function of the form $U_t = k_0 + k_1 c_t + k_2 z_t$, where k_0, k_1, and k_2 are scalar constants. In detail, derive formulas for k_0, k_1, k_2 that verify the recursion (14.7.5) (or (1) above).

b. (*Optional extra credit*)   Derive a formula for the stochastic discount factor in this economy.

Exercise 14.11   **Stochastic volatility**

Consider the following pure exchange economy with one representative consumer and an exogenous consumption process. Consumption $C_t = \exp(c_t)$ follows the stochastic process

$$c_{t+1} = \mu + c_t + \sigma_\varepsilon(s_t)\varepsilon_{t+1},$$

where $\varepsilon_{t+1} \sim \mathcal{N}(0,1)$ is an i.i.d. scalar random process, $\mu > 0$, and c_0, s_0 are initial conditions, and s_t is the time t realization of a two state Markov chain on $\{0,1\}$ with transition matrix $P = \begin{bmatrix} \pi_0 & 1 - \pi_0 \\ 1 - \pi_1 & \pi_1 \end{bmatrix}$, where $\pi_0 \in (0,1)$ and $\pi_1 \in (0,1)$. It is true that

$$\sigma_\varepsilon(s_t) = \begin{cases} \sigma_L, & \text{if } s_t = 0; \\ \sigma_H, & \text{if } s_t = 1, \end{cases}$$

where $0 < \sigma_L < \sigma_H$. At time t, the consumer observes c_t, s_t at time t. The consumer ranks consumption streams according to

$$U_0 = E_0 \sum_{t=0}^{\infty} \beta^t \left(\frac{C_t^{(1-\gamma)}}{1 - \gamma} \right), \tag{1}$$

where $\beta \in (0,1)$, E_0 denotes the mathematical expectation conditional on C_0, s_0, and $\gamma \geq 0$.

a. Define the consumer's stochastic discount factor m_{t+1}.

b. Find a formula for the consumer's stochastic discount factor m_{t+1}.

c. Compute $E_t(m_{t+1})$ and interpret it.

d. Compute $E_t(m_{t+1}m_{t+2})$ and interpret it.

e. Use your answers to parts **c** and **d** to tell how you would expect the term structure of interest rates to behave over time in this economy.

f. Describe how your approach to answering question **c** would change if, instead of observing c_t, s_t at time t, the consumer observes only current and past values $c_t, c_{t-1}, \ldots, c_0$ while having a prior distribution $s_0 \sim \tilde{\pi}_0(s_0)$.

Exercise 14.12   **Unknown** μ

Consider the following pure exchange economy with one representative consumer and an exogenous consumption process. Consumption $C_t = \exp(c_t)$ follows the stochastic process

$$c_{t+1} = \mu + z_t + c_t + \sigma_\varepsilon \varepsilon_{t+1},$$

where $\varepsilon_{t+1} \sim \mathcal{N}(0,1)$ is an i.i.d. scalar random process, $\mu > 0$, and c_0 is an initial condition. The consumer ranks consumption streams according to

$$U_0 = E_0 \sum_{t=0}^{\infty} \beta^t \log(C_t), \tag{1}$$

where $\beta \in (0,1)$, E_0 denotes the mathematical expectation conditional on C_0, and $\gamma \geq 0$. The consumer does not know μ but at time 0 believes that μ is described by a prior probability density $\mu \sim \mathcal{N}(\hat{\mu}_0, \sigma_\mu^2)$, where $\sigma_\mu > 0$. At the beginning of time $t+1$, the consumer has observed the history $\{c_{s+1} - c_s, s = 1, \ldots, t\}$.

a. Find a formula for the consumer's stochastic discount factor m_{t+1} for $t \geq 0$.

b. In this economy, how does the gross rate of return on a one-period risk free bond at time t behave through time? Can you tell whether it increases or decreases?

Exercise 14.13   **Long-run risk, III**

Consider the following pure exchange economy with one representative consumer and an exogenous consumption process. Consumption $C_t = \exp(c_t)$ follows the stochastic process

$$c_{t+1} = \mu + z_t + c_t + \sigma_\varepsilon \varepsilon_{t+1}$$

$$z_{t+1} = \rho z_t + \sigma_z \varepsilon_{t+1},$$

where $\varepsilon_{t+1} \sim \mathcal{N}(0,1)$ is an i.i.d. scalar random process, $\mu > 0$, and c_0, z_0 are initial conditions, $\rho \in (0,1)$. By setting ρ close to but less than unity and σ_z to be small, z_t becomes a slowly moving component of the conditional mean of $E_t(c_{t+1} - c_t)$. The consumer ranks consumption streams according to

$$U_0 = E_0 \sum_{t=0}^{\infty} \beta^t \log(C_t), \tag{1}$$

where $\beta \in (0,1)$, E_0 denotes the mathematical expectation conditional on C_0, and $\gamma \geq 0$. At time 0 the consumer believes that μ is described by a prior probability density $\mu \sim \mathcal{N}(\hat{\mu}_0, \sigma_\mu^2)$, where $\sigma_\mu > 0$. The consumer never observes z_t and at the start of period 0 believes that z_0 is distributed independently of μ and that $z_0 \sim \mathcal{N}(\hat{z}, \sigma_z^2)$. At the beginning of time $t+1$, the consumer has observed the history $\{c_{s+1} - c_s, s = 1, \ldots, t\}$.

a. Find a formula for the consumer's stochastic discount factor m_{t+1} for $t \geq 0$.

b. Assume the values $(.995, .005, 0, .99, .005, .00005)$ for $(\beta, \hat{\mu}_0, \hat{z}_0, \rho, \sigma_\varepsilon, \sigma_z)$. Please write a Matlab program to compute the gross rate of return on a one-period risk-free bond for $t = 0, \ldots, 10,000$. Plot it. In this economy, how does the gross rate of return on a one-period risk free bond at time t behave through time?

Exercise 14.14   **Affine term structure model**

Recall the affine term structure model of section 14.11.2

$$p_t(n) = \exp\left(\bar{A}_n + \bar{B}_n z_t\right).$$

Please verify that $(\bar{A}_n, \bar{B}_n)$ can be computed recursively from

$$\bar{A}_{n+1} = \bar{A}_n + \bar{B}'_n (\mu - C\Lambda_0) + \frac{1}{2}\bar{B}'_n CC' \bar{B}_n - \delta_0$$

$$\bar{B}'_{n+1} = \bar{B}'_n (\phi - C\Lambda_z) - \delta'_1,$$

subject to the initial conditions $\bar{A}_1 = -\delta_0, \bar{B}_1 = -\delta_1$. *Hint:* Apply the formula for the mean of a log normal random variable.

Exercise 14.15 **Reverse engineering**

An econometrician has discovered that the logarithm of consumption c_t is well described by a stochastic process

$$(0) \qquad c_{t+1} - c_t = \mu + \sigma_\varepsilon \varepsilon_{t+1}$$

where $\varepsilon_{t+1} \sim \mathcal{N}(0,1)$ is an iid scalar stochastic process. The econometrician has also discovered a stochastic discount factor of the form

$$(1) \qquad \log m_{t+1} = -\delta - (c_{t+1} - c_t) + w\varepsilon_{t+1} - \frac{1}{2}w^2,$$

where $\delta > 0, w = \sigma_\varepsilon(1 - \gamma), \gamma \geq 1$. The stochastic discount factor works well in the sense that for a set of assets $j = 1, \ldots, J$, GMM estimates of the Euler equations

$$(2) \qquad E_t\left(R_{j,t+1} m_{t+1}\right) = 1$$

are satisfied to a good approximation. In addition, the econometrician has found that the gross return $R_{j,t+1}$ on asset j is well described by

$$(3) \qquad \log\left(R_{j,t+1}\right) = \eta_{jt} + \alpha_j \varepsilon_{t+1} + \sigma_j u_{j,t+1} - \left(\frac{1}{2}\right)\left(\alpha_j^2 + \sigma_j^2\right),$$

where $u_{j,t+1} \sim \mathcal{N}(0,1)$, $E u_{j,t+1}\varepsilon_{t+1} = 0$ for all j, and $\alpha_1 = 0$.

a. Use restriction (2) to get a formula for η_{jt} as a function of the other parameters in equations (0), (1), (2), and (3).

b. Recalling that $\alpha_1 = 0$, state a formula for η_{1t} that is a special case of the formula that you derived in part **a**. Please interpret η_{1t}.

c. From your answers to parts **a** and **b**, please derive a formula that relates η_{jt} to η_{1t}. Please interpret it in terms of prices of risks.

d. Reverse engineer an economic model in which m_{t+1} described by (1) reflects the preferences of a representative consumer.

e. Describe the representative consumer's preferences derived in part **d** in the special case in which $\gamma = 1$.

f. Describe the representative consumer's preferences derived in part **d** in the general case that $\gamma > 1$. Please say what aspects of preferences the parameter γ governs (e.g., risk aversion or intertemporal substitution or yet other things).

Chapter 15
Economic Growth

15.1. Introduction

This chapter describes basic nonstochastic models of sustained economic growth. We begin by describing a benchmark exogenous growth model where sustained growth is driven by exogenous growth in labor productivity. Then we turn our attention to several endogenous growth models where sustained growth of labor productivity is somehow *chosen* by the households in the economy. We describe several models that differ in whether the equilibrium market economy matches what a benevolent planner would choose. Where the market outcome doesn't match the planner's outcome, there can be room for welfare-improving government interventions. The objective of the chapter is to shed light on the mechanisms at work in different models. We try to facilitate comparison by using the same production function for most of our discussion while changing the meaning of one of its arguments.

Paul Romer's work has been an impetus to the revived interest in the theory of economic growth. In the spirit of Arrow's (1962) model of learning by doing, Romer (1986) presents an endogenous growth model where the accumulation of capital (or knowledge) is associated with a positive externality on the available technology. The aggregate of all agents' holdings of capital is positively related to the level of technology, which in turn interacts with individual agents' savings decisions and thereby determines the economy's growth rate. Thus, the households in this economy are *choosing* how fast the economy is growing, but they do so in an unintentional way. The competitive equilibrium growth rate falls short of the socially optimal one.

Another approach to generating endogenous growth is to assume that all production factors are reproducible. Following Uzawa (1965), Lucas (1988) formulates a model with accumulation of both physical and human capital. The joint accumulation of all inputs ensures that growth will not come to a halt even though each individual factor in the final-good production function is subject to diminishing returns. In the absence of externalities, the growth rate in the competitive equilibrium coincides in this model with the social optimum.

Romer (1987) constructs a model where agents can choose to engage in research that produces technological improvements. Each invention represents a technology for producing a new type of intermediate input that can be used in the production of final goods without affecting the marginal product of existing intermediate inputs. The introduction of new inputs enables the economy to experience sustained growth even though each intermediate input taken separately is subject to diminishing returns. In a decentralized equilibrium, private agents will expend resources on research only if they are granted property rights over their inventions. Under the assumption of infinitely lived patents, Romer solves for a monopolistically competitive equilibrium that exhibits the classic tension between static and dynamic efficiency. Patents and the associated market power are necessary for there to be research and new inventions in a decentralized equilibrium, while the efficient production of existing intermediate inputs would require marginal-cost pricing, that is, the abolition of granted patents. The monopolistically competitive equilibrium is characterized by a smaller supply of each intermediate input and a lower growth rate than would be socially optimal.

Finally, we revisit the question of when nonreproducible factors may not pose an obstacle to growth. Rebelo (1991) shows that even if there are nonreproducible factors in fixed supply in a neoclassical growth model, sustained growth is possible if there is a "core" of capital goods that is produced without the direct or indirect use of the nonreproducible factors. Because of the ever-increasing relative scarcity of a nonreproducible factor, Rebelo finds that its price increases over time relative to a reproducible factor. Romer (1990) assumes that research requires the input of labor and not only goods as in his earlier model (1987). Now, if labor is in fixed supply and workers' innate productivity is constant, it follows immediately that growth must asymptotically come to an halt. To make sustained growth feasible, we can take a cue from our earlier discussion. One modeling strategy would be to introduce an externality that enhances researchers' productivity, and an alternative approach would be to assume that researchers can accumulate human capital. Romer adopts the first type of assumption, and we find it instructive to focus on its role in overcoming a barrier to growth that nonreproducible labor would otherwise pose.

15.2. The economy

The economy has a constant population of a large number of identical agents who order consumption streams $\{c_t\}_{t=0}^{\infty}$ according to

$$\sum_{t=0}^{\infty} \beta^t u\left(c_t\right), \quad \text{with} \ \ \beta \in (0,1) \ \ \text{and} \ \ u\left(c\right) = \frac{c^{1-\sigma} - 1}{1 - \sigma} \quad \text{for} \ \sigma \in [0,\infty), \quad (15.2.1)$$

and $\sigma = 1$ is taken to be logarithmic utility.[1] Lowercase letters for quantities, such as c_t for consumption, are used to denote individual variables, and uppercase letters stand for aggregate quantities.

For most part of our discussion of economic growth, the production function takes the form

$$F\left(K_t, X_t\right) = X_t f\left(\hat{K}_t\right), \quad \text{where} \ \hat{K}_t \equiv \frac{K_t}{X_t}. \tag{15.2.2}$$

That is, the production function $F(K, X)$ exhibits constant returns to scale in its two arguments, which via Euler's theorem on linearly homogeneous functions implies

$$F\left(K, X\right) = F_1\left(K, X\right) K + F_2\left(K, X\right) X, \tag{15.2.3}$$

where $F_i(K, X)$ is the derivative with respect to the ith argument (and $F_{ii}(K, X)$ will be used to denote the second derivative with respect to the ith argument). The input K_t is physical capital with a rate of depreciation equal to δ. New capital can be created by transforming one unit of output into one unit of capital. Past investments are reversible. It follows that the relative price of capital in terms of the consumption good must always be equal to 1. The second argument X_t captures the contribution of labor. Its precise meaning will differ among the various setups that we will examine.

We assume that the production function satisfies standard assumptions of positive but diminishing marginal products,

$$F_i\left(K, X\right) > 0, \quad F_{ii}\left(K, X\right) < 0, \quad \text{for} \ i = 1, 2;$$

and the Inada conditions,

$$\lim_{K \to 0} F_1\left(K, X\right) = \lim_{X \to 0} F_2\left(K, X\right) = \infty,$$
$$\lim_{K \to \infty} F_1\left(K, X\right) = \lim_{X \to \infty} F_2\left(K, X\right) = 0,$$

---

[1] By virtue of L'Hôpital's rule, the limit of $(c^{1-\sigma} - 1)/(1 - \sigma)$ is $\log(c)$ as σ goes to 1.

which imply

$$\lim_{\hat{K}\to 0} f'\left(\hat{K}\right) = \infty, \quad \lim_{\hat{K}\to\infty} f'\left(\hat{K}\right) = 0. \tag{15.2.4}$$

We will also make use of the mathematical fact that a linearly homogeneous function $F(K, X)$ has first derivatives $F_i(K, X)$ homogeneous of degree 0; thus, the first derivatives are only functions of the ratio $\hat{K}$. In particular, we have

$$F_1\left(K, X\right) = \frac{\partial X f\left(K/X\right)}{\partial K} = f'\left(\hat{K}\right), \tag{15.2.5a}$$

$$F_2\left(K, X\right) = \frac{\partial X f\left(K/X\right)}{\partial X} = f\left(\hat{K}\right) - f'\left(\hat{K}\right)\hat{K}. \tag{15.2.5b}$$

15.2.1. Balanced growth path

We seek additional technological assumptions to generate market outcomes with steady-state growth of consumption at a constant rate $1 + \mu = c_{t+1}/c_t$. The literature uses the term "balanced growth path" to denote a situation where all endogenous variables grow at constant (but possibly different) rates. Along such a steady-state growth path (and during any transition toward the steady state), the return to physical capital must be such that households are willing to hold the economy's capital stock.

In a competitive equilibrium where firms rent capital from the agents, the rental payment r_t is equal to the marginal product of capital,

$$r_t = F_1\left(K_t, X_t\right) = f'\left(\hat{K}_t\right). \tag{15.2.6}$$

Households maximize utility given by equation (15.2.1) subject to the sequence of budget constraints

$$c_t + k_{t+1} = r_t k_t + (1 - \delta) k_t + \chi_t, \tag{15.2.7}$$

where χ_t stands for labor-related budget terms. The first-order condition with respect to k_{t+1} is

$$u'\left(c_t\right) = \beta u'\left(c_{t+1}\right)\left(r_{t+1} + 1 - \delta\right). \tag{15.2.8}$$

After using equations (15.2.1) and (15.2.6) in equation (15.2.8), we arrive at the following equilibrium condition:

$$\left(\frac{c_{t+1}}{c_t}\right)^\sigma = \beta\left[f'\left(\hat{K}_{t+1}\right) + 1 - \delta\right]. \tag{15.2.9}$$

We see that a constant consumption growth rate on the left side is sustained in an equilibrium by a constant rate of return on the right side. It was also for this reason that we chose the class of utility functions in equation (15.2.1) that exhibits a constant intertemporal elasticity of substitution. These preferences allow for balanced growth paths.[2]

Equation (15.2.9) makes clear that capital accumulation alone cannot sustain steady-state consumption growth when the labor input X_t is constant over time, $X_t = L$. Given the second Inada condition in equations (15.2.4), the limit of the right side of equation (15.2.9) is $\beta(1-\delta)$ when $\hat{K}$ approaches infinity. The steady state with a constant labor input must therefore be a constant consumption level and a capital-labor ratio $\hat{K}^\star$ given by

$$f'\left(\hat{K}^\star\right) = \beta^{-1} - (1-\delta). \qquad (15.2.10)$$

In chapter 5 we derived a closed-form solution for the transition dynamics toward such a steady state in the case of logarithmic utility, a Cobb-Douglas production function, and $\delta = 1$.

15.3. Exogenous growth

As in Solow's (1956) classic article, the simplest way to ensure steady-state consumption growth is to postulate exogenous labor-augmenting technological change at the constant rate $1 + \mu \geq 1$,

$$X_t = A_t L, \qquad \text{with } A_t = (1 + \mu) A_{t-1},$$

where L is a fixed stock of labor. Our conjecture is then that both consumption and physical capital will grow at that same rate $1 + \mu$ along a balanced growth path. The same growth rate of K_t and A_t implies that the ratio $\hat{K}$ and therefore the marginal product of capital remain constant in the steady state. A time-invariant rate of return is in turn consistent with households choosing a constant growth rate of consumption, given the assumption of isoelastic preferences.

---

[2] To ensure well-defined maximization problems, a maintained assumption throughout the chapter is that parameters are such that any derived consumption growth rate $1+\mu$ yields finite lifetime utility; i.e., the implicit restriction on parameter values is that $\beta(1+\mu)^{1-\sigma} < 1$. To see that this condition is needed, substitute the consumption sequence $\{c_t\}_{t=0}^{\infty} = \{(1+\mu)^t c_0\}_{t=0}^{\infty}$ into equation (15.2.1).

Evaluating equation (15.2.9) at a steady state, the optimal ratio $\hat{K}^\star$ is given by

$$(1+\mu)^\sigma = \beta \left[f'\left(\hat{K}^\star\right) + 1 - \delta \right]. \tag{15.3.1}$$

While the steady-state consumption growth rate is exogenously given by $1+\mu$, the endogenous steady-state ratio $\hat{K}^\star$ is such that the implied rate of return on capital induces the agents to choose a consumption growth rate of $1+\mu$. As can be seen, a higher degree of patience (a larger β), a higher willingness intertemporally to substitute (a lower σ), and a more durable capital stock (a lower δ) each yield a higher ratio $\hat{K}^\star$, and therefore more output (and consumption) at a point in time, but the growth rate remains fixed at the rate of exogenous labor-augmenting technological change. It is straightforward to verify that the competitive equilibrium outcome is Pareto optimal, since the private return to capital coincides with the social return.

Physical capital is compensated according to equation (15.2.6), and labor is also paid its marginal product in a competitive equilibrium,

$$w_t = F_2\left(K_t, X_t\right) \frac{\mathrm{d}\,X_t}{\mathrm{d}\,L} = F_2\left(K_t, X_t\right) A_t. \tag{15.3.2}$$

So, by equation (15.2.3), we have

$$r_t K_t + w_t L = F\left(K_t, A_t L\right).$$

Factor payments are equal to total production, which is the standard result of a competitive equilibrium with constant-returns-to-scale technologies. However, it is interesting to note that if A_t were a separate production factor, there could not exist a competitive equilibrium, since factor payments based on marginal products would exceed total production. In other words, the dilemma would then be that the production function $F(K_t, A_t L)$ exhibits increasing returns to scale in the three "inputs" K_t, A_t, and L, which is not compatible with the existence of a competitive equilibrium. This problem is to be kept in mind as we now turn to one way to endogenize economic growth.

15.4. Externality from spillovers

Inspired by Arrow's (1962) paper on learning by doing, Romer (1986) suggests that economic growth can be endogenized by assuming that technology grows because of aggregate spillovers coming from firms' production activities. The problem alluded to in the previous section, that a competitive equilibrium fails to exist in the presence of increasing returns to scale, is avoided by letting technological advancement be external to firms.[3] As an illustration, we assume that firms face a fixed labor productivity that is proportional to the current economy-wide average of physical capital per worker.[4] In particular,

$$X_t = \bar{K}_t L, \qquad \text{where } \bar{K}_t = \frac{K_t}{L}.$$

The competitive rental rate of capital is still given by equation (15.2.6), but we now trivially have $\hat{K}_t = 1$, so equilibrium condition (15.2.9) becomes

$$\left(\frac{c_{t+1}}{c_t}\right)^\sigma = \beta \left[f'(1) + 1 - \delta\right]. \tag{15.4.1}$$

Note first that this economy has no transition dynamics toward a steady state. Regardless of the initial capital stock, equation (15.4.1) determines a time-invariant growth rate. To ensure a positive growth rate, we require the parameter restriction $\beta[f'(1) + 1 - \delta] \geq 1$. A second critical property of the model is that the economy's growth rate is now a function of preference and technology parameters.

The competitive equilibrium is no longer Pareto optimal, since the private return on capital falls short of the social rate of return, with the latter return given by

$$\frac{\mathrm{d}\,F\left(K_t, \frac{K_t}{L}L\right)}{\mathrm{d}\,K_t} = F_1(K_t, K_t) + F_2(K_t, K_t) = f(1), \tag{15.4.2}$$

---

[3] Arrow (1962) focuses on learning from experience that is assumed to get embodied in capital goods, while Romer (1986) postulates spillover effects of firms' investments in knowledge. In both analyses, the productivity of a given firm is a function of an aggregate state variable, either the economy's stock of physical capital or stock of knowledge.

[4] This specific formulation of spillovers is analyzed in a rarely cited paper by Frankel (1962).

where the last equality follows from equations (15.2.5). This higher social rate of return enters a planner's first-order condition, which then also implies a higher optimal consumption growth rate,

$$\left(\frac{c_{t+1}}{c_t}\right)^{\sigma} = \beta \left[f\left(1\right) + 1 - \delta\right]. \tag{15.4.3}$$

Let us reconsider the suboptimality of the decentralized competitive equilibrium. Since the agents and the planner share the same objective of maximizing utility, we are left with exploring differences in their constraints. For a given sequence of the spillover $\{\bar{K}_t\}_{t=0}^{\infty}$, the production function $F(k_t, \bar{K}_t l_t)$ exhibits constant returns to scale in k_t and l_t. So, once again, factor payments in a competitive equilibrium will be equal to total output, and optimal firm size is indeterminate. Therefore, we can consider a representative agent with one unit of labor endowment who runs his own production technology, taking the spillover effect as given. His resource constraint becomes

$$c_t + k_{t+1} = F\left(k_t, \bar{K}_t\right) + \left(1 - \delta\right) k_t = \bar{K}_t f\left(\frac{k_t}{\bar{K}_t}\right) + \left(1 - \delta\right) k_t,$$

and the private gross rate of return on capital is equal to $f'(k_t/\bar{K}_t) + 1 - \delta$. After invoking the equilibrium condition $k_t = \bar{K}_t$, we arrive at the competitive equilibrium return on capital $f'(1) + 1 - \delta$ that appears in equation (15.4.1). In contrast, the planner maximizes utility subject to a resource constraint where the spillover effect is internalized,

$$C_t + K_{t+1} = F\left(K_t, \frac{K_t}{L}L\right) + \left(1 - \delta\right) K_t = \left[f\left(1\right) + 1 - \delta\right] K_t.$$

15.5. All factors reproducible

15.5.1. One-sector model

An alternative approach to generating endogenous growth is to assume that all factors of production are producible. Remaining within a one-sector economy, we now assume that human capital X_t can be produced in the same way as physical capital but rates of depreciation might differ. Let δ_X and δ_K be the rates of depreciation of human capital and physical capital, respectively.

The competitive equilibrium wage is equal to the marginal product of human capital

$$w_t = F_2 \left(K_t, X_t \right). \tag{15.5.1}$$

Households maximize utility subject to budget constraint (15.2.7) where the term χ_t is now given by

$$\chi_t = w_t x_t + \left(1 - \delta_X \right) x_t - x_{t+1}.$$

The first-order condition with respect to human capital becomes

$$u' \left(c_t \right) = \beta u' \left(c_{t+1} \right) \left(w_{t+1} + 1 - \delta_X \right). \tag{15.5.2}$$

Since both equations (15.2.8) and (15.5.2) must hold, the rates of return on the two assets have to obey

$$F_1 \left(K_{t+1}, X_{t+1} \right) - \delta_K = F_2 \left(K_{t+1}, X_{t+1} \right) - \delta_X,$$

and after invoking equations (15.2.5),

$$f \left(\hat{K}_{t+1} \right) - \left(1 + \hat{K}_{t+1} \right) f' \left(\hat{K}_{t+1} \right) = \delta_X - \delta_K, \tag{15.5.3}$$

which uniquely determines a time-invariant competitive equilibrium ratio $\hat{K}^\star$, as a function solely of depreciation rates and parameters of the production function.[5]

---

[5] The left side of equation (15.5.3) is strictly increasing, since the derivative with respect to $\hat{K}$ is $-(1 + \hat{K}) f''(\hat{K}) > 0$. Thus, there can only be one solution to equation (15.5.3) and

After solving for $f'(\hat{K}^\star)$ from equation (15.5.3) and substituting into equation (15.2.9), we arrive at an expression for the equilibrium growth rate

$$\left(\frac{c_{t+1}}{c_t}\right)^\sigma = \beta\left[\frac{f\left(\hat{K}^\star\right)}{1+\hat{K}^\star} + 1 - \frac{\delta_X + \hat{K}^\star \delta_K}{1+\hat{K}^\star}\right]. \qquad (15.5.4)$$

As in the previous model with an externality, the economy here is void of any transition dynamics toward a steady state. But this implication now critically hinges on investments being reversible so that the initial stocks of physical capital and human capital are inconsequential. In contrast to the previous model, the present competitive equilibrium is Pareto optimal because there is no longer any discrepancy between private and social rates of return.[6]

The problem of optimal taxation with commitment (see chapter 16) is studied for this model of endogenous growth by Jones, Manuelli, and Rossi (1993), who adopt the assumption of irreversible investments.

---

existence is guaranteed because the left side ranges from minus infinity to plus infinity. The limit of the left side when $\hat{K}$ approaches zero is $f(0) - \lim_{\hat{K}\to 0} f'(\hat{K})$, which is equal to minus infinity by equations (15.2.4) and the fact that $f(0) = 0$. (Barro and Sala-i-Martin (1995) show that the Inada conditions and constant returns to scale imply that all production factors are essential, i.e., $f(0) = 0$.) To establish that the left side of equation (15.5.3) approaches plus infinity when $\hat{K}$ goes to infinity, we can define the function g as $F(K, X) = Kg(\hat{X})$ where $\hat{X} \equiv X/K$ and derive an alternative expression for the left side of equation (15.5.3), $(1 + \hat{X})g'(\hat{X}) - g(\hat{X})$, for which we take the limit when $\hat{X}$ goes to zero.

[6] It is instructive to compare the present model with two producible factors, $F(K, X)$, to the previous setup with one producible factor and an externality, $\tilde{F}(K, X)$ with $X = \bar{K}L$. Suppose the present technology is such that $\hat{K}^\star = 1$ and $\delta_K = \delta_X$, and the two different setups are equally productive; i.e., we assume that $F(K, X) = \tilde{F}(2K, 2X)$, which implies $f(\hat{K}) = 2\tilde{f}(\hat{K})$. We can then verify that the present competitive equilibrium growth rate in equation (15.5.4) is the same as the planner's solution for the previous setup in equation (15.4.3).

15.5.2. Two-sector model

Following Uzawa (1965), Lucas (1988) explores endogenous growth in a two-sector model with all factors being producible. The resource constraint in the goods sector is

$$C_t + K_{t+1} = K_t^\alpha \left(\phi_t X_t\right)^{1-\alpha} + \left(1 - \delta\right) K_t, \qquad (15.5.5a)$$

and the linear technology for accumulating additional human capital is

$$X_{t+1} - X_t = A \left(1 - \phi_t\right) X_t, \qquad (15.5.5b)$$

where $\phi_t \in [0, 1]$ is the fraction of human capital employed in the goods sector, and $(1 - \phi_t)$ is devoted to human capital accumulation. (Lucas provides an alternative interpretation that we will discuss later.)

We seek a balanced growth path where consumption, physical capital, and human capital grow at constant rates (but not necessarily the same ones) and the fraction ϕ stays constant over time. Let $1 + \mu$ be the growth rate of consumption, and equilibrium condition (15.2.9) becomes

$$(1 + \mu)^\sigma = \beta \left(\alpha K_t^{\alpha-1} [\phi X_t]^{1-\alpha} + 1 - \delta\right). \qquad (15.5.6)$$

That is, along the balanced growth path, the marginal product of physical capital must be constant. With the assumed Cobb-Douglas technology, the marginal product of capital is proportional to the average product, so that by dividing equation (15.5.5a) through by K_t and applying equation (15.5.6) we obtain

$$\frac{C_t}{K_t} + \frac{K_{t+1}}{K_t} = \frac{(1 + \mu)^\sigma \beta^{-1} - (1 - \alpha)(1 - \delta)}{\alpha}. \qquad (15.5.7)$$

By definition of a balanced growth path, K_{t+1}/K_t is constant, so equation (15.5.7) implies that C_t/K_t is constant; that is, the capital stock must grow at the same rate as consumption.

Substituting $K_t = (1 + \mu)K_{t-1}$ into equation (15.5.6),

$$(1 + \mu)^\sigma - \beta (1 - \delta) = \beta \alpha \left[(1 + \mu) K_{t-1}\right]^{\alpha-1} [\phi X_t]^{1-\alpha},$$

and dividing by the similarly rearranged equation (15.5.6) for period $t - 1$, we arrive at

$$1 = (1 + \mu)^{\alpha-1} \left[\frac{X_t}{X_{t-1}}\right]^{1-\alpha},$$

which directly implies that human capital must also grow at the rate $1+\mu$ along a balanced growth path. Moreover, by equation ($15.5.5b$), the growth rate is

$$1 + \mu = 1 + A\left(1 - \phi\right), \tag{15.5.8}$$

so it remains to determine the steady-state value of ϕ.

The equilibrium value of ϕ has to be such that a unit of human capital receives the same factor payment in both sectors; that is, the marginal products of human capital must be the same,

$$p_t A = \left(1 - \alpha\right) K_t^{\alpha} \left[\phi X_t\right]^{-\alpha},$$

where p_t is the relative price of human capital in terms of the composite consumption/capital good. Since the ratio K_t/X_t is constant along a balanced growth path, it follows that the price p_t must also be constant over time. Finally, the remaining equilibrium condition is that the rates of return on human and physical capital be equal,

$$\frac{p_t\left(1 + A\right)}{p_{t-1}} = \alpha K_t^{\alpha-1} \left[\phi X_t\right]^{1-\alpha} + 1 - \delta,$$

and after invoking a constant steady-state price of human capital and equilibrium condition ($15.5.6$), we obtain

$$1 + \mu = \left[\beta\left(1 + A\right)\right]^{1/\sigma}. \tag{15.5.9}$$

Thus, the growth rate is positive as long as $\beta(1+A) \geq 1$, but feasibility requires also that solution ($15.5.9$) fall below $1+A$, which is the maximum growth rate of human capital in equation ($15.5.5b$). This parameter restriction, $[\beta(1+A)]^{1/\sigma} < (1 + A)$, also ensures that the growth rate in equation ($15.5.9$) yields finite lifetime utility.

As in the one-sector model, there is no discrepancy between private and social rates of return, so the competitive equilibrium is Pareto optimal. Lucas (1988) does allow for an externality (in the spirit of our earlier section) where the economy-wide average of human capital per worker enters the production function in the goods sector, but, as he notes, the externality is not needed to generate endogenous growth.

Lucas provides an alternative interpretation of the technologies in equations ($15.5.5$). Each worker is assumed to be endowed with one unit of time. The time spent in the goods sector is denoted ϕ_t, which is multiplied by the agent's human capital x_t to arrive at the efficiency units of labor supplied. The remaining time is spent in the education sector with a constant marginal productivity of $A x_t$ additional units of human capital acquired. Even though Lucas's interpretation does introduce a nonreproducible factor in the form of a time endowment, the multiplicative specification makes the model identical to an economy with only two factors that are both reproducible. One section ahead we will study a setup with a nonreproducible factor that has some nontrivial implications.

15.6. Research and monopolistic competition

Building on Dixit and Stiglitz's (1977) formulation of the demand for differentiated goods and the extension to differentiated inputs in production by Ethier (1982), Romer (1987) studied an economy with an aggregate resource constraint of the following type:

$$C_t + \int_0^{A_{t+1}} Z_{t+1}(i) \, \mathrm{d}i + (A_{t+1} - A_t)\,\kappa = L^{1-\alpha} \int_0^{A_t} Z_t(i)^{\alpha} \, \mathrm{d}i, \qquad (15.6.1)$$

where one unit of the intermediate input $Z_{t+1}(i)$ can be produced from one unit of output at time t, and $Z_{t+1}(i)$ is used in production in the following period $t+1$. The continuous range of inputs at time t, $i \in [0, A_t]$, can be augmented for next period's production function at the constant marginal cost κ.

   In the allocations that we are about to study, the quantity of an intermediate input will be the same across all existing types, $Z_t(i) = Z_t$ for $i \in [0, A_t]$. The resource constraint $(15.6.1)$ can then be written as

$$C_t + A_{t+1} Z_{t+1} + (A_{t+1} - A_t)\,\kappa = L^{1-\alpha} A_t Z_t^{\alpha}. \qquad (15.6.2)$$

If A_t were constant over time, say, let $A_t = 1$ for all t, we would just have a parametric example of an economy yielding a no-growth steady state given by equation $(15.2.10)$ with $\delta = 1$. Hence, growth can only be sustained by allocating resources to a continuous expansion of the range of inputs. But this approach poses a barrier to the existence of a competitive equilibrium, since the production relationship $L^{1-\alpha} A_t Z_t^{\alpha}$ exhibits increasing returns to scale in its three "inputs." Following Judd's (1985a) treatment of patents in a dynamic setting of Dixit and Stiglitz's (1977) model of monopolistic competition, Romer (1987) assumes that an inventor of a new intermediate input obtains an infinitely lived patent on that design. As the sole supplier of an input, the inventor can recoup the investment cost κ by setting a price of the input above its marginal cost.

15.6.1. Monopolistic competition outcome

The final-goods sector is still assumed to be characterized by perfect competition because it exhibits constant returns to scale in the labor input L and the existing continuous range of intermediate inputs $Z_t(i)$. Thus, a competitive outcome prescribes that each input is paid its marginal product,

$$w_t = (1 - \alpha) L^{-\alpha} \int_0^{A_t} Z_t(i)^\alpha \, di, \qquad (15.6.3)$$

$$p_t(i) = \alpha L^{1-\alpha} Z_t(i)^{\alpha-1}, \qquad (15.6.4)$$

where $p_t(i)$ is the price of intermediate input i at time t in terms of the final good.

Let $1 + R_m$ be the steady-state interest rate along the balanced growth path that we are seeking. In order to find the equilibrium invention rate of new inputs, we first compute the profits from producing and selling an existing input i. The profit at time t is equal to

$$\pi_t(i) = [p_t(i) - (1 + R_m)] Z_t(i), \qquad (15.6.5)$$

where the cost of supplying one unit of the input i is one unit of the final good acquired in the previous period; that is, the cost is the intertemporal price $1 + R_m$. The first-order condition of maximizing the profit in equation (15.6.5) is the familiar expression that the monopoly price $p_t(i)$ should be set as a markup above marginal cost, $1 + R_m$, and the markup is inversely related to the absolute value of the demand elasticity of input i, $|\epsilon_t(i)|$:

$$p_t(i) = \frac{1 + R_m}{1 + \epsilon_t(i)^{-1}}, \qquad (15.6.6)$$

$$\epsilon_t(i) = \left[\frac{\partial p_t(i)}{\partial Z_t(i)} \frac{Z_t(i)}{p_t(i)} \right]^{-1} < 0.$$

The constant marginal cost, $1 + R_m$, and the constant-elasticity demand curve $(15.6.4)$, $\epsilon_t(i) = -(1-\alpha)^{-1}$, yield a time-invariant monopoly price, which, substituted into demand curve $(15.6.4)$, results in a time-invariant equilibrium quantity of input i:

$$p_t(i) = \frac{1 + R_m}{\alpha}, \tag{15.6.7a}$$

$$Z_t(i) = \left(\frac{\alpha^2}{1 + R_m}\right)^{1/(1-\alpha)} L \equiv Z_m. \tag{15.6.7b}$$

By substituting equation $(15.6.7)$ into equation $(15.6.5)$, we obtain an input producer's steady-state profit flow,

$$\pi_t(i) = (1-\alpha)\,\alpha^{1/(1-\alpha)} \left(\frac{\alpha}{1 + R_m}\right)^{\alpha/(1-\alpha)} L \equiv \Omega_m(R_m). \tag{15.6.8}$$

In an equilibrium with free entry, the cost κ of inventing a new input must be equal to the discounted stream of future profits associated with being the sole supplier of that input,

$$\sum_{t=1}^{\infty} (1 + R_m)^{-t}\, \Omega_m(R_m) = \frac{\Omega_m(R_m)}{R_m}; \tag{15.6.9}$$

that is,

$$R_m \kappa = \Omega_m(R_m). \tag{15.6.10}$$

The profit function $\Omega_m(R)$ is positive, strictly decreasing in R, and convex, as depicted in Figure 15.6.1. It follows that there exists a unique intersection between $\Omega(R)$ and $R\kappa$ that determines R_m. Using the corresponding version of equilibrium condition $(15.2.9)$, the computed interest rate R_m characterizes a balanced growth path with

$$\left(\frac{c_{t+1}}{c_t}\right)^{\sigma} = \beta(1 + R_m), \tag{15.6.11}$$

as long as $1 + R_m \geq \beta^{-1}$; that is, the technology must be sufficiently productive relative to the agents' degree of impatience.[7] It is straightforward to verify that the range of inputs must grow at the same rate as consumption in a steady state. After substituting the constant quantity Z_m into resource constraint $(15.6.2)$ and dividing by A_t, we see that a constant A_{t+1}/A_t implies that C_t/A_t stays constant; that is, the range of inputs must grow at the same rate as consumption.

---

[7] If the computed value $1 + R_m$ falls short of β^{-1}, the technology does not present sufficient private incentives for new inventions, so the range of intermediate inputs stays constant over time, and the equilibrium interest rate equals β^{-1}.

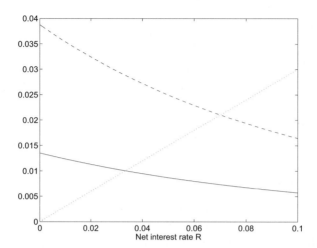

Figure 15.6.1: Interest rates in a version of Romer's (1987) model of research and monopolistic competition. The dotted line is the linear relationship κR, while the solid and dashed curves depict $\Omega_m(R)$ and $\Omega_s(R)$, respectively. The intersection between κR and $\Omega_m(R)$ [$\Omega_s(R)$] determines the interest rate along a balanced growth path for the laissez-faire economy (planner allocation), as long as $R \geq \beta^{-1} - 1$. The parameterization is $\alpha = 0.9$, $\kappa = 0.3$, and $L = 1$.

Note that the solution to equation (15.6.10) exhibits positive scale effects where a larger labor force L implies a higher interest rate and therefore a higher growth rate in equation (15.6.11). The reason is that a larger economy enables input producers to profit from a larger sales volume in equation (15.6.7b), which spurs more inventions until the discounted stream of profits of an input is driven down to the invention cost κ by means of the higher equilibrium interest rate. In other words, it is less costly for a larger economy to expand its range of inputs because the cost of an additional input is smaller in per capita terms.

15.6.2. Planner solution

Let $1 + R_s$ be the social rate of interest along an optimal balanced growth path. We analyze the planner problem in two steps. First, we establish that the socially optimal supply of an input i is the same across all existing inputs and constant over time. Second, we derive $1 + R_s$ and the implied optimal growth rate of consumption.

For a given social interest rate $1 + R_s$ and a range of inputs $[0, A_t]$, the planner would choose the quantities of intermediate inputs that maximize

$$L^{1-\alpha} \int_0^{A_t} Z_t(i)^\alpha \, \mathrm{d}i - (1 + R_s) \int_0^{A_t} Z_t(i) \, \mathrm{d}i,$$

with the following first-order condition with respect to $Z_t(i)$:

$$Z_t(i) = \left(\frac{\alpha}{1 + R_s}\right)^{1/(1-\alpha)} L \equiv Z_s. \tag{15.6.12}$$

Thus, the quantity of an intermediate input is the same across all inputs and constant over time. Hence, the planner's problem is simplified to one where utility function (15.2.1) is maximized subject to resource constraint (15.6.2) with quantities of intermediate inputs given by equation (15.6.12). The first-order condition with respect to A_{t+1} is then

$$\left(\frac{c_{t+1}}{c_t}\right)^\sigma = \beta \frac{L^{1-\alpha} Z_s^\alpha + \kappa}{Z_s + \kappa} = \beta (1 + R_s), \tag{15.6.13}$$

where the last equality merely invokes the definition of $1 + R_s$ as the social marginal rate of intertemporal substitution, $\beta^{-1}(c_{t+1}/c_t)^\sigma$. After substituting equation (15.6.12) into equation (15.6.13) and rearranging the last equality, we obtain

$$R_s \kappa = (1 - \alpha) \left(\frac{\alpha}{1 + R_s}\right)^{\alpha/(1-\alpha)} L \equiv \Omega_s(R_s). \tag{15.6.14}$$

The solution to this equation, $1 + R_s$, is depicted in Figure 15.6.1, and existence is guaranteed in the same way as in the case of $1 + R_m$.

We conclude that the social rate of return $1 + R_s$ and therefore the optimal growth rate exceed the laissez-faire outcome, since the function $\Omega_s(R)$ lies above the function $\Omega_m(R)$,

$$\Omega_m(R) = \alpha^{1/(1-\alpha)} \Omega_s(R). \tag{15.6.15}$$

We can also show that the laissez-faire supply of an input falls short of the socially optimal one:

$$Z_m < Z_s \qquad \Longleftrightarrow \qquad \alpha \frac{1 + R_s}{1 + R_m} < 1. \qquad (15.6.16)$$

To establish condition $(15.6.16)$, divide equation $(15.6.7b)$ by equation $(15.6.12)$. Thus, the laissez-faire equilibrium is characterized by a smaller supply of each intermediate input and a lower growth rate than would be socially optimal. These inefficiencies reflect the fact that suppliers of intermediate inputs do not internalize the full contribution of their inventions, and so their monopolistic pricing results in less than socially efficient quantities of inputs.

15.7. Growth in spite of nonreproducible factors

15.7.1. *"Core" of capital goods produced without nonreproducible inputs*

It is not necessary that all factors be producible in order to experience sustained growth through factor accumulation in the neoclassical framework. Instead, Rebelo (1991) shows that the critical requirement for perpetual growth is the existence of a "core" of capital goods that is produced with constant returns technologies and without the direct or indirect use of nonreproducible factors. Here we will study the simplest version of his model with a single capital good that is produced without any input of the economy's constant labor endowment. Jones and Manuelli (1990) provide a general discussion of convex models of economic growth and highlight the crucial feature that the rate of return to accumulated capital must remain bounded above the inverse of the subjective discount factor in spite of any nonreproducible factors in production.

Rebelo (1991) analyzes the competitive equilibrium for the following technology:

$$C_t = L^{1-\alpha} \left(\phi_t K_t\right)^{\alpha}, \qquad (15.7.1a)$$

$$I_t = A \left(1 - \phi_t\right) K_t, \qquad (15.7.1b)$$

$$K_{t+1} = (1 - \delta) K_t + I_t, \qquad (15.7.1c)$$

where $\phi_t \in [0, 1]$ is the fraction of capital employed in the consumption goods sector and $(1 - \phi_t)$ is employed in the linear technology producing investment goods I_t. In a competitive equilibrium, the rental price of capital r_t (in terms of consumption goods) is equal to the marginal product of capital, which then has to be the same across the two sectors (as long as they both are operating):

$$r_t = \alpha L^{1-\alpha} \left(\phi_t K_t\right)^{\alpha - 1} = p_t A, \tag{15.7.2}$$

where p_t is the relative price of capital in terms of consumption goods.

Along a steady-state growth path with a constant ϕ, we can compute the growth rate of capital by substituting equation $(15.7.1b)$ into equation $(15.7.1c)$ and dividing by K_t,

$$\frac{K_{t+1}}{K_t} = (1 - \delta) + A (1 - \phi) \equiv 1 + \rho (\phi). \tag{15.7.3}$$

Given the growth rate of capital, $1 + \rho(\phi)$, it is straightforward to compute other rates of change:

$$\frac{p_{t+1}}{p_t} = [1 + \rho (\phi)]^{\alpha - 1}, \tag{15.7.4a}$$

$$\frac{C_{t+1}}{C_t} = \frac{p_{t+1} I_{t+1}}{p_t I_t} = \frac{p_{t+1} K_{t+1}}{p_t K_t} = [1 + \rho (\phi)]^{\alpha}. \tag{15.7.4b}$$

Since the values of investment goods and the capital stock in terms of consumption goods grow at the same rate as consumption, $[1 + \rho(\phi)]^{\alpha}$, this common rate is also the steady-state growth rate of the economy's net income, measured as $C_t + p_t I_t - \delta p_t K_t$.

Agents maximize utility given by condition $(15.2.1)$ subject to budget constraint $(15.2.7)$ modified to incorporate the relative price p_t,

$$c_t + p_t k_{t+1} = r_t k_t + (1 - \delta) p_t k_t + \chi_t. \tag{15.7.5}$$

The first-order condition with respect to capital is

$$\left(\frac{c_{t+1}}{c_t}\right)^{\sigma} = \beta \frac{(1 - \delta) p_{t+1} + r_{t+1}}{p_t}. \tag{15.7.6}$$

After substituting $r_{t+1} = p_{t+1} A$ from equation $(15.7.2)$ and steady-state rates of change from equation $(15.7.4)$ into equation $(15.7.6)$, we arrive at the following equilibrium condition:

$$[1 + \rho (\phi)]^{1 - \alpha(1-\sigma)} = \beta (1 - \delta + A). \tag{15.7.7}$$

Thus, the growth rate of capital and therefore the growth rate of consumption are positive as long as

$$\beta\left(1 - \delta + A\right) \geq 1. \tag{15.7.8a}$$

Moreover, the maintained assumption of this chapter that parameters are such that derived growth rates yield finite lifetime utility, $\beta(c_{t+1}/c_t)^{1-\sigma} < 1$, imposes here the parameter restriction $\beta[\beta(1 - \delta + A)]^{\alpha(1-\sigma)/[1-\alpha(1-\sigma)]} < 1$, which can be simplified to read

$$\beta\left(1 - \delta + A\right)^{\alpha(1-\sigma)} < 1. \tag{15.7.8b}$$

Given that conditions $(15.7.8)$ are satisfied, there is a unique equilibrium value of ϕ because the left side of equation $(15.7.7)$ is monotonically decreasing in $\phi \in [0, 1]$ and it is strictly greater (smaller) than the right side for $\phi = 0$ ($\phi = 1$). The outcome is socially efficient because private and social rates of return are the same as in the previous models with all factors reproducible.

15.7.2. Research labor enjoying an externality

Romer's (1987) model includes labor as a fixed nonreproducible factor, but similar to the last section, an important assumption is that this nonreproducible factor is not used in the production of inventions that expand the input variety (which constitutes a kind of reproducible capital in that model). In his sequel, Romer (1990) assumes that the input variety A_t is expanded through the effort of researchers rather than the resource cost κ in terms of final goods. Suppose that we specify this new invention technology as

$$A_{t+1} - A_t = \eta\left(1 - \phi_t\right) L,$$

where $(1 - \phi_t)$ is the fraction of the labor force employed in the research sector (and ϕ_t is working in the final-goods sector). After dividing by A_t, it becomes clear that this formulation cannot support sustained growth, since new inventions bounded from above by ηL must become a smaller fraction of any growing range A_t. Romer solves this problem by assuming that researchers' productivity grows with the range of inputs (i.e., an externality as discussed previously):

$$A_{t+1} - A_t = \eta A_t\left(1 - \phi_t\right) L,$$

so the growth rate of A_t is

$$\frac{A_{t+1}}{A_t} = 1 + \eta \left(1 - \phi_t\right) L. \tag{15.7.9}$$

When seeking a balanced growth path with a constant ϕ, we can use the earlier derivations, since the optimization problem of monopolistic input producers is the same as before. After replacing L in equations $(15.6.7b)$ and $(15.6.8)$ by ϕL, the steady-state supply of an input and the profit flow of an input producer are

$$Z_m = \left(\frac{\alpha^2}{1 + R_m}\right)^{1/(1-\alpha)} \phi L, \tag{15.7.10a}$$

$$\Omega_m\left(R_m\right) = \left(1 - \alpha\right) \alpha^{1/(1-\alpha)} \left(\frac{\alpha}{1 + R_m}\right)^{\alpha/(1-\alpha)} \phi L. \tag{15.7.10b}$$

In an equilibrium, agents must be indifferent between earning the wage in the final-goods sector equal to the marginal product of labor and being a researcher who expands the range of inputs by ηA_t and receives the associated discounted stream of profits in equation $(15.6.9)$:

$$\left(1 - \alpha\right) \left(\phi L\right)^{-\alpha} A_t Z_m^\alpha = \eta A_t \frac{\Omega_m\left(R_m\right)}{R_m}.$$

The substitution of equation $(15.7.10)$ into this expression yields

$$\phi = \frac{R_m}{\alpha \eta L}, \tag{15.7.11}$$

which, used in equation $(15.7.9)$, determines the growth rate of the input range,

$$\frac{A_{t+1}}{A_t} = 1 + \eta L - \frac{R_m}{\alpha}. \tag{15.7.12}$$

Thus, the maximum feasible growth rate in equation $(15.7.9)$, that is, $1 + \eta L$ with $\phi = 0$, requires an interest rate $R_m = 0$, while the growth vanishes as R_m approaches $\alpha \eta L$.

As previously, we can show that both consumption and the input range must grow at the same rate along a balanced growth path. It then remains to determine which consumption growth rate given by equation (15.7.12), is supported by Euler equation (15.6.11):

$$1 + \eta L - \frac{R_m}{\alpha} = [\beta (1 + R_m)]^{1/\sigma} . \qquad (15.7.13)$$

The left side of equation (15.7.13) is monotonically decreasing in R_m, and the right side is increasing. It is also trivially true that the left side is strictly greater than the right side for $R_m = 0$. Thus, a unique solution exists as long as the technology is sufficiently productive, in the sense that $\beta(1 + \alpha\eta L) > 1$. This parameter restriction ensures that the left side of equation (15.7.13) is strictly less than the right side at the interest rate $R_m = \alpha\eta L$ corresponding to a situation with zero growth, since no labor is allocated to the research sector, $\phi = 1$.

Equation (15.7.13) shows that this alternative model of research shares the scale implications described earlier; that is, a larger economy in terms of L has a higher equilibrium interest rate and therefore a higher growth rate. It can also be shown that the laissez-faire outcome continues to produce a smaller quantity of each input and to yield a lower growth rate than what is socially optimal. An additional source of underinvestment is now that agents who invent new inputs do not take into account that their inventions will increase the productivity of all future researchers.

15.8. Concluding remarks

This chapter has focused on the mechanical workings of endogenous growth models, with only limited reference to the motivation behind assumptions. For example, we have examined how externalities might enter models to overcome the onset of diminishing returns from nonreproducible factors without referring too much to the authors' interpretation of those externalities. The formalism of models is of course silent on why the assumptions are made, but the conceptual ideas behind the models contain valuable insights. In the last setup, Paul Romer argues that input designs represent excludable factors in the monopolists' production of inputs but the input variety A is also an aggregate stock of knowledge that enters as a nonexcludable factor in the production of new inventions. That is, the patent holder of an input type has the sole right to produce and sell that particular input, but she cannot stop inventors from studying the input design and learning knowledge that helps to invent new inputs. This multiple use of an input design hints at the nonrival nature of ideas and technology (i.e., a nonrival object has the property that its use by one person in no way limits its use by another). Romer (1990, p. S75) emphasizes this fundamental nature of technology and its implication; "If a nonrival good has productive value, then output cannot be a constant-returns-to-scale function of all its inputs taken together. The standard replication argument used to justify homogeneity of degree one does not apply because it is not necessary to replicate nonrival inputs." Thus, an endogenous growth model that is driven by technological change must be one where the advancement enters the economy as an externality or the assumption of perfect competition must be abandoned. Besides technological change, an alternative approach in the endogenous growth literature is to assume that all production factors are reproducible, or that there is a "core" of capital goods produced without the direct or indirect use of nonreproducible factors.

As we have seen, much of the effort in the endogenous growth literature is geared toward finding the proper technology specification. Even though growth is an endogenous outcome in these models, its manifestation ultimately hinges on technology assumptions. In the case of the last setup, as pointed out by Romer (1990, p. S84), "Linearity in A is what makes unbounded growth possible, and in this sense, unbounded growth is more like an assumption than a result of the model." It follows that various implications of the analyses stand and fall with the assumptions on technology. For example, the preceding model of research and monopolistic competition implies that the laissez-faire economy grows at a slower rate than the social optimum, but Benassy (1998) shows how this result can be overturned if the production function for final goods on the right side of equation (15.6.1) is multiplied by the input range raised to some power ν, A_t^{ν}. It then becomes possible that the laissez-faire growth rate exceeds the socially optimal rate because the new production function disentangles input producers' market power, determined by the parameter α, and the economy's returns to specialization, which is here also related to the parameter ν.

Segerstrom, Anant, and Dinopoulos (1990), Grossman and Helpman (1991), and Aghion and Howitt (1992) provide early attempts to explore endogenous growth arising from technologies that allow for product improvements and therefore product obsolescence. These models open the possibility that the laissez-faire growth rate is excessive because of a *business-stealing* effect, where agents fail to internalize the fact that their inventions exert a negative effect on incumbent producers. As in the models of research by Romer (1987, 1990) covered in this chapter, these other technologies exhibit scale effects, so that increases in the resources devoted to research imply faster economic growth. Charles Jones (1995), Young (1998), and Segerstrom (1998) criticize this feature and propose assumptions on technology that do not give rise to scale effects.

Exercises

Exercise 15.1   **Government spending and investment**, donated by Rodolfo Manuelli

Consider the following economy. There is a representative agent who has preferences given by

$$\sum_{t=0}^{\infty} \beta^t u\left(c_t\right),$$

where the function u is differentiable, increasing, and strictly concave. The technology in this economy is given by

$$c_t + x_t + g_t \leq f\left(k_t, g_t\right),$$
$$k_{t+1} \leq (1 - \delta) k_t + x_t,$$
$$(c_t, k_{t+1}, x_t) \geq (0, 0, 0),$$

and the initial condition $k_0 > 0$, given. Here k_t and g_t are capital per worker and government spending per worker. The function f is assumed to be strictly concave, increasing in each argument, twice differentiable, and such that the partial derivative with respect to both arguments converge to zero as the quantity of them grows without bound.

a.   Describe a set of equations that characterize an interior solution to the planner's problem when the planner can choose the sequence of government spending.

b.   Describe the steady state for the "general" specification of this economy. If necessary, make assumptions to guarantee that such a steady state exists.

c. Go as far as you can describing how the steady-state levels of capital per worker and government spending per worker change as a function of the discount factor.

d. Assume that the technology level can vary. More precisely, assume that the production function is given by $f(k, g, z) = z k^\alpha g^\eta$, where $0 < \alpha < 1$, $0 < \eta < 1$, and $\alpha + \eta < 1$. Go as far as you can describing how the investment/GDP ratio and the government spending/GDP ratio vary with the technology level z at the steady state.

Exercise 15.2 **Productivity and employment**, donated by Rodolfo Manuelli

Consider a basic growth economy with one modification. Instead of assuming that the labor supply is fixed at 1, we include leisure in the utility function. To simplify, we consider the total endowment of time to be 1. With this modification, preferences and technology are given by

$$\sum_{t=0}^{\infty} \beta^t u\left(c_t, 1 - n_t\right),$$

$$c_t + x_t + g_t \le z f\left(k_t, n_t\right),$$

$$k_{t+1} \le (1 - \delta) k_t + x_t.$$

In this setting, n_t is the number of hours worked by the representative household at time t. The rest of the time, $1 - n_t$, is consumed as leisure. The functions u and f are assumed to be strictly increasing in each argument, concave, and twice differentiable. In addition, f is such that the marginal product of capital converges to zero as the capital stock goes to infinity for any given value of labor, n.

a. Describe the steady state of this economy. If necessary, make additional assumptions to guarantee that it exists and is unique. If you make additional assumptions, go as far as you can giving an economic interpretation of them.

b. Assume that $f(k, n) = k^\alpha n^{1-\alpha}$ and $u(c, 1 - n) = [c^\mu (1 - n)^{1-\mu}]^{1-\sigma}/(1 - \sigma)$. What is the effect of changes in the technology (say increases in z) on employment and output per capita?

c. Consider next an increase in g. Are there conditions under which an increase in g will result in an increase in the steady-state k/n ratio? How about an increase in the steady-state level of output per capita? Go as far as you can giving an economic interpretation of these conditions. (Try to do this for general $f(k, n)$ functions with the appropriate convexity assumptions, but if this proves too hard, use the Cobb-Douglas specification.)

Exercise 15.3   **Vintage capital and cycles**, donated by Rodolfo Manuelli

Consider a standard one-sector optimal growth model with only one difference: If k_{t+1} new units of capital are built at time t, these units remain fully productive (i.e., they do not depreciate) until time $t+2$, at which point they disappear. Thus, the technology is given by

$$c_t + k_{t+1} \leq zf(k_t + k_{t-1}).$$

a. Formulate the optimal growth problem.

b. Show that, under standard conditions, a steady state exists and is unique.

c. A researcher claims that with the unusual depreciation pattern, it is possible that the economy displays cycles. By this he means that, instead of a steady state, the economy will converge to a period two sequence like $(c^o, c^e, c^o, c^e, \dots)$ and $(k^o, k^e, k^o, k^e, \dots)$, where c^o (k^o) indicates consumption (investment) in odd periods, and c^e (k^e) indicates consumption (investment) in even periods. Go as far as you can determining whether this can happen. If it is possible, try to provide an example.

Exercise 15.4   **Excess capacity**, donated by Rodolfo Manuelli

In the standard growth model, there is no room for varying the rate of utilization of capital. In this problem, you will explore how the nature of the solution is changed when variable rates of capital utilization are allowed.

As in the standard model, there is a representative agent with preferences given by

$$\sum_{t=0}^{\infty} \beta^t u(c_t), \quad 0 < \beta < 1.$$

It is assumed that u is strictly increasing, concave, and twice differentiable. Output depends on the actual number of machines used at time t, κ_t. Thus, the aggregate resource constraint is

$$c_t + x_t \leq zf(\kappa_t),$$

where the function f is strictly increasing, concave, and twice differentiable. In addition, f is such that the marginal product of capital converges to zero as the stock goes to infinity. Capital that is not used does not depreciate. Thus, capital accumulation satisfies

$$k_{t+1} \leq (1 - \delta)\, \kappa_t + (k_t - \kappa_t) + x_t,$$

where we require that the number of machines used, κ_t, is no greater than the number of machines available, k_t, or $k_t \geq \kappa_t$. This specification captures the idea that if some machines are not used, $k_t - \kappa_t > 0$, they do not depreciate.

a. Describe the planner's problem and analyze, as thoroughly as you can, the first-order conditions. Discuss your results.

b. Describe the steady state of this economy. If necessary, make additional assumptions to guarantee that it exists and is unique. If you make additional assumptions, go as far as you can giving an economic interpretation of them.

c. What is the optimal level of capacity utilization in this economy in the steady state?

d. Is this model consistent with the view that cross-country differences in output per capita are associated with differences in capacity utilization?

Exercise 15.5 **Heterogeneity and growth**, donated by Rodolfo Manuelli

Consider an economy populated by a large number of households indexed by i. The utility function of household i is

$$\sum_{t=0}^{\infty} \beta^t u_i(c_{it}),$$

where $0 < \beta < 1$, and u_i is differentiable, increasing and strictly concave. Note that although we allow the utility function to be "household specific," all households share the same discount factor. All households are endowed with one unit of labor that is supplied inelastically.

Assume that in this economy capital markets are perfect and that households start with initial capital given by $k_{i0} > 0$. Let total capital in the economy at time t be denoted k_t and assume that total labor is normalized to 1.

Assume that there is a large number of firms that produce output using capital and labor. Each firm has a production function given by $F(k, n)$ which is increasing, differentiable, concave, and homogeneous of degree 1. Firms maximize the present discounted value of profits. Assume that initial ownership of firms is uniformly distributed across households.

a. Define a competitive equilibrium.

b. Discuss (i) and (ii) and justify your answer. Be as formal as you can.

(i) Economist A argues that the steady state of this economy is unique and independent of the u_i functions, while B says that without knowledge of the u_i functions it is impossible to calculate the steady-state interest rate.

(ii) Economist A says that if k_0 is the steady-state aggregate stock of capital, then the pattern of "consumption inequality" will mirror exactly the pattern of "initial capital inequality" (i.e., k_{i0}), even though capital markets are perfect. Economist B argues that for all k_0, in the long run, per capita consumption will be the same for all households.

c. Assume that the economy is at the steady state. Describe the effects of the following three policies.

(i) At time zero, capital is redistributed across households (i.e., some people must surrender capital and others get their capital).

(ii) Half of the households are required to pay a lump-sum tax. The proceeds of the tax are used to finance a transfer program to the other half of the population.

(iii) Two-thirds of the households are required to pay a lump-sum tax. The proceeds of the tax are used to finance the purchase of a public good, say g, which does not enter in either preferences or technology.

Exercise 15.6   **Taxes and growth**, donated by Rodolfo Manuelli

Consider a simple two-planner economy. The first planner picks "tax rates," τ_t, and makes transfers to the representative agent, v_t. The second planner takes the tax rates and the transfers as given. That is, even though we know the connection between tax rates and transfers, the second planner does not, he or she takes the sequence of tax rates and transfers as given and beyond his or her control when solving for the optimal allocation. Thus, the problem faced by the second planner (the only one we will analyze for now) is

$$\max \sum_{t=0}^{\infty} \beta^t u\left(c_t\right)$$

subject to

$$c_t + x_t + g_t - v_t \leq \left(1 - \tau_t\right) f\left(k_t\right),$$
$$k_{t+1} \leq \left(1 - \delta\right) k_t + x_t,$$
$$\left(c_t, k_{t+1}, x_t\right) \geq \left(0, 0, 0\right),$$

and the initial condition $k_0 > 0$, given. The functions u and f are assumed to be strictly increasing, concave, and twice differentiable. In addition, f is such that the marginal product of capital converges to zero as the capital stock goes to infinity.

a. Assume that $0 < \tau_t = \tau < 1$, that is, the tax rate is constant. Assume that $v_t = \tau f(k_t)$ (remember that we know this, but the planner takes v_t as given at the time he or she maximizes). Show that there exists a steady state, and that for any initial condition $k_0 > 0$ the economy converges to the steady state.

b. Assume now that the economy has reached the steady state you analyzed in a. The first planner decides to change the tax rate to $0 < \tau' < \tau$. (Of course, the first planner and we know that this will result in a change in v_t; however, the second planner, the one that maximizes, acts as if v_t is a given sequence that is independent of his or her decisions.) Describe the new steady state as well as the dynamic path followed by the economy to reach this new steady state. Be as precise as you can about consumption, investment and output.

c. Consider now a competitive economy in which households, but not firms, pay income tax at rate τ_t on both labor and capital income. In addition, each household receives a transfer, v_t, that it takes to be given and independent of its own actions. Let the aggregate per capita capital stock be k_t. Then, balanced budget on the part of the government implies $v_t = \tau_t(r_t k_t + w_t, n_t)$, where r_t and w_t are the rental prices of capital and labor, respectively. Assume that the production function is $F(k, n)$, with F homogeneous of degree 1, concave, and "nice." Go as far as you can describing the impact of the change described in b on the equilibrium interest rate.

Chapter 16
Optimal Taxation with Commitment

16.1. Introduction

This chapter formulates a dynamic optimal taxation problem called a Ramsey problem with a solution called a Ramsey plan. The government's goal is to maximize households' welfare subject to raising set revenues through distortionary taxation. When designing an optimal policy, the government takes into account the competitive equilibrium reactions by consumers and firms to the tax system. We first study a nonstochastic economy, then a stochastic economy.

The model is a competitive equilibrium version of the basic neoclassical growth model with a government that finances an exogenous stream of government purchases. In the simplest version, the production factors are raw labor and physical capital on which the government levies distorting flat-rate taxes. The problem is to determine optimal sequences for the two tax rates. In a nonstochastic economy, Chamley (1986) and Judd (1985b) show in related settings that if an equilibrium has an asymptotic steady state, then the optimal policy is eventually to set the tax rate on capital to zero. This remarkable result asserts that capital income taxation serves neither efficiency nor redistributive purposes in the long run. This conclusion is robust to whether the government can issue debt or must run a balanced budget in each period. However, if the tax system is incomplete, the limiting value of optimal capital tax can be different from zero. To illustrate this possibility, we follow Correia (1996), and study a case with an additional fixed production factor that cannot be taxed by the government.

In a stochastic version of the model with complete markets, we find indeterminacy of state-contingent debt and capital taxes. Infinitely many plans implement the same competitive equilibrium allocation. For example, two alternative extreme cases are (1) that the government issues risk-free bonds and lets the capital tax rate depend on the current state, or (2) that the government fixes the capital tax rate one period ahead and lets debt be state contingent. While the state-by-state capital tax rates cannot be pinned down, an optimal plan does determine the current market value of next period's tax payments across states of nature. Dividing by the current market value of capital income gives a measure that we call the *ex ante capital tax rate*. If there exists a stationary Ramsey allocation, Zhu (1992) shows that for some special utility functions, the Ramsey plan prescribes a zero *ex ante* capital tax rate that can be implemented by setting a zero tax on capital income. But except for those preferences, Zhu concludes that the *ex ante* capital tax rate should vary around zero, in the sense that there is a positive measure of states with positive tax rates and a positive measure of states with negative tax rates. Chari, Christiano, and Kehoe (1994) perform numerical simulations and conclude that there is a quantitative presumption that the *ex ante* capital tax rate is approximately zero.

To gain further insights into optimal taxation and debt policies, we turn to Lucas and Stokey (1983) who analyze a model without physical capital. Examples of deterministic and stochastic government expenditure streams bring out the important role of government debt in smoothing tax distortions over both time and states. State-contingent government debt is used as an "insurance policy" that allows the government to smooth taxes across states. In this complete markets model, the current value of the government's debt reflects the current and likely future path of government expenditures rather than anything about its past. This feature of an optimal debt policy is especially apparent when government expenditures follow a Markov process because then the beginning-of-period state-contingent government debt is a function of the current state only and hence there are no lingering effects of past government expenditures. Aiyagari, Marcet, Sargent, and Seppälä (2002) alter that feature of optimal policy in Lucas and Stokey's model by assuming that the government can issue only risk-free debt. Not having access to state-contingent debt constrains the government's ability to smooth taxes over states and allows past values of government expenditures to have persistent effects on both future tax rates and debt levels. Reasoning by analogy from the savings problem of chapter 17 to an optimal taxation problem, Barro (1979) asserted that tax revenues would be a martingale that is cointegrated with government debt. Barro thus predicted persistent effects of government expenditures that are absent from the Ramsey plan in Lucas and Stokey's model. Aiyagari et. al.'s suspension of complete markets in Lucas and Stokey's environment goes a long way toward rationalizing outcomes Barro had predicted.

Returning to a nonstochastic setup, Jones, Manuelli, and Rossi (1997) augment the model by allowing human capital accumulation. They make the particular assumption that the technology for human capital accumulation is linearly homogeneous in a stock of human capital and a flow of inputs coming from current output. Under this special constant returns assumption, they show that a zero limiting tax applies also to labor income; that is, the return to human capital should not be taxed in the limit. Instead, the government should resort to a consumption tax. But even this consumption tax, and therefore all taxes, should be zero in the limit for a particular class of preferences where it is optimal during a transition period for the government to amass so many claims on the private economy that the interest earnings suffice to finance government expenditures. While these successive results on optimal taxation require ever more stringent assumptions, the basic prescription for a zero *capital* tax in a nonstochastic steady state is an immediate implication of a standard constant-returns-to-scale production technology, competitive markets, and a complete set of flat-rate taxes.

Throughout the chapter we maintain the assumption that the government can commit to future tax rates.

16.2. A nonstochastic economy

An infinitely lived representative household likes consumption, leisure streams $\{c_t, \ell_t\}_{t=0}^{\infty}$ that give higher values of

$$\sum_{t=0}^{\infty} \beta^t u\left(c_t, \ell_t\right), \ \beta \in (0, 1) \tag{16.2.1}$$

where u is increasing, strictly concave, and three times continuously differentiable in consumption c and leisure ℓ. The household is endowed with one unit of time that can be used for leisure ℓ_t and labor n_t:

$$\ell_t + n_t = 1. \tag{16.2.2}$$

The single good is produced with labor n_t and capital k_t. Output can be consumed by households, used by the government, or used to augment the capital stock. The technology is

$$c_t + g_t + k_{t+1} = F\left(k_t, n_t\right) + (1 - \delta) k_t, \tag{16.2.3}$$

where $\delta \in (0,1)$ is the rate at which capital depreciates and $\{g_t\}_{t=0}^{\infty}$ is an exogenous sequence of government purchases. We assume a standard concave production function $F(k,n)$ that exhibits constant returns to scale. By Euler's theorem on homogeneous functions, linear homogeneity of F implies

$$F(k,n) = F_k k + F_n n. \tag{16.2.4}$$

Let u_c be the derivative of $u(c_t, \ell_t)$ with respect to consumption; u_ℓ is the derivative with respect to ℓ. We use $u_c(t)$ and $F_k(t)$ and so on to denote the time t values of the indicated objects, evaluated at an allocation to be understood from the context.

16.2.1. Government

The government finances its stream of purchases $\{g_t\}_{t=0}^{\infty}$ by levying flat-rate, time-varying taxes on earnings from capital at rate τ_t^k and earnings from labor at rate τ_t^n. The government can also trade one-period bonds, sequential trading of which suffices to accomplish any intertemporal trade in a world without uncertainty. Let b_t be government indebtedness to the private sector, denominated in time t-goods, maturing at the beginning of period t. The government's budget constraint is

$$g_t = \tau_t^k r_t k_t + \tau_t^n w_t n_t + \frac{b_{t+1}}{R_t} - b_t, \tag{16.2.5}$$

where r_t and w_t are the market-determined rental rate of capital and the wage rate for labor, respectively, denominated in units of time t goods, and R_t is the gross rate of return on one-period bonds held from t to $t+1$. Interest earnings on bonds are assumed to be tax exempt; this assumption is innocuous for bond exchanges between the government and the private sector.

16.2.2. Household

A representative household chooses $\{c_t, n_t, k_{t+1}, b_{t+1}\}_{t=0}^{\infty}$ to maximize expression (16.2.1) subject to the following sequence of budget constraints:

$$c_t + k_{t+1} + \frac{b_{t+1}}{R_t} = (1 - \tau_t^n)\, w_t n_t + (1 - \tau_t^k)\, r_t k_t + (1 - \delta)\, k_t + b_t, \quad (16.2.6)$$

for $t \geq 0$. With $\beta^t \lambda_t$ as the Lagrange multiplier on the time t budget constraint, the first-order conditions are

$$c_t: \quad u_c(t) = \lambda_t, \quad (16.2.7)$$

$$n_t: \quad u_\ell(t) = \lambda_t (1 - \tau_t^n)\, w_t, \quad (16.2.8)$$

$$k_{t+1}: \quad \lambda_t = \beta \lambda_{t+1} \left[(1 - \tau_{t+1}^k)\, r_{t+1} + 1 - \delta \right], \quad (16.2.9)$$

$$b_{t+1}: \quad \lambda_t \frac{1}{R_t} = \beta \lambda_{t+1}. \quad (16.2.10)$$

Substituting equation (16.2.7) into equations (16.2.8) and (16.2.9), we obtain

$$u_\ell(t) = u_c(t)\, (1 - \tau_t^n)\, w_t, \quad (16.2.11a)$$

$$u_c(t) = \beta u_c(t+1) \left[(1 - \tau_{t+1}^k)\, r_{t+1} + 1 - \delta \right]. \quad (16.2.11b)$$

Moreover, equations (16.2.9) and (16.2.10) imply

$$R_t = (1 - \tau_{t+1}^k)\, r_{t+1} + 1 - \delta, \quad (16.2.12)$$

which is a condition not involving any quantities that the household is free to adjust. Because only one financial asset is needed to accomplish all intertemporal trades in a world without uncertainty, condition (16.2.12) constitutes a no-arbitrage condition for trades in capital and bonds that ensures that these two assets have the same rate of return. This no-arbitrage condition can be obtained by consolidating two consecutive budget constraints; constraint (16.2.6) and its counterpart for time $t + 1$ can be merged by eliminating the common quantity b_{t+1} to get

$$c_t + \frac{c_{t+1}}{R_t} + \frac{k_{t+2}}{R_t} + \frac{b_{t+2}}{R_t R_{t+1}} = (1 - \tau_t^n)\, w_t n_t$$
$$+ \frac{(1 - \tau_{t+1}^n)\, w_{t+1} n_{t+1}}{R_t} + \left[\frac{(1 - \tau_{t+1}^k)\, r_{t+1} + 1 - \delta}{R_t} - 1 \right] k_{t+1}$$
$$+ (1 - \tau_t^k)\, r_t k_t + (1 - \delta)\, k_t + b_t, \quad (16.2.13)$$

where the left side is the use of funds and the right side measures the resources at the household's disposal. If the term multiplying k_{t+1} is not zero, the household can make its budget set unbounded either by buying an arbitrarily large k_{t+1} when $(1-\tau_{t+1}^k)r_{t+1}+1-\delta > R_t$, or, in the opposite case, by selling capital short to achieve an arbitrarily large negative k_{t+1}. In such arbitrage transactions, the household would finance purchases of capital or invest the proceeds from short sales in the bond market between periods t and $t+1$. Thus, to ensure the existence of a competitive equilibrium with bounded budget sets, condition (16.2.12) must hold.

If we continue the process of recursively using successive budget constraints to eliminate successive b_{t+j} terms, begun in equation (16.2.13), we arrive at the household's present-value budget constraint,

$$\sum_{t=0}^{\infty} \left(\prod_{i=0}^{t-1} R_i^{-1} \right) c_t = \sum_{t=0}^{\infty} \left(\prod_{i=0}^{t-1} R_i^{-1} \right) (1 - \tau_t^n) w_t n_t$$

$$+ \left[\left(1 - \tau_0^k \right) r_0 + 1 - \delta \right] k_0 + b_0, \qquad (16.2.14)$$

where we have imposed the transversality conditions

$$\lim_{T \to \infty} \left(\prod_{i=0}^{T-1} R_i^{-1} \right) k_{T+1} = 0, \qquad (16.2.15)$$

$$\lim_{T \to \infty} \left(\prod_{i=0}^{T-1} R_i^{-1} \right) \frac{b_{T+1}}{R_T} = 0. \qquad (16.2.16)$$

As discussed in chapter 13, the household would not like to violate these transversality conditions by choosing k_{t+1} or b_{t+1} to be larger, because alternative feasible allocations with higher consumption in finite time would yield higher lifetime utility. A consumption/savings plan that made either expression negative would not be possible because the household would not find anybody willing to be on the lending side of the implied transactions.

16.2.3. Firms

In each period, the representative firm takes (r_t, w_t) as given, rents capital and labor from households, and maximizes profits,

$$\Pi = F\left(k_t, n_t\right) - r_t k_t - w_t n_t. \tag{16.2.17}$$

The first-order conditions for this problem are

$$r_t = F_k\left(t\right), \tag{16.2.18a}$$
$$w_t = F_n\left(t\right). \tag{16.2.18b}$$

In words, inputs should be employed until the marginal product of the last unit is equal to its rental price. With constant returns to scale, we get the standard result that pure profits are zero and the size of an individual firm is indeterminate.

An alternative way of establishing the equilibrium conditions for the rental price of capital and the wage rate for labor is to substitute equation (16.2.4) into equation (16.2.17) to get

$$\Pi = \left[F_k\left(t\right) - r_t\right] k_t + \left[F_n\left(t\right) - w_t\right] n_t.$$

If the firm's profits are to be nonnegative and finite, the terms multiplying k_t and n_t must be zero; that is, condition (16.2.18) must hold. These conditions imply that in any equilibrium, $\Pi = 0$.

16.3. The Ramsey problem

We shall use symbols without subscripts to denote the one-sided infinite sequence for the corresponding variable, e.g., $c \equiv \{c_t\}_{t=0}^{\infty}$.

DEFINITION: A *feasible allocation* is a sequence (k, c, ℓ, g) that satisfies equation (16.2.3).

DEFINITION: A *price system* is a 3-tuple of nonnegative bounded sequences (w, r, R).

DEFINITION: A *government policy* is a 4-tuple of sequences (g, τ^k, τ^n, b).

DEFINITION: A *competitive equilibrium* is a feasible allocation, a price system, and a government policy such that (a) given the price system and the government policy, the allocation solves both the firm's problem and the household's problem; and (b) given the allocation and the price system, the government policy satisfies the sequence of government budget constraints (16.2.5).

There are many competitive equilibria, indexed by different government policies. This multiplicity motivates the Ramsey problem.

DEFINITION: Given k_0 and b_0, the *Ramsey problem* is to choose a competitive equilibrium that maximizes expression (16.2.1).

To make the Ramsey problem interesting, we always impose a restriction on τ_0^k, for example, by taking it as given at a small number, say, 0. This approach rules out taxing the initial capital stock via a so-called capital levy that would constitute a lump-sum tax, since k_0 is in fixed supply. One often imposes other restrictions on $\tau_t^k, t \geq 1$, namely, that they be bounded above by some arbitrarily given numbers. These bounds play an important role in shaping the near-term temporal properties of the optimal tax plan, as discussed by Chamley (1986) and explored in computational work by Jones, Manuelli, and Rossi (1993). In the analysis that follows, we shall impose the bound on τ_t^k only for $t = 0$.[1]

---

[1] According to our assumption on the technology in equation (16.2.3), capital is reversible and can be transformed back into the consumption good. Thus, the capital stock is a fixed factor for only one period at a time, so τ_0^k is the only tax that we need to restrict to ensure an interesting Ramsey problem.

16.4. Zero capital tax

Following Chamley (1986), we formulate the Ramsey problem as if the government chooses the after-tax rental rate of capital $\tilde{r}_t$, and the after-tax wage rate $\tilde{w}_t$:

$$\tilde{r}_t \equiv \left(1 - \tau_t^k\right) r_t,$$
$$\tilde{w}_t \equiv \left(1 - \tau_t^n\right) w_t.$$

Using equations $(16.2.18)$ and $(16.2.4)$, Chamley expresses government tax revenues as

$$
\begin{aligned}
\tau_t^k r_t k_t + \tau_t^n w_t n_t &= (r_t - \tilde{r}_t)\, k_t + (w_t - \tilde{w}_t)\, n_t \\
&= F_k\,(t)\, k_t + F_n\,(t)\, n_t - \tilde{r}_t k_t - \tilde{w}_t n_t \\
&= F\,(k_t, n_t) - \tilde{r}_t k_t - \tilde{w}_t n_t.
\end{aligned}
$$

Substituting this expression into equation $(16.2.5)$ consolidates the firm's first-order conditions with the government's budget constraint. The government's policy choice is also constrained by the aggregate resource constraint $(16.2.3)$ and the household's first-order conditions $(16.2.11)$. To solve the Ramsey problem, form a Lagrangian

$$
\begin{aligned}
L = \sum_{t=0}^{\infty} \beta^t \Big\{ &u\,(c_t, 1 - n_t) \\
&+ \Psi_t \left[F\,(k_t, n_t) - \tilde{r}_t k_t - \tilde{w}_t n_t + \frac{b_{t+1}}{R_t} - b_t - g_t \right] \\
&+ \theta_t \left[F\,(k_t, n_t) + (1 - \delta)\, k_t - c_t - g_t - k_{t+1} \right] \\
&+ \mu_{1t} \left[u_\ell\,(t) - u_c\,(t)\, \tilde{w}_t \right] \\
&+ \mu_{2t} \left[u_c\,(t) - \beta u_c\,(t+1)\, (\tilde{r}_{t+1} + 1 - \delta) \right] \Big\},
\end{aligned}
\qquad (16.4.1)
$$

where $R_t = \tilde{r}_{t+1} + 1 - \delta$, as given by equation $(16.2.12)$. Note that the household's budget constraint is not explicitly included because it is redundant when the government satisfies its budget constraint and the resource constraint holds.

The first-order condition for maximizing the Lagrangian $(16.4.1)$ with respect to k_{t+1} is

$$\theta_t = \beta \left\{ \Psi_{t+1} \left[F_k\,(t+1) - \tilde{r}_{t+1} \right] + \theta_{t+1} \left[F_k\,(t+1) + 1 - \delta \right] \right\}. \qquad (16.4.2)$$

The equation has a straightforward interpretation. A marginal increment of capital investment in period t increases the quantity of available goods at time $t+1$ by the amount $[F_k(t+1)+1-\delta]$, which has a social marginal value θ_{t+1}. In addition, there is an increase in tax revenues equal to $[F_k(t+1)-\tilde{r}_{t+1}]$, which enables the government to reduce its debt or other taxes by the same amount. The reduction of the "excess burden" equals $\Psi_{t+1}[F_k(t+1)-\tilde{r}_{t+1}]$. The sum of these two effects in period $t+1$ is discounted by the discount factor β and set equal to the social marginal value of the initial investment good in period t, which is given by θ_t.

Suppose that government expenditures stay constant after some period T, and assume that the solution to the Ramsey problem converges to a steady state; that is, all endogenous variables remain constant. Using equation $(16.2.18a)$, the steady-state version of equation $(16.4.2)$ is

$$\theta = \beta\left[\Psi\left(r-\tilde{r}\right)+\theta\left(r+1-\delta\right)\right]. \qquad (16.4.3)$$

Now with a constant consumption stream, the steady-state version of the household's optimality condition for the choice of capital in equation $(16.2.11b)$ is

$$1 = \beta\left(\tilde{r}+1-\delta\right). \qquad (16.4.4)$$

A substitution of equation $(16.4.4)$ into equation $(16.4.3)$ yields

$$(\theta+\Psi)\left(r-\tilde{r}\right) = 0. \qquad (16.4.5)$$

Since the marginal social value of goods θ is strictly positive and the marginal social value of reducing government debt or taxes Ψ is nonnegative, it follows that r must be equal to $\tilde{r}$, so that $\tau^k = 0$. This analysis establishes the following celebrated result, versions of which were attained by Chamley (1986) and Judd (1985b).

PROPOSITION 1: If there exists a steady-state Ramsey allocation, the associated limiting tax rate on capital is zero.

Its ability to borrow and *lend* a risk-free one period asset makes it feasible for the government to amass a stock of claims on the private economy that is so large that eventually the interest earnings suffice to finance the stream of government expenditures.[2] Then it can set *all* tax rates to zero. But this is *not* the force that underlies the above result that τ_k should be zero asymptotically. The zero-capital-tax outcome would prevail even if we were to prohibit the government from borrowing or lending by requiring it to run a balanced budget in each period. To see this, notice that if we had set b_t and b_{t+1} equal to zero in equation (16.4.1), nothing would change in our derivation of the conclusion that $\tau^k = 0$. Thus, even when the government must perpetually raise positive revenues from *some* source each period, it is optimal eventually to set τ_k to zero.

16.5. Limits to redistribution

The optimality of a limiting zero capital tax extends to an economy with heterogeneous agents, as mentioned by Chamley (1986) and explored in depth by Judd (1985b). Assume a finite number of different classes of agents, N, and for simplicity, let each class be the same size. The consumption, labor supply, and capital stock of the representative agent in class i are denoted c_t^i, n_t^i, and k_t^i, respectively. The utility function might also depend on the class, $u^i(c_t^i, 1 - n_t^i)$, but the discount factor is assumed to be identical across all agents.

The government can make positive class-specific lump-sum transfers $S_t^i \geq 0$, but there are no lump-sum taxes. As before, the government must rely on flat-rate taxes on earnings from capital and labor. We assume that the government has a social welfare function that is a positively weighted average of individual utilities with weight $\alpha^i \geq 0$ on class i. We assume that the government runs a balanced budget, which does not affect the limiting zero-tax on capital tax outcome. The Lagrangian associated with the government's optimization problem becomes

$$L = \sum_{t=0}^{\infty} \beta^t \left\{ \sum_{i=1}^{N} \alpha^i u^i \left(c_t^i, 1 - n_t^i \right) \right.$$

$$+ \Psi_t \left[F\left(k_t, n_t \right) - \tilde{r}_t k_t - \tilde{w}_t n_t - g_t - S_t \right]$$

$$+ \theta_t \left[F\left(k_t, n_t \right) + (1 - \delta) k_t - c_t - g_t - k_{t+1} \right]$$

---

[2] Below we shall describe a stochastic economy in which the government cannot issue state-contingent debt. For that economy, such a policy would actually be the optimal one.

$$+ \sum_{i=1}^{N} \epsilon_t^i \left[\tilde{w}_t n_t^i + \tilde{r}_t k_t^i + (1 - \delta) k_t^i + S_t^i - c_t^i - k_{t+1}^i \right]$$

$$+ \sum_{i=1}^{N} \mu_{1t}^i \left[u_\ell^i(t) - u_c^i(t) \tilde{w}_t \right]$$

$$+ \sum_{i=1}^{N} \mu_{2t}^i \left[u_c^i(t) - \beta u_c^i(t+1)(\tilde{r}_{t+1} + 1 - \delta) \right] \Bigg\}, \qquad (16.5.1)$$

where $x_t \equiv \sum_{i=1}^{N} x_t^i$, for $x = c, n, k, S$. Here we have to include the budget constraints and the first-order conditions for each class of agents.

The social marginal value of an increment in the capital stock depends now on whose capital stock is augmented. The Ramsey problem's first-order condition with respect to k_{t+1}^i is

$$\theta_t + \epsilon_t^i = \beta \Big\{ \Psi_{t+1} \left[F_k(t+1) - \tilde{r}_{t+1} \right] + \theta_{t+1} \left[F_k(t+1) + 1 - \delta \right]$$

$$+ \epsilon_{t+1}^i (\tilde{r}_{t+1} + 1 - \delta) \Big\}. \qquad (16.5.2)$$

If an asymptotic steady state exists in equilibrium, the time-invariant version of this condition becomes

$$\theta + \epsilon^i \left[1 - \beta(\tilde{r} + 1 - \delta) \right] = \beta \left[\Psi(r - \tilde{r}) + \theta(r + 1 - \delta) \right]. \qquad (16.5.3)$$

Since the steady-state condition (16.4.4) holds for each individual household, the term multiplying ϵ^i is zero, and we can once again deduce condition (16.4.5) asserting that the limiting capital tax must be zero in any convergent Pareto-efficient tax program.

Judd (1985b) discusses one extreme version of heterogeneity with two classes of agents. Agents of class 1 are workers who do not save, so their budget constraint is

$$c_t^1 = \tilde{w}_t n_t^1 + S_t^1.$$

Agents of class 2 are capitalists who do not work, so their budget constraint is

$$c_t^2 + k_{t+1}^2 = \tilde{r}_t k_t^2 + (1 - \delta) k_t^2 + S_t^2.$$

Since this setup is also covered by the preceding analysis, a limiting zero capital tax remains optimal if there is a steady state. This fact implies, for example, that if the government only values the welfare of workers ($\alpha^1 > \alpha^2 = 0$), there will not be any recurring redistribution in the limit. Government expenditures will be financed solely by levying wage taxes on workers.

It is important to keep in mind that the zero tax on capital result pertains only to the limiting steady state. Our analysis is silent about how much redistribution is accomplished in the transition period.

16.6. Primal approach to the Ramsey problem

In the formulation of the Ramsey problem in expression (16.4.1), Chamley reduced a pair of taxes (τ_t^k, τ_t^n) and a pair of prices (r_t, w_t) to just one pair of numbers $(\tilde{r}_t, \tilde{w}_t)$ by utilizing the firm's first-order conditions and equilibrium outcomes in factor markets. In a similar spirit, we will now eliminate all prices and taxes so that the government can be thought of as directly choosing a feasible allocation, subject to constraints that ensure the existence of prices and taxes such that the chosen allocation is consistent with the optimization behavior of households and firms. This primal approach to the Ramsey problem, as opposed to the dual approach in which tax rates are viewed as governmental decision variables, is used in Lucas and Stokey's (1983) analysis of an economy without capital. Here we will follow the setup of Jones, Manuelli, and Rossi (1997).

It is useful to compare our primal approach to the Ramsey problem with the formulation in (16.4.1). First, we will now consider only the case when the government is free to trade in the bond market. The constraints associated with Lagrange multipliers Ψ_t in (16.4.1) can therefore be replaced with a single present-value budget constraint for either the government or the representative household. (One of them is redundant, since we are also imposing the aggregate resource constraint.) The problem simplifies nicely if we choose the present-value budget constraint of the household (16.2.14), in which future capital stocks have been eliminated with the use of no-arbitrage conditions. For convenience, we repeat the household's present-value budget constraint (16.2.14) here in the form:

$$\sum_{t=0}^{\infty} q_t^0 c_t = \sum_{t=0}^{\infty} q_t^0 \left(1 - \tau_t^n\right) w_t n_t + \left[\left(1 - \tau_0^k\right) r_0 + 1 - \delta\right] k_0 + b_0 . \qquad (16.6.1)$$

In equation (16.6.1), q_t^0 is the Arrow-Debreu price

$$q_t^0 = \prod_{i=0}^{t-1} R_i^{-1}, \qquad \forall t \geq 1; \tag{16.6.2}$$

with the numeraire $q_0^0 = 1$. Second, we use two constraints in expression (16.4.1) to replace prices q_t^0 and $(1 - \tau_t^n)w_t$ in equation (16.6.1) with the household's marginal rates of substitution.

A stepwise summary of the primal approach is as follows:

1. Obtain the first-order conditions of the household's and the firm's problems, as well as any arbitrage pricing conditions. Solve these conditions for $\{q_t^0, r_t, w_t, \tau_t^k, \tau_t^n\}_{t=0}^{\infty}$ as functions of the allocation $\{c_t, n_t, k_{t+1}\}_{t=0}^{\infty}$.

2. Substitute these expressions for taxes and prices in terms of the allocation into the household's present-value budget constraint. This is an intertemporal constraint involving only the allocation.

3. Solve for the Ramsey allocation by maximizing expression (16.2.1) subject to equation (16.2.3) and the "implementability condition" derived in step 2.

4. After the Ramsey allocation is solved, use the formulas from step 1 to find taxes and prices.

16.6.1. Constructing the Ramsey plan

We now carry out the steps outlined in the preceding list of instructions.

Step 1. Let λ be a Lagrange multiplier on the household's budget constraint (16.6.1). The first-order conditions for the household's problem are

$$c_t: \quad \beta^t u_c(t) - \lambda q_t^0 = 0,$$
$$n_t: \quad -\beta^t u_\ell(t) + \lambda q_t^0 (1 - \tau_t^n) w_t = 0.$$

With the numeraire $q_0^0 = 1$, these conditions imply

$$q_t^0 = \beta^t \frac{u_c(t)}{u_c(0)}, \tag{16.6.3a}$$

$$(1 - \tau_t^n) w_t = \frac{u_\ell(t)}{u_c(t)}. \tag{16.6.3b}$$

As before, we can derive the arbitrage condition $(16.2.12)$, which now reads

$$\frac{q_t^0}{q_{t+1}^0} = \left(1 - \tau_{t+1}^k\right) r_{t+1} + 1 - \delta. \tag{16.6.4}$$

Profit maximization and factor market equilibrium imply equations $(16.2.18)$.

Step 2. Substitute equations $(16.6.3)$ and $r_0 = F_k(0)$ into equation $(16.6.1)$, so that we can write the household's budget constraint as

$$\sum_{t=0}^{\infty} \beta^t \left[u_c\left(t\right)c_t - u_\ell\left(t\right)n_t\right] - A = 0, \tag{16.6.5}$$

where A is given by

$$A = A\left(c_0, n_0, \tau_0^k, b_0\right) = u_c\left(0\right)\left\{\left[\left(1 - \tau_0^k\right)F_k\left(0\right) + 1 - \delta\right]k_0 + b_0\right\}. \tag{16.6.6}$$

Step 3. The Ramsey problem is to maximize expression $(16.2.1)$ subject to equation $(16.6.5)$ and the feasibility constraint $(16.2.3)$. As before, we proceed by assuming that government expenditures are small enough that the problem has a convex constraint set and that we can approach it using Lagrangian methods. In particular, let Φ be a Lagrange multiplier on equation $(16.6.5)$ and define

$$V\left(c_t, n_t, \Phi\right) = u\left(c_t, 1 - n_t\right) + \Phi\left[u_c\left(t\right)c_t - u_\ell\left(t\right)n_t\right]. \tag{16.6.7}$$

Then form the Lagrangian

$$\begin{aligned}
J = \sum_{t=0}^{\infty} \beta^t &\left\{V\left(c_t, n_t, \Phi\right) + \theta_t\left[F\left(k_t, n_t\right) + \left(1 - \delta\right)k_t\right.\right. \\
&\left.\left. - c_t - g_t - k_{t+1}\right]\right\} - \Phi A,
\end{aligned} \tag{16.6.8}$$

where $\{\theta_t\}_{t=0}^{\infty}$ is a sequence of Lagrange multipliers. For given k_0 and b_0, we fix τ_0^k and maximize J with respect to $\{c_t, n_t, k_{t+1}\}_{t=0}^{\infty}$. First-order conditions for this problem are[3]

$$c_t: \quad V_c(t) = \theta_t, \quad t \geq 1$$

$$n_t: \quad V_n(t) = -\theta_t F_n(t), \quad t \geq 1$$

$$k_{t+1}: \quad \theta_t = \beta \theta_{t+1} \left[F_k(t+1) + 1 - \delta \right], \quad t \geq 0$$

$$c_0: \quad V_c(0) = \theta_0 + \Phi A_c,$$

$$n_0: \quad V_n(0) = -\theta_0 F_n(0) + \Phi A_n.$$

These conditions become

$$V_c(t) = \beta V_c(t+1) \left[F_k(t+1) + 1 - \delta \right], \quad t \geq 1 \qquad (16.6.9a)$$

$$V_n(t) = -V_c(t) F_n(t), \quad t \geq 1 \qquad (16.6.9b)$$

$$V_c(0) - \Phi A_c = \beta V_c(1) \left[F_k(1) + 1 - \delta \right], \qquad (16.6.9c)$$

$$V_n(0) = \left[\Phi A_c - V_c(0) \right] F_n(0) + \Phi A_n. \qquad (16.6.9d)$$

To these we add equations $(16.2.3)$ and $(16.6.5)$, which we repeat here for convenience:

$$c_t + g_t + k_{t+1} = F(k_t, n_t) + (1 - \delta) k_t, \quad t \geq 0 \qquad (16.6.10a)$$

$$\sum_{t=0}^{\infty} \beta^t \left[u_c(t) c_t - u_\ell(t) n_t \right] - A = 0. \qquad (16.6.10b)$$

We seek an allocation $\{c_t, n_t, k_{t+1}\}_{t=0}^{\infty}$, and a multiplier Φ that satisfies the system of difference equations formed by equations $(16.6.9)$–$(16.6.10)$.[4]

Step 4: After an allocation has been found, obtain q_t^0 from equation $(16.6.3a)$, r_t from equation $(16.2.18a)$, w_t from equation $(16.2.18b)$, τ_t^n from equation $(16.6.3b)$, and finally τ_t^k from equation $(16.6.4)$.

---

[3] Comparing the first-order condition for k_{t+1} to the earlier one in equation $(16.4.2)$, obtained under Chamley's alternative formulation of the Ramsey problem, note that the Lagrange multiplier θ_t is different across formulations. Specifically, the present specification of the objective function V subsumes parts of the household's present-value budget constraint. To bring out this difference, a more informative notation would be to write $V_j(t, \Phi)$ for $j = c, n$ rather than just $V_j(t)$.

[4] This system of nonlinear equations can be solved iteratively. First, fix Φ, and solve equations $(16.6.9)$ and $(16.6.10a)$ for an allocation. Then check the implementability condition $(16.6.10b)$, and increase or decrease Φ depending on whether the budget is in deficit or surplus. Note that the multiplier Φ is nonnegative because we are facing the constraint that the left-hand side of equation $(16.6.10b)$ is *greater* than or equal to zero. That is, we are constrained by the equilibrium outcome that households fully exhaust their incomes and, hence, are not free to choose households' expenditures strictly less than their incomes.

16.6.2. *Revisiting a zero capital tax*

Consider the special case in which there is a $T \geq 0$ for which $g_t = g$ for all $t \geq T$. Assume that there exists a solution to the Ramsey problem and that it converges to a time-invariant allocation, so that c, n, and k are constant after some time. Then because $V_c(t)$ converges to a constant, the stationary version of equation (16.6.9a) implies

$$1 = \beta \left(F_k + 1 - \delta \right). \tag{16.6.11}$$

Now because c_t is constant in the limit, equation (16.6.3a) implies that $\left(q_t^0 / q_{t+1}^0 \right) \to \beta^{-1}$ as $t \to \infty$. Then the no-arbitrage condition for capital (16.6.4) becomes

$$1 = \beta \left[\left(1 - \tau^k \right) F_k + 1 - \delta \right]. \tag{16.6.12}$$

Equalities (16.6.11) and (16.6.12) imply that $\tau_k = 0$.

16.7. Taxation of initial capital

Thus far, we have set τ_0^k at zero (or some other small fixed number). Now suppose that the government is free to choose τ_0^k. The derivative of J in equation (16.6.8) with respect to τ_0^k is

$$\frac{\partial J}{\partial \tau_0^k} = \Phi u_c \left(0 \right) F_k \left(0 \right) k_0, \tag{16.7.1}$$

which is strictly positive for all τ_0^k as long as $\Phi > 0$. The nonnegative Lagrange multiplier Φ measures the utility costs of raising government revenues through distorting taxes. Without distortionary taxation, a competitive equilibrium would attain the first-best outcome for the representative household, and Φ would be equal to zero, so that the household's (or equivalently, by Walras' Law, the government's) present-value budget constraint would not exert any additional constraining effect on welfare maximization beyond what is present in the economy's technology. In contrast, when the government has to use some of the tax rates $\{\tau_t^n, \tau_{t+1}^k\}_{t=0}^{\infty}$, the multiplier Φ is strictly positive and reflects the welfare cost of the distorted margins, implicit in the present-value budget constraint (16.6.10b), which govern the household's optimization behavior.

By raising τ_0^k and thereby increasing the revenues from lump-sum taxation of k_0, the government reduces its need to rely on future distortionary taxation, and hence the value of Φ falls. In fact, the ultimate implication of condition $(16.7.1)$ is that the government should set τ_0^k high enough to drive Φ down to zero. In other words, the government should raise *all* revenues through a time 0 capital levy, then lend the proceeds to the private sector and finance government expenditures by using the interest from the loan; this would enable the government to set $\tau_t^n = 0$ for all $t \geq 0$ and $\tau_t^k = 0$ for all $t \geq 1$.[5]

16.8. Nonzero capital tax due to incomplete taxation

The result that the limiting capital tax should be zero hinges on a complete set of flat-rate taxes. The consequences of incomplete taxation are illustrated by Correia (1996), who introduces an additional production factor z_t in fixed supply $z_t = Z$ that cannot be taxed, $\tau_t^z = 0$.

The new production function $F(k_t, n_t, z_t)$ exhibits constant returns to scale in all of its inputs. Profit maximization implies that the rental price of the new factor equals its marginal product:

$$p_t^z = F_z(t).$$

The only change to the household's present-value budget constraint $(16.6.1)$ is that a stream of revenues is added to the right side:

$$\sum_{t=0}^{\infty} q_t^0 p_t^z Z.$$

---

[5] The scheme may involve $\tau_0^k > 1$ for high values of $\{g_t\}_{t=0}^{\infty}$ and b_0. However, such a scheme cannot be implemented if the household could avoid the tax liability by not renting out its capital stock at time 0. The government would then be constrained to choose $\tau_0^k \leq 1$.

In the rest of the chapter, we do not impose that $\tau_t^k \leq 1$. If we were to do so, an extra constraint in the Ramsey problem would be

$$u_c(t) \geq \beta(1 - \delta) u_c(t + 1),$$

which can be obtained by substituting equation $(16.6.3a)$ into equation $(16.6.4)$.

Following our scheme of constructing the Ramsey plan, step 2 yields the following implementability condition:

$$\sum_{t=0}^{\infty} \beta^t \left\{ u_c(t) \left[c_t - F_z(t) Z \right] - u_\ell(t) n_t \right\} - A = 0, \qquad (16.8.1)$$

where A remains defined by equation (16.6.6). In step 3 we formulate

$$V(c_t, n_t, k_t, \Phi) = u(c_t, 1 - n_t)$$
$$+ \Phi \left\{ u_c(t) \left[c_t - F_z(t) Z \right] - u_\ell(t) n_t \right\}. \qquad (16.8.2)$$

In contrast to equation (16.6.7), k_t enters now as an argument in V because of the presence of the marginal product of the factor Z (but we have chosen to suppress the quantity Z itself, since it is in fixed supply).

Except for these changes of the functions F and V, the Lagrangian of the Ramsey problem is the same as equation (16.6.8). The first-order condition with respect to k_{t+1} is

$$\theta_t = \beta V_k(t+1) + \beta \theta_{t+1} \left[F_k(t+1) + 1 - \delta \right]. \qquad (16.8.3)$$

Assuming the existence of a steady state, the stationary version of equation (16.8.3) becomes

$$1 = \beta (F_k + 1 - \delta) + \beta \frac{V_k}{\theta}. \qquad (16.8.4)$$

Condition (16.8.4) and the no-arbitrage condition for capital (16.6.12) imply an optimal value for τ^k:

$$\tau^k = \frac{-V_k}{\theta F_k} = \frac{\Phi u_c Z}{\theta F_k} F_{zk}.$$

As discussed earlier, in a second-best solution with distortionary taxation, $\Phi > 0$, so the limiting tax rate on capital is zero only if $F_{zk} = 0$. Moreover, the sign of τ^k depends on the direction of the effect of capital on the marginal product of the untaxed factor Z. If k and Z are complements, the limiting capital tax is positive, and it is negative in the case where the two factors are substitutes.

Other examples of a nonzero limiting capital tax are presented by Stiglitz (1987) and Jones, Manuelli, and Rossi (1997), who assume that two types of labor must be taxed at the same tax rate. Once again, the incompleteness of the tax system makes the optimal capital tax depend on how capital affects the marginal products of the other factors.

16.9. A stochastic economy

We now turn to optimal taxation in a stochastic version of our economy. With the notation of chapter 8, we follow the setups of Zhu (1992) and Chari, Christiano, and Kehoe (1994). The stochastic state s_t at time t determines an exogenous shock both to the production function $F(\cdot, \cdot, s_t)$ and to government purchases $g_t(s_t)$. We use the history of events s^t to define history-contingent commodities: $c_t(s^t)$, $\ell_t(s^t)$, and $n_t(s^t)$ are the household's consumption, leisure, and labor at time t given history s^t, and $k_{t+1}(s^t)$ denotes the capital stock carried over to next period $t+1$. Following our earlier convention, $u_c(s^t)$ and $F_k(s^t)$ and so on denote the values of the indicated objects at time t for history s^t, evaluated at an allocation to be understood from the context.

The household's preferences are ordered by

$$\sum_{t=0}^{\infty} \sum_{s^t} \beta^t \pi_t\left(s^t\right) u\left[c_t\left(s^t\right), \ell_t\left(s^t\right)\right]. \tag{16.9.1}$$

The production function has constant returns to scale in labor and capital. Feasibility requires that

$$
\begin{aligned}
c_t\left(s^t\right) + g_t\left(s_t\right) + k_{t+1}\left(s^t\right) & \\
= F\left[k_t\left(s^{t-1}\right), n_t\left(s^t\right), s_t\right] & + (1-\delta) k_t\left(s^{t-1}\right).
\end{aligned}
\tag{16.9.2}
$$

16.9.1. Government

Given history s^t at time t, the government finances its exogenous purchase $g_t(s_t)$ and any debt obligation by levying flat-rate taxes on earnings from capital at rate $\tau_t^k(s^t)$ and from labor at rate $\tau_t^n(s^t)$, and by issuing state-contingent debt. Let $b_{t+1}(s_{t+1}|s^t)$ be government indebtedness to the private sector at the beginning of period $t+1$ if event s_{t+1} is realized. This state-contingent asset is traded in period t at the price $p_t(s_{t+1}|s^t)$, in terms of time t goods. The government's budget constraint becomes

$$
\begin{aligned}
g_t\left(s_t\right) = & \tau_t^k\left(s^t\right) r_t\left(s^t\right) k_t\left(s^{t-1}\right) + \tau_t^n\left(s^t\right) w_t\left(s^t\right) n_t\left(s^t\right) \\
& + \sum_{s_{t+1}} p_t\left(s_{t+1}|s^t\right) b_{t+1}\left(s_{t+1}|s^t\right) - b_t\left(s_t|s^{t-1}\right),
\end{aligned}
\tag{16.9.3}
$$

where $r_t(s^t)$ and $w_t(s^t)$ are the market-determined rental rate of capital and the wage rate for labor, respectively.

16.9.2. Households

The representative household maximizes expression (16.9.1) subject to the following sequence of budget constraints:

$$c_t\left(s^t\right) + k_{t+1}\left(s^t\right) + \sum_{s_{t+1}} p_t\left(s_{t+1}|s^t\right) b_{t+1}\left(s_{t+1}|s^t\right)$$

$$= \left[1 - \tau_t^k\left(s^t\right)\right] r_t\left(s^t\right) k_t\left(s^{t-1}\right) + \left[1 - \tau_t^n\left(s^t\right)\right] w_t\left(s^t\right) n_t\left(s^t\right)$$
$$+ (1-\delta) k_t\left(s^{t-1}\right) + b_t\left(s_t|s^{t-1}\right) \quad \forall t. \tag{16.9.4}$$

The first-order conditions for this problem imply

$$\frac{u_\ell\left(s^t\right)}{u_c\left(s^t\right)} = \left[1 - \tau_t^n\left(s^t\right)\right] w_t\left(s^t\right), \tag{16.9.5a}$$

$$p_t\left(s_{t+1}|s^t\right) = \beta \frac{\pi_{t+1}\left(s^{t+1}\right)}{\pi_t\left(s^t\right)} \frac{u_c\left(s^{t+1}\right)}{u_c\left(s^t\right)}, \tag{16.9.5b}$$

$$u_c\left(s^t\right) = \beta E_t\left\{u_c\left(s^{t+1}\right)\right.$$
$$\left. \cdot \left[\left(1 - \tau_{t+1}^k\left(s^{t+1}\right)\right) r_{t+1}\left(s^{t+1}\right) + 1 - \delta\right]\right\}, \tag{16.9.5c}$$

where E_t is the mathematical expectation conditional on information available at time t, i.e., history s^t:

$$E_t x_{t+1}\left(s^{t+1}\right) = \sum_{s^{t+1}|s^t} \frac{\pi_{t+1}\left(s^{t+1}\right)}{\pi_t\left(s^t\right)} x_{t+1}\left(s^{t+1}\right)$$

$$= \sum_{s^{t+1}|s^t} \pi_{t+1}\left(s^{t+1}|s^t\right) x_{t+1}\left(s^{t+1}\right),$$

where the summation over $s^{t+1}|s^t$ means that we sum over all possible histories $\tilde{s}^{t+1}$ such that $\tilde{s}^t = s^t$.

Corresponding to the no-arbitrage condition (16.2.12) in the nonstochastic economy, conditions (16.9.5b) and (16.9.5c) imply

$$1 = \sum_{s_{t+1}} p_t\left(s_{t+1}|s^t\right) \left\{\left[1 - \tau_{t+1}^k\left(s^{t+1}\right)\right] r_{t+1}\left(s^{t+1}\right) + 1 - \delta\right\}. \tag{16.9.6}$$

And once again, this no-arbitrage condition can be obtained by consolidating the budget constraints of two consecutive periods. Multiply the time $t+1$ version of equation $(16.9.4)$ by $p_t(s_{t+1}|s^t)$ and sum over all realizations s_{t+1}. The resulting expression can be substituted into equation $(16.9.4)$ by eliminating $\sum_{s_{t+1}} p_t(s_{t+1}|s^t) b_{t+1}(s_{t+1}|s^t)$. To rule out arbitrage transactions in capital and state-contingent assets, the term multiplying $k_{t+1}(s^t)$ must be zero; this approach amounts to imposing condition $(16.9.6)$. Similar no-arbitrage arguments were made in chapters 8 and 13.

As before, by repeated substitution of one-period budget constraints, we can obtain the household's present-value budget constraint:

$$\sum_{t=0}^{\infty} \sum_{s^t} q_t^0\left(s^t\right) c_t\left(s^t\right) = \sum_{t=0}^{\infty} \sum_{s^t} q_t^0\left(s^t\right) \left[1 - \tau_t^n\left(s^t\right)\right] w_t\left(s^t\right) n_t\left(s^t\right)$$
$$+ \left[\left(1 - \tau_0^k\right) r_0 + 1 - \delta\right] k_0 + b_0, \tag{16.9.7}$$

where we denote time 0 variables by the time subscript 0. The price system $q_t^0(s^t)$ conforms to the following formula, versions of which were displayed in chapter 8:

$$q_{t+1}^0\left(s^{t+1}\right) = p_t\left(s_{t+1}|s^t\right) q_t^0\left(s^t\right) = \beta^{t+1} \pi_{t+1}\left(s^{t+1}\right) \frac{u_c\left(s^{t+1}\right)}{u_c\left(s^0\right)}. \tag{16.9.8}$$

Alternatively, equilibrium price $(16.9.8)$ can be computed from the first-order conditions for maximizing expression $(16.9.1)$ subject to equation $(16.9.7)$ (and choosing the numeraire $q_0^0 = 1$). Furthermore, the no-arbitrage condition $(16.9.6)$ can be expressed as

$$q_t^0\left(s^t\right) = \sum_{s^{t+1}|s^t} q_{t+1}^0\left(s^{t+1}\right)$$
$$\cdot \left\{\left[1 - \tau_{t+1}^k\left(s^{t+1}\right)\right] r_{t+1}\left(s^{t+1}\right) + 1 - \delta\right\}. \tag{16.9.9}$$

In deriving the present-value budget constraint $(16.9.7)$, we imposed two transversality conditions that specify that for any infinite history s^∞,

$$\lim_{t \to +\infty} q_t^0\left(s^t\right) k_{t+1}\left(s^t\right) = 0, \tag{16.9.10a}$$

$$\lim_{t \to +\infty} \sum_{s_{t+1}} q_{t+1}^0\left(\{s_{t+1}, s^t\}\right) b_{t+1}\left(s_{t+1}|s^t\right) = 0, \tag{16.9.10b}$$

where the limits are taken over sequences of histories s^t contained in the infinite history s^∞.

16.9.3. Firms

The static maximization problem of the representative firm remains the same. Thus, in a competitive equilibrium, production factors are paid their marginal products:

$$r_t\left(s^t\right) = F_k\left(s^t\right),\qquad\qquad\qquad (16.9.11a)$$

$$w_t\left(s^t\right) = F_n\left(s^t\right).\qquad\qquad\qquad (16.9.11b)$$

16.10. Indeterminacy of state-contingent debt and capital taxes

Consider a feasible government policy $\{g_t(s_t),\ \tau_t^k(s^t),\ \tau_t^n(s^t),\ b_{t+1}(s_{t+1}|s^t);\ \forall s^t,$ $s_{t+1}\}_{t\geq 0}$ with an associated competitive allocation $\{c_t(s^t), n_t(s^t), k_{t+1}(s^t); \forall s^t\}_{t\geq 0}$. Note that the labor tax is uniquely determined by equations $(16.9.5a)$ and $(16.9.11b)$. However, there are infinitely many plans for state-contingent debt and capital taxes that can implement a particular competitive allocation.

Intuition for the indeterminacy of state-contingent debt and capital taxes can be gleaned from the household's first-order condition $(16.9.5c)$, which states that capital tax rates affect the household's intertemporal allocation by changing the current market value of after-tax returns on capital. If a different set of capital taxes induces the same current market value of after-tax returns on capital, then they will also be consistent with the same competitive allocation. It remains only to verify that the change of capital tax receipts in different states can be offset by restructuring the government's issue of state-contingent debt. Zhu (1992) shows how such feasible alternative policies can be constructed.

Let $\{\epsilon_t(s^t); \forall s^t\}_{t\geq 0}$ be a random process such that

$$E_t u_c\left(s^{t+1}\right)\epsilon_{t+1}\left(s^{t+1}\right) r_{t+1}\left(s^{t+1}\right) = 0.\qquad (16.10.1)$$

We can then construct an alternative policy for capital taxes and state-contingent debt, $\{\hat\tau_t^k(s^t), \hat b_{t+1}(s_{t+1}|s^t); \forall s^t, s_{t+1}\}_{t\geq 0}$, as follows:

$$\hat\tau_0^k = \tau_0^k,\qquad\qquad\qquad (16.10.2a)$$

$$\hat{\tau}^k_{t+1}\left(s^{t+1}\right) = \tau^k_{t+1}\left(s^{t+1}\right) + \epsilon_{t+1}\left(s^{t+1}\right), \tag{16.10.2b}$$

$$\hat{b}_{t+1}\left(s_{t+1}|s^t\right) = b_{t+1}\left(s_{t+1}|s^t\right) + \epsilon_{t+1}\left(s^{t+1}\right) r_{t+1}\left(s^{t+1}\right) k_{t+1}\left(s^t\right), \tag{16.10.2c}$$

for $t \geq 0$. Compared to the original fiscal policy, we can verify that this alternative policy does not change the following:

1. The household's intertemporal consumption choice, governed by first-order condition $(16.9.5c)$.

2. The current market value of all government debt issued at time t, when discounted with the equilibrium expression for $p_t(s_{t+1}|s^t)$ in equation $(16.9.5b)$.

3. The government's revenue from capital taxation net of maturing government debt in any state s^{t+1}.

Thus, the alternative policy is feasible and leaves the competitive allocation unchanged.

Since there are infinitely many ways of constructing sequences of random variables $\{\epsilon_t(s^t)\}$ that satisfy equation $(16.10.1)$, it follows that the competitive allocation can be implemented by many different plans for capital taxes and state-contingent debt. It is instructive to consider two special cases where there is no uncertainty one period ahead about one of the two policy instruments. We first take the case of risk-free one-period bonds. In period t, the government issues bonds that promise to pay $\bar{b}_{t+1}(s^t)$ at time $t+1$ with certainty. Let the amount of bonds be such that their present market value is the same as that for the original fiscal plan,

$$\sum_{s_{t+1}} p_t\left(s_{t+1}|s^t\right) \bar{b}_{t+1}\left(s^t\right) = \sum_{s_{t+1}} p_t\left(s_{t+1}|s^t\right) b_{t+1}\left(s_{t+1}|s^t\right).$$

After invoking the equilibrium expression for prices $(16.9.5b)$, we can solve for the constant $\bar{b}_{t+1}(s^t)$

$$\bar{b}_{t+1}\left(s^t\right) = \frac{E_t u_c\left(s^{t+1}\right) b_{t+1}\left(s_{t+1}|s^t\right)}{E_t u_c\left(s^{t+1}\right)}. \tag{16.10.3}$$

The change in capital taxes needed to offset this shift to risk-free bonds is then implied by equation $(16.10.2c)$:

$$\epsilon_{t+1}\left(s^{t+1}\right) = \frac{\bar{b}_{t+1}\left(s^t\right) - b_{t+1}\left(s_{t+1}|s^t\right)}{r_{t+1}\left(s^{t+1}\right) k_{t+1}\left(s^t\right)}. \tag{16.10.4}$$

We can check that equations $(16.10.3)$ and $(16.10.4)$ describe a permissible policy by substituting these expressions into equation $(16.10.1)$ and verifying that the restriction is indeed satisfied.

Next, we examine a policy where the capital tax is not contingent on the realization of the current state but is already set in the previous period. Let $\bar{\tau}_{t+1}(s^t)$ be the capital tax rate in period $t+1$, conditional on information at time t. We choose $\bar{\tau}_{t+1}(s^t)$ so that the household's first-order condition $(16.9.5c)$ is unaffected:

$$E_t \left\{ u_c \left(s^{t+1} \right) \left[\left(1 - \bar{\tau}^k_{t+1} \left(s^t \right) \right) r_{t+1} \left(s^{t+1} \right) + 1 - \delta \right] \right\}$$
$$= E_t \left\{ u_c \left(s^{t+1} \right) \left[\left(1 - \tau^k_{t+1} \left(s^{t+1} \right) \right) r_{t+1} \left(s^{t+1} \right) + 1 - \delta \right] \right\},$$

which gives

$$\bar{\tau}^k_{t+1} \left(s^t \right) = \frac{E_t u_c \left(s^{t+1} \right) \tau^k_{t+1} \left(s^{t+1} \right) r_{t+1} \left(s^{t+1} \right)}{E_t u_c \left(s^{t+1} \right) r_{t+1} \left(s^{t+1} \right)}. \tag{16.10.5}$$

Thus, the alternative policy in equations $(16.10.2)$ with capital taxes known one period in advance is accomplished by setting

$$\epsilon_{t+1} \left(s^{t+1} \right) = \bar{\tau}^k_{t+1} \left(s^t \right) - \tau^k_{t+1} \left(s^{t+1} \right).$$

16.11. The Ramsey plan under uncertainty

We now ask what competitive allocation should be chosen by a benevolent government; that is, we solve the Ramsey problem for the stochastic economy. The computational strategy is in principle the same given in our recipe for a nonstochastic economy.

Step 1, in which we use private first-order conditions to solve for prices and taxes in terms of the allocation, has already been accomplished with equations $(16.9.5a)$, $(16.9.8)$, $(16.9.9)$, and $(16.9.11)$. In step 2, we use these expressions to eliminate prices and taxes from the household's present-value budget constraint $(16.9.7)$, which leaves us with

$$\sum_{t=0}^{\infty} \sum_{s^t} \beta^t \pi_t \left(s^t \right) \left[u_c \left(s^t \right) c_t \left(s^t \right) - u_\ell \left(s^t \right) n_t \left(s^t \right) \right] - A = 0, \tag{16.11.1}$$

where A is still given by equation $(16.6.6)$. Proceeding to step 3, we define

$$V\left[c_t\left(s^t\right), n_t\left(s^t\right), \Phi\right] = u\left[c_t\left(s^t\right), 1 - n_t\left(s^t\right)\right]$$
$$+ \Phi\left[u_c\left(s^t\right) c_t\left(s^t\right) - u_\ell\left(s^t\right) n_t\left(s^t\right)\right], \qquad (16.11.2)$$

where Φ is a Lagrange multiplier on equation $(16.11.1)$. Then form the Lagrangian

$$J = \sum_{t=0}^{\infty} \sum_{s^t} \beta^t \pi_t(s^t) \Big\{ V[c_t(s^t), n_t(s^t), \Phi]$$
$$+ \theta_t(s^t) \Big[F\left(k_t(s^{t-1}), n_t(s^t), s_t\right) + (1-\delta)k_t(s^{t-1})$$
$$- c_t(s^t) - g_t(s_t) - k_{t+1}(s^t) \Big] \Big\} - \Phi A, \qquad (16.11.3)$$

where $\{\theta_t(s^t); \forall s^t\}_{t \geq 0}$ is a sequence of Lagrange multipliers. For given k_0 and b_0, we fix τ_0^k and maximize J with respect to $\{c_t(s^t), n_t(s^t), k_{t+1}(s^t); \forall s^t\}_{t \geq 0}$.

The first-order conditions for the Ramsey problem are

$$c_t\left(s^t\right): \; V_c\left(s^t\right) = \theta_t\left(s^t\right), \qquad\qquad\qquad t \geq 1;$$
$$n_t\left(s^t\right): \; V_n\left(s^t\right) = -\theta_t\left(s^t\right) F_n\left(s^t\right), \qquad\qquad t \geq 1;$$
$$k_{t+1}\left(s^t\right): \; \theta_t\left(s^t\right) = \beta \sum_{s^{t+1}|s^t} \frac{\pi_{t+1}\left(s^{t+1}\right)}{\pi_t\left(s^t\right)} \theta_{t+1}\left(s^{t+1}\right)$$
$$\cdot \left[F_k\left(s^{t+1}\right) + 1 - \delta\right], \quad t \geq 0;$$

where we have left out the conditions for c_0 and n_0, which are different because they include terms related to the initial stocks of capital and bonds. The first-order conditions for the problem imply, for $t \geq 1$,

$$V_c\left(s^t\right) = \beta E_t V_c\left(s^{t+1}\right) \left[F_k\left(s^{t+1}\right) + 1 - \delta\right], \qquad (16.11.4a)$$
$$V_n\left(s^t\right) = -V_c\left(s^t\right) F_n\left(s^t\right). \qquad\qquad\qquad (16.11.4b)$$

These expressions reveal an interesting property of the Ramsey allocation. If the stochastic process s is Markov, equations $(16.11.4)$ suggest that the allocations from period 1 onward can be described by time-invariant allocation rules $c(s, k)$, $n(s, k)$, and $k'(s, k)$.[6]

---

[6] To emphasize that the second-best allocation depends critically on the extent to which the government has to resort to distortionary taxation, we might want to include the constant Φ as an explicit argument in $c(s, k)$, $n(s, k)$, and $k'(s, k)$.

16.12. Ex ante capital tax varies around zero

In a nonstochastic economy, we proved that if the equilibrium converges to a steady state, then the optimal limiting capital tax is zero. The counterpart to a steady state in a stochastic economy is a stationary equilibrium. Therefore, we now assume that the process on s follows a Markov process with transition probabilities $\pi(s'|s) \equiv \text{Prob}(s_{t+1} = s'|s_t = s)$. As noted in the previous section, this assumption implies that the allocation rules are time-invariant functions of (s, k). If the economy converges to a stationary equilibrium, the stochastic process $\{s_t, k_t\}$ is a stationary, ergodic Markov process on the compact set $\mathbf{S} \times [0, \bar{k}]$ where $\mathbf{S}$ is a finite set of possible realizations for s_t and $\bar{k}$ is an upper bound on the capital stock.[7]

Because of the indeterminacy of state-contingent government debt and capital taxes, it is not possible uniquely to characterize a stationary distribution of realized capital tax rates, but we can study the *ex ante capital tax rate* defined as

$$\bar{\tau}^k_{t+1}\left(s^t\right) = \frac{\sum_{s_{t+1}} p_t\left(s_{t+1}|s^t\right)\tau^k_{t+1}\left(s^{t+1}\right)r_{t+1}\left(s^{t+1}\right)}{\sum_{s_{t+1}} p_t\left(s_{t+1}|s^t\right)r_{t+1}\left(s^{t+1}\right)}. \tag{16.12.1}$$

That is, the *ex ante* capital tax rate is the ratio of current market value of taxes on capital income to the present market value of capital income. After invoking the equilibrium price of equation $(16.9.5b)$, we see that this expression is identical to equation $(16.10.5)$. Recall that equation $(16.10.5)$ resolved the indeterminacy of the Ramsey plan by pinning down a unique fixed capital tax rate for period $t + 1$ conditional on information at time t. It follows that the alternative interpretation of $\bar{\tau}^k_{t+1}(s^t)$ in equation $(16.12.1)$ as the *ex ante* capital tax rate offers a unique measure across the multiplicity of capital tax schedules under the Ramsey plan. Moreover, it is quite intuitive that one way for the government to tax away, in present value terms, a fraction $\bar{\tau}^k_{t+1}(s^t)$ of next period's capital income is to set a constant tax rate exactly equal to that number.

Let $P^\infty(\cdot)$ be the probability measure over the outcomes in such a stationary equilibrium. We now state the proposition of Zhu (1992) that the *ex ante* capital tax rate in a stationary equilibrium either equals zero or varies around zero.

---

[7] An upper bound on the capital stock can be constructed as follows:

$$\bar{k} = \max\{\bar{k}\left(s\right) : F\left[\bar{k}\left(s\right), 1, s\right] = \delta\bar{k}\left(s\right); s \in \mathbf{S}\}.$$

PROPOSITION 2: If there exists a stationary Ramsey allocation, the *ex ante* capital tax rate is such that

(a) either $P^\infty(\bar{\tau}_t^k = 0) = 1$, or $P^\infty(\bar{\tau}_t^k > 0) > 0$ and $P^\infty(\bar{\tau}_t^k < 0) > 0$;

(b) $P^\infty(\bar{\tau}_t^k = 0) = 1$ if and only if $P^\infty[V_c(c_t, n_t, \Phi)/u_c(c_t, \ell_t) = \Lambda] = 1$ for some constant Λ.

A sketch of the proof is provided in the next subsection. Let us just add here that the two possibilities with respect to the *ex ante* capital tax rate are not vacuous. One class of utilities that imply $P^\infty(\bar{\tau}_t^k = 0) = 1$ is

$$u(c_t, \ell_t) = \frac{c_t^{1-\sigma}}{1-\sigma} + v(\ell_t),$$

for which the ratio $V_c(c_t, n_t, \Phi)/u_c(c_t, \ell_t)$ is equal to $[1 + \Phi(1-\sigma)]$, which plays the role of the constant Λ required by Proposition 2. Chari, Christiano, and Kehoe (1994) solve numerically for Ramsey plans when the preferences do not satisfy this condition. In their simulations, the *ex ante* tax on capital income remains approximately equal to zero.

To revisit Chamley (1986) and Judd's (1985b) result on the optimality of a zero capital tax in a nonstochastic economy, it is trivially true that the ratio $V_c(c_t, n_t, \Phi)/u_c(c_t, \ell_t)$ is constant in a nonstochastic steady state. In a stationary equilibrium of a stochastic economy, Proposition 2 extends this result: for some utility functions, the Ramsey plan prescribes a zero *ex ante* capital tax rate that can be implemented by setting a zero tax on capital income. But except for such special classes of preferences, Proposition 2 states that the *ex ante* capital tax rate should fluctuate around zero, in the sense that $P^\infty(\bar{\tau}_t^k > 0) > 0$ and $P^\infty(\bar{\tau}_t^k < 0) > 0$.

16.12.1. Sketch of the proof of Proposition 2

Note from equation (16.12.1) that $\bar{\tau}_{t+1}^k(s^t) \geq (\leq) 0$ if and only if

$$\sum_{s_{t+1}} p_t\left(s_{t+1}|s^t\right) \tau_{t+1}^k\left(s^{t+1}\right) r_{t+1}\left(s^{t+1}\right) \geq (\leq) 0,$$

which, together with equation (16.9.6), implies

$$1 \leq (\geq) \sum_{s_{t+1}} p_t\left(s_{t+1}|s^t\right) \left[r_{t+1}\left(s^{t+1}\right) + 1 - \delta\right].$$

Substituting equations (16.9.5b) and (16.9.11a) into this expression yields

$$u_c\left(s^t\right) \leq (\geq) \beta E_t u_c\left(s^{t+1}\right) \left[F_k\left(s^{t+1}\right) + 1 - \delta\right] \qquad (16.12.2)$$

if and only if $\bar{\tau}_{t+1}^k(s^t) \geq (\leq) 0$.

Define

$$H\left(s^t\right) \equiv \frac{V_c\left(s^t\right)}{u_c\left(s^t\right)}. \qquad (16.12.3)$$

Using equation (16.11.4a), we have

$$u_c\left(s^t\right) H\left(s^t\right) = \beta E_t u_c\left(s^{t+1}\right) H\left(s^{t+1}\right) \left[F_k\left(s^{t+1}\right) + 1 - \delta\right]. \qquad (16.12.4)$$

By formulas (16.12.2) and (16.12.4), $\bar{\tau}_{t+1}^k(s^t) \geq (\leq) 0$ if and only if

$$H\left(s^t\right) \geq (\leq) \frac{E_t\omega\left(s^{t+1}\right) H\left(s^{t+1}\right)}{E_t\omega\left(s^{t+1}\right)}, \qquad (16.12.5)$$

where $\omega(s^{t+1}) \equiv u_c(s^{t+1})[F_k(s^{t+1}) + 1 - \delta]$.

Since a stationary Ramsey equilibrium has time-invariant allocation rules $c(s, k)$, $n(s, k)$, and $k'(s, k)$, it follows that $\bar{\tau}_{t+1}^k(s^t)$, $H(s^t)$, and $\omega(s^t)$ can also be expressed as functions of (s, k). The stationary version of expression (16.12.5) with transition probabilities $\pi(s'|s)$ becomes

$$\bar{\tau}^k(s, k) \geq (\leq) 0 \quad \text{if and only if}$$

$$H(s, k) \geq (\leq) \frac{\sum_{s'} \pi(s'|s)\,\omega[s', k'(s, k)]\,H[s', k'(s, k)]}{\sum_{s'} \pi(s'|s)\,\omega[s', k'(s, k)]} \qquad (16.12.6)$$

$$\equiv \Gamma H(s, k).$$

Note that the operator Γ is a weighted average of $H[s', k'(s, k)]$ and that it has the property that $\Gamma H^* = H^*$ for any constant H^*.

Under some regularity conditions, $H(s, k)$ attains a minimum H^- and a maximum H^+ in the stationary equilibrium. That is, there exist equilibrium states (s^-, k^-) and (s^+, k^+) such that

$$P^\infty \left[H\left(s, k\right) \geq H^- \right] = 1, \qquad (16.12.7a)$$

$$P^\infty \left[H\left(s, k\right) \leq H^+ \right] = 1, \qquad (16.12.7b)$$

where $H^- = H(s^-, k^-)$ and $H^+ = H(s^+, k^+)$. We will now show that if

$$P^\infty \left[H\left(s, k\right) \geq \Gamma H\left(s, k\right) \right] = 1, \qquad (16.12.8a)$$

or

$$P^\infty \left[H\left(s, k\right) \leq \Gamma H\left(s, k\right) \right] = 1, \qquad (16.12.8b)$$

then there must exist a constant H^* such that

$$P^\infty \left[H\left(s, k\right) = H^* \right] = 1. \qquad (16.12.8c)$$

First, take equation $(16.12.8a)$ and consider the state $(s, k) = (s^-, k^-)$ that is associated with a set of possible states in the next period, $\{s', k'(s, k); \forall s' \in \mathbf{S}\}$. By equation $(16.12.7a)$, $H(s', k') \geq H^-$, and since $H(s, k) = H^-$, condition $(16.12.8a)$ implies that $H(s', k') = H^-$. We can repeat the same argument for each (s', k'), and thereafter for the equilibrium states that they map into, and so on. Thus, using the ergodicity of $\{s_t, k_t\}$, we obtain equation $(16.12.8c)$ with $H^* = H^-$. A similar reasoning can be applied to equation $(16.12.8b)$, but we now use $(s, k) = (s^+, k^+)$ and equation $(16.12.7b)$ to show that equation $(16.12.8c)$ is implied.

By the correspondence in expression $(16.12.6)$ we have established part (a) of Proposition 2. Part (b) follows after recalling definition $(16.12.3)$; the constant H^* in equation $(16.12.8c)$ is the sought-after Λ.

16.13. Examples of labor tax smoothing

To gain some insight into optimal tax policies, we consider several examples of government expenditures to be financed in a model without physical capital. The technology is now described by

$$c_t\left(s^t\right) + g_t\left(s_t\right) = n_t\left(s^t\right). \tag{16.13.1}$$

Since one unit of labor yields one unit of output, the competitive equilibrium wage is $w_t(s^t) = 1$. The model is otherwise identical to the previous framework. This very model is analyzed by Lucas and Stokey (1983), who also study the time consistency of the optimal fiscal policy by allowing the government to choose taxes sequentially rather than once-and-for-all at time 0.[8]

The household's present-value budget constraint is given by equation (16.9.7) except that we delete the part involving physical capital. Prices and taxes are expressed in terms of the allocation by conditions (16.9.5a) and (16.9.8). After using these expressions to eliminate prices and taxes, the implementability condition, equation (16.11.1), becomes

$$\sum_{t=0}^{\infty} \sum_{s^t} \beta^t \pi_t\left(s^t\right) \left[u_c\left(s^t\right) c_t\left(s^t\right) - u_\ell\left(s^t\right) n_t\left(s^t\right)\right] - u_c\left(s^0\right) b_0 = 0. \tag{16.13.2}$$

We then form the Lagrangian in the same way as before. After writing out the derivatives $V_c(s^t)$ and $V_n(s^t)$, the first-order conditions of this Ramsey problem are

---

[8] The optimal tax policy is in general time inconsistent, as studied in chapter 26 and as indicated by the preceding discussion about taxation of initial capital. However, Lucas and Stokey (1983) show that the optimal tax policy in the model without physical capital can be made time consistent if the government can issue debt at all maturities (and so is not restricted to issue only one-period debt as in our formulation). There exists a period-by-period strategy for structuring a term structure of history-contingent claims that preserves the initial Ramsey allocation $\{c_t(s^t), n_t(s^t); \forall s^t\}_{t \geq 0}$ as the Ramsey allocation for the continuation economy. By induction, the argument extends to subsequent periods. Alvarez, Kehoe and Neumeyer (2004) apply the argument to the maturity structure of both real and *nominal* bonds in a monetary economy. For a class of economies where the Friedman rule of setting nominal interest rates to zero is optimal under commitment, they show that optimal monetary and fiscal policies are time consistent. When the Friedman rule is optimal, households are satiated with money balances – as if money has disappeared – so that the economy is equivalent to a real economy with one consumption good and labor.

$$c_t\left(s^t\right): \quad (1+\Phi)\, u_c\left(s^t\right) + \Phi\left[u_{cc}\left(s^t\right) c_t\left(s^t\right) - u_{\ell c}\left(s^t\right) n_t\left(s^t\right)\right]$$
$$- \theta_t\left(s^t\right) = 0, \qquad t \geq 1; \qquad (16.13.3a)$$

$$n_t\left(s^t\right): \quad -(1+\Phi)\, u_\ell\left(s^t\right) - \Phi\left[u_{c\ell}\left(s^t\right) c_t\left(s^t\right) - u_{\ell\ell}\left(s^t\right) n_t\left(s^t\right)\right]$$
$$+ \theta_t\left(s^t\right) = 0, \qquad t \geq 1; \qquad (16.13.3b)$$

$$c_0\left(s^0\right): \quad (1+\Phi)\, u_c\left(s^0\right) + \Phi\left[u_{cc}\left(s^0\right) c_0\left(s^0\right) - u_{\ell c}\left(s^0\right) n_0\left(s^0\right)\right]$$
$$- \theta_0\left(s^0\right) - \Phi u_{cc}\left(s^0\right) b_0 = 0; \qquad (16.13.3c)$$

$$n_0\left(s^0\right): \quad -(1+\Phi)\, u_\ell\left(s^0\right) - \Phi\left[u_{c\ell}\left(s^0\right) c_0\left(s^0\right) - u_{\ell\ell}\left(s^0\right) n_0\left(s^0\right)\right]$$
$$+ \theta_0\left(s^0\right) + \Phi u_{c\ell}\left(s^0\right) b_0 = 0. \qquad (16.13.3d)$$

Here we retain our assumption that the government does not set taxes sequentially but commits to a policy at time 0.

To uncover a key property of the optimal allocation for $t \geq 1$, it is instructive to merge first-order conditions $(16.13.3a)$ and $(16.13.3b)$ by substituting out for the multiplier $\theta_t(s^t)$:

$$(1+\Phi)u_c(c, 1-c-g) + \Phi\big[cu_{cc}(c, 1-c-g)$$
$$- (c+g)u_{\ell c}(c, 1-c-g)\big]$$
$$= (1+\Phi)u_\ell(c, 1-c-g) + \Phi\big[cu_{c\ell}(c, 1-c-g)$$
$$- (c+g)u_{\ell\ell}(c, 1-c-g)\big], \qquad (16.13.4)$$

where we have invoked the resource constraints $(16.13.1)$ and $\ell_t(s^t) + n_t(s^t) = 1$. We have also suppressed the time subscript and the index s^t for the quantities of consumption, leisure, and government purchases in order to highlight a key property of the optimal allocation. In particular, if the quantities of government purchases are the same after two histories s^t and $\tilde{s}^j$ for $t, j \geq 0$, i.e., $g_t(s_t) = g_j(\tilde{s}_j) = g$, then it follows from equation $(16.13.4)$ that the optimal choices of consumption and leisure, $(c_t(s^t), \ell_t(s^t))$ and $(c_j(\tilde{s}^j), \ell_j(\tilde{s}^j))$, must be identical. Hence, the optimal allocation is a function only of the current realized quantity of government purchases g and does *not* depend on the specific history leading up to that outcome. This history independence is reminiscent of and connected to the analogous history independence of the competitive equilibrium allocation with complete markets in chapter 8.

The following preliminary calculations will be useful in shedding further light on optimal tax policies for some examples of government expenditure streams. First, substitute equations $(16.9.5a)$ and $(16.13.1)$ into equation $(16.13.2)$ to get

$$\sum_{t=0}^{\infty} \sum_{s^t} \beta^t \pi_t \left(s^t\right) u_c \left(s^t\right) \left[\tau_t^n \left(s^t\right) n_t \left(s^t\right) - g_t \left(s_t\right)\right] - u_c \left(s^0\right) b_0 = 0. \quad (16.13.5)$$

Then multiplying equation $(16.13.3a)$ by $c_t(s^t)$ and equation $(16.13.3b)$ by $n_t(s^t)$ and summing, we find

$$(1 + \Phi) \left[c_t \left(s^t\right) u_c \left(s^t\right) - n_t \left(s^t\right) u_\ell \left(s^t\right)\right]$$
$$+ \Phi \left[c_t \left(s^t\right)^2 u_{cc} \left(s^t\right) - 2n_t \left(s^t\right) c_t \left(s^t\right) u_{\ell c} \left(s^t\right) + n_t \left(s^t\right)^2 u_{\ell\ell} \left(s^t\right)\right]$$
$$- \theta_t \left(s^t\right) \left[c_t \left(s^t\right) - n_t \left(s^t\right)\right] = 0, \qquad t \geq 1. \quad (16.13.6a)$$

Similarly, multiplying equation $(16.13.3c)$ by $[c_0(s^0) - b_0]$ and equations $(16.13.3d)$ by $n_0(s^0)$ and summing, we obtain

$$(1 + \Phi) \left\{\left[c_0 \left(s^0\right) - b_0\right] u_c \left(s^0\right) - n_0 \left(s^0\right) u_\ell \left(s^0\right)\right\}$$
$$+ \Phi \left\{\left[c_0 \left(s^0\right) - b_0\right]^2 u_{cc} \left(s^0\right) - 2n_0 \left(s^0\right) \left[c_0 \left(s^0\right) - b_0\right] u_{\ell c} \left(s^0\right)\right.$$
$$\left. + n_0 \left(s^0\right)^2 u_{\ell\ell} \left(s^0\right)\right\} - \theta_0 \left(s^0\right) \left[c_0 \left(s^0\right) - b_0 - n_0 \left(s^0\right)\right] = 0. \quad (16.13.6b)$$

Note that since the utility function is strictly concave, the quadratic forms multiplying Φ appearing in equations $(16.13.6a)$ and $(16.13.6b)$ are both negative.[9]

---

[9] To see that the quadratic term in equation $(16.13.6a)$ is negative, complete the square by adding and subtracting the quantity $n^2 u_{\ell c}^2 / u_{cc}$ (where we have suppressed the time subscript and the argument s^t):

$$c^2 u_{cc} - 2nc u_{\ell c} + n^2 u_{\ell\ell} + n^2 \frac{u_{\ell c}^2}{u_{cc}} - n^2 \frac{u_{\ell c}^2}{u_{cc}}$$
$$= u_{cc} \left(c^2 - 2nc \frac{u_{\ell c}}{u_{cc}} + n^2 \frac{u_{\ell c}^2}{u_{cc}^2}\right) + \left(u_{\ell\ell} - \frac{u_{\ell c}^2}{u_{cc}}\right) n^2$$
$$= u_{cc} \left(c - \frac{u_{\ell c}}{u_{cc}} n\right)^2 + \frac{u_{cc} u_{\ell\ell} - u_{\ell c}^2}{u_{cc}} n^2.$$

Since the conditions for a strictly concave u are $u_{cc} < 0$ and $u_{cc} u_{\ell\ell} - u_{\ell c}^2 > 0$, it follows immediately that the quadratic term in equation $(16.13.6a)$ is negative. The same argument applies to the quadratic term in equation $(16.13.6b)$.

Finally, multiplying equation $(16.13.6a)$ by $\beta^t \pi_t(s^t)$, summing over t and s^t, and adding equation $(16.13.6b)$, we find that

$$(1 + \Phi) \left(\sum_{t=0}^{\infty} \sum_{s^t} \beta^t \pi_t\left(s^t\right) \left[c_t\left(s^t\right) u_c\left(s^t\right) - n_t\left(s^t\right) u_\ell\left(s^t\right)\right] - u_c\left(s^0\right) b_0 \right)$$

$$+ \Phi Q - \sum_{t=0}^{\infty} \sum_{s^t} \beta^t \pi_t\left(s^t\right) \theta_t\left(s^t\right) \left[c_t\left(s^t\right) - n_t\left(s^t\right)\right] + \theta_0\left(s^0\right) b_0 = 0,$$

where Q is the sum of negative (quadratic) terms. Using equations $(16.13.2)$ and $(16.13.1)$, we arrive at

$$\Phi Q + \sum_{t=0}^{\infty} \sum_{s^t} \beta^t \pi_t\left(s^t\right) \theta_t\left(s^t\right) g_t\left(s_t\right) + \theta_0\left(s^0\right) b_0 = 0. \qquad (16.13.7)$$

Expression $(16.13.7)$ furthers our understanding of the Lagrange multiplier Φ on the household's present value budget constraint and how it relates to the shadow values associated with the economy's resource constraints $\{\theta_t(s^t); \forall s^t\}_{t \geq 0}$. Let us first examine under what circumstances the Lagrange multiplier Φ is equal to zero. Setting $\Phi = 0$ in equations $(16.13.3)$ and $(16.13.7)$ yields

$$u_c\left(s^t\right) = u_\ell\left(s^t\right) = \theta_t\left(s^t\right), \qquad t \geq 0; \qquad (16.13.8)$$

and, thus,

$$\sum_{t=0}^{\infty} \sum_{s^t} \beta^t \pi_t\left(s^t\right) u_c\left(s^t\right) g_t\left(s_t\right) + u_c\left(s^0\right) b_0 = 0.$$

Dividing this expression by $u_c(s^0)$ and using equation $(16.9.8)$, we find that

$$\sum_{t=0}^{\infty} \sum_{s^t} q_t^0\left(s^t\right) g_t\left(s_t\right) = -b_0.$$

In other words, when the government's initial claims $-b_0$ against the private sector equal the present value of all future government expenditures, the Lagrange multiplier Φ is zero; that is, the household's present-value budget does not exert any additional constraining effect on welfare maximization beyond what is already present in the economy's technology. The reason is that the government does not have to resort to any distortionary taxation, as can be seen from conditions $(16.9.5a)$ and $(16.13.8)$, which imply $\tau_t^n(s^t) = 0$. If the government's initial claims against the private sector were to exceed the present value of future government expenditures, a trivial implication would be that the government would like to return this excess financial wealth as lump-sum transfers to the households, and our argument here with $\Phi = 0$ would remain applicable. In the opposite case, when the present value of all government expenditures exceeds the value of any initial claims against the private sector, the Lagrange multiplier $\Phi > 0$. For example, suppose $b_0 = 0$ and that there is some $g_t(s_t) > 0$. After recalling that $Q < 0$ and $\theta_t(s^t) > 0$, it follows from equation (16.13.7) that $\Phi > 0$.

Following Lucas and Stokey (1983), we now exhibit some examples of government expenditure streams and how they affect optimal tax policies. Throughout we assume that $b_0 = 0$.

16.13.1. Example 1: $g_t = g$ for all $t \geq 0$

Given a constant amount of government purchases $g_t = g$, the first-order condition (16.13.4) is the same in every period, and we conclude that the optimal allocation is constant over time: $(c_t, n_t) = (\hat{c}, \hat{n})$ for $t \geq 0$. It then follows from condition (16.9.5a) that the tax rate required to implement the optimal allocation is also constant over time: $\tau_t^n = \hat{\tau}^n$, for $t \geq 0$. Consequently, equation (16.13.5) implies that the government budget is balanced in each period.

Government debt issues in this economy serve to smooth distortions over time. Because government expenditures are already smooth in this economy, they are optimally financed from contemporaneous taxes. Nothing is gained from using debt to change the timing of tax collection.

16.13.2. Example 2: $g_t = 0$ for $t \neq T$ and nonstochastic $g_T > 0$

Setting $g = 0$ in expression (16.13.4), the optimal allocation $(c_t, n_t) = (\hat{c}, \hat{n})$ is constant for $t \neq T$, and consequently, from condition (16.9.5a), the tax rate is also constant over these periods, $\tau_t^n = \hat{\tau}^n$ for $t \neq T$. Using equations (16.13.6), we can study tax revenues. Recall that $c_t - n_t = 0$ for $t \neq T$ and that $b_0 = 0$. Thus, the last term in equations (16.13.6) drops out. Since $\Phi > 0$, the second (quadratic) term is negative, so the first term must be positive. Since $(1 + \Phi) > 0$, this fact implies

$$0 < \hat{c} - \frac{u_\ell}{u_c}\hat{n} = \hat{c} - (1 - \hat{\tau}^n)\,\hat{n} = \hat{\tau}^n\hat{n},$$

where the first equality invokes condition (16.9.5a). We conclude that tax revenue is positive for $t \neq T$. For period T, the last term in equation (16.13.6), $\theta_T g_T$, is positive. Therefore, the sign of the first term is indeterminate: labor may be either taxed or subsidized in period T.

This example is a stark illustration of tax smoothing where debt is used to redistribute tax distortions over time. With the same tax revenues in all periods before and after time T, the optimal debt policy is as follows: in each period $t = 0, 1, \ldots, T - 1$, the government runs a surplus, using it to buy bonds issued by the private sector. In period T, the expenditure g_T is met by selling all of these bonds, possibly levying a tax on current labor income, and issuing new bonds that are thereafter rolled over forever. Interest payments on that constant outstanding government debt are equal to the constant tax revenue for $t \neq T$, $\hat{\tau}^n\hat{n}$. Thus, the tax distortion is the same in all periods surrounding period T, regardless of the proximity to the date T.

16.13.3. Example 3: $g_t = 0$ for $t \neq T$, and g_T is stochastic

We assume that $g_T = g > 0$ with probability $\alpha \in (0, 1)$ and $g_T = 0$ with probability $1 - \alpha$. As in the previous example, there is an optimal constant allocation $(c_t, n_t) = (\hat{c}, \hat{n})$ for all periods $t \neq T$ (although the optimum values of $\hat{c}$ and $\hat{n}$ will not, in general, be the same as in example 2). In addition, equation (16.13.4) implies that $(c_T, n_T) = (\hat{c}, \hat{n})$ if $g_T = 0$. The argument in example 2 shows that tax revenue is positive in all these states. Consequently, debt issues are as follows.

In each period $t = 0, 1, \ldots, T - 2$, the government runs a surplus, using it to buy risk-free one-period bonds issued by the private sector. A significant difference from example 2 occurs in period $T - 1$, when the government now sells all these bonds and uses the proceeds plus current labor tax revenue to buy one-period contingent bonds that pay off in the next period only if $g_T = g$ and otherwise have no value. In addition, the government *buys* more of these contingent claims in period $T - 1$ by going short in one-period noncontingent claims. And as in example 2, the noncontingent government debt will be rolled over forever with interest payments equal to $\hat{\tau}^n \hat{n}$, but here it is issued one period earlier. If $g_T = 0$ in the next period, the government clearly satisfies its intertemporal budget constraint. In the case $g_T = g$, the construction of our Ramsey equilibrium ensures that the payoff on the government's holdings of contingent claims against the private sector is equal to g plus interest payments of $\hat{\tau}^n \hat{n}$ on government debt net of any current labor tax/subsidy in period T. In periods $T + 1, T + 2, \ldots$, the situation is as in example 2, regardless of whether $g_T = 0$ or $g_T = g$.

This is another example of tax smoothing over time where the tax distortion is the same in all periods around time T. It also demonstrates the risk-spreading aspects of fiscal policy under uncertainty. In effect, the government in period $T - 1$ buys insurance from the private sector against the event that $g_T = g$.

16.14. Lessons for optimal debt policy

Lucas and Stokey (1983) draw three lessons from their analysis of the model in our previous section. The first is built into the model at the outset: budget balance in a present-value sense must be respected. In a stationary economy, fiscal policies that have occasional deficits necessarily have offsetting surpluses at other dates. Thus, in the examples with erratic government expenditures, good times are associated with budget surpluses. Second, in the face of erratic government spending, the role of government debt is to smooth tax distortions over time, and the government should not seek to balance its budget on a continual basis. Third, the contingent-claim character of government debt is important for an optimal policy. [10]

To highlight the role of an optimal state-contingent government debt policy further, we study the government's budget constraint at time t after history s^t:

$$
\begin{aligned}
b_t\left(s_t|s^{t-1}\right) &= \tau_t^n\left(s^t\right)n_t\left(s^t\right) - g_t\left(s_t\right) \\
&\quad + \sum_{j=1}^{\infty}\sum_{s^{t+j}|s^t} q_{t+j}^t\left(s^{t+j}\right)\left[\tau_{t+j}^n\left(s^{t+j}\right)n_{t+j}\left(s^{t+j}\right) - g_{t+j}\left(s_{t+j}\right)\right] \\
&= \sum_{j=0}^{\infty}\sum_{s^{t+j}|s^t} \beta^j \pi_{t+j}\left(s^{t+j}|s^t\right)\frac{u_c\left(s^{t+j}\right)}{u_c\left(s^t\right)}\left\{\left[1 - \frac{u_\ell\left(s^{t+j}\right)}{u_c\left(s^{t+j}\right)}\right]\right. \\
&\quad \left. \cdot\left[c_{t+j}\left(s^{t+j}\right) + g_{t+j}\left(s_{t+j}\right)\right] - g_{t+j}\left(s_{t+j}\right)\right\},
\end{aligned}
\tag{16.14.1}
$$

where we have invoked the resource constraint $(16.13.1)$ and conditions $(16.9.5a)$ and $(16.9.8)$ that express taxes and prices in terms of the allocation. Recall from our discussion of first-order condition $(16.13.4)$ that the optimal allocation $\{c_{t+j}(s^{t+j}), \ell_{t+j}(s^{t+j})\}$ is history independent and depends only on the present realization of government purchases in any given period. We now ask, on what aspects of history does the optimal amount of state-contingent debt

---

[10] Aiyagari, Marcet, Sargent, and Seppälä (2002) offer a qualification to the importance of state-contingent government debt in the model of Lucas and Stokey (1983). In numerical simulations, they explore Ramsey outcomes under the assumption that contingent claims cannot be traded. (Their setup is presented and analyzed in our next section.) They find that the incomplete markets Ramsey allocation is very close to the complete markets Ramsey allocation. This proximity comes from the Ramsey policy's use of self-insurance through risk-free borrowing and lending with households. Compare this outcome to our chapter 18 on heterogeneous agents and how self-insurance can soften the effects of market incompleteness.

that matures in period t after history s^t depend? Investigating the right side of expression (16.14.1), we see that history dependence would only arise because of the transition probabilities $\{\pi_{t+j}(s^{t+j}|s^t)\}$ that govern government purchases. Hence, if government purchases are governed by a Markov process, we conclude that there can be no history dependence: the beginning-of-period state-contingent government debt is a function only of the current state s_t, since everything on the right side of (16.13.1) depends solely on s_t. This is a remarkable feature of the optimal tax-debt policy. By purposefully trading in state-contingent debt markets, the government insolates its net indebtedness to the private sector from any lingering effects of past shocks to government purchases. Its beginning-of-period indebtedness is completely tailored to its present circumstances as captured by the realization of the current state s_t. In contrast, our stochastic example 3 above is a nonstationary environment where the debt policy associated with the optimal allocation depends on both calender time and past events.[11]

Finally, we take a look at the value of contingent government debt in our earlier model with physical capital. Here we cannot expect any sharp result concerning beginning-of-period debt because of our finding above on the indeterminacy of state-contingent debt and capital taxes. However, the derivations of that specific finding suggest that we instead should look at the value of outstanding debt at the end of a period. By multiplying equation (16.9.4) by $p_{t-1}(s_t|s^{t-1})$ and summing over s_t, we express the household's budget constraint for period t in terms of time $t-1$ values,

$$k_t\left(s^{t-1}\right) + \sum_{s_t} p_{t-1}\left(s_t|s^{t-1}\right) b_t\left(s_t|s^{t-1}\right)$$

$$= \sum_{s_t} p_{t-1}\left(s_t|s^{t-1}\right) \left\{ c_t\left(s^t\right) - \left[1 - \tau_t^n\left(s^t\right)\right] w_t\left(s^t\right) n_t\left(s^t\right) \right.$$

$$\left. + k_{t+1}\left(s^t\right) + \sum_{s_{t+1}} p_t\left(s_{t+1}|s^t\right) b_{t+1}\left(s_{t+1}|s^t\right) \right\}, \qquad (16.14.2)$$

where the unit coefficient on $k_t(s^{t-1})$ is obtained by invoking conditions (16.9.5b) and (16.9.5c). Expression (16.14.2) states that the household's ownership of capital and contingent debt at the end of period $t-1$ is equal to the present

---

[11] An alternative way to express this statement about example 3 is that time t must be added to the state s_t to acquire an augmented state that is Markov.

value of next period's contingent purchases of goods and financial assets net of labor earnings. We can eliminate next period's purchases of capital and state-contingent bonds by using next period's version of equation (16.14.2). After invoking transversality conditions (16.9.10), continued substitutions yield

$$
\sum_{s_t} p_{t-1}\left(s_t|s^{t-1}\right) b_t\left(s_t|s^{t-1}\right)
$$

$$
= \sum_{j=t}^{\infty} \sum_{s^j|s^{t-1}} \beta^{j+1-t} \pi_j\left(s^j|s^{t-1}\right) \frac{u_c\left(s^j\right) c_j\left(s^j\right) - u_\ell\left(s^j\right) n_j\left(s^j\right)}{u_c\left(s^{t-1}\right)}
$$

$$
- k_t\left(s^{t-1}\right), \tag{16.14.3}
$$

where we have invoked conditions (16.9.5a) and (16.9.5b). Suppose now s follows a Markov process. Then recall from earlier that the allocations from period 1 onward can be described by time-invariant allocation rules with the current state s and beginning-of-period capital stock k as arguments. Thus, equation (16.14.3) implies that the end-of-period government debt is also a function of the state vector (s, k), since the current state fully determines the end-of-period capital stock and is the only information needed to form conditional expectations of future states. Putting together the lessons of this section with earlier ones, reliance on state-contingent debt and/or state-contingent capital taxes enables the government to avoid any lingering effects on indebtedness from past shocks to government expenditures and past productivity shocks that affected labor tax revenues.

This striking lack of history dependence contradicts the extensive history-dependence of the stock of government debt that Robert Barro (1979) identified as one of the salient characteristics of his model of optimal fiscal policy. According to Barro, government debt should be cointegrated with tax revenues, which in turn should follow a random walk, with innovations that are perfectly correlated with innovations in the government expenditure process. Important aspects of such behavior of government debt seem to be observed. For example, Sargent and Velde (1995) display long series of government debt for eighteenth century Britain that more closely resembles the outcome from Barro's model than from Lucas and Stokey's. Partly inspired by those observations, Aiyaragi et al. returned to the environment of Lucas and Stokey's model and altered the market structure in a way that brought outcomes closer to Barro's. We create their model by closing almost all of the markets that Lucas and Stokey had allowed.[12]

16.15. Taxation without state-contingent debt

Returning to the model without physical capital, we follow Aiyagari, Marcet, Sargent, and Seppälä (2002) and study optimal taxation without state-contingent debt. The government's budget constraint in expression (16.9.3) has to be modified by replacing state-contingent debt by risk-free government bonds. In period t and history s^t, let $b_{t+1}(s^t)$ be the amount of government indebtedness carried over to and maturing in the next period $t+1$, denominated in time $(t+1)$ goods. The market value at time t of that government indebtedness equals $b_{t+1}(s^t)$ divided by the risk-free gross interest rate between periods t and $t+1$, denoted by $R_t(s^t)$. Thus, the government's budget constraint in period t and history s^t becomes

$$b_t\left(s^{t-1}\right) = \tau_t^n\left(s^t\right) n_t\left(s^t\right) - g_t\left(s_t\right) - T_t\left(s^t\right) + \frac{b_{t+1}\left(s^t\right)}{R\left(s^t\right)}$$

$$\equiv z\left(s^t\right) + \frac{b_{t+1}\left(s^t\right)}{R_t\left(s^t\right)}, \tag{16.15.1}$$

where $T_t(s^t)$ is a nonnegative lump-sum transfer to the representative household and $z(s^t)$ is a function for the net-of-interest government surplus. It might seem strange to include the term $T_t(s^t)$ that allows for a nonnegative lump-sum transfer to the private sector. In an optimal taxation allocation that includes the levy of distortionary taxes, why would the government ever want to hand back resources to the private sector that have been raised with distortionary taxes? Certainly that would never happen in an economy with state-contingent debt, since any such allocation could be improved by lowering distortionary taxes rather than handing out lump-sum transfers. But as we will see, without

---

12 Werning (2007) extends the Lucas-Stokey model in another interesting direction. He assumes that there are complete markets in consumption, that agents are heterogeneous in the efficiencies of their labor supplies, and that taxes can be nonlinear functions of labor earnings. For example, with affine, rather than linear taxes, he explores how distorting taxes on labor are imposed to redistribute income as well as to raise revenues for financing expenditures. Without heterogeneity of labor efficiencies, no distorting taxes on labor are imposed.

state-contingent debt there can be circumstances when a government would like
to make lump-sum transfers to the private sector. However, most of the time
we shall be able to ignore this possibility.

To rule out Ponzi schemes, we assume that the government is subject to
versions of the natural debt limits defined in chapters 8 and 18. The consumption
Euler equation for the representative household able to trade risk-free debt with
one-period gross interest rate $R_t(s^t)$ is

$$\frac{1}{R_t(s^t)} = \sum_{s_{t+1}} p_t\left(s_{t+1}|s^t\right) = \sum_{s^{t+1}|s^t} \beta \pi_{t+1}\left(s^{t+1}|s^t\right) \frac{u_c\left(s^{t+1}\right)}{u_c\left(s^t\right)}.$$

Substituting this expression into the government's budget constraint $(16.15.1)$
yields:

$$b_t\left(s^{t-1}\right) = z\left(s^t\right) + \sum_{s^{t+1}|s^t} \beta \pi_{t+1}\left(s^{t+1}|s^t\right) \frac{u_c\left(s^{t+1}\right)}{u_c\left(s^t\right)} b_{t+1}\left(s^t\right). \qquad (16.15.2)$$

Note that the constant $b_{t+1}(s^t)$ is the same for all realizations of s_{t+1}. We will
now replace that constant $b_{t+1}(s^t)$ by another expression of the same magnitude.
In fact, we have as many candidate expressions of that magnitude as there
are possible states s_{t+1}, i.e., for each state s_{t+1} there is a government budget
constraint that is the analogue to expression $(16.15.1)$ but where the time index
is moved one period forward. And all those budget constraints have a right side
that is equal to $b_{t+1}(s^t)$. Instead of picking one of these candidate expressions
to replace all occurrences of $b_{t+1}(s^t)$ in equation $(16.15.2)$, we replace $b_{t+1}(s^t)$
when the summation index in equation $(16.15.2)$ is s_{t+1} by the right side of
next period's budget constraint that is associated with that particular realization
s_{t+1}. These substitutions give rise to the following expression:

$$b_t\left(s^{t-1}\right) = z\left(s^t\right) + \sum_{s^{t+1}|s^t} \beta \pi_{t+1}\left(s^{t+1}|s^t\right) \frac{u_c\left(s^{t+1}\right)}{u_c\left(s^t\right)}$$

$$\cdot \left[z\left(s^{t+1}\right) + \frac{b_{t+2}\left(s^{t+1}\right)}{R_{t+1}\left(s^{t+1}\right)} \right].$$

After similar repeated substitutions for all future occurrences of government
indebtedness, and by invoking the natural debt limit, we arrive at a final expression:

$$b_t\left(s^{t-1}\right) = \sum_{j=0}^{\infty} \sum_{s^{t+j}|s^t} \beta^j \pi_{t+j}\left(s^{t+j}|s^t\right) \frac{u_c\left(s^{t+j}\right)}{u_c\left(s^t\right)} z\left(s^{t+j}\right)$$

$$= E_t \sum_{j=0}^{\infty} \beta^j \frac{u_c\left(s^{t+j}\right)}{u_c\left(s^t\right)} \, z\left(s^{t+j}\right). \tag{16.15.3}$$

Expression $(16.15.3)$ at time $t = 0$ and initial state s^0, constitutes an implementability condition derived from the present-value budget constraint that the government must satisfy when seeking a solution to the Ramsey taxation problem:

$$b_0\left(s^{-1}\right) = E_0 \sum_{j=0}^{\infty} \beta^j \frac{u_c\left(s^j\right)}{u_c\left(s^0\right)} \, z\left(s^j\right). \tag{16.15.4}$$

Now it is instructive to compare the present economy without state-contingent debt to the earlier economy with state-contingent debt. Suppose that the initial government debt in period 0 and state s^0 is the same across the two economies, i.e., $b_0(s^{-1}) = b_0(s_0|s^{-1})$. Implementability condition $(16.15.4)$ of the present economy is then exactly the same as the one for the economy with state-contingent debt, as given by expression $(16.14.1)$ evaluated in period $t = 0$. But while this is the only implementability condition arising from budget constraints in the complete markets economy, many more implementability conditions must be satisfied in the economy without state-contingent debt. Specifically, because the beginning-of-period indebtedness is the same across any two histories, for any two realizations s^t and $\tilde{s}^t$ that share the same history until the previous period, i.e., $s^{t-1} = \tilde{s}^{t-1}$, we must impose equality across the right sides of their respective budget constraints, as depicted in expression $(16.15.3)$.[13] Hence, the Ramsey taxation problem without state-contingent debt becomes

$$\max_{\{c_t(s^t), b_{t+1}(s^t)\}} E_0 \sum_{t=0}^{\infty} \beta^t u\left(c_t\left(s^t\right), 1 - c_t\left(s^t\right) - g_t\left(s_t\right)\right)$$

$$\text{s.t. } E_0 \sum_{j=0}^{\infty} \beta^j \frac{u_c\left(s^j\right)}{u_c\left(s^0\right)} \, z\left(s^j\right) \geq b_0\left(s^{-1}\right); \tag{16.15.5a}$$

$$E_t \sum_{j=0}^{\infty} \beta^j \frac{u_c\left(s^{t+j}\right)}{u_c\left(s^t\right)} \, z\left(s^{t+j}\right) = b_t\left(s^{t-1}\right), \quad \text{for all } s^t; \tag{16.15.5b}$$

$$\text{given } b_0\left(s^{-1}\right),$$

---

[13] Aiyagari et al. (2002) regard these conditions as imposing measurability of the right-hand side of $(16.15.3)$ with respect to s^{t-1}.

where we have substituted the resource constraint $(16.13.1)$ into the utility function. It should also be understood that we have substituted the resource constraint into the net-of-interest government surplus and used the household's first-order condition, $1 - \tau_t^n(s^t) = u_\ell(s^t)/u_c(s^t)$, to eliminate the labor tax rate. Hence, the net-of-interest government surplus now reads as

$$z\left(s^t\right) = \left[1 - \frac{u_\ell\left(s^t\right)}{u_c\left(s^t\right)}\right]\left[c_t\left(s^t\right) + g_t\left(s_t\right)\right] - g_t\left(s_t\right) - T_t\left(s^t\right) . \qquad (16.15.6)$$

Next, we compose a Lagrangian for the Ramsey problem. Let $\gamma_0(s^0)$ be the nonnegative Lagrange multiplier on constraint $(16.15.5a)$. As in the earlier economy with state-contingent debt, this multiplier is strictly positive if the government must resort to distortionary taxation, and otherwise equal to zero. The force of the assumption that markets in state-contingent securities have been shut down but that a market in a risk-free security remains is that we have to attach stochastic processes $\{\gamma_t(s^t)\}_{t=1}^\infty$ of Lagrange multipliers to the new implementability constraints $(16.15.5b)$. These multipliers might be positive or negative, depending on direction in which the constraints are binding:

$$\gamma_t\left(s^t\right) \geq (\leq) \ 0 \quad \text{if the constraint is binding in the direction}$$

$$E_t \sum_{j=0}^\infty \beta^j \frac{u_c\left(s^{t+j}\right)}{u_c\left(s^t\right)} \, z\left(s^{t+j}\right) \geq (\leq) \ b_t\left(s^{t-1}\right).$$

A negative multiplier $\gamma_t(s^t) < 0$ means that if we could relax constraint $(16.15.5b)$, we would like to *increase* the beginning-of-period indebtedness for that particular realization of history s^t, which would presumably enable us to reduce the beginning-of-period indebtedness for some other history. In particular, as we will soon see from the first-order conditions of the Ramsey problem, there would then exist another realization $\tilde{s}^t$ with the same history up until the previous period, i.e., $\tilde{s}^{t-1} = s^{t-1}$, but where the multiplier on constraint $(16.15.5b)$ takes on a positive value $\gamma_t(\tilde{s}^t) > 0$. All this is indicative of the fact that the government cannot use state-contingent debt and therefore cannot allocate its indebtedness most efficiently across future states.

We apply two transformations to the Lagrangian. We multiply constraint $(16.15.5a)$ by $u_c(s^0)$ and the constraints $(16.15.5b)$ by $\beta^t u_c(s^t)$. The Lagrangian for the Ramsey problem can then be represented as follows, where the second equality invokes the law of iterated expectations and uses Abel's summation formula:[14]

$$J = E_0 \sum_{t=0}^{\infty} \beta^t \bigg\{ u \left(c_t(s^t), 1 - c_t(s^t) - g_t(s_t) \right)$$

$$+ \gamma_t(s^t) \bigg[E_t \sum_{j=0}^{\infty} \beta^j u_c(s^{t+j}) \, z(s^{t+j}) - u_c(s^t) \, b_t(s^{t-1}) \bigg] \bigg\}$$

$$= E_0 \sum_{t=0}^{\infty} \beta^t \bigg\{ u \left(c_t(s^t), 1 - c_t(s^t) - g_t(s_t) \right)$$

$$+ \Psi_t(s^t) \, u_c(s^t) \, z(s^t) - \gamma_t(s^t) \, u_c(s^t) \, b_t(s^{t-1}) \bigg] \bigg\}, \qquad (16.15.7a)$$

where

$$\Psi_t(s^t) = \Psi_{t-1}(s^{t-1}) + \gamma_t(s^t) \qquad (16.15.7b)$$

and $\Psi_{-1}(s^{-1}) = 0$. The first-order condition with respect to $c_t(s^t)$ can be expressed as

$$u_c(s^t) - u_\ell(s^t)$$
$$+ \Psi_t(s^t) \left\{ \left[u_{cc}(s^t) - u_{c\ell}(s^t) \right] z(s^t) + u_c(s^t) \, z_c(s^t) \right\}$$
$$- \gamma_t(s^t) \left[u_{cc}(s^t) - u_{c\ell}(s^t) \right] b_t(s^{t-1}) = 0, \qquad (16.15.8a)$$

and with respect to $b_t(s^t)$,

$$E_t \left[\gamma_{t+1}(s^{t+1}) \, u_c(s^{t+1}) \right] = 0. \qquad (16.15.8b)$$

If we substitute $z(s^t)$ from equation (16.15.6) and its derivative $z_c(s^t)$ into first-order condition (16.15.8a), we will find only two differences from the corresponding condition (16.13.4) for the optimal allocation in an economy with state-contingent government debt. First, the term involving $b_t(s^{t-1})$ in first-order condition (16.15.8a) does not appear in expression (16.13.4). Once again, this term reflects the constraint that beginning-of-period government indebtedness must be the same across all realizations of next period's state, a constraint that is not present if government debt can be state contingent. Second, the Lagrange multiplier $\Psi_t(s^t)$ in first-order condition (16.15.8a) may change over time in response to realizations of the state, while the multiplier Φ in expression (16.13.4) is time invariant.

---

[14] See Apostol (1974, p. 194). For another application, see chapter 20, page 820.

Next, we are interested to learn if the optimal allocation without state-contingent government debt will eventually be characterized by an expression similar to (16.13.4), i.e., whether or not the Lagrange multiplier $\Psi_t(s^t)$ converges to a constant, so that from there on, the absence of state-contingent debt no longer binds.

16.15.1. Future values of $\{g_t\}$ become deterministic

Aiyagari et al. (2002) prove that if $\{g_t(s_t)\}$ has absorbing states in the sense that $g_t = g_{t-1}$ almost surely for t large enough, then $\Psi_t(s^t)$ converges when $g_t(s_t)$ enters an absorbing state. The optimal tail allocation for this economy without state-contingent government debt coincides with the allocation of an economy with state-contingent debt that would have occurred under the same shocks, but for different initial debt. That is, the limiting random variable Ψ_∞ will then play the role of the single multiplier in an economy with state-contingent debt because, as noted above, the first-order condition $(16.15.8a)$ will then be the same as expression $(16.13.4)$, where $\Phi = \Psi_\infty$. The value of Ψ_∞ depends on the realization of the government expenditure path. If the absorbing state is reached after many bad shocks (high values of $g_t(s_t)$), the government will have accumulated high debt, and convergence will occur to a contingent-debt economy with high initial debt and therefore a high value of the multiplier Φ.

This particular result about convergence can be stated in more general terms, i.e., $\Psi_t(s^t)$ can be shown to converge if the future path of government expenditures eventually becomes deterministic, for example, if government expenditures eventually become constant. Once uncertainty about future government expenditures ceases, the government can thereafter attain the Ramsey allocation with one-period risk-free bonds, as described at the beginning of this chapter. In the present setup, this becomes apparent from examining first-order condition $(16.15.8b)$ when there is no uncertainty: next period's nonstochastic marginal utility of consumption must be multiplied by a nonstochastic multiplier $\gamma_{t+1} = 0$ in order for that first-order condition to be satisfied under certainty. The zero value of all future multipliers $\{\gamma_t\}$ implies convergence of $\Psi_t(s^t) = \Psi_\infty$, and we are back to our earlier logic where expression $(16.13.4)$ with $\Phi_t = \Psi_\infty$ characterizes the optimal tail allocation for an economy without state-contingent government debt when there is no uncertainty.

16.15.2. Stochastic $\{g_t\}$ but special preferences

To study whether $\Psi_t(s^t)$ can converge when $g_t(s_t)$ remains stochastic forever, it is helpful to substitute expression $(16.15.7b)$ into first-order condition $(16.15.8b)$

$$E_t \left\{ \left[\Psi_{t+1}\left(s^{t+1}\right) - \Psi_t\left(s^t\right) \right] u_c\left(s^{t+1}\right) \right\} = 0,$$

which can be rewritten as

$$\Psi_t\left(s^t\right) = E_t \left[\Psi_{t+1}\left(s^{t+1}\right) \frac{u_c\left(s^{t+1}\right)}{E_t u_c\left(s^{t+1}\right)} \right]$$

$$= E_t \Psi_{t+1}\left(s^{t+1}\right) + \frac{\text{COV}_t\left(\Psi_{t+1}\left(s^{t+1}\right), u_c\left(s^{t+1}\right)\right)}{E_t u_c\left(s^{t+1}\right)}. \qquad (16.15.9)$$

Aiyagari et al. (2002) present a convergence result for a special class of preferences that make the covariance term in equation $(16.15.9)$ identically equal to zero. The household's utility is assumed to be linear in consumption and additively separable from the utility of leisure. (See the preference specification in our next subsection.) Thus, the marginal utility of consumption is constant and expression $(16.15.9)$ reduces to

$$\Psi_t\left(s^t\right) = E_t \Psi_{t+1}\left(s^{t+1}\right).$$

The stochastic process $\Psi_t(s^t)$ is evidently a nonnegative martingale. As described in equation $(16.15.7b)$, $\Psi_t(s^t)$ fluctuates over time in response to realizations of the multiplier $\gamma_t(s^t)$ that can be either positive or negative; $\gamma_t(s^t)$ measures the marginal impact of news about the present value of government expenditures on the maximum utility attained by the planner. The cumulative multiplier $\Psi_t(s^t)$ remains strictly positive so long as the government must resort to distortionary taxation either in the current period or for some realization of the state in a future period.

By a theorem of Doob (1953, p. 324), a nonnegative martingale like $\Psi_t(s^t)$ converges almost surely.[15] If the process for government expenditures is sufficiently stochastic, e.g., when $g_t(s_t)$ is stationary with a strictly positive variance, then Aiyagari et al. (2002) prove that $\Psi_t(s^t)$ converges almost surely to

---

[15] For a discussion of the martingale convergence theorem, see the appendix to chapter 17.

zero. When setting $\Psi_\infty = \gamma_\infty = 0$ in first-order condition $(16.15.8a)$, it follows that the optimal tax policy must eventually lead to a first-best allocation with $u_c(s^t) = u_\ell(s^t)$, i.e., $\tau^n_\infty = 0$. This implies that government assets converge to a level always sufficient to support government expenditures from interest earnings alone. Unspent interest earnings on government-owned assets are returned to the households as positive lump-sum transfers. Such transfers occur whenever government expenditures fall below their maximum possible level.

A proof that $\Psi_t(s^t)$ converges to zero and that government assets eventually become large enough to finance all future government expenditures can be constructed along lines that supported analogous outcomes in our chapter 17 analysis of self-insurance with incomplete markets. Like the analysis there, we can appeal to a martingale convergence theorem and use an argument based on contradictions to rule out convergence to any number other than zero. To establish a contradiction in the present setting, suppose that $\Psi_t(s^t)$ does not converge to zero but rather to a strictly positive limit, $\Psi_\infty > 0$. According to our argument above, the optimal tail allocation for this economy without state-contingent government debt will then coincide with the allocation of an economy that has state-contingent debt and a particular initial debt level. It follows that these two economies should have identical labor tax rates supporting that optimal tail allocation. But Aiyagari et al. (2002) show that a government that follows such a tax policy and has access only to risk-free bonds to absorb stochastic surpluses and deficits will with positive probability see either its debt grow without bound or its assets grow without bound, two outcomes that are inconsistent with an optimal allocation. The heuristic explanation is as follows. The government in an economy with state-contingent debt uses these debt instruments as an "insurance policy" to smooth taxes across realizations of the state. The government's lack of access to such an "insurance policy" when only risk-free bonds are available means that implementing those very same tax rates, unresponsive as they are to realizations of the state, would expose the government to a positive probability of seeing either its debt level or its asset level drift off to infinity. But that contradicts a supposition that such a tax policy would be optimal in an economy without state-contingent debt. First, it is impossible for government debt to grow without bound, because households would not be willing to lend to a government that violates its natural borrowing limit. Second, it is not optimal for the government to accumulate assets without bound, because welfare could then be increased by cutting tax rates in some periods and thereby reducing the deadweight loss of taxation.[16] Therefore, we conclude that $\Psi_t(s^t)$ cannot converge to a nonnegative limit other than zero.

For more general preferences and ample randomness in government expenditures, Aiyagari et al. (2002) cannot characterize the limiting dynamics of $\Psi_t(s^t)$ except to rule out convergence to a strictly positive number. So at least two interesting possibilities remain: $\Psi(s^t)$ may converge to zero or it may have a nondegenerate distribution in the limit.

16.15.3. *Example 3 revisited:* $g_t = 0$ *for* $t \neq T$, *and* g_T *is stochastic*

To illustrate differences in optimal tax policy between economies with and without state-contingent government debt, we revisit our third example above of government expenditures that was taken from Lucas and Stokey's (1983) analysis of an economy with state-contingent debt. Let us examine how the optimal policy changes if the government has access only to risk-free bonds.[17] We assume that the household's utility function is

$$u\left(c_t\left(s^t\right), \ell_t\left(s^t\right)\right) = c_t\left(s^t\right) + H\left(\ell_t\left(s^t\right)\right),$$

where $H_\ell > 0$, $H_{\ell\ell} < 0$ and $H_{\ell\ell\ell} > 0$. We assume that $H_\ell(0) = \infty$ and $H_\ell(1) < 1$ to guarantee that the first-best allocation without distortionary taxation has an interior solution for leisure. Given these preferences, the first-order condition $(16.15.8a)$ with respect to consumption simplifies to

$$u_c\left(s^t\right) - u_\ell\left(s^t\right) + \Psi_t\left(s^t\right) u_c\left(s^t\right) z_c\left(s^t\right) = 0,$$

which after solving for the derivatives becomes

$$\left[1 + \Psi_t\left(s^t\right)\right]\left\{1 - H_\ell\left(1 - c_t\left(s^t\right) - g_t\left(s_t\right)\right)\right\}$$

$$= -\Psi_t\left(s^t\right) H_{\ell\ell}\left(1 - c_t\left(s^t\right) - g_t\left(s_t\right)\right)\left[c_t\left(s^t\right) + g_t\left(s_t\right)\right]. \tag{16.15.10}$$

---

[16] Aiyagari et al. (2002, lemma 3) suggest that unbounded growth of government-owned assets constitutes a contradiction because it violates a lower bound on debt, or an "asset limit." But we question this argument, since a government can trivially avoid violating any asset limit by making positive lump-sum transfers to the households. A correct proof should instead be based on the existence of welfare improvements associated with cutting distortionary taxes instead of making any such lump-sum transfers to households.

[17] Our first two examples above involve no uncertainty, so the issue of state-contingent debt does not arise. Hence the optimal tax policy is unaltered in those two examples.

As in our earlier analysis of this example, we assume that $g_T = g > 0$ with probability α and $g_T = 0$ with probability $1 - \alpha$. We also retain our assumption that the government starts with no assets or debt, $b_0(s^{-1}) = 0$, so that the multiplier on constraint (16.15.5a) is strictly positive, $\gamma_0(s^0) = \Psi_0(s^0) > 0$. Since no additional information about future government expenditures is revealed in periods $t < T$, it follows that the multiplier $\Psi_t(s^t) = \Psi_0(s^0) \equiv \Psi_0 > 0$ for $t < T$. Given the multiplier Ψ_0, the optimal consumption level for $t < T$, denoted c_0, satisfies the following version of first-order condition (16.15.10):

$$[1 + \Psi_0]\{1 - H_\ell(1 - c_0)\} = -\Psi_0\, H_{\ell\ell}(1 - c_0)\, c_0. \qquad (16.15.11)$$

In period T, there are two possible values of g_T, and hence the stochastic multiplier $\gamma_T(s^T)$ can take two possible values, one negative value and one positive value, according to first-order condition (16.15.8b); $\gamma_T(s^T)$ is negative if $g_T = 0$ because that represents good news that should cause the multiplier $\Psi_T(s^T)$ to fall. In fact, the multiplier $\Psi_T(s^T)$ falls all the way to zero if $g_T = 0$ because the government would then never again have to resort to distortionary taxation. And any tax revenues raised in earlier periods and carried over as government-owned assets would then also be handed back to the households as a lump-sum transfer. If, on the other hand, $g_T = g > 0$, then $\gamma_T(s^T) \equiv \gamma_T$ is strictly positive and the optimal consumption level for $t > T$, denoted $\tilde{c}$, would satisfy the following version of first-order condition (16.15.10)

$$[1 + \Psi_0 + \gamma_T]\{1 - H_\ell(1 - \tilde{c})\} = -[\Psi_0 + \gamma_T]\, H_{\ell\ell}(1 - \tilde{c})\, \tilde{c}. \qquad (16.15.12)$$

In response to $\gamma_T > 0$, the multiplicative factors within square brackets have increased on both sides of equation (16.15.12) but proportionately more so on the right side. Because both equations (16.15.11) and (16.15.12) must hold with equality at the optimal allocation, it follows that the change from c_0 to $\tilde{c}$ has to be such that $\{1 - H_\ell(1 - c)\}$ increases proportionately more than $-\{H_{\ell\ell}(1 - c)\, c\}$. Since the former expression is decreasing in c and the latter expression is increasing in c, we can then conclude that $\tilde{c} < c_0$ and hence that the implied labor tax rate is raised for all periods $t > T$ if government expenditures turn out to be strictly positive in period T.

It is obvious from this example that a government with access only to risk-free bonds cannot smooth tax rates across different realizations of the state. Recall that the optimal tax policy with state-contingent debt prescribed a constant tax rate for all $t \neq T$ regardless of the realization of g_T. Note also that, as discussed above, the multiplier $\Psi_t(s^t)$ in the economy without state-contingent debt does converge when the future path of government expenditures becomes deterministic in period T. In our example, $\Psi_t(s^t)$ converges either to zero or to $(\Psi_0 + \gamma_T) > 0$, depending on the realization of government expenditures. Starting from period T, the optimal tail allocation coincides then with the allocation of an economy with state-contingent debt that would have occurred under the same shocks, but for different initial debt, either a zero debt level associated with $\Phi = 0$, if $g_T = 0$, or the positive debt level that would correspond to $\Phi = \Psi_0 + \gamma_T$, if $g_T = g > 0$.

Schmitt-Grohe and Uribe (2004a) and Siu (2004) analyze optimal monetary and fiscal policies in economies in which the government can issue only *nominal* risk-free debt. Unanticipated inflation makes risk-free nominal debt state contingent in real terms and provides a motive for the government to make inflation vary. Schmitt-Grohe and Uribe and Siu both focus on how price stickiness would affect the government's use of fluctuations in inflation as an indirect way of introducing state-contingent debt. They find that even a very small amount of price stickiness causes the volatility of the optimal inflation rate to become very small. Thus, the government abstains from the indirect channel for synthesizing state-contingent debt. The authors relate their finding to the aspect of Aiyagari's et al.'s calculations for an economy with no state-contingent debt, mentioned in footnote 10 of this chapter, that the Ramsey allocation in their economy without state-contingent debt closely approximates that for the economy with complete markets.

16.16. Nominal debt as state-contingent real debt

We now turn to a monetary economy of Lucas and Stokey (1983) that, under particular preferences, Chari, Christiano, and Kehoe (1996) used to study the optimality of the Friedman rule and whether equilibrium price level adjustments can transform nominal non-state-contingent debt into state-contingent real debt. In particular, Chari, Christiano and Kehoe restricted preferences to satisfy Assumption 1 below in order to show optimality of the Friedman rule. They stated no additional conditions to reach their conclusion that, under an appropriate policy, non-state-contingent nominal government debt can be transformed into state-contingent real debt. However, we find that their statements about the equivalence of allocations under these two debt structures require stronger assumptions because of a potential sign-switching problem with optimal debt across state realizations at a point in time. To obtain Chari, Christiano and Kehoe's conclusion, we add our Assumption 2 below.

Our strategy is to follow Chari, Christiano, and Kehoe by first, in subsection 16.16.2, finding a Ramsey plan and an associated Ramsey allocation for a nonmonetary economy with state-contingent government debt; then, in subsection 16.16.3, stating conditions on fundamentals that suffice to allow that same allocation to be supported by a Ramsey plan for a monetary economy without real state-contingent government debt but only nominal non-state-contingent debt.

A key outcome here is that the Ramsey equilibrium in the monetary economy makes the price level fluctuate in response to shocks to government expenditures. The price level adjusts to deliver history-contingent returns to holders of government debt required to support the Ramsey allocation. This structure includes many, if not most, components of a 'fiscal theory of the price level'.[18] That it does so in a coherent way can help us describe a set of contentious issues associated with some expositions of fiscal theories of the price level. We take up this theme in section 16.17, and again in chapter 26.

---

[18] We admit that some writers could legitimately beg to differ here because we have chosen to express the fiscal theory of the price level within the straightjacket of a rational expectations competitive equilibrium in an Arrow-Debreu complete markets economy. Some would argue that the fiscal theory of the price level can dispense with auxiliary assumptions like rational expectations and complete markets.

16.16.1. Setup and main ideas

The production technology is as in (16.13.1) except that there are now two distinct goods being produced; a 'cash good' and a 'credit good'. Let $c_{1t}(s^t)$ and $c_{2t}(s^t)$ denote the household's consumption of the cash good and the credit good, respectively, while government consumption $g_t(s_t)$ is made up of credit goods only. The feasibility condition becomes

$$c_{1t}\left(s^t\right) + c_{2t}\left(s^t\right) + g_t\left(s_t\right) = n_t\left(s^t\right).
\tag{16.16.1}$$

The household's preferences are ordered by

$$\sum_{t=0}^{\infty}\sum_{s^t} \beta^t\, \pi_t\left(s^t\right)\, u\left[c_{1t}\left(s^t\right), c_{2t}\left(s^t\right), \ell_t\left(s^t\right)\right], \quad \beta \in (0,1)
\tag{16.16.2}$$

where u is increasing, strictly concave, and twice continuously differentiable in all of its arguments. Moreover, the utility function satisfies the Inada conditions

$$\lim_{c_1\downarrow 0} u_1\left(s^t\right) = \lim_{c_2\downarrow 0} u_2\left(s^t\right) = \lim_{\ell\downarrow 0} u_3\left(s^t\right) = +\infty,$$

where u_j is the derivative of the utility function with respect to its jth argument.

In period t, households trade money, assets, and goods in particular ways. At the start of period t, after observing the current state s_t, households trade money and assets in a centralized securities market. The assets are one-period, state-noncontingent, nominal discount bonds. Let $M_{t+1}(s^t)$ and $B_{t+1}(s^t)$ denote the money and the nominal bonds held at the end of the securities market trading. Let $R_t(s^t)$ denote the gross nominal interest rate, so $1/R_t(s^t)$ is the purchase price of bonds. After securities trading, each household splits into a worker and a shopper. The shopper must use the money to purchase cash goods. To purchase credit goods, the shopper issues one-period nominal claims that are to be settled in the securities market in the next period. The worker is paid in cash at the end of the period. He adds this cash to any cash unexpended during shopping and carries it into the next period.

The household's budget constraint in the securities market is

$$M_{t+1}\left(s^t\right) + B_{t+1}\left(s^t\right)/R_t\left(s^t\right) = B_t\left(s^{t-1}\right)$$
$$+ M_t\left(s^{t-1}\right) - P_{t-1}\left(s^{t-1}\right) c_{1t-1}\left(s^{t-1}\right) - P_{t-1}\left(s^{t-1}\right) c_{2t-1}\left(s^{t-1}\right)$$
$$+ P_{t-1}\left(s^{t-1}\right) \left[1 - \tau_{t-1}^n\left(s^{t-1}\right)\right] n_{t-1}\left(s^{t-1}\right),
\tag{16.16.3}$$

where P is the nominal price of goods. The left side of ($16.16.3$) is the nominal value of assets held at the end of securities market trading. The first term on the right side is the value of nominal debt bought in the preceding period. The next two terms on the right side are the shopper's unspent cash. The fourth term is the payments for credit goods, and the last term is after-tax labor earnings. Purchases of cash goods must satisfy a cash-in-advance constraint:

$$P_t\left(s^t\right) c_{1t}\left(s^t\right) \leq M_{t+1}\left(s^t\right). \tag{16.16.4}$$

Let $\lambda_t(s^t)$ and $\mu_t(s^t)$ be the Lagrange multipliers on constraints ($16.16.3$) and ($16.16.4$), respectively. The first-order conditions for the household's problem are

$$c_{1t}\left(s^t\right): \; \beta^t \pi_t\left(s^t\right) u_1\left(s^t\right) = P_t\left(s^t\right)\left[\mu_t\left(s^t\right) + \sum_{s^{t+1}|s^t} \lambda_{t+1}\left(s^{t+1}\right)\right], \tag{16.16.5a}$$

$$c_{2t}\left(s^t\right): \; \beta^t \pi_t\left(s^t\right) u_2\left(s^t\right) = P_t\left(s^t\right) \sum_{s^{t+1}|s^t} \lambda_{t+1}\left(s^{t+1}\right), \tag{16.16.5b}$$

$$n_t\left(s^t\right): \; \beta^t \pi_t\left(s^t\right) u_3\left(s^t\right) = P_t\left(s^t\right)\left[1 - \tau_t^n\left(s^t\right)\right] \sum_{s^{t+1}|s^t} \lambda_{t+1}\left(s^{t+1}\right), \tag{16.16.5c}$$

$$M_{t+1}\left(s^t\right): \; \lambda_t\left(s^t\right) = \mu_t\left(s^t\right) + \sum_{s^{t+1}|s^t} \lambda_{t+1}\left(s^{t+1}\right), \tag{16.16.5d}$$

$$B_{t+1}\left(s^t\right): \; \lambda_t\left(s^t\right)/R_t\left(s^t\right) = \sum_{s^{t+1}|s^t} \lambda_{t+1}\left(s^{t+1}\right). \tag{16.16.5e}$$

After substituting ($16.16.5d$) into ($16.16.5a$), ($16.16.5e$) into ($16.16.5b$), and ($16.16.5e$) into ($16.16.5c$), respectively, the following conditions emerge

$$\beta^t \pi_t\left(s^t\right) u_1\left(s^t\right) = P_t\left(s^t\right) \lambda_t\left(s^t\right), \tag{16.16.6a}$$

$$\beta^t \pi_t\left(s^t\right) u_2\left(s^t\right) = P_t\left(s^t\right) \lambda_t\left(s^t\right)/R_t\left(s^t\right), \tag{16.16.6b}$$

$$\beta^t \pi_t\left(s^t\right) u_3\left(s^t\right) = P_t\left(s^t\right)\left[1 - \tau_t^n\left(s^t\right)\right] \lambda_t\left(s^t\right)/R_t\left(s^t\right). \tag{16.16.6c}$$

By ($16.16.6$), marginal rates of substitution between goods and leisure satisfy

$$\frac{u_1\left(s^t\right)}{R_t\left(s^t\right) u_3\left(s^t\right)} = \frac{u_2\left(s^t\right)}{u_3\left(s^t\right)} = \frac{1}{1 - \tau_t^n\left(s^t\right)}.$$

Hence, the marginal rate of substitution between the cash and credit good equals the nominal interest rate,

$$\frac{u_1\left(s^t\right)}{u_2\left(s^t\right)} = R_t\left(s^t\right).$$

Another expression for the nominal interest rate is obtained by solving for $\lambda_t(s^t)$ from $(16.16.6a)$ and a corresponding expression for $\lambda_{t+1}(s^{t+1})$, which are then substituted into $(16.16.5e)$ to get

$$R_t\left(s^t\right) = \frac{u_1\left(s^t\right)/P_t\left(s^t\right)}{\beta \sum_{s_{t+1}} \pi_{t+1}\left(s^{t+1}|s^t\right) u_1\left(s^{t+1}\right)/P_{t+1}\left(s^{t+1}\right)}. \tag{16.16.7}$$

In an equilibrium, the nominal interest rate must satisfy $R_t(s^t) \geq 1$ because otherwise, households could make infinite profits by buying money and selling bonds. This inequality captures the much discussed zero lower bound on the net nominal interest rate $R_t(s^t) - 1$.

Money is injected into and withdrawn from the economy through government open-market operations in the securities market. The constraint on open market operations is

$$M_{t+1}\left(s^t\right) - M_t\left(s^{t-1}\right) + B_{t+1}\left(s^t\right)/R_t\left(s^t\right) = B_t\left(s^{t-1}\right)$$
$$+ P_{t-1}\left(s^{t-1}\right) g_{t-1}\left(s_{t-1}\right) - P_{t-1}\left(s^{t-1}\right) \tau_{t-1}^n\left(s^{t-1}\right) n_{t-1}\left(s^{t-1}\right). \tag{16.16.8}$$

The terms on the left side of this equation are the assets sold by the government. The first term on the right is the payment on debt incurred in the preceding period, the second term is the payment for government consumption, and the third term is tax receipts. Recall that government consumption consists solely of credit goods.

To ensure that the Ramsey problem is nontrivial in this environment, we follow Chari et al. (1996) and assume that households hold no nominal assets at the beginning of period 0, $M_0(s^{-1}) = B_0(s^{-1}) = 0$.[19] Hence, the government constraint $(16.16.8)$ in period 0 becomes

$$M_1\left(s^0\right) + B_1\left(s^0\right)/R_0\left(s^0\right) = 0. \tag{16.16.9}$$

Starting with $(16.16.8)$, and then recursively eliminating debt issued in a preceding period, by using corresponding versions of $(16.16.8)$, continuing all the

---

[19] As in the case of our earlier restriction on the capital tax in period 0, the initial condition here restricts the government's access to nondistorting ways of financing its expenditures. In a monetary economy, if the initial stock of nominal assets held by households is positive, then welfare is maximized by increasing the initial price level to infinity. If the initial stock is negative, then welfare is maximized by setting the initial price level so low that those initial assets are sufficient to finance the government's entire stream of expenditures without ever having to levy distorting taxes.

way back to time 0 when (16.16.9) applies, we arrive at the following expression for how nominal, state-noncontingent debt evolves over time,

$$
B_{t+1}\left(s^t\right) = \sum_{j=0}^{t-1} \left(\prod_{k=j}^{t} R_k\left(s^k\right) \right) P_j\left(s^j\right) \left[g_j\left(s_j\right) - \tau_j^n\left(s^j\right) n_j\left(s^j\right) \right]
$$

$$
+ \sum_{j=1}^{t-1} \left(\prod_{k=j}^{t} R_k\left(s^k\right) \right) \left[1 - R_{j-1}\left(s^{j-1}\right) \right] M_j\left(s^{j-1}\right)
$$

$$
- R_t\left(s^t\right) M_{t+1}\left(s^t\right). \tag{16.16.10}
$$

Chari et al. (1996) maintain the following assumption:

ASSUMPTION 1: The utility function is of the form $u(c_1, c_2, \ell) = V(f(c_1, c_2), \ell)$ where f is homothetic.

Under this assumption, Chari et al. (1996) demonstrate that the Friedman rule is optimal, i.e., the Ramsey equilibrium has $R_t(s^t) = 1$ for all s^t. Despite the need to use some distorting taxes in the Ramsey equilibrium, it is not optimal to levy an inflation tax. Moreover, the Ramsey solution in our monetary economy with nominal, state-noncontingent debt yields an allocation that is identical to that in a frictionless nonmonetary economy in which the government issues (real) state-contingent debt. This finding enables us in the next section first to solve the Ramsey problem for the nonmonetary economy, and then to verify that the Ramsey allocation for the nonmonetary economy is also the Ramsey allocation for the monetary economy.

The reason that identical Ramsey allocations can prevail for the two economies is that the nominal, state-noncontingent debt in the monetary economy can in effect be transformed into state-contingent real debt by systematically varying the price level. In bad times associated with high government expenditures, it is optimal to raise the price level so that real debt payments can be made to be relatively small. In good times with low government expenditures, it is optimal to lower the price level to make real debt payments become relatively large.

The above description of an optimal policy potentially identifies a technical problem that could potentially invalidate the finding of Chari et al. (1996). Specifically, while monetary policy can be used to vary the size of real indebtedness at the beginning of a period, it can never alter the *sign* of the asset position inherited from last period. Hence, monetary policy cannot replicate a state-contingent debt policy in the nonmonetary economy if there is any period when the nonmonetary economy has positive state-contingent indebtedness under some realizations of the state and negative under others. Therefore, we must add the following assumption on primitives that will suffice to justify the finding of Chari et al. (1996).

ASSUMPTION 2: The Ramsey equilibrium in the nonmonetary economy with state-contingent debt has strictly positive indebtedness, $b_t(s^t) > 0$, in all periods $t \geq 1$ after all histories s^t.

16.16.2. *Optimal taxation in a nonmonetary economy*

Here we solve the Ramsey problem in a frictionless nonmonetary economy. Then in the next section, we will show how, under a particular monetary policy, the same optimal allocation can also be supported as an equilibrium of the monetary economy.

To have a complete set of tax instruments in the nonmonetary economy, we introduce a value-added tax on good 1, $\tau_t^{c1}(s^t)$; it becomes a subsidy when negative. The government's budget constraint at time t after history s^t becomes

$$\sum_{s_{t+1}} q_{t+1}^t \left(s_{t+1}, s^t\right) b_{t+1} \left(s_{t+1}|s^t\right) = b_t \left(s_t|s^{t-1}\right)$$

$$+ g_t \left(s_t\right) - \tau_t^n \left(s^t\right) n_t \left(s^t\right) - \tau_t^{c1} \left(s^t\right) c_{1t} \left(s^t\right), \qquad (16.16.11)$$

where as earlier, $b_{t+1}(s_{t+1}|s^t)$ is a state-contingent asset traded in period t at the price $q_{t+1}^t(s_{t+1}, s^t)$, in terms of time t goods (not including any value-added tax on good 1).

The representative household maximizes expression (16.16.2) subject to its present-value budget constraint:

$$\sum_{t=0}^{\infty} \sum_{s^t} q_t^0 \left(s^t\right) \left\{ \left[1 + \tau_t^{c1} \left(s^t\right)\right] c_{1t} \left(s^t\right) + c_{2t} \left(s^t\right) \right\}$$

$$= \sum_{t=0}^{\infty} \sum_{s^t} q_t^0 \left(s^t\right) \left[1 - \tau_t^n \left(s^t\right)\right] n_t \left(s^t\right). \qquad (16.16.12)$$

The first-order conditions for this problem imply

$$q_t^0 \left(s^t\right) = \beta^t \pi \left(s^t\right) \frac{u_2 \left(s^t\right)}{u_2 \left(s^0\right)}, \qquad (16.16.13a)$$

$$1 + \tau_t^{c1}\left(s^t\right) = \frac{u_1\left(s^t\right)}{u_2\left(s^t\right)}, \tag{16.16.13b}$$

$$1 - \tau_t^{n}\left(s^t\right) = \frac{u_3\left(s^t\right)}{u_2\left(s^t\right)}. \tag{16.16.13c}$$

After using these expressions to eliminate prices and taxes from the household's present-value budget constraint (16.16.12), the implementability condition in the Ramsey problem becomes

$$\sum_{t=0}^{\infty}\sum_{s^t}\beta^t\pi_t\left(s^t\right)\left[u_1\left(s^t\right)c_{1t}\left(s^t\right) + u_2\left(s^t\right)c_{2t}\left(s^t\right) - u_3\left(s^t\right)n_t\left(s^t\right)\right] = 0. \tag{16.16.14}$$

The first-order conditions of the Ramsey problem with respect to consumption goods $c_{it}(s^t)$, $i = 1, 2$, are

$$(1 + \Phi)\,u_i\left(s^t\right) + \Phi\left[\sum_{j=1}^{2}u_{ji}\left(s^t\right)c_{jt}\left(s^t\right) - u_{3i}\left(s^t\right)n_t\left(s^t\right)\right] = \theta_t\left(s^t\right), \tag{16.16.15}$$

where Φ and $\theta_t(s^t)$ are Lagrange multipliers on implementability condition (16.16.14) and resource constraint (16.16.1), respectively. After dividing by $u_i(s^t)$ and noting that by Assumption 1, $u_{3i}(s^t)/u_i(s^t) = V_{12}(s^t)/V_1(s^t)$, we have

$$(1 + \Phi) + \Phi\left[\sum_{j=1}^{2}\frac{u_{ji}\left(s^t\right)c_{jt}\left(s^t\right)}{u_i\left(s^t\right)} - \frac{V_{12}\left(s^t\right)}{V_1\left(s^t\right)}n_t\left(s^t\right)\right] = \frac{\theta_t\left(s^t\right)}{u_i\left(s^t\right)}, \tag{16.16.16}$$

for $i = 1, 2$. Next, note that a utility function that satisfies Assumption 1 implies that the value of the summation on the left side of expression (16.16.16) is the same for $i = 1$ and $i = 2$.[20] Thus, the value of the entire left side

---

[20] Recall that homotheticity, as in Assumption 1, implies that for any constant $\kappa > 0$,

$$\frac{u_1\left[\kappa c_{1t}\left(s^t\right), \kappa c_{2t}\left(s^t\right), n_t\left(s^t\right)\right]}{u_2\left[\kappa c_{1t}\left(s^t\right), \kappa c_{2t}\left(s^t\right), n_t\left(s^t\right)\right]} = \frac{u_1\left[c_{1t}\left(s^t\right), c_{2t}\left(s^t\right), n_t\left(s^t\right)\right]}{u_2\left[c_{1t}\left(s^t\right), c_{2t}\left(s^t\right), n_t\left(s^t\right)\right]}.$$

Differentiating this expression with respect to κ and evaluating at $\kappa = 1$ gives

$$\sum_{j=1}^{2}\frac{u_{j1}\left(s^t\right)c_{jt}\left(s^t\right)}{u_1\left(s^t\right)} = \sum_{j=1}^{2}\frac{u_{j2}\left(s^t\right)c_{jt}\left(s^t\right)}{u_2\left(s^t\right)}.$$

of expression $(16.16.16)$ is the same for $i = 1$ and $i = 2$ and consequently, so must the value of the right side be, so that it follows that $u_1(s^t) = u_2(s^t)$. From expression $(16.16.13b)$, we then conclude that the two consumption goods are taxed at the same rates, i.e., $\tau_t^{c1}(s^t) = 0$.[21]

Given $\tau_t^{c1}(s^t) = 0$, the government debt satisfies an expression analogous to expression $(16.14.1)$, namely,

$$
b_t\left(s_t|s^{t-1}\right) = \tau_t^n\left(s^t\right) n_t\left(s^t\right) - g_t\left(s_t\right)
$$
$$
+ \sum_{j=1}^{\infty} \sum_{s^{t+j}|s^t} q_{t+j}^t\left(s^{t+j}\right) \left[\tau_{t+j}^n\left(s^{t+j}\right) n_{t+j}\left(s^{t+j}\right) - g_{t+j}\left(s_{t+j}\right)\right]
$$
$$
= \sum_{j=0}^{\infty} \sum_{s^{t+j}|s^t} \beta^j \pi_{t+j}\left(s^{t+j}|s^t\right) \frac{u_2\left(s^{t+j}\right)}{u_2\left(s^t\right)} \left\{\left[1 - \frac{u_3\left(s^{t+j}\right)}{u_2\left(s^{t+j}\right)}\right]\right.
$$
$$
\left. \cdot \left[c_{1t+j}\left(s^{t+j}\right) + c_{2t+j}\left(s^{t+j}\right) + g_{t+j}\left(s_{t+j}\right)\right] - g_{t+j}\left(s_{t+j}\right)\right\}. \quad (16.16.17)
$$

16.16.3. *Optimal policy in a corresponding monetary economy*

We want to verify our conjecture that the optimal allocation of the nonmonetary economy with state-contingent debt can be attained in the monetary economy with state-noncontingent nominal debt under a monetary policy that implements the Friedman rule, i.e., $R_t(s^t) = 1$ for all s^t.

At the beginning of any period $t > 0$ and history s^t, given outstanding nominal assets $\{M_t(s^{t-1}), B_t(s^{t-1})\}$ and last period's price level $P_{t-1}(s^{t-1})$, the government targets a current price level equal to

$$
P_t\left(s^t\right) = \left\{P_{t-1}\left(s^{t-1}\right)\left[g_{t-1}\left(s_{t-1}\right) - \tau_{t-1}^n\left(s^{t-1}\right) n_{t-1}\left(s^{t-1}\right)\right]\right.
$$
$$
\left. + M_t\left(s^{t-1}\right) + B_t\left(s^{t-1}\right)\right\} / b_t\left(s_t|s^{t-1}\right), \quad (16.16.18)
$$

---

[21] Given that we use tax instruments $\tau_t^{c1}(s^t)$ and $\tau_t^n(s^t)$ to control the tax wedges between $c_{1t}(s^t)$, $c_{2t}(s^t)$ and $\ell_t(s^t)$, equal taxation of the two goods implies $\tau_t^{c1}(s^t) = 0$, and hence, the optimal tax wedge between goods consumption and leisure is attained with the labor tax. If we instead had formulated the problem with two separate value-added taxes on the two goods, $\tau_t^{c1}(s^t)$ and $\tau_t^{c2}(s^t)$, but no labor tax, the equal taxation result for goods would have meant $\tau_t^{c1}(s^t) = \tau_t^{c2}(s^t)$, where the level of that common value-added tax rate would have served the same role as the present labor tax, i.e., to establish the optimal tax wedge between goods consumption and leisure.

which is evidently a function of the beginning-of-period indebtedness of the corresponding nonmonetary economy, $b_t(s_t|s^{t-1}) > 0$. The government attains the target by conducting open-market operations to make the money supply become

$$M_{t+1}\left(s^t\right) = P_t\left(s^t\right) c_{1t}\left(s^t\right), \qquad (16.16.19)$$

and the resulting nominal debt level $B_{t+1}(s^t)$ is given by $(16.16.8)$, or equivalently, by $(16.16.10)$. Note that, if our conjecture $R_t(s^t) = 1$ is correct, policy rule $(16.16.18)$ and government budget constraint $(16.16.8)$ imply

$$\frac{M_{t+1}\left(s^t\right) + B_{t+1}\left(s^t\right)}{P_t\left(s^t\right)} = b_t\left(s_t|s^{t-1}\right). \qquad (16.16.20)$$

We now proceed as if our conjecture is correct, and verify that the gross nominal interest rate given by expression $(16.16.7)$ would indeed equal to unity under the government policy we have described. The key steps in verifying that the policy attains $R_t(s^t) = 1$ are:

$$R_t\left(s^t\right) = \frac{u_2\left(s^t\right)/P_t\left(s^t\right)}{\beta \sum_{s_{t+1}} \pi_{t+1}\left(s^{t+1}|s^t\right) u_2\left(s^{t+1}\right)/P_{t+1}\left(s^{t+1}\right)}$$

$$= \frac{1/P_t\left(s^t\right)}{\sum_{s_{t+1}} q_{t+1}^t\left(s_{t+1}, s^t\right)/P_{t+1}\left(s^{t+1}\right)}$$

$$= \frac{1/P_t\left(s^t\right)}{\sum_{s_{t+1}} q_{t+1}^t\left(s_{t+1}, s^t\right) \dfrac{b_{t+1}\left(s_{t+1}|s^t\right)}{P_t\left(s^t\right)\left[g_t\left(s_t\right) - \tau_t^n\left(s^t\right)\right] + M_{t+1}\left(s^t\right) + B_{t+1}\left(s^t\right)}}$$

$$= \frac{g_t\left(s_t\right) - \tau_t^n\left(s^t\right) + b_t\left(s_t|s^{t-1}\right)}{\sum_{s_{t+1}} q_{t+1}^t\left(s_{t+1}, s^t\right) b_{t+1}\left(s_{t+1}|s^t\right)} = 1.$$

Under our conjecture, the first equality above substitutes $u_2(s^t) = u_1(s^t)$ into expression $(16.16.7)$. The second equality uses expression $(16.16.13a)$ for the price of state-contingent claims in the nonmonetary economy. The third equality uses policy rule $(16.16.18)$ for the price level $P_{t+1}(s^{t+1})$, while the fourth equality invokes expression $(16.16.20)$. The last equality follows from government budget constraint $(16.16.11)$ in the nonmonetary economy and hence, we have confirmed that our conjecture $R_t(s^t) = 1$ is correct.

Next, even though it is redundant, it is instructive to verify that the government budget constraint of the nonmonetary economy implies the budget constraint of the monetary economy. Starting with government budget constraint

$(16.16.11)$ in the nonmonetary economy with $\tau_t^{c1}(s^t) = 0$, we use expression $(16.16.13a)$ for the price of state-contingent claims (where $u_2(s^t) = u_1(s^t)$), and eliminate all state-contingent debt by using policy rule $(16.16.18)$,

$$\sum_{s_{t+1}} \beta \pi_{t+1} \left(s^{t+1} | s^t\right) \frac{u_1 \left(s^{t+1}\right)}{u_1 \left(s^t\right)}$$

$$\cdot \frac{P_t \left(s^t\right)}{P_{t+1} \left(s^{t+1}\right)} \left\{ \left[g_t \left(s_t\right) - \tau_t^n \left(s^t\right) n_t \left(s^t\right)\right] + \frac{M_{t+1} \left(s^t\right) + B_{t+1} \left(s^t\right)}{P_t \left(s^t\right)} \right\}$$

$$= \frac{P_{t-1} \left(s^{t-1}\right)}{P_t \left(s^t\right)} \left\{ \left[g_{t-1} \left(s_{t-1}\right) - \tau_{t-1}^n \left(s^{t-1}\right) n_{t-1} \left(s^{t-1}\right)\right] + \frac{M_t \left(s^{t-1}\right) + B_t \left(s^{t-1}\right)}{P_{t-1} \left(s^{t-1}\right)} \right\}$$

$$+ g_t \left(s_t\right) - \tau_t^n \left(s^t\right) n_t \left(s^t\right).$$

Invoking equilibrium expression $(16.16.7)$ for the nominal interest rate, this expression can be written

$$\left[R_t \left(s^t\right)\right]^{-1} \left\{ \left[g_t \left(s_t\right) - \tau_t^n \left(s^t\right) n_t \left(s^t\right)\right] + \frac{M_{t+1} \left(s^t\right) + B_{t+1} \left(s^t\right)}{P_t \left(s^t\right)} \right\}$$

$$= \frac{P_{t-1} \left(s^{t-1}\right)}{P_t \left(s^t\right)} \left\{ \left[g_{t-1} \left(s_{t-1}\right) - \tau_{t-1}^n \left(s^{t-1}\right) n_{t-1} \left(s^{t-1}\right)\right] + \frac{M_t \left(s^{t-1}\right) + B_t \left(s^{t-1}\right)}{P_{t-1} \left(s^{t-1}\right)} \right\}$$

$$+ g_t \left(s_t\right) - \tau_t^n \left(s^t\right) n_t \left(s^t\right).$$

Since we have confirmed that $R_t(s^t) = 1$ under the postulated monetary policy, this expression can then be simplified to become government budget constraint $(16.16.8)$ in the monetary economy.

16.17. Relation to fiscal theories of the price level

In chapter 26, we take up monetary-fiscal theories of inflation, including one that
has been christened a 'fiscal theory of the price level'. The model of the previous
section includes components that combine to give rise to all of the forces active
in that theory. As emphasized by Niepelt (2004), accounts of that theory are
too often at best incomplete because they leave implicit aspects of an underlying
general equilibrium model. For that reason, we find it enlightening to interpret
statements about the fiscal theory of the price level within the context of a
coherent general equilibrium model like that of Chari, Christiano, and Kehoe
(1996).

16.17.1. Budget constraint versus asset pricing equation

To take a prominent example, Cochrane (2005) asserts that under a fiat money
system there exists no government budget constraint, only a *valuation equation*
for nominal government debt. To express Cochrane's point of view within the
economy of Chari, Christiano, and Kehoe (1996), solve for $b_t(s_t|s^{t-1})$ from
policy rule (16.16.18) and substitute the outcome into (16.16.17) to get

$$\frac{M_t\left(s^{t-1}\right) + B_t\left(s^{t-1}\right) + P_{t-1}\left(s^{t-1}\right)\left[g_{t-1}\left(s_{t-1}\right) - \tau_{t-1}^n\left(s^{t-1}\right)\right]}{P_t\left(s^t\right)}$$

$$= \tau_t^n\left(s^t\right) n_t\left(s^t\right) - g_t\left(s_t\right)$$

$$+ \sum_{j=1}^{\infty} \sum_{s^{t+j}|s^t} q_{t+j}^t\left(s^{t+j}\right)\left[\tau_{t+j}^n\left(s^{t+j}\right) n_{t+j}\left(s^{t+j}\right) - g_{t+j}\left(s_{t+j}\right)\right]. \quad (16.17.1)$$

The left side is the nominal liabilities of the government outstanding at the beginning of period t divided by the equilibrium price level $P_t(s^t)$. The right side
is the real value of future government surpluses. Cochrane notes the resemblance
of this equation to an asset pricing equation and interprets it as a fiscal theory
of the price level by asserting that news about the right side of (16.17.1) will
cause immediate adjustments in the price level that he regards as 'revaluations'
of the government's outstanding stock of government debt.

 Niepelt (2004) objects to isolating equation (16.17.1) and interpreting it in
this way. Niepelt's perspective can be expressed neatly within the Chari, Christiano, and Kehoe model. Here it is the stochastic process for the price level that

is an equilibrium object that adjusts to create a stochastic process of realized return on nominal government debt that does the same job as state-contingent real government debt. Within an equilibrium, individuals are never 'surprised' or 'disappointed' by rate of return realizations on nominal government debt. As the economy passes from one Arrow-Debreu node to its successor, Cochrane's favorite equation (16.17.1) of course prevails (along with other equations such as the household's intertemporal optimization conditions that together determine a continuation equilibrium), but it is incomplete to the point of being misleading to promote equation by itself (16.17.1) to the status of 'a theory of the price level' that ought to displaces earlier classic theories, such as the quantity theory of money.

16.17.2. *Disappearance of quantity theory?*

Actually, a version of the quantity theory coexists along with equation (16.17.1) in the Chari, Christiano, and Kehoe model. To see this, note that the government can set the initial price level $P_0(s^0)$ to any positive number by executing an appropriate time 0 open market operation.[22] Thus, we can regard the time 0 government budget constraint (16.16.9) as a constraint on a time 0 open market operation. The government can set the nominal money supply $M_1(s^0) > 0$ to an arbitrary positive number subject to the constraint $B_1(s^0) = -M_1(s^0)$ that holds under the Friedman rule. The government issues money to purchase nominally denominated bonds subject to $B_1(s^0) = -M_1(s^0)$, as given by government budget constraint (16.16.9) under the Friedman rule. The household is willing to issue these bonds in exchange for money. Presuming that the cash-in-advance constraint is satisfied with equality, the price level and money supply

---

[22] The argument of this subsection treats time 0 in a peculiar way because no endogenous variables inherited from the past impede independently manipulating time 0 (and all subsequent) nominal quantities. This special treatment of time 0 also characterizes many other presentations of the quantity theory of money as a 'pure units change' experiment that multiplies nominal quantities at all dates and all histories by the same positive scalar. Commenting on a paper by Robert Townsend at the Minneapolis Federal Reserve Bank in 1985, Ramon Marimon asked 'when is time 0?', thereby anticipating doubts expressed by Niepelt (2004).

then conform to $P_0(s^0)c_{10}(s^0) = M_1(s^0)$, which is a version of the quantity theory equation for the price level at time 0.[23] This equation provides the basis for a sharp statement of the quantity theory: with fiscal policy being held constant, a time 0 open market operation that increases $M_1(s^0)$ leads to a proportionate increase in the price level at all histories and dates while leaving the equilibrium allocation and real rates of return unaltered.

16.17.3. Price level indeterminacy under interest rate peg

Sargent and Wallace (1975) established that an interest rate peg leads to price level indeterminacy in an *ad hoc* macroeconomic model. We demonstrate how the same result arises in two different forms in the Chari, Christiano, and Kehoe model. First, using an argument similar to that of the quantity theory in section 16.17.2, we show that a proportionate increase in the price level at all histories and dates is consistent with a given interest rate peg. Second, we extend the indeterminacy result to an economy with ongoing sunspot uncertainty.

We simplify our analysis by setting all distortionary taxes, government consumption, and government bond issues equal to zero. We assume that the government can levy a real lump sum tax $\tau_t^h(s^t)$ on the household at time t after history s^t. A negative value of $\tau_t^h(s^t)$ means a lump sum transfer to the household. As with other taxes, the lump sum tax is payable in money and paid to the government in the securities market; when $\tau_t^h(s^t)$ is negative, the household receives the lump sum transfer as money in the securities market. The new version of government budget constraint (16.16.8) becomes

$$M_{t+1}\left(s^t\right) - M_t\left(s^{t-1}\right) = -P_t\left(s^t\right)\tau_t^h\left(s^t\right), \qquad (16.17.2)$$

where now the government uses the revenues from lump sum taxation to alter the money supply in the securities market. The new version of government

---

[23] Under the Friedman rule with a zero nominal interest rate, the household would be indifferent between holding excess balances of money above and beyond cash-in-advance constraint (16.16.4) or holding of nominal government bonds. Likewise, the government would be indifferent about whether to issue nominal indebtedness in the form of nominal bonds or money, because both liabilities carry the same cost to the government, either in the form of interest payments on bonds or open-market repurchases of money to deliver a deflation that amounts to the same real return on money as on bonds. However, while the composition of nominal government liabilities is indeterminate under the Friedman rule, the total amount of such liabilities and hence, the price level is determinate.

constraint (16.16.9) in period 0 is

$$M_1\left(s^0\right) = -P_0\left(s^0\right)\tau_0^h\left(s^0\right).$$ (16.17.3)

Under an interest rate peg, the government stands ready to accommodate any money demand that arises at the pegged interest rate. What will be the resulting equilibrium price level? Note from interest rate expression (16.16.7) that any proportionate increase in the price level at all histories and dates is consistent with a given interest rate peg. This means that the price level is indeterminate under an interest rate peg.

We now turn to a form of price level indeterminacy that can be driven by ongoing sunspot uncertainty. For a given interest rate peg $R_t(s^t)$ and a current price level $P_t(s^t)$, interest rate expression (16.16.7) implies a unique magnitude for the denominator of that equation, but not for the individual price levels $P_{t+1}(s^{t+1})$ after each history s^{t+1} in the next period. There exist equilibria in which next period's price level depend on a sunspot, so long as restriction (16.16.7) is satisfied with respect to the expected inverse of next period's price level, weighted by the marginal utilities in different states next period. Hence, when the government stands ready to accommodate any such system of expectations, an interest rate peg is associated with price level indeterminacy.

16.17.4. Monetary or fiscal theory of the price level?

Sargent and Wallace (1975) show that the price level becomes determinate under a money supply rule, and the same holds true in the Chari, Christiano, and Kehoe model. Specifically, when the equilibrium nominal interest is strictly positive, it follows that cash-in-advance constraint (16.16.4) holds with equality and thus, the price level is given by $P_t(s^t) = M_{t+1}(s^t)/c_{1t}(s^t)$. Price level determinacy also prevails under the Friedman rule with a zero nominal interest rate, as discussed in footnote 23. Thus, in this economy the government's ability to control the money supply gives it the ability to control the price level.

While the analysis of Sargent and Wallace (1975) is typically viewed as describing a monetary theory of the price level, proponents of the fiscal theory of the price level might reply that it is really fiscal policy as summarized by right side of government budget constraint (16.17.2) that determines the price level. If we ignore period 0, we can confirm that assertion. The argument goes as follows. Instead of thinking in terms of the quantity of money needed to support a price level $P_t(s^t)$, the government proceeds by calculating the implied real value of households' money balances carried over from last period, $M_t(s^{t-1})/P_t(s^t)$, and if that quantity is higher (lower) than the quantity of cash goods that will be transacted in the present period, $c_{1t}(s^t)$, the government sets a lump sum tax (transfer) equal to the difference,

$$\tau_t^h\left(s^t\right) = \frac{M_t\left(s^{t-1}\right)}{P_t\left(s^t\right)} - c_{1t}\left(s^t\right). \tag{16.17.4}$$

After using cash-in-advance constraint (16.16.4) at equality to eliminate the last term in expression (16.17.4), this is just government budget constraint (16.17.2). Hence, for the same reason that a money supply rule achieved price level determinacy so would the prescribed lump sum tax associated with the required fiscal policy.

So far, so good for the fiscal theory of the price level. But in the present economy, the fiscal theory runs into trouble for period 0. Since we have assumed that there are no outstanding money balances at the beginning of period 0, the fiscal policy according to expression (16.17.4) becomes

$$\tau_0^h\left(s^0\right) = -c_{10}\left(s^0\right). \tag{16.17.5}$$

That is, since the government supplies all money balances in period 0, the fiscal policy prescribes a lump sum transfer (negative tax) equal to the quantity of cash goods that will be transacted in period 0. But what will the equilibrium price level be? Because expression (16.17.5) lacks the first term on the right side of expression (16.17.4) in all subsequent periods, there is effectively no anchor for the price level in period 0. We conclude that, in contrast to a permissible exogenous process for the nominal money supply, a permissible exogenous process for real lump sum taxes and transfers is associated with price level indeterminacy because any positive number $P_0(s^0)$ can serve as the equilibrium price level in period 0.[24]

---

[24] A monetary model with a cash-in-advance constraint like (16.16.4), is well suited to deliver a conclusion that control of the money supply can lead to price level determinacy. Because by assumption, money must be used in exchanges and therefore, cash-in-advance constraint (16.16.4) takes on the appearance of a quantity-theory-of-money equation. But this support for a particular currency's value vanishes if there are substitutes that can provide the same transaction services, like a foreign country's currency, or if money is valued because of dynamic inefficiency in an overlapping generations model, as analyzed in chapter 9. When revisiting the fiscal theory of the price level in sections 26.3 and 26.4, we will return to these issues in the context of a two-country (two-currency) overlapping generations model.

A reader might regard this way of disarming a fiscal theory of the price level as a knife-edged result because any positive initial money balances, $M_0(s^{-1}) > 0$, would render the price level determinate. Perhaps, but this special example motivates us to return to the fiscal theory of the price level in chapter 26, and section 26.4.3 in particular, where we discuss Bassetto's (2002) criticism and reformulation of that theory. In criticizing the valuation-equation view of the government budget constraint in section 16.17.1, Bassetto argues that the fiscal theory of the price level should be cast as a game where government strategies are specified for any arbitrary outcomes, not just equilibrium outcomes. Such a complete specification of a policy scheme is crucial for determining whether the scheme implements a unique equilibrium, or whether instead it leaves room for multiple equilibria indexed by systems of private sector expectations.

16.18. Zero tax on human capital

Returning to a nonstochastic nonmonetary model, Jones, Manuelli, and Rossi (1997) show that the optimality of a limiting zero tax also applies to labor income in a model with human capital, h_t, so long as the technology for accumulating human capital displays constant returns to scale in the stock of human capital and goods used (not including raw labor).

We postulate the following human capital technology,

$$h_{t+1} = (1 - \delta_h)\, h_t + H\left(x_{ht}, h_t, n_{ht}\right), \qquad (16.18.1)$$

where $\delta_h \in (0, 1)$ is the rate at which human capital depreciates. The function H describes how new human capital is created with the input of a market good x_{ht}, the stock of human capital h_t, and raw labor n_{ht}. Human capital is in turn used to produce "efficiency units" of labor e_t,

$$e_t = M\left(x_{mt}, h_t, n_{mt}\right), \qquad (16.18.2)$$

where x_{mt} and n_{mt} are the market good and raw labor used in the process. We assume that both H and M are homogeneous of degree one in market goods $(x_{jt}, j = h, m)$ and human capital (h_t), and twice continuously differentiable with strictly decreasing (but everywhere positive) marginal products of all factors.

The number of efficiency units of labor e_t replaces our earlier argument for labor in the production function, $F(k_t, e_t)$. The household's preferences are still described by expression $(16.2.1)$, with leisure $\ell_t = 1 - n_{ht} - n_{mt}$. The economy's aggregate resource constraint is

$$c_t + g_t + k_{t+1} + x_{mt} + x_{ht}$$
$$= F\left[k_t, M\left(x_{mt}, h_t, n_{mt}\right)\right] + (1 - \delta)\, k_t. \qquad (16.18.3)$$

The household's present-value budget constraint is

$$\sum_{t=0}^{\infty} q_t^0 \left(1 + \tau_t^c\right) c_t = \sum_{t=0}^{\infty} q_t^0 \left[\left(1 - \tau_t^n\right) w_t e_t - \left(1 + \tau_t^m\right) x_{mt} - x_{ht}\right]$$
$$+ \left[\left(1 - \tau_0^k\right) r_0 + 1 - \delta\right] k_0 + b_0, \qquad (16.18.4)$$

where we have added τ_t^c and τ_t^m to the set of tax instruments, to enhance the government's ability to control various margins. Substitute equation $(16.18.2)$ into equation $(16.18.4)$, and let λ be the Lagrange multiplier on this budget constraint, while α_t denotes the Lagrange multiplier on equation $(16.18.1)$. The household's first-order conditions are then

$$c_t: \quad \beta^t u_c\left(t\right) - \lambda q_t^0 \left(1 + \tau_t^c\right) = 0, \qquad (16.18.5a)$$

$$n_{mt}: \quad -\beta^t u_\ell\left(t\right) + \lambda q_t^0 \left(1 - \tau_t^n\right) w_t M_n\left(t\right) = 0, \qquad (16.18.5b)$$

$$n_{ht}: \quad -\beta^t u_\ell\left(t\right) + \alpha_t H_n\left(t\right) = 0, \qquad (16.18.5c)$$

$$x_{mt}: \quad \lambda q_t^0 \left[\left(1 - \tau_t^n\right) w_t M_x\left(t\right) - \left(1 + \tau_t^m\right)\right] = 0, \qquad (16.18.5d)$$

$$x_{ht}: \quad -\lambda q_t^0 + \alpha_t H_x\left(t\right) = 0, \qquad (16.18.5e)$$

$$h_{t+1}: \quad -\alpha_t + \lambda q_{t+1}^0 \left(1 - \tau_{t+1}^n\right) w_{t+1} M_h\left(t + 1\right)$$
$$+ \alpha_{t+1} \left[1 - \delta_h + H_h\left(t + 1\right)\right] = 0. \qquad (16.18.5f)$$

Substituting equation $(16.18.5e)$ into equation $(16.18.5f)$ yields

$$\frac{q_t^0}{H_x(t)} = q_{t+1}^0 \Big[\frac{1 - \delta_h + H_h(t + 1)}{H_x(t + 1)}$$
$$+ (1 - \tau_{t+1}^n) w_{t+1} M_h(t + 1)\Big]. \qquad (16.18.6)$$

We now use the household's first-order conditions to simplify the sum on the right side of the present-value constraint $(16.18.4)$. First, note that homogeneity of H implies that equation $(16.18.1)$ can be written as

$$h_{t+1} = (1 - \delta_h)\, h_t + H_x\left(t\right) x_{ht} + H_h\left(t\right) h_t.$$

Solve for x_{ht} with this expression, use M from equation $(16.18.2)$ for e_t, and substitute into the sum on the right side of equation $(16.18.4)$, which then becomes

$$\sum_{t=0}^{\infty} q_t^0 \Big\{ (1 - \tau_t^n) \, w_t M_x(t) \, x_{mt} + (1 - \tau_t^n) \, w_t M_h(t) \, h_t$$

$$- (1 + \tau_t^m) \, x_{mt} - \frac{h_{t+1} - [1 - \delta_h + H_h(t)] \, h_t}{H_x(t)} \Big\}.$$

Here we have also invoked the homogeneity of M. First-order condition $(16.18.5d)$ implies that the term multiplying x_{mt} is zero, $[(1 - \tau_t^n) w_t M_x(t) - (1 + \tau_t^m)] = 0$. After rearranging, we are left with

$$\left[\frac{1 - \delta_h + H_h(0)}{H_x(0)} + (1 - \tau_0^n) \, w_0 M_h(0) \right] h_0 - \sum_{t=1}^{\infty} h_t \Big\{ \frac{q_{t-1}^0}{H_x(t-1)}$$

$$- q_t^0 \left[\frac{1 - \delta_h + H_h(t)}{H_x(t)} + (1 - \tau_t^n) \, w_t M_h(t) \right] \Big\}. \tag{16.18.7}$$

However, the term in braces is zero by first-order condition $(16.18.6)$, so the sum on the right side of equation $(16.18.4)$ simplifies to the very first term in this expression.

Following our standard scheme of constructing the Ramsey plan, a few more manipulations of the household's first-order conditions are needed to solve for prices and taxes in terms of the allocation. We first assume that $\tau_0^c = \tau_0^k = \tau_0^n = \tau_0^m = 0$. If the numeraire is $q_0^0 = 1$, then condition $(16.18.5a)$ implies

$$q_t^0 = \beta^t \frac{u_c(t)}{u_c(0)} \frac{1}{1 + \tau_t^c}. \tag{16.18.8a}$$

From equations $(16.18.5b)$ and $(16.18.8a)$ and $w_t = F_e(t)$, we obtain

$$(1 + \tau_t^c) \frac{u_\ell(t)}{u_c(t)} = (1 - \tau_t^n) \, F_e(t) \, M_n(t), \tag{16.18.8b}$$

and, by equations $(16.18.5c)$, $(16.18.5e)$, and $(16.18.8a)$,

$$(1 + \tau_t^c) \frac{u_\ell(t)}{u_c(t)} = \frac{H_n(t)}{H_x(t)}, \tag{16.18.8c}$$

and equation $(16.18.5d)$ with $w_t = F_e(t)$ yields

$$1 + \tau_t^m = (1 - \tau_t^n) \, F_e(t) \, M_x(t). \tag{16.18.8d}$$

For a given allocation, expressions $(16.18.8)$ allow us to recover prices and taxes in a recursive fashion: $(16.18.8c)$ defines τ_t^c and $(16.18.8a)$ can be used to compute q_t^0, $(16.18.8b)$ sets τ_t^n, and $(16.18.8d)$ pins down τ_t^m.

Only one task remains to complete our strategy of determining prices and taxes that achieve any allocation. The additional condition $(16.18.6)$ characterizes the household's intertemporal choice of human capital, which imposes still another constraint on the price q_t^0 and the tax τ_t^n. Our determination of τ_t^n in equation $(16.18.8b)$ can be thought of as manipulating the margin that the household faces in its static choice of supplying effective labor e_t, but the tax rate also affects the household's dynamic choice of human capital h_t. Thus, in the Ramsey problem, we will have to impose the extra constraint that the allocation is consistent with the same τ_t^n entering both equations $(16.18.8b)$ and $(16.18.6)$. To find an expression for this extra constraint, solve for $(1 - \tau_t^n)$ from equation $(16.18.8b)$ and a lagged version of equation $(16.18.6)$, which are then set equal to each other. We eliminate the price q_t^0 by using equations $(16.18.8a)$ and $(16.18.8c)$, and the final constraint becomes

$$
\begin{aligned}
u_\ell (t-1) H_n (t) = & \beta u_\ell (t) H_n (t-1) \\
& \cdot \left[1 - \delta_h + H_h (t) + H_n (t) \frac{M_h (t)}{M_n (t)} \right].
\end{aligned} \tag{16.18.9}
$$

Proceeding to step 2 in constructing the Ramsey plan, we use condition $(16.18.8a)$ to eliminate $q_t^0 (1 + \tau_t^c)$ in the household's budget constraint $(16.18.4)$. After also invoking the simplified expression $(16.18.7)$ for the sum on the right side of $(16.18.4)$, the implementability condition can be written as

$$
\sum_{t=0}^{\infty} \beta^t u_c (t) c_t - \tilde{A} = 0, \tag{16.18.10}
$$

where $\tilde{A}$ is given by

$$
\begin{aligned}
\tilde{A} = & \tilde{A} (c_0, n_{m0}, n_{h0}, x_{m0}, x_{h0}) \\
= & u_c (0) \left\{ \left[\frac{1 - \delta_h + H_h (0)}{H_x (0)} + F_e (0) M_h (0) \right] h_0 \right. \\
& \left. + \left[F_k (0) + 1 - \delta_k \right] k_0 + b_0 \right\}.
\end{aligned}
$$

In step 3, we define

$$V(c_t, n_{mt}, n_{ht}, \Phi) = u(c_t, 1 - n_{mt} - n_{ht}) + \Phi u_c(t) c_t, \qquad (16.18.11)$$

and formulate a Lagrangian,

$$
\begin{aligned}
J = \sum_{t=0}^{\infty} \beta^t \Big\{ & V(c_t, n_{mt}, n_{ht}, \Phi) \\
& + \theta_t \Big\{ F[k_t, M(x_{mt}, h_t, n_{mt})] + (1 - \delta) k_t \\
& \qquad - c_t - g_t - k_{t+1} - x_{mt} - x_{ht} \Big\} \\
& + \nu_t [(1 - \delta_h) h_t + H(x_{ht}, h_t, n_{ht}) - h_{t+1}] \Big\} - \Phi \tilde{A}. \quad (16.18.12)
\end{aligned}
$$

This formulation would correspond to the Ramsey problem if it were not for the missing constraint (16.18.9). Following Jones, Manuelli, and Rossi (1997), we will solve for the first-order conditions associated with equation (16.18.12), and when it is evaluated at a steady state, we can verify that constraint (16.18.9) is satisfied even though it has not been imposed. Thus, if both the problem in expression (16.18.12) and the proper Ramsey problem with constraint (16.18.9) converge to a unique steady state, they will converge to the same steady state.

The first-order conditions for equation (16.18.12) evaluated at the steady state are

$$c: \quad V_c = \theta \qquad (16.18.13a)$$

$$n_m: \quad V_{n_m} = -\theta F_e M_n \qquad (16.18.13b)$$

$$n_h: \quad V_{n_h} = -\nu H_n \qquad (16.18.13c)$$

$$x_m: \quad 1 = F_e M_x \qquad (16.18.13d)$$

$$x_h: \quad \theta = \nu H_x \qquad (16.18.13e)$$

$$h: \quad 1 = \beta \left(1 - \delta_h + H_h + \frac{\theta}{\nu} F_e M_h \right) \qquad (16.18.13f)$$

$$k: \quad 1 = \beta (1 - \delta_k + F_k). \qquad (16.18.13g)$$

Note that $V_{n_m} = V_{n_h}$, so by conditions (16.18.13b) and (16.18.13c),

$$\frac{\theta}{\nu} = \frac{H_n}{F_e M_n}, \qquad (16.18.14)$$

which we substitute into equation $(16.18.13g)$,

$$1 = \beta \left(1 - \delta_h + H_h + H_n \frac{M_h}{M_n} \right). \tag{16.18.15}$$

Condition $(16.18.15)$ coincides with constraint $(16.18.9)$, evaluated in a steady state. In other words, we have confirmed that the problem $(16.18.12)$ and the proper Ramsey problem with constraint $(16.18.9)$ share the same steady state, under the maintained assumption that both problems converge to a unique steady state.

What is the optimal τ^n? The substitution of equation $(16.18.13e)$ into equation $(16.18.14)$ yields

$$H_x = \frac{H_n}{F_e M_n}. \tag{16.18.16}$$

The household's first-order conditions $(16.18.8b)$ and $(16.18.8c)$ imply in a steady state that

$$(1 - \tau^n) H_x = \frac{H_n}{F_e M_n}. \tag{16.18.17}$$

It follows immediately from equations $(16.18.16)$ and $(16.18.17)$ that $\tau^n = 0$. Given $\tau^n = 0$, conditions $(16.18.8d)$ and $(16.18.13d)$ imply $\tau^m = 0$. We conclude that in the present model neither labor nor capital should be taxed in the limit.

16.19. Should all taxes be zero?

The optimal steady-state tax policy of the model in the previous section is to set $\tau^k = \tau^n = \tau^m = 0$. However, in general, this implies $\tau^c \neq 0$. To see this point, use equation $(16.18.8b)$ and $\tau^n = 0$ to get

$$1 + \tau^c = \frac{u_c}{u_\ell} F_e M_n. \tag{16.19.1}$$

From equations $(16.18.13a)$ and $(16.18.13b)$

$$F_e M_n = -\frac{V_{n_m}}{V_c} = \frac{u_\ell + \Phi u_{c\ell} c}{u_c + \Phi (u_c + u_{cc} c)}. \tag{16.19.2}$$

Hence,

$$1 + \tau^c = \frac{u_c u_\ell + \Phi u_c u_{c\ell} c}{u_c u_\ell + \Phi (u_c u_\ell + u_{cc} u_\ell c)}. \tag{16.19.3}$$

As discussed earlier, a first-best solution without distortionary taxation has $\Phi = 0$, so τ^c should trivially be set equal to zero. In a second-best solution, $\Phi > 0$ and we get $\tau^c = 0$ if and only if

$$u_c u_{c\ell} c = u_c u_\ell + u_{cc} u_\ell c, \tag{16.19.4}$$

which is in general not satisfied. However, Jones, Manuelli, and Rossi (1997) point out one interesting class of utility functions that is consistent with equation (16.19.4):

$$u\left(c,\ell\right) = \begin{cases} \dfrac{c^{1-\sigma}}{1-\sigma} v\left(\ell\right) & \text{if } \sigma > 0, \sigma \neq 1 \\ \ln\left(c\right) + v\left(\ell\right) & \text{if } \sigma = 1. \end{cases}$$

If a steady state exists, the optimal solution for these preferences is eventually to set all taxes equal to zero. It follows that the optimal plan involves collecting tax revenues in excess of expenditures in the initial periods. When the government has amassed claims against the private sector so large that the interest earnings suffice to finance g, all taxes are set equal to zero. Since the steady-state interest rate is $R = \beta^{-1}$, we can use the government's budget constraint (16.2.5) to find the corresponding value of government indebtedness

$$b = \frac{\beta}{\beta - 1} g < 0.$$

16.20. Concluding remarks

Perhaps the most startling finding of this chapter is that the optimal steady-state tax on physical capital in a nonstochastic economy is equal to *zero*. The result that capital should not be taxed in the steady state is robust to whether or not the government must balance its budget in each period and to any redistributional concerns arising from a social welfare function. As a stark illustration, Judd's (1985b) example demonstrates that the result holds when the government is constrained to run a balanced budget and when it cares only about the workers who are exogenously constrained to not hold any assets. Thus, the capital owners who are assumed not to work will be exempt from taxation in the steady state, and the government will finance its expenditures solely by levying wage taxes on the group of agents that it cares about.

It is instructive to consider Jones, Manuelli, and Rossi's (1997) extension of the no-tax result to labor income, or more precisely human capital. They ask rhetorically, Is physical capital special? We are inclined to answer yes to this question for the following reason. The zero tax on human capital is derived in a model where the production of both human capital and "efficiency units" of labor show constant, returns to scale in the stock of human capital and the use of final goods but not raw labor which otherwise enters as an input in the production functions. These assumptions explain why the stream of future labor income in the household's present-value budget constraint in equation (16.18.4) is reduced to the first term in equation (16.18.7), which is the value of the household's human capital at time 0. Thus, the functional forms have made raw labor disappear as an object for taxation in future periods. Or in the words of Jones, Manuelli, and Rossi (1997, pp. 103 and 99), "Our zero tax results are driven by zero profit conditions. Zero profits follow from the assumption of linearity in the accumulation technologies. Since the activity 'capital income' and the activity 'labor income' display constant returns to scale in reproducible factors, their 'profits' cannot enter the budget constraint in equilibrium." But for alternative production functions that make the endowment of raw labor reappear, the optimal labor tax would not be zero. It is for this reason that we think physical capital is special: because the zero-tax result arises with the minimal assumptions of the standard neoclassical growth model, while the zero-tax result on labor income requires that raw labor vanish from the agent's present-value budget constraints.[25]

The weaknesses of our optimal steady-state tax analysis are that it says nothing about how long it takes to reach the zero tax on capital income and how taxes and any redistributive transfers are set during the transition period. These questions have to be studied numerically, as was done by Chari, Christiano, and Kehoe (1994), though their paper does not involve any redistributional concerns because of the assumption of a representative agent. Domeij

---

[25] One special case of Jones, Manuelli, and Rossi's (1997) framework with its zero-tax result for labor is Lucas's (1988) endogenous growth model studied in chapter 15. Recall our alternative interpretation of that model as one without any nonreproducible raw labor but just two reproducible factors: physical and human capital. No wonder that raw labor in Lucas's model does not affect the optimal labor tax, since the model can equally well be thought of as an economy without raw labor.

and Heathcote (2000) construct a model with heterogeneous agents and incomplete insurance markets to study the welfare implications of eliminating capital income taxation. Using earnings and wealth data from the United States, they calibrate a stochastic process for labor earnings that implies a wealth distribution of asset holdings resembling the empirical one. Setting initial tax rates equal to estimates of present taxes in the United States, they study the effects of an unexpected policy reform that sets the capital tax permanently equal to zero and raises the labor tax to maintain long-run budget balance. They find that a majority of households prefers the status quo to the tax reform because of the distributional implications.

This example illustrates the importance of a well-designed tax and transfer policy in the transition to a new steady state. In addition, as shown by Aiyagari (1995), the optimal capital tax in a heterogeneous-agent model with incomplete insurance markets is actually positive, even in the long run. A positive capital tax is used to counter the tendency of such an economy to overaccumulate capital because of too much precautionary saving. We say more about these heterogeneous-agent models in chapter 18.

Golosov, Kocherlakota, and Tsyvinski (2003) pursue another way of disrupting the connection between stationary values of the two key Euler equations that underlie Chamley and Judd's zero-tax-on-capital outcome. They put the Ramsey planner in a private information environment in which it cannot observe the hidden skill levels of different households. That impels the planner to design the tax system as an optimal dynamic incentive mechanism that trades off current and continuation values in an optimal way. We discuss such mechanisms for coping with private information in chapter 20. Because the information problem alters the planner's Euler equation for the household's consumption, Chamley and Judd's result does not hold for this environment.

An assumption maintained throughout the chapter has been that the government can commit to future tax rates when solving the Ramsey problem at time 0. As noted earlier, taxing the capital stock at time 0 amounts to lump-sum taxation and therefore disposes of distortionary taxation. It follows that a government without a commitment technology would be tempted in future periods to renege on its promises and levy a confiscatory tax on capital. An interesting question arises: can the incentive to maintain a good reputation replace a commitment technology? That is, can a promised policy be sustained in an equilibrium because the government wants to preserve its reputation? Reputation involves history dependence and incentives and will be studied in chapter 26.

Exercises

Exercise 16.1   **A small open economy** (Razin and Sadka, 1995)

Consider the nonstochastic model with capital and labor in this chapter, but assume that the economy is a small open economy that cannot affect the international rental rate on capital, r_t^*. Domestic firms can rent any amount of capital at this price, and the households and the government can choose to go short or long in the international capital market at this rental price. There is no labor mobility across countries. We retain the assumption that the government levies a tax τ_t^n on each household's labor income, but households no longer have to pay taxes on their capital income. Instead, the government levies a tax $\hat{\tau}_t^k$ on domestic firms' rental payments to capital regardless of the capital's origin (domestic or foreign). Thus, a domestic firm faces a total cost of $(1 + \hat{\tau}_t^k)r_t^*$ on a unit of capital rented in period t.

a. Solve for the optimal capital tax $\hat{\tau}_t^k$.

b. Compare the optimal tax policy of this small open economy to that of the closed economy of this chapter.

Exercise 16.2   **Consumption taxes**

Consider the nonstochastic model with capital and labor in this chapter, but instead of labor and capital taxation assume that the government sets labor and consumption taxes, $\{\tau_t^n, \tau_t^c\}$. Thus, the household's present-value budget constraint is now given by

$$\sum_{t=0}^{\infty} q_t^0 \left(1 + \tau_t^c\right) c_t = \sum_{t=0}^{\infty} q_t^0 \left(1 - \tau_t^n\right) w_t n_t + \left[r_0 + 1 - \delta\right] k_0 + b_0.$$

a. Solve for the Ramsey plan.

b. Suppose that the solution to the Ramsey problem converges to a steady state. Characterize the optimal limiting sequence of consumption taxes.

c. In the case of capital taxation, we imposed an exogenous upper bound on τ_0^k. Explain why a similar exogenous restriction on τ_0^c is needed to ensure an interesting Ramsey problem. (*Hint:* Explore the implications of setting $\tau_t^c = \tau^c$ and $\tau_t^n = -\tau^c$ for all $t \geq 0$, where τ^c is a large positive number.)

Exercise 16.3   **Specific utility function** (Chamley, 1986)

Consider the nonstochastic model with capital and labor in this chapter, and assume that the period utility function in equation (16.2.1) is given by

$$u\left(c_t, \ell_t\right) = \frac{c_t^{1-\sigma}}{1 - \sigma} + v\left(\ell_t\right),$$

where $\sigma > 0$. When σ is equal to 1, the term $c_t^{1-\sigma}/(1 - \sigma)$ is replaced by $\log(c_t)$.

a. Show that the optimal tax policy in this economy is to set capital taxes equal to zero in period 2 and from there on, i.e., $\tau_t^k = 0$ for $t \geq 2$. (*Hint:* Given the preference specification, evaluate and compare equations (16.6.4) and (16.6.9a).)

b. Suppose there is uncertainty in the economy, as in the stochastic model with capital and labor in this chapter. Derive the optimal *ex ante capital tax rate* for $t \geq 2$.

Exercise 16.4   **Two labor inputs** (Jones, Manuelli, and Rossi, 1997)

Consider the nonstochastic model with capital and labor in this chapter, but assume that there are two labor inputs, n_{1t} and n_{2t}, entering the production function, $F(k_t, n_{1t}, n_{2t})$. The household's period utility function is still given by $u(c_t, \ell_t)$ where leisure is now equal to

$$\ell_t = 1 - n_{1t} - n_{2t}.$$

Let τ_{it}^n be the flat-rate tax at time t on wage earnings from labor n_{it}, for $i = 1, 2$, and τ_t^k denotes the tax on earnings from capital.

a. Solve for the Ramsey plan. What is the relationship between the optimal tax rates τ_{1t}^n and τ_{2t}^n for $t \geq 1$? Explain why your answer is different for period $t = 0$. As an example, assume that k and n_1 are complements while k and n_2 are substitutes.

We now assume that the period utility function is given by $u(c_t, \ell_{1t}, \ell_{2t})$ where

$$\ell_{1t} = 1 - n_{1t}, \qquad \text{and} \qquad \ell_{2t} = 1 - n_{2t}.$$

Further, the government is now constrained to set the same tax rate on both types of labor, i.e., $\tau_{1t}^n = \tau_{2t}^n$ for all $t \geq 0$.

b. Solve for the Ramsey plan. (*Hint:* Using the household's first-order conditions, we see that the restriction $\tau_{1t}^n = \tau_{2t}^n$ can be incorporated into the Ramsey problem by adding the constraint $u_{\ell_1}(t) F_{n_2}(t) = u_{\ell_2}(t) F_{n_1}(t)$.)

c. Suppose that the solution to the Ramsey problem converges to a steady state where the constraint that the two labor taxes should be equal is binding. Show that the limiting capital tax is not zero unless $F_{n_1} F_{n_2 k} = F_{n_2} F_{n_1 k}$.

Exercise 16.5   **Another specific utility function**

Consider the following optimal taxation problem. There is no uncertainty. There is one good that is produced by labor x_t of the representative household, and that can be divided among private consumption c_t and government consumption g_t subject to

$$c_t + g_t = 1 - x_t. \tag{0}$$

The good is produced by zero-profit competitive firms that pay the worker a pretax wage of 1 per unit of $1 - x_t$ (i.e., the wage is tied down by the linear technology). A representative consumer maximizes

$$\sum_{t=0}^{\infty} \beta^t u\left(c_t, x_t\right) \tag{1}$$

subject to the sequence of budget constraints

$$c_t + q_t b_{t+1} \leq (1 - \tau_t)(1 - x_t) + b_t \tag{2}$$

where c_t is consumption, x_t is leisure, q_t is the price of consumption at $t+1$ in units of time t consumption, and b_t is a stock of one-period IOUs owned by the household and falling due at time t. Here τ_t is a flat-rate tax on the household's labor supply $1 - x_t$. Assume that $u(c, x) = c - .5(1 - x)^2$.

a. Argue that in a competitive equilibrium, $q_t = \beta$ and $x_t = \tau_t$.

b. Argue that in a competitive equilibrium with $b_0 = 0$ and $\lim_{t \to \infty} \beta^t b_t = 0$, the sequence of budget constraints (2) imply the following single intertemporal constraint:

$$\sum_{t=0}^{\infty} \beta^t \left(c_t - (1 - x_t)(1 - \tau_t) \right) = 0.$$

Given an exogenous sequence of government purchases $\{g_t\}_{t=0}^{\infty}$, a government wants to maximize (1) subject both to the budget constraint

$$\sum_{t=0}^{\infty} \beta^t \left(g_t - \tau_t (1 - x_t) \right) = 0 \tag{3}$$

and to the household's first-order condition

$$x_t = \tau_t. \tag{4}$$

c. Consider the following government expenditure process defined for $t \geq 0$:

$$g_t = \begin{cases} 0, & \text{if } t \text{ is even;} \\ .2, & \text{if } t \text{ is odd;} \end{cases}$$

Solve the Ramsey plan. Show that the optimal tax rate is given by

$$\tau_t = \bar{\tau} \quad \forall t \geq 0.$$

Please compute the value for $\bar{\tau}$ when $\beta = .95$.

d. Consider the following government expenditure process defined for $t \geq 0$:

$$g_t = \begin{cases} .2, & \text{if } t \text{ is even;} \\ 0, & \text{if } t \text{ is odd;} \end{cases}$$

Show that $\tau_t = \bar{\tau} \ \forall t \geq 0$. Compute $\bar{\tau}$ and comment on whether it is larger or smaller than the value you computed in part (c).

e. Interpret your results in parts c and d in terms of "tax-smoothing."

f. Under what circumstances, if any, would $\bar{\tau} = 0$?

Exercise 16.6   **Yet another specific utility function**

Consider an economy with a representative household with preferences over streams of consumption c_t and labor supply n_t that are ordered by

$$\sum_{t=0}^{\infty} \beta^t \left(c_t - u_1 n_t - .5 u_2 n_t^2 \right), \quad \beta \in (0, 1) \tag{1}$$

where $u_1, u_2 > 0$. The household operates a linear technology

$$y_t = n_t, \tag{2}$$

where y_t is output. There is no uncertainty. There is a government that finances an exogenous stream of government purchases $\{g_t\}$ by a flat rate tax τ_t on labor. The feasibility condition for the economy is

$$y_t = c_t + g_t. \tag{3}$$

At time 0 there are complete markets in dated consumption goods. Let q_t be the price of a unit of consumption at date t in terms of date 0 consumption. The budget constraints for the household and the government, respectively, are

$$\sum_{t=0}^{\infty} q_t \left[(1 - \tau_t) n_t - c_t \right] = 0 \tag{4}$$

$$\sum_{t=0}^{\infty} q_t \left(\tau_t n_t - g_t \right) = 0. \tag{5}$$

Part I. Call a tax rate process $\{\tau_t\}$ budget feasible if it satisfies (5).

a. Define a competitive equilibrium with taxes.

Part II. A Ramsey planner chooses a competitive equilibrium to maximize (1).

b. Formulate the Ramsey problem. Get as far as you can in solving it for the Ramsey plan, i.e., compute the competitive equilibrium price system and tax policy under the Ramsey plan. How does the Ramsey plan pertain to "tax smoothing?"

c. Consider two possible government expenditure sequences: Sequence A: $\{g_t\} = \{0, g, 0, g, 0, g, \ldots\}$. Sequence B: $\{g_t\} = \{\beta g, 0, \beta g, 0, \beta g, 0, \ldots\}$. Please tell how the Ramsey equilibrium tax rates and interest rates differ across the two equilibria associated with sequence A and sequence B.

Exercise 16.7 **Comparison of tax systems**

Consider an economy with a representative household that orders consumption, leisure streams $\{c_t, \ell_t\}_{t=0}^{\infty}$ according to

$$\sum_{t=0}^{\infty} \beta^t u(c_t, \ell_t), \qquad \beta \in (0,1)$$

where u is increasing, strictly concave, and twice continuously differentiable in c and ℓ. The household is endowed with one unit of time that can be used for leisure ℓ_t and labor n_t; $\ell_t + n_t = 1$.

A single good is produced with labor n_t and capital k_t as inputs. The output can be consumed by households, used by the government, or used to augment the capital stock. The technology is described by

$$c_t + g_t + k_{t+1} = F(k_t, n_t) + (1 - \delta) k_t,$$

where $\delta \in (0,1)$ is the rate at which capital depreciates, and $g_t \geq 0$ is an exogenous amount of government purchases in period t. The production function $F(k,n)$ exhibits constant returns to scale.

The government finances its purchases by levying two flat-rate, time varying taxes $\{\tau_t^n, \tau_t^a\}_{t=0}^{\infty}$. τ_t^n is a tax on labor earnings and the tax revenue from this source in period t is equal to $\tau_t^n w_t^n n_t$, where w_t is the wage rate. τ_t^a is a tax on capital earnings and the asset value of the capital stock net of depreciation. That is, the tax revenue from this source in period t is equal to $\tau_t^a (r_t + 1 - \delta) k_t$, where r_t is the rental rate on capital. We assume that the tax rates in period 0 cannot be chosen by the government but must be set equal to zero, $\tau_0^n = \tau_0^a = 0$. The government can trade one-period bonds. We assume that there is no outstanding government debt at time 0.

a. Formulate the Ramsey problem, and characterize the optimal government policy using the primal approach to taxation.

b. Show that if there exists a steady state Ramsey allocation, the limiting tax rate τ_∞^a is zero.

Consider another economy with identical preferences, endowment, technology and government expenditures but where labor taxation is forbidden. Instead of a labor tax this economy must use a consumption tax $\tilde{\tau}_t^c$. (We use a tilde to distinguish outcomes in this economy as compared to the previous economy.) Hence, this economy's tax revenues in period t are equal to $\tilde{\tau}_t^c \tilde{c}_t + \tilde{\tau}_t^a (\tilde{r}_t + 1 - \delta) \tilde{k}_t$. We assume that the tax rates in period 0 cannot be chosen by the government but must be set equal to zero, $\tilde{\tau}_0^c = \tilde{\tau}_0^a = 0$. And as before, the government can trade in one-period bonds and there is no outstanding government debt at time 0.

c. Formulate the Ramsey problem, and characterize the optimal government policy using the primal approach to taxation.

Let the allocation and tax rates that solve the Ramsey problem in question **a** be given by $\Omega \equiv \{c_t, \ell_t, n_t, k_{t+1}, \tau_t^n, \tau_t^a\}_{t=0}^{\infty}$. And let the allocation and tax rates that solve the Ramsey problem in question **c** be given by $\tilde{\Omega} \equiv \{\tilde{c}_t, \tilde{\ell}_t, \tilde{n}_t, \tilde{k}_{t+1}, \tilde{\tau}_t^c, \tilde{\tau}_t^a\}_{t=0}^{\infty}$.

d. Make a careful argument for how the allocation $\{c_t, \ell_t, n_t, k_{t+1}\}_{t=0}^{\infty}$ compares to the allocation $\{\tilde{c}_t, \tilde{\ell}_t, \tilde{n}_t, \tilde{k}_{t+1}\}_{t=0}^{\infty}$.

e. Find expressions for the tax rates $\{\tilde{\tau}_t^c, \tilde{\tau}_t^a\}_{t=1}^{\infty}$ solely in terms of $\{\tau_t^n, \tau_t^a\}_{t=1}^{\infty}$.

f. Write down the government's present value budget constraint in the first economy which holds with equality for the allocation and tax rates as given by Ω. Can you manipulate this expression so that you arrive at the government's present value budget constraint in the second economy by only using your characterization of $\tilde{\Omega}$ in terms of Ω in questions **d** and **e**?

Exercise 16.8 **Taxes on capital, labor, and consumption, I**

Consider the nonstochastic version of the model with capital and labor in this chapter, but now assume that the government sets labor, capital, and consumption taxes, $\{\tau_t^n, \tau_t^k, \tau_t^c\}$. Thus, the household's present-value budget constraint is

$$\sum_{t=0}^{\infty} q_t^0 \left(1 + \tau_t^c\right) c_t = \sum_{t=0}^{\infty} q_t^0 \left(1 - \tau_t^n\right) w_t n_t + \left[r_0 \left(1 - \tau_t^k\right) + 1 - \delta\right] k_0 + b_0.$$

a. Suppose that there are upper bounds on τ_0^k and τ_0^c. Solve for the Ramsey plan. Is the Ramsey tax system unique?

b. Take an arbitrary (non-optimal) tax policy that sets $\tau_t^c = 0, \tau_t^n = \hat{\tau}^n, \tau_t^k = \hat{\tau}^k$ for all $t \geq 0$. Let the equilibrium allocation under this policy be $\{\hat{c}_t, \hat{n}_t, \hat{k}_{t+1}\}_{t=0}^{\infty}$.

Show that you can support the same (ˆ) allocation with a different tax policy, namely one that sets $\tau_t^k = 0$ for all t and time varying taxes τ_t^c, τ_t^n for $t \geq 0$. Find expressions for the time varying taxes τ_t^c, τ_t^n as functions of the original (ˆ) tax policy and the (ˆ) equilibrium allocation.

c. Interpret the time-varying taxes that you computed in part **b** in terms of the intertemporal distortions on the final goods (c_t, ℓ_t) that are in effect induced by a constant tax on capital. (By 'final goods' we refer to the goods that appear in the representative household's utility function.)

d. Interpret the outcomes in parts **b** and **c** in terms of the following advice that comes from the asymptotic properties of the Ramsey plan. Lesson 1: don't distort intertemporal margins in the limit. Lesson 2: distort intratemporal margins in the same way each period.

Exercise 16.9 **Taxes on capital, labor, and consumption, II**

Consider the nonstochastic version of the model with capital and labor in this chapter, but now assume that the government sets labor, capital, and consumption taxes, $\{\tau_t^n, \tau_t^k, \tau_t^c\}$. The household's present-value budget constraint is

$$\sum_{t=0}^{\infty} q_t^0 \left(1 + \tau_t^c\right) c_t = \sum_{t=0}^{\infty} q_t^0 \left(1 - \tau_t^n\right) w_t n_t + \left[r_0 \left(1 - \tau_t^k\right) + 1 - \delta\right] k_0 + b_0.$$

a. Suppose that there are upper bounds on τ_0^k and τ_0^c. Solve for the Ramsey plan. Is the Ramsey tax system unique?

b. Take an arbitrary (non-optimal) tax policy that involves $\tau_t^c = 0, \tau_t^n = \hat{\tau}^n, \tau_t^k = \hat{\tau}^k$ for all $t \geq 0$. Let the equilibrium allocation under this policy be $\{\hat{c}_t, \hat{n}_t, \hat{k}_{t+1}\}_{t=0}^{\infty}$.

c. Freeze the capital tax policy $\tau_t^k = \hat{\tau}^k$ for all $t \geq 0$. But now allow $\{\tau_t^c, \tau_t^n\}_{t=0}^{\infty}$ to vary through time. Can you find a tax policy within this class that solves the Ramsey problem?

d. Interpret the outcome in part **c** in terms of the following advice that comes from the asymptotic properties of the Ramsey plan. Lesson 1: don't distort intertemporal margins in the limit. Lesson 2: distort intratemporal margins in the same way each period.

Exercise 16.10 **Lucas and Stokey (1983) model**

Consider the following version of Lucas and Stokey's (1983) model of optimal taxation with complete markets. A stochastic state s_t at time t determines an exogenous shock to government purchases $g_t(s_t)$. The history of events s^t index history-contingent commodities: $c_t(s^t)$, $\ell_t(s^t)$, and $n_t(s^t)$ are the household's consumption, leisure, and labor at time t given history s^t. There is no capital.

A representative household's preferences are ordered by

$$\sum_{t=0}^{\infty} \sum_{s^t} \beta^t \pi_t \left(s^t \right) u \left[c_t \left(s^t \right), \ell_t \left(s^t \right) \right], \tag{1}$$

where $\ell_t \in [0,1]$. The household has quasi-linear utility function

$$u \left(c, \ell \right) = c + H \left(\ell \right) \tag{2}$$

where $H' > 0, H'' < 0$, and H''' is well defined. The production function is linear in the only input labor, so the feasibility condition at t, s^t is

$$c_t \left(s^t \right) + g_t \left(s^t \right) = 1 - \ell_t$$

Given history s^t at time t, the government finances its exogenous purchase $g_t(s_t)$ and any debt obligation by levying a flat-rate tax on earnings from labor at rate $\tau_t^n(s^t)$, and by issuing state-contingent debt. Let $b_{t+1}(s_{t+1}|s^t)$ be government indebtedness to the private sector at the beginning of period $t+1$ if event s_{t+1} is realized. This state-contingent asset is traded in period t at the price $p_t(s_{t+1}|s^t)$, in terms of time t goods. The government's budget constraint is

$$\begin{aligned} g_t \left(s_t \right) = & \tau_t^n \left(s^t \right) w_t \left(s^t \right) n_t \left(s^t \right) \\ & + \sum_{s_{t+1}} p_t \left(s_{t+1}|s^t \right) b_{t+1} \left(s_{t+1}|s^t \right) - b_t \left(s_t|s^{t-1} \right), \end{aligned} \tag{3}$$

where $w_t(s^t)$ is the market-determined wage rate for labor.

a. Formulate the Ramsey problem using the 'primal approach'. Get as far as you can in solving it.

b. Describe how c_t, ℓ_t, and τ_t^n behave as functions of g_t.

c. Suppose that government expenditures g_t are drawn from the following stochastic process: $g_t = 0$ for $t = 0, \ldots, T-1$; for all $t \geq T, g_t = .5$ with probability $\alpha \in (0,1)$, and for all $t \geq T, g_t = 0$ with probability $(1 - \alpha)$. All uncertainty about g_t for $t \geq T$ is resolved at time T. Describe the government's optimal tax-debt strategy as it unfolds as time passes and chance occurs.

Part IV

The savings problem and Bewley models

Part IV

The auction problem and Arrovian surplus

Chapter 17
Self-Insurance

17.1. Introduction

This chapter describes a version of what is sometimes called a savings problem (e.g., Chamberlain and Wilson, 2000). A consumer wants to maximize the expected discounted sum of a concave function of one-period consumption rates, as in chapter 8. However, the consumer is cut off from all insurance markets and almost all asset markets. The consumer can purchase only nonnegative amounts of a single risk-free asset. The absence of insurance opportunities induces the consumer to use variations over time in his asset holdings to acquire "self-insurance."

This model is interesting to us partly as a benchmark to compare with the complete markets model of chapter 8 and some of the recursive contracts models of chapters 20 and 21, where information and enforcement problems restrict allocations relative to chapter 8, but nevertheless permit more insurance than is allowed in this chapter. A version of the single-agent model of this chapter will also be an important component of the incomplete markets models of chapter 18. Finally, the chapter provides our first encounter with the powerful supermartingale convergence theorem.

To highlight the effects of uncertainty and borrowing constraints, we shall study versions of the savings problem under alternative assumptions about the stringency of the borrowing constraint and about whether the household's endowment stream is known or uncertain.

17.2. The consumer's environment

An agent orders consumption streams according to

$$E_0 \sum_{t=0}^{\infty} \beta^t u\left(c_t\right), \tag{17.2.1}$$

where $\beta \in (0,1)$, and $u(c)$ is a strictly increasing, strictly concave, twice continuously differentiable function of the consumption of a single good c. The agent is endowed with an infinite random sequence $\{y_t\}_{t=0}^{\infty}$ of the good. Each period, the endowment takes one of a finite number of values, indexed by $s \in \mathbf{S}$. In particular, the set of possible endowments is $\bar{y}_1 < \bar{y}_2 < \cdots < \bar{y}_S$. Elements of the sequence of endowments are independently and identically distributed with $\text{Prob}(y = \bar{y}_s) = \Pi_s, \Pi_s \geq 0$, and $\sum_{s \in \mathbf{S}} \Pi_s = 1$. There are no insurance markets.

The agent can hold nonnegative amounts of a single risk-free asset that has a net rate of return r, where $(1 + r)\beta = 1$. Let $a_t \geq 0$ be the agent's assets at the beginning of period t, including the current realization of the income process. (Later we shall use an alternative notation by defining $b_t = -a_t + y_t$ as the *debt* of the consumer at the beginning of period t, *excluding* the time t endowment.) We assume that $a_0 = y_0$ is drawn from the time-invariant endowment distribution $\{\Pi_s\}$. (This is equivalent to assuming that $b_0 = 0$ in the alternative notation.) The agent faces the sequence of budget constraints

$$a_{t+1} = (1 + r)\left(a_t - c_t\right) + y_{t+1}, \tag{17.2.2}$$

where $0 \leq c_t \leq a_t$, with a_0 given. That $c_t \leq a_t$ expresses the constraint that holdings of the asset at the end of the period (which evidently equal $\frac{a_{t+1} - y_{t+1}}{1+r}$) must be nonnegative. The very important constraint $c_t \geq 0$ is either imposed or comes from an Inada condition $\lim_{c \downarrow 0} u'(c) = +\infty$.

The Bellman equation for an agent with $a > 0$ is

$$V(a) = \max_c \left\{ u(c) + \sum_{s=1}^{S} \beta \, \Pi_s V\left[(1+r)(a-c) + \bar{y}_s\right] \right\} \tag{17.2.3}$$

$$\text{subject to} \quad 0 \leq c \leq a,$$

where $\bar{y}_s$ is the income realization in state $s \in \mathbf{S}$. The value function $V(a)$ inherits some properties from $u(c)$; in particular, $V(a)$ is increasing, strictly concave, and differentiable.

"Self-insurance" occurs when the agent uses savings to insure himself against income fluctuations. On the one hand, in response to low income realizations, an agent can draw down his savings and avoid temporary large drops in consumption. On the other hand, the agent can partly save high income realizations in anticipation of poor outcomes in the future. We are interested in the long-run properties of an optimal "self-insurance" scheme. Will the agent's future consumption settle down around some level $\bar{c}$?[1] Or will the agent eventually become impoverished?[2] Following the analysis of Chamberlain and Wilson (2000) and Sotomayor (1984), we will show that neither of these outcomes occurs: consumption will diverge to infinity!

Before analyzing it under uncertainty, we'll briefly consider the savings problem under a certain endowment sequence. With a nonrandom endowment that does not grow perpetually, consumption *does* converge.

17.3. Nonstochastic endowment

Without uncertainty, the question of insurance is moot. However, it is instructive to study the optimal consumption decisions of an agent with an uneven income stream who faces a borrowing constraint. We break our analysis of the nonstochastic case into two parts, depending on the stringency of the borrowing constraint. We begin with the least stringent possible borrowing constraint, namely, the natural borrowing constraint on one-period Arrow securities, which are risk free in the current context. After that, we'll arbitrarily tighten the borrowing constraint to arrive at the no-borrowing condition $a_{t+1} \geq y_{t+1}$ imposed in the statement of the problem in the previous section. With the natural borrowing constraint, the outcome is that the agent completely smooths consumption, having a constant consumption rate over time. With the more stringent no-borrowing constraint, in general the outcome will be different. Here consumption will be a monotonic increasing sequence with jumps in consumption at times when the no-borrowing constraint binds.

---

[1] As will occur in the model of social insurance without commitment, to be analyzed in chapter 20.

[2] As in the case of social insurance with asymmetric information, to be analyzed in chapter 20.

For convenience, we temporarily use our alternative notation. We let b_t be the amount of one-period *debt* that the consumer *owes* at time t; b_t is related to a_t by

$$a_t = -b_t + y_t,$$

with $b_0 = 0$. Here $-b_t$ is the consumer's asset position *before* the realization of his time t endowment. In this notation, the time t budget constraint (17.2.2) becomes

$$c_t + b_t \leq \beta b_{t+1} + y_t \tag{17.3.1}$$

where in terms of b_{t+1}, we would express a no-borrowing constraint ($a_{t+1} \geq y_{t+1}$) as

$$b_{t+1} \leq 0. \tag{17.3.2}$$

The no-borrowing constraint (17.3.2) is evidently more stringent than the natural borrowing constraint on one-period Arrow securities that we imposed in chapter 8. Under an Inada condition on $u(c)$ at $c = 0$, or alternatively when $c_t \geq 0$ is imposed, the natural borrowing constraint in this nonstochastic case is found by solving (17.3.1) forward with $c_t \equiv 0$:

$$b_t \leq \sum_{j=0}^{\infty} \beta^j y_{t+j} \equiv \bar{b}_t. \tag{17.3.3}$$

The right side is the present value of the endowment, which is the maximal amount that it is feasible to repay at time t when $c_t \geq 0$.

Solve (17.3.1) forward and impose the initial condition $b_0 = 0$ and the terminal condition $\lim_{T \to \infty} \beta^{T+1} b_{T+1} = 0$ to get

$$\sum_{t=0}^{\infty} \beta^t c_t \leq \sum_{t=0}^{\infty} \beta^t y_t. \tag{17.3.4}$$

When $c_t \geq 0$, under the natural borrowing constraints, this is the only restriction that the budget constraints (17.3.1) impose on the $\{c_t\}$ sequence. The first-order necessary conditions for maximizing (17.2.1) subject to (17.3.4) are

$$u'(c_t) \geq u'(c_{t+1}), \quad = \text{ if } b_{t+1} < \bar{b}_{t+1}, \quad t \geq 0. \tag{17.3.5}$$

It is possible to satisfy these first-order conditions by setting $c_t = \bar{c}$ for all $t \geq 0$, where $\bar{c}$ is the constant consumption level chosen to satisfy (17.3.4) at equality:

$$\frac{\bar{c}}{1 - \beta} = \sum_{t=0}^{\infty} \beta^t y_t. \tag{17.3.6}$$

At equality, the sequence of budget constraints $(17.3.1)$ implies that debt b_t satisfies the following equation:

$$\sum_{j=0}^{\infty} \beta^j c_{t+j} + b_t = \sum_{j=0}^{\infty} \beta^j y_{t+j}. \tag{17.3.7}$$

Under the particular consumption-smoothing policy $(17.3.6)$, b_t is given by

$$b_t = \beta^{-t} \sum_{j=0}^{t-1} \beta^j (\bar{c} - y_j) = \beta^{-t} \left(\frac{\bar{c}}{1-\beta} - \frac{\beta^t \bar{c}}{1-\beta} - \sum_{j=0}^{t-1} \beta^j y_j \right)$$

$$= \sum_{j=0}^{\infty} \beta^j y_{t+j} - \frac{\bar{c}}{1-\beta}$$

where the last equality invokes $(17.3.6)$. This expression for b_t is evidently less than or equal to $\bar{b}_t$ for all $t \geq 0$. Thus, under the natural borrowing constraints, we have constant consumption for $t \geq 0$, i.e., perfect consumption smoothing over time.

The natural debt limits allow b_t to be positive, provided that it is not too large. Next we shall study the more severe ad hoc debt limit that requires $-b_t \geq 0$, so that the consumer can *lend*, but not borrow. This restriction will limit consumption smoothing for households whose incomes are growing, and who therefore are naturally borrowers.[3]

17.3.1. An ad hoc borrowing constraint: nonnegative assets

We continue to assume a known endowment sequence but now impose a no-borrowing constraint $(1+r)^{-1} b_{t+1} \leq 0 \; \forall t \geq 0$. To facilitate the transition to our subsequent analysis of the problem under uncertainty, we work in terms of a definition of assets that include this period's income, $a_t = -b_t + y_t$.[4] Let (c_t^*, a_t^*) denote an optimal path. First-order necessary conditions for an optimum are

$$u'(c_t^*) \geq u'(c_{t+1}^*), \quad = \text{ if } c_t^* < a_t^* \tag{17.3.8}$$

---

[3] See exercise *17.1* for how income growth and shrinkage impinge on consumption in the presence of an ad hoc borrowing constraint.

[4] When $\{y_t\}$ is an i.i.d. process, working with a_t rather than b_t makes it possible to formulate the consumer's Bellman equation in terms of the single state variable a_t, rather than the pair b_t, y_t. We'll exploit this idea again in chapter 18.

for $t \geq 0$. Along an optimal path for $t \geq 1$, it must be true that either

    (a) $c_{t-1}^* = c_t^*$; or

    (b) $c_{t-1}^* < c_t^*$ and $c_{t-1}^* = a_{t-1}^*$, and hence $a_t^* = y_t$.

Condition (b) states that the no-borrowing constraint binds only when the consumer desires to shift consumption from the future to the present. He will desire to do that only when his endowment is growing.

    According to conditions a and b, c_{t-1} can never exceed c_t. The reason is that a declining consumption sequence can be improved by cutting a marginal unit of consumption at time $t-1$ with a utility loss of $u'(c_{t-1})$ and increasing consumption at time t by the saving plus interest with a discounted utility gain of $\beta(1+r)u'(c_t) = u'(c_t) > u'(c_{t-1})$, where the inequality follows from the strict concavity of $u(c)$ and $c_{t-1} > c_t$. A symmetric argument rules out $c_{t-1} < c_t$ as long as the nonnegativity constraint on savings is not binding; that is, an agent would choose to cut his savings to make c_{t-1} equal to c_t as in condition a. Therefore, consumption increases from one period to another as in condition b only for a constrained agent with zero savings, $a_{t-1}^* - c_{t-1}^* = 0$. It follows that next period's assets are then equal to next period's income, $a_t^* = y_t$.

    Solving the budget constraint (17.2.2) at equality forward for a_t and rearranging gives

$$\sum_{j=0}^{\infty} \beta^j c_{t+j} = a_t + \sum_{j=1}^{\infty} \beta^j y_{t+j}. \tag{17.3.9}$$

At dates $t \geq 1$ for which $a_t = y_t$, so that the no-borrowing constraint was binding at time $t-1$, (17.3.9) becomes

$$\sum_{j=0}^{\infty} \beta^j c_{t+j} = \sum_{j=0}^{\infty} \beta^j y_{t+j}. \tag{17.3.10}$$

Equations (17.3.9) and (17.3.10) contain important information about the optimal solution. Equation (17.3.9) holds for all dates $t \geq 1$ at which the consumer arrives with positive net assets $a_t - y_t > 0$. Equation (17.3.10) holds for those dates t at which net assets or savings $a_t - y_t$ are zero, i.e., when the no-borrowing constraint was binding at $t-1$. If the no-borrowing constraint is binding only finitely often, then after the *last* date $\bar{t}-1$ at which it was binding, (17.3.10) and the Euler equation (17.3.8) imply that consumption will thereafter be constant at a rate $\tilde{c}$ that satisfies $\frac{\tilde{c}}{1-\beta} = \sum_{j=0}^{\infty} \beta^j y_{\bar{t}+j}$.

In more detail, suppose that an agent arrives in period t with zero savings and knows that the borrowing constraint will never bind again. He would then find it optimal to choose the highest sustainable *constant* consumption. This is given by the annuity return on the present value of the tail of the income process starting from period t,

$$x_t \equiv \frac{r}{1+r} \sum_{j=t}^{\infty} (1+r)^{t-j} y_j. \tag{17.3.11}$$

In the optimization problem under certainty, the impact of the borrowing constraint will not vanish until the date at which the annuity return on the present value of the tail (or remainder) of the income process is maximized. We state this in the following proposition.

PROPOSITION 1: Given a borrowing constraint and a nonstochastic endowment stream, the limit of the nondecreasing optimal consumption path is

$$\bar{c} \equiv \lim_{t \to \infty} c_t^* = \sup_t x_t \equiv \bar{x}. \tag{17.3.12}$$

PROOF: We will first show that $\bar{c} \le \bar{x}$. Suppose to the contrary that $\bar{c} > \bar{x}$. Then conditions a and b imply that there is a t such that $a_t^* = y_t$ and $c_j^* > x_t$ for all $j \ge t$. Therefore, there is a τ sufficiently large that

$$0 < \sum_{j=t}^{\tau} (1+r)^{t-j} \left(c_j^* - y_j \right) = (1+r)^{t-\tau} \left(c_\tau^* - a_\tau^* \right),$$

where the equality uses $a_t^* = y_t$ and successive iterations on budget constraint (17.2.2). The implication that $c_\tau^* > a_\tau^*$ constitutes a contradiction because it violates the constraint that savings are nonnegative in optimization problem (17.2.3).

To show that $\bar{c} \ge \bar{x}$, suppose to the contrary that $\bar{c} < \bar{x}$. Then there is an x_t such that $c_j^* < x_t$ for all $j \ge t$, and hence

$$\sum_{j=t}^{\infty} (1+r)^{t-j} c_j^* < \sum_{j=t}^{\infty} (1+r)^{t-j} x_t = \sum_{j=t}^{\infty} (1+r)^{t-j} y_j$$

$$\le a_t^* + \sum_{j=t+1}^{\infty} (1+r)^{t-j} y_j,$$

where the last weak inequality uses $a_t^* \geq y_t$. Therefore, there is an $\epsilon > 0$ and $\hat{\tau} > t$ such that for all $\tau > \hat{\tau}$,

$$\sum_{j=t}^{\tau} (1+r)^{t-j} c_j^* < a_t^* + \sum_{j=t+1}^{\tau} (1+r)^{t-j} y_j - \epsilon,$$

and after invoking budget constraint $(17.2.2)$ repeatedly,

$$(1+r)^{t-\tau} c_\tau^* < (1+r)^{t-\tau} a_\tau^* - \epsilon,$$

or, equivalently,

$$c_\tau^* < a_\tau^* - (1+r)^{\tau-t} \epsilon.$$

We can then construct an alternative feasible consumption sequence $\{c_j^\epsilon\}$ such that $c_j^\epsilon = c_j^*$ for $j \neq \hat{\tau}$ and $c_j^\epsilon = c_j^* + \epsilon$ for $j = \hat{\tau}$. The fact that this alternative sequence yields higher utility establishes the contradiction. ∎

More generally, we know that at each date $t \geq 1$ for which the no-borrowing constraint is binding at date $t-1$, consumption will increase to satisfy $(17.3.10)$. The time series of consumption will thus be a discrete time step function whose jump dates $\bar{t}$ coincide with the dates at which x_t attains new highs:

$$\bar{t} = \{t : x_t > x_s, s < t\}.$$

If there is a finite last date $\bar{t}$, optimal consumption is a monotone bounded sequence that converges to a finite limit.

In summary, we have shown that under certainty, the optimal consumption sequence converges to a finite limit as long as the discounted value of future income is bounded across all starting dates t. Surprisingly enough, that result is overturned when there is uncertainty. But first, consider a simple example of a nonstochastic endowment process.

17.3.2. Example: periodic endowment process

Suppose that the endowment oscillates between one unit of the consumption good in even periods and zero units in odd periods. The annuity return on this endowment process is equal to

$$x_t\Big|_{t \text{ even}} = \frac{r}{1+r} \sum_{j=0}^{\infty} (1+r)^{-2j} = (1-\beta) \sum_{j=0}^{\infty} \beta^{2j} = \frac{1}{1+\beta}, \qquad (17.3.13a)$$

$$x_t\Big|_{t \text{ odd}} = \frac{1}{1+r} x_t\Big|_{t \text{ even}} = \frac{\beta}{1+\beta}. \qquad (17.3.13b)$$

According to Proposition 1, the limit of the optimal consumption path is then $\bar{c} = (1+\beta)^{-1}$. That is, as soon as the agent reaches the first even period in life, he sets consumption equal to $\bar{c}$ forevermore. The associated beginning-of-period assets a_t fluctuates between $(1+\beta)^{-1}$ and 1.

The exercises at the end of this chapter contain more examples.

17.4. Quadratic preferences

It is useful briefly to consider the linear quadratic permanent income model as a benchmark for the results to come. Assume as before that $\beta(1+r) = 1$ and that the household's budget constraint at t is (17.3.1). Rather than the no-borrowing constraint (17.3.2), we impose that[5]

$$E_0 \left(\lim_{t \to \infty} \beta^t b_t^2 \right) = 0. \qquad (17.4.1)$$

This constrains the asymptotic rate at which debt can grow. Subject to this constraint, solving (17.3.1) forward yields

$$b_t = \sum_{j=0}^{\infty} \beta^j \left(y_{t+j} - c_{t+j} \right). \qquad (17.4.2)$$

---

[5] The natural borrowing limit assumes that consumption is nonnegative, while the model with quadratic preferences permits consumption to be negative. When consumption can be negative, there seems to be no natural lower bound to the amount of debt that could be repaid, since more payments can always be wrung out of the consumer. Thus, with quadratic preferences and the associated possibility of negative consumption, we have to rethink the sense of a borrowing constraint. The restriction (17.4.1) allows negative consumption but limits the rate at which debt is allowed to grow in a way designed to rule out a Ponzi scheme that would have the consumer always consume bliss consumption by accumulating debt without limit.

We alter the preference specification above to make $u(c_t)$ a quadratic function $-.5(c_t - \gamma)^2$, where $\gamma > 0$ is a bliss consumption level. Marginal utility is linear in consumption: $u'(c) = \gamma - c$. We put no bounds on c; in particular, we allow consumption to be negative. We allow $\{y_t\}$ to be an arbitrary stationary stochastic process.

The weakness of constraint (17.4.1) allows the household's first-order condition to prevail with equality at all $t \geq 0$: $u'(c_t) = E_t u'(c_{t+1})$. The linearity of marginal utility in turn implies

$$E_t c_{t+1} = c_t, \tag{17.4.3}$$

which states that c_t is a martingale. Combining (17.4.3) with (17.4.2) and taking expectations conditional on time t information gives $b_t = E_t \sum_{j=0}^{\infty} \beta^j y_{t+j} - \frac{1}{1-\beta} c_t$ or

$$c_t = \frac{r}{1+r} \left[-b_t + E_t \sum_{j=0}^{\infty} \left(\frac{1}{1+r} \right)^j y_{t+j} \right]. \tag{17.4.4}$$

Equation (17.4.4) is a version of the permanent income hypothesis and tells the consumer to set his current consumption equal to the annuity return on his nonhuman $(-b_t)$ and human wealth $(E_t \sum_{j=0}^{\infty} \left(\frac{1}{1+r} \right)^j y_{t+j})$. We can substitute this consumption rule into (17.3.1) and rearrange to get

$$b_{t+1} = b_t + r E_t \sum_{j=0}^{\infty} \left(\frac{1}{1+r} \right)^j y_{t+j} - (1+r) y_t. \tag{17.4.5}$$

Equations (17.4.4) and (17.4.5) imply that under the optimal policy, c_t, b_t both have unit roots and that they are cointegrated.[6]

Consumption rule (17.4.4) has the remarkable feature of certainty equivalence: consumption c_t depends only on the first moment of the discounted value of the endowment sequence. In particular, the conditional variance of the present value of the endowment does not matter.[7] Under rule (17.4.4), consumption is a martingale and the consumer's assets b_t are a unit root process. Neither consumption nor assets converge, though at each point in time, the consumer expects his consumption not to drift in its average value.

---

[6] See section 2.11, especially page 75.

[7] This property of the consumption rule reflects the workings of the type of certainty equivalence that we discussed in chapter 5.

The next section shows that these outcomes change dramatically when we alter the specification of the utility function to rule out negative consumption.

17.5. Stochastic endowment process: i.i.d. case

With uncertain endowments, the first-order condition for the optimization problem $(17.2.3)$ is

$$u'(c) \geq \sum_{s=1}^{S} \beta(1+r)\Pi_s V'\Big[(1+r)(a-c) + \overline{y}_s\Big], \qquad (17.5.1)$$

with equality if the nonnegativity constraint on savings is not binding. The Benveniste-Scheinkman formula implies $u'(c) = V'(a)$, so the first-order condition can also be written as

$$V'(a) \geq \sum_{s=1}^{S} \beta(1+r)\,\Pi_s V'(a'_s), \qquad (17.5.2)$$

where a'_s is next period's assets if today's income shock is $\overline{y}_s$. (Recall again that $V(a)$ is increasing, strictly concave, and differentiable.) Since $\beta^{-1} = (1+r)$, $V'(a)$ is a nonnegative supermartingale. By a theorem of Doob (1953, p. 324),[8] $V'(a)$ must then converge almost surely.[9] The limiting value of $V'(a)$ must be zero because of the following argument. Suppose to the contrary that $V'(a)$ converges to a strictly positive limit. That implies that a converges to a finite positive value. But this implication is contradicted by budget constraint $(17.2.2)$, which states that assets are equal to the value of the past period's savings including interest plus a stochastic income y_s. The random nature of y_s contradicts a finite limit for a. Instead, $V'(a)$ must converge to zero, implying that assets converge to infinity. (We return to this result in chapter 18.)

Although assets diverge to infinity, they do not increase monotonically. Since assets are used for self-insurance, *low* income realizations are associated with *reductions* in assets. To show this outcome, suppose to the contrary that

---

8 See the appendix of this chapter for a statement of the theorem.
9 See footnote 29 on page 71 for an earlier encounter with this force.

even the lowest income realization $\bar{y}_1$ is associated with nondecreasing assets; that is, $(1+r)(a-c) + \bar{y}_1 \geq a$. Then we have

$$V'\left[(1+r)(a-c) + \bar{y}_1\right] \leq V'(a)$$

$$= \sum_{s=1}^{S} \Pi_s V'\left[(1+r)(a-c) + \bar{y}_s\right], \qquad (17.5.3)$$

where the last equality is first-order condition $(17.5.2)$ when the nonnegativity constraint on savings is not binding and when $\beta^{-1} = (1+r)$. Since $V'[(1+r)(a-c)+\bar{y}_s] \leq V'[(1+r)(a-c)+\bar{y}_1]$ for all $s \in \mathbf{S}$, expression $(17.5.3)$ implies that the derivatives of V evaluated at different asset values are equal to each other, an implication that is contradicted by the strict concavity of V.

The fact that assets diverge to infinity means that the individual's consumption also diverges to infinity. After invoking the Benveniste-Scheinkman formula, first-order condition $(17.5.1)$ can be rewritten as

$$u'(c) \geq \sum_{s=1}^{S} \beta (1+r) \Pi_s u'(c'_s) = \sum_{s=1}^{S} \Pi_s u'(c'_s), \qquad (17.5.4)$$

where c'_s is next period's consumption if the income shock is $\bar{y}_s$, and the last equality uses $(1+r) = \beta^{-1}$. It is important to recognize that the individual will never find it optimal to choose a time-invariant consumption level for the indefinite future. Suppose to the contrary that the individual at time t were to choose a constant consumption level for all future periods. The maximum constant consumption level that would be sustainable under all conceivable future income realizations is the annuity return on his current assets a_t and a stream of future incomes all equal to the lowest income realization. But whenever there is a future period with a higher income realization, we can use an argument similar to our section 17.3.1 construction of the sequence $\{c_j^\epsilon\}$ in the case of certainty to show that the initial time-invariant consumption level does not maximize the agent's utility. It follows that future consumption will vary with income realizations and that consumption cannot converge to a finite limit with an i.i.d. endowment process. Hence, from the martingale convergence theorem, the nonnegative supermartingale $u'(c)$ in $(17.5.4)$ must converge to zero, since any strictly positive limit would imply that consumption converges to a finite limit, which cannot be.

17.6. Stochastic endowment process: general case

The result that consumption diverges to infinity with an i.i.d. endowment process is extended by Chamberlain and Wilson (2000) to an arbitrary stationary stochastic endowment process that is sufficiently stochastic. Let I_t denote the information set at time t. Then the general version of first-order condition (17.5.4) becomes

$$u'(c_t) \geq E\Big[u'(c_{t+1})\Big|I_t\Big], \qquad (17.6.1)$$

where $E(\cdot|I_t)$ is the expectation operator conditioned upon information set I_t. Assuming a bounded utility function, Chamberlain and Wilson prove the following result, where x_t is defined in (17.3.11):

PROPOSITION 2: If there is an $\epsilon > 0$ such that for any $\alpha \in I\!\!R^+$

$$P\Big(\alpha \leq x_t \leq \alpha + \epsilon\Big|I_t\Big) < 1 - \epsilon$$

for all I_t and $t \geq 0$, then $P(\lim_{t\to\infty} c_t = \infty) = 1$.

Without providing a proof here, it is useful to make a connection to the nonstochastic case in Proposition 1. Under certainty, the limiting value of the consumption path is given by the highest annuity return on the endowment process across all starting dates t; $\bar{c} = \sup_t x_t$. Under uncertainty, Proposition 2 says that the consumption path will never converge to any finite limit if the annuity return on the endowment process is sufficiently stochastic. Instead, the optimal consumption path will converge to infinity. This stark difference between the case of certainty and uncertainty is quite remarkable.[10]

---

[10] In exercise *17.3*, you will be asked to prove that the divergence of consumption to $+\infty$ also occurs under a stochastic counterpart to the natural borrowing limits. These are less stringent than the no-borrowing condition.

17.7. Intuition

Imagine that you perturb any constant endowment stream by adding the slightest i.i.d. component. Our two propositions then say that the optimal consumption path changes from being a constant to becoming a stochastic process that goes to infinity. Beyond appealing to martingale convergence theorems, Chamberlain and Wilson (2000, p. 381) comment on the difficulty of developing economic intuition for this startling finding:

> Unfortunately, the line of argument used in the proof does not provide a very convincing economic explanation. Clearly the strict concavity of the utility function must play a role. (The result does not hold if, for instance, u is a linear function over a sufficiently large domain and (x_t) is bounded.) But to simply attribute the result to risk aversion on the grounds that uncertain future returns will cause risk-averse consumers to save more, given any initial asset level, is not a completely satisfactory explanation either. In fact, it is a bit misleading. First, that argument only explains why expected accumulated assets would tend to be larger in the limit. It does not really explain why consumption should grow without bound. Second, over any finite time horizon, the argument is not even necessarily correct.

Given a finite horizon, Chamberlain and Wilson proceed to discuss how mean-preserving spreads of future income leave current consumption unaffected when the agent's utility function is quadratic over a sufficiently large domain.

We believe that the economic intuition is to be found in the strict concavity of the utility function *and* the assumption that the marginal utility of consumption must remain positive for any arbitrarily high consumption level. This rules out quadratic utility, for example. To advance this explanation, we first focus on utility functions whose marginal utility of consumption is strictly convex, i.e., $u''' > 0$ if the function is thrice differentiable. Then, Jensen's inequality implies $\sum_s \Pi_s u'(c_s) > u'(\sum_s \Pi_s c_s)$; first-order condition (17.5.4) then implies

$$c \; < \; \sum_{s=1}^{S} \Pi_s c'_s \,, \tag{17.7.1}$$

where the strict inequality follows from our earlier argument that future consumption levels will not be constant but will vary with income realizations. In other words, when the marginal utility of consumption is strictly convex, a given absolute decline in consumption is not only more costly in utility than a gain from an identical absolute increase in consumption, but the former is also associated with a larger rise in *marginal* utility as compared to the drop in *marginal* utility of the latter. To set today's marginal utility of consumption equal to next period's expected marginal utility of consumption, the consumer must therefore balance future states with expected declines in consumption against appropriately higher expected increases in consumption for other states. Of course, when next period arrives and the consumer chooses optimal consumption (which is then on average higher than last period's consumption), the same argument applies again. That is, the process exhibits a "ratchet effect" by which consumption tends toward ever higher levels. Moreover, this on-average increasing consumption sequence cannot converge to a finite limit because of our earlier argument based on an agent's desire to exhaust all his resources while respecting his budget constraint.

This argument for the optimality of unbounded consumption growth applies to utility functions whose marginal utility of consumption is strictly convex. But even utility functions that do not have convex marginal utility globally must ultimately conform to a similar condition over long enough intervals of the positive real line, because otherwise those utility functions would eventually violate the assumptions of a strictly positive, strictly diminishing marginal utility of consumption, $u' > 0$ and $u'' < 0$. Chamberlain and Wilson's reference to a quadratic utility function illustrates the problem of how otherwise the marginal utility of consumption will turn negative at large consumption levels. Thus, our understanding of the remarkable result in Proposition 2 is aided by considering the inexorable ratchet effect on consumption implied by the first-order condition for the agent's optimal intertemporal choice.

17.8. Endogenous labor supply

Contributions by Marcet, Obiols-Homs, and Weil (2007) and Zhu (2009) studied how adding an endogenous labor supply decision would affect outcomes in the presence of a precautionary savings motive. A key insight is that a wealth effect on labor supply can instruct an infinitely lived household to accumulate sufficient wealth to retire and thereby isolate itself thereafter from non-financial income risk. That force can allow both assets and consumption to converge while work converges to zero. The Marcet, Obiols-Homs, and Weil (2007) and Zhu (2009) work provides valuable additional intuition about the structure of the divergence of assets and consumption in the Chamberlain-Wilson model.

At each date $t \geq 0$, a household chooses $c_t \geq 0$ and $h_t \in [0, 1]$, where c_t is the consumption of a single good and h_t is leisure. The household orders a stochastic process $\{c_t, h_t\}_{t=0}^{\infty}$ according to the utility functional

$$E_0 \sum_{t=0}^{\infty} \beta^t u(c_t, h_t), \quad 0 < \beta < +\infty. \tag{17.8.1}$$

Following Zhu (2009), we adopt the assumptions that $u(c, h)$ is (A1), twice continuously differentiable; (A2) strictly increasing and strictly concave in c and h, $\lim_{c \to 0} u_1(c, h) = +\infty \forall h \in [0, 1]$, and $\lim_{h \to 0} u_2(c, h) = +\infty \forall c \geq 0$; and (A3) $u(c, h) \in [0, M], M > 0$.

The consumer can hold nonnegative amounts of a single asset that bears a constant net rate of interest $r > 0$. We assume that $\beta R = 1$, where $R = (1+r)$. The consumer receives labor income $(1 - h_t)e_t w$ at time t, where w is a fixed wage and e_t is the time t realization of a productivity shock. The productivity or labor efficiency shock e_t follows a discrete state Markov process on the state space $E = [\bar{e}_1 \ \ldots \ \bar{e}_n]$ where $[0 < \bar{e}_1 < \cdots < \bar{e}_n]$; the transition density $\pi(e'|e)$ satisfies $\sum_{e'} \pi(e'|e) = 1, \pi(e'|e) > 0 \ \forall \ (e, e') \in E \times E$. The consumer's time t budget constraint is $c_t + A_{t+1} = RA_t + (1 - h_t)e_t w$ or

$$c_t + h_t e_t w = RA_t + e_t w - A_{t+1}. \tag{17.8.2}$$

We regard the left side as the consumer's total time t expenditures on consumption and leisure.

The household chooses $\{c_t, h_t, A_{t+1}\}_{t=0}^{\infty}$ to maximize (17.8.1) subject to $A_0 \geq 0$ given, (17.8.2) for all $t \geq 0$, and the information assumption that A_t, e_t are known at time t. Following Foley and Hellwig (1975) and Zhu (2009), we solve the problem in two steps. Step 1 solves an intratemporal problem and step 2 solves an intertemporal problem.

Step 1: Form the indirect utility function $J(Y, e)$ defined as

$$J(Y, e) = \max_{c,h} u(c, h) \qquad (17.8.3)$$

subject to

$$c + hew = Y, \quad 0 \le h \le 1, c \ge 0. \qquad (17.8.4)$$

Here $Y = RA + we - A'$ stands for total current expenditures on consumption an leisure. The first-order necessary condition for the problem on the right side of $(17.8.3)$ is

$$\frac{u_2(c, h)}{u_1(c, h)} \ge ew, \quad = \text{if } h < 1. \qquad (17.8.5)$$

The right side of $(17.8.3)$ is attained by policy functions $c = c(Y, e), h = h(Y, e)$. Under assumptions A1-A3, Zhu (2009) shows that the indirect utility function $J(Y, e)$ is bounded; strictly increasing and strictly concave in Y; and continuously differentiable in Y with

$$J_1(Y, e) = u_1(c(Y, e), h(Y, e)) \quad \forall Y \in (0, +\infty). \qquad (17.8.6)$$

Step 2: Solve the intertemporal maximization problem

$$V(A_0, e_0) = \max_{Y_t \ge 0} E_0 \sum_{t=0}^{\infty} \beta^t J(Y_t, e_t) \qquad (17.8.7)$$

subject to the sequence of constraints

$$Y_t + A_{t+1} = RA_t + e_t w, \quad t \ge 0 \qquad (17.8.8)$$

and given $A_0 \ge 0$. The Bellman equation associated with this problem is

$$V(A, e) = \max_{A' \in \Gamma(A, e)} \left\{ J(RA + ew - A', e) + \beta E \left[V(A', e') | e \right] \right\} \qquad (17.8.9)$$

where

$$\Gamma(A, e) = \{ A' : 0 \le A' \le RA + ew \}.$$

An optimum policy $Y = Y(A, e)$ and the implied $A' = A(A, e)$ attain

$$V(A, e) = J(Y(A, e), e) + \beta \sum_{e'} V\big(A(A, e), e'\big)\pi(e'|e).$$

Under assumptions A1-A3, Zhu shows that $V(A, e)$ is continuous, strictly increasing, strictly concave in A, continuously differentiable, and that

$$V_1(A, e) = RJ_1(Y(A, e), e) \quad \forall A \in [0, +\infty] \tag{17.8.10}$$

Zhu shows that $A(A, e)$ is continuous and weakly increasing in A and that $Y(A, e)$ is strictly increasing in A. Furthermore, the Inada conditions A2 imply that $\lim_{Y \to 0} J_1(Y, e) = +\infty \; \forall e \in E$. Assembling earlier results also tells us that

$$V_1(A, h) = RJ_1(A, e) = Ru_1(c, h). \tag{17.8.11}$$

Outcome with endogenous labor supply: The first-order necessary condition for asset choice is

$$V_1(A, e) \geq \beta RE[V_1(A', e')|e], \quad = \text{if } A' > 0. \tag{17.8.12}$$

We assume that $\beta R = 1$ as we have done throughout this chapter. When $\beta R = 1$, inequality (17.8.12) asserts that $V_1(A, h)$ and therefore $u_1(c, h)$ is a nonnegative supermartingale. Applying Doob's supermartingale convergence theorem implies that $\lim_{t \to +\infty} V_1(A_t, e_t) = \lim_{t \to +\infty} Ru_1(c_t, h_t)$ exists and is almost surely finite. This result leaves open two possibilities: Either

1. $V_1(A_t, e_t) = Ru_1(c_t, h_t) \to 0$ while h_t remains uniformly bounded away from one infinitely often with the outcomes that $\lim_{t \to \infty} A_t = +\infty$ and $\lim_{t \to \infty} c_t = +\infty$; or

2. $h_t \to 1, A_t \to \bar{A}, c_t \to r\bar{A}$ almost surely.

Case 1 is a version of our earlier result with exogenous and perpetually random non-financial income. Here a precautionary savings motive causes assets and consumption both to diverge to $+\infty$. The possibility of case 2 inspired Marcet, Obiols-Homs, and Weil (2007) and Zhu (2009) to make labor supply decision endogenous. What drives case 2 is a wealth effect that causes the household eventually to withdraw all labor from the market and thereafter consume leisure 100% of his time. That shuts down his effective exposure to the random labor productivity process and extinguishes subsequent randomness in his income process.

Whether case 1 or 2 prevails depends on the shape of the consumer's utility function $u(c, h)$. Zhu (2009) considers the following two assumptions that tilt things toward case 2. Assumption A4 asserts that $u_{12}u_1 - u_{11}u_2 > 0$ and $u_{12}u_2 - u_{22}u_1 > 0$. This assumption makes c and h both be normal goods, and also implies that $\frac{u_2}{u_1}$ is *increasing* in c and *decreasing* in h. It also implies that $c(Y, e)$ and $h(Y, e)$ are both increasing in Y. Zhu also makes the stronger assumption A4' that $u_{12} > 0$, which makes c and h be complements and implies A4. Zhu (2009) establishes the following:

PROPOSITION: Under assumptions A1-A4', (a) $c(A, Y)$ and $h(A, Y)$ are both continuous and increasing in A; (b) $h(A, e) = 1 \quad \forall e$ when A is sufficiently large.

Zhu shows how this proposition is the heart of an argument that generates sufficient conditions for case 2 to prevail. In this way, he constructs circumstances that disarm the divergence outcomes of Chamberlain and Wilson.[11]

17.9. Concluding remarks

This chapter has maintained the assumption that $\beta(1 + r) = 1$, which is a very important ingredient in delivering the divergence toward infinity of the agent's asset and consumption level. Chamberlain and Wilson (1984) study a much more general version of the model where they relax this condition.

To build some incomplete markets models, chapter 18 will put together continua of agents facing generalizations of the savings problems. The models of that chapter will determine the interest rate $1 + r$ as an equilibrium object. In these models, to define a stationary equilibrium, we want the sequence of distributions of each agent's asset holdings to converge to a well-defined invariant distribution with finite first and second moments. For there to exist a stationary

---

[11] Zhu also provides examples of preferences that push things toward case 1. An example is preferences of a type used by Greenwood, Hercowitz, and Huffman (1988): $u(c, h) = U(c - G(1 - h))$, where $U' > 0, U'' < 0, G' > 0, G'' > 0$ with U bounded above. Here the marginal rate of substitution between c and h depends only on h and labor supplied is independent of the intertemporal consumption-savings choice.

equilibrium without aggregate uncertainty, the findings of the present chapter would lead us to anticipate that the equilibrium interest rate in those models must fall short of β^{-1}. In a production economy with physical capital, that result implies that the marginal product of capital will be less than the one that would prevail in a complete markets world when the stationary interest rate would be given by β^{-1}. In other words, an incomplete markets economy is characterized by an overaccumulation of capital that drives the interest rate below β^{-1}, which serves to choke off the desire to accumulate an infinite amount of assets that agents would have had if the interest rate had been equal to β^{-1}.

Chapters 20 and 21 will consider several models in which the condition $\beta(1+r) = 1$ is maintained. There the assumption will be that a social planner has access to risk-free loans outside the economy and seeks to maximize agents' welfare subject to enforcement and/or information problems. The environment is once again assumed to be stationary without aggregate uncertainty, so in the absence of enforcement and information problems the social planner would just redistribute the economy's resources in each period without any intertemporal trade with the outside world. But when agents are free to leave the economy with their endowment streams and forever live in autarky, optimality prescribes that the planner amass sufficient outside claims so that each agent is granted a constant consumption stream in the limit, at a level that weakly dominates autarky for all realizations of an agent's endowment. In the case of asymmetric information, where the planner can induce agents to tell the truth only by manipulating promises of future utilities, we obtain a conclusion that is diametrically opposite to the self-insurance outcome of the present chapter. Instead of consumption approaching infinity in the limit, the optimal solution has all agents' consumption approaching its lower bound.

A. Supermartingale convergence theorem

This appendix states the supermartingale convergence theorem. Let the elements of the 3-tuple $(\Omega, \mathcal{F}, P)$ denote a sample space, a collection of events, and a probability measure, respectively. Let $t \in T$ index time, where T denotes the nonnegative integers. Let $\mathcal{F}_t$ denote an increasing sequence of σ-fields of $\mathcal{F}$ sets. Suppose that

(i) Z_t is measurable with respect to $\mathcal{F}_t$;

(ii) $E|Z_t| < +\infty$;

(iii) $E(Z_t|\mathcal{F}_s) = Z_s$ almost surely for all $s < t; s, t \in T$.

Then $\{Z_t, t \in T\}$ is said to be a *martingale* with respect to $\mathcal{F}_t$. If (iii) is replaced by $E(Z_t|\mathcal{F}_s) \geq Z_s$ almost surely, then $\{Z_t\}$ is said to be a *submartingale*. If (iii) is replaced by $E(Z_t|\mathcal{F}_s) \leq Z_s$ almost surely, then $\{Z_t\}$ is said to be a *supermartingale*.

We have the following important theorem.

SUPERMARTINGALE CONVERGENCE THEOREM: Let $\{Z_t, \mathcal{F}_t\}$ be a nonnegative supermartingale. Then there exists a random variable Z such that $\lim Z_t = Z$ almost surely and $E|Z| < +\infty$, i.e., Z_t converges almost surely to a finite limit.

Exercises

Exercise 17.1   A consumer has preferences over sequences of a single consumption good that are ordered by $\sum_{t=0}^{\infty} \beta^t u(c_t)$, where $\beta \in (0,1)$ and $u(\cdot)$ is strictly increasing, twice continuously differentiable, strictly concave, and satisfies the Inada condition $\lim_{c\downarrow 0} u'(c) = +\infty$. The one good is not storable. The consumer has an endowment sequence of the one good $y_t = \lambda^t, t \geq 0$, where $|\lambda\beta| < 1$. The consumer can borrow or lend at a constant and exogenous risk-free net interest rate of r that satisfies $(1+r)\beta = 1$. The consumer's budget constraint at time t is

$$b_t + c_t \leq y_t + (1+r)^{-1}b_{t+1}$$

for all $t \geq 0$, where b_t is the *debt* (if positive) or *assets* (if negative) due at t, and the consumer has initial debt $b_0 = 0$.

Part I. In this part, assume that the consumer is subject to the ad hoc borrowing constraint $b_t \leq 0 \ \forall t \geq 1$. Thus, the consumer can lend but not borrow.

a. Assume that $\lambda < 1$. Compute the household's optimal plan for $\{c_t, b_{t+1}\}_{t=0}^{\infty}$.

b. Assume that $\lambda > 1$. Compute the household's optimal plan $\{c_t, b_{t+1}\}_{t=0}^{\infty}$.

Part II. In this part, assume that the consumer is subject to the natural borrowing constraint associated with the given endowment sequence.

c. Compute the natural borrowing limits for all $t \geq 0$.

d. Assume that $\lambda < 1$. Compute the household's optimal plan for $\{c_t, b_{t+1}\}_{t=0}^{\infty}$.

e. Assume that $\lambda > 1$. Compute the household's optimal plan $\{c_t, b_{t+1}\}_{t=0}^{\infty}$.

Exercise 17.2   The household has preferences over stochastic processes of a single consumption good that are ordered by $E_0 \sum_{t=0}^{\infty} \beta^t \ln(c_t)$, where $\beta \in (0,1)$ and E_0 is the mathematical expectation with respect to the distribution of the consumption sequence of a single nonstorable good, conditional on the value of the time 0 endowment. The consumer's endowment is the following stochastic process: at times $t = 0, 1$, the household's endowment is drawn from the distribution $\text{Prob}(y_t = 2) = \pi$, $\text{Prob}(y_t = 1) = 1 - \pi$, where $\pi \in (0, 1)$. At all times $t \geq 2$, $y_t = y_{t-1}$. At each date $t \geq 0$, the household can lend, but not borrow, at an exogenous and constant risk-free one-period net interest rate of r that satisfies $(1 + r)\beta = 1$. The consumer's budget constraint at t is $a_{t+1} = (1+r)(a_t - c_t) + y_{t+1}$, subject to the initial condition $a_0 = y_0$. One-period assets carried $(a_t - c_t)$ over into period $t + 1$ from t must be nonnegative, so that the no-borrowing constraint is $a_t \geq c_t$. At time $t = 0$, after y_0 is realized, the consumer devises an optimal consumption plan.

a. Draw a tree that portrays the possible paths for the endowment sequence from date 0 onward.

b. Assume that $y_0 = 2$. Compute the consumer's optimal consumption and lending plan.

c. Assume that $y_0 = 1$. Compute the consumer's optimal consumption and lending plan.

d. Under the two assumptions on the initial condition for y_0 in the preceding two questions, compute the asymptotic distribution of the marginal utility of consumption $u'(c_t)$ (which in this case is the distribution of $u'(c_t) = V_t'(a_t)$ for $t \geq 2$), where $V_t(a)$ is the consumer's value function at date t).

e. Discuss whether your results in part d conform to Chamberlain and Wilson's application of the supermartingale convergence theorem.

Exercise 17.3 Consider the stochastic version of the savings problem under the following *natural borrowing constraints*. At each date $t \geq 0$, the consumer can issue risk-free one-period debt up to an amount that it is feasible for him to repay almost surely, given the nonnegativity constraint on consumption $c_t \geq 0$ for all $t \geq 0$.

a. Verify that the natural debt limit is $(1 + r)^{-1} b_{t+1} \leq \frac{\bar{y}_1}{r}$.

b. Show that the natural debt limit can also be expressed as $a_{t+1} - y_{t+1} \geq -\frac{(1+r)\bar{y}_1}{r}$ for all $t \geq 0$.

c. Assume that y_t is an i.i.d. process with nontrivial distribution $\{\Pi_s\}$, in the sense that at least two distinct endowments occur with positive probabilities. Prove that optimal consumption diverges to $+\infty$ under the natural borrowing limits.

d. For identical realizations of the endowment sequence, get as far as you can in comparing what would be the sequences of optimal consumption under the natural and ad hoc borrowing constraints.

Exercise 17.4 **Trade?**

A pure endowment economy consists of two households with identical preferences but different endowments. A household of type i has preferences that are ordered by

$$(1) \qquad E_0 \sum_{t=0}^{\infty} \beta^t u(c_{it}), \quad \beta \in (0, 1)$$

where c_{it} is time t consumption of a single consumption good, $u(c_{it}) = u_1 c_{it} - .5 u_2 c_{it}^2$, where $u_1, u_2 > 0$, and E_0 denotes the mathematical expectation conditioned on time 0 information. The household of type 1 has a stochastic endowment y_{1t} of the good governed by

$$(2) \qquad y_{1t+1} = y_{1t} + \sigma \epsilon_{t+1}$$

where $\sigma > 0$ and ϵ_{t+1} is an i.i.d. process Gaussian process with mean 0 and variance 1. The household of type 2 has endowment

$$(3) \qquad\qquad y_{2t+1} = y_{2t} - \sigma\epsilon_{t+1}$$

where ϵ_{t+1} is the *same* random process as in (2). At time t, y_{it} is realized before consumption at t is chosen. Assume that at time 0, $y_{10} = y_{20}$ and that y_{10} is substantially less than the bliss point u_1/u_2. To make the computation easier, please assume that there is no disposal of resources.

Part I. In this part, please assume that there are complete markets in history- and date-contingent claims.

a. Define a competitive equilibrium, being careful to specify all of the objects of which a competitive equilibrium is composed.

b. Define a Pareto problem for a fictitious planner who attaches equal weight to the two households. Find the consumption allocation that solves the Pareto (or planning) problem.

c. Compute a competitive equilibrium.

Part II. Now assume that markets are incomplete. There is only one traded asset: a one-period risk-free bond that both households can either purchase or issue. The gross rate of return on the asset between date t and date $t + 1$ is R_t. Household i's budget constraint at time t is

$$(4) \qquad\qquad c_{it} + R_t^{-1} b_{it+1} = y_{it} + b_{it}$$

where b_{it} is the value in terms of time t consumption goods of household's i holdings of one-period risk-free bonds. We require that a consumers's holdings of bonds are subject to the restriction

$$(5) \qquad\qquad \lim_{t\to+\infty} \beta^t u'(c_{it}) E b_{it+1} = 0.$$

Assume that $b_{10} = b_{20} = 0$. An incomplete markets competitive equilibrium is a gross interest rate sequence $\{R_t\}$, sequences of bond holdings $\{b_{it}\}$ for $i = 1, 2$, and feasible allocations $\{c_{it}\}, i = 1, 2$ such that given $\{R_t\}$, household $i = 1, 2$ is maximizing (1) subject to the sequence of budget constraints (4) and the given initial levels of b_{10}, b_{20}.

d. A friend of yours recommends the guess-and-verify method and offers the following guess about the equilibrium. He conjectures that there are no gains to trade: in equilibrium, each household simply consumes its endowment. Please verify or falsify this guess. If you verify it, please give formulas for the equilibrium $\{R_t\}$ and the stocks of bonds held by each household at each date.

Exercise 17.5 **Trade??**

A consumer orders consumption streams according to

$$
(1) \qquad\qquad E_0 \sum_{t=0}^{\infty} \beta^t \frac{c_t^{1-\gamma}}{1-\gamma}, \quad \beta \in (0,1)
$$

where $\gamma > 1$ and E_0 is the mathematical expectation conditional on time 0 information. The consumer can borrow or lend a one-period risk-free security that bears a fixed rate of return of $R = \beta^{-1}$. The consumer's budget constraint at time t is

$$
(2) \qquad\qquad c_t + R^{-1} b_{t+1} = y_t + b_t
$$

where b_t is the level of the asset that the consumer brings into period t. The household is subject to a "natural" borrowing limit. The household's initial asset level is $b_0 = 0$ and his endowment sequence y_t follows the process

$$
(3) \qquad\qquad y_{t+1} = y_t \exp(\sigma_\varepsilon \varepsilon_{t+1} + \mu)
$$

where ε_{t+1} is an i.i.d. Gaussian process with mean zero and variance 1, $\mu = .5\gamma\sigma_\varepsilon^2$, and $\sigma_\varepsilon > 0$. The consumer chooses a process $\{c_t, b_{t+1}\}_{t=0}^{\infty}$ to maximize (1) subject to (2), (3), and the natural borrowing limit.

a. Give a closed-form expression for the consumer's optimal consumption and asset accumulation plan.

Hint 1: If $\log x$ is $\mathcal{N}(\mu, \sigma^2)$, then $Ex = \exp(\mu + \sigma^2/2)$.

Hint 2: You could start by trying to verify the following guess: the optimal policy has $b_{t+1} = 0$ for all $t \geq 0$.

b. Discuss the solution that you obtained in part **a** in terms of Friedman's permanent income hypothesis.

c. Does the household engage in precautionary savings?

Chapter 18
Incomplete Markets Models

18.1. Introduction

In the complete markets model of chapter 8, the optimal consumption allocation is not history dependent; the allocation depends on the current value of the Markov state variable only. This outcome reflects the comprehensive opportunities to insure risks that markets provide. This chapter and chapter 20 describe settings with more impediments to exchanging risks. These reduced opportunities make allocations history dependent. In this chapter, the history dependence is encoded in the dependence of a household's consumption on the household's current asset holdings. In chapter 20, history dependence is encoded in the dependence of the consumption allocation on a continuation value promised by a planner or principal.

The present chapter describes a particular type of incomplete markets model. The models have a large number of *ex ante* identical but *ex post* heterogeneous agents who trade a single security. For most of this chapter, we study models with no aggregate uncertainty and no variation of an aggregate state variable over time (so macroeconomic time series variation is absent). But there is much uncertainty at the individual level. Households' only option is to "self-insure" by managing a stock of a single asset to buffer their consumption against adverse shocks. We study several models that differ mainly with respect to the particular asset that is the vehicle for self-insurance, for example, fiat currency or capital.

The tools for constructing these models are discrete-state discounted dynamic programming, used to formulate and solve problems of the individuals, and Markov chains, used to compute a stationary wealth distribution. The models produce a stationary wealth distribution that is determined simultaneously with various aggregates that are defined as means across corresponding individual-level variables.

We begin by recalling our discrete-state formulation of a single-agent infinite horizon savings problem. We then describe several economies in which households face some version of this infinite horizon savings problem, and where some of the prices taken parametrically in each household's problem are determined by the *average* behavior of all households.

This class of models was invented by Bewley (1977, 1980, 1983, 1986), partly to study a set of classic issues in monetary theory. The second half of this chapter joins that enterprise by using the model to represent inside and outside money, a free banking regime, a subtle limit to the scope of Friedman's optimal quantity of money, a model of international exchange rate indeterminacy, and some related issues. The chapter closes by describing work of Krusell and Smith (1998) that extended the domain of such models to include a time-varying stochastic aggregate state variable. As we shall see, this innovation makes the state of the household's problem include the time t cross-section distribution of wealth, an immense object.

Researchers have used calibrated versions of Bewley models to give quantitative answers to questions including the welfare costs of inflation (İmrohoroğlu, 1992), the risk-sharing benefits of unfunded social security systems (İmrohoroğlu, İmrohoroğlu, and Joines, 1995), the benefits of insuring unemployed people (Hansen and İmrohoroğlu, 1992), and the welfare costs of taxing capital (Aiyagari, 1995). Also see Heathcote, Storesletten, and Violante (2008), and Krueger, Perri, Pistaferri, and Violante (2010). See Kaplan and Violante (2010) for a quantitative study of how much insurance consumers seem to attain beyond the self-insurance allowed in Bewley models. Heathcote, Storesletten, and Violante (2012) combine ideas of Bewley with those of Constantinides and Duffie (1996) to build a model of partial insurance. Heathcote, Perri, and Violante (2010) present an enlightening account of recent movements in the distributions of wages, earnings, and consumption across people and across time in the U.S.

18.2. A savings problem

Recall the discrete-state savings problem described in chapters 4. The household's labor income at time t, s_t, evolves according to an m-state Markov chain with transition matrix $\mathcal{P}$. Think of initiating the process from the invariant distribution of $\mathcal{P}$ over $\bar{s}_i$'s. If the realization of the process at t is $\bar{s}_i$, then at time t the household receives labor income $w\bar{s}_i$. Thus, employment opportunities determine the labor income process. We shall sometimes assume that m is 2, and that s_t takes the value 0 in an unemployed state and 1 in an employed state.

We constrain holdings of a single asset to a grid $\mathcal{A} = [0 < \bar{a}_1 < \bar{a}_2 < \cdots < \bar{a}_n]$. For given values of (w, r) and given initial values (a_0, s_0), the household chooses a policy for $\{a_{t+1}\}_{t=0}^{\infty}$ to maximize

$$E_0 \sum_{t=0}^{\infty} \beta^t u(c_t), \tag{18.2.1}$$

subject to

$$c_t + a_{t+1} = (1 + r)a_t + w s_t$$
$$a_{t+1} \in \mathcal{A} \tag{18.2.2}$$

where $\beta \in (0, 1)$ is a discount factor; $u(c)$ is a strictly increasing, strictly concave, twice continuously differentiable one-period utility function satisfying the Inada condition $\lim_{c \downarrow 0} u'(c) = +\infty$; and $\beta(1 + r) < 1$.[1]

The Bellman equation, for each $i \in [1, \ldots, m]$ and each $h \in [1, \ldots, n]$, is

$$v(\bar{a}_h, \bar{s}_i) = \max_{a' \in \mathcal{A}} \{ u[(1 + r)\bar{a}_h + w\bar{s}_i - a'] + \beta \sum_{j=1}^{m} \mathcal{P}(i, j) v(a', \bar{s}_j) \}, \tag{18.2.3}$$

where a' is next period's value of asset holdings. Here $v(a, s)$ is the optimal value of the objective function, starting from asset-employment state (a, s). Note that the grid $\mathcal{A}$ incorporates upper and lower limits on the quantity that can be borrowed (i.e., the amount of the asset that can be issued). The upper bound on $\mathcal{A}$ is restrictive. In some of our prior theoretical discussions, especially in chapter 17, it was important to dispense with that upper bound.

---

[1] The Inada condition makes consumption nonnegative, and this fact plays a role in justifying the natural debt limit below.

In chapter 4, we described how to solve equation $(18.2.3)$ for a value function $v(a, s)$ and an associated policy function $a' = g(a, s)$ mapping this period's (a, s) pair into an optimal choice of assets to carry into next period.

18.2.1. Wealth-employment distributions

Define the unconditional distribution of (a_t, s_t) pairs, $\lambda_t(a, s) = \text{Prob}(a_t = a, s_t = s)$. The exogenous Markov transition matrix $\mathcal{P}$ on s and the optimal policy function $a' = g(a, s)$ induce a law of motion for the distribution λ_t, namely,

$$\text{Prob}(a_{t+1} = a', s_{t+1} = s') = \sum_{a_t} \sum_{s_t} \text{Prob}(a_{t+1} = a' | a_t = a, s_t = s)$$
$$\cdot \text{Prob}(s_{t+1} = s' | s_t = s) \cdot \text{Prob}(a_t = a, s_t = s),$$

or

$$\lambda_{t+1}(a', s') = \sum_{a} \sum_{s} \lambda_t(a, s) \text{Prob}(s_{t+1} = s' | s_t = s) \cdot \mathcal{I}(a', s, a),$$

where we define the indicator function $\mathcal{I}(a', a, s) = 1$ if $a' = g(a, s)$, and 0 otherwise.[2] The indicator function $\mathcal{I}(a', a, s) = 1$ identifies the time t states a, s that are sent into a' at time $t+1$. The preceding equation can be expressed as

$$\lambda_{t+1}(a', s') = \sum_{s} \sum_{\{a: a' = g(a,s)\}} \lambda_t(a, s) \mathcal{P}(s, s'). \tag{18.2.4}$$

A time-invariant probability distribution λ that solves equation $(18.2.4)$ (i.e., one for which $\lambda_{t+1} = \lambda_t$) is called a *stationary distribution*. In chapter 2, we described two ways to compute a stationary distribution for a Markov chain. One way is in effect to iterate to convergence on equation $(18.2.4)$. An alternative is to create a Markov chain that describes the solution of the optimum problem, then to compute an invariant distribution from a left eigenvector associated with a unit eigenvalue of the stochastic matrix (see chapter 2).

To deduce this Markov chain, we map the pair (a, s) of vectors into a single state vector x as follows. For $i = 1, \ldots, n$, $h = 1, \ldots, m$, let the jth

---

[2] This construction exploits the fact that the optimal policy is a deterministic function of the state, which comes from the concavity of the objective function and the convexity of the constraint set.

element of x be the *pair* (a_i, s_h), where $j = (i - 1)m + h$. Denote $x' = [(\bar{a}_1, \bar{s}_1), (\bar{a}_1, \bar{s}_2), \ldots, (\bar{a}_1, \bar{s}_m), (\bar{a}_2, \bar{s}_1), \ldots, (\bar{a}_2, \bar{s}_m), \ldots, (\bar{a}_n, \bar{s}_1), \ldots, (\bar{a}_n, \bar{s}_m)]$. The optimal policy function $a' = g(a, s)$ and the Markov chain $\mathcal{P}$ on s induce a Markov chain for x via the formula

$$\text{Prob}[(a_{t+1} = a', s_{t+1} = s')|(a_t = a, s_t = s)]$$
$$= \text{Prob}(a_{t+1} = a'|a_t = a, s_t = s) \cdot \text{Prob}(s_{t+1} = s'|s_t = s)$$
$$= \mathcal{I}(a', a, s)\mathcal{P}(s, s'),$$

where $\mathcal{I}(a', a, s) = 1$ is defined as above. This formula defines an $N \times N$ matrix P, where $N = n \cdot m$. This is the Markov chain on the household's state vector x.[3]

Suppose that the Markov chain associated with P is asymptotically stationary and has a unique invariant distribution π_∞. Typically, all states in the Markov chain will be recurrent, and the individual will occasionally revisit each state. For long samples, the distribution π_∞ tells the fraction of time that the household spends in each state. We can "unstack" the state vector x and use π_∞ to deduce the stationary probability measure $\lambda(\bar{a}_i, \bar{s}_h)$ over $(\bar{a}_i, \bar{s}_h)$ pairs, where

$$\lambda(\bar{a}_i, \bar{s}_h) = \text{Prob}(a_t = \bar{a}_i, s_t = \bar{s}_h) = \pi_\infty(j),$$

and where $\pi_\infty(j)$ is the jth component of the vector π_∞, and $j = (i-1)m+h$.

18.2.2. Reinterpretation of the distribution λ

The solution of the household's optimum savings problem induces a stationary distribution $\lambda(a, s)$ that tells the fraction of time that an infinitely lived agent spends in state (a, s). We want to reinterpret $\lambda(a, s)$. Thus, let (a, s) index the state of a particular household at a particular time period t, and assume that there is a cross-section of households distributed over states (a, s). We start the economy at time $t = 0$ with a cross-section $\lambda(a, s)$ of households that we want to repeat over time. The models in this chapter arrange the initial distribution and other things so that the cross-section distribution of agents over individual state variables (a, s) remains constant over time even though the state of the individual household is a stochastic process.

---

[3] Matlab programs to be described later in this chapter create the Markov chain for the joint (a, s) state.

In a model of this type, for a given interest rate r, the population mean

$$E(a)(r) = \sum_{a,s} \lambda(a,s)g(a,s)$$

has two interpretations. First, it is the average asset level experienced by a single household, where here the average is *across time*. Second, it is the average asset level held by the economy as a whole, where here the average is across households indexed by (a,s) pairs. The spirit of what we shall call 'Bewley models' is to make r an equilibrium object that adjusts to set $E(a)(r)$ equal to a particular value. We shall study several models of this type, where the models differ in how we formulate the value to which $E(a)(r)$ must be equated through an appropriate adjustment of r.

18.2.3. Example 1: a pure credit model

Mark Huggett (1993) studied a pure consumption loans economy. Each of a continuum of households has access to a centralized loan market in which it can borrow or lend at a constant net risk-free interest rate of r. Each household's endowment is governed by the Markov chain $(\mathcal{P}, \bar{s})$. The household can either borrow or lend at a constant risk-free rate. However, total borrowing cannot exceed $\phi > 0$, where ϕ is a parameter set by Huggett. A household's setting of next period's level of assets is restricted to the discrete set $\mathcal{A} = [\bar{a}_1, \ldots, \bar{a}_m]$, where the lower bound on assets is $\bar{a}_1 = -\phi$. Later we'll discuss alternative ways to set ϕ, and how it relates to a natural borrowing limit.[4] For now, we simply note the fact that ϕ must be set so that it is feasible for the consumer to honor his loans with probability 1. Otherwise, it is not coherent to posit that loans are risk-free.

The solution of a typical household's problem is a policy function $a' = g(a,s)$ that induces a stationary distribution $\lambda(a,s)$ over states. Huggett uses the following definition:

DEFINITION: Given a borrowing limit ϕ, a *stationary equilibrium* is an interest rate r, a policy function $g(a,s)$, and a stationary distribution $\lambda(a,s)$ for which

(a) Given r, the policy function $g(a,s)$ solves the household's optimum problem;

---

[4] For a related discussion of borrowing limits in economies with sequential trading of IOU's, see chapter 8.

(b) The probability distribution $\lambda(a, s)$ is the invariant distribution of the Markov chain on (a, s) induced by the Markov chain $(\mathcal{P}, \bar{s})$ and the optimal policy $g(a, s)$;

(c) When $\lambda(a, s)$ describes the cross-section of households at each date, the loan market clears

$$\sum_{a,s} \lambda(a, s)g(a, s) = 0.$$

18.2.4. Equilibrium computation

Huggett computed equilibria by using an iterative algorithm that adjusted r to make $\sum_{a,s} \lambda(a, s)g(a, s) = 0$. He fixed an $r = r_j$ for $j = 0$, and for that r solved the household's problem for a policy function $g_j(a, s)$ and an associated stationary distribution $\lambda_j(a, s)$. Then he checked to see whether the loan market clears at r_j by computing

$$\sum_{a,s} \lambda_j(a, s)g(a, s) = e_j^*.$$

If $e_j^* > 0$, Huggett lowered r_{j+1} below r_j and recomputed excess demand, continuing these iterations until he found an r at which excess demand for loans is zero.

18.2.5. Example 2: a model with capital

The next model was created by Rao Aiyagari (1994). He used a version of the savings problem in an economy with many agents and interpreted the single asset as homogeneous physical capital, denoted k. The capital holdings of a household evolve according to

$$k_{t+1} = (1 - \delta)k_t + x_t$$

where $\delta \in (0, 1)$ is a depreciation rate and x_t is gross investment. The household's consumption is constrained by

$$c_t + x_t = \tilde{r}k_t + ws_t,$$

where $\tilde{r}$ is the rental rate on capital and w is a competitive wage, to be determined later. The preceding two equations can be combined to become

$$c_t + k_{t+1} = (1 + \tilde{r} - \delta)k_t + ws_t,$$

which agrees with equation (18.2.2) if we take $a_t \equiv k_t$ and $r \equiv \tilde{r} - \delta$.

There is a large number of households with identical preferences (18.2.1) whose distribution across (k, s) pairs is given by $\lambda(k, s)$, and whose average behavior determines (w, r) as follows: Households are identical in their preferences, the Markov processes governing their employment opportunities, and the prices that they face. However, they differ in their histories $s_0^t = \{s_h\}_{h=0}^t$ of employment opportunities, and therefore in the capital that they have accumulated. Each household has its own history s_0^t as well as its own initial capital k_0. The productivity processes are assumed to be independent across households. The behavior of the collection of these households determines the wage and interest rate (w, r).

Assume an initial distribution *across* households of $\lambda(k, s)$. The average level of capital per household K satisfies

$$K = \sum_{k,s} \lambda(k, s)g(k, s),$$

where $k' = g(k, s)$. Assuming that we start from the invariant distribution, the average level of employment is

$$N = \xi_\infty' \bar{s},$$

where ξ_∞ is the invariant distribution associated with $\mathcal{P}$ and $\bar{s}$ is the exogenously specified vector of individual employment rates. The average employment rate is exogenous, but the average level of capital is endogenous.

There is an aggregate production function whose arguments are the average levels of capital and employment. The production function determines the rental rates on capital and labor from the first-order conditions

$$w = \partial F(K, N)/\partial N$$
$$\tilde{r} = \partial F(K, N)/\partial K,$$

where $F(K, N) = AK^\alpha N^{1-\alpha}$ and $\alpha \in (0, 1)$.

We now have identified all of the objects in terms of which a stationary equilibrium is defined.

DEFINITION OF EQUILIBRIUM: A *stationary equilibrium* is a policy function $g(k, s)$, a probability distribution $\lambda(k, s)$, and positive real numbers $(K, \tilde{r}, w)$ such that

(a) The prices (w, r) satisfy

$$
\begin{aligned}
w &= \partial F(K, N)/\partial N \\
r &= \partial F(K, N)/\partial K - \delta;
\end{aligned}
\tag{18.2.5}
$$

(b) The policy function $g(k, s)$ solves the household's optimum problem;

(c) The probability distribution $\lambda(k, s)$ is a stationary distribution associated with $[g(k, s), \mathcal{P}]$; that is, it satisfies

$$
\lambda(k', s') = \sum_s \sum_{\{k : k' = g(k, s)\}} \lambda(k, s) \mathcal{P}(s, s');
$$

(d) The cross-section average value of K is implied by the average of the households' decisions

$$
K = \sum_{k, s} \lambda(k, s) g(k, s).
$$

18.2.6. Computation of equilibrium

Aiyagari computed an equilibrium of the model by defining a mapping from $K \in \mathbb{R}$ into $\mathbb{R}$, with the property that a fixed point of the mapping is an equilibrium K. Here is an algorithm for finding a fixed point:

1. For fixed value of $K = K_j$ with $j = 0$, compute (w, r) from equation (18.2.5), then solve the household's optimum problem. Use the optimal policy $g_j(k, s)$ to deduce an associated stationary distribution $\lambda_j(k, s)$.

2. Compute the average value of capital associated with $\lambda_j(k, s)$, namely,

$$
K_j^* = \sum_{k, s} \lambda_j(k, s) g_j(k, s).
$$

3. For a fixed "relaxation parameter" $\xi \in (0, 1)$, compute a new estimate of K from method[5]

$$K_{j+1} = \xi K_j + (1 - \xi) K_j^*.$$

4. Iterate on this scheme to convergence.

Later, we shall display some computed examples of equilibria of both Huggett's model and Aiyagari's model. But first we shall analyze some features of both models more formally.

18.3.   Unification and further analysis

We can display salient features of several models by using a graphical apparatus of Aiyagari (1994). We shall show relationships among several models that have identical household sectors but make different assumptions about the single asset being traded.

For convenience, recall the basic savings problem. The household's objective is to maximize

$$E_0 \sum_{t=0}^{\infty} \beta^t u(c_t) \qquad (18.3.1a)$$

$$c_t + a_{t+1} = w s_t + (1 + r) a_t \qquad (18.3.1b)$$

subject to the borrowing constraint

$$a_{t+1} \geq -\phi. \qquad (18.3.1c)$$

We now temporarily suppose that a_{t+1} can take any real value exceeding $-\phi$. Thus, we now suppose that $a_t \in [-\phi, +\infty)$. We occasionally find it useful to express the discount factor $\beta \in (0, 1)$ in terms of a discount *rate* ρ as $\beta = \frac{1}{1+\rho}$. In equation (18.3.1b), we sometimes express w as a given function $\psi(r)$ of the net interest rate r.

---

[5] By setting $\xi < 1$, the relaxation method often converges to a fixed point in cases in which direct iteration (i.e., setting $\xi = 0$) fails to converge.

18.4. The nonstochastic savings problem when $\beta(1+r) < 1$

It is useful briefly to study the nonstochastic version of the savings problem when $\beta(1+r) < 1$. For $\beta(1+r) = 1$, we studied this problem in chapter 17. To get the nonstochastic savings problem, assume that s_t is permanently fixed at some positive level s. Associated with the household's maximum problem is the Lagrangian

$$L = \sum_{t=0}^{\infty} \beta^t \left\{ u(c_t) + \theta_t \left[(1+r)a_t + ws - c_t - a_{t+1} \right] \right\}, \tag{18.4.1}$$

where $\{\theta_t\}_{t=0}^{\infty}$ is a sequence of nonnegative Lagrange multipliers on the budget constraint. The first-order conditions for this problem are

$$u'(c_t) \geq \beta(1+r)u'(c_{t+1}), \quad = \text{if } a_{t+1} > -\phi. \tag{18.4.2}$$

When $a_{t+1} > -\phi$, the first-order condition implies

$$u'(c_{t+1}) = \frac{1}{\beta(1+r)} u'(c_t), \tag{18.4.3}$$

which because $\beta(1+r) < 1$ in turn implies that $u'(c_{t+1}) > u'(c_t)$ and $c_{t+1} < c_t$. Consumption is declining during periods when the household is not borrowing constrained, so $\{c_t\}_{t=0}^{\infty}$ is a monotone decreasing sequence. If $\{c_t\}_{t=0}^{\infty}$ is bounded below, either because of an Inada condition $\lim_{c \downarrow 0} u'(c) = +\infty$ or a nonnegativity constraint on c_t, then c_t will converge as $t \to +\infty$. When it converges, the household will be "borrowing constrained".

We can compute the steady level of consumption when the household eventually becomes permanently stuck at the borrowing constraint. Set $a_{t+1} = a_t = -\phi$. This and $(18.3.1b)$ gives

$$c_t = \bar{c} = ws - r\phi. \tag{18.4.4}$$

This is the amount of labor income remaining after paying the net interest on the debt at the borrowing limit. The household would like to shift consumption from tomorrow to today but can't.

If we solve the budget constraint $(18.3.1b)$ forward, we obtain the present-value budget constraint

$$a_0 = (1+r)^{-1} \sum_{t=0}^{\infty} (1+r)^{-t}(c_t - ws). \tag{18.4.5}$$

Thus, when $\beta(1+r) < 1$, the household's consumption plan can be found by solving equations $(18.4.5)$, $(18.4.4)$, and $(18.4.3)$ for an initial c_0 and a date T after which the debt limit is binding and c_t is constant.

If consumption is required to be nonnegative,[6] equation $(18.4.4)$ implies that the debt limit must satisfy

$$\phi \leq \frac{ws}{r}. \tag{18.4.6}$$

We call the right side the *natural debt limit*. If $\phi < \frac{ws}{r}$, we say that we have imposed an *ad hoc* debt limit.

We have deduced that when $\beta(1+r) < 1$, if a steady-state level exists, consumption is given by equation $(18.4.4)$ and assets by $a_t = -\phi$.

Now turn to the case that $\beta(1+r) = 1$. Here equation $(18.4.3)$ implies that $c_{t+1} = c_t$ and the budget constraint implies $c_t = ws + ra$ and $a_{t+1} = a_t = a_0$. So when $\beta(1+r) = 1$, *any* a_0 is a stationary value of a. It is optimal forever to roll over initial assets.

In summary, in the deterministic case, the steady-state demand for assets is $-\phi$ when $(1+r) < \beta^{-1}$ (i.e., when $r < \rho$); and it equals a_0 when $r = \rho$. Letting the steady-state level be $\bar{a}$, we have

$$\bar{a} = \begin{cases} -\phi, & \text{if } r < \rho; \\ a_0, & \text{if } r = \rho, \end{cases}$$

where $\beta = (1+\rho)^{-1}$. When $r = \rho$, we say that the steady-state asset level $\bar{a}$ is indeterminate.

---

[6] Consumption must be nonnegative, for example, if we impose the Inada condition discussed earlier.

18.5. Borrowing limits: natural and ad hoc

We return to the stochastic case and take up the issue of debt limits. Imposing $c_t \geq 0$ implies the emergence of what Aiyagari calls a natural debt limit. Thus, imposing $c_t \geq 0$ and solving equation $(18.3.1b)$ forward gives

$$a_t \geq -\frac{1}{1+r} \sum_{j=0}^{\infty} w s_{t+j}(1+r)^{-j}. \tag{18.5.1}$$

Since the right side is a random variable, not known at t, we have to supplement equation $(18.5.1)$ to obtain the borrowing constraint. One possible approach is to replace the right side of equation $(18.5.1)$ with its conditional expectation, and to require equation $(18.5.1)$ to hold in expected value. But this expected value formulation is incompatible with the notion that the loan is risk free, and that the household can repay it for sure. We want to impose a restriction that will guarantee that it is feasible for the household to repay its debt for all possible sequences of income realizations. If we insist that equation $(18.5.1)$ hold almost surely for all $t \geq 0$, then we obtain the constraint that emerges by replacing s_t with $\min s \equiv \bar{s}_1$, which yields

$$a_t \geq -\frac{\bar{s}_1 w}{r}. \tag{18.5.2}$$

Aiyagari (1994) calls this the natural debt limit. To accommodate possibly more stringent debt limits, beyond those dictated by the notion that it is feasible to repay the debt for sure, Aiyagari specifies the debt limit as

$$a_t \geq -\phi, \tag{18.5.3}$$

where

$$\phi = \min\left[b, \frac{\bar{s}_1 w}{r}\right], \tag{18.5.4}$$

and $b > 0$ is an arbitrary parameter defining an "ad hoc" debt limit.

18.5.1. A candidate for a single state variable

For the special case in which s is i.i.d., Aiyagari showed how to cast the model in terms of a single state variable to appear in the household's value function. To synthesize a single state variable, note that the "disposable resources" available to be allocated at t are $z_t = ws_t + (1+r)a_t + \phi$. Thus, z_t is the sum of the current endowment, current savings at the beginning of the period, and the maximal borrowing capacity ϕ. This can be rewritten as

$$z_t = ws_t + (1+r)\hat{a}_t - r\phi$$

where $\hat{a}_t \equiv a_t + \phi$. In terms of the single state variable z_t, the household's budget set can be represented recursively as

$$c_t + \hat{a}_{t+1} \le z_t \tag{18.5.5a}$$

$$z_{t+1} = ws_{t+1} + (1+r)\hat{a}_{t+1} - r\phi \tag{18.5.5b}$$

where we must have $\hat{a}_{t+1} \ge 0$. The Bellman equation is

$$v(z_t, s_t) = \max_{\hat{a}_{t+1} \ge 0} \left\{ u(z_t - \hat{a}_{t+1}) + \beta E v(z_{t+1}, s_{t+1}) \right\}. \tag{18.5.6}$$

Here s_t appears in the state vector purely as an information variable for predicting the employment component s_{t+1} of next period's disposable resources z_{t+1}, conditional on the choice of $\hat{a}_{t+1}$ made this period. Therefore, it disappears from both the value function and the decision rule in the i.i.d. case.

More generally, with a serially correlated state, associated with the solution of the Bellman equation is a policy function

$$\hat{a}_{t+1} = A(z_t, s_t). \tag{18.5.7}$$

18.5.2. *Supermartingale convergence again*

Let's revisit a main issue from chapter 17, but now consider the possible case $\beta(1+r) < 1$. From equation $(18.5.5a)$, optimal consumption satisfies $c_t = z_t - A(z_t, s_t)$. The optimal policy obeys the Euler inequality:

$$u'(c_t) \geq \beta(1+r)E_t u'(c_{t+1}), \quad = \text{ if } \hat{a}_{t+1} > 0. \tag{18.5.8}$$

We can use equation $(18.5.8)$ to deduce significant aspects of the limiting behavior of mean assets as a function of r. Following Chamberlain and Wilson (2000) and others, to deduce the effect of r on the mean of assets, we analyze the limiting behavior of consumption implied by the Euler inequality $(18.5.8)$. Define

$$M_t = \beta^t(1+r)^t u'(c_t) \geq 0.$$

Then $M_{t+1} - M_t = \beta^t(1+r)^t[\beta(1+r)u'(c_{t+1}) - u'(c_t)]$. Equation $(18.5.8)$ can be written

$$E_t(M_{t+1} - M_t) \leq 0, \tag{18.5.9}$$

which asserts that M_t is a supermartingale. Because M_t is nonnegative, the supermartingale convergence theorem applies. It asserts that M_t converges almost surely to a nonnegative random variable $\bar{M}$: $M_t \to_{\text{a.s.}} \bar{M}$.

It is interesting to consider three cases: (1) $\beta(1+r) > 1$; (2) $\beta(1+r) < 1$, and (3) $\beta(1+r) = 1$. In case 1, the fact that M_t converges implies that $u'(c_t)$ converges to zero almost surely. Because $u'(c_t) > 0$ and $u''(c_t) < 0$, this fact then implies that $c_t \to +\infty$ and that the consumer's asset holdings diverge to $+\infty$. Chamberlain and Wilson (2000) show that such results also characterize the borderline case (3) (see chapter 17). In case 2, convergence of M_t leaves open the possibility that $u'(c)$ does not converge almost surely. To take a simple example of nonconvergence in case 2, consider the case of a nonstochastic endowment. Under the natural borrowing constraint, the consumer chooses to drive $u'(c) \to +\infty$ as time passes and so asymptotically chooses to impoverish himself. The marginal utility $u'(c)$ diverges.

It is easier to analyze the borderline case $\beta(1+r) = 1$ in the special case that the employment process is independently and identically distributed, meaning that the stochastic matrix $\mathcal{P}$ has identical rows.[7] In this case, s_t provides no information about z_{t+1}, and so s_t can be dropped as an argument

---

[7] See chapter 17 for a closely related proof.

of both $v(\cdot)$ and $A(\cdot)$. For the case in which s_t is i. i. d., Aiyagari (1994) uses the following argument by contradiction to show that if $\beta(1 + r) = 1$, then z_t diverges to $+\infty$. Assume that there is some upper limit $z_{\max}$ such that $z_{t+1} \leq z_{\max} = ws_{\max} + (1 + r)A(z_{\max}) - r\phi$. Then when $\beta(1 + r) = 1$, the strict concavity of the value function, the Benveniste-Scheinkman formula, and equation (18.5.8) imply

$$
\begin{aligned}
v'(z_{\max}) &\geq E_t v'\big[ws_{t+1} + (1 + r)A(z_{\max}) - r\phi\big] \\
&> v'\big[ws_{\max} + (1 + r)A(z_{\max}) - r\phi\big] = v'(z_{\max}),
\end{aligned}
$$

which is a contradiction.

18.6. Average assets as a function of r

In the next several sections, we use versions of a graph of Aiyagari (1994) to analyze several models. The graph plots the average level of assets as a function of r. In the model with capital, the graph is constructed to incorporate the equilibrium dependence of the wage w on r. In models without capital, like Huggett's, the wage is fixed. We shall focus on situations where $\beta(1 + r) < 1$. We consider cases where the optimal decision rule $A(z_t, s_t)$ and the Markov chain for s induce a Markov chain jointly for assets and s that has a unique invariant distribution. For fixed r, let $Ea(r)$ denote the mean level of assets a and let $E\hat{a}(r) = Ea(r) + \phi$ be the mean level of assets plus borrowing capacity $\hat{a} = a + \phi$, where the mean is taken with respect to the invariant distribution. Here it is understood that $Ea(r)$ is a function of ϕ; when we want to make the dependence explicit we write $Ea(r; \phi)$. Also, as we have said, where the single asset is capital, it is appropriate to make the wage w a function of r. This approach incorporates the way different values of r affect average capital, the marginal product of labor, and therefore the wage.

The preceding analysis applying the supermartingale convergence theorem implies that as $\beta(1 + r)$ goes to 1 from below (i.e., r goes to ρ from below), $Ea(r)$ diverges to $+\infty$. This feature is reflected in the shape of the $Ea(r)$ curve in Figure 18.6.1.[8]

---

[8] As discussed in Aiyagari (1994), $Ea(r)$ need not be a monotonically increasing function of r, especially because w can be a function of r.

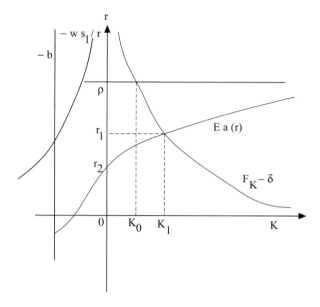

Figure 18.6.1: Demand for capital and determination of interest rate. The $Ea(r)$ curve is constructed for a fixed wage that equals the marginal product of labor at level of capital K_1. In the nonstochastic version of the model with capital, the equilibrium interest rate and capital stock are (ρ, K_0), while in the stochastic version they are (r, K_1). For a version of the model without capital in which w is fixed at this same fixed wage, the equilibrium interest rate in Huggett's pure credit economy occurs at the intersection of the $Ea(r)$ curve with the r-axis.

Figure 18.6.1 assumes that the wage w is fixed in drawing the $Ea(r)$ curve. Later, we will discuss how to draw a similar curve, making w adjust as the function of r that is induced by the marginal productivity conditions for positive values of K. For now, we just assume that w is fixed at the value equal to the marginal product of labor when $K = K_1$, the equilibrium level of capital in the model. The equilibrium interest rate is determined at the intersection of the $Ea(r)$ curve with the marginal productivity of capital curve. Notice that the equilibrium interest rate r is lower than ρ, its value in the nonstochastic version of the model, and that the equilibrium value of capital K_1 exceeds the equilibrium value K_0 (determined by the marginal productivity of capital at $r = \rho$ in the nonstochastic version of the model.)

For a pure credit version of the model like Huggett's, but the same $Ea(r)$ curve, the equilibrium interest rate is determined by the intersection of the $Ea(r)$ curve with the r-axis.

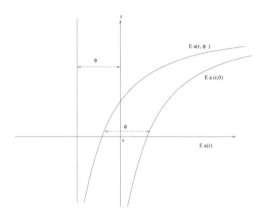

Figure 18.6.2: The effect of a shift in ϕ on the $Ea(r)$ curve. Both $Ea(r)$ curves are drawn assuming that the wage is fixed.

For the purpose of comparing some of the models that follow, it is useful to note the following aspect of the dependence of $Ea(0)$ on ϕ:

PROPOSITION 1: When $r = 0$, the optimal rule $\hat{a}_{t+1} = A(z_t, s_t)$ is independent of ϕ. This implies that for $\phi > 0$, $Ea(0; \phi) = Ea(0; 0) - \phi$.

PROOF: It is sufficient to note that when $r = 0$, ϕ disappears from the right side of equation (18.5.5b) (the consumer's budget constraint). Therefore, the optimal rule $\hat{a}_{t+1} = A(z_t, s_t)$ does not depend on ϕ when $r = 0$. More explicitly, when $r = 0$, add ϕ to both sides of the household's budget constraint to get

$$(a_{t+1} + \phi) + c_t \le (a_t + \phi) + ws_t.$$

If the household's problem with $\phi = 0$ is solved by the decision rule $a_{t+1} = g(a_t, z_t)$, then the household's problem with $\phi > 0$ is solved with the same decision rule evaluated at $a_{t+1} + \phi = g(a_t + \phi, z_t)$. ∎

Thus, it follows that at $r = 0$, an increase in ϕ displaces the $Ea(r)$ curve to the left by the same amount. See Figure 18.6.2. We shall use this result to analyze several models.

In the following sections, we use a version of Figure 18.6.1 to compute equilibria of various models. For models without capital, the figure is drawn assuming that the wage is fixed. Typically, the $Ea(r)$ curve will have the same shape as Figure 18.6.1. In Huggett's model, the equilibrium interest rate is determined by the intersection of the $Ea(r)$ curve with the r-axis, reflecting that the asset (pure consumption loans) is available in zero net supply. In some models with money, the availability of fiat currency as a perfect substitute for consumption loans creates a positive net supply.

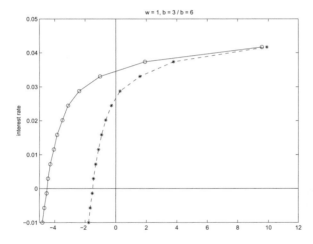

Figure 18.6.3: Two $Ea(r)$ curves, one with $b = 6$, the other with $b = 3$, with w fixed at $w = 1$. Notice that at $r = 0$, the difference between the two curves is 3, the difference in the b's.

18.7. Computed examples

We used some Matlab programs that solve discrete-state dynamic programming problems to compute some examples.[9] We discretized the space of assets from $-\phi$ to a parameter $a_{\max} = 16$ with step size .2.

The utility function is $u(c) = (1 - \mu)^{-1} c^{1-\mu}$, with $\mu = 3$. We set $\beta = .96$. We used two specifications of the Markov process for s. First, we used Tauchen's (1986) method to get a discrete-state Markov chain to approximate a first-order autoregressive process

$$\log s_t = \rho \log s_{t-1} + u_t,$$

where u_t is a sequence of i.i.d. Gaussian random variables. We set $\rho = .2$ and the standard deviation of u_t equal to $.4\sqrt{(1 - \rho)^2}$. We used Tauchen's method with $N = 7$ being the number of points in the grid for s.

For the second specification, we assumed that s is i.i.d. with mean 1.0903. For this case, we compared two settings for the variance: .22 and .68. Figures 18.6.3 and 18.7.1 plot the $Ea(r)$ curves for these various specifications. Figure 18.7.1 plots $Ea(r)$ for the first case of serially correlated s. The two $E[a(r)]$ curves correspond to two distinct settings of the ad hoc debt constraint. One is for $b = 3$, the other for $b = 6$. Figure 18.7.2 plots the invariant distribution of asset holdings for the case in which $b = 3$ and the interest rate is determined at the intersection of the $Ea(r)$ curve and the r-axis.

Figure 18.7.1 summarizes a precautionary savings experiment for the i.i.d. specification of s. Two $Ea(r)$ curves are plotted. For each, we set the ad hoc debt limit $b = 0$. The $Ea(r)$ curve further to the right is the one for the higher variance of the endowment shock s. Thus, a larger variance in the random shock causes increased savings.

Keep these graphs in mind as we turn to analyze some particular models in more detail.

---

[9] The Matlab programs used to compute the $Ea(r)$ functions are `bewley99.m`, `bewley99v2.m`, `aiyagari2.m`, `bewleyplot.m`, and `bewleyplot2.m`. The program `markovapprox.m` implements Tauchen's method for approximating a continuous autoregressive process with a Markov chain. A program `markov.m` simulates a Markov chain. The programs can be downloaded from < https://files.nyu.edu/ts43/public/books.html >.

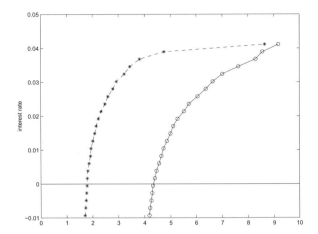

Figure 18.7.1: Two $Ea(r)$ curves when $b = 0$ and the endowment shock s is i.i.d. but with different variances; the curve with circles belongs to the economy with the higher variance.

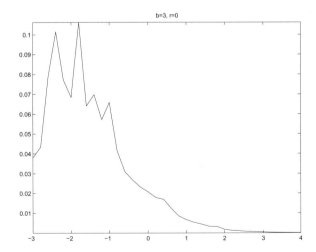

Figure 18.7.2: The invariant distribution of capital when $b = 3$.

18.8. Several Bewley models

We consider several models in which a continuum of households faces the same savings problem. Their behavior generates the asset demand function $Ea(r; \phi)$. The models share the same family of $Ea(r; \phi)$ curves as functions of ϕ, but differ in their settings of ϕ and in their interpretations of the supply of the asset. The models are (1) Aiyagari's (1994, 1995) model in which the risk-free asset is either physical capital or private IOUs, with physical capital being the net supply of the asset; (2) Huggett's model (1993), where the asset is private IOUs, available in zero net supply; (3) Bewley's model of fiat currency; (4) modifications of Bewley's model to permit an inflation tax; and (5) modifications of Bewley's model to pay interest on currency, either explicitly or implicitly through deflation.

18.8.1. Optimal stationary allocation

Because there is no aggregate risk and the aggregate endowment is constant, a stationary optimal allocation would have consumption constant over time for each household. Each household's consumption plan would have constant consumption over time. The implicit risk-free interest rate associated with such an allocation would be $r = \rho$, where recall that $\beta = (1 + \rho)^{-1}$. In the version of the model with capital, the stationary aggregate capital stock solves

$$F_K(K, N) - \delta = \rho. \qquad (18.8.1)$$

Equation (18.8.1) restricts the stationary optimal capital stock in the non-stochastic optimal growth model of Cass (1965) and Koopmans (1965). The stationary level of capital is K_0 in Figure 18.6.1, depicted as the ordinate of the intersection of the marginal productivity net of depreciation curve with a horizontal line $r = \rho$. As we saw before, the horizontal line at $r = \rho$ acts as a "long-run" demand curve for savings for a nonstochastic version of the savings problem. The stationary optimal allocation matches the one produced by a nonstochastic growth model. We shall use the risk-free interest rate $r = \rho$ as a benchmark against which to compare some alternative incomplete market allocations. Aiyagari's (1994) model replaces the horizontal line $r = \rho$ with an upward-sloping curve $Ea(r)$, causing the stationary equilibrium interest rate to fall and the capital stock to rise relative to the savings model with a risk-free endowment sequence.

18.9. A model with capital and private IOUs

Figure 18.6.1 can be used to depict the equilibrium of Aiyagari's model described above. The single asset is capital. There is an aggregate production function $Y = F(K, N)$, and $w = F_N(K, N)$, $r + \delta = F_K(K, N)$. We can invert the marginal condition for capital to deduce a downward-sloping curve $K = K(r)$. This is drawn as the curve labeled $F_K - \delta$ in Figure 18.6.1. We can use the marginal productivity conditions to deduce a factor price frontier $w = \psi(r)$. For fixed r, we use $w = \psi(r)$ as the wage in the savings problem and then deduce $Ea(r)$. We want the equilibrium r to satisfy

$$Ea(r) = K(r). \tag{18.9.1}$$

The equilibrium interest rate occurs at the intersection of $Ea(r)$ with the $F_K - \delta$ curve. See Figure 18.6.1.[10]

It follows from the shape of the curves that the equilibrium capital stock K_1 exceeds K_0, the capital stock required at the given level of total labor to make the interest rate equal ρ. There is capital overaccumulation in the stochastic version of the model.

---

[10] Recall that Figure 18.6.1 was drawn for a fixed wage w, fixed at the value equal to the marginal product of labor when $K = K_1$. Thus, the new version of Figure 18.6.1 that incorporates $w = \psi(r)$ has a new curve $Ea(r)$ that intersects the $F_K - \delta$ curve at the same point (r_1, K_1) as the old curve $Ea(r)$ with the fixed wage. Further, the new $Ea(r)$ curve would not be defined for negative values of K.

18.10. Private IOUs only

It is easy to compute the equilibrium of Mark Huggett's (1993) model with Figure 18.6.1. Recall that in Huggett's model the one asset consists of risk-free loans issued by other households. There are no "outside" assets. This fits the basic model, with a_t being the quantity of loans owed to the individual at the beginning of t. The equilibrium condition is

$$Ea(r, \phi) = 0, \qquad (18.10.1)$$

which is depicted as the intersection of the $Ea(r)$ curve in Figure 18.6.1 with the r-axis. There is a family of such curves, one for each value of the "ad hoc" debt limit. Relaxing the ad hoc debt limit (by driving $b \to +\infty$) sends the equilibrium interest rate upward toward the intersection of the furthest to the left $Ea(r)$ curve, the one that is associated with the natural debt limit, with the r-axis.

18.10.1. Limitation of what credit can achieve

The equilibrium condition (18.10.1) and $\lim_{r \nearrow \rho} Ea(r) = +\infty$ imply that the equilibrium value of r is less than ρ, for all values of the debt limit respecting the natural debt limit. This outcome supports the following conclusion:

PROPOSITION 2: (Suboptimality of equilibrium with credit) The equilibrium interest rate associated with the natural debt limit is the highest one that Huggett's model can support. This interest rate falls short of ρ, the interest rate that would prevail in a complete markets world.[11]

---

[11] Huggett used the model to study how tightening the ad hoc debt limit parameter b would reduce the risk-free rate far enough below ρ to explain the "risk-free rate" puzzle.

18.10.2. Proximity of r to ρ

Notice how in Figure 18.6.3 the equilibrium interest rate r gets closer to ρ as the borrowing constraint is relaxed. How close it can get under the natural borrowing limit depends on several key parameters of the model: (1) the discount factor β, (2) the curvature of $u(\cdot)$, (3) the persistence of the endowment process, and (4) the volatility of the innovations to the endowment process. When he selected a plausible β and $u(\cdot)$ and then calibrated the persistence and volatility of the endowment process to U.S. panel data on workers' earnings, Huggett (1993) found that under the natural borrowing limit, r is quite close to ρ and that the household can achieve substantial self-insurance.[12] We shall encounter an echo of this finding when we review Krusell and Smith's (1998) finding that under their calibration of idiosyncratic risk, a real business cycle model with complete markets does a good job of approximating the prices and the aggregate allocation of a model with identical preferences and technology but in which only a single asset, physical capital, can be traded.

18.10.3. Inside money or free banking interpretation

Huggett's can be viewed as a model of pure "inside money," or of circulating private IOUs. Every person is a "banker" in this setting, being entitled to issue "notes" or evidences of indebtedness, subject to the debt limit (18.5.3). A household has issued more IOU notes of its own than it holds of those issued by others whenever $a_{t+1} < 0$.

There are several ways to think about the "clearing" of notes imposed by equation (18.10.1). Here is one: In period t, trading occurs in subperiods as follows. First, households realize their s_t. Second, some households choose to set $a_{t+1} < a_t \leq 0$ by issuing new IOUs in the amount $-a_{t+1} + a_t$. Other households with $a_t < 0$ may decide to set $a_{t+1} \geq 0$, meaning that they want to "redeem" their outstanding notes and possibly acquire notes issued by others. Third, households go to the market and exchange goods for notes. Fourth, notes are "cleared" or "netted out" in a centralized clearinghouse: positive holdings of

---

[12] This result depends sensitively on how one specifies the left tail of the endowment distribution. Notice that if the minimum endowment $\bar{s}_1$ is set to zero, then the natural borrowing limit is zero. However, Huggett's calibration permits positive borrowing under the natural borrowing limit.

notes issued by others are used to retire possibly negative initial holdings of one's own notes. If a person holds positive amounts of notes issued by others, some of these are used to retire any of his own notes outstanding. This clearing operation leaves each person with a particular a_{t+1} to carry into the next period, with no owner of IOUs also being in the position of having some notes outstanding.

There are other ways to interpret the trading arrangement in terms of circulating notes that implement multilateral long-term lending among corresponding "banks": notes issued by individual A and owned by B are "honored" or redeemed by individual C by being exchanged for goods.[13] In a different setting, Kocherlakota (1996b) and Kocherlakota and Wallace (1998) describe such trading mechanisms.

Under the natural borrowing limit, we might think of this pure consumption loans or inside money model as a model of free banking. In the model, households' ability to issue IOUs is restrained only by the requirement that all loans be risk-free and of one period in duration. Later, we'll use the equilibrium allocation of this free banking model as a benchmark against which to judge the celebrated Friedman rule in a model with outside money and a severe borrowing limit.

We now tighten the borrowing limit enough to make room for some "outside money."

18.10.4. Bewley's basic model of fiat money

This version of the model is set up to generate a demand for fiat money, an inconvertible currency supplied in a fixed nominal amount by an entity outside the model called the government. Individuals can hold currency, but not issue it. To map the individual's problem into problem (18.3.1), we let $m_{t+1}/p = a_{t+1}, b = \phi = 0$, where m_{t+1} is the individual's holding of currency from t to $t+1$, and p is a constant price level. With a constant price level, $r = 0$. With $b = \phi = 0$, $\hat{a}_t = a_t$. Currency is the only asset that can be held. The fixed supply of currency is M. The condition for a stationary equilibrium is

$$Ea(0) = \frac{M}{p}. \tag{18.10.2}$$

---

[13] It is possible to tell versions of this story in which notes issued by one individual or group of individuals are "extinguished" by another.

This equation is to be solved for p. The equation states a version of the quantity theory of money.

Since $r = 0$, we need *some* ad hoc borrowing constraint (i.e., $b < \infty$) to make this model have a stationary equilibrium. If we relax the borrowing constraint from $b = 0$ to permit some borrowing (letting $b > 0$), the $Ea(r)$ curve shifts to the left, causing $Ea(0)$ to fall and the stationary price level to rise.

Let $\bar{m} = Ea(0, \phi = 0)$ be the solution of equation $(18.10.2)$ when $\phi = 0$. Proposition 1 tells how to construct a set of stationary equilibria, indexed by $\phi \in (0, \bar{m})$, which have identical allocations but different price levels. Given an initial stationary equilibrium with $\phi = 0$ and a price level satisfying equation $(18.10.2)$, we construct the equilibrium for $\phi \in (0, \bar{m})$ by setting $\hat{a}_t$ for the new equilibrium equal to $\hat{a}_t$ for the old equilibrium for each person for each period.

This set of equilibria highlights how expanding the amount of "inside money," by substituting for "outside" money, causes the value of outside money (currency) to fall. The construction also indicates that if we set $\phi > \bar{m}$, then there exists no stationary monetary equilibrium with a finite positive price level. For $\phi > \bar{m}$, $Ea(0) < 0$, indicating a force for the interest rate to rise and for private IOUs to dominate currency in rate of return and to drive it out of the model. This outcome leads us to consider proposals to get currency back into the model by paying interest on it. Before we do, let's consider some situations more often observed, where a government raises revenues by an inflation tax.

18.11. A model of seigniorage

The household side of the model is described in the previous section; we continue to summarize this in a stationary demand function $Ea(r)$. We suppose that $\phi = 0$, so individuals cannot borrow. But now the government augments the nominal supply of currency over time to finance a fixed aggregate flow of real purchases G. The government budget constraint at $t \geq 0$ is

$$M_{t+1} = M_t + p_t G, \tag{18.11.1}$$

which for $t \geq 1$ can be expressed

$$\frac{M_{t+1}}{p_t} = \frac{M_t}{p_{t-1}} \left(\frac{p_{t-1}}{p_t} \right) + G.$$

We shall seek a stationary equilibrium with $\frac{p_{t-1}}{p_t} = (1 + r)$ for $t \geq 1$ and $\frac{M_{t+1}}{p_t} = \bar{a}$ for $t \geq 0$. These guesses make the previous equation become

$$\bar{a} = \frac{G}{-r}. \tag{18.11.2}$$

For $G > 0$, this is a rectangular hyperbola in the southeast quadrant. A stationary equilibrium value of r is determined at the intersection of this curve with $Ea(r)$ (see Figure 18.11.1). Evidently, when $G > 0$, the equilibrium net interest rate $r < 0$; $-r$ can be regarded as an inflation tax. Notice that if there is one equilibrium value, there is typically more than one. This is a symptom of the Laffer curve present in this model. Typically if a stationary equilibrium exists, there are at least two stationary inflation rates that finance the government budget. This conclusion follows from the fact that both curves in Figure 18.11.1 have positive slopes.

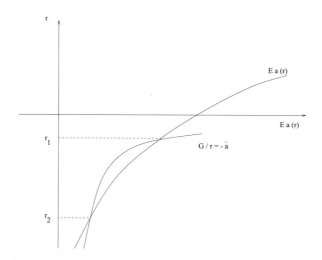

Figure 18.11.1: Two stationary equilibrium rates of return on currency that finance the constant government deficit G.

After r is determined, the initial price level can be determined by the time 0 version of the government budget constraint (18.11.1), namely,

$$\bar{a} = M_0/p_0 + G.$$

This is the version of the quantity theory of money that prevails in this model. An increase in M_0 increases p_0 and all subsequent prices proportionately.

Since there are generally multiple stationary equilibrium inflation rates, which one should we select? We recommend choosing the one with the highest rate of return to currency, that is, the lowest inflation tax. This selection gives "classical" comparative statics: increasing G causes r to fall. In distinct but related settings, Marcet and Sargent (1989) and Bruno and Fischer (1990) give learning procedures that select the same equilibrium we have recommended. Marimon and Sunder (1993) describe experiments with human subjects that they interpret as supporting this selection.

Note the effects of alterations in the debt limit ϕ on the inflation rate. Raising ϕ causes the $Ea(r)$ curve to shift to the left, and *lowers* r. It is even possible for such an increase in ϕ to cause all stationary equilibria to vanish. This experiment indicates why governments intent on raising seigniorage might want to restrict private borrowing. See Bryant and Wallace (1984) for an extensive theoretical elaboration of this and related points. See Sargent and Velde (1995) for a practical example from the French Revolution.

18.12. Exchange rate indeterminacy

We can adapt the preceding model to display a version of Kareken and Wallace's (1980) theory of exchange rate indeterminacy. Consider a model consisting of two countries, each of which is a Bewley economy with stationary money demand function $Ea_i(r)$ in country i. The same single consumption good is available in each country. Residents of both countries are free to hold the currency of either country. Households of either country are indifferent between the two currencies as long as their rates of return are equal. Let p_{it} be the price level in country i, and let $p_{1t} = e_t p_{2t}$ define the time t exchange rate e_t. The gross return on currency i between $t-1$ and t is $(1+r) = \left(\frac{p_{i,t-1}}{p_{i,t}}\right)$ for $i = 1, 2$. Equality of rates of return implies $e_t = e_{t-1}$ for all t and therefore $p_{1,t} = e p_{2,t}$ for all t, where e is a *constant* exchange rate to be determined.

Each of the two countries finances a fixed expenditure level G_i by printing its own currency. Let $\bar{a}_i$ be the stationary level of real balances in country i's currency. Stationary versions of the two countries' budget constraints are

$$\bar{a}_1 = \bar{a}_1(1+r) + G_1 \qquad\qquad (18.12.1)$$

$$\bar{a}_2 = \bar{a}_2(1+r) + G_2 \qquad\qquad (18.12.2)$$

Sum these to get

$$\bar{a}_1 + \bar{a}_2 = \frac{(G_1 + G_2)}{-r}.$$

Setting this curve against $Ea_1(r) + Ea_2(r)$ determines a stationary equilibrium rate of return r. To determine the initial price level and exchange rate, we use the time 0 budget constraints of the two governments. The time 0 budget constraint for country i is

$$\frac{M_{i,1}}{p_{i,0}} = \frac{M_{i,0}}{p_{i,0}} + G_i$$

or

$$\bar{a}_i = \frac{M_{i,0}}{p_{i,0}} + G_i. \qquad\qquad (18.12.3)$$

Add these and use $p_{1,0} = e p_{2,0}$ to get

$$(\bar{a}_1 + \bar{a}_2) - (G_1 + G_2) = \frac{M_{1,0} + e M_{2,0}}{p_{1,0}}.$$

This is one equation in two variables $(e, p_{1,0})$. If there is a solution for some $e \in (0, +\infty)$, then there is a solution for any other $e \in (0, +\infty)$. In this sense, the equilibrium exchange rate is indeterminate.

Equation (18.12.3) is a quantity theory of money stated in terms of the initial "world money supply" $M_{1,0} + e M_{2,0}$.

18.13. Interest on currency

Bewley (1980, 1983) studied whether Friedman's recommendation to pay interest on currency could improve outcomes in a stationary equilibrium, and possibly even support an optimal allocation. He found that when $\beta < 1$, Friedman's rule could improve things but could not implement an optimal allocation, for reasons we now describe.

As in the earlier fiat money model, there is one asset, fiat currency, issued by a government. Households cannot borrow ($b = 0$). The consumer's budget constraint is

$$m_{t+1} + p_t c_t \leq (1 + \tilde{r})m_t + p_t w s_t - \tau p_t$$

where $m_{t+1} \geq 0$ is currency carried over from t to $t + 1$, p_t is the price level at t, $\tilde{r}$ is nominal interest on currency paid by the government, and τ is a real lump-sum tax. This tax is used to finance the interest payments on currency. The government's budget constraint at t is

$$M_{t+1} = M_t + \tilde{r}M_t - \tau p_t,$$

where M_t is the nominal stock of currency per person at the beginning of t.

There are two versions of this model: one where the government pays explicit interest while keeping the nominal stock of currency fixed, another where the government pays no explicit interest but varies the stock of currency to pay interest through deflation.

For each setting, we can show that paying interest on currency, where currency holdings continue to obey $m_t \geq 0$, can be viewed as a device for weakening the impact of this nonnegativity constraint. We establish this point for each setting by showing that the household's problem is isomorphic with Aiyagari's problem as expressed in (18.3.1), (18.5.3), and (18.5.4).

18.13.1. Explicit interest

In the first setting, the government leaves the money supply fixed, setting $M_{t+1} = M_t \; \forall t$, and undertakes to support a constant price level. These settings make the government budget constraint imply

$$\tau = \tilde{r} M / p.$$

Substituting this into the household's budget constraint and rearranging gives

$$\frac{m_{t+1}}{p} + c_t \leq \frac{m_t}{p}(1 + \tilde{r}) + ws_t - \tilde{r}\frac{M}{p}$$

where the choice of currency is subject to $m_{t+1} \geq 0$. With appropriate transformations of variables, this matches Aiyagari's setup of expressions (18.3.1), (18.5.3), and (18.5.4). In particular, take $r = \tilde{r}$, $\phi = \frac{M}{p}$, $\frac{m_{t+1}}{p} = \hat{a}_{t+1} \geq 0$. With these choices, the solution of the savings problem of a household living in an economy with aggregate real balances of $\frac{M}{p}$ and with nominal interest $\tilde{r}$ on currency can be read from the solution of the savings problem with the real interest rate $\tilde{r}$ and a borrowing constraint parameter $\phi \equiv \frac{M}{p}$. Let the solution of this problem be given by the policy function $a_{t+1} = g(a, s; r, \phi)$. Because we have set $\frac{m_{t+1}}{p} = \hat{a}_{t+1} \equiv a_{t+1} + \frac{M}{p}$, the condition that the supply of real balances equals the demand $E\frac{m_{t+1}}{p} = \frac{M}{p}$ is equivalent with $E\hat{a}(r) = \phi$. Note that because $a_t = \hat{a}_t - \phi$, the equilibrium can also be expressed as $Ea(r) = 0$, where as usual $Ea(r)$ is the average of a computed with respect to the invariant distribution $\lambda(a, s)$.

The preceding argument shows that an equilibrium of the money economy with $m_{t+1} \geq 0$, equilibrium real balances $\frac{M}{p}$, and explicit interest on currency r therefore is isomorphic to a pure credit economy with borrowing constraint $\phi = \frac{M}{p}$. We formalize this conclusion in the following proposition:

PROPOSITION 3: A stationary equilibrium with interest on currency financed by lump-sum taxation has the same allocation and interest rate as an equilibrium of Huggett's free banking model for debt limit ϕ equaling the equilibrium real balances from the monetary economy.

To compute an equilibrium with interest on currency, we use a "back-solving" method. [14] Thus, even though the spirit of the model is that the government names $\tilde{r} = r$ and commits itself to set the lump-sum tax needed to finance

---

[14] See Sims (1989) and Diaz-Giménez, Prescott, Fitgerald, and Alvarez (1992) for an explanation and application of back-solving.

interest payments on whatever $\frac{M}{p}$ emerges, we can compute the equilibrium by naming $\frac{M}{p}$ *first*, then finding an r that makes things work. In particular, we use the following steps:

1. Set ϕ to satisfy $0 \leq \phi \leq \frac{ws_1}{r}$. (We will elaborate on the upper bound in the next section.) Compute real balances and therefore p by solving $\frac{M}{p} = \phi$.

2. Find r from $E\hat{a}(r) = \frac{M}{p}$ or $Ea(r) = 0$.

3. Compute the equilibrium tax rate from the government budget constraint $\tau = r\frac{M}{p}$.

This construction finds a constant tax that satisfies the government budget constraint and that supports a level of real balances in the interval $0 \leq \frac{M}{p} \leq \frac{ws_1}{r}$. Evidently, the largest level of real balances that can be supported in equilibrium is the one associated with the natural debt limit. The levels of interest rates that are associated with monetary equilibria are in the range $0 \leq r \leq r_{FB}$, where $Ea(r_{FB}) = 0$ and r_{FB} is the equilibrium interest rate in the pure credit economy (i.e., Huggett's model) under the natural debt limit.

18.13.2. The upper bound on $\frac{M}{p}$

To interpret the upper bound on attainable $\frac{M}{p}$, note that the government's budget constraint and the budget constraint of a household with zero real balances imply that $\tau = r\frac{M}{p} \leq ws$ for all realizations of s. Assume that the stationary distribution of real balances has a positive fraction of agents with real balances arbitrarily close to zero. Let the distribution of employment shocks s be such that a positive fraction of these low-wealth consumers receive income ws_1 at any time. Then, for it to be feasible for the lowest wealth consumers to pay their lump-sum taxes, we must have $\tau \equiv \frac{rM}{p} \leq ws_1$ or $\frac{M}{p} \leq \frac{ws_1}{r}$.

In Figure 18.6.1, the equilibrium real interest rate r can be read from the intersection of the $Ea(r)$ curve and the r-axis. Think of a graph with two $Ea(r)$ curves, one with the natural debt limit $\phi = \frac{s_1 w}{r}$, the other one with an ad hoc debt limit $\phi = \min[b, \frac{s_1 w}{r}]$ shifted to the right. The highest interest rate that can be supported by an interest on currency policy is evidently determined by the point where the $Ea(r)$ curve for the natural debt limit passes through the r-axis. This is higher than the equilibrium interest rate associated with any of the ad hoc debt limits, but must be below ρ. Note that ρ is the interest rate associated with the optimal quantity of money. Thus, we have Aiyagari's (1994) graphical version of Bewley's (1983) result that the optimal quantity of money (Friedman's rule) cannot be implemented in this setting.

We summarize this discussion with a proposition about free banking and Friedman's rule:

PROPOSITION 4:   The highest interest rate that can be supported by paying interest on currency equals that associated with the pure credit (i.e., the pure inside money) model with the natural debt limit.

If $\rho > 0$, Friedman's rule—to pay real interest on currency at the rate ρ—cannot be implemented in this model. The most that can be achieved by paying interest on currency is to eradicate the restriction that prevents households from issuing currency in competition with the government and to implement the free banking outcome.

18.13.3. A very special case

Levine and Zame (2002) have studied a special limiting case of the preceding model in which the free banking equilibrium, which we have seen is equivalent to the best stationary equilibrium with interest on currency, is optimal. They attain this special case as the limit of a sequence of economies with $\rho \downarrow 0$. Heuristically, under the natural debt limits, the $Ea(r)$ curves converge to a horizontal line at $r = 0$. At the limit $\rho = 0$, the argument leading to Proposition 4 allows for the optimal $r = \rho$ equilibrium.

18.13.4. *Implicit interest through deflation*

There is another arrangement equivalent to paying explicit interest on currency. Here the government aspires to pay interest through deflation, but abstains from paying explicit interest. This purpose is accomplished by setting $\tilde{r} = 0$ and $\tau p_t = -gM_t$, where it is intended that the outcome will be $(1+r)^{-1} = (1+g)$, with $g < 0$. The government budget constraint becomes $M_{t+1} = M_t(1+g)$. This can be written

$$\frac{M_{t+1}}{p_t} = \frac{M_t}{p_{t-1}}\frac{p_{t-1}}{p_t}(1+g).$$

We seek a steady state with constant real balances and inverse of the gross inflation rate $\frac{p_{t-1}}{p_t} = (1+r)$. Such a steady state implies that the preceding equation gives $(1+r) = (1+g)^{-1}$, as desired. The implied lump-sum tax rate is $\tau = -\frac{M_t}{p_{t-1}}(1+r)g$. Using $(1+r) = (1+g)^{-1}$, this can be expressed

$$\tau = \frac{M_t}{p_{t-1}}r.$$

The household's budget constraint with taxes set in this way becomes

$$c_t + \frac{m_{t+1}}{p_t} \leq \frac{m_t}{p_{t-1}}(1+r) + ws_t - \frac{M_t}{p_{t-1}}r \tag{18.13.1}$$

This matches Aiyagari's setup with $\frac{M_t}{p_{t-1}} = \phi$.

With these matches the steady-state equilibrium is determined just as though explicit interest were paid on currency. The intersection of the $Ea(r)$ curve with the r-axis determines the real interest rate. Given the parameter b setting the debt limit, the interest rate equals that for the economy with explicit interest on currency.

18.14. Precautionary savings

As we have seen in the production economy with idiosyncratic labor income shocks, the steady-state capital stock is larger when agents have no access to insurance markets as compared to the capital stock in a complete markets economy. The "excessive" accumulation of capital can be thought of as the economy's aggregate amount of *precautionary savings*—a point emphasized by Huggett and Ospina (2000). The precautionary demand for savings is usually described as the extra savings caused by future income being random rather than determinate. [15]

In a partial equilibrium savings problem, it has been known since Leland (1968) and Sandmo (1970) that precautionary savings in response to risk are associated with convexity of the marginal utility function, or a positive third derivative of the utility function. In a two-period model, the intuition can be obtained from the Euler equation, assuming an interior solution with respect to consumption:

$$u'[(1+r)a_0 + w_0 - a_1] = \beta(1+r)E_0 u'[(1+r)a_1 + w_1],$$

where $1+r$ is the gross interest rate, w_t is labor income (endowment) in period $t = 0, 1$, a_0 is an initial asset level, and a_1 is the optimal amount of savings between periods 0 and 1. Now compare the optimal choice of a_1 in two economies where next period's labor income w_1 is either determinate and equal to $\bar{w}_1$, or random with a mean value of $\bar{w}_1$. Let a_1^n and a_1^s denote the optimal choice of savings in the nonstochastic and stochastic economy, respectively, that satisfy the Euler equations:

$$u'[(1+r)a_0 + w_0 - a_1^n] = \beta(1+r)u'[(1+r)a_1^n + \bar{w}_1]$$

$$u'[(1+r)a_0 + w_0 - a_1^s] = \beta(1+r)E_0 u'[(1+r)a_1^s + w_1]$$

$$> \beta(1+r)u'[(1+r)a_1^s + \bar{w}_1],$$

---

[15] Neng Wang (2003) describes an analytically tractable Bewley model with exponential utility. He is able to decompose the savings of an infinitely lived agent into three pieces: (1) a part reflecting a "rainy day" motive that would also be present with quadratic preferences; (2) a part coming from a precautionary motive; and (3) a dissaving component due to impatience that reflects the relative sizes of the interest rate and the consumer's discount rate. Wang computes the equilibrium of a Bewley model by hand and shows that, at the equilibrium interest rate, the second and third components cancel, effectively leaving the consumer to behave as a permanent-income consumer having a martingale consumption policy.

where the strict inequality is implied by Jensen's inequality under the assumption that $u''' > 0$. It follows immediately from these expressions that the optimal asset level is strictly greater in the stochastic economy as compared to the nonstochastic economy, $a_1^s > a_1^n$.

Versions of precautionary savings have been analyzed by Miller (1974), Sibley (1975), Zeldes (1989), Caballero (1990), Kimball (1990, 1993), and Carroll and Kimball (1996), to mention just a few other studies in a vast literature. Using numerical methods for a finite horizon savings problem and assuming a constant relative risk aversion utility function, Zeldes (1989) found that introducing labor income uncertainty made the optimal consumption function concave in assets. That is, the marginal propensity to consume out of assets or transitory income declines with the level of assets. In contrast, without uncertainty and when $\beta(1 + r) = 1$ (as assumed by Zeldes), the marginal propensity to consume depends only on the number of periods left to live, and is neither a function of the agent's asset level nor the present value of lifetime wealth.[16] Here we briefly summarize Carroll and Kimball's (1996) analytical explanation for the concavity of the consumption function that income uncertainty seemed to induce.

In a finite horizon model where both the interest rate and endowment are stochastic processes, Carroll and Kimball cast their argument in terms of the class of hyperbolic absolute risk aversion (HARA) one-period utility functions. These are defined by $\frac{u'''u'}{u''^2} = k$ for some number k. To induce precautionary savings, it must be true that $k > 0$. Most commonly used utility functions are of the HARA class: quadratic utility has $k = 0$, constant absolute risk aversion (CARA) corresponds to $k = 1$, and constant relative risk aversion (CRRA) utility functions satisfy $k > 1$.

Carroll and Kimball show that if $k > 0$, then consumption is a concave function of wealth. Moreover, except for some special cases, they show that the consumption function is *strictly* concave; that is, the marginal propensity to consume out of wealth declines with increases in wealth. The exceptions to

---

[16] When $\beta(1 + r) = 1$ and there are T periods left to live in a nonstochastic economy, consumption smoothing prescribes a constant consumption level c given by $\sum_{t=0}^{T-1} \frac{c}{(1+r)^t} = \Omega$, which implies $c = \frac{r}{1+r} \left[1 - \frac{1}{(1+r)^T} \right]^{-1} \Omega \equiv \mathrm{MPC}_T \, \Omega$, where Ω is the agent's current assets plus the present value of her future labor income. Hence, the marginal propensity to consume out of an additional unit of assets or transitory income, MPC_T, is only a function of the time horizon T.

strict concavity include two well-known cases: CARA utility if all of the risk is to labor income (no rate-of-return risk), and CRRA utility if all of the risk is rate-of-return risk (no labor-income risk).

In the course of the proof, Carroll and Kimball generalize the result of Sibley (1975) that a positive third derivative of the utility function is inherited by the value function. For there to be precautionary savings, the third derivative of the value function with respect to assets must be *positive*; that is, the marginal utility of assets must be a convex function of assets. The case of the quadratic one-period utility is an example where there is no precautionary saving. Off corners, the value function is quadratic, and the third derivative of the value function is zero.[17]

Where precautionary saving occurs, and where the marginal utility of consumption is always positive, the consumption function becomes approximately linear for large asset levels.[18] This feature of the consumption function plays a decisive role in governing the behavior of a model of Krusell and Smith (1998), to which we now turn.

18.15. Models with fluctuating aggregate variables

That the aggregate equilibrium state variables are constant helps makes the preceding models tractable. This section describes a way to extend such models to situations with time-varying stochastic aggregate state variables.[19]

Krusell and Smith (1998) modified Aiyagari's (1994) model by adding an aggregate state variable z, a technology shock that follows a Markov process. Each household continues to receive an idiosyncratic labor-endowment shock s

---

[17] In linear-quadratic models, decision rules for consumption and asset accumulation are independent of the variances of innovations to exogenous income processes.

[18] Roughly speaking, this follows from applying the Benveniste-Scheinkman formula and noting that, where v is the value function, v'' is increasing in savings and v'' is bounded.

[19] See Duffie, Geanakoplos, Mas-Colell, and McLennan (1994) for a general formulation and equilibrium existence theorem for such models. These authors cast doubt on whether in general the current distribution of wealth is enough to serve as a complete description of the history of the aggregate state. They show that in addition to the distribution of wealth, it can be necessary to add a sunspot to the state. See Miao (2003) for a later treatment and for an interpretation of the additional state variable in terms of a distribution of continuation values. See Marcet and Singleton (1999) for a computational strategy for incomplete markets models with a finite number of heterogeneous agents.

that averages to the same constant value for each value of the aggregate shock z. The aggregate shock causes the size of the state of the economy to expand dramatically, because every household's wealth will depend on the history of the *aggregate* shock z, call it z^t, as well as the history of the household-specific shock s^t. That makes the joint histories of z^t, s^t correlated across households, which in turn makes the cross-section distribution of (k, s) vary randomly over time. Therefore, the interest rate and wage will also vary randomly over time.

One way to specify the state is to include the cross-section distribution $\lambda(k, s)$ *each period* among the state variables. Thus, the state includes a cross-section probability distribution of (capital, employment) pairs. In addition, a description of a recursive competitive equilibrium must include a law of motion mapping today's distribution $\lambda(k, s)$ into tomorrow's distribution.

18.15.1. Aiyagari's model again

To prepare the way for Krusell and Smith's way of handling such a model, we recall the structure of Aiyagari's model. The household's Bellman equation in Aiyagari's model is

$$v(k, s) = \max_{c, k'} \{ u(c) + \beta E[v(k', s')|s] \} \tag{18.15.1}$$

where the maximization is subject to

$$c + k' = \tilde{r}k + ws + (1 - \delta)k, \tag{18.15.2}$$

and the prices $\tilde{r}$ and w are fixed numbers satisfying

$$\tilde{r} = \tilde{r}(K, N) = \alpha \left(\frac{K}{N} \right)^{\alpha - 1} \tag{18.15.3a}$$

$$w = w(K, N) = (1 - \alpha) \left(\frac{K}{N} \right)^{\alpha}. \tag{18.15.3b}$$

Recall that aggregate capital and labor K, N are the average values of k, s computed from

$$K = \int k\lambda(k, s)dkds \tag{18.15.4}$$

$$N = \int s\lambda(k, s)dkds. \tag{18.15.5}$$

Here we are following Aiyagari by assuming a Cobb-Douglas aggregate production function. The definition of a stationary equilibrium requires that $\lambda(k, s)$ be the stationary distribution of (k, s) across households induced by the decision rule that attains the right side of equation (18.15.1).

18.15.2. Krusell and Smith's extension

Krusell and Smith (1998) modify Aiyagari's model by adding an aggregate productivity shock z to the price equations, emanating from the presence of z in the production function. The shock z is governed by an exogenous Markov process. Now the state must include λ and z too, so the household's Bellman equation becomes

$$v(k, s; \lambda, z) = \max_{c, k'} \{u(c) + \beta E[v(k', s'; \lambda', z')|(s, z, \lambda)]\} \qquad (18.15.6)$$

where the maximization is subject to

$$c + k' = \tilde{r}(K, N, z)k + w(K, N, z)s + (1 - \delta)k \qquad (18.15.7a)$$

$$\tilde{r} = \tilde{r}(K, N, z) = z\alpha \left(\frac{K}{N}\right)^{\alpha-1} \qquad (18.15.7b)$$

$$w = w(K, N, z) = z(1 - \alpha) \left(\frac{K}{N}\right)^{\alpha} \qquad (18.15.7c)$$

$$\lambda' = H(\lambda, z) \qquad (18.15.7d)$$

where (K, N) is a stochastic processes determined from [20]

$$K_t = \int k\lambda_t(k, s)dkds \qquad (18.15.8)$$

$$N_t = \int s\lambda_t(k, s)dkds. \qquad (18.15.9)$$

Here $\lambda_t(k, s)$ is the distribution of k, s across households at time t. The *distribution* is itself a random function disturbed by the aggregate shock z_t.

---

[20] In our simplified formulation, N is actually constant over time. But in Krusell and Smith's model, N too can be a stochastic process, because leisure is in the one-period utility function.

Krusell and Smith make the plausible guess that $\lambda_t(k, s)$ is enough to complete the description of the state.[21],[22] The Bellman equation and the pricing functions induce the household to want to forecast the average capital stock K, in order to forecast future prices. That desire makes the household want to forecast the cross-section distribution of holdings of capital. To do so it consults the law of motion ($18.15.7d$).

DEFINITION: A recursive competitive equilibrium is a pair of price functions $\tilde{r}, w$, a value function, a decision rule $k' = f(k, s; \lambda, z)$, and a law of motion H for $\lambda(k, s)$ such that

(a) given the price functions and H, the value function solves the Bellman equation ($18.15.6$) and the optimal decision rule is f;

(b) the decision rule f and the Markov processes for s and z imply that today's distribution $\lambda(k, s)$ is mapped into tomorrow's $\lambda'(k, s)$ by H.

The curse of dimensionality makes an equilibrium difficult to compute. Krusell and Smith propose a way to approximate an equilibrium using simulations.[23] First, they characterize the distribution $\lambda(k, s)$ by a finite set of moments of capital $m = (m_1, \ldots, m_I)$. They assume a parametric functional form for H mapping today's m into next period's value m'. They assume a form that can be conveniently estimated using least squares. They assume initial values for the parameters of H. Given H, they use numerical dynamic programming to solve the Bellman equation

$$v(k, s; m, z) = \max_{c, k'} \left\{ u(c) + \beta E[v(k', s'; m', z')|(s, z, m)] \right\}$$

subject to the assumed law of motion H for m. They take the solution of this problem and draw a single long realization from the Markov process for $\{z_t\}$,

---

[21] However, in general settings, this guess remains to be verified. Duffie, Geanakoplos, Mas-Colell, and McLennan (1994) give an example of an incomplete markets economy in which it is necessary to keep track of a longer history of the distribution of wealth.

[22] Loosely speaking, that the individual moves through the distribution of wealth as time passes indicates that his implicit Pareto weight is fluctuating.

[23] These simulations can be justified formally using lessons learned from the literature on convergence of least squares learning to rational expectations in self-referential environments. See footnote 5 of chapter 7, the paper by Marcet and Sargent (1989), and the book with extensions and many applications by Evans and Honkapohja (2001).

say, of length T. For that particular realization of z, they then simulate paths of $\{k_t, s_t\}$ of length T for a large number M of households. They assemble these M simulations into a history of T empirical cross-section distributions $\lambda_t(k, s)$. They use the cross section at t to compute the cross-section moments $m(t)$, thereby assembling a time series of length T of the cross-section moments $m(t)$. They use this sample and nonlinear least squares to estimate the transition function H mapping $m(t)$ into $m(t + 1)$. They return to the beginning of the procedure, use this new guess at H, and continue, iterating to convergence of the function H.

Krusell and Smith compare the aggregate time series $K_t, N_t, \tilde{r}_t, w_t$ from this model with a corresponding representative agent (or complete markets) model. They find that the statistics for the aggregate quantities and prices for the two types of models are very close. Krusell and Smith interpret this result in terms of an "approximate aggregation theorem" that follows from two properties of their parameterized model. First, consumption as a function of wealth is concave but close to linear for moderate to high wealth levels. Second, most of the saving is done by the high-wealth people. These two properties mean that fluctuations in the distribution of wealth have only a small effect on the aggregate amount saved and invested. Thus, distribution effects are small. Also, for these high-wealth people, self-insurance works quite well, so aggregate consumption is not much lower than it would be for the complete markets economy.

Krusell and Smith compare the distributions of wealth from their model to the U.S. data. Relative to the data, the model with a constant discount factor generates too few very poor people and too many rich people. Krusell and Smith modify the model by making the discount factor an exogenous stochastic process. The discount factor switches occasionally between two values. Krusell and Smith find that a modest difference between two discount factors can bring the model's wealth distribution much closer to the data. Patient people become wealthier; impatient people eventually become poorer.

18.16. Concluding remarks

The models in this chapter pursue some of the adjustments that households make when their preferences and endowments give a motive to insure but markets offer limited opportunities to do so. We have studied settings where households' saving occurs through a single risk-free asset. Households use the asset to "self-insure," by making intertemporal adjustments of the asset holdings to smooth their consumption. Their consumption rates at a given date become a function of their asset holdings, which in turn depend on the histories of their endowments. In pure exchange versions of the model, the equilibrium allocation becomes individual history specific, in contrast to the history-independence of the corresponding complete markets model.

The models of this chapter arbitrarily shut down or allow markets without explanation. The market structure is imposed, its consequences then analyzed. In chapter 20, we study a class of models for similar environments that, like the models of this chapter, make consumption allocations history dependent. But the spirit of the models in chapter 20 differs from those in this chapter in requiring that the trading structure be more firmly motivated by the environment. In particular, the models in chapter 20 posit a particular reason that complete markets do not exist, coming from enforcement or information problems, and then study how risk sharing among people can best be arranged.

Exercises

Exercise 18.1 **Random discount factor**   (Bewley-Krusell-Smith)

A household has preferences over consumption of a single good ordered by a value function defined recursively by $v(\beta_t, a_t, s_t) = u(c_t) + \beta_t E_t v(\beta_{t+1}, a_{t+1}, s_{t+1})$, where $\beta_t \in (0, 1)$ is the time t value of a discount factor, and a_t is time t holding of a single asset. Here v is the discounted utility for a consumer with asset holding a_t, discount factor β_t, and employment state s_t. The discount factor evolves according to a three-state Markov chain with transition probabilities $P_{i,j} = \text{Prob}(\beta_{t+1} = \bar{\beta}_j | \beta_t = \bar{\beta}_i)$. The discount factor and employment state at t are both known. The household faces the sequence of budget constraints

$$a_{t+1} + c_t \leq (1 + r)a_t + ws_t$$

where s_t evolves according to an n-state Markov chain with transition matrix $\mathcal{P}$. The household faces the borrowing constraint $a_{t+1} \geq -\phi$ for all t.

Formulate Bellman equations for the household's problem. Describe an algorithm for solving the Bellman equations. (*Hint:* Form three coupled Bellman equations.)

Exercise 18.2 **Mobility costs** (Bertola)

A worker seeks to maximize $E \sum_{t=0}^{\infty} \beta^t u(c_t)$, where $\beta \in (0, 1)$ and $u(c) = \frac{c^{1-\sigma}}{(1-\sigma)}$, and E is the expectation operator. Each period, the worker supplies one unit of labor inelastically (there is no unemployment) and either w^g or w^b, where $w^g > w^b$. A new "job" starts off paying w^g the first period. Thereafter, a job earns a wage governed by the two-state Markov process governing transition between good and bad wages on all jobs; the transition matrix is $\begin{bmatrix} p & (1-p) \\ (1-p) & p \end{bmatrix}$. A new (well-paying) job is always available, but the worker must pay mobility cost $m > 0$ to change jobs. The mobility cost is paid at the beginning of the period that a worker decides to move. The worker's period t budget constraint is

$$A_{t+1} + c_t + mI_t \leq RA_t + w_t,$$

where R is a gross interest rate on assets, c_t is consumption at t, $m > 0$ is moving costs, I_t is an indicator equaling 1 if the worker moves in period t, zero otherwise, and w_t is the wage. Assume that $A_0 > 0$ is given and that the worker faces the no-borrowing constraint, $A_t \geq 0$ for all t.

a. Formulate the Bellman equation for the worker.

b. Write a Matlab program to solve the worker's Bellman equation. Show the optimal decision rules computed for the following parameter values: $m = .9, p = .8, R = 1.02, \beta = .95, w^g = 1.4, w^b = 1, \sigma = 4$. Use a range of assets levels of $[0, 3]$. Describe how the decision to move depends on wealth.

c. Compute the Markov chain governing the transition of the individual's state (A, w). If it exists, compute the invariant distribution.

d. In the fashion of Bewley, use the invariant distribution computed in part c to describe the distribution of wealth across a large number of workers all facing this same optimum problem.

Exercise 18.3 **Unemployment**

There is a continuum of workers with identical probabilities λ of being fired each period when they are employed. With probability $\mu \in (0,1)$, each unemployed worker receives one offer to work at wage w drawn from the cumulative distribution function $F(w)$. If he accepts the offer, the worker receives the offered wage each period until he is fired. With probability $1 - \mu$, an unemployed worker receives no offer this period. The probability μ is determined by the function $\mu = f(U)$, where U is the unemployment rate, and $f'(U) < 0, f(0) = 1, f(1) = 0$. A worker's utility is given by $E \sum_{t=0}^{\infty} \beta^t y_t$, where $\beta \in (0,1)$ and y_t is income in period t, which equals the wage if employed and zero otherwise. There is no unemployment compensation. Each worker regards U as fixed and constant over time in making his decisions.

a. For fixed U, write the Bellman equation for the worker. Argue that his optimal policy has the reservation wage property.

b. Given the typical worker's policy (i.e., his reservation wage), display a difference equation for the unemployment rate. Show that a stationary unemployment rate must satisfy

$$\lambda(1 - U) = f(U)\big[1 - F(\bar{w})\big]U,$$

where $\bar{w}$ is the reservation wage.

c. Define a *stationary equilibrium*.

d. Describe how to compute a stationary equilibrium. You don't actually have to compute it.

Exercise 18.4 **Asset insurance**

Consider the following setup. There is a continuum of households that maximize

$$E \sum_{t=0}^{\infty} \beta^t u(c_t),$$

subject to

$$c_t + k_{t+1} + \tau \le y + \max(x_t, g)k_t^\alpha, \quad c_t \ge 0, \ k_{t+1} \ge 0, \ t \ge 0,$$

where $y > 0$ is a constant level of income not derived from capital, $\alpha \in (0,1)$, τ is a fixed lump-sum tax, k_t is the capital held at the beginning of t, $g \le 1$ is an "investment insurance" parameter set by the government, and x_t is a stochastic household-specific gross rate of return on capital. We assume that x_t is governed by a two-state Markov process with stochastic matrix $\mathcal{P}$, which takes on the two values $\bar{x}_1 > 1$ and $\bar{x}_2 < 1$. When the bad investment return occurs, $(x_t = \bar{x}_2)$, the government supplements the household's return by $\max(0, g - \bar{x}_2)$.

The household-specific randomness is distributed identically and independently across households. Except for paying taxes and possibly receiving insurance payments from the government, households have no interactions with one another; there are no markets.

Given the government policy parameters τ, g, the household's Bellman equation is

$$v(k, x) = \max_{k'}\{u\big[\max(x, g)k^\alpha - k' - \tau\big] + \beta \sum_{x'} v(k', x')\mathcal{P}(x, x')\}.$$

The solution of this problem is attained by a decision rule

$$k' = G(k, x),$$

that induces a stationary distribution $\lambda(k, x)$ of agents across states (k, x).

The average (or per capita) physical output of the economy is

$$Y = \sum_k \sum_x (x \times k^\alpha)\lambda(k, x).$$

The average return on capital to households, *including* the investment insurance, is

$$\nu = \sum_k \bar{x}_1 k^\alpha \lambda(k, x_1) + \max(g, \bar{x}_2) \sum_k k^\alpha \lambda(k, x_2),$$

which states that the government pays out insurance to all households for which $g > \bar{x}_2$.

Define a stationary equilibrium.

Exercise 18.5   **Matching and job quality**

Consider the following Bewley model, a version of which Daron Acemoglu and Robert Shimer (2000) calibrate to deduce quantitative statements about the effects of government-supplied unemployment insurance on the equilibrium level of unemployment, output, and workers' welfare. Time is discrete. Each of a continuum of *ex ante* identical workers can accumulate nonnegative amounts of a single risk-free asset bearing gross one-period rate of return R; R is exogenous and satisfies $\beta R < 1$. There are good jobs with wage w_g and bad jobs with wage $w_b < w_g$. Both wages are exogenous. Unemployed workers must decide whether to search for good jobs or bad jobs. (They cannot search for both.) If an unemployment worker devotes h units of time to search for a good job, a good job arrives with probability $m_g h$; h units of time devoted to searching for bad jobs makes a bad job arrive with probability $m_b h$. Assume that $m_g < m_b$. Good jobs terminate exogenously each period with probability δ_g, bad jobs with probability δ_b. Exogenous terminations entitle an unemployed worker to unemployment compensation of b, which is independent of the worker's lagged earnings. However, each period, an unemployed worker's entitlement to unemployment insurance is exposed to an i.i.d. probability of ϕ of expiring. Workers who quit are not entitled to unemployment insurance.

Workers choose $\{c_t, h_t\}_{t=0}^{\infty}$ to maximize

$$E_0 \sum_{t=0}^{\infty} \beta^t (1-\theta)^{-1} (c_t(\bar{h} - h_t)^\eta)^{1-\theta},$$

where $\beta \in (0, 1)$, and θ is a coefficient of relative risk aversion, subject to the asset accumulation equation

$$a_{t+1} = R(a_t + y_t - c_t)$$

and the no-borrowing condition $a_{t+1} \geq 0$; η governs the substitutability between consumption and leisure. Unemployed workers eligible for unemployment insurance receive income $y_t = b$, while those not eligible receive 0. Employed workers with good jobs receive after-tax income of $y_t = w_g h(1 - \tau)$, and those with bad jobs receive $y_t = w_b h(1 - \tau)$. In equilibrium, the flat-rate tax is set so that the government budget for unemployment insurance balances. Workers with bad jobs have the option of quitting to search for good jobs.

Define a worker's composite *state* as his asset level, together with one of four possible employment states: (1) employed in a good job, (2) employed in a bad job, (3) unemployed and eligible for unemployment insurance; (4) unemployed and ineligible for unemployment insurance.

a. Formulate value functions for the four types of employment states, and describe Bellman equations that link them.

b. In the fashion of Bewley, define a stationary stochastic equilibrium, being careful to define all of the objects composing an equilibrium.

c. Adjust the Bellman equations to accommodate the following modification. Assume that every period that a worker finds himself in a bad job, there is a probability $\delta_{upgrade}$ that the following period, the bad job is upgraded to a good job, conditional on not having been fired.

d. Acemoglu and Shimer calibrate their model to U.S. high school graduates, then perform a local analysis of the consequences of increasing the unemployment compensation rate b. For their calibration, they find that there are substantial benefits to raising the unemployment compensation rate and that this conclusion prevails despite the presence of a "moral hazard problem" associated with providing unemployment insurance benefits in their model. The reason is that too many workers choose to search for bad rather than good jobs. They calibrate β so that workers are sufficiently impatient that most workers with low assets search for bad jobs. If workers were more fully insured, more workers would search for better jobs. That would put a larger fraction of workers in good jobs and raise average productivity. In equilibrium, unemployed workers with high asset levels *do* search for good jobs, because their assets provide them with the "self-insurance" needed to support their investment in search for good jobs. Do you think that the modification suggested in part c would affect the outcomes of increasing unemployment compensation b?

Part V
Recursive contracts

Chapter 19
Dynamic Stackelberg Problems

19.1. History dependence

Previous chapters described decision problems that are recursive in what we can call "natural" state variables, i.e., state variables that describe stocks of capital, wealth, and information that helps forecast future values of prices and quantities that impinge on future utilities or profits. In problems that are recursive in the natural state variables, optimal decision rules are functions of the natural state variables.

This chapter is our first encounter with a class of problems that are not recursive in the natural state variables. Kydland and Prescott (1977), Prescott (1977), and Calvo (1978) gave macroeconomic examples of decision problems whose solutions exhibited *time inconsistency* because they are not recursive in the natural state variables. Those authors studied the decision problem of a large agent (a government) that confronts a competitive market composed of many small private agents whose decisions are influenced by their *forecasts* of the government's future actions. In such settings, the natural state variables of private agents at time t are partly shaped by past decisions that were influenced by their earlier forecasts of the government's action at time t. In a rational expectations equilibrium, the government on average confirms private agents' earlier forecasts of the government's time t actions. This requirement to confirm prior forecasts puts constraints on the government's time t decisions that prevent its problem from being recursive in natural state variables. These additional constraints make the government's decision rule at t depend on the entire history of the state from time 0 to time t.

It took some time for economists to figure out how to formulate policy problems of this type recursively. Prescott (1977) asserted that recursive optimal control theory does not apply to problems with this structure. This chapter and chapters 20 and 23 show how Prescott's pessimism about the inapplicability of optimal control theory has been overturned by more recent work.[1] An important finding is that if the natural state variables are augmented with additional state variables that measure costs in terms of the government's *current* continuation value of confirming *past* private sector expectations about its current behavior, this class of problems can be made recursive. This fact affords immense computational advantages and yields substantial insights. This chapter displays these within the tractable framework of linear quadratic problems.

19.2.   The Stackelberg problem

To exhibit the essential structure of the problems that concerned Kydland and Prescott (1977) and Calvo (1979), this chapter uses the optimal linear regulator to solve a linear quadratic version of what is known as a dynamic Stackelberg problem.[2] For now we refer to the Stackelberg leader as the government and the Stackelberg follower as the representative agent or private sector. Soon we'll give an application with another interpretation of these two players.

Let z_t be an $n_z \times 1$ vector of natural state variables, x_t an $n_x \times 1$ vector of endogenous variables free to jump at t, and u_t a vector of government instruments. The z_t vector is inherited from the past. But x_t is *not* inherited from the past. The model determines the "jump variables" x_t at time t. Included in x_t are prices and quantities that adjust instantaneously to clear markets at time t. Let $y_t = \begin{bmatrix} z_t \\ x_t \end{bmatrix}$. Define the government's one-period loss function[3]

$$r(y, u) = y'Ry + u'Qu. \qquad (19.2.1)$$

Subject to an initial condition for z_0, but not for x_0, a government wants to maximize

$$-\sum_{t=0}^{\infty} \beta^t r(y_t, u_t). \qquad (19.2.2)$$

---

[1]   The important contribution by Kydland and Prescott (1980) helped to dissipate Prescott's initial pessimism.

[2]   In some settings it is called a Ramsey problem.

[3]   The problem assumes that there are no cross products between states and controls in the return function. There is a simple transformation that converts a problem whose return function has cross products into an equivalent problem that has no cross products. For example, see Hansen and Sargent (2008, chapter 4, pp. 72-73).

The government makes policy in light of the model

$$\begin{bmatrix} I & 0 \\ G_{21} & G_{22} \end{bmatrix} \begin{bmatrix} z_{t+1} \\ x_{t+1} \end{bmatrix} = \begin{bmatrix} \hat{A}_{11} & \hat{A}_{12} \\ \hat{A}_{21} & \hat{A}_{22} \end{bmatrix} \begin{bmatrix} z_t \\ x_t \end{bmatrix} + \hat{B} u_t. \qquad (19.2.3)$$

We assume that the matrix on the left is invertible, so that we can multiply both sides of the above equation by its inverse to obtain[4]

$$\begin{bmatrix} z_{t+1} \\ x_{t+1} \end{bmatrix} = \begin{bmatrix} A_{11} & A_{12} \\ A_{21} & A_{22} \end{bmatrix} \begin{bmatrix} z_t \\ x_t \end{bmatrix} + B u_t \qquad (19.2.4)$$

or

$$y_{t+1} = A y_t + B u_t. \qquad (19.2.5)$$

The government maximizes $(19.2.2)$ by choosing sequences $\{u_t, x_t, z_{t+1}\}_{t=0}^{\infty}$ subject to $(19.2.5)$ and the initial condition for z_0.

The private sector's behavior is summarized by the second block of equations of $(19.2.3)$ or $(19.2.4)$. These typically include the first-order conditions of private agents' optimization problem (i.e., their Euler equations). They summarize the forward-looking aspect of private agents' behavior. We shall provide an example later in this chapter in which, as is typical of these problems, the last n_x equations of $(19.2.4)$ or $(19.2.5)$ constitute *implementability constraints* that are formed by the Euler equations of a competitive fringe or private sector. When combined with a stability condition to be imposed below, these Euler equations summarize the private sector's best response to the sequence of actions by the government.

The certainty equivalence principle stated in chapter 5 allows us to work with a nonstochastic model. We would attain the same decision rule if we were to replace x_{t+1} with the forecast $E_t x_{t+1}$ and to add a shock process $C \epsilon_{t+1}$ to the right side of $(19.2.4)$, where ϵ_{t+1} is an i.i.d. random vector with mean of zero and identity covariance matrix.

Let X^t denote the history of any variable X from 0 to t. Miller and Salmon (1982, 1985), Hansen, Epple, and Roberds (1985), Pearlman, Currie, and Levine (1986), Sargent (1987), Pearlman (1992), and others have all studied versions of the following problem:

---

[4] We have assumed that the matrix on the left of $(19.2.3)$ is invertible for ease of presentation. However, by appropriately using the invariant subspace methods described under step 2 below (see Appendix 19B), it is straightforward to adapt the computational method when this assumption is violated.

Problem S: The *Stackelberg problem* is to maximize (19.2.2) by choosing an x_0 and a sequence of decision rules, the time t component of which maps the time t history of the state z^t into the time t decision u_t of the Stackelberg leader. The Stackelberg leader commits to this sequence of decision rules at time 0. The maximization is subject to a given initial condition for z_0. But x_0 is among the objects to be chosen by the Stackelberg leader.

The optimal decision rule is history dependent, meaning that u_t depends not only on z_t but also on lags of z. History dependence has two sources: (a) the government's ability to commit[5] to a sequence of rules at time 0, and (b) the forward-looking behavior of the private sector embedded in the second block of equations (19.2.4). The history dependence of the government's plan is expressed in the dynamics of Lagrange multipliers μ_x on the last n_x equations of (19.2.3) or (19.2.4). These multipliers measure the costs today of honoring past government promises about current and future settings of u. It is appropriate to initialize the multipliers to zero at time $t = 0$, because then there are no past promises about u to honor. But the multipliers μ_x take nonzero values thereafter, reflecting future costs to the government of adhering to its commitment.

19.3. Solving the Stackelberg problem

This section describes a remarkable four-step algorithm for solving the Stackelberg problem.

---

[5] The government would make different choices were it to choose sequentially, that is, were it to select its time t action at time t.

19.3.1. Step 1: solve an optimal linear regulator

Step 1 seems to disregard the forward-looking aspect of the problem (step 3 will take account of that). If we temporarily ignore the fact that the x_0 component of the state $y_0 = \begin{bmatrix} z_0 \\ x_0 \end{bmatrix}$ is *not* actually part of the true state vector, then superficially the Stackelberg problem (19.2.2), (19.2.5) has the form of an optimal linear regulator problem. It can be solved by forming a Bellman equation and iterating until it converges. The optimal value function has the form $v(y) = -y'Py$, where P satisfies the Riccati equation (19.3.5). The next steps note how the value function $v(y) = -y'Py$ encodes objects that solve the Stackelberg problem, then tell how to decode them.

A reader not wanting to be reminded of the details of the Bellman equation can now move directly to step 2. For those wanting a reminder, here it is. The linear regulator is

$$v(y_0) = -y_0'Py_0 = \max_{\{u_t, y_{t+1}\}_{t=0}^{\infty}} - \sum_{t=0}^{\infty} \beta^t \left(y_t'Ry_t + u_t'Qu_t \right) \qquad (19.3.1)$$

where the maximization is subject to a fixed initial condition for y_0 and the law of motion [6]

$$y_{t+1} = Ay_t + Bu_t. \qquad (19.3.2)$$

Associated with problem (19.3.1), (19.3.2) is the Bellman equation

$$-y'Py = \max_{u, y^*} \left\{ -y'Ry - u'Qu - \beta y^{*\prime}Py^* \right\} \qquad (19.3.3)$$

where the maximization is subject to

$$y^* = Ay + Bu \qquad (19.3.4)$$

where y^* denotes next period's value of the state. Problem (19.3.3), (19.3.4) gives rise to the matrix Riccati equation

$$P = R + \beta A'PA - \beta^2 A'PB(Q + \beta B'PB)^{-1}B'PA \qquad (19.3.5)$$

and the formula for F in the decision rule $u_t = -Fy_t$

$$F = \beta(Q + \beta B'PB)^{-1}B'PA. \qquad (19.3.6)$$

---

[6] In step 4, we acknowledge that the x_0 component is *not* given but is to be chosen by the Stackelberg leader.

Thus, we can solve problem (19.2.2), (19.2.5) by iterating to convergence on the difference equation counterpart to the algebraic Riccati equation (19.3.5), or by using a faster computational method that emerges as a by-product in step 2. This method is described in Appendix 19B.

19.3.2. Step 2: use the stabilizing properties of shadow price Py_t

At this point, we decode the information in the matrix P in terms of shadow prices that are associated with a Lagrangian. We adapt a method described earlier in section 5.5 that solves a linear quadratic control problem of the form (19.2.2), (19.2.5) by attaching a sequence of Lagrange multipliers $2\beta^{t+1}\mu_{t+1}$ to the sequence of constraints (19.2.5) and then forming the Lagrangian:

$$\mathcal{L} = -\sum_{t=0}^{\infty} \beta^t \left[y_t' R y_t + u_t' Q u_t + 2\beta \mu_{t+1}'(A y_t + B u_t - y_{t+1}) \right]. \qquad (19.3.7)$$

For the Stackelberg problem, it is important to partition μ_t conformably with our partition of $y_t = \begin{bmatrix} z_t \\ x_t \end{bmatrix}$, so that $\mu_t = \begin{bmatrix} \mu_{zt} \\ \mu_{xt} \end{bmatrix}$, where μ_{xt} is an $n_x \times 1$ vector of multipliers adhering to the implementability constraints. For now, we can ignore the partitioning of μ_t, but it will be very important when we turn our attention to the specific requirements of the Stackelberg problem in step 3.

We want to maximize (19.3.7) with respect to sequences for u_t and y_{t+1}. The first-order conditions with respect to u_t, y_t, respectively, are:

$$0 = Q u_t + \beta B' \mu_{t+1} \qquad (19.3.8a)$$

$$\mu_t = R y_t + \beta A' \mu_{t+1}. \qquad (19.3.8b)$$

Solving (19.3.8a) for u_t and substituting into (19.2.5) gives

$$y_{t+1} = A y_t - \beta B Q^{-1} B' \mu_{t+1}. \qquad (19.3.9)$$

We can represent the system formed by (19.3.9) and (19.3.8b) as

$$\begin{bmatrix} I & \beta B Q^{-1} B' \\ 0 & \beta A' \end{bmatrix} \begin{bmatrix} y_{t+1} \\ \mu_{t+1} \end{bmatrix} = \begin{bmatrix} A & 0 \\ -R & I \end{bmatrix} \begin{bmatrix} y_t \\ \mu_t \end{bmatrix} \qquad (19.3.10)$$

or

$$L^* \begin{bmatrix} y_{t+1} \\ \mu_{t+1} \end{bmatrix} = N \begin{bmatrix} y_t \\ \mu_t \end{bmatrix}. \qquad (19.3.11)$$

We seek a "stabilizing" solution of (19.3.11), i.e., one that satisfies

$$\sum_{t=0}^{\infty} \beta^t y_t' y_t < +\infty.$$

19.3.3. Stabilizing solution

By the same argument used in section 5.5 of chapter 5, a stabilizing solution satisfies $\mu_0 = P y_0$, where P solves the matrix Riccati equation (19.3.5). The solution for μ_0 replicates itself over time in the sense that

$$\mu_t = P y_t. \tag{19.3.12}$$

Appendix 19A verifies that the matrix P that satisfies the Riccati equation (19.3.5) is the same P that defines the stabilizing initial conditions $(y_0, P y_0)$. In Appendix 19B, we describe how to construct P by computing generalized eigenvalues and eigenvectors.

19.3.4. Step 3: convert implementation multipliers into state variables

19.3.4.1. Key insight

We now confront the fact that the x_0 component of y_0 consists of variables that are not state variables, i.e., they are not inherited from the past but are to be determined at time t. In the optimal linear regulator problem, y_0 is a state vector inherited from the past; the multiplier μ_0 jumps at t to satisfy $\mu_0 = P y_0$ and thereby stabilize the system. But in the Stackelberg problem, pertinent components of *both* y_0 *and* μ_0 must adjust to satisfy $\mu_0 = P y_0$. In particular, partition μ_t conformably with the partition of y_t into $\begin{bmatrix} z'_t & x'_t \end{bmatrix}'$:[7]

$$\mu_t = \begin{bmatrix} \mu_{zt} \\ \mu_{xt} \end{bmatrix}.$$

For the Stackelberg problem, the first n_z elements of y_t are predetermined but the remaining components are free. And while the first n_z elements of μ_t are free to jump at t, the remaining components are not. The third step completes the solution of the Stackelberg problem by acknowledging these facts. *After* we have performed the key step of computing the matrix P that solves the Riccati equation (19.3.5), we convert the last n_x Lagrange multipliers μ_{xt} into state variables by using the following procedure

Write the last n_x equations of (19.3.12) as

$$\mu_{xt} = P_{21} z_t + P_{22} x_t, \tag{19.3.13}$$

where the partitioning of P is conformable with that of y_t into $\begin{bmatrix} z_t & x_t \end{bmatrix}'$. The vector μ_{xt} becomes part of the state at t, while x_t is free to jump at t. Therefore, we solve (19.3.13) for x_t in terms of (z_t, μ_{xt}):

$$x_t = -P_{22}^{-1} P_{21} z_t + P_{22}^{-1} \mu_{xt}. \tag{19.3.14}$$

Then we can write

$$y_t = \begin{bmatrix} z_t \\ x_t \end{bmatrix} = \begin{bmatrix} I & 0 \\ -P_{22}^{-1} P_{21} & P_{22}^{-1} \end{bmatrix} \begin{bmatrix} z_t \\ \mu_{xt} \end{bmatrix} \tag{19.3.15}$$

and from (19.3.13)

$$\mu_{xt} = \begin{bmatrix} P_{21} & P_{22} \end{bmatrix} y_t. \tag{19.3.16}$$

With these modifications, the key formulas (19.3.6) and (19.3.5) from the optimal linear regulator for F and P, respectively, continue to apply. Using (19.3.15), the optimal decision rule is

$$u_t = -F \begin{bmatrix} I & 0 \\ -P_{22}^{-1} P_{21} & P_{22}^{-1} \end{bmatrix} \begin{bmatrix} z_t \\ \mu_{xt} \end{bmatrix}. \tag{19.3.17}$$

---

[7] This argument just adapts one in Pearlman (1992). The Lagrangian associated with the Stackelberg problem remains (19.3.7), which means that the same section 5.5 logic implies that the stabilizing solution must satisfy (19.3.12). It is only in how we impose (19.3.12) that the solution diverges from that for the linear regulator.

Then we have the following complete description of the Stackelberg plan:[8]

$$\begin{bmatrix} z_{t+1} \\ \mu_{x,t+1} \end{bmatrix} = \begin{bmatrix} I & 0 \\ P_{21} & P_{22} \end{bmatrix} (A - BF) \begin{bmatrix} I & 0 \\ -P_{22}^{-1}P_{21} & P_{22}^{-1} \end{bmatrix} \begin{bmatrix} z_t \\ \mu_{xt} \end{bmatrix} \qquad (19.3.19a)$$

$$x_t = \begin{bmatrix} -P_{22}^{-1}P_{21} & P_{22}^{-1} \end{bmatrix} \begin{bmatrix} z_t \\ \mu_{xt} \end{bmatrix}. \qquad (19.3.19b)$$

The difference equation $(19.3.19a)$ is to be initialized from the given value of z_0 and a value for μ_{x0} to be determined in step 4.

19.3.5. Step 4: solve for x_0 and μ_{x0}

The value function $V(y_0)$ satisfies

$$V(y_0) = -z_0' P_{11} z_0 - 2x_0' P_{21} z_0 - x_0' P_{22} x_0. \qquad (19.3.20)$$

Now choose x_0 by equating to zero the gradient of $V(y_0)$ with respect to x_0:

$$-2P_{21}z_0 - 2P_{22}x_0 = 0,$$

which by virtue of $(19.3.13)$ is equivalent with

$$\mu_{x0} = 0. \qquad (19.3.21)$$

Then we can compute x_0 from $(19.3.14)$ to arrive at

$$x_0 = -P_{22}^{-1}P_{21}z_0. \qquad (19.3.22)$$

Setting $\mu_{x0} = 0$ means that at time 0 there are no past promises to keep.

---

[8] When a random shock $C\epsilon_{t+1}$ is present, we must add

$$\begin{bmatrix} I & 0 \\ P_{21} & P_{22} \end{bmatrix} C\epsilon_{t+1} \qquad (19.3.18)$$

to the right side of $(19.3.19a)$.

19.3.6. Summary

In summary, we solve the Stackelberg problem by formulating a particular optimal linear regulator, solving the associated matrix Riccati equation (19.3.5) for P, computing F, and then partitioning P to obtain representation (19.3.19).

19.3.7. History-dependent representation of decision rule

For some purposes, it is useful to eliminate the implementation multipliers μ_{xt} and to express the decision rule for u_t as a function of z_t, z_{t-1}, and u_{t-1}. This can be accomplished as follows.[9] First represent (19.3.19a) compactly as

$$\begin{bmatrix} z_{t+1} \\ \mu_{x,t+1} \end{bmatrix} = \begin{bmatrix} m_{11} & m_{12} \\ m_{21} & m_{22} \end{bmatrix} \begin{bmatrix} z_t \\ \mu_{xt} \end{bmatrix} \tag{19.3.23}$$

and write the feedback rule for u_t

$$u_t = f_{11} z_t + f_{12} \mu_{xt}. \tag{19.3.24}$$

Then where f_{12}^{-1} denotes the generalized inverse of f_{12}, (19.3.24) implies $\mu_{x,t} = f_{12}^{-1}(u_t - f_{11} z_t)$. Equate the right side of this expression to the right side of the second line of (19.3.23) lagged once and rearrange by using (19.3.24) lagged once to eliminate $\mu_{x,t-1}$ to get

$$u_t = f_{12} m_{22} f_{12}^{-1} u_{t-1} + f_{11} z_t + f_{12}(m_{21} - m_{22} f_{12}^{-1} f_{11}) z_{t-1} \tag{19.3.25a}$$

or

$$u_t = \rho u_{t-1} + \alpha_0 z_t + \alpha_1 z_{t-1} \tag{19.3.25b}$$

for $t \geq 1$. For $t = 0$, the initialization $\mu_{x,0} = 0$ implies that

$$u_0 = f_{11} z_0. \tag{19.3.25c}$$

By making the instrument feed back on itself, the form of (19.3.25) potentially allows for "instrument-smoothing" to emerge as an optimal rule under commitment.[10]

---

[9] Peter Von Zur Muehlen suggested this representation to us.

[10] This insight partly motivated Woodford (2003) to use his model to interpret empirical evidence about interest rate smoothing in the United States.

19.3.8. Digression on determinacy of equilibrium

Appendix 19B describes methods for solving a system of difference equations of the form (19.2.3) or (19.2.4) with an arbitrary feedback rule that expresses the decision rule for u_t as a function of current and previous values of y_t and perhaps previous values of itself. The difference equation system has a unique solution satisfying the stability condition $\sum_{t=0}^{\infty} \beta^t y_t \cdot y_t$ if the eigenvalues of the matrix (19.B.1) split, with half being greater than unity and half being less than unity in modulus. If more than half are less than unity in modulus, the equilibrium is said to be indeterminate in the sense that there are multiple equilibria starting from any initial condition.

If we choose to represent the solution of a Stackelberg or Ramsey problem in the form (19.3.25), we can substitute that representation for u_t into (19.2.4), obtain a difference equation system in y_t, u_t, and ask whether the resulting system is determinate. To answer this question, we would use the method of Appendix 19B, form system (19.B.1), then check whether the generalized eigenvalues split as required. Researchers have used this method to study the determinacy of equilibria under Stackelberg plans with representations like (19.3.25) and have discovered that sometimes an equilibrium can be indeterminate.[11] See Evans and Honkapohja (2003) for a discussion of determinacy of equilibria under commitment in a class of equilibrium monetary models and how determinacy depends on how the decision rule of the Stackelberg leader is represented. Evans and Honkapohja argue that casting a government decision rule in a way that leads to indeterminacy is a bad idea.

---

[11] The existence of a Stackelberg plan is not at issue because we know how to construct one using the method in the text.

19.4. A large firm with a competitive fringe

As an example, this section studies the equilibrium of an industry with a large firm that acts as a Stackelberg leader with respect to a competitive fringe. Sometimes the large firm is called 'the monopolist' even though there are actually many firms in the industry. The industry produces a single nonstorable homogeneous good. One large firm produces Q_t and a representative firm in a competitive fringe produces q_t. The representative firm in the competitive fringe acts as a price taker and chooses sequentially. The large firm commits to a policy at time 0, taking into account its ability to manipulate the price sequence, both directly through the effects of its quantity choices on prices, and indirectly through the responses of the competitive fringe to its forecasts of prices.[12]

The costs of production are $\mathcal{C}_t = eQ_t + .5gQ_t^2 + .5c(Q_{t+1} - Q_t)^2$ for the large firm and $\sigma_t = dq_t + .5hq_t^2 + .5c(q_{t+1} - q_t)^2$ for the competitive firm, where $d > 0, e > 0, c > 0, g > 0, h > 0$ are cost parameters. There is a linear inverse demand curve

$$p_t = A_0 - A_1(Q_t + \bar{q}_t) + v_t, \tag{19.4.1}$$

where A_0, A_1 are both positive and v_t is a disturbance to demand governed by

$$v_{t+1} = \rho v_t + C_\epsilon \check{\epsilon}_{t+1} \tag{19.4.2}$$

and where $|\rho| < 1$ and $\check{\epsilon}_{t+1}$ is an i.i.d. sequence of random variables with mean zero and variance 1. In (19.4.1), $\bar{q}_t$ is equilibrium output of the representative competitive firm. In equilibrium, $\bar{q}_t = q_t$, but we must distinguish between q_t and $\bar{q}_t$ in posing the optimum problem of a competitive firm.

---

[12] Hansen and Sargent (2008, ch. 16) use this model as a laboratory to illustrate an equilibrium concept featuring robustness in which at least one of the agents has doubts about the stochastic specification of the demand shock process.

19.4.1. The competitive fringe

The representative competitive firm regards $\{p_t\}_{t=0}^{\infty}$ as an exogenous stochastic process and chooses an output plan to maximize

$$E_0 \sum_{t=0}^{\infty} \beta^t \{p_t q_t - \sigma_t\}, \quad \beta \in (0,1) \tag{19.4.3}$$

subject to q_0 given, where $c > 0, d > 0, h > 0$ are cost parameters, and E_t is the mathematical expectation based on time t information. Let $i_t = q_{t+1} - q_t$. We regard i_t as the representative firm's control at t. The first-order conditions for maximizing (19.4.3) are

$$i_t = E_t \beta i_{t+1} - c^{-1} \beta h q_{t+1} + c^{-1} \beta E_t (p_{t+1} - d) \tag{19.4.4}$$

for $t \geq 0$. We appeal to the certainty equivalence principle stated on page 131 to justify working with a non-stochastic version of (19.4.4) formed by dropping the expectation operator and the random term $\check{\epsilon}_{t+1}$ from (19.4.2). We use a method of Sargent (1979) and Townsend (1983).[13] We shift (19.4.1) forward one period, replace conditional expectations with realized values, use (19.4.1) to substitute for p_{t+1} in (19.4.4), and set $q_t = \bar{q}_t$ for all $t \geq 0$ to get

$$i_t = \beta i_{t+1} - c^{-1} \beta h \bar{q}_{t+1} + c^{-1} \beta (A_0 - d) - c^{-1} \beta A_1 \bar{q}_{t+1} - c^{-1} \beta A_1 Q_{t+1} + c^{-1} \beta v_{t+1}. \tag{19.4.5}$$

Given sufficiently stable sequences $\{Q_t, v_t\}$, we could solve (19.4.5) and $i_t = \bar{q}_{t+1} - \bar{q}_t$ to express the competitive fringe's output sequence as a function of the (tail of the) monopolist's output sequence. The dependence of i_t on future Q_t's opens an avenue for the monopolist to influence current outcomes by committing to future actions today. It is this feature that makes the monopolist's problem fail to be recursive in the natural state variables $\bar{q}, Q$. The monopolist arrives at period $t > 0$ facing the constraint that it must confirm the expectations about its time t decision upon which the competitive fringe based its decisions at dates before t.

---

[13] They used this method to compute a rational expectations competitive equilibrium. The key step was to eliminate price and output by substituting from the inverse demand curve and the production function into the firm's first-order conditions to get a difference equation in capital.

19.4.2. The monopolist's problem

The monopolist views the competitive firm's sequence of Euler equations as constraints on its own opportunities. They are *implementability constraints* on the monopolist's choices. Including (19.4.5), we can represent the constraints in terms of the transition law impinging on the monopolist:

$$
\begin{bmatrix}
1 & 0 & 0 & 0 & 0 \\
0 & 1 & 0 & 0 & 0 \\
0 & 0 & 1 & 0 & 0 \\
0 & 0 & 0 & 1 & 0 \\
A_0 - d & 1 & -A_1 & -A_1 - h & c
\end{bmatrix}
\begin{bmatrix}
1 \\ v_{t+1} \\ Q_{t+1} \\ \bar{q}_{t+1} \\ i_{t+1}
\end{bmatrix}
=
\begin{bmatrix}
1 & 0 & 0 & 0 & 0 \\
0 & \rho & 0 & 0 & 0 \\
0 & 0 & 1 & 0 & 0 \\
0 & 0 & 0 & 1 & 1 \\
0 & 0 & 0 & 0 & \frac{c}{\beta}
\end{bmatrix}
\begin{bmatrix}
1 \\ v_t \\ Q_t \\ \bar{q}_t \\ i_t
\end{bmatrix}
$$

$$
+
\begin{bmatrix}
0 \\ 0 \\ 1 \\ 0 \\ 0
\end{bmatrix}
u_t,
$$

(19.4.6)

where $u_t = Q_{t+1} - Q_t$ is the control of the monopolist. The last row portrays the implementability constraints (19.4.5). Represent (19.4.6) as

$$
y_{t+1} = A y_t + B u_t.
\tag{19.4.7}
$$

Although we have entered i_t as a component of the "state" y_t in the monopolist's transition law (19.4.7), i_t is actually a "jump" variable. Nevertheless, the analysis in earlier sections of this chapter implies that the solution of the large firm's problem is encoded in the Riccati equation associated with (19.4.7) as the transition law. Let's decode it.

To match our general setup, we partition y_t as $y_t' = [\, z_t' \quad x_t' \,]$ where $z_t' = [1 \quad v_t \quad Q_t \quad \bar{q}_t]$ and $x_t = i_t$. The large firm's problem is

$$
\max_{\{u_t, p_t, Q_{t+1}, \bar{q}_{t+1}, i_t\}} \sum_{t=0}^{\infty} \beta^t \{ p_t Q_t - \mathcal{C}_t \}
$$

subject to the given initial condition for z_0, equations (19.4.1) and (19.4.5) and $i_t = \bar{q}_{t+1} - \bar{q}_t$, as well as the laws of motion of the natural state variables z. Notice that the monopolist in effect chooses the price sequence, as well as the quantity sequence of the competitive fringe, albeit subject to the restrictions imposed by the behavior of consumers, as summarized by the demand curve (19.4.1), and the implementability constraint (19.4.5) that summarizes the best responses of the competitive fringe.

By substituting (19.4.1) into the above objective function, the monopolist's problem can be expressed as

$$\max_{\{u_t\}} \sum_{t=0}^{\infty} \beta^t \left\{ (A_0 - A_1(\bar{q}_t + Q_t) + v_t)Q_t - eQ_t - .5gQ_t^2 - .5cu_t^2 \right\} \qquad (19.4.8)$$

subject to (19.4.7). This can be written

$$\max_{\{u_t\}} - \sum_{t=0}^{\infty} \beta^t \left\{ y_t' R y_t + u_t' Q u_t \right\} \qquad (19.4.9)$$

subject to (19.4.7) where

$$R = - \begin{bmatrix} 0 & 0 & \frac{A_0 - e}{2} & 0 & 0 \\ 0 & 0 & \frac{1}{2} & 0 & 0 \\ \frac{A_0 - e}{2} & \frac{1}{2} & -A_1 - .5g & -\frac{A_1}{2} & 0 \\ 0 & 0 & -\frac{A_1}{2} & 0 & 0 \\ 0 & 0 & 0 & 0 & 0 \end{bmatrix}$$

and $Q = \frac{c}{2}$.

19.4.3. Equilibrium representation

We can use (19.3.19) to represent the solution of the monopolist's problem (19.4.9) in the form:

$$\begin{bmatrix} z_{t+1} \\ \mu_{x,t+1} \end{bmatrix} = \begin{bmatrix} m_{11} & m_{12} \\ m_{21} & m_{22} \end{bmatrix} \begin{bmatrix} z_t \\ \mu_{x,t} \end{bmatrix} \qquad (19.4.10)$$

or

$$\begin{bmatrix} z_{t+1} \\ \mu_{x,t+1} \end{bmatrix} = m \begin{bmatrix} z_t \\ \mu_{x,t} \end{bmatrix}. \qquad (19.4.11)$$

The monopolist is constrained to set $\mu_{x,0} \leq 0$, but will find it optimal to set it to zero. Recall that $z_t = \begin{bmatrix} 1 & v_t & Q_t & \bar{q}_t \end{bmatrix}'$. Thus, (19.4.11) includes the equilibrium law of motion for the quantity $\bar{q}_t$ of the competitive fringe. By construction, $\bar{q}_t$ satisfies the Euler equation of the representative firm in the competitive fringe, as we elaborate in Appendix 19C.

19.4.4. Numerical example

We computed the optimal Stackelberg plan for parameter settings $A_0, A_1, \rho, C_\epsilon$, $c, d, e, g, h, \beta = 100, 1, .8, .2, 1, 20, 20, .2, .2, .95$.[14] For these parameter values the decision rule is

$$u_t = (Q_{t+1} - Q_t) = \begin{bmatrix} 19.78 & .19 & -.64 & -.15 & -.30 \end{bmatrix} \begin{bmatrix} z_t \\ \mu_{xt} \end{bmatrix} \tag{19.4.12}$$

which can also be represented as

$$u_t = 0.44 u_{t-1} + \begin{bmatrix} 19.7827 \\ 0.1885 \\ -0.6403 \\ -0.1510 \end{bmatrix}' z_t + \begin{bmatrix} -6.9509 \\ -0.0678 \\ 0.3030 \\ 0.0550 \end{bmatrix}' z_{t-1}. \tag{19.4.13}$$

Note how in representation $(19.4.12)$ the monopolist's decision for $u_t = Q_{t+1} - Q_t$ feeds back negatively on the implementation multiplier.[15]

19.5. Concluding remarks

This chapter is our first brush with a class of problems in which optimal decision rules are history dependent. We shall confront many more such problems in chapters 20, 21, and 23 and shall see in various contexts how history dependence can be represented recursively by appropriately augmenting the natural state variables with counterparts to our implementability multipliers. A hint at what these counterparts are is gleaned by appropriately interpreting implementability multipliers as derivatives of value functions. In chapters 20,21, and 23, we make dynamic incentive and enforcement problems recursive by augmenting the state with continuation values of other decision makers.[16]

---

[14] These calculations were performed by the Matlab program `oligopoly5.m`

[15] We also computed impulse responses to the demand innovation ϵ_t. The impulse responses show that a demand innovation pushes the implementation multiplier down and leads the monopolist to expand output while the representative competitive firm contracts output in subsequent periods. The response of price to a demand shock innovation is to rise on impact but then to decrease in subsequent periods in response to the increase in total supply $\bar{q} + Q$ engineered by the monopolist.

[16] In chapter 20, we describe Marcet and Marimon's (1992, 1999) method of constructing recursive contracts, which is closely related to the method that we have presented in this chapter.

A. The stabilizing $\mu_t = Py_t$

We verify that the P associated with the stabilizing $\mu_0 = Py_0$ satisfies the Riccati equation associated with the Bellman equation. Substituting $\mu_t = Py_t$ into (19.3.9) and (19.3.8b) gives

$$(I + \beta BQ^{-1}BP)y_{t+1} = Ay_t \qquad (19.A.1a)$$

$$\beta A'Py_{t+1} = -Ry_t + Py_t. \qquad (19.A.1b)$$

A matrix inversion identity implies

$$(I + \beta BQ^{-1}B'P)^{-1} = I - \beta B(Q + \beta B'PB)^{-1}B'P. \qquad (19.A.2)$$

Solving (19.A.1a) for y_{t+1} gives

$$y_{t+1} = (A - BF)y_t \qquad (19.A.3)$$

where

$$F = \beta(Q + \beta B'PB)^{-1}B'PA. \qquad (19.A.4)$$

Premultiplying (19.A.3) by $\beta A'P$ gives

$$\beta A'Py_{t+1} = \beta(A'PA - A'PBF)y_t. \qquad (19.A.5)$$

For the right side of (19.A.5) to agree with the right side of (19.A.1b) for any initial value of y_0 requires that

$$P = R + \beta A'PA - \beta^2 A'PB(Q + \beta B'PB)^{-1}B'PA. \qquad (19.A.6)$$

Equation (19.A.6) is the algebraic matrix Riccati equation associated with the optimal linear regulator for the system A, B, Q, R.

B. Matrix linear difference equations

This appendix generalizes some calculations from chapter 5 for solving systems of linear difference equations. Returning to system (19.3.11), let $L = L^*\beta^{-.5}$ and transform the system (19.3.11) to

$$L \begin{bmatrix} y^*_{t+1} \\ \mu^*_{t+1} \end{bmatrix} = N \begin{bmatrix} y^*_t \\ \mu^*_t \end{bmatrix}, \qquad (19.B.1)$$

where $y^*_t = \beta^{t/2} y_t, \mu^*_t = \mu_t \beta^{t/2}$. Now $\lambda L - N$ is a symplectic pencil,[17] so that the generalized eigenvalues of L, N occur in reciprocal pairs: if λ_i is an eigenvalue, then so is λ_i^{-1}.

We can use Evan Anderson's Matlab program `schurg.m` to find a stabilizing solution of system (19.B.1).[18] The program computes the ordered real generalized Schur decomposition of the matrix pencil. Thus, `schurg.m` computes matrices $\bar{L}, \bar{N}, V$ such that $\bar{L}$ is upper triangular, $\bar{N}$ is upper block triangular, and V is the matrix of right Schur vectors such that for some orthogonal matrix W, the following hold:

$$\begin{aligned} WLV &= \bar{L} \\ WNV &= \bar{N}. \end{aligned} \qquad (19.B.2)$$

Let the stable eigenvalues (those less than 1) appear first. Then the stabilizing solution is

$$\mu^*_t = P y^*_t \qquad (19.B.3)$$

where

$$P = V_{21} V_{11}^{-1},$$

V_{21} is the lower left block of V, and V_{11} is the upper left block.

If L is nonsingular, we can represent the solution of the system as[19]

$$\begin{bmatrix} y^*_{t+1} \\ \mu^*_{t+1} \end{bmatrix} = L^{-1} N \begin{bmatrix} I \\ P \end{bmatrix} y^*_t. \qquad (19.B.4)$$

---

[17] A *pencil* $\lambda L - N$ is the family of matrices indexed by the complex variable λ. A pencil is *symplectic* if $LJL' = NJN'$, where $J = \begin{bmatrix} 0 & -I \\ I & 0 \end{bmatrix}$. See Anderson, Hansen, McGratten, and Sargent (1996).

[18] This program is available at $<$http://www.math.niu.edu/~anderson$>$.

[19] The solution method in the text assumes that L is nonsingular and well conditioned. If it is not, the following method proposed by Evan Anderson will work. We want to solve for a solution of the form

$$y^*_{t+1} = A^*_o y^*_t.$$

The solution is to be initialized from $(19.B.3)$. We can use the first half and then the second half of the rows of this representation to deduce the following recursive solutions for y_{t+1}^* and μ_{t+1}^*:

$$\begin{aligned} y_{t+1}^* &= A_o^* y_t^* \\ \mu_{t+1}^* &= \psi^* y_t^*. \end{aligned} \qquad (19.B.5)$$

Now express this solution in terms of the original variables:

$$\begin{aligned} y_{t+1} &= A_o y_t \\ \mu_{t+1} &= \psi y_t, \end{aligned} \qquad (19.B.6)$$

where $A_o = A_o^* \beta^{-.5}, \psi = \psi^* \beta^{-.5}$. We also have the representation

$$\mu_t = P y_t. \qquad (19.B.7)$$

The matrix $A_o = A - BF$, where F is the matrix for the optimal decision rule.

Note that with $(19.B.3)$,
$$L[I; P] y_{t+1}^* = N[I; P] y_t^*$$

The solution A_o^* will then satisfy

$$L[I; P] A_o^* = N[I; P].$$

Thus A^{o*} can be computed via the Matlab command

$$A_o^* = (L * [I; P]) \backslash (N * [I; P]).$$

C. Forecasting formulas

The decision rule for the competitive fringe incorporates forecasts of future prices from (19.4.11) under m. Thus, the representative competitive firm uses equation (19.4.11) to forecast future values of (Q_t, q_t) in order to forecast p_t. The representative competitive firm's forecasts are generated from the jth iterate of (19.4.11):[20]

$$\begin{bmatrix} z_{t+j} \\ \mu_{x,t+j} \end{bmatrix} = m^j \begin{bmatrix} z_t \\ \mu_{x,t} \end{bmatrix}. \tag{19.C.1}$$

The following calculation verifies that the representative firm forecasts by iterating the law of motion associated with m. Write the Euler equation for i_t (19.4.4) in terms of a polynomial in the lag operator L and factor it: $(1 - (\beta^{-1} + (1 + c^{-1}h))L + \beta^{-1}L^2) = -(\beta\lambda)^{-1}L(1 - \beta\lambda L^{-1})(1 - \lambda L)$ where $\lambda \in (0, 1)$ and $\lambda = 1$ when $h = 0$.[21] By taking the nonstochastic version of (19.4.4) and solving an unstable root forward and a stable root backward using the technique of Sargent (1979 or 1987a, chap. IX), we obtain

$$i_t = (\lambda - 1)q_t + c^{-1} \sum_{j=1}^{\infty} (\beta\lambda)^j p_{t+j}, \tag{19.C.2}$$

or

$$i_t = (\lambda - 1)q_t + c^{-1} \sum_{j=1}^{\infty} (\beta\lambda)^j [(A_0 - d) - A_1(Q_{t+j} + q_{t+j}) + v_{t+j}], \tag{19.C.3}$$

This can be expressed as

$$i_t = (\lambda - 1)q_t + c^{-1} e_p \beta\lambda m (I - \beta\lambda m)^{-1} \begin{bmatrix} z_t \\ \mu_{xt} \end{bmatrix} \tag{19.C.4}$$

where $e_p = [(A_0 - d) \quad 1 \quad -A_1 \quad -A_1 \quad 0]$ is a vector that forms $p_t - d$ upon postmultiplication by $\begin{bmatrix} z_t \\ \mu_{xt} \end{bmatrix}$. It can be verified that the solution procedure builds in (19.C.4) as an identity, so that (19.C.4) agrees with

$$i_t = -P_{22}^{-1} P_{21} z_t + P_{22}^{-1} \mu_{xt}. \tag{19.C.5}$$

---

[20] The representative firm acts as though (q_t, Q_t) were exogenous to it.

[21] See Sargent (1979 or 1987a) for an account of the method we are using here.

Exercises

Exercise 19.1   There is no uncertainty. For $t \geq 0$, a monetary authority sets the growth of the (log) of money according to

(1) $$m_{t+1} = m_t + u_t$$

subject to the initial condition $m_0 > 0$ given. The demand for money is

(2) $$m_t - p_t = -\alpha(p_{t+1} - p_t), \alpha > 0,$$

where p_t is the log of the price level. Equation (2) can be interpreted as the Euler equation of the holders of money.

a. Briefly interpret how equation (2) makes the demand for real balances vary inversely with the expected rate of inflation. Temporarily (only for this part of the exercise) drop equation (1) and assume instead that $\{m_t\}$ is a given sequence satisfying $\sum_{t=0}^{\infty} m_t^2 < +\infty$. Please solve the difference equation (2) "forward" to express p_t as a function of current and future values of m_s. Note how future values of m influence the current price level.

At time 0, a monetary authority chooses a possibly history-dependent strategy for setting $\{u_t\}_{t=0}^{\infty}$. (The monetary authority commits to this strategy.) The monetary authority orders sequences $\{m_t, p_t\}_{t=0}^{\infty}$ according to

(3) $$-\sum_{t=0}^{\infty} .95^t \left[(p_t - \bar{p})^2 + u_t^2 + .00001 m_t^2 \right].$$

Assume that $m_0 = 10, \alpha = 5, \bar{p} = 1$.

b. Please briefly interpret this problem as one where the monetary authority wants to stabilize the price level, subject to costs of adjusting the money supply and some implementability constraints. (We include the term $.00001 m_t^2$ for purely technical reasons that you need not discuss.)

c. Please write and run a Matlab program to find the optimal sequence $\{u_t\}_{t=0}^{\infty}$.

d. Display the optimal decision rule for u_t as a function of u_{t-1}, m_t, m_{t-1}.

e. Compute the optimal $\{m_t, p_t\}_t$ sequence for $t = 0, \ldots, 10$.

Hint: The optimal $\{m_t\}$ sequence must satisfy $\sum_{t=0}^{\infty} (.95)^t m_t^2 < +\infty$. You are free to apply the Matlab program `olrp.m`.

Exercise 19.2   A representative consumer has quadratic utility functional

(1)
$$\sum_{t=0}^{\infty} \beta^t \left\{ -.5(b - c_t)^2 \right\}$$

where $\beta \in (0, 1)$, $b = 30$, and c_t is time t consumption. The consumer faces a sequence of budget constraints

(2)
$$c_t + a_{t+1} = (1 + r)a_t + y_t - \tau_t$$

where a_t is the household's holdings of an asset at the beginning of t, $r > 0$ is a constant net interest rate satisfying $\beta(1 + r) < 1$, and y_t is the consumer's endowment at t. The consumer's plan for (c_t, a_{t+1}) has to obey the boundary condition $\sum_{t=0}^{\infty} \beta^t a_t^2 < +\infty$. Assume that y_0, a_0 are given initial conditions and that y_t obeys

(3)
$$y_t = \rho y_{t-1}, \quad t \geq 1,$$

where $|\rho| < 1$. Assume that $a_0 = 0$, $y_0 = 3$, and $\rho = .9$.

At time 0, a planner commits to a plan for taxes $\{\tau_t\}_{t=0}^{\infty}$. The planner designs the plan to maximize

(4)
$$\sum_{t=0}^{\infty} \beta^t \left\{ -.5(c_t - b)^2 - \tau_t^2 \right\}$$

over $\{c_t, \tau_t\}_{t=0}^{\infty}$ subject to the implementability constraints (2) for $t \geq 0$ and

(5)
$$\lambda_t = \beta(1 + r)\lambda_{t+1}$$

for $t \geq 0$, where $\lambda_t \equiv (b - c_t)$.

a. Argue that (5) is the Euler equation for a consumer who maximizes (1) subject to (2), taking $\{\tau_t\}$ as a given sequence.

b. Formulate the planner's problem as a Stackelberg problem.

c. For $\beta = .95, b = 30, \beta(1 + r) = .95$, formulate an artificial optimal linear regulator problem and use it to solve the Stackelberg problem.

d. Give a recursive representation of the Stackelberg plan for τ_t.

Chapter 20
Insurance Versus Incentives

20.1. Insurance with recursive contracts

This chapter studies a planner who designs an efficient contract to supply insurance in the presence of incentive constraints imposed by his limited ability either to enforce contracts or to observe households' actions or incomes. We pursue two themes, one substantive, the other technical. The substantive theme is a tension that exists between offering insurance and providing incentives. A planner can overcome incentive problems by offering "sticks and carrots" that adjust an agent's future consumption and thereby provide less insurance. Balancing incentives against insurance shapes the evolution of distributions of wealth and consumption.

The technical theme is how memory can be encoded recursively and how incentive problems can be managed with contracts that retain memory and make promises. Contracts issue rewards that depend on the history either of publicly observable outcomes or of an agent's announcements about his privately observed outcomes. Histories are large-dimensional objects. But Spear and Srivastava (1987), Thomas and Worrall (1988), Abreu, Pearce, and Stacchetti (1990), and Phelan and Townsend (1991) discovered that the dimension can be contained by using an accounting system cast solely in terms of a "promised value," a one-dimensional object that summarizes relevant aspects of an agent's history. Working with promised values permits us to formulate the contract design problem recursively.

Three basic models are set within a single physical environment but assume different structures of information, enforcement, and storage possibilities. The first adapts a model of Thomas and Worrall (1988) and Kocherlakota (1996b) that focuses on commitment or enforcement problems and has all information being public. The second is a model of Thomas and Worrall (1990) that has an incentive problem coming from private information but that assumes away commitment and enforcement problems. Common to both of these models is that the insurance contract is assumed to be the *only* vehicle for households to transfer wealth across states of the world and over time. The third model, created by Allen (1985) and Cole and Kocherlakota (2001), extends Thomas and Worrall's (1990) model by introducing private storage that cannot be observed publicly. Ironically, because it lets households self-insure as in chapters 17 and 18, the possibility of private storage reduces *ex ante* welfare by limiting the amount of social insurance that can be attained when incentive constraints are present.

20.2. Basic environment

Imagine a village with a large number of *ex ante* identical households. Each household has preferences over consumption streams that are ordered by

$$E_{-1} \sum_{t=0}^{\infty} \beta^t u(c_t), \tag{20.2.1}$$

where $u(c)$ is an increasing, strictly concave, and twice continuously differentiable function, $\beta \in (0,1)$ is a discount factor, and E_{-1} is the mathematical expectation not conditioning on any information available at time 0 or later. Each household receives a stochastic endowment stream $\{y_t\}_{t=0}^{\infty}$, where for each $t \geq 0$, y_t is independently and identically distributed according to the discrete probability distribution $\text{Prob}(y_t = \overline{y}_s) = \Pi_s$, where $s \in \{1, 2, \ldots, S\} \equiv \mathbf{S}$ and $\overline{y}_{s+1} > \overline{y}_s$. The consumption good is not storable. At time $t \geq 1$, the household has received a history of endowments $h_t = (y_t, y_{t-1}, \ldots, y_0)$. The endowment processes are independently and identically distributed both across time and across households.

   In this setting, if there were a competitive equilibrium with complete markets as described in chapter 8, at date 0 households would trade history- and date-contingent claims before the realization of endowments. Since all households are *ex ante* identical, each household would end up consuming the per capita endowment in every period, and its lifetime utility would be

$$v_{\text{pool}} = \sum_{t=0}^{\infty} \beta^t u\left(\sum_{s=1}^{S} \Pi_s \overline{y}_s\right) = \frac{1}{1-\beta} u\left(\sum_{s=1}^{S} \Pi_s \overline{y}_s\right). \tag{20.2.2}$$

Households would thus insure away all of the risk associated with their individual endowment processes. But the incentive constraints that we are about to specify make this allocation unattainable. For each specification of incentive constraints, we shall solve a planning problem for an efficient allocation that respects those constraints.

Following a tradition started by Green (1987), we assume that a "moneylender" or "planner" is the only person in the village who has access to a risk-free loan market outside the village. The moneylender can borrow or lend at a constant risk-free gross interest rate $R = \beta^{-1}$. The households cannot borrow or lend with one another, and can trade only with the moneylender. Furthermore, we assume that the moneylender is committed to honor his promises. We will study three types of incentive constraints.

(a) Both the money lender and the household observe the household's history of endowments at each time t. Although the moneylender *can* commit to honor a contract, households *cannot* commit and at any time are free to walk away from an arrangement with the moneylender and live in perpetual autarky thereafter. They must be induced not to do so by the structure of the contract. This is a model of "one-sided commitment" in which the contract is "self-enforcing" because the household prefers to conform to it.

(b) Households *can* make commitments and enter into enduring and binding contracts with the moneylender, but they have private information about their own incomes. The moneylender can see neither their income nor their consumption. It follows that any exchanges between the moneylender and a household must be based on the household's own reports about income realizations. An incentive-compatible contract must induce a household to report its income truthfully.

(c) The environment is the same as in b except for the additional assumption that households have access to a storage technology that cannot be observed by the moneylender. Households can store nonnegative amounts of goods at a risk-free gross return of R equal to the interest rate that the moneylender faces in the outside credit market. Since the moneylender can both borrow and lend at the interest rate R outside of the village, the private storage technology does not change the economy's aggregate resource constraint, but it does substantially affect the set of incentive-compatible contracts between the moneylender and the households.

When we compute efficient allocations for each of these three environments, we shall find that the dynamics of the implied consumption allocations differ

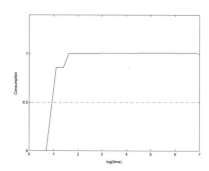

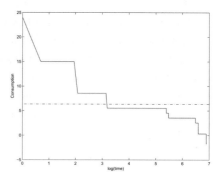

Figure 20.2.1.a: Typical consumption path in environment a.

Figure 20.2.1.b: Typical consumption path in environment b.

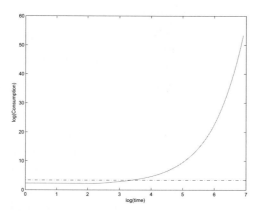

Figure 20.2.2: Typical consumption path in environment c.

dramatically. As a prelude, Figures 20.2.1 and 20.2.2 depict the different consumption streams that are associated with the *same* realization of a random endowment stream for households living in environments a, b, and c, respectively. For all three of these economies, we set $u(c) = -\gamma^{-1}\exp(-\gamma c)$ with

$\gamma = .8$, $\beta = .92$, $[\overline{y}_1, \ldots, \overline{y}_{10}] = [6, \ldots, 15]$, and $\Pi_s = \frac{1-\lambda}{1-\lambda^{10}} \lambda^{s-1}$ with $\lambda = 2/3$. As a benchmark, a horizontal dotted line in each graph depicts the constant consumption level that would be attained in a complete markets equilibrium where there are no incentive problems. In all three environments, prior to date 0, the households have entered into efficient contracts with the moneylender. The dynamics of consumption outcomes evidently differ substantially across the three environments, increasing and then flattening out in environment a, heading "south" in environment b, and heading "north" in environment c. This chapter explains why the sample paths of consumption differ so much across these three settings.

20.3. One-sided no commitment

Our first incentive problem is a lack of commitment. A moneylender is committed to honor his promises, but villagers are free to walk away from their contract with the moneylender at any time. The moneylender designs a contract that the villager wants to honor at every moment and contingency. Such a contract is said to be self-enforcing. In chapter 21, we shall study another economy in which there is no moneylender, only another villager, and when no one is able to keep prior commitments. Such a contract design problem with participation constraints on both sides of an exchange represents a problem with two-sided lack of commitment, in contrast to the problem with one-sided lack of commitment treated here.[1]

---

[1] For an earlier two-period model of a one-sided commitment problem, see Holmström (1983).

20.3.1. *Self-enforcing contract*

A moneylender can borrow or lend resources from outside the village but the villagers cannot. A *contract* is a sequence of functions $c_t = f_t(h_t)$ for $t \geq 0$, where $h_t = (y_t, \ldots, y_0)$. The sequence of functions $\{f_t\}$ assigns a history-dependent consumption stream $c_t = f_t(h_t)$ to the household. The contract specifies that each period, the villager contributes his time t endowment y_t to the moneylender who then returns c_t to the villager. From this arrangement, the moneylender earns an *ex ante* expected present value

$$P_{-1} = E_{-1} \sum_{t=0}^{\infty} \beta^t (y_t - f(h_t)). \qquad (20.3.1)$$

By plugging the associated consumption process into expression (20.2.1), we find that the contract assigns the villager an expected present value of $v = E_{-1} \sum_{t=0}^{\infty} \beta^t u\left(f_t(h_t)\right)$.

The contract must be self-enforcing. At any point in time, the household is free to walk away from the contract and thereafter consume its endowment stream. Thus, if the household walks away from the contract, it must live in autarky evermore. The *ex ante* value associated with consuming the endowment stream, to be called the autarky value, is

$$v_{\text{aut}} = E_{-1} \sum_{t=0}^{\infty} \beta^t u(y_t) = \frac{1}{1-\beta} \sum_{s=1}^{S} \Pi_s u(\overline{y}_s). \qquad (20.3.2)$$

At time t, *after* having observed its current-period endowment, the household can guarantee itself a present value of utility of $u(y_t) + \beta v_{\text{aut}}$ by consuming its own endowment. The moneylender's contract must offer the household at least this utility at every possible history and every date. Thus, the contract must satisfy

$$u[f_t(h_t)] + \beta E_t \sum_{j=1}^{\infty} \beta^{j-1} u[f_{t+j}(h_{t+j})] \geq u(y_t) + \beta v_{\text{aut}}, \qquad (20.3.3)$$

for all $t \geq 0$ and for all histories h_t. Equation (20.3.3) is called the *participation constraint* for the villager. A contract that satisfies equation (20.3.3) is said to be *sustainable*.

20.3.2. *Recursive formulation and solution*

A difficulty with constraints like equation (20.3.3) is that there are so many of them: the dimension of the argument h_t grows exponentially with t. Fortunately, there is a recursive way to describe an interesting subset of history-dependent contracts. In particular, consider the following way of representing a contract $\{f_t\}$ recursively in terms of a state variable x_t:

$$c_t = g(x_t),$$

$$x_{t+1} = \ell(x_t, y_t).$$

Here g and ℓ are time-invariant functions. Notice that by iterating the $\ell(\cdot)$ function t times starting from (x_0, y_0), one obtains

$$x_t = m_t(x_0; y_{t-1}, \ldots, y_0), \quad t \geq 1.$$

Thus, x_t summarizes histories of endowments h_{t-1}. In this sense, x_t is a "backward-looking" variable.

A remarkable fact is that the appropriate state variable x_t is a *promised expected discounted future value* $v_t = E_{t-1} \sum_{j=0}^{\infty} \beta^j u(c_{t+j})$. This "forward-looking" variable summarizes a stream of future utilities. We shall formulate the contract recursively by having the moneylender arrive at t, before y_t is realized, with a previously made promised value v_t. He delivers v_t by letting c_t and the continuation value v_{t+1} both respond to y_t. In terms of $v_t(h_{t-1})$, the participation constraint (20.3.3) becomes

$$v_t(h_{t-1}) = u(f(_t(h_t)) + \beta v_{t+1}(h_t) \geq u(y_t) + \beta v_{\text{aut}}.$$

We shall treat the promised value v as a *state* variable, then formulate a functional equation for a moneylender. The moneylender gives a prescribed value v by delivering a state-dependent current consumption c and a promised value starting tomorrow, say v', where c and v' each depend on the current endowment y and the preexisting promise v. The moneylender chooses c and v' to provide the promised value v in a way that maximizes his profits (20.3.1).

Each period, the household must be induced to surrender the time t endowment y_t to the moneylender, who possibly gives some of it to other households and invests the rest outside the village at a constant risk-free one-period gross interest rate of β^{-1}. In exchange, the moneylender delivers a state-contingent consumption stream to the household that keeps it participating in the arrangement every period and after every history. The moneylender wants to do this in the most efficient way, that is, the profit-maximizing way. Let v be the expected discounted future utility previously promised to a villager. Let $P(v)$ be the expected present value of the "profit stream" $\{y_t - c_t\}$ for a moneylender who delivers promised value v in the optimal way. The optimum value $P(v)$ obeys the functional equation

$$P(v) = \max_{\{c_s, w_s\}} \sum_{s=1}^{S} \Pi_s [(\overline{y}_s - c_s) + \beta P(w_s)] \qquad (20.3.4)$$

where the maximization is subject to the constraints

$$\sum_{s=1}^{S} \Pi_s [u(c_s) + \beta w_s] \geq v, \qquad (20.3.5)$$

$$u(c_s) + \beta w_s \geq u(\overline{y}_s) + \beta v_{\text{aut}}, \quad s = 1, \ldots, S; \qquad (20.3.6)$$

$$c_s \in [c_{\min}, c_{\max}], \qquad (20.3.7)$$

$$w_s \in [v_{\text{aut}}, \overline{v}]. \qquad (20.3.8)$$

Here w_s is the promised value with which the consumer will enter next period, given that $y = \overline{y}_s$ this period; $[c_{\min}, c_{\max}]$ is a bounded set to which we restrict the choice of c_t each period. We restrict the continuation value w_s to be in the set $[v_{\text{aut}}, \overline{v}]$, where $\overline{v}$ is a very large number. Soon we'll compute the highest value that the moneylender would ever want to set w_s. All we require now is that $\overline{v}$ exceed this value. Constraint (20.3.5) is the promise-keeping constraint. It requires that the contract deliver at least promised value v. Constraints (20.3.6), one for each state s, are the participation constraints. Evidently, P must be a decreasing function of v because the higher the consumption stream of the villager, the lower must be the profits of the moneylender.

The constraint set is convex. The one-period return function in equation (20.3.4) is concave. The value function $P(v)$ that solves equation (20.3.4) is concave. In fact, $P(v)$ is strictly concave as will become evident from our characterization of the optimal contract. Form the Lagrangian

$$L = \sum_{s=1}^{S} \Pi_s[(\overline{y}_s - c_s) + \beta P(w_s)]$$

$$+ \mu \left\{ \sum_{s=1}^{S} \Pi_s[u(c_s) + \beta w_s] - v \right\} \qquad (20.3.9)$$

$$+ \sum_{s=1}^{S} \lambda_s \left\{ u(c_s) + \beta w_s - [u(\overline{y}_s) + \beta v_{\text{aut}}] \right\}.$$

For each v and for $s = 1, \ldots, S$, the first-order conditions for maximizing L with respect to c_s, w_s, respectively, are[2]

$$(\lambda_s + \mu \Pi_s)u'(c_s) = \Pi_s, \qquad (20.3.10)$$

$$\lambda_s + \mu \Pi_s = -\Pi_s P'(w_s). \qquad (20.3.11)$$

By the envelope theorem, if P is differentiable, then $P'(v) = -\mu$. We will proceed under the assumption that P is differentiable but it will become evident that P is indeed differentiable when we understand the optimal contract.

Equations $(20.3.10)$ and $(20.3.11)$ imply the following relationship between c_s and w_s:

$$u'(c_s) = -P'(w_s)^{-1}. \qquad (20.3.12)$$

This condition states that the household's marginal rate of substitution between c_s and w_s, given by $u'(c_s)/\beta$, should equal the moneylender's marginal rate of transformation as given by $-[\beta P'(w_s)]^{-1}$. The concavity of P and u means that equation $(20.3.12)$ traces out a positively sloped curve in the c, w plane, as depicted in Figure 20.3.1. We can interpret this condition as making c_s a function of w_s. To complete the optimal contract, it will be enough to find how w_s depends on the promised value v and the income state $\overline{y}_s$.

Condition $(20.3.11)$ can be written

$$P'(w_s) = P'(v) - \lambda_s/\Pi_s. \qquad (20.3.13)$$

How w_s varies with v depends on which of two mutually exclusive and exhaustive sets of states (s, v) falls into after the realization of $\overline{y}_s$: those in which the participation constraint $(20.3.6)$ binds (i.e., states in which $\lambda_s > 0$) or those in which it does not (i.e., states in which $\lambda_s = 0$).

---

[2] Please note that the λ_s's depend on the promised value v. In particular, which λ_s's are positive and which are zero will depend on v, with more of them being zero when the promised value v is higher. See figure 20.3.1.

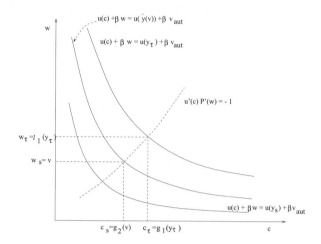

Figure 20.3.1: Determination of consumption and promised utility (c, w). Higher realizations of $\bar{y}_s$ are associated with higher indifference curves $u(c) + \beta w = u(\bar{y}_s) + \beta v_{\text{aut}}$. For a given v, there is a threshold level $\bar{y}(v)$ above which the participation constraint is binding and below which the moneylender awards a constant level of consumption, as a function of v, and maintains the same promised value $w = v$. The cutoff level $\bar{y}(v)$ is determined by the indifference curve going through the intersection of a horizontal line at level v with the "expansion path" $u'(c)P'(w) = -1$.

States where $\lambda_s > 0$

When $\lambda_s > 0$, the participation constraint (20.3.6) holds with equality. When $\lambda_s > 0$, (20.3.13) implies that $P'(w_s) < P'(v)$, which in turn implies, by the concavity of P, that $w_s > v$. Further, the participation constraint at equality implies that $c_s < \bar{y}_s$ (because $w_s > v \geq v_{\text{aut}}$). Together, these results say that when the participation constraint (20.3.6) binds, the moneylender induces the household to consume less than its endowment today by raising its continuation value.

When $\lambda_s > 0$, c_s and w_s solve the two equations

$$u(c_s) + \beta w_s = u(\bar{y}_s) + \beta v_{\text{aut}}, \tag{20.3.14}$$

$$u'(c_s) = -P'(w_s)^{-1}. \tag{20.3.15}$$

The participation constraint holds with equality. Notice that these equations are independent of v. This property is a key to understanding the form of the optimal contract. It imparts to the contract what Kocherlakota (1996b) calls *amnesia*: when incomes y_t are realized that cause the participation constraint to bind, the contract disposes of all history dependence and makes both consumption and the continuation value depend only on the current income state y_t. We portray amnesia by denoting the solutions of equations (20.3.14) and (20.3.15) by

$$c_s = g_1(\overline{y}_s), \tag{20.3.16a}$$

$$w_s = \ell_1(\overline{y}_s). \tag{20.3.16b}$$

Later, we'll exploit the amnesia property to produce a computational algorithm.

States where $\lambda_s = 0$

When the participation constraint does not bind, $\lambda_s = 0$ and first-order condition (20.3.11) imply that $P'(v) = P'(w_s)$, which implies that $w_s = v$. Therefore, from (20.3.12), we can write $u'(c_s) = -P'(v)^{-1}$, so that consumption in state s depends on promised utility v but not on the endowment in state s. Thus, when the participation constraint does not bind, the moneylender awards

$$c_s = g_2(v), \tag{20.3.17a}$$

$$w_s = v, \tag{20.3.17b}$$

where $g_2(v)$ solves $u'[g_2(v)] = -P'(v)^{-1}$.

The optimal contract

Combining the branches of the policy functions for the cases where the participation constraint does and does not bind, we obtain

$$c = \max\{g_1(y), g_2(v)\}, \tag{20.3.18}$$

$$w = \max\{\ell_1(y), v\}. \tag{20.3.19}$$

The optimal policy is displayed graphically in Figures 20.3.1 and 20.3.2. To interpret the graphs, it is useful to study equations (20.3.6) and (20.3.12) for the case in which $w_s = v$. By setting $w_s = v$, we can solve these equations for a "cutoff value," call it $\bar{y}(v)$, such that the participation constraint binds only when $\overline{y}_s \geq \bar{y}(v)$. To find $\bar{y}(v)$, we first solve equation $(20.3.12)$ for the value c_s associated with v for those states in which the participation constraint is not binding:

$$u'[g_2(v)] = -P'(v)^{-1},$$

and then substitute this value into $(20.3.6)$ at equality to solve for $\bar{y}(v)$:

$$u[\bar{y}(v)] = u[g_2(v)] + \beta(v - v_{\text{aut}}). \tag{20.3.20}$$

By the concavity of P, the cutoff value $\bar{y}(v)$ is increasing in v.

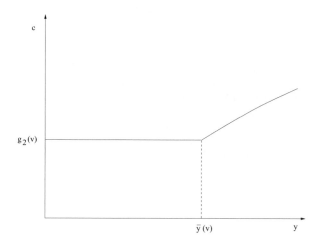

Figure 20.3.2: The shape of consumption as a function of realized endowment, when the promised initial value is v.

Associated with a given level of $v_t \in (v_{\text{aut}}, \bar{v})$, there are two numbers $g_2(v_t)$, $\bar{y}(v_t)$ such that if $y_t \leq \bar{y}(v_t)$ the moneylender offers the household $c_t = g_2(v_t)$ and leaves the promised utility unaltered, $v_{t+1} = v_t$. The moneylender is thus insuring the villager against the states $\overline{y}_s \leq \bar{y}(v_t)$ at time t. If $y_t > \bar{y}(v_t)$, the participation constraint binds, prompting the moneylender to induce the household to surrender some of its current-period endowment in exchange for a raised promised utility $v_{t+1} > v_t$. Promised values never decrease. They stay constant for low-y states $\overline{y}_s < \bar{y}(v_t)$ and increase in high-endowment states that threaten to violate the participation constraint. Consumption stays constant during periods when the participation constraint fails to bind and increases during periods when it threatens to bind. Whenever the participation binds, the household makes a net transfer to the money lender in return for a higher promised continuation utility. A household that has ever realized the highest endowment y_S is permanently awarded the highest consumption level with an associated promised value $\bar{v}$ that satisfies

$$u[g_2(\bar{v})] + \beta\bar{v} = u(\overline{y}_S) + \beta v_{\mathrm{aut}}.$$

20.3.3. Recursive computation of contract

Suppose that the initial promised value v_0 is v_{aut}. We can compute the optimal contract recursively by using the fact that the villager will ultimately receive a constant welfare level equal to $u(\overline{y}_S) + \beta v_{\mathrm{aut}}$ after ever having experienced the maximum endowment $\overline{y}_S$. We can characterize the optimal policy in terms of numbers $\{\overline{c}_s, \overline{w}_s\}_{s=1}^{S} \equiv \{g_1(\overline{y}_s), \ell_1(\overline{y}_s)\}_{s=1}^{S}$ where $g_1(\overline{y}_s)$ and $\ell_1(\overline{s})$ are given by $(20.3.16)$. These numbers can be computed recursively by working backward as follows. Start with $s = S$ and compute $(\overline{c}_S, \overline{w}_S)$ from the nonlinear equations:

$$u(\overline{c}_S) + \beta\overline{w}_S = u(\overline{y}_S) + \beta v_{\mathrm{aut}}, \qquad (20.3.21a)$$

$$\overline{w}_S = \frac{u(\overline{c}_S)}{1 - \beta}. \qquad (20.3.21b)$$

Working backward for $j = S - 1, \ldots, 1$, compute $\overline{c}_j, \overline{w}_j$ from the two nonlinear equations

$$u(\overline{c}_j) + \beta\overline{w}_j = u(\overline{y}_j) + \beta v_{\mathrm{aut}}, \qquad (20.3.22a)$$

$$\overline{w}_j = [u(\overline{c}_j) + \beta\overline{w}_j] \sum_{k=1}^{j} \Pi_k + \sum_{k=j+1}^{S} \Pi_k[u(\overline{c}_k) + \beta\overline{w}_k]. \qquad (20.3.22b)$$

These successive iterations yield the optimal contract characterized by $\{\overline{c}_s, \overline{w}_s\}_{s=1}^{S}$. *Ex ante*, before the time 0 endowment has been realized, the contract offers the household

$$v_0 = \sum_{k=1}^{S} \Pi_k[u(\overline{c}_k) + \beta\overline{w}_k] = \sum_{k=1}^{S} \Pi_k[u(\overline{y}_k) + \beta v_{\mathrm{aut}}] = v_{\mathrm{aut}}, \qquad (20.3.23)$$

where we have used $(20.3.22a)$ to verify that the contract indeed delivers $v_0 = v_{\text{aut}}$.

Some additional manipulations will enable us to express $\{\bar{c}_j\}_{j=1}^S$ solely in terms of the utility function and the endowment process. First, solve for $\overline{w}_j$ from $(20.3.22b)$,

$$\overline{w}_j = \frac{u(\bar{c}_j) \sum_{k=1}^j \Pi_k + \sum_{k=j+1}^S \Pi_k [u(\bar{y}_k) + \beta v_{\text{aut}}]}{1 - \beta \sum_{k=1}^j \Pi_k}, \qquad (20.3.24)$$

where we have invoked $(20.3.22a)$ when replacing $[u(\bar{c}_k) + \beta \overline{w}_k]$ by $[u(\bar{y}_k) + \beta v_{\text{aut}}]$. Next, substitute $(20.3.24)$ into $(20.3.22a)$ and solve for $u(\bar{c}_j)$,

$$u(\bar{c}_j) = \left[1 - \beta \sum_{k=1}^j \Pi_k\right] \left[u(\bar{y}_j) + \beta v_{\text{aut}}\right] - \beta \sum_{k=j+1}^S \Pi_k \left[u(\bar{y}_k) + \beta v_{\text{aut}}\right]$$

$$= u(\bar{y}_j) + \beta v_{\text{aut}} - \beta u(\bar{y}_j) \sum_{k=1}^j \Pi_k - \beta^2 v_{\text{aut}} - \beta \sum_{k=j+1}^S \Pi_k u(\bar{y}_k)$$

$$= u(\bar{y}_j) + \beta v_{\text{aut}} - \beta u(\bar{y}_j) \sum_{k=1}^j \Pi_k - \beta^2 v_{\text{aut}} - \beta \left[(1 - \beta) v_{\text{aut}} - \sum_{k=1}^j \Pi_k u(\bar{y}_k)\right]$$

$$= u(\bar{y}_j) - \beta \sum_{k=1}^j \Pi_k \left[u(\bar{y}_j) - u(\bar{y}_k)\right]. \qquad (20.3.25)$$

According to $(20.3.25)$, $u(\bar{c}_1) = u(\bar{y}_1)$ and $u(\bar{c}_j) < u(\bar{y}_j)$ for $j \geq 2$. That is, a household that realizes a record high endowment of $\bar{y}_j$ must surrender some of that endowment to the moneylender unless the endowment is the lowest possible value $\bar{y}_1$. Households are willing to surrender parts of their endowments in exchange for promises of insurance (i.e., future state-contingent transfers) that are encoded in the associated continuation values, $\{\overline{w}_j\}_{j=1}^S$. For those unlucky households that have so far realized only endowments equal to $\bar{y}_1$, the profit-maximizing contract prescribes that the households retain their endowment, $\bar{c}_1 = \bar{y}_1$ and by $(20.3.22a)$, the associated continuation value is $\overline{w}_1 = v_{\text{aut}}$. That is, to induce those low-endowment households to adhere to the contract, the moneylender has only to offer a contract that assures them an autarky continuation value in the next period.

Contracts when $v_0 > \overline{w}_1 = v_{\text{aut}}$

We have shown how to compute the optimal contract when $v_0 = \overline{w}_1 = v_{\text{aut}}$ by computing quantities $(\overline{c}_s, \overline{w}_s)$ for $s = 1, \ldots, S$. Now suppose that we want to construct a contract that assigns initial value $v_0 \in [\overline{w}_{k-1}, \overline{w}_k)$ for $1 < k \leq S$. Given v_0, we can deduce k, then solve for $\tilde{c}$ satisfying

$$v_0 = \left(\sum_{j=1}^{k-1} \Pi_j\right) [u(\tilde{c}) + \beta v_0] + \sum_{j=k}^{S} \Pi_j [u(\overline{c}_j) + \beta \overline{w}_j]. \tag{20.3.26}$$

The optimal contract promises $(\tilde{c}, v_0)$ so long as the maximum y_t to date is less than or equal to $\overline{y}_{k-1}$. When the maximum y_t experienced to date equals $\overline{y}_j$ for $j \geq k$, the contract offers $(\overline{c}_j, \overline{w}_j)$.

It is plausible that a higher initial expected promised value $v_0 > v_{\text{aut}}$ can be delivered in the most cost-effective way by choosing a higher consumption level $\tilde{c}$ for households that experience low endowment realizations, $\tilde{c} > \overline{c}_j$ for $j = 1, \ldots, k-1$. The reason is that those unlucky households have high marginal utilities of consumption. Therefore, transferring resources to them minimizes the resources that are needed to increase the *ex ante* promised expected utility. As for those lucky households that have received relatively high endowment realizations, the optimal contract prescribes an unchanged allocation characterized by $\{\overline{c}_j, \overline{w}_j\}_{j=k}^{S}$.

If we want to construct a contract that assigns initial value $v_0 \geq \overline{w}_S$, the efficient solution is simply to find the constant consumption level $\tilde{c}$ that delivers lifetime utility v_0:

$$v_0 = \sum_{j=1}^{S} \Pi_j [u(\tilde{c}) + \beta v_0] \qquad \Longrightarrow \qquad v_0 = \frac{u(\tilde{c})}{1 - \beta}.$$

This contract trivially satisfies all participation constraints, and a constant consumption level maximizes the expected profit of delivering v_0.

Summary of optimal contract

Define

$$s(t) = \{j : \overline{y}_j = \max\{y_0, y_1, \ldots, y_t\}\}.$$

That is, $\overline{y}_{s(t)}$ is the maximum endowment that the household has experienced up until period t.

The optimal contract has the following features. To deliver promised value $v_0 \in [v_{\mathrm{aut}}, \overline{w}_S]$ to the household, the contract offers stochastic consumption and continuation values, $\{c_t, v_{t+1}\}_{t=0}^{\infty}$, that satisfy

$$c_t = \max\{\tilde{c}, \overline{c}_{s(t)}\}, \tag{20.3.27a}$$

$$v_{t+1} = \max\{v_0, \overline{w}_{s(t)}\}, \tag{20.3.27b}$$

where $\tilde{c}$ is given by $(20.3.26)$.

20.3.4. Profits

We can use $(20.3.4)$ to compute expected profits from offering continuation value $\overline{w}_j$, $j = 1, \ldots, S$. Starting with $P(\overline{w}_S)$, we work backward to compute $P(\overline{w}_k)$, $k = S - 1, S - 2, \ldots, 1$:

$$P(\overline{w}_S) = \sum_{j=1}^{S} \Pi_j \left(\frac{\overline{y}_j - \overline{c}_S}{1 - \beta} \right), \tag{20.3.28a}$$

$$P(\overline{w}_k) = \sum_{j=1}^{k} \Pi_j (\overline{y}_j - \overline{c}_k) + \sum_{j=k+1}^{S} \Pi_j (\overline{y}_j - \overline{c}_j)$$

$$+ \beta \left[\sum_{j=1}^{k} \Pi_j P(\overline{w}_k) + \sum_{j=k+1}^{S} \Pi_j P(\overline{w}_j) \right]. \tag{20.3.28b}$$

Strictly positive profits for $v_0 = v_{\mathrm{aut}}$

We will now demonstrate that a contract that offers an initial promised value of v_{aut} is associated with strictly positive expected profits. In order to show that $P(v_{\mathrm{aut}}) > 0$, let us first examine the expected profit implications of the following limited obligation. Suppose that a household has just experienced $\overline{y}_j$ for the first time and that the limited obligation amounts to delivering $\overline{c}_j$ to the household in that period and in all future periods until the household realizes an endowment higher than $\overline{y}_j$. At the time of such a higher endowment realization in the future, the limited obligation ceases without any further transfers. Would such a limited obligation be associated with positive or negative expected profits? In the case of $\overline{y}_j = \overline{y}_1$, this would entail a deterministic profit equal to zero, since we have shown above that $\bar{c}_1 = \bar{y}_1$. But what is true for other endowment realizations?

To study the expected profit implications of such a limited obligation for any given $\bar{y}_j$, we first compute an upper bound for the obligation's consumption level $\bar{c}_j$ by using (20.3.25):

$$u(\bar{c}_j) = \left[1 - \beta \sum_{k=1}^{j} \Pi_k\right] u(\bar{y}_j) + \beta \sum_{k=1}^{j} \Pi_k u(\bar{y}_k)$$

$$\leq u\left(\left[1 - \beta \sum_{k=1}^{j} \Pi_k\right] \bar{y}_j + \beta \sum_{k=1}^{j} \Pi_k \bar{y}_k\right),$$

where the weak inequality is implied by the strict concavity of the utility function, and evidently the expression holds with strict inequality for $j > 1$. Therefore, an upper bound for $\bar{c}_j$ is

$$\bar{c}_j \leq \left[1 - \beta \sum_{k=1}^{j} \Pi_k\right] \bar{y}_j + \beta \sum_{k=1}^{j} \Pi_k \bar{y}_k. \tag{20.3.29}$$

We can sort out the financial consequences of the limited obligation by looking separately at the first period and then at all future periods. In the first period, the moneylender obtains a nonnegative profit,

$$\bar{y}_j - \bar{c}_j \geq \bar{y}_j - \left(\left[1 - \beta \sum_{k=1}^{j} \Pi_k\right] \bar{y}_j + \beta \sum_{k=1}^{j} \Pi_k \bar{y}_k\right)$$

$$= \beta \sum_{k=1}^{j} \Pi_k \left[\bar{y}_j - \bar{y}_k\right], \tag{20.3.30}$$

where we have invoked the upper bound on $\bar{c}_j$ in (20.3.29). After that first period, the moneylender must continue to deliver $\bar{c}_j$ for as long as the household does not realize an endowment greater than $\bar{y}_j$. So the probability that the household remains within the limited obligation for another t number of periods is $(\sum_{i=1}^{j} \Pi_i)^t$. Conditional on remaining within the limited obligation, the household's average endowment realization is $(\sum_{k=1}^{j} \Pi_k \bar{y}_k)/(\sum_{k=1}^{j} \Pi_k)$. Consequently, the expected discounted profit stream associated with all future periods of the limited obligation, expressed in first-period values, is

$$\sum_{t=1}^{\infty} \beta^t \left[\sum_{i=1}^{j} \Pi_i \right]^t \left[\frac{\sum_{k=1}^{j} \Pi_k \overline{y}_k}{\sum_{k=1}^{j} \Pi_k} - \overline{c}_j \right] = \frac{\left[\beta \sum_{i=1}^{j} \Pi_i \right]}{1 - \beta \sum_{i=1}^{j} \Pi_i} \left[\frac{\sum_{k=1}^{j} \Pi_k \overline{y}_k}{\sum_{k=1}^{j} \Pi_k} - \overline{c}_j \right]$$

$$\geq -\beta \sum_{k=1}^{j} \Pi_k \left[\overline{y}_j - \overline{y}_k \right], \qquad (20.3.31)$$

where the inequality is obtained after invoking the upper bound on $\overline{c}_j$ in (20.3.29). Since the sum of (20.3.30) and (20.3.31) is nonnegative, we conclude that the limited obligation at least breaks even in expectation. In fact, for $\overline{y}_j > \overline{y}_1$ we have that (20.3.30) and (20.3.31) hold with strict inequalities, and thus, each such limited obligation is associated with strictly positive profits.

Since the optimal contract with an initial promised value of v_{aut} can be viewed as a particular constellation of all of the described limited obligations, it follows immediately that $P(v_{\text{aut}}) > 0$.

Contracts with $P(v_0) = 0$

In exercise *20.2*, you will be asked to compute v_0 such that $P(v_0) = 0$. Here is a good way to do this. After computing the optimal contract for $v_0 = v_{\text{aut}}$, suppose that we can find some k satisfying $1 < k \leq S$ such that for $j \geq k, P(\overline{w}_j) \leq 0$ and for $j < k, P(\overline{w}_k) > 0$. Use a zero-profit condition to find an initial $\tilde{c}$ level:

$$0 = \sum_{j=1}^{k-1} \Pi_j (\overline{y}_j - \tilde{c}) + \sum_{j=k}^{S} \Pi_j \left[\overline{y}_j - \overline{c}_j + \beta P(\overline{w}_j) \right].$$

Given $\tilde{c}$, we can solve (20.3.26) for v_0.

However, such a k will fail to exist if $P(\overline{w}_S) > 0$. In that case, the efficient allocation associated with $P(v_0) = 0$ is a trivial one. The moneylender would simply set consumption equal to the average endowment value. This contract breaks even on average, and the household's utility is equal to the first-best unconstrained outcome, $v_0 = v_{\text{pool}}$, as given in (20.2.2).

20.3.5. $P(v)$ is strictly concave and continuously differentiable

Consider a promised value $v_0 \in [\overline{w}_{k-1}, \overline{w}_k)$ for $1 < k \leq S$. We can then use equation (20.3.26) to compute the amount of consumption $\tilde{c}(v_0)$ awarded to a household with promised value v_0, as long as the household is not experiencing an endowment greater than $\overline{y}_{k-1}$:

$$u[\tilde{c}(v_0)] = \frac{\left[1 - \beta \sum_{j=1}^{k-1} \Pi_j\right] v_0 - \sum_{j=k}^{S} \Pi_j \left[u(\overline{c}_j) + \beta \overline{w}_j\right]}{\sum_{j=1}^{k-1} \Pi_j} \equiv \Phi_k(v_0), \quad (20.3.32)$$

that is,

$$\tilde{c}(v_0) = u^{-1}\left[\Phi_k(v_0)\right]. \quad (20.3.33)$$

Since the utility function is strictly concave, it follows that $\tilde{c}(v_0)$ is strictly convex in the promised value v_0:

$$\tilde{c}'(v_0) = \frac{\left[1 - \beta \sum_{j=1}^{k-1} \Pi_j\right]}{\sum_{j=1}^{k-1} \Pi_j} u^{-1\prime}\left[\Phi_k(v_0)\right] > 0, \quad (20.3.34a)$$

$$\tilde{c}''(v_0) = \frac{\left[1 - \beta \sum_{j=1}^{k-1} \Pi_j\right]^2}{\left[\sum_{j=1}^{k-1} \Pi_j\right]^2} u^{-1\prime\prime}\left[\Phi_k(v_0)\right] > 0. \quad (20.3.34b)$$

Next, we evaluate the expression for expected profits in (20.3.4) at the optimal contract,

$$P(v_0) = \sum_{j=1}^{k-1} \Pi_j \left[\overline{y}_j - \tilde{c}(v_0) + \beta P(v_0)\right] + \sum_{j=k}^{S} \Pi_j \left[\overline{y}_j - \overline{c}_j + \beta P(\overline{w}_j)\right],$$

which can be rewritten as

$$P(v_0) = \frac{\sum_{j=1}^{k-1} \Pi_j \left[\overline{y}_j - \tilde{c}(v_0)\right] + \sum_{j=k}^{S} \Pi_j \left[\overline{y}_j - \overline{c}_j + \beta P(\overline{w}_j)\right]}{1 - \beta \sum_{j=1}^{k-1} \Pi_j}.$$

We can now verify that $P(v_0)$ is strictly concave for $v_0 \in [\overline{w}_{k-1}, \overline{w}_k)$,

$$P'(v_0) = -\frac{\sum_{j=1}^{k-1} \Pi_j}{1 - \beta \sum_{j=1}^{k-1} \Pi_j} \tilde{c}'(v_0) = -u^{-1\prime}\left[\Phi_k(v_0)\right] < 0, \quad (20.3.35a)$$

$$P''(v_0) = -\frac{\sum_{j=1}^{k-1} \Pi_j}{1 - \beta \sum_{j=1}^{k-1} \Pi_j} \tilde{c}''(v_0)$$

$$= -\frac{\left[1 - \beta \sum_{j=1}^{k-1} \Pi_j\right]}{\sum_{j=1}^{k-1} \Pi_j} u^{-1''}\left[\Phi_k(v_0)\right] < 0, \qquad (20.3.35b)$$

where we have invoked expressions $(20.3.34)$.

To shed light on the properties of the value function $P(v_0)$ around the promised value $\overline{w}_k$, we can establish that

$$\lim_{v_0 \uparrow \overline{w}_k} \Phi_k(v_0) = \Phi_k(\overline{w}_k) = \Phi_{k+1}(\overline{w}_k), \qquad (20.3.36)$$

where the first equality is a trivial limit of expression $(20.3.32)$ while the second equality can be shown to hold because a rearrangement of that equality becomes merely a restatement of a version of expression $(20.3.22b)$. On the basis of $(20.3.36)$ and $(20.3.33)$, we can conclude that the consumption level $\tilde{c}(v_0)$ is continuous in the promised value which in turn implies continuity of the value function $P(v_0)$. Moreover, expressions $(20.3.36)$ and $(20.3.35a)$ ensure that the value function $P(v_0)$ is continuously differentiable in the promised value.

20.3.6. Many households

Consider a large village in which a moneylender faces a continuum of such households. At the beginning of time $t = 0$, before the realization of y_0, the moneylender offers each household v_{aut} (or maybe just a small amount more). As time unfolds, the moneylender executes the contract for each household. A society of such households would experience a "fanning out" of the distributions of consumption and continuation values across households for a while, to be followed by an eventual "fanning in" as the cross-sectional distribution of consumption asymptotically becomes concentrated at the single point $g_2(\bar{v})$ computed earlier (i.e., the minimum c such that the participation constraint will never again be binding). Notice that early on the moneylender would on average, across villagers, be collecting money from the villagers, depositing it in the bank, and receiving the gross interest rate β^{-1} on the bank balance. Later he could be using the interest on his account outside the village to finance payments to the villagers. Eventually, the villagers are completely insured, i.e., they experience no fluctuations in their consumptions.

For a contract that offers initial promised value $v_0 \in [v_{\text{aut}}, \overline{w}_S]$, constructed as above, we can compute the dynamics of the cross-section distribution of consumption by appealing to a law of large numbers of the kind used in chapter 18. At time 0, after the time 0 endowments have been realized, the cross-section distribution of consumption is evidently

$$\text{Prob}\{c_0 = \tilde{c}\} = \left(\sum_{s=1}^{k-1} \Pi_s \right) \tag{20.3.37a}$$

$$\text{Prob}\{c_0 \leq \overline{c}_j\} = \left(\sum_{s=1}^{j} \Pi_s \right), \ j \geq k. \tag{20.3.37b}$$

After t periods,

$$\text{Prob}\{c_t = \tilde{c}\} = \left(\sum_{s=1}^{k-1} \Pi_s \right)^{t+1} \tag{20.3.38a}$$

$$\text{Prob}\{c_t \leq \overline{c}_j\} = \left(\sum_{s=1}^{j} \Pi_s \right)^{t+1}, \ j \geq k. \tag{20.3.38b}$$

From the cumulative distribution functions $(20.3.37)$ and $(20.3.38)$, it is easy to compute the corresponding densities

$$f_{j,t} = \text{Prob}(c_t = \overline{c}_j) \tag{20.3.39}$$

where here we set $\overline{c}_j = \tilde{c}$ for all $j < k$. These densities allow us to compute the evolution over time of the moneylender's bank balance. Starting with initial balance $\beta^{-1}B_{-1} = 0$ at time 0, the moneylender's balance at the bank evolves according to

$$B_t = \beta^{-1}B_{t-1} + \left(\sum_{j=1}^{S} \Pi_j \overline{y}_j - \sum_{j=1}^{S} f_{j,t}\overline{c}_j \right) \tag{20.3.40}$$

for $t \geq 0$, where B_t denotes the end-of-period balance in period t. Let $\beta^{-1} = 1 + r$. After the cross-section distribution of consumption has converged to a distribution concentrated on $\overline{c}_S$, the moneylender's bank balance will obey the difference equation

$$B_t = (1 + r)B_{t-1} + E(y) - \overline{c}_S, \tag{20.3.41}$$

where $E(y)$ is the mean of y.

A convenient formula links $P(v_0)$ to the tail behavior of B_t, in particular, to the behavior of B_t after the consumption distribution has converged to $\overline{c}_S$. Here we are once again appealing to a law of large numbers so that the expected profits $P(v_0)$ becomes a nonstochastic present value of profits associated with making a promise v_0 to a large number of households. Since the moneylender lets all surpluses and deficits accumulate in the bank account, it follows that $P(v_0)$ is equal to the present value of the sum of any future balances B_t and the continuation value of the remaining profit stream. After all households' promised values have converged to $\overline{w}_S$, the continuation value of the remaining profit stream is evidently equal to $\beta P(\overline{w}_S)$. Thus, for t such that the distribution of c has converged to $\overline{c}_s$, we deduce that

$$P(v_0) = \frac{B_t + \beta P(\overline{w}_S)}{(1+r)^t}. \qquad (20.3.42)$$

Since the term $\beta P(\overline{w}_S)/(1+r)^t$ in expression $(20.3.42)$ will vanish in the limit, the expression implies that the bank balances B_t will eventually change at the gross rate of interest. If the initial v_0 is set so that $P(v_0) > 0$ $(P(v_0) < 0)$, then the balances will eventually go to plus infinity (minus infinity) at an exponential rate. The asymptotic balances would be constant only if the initial v_0 is set so that $P(v_0) = 0$. This has the following implications. First, recall from our calculations above that there can exist an initial promised value $v_0 \in [v_{\text{aut}}, \overline{w}_S]$ such that $P(v_0) = 0$ only if it is true that $P(\overline{w}_S) \leq 0$, which by $(20.3.28a)$ implies that $E(y) \leq \overline{c}_S$. After imposing $P(v_0) = 0$ and using the expression for $P(\overline{w}_S)$ in $(20.3.28a)$, equation $(20.3.42)$ becomes $B_t = -\beta \frac{E(y)-\overline{c}_S}{1-\beta}$, or

$$B_t = \frac{\overline{c}_S - E(y)}{r} \geq 0,$$

where we have used the definition $\beta^{-1} = 1+r$. Thus, if the initial promised value v_0 is such that $P(v_0) = 0$, then the balances will converge when all households' promised values converge to $\overline{w}_S$. The interest earnings on those stationary balances will equal the one-period deficit associated with delivering $\overline{c}_S$ to every household while collecting endowments per capita equal to $E(y) \leq \overline{c}_S$.

After enough time has passed, all of the villagers will be perfectly insured because according to $(20.3.38)$, $\lim_{t\to+\infty} \text{Prob}(c_t = \overline{c}_S) = 1$. How much time it takes to converge depends on the distribution Π. Eventually, everyone will have received the highest endowment realization sometime in the past, after which his continuation value remains fixed. Thus, this is a model of temporary imperfect insurance, as indicated by the eventual "fanning in" of the distribution of continuation values.

20.3.7. An example

Figures 20.3.3 and 20.3.4 summarize aspects of the optimal contract for a version of our economy in which each household has an i.i.d. endowment process that is distributed as

$$\text{Prob}(y_t = \overline{y}_s) = \frac{1 - \lambda}{1 - \lambda^S}\lambda^{s-1}$$

where $\lambda \in (0, 1)$ and $\overline{y}_s = s + 5$ is the sth possible endowment value, $s = 1, \ldots, S$. The typical household's one-period utility function is $u(c) = (1 - \gamma)^{-1}c^{1-\gamma}$, where γ is the household's coefficient of relative risk aversion. We have assumed the parameter values $(\beta, S, \gamma, \lambda) = (.5, 20, 2, .95)$. The initial promised value v_0 is set so that $P(v_0) = 0$.

The moneylender's bank balance in Figure 20.3.3, panel d, starts at zero. The moneylender makes money at first, which he deposits in the bank. But as time passes, the moneylender's bank balance converges to the point that he is earning just enough interest on his balance to finance the extra payments he must make to pay $\overline{c}_S$ to each household each period. These interest earnings make up for the deficiency of his per capita period income $E(y)$, which is less than his per period per capita expenditures $\overline{c}_S$.

20.4. A Lagrangian method

Marcet and Marimon (1992, 1999) have proposed an approach that applies to most of the contract design problems of this chapter. They form a Lagrangian and use the Lagrange multipliers on incentive constraints to keep track of promises. Their approach extends the work of Kydland and Prescott (1980) and is related to Hansen, Epple, and Roberds' (1985) formulation for linear quadratic environments.[3] We can illustrate the method in the context of the preceding model.

---

[3] Marcet and Marimon's method is a variant of the method used to compute Stackelberg or Ramsey plans in chapter 19. See chapter 19 for a more extensive review of the history of

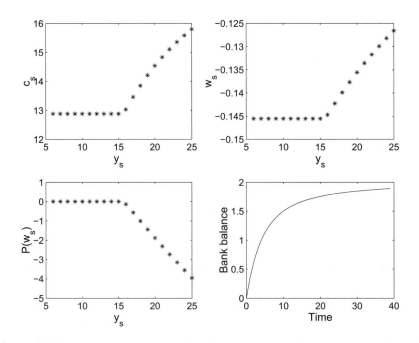

Figure 20.3.3: Optimal contract when $P(v_0) = 0$. Panel a: $\overline{c}_s$ as function of maximum $\overline{y}_s$ experienced to date. Panel b: $\overline{w}_s$ as function of maximum $\overline{y}_s$ experienced. Panel c: $P(\overline{w}_s)$ as function of maximum $\overline{y}_s$ experienced. Panel d: The moneylender's bank balance.

Marcet and Marimon's approach would be to formulate the problem directly in the space of stochastic processes (i.e., random sequences) and to form a Lagrangian for the moneylender. The contract specifies a stochastic process for consumption obeying the following constraints:

$$u(c_t) + E_t \sum_{j=1}^{\infty} \beta^j u(c_{t+j}) \geq u(y_t) + \beta v_{\text{aut}} \, , \forall t \geq 0, \qquad (20.4.1a)$$

$$E_{-1} \sum_{t=0}^{\infty} \beta^t u(c_t) \geq v, \qquad (20.4.1b)$$

the ideas underlying Marcet and Marimon's approach, in particular, some work from Great Britain in the 1980s by Miller, Salmon, Pearlman, Currie, and Levine.

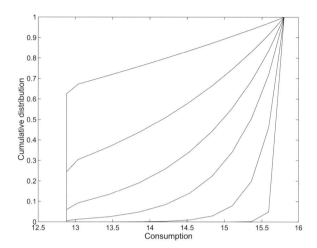

Figure 20.3.4: Cumulative distribution functions $F_t(c_t)$ for consumption for $t = 0, 2, 5, 10, 25, 100$ when $P(v_0) = 0$ (later dates have c.d.f.s shifted to right).

where $E_{-1}(\cdot)$ denotes the conditional expectation before y_0 has been realized. Here v is the initial promised value to be delivered to the villager starting in period 0. Equation $(20.4.1a)$ gives the participation constraints.

The moneylender's Lagrangian is

$$J = E_{-1} \sum_{t=0}^{\infty} \beta^t \left\{ (y_t - c_t) + \alpha_t \left[E_t \sum_{j=0}^{\infty} \beta^j u(c_{t+j}) - [u(y_t) + \beta v_{\text{aut}}] \right] \right\}$$
$$+ \phi \left[E_{-1} \sum_{t=0}^{\infty} \beta^t u(c_t) - v \right], \tag{20.4.2}$$

where $\{\alpha_t\}_{t=0}^{\infty}$ is a stochastic process of nonnegative Lagrange multipliers on the participation constraint of the villager and ϕ is the strictly positive multiplier on the initial promise-keeping constraint that states that the moneylender must deliver v. It is useful to transform the Lagrangian by making use of the following equality, which is a version of the "partial summation formula of Abel" (see Apostol, 1975, p. 194):

$$\sum_{t=0}^{\infty} \beta^t \alpha_t \sum_{j=0}^{\infty} \beta^j u(c_{t+j}) = \sum_{t=0}^{\infty} \beta^t \mu_t u(c_t), \tag{20.4.3}$$

where

$$\mu_t = \mu_{t-1} + \alpha_t, \qquad \text{with} \quad \mu_{-1} = 0. \tag{20.4.4}$$

Formula $(20.4.3)$ can be verified directly. If we substitute formula $(20.4.3)$ into formula $(20.4.2)$ and use the law of iterated expectations to justify $E_{-1}E_t(\cdot) = E_{-1}(\cdot)$, we obtain

$$J = E_{-1} \sum_{t=0}^{\infty} \beta^t \left\{ (y_t - c_t) + (\mu_t + \phi)u(c_t) \right.$$

$$\left. - (\mu_t - \mu_{t-1}) \left[u(y_t) + \beta v_{\mathrm{aut}} \right] \right\} - \phi v. \tag{20.4.5}$$

For a given value v, we seek a saddle point: a maximum with respect to $\{c_t\}$, a minimum with respect to $\{\mu_t\}$ and ϕ. The first-order condition with respect to c_t is

$$u'(c_t) = \frac{1}{\mu_t + \phi}, \tag{20.4.6a}$$

which is a version of equation $(20.3.12)$. Thus, $-(\mu_t + \phi)$ equals $P'(w)$ from the previous section, so that the multipliers encode the information contained in the derivative of the moneylender's value function. We also have the complementary slackness conditions

$$u(c_t) + E_t \sum_{j=1}^{\infty} \beta^j u(c_{t+j}) - [u(y_t) + \beta v_{\mathrm{aut}}] \geq 0, \qquad = 0 \text{ if } \alpha_t > 0; \tag{20.4.6b}$$

$$E_{-1} \sum_{t=0}^{\infty} \beta^t u(c_t) - v = 0. \tag{20.4.6c}$$

Equation $(20.4.6)$ together with the transition law $(20.4.4)$ characterizes the solution of the moneylender's maximization problem.

To explore the time profile of the optimal consumption process, we now consider some period $t \geq 0$ when (y_t, μ_{t-1}, ϕ) are known. First, we tentatively try the solution $\alpha_t = 0$ (i.e., the participation constraint is not binding). Equation $(20.4.4)$ instructs us then to set $\mu_t = \mu_{t-1}$, which by first-order condition $(20.4.6a)$ implies that $c_t = c_{t-1}$. If this outcome satisfies participation constraint $(20.4.6b)$, we have our solution for period t. If not, it signifies that the participation constraint binds. In other words, the solution has $\alpha_t > 0$ and $c_t > c_{t-1}$. Thus, equations $(20.4.4)$ and $(20.4.6a)$ immediately show us that c_t

is a nondecreasing random sequence, that c_t stays constant when the participation constraint is not binding, and that it rises when the participation constraint binds.

The numerical computation of a solution to equation $(20.4.5)$ is complicated by the fact that slackness conditions $(20.4.6b)$ and $(20.4.6c)$ involve conditional expectations of future endogenous variables $\{c_{t+j}\}$. Marcet and Marimon (1992) handle this complication by resorting to the parameterized expectation approach; that is, they replace the conditional expectation by a parameterized function of the state variables.[4] Marcet and Marimon (1992, 1999) describe a variety of other examples using the Lagrangian method. See Kehoe and Perri (2002) for an application to an international trade model.

20.5. Insurance with asymmetric information

The moneylender-villager environment of section 20.3 has a commitment problem because agents are free to choose autarky each period; but there is no information problem. We now study a contract design problem where the incentive problem comes not from a commitment problem, but instead from asymmetric information. As before, the moneylender or planner can borrow or lend outside the village at the constant risk-free gross interest rate of β^{-1}, and each household's income y_t is independently and identically distributed across time and across households. However, we now assume that both the planner and households can enter into enduring and binding contracts. At the beginning of time, let v^o be the expected lifetime utility that the planner promises to deliver to a household. The initial promise v^o could presumably not be less than v_{aut}, since a household would not accept a contract that gives a lower utility than he could attain at time 0 by choosing autarky. We defer discussing how v^o is determined until the end of the section. The other new assumption here is that households have private information about their own income, and that the planner can see neither their income nor their consumption. It follows that any transfers between the planner and a household must be based on the household's

---

[4] For details on the implementation of the parameterized expectation approach in a simple growth model, see den Haan and Marcet (1990).

own reports about income realizations. An incentive-compatible contract makes households choose to report their incomes truthfully.

Our analysis follows the work by Thomas and Worrall (1990), who make a few additional assumptions about the preferences in expression $(20.2.1)$: $u : (a, \infty) \to I\!R$ is twice continuously differentiable with $\sup u(c) < \infty$, $\inf u(c) = -\infty$, $\lim_{c \to a} u'(c) = \infty$. Thomas and Worrall also use the following special assumption:

CONDITION A: $-u''/u'$ is nonincreasing.

This is a sufficient condition to make the value function concave, as we will discuss. The roles of the other restrictions on preferences will also be revealed.

An efficient insurance contract solves a dynamic programming problem.[5] A planner maximizes expected discounted profits, $P(v)$, where v is the household's promised utility from last period. The planner's current payment to the household, denoted b (repayments from the household register as negative numbers), is a function of the state variable v and the household's reported current income y. Let b_s and w_s be the payment and continuation utility awarded to the household if it reports income $\overline{y}_s$. The optimum value function $P(v)$ obeys the functional equation

$$P(v) = \max_{\{b_s, w_s\}} \sum_{s=1}^{S} \Pi_s [-b_s + \beta P(w_s)] \tag{20.5.1}$$

where the maximization is subject to the constraints

$$\sum_{s=1}^{S} \Pi_s \left[u(\overline{y}_s + b_s) + \beta w_s \right] = v \tag{20.5.2}$$

$$C_{s,k} \equiv u(\overline{y}_s + b_s) + \beta w_s - \left[u(\overline{y}_s + b_k) + \beta w_k \right] \geq 0, \ s, k \in \mathbf{S} \times \mathbf{S} \tag{20.5.3}$$

$$b_s \in [a - \overline{y}_s, \infty], \ s \in \mathbf{S} \tag{20.5.4}$$

$$w_s \in [-\infty, v_{\max}], \ s \in \mathbf{S} \tag{20.5.5}$$

where $v_{\max} = \sup u(c)/(1 - \beta)$. Equation $(20.5.2)$ is the "promise-keeping" constraint guaranteeing that the promised utility v is delivered. Note that

---

[5] It is important that the endowment is independently distributed over time. See Fernandes and Phelan (2000) for a related analysis that shows complications that arise when the iid assumption is relaxed

our earlier weak inequality in (20.3.5) is replaced by an equality. The planner cannot award a higher utility than v because that could violate an incentive-compatibility constraint for telling the truth in earlier periods. The set of constraints (20.5.3) ensures that the households have no incentive to lie about their endowment realization in each state $s \in \mathbf{S}$. Here s indexes the actual income state, and k indexes the reported income state. We express the incentive compatibility constraints when the endowment is in state s as $C_{s,k} \geq 0$ for $k \in \mathbf{S}$. Note also that there are no "participation constraints" like expression (20.3.6) from our earlier model, an absence that reflects the assumption that both parties are committed to the contract.

It is instructive to establish bounds on the value function $P(v)$. Consider first a contract that pays a constant amount $\bar{b} = \bar{b}(v)$ in all periods, where $\bar{b}(v)$ satisfies $\sum_{s=1}^{S} \Pi_s u(\bar{y}_s + \bar{b})/(1 - \beta) = v$. It is trivially incentive compatible and delivers the promised utility v. Therefore, the discounted profits from this contract, $-\bar{b}/(1-\beta)$, provide a lower bound on $P(v)$. In addition, $P(v)$ cannot exceed the value of the unconstrained first-best contract that pays $\bar{c} - \bar{y}_s$ in all periods, where $\bar{c}$ satisfies $\sum_{s=1}^{S} \Pi_s u(\bar{c})/(1 - \beta) = v$. Thus, the value function is bounded by

$$-\bar{b}(v)/(1 - \beta) \;\leq\; P(v) \;\leq\; \sum_{s=1}^{S} \Pi_s [\bar{y}_s - \bar{c}(v)]/(1 - \beta) . \qquad (20.5.6)$$

The bounds are depicted in Figure 20.5.1, which also illustrates a few other properties of $P(v)$. Since $\lim_{c \to a} u'(c) = \infty$, it becomes very cheap for the planner to increase the promised utility when the current promise is very low, that is, $\lim_{v \to -\infty} P'(v) = 0$. The situation is different when the household's promised utility is close to the upper bound $v_{\max}$ where the household has a low marginal utility of additional consumption, which implies that both $\lim_{v \to v_{\max}} P'(v) = -\infty$ and $\lim_{v \to v_{\max}} P(v) = -\infty$.

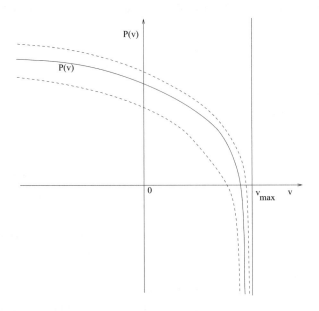

Figure 20.5.1: Value function $P(v)$ and the two dashed curves depict the bounds on the value function. The vertical solid line indicates $v_{\max} = \sup u(c)/(1-\beta)$.

20.5.1. *Efficiency implies* $b_{s-1} \geq b_s, w_{s-1} \leq w_s$

An incentive-compatible contract must satisfy $b_{s-1} \geq b_s$ (insurance) and $w_{s-1} \leq w_s$ (*partial* insurance). This can be established by adding the "downward constraint" $C_{s,s-1} \geq 0$ and the "upward constraint" $C_{s-1,s} \geq 0$ to get

$$u(\overline{y}_s + b_s) - u(\overline{y}_{s-1} + b_s) \geq u(\overline{y}_s + b_{s-1}) - u(\overline{y}_{s-1} + b_{s-1}),$$

where the concavity of $u(c)$ implies $b_s \leq b_{s-1}$. It then follows directly from $C_{s,s-1} \geq 0$ that $w_s \geq w_{s-1}$. Thus, for any v, a household reporting a lower income receives a higher transfer from the planner in exchange for a lower future utility.

20.5.2. Local upward and downward constraints are enough

Constraint set (20.5.3) can be simplified. We can show that if the local downward constraints $C_{s,s-1} \geq 0$ and upward constraints $C_{s,s+1} \geq 0$ hold for each $s \in \mathbf{S}$, then the global constraints $C_{s,k} \geq 0$ hold for each $s, k \in \mathbf{S}$. The argument goes as follows: Suppose we know that the downward constraint $C_{s,k} \geq 0$ holds for some $s > k$,

$$u(\overline{y}_s + b_s) + \beta w_s \geq u(\overline{y}_s + b_k) + \beta w_k. \tag{20.5.7}$$

From above we know that $b_s \leq b_k$, so the concavity of $u(c)$ implies

$$u(\overline{y}_{s+1} + b_s) - u(\overline{y}_s + b_s) \geq u(\overline{y}_{s+1} + b_k) - u(\overline{y}_s + b_k). \tag{20.5.8}$$

By adding expressions (20.5.7) and (20.5.8) and using the local downward constraint $C_{s+1,s} \geq 0$, we arrive at

$$u(\overline{y}_{s+1} + b_{s+1}) + \beta w_{s+1} \geq u(\overline{y}_{s+1} + b_k) + \beta w_k,$$

that is, we have shown that the downward constraint $C_{s+1,k} \geq 0$ holds. In this recursive fashion we can verify that all global downward constraints are satisfied when the local downward constraints hold. A symmetric reasoning applies to the upward constraints. Starting from any upward constraint $C_{k,s} \geq 0$ with $k < s$, we can show that the local upward constraint $C_{k-1,k} \geq 0$ implies that the upward constraint $C_{k-1,s} \geq 0$ must also hold, and so forth.

20.5.3. Concavity of P

Thus far, we have not appealed to the concavity of the value function, but henceforth we shall have to. Thomas and Worrall showed that under condition A, P is concave.

PROPOSITION: The value function $P(v)$ is concave.

We recommend just skimming the following proof on first reading:

PROOF: Let $T(P)$ be the operator associated with the right side of equation (20.5.1). We could compute the optimum value function by iterating to convergence on T. We want to show that T maps strictly concave P to strictly concave function $T(P)$. Thomas and Worrall use the following argument:

Let $P_{k-1}(v)$ be the $k-1$ iterate on T. Assume that $P_{k-1}(v)$ is strictly concave. We want to show that $P_k(v)$ is strictly concave. Consider any v^o and v' with associated contracts $(b_s^o, w_s^o)_{s \in S}, (b_s', w_s')_{s \in S}$. Let $w_s^* = \delta w_s^o + (1 - \delta)w_s'$ and define b_s^* by $u(b_s^* + \overline{y}_s) = \delta u(b_s^o + \overline{y}_s) + (1 - \delta)u(b_s' + \overline{y}_s)$ where $\delta \in (0, 1)$. Therefore, $(b_s^*, w_s^*)_{s \in S}$ gives the borrower a utility that is the weighted average of the two utilities, and gives the lender no less than the average utility $\delta P_k(v^o) + (1 - \delta)P_k(v')$. Then $C_{s,s-1}^* = \delta C_{s,s-1}^o + (1 - \delta)C_{s,s-1}' + [\delta u(b_{s-1}^o + \overline{y}_s) + (1 - \delta)u(b_{s-1}' + \overline{y}_s) - u(b_{s-1}^* + \overline{y}_s)]$. Because the downward constraints $C_{s,s-1}^o$ and $C_{s,s-1}'$ are satisfied, and because the third term is nonnegative under condition A, the downward incentive constraints $C_{s,s-1}^* \geq 0$ are satisfied. However, $(b_s^*, w_s^*)_{s \in S}$ may violate the upward incentive constraints. But Thomas and Worrall use the following argument to construct a new contract from $(b_s^*, w_s^*)_{s \in S}$ that is incentive compatible and that offers both the lender and the borrower no less utility. Thus, keep w_1 fixed and reduce w_2 until $C_{2,1} = 0$ or $w_2 = w_1$. Then reduce w_3 in the same way, and so on. Add the constant necessary to leave $\sum_s \Pi_s w_s$ constant. This step will not make the lender worse off, by the concavity of $P_{k-1}(v)$. Now if $w_2 = w_1$, which implies $b_2^* > b_1^*$, reduce b_2 until $C_{2,1} = 0$, and proceed in the same way for b_3, and so on. Since $b_s + \overline{y}_s > b_{s-1} + \overline{y}_{s-1}$, adding a constant to each b_s to leave $\sum_s \Pi_s b_s$ constant cannot make the borrowers worse off. So in this new contract, $C_{s,s-1} = 0$ and $b_{s-1} \geq b_s$. Thus, the upward constraints also hold. Strict concavity of $P_k(v)$ then follows because it is not possible to have both $b_s^o = b_s'$ and $w_s^o = w_s'$ for all $s \in S$ and $v^o \neq v'$, so the contract (b_s^*, w_s^*) yields the lender strictly more than $\delta P_k(v^o) + (1 - \delta)P_k(v')$. To complete the induction argument, note that starting from $P_0(v) = 0$, $P_1(v)$ is strictly concave. Therefore, $\lim_{k=\infty} P_k(v)$ is concave. ∎

We now turn to some properties of the optimal allocation that require strict concavity of the value function. Thomas and Worrall derive these results for the finite horizon problem with value function $P_k(v)$, which is strictly concave by the preceding proposition. In order for us to stay with the infinite horizon value function $P(v)$, we make the following assumption about $\lim_{k=\infty} P_k(v)$:[6]

ASSUMPTION: The value function $P(v)$ is strictly concave.

---

[6] To get the main result reported below that all households become impoverished in the limit, Thomas and Worrall provide a proof that requires only concavity of $P(v)$ as established in the preceding proposition.

20.5.4. Local downward constraints always bind

At the optimal solution, the local downward incentive constraints always bind, while the local upward constraints never do. That is, a household is always indifferent between reporting the truth and reporting that its income is actually a little lower than it is; but it never wants to report that its income is higher. To see that the downward constraints must bind, suppose to the contrary that $C_{k,k-1} > 0$ for some $k \in \mathbf{S}$. Since $b_k \leq b_{k-1}$, it must then be the case that $w_k > w_{k-1}$. Consider changing $\{b_s, w_s; s \in \mathbf{S}\}$ as follows. Keep w_1 fixed, and if necessary reduce w_2 until $C_{2,1} = 0$. Next reduce w_3 until $C_{3,2} = 0$, and so on, until $C_{s,s-1} = 0$ for all $s \in \mathbf{S}$. (Note that any reductions cumulate when moving up the sequence of constraints.) Thereafter, add the necessary constant to each w_s to leave the expected value of all future promises unchanged, $\sum_{s=1}^{S} \Pi_s w_s$. The new contract offers the household the same utility and is incentive compatible because $b_s \leq b_{s-1}$ and $C_{s,s-1} = 0$ together imply that the local upward constraint $C_{s-1,s} \geq 0$ does not bind. At the same time, since the mean of promised values is unchanged and the differences $(w_s - w_{s-1})$ have either been left the same or reduced, the strict concavity of the value function $P(v)$ implies that the planner's profits have increased. That is, we have engineered a mean-preserving *decrease* in the spread in the continuation values w. Because $P(v)$ is strictly concave, $\sum_{s \in S} \Pi_s P(w_s)$ rises and therefore $P(v)$ rises. Thus, the original contract with a nonbinding local downward constraint could not have been an optimal solution.

20.5.5. Coinsurance

The optimal contract is characterized by *coinsurance*, meaning that the household's utility and the planner's profits both increase with a higher income realization:

$$u(\overline{y}_s + b_s) + \beta w_s > u(\overline{y}_{s-1} + b_{s-1}) + \beta w_{s-1} \tag{20.5.9}$$

$$-b_s + \beta P(w_s) \geq -b_{s-1} + \beta P(w_{s-1}). \tag{20.5.10}$$

The higher utility of the household in expression (20.5.9) follows trivially from the downward incentive-compatibility constraint $C_{s,s-1} = 0$. Concerning the planner's profits in expression (20.5.10), suppose to the contrary that $-b_s + \beta P(w_s) < -b_{s-1} + \beta P(w_{s-1})$. Then replacing (b_s, w_s) in the contract by

(b_{s-1}, w_{s-1}) raises the planner's profits but leaves the household's utility unchanged because $C_{s,s-1} = 0$, and the change is also incentive compatible. Thus, an optimal contract must be such that the planner's profits weakly increase in the household's income realization.

20.5.6. $P'(v)$ *is a martingale*

If we let λ and μ_s, $s = 2, \ldots, S$, be Lagrange multipliers associated with the constraints (20.5.2) and $C_{s,s-1} \geq 0$, $s = 2, \ldots, S$, respectively, the first-order necessary conditions with respect to b_s and w_s, $s \in \mathbf{S}$, are

$$\Pi_s \left[1 - \lambda u'(\overline{y}_s + b_s) \right] = \mu_s u'(\overline{y}_s + b_s) - \mu_{s+1} u'(\overline{y}_{s+1} + b_s), \quad (20.5.11)$$

$$\Pi_s \left[P'(w_s) + \lambda \right] = \mu_{s+1} - \mu_s, \quad (20.5.12)$$

for $s \in \mathbf{S}$, where $\mu_1 = \mu_{S+1} = 0$. (There are no constraints corresponding to μ_1 and μ_{S+1}.) From the envelope condition,

$$P'(v) = -\lambda. \quad (20.5.13)$$

Summing equation (20.5.12) over $s \in \mathbf{S}$ and using $\sum_{s=1}^{S}(\mu_{s+1} - \mu_s) = \mu_{S+1} - \mu_1 = 0$ and equation (20.5.13) yields

$$\sum_{s=1}^{S} \Pi_s P'(w_s) = P'(v). \quad (20.5.14)$$

This equation states that P' is a martingale.

20.5.7. Comparison to model with commitment problem

In the model with a commitment problem studied in section 20.3, the efficient allocation had to satisfy equation $(20.3.12)$, i.e., $u'(\overline{y}_s + b_s) = -P'(w_s)^{-1}$. As we explained then, this condition sets the household's marginal rate of substitution equal to the planner's marginal rate of transformation with respect to transfers in the current period and continuation values in the next period. This condition fails to hold in the present framework with incentive-compatibility constraints associated with telling the truth. The efficient trade-off between current consumption and a continuation value for a household with income realization $\overline{y}_s$ can not be determined without taking into account the incentives that other households have to report $\overline{y}_s$ untruthfully in order to obtain the corresponding bundle of current and future transfers from the planner. It is instructive to note that equation $(20.3.12)$ *would* continue to hold in the present framework if the incentive-compatibility constraints for truth telling were not binding. That is, set the multipliers μ_s, $s = 2, \ldots, S$, equal to zero and substitute first-order condition $(20.5.12)$ into $(20.5.11)$ to obtain $u'(\overline{y}_s + b_s) = -P'(w_s)^{-1}$.

20.5.8. Spreading continuation values

An efficient contract requires that the promised future utility falls (rises) when the household reports the lowest (highest) income realization, that is, that $w_1 < v < w_S$. To show that $w_S > v$, suppose to the contrary that $w_S \leq v$. That this assumption leads to a contradiction is established by the following line of argument. Since $w_S \geq w_s$ for all $s \in \mathbf{S}$ and $P(v)$ is strictly concave, equation $(20.5.14)$ implies that $w_s = v$ for all $s \in \mathbf{S}$. Substitution of equation $(20.5.13)$ into equation $(20.5.12)$ then yields a zero on the left side of equation $(20.5.12)$. Moreover, the right side of equation $(20.5.12)$ is equal to μ_2 when $s = 1$ and $-\mu_S$ when $s = S$, so we can successively unravel from the constraint set $(20.5.12)$ that $\mu_s = 0$ for all $s \in \mathbf{S}$. Turning to equation $(20.5.11)$, it follows that the marginal utility of consumption is equalized across income realizations, $u'(\overline{y}_s + b_s) = \lambda^{-1}$ for all $s \in \mathbf{S}$. Such consumption smoothing requires $b_{s-1} > b_s$, but from incentive compatibility, $w_{s-1} = w_s$ implies $b_{s-1} = b_s$, a contradiction. We conclude that an efficient contract must have $w_S > v$. A symmetric argument establishes $w_1 < v$.

The planner must spread out promises to future utility because otherwise it would be impossible to provide any insurance in the form of contingent payments today. Equation (20.5.14) describes how the planner balances the delivery of utility today versus tomorrow. To understand this expression, consider having the planner increase the household's promised utility v by one unit. One way of doing so is to increase every w_s by an increment $1/\beta$ while keeping every b_s constant. Such a change preserves incentive compatibility at an expected discounted cost to the planner of $\sum_{s=1}^{S} \Pi_s P'(w_s)$. By the envelope theorem, locally this is as good a way to increase v as any other, and its cost is therefore equal to $P'(v)$; that is, we obtain expression (20.5.14). In other words, given a planner's obligation to deliver utility v to the agent, it is cost-efficient to balance today's contingent deliveries of goods, $\{b_s\}$, and the bundle of future utilities, $\{w_s\}$, so that the expected marginal cost of next period's promises, $\sum_{s=1}^{S} \Pi_s P'(w_s)$, becomes equal to the marginal cost of the current obligation, $P'(v)$. No intertemporal price affects this trade-off, since any interest earnings on postponed payments are just sufficient to compensate the agent for his own subjective rate of discounting, $(1+r) = \beta^{-1}$.

20.5.9. Martingale convergence and poverty

The martingale property (20.5.14) for $P'(v)$ has an intriguing implication for the long-run tendency of a household's promised future utility. Recall that $\lim_{v \to -\infty} P'(v) = 0$ and $\lim_{v \to v_{\max}} P'(v) = -\infty$, so $P'(v)$ in expression (20.5.14) is a nonpositive martingale. By a theorem of Doob (1953, p. 324), $P'(v)$ then converges almost surely. We can show that $P'(v)$ must converge to 0, so that v converges to $-\infty$. Suppose to the contrary that $P'(v)$ converges to a nonzero limit, which implies that v converges to a finite limit. However, this assumption contradicts our earlier result that future w_s always spread out to provide incentives. The contradiction is avoided only for v converging to $-\infty$; therefore, the limit of $P'(v)$ must be zero.

The result that all households become impoverished in the limit can be understood in terms of the concavity of $P(v)$. First, if there were no asymmetric information, the least expensive way of delivering lifetime utility v would be to assign the household a constant consumption stream, given by the upper bound on the value function in expression (20.5.6). The concavity of $P(v)$

and standard intertemporal considerations favor a time-invariant consumption stream. But the presence of asymmetric information makes it necessary for the planner to vary promises of future utility to induce truth telling, which is costly due to the concavity of $P(v)$. For example, Thomas and Worrall pointed out that if $S = 2$, the cost of spreading w_1 and w_2 an equally small amount ϵ on either side of their average value $\bar{w}$ is approximately $-0.5\epsilon^2 P''(\bar{w})$.[7] In general, we cannot say how this cost differs for any two values of $\bar{w}$, but it follows from the properties of $P(v)$ at its endpoints that $\lim_{v \to -\infty} P''(v) = 0$, and $\lim_{v \to v_{\max}} P''(v) = -\infty$. Thus, the cost of spreading promised values goes to zero at one endpoint and to infinity at the other endpoint. Therefore, the concavity of $P(v)$ and incentive compatibility considerations impart a downward drift to future utilities and, consequently, consumption. That is, with private information the ideal time-invariant consumption level without private information is abandoned in favor of random consumption paths that are expected to be tilted toward the present.

One possibility is that the initial utility level v^o is determined in competition between insurance providers. If there are no costs associated with administering contracts, v^o would then be implicitly determined by the zero-profit condition, $P(v^o) = 0$. Such a contract must be enforceable because, as we have seen, the household will almost surely eventually wish that it could revert to autarky. However, since the contract is the solution to a dynamic programming problem where the continuation of the contract is always efficient at every date, the insurer and the household will never mutually agree to renegotiate the contract.

---

[7] The expected discounted profits of providing promised values $w_1 = \bar{w} - \epsilon$ and $w_2 = \bar{w} + \epsilon$ with equal probabilities, can be approximated with a Taylor series expansion around $\bar{w}$,

$$\sum_{s=1}^{2} \tfrac{1}{2} P(w_s) \approx \sum_{s=1}^{2} \tfrac{1}{2} \left[P(\bar{w}) + (w_s - \bar{w})P'(\bar{w}) + \frac{(w_s - \bar{w})^2}{2} P''(\bar{w}) \right] = P(\bar{w}) + \frac{\epsilon^2}{2} P''(\bar{w}).$$

20.5.10. Extension to general equilibrium

Atkeson and Lucas (1992) provide examples of closed economies where the constrained efficient allocation also has each household's expected utility converging to the minimum level with probability 1. Here the planner chooses the incentive-compatible allocation for all agents subject to a constraint that the total consumption handed out in each period to the population of households cannot exceed some constant endowment level. Households are assumed to experience unobserved idiosyncratic taste shocks ϵ that are i.i.d. over time and households. The taste shock enters multiplicatively into preferences that take either the logarithmic form $u(c, \epsilon) = \epsilon \log(c)$, the constant relative risk aversion (CRRA) form $u(c, \epsilon) = \epsilon c^\gamma / \gamma$, $\gamma < 1$, $\gamma \neq 0$, or the constant absolute risk aversion (CARA) form $u(c, \epsilon) = -\epsilon \exp(-\gamma c)$, $\gamma > 0$. The assumption that the utility function belongs to one of these families greatly simplifies the analytics of the evolution of the wealth distribution. Atkeson and Lucas show that an equilibrium of this model yields an efficient allocation that assigns an ever-increasing fraction of resources to an ever-diminishing fraction of the economy's population.

20.5.11. Comparison with self-insurance

We have just seen how in the Thomas and Worrall model, the planner responds to the incentive problem created by the consumer's private information by putting a downward tilt into temporal consumption profiles. It is useful to recall how in the savings problem of chapters 17 and 18, the martingale convergence theorem was used to show that the consumption profile acquired an upward tilt coming from the motive of the consumer to self-insure.

20.6. Insurance with unobservable storage

In the spirit of an analysis of Franklin Allen (1985), we now augment the model of the previous section by assuming that households have access to a technology that enables them to store nonnegative amounts of goods at a risk-free gross return of $R > 0$. The planner cannot observe private storage. The planner can borrow and lend outside the village at a risk-free gross interest rate that also equals R, so that private and public storage yield identical rates of return. The planner retains an advantage over households of being the only one able to *borrow* outside of the village.

The outcome of our analysis will be to show that allowing households to store amounts that are not observable to the planner so impedes the planner's ability to manipulate the household's continuation valuations that no social insurance can be supplied. Instead, the planner helps households overcome the nonnegativity constraint on households' storage by in effect allowing them to engage also in private *borrowing* at the risk-free rate R, subject to natural borrowing limits. Thus, outcomes share many features of the allocations studied in chapters 17 and 18.

Our analysis partly follows Cole and Kocherlakota (2001), who assume that a household's utility function $u(\cdot)$ is strictly concave and twice continuously differentiable over $(0, \infty)$ with $\lim_{c \to 0} u'(c) = \infty$. The domain of u is the entire real line with $u(c) = -\infty$ for $c < 0$.[8] They also assume that u satisfies condition A above. This preference specification allows Cole and Kocherlakota to characterize an efficient allocation in a finite horizon model. Their extension to an infinite horizon involves a few other assumptions, including upper and lower bounds on the utility function.

We retain our earlier assumption that the planner has access to a risk-free loan market outside of the village. Cole and Kocherlakota (2001) postulate a closed economy where the planner is constrained to choose nonnegative amounts of storage. Hence, our concept of feasibility differs from theirs.

---

[8] Allowing for negative consumption while setting utility equal to $-\infty$ is a convenient device for avoiding having to deal with transfers that exceed the household's resources.

20.6.1. Feasibility

Anticipating that our characterization of efficient outcomes will be in terms of sequences of quantities, we let the history of a household's reported income enter as an argument in the function specifying the planner's transfer scheme. In period t, a household with an earlier history h_{t-1} and a currently reported income of y_t receives a transfer $b_t(\{h_{t-1}, y_t\})$ that can be either positive or negative. If all households report their incomes truthfully, the planner's time t budget constraint is

$$K_t + \sum_{h_t} \pi(h_t) b_t(h_t) \leq R K_{t-1}, \qquad (20.6.1)$$

where K_t is the planner's end-of-period savings (or, if negative, borrowing) and $\pi(h_t)$ is the unconditional probability that a household experiences history h_t, which in the planner's budget constraint equals the fraction of households that experience history h_t. Given a finite horizon with a final period T, solvency of the planner requires that $K_T \geq 0$.

We use a household's history h_t to index consumption and private storage at time t; $c_t(h_t) \geq 0$ and $k_t(h_t) \geq 0$. The household's resource constraint at history h_t at time t is

$$c_t(h_t) + k_t(h_t) \leq y_t(h_t) + R k_{t-1}(h_{t-1}) + b_t(h_t), \qquad (20.6.2)$$

where the function for current income $y_t(h_t)$ returns the tth element of the household's history h_t. We assume that the household has always reported its income truthfully, so that the transfer in period t is given by $b_t(h_t)$.

Given initial conditions $K_0 = k_0 = 0$, an allocation $(c, k, b, K) \equiv \{c_t(h_t), k_t(h_t), b_t(h_t), K_t\}$ is physically feasible if inequalities (20.6.1), (20.6.2) and $k_t(h_t) \geq 0$ are satisfied for all periods t and all histories h_t, and $K_T \geq 0$.

20.6.2. Incentive compatibility

Since income realizations and private storage are both unobservable, households are free to deviate from an allocation (c, k, b, K) in two ways. First, households can lie about their income and thereby receive the transfer payments associated with the reported but untrue income history. Second, households can choose different levels of storage. Let Ω^T be the set of reporting and storage strategies $(\hat{y}, \hat{k}) \equiv \{\hat{y}_t(h_t), \hat{k}_t(h_t); \text{for all } t, h_t\}$, where h_t denotes the household's true history.

Let $\hat{h}_t$ denote the history of reported incomes, $\hat{h}_t(h_t) = \{\hat{y}_1(h_1), \hat{y}_2(h_2), \ldots, \hat{y}_t(h_t)\}$. With some abuse of notation, we let y denote the truth-telling strategy for which $\hat{y}_t(\{h_{t-1}, y_t\}) = y_t$ for all (t, h_{t-1}), and hence for which $\hat{h}_t(h_t) = h_t$.

Given a transfer scheme b, the expected utility of following reporting and storage strategy $(\hat{y}, \hat{k})$ is

$$\Gamma(\hat{y}, \hat{k}; b) \equiv \sum_{t=1}^{T} \beta^{t-1} \sum_{h_t} \pi(h_t)$$
$$\cdot u\left(y_t(h_t) + R\hat{k}_{t-1}(h_{t-1}) + b_t(\hat{h}_t(h_t)) - \hat{k}_t(h_t)\right), \qquad (20.6.3)$$

given $k_0 = 0$. An allocation is incentive compatible if

$$\Gamma(y, k; b) = \max_{(\hat{y}, \hat{k}) \in \Omega^T} \Gamma(\hat{y}, \hat{k}; b). \qquad (20.6.4)$$

An allocation that is both incentive compatible and feasible is called an *incentive feasible* allocation. The following proposition asserts that any incentive feasible allocation with private storage can be attained with an alternative incentive feasible allocation without private storage.

PROPOSITION 1: Given any incentive feasible allocation (c, k, b, K), there exists another incentive feasible allocation $(c, 0, b^o, K^o)$.

PROOF: We claim that $(c, 0, b^o, K^o)$ is incentive feasible where

$$b_t^o(h_t) \equiv b_t(h_t) - k_t(h_t) + Rk_{t-1}(h_{t-1}), \qquad (20.6.5)$$
$$K_t^o \equiv \sum_{h_t} \pi(h_t)k_t(h_t) + K_t. \qquad (20.6.6)$$

Feasibility follows from the assumed feasibility of (c, k, b, K). Note also that $\Gamma(y, 0; b^o) = \Gamma(y, k; b)$. The proof of incentive compatibility is by contradiction.

Suppose that $(c, 0, b^o, K^o)$ is not incentive compatible, i.e., that there exists a reporting and storage strategy $(\hat{y}, \hat{k}) \in \Omega^T$ such that

$$\Gamma(\hat{y}, \hat{k}; b^o) > \Gamma(y, 0; b^o) = \Gamma(y, k; b). \tag{20.6.7}$$

After invoking expression $(20.6.5)$ for transfer payment $b_t^o(\hat{h}_t(h_t))$, the left side of inequality $(20.6.7)$ becomes

$$
\begin{aligned}
\Gamma(\hat{y}, \hat{k}; b^o) &= \sum_{t=1}^{T} \beta^{t-1} \sum_{h_t} \pi(h_t)\, u\Big(y_t(h_t) + R\hat{k}_{t-1}(h_{t-1}) - \hat{k}_t(h_t) \\
&\quad + \Big[b_t(\hat{h}_t(h_t)) - k_t(\hat{h}_t(h_t)) + Rk_{t-1}(\hat{h}_{t-1}(h_{t-1})) \Big] \Big) \\
&= \Gamma(\hat{y}, k^*; b),
\end{aligned}
$$

where we have defined $k_t^*(h_t) \equiv \hat{k}_t(h_t) + k_t(\hat{h}_t(h_t))$. Thus, inequality $(20.6.7)$ implies that

$$\Gamma(\hat{y}, k^*; b) > \Gamma(y, k; b),$$

which contradicts the assumed incentive compatibility of (c, k, b, K). ∎

20.6.3. Efficient allocation

An incentive feasible allocation that maximizes *ex ante* utility is called an efficient allocation. It solves the following problem:

$$(\text{P1}) \qquad \max_{\{c,k,b,K\}} \sum_{t=1}^{T} \beta^{t-1} \sum_{h_t} \pi(h_t) u(c_t(h_t))$$

subject to

$$
\begin{aligned}
&\Gamma(y, k; b) = \max_{(\hat{y}, \hat{k}) \in \Omega^T} \Gamma(\hat{y}, \hat{k}; b) \\
&c_t(h_t) + k_t(h_t) = y_t(h_t) + Rk_{t-1}(h_{t-1}) + b_t(h_t), \qquad \forall t, h_t \\
&K_t + \sum_{h_t} \pi(h_t) b_t(h_t) \leq RK_{t-1}, \qquad \forall t \\
&k_t(h_t) \geq 0, \qquad \forall t, h_t \\
&K_T \geq 0, \\
&K_0 = k_0 = 0.
\end{aligned}
$$

The incentive compatibility constraint with unobservable private storage makes problem (P1) exceedingly difficult to solve. To find the efficient allocation we will adopt a guess-and-verify approach. We will guess that the consumption allocation that solves (P1) coincides with the optimal consumption allocation in another economic environment. For example, we might guess that the consumption allocation that solves (P1) is the same as in a complete markets economy with complete enforcement. A better guess might be the autarkic consumption allocation where each household stores goods only for its own use, behaving according to a version of the chapter 17 model with a no-borrowing constraint. Our analysis of the model without private storage in the previous section makes the first guess doubtful. In fact, both guesses are wrong. What turns out to be true is the following.

PROPOSITION 2: An incentive feasible allocation (c, k, b, K) is efficient if and only if $c = c^*$, where c^* is the consumption allocation that solves

$$(\text{P2}) \qquad \max_{\{c\}} \sum_{t=1}^{T} \beta^{t-1} \sum_{h_t} \pi(h_t) u(c_t(h_t))$$

subject to

$$\sum_{t=1}^{T} R^{1-t} \left[y_t(h_T) - c_t(h_t(h_T)) \right] \geq 0, \qquad \forall h_T.$$

The proposition says that the consumption allocation that solves (P1) is the same as that in an economy where each household can borrow or lend outside the village at the risk-free gross interest rate R subject to a solvency requirement.[9] Below we will provide a proof for the case of two periods $(T = 2)$. We refer readers to Cole and Kocherlakota (2001) for a general proof.

Central to the proof are the first-order conditions of problem (P2), namely,

$$u'(c_t(h_t)) = \beta R \sum_{s=1}^{S} \Pi_s u' \left(c_{t+1}(\{h_t, \overline{y}_s\}) \right), \qquad \forall t, h_t \qquad (20.6.8)$$

$$\sum_{t=1}^{T} R^{1-t} \left[y_t(h_T) - c_t(h_t(h_T)) \right] = 0, \qquad \forall h_T. \qquad (20.6.9)$$

---

[9] The solvency requirement is equivalent to the *natural debt limit* discussed in chapters 17 and 18.

Given the continuous, strictly concave objective function and the compact, convex constraint set in problem (P2), the solution c^* is unique and the first-order conditions are both necessary and sufficient.

In the efficient allocation, the planner chooses transfers that in effect relax the nonnegativity constraint on a household's storage is not binding, i.e., consumption smoothing condition (20.6.8) is satisfied. However, the optimal transfer scheme offers no insurance across households because the present value of transfers is zero for any history h_T, i.e., the net-present value condition (20.6.9) is satisfied.

20.6.4. The case of two periods ($T = 2$)

In a finite horizon model, an immediate implication of the incentive constraints is that transfers in the final period T must be independent of households' reported values of y_T. In the case of two periods, we can therefore encode permissible transfer schemes as

$$b_1(\overline{y}_s) = b_s, \quad \forall s \in \mathbf{S},$$
$$b_2(\{\overline{y}_s, \overline{y}_j\}) = e_s, \quad \forall s, j \in \mathbf{S},$$

where b_s and e_s denote the transfer in the first and second period, respectively, when the household reports income $\overline{y}_s$ in the first period and income $\overline{y}_j$ in the second period.

Following Cole and Kocherlakota (2001), we will first characterize the solution to the modified planner's problem (P3) stated below. It has the same objective function as (P1) but a larger constraint set. In particular, we enlarge the constraint set by considering a smaller set of reporting strategies for the households, Ω_R^2. A household strategy $(\hat{y}, \hat{k})$ is an element of Ω_R^2 if

$$\hat{y}_1(\overline{y}_s) \in \{\overline{y}_{s-1}, \overline{y}_s\}, \quad \text{for } s = 2, 3, \ldots, S$$
$$\hat{y}_1(\overline{y}_1) = \overline{y}_1.$$

That is, a household can either tell the truth or lie downward by one notch in the grid of possible income realizations. There is no restriction on possible storage strategies.

Given $T = 2$, we state problem (P3) as follows. Choose $\{b_s, e_s\}_{s=1}^{S}$ to maximize

$$(P3) \qquad \sum_{s=1}^{S} \Pi_s \left[u(\overline{y}_s + b_s) + \beta \sum_{j=1}^{S} \Pi_j u(\overline{y}_j + e_s) \right]$$

subject to

$$\Gamma(y, 0; b) = \max_{(\hat{y}, \hat{k}) \in \Omega_R^2} \Gamma(\hat{y}, \hat{k}; b)$$

$$c_t(h_t) = y_t(h_t) + b_t(h_t), \qquad \forall t, h_t$$

$$k_t(h_t) = 0, \qquad \forall t, h_t$$

$$K_t + \sum_{h_t} \pi(h_t) b_t(h_t) \leq R K_{t-1}, \qquad \forall t$$

$$K_2 \geq 0,$$

$$\text{given } K_0 = k_0 = 0.$$

Beyond the restricted strategy space Ω_R^2, problem (P3) differs from (P1) in considering only allocations that have zero private storage. But by Proposition 1, we know that this is an innocuous restriction that does not affect the maximized value of the objective.

Here it is useful to explain why we are first studying the contrived problem (P3) rather than turning immediately to the real problem (P1). Certainly problem (P3) is easier to solve because we are exogenously restricting the households' reporting strategies to either telling the truth or making one specific lie. But how can knowledge of the solution to problem (P3) help us understand problem (P1)? Well, suppose it happens that problem (P3) has a unique solution equal to the optimal consumption allocation c^* from Proposition 2 (which will in fact turn out to be true). In that case, it follows that c^* is also the solution to problem (P1) because of the following argument. First, it is straightforward to verify that c^* is incentive compatible with respect to the unrestricted set Ω^2 of reporting strategies. Second, given that no better allocation than c^* can be supported with the restricted set Ω_R^2 of reporting strategies (telling the truth or making one specific lie), it is impossible that we can attain better outcomes by merely introducing additional ways of lying.

Let us therefore first study problem (P3). In particular, using a proof by contradiction, we now show that any allocation $(c, 0, b, K)$ that solves problem (P3) must satisfy three conditions:[10]

(i) The aggregate resource constraint $(20.6.1)$ holds with equality in both periods and $K_2 = 0$;

(ii) $u'(c_1(\overline{y}_s)) = \beta R \sum_{j=1}^{S} \Pi_j u'\left(c_2(\{\overline{y}_s, \overline{y}_j\})\right), \quad \forall s;$

(iii) $b_s + R^{-1}e_s = 0, \quad \forall s.$

Condition (i) is easy to establish given the restricted strategy space Ω_R^2. Suppose that condition (i) is violated and hence, some aggregate resources have not been transferred to the households. In that case, the planner should store all unused resources until period 2 and give them to any household who reported the highest income in period 1. Given strategy space Ω_R^2, households are only allowed to lie downward so the proposed allocation cannot violate the incentive constraints for truthful reporting. Also, transferring more consumption in the last period will not lead to any private storage. We conclude that condition (i) must hold for any solution to problem (P3).

Next, suppose that condition (ii) is violated, i.e., for some $i \in \mathbf{S}$,

$$u'(c_1(\overline{y}_i)) > \beta R \sum_{s=1}^{S} \Pi_s u'\left(c_2(\{\overline{y}_i, \overline{y}_s\})\right). \qquad (20.6.10)$$

(The reverse inequality is obviously inconsistent with the incentive constraints since households are free to store goods between periods.) We can then construct an alternative incentive feasible allocation that yields higher *ex ante* utility as follows. Set $K_1^o = K_1 - \epsilon \Pi_i$, $b_i^o = b_i + \epsilon$, $e_i^o = e_i - \delta$, and choose (ϵ, δ) such that

$$u(\overline{y}_i + b_i + \epsilon) + \beta \sum_{s=1}^{S} \Pi_s u\left(\overline{y}_s + e_i - \delta\right)$$

$$= u(\overline{y}_i + b_i) + \beta \sum_{s=1}^{S} \Pi_s u\left(\overline{y}_s + e_i\right), \qquad (20.6.11)$$

---

[10] The proof by contradiction goes as follows. Suppose that an allocation $(c, 0, b, K)$ solves problem (P3) but violates one of our conditions. Then we can show either that $(c, 0, b, K)$ cannot be incentive feasible with respect to (P3) or that there exists another incentive feasible allocation $(c^o, 0, b^o, K^o)$ that yields an even higher *ex ante* utility than $(c, 0, b, K)$.

$$u'(\overline{y}_i + b_i + \epsilon) \geq \beta R \sum_{s=1}^{S} \Pi_s u' \left(\overline{y}_s + e_i - \delta \right). \qquad (20.6.12)$$

By the envelope condition, $(20.6.10)$ implies that $\delta > R\epsilon$, so this alternative allocation frees up resources that can be used to improve *ex ante* utility. But we have to check that the incentive constraints are respected. For households experiencing $\overline{y}_i$, the proposed allocation is clearly incentive compatible, since their payoffs from reporting truthfully or lying are unchanged, and condition $(20.6.12)$ ensures that they are not deviating from zero private storage. It can be verified that a household with the next higher income shock $\overline{y}_{i+1}$ would not want to lie downward because a household with a higher income $\overline{y}_{i+1}$ would not want the proposed loan against the future at the implied interest rate, $\delta/\epsilon > R$, at which the lower-income household is indifferent to the transaction. The following lemma shows this formally.

LEMMA: Let ϵ, $\delta > 0$ satisfy $\delta > R\epsilon$, and define

$$Z(m) \equiv \max_{k \geq 0} \Big[u(m - k) + \beta E_y u(y + Rk) \Big]$$

$$W(m) \equiv \max_{k \geq 0} \Big[u(m - k + \epsilon) + \beta E_y u(y + Rk - \delta) \Big],$$

where u is a strictly concave function and the expectation E_y is taken with respect to a random second-period income y. If $Z(m_a) = W(m_a)$ and $m_b > m_a$, then $Z(m_b) > W(m_b)$.

PROOF: Let the unique, weakly increasing sequence of maximizers of the savings problems Z and W be denoted $k_Z(m)$ and $k_W(m)$, respectively, which are guaranteed to exist by the strict concavity of u. The proof of the lemma proceeds by contradiction. Suppose that $Z(m_b) \leq W(m_b)$. Then by the mean value theorem, there exists $m_c \in (m_a, m_b)$ such that $Z'(m_c) \leq W'(m_c)$. This implies that

$$u'(m_c - k_Z(m_c)) \leq u'(m_c - k_W(m_c) + \epsilon).$$

The concavity of u implies that $0 \leq k_Z(m_c) \leq k_W(m_c) - \epsilon$. The weak monotonicity of k_W implies that $k_W(m_b) \geq k_W(m_c)$, so we know that $0 \leq k_W(m_b) - \epsilon$ and we can write

$$Z(m_b) \geq u(m_b - k_W(m_b) + \epsilon) + \beta E_y u(y + Rk_w(m_b) - R\epsilon)$$

$$> u(m_b - k_W(m_b) + \epsilon) + \beta E_y u(y + Rk_w(m_b) - \delta) = W(m_b),$$

which is a contradiction.   ∎

Finally, suppose that condition (iii) is violated, i.e., for some $i \in \mathbf{S}$,

$$\Psi_s \equiv b_s + R^{-1}e_s \neq b_{s-1} + R^{-1}e_{s-1} \equiv \Psi_{s-1}.$$

First, we can rule out $\Psi_s < \Psi_{s-1}$ because it would compel households with income shock $\overline{y}_s$ in the first period to lie downward. This is so because our condition (ii) implies that the nonnegative storage constraint binds for neither these households nor the households with the lower income shock $\overline{y}_{s-1}$. Hence, households with income shock $\overline{y}_s$ will only report truthfully if $Z(\overline{y}_s + \Psi_s) \geq Z(\overline{y}_s + \Psi_{s-1})$, where $Z(\cdot)$ is the value of the first savings problem defined in the lemma above. Thus, we conclude that $\Psi_s \geq \Psi_{s-1}$.

Second, we can rule out $\Psi_s > \Psi_{s-1}$ by constructing an alternative incentive feasible allocation that attains a higher *ex ante* utility. Compute the certainty equivalent $\tilde{\Psi}$ such that

$$\Pi_s Z(\overline{y}_s + \tilde{\Psi}) + \Pi_{s-1} Z(\overline{y}_{s-1} + \tilde{\Psi}) = \Pi_s Z(\overline{y}_s + \Psi_s) + \Pi_{s-1} Z(\overline{y}_{s-1} + \Psi_{s-1}).$$

Then change the transfer scheme so that households reporting $\overline{y}_s$ or $\overline{y}_{s-1}$ get the same present value of transfers equal to $\tilde{\Psi}$. Because of the strict concavity of the utility function, the new scheme frees up resources that can be used to improve *ex ante* utility. Also, the new scheme does not violate any incentive constraints. Households with income shock $\overline{y}_{s-1}$ are now better off when reporting truthfully, households with income shock $\overline{y}_s$ are indifferent to telling the truth, and households with income shock $\overline{y}_{s+1}$ will not lie because the present value of the transfers associated with lying has gone down. Since the planner satisfies the aggregate resource constraint at equality in our condition (i), we conclude that all households receive the same present value of transfers equal to zero.

By establishing conditions (i)–(iii), we have in effect shown that any solution to (P3) must satisfy equations (20.6.8) and (20.6.9). Thus, problem (P3) has a unique solution $(c^*, 0, b^*, K^*)$, where c^* is given by Proposition 2 and

$$b_t^*(h_t) = c_t^*(h_t) - y_t(h_t),$$

$$K_t^* = -\sum_{h_t} \pi(h_t) \sum_{j=1}^{t} R^{t-1} b_j^*(h_j(h_t)).$$

Moreover, $(c^*, 0, b^*, K^*)$ is incentive compatible with respect to the unrestricted strategy set Ω^2. If a household tells the truth, its consumption is optimally smoothed. Hence, households weakly prefer to tell the truth and not store.

The proof of Proposition 2 for $T = 2$ is completed by noting that by construction, if some allocation $(c^*, 0, b^*, K^*)$ solves (P3), and $(c^*, 0, b^*, K^*)$ is incentive compatible with respect to Ω^2, then $(c^*, 0, b^*, K^*)$ solves (P1). Also, since equations (20.6.8) and (20.6.9) fully characterize the consumption allocation c^*, we have uniqueness with respect to c^* (but there exists a multitude of storage and transfer schemes that the planner can use to implement c^* in problem (P1)).

20.6.5. Role of the planner

Proposition 2 states that any allocation (c, k, b, K) that solves the planner's problem (P1) has the same consumption outcome $c = c^*$ as the solution to (P2), i.e., the market outcome when each household can lend *or* borrow at the risk-free interest rate R. This result has both positive and negative messages about the role of the planner. Because households have access only to a storage technology, the planner implements the efficient allocation by designing an elaborate transfer scheme that effectively undoes each household's nonnegativity constraint on storage while respecting solvency requirements. In this sense, the planner has an important role to play. However, the optimal transfer scheme offers no insurance across households and implements only a self-insurance scheme tantamount to a borrowing-and-lending outcome for each household. Thus, the planner's accomplishments as an insurance provider are very limited.

If we had assumed that households themselves have direct access to the credit market outside of the village, it would follow immediately that the planner would be irrelevant, since the households could then implement the efficient allocation themselves. Allen (1985) first made this observation. Given any transfer scheme, he showed that all households would choose to report the income that yields the highest present value of transfers regardless of what the actual income is. In our setting where the planner has no resources of his own, we get the zero net present value condition for the stream of transfers to any individual household.

20.6.6. Decentralization in a closed economy

Suppose that consumption allocation c^* in Proposition 2 satisfies

$$\sum_{h_t} \pi(h_t) \sum_{j=1}^{t} R^{t-j} \left[y_j(h_t) - c_j^*(h_j(h_t)) \right] \geq 0, \quad \forall t. \tag{20.6.13}$$

That is, aggregate storage is nonnegative at all dates. It follows that the efficient allocation in Proposition 2 would then also be the solution to a closed system where the planner has no access to outside borrowing. Moreover, c^* can then be decentralized as the equilibrium outcome in an incomplete markets economy where households competitively trade consumption and risk-free one-period bonds that are available in zero net supply in each period. Here we are assuming complete enforcement so that households must pay off their debts in every state of the world, and they cannot end their lives in debt.

In the decentralized equilibrium, let $a_t(h_t)$ and $k_t^d(h_t)$ denote bond holdings and storage, respectively, of a household indexed by its history h_t. The gross interest rate on bonds between periods t and $t+1$ is denoted $1 + r_t$. We claim that the efficient allocation $(c^*, 0, b^*, K^*)$ can be decentralized by recursively defining

$$r_t \equiv R - 1, \tag{20.6.14}$$

$$k_t^d(h_t) \equiv K_t^*, \tag{20.6.15}$$

$$a_t(h_t) \equiv y_t(h_t) - c_t^*(h_t) - K_t^* + RK_{t-1}^* + Ra_{t-1}(h_{t-1}), \tag{20.6.16}$$

with $a_0 = 0$. First, we verify that households are behaving optimally. Note that we have chosen the interest rate so that households are indifferent between lending and storing. Because we also know that the household's consumption is smoothed at c^*, we need only to check that households' budget constraints hold with equality. By substituting (20.6.15) into (20.6.16), we obtain the household's one-period budget constraint. The consolidation of all one-period budget constraints yields

$$a_T(h_T) = - k_T^d(h_T) + \sum_{t=1}^{T} R^{T-t} \left[y_t(h_T) - c_t^*(h_t(h_T)) \right]$$
$$+ R^{T-1}(k_0^d + a_0) = 0$$

where the last equality is implied by $K_T^* = K_0 = a_0 = 0$ and $(20.6.9)$. Second, we verify that the bond market clears by summing all households' one-period budget constraints,

$$\sum_{h_t} \pi(h_t) a_t(h_t) = \sum_{h_t} \pi(h_t) \Big[y_t(h_t) - c_t^*(h_t) - k_t^d(h_t)$$
$$+ Rk_{t-1}^d(h_{t-1}(h_t)) + Ra_{t-1}(h_{t-1}(h_t)) \Big].$$

After invoking $(20.6.15)$ and the fact that $b_t^*(h_t) = c_t^*(h_t) - y_t(h_t)$, we can rewrite this expression as

$$\sum_{h_t} \pi(h_t) a_t(h_t) = - K_t^* + RK_{t-1}^*$$
$$- \sum_{h_t} \pi(h_t) \Big[b_t^*(h_t) - Ra_{t-1}(h_{t-1}(h_t)) \Big]$$
$$= R \sum_{h_{t-1}} \pi(h_{t-1}) a_{t-1}(h_{t-1}) = 0 \,,$$

where the second equality is implied by $(20.6.1)$ holding with equality at the allocation $(c^*, 0, b^*, K^*)$, and the last equality follows from successive substitutions leading back to the initial condition $a_0 = 0$.

It is straightforward to make the reverse argument and show that if $1 + r_t = R$ for all t in our incomplete markets equilibrium, then the equilibrium consumption allocation is efficient and equal to c^*, as given in Proposition 2.

Cole and Kocherlakota note that seemingly ad hoc restrictions on the securities available for trade are consistent with the implementation of the efficient allocation in this setting, and they argue that their framework provides an explicit micro foundation for incomplete markets models such as Aiyagari's (1994) model that we studied in chapter 18.

20.7. Concluding remarks

The idea of using promised values as a state variable has made it possible to use dynamic programming to study problems with history dependence. In this chapter we have studied how using a promised value as a state variable helps to study optimal risk-sharing arrangements when there are incentive problems coming from limited enforcement or limited information. The next several chapters apply and extend this idea in other contexts. Chapter 21 discusses how to build a closed-economy, or general equilibrium, version of our model with imperfect enforcement. Chapter 22 discusses ways of designing unemployment insurance that optimally compromise between supplying insurance and providing incentives for unemployed workers to search diligently. Chapter 23 uses a continuation value as a state variable to encode a government's reputation. Chapter 25 discusses some models of contracts and government policies that have been applied to some enforcement problems in international trade.

A. Historical development

20.A.1. Spear and Srivastava

Spear and Srivastava (1987) introduced the following recursive formulation of an infinitely repeated, discounted repeated principal-agent problem: A *principal* owns a technology that produces output q_t at time t, where q_t is determined by a family of c.d.f.'s $F(q_t|a_t)$, and a_t is an action taken at the beginning of t by an *agent* who operates the technology. The principal has access to an outside loan market with constant risk-free gross interest rate β^{-1}. The agent has preferences over consumption streams ordered by $E_0 \sum_{t=0}^{\infty} \beta^t u(c_t, a_t)$. The principal is risk neutral and offers a contract to the agent designed to maximize $E_0 \sum_{t=0}^{\infty} \beta^t \{q_t - c_t\}$ where c_t is the principal's payment to the agent at t.

20.A.2. Timing

Let w denote the discounted utility promised to the agent at the beginning of the period. Given w, the principal selects three functions $a(w)$, $c(w,q)$, and $\tilde{w}(w,q)$ determining the current action $a_t = a(w_t)$, the current consumption $c = c(w_t, q_t)$, and a promised utility $w_{t+1} = \tilde{w}(w_t, q_t)$. The choice of the three functions $a(w)$, $c(w,q)$, and $\tilde{w}(w,q)$ must satisfy the following two sets of constraints:

$$w = \int \{u[c(w,q), a(w)] + \beta \tilde{w}(w,q)\} \, dF[q|a(w)] \qquad (20.A.1)$$

and

$$\int \{u[c(w,q), a(w)] + \beta \tilde{w}(w,q)\} \, dF[q|a(w)]$$
$$\geq \int \{u[c(w,q), \hat{a}] + \beta \tilde{w}(w,q)\} dF(q|\hat{a}), \qquad \forall \, \hat{a} \in A. \qquad (20.A.2)$$

Equation $(20.A.1)$ requires the contract to deliver the promised level of discounted utility. Equation $(20.A.2)$ is the *incentive compatibility* constraint requiring the agent to want to deliver the amount of effort called for in the contract. Let $v(w)$ be the value to the principal associated with promising discounted utility w to the agent. The principal's Bellman equation is

$$v(w) = \max_{a,c,\tilde{w}} \{q - c(w,q) + \beta \, v[\tilde{w}(w,q)]\} \, dF[q|a(w)] \qquad (20.A.3)$$

where the maximization is over functions $a(w)$, $c(w,q)$, and $\tilde{w}(w,q)$ and is subject to the constraints $(20.A.1)$ and $(20.A.2)$. This value function $v(w)$ and the associated optimum policy functions are to be solved by iterating on the Bellman equation $(20.A.3)$.

20.A.3. Use of lotteries

In various implementations of this approach, a difficulty can be that the constraint set fails to be convex as a consequence of the structure of the incentive constraints. This problem has been overcome by Phelan and Townsend (1991) by convexifying the constraint set through randomization. Thus, Phelan and Townsend simplify the problem by extending the principal's choice to the space of lotteries over actions a and outcomes c, w'. To introduce Phelan and Townsend's formulation, let $P(q|a)$ be a family of discrete probability distributions over discrete spaces of outputs and actions Q, A, and imagine that consumption and values are also constrained to lie in discrete spaces C, W, respectively. Phelan and Townsend instruct the principal to choose a probability distribution $\Pi(a, q, c, w')$ subject first to the constraint that for all fixed $(\bar{a}, \bar{q})$

$$\sum_{C \times W} \Pi(\bar{a}, \bar{q}, c, w') = P(\bar{q}|\bar{a}) \sum_{Q \times C \times W} \Pi(\bar{a}, q, c, w') \qquad (20.A.4a)$$

$$\Pi(a, q, c, w') \geq 0 \qquad (20.A.4b)$$

$$\sum_{A \times Q \times C \times W} \Pi(a, q, c, w') = 1. \qquad (20.A.4c)$$

Equation $(20.A.4a)$ simply states that $\mathrm{Prob}(\bar{a}, \bar{q}) = \mathrm{Prob}(\bar{q}|\bar{a})\mathrm{Prob}(\bar{a})$. The remaining pieces of $(20.A.4)$ just require that "probabilities are probabilities." The counterpart of Spear-Srivastava's equation $(20.A.1)$ is

$$w = \sum_{A \times Q \times C \times W} \{u(c, a) + \beta w'\} \, \Pi(a, q, c, w'). \qquad (20.A.5)$$

The counterpart to Spear-Srivastava's equation $(20.A.2)$ for each $a, \hat{a}$ is

$$\sum_{Q \times C \times W} \{u(c, a) + \beta w'\} \, \Pi(c, w'|q, a)P(q|a)$$

$$\geq \sum_{Q \times C \times W} \{u(c, \hat{a}) + \beta w'\} \, \Pi(c, w'|q, a)P(q|\hat{a}).$$

Here $\Pi(c, w'|q, a)P(q|\hat{a})$ is the probability of (c, w', q) if the agent claims to be working a but is actually working $\hat{a}$. Express

$$\Pi(c, w'|q, a)P(q|\hat{a}) =$$

$$\Pi(c, w'|q, a)P(q|a) \frac{P(q|\hat{a})}{P(q|a)} = \Pi(c, w', q|a) \cdot \frac{P(q|\hat{a})}{P(q|a)}.$$

To write the incentive constraint as

$$\sum_{Q \times C \times W} \{u(c, a) + \beta w'\} \Pi(c, w', q|a)$$

$$\geq \sum_{Q \times C \times W} \{u(c, \hat{a}) + \beta w'\} \; \Pi(c, w', q|\hat{a}) \; \cdot \; \frac{P(q|\hat{a})}{P(q|a)}.$$

Multiplying both sides by the unconditional probability $P(a)$ gives expression $(20.A.6)$.

$$\sum_{Q \times C \times W} \{u(c, a) + \beta w'\} \; \Pi(a, q, c, w')$$

$$\geq \sum_{Q \times C \times W} \{u(c, \hat{a}) + \beta w'\} \; \frac{P(q|\hat{a})}{P(q|a)} \; \Pi(a, q, c, w') \qquad (20.A.6)$$

The Bellman equation for the principal's problem is

$$v(w) = \max_{\Pi}\{(q - c) + \beta v(w')\}\Pi(a, q, c, w'), \qquad (20.A.7)$$

where the maximization is over the probabilities $\Pi(a, q, c, w')$ subject to equations $(20.A.4)$, $(20.A.5)$, and $(20.A.6)$. The problem on the right side of equation $(20.A.7)$ is a linear programming problem. Think of each of (a, q, c, w') being constrained to a discrete grid of points. Then, for example, the term $(q - c) + \beta v(w')$ on the right side of equation $(20.A.7)$ can be represented as a *fixed* vector that multiplies a vectorized version of the probabilities $\Pi(a, q, c, w')$. Similarly, each of the constraints $(20.A.4)$, $(20.A.5)$, and $(20.A.6)$ can be represented as a linear inequality in the choice variables, the probabilities Π. Phelan and Townsend compute solutions of these linear programs to iterate on the Bellman equation $(20.A.7)$. Note that at each step of the iteration on the Bellman equation, there is one linear program to be solved for each point w in the space of grid values for W.

In practice, Phelan and Townsend have found that lotteries are often redundant in the sense that most of the $\Pi(a, q, c, w')$'s are zero, and a few are 1.

Exercises

Exercise 20.1   **Thomas and Worrall meet Markov**

A household orders sequences $\{c_t\}_{t=0}^{\infty}$ by

$$E \sum_{t=0}^{\infty} \beta^t u(c_t), \qquad \beta \in (0,1)$$

where u is strictly increasing, twice continuously differentiable, and strictly concave with $u'(0) = +\infty$. The good is nondurable. The household receives an endowment of the consumption good of y_t that obeys a discrete-state Markov chain with $P_{ij} = \text{Prob}(y_{t+1} = \bar{y}_j | y_t = \bar{y}_i)$, where the endowment y_t can take one of the I values $[\bar{y}_1, \ldots, \bar{y}_I]$.

a. Conditional on having observed the time t value of the household's endowment, a social insurer wants to deliver expected discounted utility v to the household in the least costly way. The insurer observes y_t at the beginning of every period, and contingent on the observed history of those endowments, can make a transfer τ_t to the household. The transfer can be positive or negative and can be enforced without cost. Let $C(v, i)$ be the minimum expected discounted cost to the insurance agency of delivering promised discounted utility v when the household has just received endowment $\bar{y}_i$. (Let the insurer discount with factor β.) Write a Bellman equation for $C(v, i)$.

b. Characterize the consumption plan and the transfer plan that attains $C(v, i)$; find an associated law of motion for promised discounted value.

c. Now assume that the household is isolated and has no access to insurance. Let $v^a(i)$ be the expected discounted value of utility for a household in autarky, conditional on current income being $\bar{y}_i$. Formulate Bellman equations for $v^a(i), i = 1, \ldots, I$.

d. Now return to the problem of the insurer mentioned in part b, but assume that the insurer cannot enforce transfers because each period the consumer is free to walk away from the insurer and live in autarky thereafter. The insurer must structure a history-dependent transfer scheme that prevents the household from ever exercising the option to revert to autarky. Again, let $C(v, i)$ be the minimum cost for an insurer that wants to deliver promised discounted utility v to a household with current endowment i. Formulate Bellman equations for $C(v, i), i = 1, \ldots, I$. Briefly discuss the form of the law of motion for v associated with the minimum cost insurance scheme.

Exercise 20.2   Wealth dynamics in moneylender model

Consider the model in the text of the village with a moneylender. The village consists of a large number (e.g., a continuum) of households, each of which has an i.i.d. endowment process that is distributed as

$$\text{Prob}(y_t = \overline{y}_s) = \frac{1 - \lambda}{1 - \lambda^S} \lambda^{s-1}$$

where $\lambda \in (0, 1)$ and $\overline{y}_s = s + 5$ is the sth possible endowment value, $s = 1, \ldots, S$. Let $\beta \in (0, 1)$ be the discount factor and β^{-1} the gross rate of return at which the moneylender can borrow or lend. The typical household's one-period utility function is $u(c) = (1 - \gamma)^{-1} c^{1-\gamma}$, where γ is the household's coefficient of relative risk aversion. Assume the parameter values $(\beta, S, \gamma, \lambda) = (.5, 20, 2, .95)$.

Hint: The formulas given in the section 20.3.3 will be helpful in answering the following questions.

a. Using Matlab, compute the optimal contract that the moneylender offers a villager, assuming that the contract leaves the villager indifferent between refusing and accepting the contract.

b. Compute the expected profits that the moneylender earns by offering this contract for an initial discounted utility that equals the one that the household would receive in autarky.

c. Let the cross-section distribution of consumption at time $t \geq 0$ be given by the c.d.f. $\text{Prob}(c_t \leq \overline{C}) = F_t(\overline{C})$. Compute F_t. Plot it for $t = 0$, $t = 5$, $t = 10$, $t = 500$.

d. Compute the moneylender's savings for $t \geq 0$ and plot it for $t = 0, \ldots, 100$.

e. Now adapt your program to find the initial level of promised utility $v > v_{\text{aut}}$ that would set $P(v) = 0$.

Exercise 20.3   Thomas and Worrall (1988)

There is a competitive spot market for labor always available to each of a continuum of workers. Each worker is endowed with one unit of labor each period that he supplies inelastically to work either permanently for "the company" or each period in a new one-period job in the spot labor market. The worker's productivity in either the spot labor market or with the company is an i.i.d. endowment process that is distributed as

$$\text{Prob}(w_t = \overline{w}_s) = \frac{1 - \lambda}{1 - \lambda^S} \lambda^{s-1}$$

where $\lambda \in (0, 1)$ and $\overline{w}_s = s + 5$ is the sth possible marginal product realization, $s = 1, \ldots, S$. In the spot market, the worker is paid w_t. In the company, the worker is offered a history-dependent payment $\omega_t = f_t(h_t)$ where $h_t = w_t, \ldots, w_0$. Let $\beta \in (0, 1)$ be the discount factor and β^{-1} the gross rate of return at which the company can borrow or lend. The worker cannot borrow or lend. The worker's one-period utility function is $u(\omega) = (1 - \gamma)^{-1} w^{1-\gamma}$ where ω is the period wage from the company, which equals consumption, and γ is the worker's coefficient of relative risk aversion. Assume the parameter values $(\beta, S, \gamma, \lambda) = (.5, 20, 2, .95)$.

The company's discounted expected profits are

$$E \sum_{t=0}^{\infty} \beta^t (w_t - \omega_t).$$

The worker is free to walk away from the company at the start of any period, but must then stay in the spot labor market forever. In the spot labor market, the worker receives continuation value

$$v_{\text{spot}} = \frac{Eu(w)}{1 - \beta}.$$

The company designs a history-dependent compensation contract that must be sustainable (i.e., self-enforcing) in the face of the worker's freedom to enter the spot labor market at the beginning of period t *after* he has observed w_t but before he receives the t period wage.

Hint: Do these questions ring a bell? See exercise *20.2*.

a. Using Matlab, compute the optimal contract that the company offers the worker, assuming that the contract leaves the worker indifferent between refusing and accepting the contract.

b. Compute the expected profits that the firm earns by offering this contract for an initial discounted utility that equals the one that the worker would receive by remaining forever in the spot market.

c. Let the distribution of wages that the firm offers to its workers at time $t \geq 0$ be given by the c.d.f. $\text{Prob}(\omega_t \leq \overline{w}) = F_t(\overline{w})$. Compute F_t. Plot it for $t = 0$, $t = 5$, $t = 10$, $t = 500$.

d. Plot an expected wage-tenure profile for a new worker.

e. Now assume that there is competition among companies and free entry. New companies enter by competing for workers by raising initial promised utility with the company. Adapt your program to find the initial level of promised utility $v > v_{\text{spot}}$ that would set expected profits from the average worker $P(v) = 0$.

Exercise 20.4 **Thomas-Worrall meet Phelan-Townsend**

Consider the Thomas Worrall environment and denote $\Pi(y)$ the density of the i.i.d. endowment process, where y belongs to the discrete set of endowment levels $Y = [\overline{y}_1, \ldots, \overline{y}_S]$. The one-period utility function is $u(c) = (1 - \gamma)^{-1}(c - a)^{1-\gamma}$ where $\gamma > 1$ and $\overline{y}_S > a > 0$.

Discretize the set of transfers B and the set of continuation values W. We assume that the discrete set $B \subset (a - \overline{y}_S, \overline{b}]$. Notice that with the one-period utility function above, the planner could never extract more than $a - \overline{y}_S$ from the agent. Denote $\Pi^v(b, w|y)$ the joint density over (b, w) that the planner offers the agent who reports y and to whom he has offered beginning-of-period promised value v. For each $y \in Y$ and each $v \in W$, the planner chooses a set of conditional probabilities $\Pi^v(b, w|y)$ to satisfy the Bellman equation

$$P(v) = \max_{\Pi^v(b,w,y)} \sum_{B \times W \times Y} [-b + \beta P(w)] \, \Pi^v(b, w, y) \tag{1}$$

subject to the following constraints:

$$v = \sum_{B \times W \times Y} [u(y + b) + \beta w] \, \Pi^v(b, w, y) \tag{2}$$

$$\sum_{B \times W} [u(y + b) + \beta w] \, \Pi^v(b, w|y) \geq \sum_{B \times W} [u(y + b) + \beta w] \, \Pi^v(b, w|\tilde{y})$$

$$\forall (y, \tilde{y}) \in Y \times Y \tag{3}$$

$$\Pi^v(b, w, y) = \Pi(y)\Pi^v(b, w|y) \qquad \forall (b, w, y) \in B \times W \times Y \tag{4}$$

$$\sum_{B \times W \times Y} \Pi^v(b, w, y) = 1. \tag{5}$$

Here (2) is the promise-keeping constraint, (3) are the truth-telling constraints, and (4), (5) are restrictions imposed by the laws of probability.

a. Verify that given $P(w)$, one step on the Bellman equation is a linear programming problem.

b. Set $\beta = .94, a = 5, \gamma = 3$. Let S, N_B, N_W be the number of points in the grids for Y, B, W, respectively. Set $S = 10$, $N_B = N_W = 25$. Set $Y = [6 \quad 7 \quad \dots 15]$, $\text{Prob}(y_t = \bar{y}_s) = S^{-1}$. Set $W = [w_{\min}, \dots, w_{\max}]$ and $B = [b_{\min}, \dots, b_{\max}]$, where the intermediate points in W and B, respectively, are equally spaced. Please set $w_{\min} = \frac{1}{1-\beta} \frac{1}{1-\gamma} (y_{\min} - a)^{1-\gamma}$ and $w_{\max} = w_{\min}/20$ (these are negative numbers, so $w_{\min} < w_{\max}$). Also set $b_{\min} = (1 - y_{\max} + .33)$ and $b_{\max} = y_{\max} - y_{\min}$.

For these parameter values, compute the optimal contract by formulating a linear program for one step on the Bellman equation, then iterating to convergence on it.

c. Notice the following probability laws:

$$\text{Prob}(b_t, w_{t+1}, y_t | w_t) \equiv \Pi^{w_t}(b_t, w_{t+1}, y_t)$$
$$\text{Prob}(w_{t+1} | w_t) = \sum_{b \in B, y \in Y} \Pi^{w_t}(b, w_{t+1}, y)$$
$$\text{Prob}(b_t, y_t | w_t) = \sum_{w_{t+1} \in W} \Pi^{w_t}(b_t, w_{t+1}, y_t).$$

Please use these and other probability laws to compute $\text{Prob}(w_{t+1} | w_t)$. Show how to compute $\text{Prob}(c_t)$, assuming a given initial promised value w_0.

d. Assume that $w_0 \approx -2$. Compute and plot $F_t(c) = \text{Prob}(c_t \leq c)$ for $t = 1, 5, 10, 100$. Qualitatively, how do these distributions compare with those for the simple village and moneylender model with no information problem and one-sided lack of commitment?

Exercise 20.5 **The IMF**

Consider the problem of a government of a small country that has to finance an exogenous stream of expenditures $\{g_t\}$. For time $t \geq 0$, g_t is i.i.d. with $\text{Prob}(g_t = \bar{g}_s) = \pi_s$ where $\pi_s > 0, \sum_{s=1}^{S} \pi_s = 1$ and $0 < \bar{g}_1 < \cdots < \bar{g}_S$. Raising revenues by taxation is distorting. In fact, the government confronts a deadweight loss function $W(T_t)$ that measures the distortion at time t. Assume that

W is an increasing, twice continuously differentiable, strictly convex function that satisfies $W(0) = 0, W'(0) = 0, W'(T) > 0$ for $T > 0$ and $W''(T) > 0$ for $T \geq 0$. The government's intertemporal loss function for taxes is such that it wants to minimize

$$E_{-1} \sum_{t=0}^{\infty} \beta^t W(T_t), \quad \beta \in (0, 1)$$

where E_{-1} is the mathematical expectation before g_0 is realized. If it cannot borrow or lend, the government's budget constraint is $g_t = T_t$. In fact, the government is unable to borrow and lend *except* through an international coalition of lenders called the IMF. If it does not have an arrangement with the IMF, the country is in autarky and the government's loss is the value

$$v_{\text{aut}} = E \sum_{t=0}^{\infty} \beta^t W(g_t).$$

The IMF itself is able to borrow and lend at a constant risk-free gross rate of interest of $R = \beta^{-1}$. The IMF offers the country a contract that gives the country a net transfer of $g_t - T_t$. A *contract* is a sequence of functions for $t \geq 0$, the time t component of which maps the history g^t into a net transfer $g - T_t$. The IMF has the ability to commit to the contract. However, the country cannot commit to honor the contract. Instead, at the beginning of each period, after g_t has been realized but before the net transfer $g_t - T_t$ has been received, the government can default on the contract, in which case it receives loss $W(g_t)$ this period and the autarky value ever after. A contract is said to be *sustainable* if it is immune to the threat of repudiation, i.e., if it provides the country with the incentive not to leave the arrangement with the IMF. The present value of the contract to the IMF is

$$E \sum_{t=0}^{\infty} \beta^t (T_t - g_t).$$

a. Write a Bellman equation that can be used to find an optimal sustainable contract.

b. Characterize an optimal sustainable contract that delivers initial promised value v_{aut} to the country (i.e., a contract that renders the country indifferent between accepting and not accepting the IMF contract starting from autarky).

c. Can you say anything about a typical pattern of government tax collections T_t and distortions $W(T_t)$ over time for a country in an optimal sustainable contract with the IMF? What about the average pattern of government surpluses $T_t - g_t$ across a panel of countries with identical g_t processes and W functions? Would there be a "cohort" effect in such a panel (i.e., would the calendar date when the country signed up with the IMF matter)?

d. If the optimal sustainable contract gives the country value v_{aut}, can the IMF expect to earn anything from the contract?

Chapter 21
Equilibrium without Commitment

21.1. Two-sided lack of commitment

In section 20.3 of the previous chapter, we studied insurance without commitment. That was a partial equilibrium analysis since the moneylender could borrow or lend resources outside of the village at a given interest rate. Recall also the asymmetry in the environment where villagers could not make any commitments while the moneylender was assumed to be able to commit. We will now study a closed system without access to an outside credit market. Any household's consumption in excess of its own endowment must then come from the endowments of the other households in the economy. We will also adopt the symmetric assumption that no one is able to make commitments. That is, any contract prescribing an exchange of goods today in anticipation of future exchanges of goods represents a sustainable allocation only if current and future exchanges satisfy participation constraints for all households involved in the contractual arrangement. Households are free to walk away from the arrangement at any point in time and thereafter to live in autarky. Such a contract design problem with participation constraints on both sides of an exchange represents a problem with two-sided lack of commitment, as compared to the problem with one-sided lack of commitment in section 20.3.

This chapter draws on the work of Thomas and Worrall (1988, 1994) and Kocherlakota (1996b). At the end of the chapter, we also discuss market arrangements for decentralizing the constrained Pareto optimal allocation, as studied by Kehoe and Levine (1993) and Alvarez and Jermann (2000).

21.2. A closed system

Thomas and Worrall's (1988) model of self-enforcing wage contracts is an antecedent to our villager-moneylender environment. The counterpart to our moneylender in their model is a risk-neutral firm that forms a long-term relationship with a risk-averse worker. In their model, there is also a competitive spot market for labor where a worker is paid y_t at time t. The worker is always free to walk away from the firm and work in that spot market. But if he does, he can never again enter into a long-term relationship with another firm. The firm seeks to maximize the discounted stream of expected future profits by designing a long-term wage contract that is self-enforcing in the sense that it never gives the worker an incentive to quit. In a contract that stipulates a wage c_t at time t, the firm earns time t profits of $y_t - c_t$ (as compared to hiring a worker in the spot market for labor). If Thomas and Worrall had assumed a commitment problem only on the part of the worker, their model would be formally identical to our villager-moneylender environment. However, Thomas and Worrall also assume that the firm itself can renege on a wage contract and buy labor at the random spot market wage. Hence, they require that a self-enforcing wage contract be one in which neither party ever has an incentive to renege.

Kocherlakota (1996b) studies a model that has some valuable features in common with Thomas and Worrall's.[1] Kocherlakota's counterpart to Thomas and Worrall's firm is a risk-averse second household. In Kocherlakota's model, two households receive stochastic endowments. The contract design problem is to find an insurance/transfer arrangement that reduces consumption risk while respecting participation constraints for both households: both households must be induced each period not to walk away from the arrangement to live in autarky. Kocherlakota uses his model in an interesting way to help interpret empirically estimated conditional consumption-income covariances that seem to violate the hypothesis of complete risk sharing. Kocherlakota investigates the extent to which those failures reflect impediments to enforcement that are captured by his participation constraints.

For the purpose of studying those conditional covariances in a stationary stochastic environment, Kocherlakota's use of an environment with two-sided

---

[1] The working paper of Thomas and Worrall (1994) also analyzed a multiple agent closed model like Kocherlakota's. Thomas and Worrall's (1994) analysis evolved into an article by Ligon, Thomas, and Worrall (2002) that we discuss in section 21.13.

lack of commitment is important. In our model of villagers facing a moneylender in section 20.3, imperfect risk sharing is temporary and so would not prevail in a stochastic steady state. In Kocherlakota's model, imperfect risk sharing can be perpetual. There are equal numbers of two types of households in the village. Each of the households has the preferences, endowments, and autarkic utility possibilities described in chapter 20. Here we assume that the endowments of the two types of households are perfectly negatively correlated. Whenever a household of type 1 receives $\overline{y}_s$, a household of type 2 receives $1 - \overline{y}_s$. We assume that y_t is independently and identically distributed according to the discrete probability distribution $\text{Prob}(y_t = \overline{y}_s) = \Pi_s$, where we assume that $\overline{y}_s \in [0, 1]$. We also assume that the Π_s's are such that the distribution of y_t is identical to that of $1 - y_t$. Also, now the planner has access to neither borrowing nor lending opportunities, and is confined to reallocating consumption goods between the two types of households. This limitation leads to two participation constraints. At time t, the type 1 household receives endowment y_t and consumption c_t, while the type 2 household receives $1 - y_t$ and $1 - c_t$.

In this setting, an allocation is said to be *sustainable*[2] if for all $t \geq 0$ and for all histories h_t

$$u(c_t) - u(y_t) + \beta E_t \sum_{j=1}^{\infty} \beta^{j-1} \left[u(c_{t+j}) - u(y_{t+j}) \right] \geq 0, \qquad (21.2.1a)$$

$$u(1 - c_t) - u(1 - y_t) + \beta E_t \sum_{j=1}^{\infty} \beta^{j-1} \left[u(1 - c_{t+j}) - u(1 - y_{t+j}) \right] \geq 0. \quad (21.2.1b)$$

Let Γ denote the set of sustainable allocations. We seek the following function:

$$Q(\triangle) = \max_{\{c_t\}} E_{-1} \sum_{t=0}^{\infty} \beta^t \left[u(1 - c_t) - u(1 - y_t) \right] \qquad (21.2.2a)$$

subject to

$$\{c_t\} \in \Gamma, \qquad (21.2.2b)$$

$$E_{-1} \sum_{t=0}^{\infty} \beta^t \left[u(c_t) - u(y_t) \right] \geq \triangle. \qquad (21.2.2c)$$

The function $Q(\triangle)$ depicts a (constrained) Pareto frontier by portraying the maximized value of the expected lifetime utility of the type 2 household subject to the type 1 household receiving an expected lifetime utility that exceeds

---

[2] Kocherlakota says *subgame perfect* rather than *sustainable*.

its autarkic welfare level by at least $\triangle$ utils. To find this Pareto frontier, we first solve for the consumption dynamics that characterize all efficient contracts. From these optimal consumption dynamics, it will be straightforward to compute the *ex ante* division of gains from an efficient contract.

21.3. Recursive formulation

We choose to study Kocherlakota's model using the approach proposed by Thomas and Worrall.[3] Thomas and Worrall (1988) formulate the contract design problem as a dynamic program, where the state of the system *prior* to the current period's endowment realization is given by a vector $[x_1 \; x_2 \; \ldots \; x_s \; \ldots \; x_S]$. Here x_s is the value of the expression on the left side of $(21.2.1a)$ that is promised to a type 1 agent conditional on the current period's endowment realization being $\overline{y}_s$. Let $Q_s(x_s)$ then denote the corresponding value of expression $(21.2.1b)$ that is promised to a type 2 agent.[4] When the endowment realization $\overline{y}_s$ is associated with a promise to a type 1 agent equal to $x_s = x$, we can write the Bellman equation as

$$Q_s(x) = \max_{c, \{\chi_j\}_{j=1}^S} \left\{ u(1-c) - u(1-\overline{y}_s) + \beta \sum_{j=1}^S \Pi_j Q_j(\chi_j) \right\} \qquad (21.3.1a)$$

subject to

$$u(c) - u(\overline{y}_s) + \beta \sum_{j=1}^S \Pi_j \chi_j \geq x, \qquad (21.3.1b)$$

$$\chi_j \geq 0, \qquad\qquad j = 1, \ldots, S; \qquad (21.3.1c)$$

$$Q_j(\chi_j) \geq 0, \qquad\quad j = 1, \ldots, S; \qquad (21.3.1d)$$

$$c \in [0,1], \qquad\qquad\qquad\qquad (21.3.1e)$$

---

[3] Kocherlakota instead extended the approach that we used in the villager-moneylender model of section 20.3 to an environment with two-sided lack of commitment. We followed Kocherlakota in chapter 15 of the first edition of this book. However, when applied to problems with two-sided lack of commitment, this approach encounters a technical difficulty associated with possible kinks in the Pareto frontier. (We first encountered this difficulty when we assigned a version of exercise *21.3* to our students.) Thomas and Worrall's approach avoids this nondifferentiability problem by using conditional Pareto frontiers, one for each realization of the endowment.

[4] $Q_s(\cdot)$ is a Pareto frontier conditional on the endowment realization $\overline{y}_s$, while $Q(\cdot)$ in $(21.2.2a)$ is an *ex ante* Pareto frontier before observing any endowment realization.

where expression $(21.3.1b)$ is the promise-keeping constraint, expression $(21.3.1c)$ is the participation constraint for the type 1 agent, and expression $(21.3.1d)$ is the participation constraint for the type 2 agent. The set of feasible c is given by expression $(21.3.1e)$.

Thomas and Worrall prove the existence of a compact interval that contains all permissible continuation values χ_j:

$$\chi_j \in [0, \overline{x}_j] \text{ for } j = 1, 2, \ldots, S. \qquad (21.3.1f)$$

Thomas and Worrall also show that the Pareto-frontier $Q_j(\cdot)$ is decreasing, strictly concave, and continuously differentiable on $[0, \overline{x}_j]$. The bounds on χ_j are motivated as follows. The contract cannot award the type 1 agent a value of χ_j less than zero because that would correspond to an expected future lifetime utility below the agent's autarky level. There exists an upper bound $\overline{x}_j$ above which the planner would never find it optimal to award the type 1 agent a continuation value conditional on next period's endowment realization being $\overline{y}_j$. It would simply be impossible to deliver a higher continuation value because of the participation constraints. In particular, the upper bound $\overline{x}_j$ is such that

$$Q_j(\overline{x}_j) = 0. \qquad (21.3.2)$$

Here a type 2 agent receives an expected lifetime utility equal to his autarky level if the next period's endowment realization is $\overline{y}_j$ and a type 1 agent is promised the upper bound $\overline{x}_j$. Our two- and three-state examples in sections 21.10 and 21.11 illustrate what determines $\overline{x}_j$.

Attach Lagrange multipliers μ, $\beta\Pi_j\lambda_j$, and $\beta\Pi_j\theta_j$ to expressions $(21.3.1b)$, $(21.3.1c)$, and $(21.3.1d)$, then get the following first-order conditions for c and χ_j:[5]

$$c: \quad -u'(1-c) + \mu u'(c) = 0, \qquad (21.3.3a)$$

$$\chi_j: \quad \beta\Pi_j Q'_j(\chi_j) + \mu\beta\Pi_j + \beta\Pi_j\lambda_j + \beta\Pi_j\theta_j Q'_j(\chi_j) = 0. \qquad (21.3.3b)$$

By the envelope theorem,

$$Q'_s(x) = -\mu. \qquad (21.3.4)$$

---

[5] Here we are proceeding under the conjecture that the nonnegativity constraints on consumption in $(21.3.1e)$, $c \geq 0$ and $1 - c \geq 0$, are not binding. This conjecture is confirmed below when it is shown that optimal consumption levels satisfy $c \in [\overline{y}_1, \overline{y}_S]$.

After substituting $(21.3.4)$ into $(21.3.3a)$ and $(21.3.3b)$, respectively, the optimal choices of c and χ_j satisfy

$$Q'_s(x) = -\frac{u'(1-c)}{u'(c)}, \tag{21.3.5a}$$

$$Q'_s(x) = (1+\theta_j)Q'_j(\chi_j) + \lambda_j. \tag{21.3.5b}$$

21.4. Equilibrium consumption

21.4.1. Consumption dynamics

From equation $(21.3.5a)$, the consumption c of a type 1 agent is an increasing function of the promised value x. The properties of the Pareto frontier $Q_s(x)$ imply that c is a differentiable function of x on $[0, \overline{x}_s]$. Since $x \in [0, \overline{x}_s]$, c is contained in the nonempty compact interval $[\underline{c}_s, \overline{c}_s]$, where

$$Q'_s(0) = -\frac{u'(1-\underline{c}_s)}{u'(\underline{c}_s)} \qquad \text{and} \qquad Q'_s(\overline{x}_s) = -\frac{u'(1-\overline{c}_s)}{u'(\overline{c}_s)}.$$

Thus, if $c = \underline{c}_s$, $x = 0$, so that a type 1 agent gets no gain from the contract from then on. If $c = \overline{c}_s$, $Q_s(x) = Q_s(\overline{x}_s) = 0$, so that a type 2 agent gets no gain.

Equation $(21.3.5a)$ can be expressed as

$$c = g(Q'_s(x)), \tag{21.4.1}$$

where g is a continuously and strictly decreasing function. By substituting the inverse of that function into equation $(21.3.5b)$, we obtain the expression

$$g^{-1}(c) = (1+\theta_j)\,g^{-1}(c_j) + \lambda_j, \tag{21.4.2}$$

where c is again the current consumption of a type 1 agent and c_j is his next period's consumption when next period's endowment realization is $\overline{y}_j$. The optimal consumption dynamics implied by an efficient contract are evidently governed by whether or not agents' participation constraints are binding. For any given endowment realization $\overline{y}_j$ next period, only one of the participation constraints in $(21.3.1c)$ and $(21.3.1d)$ can bind. Hence, there are three regions of interest for any given realization $\bar{y}_j$:

1. Neither participation constraint binds. When $\lambda_j = \theta_j = 0$, the consumption dynamics in $(21.4.2)$ satisfy

$$g^{-1}(c) = g^{-1}(c_j) \qquad \Longrightarrow \qquad c = c_j,$$

where $c = c_j$ follows from the fact that $g^{-1}(\cdot)$ is a strictly decreasing function. Hence, consumption is independent of the endowment and the agents are offered full insurance against endowment realizations so long as there are no binding participation constraints. The constant consumption allocation is determined by the "temporary relative Pareto weight" μ in equation $(21.3.3a)$.

2. The participation constraint of a type 1 person binds $(\lambda_j > 0)$, but $\theta_j = 0$. Thus, condition $(21.4.2)$ becomes

$$g^{-1}(c) = g^{-1}(c_j) + \lambda_j \quad \Longrightarrow \quad g^{-1}(c) > g^{-1}(c_j) \quad \Longrightarrow \quad c < c_j.$$

The planner raises the consumption of the type 1 agent in order to satisfy his participation constraint. The strictly positive Lagrange multiplier, $\lambda_j > 0$, implies that $(21.3.1c)$ holds with equality, $\chi_j = 0$. That is, the planner raises the welfare of a type 1 agent just enough to make her indifferent between choosing autarky and staying with the optimal insurance contract. In effect, the planner minimizes the change in last period's relative welfare distribution that is needed to induce the type 1 agent not to abandon the contract. The welfare of the type 1 agent is raised both through the mentioned higher consumption $c_j > c$ and through the expected higher future consumption. Recall our earlier finding that implies that the new higher consumption level will remain unchanged so long as there are no binding participation constraints. It follows that the contract for agent 1 displays *amnesia* when agent 1's participation constraint is binding, because the previously promised value x becomes irrelevant for the consumption allocated to agent 1 from now on.

3. The participation constraint of a type 2 person binds $(\theta_j > 0)$, but $\lambda_j = 0$. Thus, condition $(21.4.2)$ becomes

$$g^{-1}(c) = (1 + \theta_j) g^{-1}(c_j) \quad \Longrightarrow \quad g^{-1}(c) < g^{-1}(c_j) \quad \Longrightarrow \quad c > c_j,$$

where we have used the fact that $g^{-1}(\cdot)$ is a negative number. This situation is the mirror image of the previous case. When the participation constraint of the type 2 agent binds, the planner induces the agent to remain with the optimal contract by increasing her consumption $(1 - c_j) > (1 - c)$ but only by enough that she remains indifferent to the alternative of choosing autarky, $Q_j(\chi_j) = 0$. And once again, the change in the welfare distribution persists in the sense that the new consumption level will remain unchanged so long as there are no binding participation constraints. The amnesia property prevails again.

We can assemble these results to characterize the contract by arguing that when $c < \underline{c}_j$, the participation constraint of the type 1 agent binds, while if $c > \overline{c}_j$, the participation constraint of a type 2 agent binds. Thus, assume that $c < \underline{c}_j$. Then since it must be that $c_j \geq \underline{c}_j$, it follows that $c \leq \underline{c}_j$, so we must be in the region where the participation constraint of the type 1 agent binds, which in turn implies that $c_j = \underline{c}_j$. A symmetric argument that applies when the participation constraint of a type 2 argument applies, allowing us to summarize the consumption dynamics of an efficient contract as follows. Given the current consumption c of the type 1 agent, next period's consumption conditional on next period's endowment realization $\overline{y}_j$ satisfies

$$
c_j = \begin{cases} \underline{c}_j & \text{if } c < \underline{c}_j & \text{(p.c. of type 1 binds)}, \\ c & \text{if } c \in [\underline{c}_j, \overline{c}_j] & \text{(p.c. of neither type binds)}, \\ \overline{c}_j & \text{if } c > \overline{c}_j & \text{(p.c. of type 2 binds)}. \end{cases} \tag{21.4.3}
$$

21.4.2. Consumption intervals cannot contain each other

We will show that

$$
\overline{y}_k > \overline{y}_q \quad \Longrightarrow \quad \overline{c}_k > \overline{c}_q \text{ and } \underline{c}_k > \underline{c}_q. \tag{21.4.4}
$$

Hence, no consumption interval can contain another. Depending on parameter values, the consumption intervals can be either overlapping or disjoint.

As an intermediate step, it is useful to first verify that the following assertion is correct for any $k, q = 1, 2, \ldots, S$, and for any $x \in [0, \overline{x}_q]$:

$$
Q_k\big(x + u(\overline{y}_q) - u(\overline{y}_k)\big) = Q_q(x) + u(1 - \overline{y}_q) - u(1 - \overline{y}_k). \tag{21.4.5}
$$

After invoking functional equation $(21.3.1)$, the left side of $(21.4.5)$ is equal to

$$Q_k\big(x + u(\overline{y}_q) - u(\overline{y}_k)\big) = \max_{c,\,\{\chi_j\}_{j=1}^S} \Big\{u(1-c) - u(1-\overline{y}_k) + \beta \sum_{j=1}^S \Pi_j Q_j(\chi_j)\Big\}$$

subject to

$$u(c) - u(\overline{y}_k) + \beta \sum_{j=1}^S \Pi_j \chi_j \geq x + u(\overline{y}_q) - u(\overline{y}_k)$$

and $(21.3.1c) - (21.3.1e)$; and the right side of $(21.4.5)$ is equal to

$$Q_q(x) + u(1 - \overline{y}_q) - u(1 - \overline{y}_k)$$

$$= \max_{c,\,\{\chi_j\}_{j=1}^S} \Big\{u(1-c) - u(1 - \overline{y}_q) + \beta \sum_{j=1}^S \Pi_j Q_j(\chi_j)\Big\} + u(1 - \overline{y}_q) - u(1 - \overline{y}_k)$$

subject to

$$u(c) - u(\overline{y}_q) + \beta \sum_{j=1}^S \Pi_j \chi_j \geq x$$

and $(21.3.1c) - (21.3.1e)$. We can then verify $(21.4.5)$.[6] And after differentiating that expression with respect to x,

$$Q_k'\big(x + u(\overline{y}_q) - u(\overline{y}_k)\big) = Q_q'(x). \tag{21.4.6}$$

To show that $\overline{y}_k > \overline{y}_q$ implies $\overline{c}_k > \overline{c}_q$, set $x = \overline{x}_q$ in expression $(21.4.5)$,

$$Q_k\big(\overline{x}_q + u(\overline{y}_q) - u(\overline{y}_k)\big) = u(1 - \overline{y}_q) - u(1 - \overline{y}_k) > 0, \tag{21.4.7}$$

where we have used $Q_q(\overline{x}_q) = 0$. After also invoking $Q_k(\overline{x}_k) = 0$ and the fact that $Q_k(\cdot)$ is decreasing, it follows from $Q_k\big(\overline{x}_q + u(\overline{y}_q) - u(\overline{y}_k)\big) > 0$ that

$$\overline{x}_k > \overline{x}_q + u(\overline{y}_q) - u(\overline{y}_k).$$

---

[6] The two optimization problems on the left and the right sides of expression $(21.4.5)$ share the common objective of maximizing the expected utility of the type 2 agent, minus an identical constant. The optimization is subject to the same constraints, $u(c) - u(\overline{y}_q) + \beta \sum_{j=1}^S \Pi_j \chi_j \geq x$ and $(21.3.1c) - (21.3.1e)$. Hence, they are identical well-defined optimization problems. The observant reader should not be concerned with the fact that $Q_k(\cdot)$ on the left side of $(21.4.5)$ might be evaluated at a promised value outside of the range $[0, \overline{x}_k]$. This causes no problem because the optimization problem imposes no participation constraint in the current period, in contrast to the restrictions on future continuation values in $(21.3.1c)$ and $(21.3.1d)$.

So by the strict concavity of $Q_k(\cdot)$, we have

$$Q'_k(\overline{x}_k) < Q'_k\big(\overline{x}_q + u(\overline{y}_q) - u(\overline{y}_k)\big) = Q'_q(\overline{x}_q), \qquad (21.4.8)$$

where the equality is given by $(21.4.6)$. Finally, by using function $(21.4.1)$ and the present finding that $Q'_k(\overline{x}_k) < Q'_q(\overline{x}_q)$, we can verify our assertion that

$$\overline{c}_k = g(Q'_k(\overline{x}_k)) > g(Q'_q(\overline{x}_q)) = \overline{c}_q.$$

We leave it to the reader as an exercise to construct a symmetric argument to show that $\overline{y}_k > \overline{y}_q$ implies $\underline{c}_k > \underline{c}_q$.

21.4.3. Endowments are contained in the consumption intervals

We will show that

$$\overline{y}_s \in [\underline{c}_s, \overline{c}_s], \quad \forall s; \quad \text{and} \quad \overline{y}_1 = \underline{c}_1 \text{ and } \overline{y}_S = \overline{c}_S. \qquad (21.4.9)$$

First, we show that $\overline{y}_s \le \overline{c}_s$ for all s; and $\overline{y}_S = \overline{c}_S$. Let $x = \overline{x}_s$ in the functional equation $(21.3.1)$, then $c = \overline{c}_s$ and

$$u(1 - \overline{c}_s) - u(1 - \overline{y}_s) + \beta \sum_{j=1}^{S} \Pi_j Q_j(\chi_j) = 0 \qquad (21.4.10)$$

with $\{\chi_j\}_{j=1}^{S}$ being optimally chosen. Since $Q_j(\chi_j) \ge 0$, it follows immediately that

$$u(1 - \overline{c}_s) - u(1 - \overline{y}_s) \le 0 \quad \implies \quad \overline{y}_s \le \overline{c}_s.$$

To establish strict equality for $s = S$, we note that

$$Q'_j(\chi_j) \ge Q'_j(\overline{x}_j) \ge Q'_S(\overline{x}_S),$$

where the first weak inequality follows from the fact that all permissible $\chi_j \le \overline{x}_j$ and $Q_j(\cdot)$ is strictly concave, and the second weak inequality is given by $(21.4.8)$. In fact, we showed above that the second inequality holds strictly for $j < S$ and therefore, by the condition for optimality in $(21.3.5b)$,

$$Q'_S(\overline{x}_S) = (1 + \theta_j)Q'_j(\chi_j) \quad \text{with} \quad \theta_j > 0, \text{ for } j < S; \quad \text{and} \quad \theta_S = 0,$$

which imply $\chi_j = \overline{x}_j$ for all j. After also invoking the corresponding expression $(21.4.10)$ for $s = S$, we can complete the argument:

$$\beta \sum_{j=1}^{S} \Pi_j Q_j(\overline{x}_j) = 0 \quad \implies \quad u(1 - \overline{c}_S) - u(1 - \overline{y}_S) = 0 \quad \implies \quad \overline{y}_S = \overline{c}_S.$$

We leave it as an exercise for the reader to construct a symmetric argument showing that $\overline{y}_s \geq \underline{c}_s$ for all s; and $\overline{y}_1 = \underline{c}_1$.

21.4.4. All consumption intervals are nondegenerate (unless autarky is the only sustainable allocation)

Suppose that the consumption interval associated with endowment realization $\overline{y}_k$ is degenerate, i.e., $\underline{c}_k = \overline{c}_k = \overline{y}_k$. (The last inequality follows from section 21.4.3, where we established that the endowment is contained in the consumption interval.) Since the consumption interval is degenerate, it follows that the range of permissible continuation values associated with endowment realization $\overline{y}_k$ is also degenerate, i.e., $\chi_k \in [0, \overline{x}_k] = \{0\}$. Recall that χ_k is the number of utils awarded to the type 1 household over and above its autarkic welfare level, given endowment realization $\overline{y}_k$:

$$0 = \chi_k = u(\overline{y}_k) - u(\overline{y}_k) + \beta \sum_{j=1}^{S} \Pi_j \chi_j,$$

where we have invoked the degenerate consumption interval, $c = \overline{y}_k$, and where χ_j are optimally chosen subject to the constraints $\chi_j \geq 0$ for all $j \in \mathbf{S}$. It follows immediately that $\chi_j = 0$ for all $j \in \mathbf{S}$, given the current endowment realization $\overline{y}_k$.

Due to the degenerate range of continuation values associated with endowment realization $\overline{y}_k$, i.e., $\chi_k \in \{0\}$, it must be the case that the type 2 household also receives its autarkic welfare level, given endowment realization $\overline{y}_k$:[7]

$$0 = Q_k(\chi_k) = u(1 - \overline{y}_k) - u(1 - \overline{y}_k) + \beta \sum_{j=1}^{S} \Pi_j Q_j(\chi_j)$$

---

[7] Obviously, there would be a contradiction if the type 2 household were to receive $Q_k(\chi_k) > 0$. The reason is that then it would be possible to transfer current consumption from the type 2 household to the type 1 household without violating the type 2 household's participation constraint, and hence the consumption interval associated with endowment realization $\overline{y}_k$ could not be degenerate.

where we have invoked the degenerate consumption interval, and where continuation values $Q_j(\chi_j)$ are subject to the constraints $Q_j(\chi_j) \geq 0$ for all $j \in \mathbf{S}$. It follows immediately that $Q_j(\chi_j) = 0$ for all $j \in \mathbf{S}$, given the current endowment realization $\overline{y}_k$. Hence, given endowment realization $\overline{y}_k$, we have continuation values of the type 2 household satisfying $Q_j(\chi_j) = 0$, so it must be the case that the optimally chosen χ_j are set at their maximum permissible values, $\chi_j = \overline{x}_j$. Moreover, we know from above that the optimal values also satisfy $\chi_j = 0$, and therefore we can conclude that $\overline{x}_j = 0$ for all $j \in \mathbf{S}$.

We have shown that if one consumption interval is degenerate, then all consumption intervals must be degenerate, i.e., $\underline{c}_s = \overline{c}_s = \overline{y}_s$ for all $s \in \mathbf{S}$. This finding seems rather intuitive. A degenerate consumption interval associated with any endowment realization $\overline{y}_k$ implies that, given the realization of $\overline{y}_k$, none of the households has anything to gain from the optimal contract, neither from current transfers nor from future risk sharing. That can only happen if autarky is the only sustainable allocation.

21.5.   Pareto frontier and ex ante division of the gains

We have characterized the optimal consumption dynamics of any efficient contract. The consumption intervals $\{[\underline{c}_j, \overline{c}_j]\}_{j=1}^{S}$ and the updating rules in $(21.4.3)$ are identical for all efficient contracts. The *ex ante* division of gains from an efficient contract can be viewed as being determined by an implicit past consumption level, $c_\triangle \in [\underline{c}_1, \overline{c}_S]$: by $(21.4.9)$, this can also be written as $c_\triangle \in [\overline{y}_1, \overline{y}_S])$. A contract with an implicit past consumption level $c_\triangle = \underline{c}_1$ gives all of the surplus to the type 2 agent and none to the type 1 agent. This follows immediately from the updating rules in $(21.4.3)$ that prescribe a first-period consumption level equal to $\underline{c}_j$ if the endowment realization is $\overline{y}_j$. The corresponding promised value to the type 1 agent, conditional on endowment realization $\overline{y}_j$, is $\chi_j = 0$. Thus, the *ex ante* gain to the type 1 agent in expression $(21.2.2c)$ becomes

$$\triangle\Big|_{c_\triangle = \underline{c}_1} = \sum_{j=1}^{S} \Pi_j \chi_j \Big|_{c_\triangle = \underline{c}_1} = 0.$$

Similarly, we can show that a contract with an implicit consumption level $c_\triangle = \overline{c}_S$ gives all of the surplus to the type 1 agent and none to the type 2 agent. The updating rules in (21.4.3) will then prescribe a first-period consumption level equal to $\bar{c}_j$ if the endowment realization is $\bar{y}_j$ with a corresponding promised value of $\chi_j = \bar{x}_j$. We can compute the *ex ante* gain to the type 1 agent as

$$\triangle\Big|_{c_\triangle = \bar{c}_S} = \sum_{j=1}^{S} \Pi_j \bar{x}_j \equiv \triangle_{\max}.$$

For these two endpoints of the interval $c_\triangle \in [\underline{c}_1, \bar{c}_S]$, the *ex ante* gains attained by the type 2 agent in expression (21.2.2a) become

$$Q(\triangle)\Big|_{c_\triangle = \underline{c}_1} = Q(0) = \sum_{j=1}^{S} \Pi_j Q_j(0) = \triangle_{\max},$$

$$Q(\triangle)\Big|_{c_\triangle = \bar{c}_S} = Q(\triangle_{\max}) = \sum_{j=1}^{S} \Pi_j Q_j(\bar{x}_j) = 0,$$

where the equality $Q(0) = \triangle_{\max}$ follows from the symmetry of the environment with respect to the type 1 and type 2 agents' preferences and endowment processes.

21.6. Consumption distribution

21.6.1. Asymptotic distribution

The asymptotic consumption distribution depends sensitively on whether there exists a first-best sustainable allocation. We say that a sustainable allocation is *first best* if the participation constraint of neither agent ever binds. As we have seen, nonbinding participation constraints imply that consumption remains constant over time. Thus, a first-best sustainable allocation can exist only if the intersection of all the consumption intervals $\{[\underline{c}_j, \bar{c}_j]\}_{j=1}^{S}$ is nonempty. Define the following two critical numbers

$$\bar{c}_{\min} \equiv \min\{\bar{c}_j\}_{j=1}^{S} = \bar{c}_1, \tag{21.6.1a}$$

$$\underline{c}_{\max} \equiv \max\{\underline{c}_j\}_{j=1}^{S} = \underline{c}_S, \tag{21.6.1b}$$

where the two equalities are implied by (21.4.4). A necessary and sufficient condition for the existence of a first-best sustainable allocation is that $\bar{c}_{\min} \geq \underline{c}_{\max}$. Within a first-best sustainable allocation, there is complete risk sharing.

For high enough values of β, sufficient endowment risk, and enough curvature of $u(\cdot)$, there will exist a set of first-best sustainable allocations, i.e., $\bar{c}_{\min} \geq \underline{c}_{\max}$. If the *ex ante* division of the gains is then given by an implicit initial consumption level $c_{\triangle} \in [\underline{c}_{\max}, \bar{c}_{\min}]$, it follows by the updating rules in (21.4.3) that consumption remains unchanged forever, and therefore the asymptotic consumption distribution is degenerate.

But what happens if the *ex ante* division of gains is associated with an implicit initial consumption level outside of this range, or if there does not exist any first-best sustainable allocation ($\bar{c}_{\min} < \underline{c}_{\max}$)? To understand the convergence of consumption to an asymptotic distribution in general, we make the following observations. According to the updating rules in (21.4.3), any increase in the consumption of a type 1 person between two consecutive periods has consumption attaining the lower bound of some consumption interval. It follows that in periods of increasing consumption, the consumption level is bounded above by $\underline{c}_{\max}$ ($= \underline{c}_S$) and hence increases can occur only if the initial consumption level is less than $\underline{c}_{\max}$. Similarly, any decrease in consumption between two consecutive periods has consumption attain the upper bound of some consumption interval. It follows that in periods of decreasing consumption, consumption is bounded below by $\bar{c}_{\min}$ ($= \bar{c}_1$) and hence decreases can only occur if initial consumption is higher than $\bar{c}_{\min}$. Given a current consumption level c, we can then summarize the permissible range for next-period consumption c' as follows:

$$\text{if } \ c \leq \underline{c}_{\max} \ \text{ then } \quad c' \in [\min\{c, \bar{c}_{\min}\}, \ \underline{c}_{\max}], \qquad (21.6.2a)$$

$$\text{if } \ c \geq \bar{c}_{\min} \ \text{ then } \quad c' \in [\bar{c}_{\min}, \ \max\{c, \underline{c}_{\max}\}]. \qquad (21.6.2b)$$

21.6.2. Temporary imperfect risk sharing

We now return to the case that there exist first-best sustainable allocations, $\bar{c}_{\min} \geq \underline{c}_{\max}$, but we let the *ex ante* division of gains be given by an implicit initial consumption level $c_\triangle \notin [\underline{c}_{\max}, \bar{c}_{\min}]$. The permissible range for next-period consumption, as given in (21.6.2), and the support of the asymptotic consumption becomes

$$\text{if } c \leq \underline{c}_{\max} \quad \text{then} \quad c' \in [c, \underline{c}_{\max}] \quad \text{and} \quad \lim_{t \to \infty} c_t = \underline{c}_{\max} = \underline{c}_S, \quad (21.6.3a)$$

$$\text{if } c \geq \bar{c}_{\min} \quad \text{then} \quad c' \in [\bar{c}_{\min}, c] \quad \text{and} \quad \lim_{t \to \infty} c_t = \bar{c}_{\min} = \bar{c}_1. \quad (21.6.3b)$$

We have monotone convergence in (21.6.3a) for two reasons. First, consumption is bounded from above by $\underline{c}_{\max}$. Second, consumption cannot decrease when $c \leq \bar{c}_{\min}$ and by assumption $\bar{c}_{\min} \geq \underline{c}_{\max}$, so consumption cannot decrease when $c \leq \underline{c}_{\max}$. It follows immediately that $\underline{c}_{\max}$ is an absorbing point that is attained as soon as the endowment $\bar{y}_S$ is realized with its consumption level $\underline{c}_S = \underline{c}_{\max}$. Similarly, the explanation for monotone convergence in (21.6.3b) goes as follows. First, consumption is bounded from below by $\bar{c}_{\min}$. Second, consumption cannot increase when $c \geq \underline{c}_{\max}$ and by assumption $\bar{c}_{\min} \geq \underline{c}_{\max}$, so consumption cannot increase when $c \geq \bar{c}_{\min}$. It follows immediately that $\bar{c}_{\min}$ is an absorbing point that is attained as soon as the endowment $\bar{y}_1$ is realized with its consumption level $\bar{c}_1 = \bar{c}_{\min}$.

These convergence results assert that imperfect risk sharing is at most temporary if the set of first-best sustainable allocations is nonempty. Notice that when an economy begins with an implicit initial consumption outside of the interval of sustainable constant consumption levels, the subsequent monotone convergence to the closest endpoint of that interval is reminiscent of our earlier analysis of the moneylender and the villagers with one-sided lack of commitment in section 20.3. In the current setting, the agent who is relatively disadvantaged under the initial welfare assignment will see her consumption weakly increase over time until she has experienced the endowment realization that is most favorable to her. From there on, the consumption level remains constant forever, and the participation constraints will never bind again.

21.6.3. Permanent imperfect risk sharing

If the set of first-best sustainable allocations is empty ($\bar{c}_{min} < \underline{c}_{max}$), it breaks the monotone convergence to a constant consumption level. The updating rules in (21.4.3) imply that the permissible range for next-period consumption in (21.6.2) will ultimately shrink to $[\bar{c}_{min}, \underline{c}_{max}]$, regardless of the initial welfare assignment. If the implicit initial consumption lies outside of that set, consumption is bound to converge to it, again because of the monotonicity of consumption when $c \leq \bar{c}_{min}$ or $c \geq \underline{c}_{max}$. And as soon as there is a binding participation constraint with an associated consumption level that falls inside of the interval $[\bar{c}_{min}, \underline{c}_{max}]$, the updating rules in (21.4.3) will never take us outside of this interval again. Thereafter, the only observed consumption levels belong to the ergodic set

$$\left\{ [\bar{c}_{min}, \underline{c}_{max}] \bigcap \{\underline{c}_j, \bar{c}_j\}_{j=1}^{S} \right\}, \tag{21.6.4}$$

with a unique asymptotic distribution. Within this invariant set, the participation constraints of both agents occasionally bind, reflecting imperfections in risk sharing.

If autarky is the only sustainable allocation, then each consumption interval is degenerate with $\underline{c}_j = \bar{c}_j = \bar{y}_j$ for all $j \in \mathbf{S}$, as discussed in section 21.4.4. Hence, the ergodic consumption set in (21.6.4) is then trivially equal to the set of endowment levels, $\{\bar{y}_j\}_{j=1}^{S}$.

21.7. Alternative recursive formulation

Kocherlakota (1996b) used an alternative recursive formulation of the contract design problem, one that more closely resembles our treatment of the moneylender villager economy of section 20.3. After replacing the argument in the function of (21.2.2a) by the expected utility of the type 1 agent, Kocherlakota writes the Bellman equation as

$$P(v) = \max_{\{c_s, w_s\}_{s=1}^{S}} \sum_{s=1}^{S} \Pi_s \{u(1 - c_s) + \beta P(w_s)\} \tag{21.7.1a}$$

subject to

$$\sum_{s=1}^{S} \Pi_s [u(c_s) + \beta w_s] \geq v, \tag{21.7.1b}$$

$$u(c_s) + \beta w_s \geq u(\overline{y}_s) + \beta v_{\mathrm{aut}}, \qquad\qquad s = 1, \ldots, S; \qquad (21.7.1c)$$

$$u(1 - c_s) + \beta P(w_s) \geq u(1 - \overline{y}_s) + \beta v_{\mathrm{aut}}, \quad s = 1, \ldots, S; \qquad (21.7.1d)$$

$$c_s \in [0, 1], \qquad\qquad\qquad\qquad (21.7.1e)$$

$$w_s \in [v_{\mathrm{aut}}, v_{\mathrm{max}}]. \qquad\qquad\qquad\qquad (21.7.1f)$$

Here the planner comes into a period with a state variable v that is a promised expected utility to the type 1 agent. Before observing the current endowment realization, the planner chooses a consumption level c_s and a continuation value w_s for each possible realization of the current endowment. This state-contingent portfolio $\{c_s, w_s\}_{s=1}^{S}$ must deliver at least the promised value v to the type 1 agent, as stated in $(21.7.1b)$, and must also be consistent with the agents' participation constraints in $(21.7.1c)$ and $(21.7.1d)$.

Notice the difference in timing with our presentation, which we have based on Thomas and Worrall's (1988) analysis. Kocherlakota's planner leaves the current period with only one continuation value w_s and postpones the question of how to deliver that promised value across future states until the beginning of next period but *before* observing next period's endowment. In contrast, in our setting, in the current period the planner chooses a state-contingent set of continuation values for the next period, $\{\chi_j\}_{j=1}^{S}$, where χ_j is the number of utils that the type 1 agent's expected utility should exceed her autarky level in the next period if that period's endowment is $\overline{y}_j$. We can evidently express Kocherlakota's one state variable in terms of our state vector,

$$w_s = \sum_{j=1}^{S} \Pi_j \left[\chi_j + u(\overline{y}_j) + \beta v_{\mathrm{aut}} \right] = v_{\mathrm{aut}} + \sum_{j=1}^{S} \Pi_j \chi_j,$$

where v_{aut} is the *ex ante* welfare level in autarky as given by $(20.3.2)$. Similarly, Kocherlakota's upper bound on permissible values of next period's continuation value in $(21.7.1f)$ is related to our upper bounds $\{\overline{x}_j\}_{j=1}^{S}$,

$$v_{\mathrm{max}} = v_{\mathrm{aut}} + \sum_{j=1}^{S} \Pi_j \overline{x}_j = v_{\mathrm{aut}} + \triangle_{\mathrm{max}}.$$

21.8. Pareto frontier revisited

Given our earlier characterization of the optimal solution, we can map Kocherlakota's promised value v into an implicit promised consumption level $c_\triangle \in [\underline{c}_1, \bar{c}_S] = [\underline{y}_1, \bar{y}_S]$. Let that mapping be encoded in the function $v(c_\triangle)$. Hence, given a promised a value $v(c_\triangle)$, the optimal consumption dynamics in section 21.4.1 instruct us to set Kocherlakota's choice variables, $\{c_s, w_s\}_{s=1}^{S}$, as follows:

$$c_s = c_\triangle + \max\{0, \underline{c}_s - c_\triangle\} - \max\{0, c_\triangle - \bar{c}_s\}, \qquad (21.8.1a)$$

$$w_s = v(c_s). \qquad (21.8.1b)$$

For a given value of $c_\triangle$, we define the following three sets that partition the set $\mathbf{S}$ of endowment realizations:

$$\mathbf{S}^o(c_\triangle) \equiv \Big\{ j \in \mathbf{S} : c_\triangle \in (\underline{c}_j, \bar{c}_j) \Big\}, \qquad (21.8.2a)$$

$$\mathbf{S}^-(c_\triangle) \equiv \Big\{ j \in \mathbf{S} : c_\triangle \geq \bar{c}_j \Big\}, \qquad (21.8.2b)$$

$$\mathbf{S}^+(c_\triangle) \equiv \Big\{ j \in \mathbf{S} : c_\triangle \leq \underline{c}_j \Big\}. \qquad (21.8.2c)$$

According to our characterization of consumption intervals in (21.4.4), these three sets are mutually exclusive and their union is equal to $\mathbf{S}$. $\mathbf{S}^o(c_\triangle)$ is the set of states, i.e., endowment realizations, for which the optimal consumption level is $c_s = c_\triangle$. But if the endowment realization falls outside of $\mathbf{S}^o(c_\triangle)$, the optimal consumption c_s is determined by either the upper or lower bound of the consumption interval associated with that endowment realization. In particular, for $s \in \mathbf{S}^-(c_\triangle)$, consumption should drop to the upper bound of the consumption interval, $c_s = \bar{c}_s$; and for $s \in \mathbf{S}^+(c_\triangle)$, consumption should increase to the lower bound of the consumption interval, $c_s = \underline{c}_s$.

The continuation value $v(c_\triangle)$ can then be expressed as

$$v(c_\triangle) = \sum_{s \in \mathbf{S}^o(c_\triangle)} \Pi_s \Big[u(c_\triangle) + \beta v(c_\triangle) \Big] + \sum_{s \in \mathbf{S}^-(c_\triangle)} \Pi_s \Big[u(\bar{c}_s) + \beta v(\bar{c}_s) \Big]$$

$$+ \sum_{s \in \mathbf{S}^+(c_\triangle)} \Pi_s \Big[u(\underline{c}_s) + \beta v(\underline{c}_s) \Big] \qquad (21.8.3)$$

which can be rewritten as

$$v(c_\triangle) = \Big(1 - \beta \sum_{s \in \mathbf{S}^o(c_\triangle)} \Pi_s \Big)^{-1} \Big\{ u(c_\triangle) \sum_{s \in \mathbf{S}^o(c_\triangle)} \Pi_s$$

$$+ \sum_{s \in \mathbf{S}^-(c_\triangle)} \Pi_s \Big[u(\bar{c}_s) + \beta v(\bar{c}_s) \Big] + \sum_{s \in \mathbf{S}^+(c_\triangle)} \Pi_s \Big[u(\underline{c}_s) + \beta v(\underline{c}_s) \Big] \bigg\}. \quad (21.8.4)$$

Similarly, the promised value to the type 2 household can be expressed as

$$P(v(c_\triangle)) = \bigg(1 - \beta \sum_{s \in \mathbf{S}^o(c_\triangle)} \Pi_s \bigg)^{-1} \bigg\{ u(1 - c_\triangle) \sum_{s \in \mathbf{S}^o(c_\triangle)} \Pi_s$$

$$+ \sum_{s \in \mathbf{S}^-(c_\triangle)} \Pi_s \Big[u(1 - \bar{c}_s) + \beta P(v(\bar{c}_s)) \Big]$$

$$+ \sum_{s \in \mathbf{S}^+(c_\triangle)} \Pi_s \Big[u(1 - \underline{c}_s) + \beta P(v(\underline{c}_s)) \Big] \bigg\}. \quad (21.8.5)$$

21.8.1. Values are continuous in implicit consumption

Both $v(c_\triangle)$ and $P(v(c_\triangle))$ are continuous in the implicit consumption level $c_\triangle$. From (21.8.4) and (21.8.5) this is trivially true when variations in $c_\triangle$ do not change the partition of states given by the sets $\mathbf{S}^o(\cdot)$, $\mathbf{S}^-(\cdot)$ and $\mathbf{S}^+(\cdot)$. It can also be shown to be true when variations in $c_\triangle$ do involve changes in the partition of states. As an illustration, let us compute the limiting values of $v(c_\triangle)$ when $c_\triangle$ approaches $\bar{c}_k$ from below and from above, respectively, where we recall that $\bar{c}_k$ is the upper bound of the consumption interval associated with endowment $\bar{y}_k$.

We can choose a sufficiently small $\epsilon > 0$ such that

$$\{\underline{c}_s, \bar{c}_s\}_{s=1}^S \bigcap [\bar{c}_k - \epsilon, \bar{c}_k + \epsilon] = \bar{c}_k.$$

In particular, the findings in (21.4.4) ensure that we can choose a sufficiently small ϵ so that this intersection contains no upper bounds on consumption intervals other than $\bar{c}_k$. Similarly, ϵ can be chosen sufficiently small that the intersection does not contain any lower bound on consumption intervals *unless* there exists a consumption interval with a lower bound that is exactly equal to $\bar{c}_k$, i.e., if for some $j \geq 1$, $\underline{c}_{k+j} = \bar{c}_k$. We will have to keep this possibility in mind as we proceed in our characterization of the sets $\mathbf{S}^o(\cdot)$, $\mathbf{S}^-(\cdot)$ and $\mathbf{S}^+(\cdot)$.

All three sets are constant for an implicit consumption $c_\triangle \in [\bar{c}_k - \epsilon, \bar{c}_k)$ with $\max\{\mathbf{S}^-(c_\triangle)\} = k - 1$. For an implicit consumption $c_\triangle \in [\bar{c}_k, \bar{c}_k + \epsilon]$, the set $\mathbf{S}^-(c_\triangle)$ is constant with $\max\{\mathbf{S}^-(c_\triangle)\} = k$, while the configuration of the other two sets depends on which one of the following two possible cases applies.

Case a: $\bar{c}_k \ne \underline{c}_s$ for all $s \in \mathbf{S}$. Here it follows that the set $\mathbf{S}^+(c_\triangle)$ is constant for any implicit consumption $c_\triangle \in [\bar{c}_k - \epsilon, \bar{c}_k + \epsilon]$. Using $(21.8.3)$, the limiting values of $v(c_\triangle)$ when $c_\triangle$ approaches $\bar{c}_k$ from below and from above, respectively, are then equal to

$$
\lim_{c_\triangle \uparrow \bar{c}_k} v(c_\triangle) = \sum_{s \in \mathbf{S}^o(\bar{c}_k - \epsilon)} \Pi_s \Big[u(\bar{c}_k) + \beta v(\bar{c}_k) \Big] + \sum_{s=1}^{k-1} \Pi_s \Big[u(\bar{c}_s) + \beta v(\bar{c}_s) \Big]
$$

$$
+ \sum_{s \in \mathbf{S}^+(\bar{c}_k - \epsilon)} \Pi_s \Big[u(\underline{c}_s) + \beta v(\underline{c}_s) \Big]
$$

$$
= \sum_{s \in \mathbf{S}^o(\bar{c}_k + \epsilon)} \Pi_s \Big[u(\bar{c}_k) + \beta v(\bar{c}_k) \Big] + \sum_{s=1}^{k} \Pi_s \Big[u(\bar{c}_s) + \beta v(\bar{c}_s) \Big]
$$

$$
+ \sum_{s \in \mathbf{S}^+(\bar{c}_k + \epsilon)} \Pi_s \Big[u(\underline{c}_s) + \beta v(\underline{c}_s) \Big] = \lim_{c_\triangle \downarrow \bar{c}_k} v(c_\triangle). \tag{21.8.6}
$$

Case b: $\bar{c}_k = \underline{c}_{k+j}$ for some $j \ge 1$. Here it follows that the set $\mathbf{S}^+(c_\triangle)$ is constant with $\min\{\mathbf{S}^+(c_\triangle)\} = k + j$ for any implicit consumption $c_\triangle \in [\bar{c}_k - \epsilon, \bar{c}_k]$; and $\mathbf{S}^+(c_\triangle)$ is constant with $\min\{\mathbf{S}^+(c_\triangle)\} = k + j + 1$ for any implicit consumption $c_\triangle \in (\bar{c}_k, \bar{c}_k + \epsilon]$. Using $(21.8.3)$ and invoking the fact that $\underline{c}_{k+j} = \bar{c}_k$, the limiting values of $v(c_\triangle)$ when $c_\triangle$ approaches $\bar{c}_k$ from below and from above, respectively, are then equal to

$$
\lim_{c_\triangle \uparrow \bar{c}_k} v(c_\triangle) = \sum_{s=k}^{k+j-1} \Pi_s \Big[u(\bar{c}_k) + \beta v(\bar{c}_k) \Big] + \sum_{s=1}^{k-1} \Pi_s \Big[u(\bar{c}_s) + \beta v(\bar{c}_s) \Big]
$$

$$
+ \sum_{s=k+j}^{S} \Pi_s \Big[u(\underline{c}_s) + \beta v(\underline{c}_s) \Big]
$$

$$
= \sum_{s=k+1}^{k+j} \Pi_s \Big[u(\bar{c}_k) + \beta v(\bar{c}_k) \Big] + \sum_{s=1}^{k} \Pi_s \Big[u(\bar{c}_s) + \beta v(\bar{c}_s) \Big]
$$

$$
+ \sum_{s=k+j+1}^{S} \Pi_s \Big[u(\underline{c}_s) + \beta v(\underline{c}_s) \Big] = \lim_{c_\triangle \downarrow \bar{c}_k} v(c_\triangle). \tag{21.8.7}
$$

We have shown that $v(c_\triangle)$ is continuous at the upper bound of any consumption interval even though the partition of states changes at such a point. Similarly, we can show that $v(c_\triangle)$ is continuous at the lower bound of any consumption interval. And in the same manner, we can also establish that $P(v(c_\triangle))$ is continuous in the implicit consumption $c_\triangle$.

21.8.2. Differentiability of the Pareto frontier

Consider an implicit consumption level $c_\triangle \in [\overline{y}_1, \overline{y}_S]$ that falls strictly inside at least one consumption interval. We can then use expressions (21.8.4) and (21.8.5) to compute the derivative of the Pareto frontier at $v(c_\triangle)$ by differentiating with respect to $c_\triangle$:

$$ P'(v(c_\triangle)) = \frac{\dfrac{dP(v(c_\triangle))}{dc_\triangle}}{\dfrac{dv(c_\triangle)}{dc_\triangle}} = -\frac{u'(1 - c_\triangle)}{u'(c_\triangle)}. \tag{21.8.8} $$

It can be verified that (21.8.8) is the derivative of the Pareto frontier so long as the set $\mathbf{S}^o(c_\triangle)$ remains nonempty. That is, changes in the set $\mathbf{S}^o(c_\triangle)$ induced by varying $c_\triangle$ do not affect the expression for the derivative in (21.8.8). This follows from the fact that the derivatives are the same to the left and to the right of an implicit consumption level where the set $\mathbf{S}^o(c_\triangle)$ changes, and the fact that $v(c_\triangle)$ and $P(v(c_\triangle))$ are continuous in the implicit consumption level, as shown in section 21.8.1. It can also be verified that the derivative in (21.8.8) exists in the knife-edged case that occurs when $\mathbf{S}^o(c_\triangle)$ becomes empty at a single point because two adjacent consumption intervals share only one point, i.e., when $\overline{c}_k = \underline{c}_{k+1}$, which implies that $\mathbf{S}^o(\overline{c}_k) = \mathbf{S}^o(\underline{c}_{k+1}) = \emptyset$.

The Pareto frontier becomes nondifferentiable when two adjacent consumption intervals are disjoint. Consider such a situation where an implicit consumption level $c_\triangle \in [\overline{y}_1, \overline{y}_S]$ does not fall inside any consumption interval, which implies that the set $\mathbf{S}^o(c_\triangle)$ is empty. Let $\overline{y}_k$ and $\overline{y}_{k+1}$ be the endowment realizations associated with the consumption interval to the left and to the right of $c_\triangle$, respectively. That is,

$$ \overline{c}_k < c_\triangle < \underline{c}_{k+1}. $$

According to (21.8.3), the continuation value for any implicit consumption level $c \in [\bar{c}_k, \underline{c}_{k+1}]$ is then constant and equal to

$$\hat{v} = \sum_{s \in \mathbf{S}^-(c_\triangle)} \Pi_s \Big[u(\bar{c}_s) + \beta v(\bar{c}_s) \Big] + \sum_{s \in \mathbf{S}^+(c_\triangle)} \Pi_s \Big[u(\underline{c}_s) + \beta v(\underline{c}_s) \Big]. \qquad (21.8.9)$$

By using expression (21.8.8), we can compute the derivative of the Pareto frontier on the left side and the right side of $\hat{v}$,

$$\lim_{v \uparrow \hat{v}} P'(v) = \lim_{c \uparrow \bar{c}_k} \frac{\dfrac{dP(v(c))}{dc}}{\dfrac{dv(c)}{dc}} = -\frac{u'(1 - \bar{c}_k)}{u'(\bar{c}_k)},$$

$$\lim_{v \downarrow \hat{v}} P'(v) = \lim_{c \downarrow \underline{c}_{k+1}} \frac{\dfrac{dP(v(c))}{dc}}{\dfrac{dv(c)}{dc}} = -\frac{u'(1 - \underline{c}_{k+1})}{u'(\underline{c}_{k+1})}.$$

Since $\bar{c}_k < \underline{c}_{k+1}$, it follows that

$$\lim_{v \uparrow \hat{v}} P'(v) > \lim_{v \downarrow \hat{v}} P'(v)$$

and hence, the Pareto frontier is not differentiable at $\hat{v}$.[8]

---

[8] Kocherlakota (1996b) prematurely assumed that Thomas and Worrall's (1988) demonstration of the differentiability of the Pareto frontier $Q_s(\cdot)$ would imply that his conceptually different frontier $P(\cdot)$ would be differentiable. Koeppl (2003) uses the approach of Benveniste and Scheinkman (1979) to establish a sufficient condition for differentiability of the Pareto frontier $P(v)$. For a given value of v, the sufficient condition is that there exists at least one realization of the endowment such that the participation constraints are not binding for any household in that state, i.e., our set $\mathbf{S}^o(c_\triangle)$ should be nonempty for the implicit consumption level $c_\triangle$ associated with that particular value of v. That condition is sufficient but not necessary, since we have seen above that $P(v)$ is also differentiable at a knife-edged case with $c_\triangle = \bar{c}_k = \underline{c}_{k+1}$, even though the set $\mathbf{S}^o(c_\triangle)$ would then be empty.

21.9. Continuation values à la Kocherlakota

21.9.1. *Asymptotic distribution is nondegenerate for imperfect risk sharing (except when $S = 2$)*

Here we assume that there exist sustainable allocations other than autarky but that first-best outcomes are not attainable, i.e., there exist sustainable allocations with imperfect risk sharing. Kocherlakota (1996b, Proposition 4.2) states that the continuation values will then converge to a unique nondegenerate distribution. Here we will verify that the claim of a nondegenerate asymptotic distribution is correct except for when there are only two states ($S = 2$).

The assumption that the distribution of y_t is identical to that of $1 - y_t$ means that

$$\Pi_j = \Pi_{S+1-j}, \qquad\qquad (21.9.1a)$$

$$\overline{y}_j = 1 - \overline{y}_{S+1-j}, \qquad\qquad (21.9.1b)$$

for all $j \in \mathbf{S}$. The symmetric environment bestows symmetry on the consumption intervals of section 21.4

$$\overline{c}_j = 1 - \underline{c}_{S+1-j}, \qquad\qquad (21.9.1c)$$

for all $j \in \mathbf{S}$, and symmetry on the continuation values of the type 1 and type 2 household

$$v(c_\triangle) = P(v(1 - c_\triangle)). \qquad\qquad (21.9.1d)$$

As discussed in section 21.6.1, the condition for the nonexistence of first-best sustainable allocations is that $\overline{c}_{\min} < \underline{c}_{\max}$, which by $(21.6.1)$ is the same as

$$\overline{c}_1 < \underline{c}_S \qquad \Longrightarrow \qquad \underline{c}_S > 0.5 \qquad\qquad (21.9.2)$$

where the implication follows from using $\overline{c}_1 = 1 - \underline{c}_S$ as given by $(21.9.1c)$. It is quite intuitive that the consumption interval $[\underline{c}_S, \overline{c}_S]$ associated with the highest endowment realization $\overline{y}_S$ cannot contain the average value of the stochastic endowment, $\sum_{i=1}^{S} \Pi_i \overline{y}_i = 0.5$. Otherwise, there would certainly exist first-best sustainable allocations, a contradiction.

To prove the existence of a nondegenerate asymptotic distribution of continuation values, it is sufficient to show that the continuation value of an agent experiencing the highest endowment, say, the type 1 household, exceeds the continuation value of the other agent who is then experiencing the lowest endowment, say, the type 2 household. Given her current realization of the highest endowment $\overline{y}_S$, the type 1 household is awarded the highest consumption level $\underline{c}_S$ ($=\underline{c}_{\max}$) in the ergodic consumption set of (21.6.4). Conditional on next period's endowment realization $\overline{y}_i$, the type 1 household's consumption $\hat{c}_i$ in the next period is determined by (21.8.1a), where $c_\triangle = \underline{c}_S$. From (21.4.4) we know that $\underline{c}_S \geq \underline{c}_i$ for all $i \in \mathbf{S}$, so next period's consumption of the type 1 household as determined by (21.8.1a) can be written as

$$\hat{c}_i = \min\{\overline{c}_i, \underline{c}_S\}. \tag{21.9.3}$$

Given the vector $\{\hat{c}_i\}_{i=1}^S$ for next period's consumption, we can use (21.8.3) to compute the type 1 household's outgoing continuation value in the current period,

$$v(\underline{c}_S) = \sum_{i=1}^S \Pi_i \Big[u(\hat{c}_i) + \beta v(\hat{c}_i) \Big].$$

Next, we are interested in computing the difference between the continuation values of the type 1 and the type 2 households,

$$v(\underline{c}_S) - P(v(\underline{c}_S))) = \sum_{i=1}^S \Pi_i \Big[u(\hat{c}_i) + \beta v(\hat{c}_i) - u(1 - \hat{c}_i) - \beta P(v(\hat{c}_i)) \Big]$$

$$= \sum_{i=1}^S \Pi_i \Big[u(\hat{c}_i) + \beta v(\hat{c}_i) - u(1 - \hat{c}_{S+1-i}) - \beta P(v(\hat{c}_{S+1-i})) \Big]$$

$$= \sum_{i=1}^S \Pi_i \Big[u(\hat{c}_i) + \beta v(\hat{c}_i) - u(1 - \hat{c}_{S+1-i}) - \beta v(1 - \hat{c}_{S+1-i}) \Big], \tag{21.9.4}$$

where the second equality emerges from the symmetric probabilities in (21.9.1a), and the third equality follows from (21.9.1d). To establish that the difference in (21.9.4) is strictly positive, it is sufficient to show that

$$\hat{c}_i \geq 1 - \hat{c}_{S+1-i},$$

or, by using (21.9.3),

$$\min\{\overline{c}_i, \underline{c}_S\} \geq 1 - \min\{\overline{c}_{S+1-i}, \underline{c}_S\} \tag{21.9.5}$$

for all $i \in \mathbf{S}$, and at least one of them holds with strict inequality. The proof proceeds by considering four possible cases for each $i \in \mathbf{S}$.

Case a: $\bar{c}_i \leq \underline{c}_S$ and $\bar{c}_{S+1-i} \leq \underline{c}_S$. According to $(21.9.1c)$ $\bar{c}_{S+1-i} = 1 - \underline{c}_i$, so inequality $(21.9.5)$ can then be written as

$$\bar{c}_i \geq 1 - \bar{c}_{S+1-i} = \underline{c}_i \text{ which is true since } \bar{c}_i > \underline{c}_i \quad \text{for all} \quad i \in \mathbf{S},$$

as established in section 21.4.4.

Case b: $\bar{c}_i \leq \underline{c}_S$ and $\bar{c}_{S+1-i} > \underline{c}_S$. According to $(21.9.1c)$ $\bar{c}_i = 1 - \underline{c}_{S+1-i}$, so inequality $(21.9.5)$ can then be written as

$$\bar{c}_i = 1 - \underline{c}_{S+1-i} \geq 1 - \underline{c}_S \text{ which is true since } \underline{c}_{S+1-i} < \underline{c}_S \quad \text{for all} \quad i \neq 1,$$

as established in section 21.4.2.

Case c: $\bar{c}_i > \underline{c}_S$ and $\bar{c}_{S+1-i} \leq \underline{c}_S$. According to $(21.9.1c)$ $\bar{c}_{S+1-i} = 1 - \underline{c}_i$, so inequality $(21.9.5)$ can then be written as

$$\underline{c}_S \geq 1 - \bar{c}_{S+1-i} = \underline{c}_i \text{ which is true since } \underline{c}_S > \underline{c}_i \quad \text{for all} \quad i \neq S,$$

as established in section 21.4.2.

Case d: $\bar{c}_i > \underline{c}_S$ and $\bar{c}_{S+1-i} > \underline{c}_S$. The inequality $(21.9.5)$ can then be written as

$$\underline{c}_S \geq 1 - \underline{c}_S \text{ which is true since } \underline{c}_S > 0.5$$

as established in $(21.9.2)$.

We can conclude that the inequality $(21.9.5)$ holds with *strict* inequality with only two exceptions: (1) when $i = 1$ and case b applies; and (2) when $i = S$ and case c applies. It follows that the difference in $(21.9.4)$ is definitely strictly positive if there are more than two states, and hence the asymptotic distribution of continuation values is nondegenerate. But what about when there are only two states ($S = 2$)? Since $\bar{c}_1 < \underline{c}_S$ by $(21.9.2)$ and $\bar{c}_S > \underline{c}_S$, it follows that case b applies when $i = 1$ and case c applies when $i = 2 = S$. Therefore, the difference in $(21.9.4)$ is zero, and thus the continuation value of an agent experiencing the highest endowment is equal to that of the other agent who is then experiencing the lowest endowment. Since there are no other continuation values in an economy with only two possible endowment realizations, it follows that the asymptotic distribution of continuation values is degenerate when there are only two states ($S = 2$).

A two-state example in section 21.10 illustrates our findings. The intuition for the degenerate asymptotic distribution of continuation values is straightforward. On the one hand, the planner would like to vary continuation values and thereby avoid large changes in current consumption that would otherwise be needed to satisfy binding participation constraints. But, on the other hand, different continuation values presuppose that there exist "intermediate" states in which a higher continuation value can be awarded. In our two-state example, the participation constraint of either one or the other type of agent always binds, and the asymptotic distribution is degenerate with only one continuation value.

21.9.2. *Continuation values do not always respond to binding participation constraints*

Evidently, continuation values will eventually not respond to binding participation constraints in a two-state economy, since we have just shown that the asymptotic distribution is degenerate with only one continuation value. But the outcome that continuation values might not respond to binding participation constraints occurs even with more states when endowments are i.i.d. In fact, it is present whenever the consumption intervals of two adjacent endowment realizations, $\overline{y}_k$ and $\overline{y}_{k+1}$, do not overlap, i.e., when $\overline{c}_k < \underline{c}_{k+1}$. Here is how the argument goes.

Since $\overline{c}_k < \underline{c}_{k+1}$ it follows from (21.6.4) that both $\overline{c}_k$ and $\underline{c}_{k+1}$ belong to the ergodic set of consumption. Moreover, (21.4.4) implies that $\mathbf{S}^o(\overline{c}_k) = \mathbf{S}^o(\underline{c}_{k+1}) = \emptyset$, where $\mathbf{S}^o(\cdot)$ is defined in (21.8.2a). Using expression (21.8.3), we can compute a common continuation value $v(\overline{c}_k) = v(\underline{c}_{k+1}) = \hat{v}$, where $\hat{v}$ is given by (21.8.9) when that expression is evaluated for any $c_\triangle \in [\overline{c}_k, \underline{c}_{k+1}]$. Given this identical continuation value, it follows that there are situations where households' continuation values will not respond to binding participation constraints.

As an example, let the current consumption and continuation value of the type 1 household be $\overline{c}_k$ and $v(\overline{c}_k) = \hat{v}$, and suppose that the household next period realizes the endowment $\overline{y}_{k+1}$. It follows that the participation constraint of the type 1 household is binding and that the optimal solution in (21.8.1) is to award the household a consumption level $\underline{c}_{k+1}$ and continuation value $v(\underline{c}_{k+1})$. That is, the household is induced not to defect into autarky by increasing its consumption, $\underline{c}_{k+1} > \overline{c}_k$, but its continuation value is kept unchanged, $v(\underline{c}_{k+1}) = \hat{v}$. Suppose next that the type 1 household experiences $\overline{y}_k$ in the following period. This time it will be the participation constraint of the type 2 households that binds and the optimal solution in (21.8.1) prescribes that the type 1 household is awarded consumption $\overline{c}_k$ and continuation value $v(\overline{c}_k) = \hat{v}$. Hence, only consumption levels but not continuation values are adjusted in these two realizations with alternating binding participation constraints.

We use a three-state example in section 21.11 to elaborate on the point that even though an incoming continuation value lies in the interior of the range of permissible continuation values in (21.7.1f), a binding participation constraint still might not trigger a change in the outgoing continuation value because there may not exist any efficient way to deliver a changed continuation value. Continuation values that do not respond to binding participation constraints are a manifestation of the possibility that the Pareto frontier $P(\cdot)$ need not be differentiable everywhere on the interval $[v_{\mathrm{aut}}, v_{\mathrm{max}}]$, as shown in section 21.8.2.

21.10. A two-state example: amnesia overwhelms memory

In this example and the three-state example of the following section, we use the term "continuation value" to denote the state variable of Kocherlakota (1996b) as described in the preceding section.[9] That is, at the end of a period, the continuation value v is the promised expected utility to the type 1 agent that will be delivered at the start of the next period.

Assume that there are only two possible endowment realizations, $S = 2$, with $\{\overline{y}_1, \overline{y}_2\} = \{1 - \overline{y}, \overline{y}\}$, where $\overline{y} \in (.5, 1)$. Each endowment realization is equally likely to occur, $\{\Pi_1, \Pi_2\} = \{0.5, 0.5\}$. Hence, the two types of agents

---

[9] See Krueger and Perri (2003b) for another analysis of a two-state example.

face the same *ex ante* welfare level in autarky,

$$v_{\text{aut}} = \frac{.5}{1 - \beta} \left[u(\overline{y}) + u(1 - \overline{y}) \right].$$

We will focus on parameterizations for which there exist no first-best sustainable allocations (i.e., $\overline{c}_{\min} < \underline{c}_{\max}$, which here amounts to $\overline{c}_1 < \underline{c}_2$). An efficient allocation will then asymptotically enter the ergodic consumption set in (21.6.4) that here is given by two points, $\{\overline{c}_1, \underline{c}_2\}$. Because of the symmetry in preferences and endowments, it must be true that $\underline{c}_2 = 1 - \overline{c}_1 \equiv \overline{c}$, where we let $\overline{c}$ denote the consumption allocated to an agent whose participation constraint is binding and $1 - \overline{c}$ be the consumption allocated to the other agent.

Before determining the optimal values $\{1-\overline{c}, \overline{c}\}$, we will first verify that any such stationary allocation delivers the same continuation value to both types of agent. Let v^+ be the continuation value for the consumer who last received a high endowment and let v^- be the continuation value for the consumer who last received a low endowment. The promise-keeping constraint for v^+ is

$$v^+ = .5[u(\overline{c}) + \beta v^+] + .5[u(1 - \overline{c}) + \beta v^-]$$

and the promise-keeping constraint for v^- is

$$v^- = .5[u(\overline{c}) + \beta v^+] + .5[u(1 - \overline{c}) + \beta v^-].$$

Notice that the promise-keeping constraints make v^+ and v^- identical. Therefore, there is a unique stationary continuation value $\overline{v} \equiv v^+ = v^-$ that is independent of the current period endowment, as established in section 21.9.1 for $S = 2$. Setting $v^+ = v^- = \overline{v}$ in one of the two equations above and solving gives the stationary continuation value:

$$\overline{v} = \frac{.5}{1 - \beta} \left[u(\overline{c}) + u(1 - \overline{c}) \right]. \tag{21.10.1}$$

To determine the optimal $\overline{c}$ in this two-state example, we use the following two facts. First, $\overline{c}$ is the lower bound of the consumption interval $[\underline{c}_2, \overline{c}_2]$; $\overline{c}$ is the consumption level that should be awarded to the type 1 agent when she experiences the highest endowment $\overline{y}_2 = \overline{y}$ and we want to maximize the welfare of the type 2 agent subject to the type 1 agent's participation constraint. Second, $\overline{c}$ belongs also to the ergodic set $\{\overline{c}_1, \underline{c}_2\}$ that characterizes the stationary

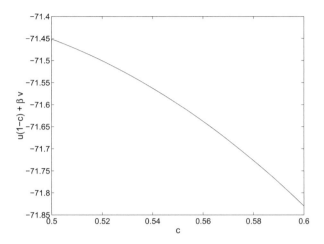

Figure 21.10.1: Welfare of the agent with low endowment as a function of $\bar{c}$.

efficient allocation, and we know that the associated efficient continuation values are then the same for all agents and given by $\bar{v}$ in (21.10.1). The maximization problem above can therefore be written as

$$\max_{\bar{c}} \ u(1 - \bar{c}) + \beta\bar{v} \tag{21.10.2a}$$

$$\text{subject to} \quad u(\bar{c}) + \beta\bar{v} - [u(\bar{y}) + \beta v_{\text{aut}}] \geq 0, \tag{21.10.2b}$$

where $\bar{v}$ is given by (21.10.1). We graphically illustrate how $\bar{c}$ is chosen in order to maximize (21.10.2a) subject to (21.10.2b) in Figures 21.10.1 and 21.10.2 for utility function $(1 - \gamma)^{-1}c^{1-\gamma}$ and parameter values $(\beta, \gamma, \bar{y}) = (.85, 1.1, .6)$. It can be verified numerically that $\bar{c} = .536$. Figure 21.10.1 shows (21.10.2a) as a decreasing function of $\bar{c}$ in the interval $[.5, .6]$. Figure 21.10.2 plots the left side of (21.10.2b) as a function of $\bar{c}$. Values of $\bar{c}$ for which the expression is negative are not sustainable (i.e., values less than .536). Values of $\bar{c}$ for which the expression is nonnegative are sustainable. Since the welfare of the agent with a low endowment realization in (21.10.2a) is decreasing as a function of $\bar{c}$ in the interval $[.5, .6]$, the best sustainable value of $\bar{c}$ is the lowest value for which the expression in (21.10.2b) is nonnegative. This value for $\bar{c}$ gives the most risk sharing that is compatible with the participation constraints.

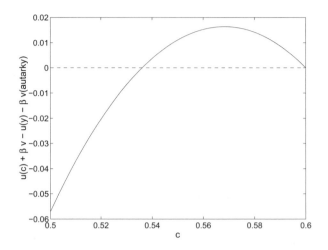

Figure 21.10.2: The participation constraint is satisfied for values of $\bar{c}$ for which the difference $u(\bar{c}) + \beta\bar{v} - [u(\bar{y}) + \beta v_{\text{aut}}]$ plotted here is positive.

21.10.1. Pareto frontier

It is instructive to find the entire set of sustainable values V. In addition to the value $\bar{v}$ above associated with a stationary sustainable allocation, other values can be sustained, for example, by promising a value $\hat{v} > \bar{v}$ to a type 1 agent who has yet to receive a low endowment realization. Thus, let $\hat{v}$ be a promised value to such a consumer and let c^+ be the consumption assigned to that consumer in the event that his endowment is high. Then promise keeping for the two types of agents requires

$$\hat{v} = .5[u(c^+) + \beta\hat{v}] + .5[u(1 - \bar{c}) + \beta\bar{v}], \qquad (21.10.3a)$$

$$P(\hat{v}) = .5[u(1 - c^+) + \beta P(\hat{v})] + .5[u(\bar{c}) + \beta\bar{v}]. \qquad (21.10.3b)$$

If the type 1 consumer receives the high endowment, sustainability of the allocation requires

$$u(c^+) + \beta\hat{v} \geq u(\bar{y}) + \beta v_{\text{aut}}, \qquad (21.10.4a)$$

$$u(1 - c^+) + \beta P(\hat{v}) \geq u(1 - \bar{y}) + \beta v_{\text{aut}}. \qquad (21.10.4b)$$

If the type 2 consumer receives the high endowment, awarding him $\overline{c}, \overline{v}$ automatically satisfies the sustainability requirements because these are already built into the construction of the stationary sustainable value $\overline{v}$.

Let's solve for the *highest* sustainable initial value of $\hat{v}$, namely, $v_{\max}$. To do so, we must solve the three equations formed by the promise-keeping constraints (21.10.3a) and (21.10.3b) and the participation constraint (21.10.4b) of a type 2 agent when it receives $1 - \overline{y}$ at equality:

$$u(1 - c^+) + \beta P(\hat{v}) = u(1 - \overline{y}) + \beta v_{\text{aut}}. \tag{21.10.5}$$

Equation (21.10.3b) and (21.10.5) are two equations in $(c^+, P(v_{\max}))$. After solving them, we can solve (21.10.3a) for $v_{\max}$. Substituting (21.10.5) into (21.10.3b) gives

$$P(v_{\max}) = .5[u(1 - \overline{y}) + \beta v_{\text{aut}}] + .5[u(\overline{c}) + \beta \overline{v}]. \tag{21.10.6}$$

But from the participation constraint of a high endowment household in a stationary allocation, recall that $u(\overline{c}) + \beta \overline{v} = u(\overline{y}) + \beta v_{\text{aut}}$. Substituting this into (21.10.6) and rearranging gives

$$P(v_{\max}) = v_{\text{aut}}$$

and therefore by (21.10.5), $c^+ = \overline{y}$.[10] Solving (21.10.3a) for $v_{\max}$ we find

$$v_{\max} = \frac{1}{2 - \beta}[u(\overline{y}) + u(1 - \overline{c}) + \beta \overline{v}]. \tag{21.10.7}$$

Now let us study what happens when we set $v \in (\overline{v}, v_{\max})$ and drive v toward $\overline{v}$ from above. Totally differentiating (21.10.3a) and (21.10.3b), we find

$$\frac{dP(\hat{v})}{d\hat{v}} = -\frac{u'(1 - c^+)}{u'(c^+)}.$$

Evidently

$$\lim_{v \downarrow \overline{v}} \frac{dP(v)}{dv} = -\frac{u'(1 - \overline{c})}{u'(\overline{c})} < -1.$$

---

[10] According to our general characterization of the *ex ante* division of the gains of an efficient contract in section 21.5, it can be viewed as determined by an implicit initial consumption level $c_\triangle \in [\overline{y}_1, \overline{y}_S]$. Notice that the present calculations have correctly computed the upper bound of that interval for our two-state example, $\overline{y}_S = \overline{y}_2 \equiv \overline{y}$.

By symmetry,

$$\lim_{v \uparrow \overline{v}} \frac{dP(v)}{dv} = -\frac{u'(\overline{c})}{u'(1 - \overline{c})} > -1.$$

Thus, there is a kink in the value function $P(v)$ at $v = \overline{v}$. At $\overline{v}$, the value function is not differentiable as established in section 21.8.2 when two adjacent consumption intervals are disjoint. At $\overline{v}$, $P'(v)$ exists only in the sense of a subgradient in the interval $[-u'(1 - \overline{c})/u'(\overline{c}), \ -u'(\overline{c})/u'(1 - \overline{c})]$. Figure 21.10.3 depicts the kink in $P(v)$.

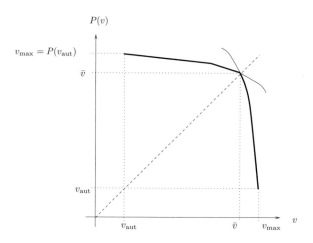

Figure 21.10.3: The kink in $P(v)$ at the stationary value of v for the two-state symmetric example.

21.10.2. *Interpretation*

Recall our characterization of the optimal consumption dynamics in (21.4.3). Consumption remains unchanged between periods when neither participation constraint binds, and hence the efficient contract displays memory or history dependence. When either of the participation constraints binds, history dependence is limited to selecting either the lower or the upper bound of a consumption range $[\underline{c}_j, \overline{c}_j]$, where the range and its bounds are functions of the current endowment realization $\overline{y}_j$. After someone's participation constraint has once been binding, history becomes irrelevant, because past consumption has no additional impact on the level of current consumption.

Now, in the case of our two-state example, there are only two consumption ranges, $[\underline{c}_1, \overline{c}_1]$ and $[\underline{c}_2, \overline{c}_2]$. And as a consequence, the asymptotic consumption distribution has only two points, $\overline{c}_1$ and $\underline{c}_2$ (or in our notation, $1 - \overline{c}$ and $\overline{c}$). It follows that history becomes irrelevant because consumption is then determined by the endowment realization. Thus, it can be said that "amnesia overwhelms memory" in this example, and the asymptotic distribution of continuation values becomes degenerate with a single point $\overline{v}$.[11]

We further explore the variation or the lack of variation in continuation values in the three-state example of the following section.

21.11.   A three-state example

As the two-state example stresses, any variation of continuation values in an efficient allocation requires that the environment be such that when a household's participation constraint is binding, the planner has room to increase both the current consumption *and* the continuation value of that household. In the stationary allocation in the two-state example, there is no room to adjust the continuation value because of the restrictions that promise keeping imposes. We now analyze the stationary allocation of a three-state ($S = 3$) example in

---

[11] If we adopt the recursive formulation of Thomas and Worrall in (21.3.1), amnesia manifests itself as a time-invariant state vector $[x_1, x_2]$ where $x_1 = u(1 - \overline{c}) - u(1 - \overline{y}) + \beta[\overline{v} - v_{\text{aut}}]$ and $x_2 = u(\overline{c}) - u(\overline{y}) + \beta[\overline{v} - v_{\text{aut}}]$.

which the environment still limits the planner's ability to manipulate continuation values, but nevertheless sometimes allows adjustments in the continuation value.

Thus, consider an environment in which $S = 3$. We assume that the distributions of y_t and $1 - y_t$ are identical. In particular, we let $\{\bar{y}_1, \bar{y}_2, \bar{y}_3\} = \{1 - \bar{y}, 0.5, \bar{y}\}$ and $\{\Pi_1, \Pi_2, \Pi_3\} = \{\Pi/2, 1 - \Pi, \Pi/2\}$ where $\bar{y} \in (.5, 1]$ and $\Pi \in [0, 1]$. Given parameter values such that there is no first-best sustainable allocation (i.e., $\bar{c}_1 < \underline{c}_3$), we will study the efficient allocation that is attained asymptotically. According to (21.6.4), this ergodic consumption set is given by

$$\left\{ [\bar{c}_1, \underline{c}_3] \bigcap \{\bar{c}_1, \underline{c}_2, \bar{c}_2, \underline{c}_3\} \right\}, \qquad (21.11.1)$$

which contains at least two points ($\bar{c}_1$, $\underline{c}_3$) and maybe two additional points ($\underline{c}_2$, $\bar{c}_2$).

When there are no first-best sustainable allocations, the efficient stationary allocation must be such that the participation constraints of a type 1 person and a type 2 person bind in state 3 and state 1, respectively. Let $\bar{c} \in [0.5, 1]$ and $\bar{w}^+$ be the consumption and continuation value allocated to the agent whose participation constraint is binding because his endowment is equal to $\bar{y}$:

$$u(\bar{c}) + \beta \bar{w}^+ = u(\bar{y}) + \beta v_{\text{aut}}. \qquad (21.11.2)$$

In such a state, the agent whose participation constraint is *not* binding consumes $1 - \bar{c}$ and is assigned continuation value $\bar{w}^-$. Because of the assumed symmetries with respect to preferences and endowments, we have $\bar{c} = \underline{c}_3 = 1 - \bar{c}_1$.

The consumption allocation in state 2 depends on the different promised continuation values with which agents enter a period. The symmetry in our environment and the existence of only three states imply that there is a single consumption level $\hat{c}$ that is granted to the type of person that last realized the highest endowment $\bar{y}$. Let $\hat{w}^+$ be the continuation value that in state 2 is allocated to the type of person that last received endowment $\bar{y}$. According to our earlier characterization of an efficient allocation, the agents who realize the highest endowment $\bar{y}$ are induced not to defect into autarky by granting them both higher current consumption and a higher continuation value. Hence, state 2 is "payback time" for the agents who were promised a higher continuation value and it must be true that $\hat{c} \in [0.5, 1]$. In state 2, the type of person that did not last receive $\bar{y}$ is allocated consumption $1 - \hat{c}$ and continuation value

$\hat{w}^-$. The participation constraint of this type of person might conceivably be binding in state 2,

$$u(1 - \hat{c}) + \beta\hat{w}^- \geq u(0.5) + \beta v_{\text{aut}}. \tag{21.11.3}$$

According to the optimal consumption dynamics in (21.4.3), we know that $\hat{c} = \min\{\bar{c}, \bar{c}_2\}$. That is, a person who had the highest endowment realization $\bar{y}$ with associated consumption level $\bar{c}$ will retain that consumption level when moving into state 2 ($\hat{c} = \bar{c}$) unless the participation constraint of the other agent becomes binding in state 2. In the latter case, the person who had the highest endowment realization is awarded consumption $\hat{c} = \bar{c}_2$ in state 2 and the participation constraint for the other person in (21.11.3) will hold with strict equality.

While there can exist four different consumption levels in the efficient stationary allocation, $\{1 - \bar{c}, 1 - \hat{c}, \hat{c}, \bar{c}\}$, it is possible to have at most two distinct continuation values:

$$\bar{w}^+ = \hat{w}^+ = (\Pi/2)\left[u(\bar{c}) + \beta\bar{w}^+\right] + (1 - \Pi)\left[u(\hat{c}) + \beta\hat{w}^+\right]$$
$$+ (\Pi/2)\left[u(1 - \bar{c}) + \beta\bar{w}^-\right], \tag{21.11.4a}$$

$$\bar{w}^- = \hat{w}^- = (\Pi/2)\left[u(\bar{c}) + \beta\bar{w}^+\right] + (1 - \Pi)\left[u(1 - \hat{c}) + \beta\hat{w}^-\right]$$
$$+ (\Pi/2)\left[u(1 - \bar{c}) + \beta\bar{w}^-\right]. \tag{21.11.4b}$$

As can be seen on the right side of (21.11.4a), the expressions for $\bar{w}^+$ and $\hat{w}^+$ are the same, and so $\bar{w}^+ = \hat{w}^+ \equiv w^+$. The same holds true for $\bar{w}^-$ and $\hat{w}^-$ in (21.11.4b), and hence $\bar{w}^- = \hat{w}^- \equiv w^-$. By manipulating equations (21.11.4), we can express the two continuation values in terms of $(\bar{c}, \hat{c})$:

$$w^+ = \left\{ (\Pi/2)\left[1 + \beta\kappa\Pi/2\right]\left[u(\bar{c}) + u(1 - \bar{c})\right] \right.$$
$$+ (1 - \Pi)\left[u(\hat{c}) + \beta\kappa\Pi u(1 - \hat{c})/2\right] \Big\}$$
$$\cdot \left\{ \left[1 - \beta(1 - \Pi)\right](1 - \beta)\kappa \right\}^{-1} \tag{21.11.5a}$$

$$w^- = w^+ - \frac{1 - \Pi}{1 - \beta(1 - \Pi)}\left[u(\hat{c}) - u(1 - \hat{c})\right], \tag{21.11.5b}$$

where $\kappa = \left[1 - (1 - \Pi/2)\beta\right]^{-1}$.

To determine the optimal $\{\bar{c}, \hat{c}\}$ in this three-state example, it is helpful to focus on a state in which the agents realize different endowments, say, state

3 in which the type 1 agent realizes the highest endowment $\overline{y}$ and is awarded consumption level $\overline{c}$. We can then exploit the following two facts. First, $\overline{c}$ is the lower bound of the consumption interval $[\underline{c}_3, \overline{c}_3]$, so $\overline{c}$ is the consumption level that should be awarded to the type 1 agent when she experiences the highest endowment $\overline{y}_3 = \overline{y}$ and we want to maximize the welfare of the type 2 agent subject to the type 1 agent's participation constraint. Second, $\overline{c}$ belongs also to the ergodic set in $(21.11.1)$ that characterizes the stationary efficient allocation, and we know that the associated efficient continuation values are w^+ for the agents with high endowment and w^- for the other agents. By invoking functions $(21.11.5)$ that express these continuation values in terms of $\{\overline{c}, \hat{c}\}$ and by using participation constraint $(21.11.2)$ that determines permissible values of $\hat{c}$, the optimization problem above becomes:

$$\max_{\overline{c}, \hat{c}} \; u(1 - \overline{c}) + \beta w^- \tag{21.11.6a}$$

$$\text{subject to} \quad u(\overline{c}) + \beta w^+ - [u(\overline{y}) + \beta v_{\text{aut}}] \geq 0 \tag{21.11.6b}$$

$$u(1 - \hat{c}) + \beta w^- - [u(0.5) + \beta v_{\text{aut}}] \geq 0, \tag{21.11.6c}$$

where w^- and w^+ are given by $(21.11.5)$.

To illustrate graphically how an efficient stationary allocation $\{\overline{c}, \hat{c}\}$ can be computed from optimization problem $(21.11.6)$, we assume a utility function $c^{1-\gamma}/(1 - \gamma)$ and parameter values $(\beta, \gamma, \Pi, \overline{y}) = (0.7, 1.1, 0.6, 0.7)$. It should now be evident that we can restrict attention to consumption levels $\overline{c} \in [0.5, \overline{y}]$ and $\hat{c} \in [0.5, \overline{c}]$. Figures 21.11.1a and 21.11.1b show the sets $(\overline{c}, \hat{c}) \in [0.5, \overline{y}] \times [0.5, \overline{c}]$ that satisfy participation constraint $(21.11.6b)$ and $(21.11.6c)$, respectively. The intersection of these sets is depicted in Figure 21.11.2 where the circle indicates the efficient stationary allocation that maximizes $(21.11.6a)$.

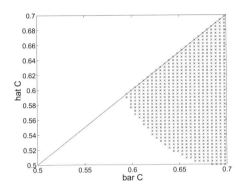

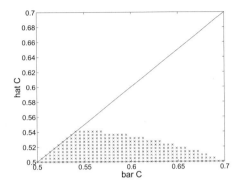

Figure 21.11.1a: Pairs of $(\overline{c}, \hat{c})$ that satisfy $u(\overline{c}) + \beta w^+ \geq u(\overline{y}) + \beta v_{\text{aut}}$.

Figure 21.11.1b: Pairs of $(\overline{c}, \hat{c})$ that satisfy $u(1 - \hat{c}) + \beta w^- \geq u(0.5) + \beta v_{\text{aut}}$.

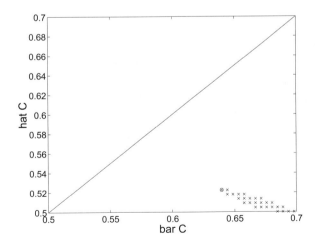

Figure 21.11.2: Pairs of $(\overline{c}, \hat{c})$ that satisfy $u(\overline{c}) + \beta w^+ \geq u(\overline{y}) + \beta v_{\text{aut}}$ and $u(1 - \hat{c}) + \beta w^- \geq u(0.5) + \beta v_{\text{aut}}$. The efficient stationary allocation within this set is marked with a circle.

21.11.1. Perturbation of parameter values

We also compute efficient stationary allocations for different values of $\Pi \in [0, 1]$ while retaining all other parameter values. As a function of Π, Figures 21.11.3a and 21.11.3b depict consumption levels and continuation values, respectively. For low values of Π, we see that there cannot be any risk sharing among the agents, so that autarky is the only sustainable allocation. The explanation for this is as follows. Given a low value of Π, an agent who has realized the high endowment $\overline{y}$ is heavily discounting the insurance value of any transfer in a future state when her endowment might drop to $1 - \overline{y}$ because such a state occurs only with a small probability equal to $\Pi/2$. Hence, in order for that agent to surrender some of her endowment in the current period, she must be promised a significant combined payoff in that unlikely event of a low endowment in the future and a positive transfer in the most common state 2. But such promises are difficult to make compatible with participation constraints, because all agents will be discounting the value of any insurance arrangement as soon as the common state 2 is realized since then there is once again only a small probability of experiencing anything else.

When the probability of experiencing extreme values of the endowment realization is set sufficiently high, there exist efficient allocations that deliver risk sharing. When Π exceeds 0.4 in Figure 21.11.3a, the lucky agent is persuaded to surrender some of her endowment, and her consumption becomes $\overline{c} < \overline{y}$. The lucky agent is compensated for her sacrifice not only through the insurance value of being entitled to an equivalent transfer in the future when she herself might realize the low endowment $1 - \overline{y}$ but also through a higher consumption level in state 2, $\hat{c} > 0.5$.[12] In fact, if the consumption smoothing motive could operate unhindered in this situation, the lucky agent's consumption would indeed by equalized across states. But what hinders such an outcome is the participation constraint of the unlucky agent when entering state 2. It must be incentive compatible for that earlier unlucky agent to give up parts of her endowment in state 2 when both agents now have the same endowment and the value of the insurance arrangement lies in the future. Notice that this participation constraint of the earlier unlucky agent is no longer binding in our example when

---

[12] Recall that we established in section 21.4.4 that *all* consumption intervals are nondegenerate if there is risk sharing. We can use this fact to prove that as soon as the parameter value for Π exceeds the critical value where risk sharing becomes viable, it follows that $\hat{c} = \overline{c}_2 > \overline{y}_2 = 0.5$.

Π is greater than 0.94, because the efficient allocation prescribes $\hat{c} = \bar{c}$. In terms of Thomas and Worrall's characterization of the optimal consumption dynamics, the parameterization is then such that $\bar{c}_2 > \underline{c}_3$ and the ergodic set in $(21.11.1)$ is given by $\{\bar{c}_1, \underline{c}_3\}$ or, in our notation, by $\{1 - \bar{c}, \bar{c}\}$.

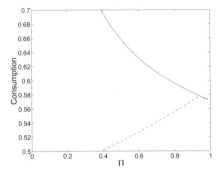

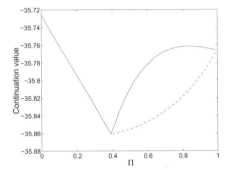

Figure 21.11.3a: Consumption levels as a function of Π. The solid line depicts $\bar{c}$, i.e., consumption in states 1 and 3 of a person who realizes the highest endowment $\bar{y}$. The dashed line depicts $\hat{c}$, i.e., consumption in state 2 of the type of person that was the last one to have received $\bar{y}$.

Figure 21.11.3b: Continuation values as a function of Π. The solid line depicts w^+, i.e., continuation value of the type of person that was the last one to have received $\bar{y}$. The dashed line is the continuation value of the other type of person, i.e., w^-.

The fact that the efficient allocation raises the consumption of the lucky agent in future realizations of state 2 is reflected in the spread of continuation values in Figure 21.11.3b. The spread vanishes only in the limit when $\Pi = 1$ because then the three-state example turns into our two-state example of the preceding section where there is only a single continuation value. But while the planner is able to vary continuation values in the three-state example, there remains an important limitation to when those continuation values can be varied. Consider a parameterization with $\Pi \in (0.4, 0.94)$ for which we know that $\hat{c} < \bar{c}$ in Figure 21.11.3a. The agent who last experienced the highest endowment $\bar{y}$

is consuming $\hat{c}$ in state 2 in the efficient stationary allocation, and is awarded continuation value w^+. Suppose now that agent once again realizes the highest endowment $\bar{y}$ and his participation constraint becomes binding. To prevent him from defecting to autarky, the planner responds by raising his consumption to $\bar{c}$ $(> \hat{c})$ *but* keeps his continuation value unchanged at w^+. In other words, the optimal consumption dynamics in the efficient stationary allocation leaves no room for increasing the continuation value further. The unchanging continuation value is a reflection of the nondifferentiability of the Pareto frontier at $v = w^+$.

21.11.2. Pareto frontier

As described in section 21.8, the *ex ante* division of the gains from an efficient contract can be viewed as determined by an implicit initial consumption level, $c_\triangle \in [\bar{y}_1, \bar{y}_S]$. In our symmetric environment, it is sufficient to focus on half of this range because the other half will just be the mirror image of those computations. Let us therefore compute the Pareto frontier for $c_\triangle \in [0.5, \bar{y}_3] \equiv [0.5, \bar{y}]$. We assume a parameterization such that the consumption intervals, $\{[\underline{c}_j, \bar{c}_j]\}_{j=1}^3$, are disjoint, i.e., the parameterization is such that $\hat{c} < \bar{c}$, which corresponds to a parameterization with $\Pi \in (0.4, 0.94)$ in Figure 21.11.3a.

First, we study the formulas for computing v and $P(v)$ in the range $c_\triangle \in [0.5, \hat{c}]$:

$$v = \frac{\Pi}{2}\Big\{u(1-\bar{c}) + \beta w^-\Big\} + (1-\Pi)\Big\{u(c_\triangle) + \beta v\Big\} + \frac{\Pi}{2}\Big\{u(\bar{c}) + \beta w^+\Big\},$$

$$P(v) = \frac{\Pi}{2}\Big\{u(\bar{c}) + \beta w^+\Big\} + (1-\Pi)\Big\{u(1-c_\triangle) + \beta P(v)\Big\}$$
$$+ \frac{\Pi}{2}\Big\{u(1-\bar{c}) + \beta w^-\Big\}.$$

When the type 1 agent is assigned an implicit initial consumption level $c_\triangle \in [0.5, \hat{c}]$, her consumption is indeed equal to $c_\triangle$ for any initial uninterrupted string of realizations of state 2 with some continuation value v, and the corresponding consumption of the type 2 agent is $1 - c_\triangle$ with an associated continuation value $P(v)$. But as soon as either state 1 or state 3 is realized for the first time, the updating rules in (21.4.3) imply that the economy enters the ergodic set of the efficient stationary allocation. In particular, if state 1 is realized and the participation constraint of the type 2 agent becomes binding, the type 1 agent is awarded consumption $\bar{c}_1 \equiv 1 - \bar{c}$ and continuation value w^- while the type 2 agent consumes $1 - \bar{c}_1 \equiv \bar{c}$ with continuation value $P(w^-) = w^+$. But if state 3 is realized and the participation constraint of the type 1 agent becomes binding, the type 1 agent is awarded consumption $\underline{c}_3 \equiv \bar{c}$ and continuation value w^+ while the type 2 agent consumes $1 - \underline{c}_3 \equiv 1 - \bar{c}$ with continuation value $P(w^+) = w^-$. Given the implicit initial consumption level $c_\triangle$, it is straightforward to solve for the initial welfare assignment $\{v, P(v)\}$ from the equations above.

Similarly, we can use the updating rules (21.4.3) to get formulas for computing v and $P(v)$ in the range $c_\triangle \in [\bar{c}, \bar{y}]$:

$$v = \frac{\Pi}{2}\Big\{u(1 - \bar{c}) + \beta w^-\Big\} + (1 - \Pi)\Big\{u(\hat{c}) + \beta w^+\Big\} + \frac{\Pi}{2}\Big\{u(c_\triangle) + \beta v\Big\},$$

$$P(v) = \frac{\Pi}{2}\Big\{u(\bar{c}) + \beta w^+\Big\} + (1 - \Pi)\Big\{u(1 - \bar{c}) + \beta w^-\Big\}$$
$$+ \frac{\Pi}{2}\Big\{u(1 - c_\triangle) + \beta P(v)\Big\}.$$

Concerning the remaining range of implicit initial consumption levels $c_\triangle \in (\hat{c}, \bar{c})$, we can immediately verify that either pair of equations above can be used when setting $c_\triangle = \hat{c}$ in the first pair of equations or $c_\triangle = \bar{c}$ in the second pair of equations. Hence, the initial welfare assignment is the same for implicit initial consumption $c_\triangle \in [\hat{c}, \bar{c}]$, and it is given by $\{w^+, P(w^+)\}$. At this point in Figure 21.11.4 the Pareto frontier becomes nondifferentiable.

21.12. Empirical motivation

Kocherlakota was interested in the case of perpetual imperfect risk sharing because he wanted to use his model to think about the empirical findings from panel studies by Mace (1991), Cochrane (1991), and Townsend (1994). Those studies found that, after conditioning on aggregate income, individual consumption and earnings are positively correlated, belying the risk-sharing implications of the complete markets models with recursive utility of the type we studied in chapter 8. So long as no first-best allocation is sustainable, the action of the occasionally binding participation constraints lets the model with two-sided lack of commitment reproduce that positive conditional covariation. In recent work,

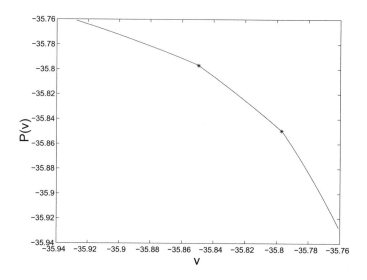

Figure 21.11.4: Pareto frontier $P(v)$ for the three-state symmetric example. Kinks occur at coordinates $(v, P(v)) = (w^-, w^+)$ and $(v, P(v)) = (w^+, w^-)$.

Albarran and Attanasio (2003) and Kehoe and Perri (2003a, 2003b) pursue more implications of models like Kocherlakota's.

21.13. Generalization

Our formal analysis has followed the approach taken by Thomas and Worrall (1988). We have converted the risk-neutral firm into a risk-averse household, as suggested by Kocherlakota (1996b). Another difference is that our analysis is cast in a general equilibrium setting while Thomas and Worrall formulate a partial equilibrium model where the firm implicitly has access to an outside credit market with a given gross interest rate of β^{-1} when maximizing the expected present value of profits. However, this difference is not material, since an efficient contract is such that wages never exceed output.[13] Hence, Thomas and

---

[13] The outcome that efficient wages do not exceed output in Thomas and Worrall's (1988) analysis is related to our ability to solve optimization problem (21.3.1) without imposing nonnegativity constraints on consumption. See footnote 5.

Worrall's (1988) analysis can equally well be thought of as a general equilibrium analysis.

Ligon, Thomas, and Worrall (2002) further generalize the environment by assuming that the endowment follows a Markov process. This allows for the possibility of both aggregate and idiosyncratic risk and serial correlation. The efficient contract is characterized by an updating rule for the ratio of the marginal utilities of the two households that resembles our updating rule for consumption in (21.4.3). Each state of nature is associated with a particular interval of permissible ratios of marginal utilities. Given the current state and the previous period's ratio of marginal utilities, the new ratio lies within the interval associated with the current state, such that the change is minimized. That is, if last period's ratio falls outside of the current interval, then the ratio must change to an endpoint of the current interval, and one of the households will be constrained. But whenever possible, the ratio is kept constant over time. This is consistent with our analysis in chapter 8 of competitive equilibria with complete markets (and full commitment). Expression (8.5.5) states that these unconstrained first-best allocations are such that ratios of marginal utilities between pairs of agents are constant across all histories and dates.

21.14. Decentralization

By imposing constraints on each household's budget sets above and beyond those imposed by the standard household's budget constraint, Kehoe and Levine (1993) describe how to decentralize the optimal allocation in an economy like Kocherlakota's with complete competitive markets at time 0. Thus, let $q_t^0(h_t)$ be the Arrow-Debreu time 0 price of a unit of time t consumption after history h_t. The two households' budget constraints are

$$\sum_{t=0}^{\infty} \sum_{h_t} q_t^0(h_t) c_t(h_t) \leq \sum_{t=0}^{\infty} \sum_{h_t} q_t^0(h_t) y_t \qquad (21.14.1a)$$

$$\sum_{t=0}^{\infty} \sum_{h_t} q_t^0(h_t)(1 - c_t(h_t)) \leq \sum_{t=0}^{\infty} \sum_{h_t} q_t^0(h_t)(1 - y_t). \qquad (21.14.1b)$$

Kehoe and Levine augment these standard budget constraints with what were the planner's "participation constraints" $(21.2.1a)$, $(21.2.1b)$, but which now have to be interpreted as exogenous restrictions on the households' budget sets, one restriction for each consumer for each $t \geq 0$ for each history h_t.

Adding those restrictions leaves the household's budget sets convex. That allows all of the assumptions of the second welfare theorem to be fulfilled. That then implies that a competitive equilibrium (defined in the standard way to include optimization and market clearing, but with household budget sets being further restricted by $(21.2.1)$) will implement the planner's optimal allocation.

Although mechanically this decentralization works like a charm, it can nevertheless be argued that it conflicts with the spirit of a competitive equilibrium in which agents take prices as given and budget constraints are the only restrictions on agents' consumption sets. In contrast, participation constraints $(21.2.1a)$ and $(21.2.1b)$ are now modelled as direct restrictions on agents' consumption possibility sets. Partly because of this controversial feature of the Kehoe-Levine decentralization, Alvarez and Jermann use another decentralization, one that imposes portfolio/solvency constraints and is cast in terms of sequential trading of Arrow securities. The endogenously determined solvency constraints are agent and state specific and ensure that the participation constraints are satisfied. We turn to the Alvarez-Jermann decentralization in the next section.

In all fairness, one could argue that the alternative decentralization solely converts one set of participation constraints into another one. For both specifications we have a substantial departure from a decentralized equilibrium under full commitment. When removing the assumption of commitment, we are assigning a very demanding task to the "invisible hand" who must not only look for market-clearing prices but also check participation/solvency constraints for all agents and all states of the world.

21.15. Endogenous borrowing constraints

Alvarez and Jermann (2000) alter Kehoe and Levine's decentralization to attain a model with sequentially complete markets in which households face what can be interpreted as endogenous borrowing constraints. Essentially, they accomplish this by showing how the standard quantity constraints on Arrow securities (see chapter 8) can be appropriately tightened to implement the optimal allocation as constrained by the participation constraints. Their idea is to find borrowing constraints tight enough to make the highest endowment agents adhere to the allocation, while letting prices alone prompt lower endowment agents to go along with it.

For expositional simplicity, we let $y^i(y)$ denote the endowment of a household of type i when a representative household of type 1 receives y. Recall the earlier assumption that $[y^1(y), y^2(y)] = (y, 1 - y)$. The state of the economy is the current endowment realization y and the beginning-of-period asset holdings $A = (A_1, A_2)$, where A_i is the asset holding of a household of type i and $A_1 + A_2 = 0$. Because asset holdings add to zero, it is sufficient to use A_1 to characterize the wealth distribution. Define the *state* of the economy as $X = [y \quad A_1]'$. There is a complete set of markets in one-period Arrow securities. In particular, let $Q(X'|X)$ be the price of one unit of consumption in state X' tomorrow given state X today. A household of type i with beginning-of-period assets a can purchase and sell these securities subject to the budget constraint

$$c + \sum_{X'} Q(X'|X)a(X') \leq y^i(y) + a, \qquad (21.15.1)$$

where $a(X')$ is the quantity purchased (if positive) or sold (if negative) of Arrow securities that pay one unit of consumption tomorrow if X' is realized, and also subject to the borrowing constraints

$$a(X') \geq B^i(X'). \qquad (21.15.2)$$

Notice that there is one constraint for each next period state X' and that the borrowing constraints reflect history dependence through the presence of A'.

The Bellman equation for the household in the decentralized economy is

$$V^i(a, X) = \max_{c, \{a(X')\}_{X' \in \mathbf{X}}} \left\{ u(c) + \beta \sum_{X'} V^i[a(X'), X']\Pi(X'|X) \right\}$$

subject to the budget constraint (21.15.1) and borrowing constraints (21.15.2). The equilibrium law of motion for the asset distribution, A_1 is embedded in the conditional distribution $\Pi(X'|X)$.

Alvarez and Jermann define a competitive equilibrium with borrowing constraints in a standard way, with the qualification that among the equilibrium objects are the borrowing constraints $B^i(X')$, functions that the households take as given. Alvarez and Jermann show how to choose the borrowing constraints to make the allocation that solves the planning problem be an equilibrium allocation. They do so by construction, identifying the elements of the borrowing constraints that are binding from having identified the states in the planning problem where one or another agent's participation constraint is binding.

It is easy for Alvarez and Jermann to compute the equilibrium pricing kernel from the allocation that solves the planning problem. The pricing kernel satisfies

$$q(X'|X) = \max_{i=1,2} \beta \frac{u'[c^i(a', X')]}{u'[c^i(a, X)]} \Pi(X'|X), \qquad (21.15.3)$$

where $c^i(a, X)$ is the consumption decision rule of a household of type i with beginning-of-period assets a.[14] People with the highest valuation of an asset *buy* it. Buyers of state-contingent securities are unconstrained, so they equate their marginal rate of substitution to the price of the asset. At equilibrium prices, sellers of state-contingent securities will occasionally like to issue more, but are constrained from doing so by state-by-state restrictions on the amounts that they can sell. Thus, the intertemporal marginal rate of substitution of an agent whose participation constraint (or borrowing constraint) is *not* binding determines the pricing kernel. A binding *participation* constraint translates into a binding *borrowing* constraint in the previous period. A participation constraint for some state at t restricts the amount of state-contingent debts that can be issued for that state at $t-1$. In effect, constrained and unconstrained agents have their own "personal interest rates" at which they are just indifferent between borrowing or lending a infinitesimally more. A constrained agent wants to consume *more* today at equilibrium prices (i.e., at the shadow prices (21.15.3) evaluated at the solution of the planning problem), and thus has a high personal interest rate. He would like to sell more of the state-contingent security than

---

[14] For the two-state example with $\beta = .85, \gamma = 1.1, \bar{y} = .6$, described in Figure 21.10.1, we computed that $\bar{c} = .536$, which implies that the risk-free interest rate is 1.0146. Note that with complete markets the risk-free claim would be $\beta^{-1} = 1.1765$.

he is allowed to at the equilibrium state-date prices. An agent would like to *sell* state-contingent claims on consumption tomorrow in those states in which he will be well endowed tomorrow. But those high endowment states are also the ones in which he will have an incentive to default. He must be restrained from doing so by limiting the volume of debt that he is able to carry into those high endowment states. This limits his ability to smooth consumption across high and low endowment states. Thus, his consumption and continuation value increases when he enters one of those high endowment states precisely because he has been prevented from selling enough claims to smooth his consumption over time and across states.

From a general equilibrium perspective, when sellers of a state-contingent security are constrained with respect to the quantities that they can issue, it follows that the price is bid up when unconstrained buyers are competing for a smaller volume of that security. This tendency of lowering the yield on individual Arrow securities explains Alvarez and Jermann's result that interest rates are lower, when compared to a corresponding complete markets economy, a property shared with the Bewley economies studied in chapter 18. [15]

Alvarez and Jermann study how the state-contingent prices (21.15.3) behave as they vary the discount factor and the stochastic process for y. They use the additional fluctuation in the stochastic discount factor injected by the participation constraints to explain some asset pricing puzzles. See Zhang (1997) and Lustig (2000, 2003) for further work along these lines.

---

[15] In exercise *21.4*, we ask the reader to compute the allocation and interest rate in such an economy.

21.16. Concluding remarks

The model in this chapter assumes that the economy reverts to an autarkic allocation in the event that a household chooses to deviate from the allocation assigned in the contract. Of course, assigning autarky continuation values to *everyone* puts us inside the Pareto frontier and so is inefficient. In terms of sustaining an allocation, the important feature of the autarky allocation is just the continuation value that it assigns to an agent who is tempted to default, i.e., an agent whose participation constraint binds. Kletzer and Wright (2000) recognize that it can be possible to promise an agent who is tempted to default an autarky continuation value while giving those agents whose participation constraints aren't binding enough to stay on the Pareto frontier. Continuation values that lie on the Pareto frontier are said to be "renegotiation proof".

Further thought about how to model the consequences of default in these settings is likely to be fruitful. By permitting coalitions of consumers to break away and thereafter share risks among themselves, Genicot and Ray (2003) refine a notion of sustainability in a multi-consumer economy.

Exercises

Exercise 21.1 **Lagrangian method with two-sided no commitment**

Consider the model of Kocherlakota with two-sided lack of commitment. Two consumers each have preferences $E_0 \sum_{t=0}^{\infty} \beta^t u[c_i(t)]$, where u is increasing, twice differentiable, and strictly concave, and where $c_i(t)$ is the consumption of consumer i. The good is not storable, and the consumption allocation must satisfy $c_1(t) + c_2(t) \leq 1$. In period t, consumer 1 receives an endowment of $y_t \in [0,1]$, and consumer 2 receives an endowment of $1 - y_t$. Assume that y_t is i.i.d. over time and is distributed according to the discrete distribution $\text{Prob}(y_t = y_s) = \Pi_s$. At the start of each period, after the realization of y_s but before consumption has occurred, each consumer is free to walk away from the loan contract.

a. Find expressions for the expected value of autarky, before the state y_s is revealed, for consumers of each type. (*Note:* These need not be equal.)

b. Using the Lagrangian method, formulate the contract design problem of finding an optimal allocation that for each history respects feasibility and the participation constraints of the two types of consumers.

c. Use the Lagrangian method to characterize the optimal contract as completely as you can.

Exercise 21.2 **A model of Dixit, Grossman, and Gul (2000)**

For each date $t \geq 0$, two political parties divide a "pie" of fixed size 1. Party 1 receives a sequence of shares $y = \{y_t\}_{t \geq 0}$ and has utility function $E \sum_{t=0}^{\infty} \beta^t U(y_t)$, where $\beta \in (0,1)$, E is the mathematical expectation operator, and $U(\cdot)$ is an increasing, strictly concave, twice differentiable period utility function. Party 2 receives share $1 - y_t$ and has utility function $E \sum_{t=0}^{\infty} \beta^t U(1 - y_t)$. A state variable X_t is governed by a Markov process; X resides in one of K states. There is a partition S_1, S_2 of the state space. If $X_t \in S_1$, party 1 chooses the division $y_t, 1 - y_t$, where y_t is the share of party 1. If $X_t \in S_2$, party 2 chooses the division. At each point in time, each party has the option of choosing "autarky," in which case its share is 1 when it is in power and zero when it is not in power.

Formulate the optimal history-dependent sharing rule as a recursive contract. Formulate the Bellman equation. (*Hint:* Let $V[u_0(x), x]$ be the optimal value for party 1 in state x when party 2 is promised value $u_0(x)$.)

Exercise 21.3 **Two-state numerical example of social insurance**

Consider an endowment economy populated by a large number of individuals with identical preferences,

$$E \sum_{t=0}^{\infty} \beta^t u(c_t) = E \sum_{t=0}^{\infty} \beta^t \left(4 c_t - \frac{c_t^2}{2} \right), \qquad \text{with } \beta = 0.8.$$

With respect to endowments, the individuals are divided into two types of equal size. All individuals of a particular type receive zero goods with probability 0.5 and two goods with probability 0.5 in any given period. The endowments of the two types of individuals are perfectly negatively correlated so that the per capita endowment is always one good in every period.

The planner attaches the same welfare weight to all individuals. Without access to outside funds or borrowing and lending opportunities, the planner seeks to provide insurance by simply reallocating goods between the two types of individuals. The design of the social insurance contract is constrained by a lack of commitment on behalf of the individuals. The individuals are free to walk away from any social arrangement, but they must then live in autarky evermore.

a. Compute the optimal insurance contract when the planner lacks memory; that is, transfers in any given period can be a function only of the current endowment realization.

b. Can the insurance contract in part a be improved if we allow for history-dependent transfers?

c. Explain how the optimal contract changes when the parameter β goes to 1. Explain how the optimal contract changes when the parameter β goes to zero.

Exercise 21.4   **Kehoe-Levine without risk**

Consider an economy in which each of two types of households has preferences over streams of a single good that are ordered by $v = \sum_{t=0}^{\infty} \beta^t u(c_t)$, where $u(c) = (1 - \gamma)^{-1}(c + b)^{1-\gamma}$ for $\gamma \geq 1$ and $\beta \in (0, 1)$, and $b > 0$. For $\epsilon > 0$ and $t \geq 0$, households of type 1 are endowed with an endowment stream $y_{1,t} = 1 + \epsilon$ in even-numbered periods and $y_{1,t} = 1 - \epsilon$ in odd-numbered periods. Households of type 2 own an endowment stream of $y_{2,t}$ that equals $1 - \epsilon$ in even periods and $1 + \epsilon$ in odd periods. There are equal numbers of the two types of household. For convenience, you can assume that there is one of each type of household. Assume that $\beta = .8$, $b = 5$, $\gamma = 2$, and $\epsilon = .5$.

a. Compute autarky levels of discounted utility v for the two types of households. Call them $v_{\text{aut},h}$ and $v_{\text{aut},\ell}$.

b. Compute the competitive equilibrium allocation and prices. Here assume that there are no enforcement problems.

c. Compute the discounted utility to each household for the competitive equilibrium allocation. Denote them v_i^{CE} for $i = 1, 2$.

d. Verify that the competitive equilibrium allocation is not self-enforcing in the sense that at each $t > 0$, some households would prefer autarky to the competitive equilibrium allocation.

e. Now assume that there are enforcement problems because at the beginning of each period, each household can renege on contracts and other social arrangements with the consequence that it receives the autarkic allocation from that period on. Let v_i be the discounted utility at time 0 of consumer i. Formulate the consumption smoothing problem of a planner who wants to maximize v_1 subject to $v_2 \geq \tilde{v}_2$, and constraints that make the allocation self-enforcing.

f. Find an efficient self-enforcing allocation of the periodic form $c_{1,t} = \check{c}, 2 - \check{c}, \check{c}, \ldots$ and $c_{2,t} = 2 - \check{c}, \check{c}, 2 - \check{c}, \ldots$, where continuation utilities of the two agents oscillate between two values v_h and v_ℓ. Compute $\check{c}$. Compute discounted utilities v_h for the agent who receives $1 + \epsilon$ in the period and v_ℓ for the agent who receives $1 - \epsilon$ in the period.

Plot consumption paths for the two agents for (i) autarky, (ii) complete markets without enforcement problems, and (iii) complete markets with the enforcement constraint. Plot continuation utilities for the two agents for the same three allocations. Comment on them.

g. Compute one-period gross interest rates in the complete markets economies with and without enforcement constraints. Plot them over time. In which economy is the interest rate higher? Explain.

h. Keep all parameters the same, but gradually increase the discount factor. As you raise β toward 1, compute interest rates as in part g. At what value of β do interest rates in the two economies become equal? At that value of β is either participation constraint ever binding?

Exercise 21.5 **The kink**

A pure endowment economy consists of two *ex ante* identical consumers each of whom values streams of a single nondurable consumption good according to the utility functional

$$v = E \sum_{t=0}^{\infty} \beta^t u(c_t), \quad \beta \in (0, 1)$$

where E is the mathematical expectation operator and $u(\cdot)$ is a strictly concave, increasing, and twice continuously differentiable function. The endowment sequence of consumer 1 is an i.i.d. process with $\text{Prob}(y_t = \bar{y}) = .5$ and $\text{Prob}(y_t = 1 - \bar{y}) = .5$ where $\bar{y} \in [.5, 1)$. The endowment sequence of consumer 2 is identically distributed with that of consumer 1, but perfectly negatively correlated with it: whenever consumer 1 receives $\bar{y}$, consumer 2 receives $1 - \bar{y}$.

Part I. (Complete markets)

In this part, please assume that there are no enforcement (or commitment) problems.

a. Solve the Pareto problem for this economy, attaching equal weights to the two types of consumer.

b. Show how to decentralize the allocation that solves the Pareto problem with a competitive equilibrium with *ex ante* (i.e., before time 0) trading of a complete set of history-contingent commodities. Please calculate the price of a one-period risk-free security.

Part II. (Enforcement problems)

In this part, assume that there are enforcement problems. In particular, assume that there is two-sided lack of commitment.

c. Pose an *ex ante* Pareto problem in which, after having observed its current endowment but before receiving his allocation from the Pareto planner, each consumer is free at any time to defect from the social contract and live thereafter in autarky. Show how to compute the value of autarky for each type of consumer.

d. Call an allocation *sustainable* if neither household would ever choose to defect to autarky. Formulate the enforcement-constrained Pareto problem recursively. That is, please write a programming problem that can be used to compute an optimal sustainable allocation.

e. Under what circumstance will the allocation that you found in part I solve the enforcement-constrained Pareto problem in part d? I.e., state conditions on $u, \beta, \overline{y}$ that are sufficient to make the enforcement constraints never bind.

Some useful background: For the remainder of this problem, please assume that $u, \beta, \overline{y}$, are such that the allocation computed in part I is not sustainable. Recall that the amnesia property implies that the consumption allocated to an agent whose participation constraint is binding is independent of the *ex ante* promised value with which he enters the period. With the present i.i.d., two-state, symmetric endowment pattern, *ex ante*, each period each of our two agents has an equal chance that it is his participation constraint that is binding. In a symmetric sustainable allocation, let each agent *enter* the period with the *same ex ante* promised value v, and let $\overline{c}$ be the consumption allocated to the high endowment agent whose participation constraint is binding and let $1 - \bar{c}$ be the consumption allocated to the low endowment agent whose participation constraint is not binding. By the above argument, $\bar{c}$ is independent of the promised value v that an agent enters the period with, which means that the current allocation to *both* types of agent does not depend on the promised value with which they entered the period. And in a symmetric stationary sustainable allocation, both consumers enter each period with the same promised value v.

f. Please give a formula for the promised value v within a symmetric stationary sustainable allocation.

g. Use a graphical argument to show how to determine the $v, \bar{c}$ that are associated with an optimal stationary symmetric allocation.

h. In the optimal stationary sustainable allocation that you computed in part g, why doesn't the planner adjust the continuation value of the consumer whose participation constraint is binding?

i. Alvarez and Jermann showed that, provided that the usual constraints on issuing Arrow securities are tightened enough, the optimal sustainable allocation can be decentralized by trading in a complete set of Arrow securities with price

$$q(y'|y) = \max_{i=1,2} \beta \frac{u'(c_{t+1}^i(y'))}{u'(c_t^i(y))} .5,$$

where $q(y'|y)$ is the price of one unit of consumption tomorrow, contingent on tomorrow's endowment of the type 1 person being y' when it is y today. This formula has each Arrow security being priced by the agent whose participation constraint is *not* binding. Heuristically, the agent who wants to *buy* the state-contingent security determines its price because the agent who wants to sell it is constrained from selling more by a limitation on the quantity of Arrow securities that he can promise to deliver in that future state. Evidently the gross rate of interest on a one-period risk-free security is

$$R(y) = \frac{1}{\sum_{y'} q(y'|y)},$$

for $y = \bar{y}$ and $y = 1 - \bar{y}$.

For the case in which the parameters are such that the allocation computed in part I is *not* sustainable (so that the participation constraints bind), please compute the risk-free rate of interest. Is it higher or lower than that for the complete markets economy without enforcement problems that you analyzed in part I?

Chapter 22
Optimal Unemployment Insurance

22.1. History-dependent unemployment insurance

This chapter applies the recursive contract machinery studied in chapters 20, 21, and 23 in contexts that are simple enough that we can go a long way toward computing optimal contracts by hand. The contracts encode history dependence by mapping an initial promised value and a random time t observation into a time t consumption allocation and a continuation value to bring into next period. We use recursive contracts to study good ways of providing consumption insurance when incentive problems come from the insurance authority's inability to observe the effort that an unemployed person exerts searching for a job. We begin by studying a setup of Shavell and Weiss (1979) and Hopenhayn and Nicolini (1997) that focuses on a single isolated spell of unemployment followed by permanent employment. Later we take up settings of Wang and Williamson (1996) and Zhao (2001) with alternating spells of employment and unemployment in which the planner has limited information about a worker's effort while he is on the job, in addition to not observing his search effort while he is unemployed. Here history dependence manifests itself in an optimal contract with intertemporal tie-ins across these spells. Zhao uses her model to rationalize unemployment compensation that replaces a fraction of a worker's earnings on his or her previous job.

22.2. A one-spell model

This section describes a model of optimal unemployment compensation along
the lines of Shavell and Weiss (1979) and Hopenhayn and Nicolini (1997). We
shall use the techniques of Hopenhayn and Nicolini to analyze a model closer
to Shavell and Weiss's. An unemployed worker orders stochastic processes of
consumption and search effort $\{c_t, a_t\}_{t=0}^{\infty}$ according to

$$E \sum_{t=0}^{\infty} \beta^t \left[u(c_t) - a_t \right] \qquad (22.2.1)$$

where $\beta \in (0, 1)$ and $u(c)$ is strictly increasing, twice differentiable, and strictly
concave. We assume that $u(0)$ is well defined. We require that $c_t \geq 0$ and
$a_t \geq 0$. All jobs are alike and pay wage $w > 0$ units of the consumption good
each period forever. An unemployed worker searches with effort a and with
probability $p(a)$ receives a permanent job at the beginning of the next period.
Once a worker has found a job, he is beyond the grasp of the unemployment
insurance agency.[1] Furthermore, $a = 0$ when the worker is employed. The
probability of finding a job is $p(a)$ where p is an increasing and strictly concave
and twice differentiable function of a, satisfying $p(a) \in [0, 1]$ for $a \geq 0$, $p(0) =
0$. The consumption good is nonstorable. The unemployed worker has no savings
and cannot borrow or lend. The insurance agency is the unemployed worker's
only source of consumption smoothing over time and across states.

---

[1] This is Shavell and Weiss's assumption, but not Hopenhayn and Nicolini's. Hopenhayn
and Nicolini allow the unemployment insurance agency to impose history-dependent taxes
on previously unemployed workers. Since there is no incentive problem after the worker has
found a job, it is optimal for the agency to provide an employed worker with a constant level
of consumption, and hence, the agency imposes a permanent per-period history-dependent
tax on a previously unemployed worker. See exercise 22.2.

22.2.1. The autarky problem

As a benchmark, we first study the fate of the unemployed worker who has no access to unemployment insurance. Because employment is an absorbing state for the worker, we work backward from that state. Let V^e be the expected sum of discounted one-period utilities of an employed worker. Once the worker is employed, $a = 0$, making his period utility be $u(c) - a = u(w)$ forever. Therefore,

$$V^e = \frac{u(w)}{(1-\beta)}. \tag{22.2.2}$$

Now let V^u be the expected present value of utility for an unemployed worker who chooses the current period pair (c, a) optimally. The Bellman equation for V^u is

$$V^u = \max_{a \geq 0} \left\{ u(0) - a + \beta \left[p(a)V^e + (1 - p(a))V^u \right] \right\}. \tag{22.2.3}$$

The first-order condition for this problem is

$$\beta p'(a) \left[V^e - V^u \right] \leq 1, \tag{22.2.4}$$

with equality if $a > 0$. Since there is no state variable in this infinite horizon problem, there is a time-invariant optimal search intensity a and an associated value of being unemployed V^u. Let $V_{\text{aut}} = V^u$ denote the solution of Bellman equation (22.2.3).

Equations (22.2.3) and (22.2.4) form the basis for an iterative algorithm for computing $V^u = V_{\text{aut}}$. Let V_j^u be the estimate of V_{aut} at the jth iteration. Use this value in equation (22.2.4) and solve for an estimate of effort a_j. Use this value in a version of equation (22.2.3) with V_j^u on the right side to compute V_{j+1}^u. Iterate to convergence.

22.2.2. Unemployment insurance with full information

As another benchmark, we study the provision of insurance with full information. An insurance agency can observe and control the unemployed person's consumption and search effort. The agency wants to design an unemployment insurance contract to give the unemployed worker expected discounted utility $V > V_{\text{aut}}$. The planner wants to deliver value V in the most efficient way, meaning the way that minimizes expected discounted cost, using β as the discount factor. We formulate the optimal insurance problem recursively. Let $C(V)$ be the expected discounted cost of giving the worker expected discounted utility V. The cost function is strictly convex because a higher V implies a lower marginal utility of the worker; that is, additional expected "utils" can be awarded to the worker only at an increasing marginal cost in terms of the consumption good. Given V, the planner assigns first-period pair (c, a) and promised continuation value V^u, should the worker be unlucky and not find a job; (c, a, V^u) will all be chosen to be functions of V and to satisfy the Bellman equation

$$C(V) = \min_{c, a, V^u} \left\{ c + \beta[1 - p(a)]C(V^u) \right\},\tag{22.2.5}$$

where the minimization is subject to the promise-keeping constraint

$$V \le u(c) - a + \beta\left\{p(a)V^e + [1 - p(a)]V^u\right\}.\tag{22.2.6}$$

Here V^e is given by equation (22.2.2), which reflects the assumption that once the worker is employed, he is beyond the reach of the unemployment insurance agency. The right side of Bellman equation (22.2.5) is attained by policy functions $c = c(V), a = a(V)$, and $V^u = V^u(V)$. The promise-keeping constraint, equation (22.2.6), asserts that the 3-tuple (c, a, V^u) attains at least V. Let θ be the Lagrange multiplier on constraint (22.2.6). At an interior solution, the first-order conditions with respect to c, a, and V^u, respectively, are

$$\theta = \frac{1}{u'(c)},\tag{22.2.7a}$$

$$C(V^u) = \theta\left[\frac{1}{\beta p'(a)} - (V^e - V^u)\right],\tag{22.2.7b}$$

$$C'(V^u) = \theta.\tag{22.2.7c}$$

The envelope condition $C'(V) = \theta$ and equation (22.2.7c) imply that $C'(V^u) = C'(V)$. Strict convexity of C then implies that $V^u = V$. Applied repeatedly over time, $V^u = V$ makes the continuation value remain constant during the entire spell of unemployment. Equation $(22.2.7a)$ determines c, and equation $(22.2.7b)$ determines a, both as functions of the promised V. That $V^u = V$ then implies that c and a are held constant during the unemployment spell. Thus, the unemployed worker's consumption c and search effort a are both "fully smoothed" during the unemployment spell. But the worker's consumption is not smoothed across states of employment and unemployment unless $V = V^e$.

22.2.3. The incentive problem

The preceding efficient insurance scheme requires that the insurance agency control both c and a. It will not do for the insurance agency simply to announce c and then allow the worker to choose a. Here is why. The agency delivers a value V^u higher than the autarky value V_{aut} by doing two things. It *increases* the unemployed worker's consumption c and *decreases* his search effort a. But the prescribed search effort is *higher* than what the worker would choose if he were to be guaranteed consumption level c while he remains unemployed. This follows from equations $(22.2.7a)$ and $(22.2.7b)$ and the fact that the insurance scheme is costly, $C(V^u) > 0$, which imply $[\beta p'(a)]^{-1} > (V^e - V^u)$. But look at the worker's first-order condition $(22.2.4)$ under autarky. It implies that if search effort $a > 0$, then $[\beta p'(a)]^{-1} = [V^e - V^u]$, which is inconsistent with the preceding inequality $[\beta p'(a)]^{-1} > (V^e - V^u)$ that prevails when $a > 0$ under the social insurance arrangement. If he were free to choose a, the worker would therefore want to fulfill $(22.2.4)$, either at equality so long as $a > 0$, or by setting $a = 0$ otherwise. Starting from the a associated with the social insurance scheme, he would establish the desired equality in $(22.2.4)$ by *lowering* a, thereby decreasing the term $[\beta p'(a)]^{-1}$ (which also lowers $(V^e - V^u)$ when the value of being unemployed V^u increases]). If an equality can be established before a reaches zero, this would be the worker's preferred search effort; otherwise the worker would find it optimal to accept the insurance payment, set $a = 0$, and never work again. Thus, since the worker does not take the cost of the insurance scheme into account, he would choose a search effort below the socially optimal one. The efficient contract exploits the agency's ability to control *both* the unemployed worker's consumption *and* his search effort.

22.2.4. Unemployment insurance with asymmetric information

Following Shavell and Weiss (1979) and Hopenhayn and Nicolini (1997), now assume that the unemployment insurance agency cannot observe or enforce a, though it can observe and control c. The worker is free to choose a, which puts expression (22.2.4), the worker's first-order condition under autarky, back in the picture.[2] Given any contract, the individual will choose search effort according to the first-order condition (22.2.4). This fact leads the insurance agency to design the unemployment insurance contract to respect this restriction. Thus, the recursive contract design problem is now to minimize the right side of equation (22.2.5) subject to expression (22.2.6) and the incentive constraint (22.2.4).

Since the restrictions (22.2.4) and (22.2.6) are not linear and generally do not define a convex set, it becomes difficult to provide conditions under which the solution to the dynamic programming problem results in a convex function $C(V)$. As discussed in Appendix A of chapter 20, this complication can be handled by convexifying the constraint set through the introduction of lotteries. However, a common finding is that optimal plans do not involve lotteries, because convexity of the constraint set is a sufficient but not necessary condition for convexity of the cost function. Following Hopenhayn and Nicolini (1997), we therefore proceed under the assumption that $C(V)$ is strictly convex in order to characterize the optimal solution.

Let η be the multiplier on constraint (22.2.4), while θ continues to denote the multiplier on constraint (22.2.6). But now we replace the weak inequality in (22.2.6) by an equality. The unemployment insurance agency cannot award a higher utility than V because that might violate an incentive-compatibility constraint for exerting the proper search effort in earlier periods. At an interior solution, the first-order conditions with respect to c, a, and V^u, respectively, are[3]

$$\theta = \frac{1}{u'(c)}, \tag{22.2.8a}$$

$$C(V^u) = \theta \left[\frac{1}{\beta p'(a)} - (V^e - V^u) \right] - \eta \frac{p''(a)}{p'(a)} (V^e - V^u)$$

---

[2] We are assuming that the worker's best response to the unemployment insurance arrangement is completely characterized by the first-order condition (22.2.4), an instance of the so-called "first-order" approach to incentive problems.

[3] Hopenhayn and Nicolini let the insurance agency also choose V^e, the continuation value from V, if the worker finds a job. This approach reflects their assumption that the agency can tax a previously unemployed worker after he becomes employed. See exercise 22.2.

$$= -\eta \frac{p''(a)}{p'(a)} (V^e - V^u), \qquad (22.2.8b)$$

$$C'(V^u) = \theta - \eta \frac{p'(a)}{1 - p(a)}, \qquad (22.2.8c)$$

where the second equality in equation $(22.2.8b)$ follows from strict equality of the incentive constraint $(22.2.4)$ when $a > 0$. As long as the insurance scheme is associated with costs, so that $C(V^u) > 0$, first-order condition $(22.2.8b)$ implies that the multiplier η is strictly positive. The first-order condition $(22.2.8c)$ and the envelope condition $C'(V) = \theta$ together allow us to conclude that $C'(V^u) < C'(V)$. Convexity of C then implies that $V^u < V$. After we have also used equation $(22.2.8a)$, it follows that in order to provide the proper incentives, the consumption of the unemployed worker must decrease as the duration of the unemployment spell lengthens. It also follows from $(22.2.4)$ at equality that search effort a rises as V^u falls, i.e., it rises with the duration of unemployment.

The duration dependence of benefits is designed to provide incentives to search. To see this, from $(22.2.8c)$, notice how the conclusion that consumption falls with the duration of unemployment depends on the assumption that more search effort raises the prospect of finding a job, i.e., that $p'(a) > 0$. If $p'(a) = 0$, then $(22.2.8c)$ and the strict convexity of C imply that $V^u = V$. Thus, when $p'(a) = 0$, there is no reason for the planner to make consumption fall with the duration of unemployment.

22.2.5. Computed example

For parameters chosen by Hopenhayn and Nicolini, Figure 22.2.1 displays the replacement ratio c/w as a function of the duration of the unemployment spell.[4] This schedule was computed by finding the optimal policy functions

$$V_{t+1}^u = f(V_t^u)$$

$$c_t = g(V_t^u).$$

and iterating on them, starting from some initial $V_0^u > V_{\mathrm{aut}}$, where V_{aut} is the autarky level for an unemployed worker. Notice how the replacement ratio

---

[4] This figure was computed using the Matlab programs `hugo.m`, `hugo1a.m`, `hugofoc1.m`, `valhugo.m`. These are available in the subdirectory `hugo`, which contains a readme file. These programs were composed by various members of Economics 233 at Stanford in 1998, especially Eva Nagypal, Laura Veldkamp, and Chao Wei.

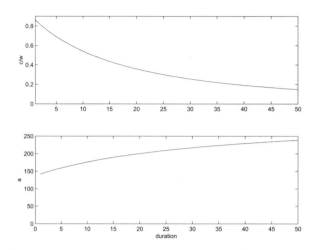

Figure 22.2.1: Top panel: replacement ratio c/w as a function of duration of unemployment in the Shavell-Weiss model. Bottom panel: effort a as a function of duration.

declines with duration. Figure 22.2.1 sets V_0^u at 16,942, a number that has to be interpreted in the context of Hopenhayn and Nicolini's parameter settings.

We computed these numbers using the parametric version studied by Hopenhayn and Nicolini.[5] Hopenhayn and Nicolini chose parameterizations and parameters as follows: They interpreted one period as one week, which led them to set $\beta = .999$. They took $u(c) = \frac{c^{(1-\sigma)}}{1-\sigma}$ and set $\sigma = .5$. They set the wage $w = 100$ and specified the hazard function to be $p(a) = 1 - \exp(-ra)$, with r chosen to give a hazard rate $p(a^*) = .1$, where a^* is the optimal search effort under autarky. To compute the numbers in Figure 22.2.1 we used these same settings.

---

[5] In section 4.7.3, we described a computational strategy of iterating to convergence on the Bellman equation (22.2.5), subject to expressions (22.2.6) at equality, and (22.2.4).

22.2.6. *Computational details*

Exercise *22.1* asks the reader to solve the Bellman equation numerically. In doing so, it is useful to note that there are natural lower and upper bounds to the set of continuation values V^u. The lower bound is the expected lifetime utility in autarky, V_{aut}. To compute the upper bound, represent condition (22.2.4) as

$$V^u \geq V^e - [\beta p'(a)]^{-1},$$

with equality if $a > 0$. If there is zero search effort, then $V^u \geq V^e - [\beta p'(0)]^{-1}$. Therefore, to rule out zero search effort we require

$$V^u < V^e - [\beta p'(0)]^{-1}.$$

(Remember that $p''(a) < 0$.) This step gives our upper bound for V^u.

   To formulate the Bellman equation numerically, we suggest using the constraints to eliminate c and a as choice variables, thereby reducing the Bellman equation to a minimization over the one choice variable V^u. First express the promise-keeping constraint (22.2.6) as $u(c) = V + a - \beta\{p(a)V^e + [1-p(a)]V^u\}$. That is, consumption is equal to

$$c = u^{-1}\left(V + a - \beta[p(a)V^e + (1 - p(a))V^u]\right). \tag{22.2.9}$$

Similarly, solving the inequality (22.2.4) for a and using the assumed functional form for $p(a)$ leads to

$$a = \max\left\{0, \frac{\log[r\beta(V^e - V^u)]}{r}\right\}. \tag{22.2.10}$$

Formulas (22.2.9) and (22.2.10) express (c, a) as functions of V and the continuation value V^u. Using these functions allows us to write the Bellman equation in $C(V)$ as

$$C(V) = \min_{V^u}\{c + \beta[1 - p(a)]C(V^u)\} \tag{22.2.11}$$

where c and a are given by equations (22.2.9) and (22.2.10).

22.2.7. Interpretations

The substantial downward slope in the replacement ratio in Figure 22.2.1 comes entirely from the incentive constraints facing the planner. We saw earlier that without private information, the planner would smooth consumption over the unemployment spell by keeping the replacement ratio constant. In the situation depicted in Figure 22.2.1, the planner can't observe the worker's search effort and therefore makes the replacement ratio fall and search effort rise as the duration of unemployment increases, especially early in an unemployment spell. There is a "carrot-and-stick" aspect to the replacement rate and search effort schedules: the "carrot" occurs in the forms of high compensation and low search effort early in an unemployment spell. The "stick" occurs in the low compensation and high effort later in the spell. We shall see this carrot-and-stick feature in some of the credible government policies analyzed in chapters 23, 24, and 25.

The planner offers declining benefits and asks for increased search effort as the duration of an unemployment spell rises in order to provide unemployed workers with proper incentives, not to punish an unlucky worker who has been unemployed for a long time. The planner believes that a worker who has been unemployed a long time is unlucky, not that he has done anything wrong (i.e., not lived up to the contract). Indeed, the contract is designed to induce the unemployed workers to search in the way the planner expects. The falling consumption and rising search effort of the unlucky ones with long unemployment spells are simply the prices that have to be paid for the common good of providing proper incentives.

22.2.8. Extension: an on-the-job tax

Hopenhayn and Nicolini allow the planner to tax the worker *after* he becomes employed, and they let the tax depend on the duration of unemployment. Giving the planner this additional instrument substantially decreases the rate at which the replacement ratio falls during a spell of unemployment. Instead, the planner makes use of a more powerful tool: a *permanent* bonus or tax after the worker becomes employed. Because it endures, this tax or bonus is especially potent when the discount factor is high. In exercise *22.2*, we ask the reader to set up the functional equation for Hopenhayn and Nicolini's model.

22.2.9. Extension: intermittent unemployment spells

In Hopenhayn and Nicolini's model, employment is an absorbing state and there are no incentive problems after a job is found. There are not multiple spells of unemployment. Wang and Williamson (1996) built a model in which there can be multiple unemployment spells, and in which there is also an incentive problem on the job. As in Hopenhayn and Nicolini's model, search effort affects the probability of finding a job. In addition, while on a job, effort affects the probability that the job ends and that the worker becomes unemployed again. Each job pays the same wage. In Wang and Williamson's setup, the promised value keeps track of the duration and number of spells of employment as well as of the number and duration of spells of unemployment. One contract transcends employment and unemployment.

22.3. A multiple-spell model with lifetime contracts

Rui Zhao (2001) modifies and extends features of Wang and Williamson's model. In her model, effort on the job affects output as well as the probability that the job will end. In Zhao's model, jobs randomly end, recurrently returning a worker to the state of unemployment. The probability that a job ends depends directly or indirectly on the effort that workers expend on the job. A planner observes the worker's output and employment status, but never his effort, and wants to insure the worker. Using recursive methods, Zhao designs a history-dependent assignment of unemployment benefits, if unemployed, and wages, if employed, that balance a planner's desire to insure the worker with the need to provide incentives to supply effort in work and search. The planner uses history dependence to tie compensation while unemployed (or employed) to earlier outcomes that partially inform the planner about the workers' efforts while employed (or unemployed). These intertemporal tie-ins give rise to what Zhao interprets broadly as a "replacement rate" feature that we seem to observe in unemployment compensation systems.

22.3.1. The setup

In a special case of Zhao's model, there are two effort levels. Where $a \in \{a_L, a_H\}$ is a worker's effort and $\overline{y}_i > \overline{y}_{i-1}$, an employed worker produces $y_t \in [\overline{y}_1, \cdots, \overline{y}_n]$ with probability

$$\mathrm{Prob}(y_t = \overline{y}_i) = p(\overline{y}_i; a).$$

Zhao assumes:

ASSUMPTION 1: $p(\overline{y}_i; a)$ satisfies the *monotone likelihood ratio* property: $\frac{p(\overline{y}_i; a_H)}{p(\overline{y}_i; a_L)}$ increases as $\overline{y}_i$ increases.

At the end of each period, jobs end with probability π_{eu}. Zhao embraces one of two alternative assumptions about the job separation rate π_{eu}, allowing it to depend on either current output y or current work effort a. She assumes:

ASSUMPTION 2: Either $\pi_{eu}(y)$ decreases with y or $\pi_{eu}(a)$ decreases with a.

Unemployed workers produce nothing and search for a job subject to the following assumption about the job finding rate $\pi_{ue}(a)$:

ASSUMPTION 3: $\pi_{ue}(a)$ increases with a.

The worker's one-period utility function is $U(c, a) = u(c) - \phi(a)$ where $u(\cdot)$ is continuously differentiable, strictly increasing and strictly concave, and $\phi(a)$ is continuous, strictly increasing, and strictly convex. The worker orders random $\{c_t, a_t\}_{t=0}^{\infty}$ sequences according to

$$E \sum_{t=0}^{\infty} \beta^t U(c_t, a_t), \quad \beta \in (0, 1). \qquad (22.3.1)$$

We shall regard a planner as being a coalition of firms united with an unemployment insurance agency. The planner is risk neutral and can borrow and lend at a constant risk-free gross one-period interest rate of $R = \beta^{-1}$.

Let the worker's employment state be $s_t \in S = \{e, u\}$ where e denotes employed, u unemployed. The worker's output at t is

$$z_t = \begin{cases} 0 & \text{if } s_t = u, \\ y_t & \text{if } s_t = e. \end{cases}$$

For $t \geq 1$, the time t component of the publicly observed information is

$$x_t = (z_{t-1}, s_t),$$

and $x_0 = s_0$. At time t, the planner observes the history x^t and the worker observes (x^t, a^t).

The transition probability for $x_{t+1} \equiv (z_t, s_{t+1})$ can be factored as follows:

$$\pi(x_{t+1}|s_t, a_t) = \pi_z(z_t; s_t, a_t)\pi_s(s_{t+1}; z_t, s_t, a_t) \qquad (22.3.2)$$

where π_z is the distribution of output conditioned on the state and the action, and π_s encodes the transition probabilities of employment status conditional on output, current employment status, and effort. In particular, Zhao assumes that

$$\begin{aligned}
\pi_s(u; 0, u, a) &= 1 - \pi_{ue}(a) \\
\pi_s(e; 0, u, a) &= \pi_{ue}(a) \\
\pi_s(u; y, e, a) &= \pi_{eu}(y, a) \\
\pi_s(e; y, e, a) &= 1 - \pi_{eu}(y, a).
\end{aligned} \qquad (22.3.3)$$

22.3.2. A recursive lifetime contract

Consider a worker with beginning-of-period employment status s and promised value v. For given (v, s), let $w(z, s')$ be the continuation value of promised utility (22.3.1) for next period when today's output is z and tomorrow's employment state is s'. At the beginning of next period, (z, s') will be the labor market outcome most recently observed by the planner. Let $W = \{W_s\}_{s \in \{u,e\}}$ be two compact sets of continuation values, one set for $s = u$ and another for $s = e$. For each (v, s), a *recursive contract* specifies a recommended effort level a today, an output-contingent consumption level $c(z)$ today, and continuation values $w(z, s')$ to be used to reset v tomorrow.

For each (v, s), the contract $(a, c(z), w(z, s'))$ must satisfy:

$$\sum_z \pi_z(z; s, a)\left(u(c(z)) + \beta \sum_{s'} \pi_s(s'; z, s, a)w(z, s')\right) - \phi(a) = v \qquad (22.3.4)$$

and

$$\begin{aligned}
&\sum_z \pi_z(z; s, a)\left(u(c(z)) + \beta \sum_{s'} \pi_s(s'; z, s, a)w(z, s')\right) - \phi(a) \geq \\
&\sum_z \pi_z(z; s, \tilde{a})\left(u(c(z)) + \beta \sum_{s'} \pi_s(s'; z, s, \tilde{a})w(z, s')\right) - \phi(\tilde{a}) \quad \forall \tilde{a}.
\end{aligned} \qquad (22.3.5)$$

Constraint $(22.3.4)$ entails *promise keeping*, while $(22.3.5)$ are the incentive-compatibility or "effort-inducing" constraints. In addition, a contract has to satisfy $\underline{c} \leq c(z) \leq \bar{c}$ for all z and $w(z, s') \in W_{s'}$ for all (z, s'). A contract is said to be *incentive compatible* if it satisfies the incentive compatibility constraints $(22.3.5)$.[6]

DEFINITION: A recursive contract $(a, c(z), w(z, s'))$ is said to be *feasible with respect to* W for a given (v, s) pair if it is incentive compatible in state s, delivers promised value v, and $w(z, s') \in W_{s'}$ for all (z, s').

Let $C(v, s)$ be the minimum cost to the planner of delivering promised value v to a worker in employment state s. We can represent the Bellman equation for $C(v, s)$ in terms of the following two-part optimization:

$$\Psi(v, s, a) = \min_{c(z), w(z, s')} \left\{ \sum_z \pi_z(z; s, a)\Big(-z + c(z) \right.$$

$$\left. + \beta \sum_{s'} \pi_s(s'; z, s, a) C(w(z, s'), s')\Big) \right\} \quad (22.3.6a)$$

subject to constraints $(22.3.4)$ and $(22.3.5)$, and

$$C(v, s) = \min_{a \in [a_L, a_H]} \Psi(v, s, a). \quad (22.3.6b)$$

The function $\Psi(v, s, a)$ assumes that the worker exerts effort level a. Later, we shall typically assume that parameters are such that $C(v, s) = \Psi(v, s, a_H)$, so that the planner finds it optimal always to induce high effort. Put a Lagrange multiplier $\lambda(v, s, a)$ on the promise-keeping constraint $(22.3.4)$ and another multiplier $\nu(v, s, a)$ on the effort-inducing constraint $(22.3.5)$ given a, and form the

---

[6] We assume two-sided commitment to the contract and therefore ignore the participation constraints that Zhao imposes on the contract. She requires that continuation values $w(z, s')$ be at least as great as the autarky values $V_{s', \text{aut}}$ for each (z, s').

Lagrangian:

$$
L = \sum_z \pi_z(z; s, a) \left\{ -z + c(z) + \beta \sum_{s'} \pi_s(s'; z, s, a) C(w(z, s'), s') \right.
$$

$$
- \lambda(v, s, a) \left[u(c(z)) + \beta \sum_{s'} \pi_s(s'; z, s, a) w(z, s')) - \phi(a) - v \right]
$$

$$
- \nu(v, s, a) \left[u(c(z)) + \beta \sum_{s'} \pi_s(s'; z, s, a) w(z, s') - \phi(a) \right.
$$

$$
\left. \left. - \frac{\pi_z(z; s, \tilde{a})}{\pi_z(z; s, a)} \left(u(c(z)) + \beta \sum_{s'} \pi_s(s'; z, s, \tilde{a}) w(z, s') - \phi(\tilde{a}) \right) \right] \right\},
$$

where $\tilde{a} \in \{a_L, a_H\}$ and $\tilde{a} \neq a$. First-order conditions for $c(z)$ and $w(z, s')$, respectively, are

$$
\frac{1}{u'(c(z))} = \lambda(v, s, a) + \nu(v, s, a) \left(1 - \frac{\pi_z(z; s, \tilde{a})}{\pi_z(z; s, a)} \right) \tag{22.3.7a}
$$

$$
C_v(w(z, s'), s') = \lambda(v, s, a)
$$
$$
+ \nu(v, s, a) \left[1 - \frac{\pi_z(z; s, \tilde{a})}{\pi_z(z; s, a)} \frac{\pi_s(s'; s, z, \tilde{a})}{\pi_s(s'; z, s, a)} \right]. \tag{22.3.7b}
$$

The envelope conditions are

$$
\Psi_v(v, s, a) = \lambda(v, s, a) \tag{22.3.8a}
$$
$$
C_v(v, s) = \Psi_v(v, s, a^*) \tag{22.3.8b}
$$

where a^* is the planner's optimal choice of a.

To deduce the dynamics of compensation, Zhao's strategy is to study the first-order conditions $(22.3.7)$ and envelope conditions $(22.3.8)$ under two cases, $s = u$ and $s = e$.

22.3.3. Compensation dynamics when unemployed

In the unemployed state $(s = u)$, the first-order conditions become

$$\frac{1}{u'(c)} = \lambda(v, u, a) \tag{22.3.9a}$$

$$C_v(w(0, u), u) = \lambda(v, u, a) + \nu(v, u, a)\left[1 - \frac{1 - \pi_{ue}(\tilde{a})}{1 - \pi_{ue}(a)}\right] \tag{22.3.9b}$$

$$C_v(w(0, e), e) = \lambda(v, u, a) + \nu(v, u, a)\left[1 - \frac{\pi_{ue}(\tilde{a})}{\pi_{ue}(a)}\right]. \tag{22.3.9c}$$

The effort-inducing constraint $(22.3.5)$ can be rearranged to become

$$\beta(\pi_{ue}(a) - \pi_{ue}(\tilde{a}))(w(0, e) - w(0, u)) \geq \phi(a) - \phi(\tilde{a}).$$

Like Hopenhayn and Nicolini, Zhao describes how compensation and effort depend on the duration of unemployment:

PROPOSITION: To induce high search effort, unemployment benefits must *fall* over an unemployment spell.

PROOF: When search effort is high, the effort-inducing constraint binds. By assumption 3,

$$\frac{1 - \pi_{ue}(a_L)}{1 - \pi_{ue}(a_H)} > 1 > \frac{\pi_{ue}(a_L)}{\pi_{ue}(a_H)}.$$

These inequalities and the first-order condition $(22.3.9)$ then imply

$$C_v(w(0, e), e) > \Psi_v(v, u, a_H) > C_v(w(0, u), u). \tag{22.3.10}$$

Let $c_u(t), v_u(t)$, respectively, be consumption and the continuation value for an unemployed worker. Equations $(22.3.9)$ and the envelope conditions imply

$$\frac{1}{u'(c_u(t))} = \Psi_v(v_u(t), u, a_H) > C_v(v_u(t+1), u) = \frac{1}{u'(c_u(t+1))}. \tag{22.3.11}$$

Concavity of u then implies that $c_u(t) > c_u(t+1)$. In addition, notice that

$$C_v(w(0, u), u) - C_v(v, u) = \nu(v, u, a_H)\left(1 - \frac{1 - \pi_{ue}(a_L)}{1 - \pi_{ue}(a_H)}\right), \tag{22.3.12}$$

which follows from the first-order conditions $(22.3.9)$ and the envelope conditions. Equation $(22.3.12)$ implies that continuation values fall with the duration of unemployment. ∎

22.3.4. Compensation dynamics while employed

When the worker is employed, for each promised value v, the contract specifies output-contingent consumption and continuation values a $c(y), w(y, s')$. When $s = e$, the first-order conditions $(22.3.7)$ become

$$\frac{1}{u'(c(y))} = \lambda(v, e, a) + \nu(v, e, a) \left(1 - \frac{p(y; \tilde{a})}{p(y; a)}\right) \tag{22.3.13a}$$

$$C_v(w(y, u), u) = \lambda(v, e, a) + \nu(v, e, a) \left(1 - \frac{p(y; \tilde{a})}{p(y; a)} \frac{\pi_{eu}(y, \tilde{a})}{\pi_{eu}(y, a)}\right) \tag{22.3.13b}$$

$$C_v(w(y, e), e) = \lambda(v, e, a) + \nu(v, e, a) \left(1 - \frac{p(y; \tilde{a})}{p(y; a)} \frac{1 - \pi_{eu}(y, \tilde{a})}{1 - \pi_{eu}(y, a)}\right). \tag{22.3.13b}$$

Zhao uses these first-order conditions to characterize how compensation depends on output:

PROPOSITION: To induce high work effort, wages and continuation values increase with current output.

PROOF: For any $y > \tilde{y}$, let $d = \frac{p(\tilde{y}; a_L)}{p(\tilde{y}; a_H)} - \frac{p(y; a_L)}{p(y; a_H)}$. Assumption 1 about $p(y; a)$ implies that $d > 0$. The first-order conditions $(22.3.13)$ imply that

$$\frac{1}{u'(c(y))} - \frac{1}{u'(c(\tilde{y}))} = \nu(v, e, a)d > 0, \tag{22.3.14a}$$

$$C_v(w(y, u), u) - C_v(w(\tilde{y}, u), u) \propto \nu(v, e, a)d > 0, \tag{22.3.14b}$$

$$C_v(w(y, e), e) - C_v(w(\tilde{y}, e), e) \propto \nu(v, e, a)d > 0. \tag{22.3.14c}$$

Concavity of u and convexity of C give the result. ∎

In the following proposition, Zhao shows how continuation values at the start of unemployment spells should depend on the history of the worker's outcomes during previous employment and unemployment spells.

PROPOSITION: If the job separation rate depends on current output, then the replacement rate immediately after a worker loses a job is 100%. If the job separation rate depends on work effort, then the replacement rate is less than 100%.

PROOF: If the job separation rate depends on *output*, the first-order conditions $(22.3.13)$ imply

$$\frac{1}{u'(c(y))} = C_v(w(y, u), u) = C_v(w(y, e), e)). \tag{22.3.15}$$

This is because $\pi_{eu}(y, \tilde{a}) = \pi_{eu}(y, a)$ when the job separation rate depends on output. Let $c_e(t), c_u(t)$ be consumption of employed and unemployed workers, and let $v_e(t), v_u(t)$ be the assigned promised values at t. Then

$$\frac{1}{u'(c_e(t))} = C_v(v_{u,t+1}, u) = \frac{1}{c_u(t+1)}$$

where the first equality follows from $(22.3.15)$ and the second from the envelope condition. If the job separation rate depends on *work effort*, then the first-order conditions $(22.3.13)$ imply

$$\frac{1}{u'(c(y))} - C_v(w(y, u), u) = \nu(v, e, a)\frac{p(y; a_L)}{p(y; a_H)}\left(\frac{\pi_{eu}(a_L)}{\pi_{eu}(a_H)} - 1\right). \qquad (22.3.16)$$

Assumption 2 implies that the right side of $(22.3.16)$ is positive, which implies that

$$\frac{1}{u'(c_e(t))} > C_v(v_u(t+1), u) = \frac{1}{u'(c_u(t+1))}.$$

∎

22.3.5. Summary

A worker in Zhao's model enters a lifetime contract that makes compensation respond to the history of outputs on the current and past jobs, as well as on the durations of all previous spells of unemployment.[7] Her model has the outcome that compensation at the beginning of an unemployment spell varies directly with the compensation attained on the previous job. This aspect of her model offers a possible explanation for why unemployment insurance systems often feature a "replacement rate" that gives more unemployment insurance payments to workers who had higher wages in their prior jobs.

---

[7] We have analyzed a version of Zhao's model in which the worker is committed to obey the contract. Zhao incorporates an enforcement problem in her model by allowing the worker to accept an outside option each period.

22.4.   Concluding remarks

The models that we have studied in this chapter isolate the worker from capital markets so that the worker cannot transfer consumption across time or states except by adhering to the contract offered by the planner. If the worker in the models of this chapter were allowed to save or issue a risk-free asset bearing a gross one-period rate of return approaching β^{-1}, it would interfere substantially with the planner's ability to provide incentives by manipulating the worker's continuation value in response to observed current outcomes. In particular, forces identical to those analyzed in the Cole and Kocherlakota setup that we analyzed at length in chapter 20 would circumscribe the planner's ability to supply insurance. In the context of unemployment insurance models like that of this chapter, this point has been studied in detail in papers by Ivan Werning (2002) and Kocherlakota (2004).

Pavoni and Violante (2007) substantially extended models like those in this chapter to perform positive and normative analysis of a sequence of government programs that try efficiently to provide insurance, training, and proper incentives for unemployed and undertrained workers to reenter employment.

Exercises

Exercise 22.1   **Optimal unemployment compensation**

a. Write a program to compute the autarky solution, and use it to reproduce Hopenhayn and Nicolini's calibration of r, as described in text.

b. Use your calibration from part a. Write a program to compute the optimum value function $C(V)$ for the insurance design problem with incomplete information. Use the program to form versions of Hopenhayn and Nicolini's table 1, column 4 for three different initial values of V, chosen by you to belong to the set (V_{aut}, V^e).

Exercise 22.2   **Taxation after employment**

Show how the functional equation (22.2.5), (22.2.6) would be modified if the planner were permitted to tax workers after they became employed.

Exercise 22.3 **Optimal unemployment compensation with unobservable wage offers**

Consider an unemployed person with preferences given by

$$E \sum_{t=0}^{\infty} \beta^t u(c_t) \,,$$

where $\beta \in (0,1)$ is a subjective discount factor, $c_t \geq 0$ is consumption at time t, and the utility function $u(c)$ is strictly increasing, twice differentiable, and strictly concave. Each period the worker draws one offer w from a uniform wage distribution on the domain $[w_L, w_H]$ with $0 \leq w_L < w_H < \infty$. Let the cumulative density function be denoted $F(x) = \text{Prob}\{w \leq x\}$, and denote its density by f, which is constant on the domain $[w_L, w_H]$. After the worker has accepted a wage offer w, he receives the wage w per period forever. He is then beyond the grasp of the unemployment insurance agency. During the unemployment spell, any consumption smoothing has to be done through the unemployment insurance agency because the worker holds no assets and cannot borrow or lend.

a. Characterize the worker's optimal reservation wage when he is entitled to a time-invariant unemployment compensation b of indefinite duration.

b. Characterize the optimal unemployment compensation scheme under full information. That is, we assume that the insurance agency can observe and control the unemployed worker's consumption and reservation wage.

c. Characterize the optimal unemployment compensation scheme under asymmetric information where the insurance agency cannot observe wage offers, though it can observe and control the unemployed worker's consumption. Discuss the optimal time profile of the unemployed worker's consumption level.

Exercise 22.4 **Full unemployment insurance**

An unemployed worker orders stochastic processes of consumption, search effort $\{c_t, a_t\}_{t=0}^{\infty}$ according to

$$E \sum_{t=0}^{\infty} \beta^t \left[u(c_t) - a_t \right]$$

where $\beta \in (0,1)$ and $u(c)$ is strictly increasing, twice differentiable, and strictly concave. It is required that $c_t \geq 0$ and $a_t \geq 0$. All jobs are alike and pay wage

$w > 0$ units of the consumption good each period forever. After a worker has found a job, the unemployment insurance agency can tax the employed worker at a rate τ consumption goods per period. The unemployment agency can make τ depend on the worker's unemployment history. The probability of finding a job is $p(a)$, where p is an increasing and strictly concave and twice differentiable function of a, satisfying $p(a) \in [0,1]$ for $a \geq 0$, $p(0) = 0$. The consumption good is nonstorable. The unemployed person cannot borrow or lend and holds no assets. If the unemployed worker is to do any consumption smoothing, it has to be through the unemployment insurance agency. The insurance agency can observe the worker's search effort and can control his consumption. An employed worker's consumption is $w - \tau$ per period.

a. Let V_{aut} be the value of an unemployed worker's expected discounted utility when he has no access to unemployment insurance. An unemployment insurance agency wants to insure unemployed workers and to deliver expected discounted utility $V > V_{\text{aut}}$ at minimum expected discounted cost $C(V)$. The insurance agency also uses the discount factor β. The insurance agency controls c, a, τ, where c is consumption of an unemployed worker. The worker pays the tax τ only after he becomes employed. Formulate the Bellman equation for $C(V)$.

Exercise 22.5   **Two effort levels**

An unemployment insurance agency wants to insure unemployed workers in the most efficient way. An unemployed worker receives no income and chooses a sequence of search intensities $a_t \in \{0, a\}$ to maximize the utility functional

$$(1) \qquad E_0 \sum_{t=0}^{\infty} \beta^t \left\{ u(c_t) - a_t \right\}, \quad \beta \in (0, 1)$$

where $u(c)$ is an increasing, strictly concave, and twice continuously differentiable function of consumption of a single good. There are two values of the search intensity, 0 and a. The probability of finding a job at the beginning of period $t + 1$ is

$$(2) \qquad \pi(a_t) = \begin{cases} \pi(a), & \text{if } a_t = a; \\ \pi(0) < \pi(a), & \text{if } a_t = 0, \end{cases}$$

where we assume that $a > 0$. Note that the worker exerts search effort in period t and possibly receives a job at the beginning of period $t + 1$. Once the worker finds a job, he receives a fixed wage w forever, sets $a = 0$, and has continuation utility $V_e = \frac{u(w)}{1-\beta}$. The consumption good is not storable and workers can neither borrow nor lend. The unemployment agency can borrow and lend at a constant one-period risk-free gross interest rate of $R = \beta^{-1}$. The unemployment agency cannot observe the worker's effort level.

Subproblem A

a. Let V be the value of (1) that the unemployment agency has promised an unemployed worker at the start of a period (before he has made his search decision). Let $C(V)$ be the minimum cost to the unemployment insurance agency of delivering promised value V. Assume that the unemployment insurance agency wants the unemployed worker to set $a_t = a$ for as long as he is unemployed (i.e., it wants to promote high search effort). Formulate a Bellman equation for $C(V)$, being careful to specify any promise-keeping and incentive constraints. (Assume that there are no participation constraints: the unemployed worker must participate in the program.)

b. Show that if the incentive constraint binds, then the unemployment agency offers the worker benefits that decline as the duration of unemployment grows.

c. Now alter assumption (2) so that $\pi(a) = \pi(0)$. Do benefits still decline with increases in the duration of unemployment? Explain.

Subproblem B

d. Now assume that the unemployment insurance agency can tax the worker after he has found a job, so that his continuation utility upon entering a state of employment is $\frac{u(w-\tau)}{1-\beta}$, where τ is a tax that is permitted to depend on the duration of the unemployment spell. Defining V as above, formulate the Bellman equation for $C(V)$.

e. Show how the tax τ responds to the duration of unemployment.

Exercise 22.6   **Partially observed search effort**

Consider the following modification of a model of Hopenhayn and Nicolini. An insurance agency wants to insure an infinitely lived unemployed worker against the risk that he will not find a job. With probability $p(a)$, an unemployed worker who searches with effort a this period will find a job that earns wage w in consumption units per period. That job will start next period, last forever, and the worker will never quit it. With probability $1 - p(a)$ he will find himself unemployed again at the beginning of next period. We assume that $p(a)$ is an increasing and strictly concave and twice differentiable function of a with $p(a) \in [0, 1]$ for $a \geq 0$ and $p(0) = 0$. The insurance agency is the worker's only source of consumption (there is no storage or saving available to the worker). The worker values consumption according to a twice continuously differentiable and strictly concave utility function $u(c)$ where $u(0)$ is finite. While unemployed, the worker's utility is $u(c) - a$; when he is employed it is $u(w)$ (no effort a need be applied when he is working).

With exogenous probability $d \in (0, 1)$ the insurance agency observes the search effort of a worker who searched last period but did not find a job. With probability $1 - d$, the insurance agency does not observe the last-period search intensity of an unemployed worker who was not successful in finding a job period.

Let V be the expected discounted utility of an unemployed worker who is searching for work this period. Let $C(V)$ be the minimum cost to the unemployment insurance agency of delivering V to the unemployed worker.

a. Formulate a Bellman equation for $C(V)$.

b. Get as far as you can in analyzing how the unemployment compensation contract offered to the worker depends on the duration of unemployment and the history of observed search efforts that are detected by the UI agency.

Hint: you might want to allow the continuation value when unemployed to depend on last period's search effort when it is observed.

Chapter 23
Credible Government Policies, I

23.1. Introduction

Kydland and Prescott (1977) opened the modern discussion of time consistency in macroeconomics with some examples that show how outcomes differ in otherwise identical economies when the assumptions about the timing of government policy choices are altered.[1] In particular, they compared a timing protocol in which a government chooses its (possibly history-contingent) policies once and for all at the beginning of time with one in which the government chooses sequentially. Because outcomes are worse when the government chooses sequentially, Kydland and Prescott's examples illustrate the value to a government of having a "commitment technology" that requires it not to choose sequentially.

Subsequent work on time consistency focused on how a reputation can substitute for a commitment technology when the government chooses sequentially.[2] The issue is whether constraints confronting the government and private sector expectations can be arranged so that a government adheres to an expected pattern of behavior because it would worsen its reputation if it did not.

A "folk theorem" from game theory states that if there is no discounting of future payoffs, then virtually any first-period payoff can be sustained as a reputational equilibrium. A main purpose of this chapter is to study how discounting might shrink the set of outcomes that are attainable with a reputational mechanism.

Modern formulations of reputational models of government policy use and extend ideas from dynamic programming. Each period, a government faces choices whose consequences include a first-period return and a reputation to

---

[1] Consider two extensive-form versions of the "battle of the sexes" game described by Kreps (1990), one in which the man chooses first, the other in which the woman chooses first. Backward induction recovers different outcomes in these two different games. Though they share the same choice sets and payoffs, these are different games.

[2] Barro and Gordon (1983a, 1983b) are early contributors to this literature. See Kenneth Rogoff (1989) for a survey.

pass on to next period. Under rational expectations, any reputation that the government carries into next period must be one that it will want to confirm. We shall study the set of possible values that the government can attain with reputations that it could conceivably want to confirm.

This and the following chapter apply an apparatus of Abreu, Pearce, and Stacchetti (1986, 1990) (APS) to reputational equilibria in a class of macroeconomic models. APS use ideas from dynamic programming.[3] Their work exploits the insight that it is more convenient to work with the set of continuation values associated with equilibrium strategies than it is to work directly with the set of equilibrium strategies. We use an economic model like those of Chari, Kehoe, and Prescott (1989) and Stokey (1989, 1991) to exhibit what Chari and Kehoe (1990) call sustainable government policies and what Stokey calls credible public policies.   The literature on sustainable or credible government policies in macroeconomics adapts ideas from the literature on repeated games so that they can be applied in contexts in which a single agent (a government) behaves strategically, and in which all other agents' behavior can be summarized in terms of a system of expectations about government actions together with competitive equilibrium outcomes that respond to the government's choices.[4]

---

[3] This chapter closely follow Stacchetti (1991), who applies Abreu, Pearce, and Stacchetti (1986, 1990) to a more general class of models than that treated here. Stacchetti also studies a class of setups in which the private sector observes only a noise-ridden signal of the government's actions.

[4] For descriptions of theories of credible government policy, see Chari and Kehoe (1990), Stokey (1989, 1991), Rogoff (1989), and Chari, Kehoe, and Prescott (1989). For applications of the framework of Abreu, Pearce, and Stacchetti, see Chang (1998), and Phelan and Stacchetti (1999).

23.1.1. Diverse sources of history dependence

The theory of credible government policy uses particular kinds of history dependence to render credible a sequence of actions chosen by a *sequence* of policy makers. Here *credible* means an action that a government decision maker *wants* to implement. By way of contrast, in chapter 19, we encountered a distinct source of history dependence in the policy of a Ramsey planner or Stackelberg leader. There history dependence came from the requirement that it is necessary properly to account for constraints that dynamic aspects of private sector behavior put on the time t choices of a Ramsey planner or Stackelberg leader that at time 0 makes once-and-for-all choices of intertemporal sequences. In this context, history dependence emerges from the requirement that at time t the Ramsey planner must confirm private sector expectations about those time t actions that at time 0 were partly designed to influence private sector outcomes in periods $0, \ldots, t-1$.

In settings in which private agents face genuinely dynamic decision problems having their own endogenous state variables like various forms of physical and human capital, *both* of these sources of history dependence influence a credible policy. It can be subtle to disentangle the economic forces contributing to history dependence in government policies in such settings. However, for a special classes of examples with private agents' choices sufficiently simplified that they deprive private agents' decision problem of any 'natural' state variables, we can isolate the source of history dependence coming from the requirement that a government policy be must be credible. We consider such a special class in this chapter for the avowed purpose of isolating the source of history dependence coming from credibility considerations and distinguishing it from the chapter 19 source that instead comes from the need to respect substantial dynamics coming from equilibrium private sector behavior. Having isolated one source of history dependence in chapter 19 and another in the present chapter, we proceed in chapter 24 to activate both sources of history dependence and then seek a recursive representation for a credible government policy in that more complicated setting.

23.2.   The one-period economy

There is a continuum of households, each of which chooses an action $\xi \in X$. A government chooses an action $y \in Y$. The sets X and Y are compact. The average level of ξ across households is denoted $x \in X$. The utility of a particular household is $u(\xi, x, y)$ when it chooses ξ, when the average household's choice is x, and when the government chooses y. The payoff function $u(\xi, x, y)$ is strictly concave and continuously differentiable in ξ and y.[5]

23.2.1. Competitive equilibrium

For given levels of y and x, the representative household faces the problem $\max_{\xi \in X} u(\xi, x, y)$. Let the solution be a function $\xi = f(x, y)$. When a household believes that the government's choice is y and that the average level of other households' choices is x, it acts to set $\xi = f(x, y)$. Because all households are alike, this fact implies that the actual level of x is $f(x, y)$. For the representative household's expectations about the average to be consistent with the average outcome, we require that $\xi = x$, or $x = f(x, y)$. This makes the representative agent representative. We use the following:[6]

DEFINITION 1: A *competitive equilibrium* or a *rational expectations equilibrium* is an $x \in X$ that satisfies $x = f(x, y)$.

A competitive equilibrium satisfies $u(x, x, y) = \max_{\xi \in X} u(\xi, x, y)$.

For each $y \in Y$, let $x = h(y)$ denote the corresponding competitive equilibrium. We adopt:

DEFINITION 2: The set of competitive equilibria is $C = \{(x, y) \mid u(x, x, y) = \max_{\xi \in X} u(\xi, x, y)\}$, or equivalently $C = \{(x, y) \mid x = h(y)\}$.

---

[5] The discrete-choice examples given later violate some of these assumptions in non essential ways.

[6] See the definition of a rational expectations equilibrium in chapter 7.

23.2.2. *The Ramsey problem*

The following timing of actions underlies a *Ramsey plan*. First, the government selects a $y \in Y$. Then, knowing the government's choice of y, the aggregate of households responds with a competitive equilibrium. The government evaluates policies $y \in Y$ with the payoff function $u(x, x, y)$; that is, the government is benevolent.

In choosing y, the government has to forecast how the economy will respond. We assume that the government correctly forecasts that the economy will respond to y with a competitive equilibrium, $x = h(y)$. We use these definitions:

DEFINITION 3: The *Ramsey problem* is $\max_{y \in Y} u[h(y), h(y), y]$, or equivalently $\max_{(x,y) \in C} u(x, x, y)$.

DEFINITION 4: The policy that attains the maximum for the Ramsey problem is denoted y^R. Let $x^R = h(y^R)$. Then (y^R, x^R) is called the *Ramsey outcome* or *Ramsey plan.*

Two remarks about the Ramsey problem are in order. First, the Ramsey outcome is typically inferior to the "dictatorial outcome" that solves the unrestricted problem $\max_{x \in X,\, y \in Y} u(x, x, y)$, because the restriction $(x, y) \in C$ is in general binding. Second, the timing of actions is important. The Ramsey problem assumes that the government chooses first and must stick with its choice regardless of how private agents subsequently choose $x \in X$.

If the government were granted the opportunity to reconsider its plan *after* households had chosen $x = x^R$, the government would in general want to deviate from y^R because often there exists an $\alpha \neq y^R$ for which $u(x^R, x^R, \alpha) > u(x^R, x^R, y^R)$. The "time consistency problem" is the incentive the government would have to deviate from the Ramsey plan if it were allowed to react *after* households had set $x = x^R$. In this one-period setting, to support the Ramsey plan requires a timing protocol that forces the government to choose first.

23.2.3. Nash equilibrium

Consider an alternative timing protocol that confronts households with a forecasting problem because the government chooses after or simultaneously with the households. Assume that households forecast that, given x, the government will set y to solve $\max_{y \in Y} u(x,x,y)$. We use:

DEFINITION 5: A *Nash equilibrium* (x^N, y^N) satisfies

(1) $(x^N, y^N) \in C$;

(2) Given x^N, $u(x^N, x^N, y^N) = \max_{\eta \in Y} u(x^N, x^N, \eta)$.

Condition (1) asserts that $x^N = h(y^N)$, or that the economy responds to y^N with a competitive equilibrium. Thus, condition (1) says that given (x^N, y^N), each individual household wants to set $\xi = x^N$; that is, the representative household has no incentive to deviate from x^N. Condition (2) asserts that given x^N, the government chooses a policy y^N from which it has no incentive to deviate. [7]

We can use the solution of the problem in condition (2) to define the government's *best response* function $y = H(x)$. The definition of a Nash equilibrium can be phrased as a pair $(x, y) \in C$ such that $y = H(x)$.

There are two timings of choices for which a Nash equilibrium is a natural equilibrium concept. One is where households choose first, forecasting that the government will respond to the aggregate outcome x by setting $y = H(x)$. Another is where the government and households choose simultaneously, in which case a Nash equilibrium (x^N, y^N) depicts a situation in which everyone has rational expectations: given that each household expects the aggregate variables to be (x^N, y^N), each household responds in a way to make $x = x^N$, and given that the government expects that $x = x^N$, it responds by setting $y = y^N$.

We let the values attained by the government under the Nash and Ramsey outcomes, respectively, be denoted $v^N = u(x^N, x^N, y^N)$ and $v^R = u(x^R, x^R, y^R)$. Because of the additional constraint embedded in the Nash equilibrium, outcomes are ordered according to

$$v^N = \max_{\{(x,y) \in C: y = H(x)\}} u(x,x,y) \leq \max_{(x,y) \in C} u(x,x,y) = v^R .$$

---

[7] Much of the language of this chapter is borrowed from game theory, but the object under study is not a game, because we do not specify all of the objects that formally define a game. In particular, we do not specify the payoffs to all agents for all feasible choices. We only specify the payoffs $u(\xi, x, y)$ where each private agent chooses the *same* value of ξ.

23.3. Nash and Ramsey outcomes

To illustrate these concepts, we consider two examples: taxation within a fully specified economy, and a black-box model with discrete choice sets.

23.3.1. Taxation example

Each of a continuum of households has preferences over leisure ℓ, private consumption c, and per capita government expenditures g. The utility function is

$$U(\ell, c, g) = \ell + \log(\alpha + c) + \log(\alpha + g), \qquad \alpha \in (0, \tfrac{1}{2}).$$

Each household is endowed with one unit of time that can be devoted to leisure or labor. The production technology is linear in labor, and the economy's resource constraint is

$$\bar{c} + g = 1 - \bar{\ell},$$

where $\bar{c}$ and $\bar{\ell}$ are the average levels of private consumption and leisure, respectively.

A benevolent government wants to maximize the utility of the representative household. A benevolent government that is subject only to the constraint imposed by the technology and would choose $\ell = 0$ and $c = g = \tfrac{1}{2}$. This "dictatorial outcome" yields welfare $W^d = 2 \log(\alpha + \tfrac{1}{2})$.

Competitive equilibrium in general imposes more restrictions on the allocations attainable by a benevolent government. Here we will focus on competitive equilibria where the government finances its expenditures by levying a flat-rate tax τ on labor income. The household's budget constraint at equality is $c = (1 - \tau)(1 - \ell)$. Given a government policy (τ, g), an individual household's optimal decision rule for leisure is

$$\ell(\tau) = \begin{cases} \dfrac{\alpha}{1 - \tau} & \text{if } \tau \in [0, 1 - \alpha]; \\ 1 & \text{if } \tau > 1 - \alpha. \end{cases}$$

Due to the linear technology and the fact that government expenditures enter additively in the utility function, the household's decision rule $\ell(\tau)$ is also the equilibrium value of individual leisure at a given tax rate τ. Imposing government budget balance, $g = \tau(1 - \ell)$, the representative household's welfare in a competitive equilibrium can be expressed as a function of τ and is equal to

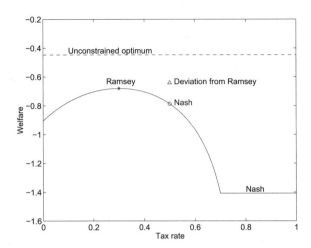

Figure 23.3.1: Welfare outcomes in the taxation example. The solid curve depicts the welfare associated with the set of competitive equilibria, $W^c(\tau)$. The set of Nash equilibria is the horizontal portion of the solid curve and the equilibrium at $\tau = 1/2$. The Ramsey outcome is marked with an asterisk. The "time inconsistency problem" is indicated with the triangle showing the outcome if the government were able to reset τ after households had chosen the Ramsey labor supply. The dashed line describes the welfare level at the unconstrained optimum, W^d. The graph sets $\alpha = 0.3$.

$$W^c(\tau) = \ell(\tau) + \log\{\alpha + (1-\tau)[1 - \ell(\tau)]\} + \log\{\alpha + \tau[1 - \ell(\tau)]\}.$$

The Ramsey tax rate and allocation are determined by the solution to $\max_\tau W^c(\tau)$. It can be verified that because $\alpha \in (0, .5)$, the Ramsey plan sets $\tau < .5$, which produces an allocation in which c, g, and $1 - \ell$ are all positive.

By way of contrast, the government's problem in a Nash equilibrium is $\max_\tau \{\ell + \log[\alpha + (1-\tau)(1-\ell)] + \log[\alpha + \tau(1-\ell)]\}$. If $\ell < 1$, the optimizer is $\tau = .5$. There is a continuum of Nash equilibria indexed by $\tau \in [1-\alpha, 1]$ where agents choose not to work, and consequently $c = g = 0$. The only Nash equilibrium with production is $\tau = 1/2$ with welfare level $W^c(1/2)$. This conclusion follows directly from the fact that the government's best response is $\tau = 1/2$ for any $\ell < 1$. These outcomes are illustrated numerically in Figure 23.3.1. Here the time inconsistency problem surfaces in the government's incentive, if offered the choice, to reset the tax rate τ, after the household has set its labor supply.

The objects of the general setup in the preceding section can be mapped into the present taxation example as follows: $\xi = \ell$, $x = \bar{\ell}$, $X = [0,1]$, $y = \tau$, $Y = [0,1]$, $u(\xi, x, y) = \xi + \log[\alpha + (1-y)(1-\xi)] + \log[\alpha + y(1-x)]$, $f(x,y) = \ell(y)$, $h(y) = \ell(y)$, and $H(x) = \frac{1}{2}$ if $x < 1$; and $H(x) \in [0,1]$ if $x = 1$.

23.3.2. Black-box example with discrete choice sets

Consider a black box example with $X = \{x_L, x_H\}$ and $Y = \{y_L, y_H\}$, in which $u(x, x, y)$ assume the values given in Table 23.3.1. Assume that values of $u(\xi, x, y)$ for $\xi \neq x$ are such that the values with asterisks for $\xi = x$ are competitive equilibria. In particular, we might assume that

$$u(\xi, x_i, y_j) = 0 \quad \text{when } \xi \neq x_i \text{ and } i = j,$$
$$u(\xi, x_i, y_j) = 20 \quad \text{when } \xi \neq x_i \text{ and } i \neq j.$$

These payoffs imply that $u(x_L, x_L, y_L) > u(x_H, x_L, y_L)$ (i.e., $3 > 0$), and $u(x_H, x_H, y_H) > u(x_L, x_H, y_H)$ (i.e., $10 > 0$). Therefore, (x_L, x_L, y_L) and (x_H, x_H, y_H) are competitive equilibria. Also, $u(x_H, x_H, y_L) < u(x_L, x_H, y_L)$ (i.e., $12 < 20$), so the dictatorial outcome cannot be supported as a competitive equilibrium.

	x_L	x_H
y_L	3*	12
y_H	1	10*

Table 23.3.1: One-period payoffs $u(x_i, x_i, y_j)$; * denotes $(x, y) \in C$; the Ramsey outcome is (x_H, y_H) and the Nash equilibrium outcome is (x_L, y_L).

Figure 23.3.2 depicts a timing of choices that supports the Ramsey outcome for this example. The government chooses first, then walks away. The Ramsey outcome (x_H, y_H) is the competitive equilibrium yielding the highest value of $u(x, x, y)$.

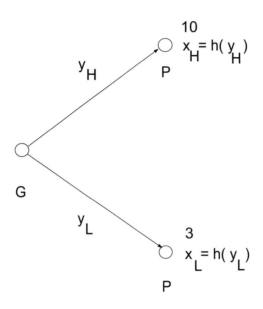

Figure 23.3.2: Timing of choices that supports Ramsey outcome. Here P and G denote nodes at which the public and the government, respectively, choose. The government has a commitment technology that binds it to "choose first." The government chooses the $y \in Y$ that maximizes $u[h(y), h(y), y]$, where $x = h(y)$ is the function mapping government actions into equilibrium values of x.

Figure 23.3.3 diagrams a timing of choices that supports the Nash equilibrium. Recall that by definition, every Nash equilibrium outcome has to be a competitive equilibrium outcome. We denote competitive equilibrium pairs (x, y) with asterisks. The government sector chooses after knowing that the private sector has set x, and chooses y to maximize $u(x, x, y)$. With this timing, if the private sector chooses $x = x_H$, the government has an incentive to set $y = y_L$, a setting of y that does not support x_H as a Nash equilibrium. The unique Nash equilibrium is (x_L, y_L), which gives a lower utility $u(x, x, y)$ than does the competitive equilibrium (x_H, y_H).

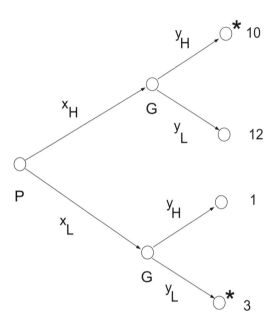

Figure 23.3.3: Timing of actions in a Nash equilibrium in which the private sector acts first. Here G denotes a node at which the government chooses and P denotes a node at which the public chooses. The private sector sets $x \in X$ before knowing the government's setting of $y \in Y$. Competitive equilibrium pairs (x, y) are denoted with an asterisk. The unique Nash equilibrium is (x_L, y_L).

23.4. Reputational mechanisms: general idea

In a finitely repeated economy, the government will certainly behave opportunistically the last period, implying that nothing better than a Nash outcome can be supported the last period. In a finite horizon economy with a unique Nash equilibrium, we won't be able to sustain anything better than a Nash equilibrium outcome in *any* earlier period. [8.]

We want to study situations in which a government might sustain a Ramsey outcome. Therefore, we shall study economies repeated an infinite number of times. Here a system of history-dependent expectations interpretable as a government reputation might be arranged to sustain something better than the Nash outcome. We strive to set things up so that the government so dearly wants to confirm a good reputation that it will not submit to the temptation to behave opportunistically. A reputation is said to be *sustainable* if it is always in the government's interests to confirm it.

A state variable that is capable of encoding a "reputation" is peculiar because it is both "backward looking" and "forward looking." It is backward looking because it remembers salient features of past behavior. It is forward-looking behavior because it measures something about what private agents expect the government to do in the future. We are about to study the ingenious machinery of Abreu, Pearce, and Stacchetti that astutely exploits these aspects of a reputational variable by recognizing that the ideal reputational state variable is a "promised value."

---

[8] If there are multiple Nash equilibria, it is sometimes possible to sustain a better-than-Nash equilibrium outcome for a while in a finite horizon economy. See exercise *23.1*, which uses an idea of Benoit and Krishna (1985).

23.4.1. *Dynamic programming squared*

A sustainable reputation for the government is one that (a) the public, having rational expectations, wants to believe, and (b) the government wants to sustain. Rather than finding all possible sustainable reputations, Abreu, Pearce, and Stacchetti (henceforth APS) (1986, 1990) used dynamic programming to characterize all *values* for the government that are attainable with sustainable reputations. This section briefly describes their main ideas, while later sections fill in many details.

First we need some language. A *strategy profile* is a pair of plans, one each for the private sector and the government. The time t components of the pair of plans maps the observed history of the economy into current-period outcomes (x, y). A *subgame perfect equilibrium* (SPE) strategy profile has a current-period outcome being a competitive equilibrium (x_t, y_t) whose y_t component the government would want to confirm at each $t \geq 1$ and for every possible history of the economy.

To characterize SPE, the method of APS is to formulate a Bellman equation that describes the value to the government of a strategy profile and that portrays the idea that the government wants to confirm the private sector's beliefs about y. For each $t \geq 1$, the government's strategy describes its first-period action $y \in Y$, which, because the public had expected it, determines an associated first-period competitive equilibrium $(x, y) \in C$. Furthermore, the strategy implies two continuation values for the government at the beginning of next period, a continuation value v_1 if it carries out the first-period choice y, and another continuation value v_2 if for any reason the government deviates from the expected first-period choice y. Associated with the government's strategy is a current value v that obeys the Bellman equation

$$v = (1 - \delta)u(x, x, y) + \delta v_1, \qquad (23.4.1a)$$

where $\delta \in (0, 1)$ is a discount factor, $(x, y) \in C$, v_1 is a continuation value awarded for confirming the private sector's expectation that the government will choose action y in the current period, and (y, v_1) are constrained to satisfy the incentive constraint

$$v \geq (1 - \delta)u(x, x, \eta) + \delta v_2, \quad \forall \eta \in Y, \qquad (23.4.1b)$$

or equivalently

$$v \geq (1 - \delta)u\big[x, x, H(x)\big] + \delta v_2,$$

where $H(x) = \arg\max_y u(x, x, y)$ is the government's opportunistic one period best policy in response to x. Here v_2 is the continuation value awarded to the government if it fails to confirm the private sector's expectation that $\eta = y$ this period. Because it receives the same continuation value v_2 for *any* deviation from y, if it does deviate, the government will choose the most rewarding action, which is to set $\eta = H(x)$.

Inequalities (23.4.1) define a Bellman equation that maps a *pair* of continuation values (v_1, v_2) into a value v and first-period outcomes (x, y). Figure 23.4.1 illustrates this mapping for the infinitely repeated version of the taxation example. Given a pair (v_1, v_2), the solid curve depicts v in equation (23.4.1a), and the dashed curve describes the right side of the incentive constraint (23.4.1b). The region in which the solid curve is above the dashed curve identifies tax rates and competitive equilibria that satisfy (23.4.1b) at the prescribed continuation values (v_1, v_2). As can be seen, when $\delta = .8$, tax rates below 18 percent cannot be sustained for the particular (v_1, v_2) pair we have chosen.

APS calculate the *set* of equilibrium values by iterating on the mapping defined by the Bellman equation (23.4.1). Let W be a set of candidate continuation values. As we vary $(v_1, v_2) \in W \times W$, the Bellman equation traces out a *set* of values, say, $v \in B(W)$. Thus, the Bellman equation maps *sets* of continuation values W (from which we can draw a *pair* of continuation values $(v_1, v_2) \in W \times W$) into sets of current values $v \in B(W)$. To qualify as SPE values, we require that $W \subset B(W)$, i.e., the *continuation* values drawn from W must themselves be *values* that are in turn supported by continuation values drawn from the same set W. APS reason that the largest set for which $W = B(W)$ *is* the set of all SPE values. APS show how iterations on the Bellman equation can determine the set of equilibrium values, provided that one starts with a big enough but bounded initial set of candidate continuation values. Furthermore, after that set of values has been found, APS show how to find a strategy that attains any equilibrium value in the set. The remainder of the chapter describes details of APS's formulation as applied in our setting. We shall see why APS want to get their hands on the entire set of equilibrium values.

Why do we call it 'dynamic programming squared'? There are two reasons.

1. The construction works by mapping *two* continuation values into one, in contrast to ordinary dynamic programming, which maps *one* continuation value tomorrow into one value function today.

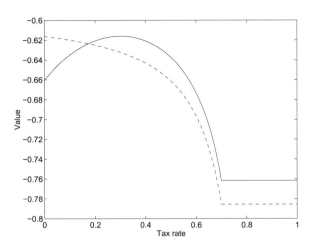

Figure 23.4.1: Mapping of continuation values (v_1, v_2) into values v in the infinitely repeated version of the taxation example. The solid curve depicts $v = (1 - \delta)u[\ell(\tau), \ell(\tau), \tau] + \delta v_1$. The dashed curve is the right side of the incentive constraint, $v \geq (1 - \delta)u\{\ell(\tau), \ell(\tau), H[\ell(\tau)]\} + \delta v_2$, where H is the government's best response function. The part of the solid curve that is above the dashed curve shows competitive equilibrium values that are sustainable for continuation values (v_1, v_2). The parameterization is $\alpha = 0.3$ and $\delta = 0.8$, and the continuation values are set as $(v_1, v_2) = (-0.6, -0.63)$.

2. A continuation value plays a double role, one as a promised value that summarizes expectations of the rewards associated with future outcomes, another as a state variable that summarizes the history of past outcomes. In the present setting, a subgame perfect equilibrium strategy profile can be represented recursively in terms of an initial value $v_1 \in I\!R$ and the following 3-tuple of functions:

$$x_t = z^g(v_t)$$
$$y_t = z^g(v_t)$$
$$v_{t+1} = \mathcal{V}(v_t, x_t, y_t)$$

the first two of which maps a promised value into a private sector decision and a government action, while the third maps a promised value and an action pair into a promised value to carry into tomorrow. By iterating these equations, we can deduce that the the triple of functions $z^h, z^g, \mathcal{V}$ induces a strategy profile that maps histories of outcomes into sequences of outcomes. The capacity to represent a subgame perfect equilibrium recursively affords immense simplifications in terms of the number of functions we must carry.

23.5. The infinitely repeated economy

Consider repeating our one-period economy forever. At each $t \geq 1$, each household chooses $\xi_t \in X$, with the result that the average $x_t \in X$; the government chooses $y_t \in Y$. We use the notation $(\vec{x}, \vec{y}) = \{(x_t, y_t)\}_{t=1}^{\infty}$, $\vec{\xi} = \{\xi_t\}_{t=1}^{\infty}$. To denote a *history* of (x_t, y_t) up to t, we use the notation $x^t = \{x_s\}_{s=1}^{t}$, $y^t = \{y_s\}_{s=1}^{t}$. These histories live in the spaces X^t and Y^t, respectively, where $X^t = X \times \cdots \times X$, the Cartesian product of X taken t times, and Y^t is the Cartesian product of Y taken t times. [9]

For the repeated economy, the government evaluates paths $(\vec{x}, \vec{y})$ according to

$$V_g(\vec{x}, \vec{y}) = \frac{(1 - \delta)}{\delta} \sum_{t=1}^{\infty} \delta^t \, r(x_t, y_t), \tag{23.5.1}$$

where $r(x_t, y_t) \equiv u(x_t, x_t, y_t)$ and $0 < \delta < 1$. [10] A *pure strategy* is defined as a sequence of functions, the tth element of which maps the history (x^{t-1}, y^{t-1}) observed at the beginning of t into an action at t. In particular, for the aggregate of households, a strategy is a sequence $\sigma^h = \{\sigma_t^h\}_{t=1}^{\infty}$ such that

$$\sigma_1^h \in X$$
$$\sigma_t^h : X^{t-1} \times Y^{t-1} \to X \qquad \text{for each} \quad t \geq 2 \, .$$

---

[9] Marco Bassetto's work (2002, 2005) shows that this specification, which is common in the literature, excludes some interesting applications. In particular, it rules out contexts in which the set of time t actions available to the government is influenced by past actions taken by households. Such excluded examples prevail, for example, in the fiscal theory of the price level. To construct sustainable plans in those interesting environments, Bassetto (2002, 2005) refines the notion of sustainability to include a more complete theory of the government's behavior off an equilibrium path.

[10] Note that we have not defined the government's payoff when $\xi_t \neq x_t$. See footnote 7.

Similarly, for the government, a strategy $\sigma^g = \{\sigma_t^g\}_{t=1}^{\infty}$ is a sequence such that

$$\sigma_1^g \in Y$$
$$\sigma_t^g : X^{t-1} \times Y^{t-1} \to Y \qquad \text{for each} \quad t \geq 2.$$

We call $\sigma = (\sigma^h, \sigma^g)$ a *strategy profile*. We let $\sigma_t = (\sigma_t^h, \sigma_t^g)$ be the tth component of the strategy profile.

23.5.1. A strategy profile implies a history and a value

APS begin with the insight that a strategy profile $\sigma = (\sigma^g, \sigma^h)$ evidently recursively generates a trajectory of outcomes $\{[x(\sigma)_t, y(\sigma)_t]\}_{t=1}^{\infty}$:

$$[x(\sigma)_1, y(\sigma)_1] = (\sigma_1^h, \sigma_1^g)$$
$$[x(\sigma)_t, y(\sigma)_t] = \sigma_t[x(\sigma)^{t-1}, y(\sigma)^{t-1}].$$

Therefore, a strategy profile also generates a pair of values for the government and the representative private agent. In particular, the value for the government of a strategy profile $\sigma = (\sigma^h, \sigma^g)$ is the value of the trajectory that it generates

$$V_g(\sigma) = V_g[\vec{x}(\sigma), \vec{y}(\sigma)].$$

23.5.2. Recursive formulation

A key step toward APS's recursive formulation comes from defining *continuation stategies* and their associated *continuation values*. Since the value of a path $(\vec{x}, \vec{y})$ in equation (23.5.1) is additively separable in its one-period returns, we can express the value recursively in terms of a one-period economy and a continuation economy. In particular, the value to the government of an outcome sequence $(\vec{x}, \vec{y})$ can be represented

$$V_g(\vec{x}, \vec{y}) = (1 - \delta) \, r(x_1, y_1) + \delta V_g(\{x_t\}_{t=2}^{\infty}, \{y_t\}_{t=2}^{\infty}) \qquad (23.5.2)$$

and the value for a household can also be represented recursively. Notice how a strategy profile σ induces a strategy profile for the continuation economy, as follows. Let $\sigma|_{(x^t,y^t)}$ denote the strategy profile for a continuation economy whose first period is $t+1$ and that is initiated after history (x^t, y^t) has been observed; here $(\sigma|_{(x^t,y^t)})_s$ is the sth component of $(\sigma|_{(x^t,y^t)})$, which for $s \geq 2$ is a function that maps $X^{s-1} \times Y^{s-1}$ into $X \times Y$, and for $s = 1$ is a point in $X \times Y$. Thus, after a first-period outcome pair (x_1, y_1), strategy σ induces the continuation strategy

$$(\sigma|_{(x_1,y_1)})_{s+1}(\nu^s, \eta^s) = \sigma_{s+2}(x_1, \nu_1, \ldots, \nu_s, y_1, \eta_1, \ldots, \eta_s)$$

$$\text{for all } (\nu^s, \eta^s) \in X^s \times Y^s, \quad \forall s \geq 0.$$

It might be helpful to write out a few terms for $s = 0, 1, \ldots$:

$$(\sigma|_{(x_1,y_1)})_1 = \sigma_2(x_1, y_1) = (\nu_1, \eta_1)$$

$$(\sigma|_{(x_1,y_1)})_2(\nu_1, \eta_1) = \sigma_3(x_1, \nu_1, y_1, \eta_1) = (\nu_2, \eta_2)$$

$$(\sigma|_{(x_1,y_1)})_3(\nu_1, \nu_2, \eta_1, \eta_2) = \sigma_4(x_1, \nu_1, \nu_2, y_1, \eta_1, \eta_2) = (\nu_3, \eta_3).$$

More generally, define the continuation strategy

$$(\sigma|_{(x^t,y^t)})_1 = \sigma_{t+1}(x^t, y^t)$$

$$(\sigma|_{(x^t,y^t)})_{s+1}(\nu^s, \eta^s) = \sigma_{t+s+1}(x_1, \ldots, x_t, \nu_1, \ldots, \nu_s; y_1, \ldots, y_t, \eta_1, \ldots, \eta_s)$$

$$\text{for all } s \geq 1 \quad \text{and all} \quad (\nu^s, \eta^s) \in X^s \times Y^s.$$

Here $(\sigma|_{(x^t,y^t)})_{s+1}(\nu^s, \eta^s)$ is the induced strategy pair to apply in the $(s+1)$th period of the continuation economy. We attain this strategy by shifting the original strategy forward t periods and evaluating it at history $(x_1, \ldots, x_t, \nu_1, \ldots, \nu_s; y_1, \ldots, y_t, \eta_1, \ldots, \eta_s)$ for the *original* economy.

In terms of the continuation strategy $\sigma|_{(x_1,y_1)}$, from equation $(23.5.2)$ we know that $V_g(\sigma)$ can be represented as

$$V_g(\sigma) = (1 - \delta)r(x_1, y_1) + \delta V_g(\sigma|_{(x_1,y_1)}). \tag{23.5.3}$$

Representation $(23.5.3)$ decomposes the value to the government of strategy profile σ into a one-period return and the continuation value $V_g(\sigma|_{(x_1,y_1)})$ associated with the continuation strategy $\sigma|_{(x_1,y_1)}$.

Any sequence $(\vec{x}, \vec{y})$ in equation $(23.5.2)$ or any strategy profile σ in equation $(23.5.3)$ can be assigned a value. We want a notion of an *equilibrium* strategy profile.

23.6. Subgame perfect equilibrium (SPE)

DEFINITION 6: A strategy profile $\sigma = (\sigma^h, \sigma^g)$ is a *subgame perfect equilibrium* (SPE) of the infinitely repeated economy if for each $t \geq 1$ and each history $(x^{t-1}, y^{t-1}) \in X^{t-1} \times Y^{t-1}$

(a) The private sector outcome $x_t = \sigma_t^h (x^{t-1}, y^{t-1})$ is consistent with competitive equilibrium when $y_t = \sigma_t^g (x^{t-1}, y^{t-1})$;

(b) For each possible government action $\eta \in Y$

$$(1 - \delta)r(x_t, y_t) + \delta V_g(\sigma|_{(x^t, y^t)}) \geq (1 - \delta)\, r(x_t, \eta) + \delta V_g(\sigma|_{(x^t; y^{t-1}, \eta)}).$$

Requirement a says two things. It attributes a theory of forecasting government behavior to members of the public, in particular, that they use the time t component σ_t^g of the government's strategy and information available at the end of period $t - 1$ to forecast the government's behavior at t. Condition a also asserts that a competitive equilibrium appropriate to the public's forecast value for y_t is the outcome at time t. Requirement b says that at each point in time and following each history, the government has no incentive to deviate from the first-period action called for by its strategy σ^g; that is, the government always wants to behave as the public expects. Notice how in condition b, the government *contemplates* setting its time t choice η_t at something other than the value forecast by the public, but confronts consequences that deter it from choosing an η_t that fails to confirm the public's expectations of it.

In section 23.15 we'll discuss the following question: who *chooses* σ^g, the government or the public? This question arises naturally because σ^g is *both* the government's sequence of policy functions *and* the private sector's rule for forecasting government behavior. Condition b of definition 6 says that the government chooses to confirm the public's forecasts.

Definition 6 implies that for each $t \geq 2$ and each $(x^{t-1}, y^{t-1}) \in X^{t-1} \times Y^{t-1}$, the continuation strategy $\sigma|_{(x^{t-1}, y^{t-1})}$ is itself an SPE. We state this formally for $t = 2$.

PROPOSITION 1: Assume that σ is an SPE. Then for all $(\nu, \eta) \in X \times Y$, $\sigma|_{(\nu, \eta)}$ is an SPE.

PROOF: Write out requirements a and b that Definition 6 asserts that the continuation strategy $\sigma|_{(\nu,\eta)}$ must satisfy to qualify as an SPE. In particular, for all $s \geq 1$ and for all $(x^{s-1}, y^{s-1}) \in X^{s-1} \times Y^{s-1}$, we require

$$(x_s, y_s) \in C, \tag{23.6.1}$$

where $x_s = \sigma^h|_{(\nu,\eta)}(x^{s-1}, y^{s-1}), y_s = \sigma^g|_{(\nu,\eta)}(x^{s-1}, y^{s-1})$. We also require that for all $\tilde{\eta} \in Y$,

$$(1-\delta)r(x_s, y_s) + \delta V_g(\sigma|_{(\eta,x^s;\nu,y^s)}) \geq (1-\delta)r(x_s, \tilde{\eta}) + \delta V_g(\sigma|_{(\nu,x^s;\eta,y^{s-1},\tilde{\eta})}) \tag{23.6.2}$$

Notice that requirements a and b of Definition 6 for $t = 2, 3, \ldots$ imply expressions $(23.6.1)$ and $(23.6.2)$ for $s = 1, 2, \ldots$. ∎

The statement that $\sigma|_{(\nu,\eta)}$ is an SPE for all $(\nu, \eta) \in X \times Y$ ensures that σ is *almost* an SPE. If we know that $\sigma|_{(\nu,\eta)}$ is an SPE for all $(\nu, \eta) \in (X \times Y)$, we must only add two requirements to ensure that σ is an SPE: first, that the $t = 1$ outcome pair (x_1, y_1) is a competitive equilibrium, and second, that the government's choice of y_1 satisfies the time 1 version of the incentive constraint b in Definition 6.

This reasoning leads to the following lemma that is at the heart of the APS analysis:

LEMMA: Consider a strategy profile σ, and let the associated first-period outcome be given by $x = \sigma_1^h, y = \sigma_1^g$. The profile σ is an SPE if and only if

(1) for each $(\nu, \eta) \in X \times Y$, $\sigma|_{(\nu,\eta)}$ is an SPE;

(2) (x, y) is a competitive equilibrium;

(3) $\forall\, \eta \in Y$, $(1 - \delta)\, r(x, y) + \delta\, V_g(\sigma|_{(x,y)}) \geq (1 - \delta)\, r(x, \eta) + \delta V_g(\sigma|_{(x,\eta)})$.

PROOF: First, prove the "if" part. Property a of the lemma and properties $(23.6.1)$ and $(23.6.2)$ of Proposition 1 show that requirements a and b of Definition 6 are satisfied for $t \geq 2$. Properties (2) and (3) of the lemma imply that requirements a and b of Definition 6 hold for $t = 1$.

Second, prove the "only if" part. Part (1) of the lemma follows from Proposition 1. Parts (2) and (3) of the lemma follow from requirements a and b of Definition 6 for $t = 1$. ∎

The lemma is very important because it characterizes SPEs in terms of a first-period competitive equilibrium outcome pair (x, y), and a *pair* of continuation values: a value $V_g(\sigma|_{(x,y)})$ to be awarded to the government next period if it adheres to the y component of the first-period pair (x, y), and a value $V_g(\sigma|_{(x,\eta)})$, $\eta \neq y$, to be awarded to the government if it deviates from the expected y component. Each of these values has to be selected from a set of values $V_g(\sigma)$ that are associated with some SPE σ.

23.7. Examples of SPE

23.7.1. Infinite repetition of one-period Nash equilibrium

It is easy to verify that the following strategy profile $\sigma^N = (\sigma^h, \sigma^g)$ forms an SPE: $\sigma_1^h = x^N, \sigma_1^g = y^N$ and for $t \geq 2$

$$\sigma_t^h = x^N \qquad \forall\, t, \quad \forall\, (x^{t-1}, y^{t-1});$$
$$\sigma_t^g = y^N \qquad \forall\, t, \quad \forall\, (x^{t-1}, y^{t-1}).$$

These strategies instruct the households and the government to choose the static Nash equilibrium outcomes for all periods for all histories. Evidently, for these strategies, $V_g(\sigma^N) = v^N = r(x^N, y^N)$. Furthermore, for these strategies the continuation value $V_g(\sigma|_{(x^t; y^{t-1}, \eta)}) = v^N$ for all outcomes $\eta \in Y$. These strategies satisfy requirement a of Definition 6 because (x^N, y^N) is a competitive equilibrium. The strategies satisfy requirement b because $r(x^N, y^N) = \max_{y \in Y} r(x^N, y)$ and because the continuation value $V_g(\sigma) = v^N$ is independent of the action chosen by the government in the first period. In this SPE, $\sigma_t^N = \{\sigma_t^h, \sigma_t^g\} = (x^N, y^N)$ for all t and for all (x^{t-1}, y^{t-1}), and the value $V_g(\sigma^N)$ and the continuation values $V_g(\sigma^N|_{(x^t, y^t)})$ for each history (x^t, y^t) equal v^N.

It is useful to think about this SPE in terms of the lemma. To verify that σ^N is a SPE, we work with the first-period outcome pair (x^N, y^N) and the pair of values $V_g(\sigma|_{(x^N, y^N)}) = v^N, V_g(\sigma|_{(x,\eta)}) = v^N$, where $v^N = r(x^N, y^N)$. With these settings, we can verify that (x^N, y^N) and v^N satisfy requirements (1), (2), and (3) of the lemma.

23.7.2. Supporting better outcomes with trigger strategies

The public can have a system of expectations about the government's behavior that induces the government to choose a better-than-Nash outcome $(\tilde{x}, \tilde{y}) \in C$. Thus, suppose that the public expects that so long as the government chooses $\tilde{y}$, it will continue to do so in the future, but if ever the government deviates from this choice, thereafter the public expects that the government will choose y^N, prompting the public (or what we can call "the market") to react with $x^N = h(y^N)$. This system of expectations confronts the government with the prospect of being "punished by the market's expectations" if it chooses to deviate from $\tilde{y}$.

To formalize this idea, we shall use the SPE σ^N as a continuation strategy and the value v^N as a continuation value on the right side of part (b) of Definition 6 of an SPE (for $\eta \neq y_t$); then by working backward one step, we shall try to construct *another* SPE $\tilde{\sigma}$ with first-period outcome $(\tilde{x}, \tilde{y}) \neq (x^N, y^N)$. In particular, for our new SPE $\tilde{\sigma}$ we propose to set

$$\tilde{\sigma}_1 = (\tilde{x}, \tilde{y})$$

$$\tilde{\sigma}|_{(x,y)} = \begin{cases} \tilde{\sigma} & \text{if } (x, y) = (\tilde{x}, \tilde{y}) \\ \sigma^N & \text{if } (x, y) \neq (\tilde{x}, \tilde{y}) \end{cases} \tag{23.7.1}$$

where $(\tilde{x}, \tilde{y}) \in C$ is a competitive equilibrium that satisfies the following particular case of part b of Definition 6:

$$\tilde{v} = (1 - \delta)\, r(\tilde{x}, \tilde{y}) + \delta \tilde{v} \geq (1 - \delta)\, r(\tilde{x}, \eta) + \delta v^N, \tag{23.7.2}$$

for all $\eta \in Y$. Inequality $(23.7.2)$ is equivalent with

$$\max_{\eta \in Y} r(\tilde{x}, \eta) - r(\tilde{x}, \tilde{y}) \leq \frac{\delta}{1 - \delta}\,(\tilde{v} - v^N). \tag{23.7.3}$$

For any $(\tilde{x}, \tilde{y}) \in C$ that satisfies expression $(23.7.3)$ with $\tilde{v} = r(\tilde{x}, \tilde{y})$, strategy $(23.7.1)$ is an SPE with value $\tilde{v}$.

If $(\tilde{x}, \tilde{y}) = (x^R, y^R)$ satisfies inequality $(23.7.3)$ with $\tilde{v} = r(x^R, y^R)$, then repetition of the Ramsey outcome (x^R, y^R) is supportable by a subgame perfect equilibrium of the form $(23.7.1)$.

This construction uses the following objects:

1. A proposed first-period competitive equilibrium $(\tilde{x}, \tilde{y}) \in C$.

2. An SPE σ^2 with value $V_g(\sigma^2)$ that is used as the continuation strategy in the event that the first-period outcome does not equal $(\tilde{x}, \tilde{y})$, so that $\tilde{\sigma}|_{(x,y)} = \sigma^2$, if $(x, y) \neq (\tilde{x}, \tilde{y})$. In the example, $\sigma^2 = \sigma^N$ and $V_g(\sigma^2) = v^N$.

3. An SPE σ^1, with value $V_g(\sigma^1)$, used to define the continuation value to be assigned after first-period outcome $(\tilde{x}, \tilde{y})$; and an associated continuation strategy $\tilde{\sigma}|_{(\tilde{x},\tilde{y})} = \sigma^1$. In the example, $\sigma^1 = \tilde{\sigma}$, which is defined recursively (and self-referentially) via equation (23.7.1).

4. A candidate for a new SPE $\tilde{\sigma}$ and a corresponding value $V_g(\tilde{\sigma})$. In the example, $V_g(\tilde{\sigma}) = r(\tilde{x}, \tilde{y})$.

In the example, objects 3 and 4 are equated.

Note how we have used the lemma in verifying that $\tilde{\sigma}$ is an SPE. We start with the SPE σ^N with associated value v^N. We guess a first-period outcome pair $(\tilde{x}, \tilde{y})$ and a value $\tilde{v}$ for a new SPE, where $\tilde{v} = r(\tilde{x}, \tilde{y})$. Then we verify requirements (2) and (3) of the lemma with $(\tilde{v}, v^N)$ as continuation values and $(\tilde{x}, \tilde{y})$ as first-period outcomes.

23.7.3. *When reversion to Nash is not bad enough*

For discount factors δ sufficiently close to one, it is typically possible to support repetition of the Ramsey outcome (x^R, y^R) with a section 23.7.2 trigger strategy of form (23.7.1). This finding conforms with a version of the folk theorem about repeated games. However, there exist discount factors δ so small that the continuation value associated with infinite repetition of the one-period Nash outcome is not low enough to support repetition of Ramsey. In that case, anticipating that it will revert to repetition of Nash after a deviation can at best support a value for the government that is less than that associated with repetition of Ramsey outcome, although perhaps better than repetition of the Nash outcome.

It is natural at this point to ask whether in this circumstance there is a better SPE? This question inspired APS to find the set of values associated with *all* SPEs. To support something *better* evidently requires finding an SPE that has a value *worse* than that associated with repetition of the one-period Nash outcome. Following APS, we shall soon see that the best and worst equilibrium values are linked.

23.8. Values of all SPEs

The role played by the lemma in analyzing our two examples hints at the central role that it plays in methods that APS developed for describing and computing values for *all* the subgame perfect equilibria. APS build on the way that the lemma characterizes SPE values in terms of a first-period competitive equilibrium outcome, along with a pair of continuation values, each element of which is itself a value associated with some SPE. The lemma directs APS's attention away from a *set of strategy profiles* σ and toward a *set of values* $V_g(\sigma)$ associated with those profiles. They define the set V of values associated with subgame perfect equilibria:

$$V = \{V_g\left(\sigma\right) \mid \sigma \text{ is an SPE}\}.$$

Evidently, $V \subset \mathbb{R}$. From the lemma, for a given competitive equilibrium $(x, y) \in C$, there exists an SPE σ for which $x = \sigma_1^h, y = \sigma_1^g$ if and only if there exist two values $(v_1, v_2) \in V \times V$ such that

$$(1 - \delta)\, r(x, y) + \delta v_1 \geq (1 - \delta)\, r(x, \eta) + \delta v_2 \quad \forall\ \eta \in Y. \tag{23.8.1}$$

Let σ^1 and σ^2 be subgame perfect equilibria for which $v_1 = V_g(\sigma^1)$, $v_2 = V_g(\sigma^2)$. The SPE σ that supports $(x, y) = (\sigma_1^h, \sigma_1^g)$ is completed by specifying the continuation strategies $\sigma|_{(x,y)} = \sigma^1$ and $\sigma|_{(\nu,\eta)} = \sigma^2$ if $(\nu, \eta) \neq (x, y)$.

This construction uses two continuation values $(v_1, v_2) \in V \times V$ to create an SPE σ with value $v \in V$ given by

$$v = (1 - \delta)\, r(x, y) + \delta v_1 \ .$$

Thus, the construction maps *pairs* of continuation values (v_1, v_2) into a strategy profile σ with first-period competitive equilibrium outcome (x, y) and a value $v = V_g(\sigma)$.

APS characterize subgame perfect equilibria by studying a mapping from pairs of continuation values $(v_1, v_2) \in V \times V$ into values $v \in V$. They use the following definitions:

DEFINITION 7: Let $W \subset \mathbb{R}$. A 4-tuple (x, y, w_1, w_2) is said to be *admissible with respect to* W if $(x, y) \in C$, $(w_1, w_2) \in W \times W$, and

$$(1 - \delta)\, r(x, y) + \delta w_1 \geq (1 - \delta)\, r(x, \eta) + \delta w_2 , \quad \forall\ \eta \in Y. \tag{23.8.2}$$

Notice that when $W \subset V$, the admissible 4-tuple (x, y, w_1, w_2) determines an SPE with strategy profile

$$\sigma_1 = (x, y), \ \sigma|_{(x,y)} = \sigma^1, \ \sigma|_{(\nu,\eta)} = \sigma^2 \ \text{for} \ (\nu, \eta) \neq (x, y)$$

where σ_1 is a continuation strategy that yields value $w_1 = V_g(\sigma^1)$ and σ_2 is a strategy that yields continuation value $w_2 = V_g(\sigma^2)$. The value of the SPE is $V_g(\sigma) = w = (1 - \delta) r(x, y) + \delta w_1$.

We want to find the set V.

23.8.1. The basic idea of dynamic programming squared

In Definition 7, W serves as a set of candidate continuation values. The idea is to pick an $(x, y) \in C$, then to check whether we can find $(w_1, w_2) \in W \times W$ that would make the government want to adhere to the y component if w_1 and w_2 could be used as continuation values for adhering to and deviating from y, respectively. If the answer is yes, we say that the 4-tuple (x, y, w_1, w_2) is "admissible with respect to W". Because we have verified that the incentive constraints are satisfied, a yes answer allows us to calculate the *value* (i.e., the left side of (23.8.2)) that can be supported with w_1, w_2 as continuation values. Thus, the idea is to use (23.8.2) to define a mapping from values tomorrow to values today, like that used in dynamic programming. In the next section, we'll define $B(W)$ as the set of possible values attained with admissible pairs of continuation values drawn from $W \times W$. Then we'll view B as an operator mapping sets of continuation values W into sets of values $B(W)$. This operator is the counterpart to the T operator associated with ordinary dynamic programming.

To pursue this analogy, recall the Bellman equation associated with the Mc-Call model of chapter 6:

$$Q = \int \max \left\{ \frac{w}{1 - \beta}, c + \beta Q \right\} d F(w).$$

Here $Q \in I\!R$ is the expected discounted value of an unemployed worker's income *before* he has drawn a wage offer. The right side defines an operator $T(Q)$, so that the Bellman equation is

$$Q = T(Q). \tag{23.8.3}$$

This equation can be solved by iterating to convergence starting from any initial Q.

Just as the right side of $(23.8.3)$ takes a candidate continuation value Q for tomorrow and maps it into a value $T(Q)$ for today, APS define a mapping $B(W)$ that, by considering only admissible 4-tuples, maps a *set* of values W tomorrow into a new *set* $B(W)$ of values today. Thus, APS use admissible 4-tuples to map candidate continuation values tomorrow into new candidate values today. In the next section, we'll iterate to convergence on $B(W)$, but as we'll see, it won't work to start from just any initial set W. We have to start from a big enough set.

23.9. The APS machinery

DEFINITION 8: For each set $W \subset \mathbb{R}$, let $B(W)$ be the set of possible values $w = (1 - \delta)\, r(x, y) + \delta w_1$ associated with admissible tuples (x, y, w_1, w_2).

Think of W as a set of potential continuation values and $B(W)$ as the set of values that they support. From the definition of admissibility it immediately follows that the operator B is *monotone*.

PROPERTY (monotonicity of B): If $W \subseteq W' \subseteq R$, then $B(W) \subseteq B(W')$.

PROOF: It can be verified directly from the definition of admissible 4-tuples that if $w \in B(W)$, then $w \in B(W')$: simply use the (w_1, w_2) pair that supports $w \in B(W)$ to support $w \in B(W')$. ∎

It can also be verified that $B(\cdot)$ maps compact sets W into compact sets $B(W)$.

The self-supporting character of subgame perfect equilibria is referred to in the following definition:

DEFINITION 9: The set W is said to be *self-generating* if $W \subseteq B(W)$.

Thus, a set of continuation values W is said to be self-generating if it is contained in the set of *values* $B(W)$ that are generated by pairs of continuation values selected from W. This description makes us suspect that if a set of values is self-generating, it must be a set of SPE values. Indeed, notice that by virtue of the lemma, the set V of SPE values $V_g(\sigma)$ is self-generating. Thus, we can write $V \subseteq B(V)$. APS show that V is the *largest* self-generating set. The key to showing this point is the following theorem:[11]

THEOREM 1 (A self generating set is a subset of V): If $W \subset \mathbb{R}$ is bounded and self-generating, then $B(W) \subseteq V$.

The proof is based on "forward induction" and proceeds by taking a point $w \in B(W)$ and constructing an SPE with value w.

PROOF: Assume $W \subseteq B(W)$. Choose an element $w \in B(W)$ and transform it as follows into a subgame perfect equilibrium:

Step 1. Because $w \in B(W)$, we know that there exist outcomes (x, y) and values w_1 and w_2 that satisfy

$$w = (1 - \delta)\, r(x, y) + \delta w_1 \geq (1 - \delta)\, r(x, \eta) + \delta w_2 \quad \forall \eta \in Y$$
$$(x, y) \in C$$
$$w_1, w_2 \in W \times W.$$

Set $\sigma_1 = (x, y)$.

Step 2. Since $w_1 \in W \subseteq B(W)$, there exist outcomes $(\tilde{x}, \tilde{y})$ and values $(\tilde{w}_1, \tilde{w}_2) \in W$ that satisfy

$$w_1 = (1 - \delta)\, r(\tilde{x}, \tilde{y}) + \delta \tilde{w}_1 \geq (1 - \delta)\, r(\tilde{x}, \eta) + \delta \tilde{w}_2, \quad \forall\, \eta \in Y$$
$$(\tilde{x}, \tilde{y}) \in C.$$

Set the first-period outcome in period 2 (the outcome to occur *given* that y was chosen in period 1) equal to $(\tilde{x}, \tilde{y})$; that is, set $(\sigma|_{(x,y)})_1 = (\tilde{x}, \tilde{y})$.

Continuing in this way, for each $w \in B(W)$, we can create a sequence of continuation values $w_1, \tilde{w}_1, \tilde{\tilde{w}}_1, \ldots$ and a corresponding sequence of first-period outcomes $(x, y), (\tilde{x}, \tilde{y}), (\tilde{\tilde{x}}, \tilde{\tilde{y}})$.

At each stage in this construction, policies are *unimprovable*, which means that given the continuation values, one-period deviations from the prescribed policies are not optimal. It follows that the strategy profile is optimal. By construction $V_g(\sigma) = w$. ∎

Collecting results, we know that

---

[11] The *unbounded* set $\mathbb{R}$ (the extended real line) is self-generating but not meaningful. It is self-generating because any value $v \in \mathbb{R}$ can be supported if there are no limits on the continuation values. It is not meaningful because most points in $\mathbb{R}$ are values that cannot be attained with *any* strategy profile.

1. $V \subseteq B(V)$ (by the lemma).

2. If $W \subseteq B(W)$, then $B(W) \subseteq V$ (by theorem 1).

3. B is monotone and maps compact sets into compact sets.

Facts 1 and 2 imply that $V = B(V)$, so that the set of equilibrium values is a "fixed point" of B, in particular, the *largest* bounded fixed point.

Monotonicity of B and the fact that it maps compact sets into compact sets provides an algorithm for computing the set V, namely, to start with a set W_0 for which $V \subseteq B(W_0) \subseteq W_0$, and to iterate on B. In more detail, we use the following steps:

1. Start with a set $W_0 = [\underline{w}_0, \overline{w}_0]$ that we know is bigger than V, and for which $B(W_0) \subseteq W_0$. It will always work to set $\overline{w}_0 = \max_{(x,y) \in C} r(x, y)$, $\underline{w}_0 = \min_{(x,y) \in C} r(x, y)$.

2. Compute the boundaries of the set $B(W_0) = [\underline{w}_1, \overline{w}_1]$. The value $\overline{w}_1$ solves the problem

$$\overline{w}_1 = \max_{(x,y) \in C} (1 - \delta) r(x, y) + \delta \overline{w}_0$$

subject to

$$(1 - \delta) r(x, y) + \delta \overline{w}_0 \geq (1 - \delta) r(x, \eta) + \delta \underline{w}_0 \quad \text{for all} \quad \eta \in Y.$$

The value $\underline{w}_1$ solves the problem

$$\underline{w}_1 = \min_{(x,y) \in C; \, (w_1, w_2) \in [\underline{w}_0, \overline{w}_0]^2} (1 - \delta) r(x, y) + \delta w_1$$

subject to

$$(1 - \delta) r(x, y) + \delta w_1 \geq (1 - \delta) r(x, \eta) + \delta w_2 \quad \forall \, \eta \in Y.$$

With $(\underline{w}_0, \overline{w}_0)$ chosen as before, it will be true that $B(W_0) \subseteq W_0$.

3. Having constructed $W_1 = B(W_0) \subseteq W_0$, continue to iterate, producing a decreasing sequence of compact sets $W_{j+1} = B(W_j) \subseteq W_j$. Iterate until the sets converge.

In section 23.13, we will present a direct way to compute the best and worst SPE values, one that evades having to iterate on the B operator.

23.10. Self-enforcing SPE

A subgame perfect equilibrium with a *worst* value $v \in V$ has the remarkable property that it is "self-enforcing." We use the following definition:

DEFINITION 10: A subgame perfect equilibrium σ with first-period outcome $(\tilde{x}, \tilde{y}) \in C$ is said to be *self-enforcing* if

$$\sigma|_{(x,y)} = \sigma \qquad \text{if } (x,y) \neq (\tilde{x}, \tilde{y}). \tag{23.10.1}$$

A strategy profile satisfying equation $(23.10.1)$ is called self-enforcing because after a one-shot deviation the consequence is simply to restart the equilibrium.

Recall our earlier characterization of a competitive equilibrium as a pair $(h(y), y)$, where $x = h(y)$ is the mapping from the government's action y to the private sector's equilibrium response. The value $\underline{v}$ associated with the worst subgame perfect equilibrium σ satisfies

$$\underline{v} = \min_{y,v} \left\{ (1 - \delta)\, r(h(y), y) + \delta v \right\} = (1 - \delta) r(h(\tilde{y}), \tilde{y}) + \delta \tilde{v}, \tag{23.10.2}$$

where the minimization is subject to $y \in Y$, $v \in V$, and the incentive constraint

$$(1 - \delta)\, r(h(y), y) + \delta v \geq (1 - \delta)\, r(h(y), \eta) + \delta \underline{v} \quad \text{for all } \eta \in Y. \tag{23.10.3}$$

Let $\tilde{v}$ be a continuation value that attains the right side of equation $(23.10.2)$, and let $\sigma_{\tilde{v}}$ be a subgame perfect equilibrium that supports continuation value $\tilde{v}$. Let $(\tilde{x}, \tilde{y})$ be the first-period outcome that attains the right side of equation $(23.10.2)$. Thus, $\underline{v} = (1 - \delta) r(\tilde{x}, \tilde{y}) + \delta \tilde{v}$. Since $\underline{v}$ is both the continuation value when first-period outcome $(x,y) \neq (\tilde{x}, \tilde{y})$ *and* the value associated with subgame perfect equilibrium σ, it follows that

$$\begin{aligned} \sigma_1 &= (\tilde{x}, \tilde{y}) \\ \sigma|_{(x,y)} &= \begin{cases} \sigma_{\tilde{v}} & \text{if } (x,y) = (\tilde{x}, \tilde{y}) \\ \sigma & \text{if } (x,y) \neq (\tilde{x}, \tilde{y}). \end{cases} \end{aligned} \tag{23.10.4}$$

Because of the double role played by $\underline{v}$, i.e., $\underline{v}$ is both the value of equilibrium σ and the "punishment" continuation value of the right side of the incentive constraint $(23.10.3)$, an equilibrium strategy σ that supports $\underline{v}$ is self-enforcing.[12]

---

[12] As we show below, the structure of the programming problem, with the double role played by $\underline{v}$, makes it possible to compute the worst value directly.

The preceding argument thus establishes

PROPOSITION 2: A subgame perfect equilibrium σ associated with $\underline{v} = \min\{v : v \in V\}$ is self-enforcing.

23.10.1. The quest for something worse than repetition of Nash outcome

Notice that the first subgame perfect equilibrium that we computed, whose outcome was infinite repetition of the one-period Nash equilibrium, is a self-enforcing equilibrium. However, in general, the infinite repetition of the one-period Nash equilibrium is not the *worst* subgame perfect equilibrium. This fact opens the possibility that even when reversion to Nash after a deviation is *not* able to support repetition of Ramsey as an SPE, we might still support repetition of the Ramsey outcome by reverting to a SPE with a value worse than that associated with repetition of the Nash outcome whenever the government deviates from an expected one-period choice.

23.11. Recursive strategies

This section emphasizes similarities between credible government policies and the recursive contracts appearing in chapter 20. We will study situations where the household's and the government's strategies have recursive representations. This approach substantially restricts the space of strategies because most history-dependent strategies cannot be represented recursively. Nevertheless, this class of strategies excludes no equilibrium payoffs $v \in V$. We use the following definitions:

DEFINITION 11: Households and the government follow *recursive strategies* if there is a 3-tuple of functions $\phi = (z^h, z^g, \mathcal{V})$ and an initial condition v_1 with the following structure:

$$
\begin{aligned}
v_1 &\in I\!R \text{ is given} \\
x_t &= z^h(v_t) \\
y_t &= z^g(v_t) \\
v_{t+1} &= \mathcal{V}(v_t, x_t, y_t),
\end{aligned}
\tag{23.11.1}
$$

where v_t is a state variable designed to summarize the history of outcomes before t.

This recursive form of strategies operates much like an autoregression to let time t actions (x_t, y_t) depend on the history $\{y_s, x_s\}_{s=1}^{t-1}$, as mediated through the state variable v_t. Representation $(23.11.1)$ induces history-dependent government policies, and thereby allows for reputation. We shall soon see that beyond its role in keeping track of histories, v_t also summarizes the future.[13]

A strategy (ϕ, v) recursively generates an outcome path expressed as $(\vec{x}, \vec{y}) = (\vec{x}, \vec{y})(\phi, v)$. By substituting the outcome path into equation $(23.5.3)$, we find that (ϕ, v) induces a value for the government, which we write as

$$V^g\big[(\vec{x}, \vec{y})(\phi, v)\big] = (1 - \delta)\, r\big[z^h(v), z^g(v)\big]$$
$$+ \delta\, V^g\Big((\vec{x}, \vec{y})\big\{\phi, \mathcal{V}[v, z^h(v), z^g(v)]\big\}\Big). \quad (23.11.2)$$

So far, we have not interpreted the state variable v, except as a particular measure of the history of outcomes. The theory of credible policy ties past and future together by making the state variable v a promised value, an outcome to be expressed

$$v = V^g\big[(\vec{x}, \vec{y})(\phi, v)\big]. \quad (23.11.3)$$

Equations $(23.11.1)$, $(23.11.2)$, and $(23.11.3)$ assert a dual role for v. In equation $(23.11.1)$, v accounts for past outcomes. In equations $(23.11.2)$ and $(23.11.3)$, v looks forward. The state v_t is a discounted future value with which the government enters time t based on past outcomes. Depending on the outcome (x, y) and the entering promised value v, $\mathcal{V}$ updates the promised value with which the government leaves the period. In section 23.15, we shall struggle with which of two valid interpretations of the government's strategy should be emphasized: something chosen by the government, or a description of a system of public expectations to which the government conforms.

Evidently, we have the following:

---

[13] By iterating equations $(23.11.1)$, we can construct a pair of sequences of functions indexed by $t \geq 1$ $\{Z_t^h(I_t), Z_t^g(I_t)\}$, mapping histories that are augmented by initial conditions $I_t = (\{x_s, y_s\}_{s=1}^{t-1}, v_1)$ into time t actions $(x_t, y_t) \in X \times Y$. Strategies for the repeated economy are a pair of sequences of such functions without the restriction that they have a recursive representation.

DEFINITION 12: Let V be the set of SPE values. A recursive strategy (ϕ, v) in equation (23.11.1) is a *subgame perfect equilibrium* (SPE) if and only if $v \in V$ and

(1) The outcome $x = z^h(v)$ is a competitive equilibrium, given that $y = z^g(v)$.

(2) For each $\eta \in Y$, $\mathcal{V}(v, z^h(v), \eta) \in V$.

(3) For each $\eta \in Y$,

$$
\begin{aligned}
v &= (1 - \delta)r\big[z^h(v), z^g(v)\big] + \delta\mathcal{V}\big[v, z^h(v), z^g(v)\big] \\
&\geq (1 - \delta)r\big[z^h(v), \eta\big] + \delta\mathcal{V}\big[v, z^h(v), \eta\big].
\end{aligned}
\tag{23.11.4}
$$

Condition (1) asserts that the first-period outcome pair (x, y) is a competitive equilibrium. Each member of the private sector forms an expectation about the government's action according to $y_t = z^g(v_t)$, and the "market" responds with a competitive equilibrium x_t,

$$
x_t = h(y_t) = h\big[z^g(v_t)\big] \equiv z^h(v_t).
\tag{23.11.5}
$$

This construction builds in rational expectations, because the private sector knows both the state variable v_t and the government's decision rule z^g.

Besides the first-period outcome (x, y), conditions (2) and (3) associate with a subgame perfect equilibrium three additional objects: a promised value v, a continuation value $v' = \mathcal{V}[v, z^h(v), z^g(v)]$ if the required first-period outcome is observed, and another continuation value $\tilde{v}(\eta) = \mathcal{V}[v, z^h(v), \eta]$ if the required first-period outcome is not observed but rather some pair (x, η). All of the continuation values must themselves be attained as subgame perfect equilibria. In terms of these objects, condition (3) is an incentive constraint inspiring the government to adhere to the equilibrium

$$
\begin{aligned}
v &= (1 - \delta)r(x, y) + \delta v' \\
&\geq (1 - \delta)r(x, \eta) + \delta\tilde{v}(\eta), \quad \forall \eta \in Y.
\end{aligned}
$$

This formula states that the government receives more if it adheres to an action called for by its strategy than if it departs. To ensure that these values constitute "credible expectations," part (2) of Definition 12 requires that the continuation values be values for subgame perfect equilibria. The definition is circular because members of the same class of objects, namely, equilibrium values v, occur on each side of expression (23.11.4). Circularity comes with recursivity.

One implication of the work of APS (1986, 1990) is that recursive equilibria of form (23.11.1) can attain *all* subgame perfect equilibrium values. As we have seen, APS's innovation was to shift the focus away from the set of equilibrium strategies and toward the set of values V attainable with subgame perfect equilibrium strategies.

23.12. Examples of SPE with recursive strategies

Our two earlier examples of subgame perfect equilibria were already of a recursive nature. But to highlight this property, we recast those SPE in the present notation for recursive strategies. Equilibria are constructed by using a guess-and-verify technique. First, guess $(v_1, z^h, z^g, \mathcal{V})$ in equations (23.11.1), then verify parts (1), (2), and (3) of Definition 12.

The examples parallel the historical development of the theory. (1) The first example is infinite repetition of a one-period Nash outcome, which was Kydland and Prescott's (1977) time-consistent equilibrium. (2) Barro and Gordon (1983a, 1983b) and Stokey (1989) used the value from infinite repetition of the Nash outcome as a continuation value to deter deviation from the Ramsey outcome. For sufficiently high discount factors, the continuation value associated with repetition of the Nash outcome can deter the government from deviating from infinite repetition of the Ramsey outcome. This is not possible for low discount factors. (3) Abreu (1988) and Stokey (1991) showed that Abreu's "stick-and-carrot" strategy induces more severe consequences than repetition of the Nash outcome.

23.12.1. Infinite repetition of Nash outcome

It is easy to construct an equilibrium whose outcome path forever repeats the one-period Nash outcome. Let $v^N = r(x^N, y^N)$. The proposed equilibrium is

$$v_1 = v^N,$$
$$z^h(v) = x^N \ \forall \ v,$$
$$z^g(v) = y^N \ \forall \ v, \text{ and}$$
$$\mathcal{V}(v, x, y) = v^N, \ \forall \ (v, x, y).$$

Here v^N plays the roles of all *three* values in condition (3) of Definition 12. Conditions (1) and (2) are satisfied by construction, and condition (3) collapses to

$$r(x^N, y^N) \ge r[x^N, H(x^N)],$$

which is satisfied at equality by the definition of a best response function.

23.12.2. Infinite repetition of a better-than-Nash outcome

Let v^b be a value associated with outcome (x^b, y^b) such that $v^b = r(x^b, y^b) > v^N$, and assume that (x^b, y^b) constitutes a competitive equilibrium. Suppose further that

$$r[x^b, H(x^b)] - r(x^b, y^b) \le \frac{\delta}{1 - \delta}(v^b - v^N). \qquad (23.12.1)$$

The left side is the one-period return to the government from deviating from y^b; it is the gain from deviating. The right side is the difference in present values associated with conforming to the plan versus reverting forever to the Nash equilibrium; it is the cost of deviating. When the inequality is satisfied, the equilibrium presents the government with an incentive not to deviate from y^b. Then an SPE is

$$v_1 = v^b$$

$$z^h(v) = \begin{cases} x^b & \text{if } v = v^b; \\ x^N & \text{otherwise}; \end{cases}$$

$$z^g(v) = \begin{cases} y^b & \text{if } v = v^b; \\ y^N & \text{otherwise}; \end{cases}$$

$$\mathcal{V}(v, x, y) = \begin{cases} v^b & \text{if } (v, x, y) = (v^b, x^b, y^b); \\ v^N & \text{otherwise}. \end{cases}$$

This strategy specifies outcome (x^b, y^b) and continuation value v^b as long as v^b is the value promised at the beginning of the period. Any deviation from y^b generates continuation value v^N. Inequality (23.12.1) validates condition (3) of Definition 12.

Barro and Gordon (1983a) considered a version of this equilibrium in which inequality (23.12.1) is satisfied with $(v^b, x^b, y^b) = (v^R, x^R, y^R)$. In this case, anticipated reversion to Nash supports the Ramsey outcome forever. When inequality (23.12.1) is *not* satisfied for $(v^b, x^b, y^b) = (v^R, x^R, y^R)$, we can solve for the best SPE value v^b, with associated actions (x^b, y^b), supportable by infinite reversion to Nash from

$$v^b = r(x^b, y^b) = (1 - \delta)r[x^b, H(x^b)] + \delta v^N > v^N. \tag{23.12.2}$$

The payoff from following the strategy equals that from deviating and reverting to Nash. Any value lower than this can be supported, but none higher.

When $v^b < v^R$, Abreu (1988) searched for a way to support something better than v^b. First, one must construct an equilibrium that yields a value *worse* than permanent repetition of the Nash outcome. The expectation of reverting to this equilibrium supports something better than v^b in equation (23.12.2).

Somehow the government must be induced temporarily to take an action $y^\#$ that yields a worse period-by-period return than the Nash outcome, meaning that the government in general would be tempted to deviate. An equilibrium system of expectations has to be constructed that makes the government expect to do better in the future only by conforming to expectations that it temporarily adheres to the bad policy $y^\#$.

23.12.3. *Something worse: a stick-and-carrot strategy*

To get something worse than repetition of the one-period Nash outcome, Abreu (1988) proposed a "stick-and-carrot punishment." The "stick" part is an outcome $(x^\#, y^\#) \in C$, which relative to (x^N, y^N) is a bad competitive equilibrium from the government's viewpoint. The "carrot" part is the Ramsey outcome (x^R, y^R), which the government attains forever after it has accepted the stick in the first period of its punishment.

We want a continuation value v^* for deviating to support the first-period outcome $(x^\#, y^\#) \in C$ and attain the value[14]

$$\tilde{v} = (1-\delta)r(x^\#, y^\#) + \delta \, v^R \geq (1-\delta)r\left[x^\#, H(x^\#)\right] + \delta \, v^*. \qquad (23.12.3)$$

Abreu proposed to set $v^* = \tilde{v}$ so that the continuation value from deviating from the first-period action equals the original value. If the stick part is severe enough, the associated strategy attains a value worse than infinite repetition of Nash. The strategy induces the government to accept the temporarily bad outcome by promising a high continuation value.

An SPE featuring stick-and-carrot punishments that attains $\tilde{v}$ is

$$
\begin{aligned}
v_1 &= \tilde{v} \\
z^h(v) &= \begin{cases} x^R & \text{if } v = v^R; \\ x^\# & \text{otherwise}; \end{cases} \\
z^g(v) &= \begin{cases} y^R & \text{if } v = v^R; \\ y^\# & \text{otherwise}; \end{cases} \qquad (23.12.4) \\
\mathcal{V}(v, x, y) &= \begin{cases} v^R & \text{if } (x,y) = [z^h(v), z^g(v)]\,; \\ \tilde{v} & \text{otherwise}. \end{cases}
\end{aligned}
$$

When the government deviates from the bad prescribed first-period action $y^\#$, the consequence is to restart the equilibrium. This means that the equilibrium is self-enforcing.

---

[14] This is a "one-period stick." The worst SPE can require more than one period of a worse-than-one-period Nash outcome.

23.13. The best and the worst SPE values

The value associated with Abreu's stick-and-carrot strategy might still not be bad enough to deter the government from deviating from repetition of the Ramsey outcome. We are therefore interested in finding the worst SPE value. We now display a pair of simple programming problems to find the best and worst SPE values. APS (1990) showed how to find the entire set of equilibrium values V. In the current setting, their ideas imply the following:

1. The set of equilibrium values V attainable by the government is a compact subset $[\underline{v}, \overline{v}]$ of $[\min_{(x,y) \in C} r(x, y), r(x^R, y^R)]$.

2. The worst equilibrium value $\underline{v}$ can be computed from a simple programming problem.

3. Given the worst equilibrium value $\underline{v}$, the best equilibrium value $\overline{v}$ can be computed from a programming problem.

4. Given a $v \in [\underline{v}, \overline{v}]$, it is easy to construct an equilibrium that attains it.

Recall from Proposition 2 that the worst equilibrium is self-enforcing, and here we repeat versions of equations $(23.10.2)$ and $(23.10.3)$,

$$\underline{v} = \min_{y \in Y, \, v_1 \in V} \left\{ (1 - \delta) \, r\big[h(y), y\big] + \delta v_1 \right\} \tag{23.13.1}$$

where the minimization is subject to the incentive constraint

$$(1 - \delta) \, r[h(y), y] + \delta v_1 \geq (1 - \delta) \, r\big\{h(y), H[h(y)]\big\} + \delta \underline{v}. \tag{23.13.2}$$

In expression $(23.13.2)$, we use the worst SPE as the continuation value in the event of a deviation. The minimum will be attained when the constraint is binding, which implies that $\underline{v} = r\{h(y), H[h(y)]\}$ for some government action y.[15] Thus, the problem of finding the worst SPE reduces to solving

$$\underline{v} = \min_{y \in Y} r\big\{h(y), H[h(y)]\big\},$$

then computing v_1 from $(1 - \delta) r[h(\underline{y}), \underline{y}] + \delta v_1 = \underline{v}$, where $\underline{y} = \arg \min r\{h(y), H[h(y)]\}$, and finally checking that v_1 is itself a value associated with an SPE. To check this condition, we need to know $\overline{v}$.

---

[15] An equivalent way to express $\underline{v}$ is $\underline{v} = \min_{y \in Y} \max_{\eta \in Y} r(h(y), \eta)$.

The computation of $\overline{v}$ utilizes the fact that the best SPE is self-rewarding; that is, the best SPE has continuation value $\overline{v}$ when the government follows the prescribed equilibrium strategy. Thus, after we have computed a candidate for the worst SPE value $\underline{v}$, we can compute a candidate for the *best* value $\overline{v}$ by solving the programming problem

$$\overline{v} = \max_{y \in Y} \; r\big[h(y), y\big]$$

$$\text{subject to} \quad r\big[h(y), y\big] \geq (1 - \delta) r\{h(y), H[h(y)]\} + \delta \underline{v}.$$

Here we are using the fact that $\overline{v}$ is the maximum continuation value available to reward adherence to the policy, so that $\overline{v} = (1 - \delta) r[h(y), y] + \delta \overline{v}$. Let y^b be the maximizing value of y. Once we have computed $\overline{v}$, we can check that the continuation value v_1 for supporting the worst value is within our candidate set $[\underline{v}, \overline{v}]$. If it is, we have succeeded in constructing V.

23.13.1. *When v_1 is outside the candidate set*

If our candidate v_1 is not within our candidate set $[\underline{v}, \overline{v}]$, we have to seek a smaller set. We could find this set by pursuing the following line of reasoning. We know that

$$\underline{v} = r\left\{h(\underline{y}), H[h(\underline{y})]\right\} \tag{23.13.3}$$

for *some* $\underline{y}$, and that for $\underline{y}$ the continuation value v_1 satisfies

$$(1 - \delta) r[h(\underline{y}), \underline{y}] + \delta v_1 = (1 - \delta) r\left\{h(\underline{y}), H[h(\underline{y})]\right\} + \delta \underline{v}.$$

Solving this equation for v_1 gives

$$v_1 = \frac{1 - \delta}{\delta}\left(r\left\{h(\underline{y}), H[h(\underline{y})]\right\} - r[h(\underline{y}), \underline{y}]\right) + r\left\{h(\underline{y}), H[h(\underline{y})]\right\} \tag{23.13.4}$$

The term in large parentheses on the right measures the one-period temptation to deviate from $\underline{y}$. It is multiplied by $\frac{1-\delta}{\delta}$, which approaches $+\infty$ as $\delta \searrow 0$. Therefore, as $\delta \searrow 0$, it is necessary that the term in braces approach 0, which means that the required $\underline{y}$ must approach y^N.

For discount factors that are so small that v_1 is outside the region of values proposed in the previous subsection because the implied v_1 exceeds the candidate $\overline{v}$, we can proceed in the spirit of Abreu's stick-and-carrot policy, but instead of using v^R as the continuation value to reward adherence (because that is too much to hope for here), we can simply reward adherence to the worst with $\overline{v}$, which we must solve for. Using $\overline{v} = v_1$ as the continuation value for adherence to the worst leads to the following four equations to be solved for $\overline{v}, \underline{v}, \overline{y}, \underline{y}$:

$$\underline{v} = r\left\{h(\underline{y}), H[h(\underline{y})]\right\} \tag{23.13.5}$$

$$\overline{v} = \frac{1-\delta}{\delta}\left(r\left\{h(\underline{y}), H[h(\underline{y})]\right\} - r[h(\underline{y}), \underline{y}]\right)$$
$$+ r\left\{h(\underline{y}), H[h(\underline{y})]\right\} \tag{23.13.6}$$

$$\overline{v} = r[h(\overline{y}), \overline{y}] \tag{23.13.7}$$

$$\overline{v} = (1-\delta)r\left\{h(\overline{y}), H[h(\overline{y})]\right\} + \delta\underline{v}. \tag{23.13.8}$$

In exercise *23.3*, we ask the reader to solve these equations for a particular example.

23.14. Examples: alternative ways to achieve the worst

We return to the situation envisioned before the last subsection, so that the candidate v_1 belongs to the required candidate set $[\underline{v}, \overline{v}]$. We describe examples of some equilibria that attain value $\underline{v}$.

23.14.1. Attaining the worst, method 1

We have seen that to evaluate the best sustainable value $\overline{v}$, we want to find the worst value $\underline{v}$. Many SPEs attain the worst value $\underline{v}$. To compute one such SPE strategy, we can use the following recursive procedure:

1. Set the first-period promised value $v_0 = \underline{v} = r\{h(y^{\#}), H[h(y^{\#})]\}$, where $y^{\#} = \arg\min r\{h(y), H[h(y)]\}$. The competitive equilibrium with the worst one-period value gives value $r[h(y^{\#}), y^{\#}]$. Given expectations $x^{\#} = h(y^{\#})$, the government is tempted toward $H(x^{\#})$, which yields one-period utility to the government of $r\{h(y^{\#}), H[h(y^{\#})]\}$. Then use $\underline{v}$ as continuation value in the event of a deviation, and construct an increasing sequence of continuation values to reward adherence, as follows:

2. Solve $\underline{v} = (1 - \delta)r[h(y^\#), y^\#] + \delta v_2$ for continuation value v_1.

3. For $j = 1, 2, \cdots$, continue solving $v_j = (1 - \delta)r[h(y^\#), y^\#] + \delta v_{j+1}$ for the continuation values v_{j+1} as long as $v_{j+1} \leq \overline{v}$. If v_{j+1} threatens to violate this constraint at step $j = \overline{j}$, then go to step 4.

4. Use $\overline{v}$ as the continuation value, and solve $v_j = (1 - \delta)r[h(\tilde{y}), \tilde{y}] + \delta\overline{v}$ for the prescription $\tilde{y}$ to be followed if promised value v_j is encountered.

5. Set $v_{j+s} = \overline{v}$ for $s \geq 1$.

23.14.2. Attaining the worst, method 2

To construct another equilibrium supporting the worst SPE value, follow steps 1 and 2, and follow step 3 also, except that we continue solving $v_j = (1 - \delta)r[h(y^\#), y^\#] + \delta v_{j+1}$ for the continuation values v_{j+1} only so long as $v_{j+1} < v^N$. As soon as $v_{j+1} = v^{**} > v^N$, we use v^{**} as both the promised value and the continuation value thereafter. In terms of our recursive strategy notation, whenever $v^{**} = r[h(y^{**}), y^{**}]$ is the promised value, $z^h(v^{**}) = h(y^{**})$, $z^g(v^{**}) = y^{**}$, and $v'[v^{**}, z^h(v^{**}), z^g(v^{**})] = v^{**}$.

23.14.3. Attaining the worst, method 3

Here is another subgame perfect equilibrium that supports $\underline{v}$. Proceed as in step 1 to find the initial continuation value v_1. Now set all subsequent values and continuation values to v_1, with associated first-period outcome $\tilde{y}$ that solves $v_1 = r[h(\tilde{y}), \tilde{y}]$. It can be checked that the incentive constraint is satisfied with $\underline{v}$ the continuation value in the event of a deviation.

23.14.4. Numerical example

We now illustrate the concepts and arguments using the infinitely repeated version of the taxation example. To make the problem of finding $\underline{v}$ nontrivial, we impose an upper bound on admissible tax rates given by $\bar{\tau} = 1 - \alpha - \epsilon$, where $\epsilon \in (0, 0.5 - \alpha)$. Given $\tau \in Y \equiv [0, \bar{\tau}]$, the model exhibits a unique Nash equilibrium with $\tau = 0.5$. For a sufficiently small ϵ, the worst one-period competitive equilibrium is $[\ell(\bar{\tau}), \bar{\tau}]$.

Set $[\alpha \quad \delta \quad \bar{\tau}] = [0.3 \quad 0.8 \quad 0.6]$. Compute

$$[\tau^R \quad \tau^N] = [0.3013 \quad 0.5000],$$
$$[v^R \quad v^N \quad \underline{v} \quad v_{\text{abreu}}] = [-0.6801 \quad -0.7863 \quad -0.9613 \quad -0.7370].$$

In this numerical example, Abreu's "stick-and-carrot" strategy fails to attain a value lower than the repeated Nash outcome. The reason is that the upper bound on tax rates makes the least favorable one-period return (the "stick") not so bad.

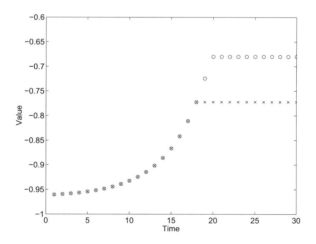

Figure 23.14.1: Continuation values (on coordinate axis) of two SPE that attain $\underline{v}$.

Figure 23.14.1 describes two SPEs that attain the worst SPE value $\underline{v}$ with the depicted sequences of time t (promised value, tax rate) pairs. The circles represent the worst SPE attained with method 1, and the x-marks correspond to method 2. By construction, the continuation values of method 2 are less than or equal to the continuation values of method 1. Since both SPEs attain the same promised value $\underline{v}$, it follows that method 2 must be associated with higher one-period returns in some periods. Figure 23.14.2 indicates that method 2 delivers those higher one-period returns around period 20 when the prescribed tax rates are closer to the Ramsey outcome $\tau^R = 0.3013$.

When varying the discount factor, we find that the cutoff value of δ below which reversion to Nash fails to support Ramsey forever is 0.2194.

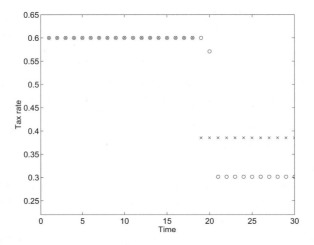

Figure 23.14.2: Tax rates associated with the continuation values of Figure 23.14.1.

23.15. Interpretations

The notion of credibility or sustainability emerges from a ruthless and complete application of two principles: rational expectations and self-interest. At each moment and for each possible history, individuals and the government act in their own best interests while expecting everyone else always to act in their best interests. A credible government policy is one that it is in the interest of the government to implement on every occasion.

The structures that we have studied have multiple equilibria that are indexed by different systems of rational expectations. Multiple equilibria are essential because what sustains a good equilibrium is a system of expectations that raises the prospect of reverting to a bad equilibrium if the government chooses to deviate from the good equilibrium. For reversion to the bad equilibrium to be credible – meaning that it is something that the private agents can expect because the government will want to act accordingly – the bad equilibrium must itself be an equilibrium. It must always be in the self-interest of all agents to behave as they are expected to. Supporting a Ramsey outcome hinges on finding an equilibrium with outcomes bad enough to deter the government from surrendering to a temporary temptation to deviate.[16]

Is the multiplicity of equilibria a strength or a weakness of such theories? Here descriptions of preferences and technologies, supplemented by the restriction of rational expectations, don't pin down outcomes. There is an independent role for expectations not based solely on fundamentals. The theory is silent about which equilibrium will prevail; the theory contains no sense in which the government *chooses* among equilibria.

Depending on the purpose, the multiplicity of equilibria can be regarded either as a strength or as a weakness of these theories. In inferior equilibria, the government is caught in an "expectations trap,"[17] an aspect of the theory that highlights how the government can be regarded as simply resigning itself to affirm the public's expectations about it. Within the theory, the government's

---

[16] This statement means that an equilibrium is supported by beliefs about behavior at prospective histories of the economy that might never be attained or observed. Part of the literature on learning in games and dynamic economies studies situations in which it is not reasonable to expect "adaptive" agents to learn so much. See Fudenberg and Kreps (1993), Kreps (1990), and Fudenberg and Levine (1998). See Sargent (1999, 2008) for macroeconomic counterparts.

[17] See Chari, Christiano, and Eichenbaum (1998).

strategy plays a dual role, as it does in any rational expectations model: one summarizing the government's choices, the other describing the public's rule for forecasting the government's behavior. In inferior equilibria, the government wishes that it could use a different strategy but nevertheless affirms the public's expectation that it will adhere to an inferior rule.

23.16. Extensions

In chapter 24, we shall describe how Chang (1998) and Phelan and Stacchetti (2001) extended the machinery of this chapter to settings in which private agents' problems have natural state variables like stocks of real balances or physical capital so that their best responses to government policies require that Euler equations (or costate equations) be satisfied. This will activate an additional source of history dependence. The approach of chapter 24 merges aspects of the method described in chapter 19 with those of this chapter.

Exercises

Exercise 23.1   Consider the following one-period economy. Let (ξ, x, y) be the choice variables available to a representative agent, the market as a whole, and a benevolent government, respectively. In a rational expectations equilibrium or competitive equilibrium, $\xi = x = h(y)$, where $h(\cdot)$ is the "equilibrium response" correspondence that gives competitive equilibrium values of x as a function of y; that is, $[h(y), y]$ is a competitive equilibrium. Let C be the set of competitive equilibria.

Let $X = \{x_M, x_H\}, Y = \{y_M, y_H\}$. For the one-period economy, when $\xi_i = x_i$, the payoffs to the government and household are given by the values of $u(x_i, x_i, y_j)$ entered in the following table:

One-period payoffs $u(x_i, x_i, y_j)$

	x_M	x_H
y_M	10*	20
y_H	4	15*

* Denotes $(x, y) \in C$.

The values of $u(\xi_k, x_i, y_j)$ not reported in the table are such that the competitive equilibria are the outcome pairs denoted by an asterisk (*).

a. Find the *Nash equilibrium* (in pure strategies) and *Ramsey outcome* for the one-period economy.

b. Suppose that this economy is repeated twice. Is it possible to support the Ramsey outcome in the first period by reverting to the Nash outcome in the second period in case of a deviation?

c. Suppose that this economy is repeated three times. Is it possible to support the Ramsey outcome in the first period? In the second period?

Consider the following expanded version of the preceding economy. $Y = \{y_L, y_M, y_H\}$, $X = \{x_L, x_M, x_H\}$. When $\xi_i = x_i$, the payoffs are given by $u(x_i, x_i, y_j)$ entered here:

One-period payoffs $u(x_i, x_i, y_j)$

	x_L	x_M	x_H
y_L	3*	7	9
y_M	1	10*	20
y_H	0	4	15*

* Denotes $(x, y) \in C$.

d. What are Nash equilibria in this one-period economy?

e. Suppose that this economy is repeated twice. Find a subgame perfect equilibrium that supports the Ramsey outcome in the first period. For what values of δ will this equilibrium work?

f. Suppose that this economy is repeated three times. Find an SPE that supports the Ramsey outcome in the first two periods (assume $\delta = 0.8$). Is it unique?

Exercise 23.2 Consider a version of the setting studied by Stokey (1989). Let (ξ, x, y) be the choice variables available to a representative agent, the market as a whole, and a benevolent government, respectively. In a rational expectations or competitive equilibrium, $\xi = x = h(y)$, where $h(\cdot)$ is the "equilibrium response" correspondence that gives competitive equilibrium values of x as a function of y; that is, $[h(y), y]$ is a competitive equilibrium. Let C be the set of competitive equilibria.

Consider the following special case. Let $X = \{x_L, x_H\}$ and $Y = \{y_L, y_H\}$. For the one-period economy, when $\xi_i = x_i$, the payoffs to the government are given by the values of $u(x_i, x_i, y_j)$ entered in the following table:

One-period payoffs $u(x_i, x_i, y_j)$

	x_L	x_H
y_L	0*	20
y_H	1	10*

* Denotes $(x, y) \in C$.

The values of $u(\xi_k, x_i, y_j)$ not reported in the table are such that the competitive equilibria are the outcome pairs denoted by an asterisk (*).

a. Define a *Ramsey plan* and a *Ramsey outcome* for the one-period economy. Find the Ramsey outcome.

b. Define a *Nash equilibrium* (in pure strategies) for the one-period economy.

c. Show that there exists no Nash equilibrium (in pure strategies) for the one-period economy.

d. Consider the infinitely repeated version of this economy, starting with $t = 1$ and continuing forever. Define a *subgame perfect equilibrium*.

e. Find the value to the government associated with the *worst* subgame perfect equilibrium.

f. Assume that the discount factor is $\delta = .8913 = (1/10)^{1/20} = .1^{.05}$. Determine whether infinite repetition of the Ramsey outcome is sustainable as an SPE. If it is, display the associated subgame perfect equilibrium.

g. Find the value to the government associated with the *best* subgame perfect equilibrium.

h. Find the outcome path associated with the *worst* subgame perfect equilibrium.

i. Find the one-period continuation value v_1 and the outcome path associated with the one-period continuation strategy σ^1 that induces adherence to the worst subgame perfect equilibrium.

j. Find the one-period continuation value v_2 and the outcome path associated with the one-period continuation strategy σ^2 that induces adherence to the first-period outcome of the σ^1 that you found in part i.

k. Proceeding recursively, define v_j and σ^j, respectively, as the one-period continuation value and the continuation strategy that induces adherence to the first-period outcome of σ^{j-1}, where (v_1, σ^1) were defined in part i. Find v_j for $j = 1, 2, \ldots$, and find the associated outcome paths.

l. Find the lowest value for the discount factor for which repetition of the Ramsey outcome is an SPE.

Exercise 23.3 **Finding the worst and best SPEs**

Consider the following model of Kydland and Prescott (1977). A government chooses the inflation rate y from a closed interval $[0, 10]$. There is a family of Phillips curves indexed by the public's expectation of inflation x:

$$(1) \qquad U = U^* - \theta(y - x)$$

where U is the unemployment rate, y is the inflation rate set by the government, and $U^* > 0$ is the natural rate of unemployment and $\theta > 0$ is the slope of the Phillips curve, and where x is the average of private agents' setting of a forecast of y, called ξ. Private agents' only decision in this model is to forecast inflation. They choose their forecast ξ to maximize

$$(2) \qquad -.5(y - \xi)^2.$$

Thus, if they know y, private agents set $\xi = y$. All agents choose the same ξ, so that $x = \xi$ in a rational expectations equilibrium. The government has one-period return function

$$(3) \qquad r(x, y) = -.5(U^2 + y^2) = -.5[(U^* - \theta(y - x))^2 + y^2].$$

Define a *competitive equilibrium* as a 3-tuple U, x, y such that given y, private agents solve their forecasting problem and (1) is satisfied.

a. Verify that in a competitive equilibrium, $x = y$ and $U = U^*$.

b. Define the government best response function in the one-period economy. Compute it.

c. Define a Nash equilibrium (in the spirit of Stokey (1989) or the text of this chapter). Compute one.

d. Define the Ramsey problem for the one-period economy. Define the Ramsey outcome. Compute it.

e. Verify that the Ramsey outcome is better than the Nash outcome.

Now consider the repeated economy where the government cares about

$$(4) \qquad\qquad (1 - \delta) \sum_{t=1}^{\infty} \delta^{t-1} r(x_t, y_t),$$

where $\delta \in (0, 1)$.

f. Define a *subgame perfect equilibrium*.

g. Define a *recursive* subgame perfect equilibrium.

h. Find a recursive subgame perfect equilibrium that sustains infinite repetition of the one-period Nash equilibrium outcome.

i. For $\delta = .95$, $U^* = 5, \theta = 1$, find the value of (4) associated with the worst subgame perfect equilibrium. Carefully and completely show your method for computing the worst subgame perfect equilibrium value. Also, compute the values associated with the repeated Ramsey outcome, the Nash equilibrium, and Abreu's simple stick-and-carrot strategy.

j. Compute a recursive subgame perfect equilibrium that attains the worst subgame perfect equilibrium value (4) for the parameter values in part i.

k. For $U^* = 5, \theta = 1$, find the cutoff value δ_c of the discount factor δ below which the Ramsey value v^R cannot be sustained by reverting to repetition of v^N as a consequence of deviation from the Ramsey y.

l. For the same parameter values as in part k, find another cut off value $\tilde{\delta}_c$ for δ below which Ramsey cannot be sustained by reverting after a deviation to an equilibrium attaining the worst subgame perfect equilibrium value. Compute the worst subgame perfect equilibrium value for $\tilde{\delta}_c$.

m. For $\delta = .08$, compute values associated with the best and worst subgame perfect equilibrium strategies.

Chapter 24
Credible Government Policies, II

24.1. Sources of history-dependent government policies

Chapter 23 adopted a simple setting designed to isolate opportunities that confront the government when the private sector's forecasting problem is the *only* source of dynamics. We studied dynamics that come exclusively from a benevolent government's incentives to confirm or disappoint private agents' forecasts of time t government actions on the basis of histories of outcomes observed through time $t-1$. To focus attention solely on the government's incentives to confirm or disappoint expectations, we analyzed credible public policies in simplified settings with competitive equilibria in which households and firms face a sequence of static problems. In particular, we began with a static setting in which a competitive equilibrium could be summarized as a pair of actions (x, y) that belongs to a set $C \in I\!\!R^2$. We formed a dynamic economy by infinitely repeating the static economy for $t = 1, 2, \ldots$, so that a competitive equilibrium for the repeated economy was simply an arbitrary sequence $\{x_t, y_t\}_{t=0}^{\infty}$ with $(x_t, y_t) \in C$ of competitive equilibria for the static economy.

Of course, in more general settings, private agents' decision problems contribute additional sources of dynamics. This typically has the consequence that a competitive equilibrium for an infinite horizon economy is itself a sequence having dynamics coming from the intertemporal optimization of competitive agents. In this more general setting, the internal dynamics within the competitive equilibrium sequence become confounded with dynamics associated with choices of government policy.

In this chapter, we describe how Chang (1998) and Phelan and Stacchetti (2001) studied credible public policies for such economies, first, by characterizing a competitive equilibrium recursively (see Kydland and Prescott (1980)) in a way reminiscent of a closely related method that we used in chapter 19 to pose Stackelberg problems in linear economies, and, second, by appropriately adapting arguments of Abreu, Pearce, and Stachetti (APS). So methodologically, this chapter combines sources of history-dependent government policies and ways of comprehending them that we have encountered earlier in chapters 19 and 23. The chapter 19 history dependence in government policies is captured through the dynamics of a vector of Lagrange multipliers on implementability constraints in the form of private sector Euler equations that summarize restrictions that competitive equilibrium imposes on outcomes. This chapter 19 history dependence concisely accounts for contributions to time t government actions of promises that a time 0 Ramsey planner makes in order to foster good outcomes at dates $s < t$.

By way of contrast, in chapter 23, no internal dynamics come from private agents' intertemporal problems. All dynamics come from private agents' rules for forecasting government policy actions. There is a sequence of government 'administrations' with the time t administration choosing only a time t government action in light of its forecasts of how future governments will act. This timing completely disarms dynamics coming from the chapter 19 multipliers. The remaining chapter 23 history dependence instead reflects constraints that a system of private agents' expectations imposes on a time t government's opportunities. Here private agents expect that a time t government will take an action only if it is in its interest to do so.

The structures of these diverse sources of history dependence differ. In this chapter, we describe how recursive methods can be used to analyze *both* sources of history dependence. A key message is that to represent credible problems recursively, it is necessary to expand the dimension of the state beyond those used in either chapter 19 or in chapter 23. We choose Roberto Chang's (1998) model as a convenient vehicle for revealing essential forces.

24.2. The setting

First, we introduce some notation. For a sequence of scalars $\vec{z} \equiv \{z_t\}_{t=0}^{\infty}$, let $\vec{z}^t = (z_0, \ldots, z_t)$, $\vec{z}_t = (z_t, z_{t+1}, \ldots)$. An infinitely lived representative agent and an infinitely lived government live at dates $t = 0, 1, \ldots$. The objects in play are an initial condition M_{-1} of nominal money holdings, a sequence of inverse money growth rates $\vec{h}$ and a consequent sequence of nominal money holdings $\vec{M}$, a sequence of values of money $\vec{q}$, a sequence of real money holdings $\vec{m}$, a sequence of total tax collections $\vec{x}$, a sequence of per capita rates of consumption $\vec{c}$, and a sequence of per capita incomes $\vec{y}$. A benevolent government chooses sequences $(\vec{M}, \vec{h}, \vec{x})$ subject to a sequence of budget constraints and constraints imposed by the requirement that a competitive equilibrium prevails. A representative household chooses sequences $(\vec{c}, \vec{m})$. In equilibrium, the price of money sequence $\vec{q}$ adjusts to clear markets in response to decisions of the government and the representative household.

Chang adopts a version of a model that Calvo (1978) designed to exhibit time-inconsistency of a Ramsey policy in a simple and transparent setting.[1] Through the household's expectations, government actions at time t affect components of household utilities for periods s before t. The source of time consistency is that a time 0 Ramsey planner with the ability to commit takes these effects into account in designing a plan of government actions for $t \geq 0$; but without commitment, i.e., with a sequence of governments each choosing just one-period action, there is no incentive for the time t government decision maker to take them into account. Under commitment, when setting a path for monetary expansion rates, the government takes into account the effects of the household's anticipations on current household decisions. But without commitment, the government in any period has no incentive to remember or value the effects of its *current* rate of money expansion on the household's *past* decisions about its holdings of real balances.

---

[1] Kydland and Prescott (1977) had described the time inconsistency of a Ramsey plan in other contexts.

24.2.1. The household's problem

A representative household faces a nonnegative value of money sequence $\vec{q}$ and sequences $\vec{y}, \vec{x}$ of income and total tax collections, respectively. The household chooses nonnegative sequences $\vec{c}, \vec{M}$ of consumption and nominal balances, respectively, to maximize

$$\sum_{t=0}^{\infty} \beta^t \left[u(c_t) + v(q_t M_t) \right] \tag{24.2.1}$$

subject to

$$q_t M_t \leq y_t + q_t M_{t-1} - c_t - x_t \tag{24.2.2}$$

$$q_t M_t \leq \bar{m}. \tag{24.2.3}$$

Here q_t is the reciprocal of the price level at t, also known as the value of money.

Chang assumes that $u : \mathbb{R}_+ \to \mathbb{R}$ is twice continuously differentiable, strictly concave, and strictly increasing; that $v : \mathbb{R}_+ \to \mathbb{R}$ is twice continuously differentiable and strictly concave; that $\lim_{c \to 0} u'(c) = \lim_{m \to 0} v'(m) = +\infty$; and that there is a finite level $m = m^f$ such that $v'(m^f) = 0$.

Real balances are given by $m_t = q_t M_t$. Inequality (24.2.2) is the household's time t budget constraint. It tells how real balances $q_t M_t$ carried out of period t depend on income, consumption, taxes, and real balances $q_t M_{t-1}$ carried into the period. Equation (24.2.3) imposes an exogenous upper bound $\bar{m}$ on the choice of real balances, where $\bar{m} \geq m^f$.

24.2.2. Government

The government chooses a sequence of inverse money growth rates with time t component $h_t \equiv \frac{M_{t-1}}{M_t} \in \Pi \equiv [\underline{\pi}, \bar{\pi}]$, where $0 < \underline{\pi} < 1 < \frac{1}{\beta} \leq \bar{\pi}$. The government faces a sequence of budget constraints with time t component

$$-x_t = q_t(M_t - M_{t-1}),$$

which by using the definitions of m_t and h_t can also be expressed as

$$-x_t = m_t(1 - h_t). \tag{24.2.4}$$

The restrictions $m_t \in [0, \bar{m}]$ and $h_t \in \Pi$ evidently imply that $x_t \in X \equiv [(\underline{\pi} - 1)\bar{m}, (\bar{\pi} - 1)\bar{m}]$. We define the set $E \equiv [0, \bar{m}] \times \Pi \times X$, so that we require that $(m, h, x) \in E$.

To represent the idea that taxes are distorting, Chang makes the following assumption about outcomes for per capita output:

$$y_t = f(x_t), \qquad (24.2.5)$$

where $f : I\!R \rightarrow I\!R$ satisfies $f(x) > 0$, is twice continuously differentiable, $f''(x) < 0$, and $f(x) = f(-x)$ for all $x \in I\!R$, so that subsidies and taxes are equally distorting. The point is not to model the tax distortions in any detail but simply to summarize the *outcome* of those distortions via the function $f(x)$. A key part of the specification is that tax distortions are increasing in the absolute value of tax revenues.

The government is benevolent and chooses a competitive equilibrium that maximizes (24.2.1). The within-period timing of decisions is as follows: first, the government chooses h_t and x_t; then given $\vec{q}$ and its expectations of future values of x and y's, the household chooses M_t and therefore m_t because $m_t = q_t M_t$; then output $y_t = f(x_t)$ is realized; and finally $c_t = y_t$. It is important to remember this within-period timing because it opens possibilities for the private sector to confront the government with incentives that flow from how the private sector responds whenever the government takes time t actions that differ from what the private sector had expected. This consideration will be important when we study credible government policies.

The model is designed to focus on the intertemporal trade-offs between the welfare benefits of deflation and the welfare costs associated with the high tax collections required to retire money at a rate that delivers deflation. A benevolent time 0 government can promote utility generating increases in real balances only by imposing an infinite sequence of sufficiently large distorting tax collections. To promote the welfare increasing effects of high real balances, the government wants to induce a gradual deflationary process.

24.2.3. Solution of household's problem

Formulate the Lagrangian for the household problem:

$$\mathcal{L} = \max_{\vec{c},\vec{M}} \min_{\vec{\lambda},\vec{\mu}} \sum_{t=0}^{\infty} \beta^t \Big\{ u(c_t) + v(m_t) + \lambda_t [y_t - c_t - x_t + q_t M_{t-1} - q_t M_t] \\ + \mu_t [\bar{m} - q_t M_t] \Big\}$$

First-order conditions with respect to c_t and M_t, respectively, are [2]

$$u'(c_t) = \lambda_t$$
$$q_t [u'(c_t) - v'(m_t)] \leq \beta u'(c_{t+1}) q_{t+1}, \quad = 0 \text{ if } m_t < \bar{m}$$

Using $h_t = \frac{M_{t-1}}{M_t}$ and $q_t = \frac{m_t}{M_t}$ in these first-order conditions and rearranging implies

$$m_t [u'(c_t) - v'(m_t)] \leq \beta u'(f(x_{t+1})) m_{t+1} h_{t+1}, \quad = \text{ if } m_t < \bar{m}. \qquad (24.2.6)$$

Define the following key variable

$$\theta_{t+1} \equiv u'(f(x_{t+1})) m_{t+1} h_{t+1}. \qquad (24.2.7)$$

This is the marginal utility of time $t+1$ real money balances. From the standpoint of the household at time t, equation (24.2.6) shows that θ_{t+1} intermediates the influences of $(\vec{x}_{t+1}, \vec{m}_{t+1})$ on the household's choice of real balances m_t. That is, the future paths $(\vec{x}_{t+1}, \vec{m}_{t+1})$ influence m_t entirely through their effects on the scalar θ_{t+1}. The observation that the one dimensional promised marginal utility of real balances θ_{t+1} functions in this way is an important step in constructing a class of competitive equilibria that have a recursive representation. A closely related observation pervaded the equilibrium representation in terms of Lagrange multipliers used in chapter 19.

---

[2] The special conditions imposed by Chang assure that these first-order conditions are both necessary and sufficient within a competitive equilibrium. Discounting and a bounded marginal utility of consumption assure satisfaction of a transversality condition.

24.2.4. Competitive equilibrium

DEFINITION: A *government policy* is a pair of sequences $(\vec{h}, \vec{x})$ where $h_t \in \Pi$ $\forall t \geq 0$. A *price system* is a nonnegative value of money sequence $\vec{q}$. An *allocation* is a triple of nonnegative sequences $(\vec{c}, \vec{m}, \vec{y})$.

It is required that time t components $(m_t, x_t, h_t) \in E$.

DEFINITION: Given M_{-1}, a government policy $(\vec{h}, \vec{x})$, price system $\vec{q}$, and allocation $(\vec{c}, \vec{m}, \vec{y})$ are said to be a *competitive equilibrium* if

  i. $m_t = q_t M_t$ and $y_t = f(x_t)$.

 ii. The government budget constraint (24.2.4) is satisfied.

iii. Given $\vec{q}, \vec{x}, \vec{y}$, $(\vec{c}, \vec{m})$ solves the household's problem.

24.3. Inventory of key objects

Chang constructs the following objects.

1. A set Ω of initial marginal utilities of money θ_0.

Let Ω denote the set of initial promised marginal utilities of money θ_0 associated with competitive equilibria. Chang exploits the fact that a competitive equilibrium consists of a first period outcome (h_0, m_0, x_0) and a continuation competitive equilibrium with marginal utility of money $\theta_1 \in \Omega$.

2. Competitive equilibria that have a recursive representation.

A competitive equilibrium with a recursive representation consists of an initial θ_0 and a four-tuple of functions (h, m, x, Ψ) mapping θ into this period's (h, m, x) and next period's θ, respectively.[3] A competitive equilibrium can be

---

[3] Proposition 3 of Chang (1998) establishes that there exists a competitive equilibrium with such a recursive representation that solves the Ramsey problem. The proposition is silent about whether there exist additional competitive equilibria not having recursive representations that also solve the Ramsey problem.

represented recursively by iterating on

$$h_t = h(\theta_t)$$
$$m_t = m(\theta_t)$$
$$x_t = x(\theta_t) \qquad\qquad (24.3.1)$$
$$\theta_{t+1} = \Psi(\theta_t)$$

starting from θ_0. The range and domain of $\Psi(\cdot)$ are both Ω.

3. A recursive representation of a Ramsey plan.

A recursive representation of a Ramsey plan is a recursive competitive equilibrium $\theta_0, (h, m, x, \Psi)$ that, among all recursive competitive equilibria, maximizes $\sum_{t=0}^{\infty} \beta^t [u(c_t) + v(q_t M_t)]$. The Ramsey planner chooses $\theta_0, (h, m, x, \Psi)$ from among the set of recursive competitive equilibria at time 0. Iterations on the function Ψ determine subsequent θ_t's that summarize the aspects of the continuation competitive equilibria that influence the household's decisions. At time 0, the Ramsey planner commits to this implied sequence $\{\theta_t\}_{t=0}^{\infty}$ and therefore to an associated sequence of continuation competitive equilibria.[4]

4. A characterization of time-inconsistency of a Ramsey plan.

If after a 'revolution' at time $t \geq 1$, a new Ramsey planner were to be given the opportunity to ignore history and solve a brand new Ramsey plan, this new planner would want to discard the θ_t implied by the original Ramsey plan and instead reset it to the θ_0 associated with the original Ramsey plan.[5] The incentive to reinitialize θ_t associated with this revolutionary experiment indicates the time-inconsistency of the Ramsey plan. By resetting θ to θ_0, the new planner

---

[4] In several contexts, Atkeson, Chari, and Kehoe (2010) show that to *implement* a Ramsey plan uniquely, it can be important to add m_t to the right side of the arguments of the functions x and Ψ determining x_t and θ_{t+1} in (24.3.1). Adding (what in Atkeson, Chari, and Kehoe's models amounts to a counterpart to) m_t to these functions allows Atkeson, Chari, and Kehoe to express how the government would respond if the representative household were to deviate from the outcome $\hat{m}_t = m(\theta_t)$ prescribed by the Ramsey plan. They construct these augmented functions to provide incentives that deter private sector deviations from the Ramsey plan.

[5] θ_t is the only state variable in Chang's model. If there were other state variables, say an exogenous Markov shock s_t, then the post-revolutionary Ramsey planner would want to set θ_t to the θ_0 that would have been appropriate if the time 0 exogenous Markov state had been s_t.

avoids the costs at time t that the original Ramsey planner must pay to reap the beneficial effects that the original Ramsey plan for $s \geq t$ had achieved via its influence on the household's decisions for $s = 0, \ldots, t-1$.

5. *A credible government policy with a recursive representation.*

Here there is no time 0 Ramsey planner. Instead there is a sequence of governments, one for each t, that choose time t government actions after forecasting what future governments will do. Let $w = \sum_{t=0}^{\infty} \beta^t \left[u(c_t) + v(q_t M_t) \right]$ be a value associated with a particular competitive equilibrium. A recursive representation of a credible government policy is a pair of initial conditions (w_0, θ_0) and a five-tuple of functions[6]

$$h(w_t, \theta_t), m(h_t, w_t, \theta_t), x(h_t, w_t, \theta_t), \chi(h_t, w_t, \theta_t), \Psi(h_t, w_t, \theta_t)$$

mapping w_t, θ_t and in some cases h_t into $\hat{h}_t, m_t, x_t, w_{t+1}$, and θ_{t+1}, respectively. Starting from initial condition (w_0, θ_0), a credible government policy can be constructed by iterating on these functions in the following order that respects the within-period timing:

$$
\begin{aligned}
\hat{h}_t &= h(w_t, \theta_t) \\
m_t &= m(h_t, w_t, \theta_t) \\
x_t &= x(h_t, w_t, \theta_t) \\
w_{t+1} &= \chi(h_t, w_t, \theta_t) \\
\theta_{t+1} &= \Psi(h_t, w_t, \theta_t).
\end{aligned}
\tag{24.3.2}
$$

Here it is to be understood that $\hat{h}_t$ is the action that the government policy instructs the government to take, while h_t possibly not equal to $\hat{h}_t$ is some other action that the government is free to take at time t. The plan is credible if it is in the time t government's interest to execute it. Credibility requires that the plan be such that for all possible choices of h_t that are consistent with competitive equilibria,

$$
\begin{aligned}
&u(f(x(\hat{h}_t, w_t, \theta_t))) + v(m(\hat{h}_t, w_t, \theta_t)) + \beta \chi(\hat{h}_t, w_t, \theta_t) \\
&\geq u(f(x(h_t, w_t, \theta_t))) + v(m(h_t, w_t, \theta_t)) + \beta \chi(h_t, w_t, \theta_t),
\end{aligned}
$$

---

[6] Again, in the spirit of footnote 3, there can exist additional credible plans that do not have a recursive representation.

so that at each instance and circumstance of choice, a government attains a weakly higher lifetime utility with continuation value $w_{t+1} = \Psi(h_t, w_t, \theta_t)$ by adhering to the plan and confirming the associated time t action $\hat{h}_t$ that the public had expected earlier.[7]

Please note the subtle change in arguments of the functions used to represent a competitive equilibrium and a Ramsey plan, on the one hand, and a credible government plan, on the other hand. The extra arguments appearing in the functions used to represent a credible plan come from allowing the government to contemplate disappointing the private sector's expectation about its time t choice $\hat{h}_t$. A credible plan induces the government to confirm the private sector's expectation. The recursive representation of the plan uses the evolution of continuation values to deter the government from wanting to disappoint the private sector's expectations.

Technically, both the Ramsey plan and the credible plan incorporate history dependence. For the Ramsey plan, this is encoded in the dynamics of the state variable θ_t, a promised marginal utility that the Ramsey plan delivers to the private sector. For a credible government plan, we require the two-dimensional state vector (w_t, θ_t) to encode the history dependence.

24.4. Formal analysis

PROPOSITION: A competitive equilibrium is characterized by a triple of sequences $(\vec{m}, \vec{x}, \vec{h}) \in E^\infty$ that satisfies (24.2.4) and (24.2.6).

DEFINITION: $CE = \{(\vec{m}, \vec{x}, \vec{h}) \in E^\infty$ such that (24.2.4) and (24.2.6) are satisfied.$\}$

CE is not empty because there exists a competitive equilibrium with $h_t = 1$ for all $t \geq 1$, namely, an equilibrium with a constant money supply and constant price level. Chang establishes that CE is also compact. Chang makes the following key observation that combines ideas of APS with insights of Kydland and Prescott (1980).

---

[7] The incentive constraint only deters the government from wanting to take one-period deviations. But this is a context in which checking for the absence of desirable one-period deviations is enough to assure us that there are no desirable deviations of any duration.

PROPOSITION: The continuation of a competitive equilibrium is a competitive equilibrium. That is, $(\vec{m}, \vec{x}, \vec{h}) \in CE$ implies that $(\vec{m}_t, \vec{x}_t, \vec{h}_t) \in CE \; \forall \; t \geq 1$.

RAMSEY PROBLEM:

$$\max_{(\vec{m}, \vec{x}, \vec{h}) \in E^\infty} \sum_{t=0}^\infty \beta^t \left[u(c_t) + v(m_t) \right]$$

subject to (24.2.4), (24.2.6), and (24.2.5). Evidently, associated with any competitive equilibrium (m_0, x_0) is an implied value of $\theta_0 = u'(f(x_0))(m_0 + x_0)$.

To bring out a recursive structure inherent in the Ramsey problem, Chang defines the set

$$\Omega = \left\{ \theta \in I\!R \text{ such that } \theta = u'(f(x_0))(m_0 + x_0) \text{ for some } (\vec{m}, \vec{x}, \vec{h}) \in CE \right\}.$$

Equation (24.2.6) inherits from the household's Euler equation for money holdings the property that the value of m_0 consistent with the representative household's choices depends on $(\vec{h}_1, \vec{m}_1)$. This dependence is captured in the definition above by making Ω be the set of first period values of θ_0 satisfying $\theta_0 = u'(f(x_0))(m_0 + x_0)$ for first period component (m_0, h_0) of competitive equilibrium sequences $(\vec{m}, \vec{x}, \vec{h})$. Chang establishes that Ω is a nonempty and compact subset of $I\!R_+$.

Next Chang advances:

DEFINITION: $\Gamma(\theta) = \left\{ (\vec{m}, \vec{x}, \vec{h}) \in CE | \theta = u'(f(x_0))(m_0 + x_0) \right\}.$

Thus, $\Gamma(\theta)$ is the set of competitive equilibrium sequences $(\vec{m}, \vec{x}, \vec{h})$ whose first period components (m_0, h_0) deliver the prescribed value θ for first period marginal utility.

If we knew the sets $\Omega, \Gamma(\theta)$, we could use the following two-step procedure to find at least the *value* of the Ramsey outcome to the representative household.
1. Find the indirect value function $w(\theta)$ defined as

$$w(\theta) = \max_{\vec{m}, \vec{x}, \vec{h}} \sum_{t=0}^\infty \beta^t \left[u(f(x_t)) + v(m_t) \right]$$

where the maximization is over elements of the set $(\vec{m}, \vec{x}, \vec{h}) \in \Gamma(\theta)$.
2. Compute the value of the Ramsey outcome by solving $\max_{\theta \in \Omega} w(\theta)$.

Next, Chang states

PROPOSITION: $w(\theta)$ satisfies the Bellman equation

$$w(\theta) = \max_{x,m,h,\theta'} \{u(f(x)) + v(m) + \beta w(\theta')\} \tag{24.4.1}$$

where the maximization is subject to

$$(m, x, h) \in E \text{ and } \theta' \in \Omega \tag{24.4.2}$$

$$\theta = u'(f(x))(m + x) \tag{24.4.3}$$

$$-x = m(1 - h) \tag{24.4.4}$$

$$m \cdot [u'(f(x)) - v'(m)] \leq \beta\theta', \quad = \text{ if } m < \bar{m}. \tag{24.4.5}$$

Before we use this proposition to recover a recursive representation of the Ramsey plan, note that the proposition relies on knowing the set Ω. Chang uses the insights of Kydland and Prescott (1980) together with a method reminiscent of chapter 23's APS iteration to convergence on an operator B that takes continuation values into values. We want an operator that takes a continuation θ into a current θ. Chang lets Q be a nonempty, bounded subset of $\mathbb{R}$. Elements of the set Q are taken to be candidate values for continuation marginal utilities. Chang defines an operator

$$B(Q) = \{\theta \in \mathbb{R} : \text{there is } (m, x, h, \theta') \in E \times Q \text{ such that}$$

$$(24.4.3), (24.4.4), \text{ and } (24.4.5) \text{ hold.}\}$$

Thus, $B(Q)$ is the set of first period θ's attainable with $(m, x, h) \in E$ and some $\theta' \in Q$.

PROPOSITION:
  i. $Q \subset B(Q)$ implies $B(Q) \subset \Omega$. ('self-generation').
 ii. $\Omega = B(\Omega)$. ('factorization')

The proposition characterizes Ω as the largest fixed point of B. It is easy to establish that $B(Q)$ is a monotone operator. This property allows Chang to compute Ω as the limit of iterations on B provided that iterations begin from a sufficiently large initial set.

For convenience, we choose to introduce another operator at this point.[8] Before we do, we want some additional notation.

24.4.1. Some useful notation

Let $\vec{h}^t = (h_0, h_1, \ldots, h_t)$ denote a history of inverse money creation rates with time t component $h_t \in \Pi$. A *government strategy* $\sigma = \{\sigma_t\}_{t=0}^\infty$ is a $\sigma_0 \in \Pi$ and for $t \geq 1$ a sequence of functions $\sigma_t : \Pi^{t-1} \to \Pi$. Chang restricts the government's choice of strategies to the following space:

$$CE_\pi = \{\vec{h} \in \Pi^\infty : \text{there is some } (\vec{m}, \vec{x}) \text{ such that } (\vec{m}, \vec{x}, \vec{h}) \in CE\}.$$

In words, CE_π is the set of money growth sequences consistent with the existence of competitive equilibria. Chang observes that CE_π is nonempty and compact.

DEFINITION: σ is said to be *admissible* if for all $t \geq 1$ and after any history $\vec{h}^{t-1}$, the continuation $\vec{h}_t$ implied by σ belongs to CE_π.

Admissibility of σ means that anticipated policy choices associated with σ are consistent with the existence of competitive equilibria after each possible subsequent history. After any history $\vec{h}^{t-1}$, admissibility restricts the government's choice in period t to the set

$$CE_\pi^0 = \{h \in \Pi : \text{there is } \vec{h} \in CE_\pi \text{ with } h = h_0\}.$$

In words, CE_π^0 is the set of all first period money growth rates $h = h_0$, each of which consistent with the existence of a sequence of money growth rates $\vec{h}$ starting from h_0 in the initial period and for which a competitive equilibrium exists.

REMARK: $CE_\pi^0 = \{h \in \Pi : \text{there is } (m, \theta') \in [0, \bar{m}] \times \Omega \text{ such that } mu'[f((h - 1)m) - v'(m)] \leq \beta\theta' \text{ with equality if } m < \bar{m}.\}$

DEFINITION: An *allocation rule* is a sequence of functions $\vec{\alpha} = \{\alpha_t\}_{t=0}^\infty$ such that $\alpha_t : \Pi^t \to [0, \bar{m}] \times X$.

---

[8] It is a version of an operator $D(Z)$ to be constructed in section 24.5.

Thus, the time t component of $\alpha_t(h^t)$ is a pair of functions $(m_t(h^t), x_t(h^t))$.

DEFINITION: Given an admissible government strategy σ, an allocation rule α is called *competitive* if given any history $\vec{h}^{t-1}$ and $h_t \in CE_\pi^0$, the continuation of σ and α after $(\vec{h}^{t-1}, h_t)$ induce a competitive equilibrium sequence.

24.4.2. Another convenient operator

Let S be the set of all pairs (w, θ) of competitive equilibrium values and associated initial marginal utilities. Let W be a bounded set of *values* in $\mathrm{I\!R}$. Let Z be a nonempty subset of $W \times \Omega$ and think of using pairs (w', θ') drawn from Z as candidate continuation value, θ pairs. Define the operator

$$D(Z) = \Big\{(w, \theta) : \text{there is } h \in CE_\pi^0$$

$$\text{and a four}-\text{tuple } (m(h), x(h), w'(h), \theta'(h)) \in [0, \bar{m}] \times X \times Z$$

$$\text{such that}$$

$$w = u(f(x(h))) + v(m(h)) + \beta w'(h) \tag{24.4.6}$$

$$\theta = u'(f(x(h)))(m(h) + x(h)) \tag{24.4.7}$$

$$x(h) = m(h)(h - 1) \tag{24.4.8}$$

$$m(h)(u'(f(x(h))) - v'(m(h))) \le \beta\theta'(h) \tag{24.4.9}$$

$$\text{with equality if } m(h) < \bar{m}.\Big\}$$

It is possible to establish

PROPOSITION:
i. If $Z \subset D(Z)$, then $D(Z) \subset S$. ('self-generation')
ii. $S = D(S)$. ('factorization')

PROPOSITION:
i. Monotonicity of D: $Z \subset Z'$ implies $D(Z) \subset D(Z')$.
ii. Z compact implies that $D(Z)$ is compact.

It can be shown that S is compact and that therefore there exists a (w, θ) pair within this set that attains the highest possible value. This (w, θ) pair

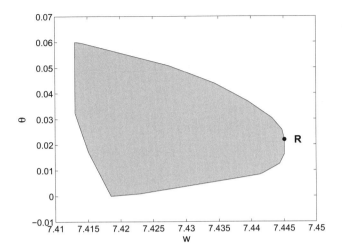

Figure 24.4.1: Set of (w, θ) pairs associated with competitive equilibria. The Ramsey is associated with the (w, θ) pair denoted R, which among points in the set maximizes w. The appropriate choice of initial θ is the projection of R onto the vertical axis.

is associated with a Ramsey plan. Further, we can compute S by iterating to convergence on D provided that one begins with a sufficiently large initial set S_0.

Figure 24.4.1 reports the set of (w, θ) pairs associated with competitive equilibria for the following parameterization: $\beta = .3, u(c) = \log(c), v(m) = \frac{1}{2000}(m\bar{m} - .5m^2)^{.5}, f(x) = 180 - (.4x)^2$. [9] The largest value associated with a competitive equilibrium is the one attained by the Ramsey plan. The (w, θ) pair associated with a Ramsey plan is denoted by an R in the figure.

REMARK: As a very useful by-product, the algorithm that finds the largest fixed point $S = D(S)$ also produces credible policies and competitive equilibrium allocations.

---

[9] We thank Ignacio Presno for computing this set and also the one in figure 24.5.1. Presno used the outer approximation method of Judd, Yeltekin, and Conklin (2003) to compute this set. A public randomization device is implicitly introduced to convexify the set of equilibrium values.

24.5. Sustainable plans

DEFINITION: A government strategy σ and an allocation rule α are said to constitute a *sustainable plan* (SP) if

i. σ is admissible.

ii. Given σ, α is competitive.

iii. After any history $\vec{h}^{t-1}$, the continuation of σ is optimal for the government; i.e., the sequence $\vec{h}_t$ induced by σ after $\vec{h}^{t-1}$ maximizes $(24.2.1)$ over CE_π given α.

REMARK: Given any history $\vec{h}^{t-1}$, the continuation of a sustainable plan is a sustainable plan.

DEFINITION: Let $\Theta = \{(\vec{m}, \vec{x}, \vec{h}) \in CE : \text{there is an SP whose outcome is } (\vec{m}, \vec{x}, \vec{h})\}$.

Sustainable outcomes are elements of Θ.

Now consider the space

$$S = \Big\{ (w, \theta) : \text{there is a sustainable outcome } (\vec{m}, \vec{x}, \vec{h}) \in \Theta \text{ with value}$$

$$w = \sum_{t=0}^{\infty} \beta^t [u(f(x_t)) + v(m_t)] \text{ and such that } u'(f(x_0))(m_0 + x_0) = \theta \Big\}$$

The space S is a compact subset of $W \times \Omega$ where $W = [\underline{w}, \overline{w}]$ is the space of values associated with sustainable plans. Here $\underline{w}$ and $\overline{w}$ are finite bounds on the set of values. Because there is at least one sustainable plan, S is nonempty.

Now recall the within-period timing protocol, which we can depict $(h, x) \to m = qM \to y = c$. With this timing protocol in mind, the time 0 component of an SP has the following components:

i. A period 0 action $\hat{h} \in \Pi$ that the public expects the government to take, together with subsequent within-period consequences $m(\hat{h}), x(\hat{h})$ when the government acts as expected.

ii. For any first period action $h \neq \hat{h}$ with $h \in CE_\pi^0$, a pair of within-period consequences $m(h), x(h)$ when the government does not act as had been expected.

iii. For every $h \in \Pi$, a pair $(w'(h), \theta'(h)) \in S$ to carry into next period.

These components must be such that it is optimal for the government to choose $\hat{h}$ as expected; and for every possible $h \in \Pi$, the government budget constraint and the household's Euler equation must hold with continuation θ being $\theta'(h)$.

Given the timing protocol within the model, the representative household's response to a government deviation from a prescribed h consists of a first period action $m(h)$ and associated subsequent actions, together with future equilibrium prices, captured by $(w'(h), \theta'(h))$.

At this point, Chang introduces an idea in the spirit of APS. Let Z be a nonempty subset of $W \times \Omega$ and think of using pairs (w', θ') drawn from Z as candidate continuation value, promised marginal utility pairs. We define the following operator:

$$\tilde{D}(Z) = \Big\{ (w, \theta) : \text{there is } \hat{h} \in CE_\pi^0 \text{ and for each } h \in CE_\pi^0$$

$$\text{a four} - \text{tuple } (m(h), x(h), w'(h), \theta'(h)) \in [0, \bar{m}] \times X \times Z$$

such that

$$w = u(f(x(\hat{h}))) + v(m(\hat{h})) + \beta w'(\hat{h}) \tag{24.5.1}$$

$$\theta = u'(f(x(\hat{h})))(m(\hat{h}) + x(\hat{h})) \tag{24.5.2}$$

and for all $h \in CE_\pi^0$

$$w \geq u(f(x(h))) + v(m(h)) + \beta w'(h) \tag{24.5.3}$$

$$x(h) = m(h)(h - 1) \tag{24.5.4}$$

and

$$m(h)(u'(f(x(h))) - v'(m(h))) \leq \beta \theta'(h) \tag{24.5.5}$$

$$\text{with equality if } m(h) < \bar{m}. \Big\}$$

This operator adds the key incentive constraint $(24.5.3)$ to the conditions that had defined the earlier $D(Z)$ operator from subsection 24.4.2. Condition $(24.5.3)$ requires that the plan deter the government from wanting to take one-shot deviations when candidate continuation values are drawn from Z.

PROPOSITION:
  i. If $Z \subset \tilde{D}(Z)$, then $\tilde{D}(Z) \subset S$. ('self-generation')
  ii. $S = \tilde{D}(S)$. ('factorization')

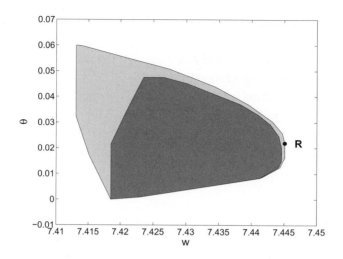

Figure 24.5.1: Sets of (w, θ) pairs associated with competitive equilibria (the larger set) and with sustainable plans (the smaller set). The Ramsey plan is associated with the (w, θ) pair denoted R.

PROPOSITION:

i. Monotonicity of $\tilde{D}$: $Z \subset Z'$ implies $\tilde{D}(Z) \subset \tilde{D}(Z')$.

ii. Z compact implies that $\tilde{D}(Z)$ is compact.

Chang establishes that S is compact and that therefore there exists a highest value SP and a lowest value SP. Further, the preceding structure allows Chang to compute S by iterating to convergence on $\tilde{D}$ provided that one begins with a sufficiently large initial set S_0.

Figure 24.5.1 adds to figure 24.4.1 the set of (w, θ) pairs associated with sustainable plans set for the same parameterization described above for figure 24.4.1. This set is smaller than the set of (w, θ) pairs associated with competitive equilibria, since the additional constraints associated with sustainability must be satisfied for this set. We have verified in separate calculations that raising β expands the set associated with sustainable plans, and that for a sufficiently large $\beta < 1$, it is possible to attain a (w, θ) pair associated with the Ramsey plan.

This structure delivers the following recursive representation of a sustainable outcome: (1) choose an initial $(w_0, \theta_0) \in S$; (2) generate a sustainable outcome recursively by iterating on (24.3.2).

Chapter 25
Two Topics in International Trade

25.1. Two dynamic contracting problems

This chapter studies two models in which recursive contracts are used to overcome incentive problems commonly thought to occur in international trade. The first is Andrew Atkeson's model of lending in the context of a dynamic setting that contains both a moral hazard problem due to asymmetric information *and* an enforcement problem due to borrowers' option to disregard the contract. It is a considerable technical achievement that Atkeson managed to include both of these elements in his contract design problem. But this substantial technical accomplishment is not just showing off. As we shall see, *both* the moral hazard *and* the self-enforcement requirement for the contract are required in order to explain the feature of observed repayments that Atkeson was after: that the occurrence of especially low output realizations prompt the contract to call for net repayments from the borrower to the lender, exactly the occasions when an unhampered insurance scheme would have lenders extend credit to borrowers.

The second is Bond and Park's model of a recursive contract that induces two countries to adopt free trade when they begin with a pair of promised values that implicitly determine the distribution of eventual welfare gains from trade liberalization. The new policy is accomplished by a gradual relaxation of tariffs, accompanied by trade concessions. Bond and Park's model of gradualism is all about the dynamics of promised values that are used optimally to manage participation constraints.

25.2. Lending with moral hazard and difficult enforcement

Andrew Atkeson (1991) designed a model to explain how, in defiance of the pattern predicted by complete markets models, low output realizations in various countries in the mid-1980s prompted international lenders to ask those countries for net repayments. A complete markets model would have net flows to a borrower during periods of bad endowment shocks. Atkeson's idea was that information and enforcement problems could produce the observed outcome. Thus, Atkeson's model combines two features of the models we have seen in chapter 20: incentive problems from private information and participation constraints coming from enforcement problems.

Atkeson showed that the optimal contract handles enforcement and information problems through the shape of the repayment schedule, thereby indirectly manipulating continuation values. Continuation values respond only by updating a single state variable, a measure of resources available to the borrower, that appears in the optimum value function, which in turn is affected only through the repayment schedule. Once this state variable is taken into account, promised values do not appear as independently manipulated state variables.[1]

Atkeson's model brings together several features. He studies a "borrower" who by himself is situated like a planner in a stochastic growth model, with the only vehicle for saving being a stochastic investment technology. Atkeson adds the possibility that the planner can also borrow subject to both participation and information constraints.

A borrower lives for $t = 0, 1, 2, \ldots$. He begins life with Q_0 units of a single good. At each date $t \geq 0$, the borrower has access to an investment technology. If $I_t \geq 0$ units of the good are invested at t, $Y_{t+1} = f(I_t, \varepsilon_{t+1})$ units of time $t+1$ goods are available, where ε_{t+1} is an i.i.d. random variable. Let $g(Y_{t+1}, I_t)$ be the probability density of Y_{t+1} conditioned on I_t. It is assumed that increased investment shifts the distribution of returns toward higher returns.

The borrower has preferences over consumption streams ordered by

$$(1 - \delta) E_0 \sum_{t=0}^{\infty} \delta^t u(c_t) \tag{25.2.1}$$

---

[1] To understand how Atkeson achieves this outcome, the reader should also digest the approach described in chapter 23.

where $\delta \in (0,1)$ and $u(\cdot)$ is increasing, strictly concave, twice continuously differentiable, and $u'(0) = +\infty$.

Atkeson used various technical conditions to render his model tractable. He assumed that for each investment I, $g(Y, I)$ has finite support $(Y_1, \ldots, Y_n)$, with $Y_n > Y_{n-1} > \ldots > Y_1$. He assumed that $g(Y_i, I) > 0$ for all values of I and all states Y_i, making it impossible precisely to infer I from Y. He further assumed that the distribution $g(Y, I)$ is given by the convex combination of two underlying distributions $g_0(Y)$ and $g_1(Y)$ as follows:

$$g(Y, I) = \lambda(I)g_0(Y) + [1 - \lambda(I)]g_1(Y), \qquad (25.2.2)$$

where $g_0(Y_i)/g_1(Y_i)$ is monotone and increasing in i, $0 \leq \lambda(I) \leq 1$, $\lambda'(I) > 0$, and $\lambda''(I) \leq 0$ for all I. Note that

$$g_I(Y, I) = \lambda'(I)[g_0(Y) - g_1(Y)], \qquad (25.2.3)$$

where g_I denotes the derivative with respect to I. Moreover, the assumption that increased investment shifts the distribution of returns toward higher returns implies

$$\sum_i Y_i \left[g_0(Y_i) - g_1(Y_i)\right] > 0. \qquad (25.2.4)$$

We shall consider the borrower's choices in three environments: (1) autarky, (2) lending from risk-neutral lenders under complete observability of the borrower's choices and complete enforcement, and (3) lending under incomplete observability and limited enforcement. Environment 3 is Atkeson's. We can use environments 1 and 2 to construct bounds on the value function for performing computations described in an appendix.

25.2.1. Autarky

Suppose that there are no lenders. Thus, the "borrower" is just an isolated household endowed with the technology. The household chooses (c_t, I_t) to maximize expression $(25.2.1)$ subject to

$$c_t + I_t \leq Q_t$$
$$Q_{t+1} = Y_{t+1}.$$

The optimal value function $U(Q)$ for this problem satisfies the Bellman equation

$$U(Q) = \max_{Q \geq I \geq 0} \left\{ (1 - \delta)\, u(Q - I) + \delta \sum_{Q'} U(Q') g(Q', I) \right\}. \qquad (25.2.5)$$

The first-order condition for I is

$$-(1 - \delta) u'(Q - I) + \delta \sum_{Q'} U(Q') g_I(Q', I) \leq 0, \qquad = 0 \text{ if } I > 0. \qquad (25.2.6)$$

This first-order condition implicitly defines a rule for accumulating capital under autarky.

25.2.2. Investment with full insurance

We now consider an environment in which in addition to investing I in the technology, the borrower can issue Arrow securities at a vector of prices $q(Y', I)$, where we let $'$ denote next period's values, and $d(Y')$ the quantity of one-period Arrow securities issued by the borrower; $d(Y')$ is the number of units of next period's consumption good that the borrower promises to deliver. Lenders observe the level of investment I, and so the pricing kernel $q(Y', I)$ depends explicitly on I. Thus, for a promise to pay one unit of output next period contingent on next-period output realization Y', for each level of I, the borrower faces a different price. (As we shall soon see, in Atkeson's model lenders cannot observe I, making it impossible to condition the price on I.) We shall assume that the Arrow securities are priced by risk-neutral investors who also have one-period discount factor δ. This implies that the price of Arrow securities is given by

$$q(Y', I) = \delta g(Y', I), \qquad (25.2.7)$$

which in turn implies that the gross one-period risk-free interest rate is δ^{-1}.

In a complete markets world where there is no problem with information or enforcement, the borrower's optimal investment decision is not a function of the borrower's own holdings of the good. Instead, the optimal investment level maximizes the project's present value when evaluated at the prices for Arrow securities:

$$\max_{I \geq 0} \left\{ -I + \sum_{Y'} Y' q(Y', I) \right\}, \tag{25.2.8}$$

and after imposing expression $(25.2.7)$

$$\max_{I \geq 0} \left\{ -I + \delta \sum_{Y'} Y' g(Y', I) \right\}.$$

Hence, when the prices for Arrow securities are determined by risk-neutral investors, the optimal investment level maximizes the project's expected payoffs discounted at the risk-free interest rate δ^{-1}. The first-order condition for I is

$$\sum_{Y'} Y' g_I(Y', I) \leq \delta^{-1}, \qquad = \delta^{-1} \text{ if } I > 0; \tag{25.2.9}$$

and after invoking equation $(25.2.3)$

$$\lambda'(I) \sum_{Y'} Y' [g_0(Y') - g_1(Y')] \leq \delta^{-1}, \qquad = \delta^{-1} \text{ if } I > 0.$$

This condition uniquely determines the investment level I, since the left side is decreasing in I and must eventually approach zero because of the upper bound on $\lambda(I)$.

As in chapter 8, we formulate the borrower's budget constraints recursively as

$$c - \sum_{Y'} q(Y', I^*) d(Y') + I^* \leq Q \tag{25.2.10a}$$

$$Q' = Y' - d(Y'), \tag{25.2.10b}$$

where I^* is the solution to investment problem $(25.2.8)$. Let $W(Q)$ be the optimal value for a borrower with goods Q. The borrower's Bellman equation is

$$W(Q) = \max_{c, d(Y')} \left\{ (1 - \delta) u(c) + \delta \sum_{Y'} W[Y' - d(Y')] g(Y', I^*) \right.$$

$$\left. + \mu \left[Q - c + \sum_{Y'} q(Y', I^*) d(Y') - I^* \right] \right\}, \tag{25.2.11}$$

where μ is a Lagrange multiplier on expression $(25.2.10a)$. First-order conditions with respect to $c, d(Y')$, respectively, are

$$c: \quad (1 - \delta)u'(c) - \mu = 0, \tag{25.2.12a}$$

$$d(Y'): \quad -\delta W'[Y' - d(Y')]g(Y', I^*) + \mu q(Y', I^*) = 0. \tag{25.2.12b}$$

By substituting $(25.2.7)$ and $(25.2.12a)$ into first-order condition $(25.2.12b)$, we obtain

$$-W'[Y' - d(Y')] + (1 - \delta)u'(c) = 0,$$

and after invoking the Benveniste-Scheinkman condition, $W'(Q') = (1-\delta)u'(c')$, we arrive at the consumption-smoothing result $c' = c$. This in turn implies, via the status of Q as the state variable in the Bellman equation, that $Q' = Q = Q_0$. Thus, the solution has I constant over time at a level I^* determined by equation $(25.2.9)$, and c and the functions $d(Y')$ satisfying

$$c + I^* = Q_0 + \sum_{Y'} q(Y', I^*)d(Y') \tag{25.2.13a}$$

$$d(Y') = Y' - Q_0. \tag{25.2.13b}$$

The borrower borrows a constant $\sum_{Y'} q(Y', I^*)d(Y')$ each period, invests the same I^* each period, and makes high repayments when Y' is high and low repayments when Y' is low. This is the standard full-insurance solution.

We now turn to Atkeson's setting where the borrower does better than under autarky but worse than with the loan contract under perfect enforcement and observable investment. Atkeson found a contract with value $V(Q)$ for which $U(Q) \leq V(Q) \leq W(Q)$. We shall want to compute $W(Q)$ and $U(Q)$ in order to compute the value of the borrower under the more restricted contract.

25.2.3. Limited commitment and unobserved investment

Atkeson designed an optimal recursive contract that copes with two impediments to risk sharing: (1) moral hazard, that is, hidden action: the lender cannot observe the borrower's action I_t that affects the probability distribution of returns Y_{t+1}; and (2) one-sided limited commitment: the borrower is free to default on the contract and can choose to revert to autarky at any state.

Each period, the borrower confronts a two-period-lived, risk-neutral lender who is endowed with $M > 0$ in each period of his life. Each lender can lend or borrow at a risk-free gross interest rate of δ^{-1} and must earn an expected return of at least δ^{-1} if he is to lend to the borrower. The lender is also willing to *borrow* at this same expected rate of return. The lender can lend up to M units of consumption to the borrower in the first period of his life, and could *repay* (if the borrower lends) up to M units of consumption in the second period of his life. The lender lends $b_t \leq M$ units to the borrower and gets a state-contingent repayment $d(Y_{t+1})$, where $-M \leq d(Y_{t+1})$, in the second period of his life. That the repayment is state contingent lets the lender insure the borrower.

A lender is willing to make a one-period loan to the borrower, but only if the loan contract ensures repayment. The borrower will fulfill the contract only if he wants. The lender observes Q, but observes neither C nor I. Next period, the lender can observe Y_{t+1}. He bases the repayment on that observation. Where $c_t + I_t - b_t = Q_t$, Atkeson's optimal recursive contract takes the form

$$d_{t+1} = d\left(Y_{t+1}, Q_t\right) \qquad (25.2.14a)$$

$$Q_{t+1} = Y_{t+1} - d_{t+1} \qquad (25.2.14b)$$

$$b_t = b(Q_t). \qquad (25.2.14c)$$

The repayment schedule $d(Y_{t+1}, Q_t)$ depends only on observables and is designed to recognize the limited commitment and moral hazard problems.

Notice how Q_t is the only state variable in the contract. Atkeson uses the apparatus of Abreu, Pearce, and Stacchetti (1990), discussed in chapter 23, to show that the state can be taken to be Q_t, and that it is not necessary to keep track of the history of past Q's. Atkeson obtains the following Bellman equation. Let $V(Q)$ be the optimum value of a borrower in state Q under the optimal contract. Let $A = (c, I, b, d(Y'))$, all to be chosen as functions of Q. The Bellman equation is

$$V(Q) = \max_A\left\{(1-\delta)\, u\left(c\right) + \delta \sum_{Y'} V\left[Y' - d\left(Y', Q\right)\right] g\left(Y', I\right)\right\} \qquad (25.2.15a)$$

subject to

$$c + I - b \leq Q, \quad b \leq M, \quad -d(Y', Q) \leq M, \quad c \geq 0, \quad I \geq 0 \qquad (25.2.15b)$$

$$b \leq \delta \sum_{Y'} d(Y') \, g(Y', I) \qquad (25.2.15c)$$

$$V[Y' - d(Y')] \geq U(Y') \qquad (25.2.15d)$$

$$I = \underset{\tilde{I} \,\epsilon\,[0, Q+b]}{\arg\max} \left\{ (1-\delta) \, u \left(Q + b - \tilde{I} \right) + \delta \sum_{Y'} V\left[Y' - d(Y', Q) \right] g(Y', \tilde{I}) \right\}. \qquad (25.2.15e)$$

Condition $(25.2.15b)$ is feasibility. Condition $(25.2.15c)$ is a rationality constraint for lenders: it requires that the gross return from lending to the borrower be at least as great as the alternative yield available to lenders, namely, the risk-free gross interest rate δ^{-1}. Condition $(25.2.15d)$ says that in every state tomorrow, the borrower must want to comply with the contract; thus the value of affirming the contract (the left side) must be at least as great as the value of autarky. Condition $(25.2.15e)$ states that the borrower chooses I to maximize his expected utility under the contract.

There are many value functions $V(Q)$ and associated contracts $b(Q), d(Y', Q)$ that satisfy conditions $(25.2.15)$. Because we want the optimal contract, we want the $V(Q)$ that is the largest (hopefully, pointwise). The usual strategy of iterating on the Bellman equation, starting from an arbitrary guess $V^0(Q)$, say, 0, will not work in this case because high candidate continuation values $V(Q')$ are needed to support good current-period outcomes. But a modified version of the usual iterative strategy does work, which is to make sure that we start with a large enough initial guess at the continuation value function $V^0(Q')$. Atkeson (1988, 1991) verified that the optimal contract can be constructed by iterating to convergence on conditions $(25.2.15)$, provided that the iterations begin from a large enough initial value function $V^0(Q)$. (See the appendix for a computational exercise using Atkeson's iterative strategy.) He adapted ideas from Abreu, Pearce, and Stacchetti (1990) to show this result.[2] In the next subsection, we shall form a Lagrangian in which the role of continuation values is explicitly accounted for.

---

[2] See chapter 23 for some work with the Abreu, Pearce, and Stacchetti structure, and for how, with history dependence, dynamic programming principles direct attention to *sets* of continuation value functions. The need to handle a set of continuation values appropriately is why Atkeson must initiate his iterations from a sufficiently high initial value function.

Binding participation constraint

Atkeson motivated his work as an effort to explain why countries often experience capital outflows in the very-low-income periods in which they would be borrowing *more* in a complete markets setting. The optimal contract associated with conditions (25.2.15) has the feature that Atkeson sought: the borrower makes net repayments $d_t > b_t$ in states with low output realizations.

Atkeson establishes this property using the following argument. First, to permit him to capture the borrower's best response with a first-order condition, he assumes the following conditions about the outcomes:[3]

ASSUMPTIONS: For the optimum contract

$$\sum_i d_i \big[g_0(Y_i) - g_1(Y_i)\big] \geq 0. \qquad (25.2.16)$$

This makes the value of repayments increasing in investment. In addition, assume that the borrower's constrained optimal investment level is interior.

Atkeson assumes conditions (25.2.16) and (25.2.2) to justify using the first-order condition for the right side of equation (25.2.15e) to characterize the investment decision. The first-order condition for investment is

$$-(1 - \delta)u'(Q + b - I) + \delta \sum_i V(Y_i - d_i)g_I(Y_i, I) = 0.$$

---

[3] The first assumption makes the lender prefer that the borrower would make larger rather than smaller investments. See Rogerson (1985b) for conditions needed to validate the first-order approach to incentive problems.

25.2.4. *Optimal capital outflows under distress*

To deduce a key property of the repayment schedule, we will follow Atkeson by introducing a continuation value $\tilde{V}$ as an additional choice variable in a programming problem that represents a form of the contract design problem. Atkeson shows how (25.2.15) can be viewed as the outcome of a more elementary programming problem in which the contract designer chooses the continuation value function from a set of permissible values.[4] Following Atkeson, let $U_d(Y_i) \equiv \tilde{V}(Y_i - d(Y_i))$ where $\tilde{V}(Y_i - d(Y_i))$ is a continuation value function to be chosen by the author of the contract. Atkeson shows that we can regard the contract author as choosing a continuation value function along with the elements of A, but that in the end it will be optimal for him to choose the continuation values to satisfy the Bellman equation (25.2.15a).

We follow Atkeson and regard the $U_d(Y_i)$'s as choice variables. They must satisfy $U_d(Y_i) \leq V(Y_i - d_i)$, where $V(Y_i - d_i)$ satisfies the Bellman equation (25.2.15). Form the Lagrangian

$$
\begin{aligned}
J(A, U_d, \mu) =& (1 - \delta)u(c) + \delta \sum_i U_d(Y_i)g(Y_i, I) \\
& + \mu_1(Q + b - c - I) \\
& + \mu_2\Big[\delta \sum_i d_i g(Y_i, I) - b\Big] \\
& + \delta \sum_i \mu_3(Y_i)g(Y_i, I)\big[U_d(Y_i) - U(Y_i)\big] \\
& + \mu_4\Big[-(1 - \delta)u'(Q + b - I) + \delta \sum_i U_d(Y_i)g_I(Y_i, I)\Big] \\
& + \delta \sum_i \mu_5(Y_i)g(Y_i, I)\big[V(Y_i - d_i) - U_d(Y_i)\big],
\end{aligned}
\tag{25.2.17}
$$

where the μ_j's are nonnegative Lagrange multipliers. To investigate the consequences of a binding participation constraint, rearrange the first-order condition with respect to $U_d(Y_i)$ to get

$$
1 + \mu_4 \frac{g_I(Y_i, I)}{g(Y_i, I)} = \mu_5(Y_i) - \mu_3(Y_i),
\tag{25.2.18}
$$

where $g_I/g = \lambda'(I)\big[\frac{g_0(Y_i) - g_1(Y_i)}{g(Y, I)}\big]$, which is negative for low Y_i and positive for high Y_i. All the multipliers are nonnegative. Then evidently when the left side

---

[4] See Atkeson (1991) and chapter 23.

of equation $(25.2.18)$ is *negative*, we must have $\mu_3(Y_i) > 0$, so that condition $(25.2.15d)$ is binding and $U_d(Y_i) = U(Y_i)$. Therefore, $V(Y_i - d_i) = U(Y_i)$ for states with $\mu_3(Y_i) > 0$. Atkeson uses this finding to show that in states Y_i where $\mu_3(Y_i) > 0$, new loans b' cannot exceed repayments $d_i = d(Y_i)$. This conclusion follows from the following argument. The optimality condition $(25.2.15e)$ implies that $V(Q)$ will satisfy

$$V(Q) = \max_{I \in [0, Q+b]} u(Q + b - I) + \delta \sum_{Y'} V(Y' - d(Y'))g(Y', I). \qquad (25.2.19)$$

Using the participation constraint $(25.2.15d)$ on the right side of $(25.2.19)$ implies

$$V(Q) \geq \max_{I \in [0, Q+b]} \left\{ u(Q + b - I) + \delta \sum_{Y'} U(Y_i')g(Y', I) \right\} \equiv U(Q + b) \quad (25.2.20)$$

where U is the value function for the autarky problem $(25.2.5)$. In states in which $\mu_3 > 0$, we know that, first, $V(Q) = U(Y)$, and, second, that by $(25.2.20)$ $V(Q) \geq U(Y + (b - d))$. But we also know that U is increasing. Therefore, we must have that $(b - d) \leq 0$, for otherwise U being increasing induces a contradiction. We conclude that for those low-Y_i states for which $\mu_3 > 0$, $b \leq d(Y_i)$, meaning that there are no capital inflows for these states.[5]

Capital outflows in bad times provide good incentives because they occur only at output realizations so low that they are more likely to occur when the borrower has undertaken too little investment. Their role is to provide incentives for the borrower to invest enough to make it unlikely that those low-output states will occur. The occurrence of capital outflows at low outputs is not called for by the complete markets contract $(25.2.13b)$. On the contrary, the complete markets contract provides a "capital inflow" to the lender in low-output states. That the pair of functions $b_t = b(Q_t)$, $d_t = d(Y_t, Q_{t-1})$ forming the optimal contract specifies repayments in those distressed states is how the contract provides incentives for the borrower to make investment decisions that reduce the likelihood that combinations of (Y_t, Q_t, Q_{t-1}) will occur that trigger capital outflows under distress.

We remind the reader of the remarkable feature of Atkeson's contract that the repayment schedule and the state variable Q "do all the work." Atkeson's

---

[5] This argument highlights the important role of limited enforcement in producing capital outflows at low output realizations.

contract manages to encode all history dependence in an extremely economical fashion. In the end, there is no need, as occurred in the problems that we studied in chapter 20, to add a promised value as an independent state variable.

25.3. Gradualism in trade policy

We now describe a version of Bond and Park's (2001) analysis of gradualism in bilateral agreements to liberalize international trade. Bond and Park cite examples in which a large country extracts a possibly rising sequence of transfers from a small country in exchange for a gradual lowering of tariffs in the large country. Bond and Park interpret gradualism in terms of the history-dependent policies that vary the continuation value of the large country in a way that induces it gradually to reduce its distortions from tariffs while still gaining from a move toward free trade. They interpret the transfers as trade concessions.[6]

We begin by laying out a simple general equilibrium model of trade between two countries.[7] The outcome of this theorizing will be a pair of indirect utility functions r_L and r_S that give the welfare of a large and small country, respectively, both as functions of a tariff t_L that the large country imposes on the small country, and a transfer e_S that the small country voluntarily offers to the large country.

---

[6] Bond and Park say that in practice, the trade concessions take the form of reforms of policies in the small country about protecting intellectual property, protecting rights of foreign investors, and managing the domestic economy. They do not claim explicitly to model these features.

[7] Bond and Park (2001) work in terms of a partial equilibrium model that differs in details but shares the spirit of our model.

25.3.1. Closed-economy model

First, we describe a one-country model. The country consists of a fixed number of identical households. A typical household has preferences

$$u(c, \ell) = c + \ell - 0.5 \, \ell^2, \tag{25.3.1}$$

where c and ℓ are consumption of a single consumption good and leisure, respectively. The household is endowed with a quantity $\bar{y}$ of the consumption good and one unit of time that can be used for either leisure or work,

$$1 = \ell + n_1 + n_2, \tag{25.3.2}$$

where n_j is the labor input in the production of intermediate good x_j, for $j = 1, 2$. The two intermediate goods can be combined to produce additional units of the final consumption good. The technology is as follows:

$$x_1 = n_1, \tag{25.3.3a}$$

$$x_2 = \gamma \, n_2, \qquad \gamma \in [0, 1], \tag{25.3.3b}$$

$$y = 2 \, \min\{x_1, x_2\}, \tag{25.3.3c}$$

$$c = y + \bar{y}, \tag{25.3.3d}$$

where consumption c is the sum of production y and the endowment $\bar{y}$.

Because of the Leontief production function for the final consumption good, a closed economy will produce the same quantity of each intermediate good. For a given production parameter γ, let $\tilde{\chi}(\gamma)$ be the identical amount of each intermediate good that would be produced per unit of labor input. That is, a fraction $\tilde{\chi}(\gamma)$ of one unit of labor input would be spent on producing $\tilde{\chi}(\gamma)$ units of intermediate good 1 and another fraction $\tilde{\chi}(\gamma)/\gamma$ of the labor input would be devoted to producing the same amount of intermediate good 2:

$$\tilde{\chi}(\gamma) + \frac{\tilde{\chi}(\gamma)}{\gamma} = 1 \quad \Longrightarrow \quad \tilde{\chi}(\gamma) = \frac{\gamma}{1 + \gamma}. \tag{25.3.4}$$

The linear technology implies a competitively determined wage at which all output is paid out as labor compensation. The optimal choice of leisure makes the marginal utility of consumption from an extra unit of labor input equal to the marginal utility of an extra unit of leisure: $2 \, \min\{\tilde{\chi}(\gamma), \tilde{\chi}(\gamma)\} = \frac{d}{d\ell}\left[\ell - 0.5 \, \ell^2\right]$.

Substituting for $\tilde{\chi}(\gamma)$ from $(25.3.4)$ gives $\frac{2\gamma}{1+\gamma} = 1 - \ell$, which can be rearranged to become

$$\ell = \mathcal{L}(\gamma) = \frac{1-\gamma}{1+\gamma}. \tag{25.3.5}$$

It follows that per capita, the equilibrium quantity of each intermediate good is given by

$$x_1 = x_2 = \chi(\gamma) \equiv \tilde{\chi}(\gamma)[1 - \mathcal{L}(\gamma)] = \frac{2\gamma^2}{(1+\gamma)^2}. \tag{25.3.6}$$

Two countries under autarky

Suppose that there are two countries named L and S (denoting large and small). Country L consists of $N \geq 1$ identical consumers, while country S consists of one household. All households have the same preferences $(25.3.1)$, but technologies differ across countries. Specifically, country L has production parameter $\gamma = 1$ while country S has $\gamma = \gamma_S < 1$.

Under no trade or *autarky*, each country is a closed economy whose allocations are given by $(25.3.5)$, $(25.3.6)$, and $(25.3.3)$. Evaluating these expressions, we obtain

$$\{\ell_L, n_{1L}, n_{2L}, c_L\} = \{0,\ 0.5,\ 0.5,\ \bar{y} + 1\},$$
$$\{\ell_S, n_{1S}, n_{2S}, c_S\} = \{\mathcal{L}(\gamma_S),\ \chi(\gamma_S),\ \chi(\gamma_S)/\gamma_S,\ \bar{y} + 2\,\chi(\gamma_S)\}.$$

The relative price between the two intermediate goods is 1 in country L while for country S, intermediate good 2 trades at a price γ_S^{-1} in terms of intermediate good 1. The difference in relative prices across countries implies gains from trade.

25.3.2. A Ricardian model of two countries under free trade

Under free trade, country L is large enough to meet both countries' demands for intermediate good 2 at a relative price of 1 and hence country S will specialize in the production of intermediate good 1 with $n_{1S} = 1$. To find the time n_{1L} that a worker in country L devotes to the production of intermediate good 1, note that the world demand at a relative price of 1 is equal to $0.5(N+1)$ and, after imposing market clearing, that

$$N n_{1L} + 1 = 0.5 (N + 1)$$
$$n_{1L} = \frac{N - 1}{2N}.$$

The free-trade allocation becomes

$$\{\ell_L, n_{1L}, n_{2L}, c_L\} = \{0, (N-1)/(2N), (N+1)/(2N), \bar{y} + 1\},$$
$$\{\ell_S, n_{1S}, n_{2S}, c_S\} = \{0, 1, 0, \bar{y} + 1\}.$$

Notice that the welfare of a household in country L is the same as under autarky because we have $\ell_L = 0$, $c_L = \bar{y} + 1$. The invariance of country L's allocation to opening trade is an immediate implication of the fact that the equilibrium prices under free trade are the same as those in country L under autarky. Only country S stands to gain from free trade.

25.3.3. Trade with a tariff

Although country L has nothing to gain from free trade, it can gain from trade if it is accompanied by a distortion to the terms of trade that is implemented through a tariff on country L's imports. Thus, assume that country L imposes a tariff of $t_L \geq 0$ on all imports into L. For any quantity of intermediate or final goods imported into country L, country L collects a fraction t_L of those goods by levying the tariff. A necessary condition for the existence of an equilibrium with trade is that the tariff does not exceed $(1 - \gamma_S)$, because otherwise country S would choose to produce intermediate good 2 rather than import it from country L.

Given that $t_L \leq 1 - \gamma_S$, we can find the equilibrium with trade as follows. From the perspective of country S, $(1 - t_L)$ acts like the production parameter

γ, i.e., it determines the cost of obtaining one unit of intermediate good 2 in terms of foregone production of intermediate good 1. Under autarky that price was γ^{-1}; with trade and a tariff t_L, that price becomes $(1-t_L)^{-1}$. For country S, we can therefore draw upon the analysis of a closed economy and just replace γ by $1-t_L$. The allocation with trade for country S becomes

$$\{\ell_S, n_{1S}, n_{2S}, c_S\} = \{\mathcal{L}(1-t_L),\ 1-\mathcal{L}(1-t_L),\ 0,\ \bar{y} + 2\chi(1-t_L)\}. \quad (25.3.7)$$

In contrast to the equilibrium under autarky, country S now allocates all labor input $1 - \mathcal{L}(1 - t_L)$ to the production of intermediate good 1 but retains only a quantity $\chi(1 - t_L)$ of total production for its own use, and exports the rest $\chi(1 - t_L)/(1 - t_L)$ to country L. After paying tariffs, country S purchases an amount $\chi(1-t_L)$ of intermediate good 2 from country L. Since this quantity of intermediate good 2 exactly equals the amount of intermediate good 1 retained in country S, production of the final consumption good given by $(25.3.3c)$ equals $2\chi(1 - t_L)$.

Country L receives a quantity $\chi(1 - t_L)/(1 - t_L)$ of intermediate good 1 from country S, partly as tariff revenue $t_L \chi(1 - t_L)/(1 - t_L)$ and partly as payments for its exports of intermediate good 2, $\chi(1 - t_L)$. In response to the inflow of intermediate good 1, an aggregate quantity of labor equal to $\chi(1 - t_L) + 0.5\, t_L\, \chi(1 - t_L)/(1 - t_L)$ is reallocated in country L from the production of intermediate good 1 to the production of intermediate good 2. This allows country L to meet the demand for intermediate good 2 from country S and at the same time increase its own use of each intermediate good by $0.5\, t_L\, \chi(1 - t_L)/(1 - t_L)$. The per capita trade allocation for country L becomes

$$\{\ell_L, n_{1L}, n_{2L}, c_L\} = \left\{ 0,\ 0.5 - \frac{(1 - 0.5t_L)\,\chi(1 - t_L)}{(1 - t_L)N}, \right.$$
$$\left. 0.5 + \frac{(1 - 0.5t_L)\,\chi(1 - t_L)}{(1 - t_L)N},\ \bar{y} + 1 + t_L\frac{\chi(1 - t_L)}{(1 - t_L)N} \right\}. \quad (25.3.8)$$

25.3.4. Welfare and Nash tariff

For a given tariff $t_L \leq 1 - \gamma_S$, we can compute the welfare levels in a trade equilibrium. Let $u_S(t_L)$ and $u_L(t_L)$ be the indirect utility of country S and country L, respectively, when the tariff is t_L. After substituting the equilibrium allocation $(25.3.7)$ and $(25.3.8)$ into the utility function of $(25.3.1)$, we obtain

$$
\begin{aligned}
u_S(t_L) &= u(c_S, \ell_S) \\
&= \bar{y} + 2\,\chi(1 - t_L) + \mathcal{L}(1 - t_L) - 0.5\,\mathcal{L}(1 - t_L)^2, \\
u_L(t_L) &= N\,u(c_L, \ell_L) = N\,(\bar{y} + 1) + t_L\frac{\chi(1 - t_L)}{1 - t_L},
\end{aligned}
\tag{25.3.9}
$$

where we multiply the utility function of the representative agent in country L by N because we are aggregating over all agents in a country. We now invoke equilibrium expressions $(25.3.5)$ and $(25.3.6)$, and take derivatives with respect to t_L. As expected, the welfare of country S decreases with the tariff while the welfare of country L is a strictly concave function that initially increases with the tariff:

$$
\frac{du_S(t_L)}{dt_L} = -\frac{4\,(1 - t_L)}{(2 - t_L)^3} < 0,
\tag{25.3.10a}
$$

$$
\frac{d\,u_L(t_L)}{d\,t_L} = \frac{2\,(2 - 3t_L)}{(2 - t_L)^3}
\begin{cases}
> 0 & \text{for } t_L < 2/3 \\
\leq 0 & \text{for } t_L \geq 2/3
\end{cases}
\tag{25.3.10b}
$$

and

$$
\frac{d^2 u_L(t_L)}{d\,t_L^2} = -\frac{12 t_L}{(2 - t_L)^4} \leq 0,
\tag{25.3.10c}
$$

where it is understood that the expressions are evaluated for $t_L \leq 1 - \gamma_S$.

The tariff enables country L to reap some of the benefits from trade. In our model, country L prefers a tariff t_L that maximizes its tariff revenues.

DEFINITION: In a one-period *Nash equilibrium*, the government of country L imposes a tariff rate that satisfies

$$
t_L^N = \min\Big\{\underset{t_L}{\arg\max}\,u_L(t_L),\; 1 - \gamma_S\Big\}.
\tag{25.3.11}
$$

From expression $(25.3.10b)$, we have $t_L^N = \min\{2/3,\, 1 - \gamma_S\}$.

REMARK: At the Nash tariff, country S gains from trade if $2/3 < 1 - \gamma_S$. Country S gets no gains from trade if $1 - \gamma_S \leq 2/3$.

Measure world welfare by $u_W(t_L) \equiv u_S(t_L) + u_L(t_L)$. This measure of world welfare satisfies

$$\frac{d\, u_W(t_L)}{d\, t_L} = -\frac{2\, t_L}{(2 - t_L)^3} \leq 0, \qquad (25.3.12a)$$

and

$$\frac{d^2 u_W(t_L)}{d\, t_L^2} = -\frac{4\,(1 + t_L)}{(2 - t_L)^4} < 0. \qquad (25.3.12b)$$

We summarize our findings:

PROPOSITION 1: World welfare $u_W(t_L)$ is strictly concave, is decreasing in $t_L \geq 0$, and is maximized by setting $t_L = 0$. But $u_L(t_L)$ is strictly concave in t_L and is maximized at $t_L^N > 0$. Therefore, $u_L(t_L^N) > u_L(0)$.

A consequence of this proposition is that country L prefers the Nash equilibrium to free trade, but country S prefers free trade. To induce country L to accept free trade, country S will have to transfer resources to it. We now study how country S can do that efficiently in an intertemporal version of the model.

25.3.5. Trade concessions

To get a model in the spirit of Bond and Park (2001), we now assume that the two countries can make trade concessions that take the form of a direct transfer of the consumption good between them. We augment utility functions u_L, u_S of the form (25.3.1) with these transfers to obtain the payoff functions

$$r_L(t_L, e_S) = u_L(t_L) + e_S \qquad (25.3.13a)$$
$$r_S(t_L, e_S) = u_S(t_L) - e_S, \qquad (25.3.13b)$$

where $t_L \geq 0$ is a tariff on the imports of country L, $e_S \geq 0$ is a transfer from country S to country L. These definitions make sense because the indirect utility functions (25.3.9) are linear in consumption of the final consumption good, so that by transferring the final consumption good, the small country transfers utility. The transfers e_S are to be voluntary and must be nonnegative (i.e., the country cannot extract transfers from the large country). We have already seen that $u_L(t_L)$ is strictly concave and twice continuously differentiable with $u_L'(0) > 0$ and that $u_W(t_L) \equiv u_S(t_L) + u_L(t_L)$ is strictly concave and twice continuously differentiable with $u'_W(0) = 0$. We call *free trade* a situation in which $t_L = 0$. We let (t_L^N, e_S^N) be the Nash equilibrium tariff rate and transfer for a one-period, simultaneous-move game in which the two countries have payoffs $(25.3.13a)$ and $(25.3.13b)$. Under Proposition 1, $t_L^N > 0, e_S^N = 0$. Also, $u_L(t_L^N) > u_L(0)$ and $u_S(0) > u_S(t_L^N)$, so that country S gains and country L loses in moving from the Nash equilibrium to free trade with $e_S = 0$.

25.3.6. A repeated tariff game

We now suppose that the economy repeats itself infinitely. for $t \geq 0$. Denote the pair of time t actions of the two countries by $\rho_t = (t_{Lt}, e_{St})$. For $t \geq 1$, denote the history of actions up to time $t - 1$ as $\rho^{t-1} = [\rho_{t-1}, \ldots, \rho_0]$. A policy σ_S for country S is an initial e_{S0} and for $t \geq 1$ a sequence of functions expressing $e_{St} = \sigma_{St}(\rho^{t-1})$. A policy σ_L for country L is an initial t_{L0} and for $t \geq 1$ a sequence of functions expressing $t_{Lt} = \sigma_{Lt}(\rho^{t-1})$. Let σ denote the pair of policies (σ_L, σ_S). The *policy* or *strategy profile* σ induces time t payoff $r_i(\sigma_t)$ for country i at time t, where σ_t is the time t component of σ. We measure country i's present discounted value by

$$v_i(\sigma) = \sum_{t=0}^{\infty} \beta^t r_i(\sigma_t) \tag{25.3.14}$$

where σ affects r_i through its effect on c_i. Define $\sigma|_{\rho^{t-1}}$ as the continuation of σ starting at t after history ρ^{t-1}. Define the continuation value of i at time t as

$$v_{it} = v_i(\sigma|_{\rho^{t-1}}) = \sum_{j=0}^{\infty} \beta^j r_i(\sigma_j|_{\rho^{t-1}}).$$

We use the following standard definition:

DEFINITION: A *subgame perfect equilibrium* is a strategy profile σ such that for all $t \geq 0$ and all histories ρ^t, country L maximizes its continuation value starting from t, given σ_S, and country S maximizes its continuation value starting from t, given σ_L.

It is easy to verify that a strategy that forever repeats the static Nash equilibrium outcome $(t_L, e_S) = (t_L^N, 0)$ is a subgame perfect equilibrium.

25.3.7. Time-invariant transfers

We first study circumstances under which there exists a time-invariant transfer $e_S > 0$ that will induce country L to move to free trade.

Let $v_i^N = \frac{u_i(t_L^N)}{1-\beta}$ be the present discounted value of country i when the static Nash equilibrium is repeated forever. If both countries are to prefer free trade with a time-invariant transfer level $e_S > 0$, the following two participation constraints must hold:

$$v_L \equiv \frac{u_L(0) + e_S}{1-\beta} \geq u_L(t_L^N) + e_S + \beta v_L^N \qquad (25.3.15)$$

$$v_S \equiv \frac{u_S(0) - e_S}{1-\beta} \geq u_S(0) + \beta v_S^N. \qquad (25.3.16)$$

The timing here articulates what it means for L and S to choose simultaneously: when L defects from $(0, e_S)$, L retains the transfer e_S for that period. Symmetrically, if S defects, it enjoys the zero tariff for that one period. These temporary gains provide the temptations to defect. Inequalities $(25.3.15)$ and $(25.3.16)$ say that countries L and S both get higher continuation values from remaining in free trade with the transfer e_S than they get in the repeated static Nash equilibrium. Inequalities $(25.3.15)$ and $(25.3.16)$ invite us to study strategies that have each country respond to any departure from what it had expected the other country to do this period by forever after choosing the Nash equilibrium actions $t_L = t_L^N$ for country L and $e_S = 0$ for country S. Thus, the response to any deviation from anticipated behavior is to revert to the repeated static Nash equilibrium, itself a subgame perfect equilibrium.[8]

Inequality $(25.3.15)$ (the participation constraint for L) and the definition of v_L^N can be rearranged to get

$$e_S \geq \frac{u_L(t_L^N) - u_L(0)}{\beta}. \qquad (25.3.17)$$

Time-invariant transfers e_S that satisfy inequality $(25.3.17)$ are sufficient to induce L to abandon the Nash equilibrium and set its tariff to zero. The *minimum* time-invariant transfer that will induce L to accept free trade is then

$$e_{S\min} = \frac{u_L(t_L^N) - u_L(0)}{\beta}. \qquad (25.3.18)$$

---

[8] In chapter 23, we study the consequences of reverting to a subgame perfect equilibrium that gives *worse* payoffs to both S and L and how the worst subgame perfect equilibrium payoffs and strategies can be constructed.

Inequality $(25.3.16)$ (the participation constraint for S) and the definition of v_S^N yield

$$e_S \leq \beta(u_S(0) - u_S(t_L^N)), \tag{25.3.19}$$

which restricts the time-invariant transfer that S is willing to make to move to free trade by setting $t_L = 0$. Evidently, the *largest* time-invariant transfer that S is willing to pay is

$$e_{S\text{max}} = \beta(u_S(0) - u_S(t_L^N)). \tag{25.3.20}$$

If we substitute $e_S = e_{S\text{min}}$ into the definition of v_L in $(25.3.15)$, we find that the *lowest* continuation value v_L for country L that can be supported by a stationary transfer is

$$v_L^* = \beta^{-1}(v_L^N - u_L(0)). \tag{25.3.21}$$

If we substitute $e_S = e_{S\text{max}}$ into the definition of v_L we can conclude that the *highest* v_L that can be sustained by a stationary transfer is

$$v_L^{**} = \frac{u_L(0) + \beta(u_S(0) - u_S(t_L^N))}{1 - \beta}. \tag{25.3.22}$$

For there to exist a time-invariant transfer e_S that induces both countries to accept free trade, we require that $v_L^* < v_L^{**}$ so that $[v_L^*, v_L^{**}]$ is nonempty. For a class of world economies differing only in their discount factors, we can compute a discount factor β that makes $v_L^* = v_L^{**}$. This is the critical value for the discount factor below which the interval $[v_L^*, v_L^{**}]$ is empty. Thus, equating the right sides of $(25.3.21)$ and $(25.3.22)$ and solving for β gives the critical value

$$\beta_c \equiv \sqrt{\frac{u_L(t_L^N) - u_L(0)}{u_S(0) - u_S(t_t^N)}}. \tag{25.3.23}$$

We know that the numerator under the square root is positive and that it is less than the denominator (because S gains by moving to free trade more than L loses, i.e., $u_W(t_L)$ is maximized at $t_L = 0$). Thus, $(25.3.23)$ has a solution $\beta_c \in (0, 1)$. For $\beta > \beta_c$, there is a nontrivial interval $[v_L^*, v_L^{**}]$. For $\beta < \beta_c$, the interval is empty.

Now consider the utility possibility frontier *without* the participation constraints $(25.3.15)$, $(25.3.16)$, namely,

$$v_S = \frac{u_W(0)}{1 - \beta} - v_L. \tag{25.3.24}$$

Then we have the following:

PROPOSITION 2: There is a critical value β_c such that for $\beta > \beta_c$, the interval $[v_L^*, v_L^{**}]$ is nonempty. For $v_L \in [v_L^*, v_L^{**}]$, a pair (v_L, v_S) on the unconstrained utility possibility frontier (25.3.24) can be attained by a time-invariant policy $(0, e_s)$ with transfer $e_S > 0$ from S to L. The policy is supported by a trigger strategy profile that reverts forever to $(t_L, 0)$ if expectations are ever disappointed.

25.3.8. Gradualism: time-varying trade policies

From now on, we assume that $\beta > \beta_c$, so that $[v_L^*, v_L^{**}]$ is nonempty. We make this assumption because we want to study settings in which the two countries eventually move to free trade even if they don't start there. Notice from expression (25.3.21) that

$$
\begin{aligned}
v_L^* &= \beta^{-1} \left(\left[u_L(t_L^N) + \beta v_L^N \right] - u_L(0) \right) \\
&= v_L^N + \beta^{-1} \left(u_L(t_L^N) - u_L(0) \right) > v_L^N.
\end{aligned}
\tag{25.3.25}
$$

Thus, even when $[v_L^*, v_L^{**}]$ is nonempty, there is an interval of continuation values $[v_L^N, v_L^*)$ that cannot be sustained by a time-invariant transfer scheme. Values $v_L > v_L^{**}$ also fail to be sustainable by a time-invariant transfer because the required e_S is too high. For initial values $v_L < v_L^*$ or $v_L > v_L^{**}$, Bond and Park construct time-varying tariff and transfer schemes that sustain continuation value v_L. They proceed by designing a recursive contract similar to ones constructed by Thomas and Worrall (1988) and again by Kocherlakota (1996a).

Let $v_L(\sigma), v_S(\sigma)$ be the discounted present values delivered to countries L and S under policy σ. For a given initial promised value v_L for country L, let $P(v_L)$ be the maximal continuation value v_S for country S, associated with a possibly time-varying trade policy. The value function $P(v_L)$ satisfies the functional equation

$$
P(v_L) = \sup_{t_L, e_S, y} \left\{ u_S(t_L) - e_S + \beta P(y) \right\},
\tag{25.3.26}
$$

where the maximization is subject to $t_L \geq 0, e_S \geq 0$ and

$$
u_L(t_L) + e_S + \beta y \geq v_L
\tag{25.3.27a}
$$

$$u_L(t_L) + e_S + \beta y \geq u_L(t_L^N) + e_S + \beta v_L^N \qquad (25.3.27b)$$

$$u_S(t_L) - e_S + \beta P(y) \geq u_S(t_L) + \beta v_S^N. \qquad (25.3.27c)$$

Here, y is the continuation value for L, meaning next period's value of v_L. Constraint $(25.3.27a)$ is the promise-keeping constraint, while $(25.3.27b)$ and $(25.3.27c)$ are the participation constraints for countries L and S, respectively. The constraint set is convex and the objective is concave, so $P(v_L)$ is concave (though not strictly concave, an important qualification, as we shall see).

As with our study of Thomas and Worrall's and Kocherlakota's model, we place nonnegative multipliers θ on $(25.3.27a)$ and μ_L, μ_S on $(25.3.27b)$ and $(25.3.27c)$, respectively, form a Lagrangian, and obtain the following first-order necessary conditions for a saddlepoint:

$$t_L : \quad u_S'(t_L) + (\theta + \mu_L)u_L'(t_L) \leq 0, \quad = 0 \text{ if } t_L > 0 \qquad (25.3.28a)$$

$$y : \quad P'(y)(1 + \mu_S) + (\theta + \mu_L) = 0 \qquad (25.3.28b)$$

$$e_S : \quad -1 + \theta - \mu_S \leq 0, \ = 0 \text{ if } e_S > 0. \qquad (25.3.28c)$$

We analyze the consequences of these first-order conditions for the optimal contract in three regions delineated by the continuation values v_L^*, v_L^{**}.

We break our analysis into two parts. We begin by displaying particular policies that attain initial values on the constrained Pareto frontier. Later, we show that there can be many additional policies that attain the same values, which as we shall see is a consequence of a flat interval in the constrained Pareto frontier.

25.3.9. Baseline policies

Region I: $v_L \in [v_L^*, v_L^{**}]$ (neither PC binds)

When the initial value is in this interval, the continuation value stays in this interval. From the envelope property, $P'(v_L) = -\theta$. If $v_L \in [v_L^*, v_L^{**}]$, neither participation constraint binds, and we have $\mu_S = \mu_L = 0$. Then $(25.3.28b)$ implies

$$P'(y) = P'(v_L).$$

This can be satisfied by setting $y = v_L$. Then $y = v_L$ and the always binding promise-keeping constraint in $(25.3.27a)$ imply that

$$v_L = y = \frac{u_L(t_L) + e_S}{1 - \beta} \geq v_L^* > v_L^N \equiv \frac{u_L(t_L^N)}{1 - \beta}, \qquad (25.3.29)$$

where the weak inequality states that v_L trivially satisfies the lower bound of region I, which in turn is strictly greater than the Nash value v_L^N according to expression (25.3.25). Because $u_L(t)$ is maximized at t_L^N, the strict inequality in expression (25.3.29) holds only if $e_S > 0$. Then inequality (25.3.28c) and $e_S > 0$ imply that $\theta = 1$. Rewrite (25.3.28a) as

$$u_W'(t_L) \le 0, \quad = 0 \text{ if } t_L > 0.$$

By Proposition 1, this implies that $t_L = 0$. We can solve for e_S from

$$v_L = \frac{u_L(0) + e_S}{1 - \beta} \tag{25.3.30}$$

and then obtain $P(v_L)$ from $\frac{u_S(0) - e_S}{1 - \beta}$.

Before turning to region II with $v_L > v_L^{**}$, we shall first establish that there indeed exist such high continuation values for the large country which cannot be sustained by a time-invariant transfer scheme. This is done by showing that $P(v_L^{**}) > v_S^N$. That is, there is scope for further increasing the continuation value of the large country beyond v_L^{**} before the associated continuation value of the small country is reduced to v_S^N. The argument goes as follows:

$$
\begin{aligned}
P(v_L^{**}) &= \frac{u_W(0)}{1 - \beta} - v_L^{**} \\
&= \frac{u_L(0) + u_S(0)}{1 - \beta} - \frac{u_L(0) + \beta(u_S(0) - u_S(t_L^N))}{1 - \beta} \\
&= u_S(0) + \beta \frac{u_S(t_L^N)}{1 - \beta} > u_S(t_L^N) + \beta \frac{u_S(t_L^N)}{1 - \beta} \equiv v_S^N,
\end{aligned}
\tag{25.3.31}
$$

where the first equality uses the fact that the continuation value v_L^{**} lies on the unconstrained Pareto frontier whose slope is -1 and the second equality invokes expression (25.3.22). It then follows that $P(v_L^{**}) > v_S^N$.

Region II: $v_L > v_L^{**}$ (PC_S binds)

We shall verify that in region II, there is a solution to the first-period first order necessary conditions with $\mu_S > 0$ and $e_S > 0$. When $v_L > v_L^{**}$, $\mu_S \ge 0$ and $\mu_L = 0$. When $\mu_S > 0$, inequality (25.3.28c) and $e_S > 0$ imply

$$\theta = 1 + \mu_S > 1. \tag{25.3.32}$$

Express $(25.3.28a)$ as

$$u'_W(t_L) + (\theta - 1)u'_L(t_L) \leq 0, \quad = 0 \text{ if } t_L > 0. \qquad (25.3.33)$$

Because $u'_W(0) = 0$ and $u'_L(0) > 0$, this inequality can be satisfied only if $t_L > 0$. Equation $(25.3.28b)$ implies that

$$P'(y) = -(1 + \mu_S)^{-1}\theta = -1,$$

where the second inequality invokes $(25.3.32)$. Therefore, $y \in [v^*_L, v^{**}_L]$, the region of the Pareto frontier whose slope is -1 and in which neither participation constraint binds. We can solve for the required transfer from $t = 1$ onward from the following version of $(25.3.30)$:

$$y = \frac{u_L(0) + e'_S}{1 - \beta}, \qquad (25.3.34)$$

where e'_S denotes the value of e_S for $t \geq 1$, because once we move into region I, we stay there, having a time invariant $e'_S > 0$ with $t'_L = 0$, as our analysis of region I indicated. We can solve for t_L, e_S for period zero as follows. For a given $\theta > 1$, solve the following equations for $y, P(y), t_L, e_S, P(v_L)$:

$$u'_S(t_L) + \theta u'_L(t_L) = 0 \qquad (25.3.35a)$$

$$v_L = u_L(t_L) + e_S + \beta y \qquad (25.3.35b)$$

$$- e_S + \beta P(y) = \beta v^N_S \qquad (25.3.35c)$$

$$P(v_L) = u_S(t_L) - e_S + \beta P(y) \qquad (25.3.35d)$$

$$y + P(y) = \frac{u_W(0)}{1 - \beta}. \qquad (25.3.35e)$$

To find the maximized value $P(v_L)$, we must search over solutions of $(25.3.35)$ for the $\theta > 1$ that corresponds to the specified initial continuation value v_L, (i.e., we are performing the minimization over μ_S entailed in finding the saddlepoint of the Lagrangian).

Using expression $(25.3.35c)$, we can show that the transfer e_S in period zero is also strictly positive,

$$e_S = \beta[P(y) - v^N_S] \geq \beta \left[P(v^{**}_L) - v^N_S \right] > 0,$$

where we have used the fact that $y \leq v^{**}_L$ and invoked the finding in expression $(25.3.31)$ that $P(v^{**}_L) > v^N_S$. Concerning the relative size of e_S at $t = 0$

compared to the transfer e'_S that the small country pays in period $t = 1$ and forever afterwards, we notice that e'_S is also subject to a participation constraint (25.3.27c) with the very same continuation value $P(y)$ (but where $u_S(t'_L) = u_S(0)$). Hence, we can express (25.3.27c) for all periods $t \geq 1$, given a time-invariant continuation value $P(y)$ determined by (25.3.35), as

$$e'_S \leq \beta[P(y) - v_S^N] = e_S,$$

where the equality sign follows from (25.3.35c). We conclude that the transfer is nonincreasing over time for our solution to an initial continuation value in region II.

Thus, in region II, $t_L > 0$ in period 0, followed by $t'_L = 0$ thereafter. Moreover, the initial promised value to the large country $v_L > v_L^{**}$ is followed by a lower time-invariant continuation value $y \leq v_L^{**}$. Subtracting (25.3.35b) from (25.3.34) gives

$$y = v_L + (u_L(0) - u_L(t_L)) + (e'_S - e_S).$$

The contract sets the continuation value $y < v_L$ by making $t_L > 0$ (thereby making $u_L(0) - u_L(t_L) < 0$) and also possibly letting $e'_S - e_S < 0$, so that transfers can fall between periods 0 and 1. In region II, country L induces S to accept free trade by a two-stage lowering of the tariff from the Nash level, so that $0 < t_L < t_L^N$ in period 0, with $t'_L = 0$ for $t \geq 1$; in return, it gets period 0 transfers of $e_S > 0$ and constant transfers $e'_S > 0$ thereafter.

Region III: $v_L \in [v_L^N, v_L^*)$ $(PC_L$ binds)

The analysis of region III is subtle.[9] It is natural to expect that $\mu_S = 0, \mu_L > 0$ in this region. However, assuming that $\mu_L > 0$ can be shown to lead to a contradiction, implying that the pair $v_L, P(v_L)$ both is and is not on the unconstrained Pareto frontier.[10]

We can avoid the contradiction by assuming that $\mu_L = 0$, so that the participation constraint for country L is barely binding. We shall construct a solution to (25.3.28) and (25.3.27) with period 0 transfer $e_S > 0$. Note that (25.3.28c) with $e_S > 0$ implies $\theta = 1$, which from the envelope property $P'(v_L) = -\theta$

---

[9] The findings of this section reproduce ones summarized in Bond and Park's (2001) corollary to their Proposition 2.

[10] Please show this in exercise *25.2*.

implies that $(v_L, P(v_L))$ is actually on the *un*constrained Pareto frontier, a reflection of the participation constraint for country L barely binding. With $\theta = 1$ and $\mu_L = 0$, (25.3.28a) implies that $t_L = 0$, which confirms $(v_L, P(v_L))$ being on the Pareto frontier. We can then solve the following equations for $P(v_L), e_S, y, P(y)$:

$$P(v_L) + v_L = \frac{u_W(0)}{1 - \beta} \tag{25.3.36a}$$

$$v_L = u_L(0) + e_S + \beta y \tag{25.3.36b}$$

$$u_L(0) + e_S + \beta y = u_L(t_L^N) + e_S + \beta v_L^N \tag{25.3.36c}$$

$$P(v_L) = u_S(0) - e_S + \beta P(y) \tag{25.3.36d}$$

$$P(y) + y = \frac{u_W(0)}{1 - \beta}. \tag{25.3.36e}$$

We shall soon see that these constitute only four linearly independent equations. Equations (25.3.36a) and (25.3.36e) impose that both $(v_L, P(v_L))$ and $(y, P(y))$ lie on the unconstrained Pareto frontier. We can solve these equations recursively. First, solve for y from (25.3.36c). Then solve for $P(y)$ from (25.3.36e). Next, solve for $P(v_L)$ from (25.3.36a). Get e_S from (25.3.36b). Finally, equations (25.3.36a), (25.3.36b), and (25.3.36d) imply that equation (25.3.36e) holds, which establishes the reduced rank of the system of equations.

We can use (25.3.34) to compute e_S', the transfer from period 1 onward. In particular, e_S' satisfies $y = u_L(0) + e_S' + \beta y$. Subtracting (25.3.36b) from this equation gives

$$y - v_L = e_S' - e_S > 0.$$

Thus, when $v_L < v_L^*$, country S induces country L immediately to reduce its tariff to zero by paying transfers that rise between period 0 and period 1 and that thereafter remain constant. That the initial tariff is zero means that we are immediately on the unconstrained Pareto frontier. It just takes time-varying transfers to put us there.

Interpretations

For values of v_L within regions II and III, time-invariant transfers e_S from country S to country L are not capable of sustaining immediate and enduring free trade. But patterns of time-varying transfers and tariff reductions are able to induce both countries to move permanently to free trade after a one-period

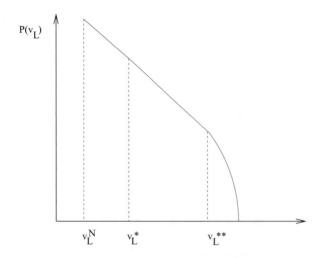

Figure 25.3.1: The constrained Pareto frontier $v_S = P(v_L)$ in the Bond-Park model.

transition. There is an asymmetry between regions II and III, revealed in Figure 25.3.1 and in our finding that $t_L = 0$ in region III, so that the move to free trade is immediate. The asymmetry emerges from a difference in the quality of instruments that the unconstrained country (L in region II, S in region III) has to induce the constrained country eventually to accept free trade by moving those instruments over time appropriately to manipulate the continuation values of the constrained country to gain its assent. In region II, where S is constrained, all that L can do is manipulate the time path of t_L, a relatively inefficient instrument because it is a distorting tax. By lowering t_L gradually, L succeeds in raising the continuation values of S gradually, but at the cost of imposing a distorting tax, thereby keeping $(v_L, P(v_L))$ inside the Pareto frontier. In region III, where L is constrained, S has at its disposal a nondistorting instrument for raising country L's continuation value by increasing the transfer e_S after period 0.

The basic principle at work is to make the continuation value rise for the country whose participation constraint is binding.

25.3.10. Multiplicity of payoffs and continuation values

We now find more equilibrium policies that support values in our three regions. The unconstrained Pareto frontier is a straight line in the space (v_L, v_S) with a slope of -1:

$$v_L + v_S = \frac{u_W(0)}{1 - \beta} \equiv W.$$

This reflects the fact that utility is perfectly transferable between the two countries. As a result, there is a continuum of ways to pick current payoffs $\{r_i; i = L, S\}$ and continuation values $\{v_i'; i = L, S\}$ that deliver the promised values v_L and v_S to country L and S, respectively. For example, each country could receive a current payoff equal to the annuity value of its promised value, $r_i = (1 - \beta)v_i$, and retain its promised value as a continuation value, $v_i' = v_i$. That would clearly deliver the promised value to each country,

$$r_i + \beta v_i' = (1 - \beta)v_i + \beta v_i = v_i.$$

Another example would reduce the prescribed current payoff to country S by $\triangle_S > 0$ and increase the prescribed payoff to country L by the same amount. Continuation values (v_S', v_L') would then have to be set such that

$$(1 - \beta)v_S - \triangle_S + \beta v_S' = v_S,$$
$$(1 - \beta)v_L + \triangle_S + \beta v_L' = v_L.$$

Solving from these equations, we get

$$\triangle_S = \beta(v_S' - v_S) = -\beta(v_L' - v_L).$$

Here country S is compensated for the reduction in current payoff by an equivalent increase in the discounted continuation value, while country L receives corresponding changes of opposite signs.

Since the constrained Pareto frontier coincides with the unconstrained Pareto frontier in regions I and III, we would expect that the tariff games would also be characterized by multiplicities of payoffs and continuation values. We will now examine how the participation constraints shape the range of admissible equilibrium values.

Region I (revisited): $v_L \in [v_L^*, v_L^{**}]$

From our earlier analysis, an equilibrium in region I satisfies

$$u_L(0) + e_S + \beta y = v_L, \qquad (25.3.37a)$$

$$u_L(0) + e_S + \beta y \geq u_L(t_L^N) + e_S + \beta v_L^N, \qquad (25.3.37b)$$

$$u_S(0) - e_S + \beta(W - y) \geq u_S(0) + \beta v_S^N, \qquad (25.3.37c)$$

where we have invoked that $P(y) = W - y$ in regions I and III. We consider only $y \in [v_L^N, v_L^{**}]$ because our earlier analysis ruled out any transitions from region I to region II.

Equation $(25.3.37a)$ determines the transfer and continuation value needed to deliver the promised value v_L to country L under free trade:

$$e_S + \beta y = v_L - u_L(0).$$

The participation constraint for country S requires that inequality $(25.3.37c)$ be satisfied, which can be rewritten as

$$e_S + \beta y \leq \beta(W - v_S^N). \qquad (25.3.38)$$

Since we are postulating that we are in region I with no binding participation constraints, this condition is indeed satisfied. Notice that incentive compatibility on behalf of country S does not impose any restrictions on the mixture of transfer and continuation value that deliver $e_S + \beta y$ to country L beyond our restriction above that $y \leq v_L^{**}$.

Turning to the participation constraint for country L, we can rearrange inequality $(25.3.37b)$ to become

$$y \geq \beta^{-1}\left(u_L(t_L^N) + \beta v_L^N - u_L(0)\right) = \beta^{-1}\left(v_L^N - u_L(0)\right) = v_L^*.$$

Thus, there cannot be a transition from region I to region III, a result to be interpreted as follows. We showed earlier that free trade is not incentive compatible with a time-invariant transfer when the promised value of country L lies in region III. In other words, an initial promised value in region III cannot by itself serve as a continuation value to support free trade. Now we are trying to attain free trade by offering country L a continuation value in that very region III together with a transfer that is even larger than the time-invariant transfer considered earlier. (The transfer is larger than the earlier time-invariant transfer because the initial promised value v_L is now assumed to lie in region I.) Since that continuation value in region III was not incentive compatible for country L at a smaller transfer from country S, it will certainly not be incentive compatible now when the transfer is larger.

We conclude that there is a multiplicity of current payoffs and continuation values in region I. Specifically, admissible equilibrium continuation values are

$$y \in \left[v_L^*, \; \min\left\{\beta^{-1}\left(v_L - u_L(0)\right), v_L^{**}\right\}\right], \tag{25.3.39}$$

where the upper bound incorporates our nonnegativity constraint on transfers from country S to country L, i.e., imposing $e_S \geq 0$ in equation $(25.3.37a)$.

Region II (revisited): $v_L > v_L^{**}$

From our earlier analysis, an equilibrium in region II satisfies:

$$u_L(t_L) + e_S + \beta y = v_L \tag{25.3.40a}$$

$$u_L(t_L) + e_S + \beta y \geq u_L(t_L^N) + e_S + \beta v_L^N \tag{25.3.40b}$$

$$u_S(t_L) - e_S + \beta(W - y) = u_S(t_L) + \beta v_S^N, \tag{25.3.40c}$$

where $0 < t_L < t_L^N$ and we have used our earlier finding that the continuation value y will be in the region of the constrained Pareto frontier whose slope is -1, i.e., $y \in [v_L^N, v_L^{**}]$ for which $P(y) = W - y$.

Equation $(25.3.40c)$ determines the combination of the transfer and continuation value received by country L:

$$e_S + \beta y = \beta(W - v_S^N).$$

Once again, this participation constraint for country S does not impose any restrictions on the relative composition of the transfer versus the continuation value assigned to country L (besides our restriction above that $y \leq v_L^{**}$). For region II, we have already shown that the combined value of $e_S + \beta y$ is not sufficient to support free trade, and that the necessary tariff in period 0 can then be computed from equation $(25.3.40a)$.

Finally, the participation constraint $(25.3.40b)$ for country L does impose a restriction on admissible equilibrium continuation values y,

$$y \geq \beta^{-1}\left(u_L(t_L^N) + \beta v_L^N - u_L(t_L)\right) = \beta^{-1}\left(v_L^N - u_L(t_L)\right).$$

Notice that this lower bound on admissible values of y lies inside region III,

$$\beta^{-1}\left(v_L^N - u_L(t_L)\right) \begin{cases} > \beta^{-1}\left(v_L^N - u_L(t_L^N)\right) = v_L^N, \\ < \beta^{-1}\left(v_L^N - u_L(0)\right) = v_L^*. \end{cases}$$

In contrast to our analysis of region I, a transition into region III is possible when the initial promised value belongs to region II. The reason is that the constrained efficient tariff is then strictly positive in period 0, which relaxes the participation constraint for country L. Hence, the range of admissible continuation values in region II becomes

$$y \in \left[\beta^{-1}\left(v_L^N - u_L(t_L)\right), \; v_L^{**}\right].$$

Region III (revisited): $v_L \in [v_L^N, v_L^*)$

The study of multiplicity of current payoffs and continuation values in region III exactly parallels our analysis of region I. The range of admissible continuation values is once again given by $(25.3.39)$. The lower bound of v_L^* is pinned down by the participation constraint $(25.3.37b)$ for country L and this implies an immediate transition out of region III into region I.

For the lowest possible promised value $v_L = v_L^N$, the range of continuation values in $(25.3.39)$ becomes degenerate, with only one admissible value of $y = v_L^*$. From equation $(25.3.37a)$, we can verify that the pair $(v_L, y) = (v_L^N, v_L^*)$ implies an equilibrium transfer that is zero, $e_S = 0$. For any other promised value in region III, $v_L \in (v_L^N, v_L^*]$, there is a multiplicity of current payoffs and continuation values. We can then pick a continuation value $y > v_L^*$ that implies that the participation constraint for country L is not binding. Without any binding participation constraints, it becomes apparent why our analysis of multiplicity in region I is also valid for region III.

25.4. Another model

Fuchs and Lippi (2006) are motivated by a vision about the nature of monetary unions that was not well captured by work in the previous literature. In particular, earlier work (1) assumed away commitment problems between members that would occur within an ongoing currency union, and (2) modeled the consequences of abandoning a currency as reversion to a worst case outcome of a repeated game played by independent monetary-fiscal authorities. The Fuchs-Lippi paper repairs both of these deficiencies by (1) imposing participation constraints each period for each member within a currency union, and (2) assuming that the consequence of a breakup is to move to the *best* outcome of the game played by independent monetary-fiscal authorities.

Here is the setup. Two countries have ideal levels of a policy setting (e.g., an interest rate) that are each hit by country-specific idiosyncratic shocks. The history of these shocks is common knowledge. When not in a union, the countries play a repeated game. The best equilibrium outcome is the point to which the countries revert after a breakup. When in a union, the two countries play another repeated game. The authors model the benefit of being in the union as making it harder to effect a surprise change than it is outside it, thereby making it easier to abstain from opportunistic monetary policy that, e.g., exploits the Phillips curve to get short run benefits in exchange for long-run costs.

The authors use 'dynamic programming squared' to express equilibrium strategies within the currency union game in terms of the current observed shock vector and continuation values. The union chooses a 'public good', namely, the common policy each period. It is a weighted average of the ideal points for the two individual countries, with the weights being tilted a country whose participation constraint is binding that period. The authors show that there are three possible cases: (1) the shocks and initial continuation values are such that only country A's constraint is binding, in which case the policy tilts toward country A's ideal point; (2) only country B's participation constraint is binding, in which case the policy tilts toward country B's ideal point; (3) the continuation values and shocks are such that both countries' participation constraints are binding. In case (3), the currency union breaks up.

Depending on the specification of functional forms, preferences, and the joint distribution of shocks, case (3) may or may not be possible. When it is, one can use the model to calculate waiting times to breakup of a union. Many currency unions have broken up in the past, an observation that could be used to help reverse engineer parameter values – something that the authors don't do.

It is interesting to compare this model with an earlier risk-sharing model of Thomas and Worrall and Kocherlakota that we studied in chapter 20. In that model, there was no case 3 and the analogue of the union, the relationship between the firm and the worker in Thomas and Worrall or between two consumers in Kocherlakota, lasts forever. One never observes defaults along the equilibrium path. What is the source of the different outcome in the Fuchs-Lippi model? The answer hinges on the part of the payoff structure of the Fuchs-Lippi model that captures the 'public good' aspect of the monetary union policy choice. Both countries have to live with the same setting of a policy instrument and what one gains the other does not necessarily lose. In the chapter 20 model, each period when one person gets more, the other necessarily gets less, creating a symmetry in the participation constraints that prevents them from binding simultaneously.

25.5. Concluding remarks

Although the substantive application differs, mechanically the models of this chapter work much like models that we studied in chapters 19, 20, and 23. The key idea is to cope with binding incentive constraints (in this case, participation constraints), partly by changing the continuation values for those agents whose incentive constraints are binding. For example, that creates "intertemporal tie-ins" that Bond and Park interpret as "gradualism."

A. Computations for Atkeson's model

It is instructive to compute a numerical example of the optimal contract for Atkeson's (1988) model. Following Atkseson, we work with the following numerical example. Assume $u(c) = 2c^{.5}, \lambda(I) = \left(\frac{I}{Y_n + 2M}\right)^{.5}, g_i(Y_j) = \frac{\exp^{-\alpha_i Y_j}}{\sum_{k=1}^{n} \exp^{-\alpha_i Y_k}}$ with $n = 5, Y_1 = 100, Y_n = 200, M = 100, \alpha_1 = \alpha_2 = -.5, \delta = .9$. Here is a version of Atkeson's numerical algorithm:

1. First, solve the Bellman equation $(25.2.5)$ and $(25.2.6)$ for the autarky value $U(Q)$. Use a polynomial for the value function.[11]

2. Solve the Bellman equation for the full-insurance setting for the value function $W(Q)$ as follows. First, solve equation $(25.2.9)$ for I. Then solve equation $(25.2.13b)$ for $d(Y') = Y' - Q$ and compute $c = c(Q)$ from $(25.2.13a)$. Since c is constant, $W(Q) = u[c(Q)]$.

Now, solve the Bellman equation for the contract with limited commitment and unobserved action. First, approximate $V(Q)$ by a polynomial, using the method described in chapter 4. Next, iterate on the Bellman equation, starting from initial value function $V^0(Q) = W(Q)$ computed earlier. As Atkeson shows, it is important to start with a value function *above* $V(Q)$. We know that $W(Q) \geq V(Q)$.

Use the following steps:

1. Let $V^j(Q)$ be the value function at the jth iteration. Let d be the vector $\begin{bmatrix} d_1 & \cdots & d_n \end{bmatrix}'$. Define

$$X(d) = \sum_i V^j(Y_i - d_i)[g_0(Y_i) - g_1(Y_i)]. \qquad (25.A.1)$$

The first-order condition for the borrower's problem $(25.2.15e)$ is

$$-(1-\delta)u'(Q+b-I) + \delta\lambda'(I)X \geq 0, \quad = 0 \text{ if } I > 0.$$

Given a candidate continuation value function V^j, a value Q, and $b, d_1, \ldots, d_n$, solve the borrower's first-order condition for a function

$$I = f(b, d_1, \ldots, d_n; Q).$$

---

[11] We recommend the Schumaker shape-preserving spline mentioned in chapter 4 and described by Judd (1998).

Evidently, when $X(d) < 0$, $I = 0$. From equation $(25.A.1)$ and the particular example,

$$I = f(b, d; Q) = \frac{\delta^2(Y_n + 2M)X(d)^2}{4(1 - \delta)^2 + \delta^2(Y_n + 2M)X(d)^2}(Q + b). \qquad (25.A.2)$$

Summarize this equation in a Matlab function.

2. Use equation $(25.A.2)$ and the constraint $(25.2.15c)$ at equality to form

$$b = \delta \sum_i d_i g[Y_i, f(b, d)].$$

Solve this equation for a new function

$$b = m(d). \qquad (25.A.3)$$

3. Write one step on the Bellman equation as

$$
\begin{aligned}
V^{j+1}(Q) = \max_d \Big\{ & (1 - \delta)u\big[Q + m(d) - f(m(d), d)\big] \\
& + \delta \sum_i V^j(Y_i - d_i)g\big[Y_i, f(m(d), d)\big] \\
& - \sum_i \theta_i \Big[\max\big(0, U(Y_i) - V^j(Y_i - d_i)\big)\Big] \\
& - \sum_i \eta_i \max[0, -d_i - M] - \eta_0 \max[0, m(d) - M] \Big\},
\end{aligned}
\qquad (25.A.4)
$$

where $V^j(Q)$ is the value function at the jth iteration, and $\theta_i > 0, \eta_i$ are positive penalty parameters designed to enforce the participation constraints $(25.2.15d)$ and the restrictions on the size of borrowing and repayments. The idea is to set the θ_i's and η_i's large enough to assure that d is set so that constraint $(25.2.15d)$ is satisfied for all i.

Exercises

Exercise 25.1

Consider a version of Bond and Park's model with $\gamma_S = .4$ and the payoff functions $(25.3.13a)$ and $(25.3.13b)$, with

$$u_L(t_L) = -.5(t_L - .5)^2$$
$$u_W(t_L) = -.5t_L^2,$$

where $u_W(t_L) = u_L(t_L) + u_S(t_L)$.

a. Compute the cutoff value β_c from $(25.3.23)$. For $\beta \in (\beta_c, 1)$, compute v_L^*, v_L^{**}.

b. Compute the constrained Pareto frontier. (*Hint:* In region II, use $(25.3.35)$ for a grid of values v_L satisfying $v_L > v_L^{**}$.)

c. For a given $v_L \in (v_L^N, v^*)$, compute e_S, e_S', y.

Exercise 25.2

Consider the Bond-Park model analyzed above. Assume that in region III, $\mu_L > 0, \mu_S = 0$. Show that this leads to a contradiction.

Part VI

Classical monetary and labor economics

Chapter 26
Fiscal-Monetary Theories of Inflation

26.1. The issues

This chapter introduces some issues in monetary theory that mostly revolve around coordinating monetary and fiscal policies. We start from the observation that complete markets models have no role for inconvertible currency, and therefore assign zero value to it.[1] We describe one way to alter a complete markets economy so that a positive value is assigned to an inconvertible currency: we impose a transaction technology with shopping time and real money balances as inputs.[2] We use the model to illustrate 10 doctrines in monetary economics. Most of these doctrines transcend many of the details of the model. The important thing about the transactions technology is that it makes demand for currency a decreasing function of the rate of return on currency. Our monetary doctrines mainly emerge from manipulating that demand function and the government's intertemporal budget constraint under alternative assumptions about government monetary and fiscal policy.[3]

---

[1] In complete markets models, money holdings would only serve as a store of value. The following transversality condition would hold in a nonstochastic economy:

$$\lim_{T \to \infty} \prod_{t=0}^{T-1} R_t^{-1} \frac{m_{T+1}}{p_T} = 0.$$

The real return on money, p_t/p_{t+1}, would have to equal the return R_t on other assets, which, substituted into the transversality condition, yields

$$\lim_{T \to \infty} \prod_{t=0}^{T-1} \frac{p_{t+1}}{p_t} \frac{m_{T+1}}{p_T} = \lim_{T \to \infty} \frac{m_{T+1}}{p_0} = 0.$$

That is, an inconvertible money (i.e., one for which $\lim_{T \to \infty} m_{T+1} > 0$) must be valueless, $p_0 = \infty$.

[2] See Bennett McCallum (1983) for an early shopping time specification.

[3] Many of the doctrines were originally developed in setups differing in details from the one in this chapter.

After describing our 10 doctrines, we use the model to analyze two important issues: the validity of Friedman's rule in the presence of distorting taxation, and its sustainability in the face of a time consistency problem. Here, we use the methods for solving an optimal taxation problem with commitment in chapter 16, and for characterizing a credible government policy in chapter 23.

26.2. A shopping time monetary economy

Consider an endowment economy with no uncertainty. A representative household has one unit of time. There is a single good of constant amount $y > 0$ each period $t \geq 0$. The good can be divided between private consumption $\{c_t\}_{t=0}^{\infty}$ and government purchases $\{g_t\}_{t=0}^{\infty}$, subject to

$$c_t + g_t = y. \tag{26.2.1}$$

The preferences of the household are ordered by

$$\sum_{t=0}^{\infty} \beta^t u(c_t, \ell_t), \tag{26.2.2}$$

where $\beta \in (0,1)$, $c_t \geq 0$ and $\ell_t \geq 0$ are consumption and leisure at time t, respectively, and u_c, $u_\ell > 0$, u_{cc}, $u_{\ell\ell} < 0$, and $u_{c\ell} \geq 0$. With one unit of time per period, the household's time constraint becomes

$$1 = \ell_t + s_t. \tag{26.2.3}$$

We use $u_c(t)$ and so on to denote the time t values of the indicated objects, evaluated at an allocation to be understood from the context.

To acquire the consumption good, the household allocates time to shopping. The amount of shopping time s_t needed to purchase a particular level of consumption c_t is negatively related to the household's holdings of real money balances m_{t+1}/p_t. Specifically, the shopping or transaction technology is

$$s_t = H\left(c_t, \frac{m_{t+1}}{p_t}\right), \tag{26.2.4}$$

where H, H_c, H_{cc}, $H_{m/p,m/p} \geq 0$, $H_{m/p}$, $H_{c,m/p} \leq 0$. A parametric example of this transaction technology is

$$H\left(c_t, \frac{m_{t+1}}{p_t}\right) = \frac{c_t}{m_{t+1}/p_t} \epsilon, \tag{26.2.5}$$

where $\epsilon > 0$. This corresponds to a transaction cost that would arise in the frameworks of Baumol (1952) and Tobin (1956). When a household spends money holdings for consumption purchases at a constant rate c_t per unit of time, $c_t(m_{t+1}/p_t)^{-1}$ is the number of trips to the bank, and ϵ is the time cost per trip to the bank.

26.2.1. Households

The household maximizes expression (26.2.2) subject to the transaction technology (26.2.4) and the sequence of budget constraints

$$c_t + \frac{b_{t+1}}{R_t} + \frac{m_{t+1}}{p_t} = y - \tau_t + b_t + \frac{m_t}{p_t}. \qquad (26.2.6)$$

Here, m_{t+1} is nominal balances held between times t and $t+1$; p_t is the price level; b_t is the real value of one-period government bond holdings that mature at the beginning of period t, denominated in units of time t consumption; τ_t is a lump-sum tax at t; and R_t is the real gross rate of return on one-period bonds held from t to $t+1$. Maximization of expression (26.2.2) is subject to $m_{t+1} \geq 0$ for all $t \geq 0$,[4] no restriction on the sign of b_{t+1} for all $t \geq 0$, and given initial stocks m_0, b_0.

After consolidating two consecutive budget constraints given by equation (26.2.6), we arrive at

$$c_t + \frac{c_{t+1}}{R_t} + \left(1 - \frac{p_t}{p_{t+1}} \frac{1}{R_t}\right) \frac{m_{t+1}}{p_t} + \frac{b_{t+2}}{R_t R_{t+1}} + \frac{m_{t+2}/p_{t+1}}{R_t}$$

$$= y - \tau_t + \frac{y - \tau_{t+1}}{R_t} + b_t + \frac{m_t}{p_t}. \qquad (26.2.7)$$

To ensure a bounded budget set, the expression in parentheses multiplying non-negative holdings of real balances must be greater than or equal to zero. Thus, we have the arbitrage condition,

$$1 - \frac{p_t}{p_{t+1}} \frac{1}{R_t} = 1 - \frac{R_{mt}}{R_t} = \frac{i_t}{1 + i_t} \geq 0, \qquad (26.2.8)$$

where $R_{mt} \equiv p_t/p_{t+1}$ is the real gross return on money held from t to $t+1$, that is, the inverse of the inflation rate, and $1 + i_t \equiv R_t/R_{mt}$ is the gross nominal

---

[4] Households cannot issue money.

interest rate. The real return on money R_{mt} must be less than or equal to the return on bonds R_t, because otherwise agents would be able to make arbitrarily large profits by choosing arbitrarily large money holdings financed by issuing bonds. In other words, the net nominal interest rate i_t cannot be negative.

The Lagrangian for the household's optimization problem is

$$\sum_{t=0}^{\infty} \beta^t \left\{ u(c_t, \ell_t) + \lambda_t \left(y - \tau_t + b_t + \frac{m_t}{p_t} - c_t - \frac{b_{t+1}}{R_t} - \frac{m_{t+1}}{p_t} \right) \right.$$
$$\left. + \mu_t \left[1 - \ell_t - H\left(c_t, \frac{m_{t+1}}{p_t} \right) \right] \right\}.$$

At an interior solution, the first-order conditions with respect to c_t, ℓ_t, b_{t+1}, and m_{t+1} are

$$u_c(t) - \lambda_t - \mu_t H_c(t) = 0, \tag{26.2.9}$$

$$u_\ell(t) - \mu_t = 0, \tag{26.2.10}$$

$$-\lambda_t \frac{1}{R_t} + \beta \lambda_{t+1} = 0, \tag{26.2.11}$$

$$-\lambda_t \frac{1}{p_t} - \mu_t H_{m/p}(t) \frac{1}{p_t} + \beta \lambda_{t+1} \frac{1}{p_{t+1}} = 0. \tag{26.2.12}$$

From equations (26.2.9) and (26.2.10),

$$\lambda_t = u_c(t) - u_\ell(t) H_c(t). \tag{26.2.13}$$

The Lagrange multiplier on the budget constraint is equal to the marginal utility of consumption reduced by the marginal disutility of having to shop for that increment in consumption. By substituting equation (26.2.13) into equation (26.2.11), we obtain an expression for the real interest rate,

$$R_t = \frac{1}{\beta} \frac{u_c(t) - u_\ell(t) H_c(t)}{u_c(t+1) - u_\ell(t+1) H_c(t+1)}. \tag{26.2.14}$$

The combination of equations (26.2.11) and (26.2.12) yields

$$\frac{R_t - R_{mt}}{R_t} \lambda_t = -\mu_t H_{m/p}(t), \tag{26.2.15}$$

which sets the cost equal to the benefit of the marginal unit of real money balances held from t to $t+1$, all expressed in time t utility. The cost of holding money balances instead of bonds is lost interest earnings $(R_t - R_{mt})$ discounted at the rate R_t and expressed in time t utility when multiplied by the shadow price λ_t. The benefit of an additional unit of real money balances is the savings in shopping time $-H_{m/p}(t)$ evaluated at the shadow price μ_t. By substituting equations $(26.2.10)$ and $(26.2.13)$ into equation $(26.2.15)$, we get

$$\left(1 - \frac{R_{mt}}{R_t}\right)\left[\frac{u_c(t)}{u_\ell(t)} - H_c(t)\right] + H_{m/p}(t) = 0, \qquad (26.2.16)$$

with $u_c(t)$ and $u_\ell(t)$ evaluated at $\ell_t = 1 - H(c_t, m_{t+1}/p_t)$. Equation $(26.2.16)$ implicitly defines a money demand function

$$\frac{m_{t+1}}{p_t} = F(c_t, R_{mt}/R_t), \qquad (26.2.17)$$

which is increasing in both of its arguments, as can be shown by applying the implicit function rule to expression $(26.2.16)$.

26.2.2. Government

The government finances the purchase of the stream $\{g_t\}_{t=0}^{\infty}$ subject to the sequence of budget constraints

$$g_t = \tau_t + \frac{B_{t+1}}{R_t} - B_t + \frac{M_{t+1} - M_t}{p_t}, \qquad (26.2.18)$$

where B_0 and M_0 are given. Here B_t is government indebtedness to the private sector, denominated in time t goods, maturing at the beginning of period t, and M_t is the stock of currency that the government has issued as of the beginning of period t.

26.2.3. Equilibrium

We use the following definitions:

DEFINITION: A *price system* is a pair of positive sequences $\{R_t, p_t\}_{t=0}^{\infty}$.

DEFINITION: We take as exogenous sequences $\{g_t, \tau_t\}_{t=0}^{\infty}$. We also take $B_0 = b_0$ and $M_0 = m_0 > 0$ as given. An *equilibrium* is a price system, a consumption sequence $\{c_t\}_{t=0}^{\infty}$, a sequence for government indebtedness $\{B_t\}_{t=1}^{\infty}$, and a positive sequence for the money supply $\{M_t\}_{t=1}^{\infty}$ for which the following statements are true: (a) given the price system and taxes, the household's optimum problem is solved with $b_t = B_t$ and $m_t = M_t$; (b) the government's budget constraint is satisfied for all $t \geq 0$; and (c) $c_t + g_t = y$.

26.2.4. "Short run" versus "long run"

We shall study government policies designed to ascribe a definite meaning to a distinction between outcomes in the "short run" (initial date) and the "long run" (stationary equilibrium). We assume

$$
\begin{aligned}
g_t &= g, \quad \forall t \geq 0; \\
\tau_t &= \tau, \quad \forall t \geq 1; \\
B_t &= B, \quad \forall t \geq 1.
\end{aligned}
\tag{26.2.19}
$$

We permit $\tau_0 \neq \tau$ and $B_0 \neq B$.

These settings of policy variables are designed to let us study circumstances in which the economy is in a stationary equilibrium for $t \geq 1$, but starts from some other position at $t = 0$. We have enough free policy variables to discuss two alternative meanings that the theoretical literature has attached to the phrase "open market operations."

26.2.5. Stationary equilibrium

We seek an equilibrium for which

$$
\begin{aligned}
p_t/p_{t+1} &= R_m, \quad \forall t \geq 0; \\
R_t &= R, \quad \forall t \geq 0; \\
c_t &= c, \quad \forall t \geq 0; \\
s_t &= s, \quad \forall t \geq 0.
\end{aligned}
\tag{26.2.20}
$$

Substituting equations $(26.2.20)$ into equations $(26.2.14)$ and $(26.2.17)$ yields

$$
\begin{aligned}
R &= \beta^{-1}, \\
\frac{m_{t+1}}{p_t} &= f(R_m),
\end{aligned}
\tag{26.2.21}
$$

where we define $f(R_m) \equiv F(c, R_m/R)$ and we have suppressed the constants c and R in the money demand function $f(R_m)$ in a stationary equilibrium. Notice that $f'(R_m) \geq 0$, an inequality that plays an important role below.

Substituting equations $(26.2.19)$, $(26.2.20)$, and $(26.2.21)$ into the government budget constraint $(26.2.18)$, using the equilibrium condition $M_t = m_t$, and rearranging gives

$$
g - \tau + B(R-1)/R = f(R_m)(1 - R_m), \quad \forall t \geq 1. \tag{26.2.22}
$$

Given the policy variables (g, τ, B), equation $(26.2.22)$ determines the stationary rate of return on currency R_m. In $(26.2.22)$, $g - \tau$ is the net of interest deficit, sometimes called the operational deficit; $g - \tau + B(R-1)/R$ is the gross of interest government deficit; and $f(R_m)(1 - R_m)$ is the rate of seigniorage revenues from printing currency.[5] The inflation tax rate is $(1 - R_m)$ and the quantity of real balances $f(R_m)$ is the base of the inflation tax.

---

[5] The stationary value of seigniorage per period is given by

$$
\frac{M_{t+1} - M_t}{p_t} = \frac{M_{t+1}}{p_t} - \frac{M_t}{p_{t-1}}\frac{p_{t-1}}{p_t} = f(R_m)(1 - R_m).
$$

26.2.6. Initial date (time 0)

Because $M_1/p_0 = f(R_m)$, the government budget constraint at $t = 0$ can be written

$$M_0/p_0 = f(R_m) - (g + B_0 - \tau_0) + B/R. \tag{26.2.23}$$

26.2.7. Equilibrium determination

Given the policy parameters (g, τ, τ_0, B), the initial stocks B_0 and M_0, and the equilibrium gross real interest rate $R = \beta^{-1}$, equations $(26.2.22)$ and $(26.2.23)$ determine (R_m, p_0). The two equations are recursive: equation $(26.2.22)$ determines R_m, then equation $(26.2.23)$ determines p_0.

It is useful to illustrate the determination of an equilibrium with a parametric example. Let the utility function and the transaction technology be given by

$$u(c_t, l_t) = \frac{c_t^{1-\delta}}{1-\delta} + \frac{l_t^{1-\alpha}}{1-\alpha},$$

$$H(c_t, m_{t+1}/p_t) = \frac{c_t}{1 + m_{t+1}/p_t},$$

where the latter is a modified version of equation $(26.2.5)$, so that transactions can be carried out even in the absence of money.

For parameter values $(\beta, \delta, \alpha, c) = (0.96, 0.7, 0.5, 0.4)$, Figure 26.2.1 displays the stationary gross of interest deficit $g - \tau + B(R-1)/R$ and the stationary seigniorage $f(R_m)(1-R_m)$;[6] Figure 26.2.2 shows $f(R_m) - (g + B_0 - \tau_0) + B/R$. Stationary equilibrium is determined as follows: name constant values $\{g, \tau, B\}$ which imply a stationary gross of interest deficit $g - \tau + B(R-1)/R$, then read an associated stationary value R_m from Figure 26.2.1 that satisfies equation $(26.2.22)$; for this value of R_m, find the value of $f(R_m) - (g + B_0 - \tau_0) + B/R$ in Figure 26.2.2 which is equal to M_0/p_0 by equation $(26.2.23)$. Thus, the initial price level p_0 is determined because M_0 is given in period 0.

---

[6] For our parameterization in Figure 26.2.1, households choose to hold zero money balances for $R_m < 0.15$, so at these rates there is no seigniorage collected. Seigniorage turns negative for $R_m > 1$ because the government is then continuously withdrawing money from circulation to raise the real return on money above 1.

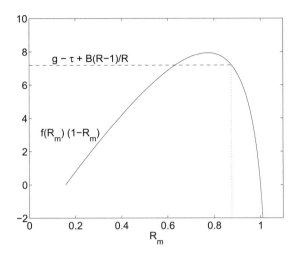

Figure 26.2.1: The stationary rate of return on currency, R_m, is determined by the intersection between the stationary gross of interest deficit $g - \tau + B(R - 1)/R$ and the stationary seigniorage $f(R_m)(1 - R_m)$.

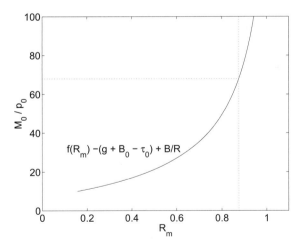

Figure 26.2.2: Given R_m, the real value of initial money balances M_0/p_0 is determined by $f(R_m) - (g + B_0 - \tau_0) + B/R$. Thus, the price level p_0 is determined because M_0 is given.

26.3. Ten monetary doctrines

We now use equations $(26.2.22)$ and $(26.2.23)$ to explain some important doctrines about money and government finance.

26.3.1. Quantity theory of money

The classic "quantity theory of money" experiment is to increase M_0 by some factor $\lambda > 1$ (a "helicopter drop" of money), leaving all of the other parameters of the model fixed (including the fiscal policy parameters (τ_0, τ, g, B)). The effect is to multiply the initial equilibrium price and money supply sequences by λ and to leave all other variables unaltered.

26.3.2. Sustained deficits cause inflation

The parameterization in Figures 26.2.1 and 26.2.2 shows that there can be multiple values of R_m that solve equation $(26.2.22)$. As can be seen in Figure 26.2.1, some values of the gross-of-interest deficit $g - \tau + B(R - 1)/R$ can be financed with either a low or high rate of return on money. The tax rate on real money balances is $(1 - R_m)$ in a stationary equilibrium, so the higher R_m that solves equation $(26.2.22)$ is on the good side of a "Laffer curve" in the inflation tax rate.

   If there are multiple values of R_m that solve equation $(26.2.22)$, we shall always select the highest one for the purposes of doing our comparative dynamic exercises.[7] The stationary equilibrium with the higher rate of return on currency is associated with classical comparative dynamics: an increase in the stationary gross-of-interest government budget deficit causes a *decrease* in the rate of return on currency (i.e., an increase in the inflation rate). Notice how the stationary equilibrium associated with the lower rate of return on currency has "perverse" comparative dynamics, from the point of view of the classical doctrine that sustained government deficits cause inflation.

---

[7] In chapter 9, we studied the perfect-foresight dynamics of a closely related system and saw that the stationary equilibrium selected here was *not* the limit point of those dynamics. Our selection of the higher rate of return equilibrium can be defended by appealing to various forms of "adaptive" (nonrational) dynamics. See Bruno and Fischer (1990), Marcet and Sargent (1989), and Marimon and Sunder (1993). Also, see exercise *26.2*.

26.3.3. Fiscal prerequisites of zero inflation policy

Equation (26.2.22) implies a restriction on fiscal policy that is necessary and sufficient to sustain a zero inflation ($R_m = 1$) equilibrium:

$$g - \tau + B(R - 1)/R = 0,$$

or

$$B = \frac{R}{R - 1}(\tau - g) = \sum_{t=0}^{\infty} R^{-t}(\tau - g).$$

This equation states that the real value of interest-bearing government indebtedness equals the present value of the net-of-interest government *surplus*, with zero revenues being contributed by an inflation tax. In this case, increased government debt implies a flow of future government surpluses, with complete abstention from the inflation tax.

26.3.4. Unpleasant monetarist arithmetic

This doctrine describes the paradoxical effects of an open market operation defined in the standard way that withholds from the monetary authority the ability to alter taxes or expenditures. Consider an open market sale of bonds at time 0, defined as a *decrease* in M_1 accompanied by an *increase* in B, with all other government fiscal policy variables constant, including (τ_0, τ). This policy can be analyzed by increasing B in equations (26.2.22) and (26.2.23). The effect of the policy is to shift the permanent gross-of-interest deficit *upward* by $(R - 1)/R$ times the increase in B, which *decreases* the real return on money R_m in Figure 26.2.1. That is, the effect is unambiguously to *increase* the stationary inflation rate (the inverse of R_m). However, the effect on the initial price level p_0 can go either way, depending on the slope of the revenue curve $f(R_m)(1 - R_m)$; the decrease in R_m reduces the right-hand side of equation (26.2.23), $f(R_m) - (g + B_0 - \tau_0) + B/R$, while the increase in B raises the value. Thus, the upward shift of the curve in Figure 26.2.2 due to the higher value of B, and the downward movement along that new curve due to the lower equilibrium value of R_m, can cause M_0/p_0 to move up or down, that is, a decrease or an increase in the initial price level p_0.

The effect of a decrease in the money supply M_1 accomplished through such an open market operation is at best temporarily to drive the price level downward, at the cost of causing the inflation rate to be permanently higher. Sargent and Wallace (1981) called this "unpleasant monetarist arithmetic."

26.3.5. An "open market" operation delivering neutrality

We now alter the definition of open market operations to be different than that used in the unpleasant monetarist arithmetic. We supplement the fiscal powers of the monetary authority in a way that lets open market operations have effects like those in the quantity theory experiment. Let there be an initial equilibrium with policy values denoted by bars over variables. Consider an open market sale or purchase defined as a decrease in M_1 and simultaneous increases in B and τ sufficient to satisfy

$$(1 - 1/R)(\hat{B} - \bar{B}) = \hat{\tau} - \bar{\tau}, \tag{26.3.1}$$

where variables with hats denote the new values of the corresponding variables. We assume that $\hat{\tau}_0 = \bar{\tau}_0$.

As long as the tax rate from time 1 on is adjusted according to equation (26.3.1), equation (26.2.22) will be satisfied at the initial value of R_m. Equation (26.3.1) imposes a requirement that the lump-sum tax τ be adjusted by just enough to service whatever additional interest payments are associated with the alteration in B resulting from the exchange of M_1 for B.[8] Under this definition of an open market operation, reductions in M_1 achieved by increases in B and the taxes needed to service B cause proportionate decreases in the paths of the money supply and the price level, leave R_m unaltered, and fulfill the pure quantity theory of money.

---

[8] This definition of an "open market" operation imputes unrealistic power to a monetary authority: on earth, central banks don't set tax rates.

26.3.6. The "optimum quantity" of money

Friedman's (1969) ideas about the optimum quantity of money can be represented in Figures 26.2.1 and 26.2.2. Friedman noted that, given the stationary levels of (g, B), the representative household prefers stationary equilibria with higher rates of return on currency. In particular, the higher the stationary level of real balances, the better the household likes it. By running a sufficiently large gross-of-interest surplus, that is, a negative value of $g - \tau + B(R-1)/R$, the government can attain any value of $R_m \in (1, \beta^{-1})$. Given (g, B) and the target value of R_m in this interval, a tax rate τ can be chosen to assure the required surplus. The proceeds of the tax are used to retire currency from circulation, thereby generating a deflation that makes the rate of return on currency equal to the target value of R_m. According to Friedman, the optimal policy is to satiate the system with real balances, insofar as it is possible to do so.

The social value of real money balances in our model is that they reduce households' shopping time. The optimum quantity of money is the one that minimizes the time allocated to shopping. For the sake of argument, suppose there is a satiation point in real balances $\psi(c)$ for any consumption level c, that is, $H_{m/p}(c, m_{t+1}/p_t) = 0$ for $m_{t+1}/p_t \geq \psi(c)$. According to condition (26.2.15), the government can attain this optimal allocation only by choosing $R_m = R$, since $\lambda_t, \mu_t > 0$. (Utility is assumed to be strictly increasing in both consumption and leisure.) Thus, welfare is at a maximum when the economy is satiated with real balances. For the transaction technology given by equation (26.2.5), the Friedman rule can only be approximately attained because money demand is insatiable.

26.3.7. Legal restrictions to boost demand for currency

If the government can somehow force households to increase their real money balances to $\tilde{f}(R_m) > f(R_m)$, it can finance a given stationary gross of interest deficit $g - \tau + B(R-1)/R$ at a higher stationary rate of return on currency R_m. The increased demand for money balances shifts the seigniorage curve in Figure 26.2.1 upward to $\tilde{f}(R_m)(1 - R_m)$, thereby increasing the higher of the two intersections of the curve $\tilde{f}(R_m)(1 - R_m)$ with the gross-of-interest deficit line in Figure 26.2.1. By increasing the base of the inflation tax, the rate $(1 - R_m)$ of inflation taxation can be diminished. Examples of legal restrictions to increase the demand for government issued currency include (a) restrictions on the rights of banks and other intermediaries to issue bank notes or other close substitutes for government issued currency,[9] (b) arbitrary limitations on trading other assets that are close substitutes with currency, and (c) reserve requirements.

Governments intent on raising revenues through the inflation tax have frequently resorted to legal restrictions and threats designed to promote the demand for its currency. In chapter 27, we shall study a version of Bryant and Wallace's (1984) theory of some of those restrictions. Sargent and Velde (1995) recount such restrictions in the Terror during the French Revolution, and the sharp tools used to enforce them.

To assess the welfare effects of policies forcing households to hold higher real balances, we must go beyond the incompletely articulated transaction process underlying equation (26.2.4). We need an explicit model of how money facilitates transactions and how the government interferes with markets to increase the demand for real balances. In such a model, there would be opposing effects on social welfare. On the one hand, our discussion of the optimum quantity of money says that a higher real return on money R_m tends to improve welfare. On the other hand, the imposition of legal restrictions aimed at forcing households to hold higher real balances might elicit socially wasteful activities from the private economy trying to evade precisely those restrictions.

26.3.8. One big open market operation

Lucas (1986) and Wallace (1989) describe a large open market purchase of private indebtedness at time 0. The purpose of the operation is to provide the government with a portfolio of interest-earning claims on the private sector, one that is sufficient to permit it to run a gross-of-interest surplus. The government uses the surplus to reduce the money supply each period, thereby engineering a deflation that raises the rate of return on money above 1. That is, the government uses its own lending to reduce the gap in rates of return between its money and higher-yield bonds. As we know from our discussion of the optimum

---

[9] In the U.S. Civil War, the U.S. Congress taxed out of existence the notes that state-chartered banks had issued, which before the war had been the country's paper currency.

quantity of money, the increase in the real return on money R_m will lead to higher welfare. [10]

To highlight the effects of the described open market policy, we impose a nonnegative net-of-interest deficit, $g - \tau \geq 0$, which prevents financing deflation by direct taxation. The proposed operation is then to increase M_1 and decrease B, with $B < 0$ indicating private indebtedness to the government. We generate a candidate policy as follows: Given values of (g, τ), use equation $(26.2.22)$ to pick a value of B that solves equation $(26.2.22)$ for a desired level of R_m, with $1 < R_m \leq \beta^{-1}$. Notice that a negative level of B will be required, since $g - \tau \geq 0$. Substituting equation $(26.2.23)$ into equation $(26.2.22)$ [by eliminating $f(R_m)$] and rearranging gives

$$M_0/p_0 = \left(\frac{R - R_m}{1 - R_m} \right) \frac{B}{R} + \left(\frac{1}{1 - R_m} \right) (g - \tau) - (g + B_0 - \tau_0). \quad (26.3.2)$$

The first term on the right side is positive, while the remainder may be positive or negative. The candidate policy is consistent with an equilibrium only if g, τ, τ_0, and B_0 assume values for which the entire right side is positive. In this case, there exists a positive price level p_0 that solves equation $(26.3.2)$.

As an example, assume that $g - \tau = 0$ and that $g + B_0 - \tau_0 = 0$, so that the government budget net of interest is balanced from time $t = 1$ onward. Then we know that the right-hand side of equation $(26.3.2)$ is positive. In this case it is feasible to operate a scheme like this to support any return on currency $1 < R_m < 1/\beta$. In the limit, when conducting an arbitrarily large open market operation, the stationary return on money R_m would approach $1/\beta = R$ and, hence, M_0/p_0 in equation $(26.3.2)$ would approach zero. This means that the government is engineering a hyperinflation in period 0 that makes the initial nominal money stock M_0 practically worthless. But how is it that the government after such a hyperinflation in period 0, can support a stationary return on money of R for the indefinite future? The explanation is as follows. Since the hyperinflation in period 0 has made the initial money holdings almost

---

[10] Beatrix Paal (2000) describes how the stabilization of the second Hungarian hyperinflation had some features of "one big open market operation." After the stabilization the government lent the one-time seigniorage revenues gathered from remonetizing the economy. The severe hyperinflation (about 4×10^{24} in the previous year) had reduced real balances of fiat currency virtually to zero. Paal argues that the fiscal aspects of the stabilization, dependent as they were on those one-time seigniorage revenues, were foreseen and shaped the dynamics of the preceding hyperinflation.

worthless, the private sector's real balances at the end of period 0, M_1/p_0, come almost entirely from that period's open-market operation. The government is injecting that money stock into the economy in exchange for interest-earnings claims on the private sector, $B/R \approx -M_1/p_0$. In future periods, the government keeps those bond holdings constant while using the net interest earnings to reduce the money supply in each future period. The government is essentially passing on the interest earnings to money holders by engineering a deflation that yields a return on money equal to $R_m \approx R$.

26.3.9. A fiscal theory of the price level

The preceding sections have illustrated what might be called a fiscal theory of *inflation*. This theory assumes a particular specification of exogenous variables that are chosen and committed to by the government. In particular, it is assumed that the government sets g, τ_0, τ, and B, that B_0 and M_0 are inherited from the past, and that the model then determines R_m and p_0 via equations (26.2.22) and (26.2.23). In particular, the system is recursive: given g, τ, and B, equation (26.2.22) determines the rate of return on currency R_m; then, given g, τ, B, and R_m, equation (26.2.23) determines p_0. After p_0 is determined, M_1 is determined from $M_1/p_0 = f(R_m)$. In this setting, the government commits to a long-run gross-of-interest government deficit $g - \tau + B(R-1)/R$, and then the market determines p_0, R_m.

Woodford (1995) and Sims (1994) have converted a version of the same model into a fiscal theory of the *price level* by altering the assumptions about the variables that the government sets. Rather than assuming that the government sets B, and thereby the gross-of-interest government deficit, Woodford assumes that B is endogenous and that instead the government sets in advance a present value of seigniorage $f(R_m)(1 - R_m)/(R - 1)$. This assumption is equivalent to saying that the government is able to commit to fix either the nominal interest rate or the gross rate of inflation R_m^{-1}. Woodford emphasizes that in the present setting, such a nominal interest rate peg leaves the equilibrium price level process determinate.[11] To illustrate Woodford's argument in our setting,

---

[11] Woodford (1995) interprets this finding against the background of a literature that occasionally asserted a different result, namely, that interest rate pegging led to price level

rearrange equation (26.2.22) to obtain

$$
\begin{aligned}
B/R &= \frac{1}{R-1}\big[(\tau - g) + f(R_m)(1 - R_m)\big] \\
&= \sum_{t=1}^{\infty} R^{-t}(\tau - g) + f(R_m)\frac{1 - R_m}{R - 1},
\end{aligned}
\tag{26.3.3}
$$

which when substituted into equation (26.2.23) yields

$$
\begin{aligned}
\frac{M_0}{p_0} + B_0 &= \sum_{t=0}^{\infty} R^{-t}(\tau_t - g_t) + f(R_m)\Big(1 + \frac{1 - R_m}{R - 1}\Big) \\
&= \sum_{t=0}^{\infty} R^{-t}(\tau_t - g_t) + \sum_{t=1}^{\infty} R^{-t} f(R_m)(R - R_m).
\end{aligned}
\tag{26.3.4}
$$

In a stationary equilibrium, the real interest rate is equal to $1/\beta$, so by multiplying the nominal interest rate by β we obtain the inverse of the corresponding value for R_m. Thus, pegging a nominal rate is equivalent to pegging the inflation rate and the steady-state flow of seigniorage $f(R_m)(1 - R_m)$. Woodford uses such equations as follows: The government chooses g, τ, τ_0, and R_m (or equivalently, $f(R_m)(1 - R_m)$). Then equation (26.3.3) determines B as the present value of the government surplus from time 1 on, including seigniorage revenues. Equation (26.3.4) then determines p_0. Equation (26.3.4) says that the price level is set to equate the real value of *total* initial government indebtedness to the present value of the net-of-interest government surplus, including seigniorage revenues. Finally, the endogenous quantity of money is determined by the demand function for money (26.2.17),

$$
M_1/p_0 = f(R_m).
\tag{26.3.5}
$$

Woodford uses this experiment to emphasize that without saying much more, the mere presence of a "quantity theory" equation of the form (26.3.5) does *not* imply the "monetarist" conclusion that it is necessary to make the money supply exogenous in order to determine the path of the price level.

---

indeterminacy because of the associated money supply endogeneity. That other literature focused on the homogeneity properties of conditions (26.2.14) and (26.2.16): the only ways in which the price level enters are as ratios to the money supply or to the price level at another date. This property suggested that a policy regime that leaves the money supply, as well as the price level, endogenous will not be able to determine the level of either.

Several commentators have remarked that the Sims-Woodford use of these equations puts the government on a different setting than the private agents.[12] Private agents' demand curves are constructed by requiring their budget constraints to hold for *all* hypothetical price processes, not just the equilibrium one. However, under Woodford's assumptions about what the government has already chosen *regardless* of the (p_0, R_m) it faces, the only way an equilibrium can exist is if p_0 adjusts to make equation (26.3.4) satisfied. The government budget constraint would not be satisfied unless p_0 adjusts to satisfy (26.3.4).

By way of contrast, in the fiscal theory of *inflation* described by Sargent and Wallace (1981) and Sargent (1992), embodied in our description of unpleasant monetarist arithmetic, the focus is on how the one tax rate that is assumed to be free to adjust, the inflation tax, responds to fiscal conditions that the government inherits. Sims and Woodford forbid the inflation tax from adjusting, having set it once and all for by pegging the nominal interest rate. They thereby force other aspects of fiscal policy and the price system to adjust.

26.3.10. Exchange rate indeterminacy

Kareken and Wallace's (1981) exchange rate indeterminacy result provides a good laboratory for putting the fiscal theory of the price level to work. First, we will describe a version of Kareken and Wallace's result. Then, we will show how it can be overturned by changing the assumptions about policy to ones like Woodford's.

To describe the theory of exchange rate indeterminacy, we change the preceding model so that there are two countries with identical technologies and preferences. Let y_i and g_i be the endowment of the good and government purchases for country $i = 1, 2$; where $y_1 + y_2 = y$ and $g_1 + g_2 = g$. Under the assumption of complete markets, equilibrium consumption c_i in country i is constant over time and $c_1 + c_2 = c$.

Each country issues currency. The government of country i has M_{it+1} units of its currency outstanding at the end of period t. The price level in terms of currency i is p_{it}, and the exchange rate e_t satisfies the purchasing power parity condition $p_{1t} = e_t p_{2t}$. The household is indifferent about which currency to use so long as both currencies bear the same rate of return, and will not hold one

---

[12] See Buiter (2002) and McCallum (2001).

with an inferior rate of return. This fact implies that $p_{1t}/p_{1t+1} = p_{2t}/p_{2t+1}$, which in turn implies that $e_{t+1} = e_t = e$. Thus, the exchange rate is constant in a nonstochastic equilibrium with two currencies being valued. We let $M_{t+1} = M_{1t+1} + eM_{2t+1}$. For simplicity, we assume that the money demand function is linear in the transaction volume, $F(c, R_m/R) = c\hat{F}(R_m/R)$. It then follows that the equilibrium condition in the world money market is

$$\frac{M_{t+1}}{p_{1t}} = f(R_m). \tag{26.3.6}$$

In order to study stationary equilibria where all real variables remain constant over time, we restrict attention to identical monetary growth rates in the two countries, $M_{it+1}/M_{it} = 1+\epsilon$ for $i = 1, 2$. We let τ_i and B_i denote constant steady-state values for lump-sum taxes, and real government indebtedness for government i. The budget constraint of government i is

$$\tau_i = g_i - B_i \frac{(1-R)}{R} - \frac{M_{it+1} - M_{it}}{p_{it}}. \tag{26.3.7}$$

Here is a version of Kareken and Wallace's exchange rate indeterminacy result: Assume that the governments of each country set g_i, B_i, and $M_{it+1} = (1 + \epsilon)M_{it}$, planning to adjust the lump-sum tax τ_i to raise whatever revenues are needed to finance their budgets. Then the constant monetary growth rate implies $R_m = (1+\epsilon)^{-1}$ and equation (26.3.6) determines the worldwide demand for real balances. But the exchange rate is not determined under these policies. Specifically, the market clearing condition for the money market at time 0 holds for *any* positive e with a price level p_{10} given by

$$\frac{M_{11} + eM_{21}}{p_{10}} = f(R_m). \tag{26.3.8}$$

For any such pair (e, p_{10}) that satisfies equation (26.3.8) with an associated value for $p_{20} = p_{10}/e$, governments' budgets are financed by setting lump-sum taxes according to (26.3.7). Kareken and Wallace conclude that under such settings for government policy variables, something more is needed to determine the exchange rate. With policy as specified here, the exchange rate is indeterminate.[13]

---

[13] See Sargent and Velde (1990) for an application of this theory to events surrounding German monetary unification.

26.3.11. Determinacy of the exchange rate retrieved

A version of Woodford's assumptions about the variables that governments choose can render the exchange rate determinate. Thus, suppose that each government sets a real level of seigniorage $x_i = (M_{it+1} - M_{it})/p_{it}$ for all $t \geq 1$. The budget constraint of government i is then

$$\tau_i = g_i - B_i \frac{(1 - R)}{R} - x_i. \tag{26.3.9}$$

In order to study stationary equilibria where all real variables remain constant over time, we allow for three cases with respect to x_1 and x_2: they are both strictly positive, strictly negative, or equal to zero.

To retrieve exchange rate determinacy, we assume that the governments of each country set g_i, B_i, x_i, and τ_i so that budgets are financed according to (26.3.9). Hence, the endogenous inflation rate is pegged to deliver the targeted levels of seigniorage,

$$x_1 + x_2 = f(R_m)(1 - R_m). \tag{26.3.10}$$

The implied return on money R_m determines the endogenous monetary growth rates in a stationary equilibrium,

$$R_m^{-1} = \frac{M_{it+1}}{M_{it}} \equiv 1 + \epsilon, \qquad \text{for } i = 1, 2. \tag{26.3.11}$$

That is, nominal supplies of both monies grow at the rate of inflation so that real money supplies remain constant over time. The levels of those real money supplies satisfy the equilibrium condition that the real value of net monetary growth is equal to the real seigniorage chosen by the government,

$$\frac{\epsilon M_{it}}{p_{it}} = x_i, \qquad \text{for } i = 1, 2. \tag{26.3.12}$$

Equations (26.3.12) determine the price levels in the two countries so long as the chosen amounts of seigniorage are not equal to zero, which in turn determine a unique exchange rate,

$$e = \frac{p_{1t}}{p_{2t}} = \frac{M_{1t}}{M_{2t}} \frac{x_2}{x_1} = \frac{(1 + \epsilon)^t M_{10}}{(1 + \epsilon)^t M_{20}} \frac{x_2}{x_1} = \frac{M_{10}}{M_{20}} \frac{x_2}{x_1}.$$

Thus, with this Sims-Woodford structure of government commitments (i.e., setting of exogenous variables), the exchange rate is determinate. It is only the third case of stationary equilibria with x_1 and x_2 equal to zero where the exchange rate is indeterminate, because then there is no relative measure of seigniorage levels that is needed to pin down the denomination of the world *real* money supply for the purpose of financing governments' budgets.

26.4. An example of exchange rate (in)determinacy

As an illustration of the Kareken-Wallace exchange rate indeterminacy and the Sims-Woodford fiscal theory of the price level, consider the following version of the two-country environment in section 26.3.10:

$$y_1 = y_2 = y/2, \qquad (26.4.1a)$$

$$g_1 = g_2 = 0, \qquad (26.4.1b)$$

$$B_1 = B_2 = 0, \qquad (26.4.1c)$$

$$M_{10} = M_{20}, \qquad (26.4.1d)$$

$$\frac{M_{1t+1}}{M_{1t}} = \frac{M_{2t+1}}{M_{2t}} = 1 + \epsilon > 1, \qquad \forall t \geq 0. \qquad (26.4.1e)$$

The governments in the two countries have no purchases to finance and no bond holdings. The seigniorage raised by printing money is handed over as lump-sum transfers to the households in each country, respectively. The budget constraint of government i is

$$-\tau_i = \frac{M_{it+1} - M_{it}}{p_{it}} = x_i, \qquad (26.4.2)$$

where the negative lump-sum tax, $-\tau_i$, is equal to the real value of the country's seigniorage, x_i.

To operationalize the concept of exchange rate indeterminacy, we assume that there is a 'sunspot' variable that can take on three values at the start of the economy. [14] Each realization of the sunspot variable is associated with a particular belief about the equilibrium value of the exchange rate $e \in \{0, 1, \infty\}$ that will prevail in period 0 and forever thereafter. That is, depending on the sunspot realization, all households will coordinate on one of the following three beliefs about the equilibrium outcome in the world money market:

---

[14] Sunspots were introduced by Cass and Shell (1983) to explain "excess market volatility." Sunspots represent extrinsic uncertainty not related to the fundamentals of the economy.

   i)   the currency of country 2 is worthless ($e = 0$ and $p_{2t} = \infty, \forall t \geq 0$);

  ii) the two currencies are traded one for one ($e = 1$ and $p_{1t} = p_{2t}, \forall t \geq 0$);

 iii) the currency of country 1 is worthless ($e = \infty$ and $p_{1t} = \infty, \forall t \geq 0$).

We assume that all households share the same belief about the sunspot process, and that each sunspot realization is perceived to occur with the same probability equal to $1/3$.

We also postulate that all households are risk-averse with identical preferences and as stated in $(26.4.1a)$ that they have the same constant endowment stream. As initial conditions, the representative household in country i owns the beginning-of-period money stock M_{i0} of its country.

26.4.1. Trading before sunspot realization

The equilibrium allocation in this economy will depend on whether or not households can trade before observing the sunspot realization. In chapter 8, we assumed that all trade took place after any uncertainty had been resolved in the first period. In our current setting, this would translate into households trading after the sunspot realization, i.e., after the agents have seen the sunspot and therefore after the coordination of beliefs about the equilibrium value of the exchange rate. In cases *i)* and *iii)*, this implies that the households in the country with a valued currency will be better off because their initial money holdings are valuable and they will receive lump-sum transfers equal to their government's revenue from seigniorage in each period. In case *ii)*, all households are equally well off in the world economy because of identical budget constraints.

Alternatively, we can assume that households can trade in markets before the sunspot realization. In a complete market world, agents would be able to trade in contingent claims with payoffs conditional on the sunspot realization. Given the symmetries in the environment with respect to preferences, endowment and expected asset/transfer outcomes associated with the sunspot process, the equilibrium allocation will be one of perfect pooling with each household consuming $y/2$ in every period.[15] Hence, the households will use security markets to pool the risks associated with the sunspot process. Given the *ex ante* symmetry in

---

[15] See Lucas (1982) for a perfect pooling equilibrium in a two-country world with two currencies. However, Lucas considers only *intrinsic* uncertainty arising from stochastic endowment streams.

possible sunspot realizations, it follows that equilibrium contingent-claim prices will be such that a household in country i can afford to trade half of its initial money holdings, $M_{i0}/2$, and half of the entitlement to its future stream of lump-sum transfers, $x_i/2$, in exchange for the corresponding quantities from a household in the other country. As a result, these diversified portfolios enable each household to finance a smooth consumption stream equal to $y/2$ in every period regardless of the sunspot realization.

We have constructed a rational expectations equilibrium where the equilibrium exchange rate is influenced by a sunspot process. But even though the exchange rate can take on three different values in this example, the households are insulated from any real effects because of their trades in complete markets prior to the sunspot realization. In this world, each government is assumed to print more of its currency each period at the net rate $\epsilon > 0$ and hand over the newly printed money to its households as lump-sum transfers. The households in turn have entered into contingent-claim contracts that oblige them to hand over half of this newly printed currency to a household in the other country, while receiving half of that other household's government transfer. Given a sunspot realization that is associated with either case *i)* or case *iii)* above, it follows that these deliveries of newly printed currencies between households are valuable in one direction but not in the other direction.

26.4.2. Fiscal theory of the price level

How can a fiscal theory of the price level overcome this indeterminacy of the exchange rate? In the spirit of section 26.3.11, suppose that each government sets a real level of seigniorage given by

$$x_1 = x_2 = 0.5 \cdot f\left(\frac{1}{1+\epsilon}\right)\left[1 - \frac{1}{1+\epsilon}\right].$$

From equations (26.3.10) and (26.3.11), we see that the governments split the total world seigniorage associated with a gross money growth rate equal to $1+\epsilon$. Given such policies, both governments can satisfy their budget constraints only if the equilibrium exchange rate is indeed $e = 1$. Hence, the fiscal theory of the price level here would claim that case *ii)* is the only viable rational expectations equilibrium. In the words of Kocherlakota and Phelan (1999), "the fiscal theory of the price level is, at its core, a device for selecting equilibria from the continuum which can exist in monetary models."

Kocherlakota and Phelan (1999) are skeptical about this recommendation for selecting an equilibrium. The fiscal theory proposes to rule out other equilibria by specifying government policies in such a way that government budget constraints hold only for one particular exchange rate. But what would happen if the sunspot realization signals case *i)* or case *iii)* to the households so that they actually abandon one currency, making it worthless? The fiscal theory formulated by Sims and Woodford contains no answer to this question. Critics of the fiscal theory of the price level instead prefer to specify government policies so that a government's budget constraint is satisfied for *all* hypothetical outcomes, including $e \in \{0, \infty\}$. For example, a government that finds itself issuing a worthless currency could surrender its aspiration to make lump-sum transfers with strictly positive value to its citizens, while the other government would accept that the value of the transfer of newly printed money to its citizens has doubled in real terms. But of course, this remedy to the puzzle would refute the fiscal theory of the price level and once again render the exchange rate indeterminate.

26.4.3. A game theoretic view of the fiscal theory of the price level

Bassetto (2002) agrees with criticisms of the fiscal theory of the price level that question how the government can adopt a fiscal policy without being concerned about outcomes that could make the policy infeasible. Bassetto reformulates the fiscal theory of the price level in terms of a game. The essence of his argument is that in order to select an equilibrium, a government must specify strategies for all arbitrary outcomes so that its desired outcome is the only one that can be supported as an equilibrium outcome, merely on the basis of individual rationality of private actors.

Bassetto (2002) studies a government that seeks to finance occasional deficits by issuing debt in a model with 'trading posts.' In such a trading environment it might happen that not all government debt can be sold because private agents fail to submit enough bids. What would the equilibrium outcome be then? The fiscal theory formulated by Sims and Woodford contains no answer since it presupposes that the government budget constraint will be satisfied for the specified fiscal policy. Bassetto provides an answer by arguing that the government should formulate a strategy for that and all other arbitrary outcomes. Specifically, the following government strategy supports the desired fiscal policy as a unique equilibrium outcome. If some debt cannot be sold, the government responds by increasing taxes to make up for the present shortfall, but without altering future taxes. Thus, "the onset of a debt crisis would be accompanied by an increase in the amount of resources that are offered in repayment of debt and hence an increase in the rate of return of government debt. As a consequence, any rational household would respond to a debt crisis by lending the government *more*, rather than less, which ensures that no such crisis can occur in an equilibrium." [16]

Because Bassetto's argument works equally well in a real economy, the preceding paragraph did not mention money or nominal prices. Moreover, our omission of money seems appropriate since Bassetto studies a cashless economy where the relative price of goods and nominal bonds merely determines the value of the unit of account (the 'dollar'). Atkeson et al. (2010) extend the analysis to a monetary economy, and follow the same approach to multiplicity of equilibria that we took in the cash-in-advance model in section 16.17.3. While theirs is a new-Keynesian model, they analyze sunspot equilibria that satisfy a constraint similar to ours when we imposed an unchanged value for the denominator of equation (16.16.7), without any constraint on each individual next-period price level. In the analysis of Atkeson et al. (2010), the corresponding restriction on sunspot equilibria is that the expected inflation is unchanged when perturbing the sunspot-driven uncertainty in next period's price level.

Note that different versions of the fiscal theory of the price level share the same key assumption that a government can fully commit to its policy or strategy. In chapters 23 and 24, we study credible government policies – policies that a government would like to enact under all circumstances.

---

[16] A similar strategy would establish Bassetto's version of the fiscal theory of the price level in section 26.4.2. For example, suppose that each government promises to increase taxation in order to purchase its currency if it turns worthless, say, at the price level that would have prevailed in case *ii)*. Such strategies can effectively rule out cases *i)* and *iii)* as equilibrium outcomes, and make exchange rate $e = 1$ the only possible equilibrium.

26.5. Optimal inflation tax: the Friedman rule

Given lump-sum taxation, the sixth monetary doctrine (about the "optimum quantity" of money) establishes the optimality of the Friedman rule. The optimal policy is to satiate the economy with real balances by generating a deflation that drives the net nominal interest rate to zero. In a stationary economy, there can be deflation only if the government retires currency with a government surplus. We now ask if such a costly scheme remains optimal when all government revenues must be raised through distortionary taxation. Or would the Ramsey plan then include an inflation tax on money holdings whose rate depends on the interest elasticity of money demand?

Following Correia and Teles (1996), we show that even with distortionary taxation the Friedman rule is the optimal policy under a transaction technology (26.2.4) that satisfies a homogeneity condition.

Earlier analyses of the optimal tax on money in models with transaction technologies include Kimbrough (1986), Faig (1988), and Guidotti and Vegh (1993). Chari, Christiano, and Kehoe (1996) also develop conditions for the optimality of the Friedman rule in models with cash and credit goods (see section 16.16), and money in the utility function.

26.5.1. Economic environment

We convert our shopping time monetary economy into a production economy with labor n_t as the only input in a linear technology:

$$c_t + g_t = n_t. \qquad (26.5.1)$$

The household's time constraint becomes

$$1 = \ell_t + s_t + n_t. \qquad (26.5.2)$$

The shopping technology is now assumed to be homogeneous of degree $\nu \geq 0$ in consumption c_t and real money balances $\hat{m}_{t+1} \equiv m_{t+1}/p_t$;

$$s_t = H(c_t, \hat{m}_{t+1}) = c_t^{\nu} H\left(1, \frac{\hat{m}_{t+1}}{c_t}\right), \qquad \text{for } c_t > 0. \qquad (26.5.3)$$

By Euler's theorem we have

$$H_c(c, \hat{m})c + H_{\hat{m}}(c, \hat{m})\hat{m} = \nu H(c, \hat{m}). \qquad (26.5.4)$$

For any consumption level c, we also assume a point of satiation in real money balances ψc such that

$$H_{\hat{m}}(c, \hat{m}) = H(c, \hat{m}) = 0, \qquad \text{for} \ \hat{m} \geq \psi c. \qquad (26.5.5)$$

26.5.2. Household's optimization problem

After replacing net income $(y - \tau_t)$ in equation (26.2.7) by $(1 - \tau_t)(1 - \ell_t - s_t)$, consolidation of budget constraints yields the household's present-value budget constraint

$$\sum_{t=0}^{\infty} q_t^0 \left(c_t + \frac{i_t}{1 + i_t} \hat{m}_{t+1} \right) = \sum_{t=0}^{\infty} q_t^0 (1 - \tau_t)(1 - \ell_t - s_t) + b_0 + \frac{m_0}{p_0}, \qquad (26.5.6)$$

where we have used equation (26.2.8), and q_t^0 is the Arrow-Debreu price

$$q_t^0 = \prod_{i=0}^{t-1} R_i^{-1}$$

with the numeraire $q_0^0 = 1$. We have also imposed the transversality conditions,

$$\lim_{T \to \infty} q_T^0 \frac{b_{T+1}}{R_T} = 0, \qquad (26.5.7a)$$

$$\lim_{T \to \infty} q_T^0 \hat{m}_{T+1} = 0. \qquad (26.5.7b)$$

Given the satiation point in equation (26.5.5), real money balances held for transaction purposes are bounded from above by ψ. Real balances may also be held purely for savings purposes if money is not dominated in rate of return by bonds, but an agent would never find it optimal to accumulate balances that violate the transversality condition. Thus, for whatever reason money is being held, condition (26.5.7b) must hold in an equilibrium.

Substitute $s_t = H(c_t, \hat{m}_{t+1})$ into equation (26.5.6), and let λ be the Lagrange multiplier on this present-value budget constraint. At an interior solution, the first-order conditions of the household's optimization problem become

$$c_t: \quad \beta^t u_c(t) - \lambda q_t^0 \left[(1 - \tau_t) H_c(t) + 1 \right] = 0, \qquad (26.5.8a)$$

$$\ell_t: \quad \beta^t u_\ell(t) - \lambda q_t^0 (1 - \tau_t) = 0, \qquad (26.5.8b)$$

$$\hat{m}_{t+1}: \quad -\lambda q_t^0 \left[(1 - \tau_t) H_{\hat{m}}(t) + \frac{i_t}{1 + i_t} \right] = 0. \qquad (26.5.8c)$$

From conditions $(26.5.8a)$ and $(26.5.8b)$, we obtain

$$\frac{u_\ell(t)}{1 - \tau_t} = u_c(t) - u_\ell(t)H_c(t). \qquad (26.5.9)$$

The left side of equation $(26.5.9)$ is the utility of extra leisure obtained from giving up one unit of disposable labor income, which at the optimum should equal the marginal utility of consumption reduced by the disutility of shopping for the marginal unit of consumption, given by the right side of equation $(26.5.9)$. Using condition $(26.5.8b)$ and the corresponding expression for $t = 0$ with the numeraire $q_0^0 = 1$, the Arrow-Debreu price q_t^0 can be expressed as

$$q_t^0 = \beta^t \frac{u_\ell(t)}{u_\ell(0)} \frac{1 - \tau_0}{1 - \tau_t}; \qquad (26.5.10)$$

and by condition $(26.5.8c)$,

$$\frac{i_t}{1 + i_t} = -(1 - \tau_t)H_{\hat{m}}(t). \qquad (26.5.11)$$

This last condition equalizes the cost of holding one unit of real balances (the left side) with the opportunity value of the shopping time that is released by an additional unit of real balances, measured on the right side by the extra after-tax labor income that can be generated.

26.5.3. Ramsey plan

Following the method for solving a Ramsey problem in chapter 16, we use the household's first-order conditions to eliminate prices and taxes from its present-value budget constraint. Specifically, we substitute equations $(26.5.10)$ and $(26.5.11)$ into equation $(26.5.6)$, and then multiply by $u_\ell(0)/(1 - \tau_0)$. After also using equation $(26.5.9)$, the implementability condition becomes

$$\sum_{t=0}^{\infty} \beta^t \left\{ \left[u_c(t) - u_\ell(t)H_c(t)\right]c_t - u_\ell(t)H_{\hat{m}}(t)\hat{m}_{t+1} - u_\ell(t)(1 - \ell_t - s_t) \right\} = 0,$$

where we have assumed zero initial assets, $b_0 = m_0 = 0$. Finally, we substitute $s_t = H(c_t, \hat{m}_{t+1})$ into this expression and invoke Euler's theorem $(26.5.4)$, to arrive at

$$\sum_{t=0}^{\infty} \beta^t \left\{ u_c(t)c_t - u_\ell(t)\left[1 - \ell_t - (1 - \nu)H(c_t, \hat{m}_{t+1})\right] \right\} = 0. \qquad (26.5.12)$$

The Ramsey problem is to maximize expression $(26.2.2)$ subject to equation $(26.5.12)$ and a feasibility constraint that combines equations $(26.5.1)$ through $(26.5.3)$:

$$1 - \ell_t - H(c_t, \hat{m}_{t+1}) - c_t - g_t = 0. \tag{26.5.13}$$

Let Φ and $\{\theta_t\}_{t=0}^{\infty}$ be a Lagrange multiplier on equation $(26.5.12)$ and a sequence of Lagrange multipliers on equation $(26.5.13)$, respectively. First-order conditions for this problem are

$$
\begin{aligned}
c_t: \quad & u_c(t) + \Phi \left\{ u_{cc}(t)c_t + u_c(t) \right. \\
& - u_{\ell c}(t) \left[1 - \ell_t - (1-\nu)H(c_t, \hat{m}_{t+1}) \right] \\
& \left. + (1-\nu)u_\ell(t)H_c(t) \right\} - \theta_t \left[H_c(t) + 1 \right] = 0, \tag{26.5.14a}
\end{aligned}
$$

$$
\begin{aligned}
\ell_t: \quad & u_\ell(t) + \Phi \left\{ u_{c\ell}(t)c_t + u_\ell(t) \right. \\
& \left. - u_{\ell\ell}(t) \left[1 - \ell_t - (1-\nu)H(c_t, \hat{m}_{t+1}) \right] \right\} = -\theta_t, \tag{26.5.14b}
\end{aligned}
$$

$$
\hat{m}_{t+1}: \quad H_{\hat{m}}(t) \left[\Phi(1-\nu)u_\ell(t) - \theta_t \right] = 0. \tag{26.5.14c}
$$

The first-order condition for real money balances $(26.5.14c)$ is satisfied when either $H_{\hat{m}}(t) = 0$ or

$$\theta_t = \Phi(1-\nu)u_\ell(t). \tag{26.5.15}$$

We now show that equation $(26.5.15)$ cannot be a solution of the problem. Notice that when $\nu > 1$, equation $(26.5.15)$ implies that the multipliers Φ and θ_t will either be zero or have opposite signs. Such a solution is excluded because Φ is nonnegative, while the insatiable utility function implies that θ_t is strictly positive. When $\nu = 1$, a strictly positive θ_t also excludes equation $(26.5.15)$ as a solution. To reject equation $(26.5.15)$ for $\nu \in [0,1)$, we substitute equation $(26.5.15)$ into equation $(26.5.14b)$,

$$u_\ell(t) + \Phi \left\{ u_{c\ell}(t)c_t + \nu u_\ell(t) - u_{\ell\ell}(t) \left[1 - \ell_t - (1-\nu)H(c_t, \hat{m}_{t+1}) \right] \right\} = 0,$$

which is a contradiction because the left side is strictly positive, given our assumption that $u_{c\ell}(t) \geq 0$. We conclude that equation $(26.5.15)$ cannot characterize the solution of the Ramsey problem when the transaction technology is homogeneous of degree $\nu \geq 0$, so the solution has to be $H_{\hat{m}}(t) = 0$. In other words, the social planner follows the Friedman rule and satiates the economy with real balances. According to condition $(26.5.8c)$, this aim can be accomplished with a monetary policy that sustains a zero net nominal interest rate.

As an illustration of how the Ramsey plan is implemented, suppose that $g_t = g$ in all periods. Example 1 of chapter 16 presents the Ramsey plan for this case if there were no transaction technology and no money in the model. The optimal outcome is characterized by a constant allocation $(\hat{c}, \hat{n})$ and a constant tax rate $\hat{\tau}$ that supports a balanced government budget. We conjecture that the Ramsey solution to the present monetary economy shares that real allocation. But how can it do so in the present economy with its additional constraint in the form of a transaction technology? First, notice that the preceding Ramsey solution calls for satiating the economy with real balances, so there will be no time allocated to shopping in the Ramsey outcome. Second, the real balances needed to satiate the economy are constant over time and equal to

$$\frac{M_{t+1}}{p_t} = \psi \hat{c}, \qquad \forall t \geq 0, \qquad (26.5.16)$$

and the real return on money is equal to the constant real interest rate,

$$\frac{p_t}{p_{t+1}} = R, \qquad \forall t \geq 0. \qquad (26.5.17)$$

Third, the real balances in equation $(26.5.16)$ also equal the real value of assets acquired by the government in period 0 from selling the money supply M_1 to the households. These government assets earn a net real return in each future period equal to

$$(R - 1)\psi \hat{c} = R \frac{M_t}{p_{t-1}} - \frac{M_{t+1}}{p_t} = \frac{p_{t-1}}{p_t} \frac{M_t}{p_{t-1}} - \frac{M_{t+1}}{p_t} = \frac{M_t - M_{t+1}}{p_t},$$

where we have invoked equations $(26.5.16)$ and $(26.5.17)$ to show that the interest earnings just equal the funds for retiring currency from circulation in all future periods needed to sustain an equilibrium in the money market with a zero net nominal interest rate. It is straightforward to verify that households would be happy to incur the indebtedness of the initial period. They use the borrowed funds to acquire money balances and meet future interest payments by surrendering some of these money balances. Yet their real money balances are unchanged over time because of the falling price level. In this way, money holdings are costless to the households, and their optimal decisions with respect to consumption and labor are the same as in the nonmonetary version of this economy.

26.6. Time consistency of monetary policy

The optimality of the Friedman rule was derived in the previous section under the assumption that the government can commit to a plan for its future actions. The Ramsey plan is not time consistent and requires that the government have a technology to bind itself to it. In each period along the Ramsey plan, the government is tempted to levy an unannounced inflation tax in order to reduce future distortionary labor taxes. Rather than examine this time consistency problem due to distortionary taxation, we now turn to another time consistency problem arising from a situation where surprise inflation can reduce unemployment.

Kydland and Prescott (1977) and Barro and Gordon (1983a, 1983b) study the time consistency problem and credible monetary policies in reduced-form models with a trade-off between surprise inflation and unemployment. In their spirit, Ireland (1997) proposes a model with microeconomic foundations that gives rise to such a trade-off because monopolistically competitive firms set nominal goods prices before the government sets monetary policy.[17] The government is here tempted to create surprise inflation that erodes firms' markups and stimulates employment above a suboptimally low level. But any anticipated inflation has negative welfare effects that arise as a result of a postulated cash-in-advance constraint. More specifically, anticipated inflation reduces the real value of nominal labor income that can be spent or invested first in the next period, thereby distorting incentives to work.

The following setup modifies Ireland's model and assumes that each household has some market power with respect to its labor supply while a single good is produced by perfectly competitive firms.

---

[17] Ireland's model takes most of its structure from those developed by Svensson (1986) and Rotemberg (1987). See Rotemberg and Woodford (1997) and King and Wolman (1999) for empirical implementations of related models.

26.6.1. Model with monopolistically competitive wage setting

There is a continuum of households indexed on the unit interval, $i \in [0,1]$. At time t, household i consumes c_{it} of a single consumption good and supplies labor $n_{it} \geq 0$.[18] The preferences of the household are

$$\sum_{t=0}^{\infty} \beta^t \left(\frac{c_{it}^{\gamma}}{\gamma} - n_{it} \right), \tag{26.6.1}$$

where $\beta \in (0,1)$ and $\gamma \in (0,1)$. The parameter restriction on γ ensures that the household's utility is well defined at zero consumption.

The technology for producing the single consumption good is

$$y_t = \left(\int_0^1 n_{it}^{\frac{1-\alpha}{1+\alpha}} \, \mathrm{d}i \right)^{\frac{1+\alpha}{1-\alpha}}, \tag{26.6.2}$$

where y_t is per capita output and $\alpha \in (0,1)$. The technology has constant returns to scale in labor inputs, and if all types of labor are supplied in the same quantity n_t, we have $y_t = n_t$. The marginal product of labor of type i is

$$\frac{\partial y_t}{\partial n_{it}} = \left(\int_0^1 n_{it}^{\frac{1-\alpha}{1+\alpha}} \, \mathrm{d}i \right)^{\frac{2\alpha}{1-\alpha}} n_{it}^{\frac{-2\alpha}{1+\alpha}} = \left(\frac{y_t}{n_{it}} \right)^{\frac{2\alpha}{1+\alpha}} \equiv \hat{w}(y_t, n_{it}). \tag{26.6.3}$$

The single good is produced by a large number of competitive firms that are willing to pay a real wage to labor of type i equal to the marginal product in equation (26.6.3).

The definition of the function $\hat{w}(y_t, n_{it})$ with its two arguments y_t and n_{it} is motivated by the first of the following two assumptions on households' labor-supply behavior.[19]

1. When maximizing the rent of its labor supply, household i perceives that it can affect the marginal product $\hat{w}(y_t, n_{it})$ through the second argument, while y_t is taken as given.

---

[18] For analytical simplicity, we assume that the households can supply any nonnegative amount of labor. When we imposed a finite time endowment in the first edition of this book, we had to confront the issue of labor rationing across firms along some equilibrium paths.

[19] Analogous assumptions are made implicitly by Ireland (1997), who takes the aggregate price index as given in the monopolistically competitive firms' profit maximization problem, and disregards firms' profitability when computing the output effect of a monetary policy deviation.

2. The nominal wage for labor of type i at time t is chosen by household i at the very beginning of period t. Given the nominal wage w_{it}, household i is obliged to deliver any amount of labor n_{it} that is demanded in the economy.

The government's only task is to increase or decrease the money supply by making lump-sum transfers $(x_t - 1)M_t$ to the households, where M_t is the per capita money supply at the beginning of period t and x_t is the gross growth rate of money in period t:

$$M_{t+1} = x_t M_t. \tag{26.6.4}$$

Following Ireland (1997), we assume that $x_t \in [\beta, \bar{x}]$. These bounds on money growth ensure the existence of a monetary equilibrium. The lower bound will be shown to yield a zero net nominal interest rate in a stationary equilibrium, whereas the upper bound $\bar{x} < \infty$ guarantees that households never abandon the use of money altogether.

During each period t, events unfold as follows for household i: The household starts period t with money m_{it} and real private bonds b_{it}, and the household sets the nominal wage w_{it} for its type of labor. After the wage is determined, the government chooses a nominal transfer $(x_t - 1)M_t$ to be handed over to the household. Thereafter, the household enters the asset market to settle maturing bonds b_{it} and to pick a new portfolio composition with money and real bonds $b_{i,t+1}$. After the asset market has closed, the household splits into a shopper and a worker.[20] During period t, the shopper purchases c_{it} units of the single good subject to the cash-in-advance constraint,

$$\frac{m_{it}}{p_t} + \frac{(x_t - 1)M_t}{p_t} + b_{it} - \frac{b_{i,t+1}}{R_t} \geq c_{it}, \tag{26.6.5}$$

where p_t and R_t are the price level and the real interest rate, respectively. Given the household's predetermined nominal wage w_{it}, the worker supplies all the labor n_{it} demanded by firms. At the end of period t when the goods market has closed, the shopper and the worker reunite, and the household's money holdings $m_{i,t+1}$ now equal the worker's labor income $w_{it} n_{it}$ plus any

---

[20] The interpretation that the household splits into a shopper and a worker follows Lucas's (1980b) cash-in-advance framework. It embodies the constraint on transactions recommended by Clower (1967).

unspent cash from the shopping round. Thus, the budget constraint of the household becomes[21]

$$\frac{m_{it}}{p_t} + \frac{(x_t - 1)M_t}{p_t} + b_{it} + \frac{w_{it}}{p_t}n_{it} = c_{it} + \frac{b_{i,t+1}}{R_t} + \frac{m_{i,t+1}}{p_t}. \tag{26.6.6}$$

26.6.2. Perfect foresight equilibrium

We first study household i's optimization problem under perfect foresight. Given initial assets (m_{i0}, b_{i0}) and sequences of prices $\{p_t\}_{t=0}^{\infty}$, real interest rates $\{R_t\}_{t=0}^{\infty}$, output levels $\{y_t\}_{t=0}^{\infty}$, and nominal transfers $\{(x_t - 1)M_t\}_{t=0}^{\infty}$, the household maximizes expression (26.6.1) by choosing sequences of consumption $\{c_{it}\}_{t=0}^{\infty}$, labor supply $\{n_{it}\}_{t=0}^{\infty}$, money holdings $\{m_{i,t+1}\}_{t=0}^{\infty}$, real bond holdings $\{b_{i,t+1}\}_{t=0}^{\infty}$, and nominal wages $\{w_{it}\}_{t=0}^{\infty}$ that satisfy cash-in-advance constraints (26.6.5) and budget constraints (26.6.6), with the real wage equaling the marginal product of labor of type i at each point in time, $w_{it}/p_t = \hat{w}(y_t, n_{it})$. The last constraint ensures that the household's choices of n_{it} and w_{it} are consistent with competitive firms' demand for labor of type i. Let us incorporate this constraint into budget constraint (26.6.6) by replacing the real wage w_{it}/p_t by the marginal product $\hat{w}(y_t, n_{it})$. With $\beta^t \mu_{it}$ and $\beta^t \lambda_{it}$ as the Lagrange multipliers on the time t cash-in-advance constraint and budget constraint, respectively, the first-order conditions at an interior solution are

$$c_{it}: \quad c_{it}^{\gamma-1} - \mu_{it} - \lambda_{it} = 0, \tag{26.6.7a}$$

$$n_{it}: \quad -1 + \lambda_{it}\left[\frac{\partial \hat{w}(y_t, n_{it})}{\partial n_{it}}n_{it} + \hat{w}(y_t, n_{it})\right] = 0, \tag{26.6.7b}$$

$$m_{i,t+1}: \quad -\lambda_{it}\frac{1}{p_t} + \beta\left(\lambda_{i,t+1} + \mu_{i,t+1}\right)\frac{1}{p_{t+1}} = 0, \tag{26.6.7c}$$

$$b_{i,t+1}: \quad -(\lambda_{it} + \mu_{it})\frac{1}{R_t} + \beta\left(\lambda_{i,t+1} + \mu_{i,t+1}\right) = 0. \tag{26.6.7d}$$

The first-order condition (26.6.7b) for the rent-maximizing labor supply n_{it} can be rearranged to read

$$\hat{w}(y_t, n_{it}) = \frac{\lambda_{it}^{-1}}{1 + \epsilon_{it}^{-1}} = \frac{1 + \alpha}{1 - \alpha}\lambda_{it}^{-1}, \tag{26.6.8}$$

---

[21] The assumptions of constant returns to scale and perfect competition in the goods market imply that profits of firms are zero.

$$\text{where} \quad \epsilon_{it} = \left[\frac{\partial\,\hat{w}(y_t, n_{it})}{\partial\,n_{it}}\frac{n_{it}}{\hat{w}(y_t, n_{it})}\right]^{-1} = -\frac{1+\alpha}{2\alpha} < 0.$$

The Lagrange multiplier λ_{it} is the shadow value of relaxing the budget constraint in period t by one unit, measured in "utils" at time t. Since preferences $(26.6.1)$ are linear in the disutility of labor, λ_{it}^{-1} is the value of leisure in period t in terms of the units of the budget constraint at time t. Equation $(26.6.8)$ is then the familiar expression that the monopoly price $\hat{w}(y_t, n_{it})$ should be set as a markup above marginal cost λ_{it}^{-1}, and the markup is inversely related to the absolute value of the demand elasticity of labor type i, $|\epsilon_{it}|$.

First-order conditions $(26.6.7c)$ and $(26.6.7d)$ for asset decisions can be used to solve for rates of return,

$$\frac{p_t}{p_{t+1}} = \frac{\lambda_{it}}{\beta\left(\lambda_{i,t+1} + \mu_{i,t+1}\right)}, \tag{26.6.9a}$$

$$R_t = \frac{\lambda_{it} + \mu_{it}}{\beta\left(\lambda_{i,t+1} + \mu_{i,t+1}\right)}. \tag{26.6.9b}$$

Whenever the Lagrange multiplier μ_{it} on the cash-in-advance constraint is strictly positive, money has a lower rate of return than bonds, or, equivalently, the net nominal interest rate is strictly positive, as shown in equation $(26.2.8)$.

Given initial conditions $m_{i0} = M_0$ and $b_{i0} = 0$, we now turn to characterizing an equilibrium under the additional assumption that the cash-in-advance constraint $(26.6.5)$ holds with equality, even when it does not bind. Since all households are perfectly symmetric, they will make identical consumption and labor decisions, $c_{it} = c_t$ and $n_{it} = n_t$, so by goods market clearing and the constant-returns-to-scale technology $(26.6.2)$, we have

$$c_t = y_t = n_t, \tag{26.6.10a}$$

and from the expression for the marginal product of labor in equation $(26.6.3)$,

$$\hat{w}(y_t, n_t) = 1. \tag{26.6.10b}$$

Equilibrium asset holdings satisfy $m_{i,t+1} = M_{t+1}$ and $b_{i,t+1} = 0$. The substitution of equilibrium quantities into the cash-in-advance constraint $(26.6.5)$ at equality yields

$$\frac{M_{t+1}}{p_t} = c_t, \tag{26.6.10c}$$

where a version of the "quantity theory of money" determines the price level, $p_t = M_{t+1}/c_t$. We now substitute this expression and conditions $(26.6.7a)$ and $(26.6.8)$ into equation $(26.6.9a)$:

$$\frac{M_{t+1}/c_t}{M_{t+2}/c_{t+1}} = \frac{\left[\frac{1-\alpha}{1+\alpha}\,\hat{w}(y_t, n_t)\right]^{-1}}{\beta\,c_{t+1}^{\gamma-1}},$$

which can be rearranged to read

$$c_t = \frac{1-\alpha}{1+\alpha}\,\frac{\beta}{x_{t+1}}\,c_{t+1}^{\gamma},$$

where we have used equations $(26.6.4)$ and $(26.6.10b)$. After taking the logarithm of this expression, we get

$$\log(c_t) = \log\left(\frac{1-\alpha}{1+\alpha}\,\beta\right) + \gamma\log(c_{t+1}) - \log(x_{t+1}).$$

Since $0 < \gamma < 1$ and x_{t+1} is bounded, this linear difference equation in $\log(c_t)$ can be solved forward to obtain

$$\log(c_t) = \frac{\log\left(\frac{1-\alpha}{1+\alpha}\,\beta\right)}{1-\gamma} - \sum_{j=0}^{\infty}\gamma^j\log(x_{t+1+j}), \tag{26.6.11}$$

where equilibrium considerations have prompted us to choose the particular solution that yields a bounded sequence.[22]

---

[22] See the appendix to chapter 2 for the solution of scalar linear difference equations.

26.6.3. Ramsey plan

The Ramsey problem is to choose a sequence of monetary growth rates $\{x_t\}_{t=0}^{\infty}$ that supports the perfect foresight equilibrium with the highest possible welfare; that is, the optimal choice of $\{x_t\}_{t=0}^{\infty}$ maximizes the representative household's utility in expression (26.6.1) subject to expression (26.6.11) and $n_t = c_t$. From the expression (26.6.11) it is apparent that the constraints on money growth, $x_t \in [\beta, \bar{x}]$, translate into lower and upper bounds on consumption, $c_t \in [\underline{c}, \bar{c}]$, where

$$\underline{c} = \left(\frac{\beta}{\bar{x}} \frac{1-\alpha}{1+\alpha} \right)^{\frac{1}{1-\gamma}}, \qquad \text{and} \qquad \bar{c} = \left(\frac{1-\alpha}{1+\alpha} \right)^{\frac{1}{1-\gamma}} < 1. \qquad (26.6.12)$$

The Ramsey plan then follows directly from inspecting the one-period return of the Ramsey optimization problem,

$$\frac{c_t^{\gamma}}{\gamma} - c_t, \qquad (26.6.13)$$

which is strictly concave and reaches a maximum at $c = 1$. Thus, the Ramsey solution calls for $x_{t+1} = \beta$ for $t \geq 0$ in order to support $c_t = \bar{c}$ for $t \geq 0$. Notice that the Ramsey outcome can be supported by any initial money growth x_0. It is only future money growth rates that must be equal to β in order to eliminate labor supply distortions that would otherwise arise from the cash-in-advance constraint if the return on money were to fall short of the return on bonds. The Ramsey outcome equalizes the returns on money and bonds; that is, it implements the Friedman rule with a zero net nominal interest rate.

It is instructive to highlight the inability of the Ramsey monetary policy to remove the distortions coming from monopolistic wage setting. Using the fact that the equilibrium real wage is unity, we solve for λ_{it} from equation (26.6.8) and substitute into equation (26.6.7a),

$$c_{it}^{\gamma-1} = \mu_{it} + \frac{1+\alpha}{1-\alpha} > 1. \qquad (26.6.14)$$

The left side of equation (26.6.14) is the marginal utility of consumption. Since technology (26.6.2) is linear in labor, the marginal utility of consumption should equal the marginal utility of leisure in a first-best allocation. But the right side of equation (26.6.14) exceeds unity, which is the marginal utility of leisure given preferences (26.6.1). While the Ramsey monetary policy succeeds in removing distortions from the cash-in-advance constraint by setting the Lagrange multiplier μ_{it} equal to zero, the policy cannot undo the distortion of monopolistic wage setting manifested in the "markup" $(1 + \alpha)/(1 - \alpha)$.[23] Notice that the Ramsey solution converges to the first-best allocation when the parameter α goes to zero, that is, when households' market power goes to zero.

To illustrate the time consistency problem, we now solve for the Ramsey plan when the initial nominal wages are taken as given, $w_{i0} = w_0 \in [\beta M_0, \bar{x} M_0]$. First, setting the initial period 0 aside, it is straightforward to show that the solution for $t \geq 1$ is the same as before. That is, the optimal policy calls for $x_{t+1} = \beta$ for $t \geq 1$ in order to support $c_t = \bar{c}$ for $t \geq 1$. Second, given w_0, the first-best outcome $c_0 = 1$ can be attained in the initial period by choosing $x_0 = w_0/M_0$. The resulting money supply $M_1 = w_0$ will then serve to transact $c_0 = 1$ at the equilibrium price $p_0 = w_0$. Specifically, firms are happy to hire any number of workers at the wage w_0 when the price of the good is $p_0 = w_0$. At the price $p_0 = w_0$, the goods market clears at full employment, since shoppers seek to spend their real balances $M_1/p_0 = 1$. The labor market also clears because workers are obliged to deliver the demanded $n_0 = 1$. Finally, money growth x_1 can be chosen freely and does not affect the real allocation of the Ramsey solution. The reason is that, because of the preset wage w_0, there cannot be any labor supply distortions at time 0 arising from a low return on money holdings between periods 0 and 1.

26.6.4. Credibility of the Friedman rule

Our comparison of the Ramsey equilibria with or without a preset initial wage w_0 hints at the government's temptation to create positive monetary surprises that will increase employment. We now ask if the Friedman rule is credible when the government lacks the commitment technology implicit in the Ramsey optimization problem. Can the Friedman rule be supported with a trigger strategy where a government deviation causes the economy to revert to the worst possible subgame perfect equilibrium?

Using the concepts and notation of chapter 23, we specify the objects of a strategy profile and state the definition of a subgame perfect equilibrium (SPE).

---

[23] The government would need to use fiscal instruments, that is, subsidies and taxation, to correct the distortion from monopolistically competitive wage setting.

Even though households possess market power with respect to their labor type, they remain atomistic vis-à-vis the government. We therefore stay within the framework of chapter 23 where the government behaves strategically, and the households' behavior can now be summarized as a "monopolistically competitive equilibrium" that responds nonstrategically to the government's choices. At every date t for all possible histories, a strategy of the households σ^h and a strategy of the government σ^g specify actions $\tilde{w}_t \in \tilde{W}$ and $x_t \in X \equiv [\beta, \bar{x}]$, respectively, where

$$\tilde{w}_t = \frac{w_t}{M_t}, \qquad \text{and} \qquad x_t = \frac{M_{t+1}}{M_t}.$$

That is, the actions multiplied by the beginning-of-period money supply M_t produce a nominal wage and a nominal money supply. (This scaling of nominal variables is used by Ireland, 1997, throughout his analysis, since the size of the nominal money supply at the beginning of a period has no significance *per se*.)

DEFINITION: A strategy profile $\sigma = (\sigma^h, \sigma^g)$ is a *subgame perfect equilibrium* if, for each $t \geq 0$ and each history $(\tilde{w}^{t-1}, x^{t-1}) \in \tilde{W}^t \times X^t$,

(1) Given the trajectory of money growth rates $\{x_{t-1+j} = x(\sigma|_{(\tilde{w}^{t-1}, x^{t-1})})_j\}_{j=1}^{\infty}$, the wage-setting outcome $\tilde{w}_t = \sigma_t^h (\tilde{w}^{t-1}, x^{t-1})$ constitutes a monopolistically competitive equilibrium.

(2) The government cannot strictly improve the households' welfare by deviating from $x_t = \sigma_t^g (\tilde{w}^{t-1}, x^{t-1})$, that is, by choosing some other money growth rate $\eta \in X$ with the implied continuation strategy profile $\sigma|_{(\tilde{w}^t; x^{t-1}, \eta)}$.

Besides changing to a "monopolistically competitive equilibrium," the main difference from Definition 6 of chapter 23 lies in requirement (1). The equilibrium in period t can no longer be stated in terms of an isolated government action at time t but requires the trajectory of the current and all future money growth rates, generated by the strategy profile $\sigma|_{(\tilde{w}^{t-1}, x^{t-1})}$. The monopolistically competitive equilibrium in requirement (1) is understood to be the perfect foresight equilibrium described previously. When the government is contemplating a deviation in requirement (2), the equilibrium is constructed as follows: In period t when the deviation takes place, equilibrium consumption c_t is a function of η and $\tilde{w}_t$ as implied by the cash-in-advance constraint at equality,

$$c_t = \frac{\eta M_t}{p_t} = \frac{\eta M_t}{w_t} = \frac{\eta}{\tilde{w}_t}, \tag{26.6.15}$$

where we use the equilibrium condition $p_t = w_t$. Starting in period $t+1$, the deviation has triggered a switch to a new perfect foresight equilibrium with a trajectory of money growth rates given by $\{x_{t+j} = x(\sigma|_{(\tilde{w}^t;x^{t-1},\eta)})_j\}_{j=1}^\infty$.

We conjecture that the worst SPE has $c_t = \underline{c}$ for all periods, and the candidate strategy profile $\hat{\sigma}$ is

$$\hat{\sigma}_t^h = \frac{\bar{x}}{\underline{c}} \quad \forall\, t\,, \quad \forall\, (\tilde{w}^{t-1}, x^{t-1});$$

$$\hat{\sigma}_t^g = \bar{x} \quad \forall\, t\,, \quad \forall\, (\tilde{w}^{t-1}, x^{t-1}).$$

The strategy profile instructs the government to choose the highest permissible money growth rate $\bar{x}$ for all periods and for all histories. Similarly, the households are instructed to set the nominal wages that would constitute a perfect foresight equilibrium when money growth will always be at its maximum. Thus, requirement (1) of an SPE is clearly satisfied. It remains to show that the government has no incentive to deviate. Since the continuation strategy profile is $\hat{\sigma}$ regardless of the history, the government needs only to find the best response in terms of the one-period return (26.6.13). After substituting the household's action $\tilde{w}_t = \bar{x}/\underline{c}$ into equation (26.6.15), we get $c_t = \underline{c}\eta/\bar{x}$, so the best response of the government is to follow the proposed strategy $\bar{x}$. We conclude that the strategy profile $\hat{\sigma}$ is indeed an SPE, and it is the worst, since $\underline{c}$ is the lower bound on consumption in any perfect foresight equilibrium.

We are now ready to address the credibility of the Friedman rule. The best chance for the Friedman rule to be credible is if a deviation triggers a reversion to the worst possible subgame perfect equilibrium given by $\hat{\sigma}$. The condition for credibility becomes

$$\frac{\frac{\bar{c}^\gamma}{\gamma} - \bar{c}}{1 - \beta} \geq \left(\frac{1}{\gamma} - 1\right) + \beta \frac{\frac{\underline{c}^\gamma}{\gamma} - \underline{c}}{1 - \beta}. \tag{26.6.16}$$

By following the Friedman rule, the government removes the labor supply distortion coming from a binding cash-in-advance constraint and keeps output at $\bar{c}$. By deviating from the Friedman rule, the government creates a positive monetary surprise that increases output to its efficient level of unity, thereby eliminating the distortion caused by monopolistically competitive wage setting as well. However, this deviation destroys the government's reputation, and the economy reverts to an equilibrium that induces the government to inflate at the highest possible rate thereafter, and output falls to $\underline{c}$. Hence, the Friedman rule is credible if and only if equation (26.6.16) holds.

The Friedman rule is the more likely to be credible, the higher is the exogenous upper bound on money growth $\bar{x}$, since $\underline{c}$ depends negatively on $\bar{x}$. In other words, a higher $\bar{x}$ translates into a larger penalty for deviating, so the government becomes more willing to adhere to the Friedman rule to avoid this penalty. In the limit when $\bar{x}$ becomes arbitrarily large, $\underline{c}$ approaches zero and condition (26.6.16) reduces to

$$\left(\frac{1-\alpha}{1+\alpha}\right)^{\frac{\gamma}{1-\gamma}} \left(\frac{1}{\gamma} - \frac{1-\alpha}{1+\alpha}\right) \geq (1-\beta)\left(\frac{1}{\gamma} - 1\right),$$

where we have used the expression for $\bar{c}$ in equations (26.6.12). The Friedman rule can be sustained for a sufficiently large value of β. The government has less incentive to deviate when households are patient and put a high weight on future outcomes. Moreover, the Friedman rule is credible for a sufficiently small value of α, which is equivalent to households having little market power. The associated small distortion from monopolistically competitive wage setting means that the potential welfare gain of a monetary surprise is also small, so the government is less tempted to deviate from the Friedman rule.

26.7. Concluding remarks

Besides shedding light on a number of monetary doctrines, this chapter has brought out the special importance of the initial date $t = 0$ in the analysis. This point is especially pronounced in Woodford's (1995) model where the initial interest-bearing government debt B_0 is not indexed but rather denominated in nominal terms. So, although the construction of a perfect foresight equilibrium ensures that all future issues of nominal bonds will *ex post* yield the real rates of return that are needed to entice the households to hold these bonds, the realized real return on the initial nominal bonds can be anything, depending on the price level p_0. Activities at the initial date were also important when we considered dynamic optimal taxation in chapter 16.

Monetary issues are also discussed in other chapters of the book. Chapters 9 and 18 study money in overlapping generations models and Bewley models, respectively. Chapters 27 and 28 present other explicit environments that give rise to a positive value of fiat money: Townsend's turnpike model and the Kiyotaki-Wright search model.

Exercises

Exercise 26.1 **Why deficits in Italy and Brazil were once extraordinary proportions of GDP**

The government's budget constraint can be written as

$$g_t - \tau_t + \frac{b_t}{R_{t-1}}(R_{t-1} - 1) = \frac{b_{t+1}}{R_t} - \frac{b_t}{R_{t-1}} + \frac{M_{t+1}}{p_t} - \frac{M_t}{p_t}. \tag{1}$$

The left side is the real gross-of-interest government deficit; the right side is change in the real value of government liabilities between $t-1$ and t.

Government budgets often report the *nominal* gross-of-interest government deficit, defined as

$$p_t(g_t - \tau_t) + p_t b_t \left(1 - \frac{1}{R_{t-1}p_t/p_{t-1}}\right),$$

and their ratio to nominal GNP, $p_t y_t$, namely,

$$\left[(g_t - \tau_t) + b_t\left(1 - \frac{1}{R_{t-1}p_t/p_{t-1}}\right)\right] / y_t.$$

For countries with a large b_t (e.g., Italy), this number can be very big even with a moderate rate of inflation. For countries with a rapid inflation rate, like Brazil in 1993, this number sometimes comes in at 30 percent of GDP. Fortunately, this number overstates the magnitude of the government's "deficit problem," and there is a simple adjustment to the interest component of the deficit that renders a more accurate picture of the problem. In particular, notice that the real values of the interest component of the real and nominal deficits are related by

$$b_t\left(1 - \frac{1}{R_{t-1}}\right) = \alpha_t b_t \left(1 - \frac{1}{R_{t-1}p_t/p_{t-1}}\right),$$

where

$$\alpha_t = \frac{R_{t-1} - 1}{R_{t-1} - p_{t-1}/p_t}.$$

Thus, we should multiply the real value of nominal interest payments $b_t[1 - p_{t-1}/(R_{t-1}p_t)]$ by α_t to get the real interest component of the debt that appears on the left side of equation (1).

a. Compute α_t for a country that has a b_t/y ratio of .5, a gross real interest rate of 1.02, and a zero net inflation rate.

b. Compute α for a country that has a b_t/y ratio of .5, a gross real interest rate of 1.02, and a 100 percent per year net inflation rate.

Exercise 26.2 **A strange example of Brock (1974)**

Consider an economy consisting of a government and a representative household. There is one consumption good, which is not produced and not storable. The exogenous supply of the good at time $t \geq 0$ is $y_t = y > 0$. The household owns the good. At time t the representative household's preferences are ordered by

$$\sum_{t=0}^{\infty} \beta^t \{\ln c_t + \gamma \ln(m_{t+1}/p_t)\}, \tag{1}$$

where c_t is the household's consumption at t, p_t is the price level at t, and m_{t+1}/p_t is the real balances that the household carries over from time t to $t+1$. Assume that $\beta \in (0,1)$ and $\gamma > 0$. The household maximizes equation (1) over choices of $\{c_t, m_{t+1}\}$ subject to the sequence of budget constraints

$$c_t + m_{t+1}/p_t = y_t - \tau_t + m_t/p_t, \quad t \geq 0, \tag{2}$$

where τ_t is a lump-sum tax due at t. The household faces the price sequence $\{p_t\}$ as a price taker and has given initial value of nominal balances m_0.

At time t the government faces the budget constraint

$$g_t = \tau_t + (M_{t+1} - M_t)/p_t, \quad t \geq 0, \tag{3}$$

where M_t is the amount of currency that the government has outstanding at the beginning of time t and g_t is government expenditures at time t. In equilibrium, we require that $M_t = m_t$ for all $t \geq 0$. The government chooses sequences of $\{g_t, \tau_t, M_{t+1}\}_{t=0}^{\infty}$ subject to the budget constraints (3) being satisfied for all $t \geq 0$ and subject to the given initial value $M_0 = m_0$.

a. Define a *competitive equilibrium*.

For the remainder of this problem assume that $g_t = g < y$ for all $t \geq 0$, and that $\tau_t = \tau$ for all $t \geq 0$. Define a *stationary equilibrium* as an equilibrium in which the rate of return on currency is constant for all $t \geq 0$.

b. Find conditions under which there exists a stationary equilibrium for which $p_t > 0$ for all $t \geq 0$. Derive formulas for real balances and the rate of return on currency in that equilibrium, given that it exists. Is the stationary equilibrium unique?

c. Find a first-order difference equation in the equilibrium level of real balances $h_t = M_{t+1}/p_t$ whose satisfaction ensures equilibrium (possibly nonstationary).

d. Show that there is a fixed point of this difference equation with positive real balances, provided that the condition that you derived in part b is satisfied. Show that this fixed point agrees with the level of real balances that you computed in part b.

Exercise 26.3 **Optimal inflation tax in a cash-in-advance model**

Consider the version of Ireland's (1997) model described in the text, but assume perfect competition (i.e., $\alpha = 0$) with flexible market-clearing wages. Suppose now that the government must finance a constant amount of purchases g in each period by levying flat-rate labor taxes and raising seigniorage. Solve the optimal taxation problem under commitment.

Exercise 26.4 **Deficits, inflation, and anticipated monetary shocks**, donated by Rodolfo Manuelli

Consider an economy populated by a large number of identical individuals. Preferences over consumption and leisure are given by

$$\sum_{t=0}^{\infty} \beta^t c_t^\alpha \ell_t^{1-\alpha},$$

where $0 < \alpha < 1$. Assume that leisure is positively related – this is just a reduced form of a shopping-time model – to the stock of real money balances, and negatively related to a measure of transactions:

$$\ell_t = A(m_{t+1}/p_t)/c_t^\eta, \quad A > 0,$$

and $\alpha - \eta(1 - \alpha) > 0$. Each individual owns a tree that drops y units of consumption per period (dividends). There is a government that issues one-period real bonds, money, and collects taxes (lump-sum) to finance spending.

Per capita spending is equal to g. Thus, consumption equals $c = y - g$. The government's budget constraint is:

$$g_t + B_t = \tau_t + B_{t+1}/R_t + (M_{t+1} - M_t)/p_t.$$

Let the rate of return on money be $R_{mt} = p_t/p_{t+1}$. Let the nominal interest rate at time t be $1 + i_t = R_t p_{t+1}/p_t = R_t \pi_t$.

a. Derive the demand for money, and show that it decreases with the nominal interest rate.

b. Suppose that the government policy is such that $g_t = g$, $B_t = B$ and $\tau_t = \tau$. Prove that the real interest rate, R, is constant and equal to the inverse of the discount factor.

c. Define the deficit as d, where $d = g + (B/R)(R-1) - \tau$. What is the highest possible deficit that can be financed in this economy? An economist claims that increases in d, which leave g unchanged, will result in increases in the inflation rate. Discuss this view.

d. Suppose that the economy is open to international capital flows and that the world interest rate is $R^* = \beta^{-1}$. Assume that $d = 0$, and that $M_t = M$. At $t = T$, the government increases the money supply to $M' = (1 + \mu)M$. This increase in the money supply is used to purchase (government) bonds. This, of course, results in a smaller deficit at $t > T$. (In this case, it will result in a surplus.) However, the government also announces its intention to cut taxes (starting at $T+1$) to bring the deficit back to zero. Argue that this open market operation will have the effect of increasing prices at $t = T$ by μ; $p' = (1 + \mu)p$, where p is the price level from $t = 0$ to $t = T - 1$.

e. Consider the same setting as in d. Suppose now that the open market operation is announced at $t = 0$ (it still takes place at $t = T$). Argue that prices will increase at $t = 0$ and, in particular, that the rate of inflation between $T - 1$ and T will be less than $1 + \mu$.

Exercise 26.5 **Interest elasticity of the demand for money**, donated by Rodolfo Manuelli

Consider an economy in which the demand for money satisfies

$$m_{t+1}/p_t = F(c_t, R_{mt}/R_t),$$

where $R_{mt} = p_t/p_{t+1}$ and R_t is the one-period interest rate. Consider the following open market operation: At $t = 0$, the government sells bonds and "destroys" the money it receives in exchange for those bonds. No other real variables, e.g., government spending or taxes, are changed. Find conditions on the income elasticity of the demand for money such that the decrease in money balances at $t = 0$ results in an increase in the price level at $t = 0$.

Exercise 26.6 **Dollarization**, donated by Rodolfo Manuelli

In recent years, several countries, e.g., Argentina and countries hit by the Asian crisis, have considered the possibility of giving up their currencies in favor of the U.S. dollar. Consider a country, say A, with deficit d and inflation rate $\pi = 1/R_m$. Output and consumption are constant, and hence the real interest rate is fixed, with $R = \beta^{-1}$. The (gross-of-interest-payments) deficit is d, with

$$d = g - \tau + (B/R)(R - 1).$$

Let the demand for money be $m_{t+1}/p_t = F(c_t, R_{mt}/R_t)$, and assume that $c_t = y - g$. Thus, the steady-state government budget constraint is

$$d = F(y - g, \beta R_m)(1 - R_m) > 0.$$

Assume that the country is considering, at $t = 0$, the retirement of its money in exchange for dollars. The government promises to give to each person who brings a "peso" to the Central Bank $1/e$ dollars, where e is the exchange rate (in pesos per dollar) between the country's currency and the U.S. dollar. Assume that the U.S. inflation rate (before and after the switch) is given and equal to $\pi^* = 1/R_m^* < \pi$, and that the country is on the "good" part of the Laffer curve.

a. If you are advising the government of A, how much would you say that it should demand from the U.S. government to make the switch? Why?

b. After the dollarization takes place, the government understands that it needs to raise taxes. Economist 1 argues that the increase in taxes (on a per period basis) will equal the loss of revenue from inflation – $F(y - g, \beta R_m)(1 - R_m)$ – while Economist 2 claims that this is an overestimate. More precisely, he or she claims that if the government is a good negotiator vis-à-vis the U.S. government, taxes need only increase by $F(y - g, \beta R_m)(1 - R_m) - F(y - g, \beta R_m^*)(1 - R_m^*)$ per period. Discuss these two views.

Exercise 26.7 **Currency boards**, donated by Rodolfo Manuelli

In the last few years, several countries, e.g., Argentina (1991), Estonia (1992), Lithuania (1994), Bosnia (1997) and Bulgaria (1997), have adopted the currency board model of monetary policy. In a nutshell, a currency board is a commitment on the part of the country to fully back its domestic currency with foreign-denominated assets. For simplicity, assume that the foreign asset is the U.S. dollar.

The government's budget constraint is given by

$$g_t + B_t + B_{t+1}^* e/(Rp_t) = \tau_t + B_{t+1}/R + B_t^* e/p_t + (M_{t+1} - M_t)/p_t,$$

where B_t^* is the stock of one-period bonds, denominated in dollars, held by this country, e is the exchange rate (pesos per dollar), and $1/R$ is the price of one-period bonds (both domestic and dollar denominated). Note that the budget constraint equates the real value of income and liabilities in units of consumption goods.

The currency board "contract" requires that the money supply be fully backed. One interpretation of this rule is that the domestic money supply is

$$M_t = eB_t^*.$$

Thus, the right side is the local currency value of foreign reserves (in bonds) held by the government, while the left side is the stock of money. Finally, let the law of one price hold: $p_t = ep_t^*$, where p_t^* is the foreign (U.S.) price level.

a. Assume that $B_t = B$, and that foreign inflation is zero, $p_t^* = p^*$. Show that even in this case, the properties of the demand for money – which you may take to be given by $F(y - g, \beta R_m)$ – are important in determining total revenue. In particular, explain how a permanent increase in y, income per capita, allows the government to lower taxes (permanently).

b. Assume that $B_t = B$. Let foreign inflation be positive, that is, $\pi^* > 1$. In this case, the price in dollars of a one-period dollar-denominated bond is $1/(R\pi^*)$. Go as far as you can describing the impact of foreign inflation on domestic inflation, and on per capita taxes, τ.

c. Assume that $B_t = B$. Go as far as you can describing the effects of a once-and-for-all surprise devaluation, i.e., an unexpected and permanent increase in e, on the level of per capita taxes.

Exercise 26.8   **Growth and inflation**, donated by Rodolfo Manuelli

Consider an economy populated by identical individuals with instantaneous utility function given by

$$u(c, \ell) = [c^\varphi \ell^{1-\varphi}]^{(1-\sigma)}/(1-\sigma).$$

Assume that shopping time is given by $s_t = \psi c_t/(m_{t+1}/p_t)$. Assume that in this economy, income grows exogenously at the rate $\gamma > 1$. Thus, at time t, $y_t = \gamma^t y$. Assume that government spending also grows at the same rate, $g_t = \gamma^t g$. Finally, $c_t = y_t - g_t$.

a. Show that for this specification, if the demand for money at t is $x = m_{t+1}/p_t$, then the demand at $t + 1$ is γx. Thus, the demand for money grows at the same rate as the economy.

b. Show that the real rate of interest depends on the growth rate. (You may assume that ℓ is constant for this calculation.)

c. Argue that even for monetary policies that keep the price level constant, that is, $p_t = p$ for all t, the government raises positive amounts of revenue from printing money. Explain.

d. Use your finding in c to discuss why, following monetary reforms that generate big growth spurts, many countries manage to "monetize" their economies (this is just jargon for increases in the money supply) without generating inflation.

Chapter 27
Credit and Currency

27.1. Credit and currency with long-lived agents

This chapter describes Townsend's (1980) turnpike model of money and puts it to work. The model uses a particular pattern of heterogeneity of endowments and locations to create a demand for currency. The model is more primitive than the shopping time model of chapter 26. As with the overlapping generations model, the turnpike model starts from a setting in which diverse intertemporal endowment patterns across agents prompt borrowing and lending. If something prevents loan markets from operating, it is possible that an unbacked currency can play a role in helping agents smooth their consumption over time. Following Townsend, we shall eventually appeal to locational heterogeneity as the force that causes loan markets to fail in this way.

The turnpike model can be viewed as a simplified version of the stochastic model proposed by Truman Bewley (1980). We use the model to study a number of interrelated issues and theories, including (1) a permanent income model of consumption, (2) a Ricardian doctrine that government borrowing and taxes have equivalent economic effects, (3) some restrictions on the operation of private loan markets needed in order that unbacked currency be valued, (4) a theory of inflationary finance, (5) a theory of the optimal inflation rate and the optimal behavior of the currency stock over time, (6) a "legal restrictions" theory of inflationary finance, and (7) a theory of exchange rate indeterminacy.[1]

---

[1] Some of the analysis in this chapter follows Manuelli and Sargent (2002). Also see Chatterjee and Corbae (1996) and Ireland (1994) for analyses of policies within a turnpike environment.

27.2. Preferences and endowments

There is one consumption good. It cannot be produced or stored. The total amount of goods available each period is constant at N. There are $2N$ households, divided into equal numbers N of two types, according to their endowment sequences. The two types of households, dubbed *odd* and *even*, have endowment sequences

$$\{y_t^o\}_{t=0}^\infty = \{1, 0, 1, 0, \ldots\},$$
$$\{y_t^e\}_{t=0}^\infty = \{0, 1, 0, 1, \ldots\}.$$

Households of both types order consumption sequences $\{c_t^h\}$ according to the common utility function

$$U = \sum_{t=0}^\infty \beta^t u(c_t^h),$$

where $\beta \in (0, 1)$, and $u(\cdot)$ is twice continuously differentiable, increasing, and strictly concave, and satisfies

$$\lim_{c \downarrow 0} u'(c) = +\infty. \qquad (27.2.1)$$

27.3. Complete markets

As a benchmark, we study a version of the economy with complete markets. Later, we shall more or less arbitrarily shut down many of the markets to make room for money.

27.3.1. A Pareto problem

Consider the following Pareto problem: Let $\theta \in [0,1]$ be a weight indexing how much a social planner likes odd agents. The problem is to choose consumption sequences $\{c_t^o, c_t^e\}_{t=0}^{\infty}$ to maximize

$$\theta \sum_{t=0}^{\infty} \beta^t u(c_t^o) + (1-\theta) \sum_{t=0}^{\infty} \beta^t u(c_t^e), \tag{27.3.1}$$

subject to

$$c_t^e + c_t^o = 1, \quad t \geq 0. \tag{27.3.2}$$

The first-order conditions are

$$\theta u'(c_t^o) - (1-\theta)u'(c_t^e) = 0.$$

Substituting the constraint $(27.3.2)$ into this first-order condition and rearranging gives the condition

$$\frac{u'(c_t^o)}{u'(1-c_t^o)} = \frac{1-\theta}{\theta}. \tag{27.3.3}$$

Since the right side is independent of time, the left must be also, so that condition $(27.3.3)$ determines the one-parameter family of optimal allocations

$$c_t^o = c^o(\theta), \quad c_t^e = 1 - c^o(\theta).$$

27.3.2. A complete markets equilibrium

A household takes the price sequence $\{q_t^0\}$ as given and chooses a consumption sequence to maximize $\sum_{t=0}^{\infty} \beta^t u(c_t)$ subject to the budget constraint

$$\sum_{t=0}^{\infty} q_t^0 c_t \leq \sum_{t=0}^{\infty} q_t^0 y_t.$$

The household's Lagrangian is

$$L = \sum_{t=0}^{\infty} \beta^t u(c_t) + \mu \sum_{t=0}^{\infty} q_t^0 (y_t - c_t),$$

where μ is a nonnegative Lagrange multiplier. The first-order conditions for the household's problem are

$$\beta^t u'(c_t) \leq \mu q_t^0, \quad = \text{ if } c_t > 0.$$

DEFINITION 1: A *competitive equilibrium* is a price sequence $\{q_t^0\}_{t=0}^{\infty}$ and an allocation $\{c_t^o, c_t^e\}_{t=0}^{\infty}$ that have the property that (a) given the price sequence, the allocation solves the optimum problem of households of both types; and (b) $c_t^o + c_t^e = 1$ for all $t \geq 0$.

To find an equilibrium, we have to produce an allocation and a price system for which we can verify that the first-order conditions of both households are satisfied. We start with a guess inspired by the constant-consumption property of the Pareto optimal allocation. We guess that $c_t^o = c^o, c_t^e = c^e$ $\forall t$, where $c^e + c^o = 1$. This guess and the first-order condition for the odd agents imply

$$q_t^0 = \frac{\beta^t u'(c^o)}{\mu^o},$$

or

$$q_t^0 = q_0^0 \beta^t, \tag{27.3.4}$$

where we are free to normalize by setting $q_0^0 = 1$. For odd agents, the right side of the budget constraint evaluated at the prices given in equation (27.3.4) is then

$$\frac{1}{1 - \beta^2},$$

and for even households it is

$$\frac{\beta}{1 - \beta^2}.$$

The left side of the budget constraint evaluated at these prices is

$$\frac{c^i}{1 - \beta}, \quad i = o, e.$$

For both of the budget constraints to be satisfied with equality, we evidently require that

$$c^o = \frac{1}{\beta + 1}$$
$$c^e = \frac{\beta}{\beta + 1}. \tag{27.3.5}$$

The price system given by equation (27.3.4) and the constant-over-time allocations given by equations (27.3.5) are a competitive equilibrium.

Notice that the competitive equilibrium allocation corresponds to a particular Pareto optimal allocation.

27.3.3. Ricardian proposition

We temporarily add a government to the model. The government levies lump-
sum taxes on agents of type $i = o, e$ at time t of τ_t^i. The government uses the
proceeds to finance a constant level of government purchases of $G \in (0, 1)$ each
period t. Consumer i's budget constraint is

$$\sum_{t=0}^{\infty} q_t^0 c_t^i \leq \sum_{t=0}^{\infty} q_t^0 (y_t^i - \tau_t^i).$$

The government's budget constraint is

$$\sum_{t=0}^{\infty} q_t^0 G = \sum_{i=o,e} \sum_{t=0}^{\infty} q_t^0 \tau_t^i.$$

We modify Definition 1 as follows:

DEFINITION 2: A *competitive equilibrium* is a price sequence $\{q_t^0\}_{t=0}^{\infty}$, a tax
system $\{\tau_t^o, \tau_t^e\}_{t=0}^{\infty}$, and an allocation $\{c_t^o, c_t^e, G_t\}_{t=0}^{\infty}$ such that given the price
system and the tax system the following conditions hold: (a) the allocation
solves each consumer's optimum problem; (b) the government budget constraint
is satisfied for all $t \geq 0$; and (c) $N(c_t^o + c_t^e) + G_t = N$ for all $t \geq 0$.

Let the present value of the taxes imposed on consumer i be $\tau^i \equiv \sum_{t=0}^{\infty} q_t^0 \tau_t^i$.
Then it is straightforward to verify that the equilibrium price system is still
equation (27.3.4) and that equilibrium allocations are

$$c^o = \frac{1}{\beta + 1} - \tau^o (1 - \beta)$$

$$c^e = \frac{\beta}{\beta + 1} - \tau^e (1 - \beta).$$

This equilibrium features a "Ricardian proposition":

RICARDIAN PROPOSITION: The equilibrium is invariant to changes in the *timing* of tax collections that leave unaltered the present value of lump-sum taxes assigned to each agent.

27.3.4. Loan market interpretation

Define total time t tax collections as $\tau_t = \sum_{i=o,e} \tau_t^i$, and write the government's budget constraint as

$$(G_0 - \tau_0) = \sum_{t=1}^{\infty} \frac{q_t^0}{q_0^0} (\tau_t - G_t) \equiv B_1,$$

where B_1 can be interpreted as government debt issued at time 0 and due at time 1. Notice that B_1 equals the present value of the future (i.e., from time 1 onward) government *surpluses* $(\tau_t - G_t)$. The government's budget constraint can also be represented as

$$\frac{q_0^0}{q_1^0}(G_0 - \tau_0) + (G_1 - \tau_1) = \sum_{t=2}^{\infty} \frac{q_t^0}{q_1^0}(\tau_t - G_t) \equiv B_2,$$

or

$$R_1 B_1 + (G_1 - \tau_1) = B_2,$$

where $R_1 = \frac{q_0^0}{q_1^0}$ is the gross rate of return between time 0 and time 1, measured in time 1 consumption goods per unit of time 0 consumption good. More generally, we can represent the government's budget constraint by the sequence of budget constraints

$$R_t B_t + (G_t - \tau_t) = B_{t+1}, \quad t \geq 0,$$

subject to the boundary condition $B_0 = 0$. In the equilibrium computed here, $R_t = \beta^{-1}$ for all $t \geq 1$.

Similar manipulations of consumers' budget constraints can be used to express them in terms of sequences of one-period budget constraints. That no opportunities are lost to the government or the consumers by representing the budget sets in this way lies behind the following fact: The Arrow-Debreu allocation in this economy can be implemented with a sequence of one-period loan markets.

In the following section, we shut down *all* loan markets, and also set government expenditures $G = 0$.

27.4. A monetary economy

We keep preferences and endowment patterns as they were in the preceding economy, but we rule out all intertemporal trades achieved through borrowing and lending or trading of future-dated consumptions. We replace complete markets with a fiat money mechanism. At time 0, the government endows each of the N even agents with M/N units of an unbacked or inconvertible currency. Odd agents are initially endowed with zero units of the currency. Let p_t be the time t price level, denominated in dollars per time t consumption good. We seek an equilibrium in which currency is valued ($p_t < +\infty \ \forall t \geq 0$) and in which each period agents not endowed with goods pass currency to agents who are endowed with goods. Contemporaneous exchanges of currency for goods are the only exchanges that we, the model builders, permit. (Later, Townsend will give us a defense or reinterpretation of this high-handed shutting down of markets.)

Given the sequence of prices $\{p_t\}_{t=0}^{\infty}$, the household's problem is to choose nonnegative sequences $\{c_t, m_t\}_{t=0}^{\infty}$ to maximize $\sum_{t=0}^{\infty} \beta^t u(c_t)$ subject to

$$m_t + p_t c_t \leq p_t y_t + m_{t-1}, \quad t \geq 0, \tag{27.4.1}$$

where m_t is currency held from t to $t+1$. Form the household's Lagrangian

$$L = \sum_{t=0}^{\infty} \beta^t \{u(c_t) + \lambda_t (p_t y_t + m_{t-1} - m_t - p_t c_t)\},$$

where $\{\lambda_t\}$ is a sequence of nonnegative Lagrange multipliers. The household's first-order conditions for c_t and m_t, respectively, are

$$u'(c_t) \leq \lambda_t p_t, \quad = \text{ if } c_t > 0,$$

$$-\lambda_t + \beta \lambda_{t+1} \leq 0, \quad = \text{ if } m_t > 0.$$

Substituting the first condition at equality into the second gives

$$\frac{\beta u'(c_{t+1})}{p_{t+1}} \leq \frac{u'(c_t)}{p_t}, \quad = \text{ if } m_t > 0. \tag{27.4.2}$$

DEFINITION 3: A *competitive equilibrium* is an allocation $\{c_t^o, c_t^e\}_{t=0}^\infty$, nonnegative money holdings $\{m_t^o, m_t^e\}_{t=-1}^\infty$, and a nonnegative price level sequence $\{p_t\}_{t=0}^\infty$ such that (a) given the price level sequence and (m_{-1}^o, m_{-1}^e); the allocation solves the optimum problems of both types of households; and (b) $c_t^o + c_t^e = 1$, $m_{t-1}^o + m_{t-1}^e = M/N$, for all $t \geq 0$.

The periodic nature of the endowment sequences prompts us to guess the following two-parameter form of stationary equilibrium:

$$\{c_t^o\}_{t=0}^\infty = \{c_0, 1 - c_0, c_0, 1 - c_0, \ldots\},$$

$$\{c_t^e\}_{t=0}^\infty = \{1 - c_0, c_0, 1 - c_0, c_0, \ldots\},$$

$$(27.4.3)$$

and $p_t = p$ for all $t \geq 0$. To determine the two undetermined parameters (c_0, p), we use the first-order conditions and budget constraint of the odd agent at time 0. His endowment sequence for periods 0 and 1, $(y_0^o, y_1^o) = (1, 0)$, and the Inada condition (27.2.1) ensure that both of his first-order conditions at time 0 will hold with equality. That is, his desire to set $c_0^o > 0$ can be met by consuming some of the endowment y_0^o, and the only way for him to secure consumption in the following period 1 is to hold strictly positive money holdings $m_0^o > 0$. From his first-order conditions at equality, we obtain

$$\frac{\beta u'(1 - c_0)}{p} = \frac{u'(c_0)}{p},$$

which implies that c_0 is to be determined as the root of

$$\beta - \frac{u'(c_0)}{u'(1 - c_0)} = 0. \qquad (27.4.4)$$

Because $\beta < 1$, it follows that $c_0 \in (\frac{1}{2}, 1)$. To determine the price level, we use the odd agent's budget constraint at $t = 0$, evaluated at $m_{-1}^o = 0$ and $m_0^o = M/N$, to get

$$pc_0 + M/N = p \cdot 1,$$

or

$$p = \frac{M}{N(1 - c_0)}. \qquad (27.4.5)$$

See Figure 27.4.1 for a graphical determination of c_0.

From equation (27.4.4), it follows that for $\beta < 1$, $c_0 > .5$ and $1 - c_0 < .5$. Thus, both types of agents experience fluctuations in their consumption sequences in this monetary equilibrium. Because Pareto optimal allocations have constant consumption sequences for each type of agent, this equilibrium allocation is not Pareto optimal.

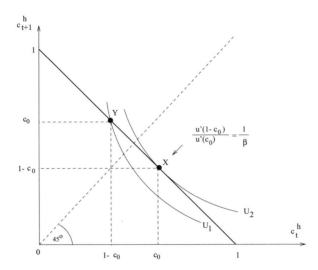

Figure 27.4.1: The trade-off between time t and time $(t+1)$ consumption faced by agent $o(e)$ in equilibrium for t even (odd). For t even, $c^o_t = c_0$, $c^o_{t+1} = 1 - c_0$, $m^o_t = p(1 - c_0)$, and $m^o_{t+1} = 0$. The slope of the indifference curve at X is $-u'(c^h_t)/\beta u'(c^h_{t+1}) = -u'(c_0)/\beta u'(1 - c_0) = -1$, and the slope of the indifference curve at Y is $-u'(1 - c_0)/\beta u'(c_0) = -1/\beta^2$.

27.5. Townsend's "turnpike" interpretation

The preceding analysis of currency is artificial in the sense that it depends entirely on our having arbitrarily ruled out the existence of markets for private loans. The physical setup of the model itself provided no reason for those loan markets not to exist, and indeed good reasons for them to exist. In addition, for many questions that we want to analyze, we want a model in which private loans and currency coexist, with currency being valued.[2]

Robert Townsend has proposed a model whose mathematical structure is identical with the preceding model, but in which a global market in private loans cannot emerge because agents are spatially separated. Townsend's setup

---

[2] In the United States today, for example, M_1 consists of the sum of demand deposits (a part of which is backed by commercial loans and another, smaller part of which is backed by reserves or currency) and currency held by the public. Thus, M_1 is not interpretable as the m in our model.

can accommodate local markets for private loans, so that it meets the objections to the model that we have expressed. But first we will focus on a version of Townsend's model where local credit markets cannot emerge, which will be mathematically equivalent to our model above.

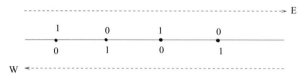

Figure 27.5.1: Endowment pattern along a Townsend turnpike. The turnpike is of infinite extent in each direction, and has equidistant trading posts. Each trading post has equal numbers of east-heading and west-heading agents. At each trading post (the black dots) each period, for each east-heading agent there is a west-heading agent with whom he would like to borrow or lend. But itineraries rule out the possibility of repayment.

The economy starts at time $t = 0$, with N east-heading migrants and N west-heading migrants physically located at each of the integers along a "turnpike" of infinite length extending in both directions. Each of the integers $n = 0, \pm 1, \pm 2, \ldots$ is a trading post number. Agents can trade the one good only with agents at the trading post at which they find themselves at a given date. An east-heading agent at an even-numbered trading post is endowed with one unit of the consumption good, and an odd-numbered trading post has an endowment of zero units (see Figure 27.5.1). A west-heading agent is endowed with zero units at an even-numbered trading post and with one unit of the consumption good at an odd-numbered trading post. Finally, at the end of each period, each east-heading agent moves one trading post to the east, whereas each west-heading agent moves one trading post to the west. The turnpike along which the trading posts are located is of infinite length in each direction, implying that the east-heading and west-heading agents who are paired at time t will never meet again. This feature means that there can be no private debt between agents moving in opposite directions. An IOU between agents moving in opposite directions can never be collected because a potential lender never meets the potential borrower again, nor does the lender meet anyone who ever meets the potential borrower, and so on, ad infinitum.

Let an agent who is endowed with one unit of the good $t = 0$ be called an agent of type o and an agent who is endowed with zero units of the good at $t = 0$ be called an agent of type e. Agents of type h have preferences summarized by $\sum_{t=0}^{\infty} \beta^t u(c_t^h)$. Finally, start the economy at time 0 by having each agent of type e endowed with $m_{-1}^e = m$ units of unbacked currency and each agent of type o endowed with $m_{-1}^o = 0$ units of unbacked currency.

With the symbols thus reinterpreted, this model involves precisely the same mathematics as that which was analyzed earlier. Agents' spatial separation and their movements along the turnpike have been set up to produce a physical reason that a global market in private loans cannot exist. The various propositions about the equilibria of the model and their optimality that were already proved apply equally to the turnpike version.[3],[4] Thus, in Townsend's version of the model, spatial separation is the "friction" that provides a potential social role for a valued unbacked currency. The spatial separation of agents and their endowment patterns give a setting in which private loan markets are limited by the need for people who trade IOUs to be linked together, if only indirectly, recurrently over time and space.

---

[3] A version of the model could be constructed in which local private markets for loans coexist with valued unbacked currency. To build such a model, one would assume some heterogeneity in the time patterns of the endowment of agents who are located at the same trading post and are headed in the same direction. If half of the east-headed agents located at trading post i at time t have present and future endowment pattern $y_t^h = (\alpha, \gamma, \alpha, \gamma \ldots)$, for example, whereas the other half of the east-headed agents have $(\gamma, \alpha, \gamma, \alpha, \ldots)$ with $\gamma \neq \alpha$, then there is room for local private loans among this cohort of east-headed agents. Whether or not there exists an equilibrium with valued currency depends on how nearly Pareto optimal the equilibrium with local loan markets is.

[4] Narayana Kocherlakota (1998) has analyzed the frictions in the Townsend turnpike and overlapping generations model. By permitting agents to use history-dependent decision rules, he has been able to support optimal allocations with the equilibrium of a gift-giving game. Those equilibria leave no room for valued fiat currency. Thus, Kocherlakota's view is that the frictions that give valued currency in the Townsend turnpike must include the restrictions on the strategy space that Townsend implicitly imposed.

27.6. The Friedman rule

Friedman's proposal to pay interest on currency by engineering a deflation can be used to solve for a Pareto optimal allocation in this economy. Friedman's proposal is to decrease the currency stock by means of lump-sum taxes at a properly chosen rate. Let the government's budget constraint be

$$M_t = (1 + \tau)M_{t-1}.$$

There are N households of each type. At time t, the government transfers or taxes nominal balances in amount $\tau M_{t-1}/(2N)$ to each household of each type. The total transfer at time t is thus τM_{t-1}, because there are $2N$ households receiving transfers.

The household's time t budget constraint becomes

$$p_t c_t + m_t \leq p_t y_t + \frac{\tau}{2} \frac{M_{t-1}}{N} + m_{t-1}.$$

We guess an equilibrium allocation of the same periodic pattern (27.4.3). For the price level, we make the "quantity theory" guess $M_t/p_t = k$, where k is a constant. Substituting this guess into the government's budget constraint gives

$$\frac{M_t}{p_t} = (1 + \tau)\frac{M_{t-1}}{p_{t-1}}\frac{p_{t-1}}{p_t}$$

or

$$k = (1 + \tau)k\frac{p_{t-1}}{p_t},$$

or

$$p_t = (1 + \tau)p_{t-1}, \tag{27.6.1}$$

which is our guess for the price level.

Substituting the price level guess and the allocation guess into the odd agent's first-order condition (27.4.2) at $t = 0$ and rearranging shows that c_0 is now the root of

$$\frac{1}{(1 + \tau)} - \frac{u'(c_0)}{\beta u'(1 - c_0)} = 0. \tag{27.6.2}$$

The price level at time $t = 0$ can be determined by evaluating the odd agent's time 0 budget constraint at $m^o_{-1} = 0$ and $m^o_0 = M_0/N = (1+\tau)M_{-1}/N$, with the result that

$$(1 - c_0)p_0 = \frac{M_{-1}}{N}\left(1 + \frac{\tau}{2}\right).$$

Finally, the allocation guess must also satisfy the even agent's first-order condition $(27.4.2)$ at $t = 0$ but not necessarily with equality, since the stationary equilibrium has $m_0^e = 0$. After substituting $(c_0^e, c_1^e) = (1 - c_0, c_0)$ and $(27.6.1)$ into $(27.4.2)$, we have

$$\frac{1}{1 + \tau} \leq \frac{u'(1 - c_0)}{\beta u'(c_0)}. \tag{27.6.3}$$

The substitution of $(27.6.2)$ into $(27.6.3)$ yields a restriction on the set of periodic allocations of type $(27.4.3)$ that can be supported as one of our stationary monetary equilibria,

$$\left[\frac{u'(c_0)}{u'(1 - c_0)}\right]^2 \leq 1 \quad \Longrightarrow \quad c_0 \geq 0.5.$$

This restriction on c_0, together with $(27.6.2)$, implies a corresponding restriction on the set of permissible monetary/fiscal policies, $1 + \tau \geq \beta$.

27.6.1. Welfare

For allocations of the class $(27.4.3)$, the utility functionals of odd and even agents, respectively, take values that are functions of the single parameter c_0, namely,

$$U^o(c_0) = \frac{u(c_0) + \beta u(1 - c_0)}{1 - \beta^2},$$

$$U^e(c_0) = \frac{u(1 - c_0) + \beta u(c_0)}{1 - \beta^2}.$$

Both expressions are strictly concave in c_0, with derivatives

$$U^{o\prime}(c_0) = \frac{u'(c_0) - \beta u'(1 - c_0)}{1 - \beta^2},$$

$$U^{e\prime}(c_0) = \frac{-u'(1 - c_0) + \beta u'(c_0)}{1 - \beta^2}.$$

The Inada condition $(27.2.1)$ ensures strictly interior maxima with respect to c_0. For the odd agents, the preferred c_0 satisfies $U^{o\prime}(c_0) = 0$, or

$$\frac{u'(c_0)}{\beta u'(1 - c_0)} = 1, \tag{27.6.4}$$

which by $(27.6.2)$ is the zero-inflation equilibrium, $\tau = 0$. For the even agents, the preferred allocation given by $U^{e\prime}(c_0) = 0$ implies $c_0 < 0.5$, and can therefore not be implemented as a monetary equilibrium above. Hence, the even agents' preferred stationary monetary equilibrium is the one with the smallest permissible c_0, i.e., $c_0 = 0.5$. According to $(27.6.2)$, this allocation can be supported by choosing money growth rate $1 + \tau = \beta$, which is then also the equilibrium gross rate of deflation. Notice that all agents, both odd and even, are in agreement that they prefer no inflation to positive inflation, that is, they prefer c_0 determined by $(27.6.4)$ to any higher value of c_0.

To abstract from the described conflict of interest between odd and even agents, suppose that the agents must pick their preferred monetary policy under a "veil of ignorance," before knowing their true identity. Since there are equal numbers of each type of agent, an individual faces a fifty-fifty chance of her identity being an odd or an even agent. Hence, prior to knowing one's identity, the expected lifetime utility of an agent is

$$\bar{U}(c_0) \equiv \frac{1}{2}U^o(c_0) + \frac{1}{2}U^e(c_0) = \frac{u(c_0) + u(1 - c_0)}{2(1 - \beta)}.$$

The *ex ante* preferred allocation c_0 is determined by the first-order condition $\bar{U}'(c_0) = 0$, which has the solution $c_0 = 0.5$. Collecting equations $(27.6.1)$, $(27.6.2)$, and $(27.6.3)$, this preferred policy is characterized by

$$\frac{p_t}{p_{t+1}} = \frac{1}{1 + \tau} = \frac{u'(c_t^o)}{\beta u'(c_{t+1}^o)} = \frac{u'(c_t^e)}{\beta u'(c_{t+1}^e)} = \frac{1}{\beta}, \qquad \forall t \geq 0,$$

where $c_j^i = 0.5$ for all $j \geq 0$ and $i \in \{o, e\}$. Thus, the real return on money, p_t/p_{t+1}, equals a common marginal rate of intertemporal substitution, β^{-1}, and this return would therefore also constitute the real interest rate if there were a credit market. Moreover, since the gross real return on money is the inverse of the gross inflation rate, it follows that the gross real interest rate β^{-1} multiplied by the gross rate of inflation is unity, or the net nominal interest rate is zero. In other words, all agents are *ex ante* in favor of Friedman's rule.

Figure 27.6.1 shows the "utility possibility frontier" associated with this economy. Except for the allocation associated with Friedman's rule, the allocations associated with stationary monetary equilibria lie inside the utility possibility frontier.

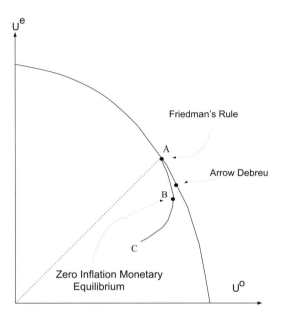

Figure 27.6.1: Utility possibility frontier on the Townsend turn-pike. The locus of points ABC denotes allocations attainable in stationary monetary equilibria. Point B is the allocation associated with the zero-inflation monetary equilibrium. Point A is associated with Friedman's rule, while points between B and C correspond to stationary monetary equilibria with inflation.

27.7. Inflationary finance

The government prints new currency in total amount $M_t - M_{t-1}$ in period t and uses it to purchase a constant amount G of goods in period t. The government's time t budget constraint is

$$M_t - M_{t-1} = p_t G, \quad t \geq 0. \tag{27.7.1}$$

Preferences and endowment patterns of odd and even agents are as specified previously. We now use the following definition:

DEFINITION 4: A competitive equilibrium is a price level sequence $\{p_t\}_{t=0}^{\infty}$, a money supply process $\{M_t\}_{t=-1}^{\infty}$, an allocation $\{c_t^o, c_t^e, G_t\}_{t=0}^{\infty}$ and nonnegative money holdings $\{m_t^o, m_t^e\}_{t=-1}^{\infty}$ such that (a) given the price sequence and (m_{-1}^o, m_{-1}^e), the allocation solves the optimum problems of households of both types; (b) the government's budget constraint is satisfied for all $t \geq 0$; and (c) $N(c_t^o + c_t^e) + G_t = N$, for all $t \geq 0$; and $m_t^o + m_t^e = M_t/N$, for all $t \geq -1$.

For $t \geq 1$, write the government's budget constraint as

$$\frac{M_t}{N p_t} = \frac{p_{t-1}}{p_t} \frac{M_{t-1}}{N p_{t-1}} + \frac{G}{N},$$

or

$$\tilde{m}_t = R_{t-1} \tilde{m}_{t-1} + g, \tag{27.7.2}$$

where $g = G/N$, $\tilde{m}_t = M_t/(N p_t)$ is per-odd-person real balances, and $R_{t-1} = p_{t-1}/p_t$ is the rate of return on currency from $t-1$ to t.

To compute an equilibrium, we guess an allocation of the periodic form

$$
\begin{aligned}
\{c_t^o\}_{t=0}^{\infty} &= \{c_0, 1 - c_0 - g, c_0, 1 - c_0 - g, \ldots\}, \\
\{c_t^e\}_{t=0}^{\infty} &= \{1 - c_0 - g, c_0, 1 - c_0 - g, c_0, \ldots\}.
\end{aligned}
\tag{27.7.3}
$$

We guess that $R_t = R$ for all $t \geq 0$, and again guess a "quantity theory" outcome

$$\tilde{m}_t = \tilde{m} \quad \forall t \geq 0.$$

Evaluating the odd household's time 0 first-order condition for currency at equality gives

$$\beta R = \frac{u'(c_0)}{u'(1 - c_0 - g)}. \tag{27.7.4}$$

With our guess, real balances held by each odd agent at the end of period 0, m_0^o/p_0, equal $1 - c_0$, and time 1 consumption, which also is R times the value of these real balances held from 0 to 1, is $1 - c_0 - g$. Thus, $(1 - c_0)R = (1 - c_0 - g)$, or

$$R = \frac{1 - c_0 - g}{1 - c_0}. \tag{27.7.5}$$

Equations (27.7.4) and (27.7.5) are two simultaneous equations that we want to solve for (c_0, R).

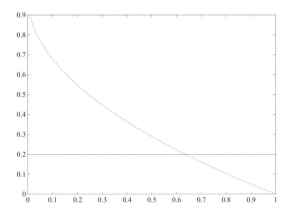

Figure 27.7.1: Revenue from inflation tax $m(R)(1 - R)$ and deficit for $\beta = .95, \delta = 2, g = .2$. The gross rate of return on currency is on the x-axis; g and the revenue from inflation are on the y-axis.

Use equation $(27.7.5)$ to eliminate $(1 - c_0 - g)$ from equation $(27.7.4)$ to get

$$\beta R = \frac{u'(c_0)}{u'[R(1 - c_0)]}.$$

Recalling that $(1 - c_0) = m_0$, this can be written

$$\beta R = \frac{u'(1 - m_0)}{u'(Rm_0)}. \tag{27.7.6}$$

For the power utility function $u(c) = \frac{c^{1-\delta}}{1-\delta}$, this equation can be solved for m_0 to get the demand function for currency

$$m_0 = \tilde{m}(R) \equiv \frac{(\beta R^{1-\delta})^{1/\delta}}{1 + (\beta R^{1-\delta})^{1/\delta}}. \tag{27.7.7}$$

Substituting this into the government budget constraint $(27.7.2)$ gives

$$\tilde{m}(R)(1 - R) = g. \tag{27.7.8}$$

This equation equates the revenue from the inflation tax, namely, $\tilde{m}(R)(1 - R)$ to the government deficit, g. The revenue from the inflation tax is the product of real balances and the inflation tax rate $1 - R$. The equilibrium value of R solves equation $(27.7.8)$.

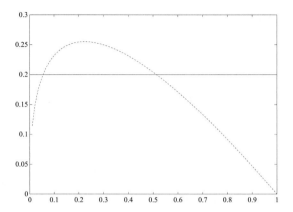

Figure 27.7.2: Revenue from inflation tax $m(R)(1 - R)$ and deficit for $\beta = .95, \delta = .7, g = .2$. The rate of return on currency is on the x-axis; g and the revenue from inflation are on the y-axis. Here there is a Laffer curve.

Figures 27.7.1 and 27.7.2 depict the determination of the stationary equilibrium value of R for two sets of parameter values. For the case $\delta = 2$, shown in Figure 27.7.1, there is a unique equilibrium R; there is a unique equilibrium for every $\delta \geq 1$. For $\delta \geq 1$, the demand function for currency slopes upward as a function of R, as for the example in Figure 27.7.3. For $\delta < 1$, there can occur multiple stationary equilibria, as for the example in Figure 27.7.2. In such cases, there is a Laffer curve in the revenue from the inflation tax. Notice that the demand for real balances is downward sloping as a function of R when $\delta < 1$.

The initial price level is determined by the time 0 budget constraint of the government, evaluated at equilibrium time 0 real balances. In particular, the time 0 government budget constraint can be written

$$\frac{M_0}{Np_0} - \frac{M_{-1}}{Np_0} = g,$$

or

$$\tilde{m} - g = \frac{M_{-1}}{Np_0}.$$

Equating $\tilde{m}$ to its equilibrium value $1 - c_0$ and solving for p_0 gives

$$p_0 = \frac{M_{-1}}{N(1 - c_0 - g)}.$$

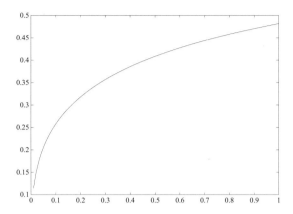

Figure 27.7.3: Demand for real balances on the y-axis as a function of the gross rate of return on currency on the x-axis when $\beta = .95, \delta = 2$.

27.8. Legal restrictions

This section adapts ideas of Bryant and Wallace (1984) and Villamil (1988) to the turnpike environment. Those authors analyzed situations in which the government could make all savers better off by introducing a price discrimination scheme for marketing its debt. The analysis formalizes some ideas mentioned by John Maynard Keynes (1940).

Figure 27.8.1 depicts the terms on which an odd agent at $t = 0$ can transfer consumption between 0 and 1 in an equilibrium with inflationary finance. The agent is endowed at the point $(1, 0)$. The monetary mechanism allows him to transfer consumption between periods on the terms $c_1 = R(1 - c_0)$, depicted by the budget line connecting 1 on the c_t-axis with the point B on the c_{t+1}-axis. The government insists on raising revenues in the amount g for each pair of an odd and an even agent, which means that R must be set so that the tangency between the agent's indifference curve and the budget line $c_1 = R(1 - c_0)$ occurs at the intersection of the budget line and the straight line connecting $1 - g$ on the c_t-axis with the point $1 - g$ on the c_{t+1}-axis. At this point, the marginal rate of substitution for odd agents is

$$\frac{u'(c_0)}{\beta u'(1 - c_0 - g)} = R,$$

(because currency holdings are positive). For even agents, the marginal rate of substitution is

$$\frac{u'(1 - c_0 - g)}{\beta u'(c_0)} = \frac{1}{\beta^2 R} > 1,$$

where the inequality follows from the fact that $R < 1$ under inflationary finance.

The fact that the odd agent's indifference curve intersects the solid line connecting $(1 - g)$ on the two axes indicates that the government could improve the welfare of the odd agent by offering him a higher rate of return subject to a minimal real balance constraint. The higher rate of return is used to send the line $c_1 = (1 - R)c_0$ into the lens-shaped area in Figure 27.8.1 onto a higher indifference curve. The minimal real balance constraint is designed to force the agent onto the "postgovernment share" feasibility line connecting the points $1 - g$ on the two axes.

Thus, notice that in Figure 27.8.1, the government can raise the same revenue by offering odd agents the *higher* rate of return associated with the line connecting 1 on the c_t axis with the point H on the c_{t+1} axis, provided that the agent is required to save at least F, if he saves at all. This minimum saving requirement would make the household's budget set the point $(1, 0)$ together with the heavy segment DH. With the setting of F, R associated with the line DH in Figure 27.8.1, odd households have the same two-period utility as without this scheme. (Points D and A lie on the same indifference curve.) However, it is apparent that there is room to lower F and lower R a bit, and thereby move the odd household into the lens-shaped area. See Figure 27.8.2.

The marginal rates of substitution that we computed earlier indicate that this scheme makes both odd and even agents better off relative to the original equilibrium. The odd agents are better off because they move into the lens-shaped area in Figure 27.8.1. The even agents are better off because relative to the original equilibrium, they are being permitted to "borrow" at a gross rate of interest of 1. Since their marginal rate of substitution at the original equilibrium is $1/(\beta^2 R) > 1$, this ability to borrow makes them better off.

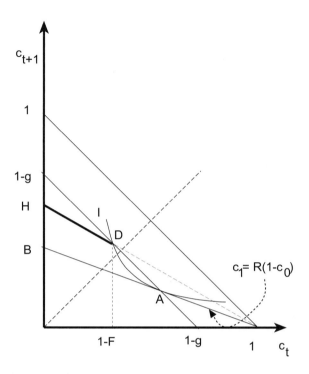

Figure 27.8.1: The budget line starting at $(1,0)$ and ending at the point B describes an odd agent's time 0 opportunities in an equilibrium with inflationary finance. Because this equilibrium has the "private consumption feasibility menu" intersecting the odd agent's indifference curve, a "forced saving" legal restriction can be used to put the odd agent onto a higher indifference curve than I, while leaving even agents better off and the government with revenue g. If the individual is confronted with a minimum denomination F at the rate of return associated with the budget line ending at H, he would choose to consume $1 - F$.

27.9. A two-money model

There are two types of currency being issued, in amounts $M_{it}, i = 1, 2$, by each of two countries. The currencies are issued according to the rules

$$M_{it} - M_{it-1} = p_{it}G_{it}, \quad i = 1, 2, \tag{27.9.1}$$

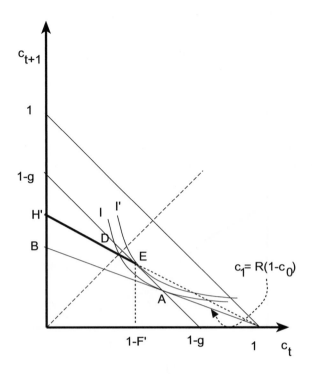

Figure 27.8.2: The minimum denomination F and the return on money can be lowered *vis-à-vis* their setting associated with line DH in Figure 27.8.1 to make the odd household better off, raise the same revenues for the government, and leave even households better off (as compared to no government intervention). The lower value of F puts the odd household at E, which leaves him at the higher indifference curve I'. The minimum denomination F and the return on money can be lowered *vis-à-vis* their setting associated with line DH in Figure 27.8.1 to make the odd household better off, raise the same revenues for the government, and leave even households better off (as compared to no government intervention). The lower value of F puts the odd household at E, which leaves him at the higher indifference curve I'.

where G_{it} is total purchases of time t goods by the government issuing currency i, and p_{it} is the time t price level denominated in units of currency i. We assume that currencies of both types are initially equally distributed among the even agents at time 0. Odd agents start out with no currency.

Household h's optimum problem becomes to maximize $\sum_{t=0}^{\infty} \beta^t u(c_t^h)$ subject to the sequence of budget constraints

$$c_t^h + \frac{m_{1t}^h}{p_{1t}} + \frac{m_{2t}^h}{p_{2t}} \leq y_t^h + \frac{m_{1t-1}^h}{p_{1t}} + \frac{m_{2t-1}^h}{p_{2t}},$$

where m_{jt-1}^h are nominal holdings of country j's currency by household h. Currency holdings of each type must be nonnegative. The first-order conditions for the household's problem with respect to m_{jt}^h for $j = 1, 2$ are

$$\frac{\beta u'(c_{t+1}^h)}{p_{1t+1}} \leq \frac{u'(c_t^h)}{p_{1t}}, \quad = \text{ if } m_{1t}^h > 0,$$

$$\frac{\beta u'(c_{t+1}^h)}{p_{2t+1}} \leq \frac{u'(c_t^h)}{p_{2t}}, \quad = \text{ if } m_{2t}^h > 0.$$

If agent h chooses to hold both currencies from t to $t + 1$, these first-order conditions imply that

$$\frac{p_{2t}}{p_{1t}} = \frac{p_{2t+1}}{p_{1t+1}},$$

or

$$p_{1t} = e p_{2t}, \quad \forall t \geq 0, \tag{27.9.2}$$

for some constant $e > 0$.[5] This equation states that if in each period there is some household that chooses to hold positive amounts of both types of currency, the rate of return from t to $t+1$ must be equal for the two types of currencies, meaning that the exchange rate must be constant over time.[6]

We use the following definition:

DEFINITION 5: A *competitive equilibrium* with two valued fiat currencies is an allocation $\{c_t^o, c_t^e, G_{1t}, G_{2t}\}_{t=0}^{\infty}$, nonnegative money holdings $\{m_{1t}^o, m_{1t}^e, m_{2t}^o, m_{2t}^e\}_{t=-1}^{\infty}$, a pair of finite price level sequences $\{p_{1t}, p_{2t}\}_{t=0}^{\infty}$ and currency supply sequences $\{M_{1t}, M_{2t}\}_{t=-1}^{\infty}$ such that (a) given the price level sequences and $(m_{1,-1}^o, m_{1,-1}^e, m_{2,-1}^o, m_{2,-1}^e)$, the allocation solves the households' problems; (b) the budget

---

[5] Evaluate both of the first-order conditions at equality, then divide one by the other to obtain this result.

[6] As long as we restrict ourselves to nonstochastic equilibria.

constraints of the governments are satisfied for all $t \geq 0$; and (c) $N(c_t^o + c_t^e) + G_{1t} + G_{2t} = N$, for all $t \geq 0$; and $m_{jt}^o + m_{jt}^e = M_{jt}/N$, for $j = 1, 2$ and all $t \geq -1$.

In the case of constant government expenditures $(G_{1t}, G_{2t}) = (Ng_1, Ng_2)$ for all $t \geq 0$, we guess an equilibrium allocation of the form $(27.7.3)$, where we reinterpret g to be $g = g_1 + g_2$. We also guess an equilibrium with a constant real value of the "world money supply," that is,

$$\tilde{m} = \frac{M_{1t}}{Np_{1t}} + \frac{M_{2t}}{Np_{2t}},$$

and a constant exchange rate, so that we impose condition $(27.9.2)$. We let $R = p_{1t}/p_{1t+1} = p_{2t}/p_{2t+1}$ be the constant common value of the rate of return on the two currencies.

With these guesses, the sum of the two countries' budget constraints for $t \geq 1$ and the conjectured form of the equilibrium allocation imply an equation of the form $(27.7.8)$, where now

$$\tilde{m}(R) = \frac{M_{1t}}{p_{1t}N} + \frac{M_{2t}}{p_{2t}N}.$$

Equation $(27.7.8)$ can be solved for R in the fashion described earlier. Once R has been determined, so has the constant real value of the world currency supply, $\tilde{m}$. To determine the time t price levels, we add the time 0 budget constraints of the two governments to get

$$\frac{M_{10}}{Np_{10}} + \frac{M_{20}}{Np_{20}} = \frac{M_{1,-1} + eM_{2,-1}}{Np_{10}} + (g_1 + g_2),$$

or

$$\tilde{m} - g = \frac{M_{1,-1} + eM_{2,-1}}{Np_{10}}.$$

In the conjectured allocation, $\tilde{m} = (1 - c_0)$, so this equation becomes

$$\frac{M_{1,-1} + eM_{2,-1}}{Np_{10}} = 1 - c_0 - g, \tag{27.9.3}$$

which, given any $e > 0$, has a positive solution for the initial country 1 price level. Given the solution p_{10} and any $e \in (0, \infty)$, the price level sequences for the two countries are determined by the constant rate of return on currency R.

To determine the values of the nominal currency stocks of the two countries, we use the government budget constraints (27.9.1).

Our findings are a special case of the following remarkable proposition:

PROPOSITION (EXCHANGE RATE INDETERMINACY): Given the initial stocks of currencies $(M_{1,-1}, M_{2,-1})$ that are equally distributed among the even agents at time 0, if there is an equilibrium for one constant exchange rate $e \in (0, \infty)$, then there exists an equilibrium for any $\hat{e} \in (0, \infty)$ with the same consumption allocation but different currency supply sequences.

PROOF: Let p_{10} be the country 1 price level at time zero in the equilibrium that is assumed to exist with exchange rate e. For the conjectured equilibrium with exchange rate $\hat{e}$, we guess that the corresponding price level is

$$\hat{p}_{10} = p_{10} \frac{M_{1,-1} + \hat{e} M_{2,-1}}{M_{1,-1} + e M_{2,-1}}.$$

After substituting this expression into (27.9.3), we can verify that the real value at time 0 of the initial "world money supply" is the same across equilibria. Next, we guess that the conjectured equilibrium shares the same rate of return on currency, R, and constant end-of-period real value of the "world money supply", $\tilde{m}$, as the the original equilibrium. By construction from the original equilibrium, we know that this setting of the world money supply process guarantees that the consolidated budget constraint of the two governments is satisfied in each period. To determine the values of each country's prices and nominal money supplies, we proceed as above. That is, given $\hat{p}_{10}$ and $\hat{e}$, the price level sequences for the two countries are determined by the constant rate of return on currency R. The evolution of the nominal money stocks of the two countries is governed by government budget constraints (27.9.1). ∎

Versions of this proposition were stated by Kareken and Wallace (1980). See chapter 26 for a discussion of a possible way to alter assumptions to make the exchange rate determinate.

27.10.   A model of commodity money

Consider the following "small-country" model.[7]   There are now two goods, the consumption good and a durable good, silver. Silver has a gross physical rate of return of 1: storing one unit of silver this period yields one unit of silver next period. Silver is not valued domestically, but it can be exchanged abroad at a fixed price of v units of the consumption good per unit of silver; v is constant over time and is independent of the amount of silver imported or exported from this country. There are equal numbers N of odd and even households, endowed with consumption good sequences

$$\{y_t^o\}_{t=0}^{\infty} = \{1, 0, 1, 0, \ldots\},$$

$$\{y_t^e\}_{t=0}^{\infty} = \{0, 1, 0, 1, \ldots\}.$$

Preferences continue to be ordered by $\sum_{t=0}^{\infty} \beta^t u(c_t^i)$ for each type of person, where c_t is consumption of the consumption good.

Each *even* person is initially endowed with S units of silver at time 0. Odd agents own no silver at $t = 0$.

Households are prohibited from borrowing or lending with each other, or with foreigners. However, they can exchange silver with each other and with foreigners. At time t, a household of type i faces the budget constraint

$$c_t^i + m_t^i v \leq y_t^i + m_{t-1}^i v,$$

subject to $m_t^i \geq 0$, where m_t^i is the amount of silver stored from time t to time $t + 1$ by agent i.

---

[7]   See Sargent and Wallace (1983), Sargent and Smith (1997), and Sargent and Velde (1999) for alternative models of commodity money.

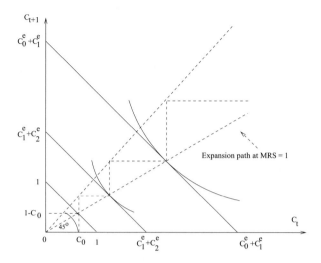

Figure 27.10.1: Determination of equilibrium when $u'(vS) < \beta u'(c_0)$. For as long as it is feasible, the even agent sets $u'(c_{t+1}^e)/u'(c_t^e) = \beta$ by running down his silver holdings. This implies that $c_{t+1}^e < c_t^e$ during the run-down period. Eventually, the even agent runs out of silver, so that the tail of his allocation is $\{c_0, 1 - c_0, c_0, 1 - c_0, \ldots\}$, determined as before. The figure depicts how the spending of silver pushes the agent onto lower and lower two-period budget sets.

27.10.1. Equilibrium

DEFINITION 6: A *competitive equilibrium* is an allocation $\{c_t^o, c_t^e\}_{t=0}^{\infty}$ and non-negative asset holdings $\{m_t^o, m_t^e\}_{t=-1}^{\infty}$ such that, given (m_{-1}^o, m_{-1}^e), the allocation solves each agent's optimum problem.

Adding the budget constraints of the two types of agents with equality at time t gives

$$c_t^o + c_t^e = 1 + v(S_{t-1} - S_t), \tag{27.10.1}$$

where $S_t = m_t^o + m_t^e$ is the total (per odd person) stock of silver in the country at time t. Equation $(27.10.1)$ asserts that total domestic consumption at time t is the sum of the country's endowment plus its imports of goods, where the latter equals its exports of silver, $v(S_{t-1} - S_t)$.

Given the opportunity to choose nonnegative asset holdings with a gross rate of return equal to 1, the equilibrium allocation to the odd agent is $\{c_t^o\}_{t=0}^{\infty} =$

$\{c_0, 1 - c_0, c_0, 1 - c_0, \ldots, \}$, where c_0 is the solution to equation $(27.4.4)$. Thus, the odd agent holds $(1 - c_0)$ units of silver from time 0 to time 1. He gets this silver either from even agents or from abroad.

Concerning the allocation to even agents, two types of equilibria are possible, depending on the value of vS relative to the value c_0 that solves equation $(27.4.4)$. If $u'(vS) \geq \beta u'(c_0)$, the equilibrium allocation to the even agent is $\{c_t^e\}_{t=0}^\infty = \{c_0^e, c_0, 1 - c_0, c_0, 1 - c_0, \ldots\}$, where $c_0^e = vS$. In this equilibrium, the even agent at time 0 sells all of his silver to support time 0 consumption. Net *exports* of silver for the country at time 0 are $S - (1 - c_0)/v$, i.e., summing up the transactions of an even and an odd agent. For $t \geq 1$, the country's allocation and trade pattern is exactly as in the original model (with a stationary fiat money equilibrium).

If the solution c_0 to equation $(27.4.4)$ and vS are such that $u'(vS) < \beta u'(c_0)$, the equilibrium allocation to the odd agents remains the same, but the allocation to the even agents is different. The situation is depicted in Figure 27.10.1. Even agents have so much silver at time 0 that they want to carry over positive amounts of silver into time 1 and maybe beyond. As long as they are carrying over positive amounts of silver from $t - 1$ to t, the allocation to even agents has to satisfy

$$\frac{u'(c_{t-1}^e)}{\beta u'(c_t^e)} = 1, \tag{27.10.2}$$

which implies that $c_t^e < c_{t-1}^e$. Also, as long as they are carrying over positive amounts of silver, their first T budget constraints can be used to deduce an intertemporal budget constraint

$$\sum_{t=0}^{T} c_t^e \leq \begin{cases} vS + (T + 1)/2, & \text{if } T \text{ odd}; \\ vS + T/2, & \text{if } T \text{ even}. \end{cases} \tag{27.10.3}$$

The even agent finds the largest horizon T over which he satisfies both $(27.10.2)$ and $(27.10.3)$ at equality with nonnegative carryover of silver for each period. This largest horizon T will occur on an even date.[8] The equilibrium allocation

---

[8] Suppose to the contrary that the largest horizon T is an odd date. That is, up until date T, both $(27.10.2)$ and $(27.10.3)$ are satisfied with nonnegative savings for each period. Now, let us examine what happens if we add one additional period and the horizon becomes $T + 1$. Since that additional period is an even date, the right side of budget constraint $(27.10.3)$ is unchanged. Therefore, condition $(27.10.2)$ implies that the extra period induces the agent to reduce consumption in all periods $t \leq T$, in order to save for consumption in period $T + 1$.

to the even agents is determined by "gluing" this initial piece with declining consumption onto a "tail" of the allocation assigned to even agents in the original model, starting on an odd date, $\{c_t\}_{t=T+1}^{\infty} = \{c_0, 1 - c_0, c_0, 1 - c_0, \ldots\}$.[9]

27.10.2. Virtue of fiat money

This is a model with an exogenous price level and an endogenous stock of currency. The model can be used to express a version of Friedman's and Keynes's condemnation of commodity money systems: the equilibrium allocation can be Pareto dominated by the allocation in a fiat money equilibrium in which, in addition to the stock of silver at time 0, the even agents are endowed with M units of an unbacked fiat currency. We can then show that there exists a monetary equilibrium with a constant price level p satisfying $(27.4.5)$,

$$ p = \frac{M}{N(1 - c_0)}. $$

In effect, the time 0 endowment of the even agents is increased by $1 - c_0$ units of consumption good. Fiat money creates wealth by removing commodity money from circulation, which instead can be transformed into consumption.

Since the initial horizon T satisfied $(27.10.2)$ and $(27.10.3)$ with nonnegative savings, it follows that so must also horizon $T + 1$. Therefore, the largest horizon T must occur on an even date.

[9] Is the equilibrium with $u'(vS) < \beta u'(c_0)$, a stylized model of Spain in the sixteenth century? At the beginning of the sixteenth century, Spain suddenly received a large claim on silver and gold from the New World. During the century, Spain exported gold and silver to the rest of Europe to finance government and private purchases.

27.11. Concluding remarks

The model of this chapter is basically a "nonstochastic incomplete markets model," a special case of the stochastic incomplete markets models of chapter 18. The virtue of the model is that we can work out many things by hand. The limitation on markets in private loans leaves room for a consumption-smoothing role to be performed by a valued fiat currency. The reader might note how some of the monetary doctrines worked out precisely in this chapter have counterparts in the stochastic incomplete markets models of chapter 18.

Exercises

Exercise 27.1 **Arrow-Debreu**

Consider an environment with equal numbers N of two types of agents, odd and even, who have endowment sequences

$$\{y_t^o\}_{t=0}^\infty = \{1,1,0,1,1,0,\ldots\}$$

$$\{y_t^e\}_{t=0}^\infty = \{0,0,1,0,0,1,\ldots\}.$$

Households of type h order consumption sequences by $\sum_{t=0}^\infty \beta^t u(c_t^h)$. Compute the Arrow-Debreu equilibrium for this economy.

Exercise 27.2 **One-period consumption loans**

Consider an environment with equal numbers N of two types of agents, odd and even, who have endowment sequences

$$\{y_t^o\}_{t=0}^\infty = \{1,0,1,0,\ldots\}$$

$$\{y_t^e\}_{t=0}^\infty = \{0,1,0,1,\ldots\}.$$

Households of type h order consumption sequences by $\sum_{t=0}^\infty \beta^t u(c_t^h)$. The only market that exists is for one-period loans. The budget constraints of household h are

$$c_t^h + b_t^h \le y_t^h + R_{t-1}b_{t-1}^h, \quad t \ge 0,$$

where $b_{-1}^h = 0, h = o, e$. Here b_t^h is agent h's lending (if positive) or borrowing (if negative) from t to $t+1$, and R_{t-1} is the gross real rate of interest on consumption loans from $t-1$ to t.

a. Define a competitive equilibrium with one-period consumption loans.

b. Compute a competitive equilibrium with one-period consumption loans.

c. Is the equilibrium allocation Pareto optimal? Compare the equilibrium allocation with that for the corresponding Arrow-Debreu equilibrium for an economy with identical endowment and preference structure.

Exercise 27.3 **Stock market**

Consider a "stock market" version of an economy with endowment and preference structure identical to the one in the previous economy. Now odd and even agents begin life owning one of two types of "trees." Odd agents own the "odd" tree, which is a perpetual claim to a dividend sequence

$$\{y_t^o\}_{t=0}^\infty = \{1, 0, 1, 0, \ldots\},$$

while even agents initially own the "even" tree, which entitles them to a perpetual claim on dividend sequence

$$\{y_t^e\}_{t=0}^\infty = \{0, 1, 0, 1, \ldots\}.$$

Each period, there is a stock market in which people can trade the two types of trees. These are the only two markets open each period. The time t price of type j trees is $a_t^j, j = o, e$. The time t budget constraint of agent h is

$$c_t^h + a_t^o s_t^{ho} + a_t^e s_t^{he} \leq (a_t^o + y_t^o)s_{t-1}^{ho} + (a_t^e + y_t^e)s_{t-1}^{he},$$

where s_t^{hj} is the number of shares of stock in tree j held by agent h from t to $t+1$. We assume that $s_{-1}^{oo} = 1, s_{-1}^{ee} = 1, s_{-1}^{jk} = 0$ for $j \neq k$.

a. Define an equilibrium of the stock market economy.

b. Compute an equilibrium of the stock market economy.

c. Compare the allocation of the stock market economy with that of the corresponding Arrow-Debreu economy.

Exercise 27.4 **Inflation**

Consider a Townsend turnpike model in which there are N odd agents and N even agents who have endowment sequences, respectively, of

$$\{y_t^o\}_{t=0}^\infty = \{1, 0, 1, 0, \ldots\}$$
$$\{y_t^e\}_{t=0}^\infty = \{0, 1, 0, 1, \ldots\}.$$

Households of each type order consumption sequences by $\sum_{t=0}^{\infty} \beta^t u(c_t)$. The government makes the stock of currency move according to

$$M_t = z M_{t-1}, \quad t \geq 0.$$

At the beginning of period t, the government hands out $(z - 1)m_{t-1}^h$ to each type h agent who held m_{t-1}^h units of currency from $t - 1$ to t. Households of type $h = o, e$ have time t budget constraint of

$$p_t c_t^h + m_t^h \leq p_t y_t^h + m_{t-1}^h + (z - 1)m_{t-1}^h.$$

a. Guess that an equilibrium endowment sequence of the periodic form $(27.4.3)$ exists. Make a guess at an equilibrium price sequence $\{p_t\}$ and compute the equilibrium values of $(c_0, \{p_t\})$. (*Hint:* Make a "quantity theory" guess for the price level.)

b. How does the allocation vary with the rate of inflation? Is inflation "good" or "bad"? Describe odd and even agents' attitudes toward living in economies with different values of z.

Exercise 27.5 **A Friedman-like scheme**

Consider Friedman's scheme to improve welfare by generating a deflation. Suppose that the government tries to boost the rate of return on currency above β^{-1} by setting $\beta > (1 + \tau)$. Show that there exists no equilibrium with an allocation of the class $(27.4.3)$ and a price-level path satisfying $p_t = (1 + \tau)p_{t-1}$, with odd agents holding $m_0^o > 0$. [(That is, the piece of the "restricted Pareto optimality frontier" does not extend above the allocation $(.5, .5)$ in Figure 27.6.1.)

Exercise 27.6 **Distribution of currency**

Consider an economy consisting of large and equal numbers of two types of infinitely lived agents. There is one kind of consumption good, which is nonstorable. "Odd" agents have period 2 endowment pattern $\{y_t^o\}_{t=0}^{\infty}$, while "even" agents have period 2 endowment pattern $\{y_t^e\}_{t=0}^{\infty}$. Agents of both types have preferences that are ordered by the utility functional

$$\sum_{t=0}^{\infty} \beta^t \ln(c_t^i), \quad i = o, e, \quad 0 < \beta < 1,$$

where c_t^i is the time t consumption of the single good by an agent of type i.

Assume the following endowment pattern:

$$y_t^o = \{1, 0, 1, 0, 1, 0, \ldots\}$$

$$y_t^e = \{0, 1, 0, 1, 0, 1, \ldots\}.$$

Now assume that all borrowing and lending is prohibited, either ex cathedra through legal restrictions or by virtue of traveling and locational restrictions of the kind introduced by Robert Townsend. At time $t = 0$, all odd agents are endowed with αH units of an unbacked, inconvertible currency, and all even units are endowed with $(1 - \alpha)H$ units of currency, where $\alpha \in [0, 1]$. The currency is denominated in dollars and is perfectly durable. Currency is the only object that agents are permitted to carry over from one period to the next. Let p_t be the price level at time t, denominated in units of dollars per time t consumption good.

a. Define an *equilibrium with valued fiat currency*.

b. Let an "eventually stationary" equilibrium with valued fiat currency be one in which there exists a $\bar{t}$ such that for $t \geq \bar{t}$, the equilibrium allocation to each type of agent is of period 2 (i.e., for each type of agent, the allocation is a periodic sequence that oscillates between two values). Show that for each value of $\alpha \in [0, 1]$, there exists such an equilibrium. Compute this equilibrium.

Exercise 27.7 **Capital overaccumulation**

Consider an environment with equal numbers N of two types of agents, odd and even, who have endowment sequences

$$\{y_t^o\}_{t=0}^\infty = \{1 - \varepsilon, \varepsilon, 1 - \varepsilon, \varepsilon, \ldots\}$$
$$\{y_t^e\}_{t=0}^\infty = \{\varepsilon, 1 - \varepsilon, \varepsilon, 1 - \varepsilon, \ldots\}.$$

Here, ε is a small positive number that is very close to zero. Households of each type h order consumption sequences by $\sum_{t=0}^\infty \beta^t \ln(c_t^h)$ where $\beta \in (0, 1)$. The one good in the model is storable. If a nonnegative amount k_t of the good is stored at time t, the outcome is that δk_t of the good is carried into period $t+1$, where $\delta \in (0, 1)$. Households are free to store nonnegative amounts of the good.

a. Assume that there are no markets. Households are on their own. Find the autarkic consumption allocations and storage sequences for the two types of agents. What is the total per-period storage in this economy?

b. Now assume that there exists a fiat currency, available in fixed supply of M, all of which is initially equally distributed among the even agents. Define an equilibrium with valued fiat currency. Compute a stationary equilibrium with valued fiat currency. Show that the associated allocation Pareto dominates the one you computed in part a.

c. Suppose that in the storage technology $\delta = 1$ (no depreciation) and that there is a fixed supply of fiat currency, initially distributed as in part b. Define an "eventually stationary" equilibrium. Show that there is a continuum of eventually stationary equilibrium price levels and allocations.

Exercise 27.8 **Altered endowments**

Consider a Bewley model identical to the one in the text, except that now the odd and even agents are endowed with the sequences

$$y_t^0 = \{1 - F, F, 1 - F, F, \ldots\}$$
$$y_t^e = \{F, 1 - F, F, 1 - F, \ldots\},$$

where $0 < F < (1 - c^o)$, where c^o is the solution of equation $(27.4.4)$.

Compute the equilibrium allocation and price level. How do these objects vary across economies with different levels of F? For what values of F does a stationary equilibrium with valued fiat currency exist?

Exercise 27.9 **Inside money**

Consider an environment with equal numbers N of two types of households, odd and even, who have endowment sequences

$$\{y_t^o\}_{t=0}^\infty = \{1, 0, 1, 0, \ldots\}$$

$$\{y_t^e\}_{t=0}^\infty = \{0, 1, 0, 1, \ldots\}.$$

Households of type h order consumption sequences by $\sum_{t=0}^\infty \beta^t u(c_t^h)$. At the beginning of time 0, each even agent is endowed with M units of an unbacked fiat currency and owes F units of consumption goods; each odd agent is owed F units of consumption goods and owns 0 units of currency. At time $t \geq 0$, a household of type h chooses to carry over $m_t^h \geq 0$ of currency from time t to $t+1$. (We start households out with these debts or assets at time 0 to support a stationary equilibrium.) Each period $t \geq 0$, households can issue indexed one-period debt in amount b_t, promising to pay off $b_t R_t$ at $t+1$, subject to the constraint that $b_t \geq -F/R_t$, where $F > 0$ is a parameter characterizing the borrowing constraint and R_t is the rate of return on these loans between time t and $t+1$. (When $F = 0$, we get the Bewley-Townsend model.) A household's period t budget constraint is

$$c_t + m_t/p_t + b_t = y_t + m_{t-1}/p_t + b_{t-1}R_{t-1},$$

where R_{t-1} is the gross real rate of return on indexed debt between time $t - 1$ and t. If $b_t < 0$, the household is borrowing at t, and if $b_t > 0$, the household is lending at t.

a. Define a competitive equilibrium in which valued fiat currency and private loans coexist.

b. Argue that, in the equilibrium defined in part a, the real rates of return on currency and indexed debt must be equal.

c. Assume that $0 < F < (1-c^o)/2$, where c^o is the solution of equation $(27.4.4)$. Show that there exists a stationary equilibrium with a constant price level and that the allocation equals that associated with the stationary equilibrium of the $F = 0$ version of the model. How does F affect the price level? Explain.

d. Suppose that $F = (1 - c^o)/2$. Show that there is a stationary equilibrium with private loans but that fiat currency is valueless in that equilibrium.

e. Suppose that $F = \frac{\beta}{1+\beta}$. For a stationary equilibrium, find an equilibrium allocation and interest rate.

f. Suppose that $F \in [(1 - c^o)/2, \frac{\beta}{1+\beta}]$. Argue that there is a stationary equilibrium (without valued currency) in which the real rate of return on debt is $R \in (1, \beta^{-1})$.

Exercise 27.10 **Initial conditions and inside money**

Consider a version of the preceding model in which each odd person is initially endowed with no currency and no IOUs, and each even person is initially endowed with M/N units of currency but no IOUs. At every time $t \geq 0$, each agent can issue one-period IOUs promising to pay off F/R_t units of consumption in period $t+1$, where R_t is the gross real rate of return on currency or IOUs between periods t and $t+1$. The parameter F obeys the same restrictions imposed in exercise *27.9*.

a. Find an equilibrium with valued fiat currency in which the tail of the allocation for $t \geq 1$ and the tail of the price level sequence, respectively, are identical with that found in exercise *27.9*.

b. Find the price level, the allocation, and the rate of return on currency and consumption loans at period 0.

Exercise 27.11 **Real bills experiment**

Consider a version of exercise *27.9*. The initial conditions and restrictions on borrowing are as described in exercise *27.9*. However, now the government augments the currency stock by an "open market operation" as follows: In period 0, the government issues $\bar{M} - M$ units per each odd agent for the purpose of purchasing Δ units of IOUs issued at time 0 by the even agents. Assume that $0 < \Delta < F$. At each time $t \geq 1$, the government uses any net real interest payments from its stock IOUs from the private sector to decrease the outstanding stock of currency. Thus, the government's budget constraint sequence is

$$\frac{\bar{M} - M}{p_0} = \Delta, \quad t = 0,$$

$$\frac{\bar{M}_t - \bar{M}_{t-1}}{p_t} = -(R_{t-1} - 1)\Delta \quad t \geq 1.$$

Here, R_{t-1} is the gross rate of return on consumption loans from $t-1$ to t, and $\bar{M}_t$ is the total stock of currency outstanding at the end of time t.

a. Verify that there exists a stationary equilibrium with valued fiat currency in which the allocation has the form $(27.4.3)$ where c_0 solves equation $(27.4.4)$.

b. Find a formula for the price level in this stationary equilibrium. Describe how the price level varies with the value of Δ.

c. Does the "quantity theory of money" hold in this example?

Chapter 28
Equilibrium Search and Matching

28.1. Introduction

This chapter presents various equilibrium models of search and matching. We describe (1) Lucas and Prescott's version of an island model; (2) some matching models in the style of Mortensen, Pissarides, and Diamond; and (3) a search model of money along the lines of Kiyotaki and Wright.

Chapter 6 studied the optimization problem of a single unemployed agent who searched for a job by drawing from an exogenous wage offer distribution. We now turn to a model with a continuum of agents who interact across a large number of spatially separated labor markets. Phelps (1970, introductory chapter) describes such an "island economy," and a formal framework is analyzed by Lucas and Prescott (1974). The agents on an island can choose to work at the market-clearing wage in their own labor market or seek their fortune by moving to another island and its labor market. In an equilibrium, agents tend to move to islands that experience good productivity shocks, while an island with bad productivity may see some of its labor force depart. Frictional unemployment arises because moves between labor markets take time.

Another approach to model unemployment is the matching framework described by Diamond (1982), Mortensen (1982), and Pissarides (1990). These models postulate the existence of a matching function that maps measures of unemployment and vacancies into a measure of matches. A match pairs a worker and a firm, who then have to bargain about how to share the "match surplus," that is, the value that will be lost if the two parties cannot agree and break the match. In contrast to the island model with price-taking behavior and no externalities, the decentralized outcome in the matching framework is in general not efficient. Unless parameter values satisfy a knife-edge restriction, there will be either too many or too few vacancies posted in an equilibrium. The efficiency problem is further exacerbated if it is assumed that heterogeneous jobs must be created via a single matching function. This assumption creates a tension between getting an efficient mix of jobs and an efficient total supply of jobs.

As a reference point to models with search and matching frictions, we also study a frictionless aggregate labor market but assume that labor is indivisible. For example, agents are constrained to work either full time or not at all. This kind of assumption has been used in the real business cycle literature to generate unemployment. If markets for contingent claims exist, Hansen (1985) and Rogerson (1988) show that employment lotteries can be welfare enhancing and that they imply that only a fraction of agents will be employed in an equilibrium. Using this model and the other two frameworks that we have mentioned, we analyze how layoff taxes affect an economy's employment level. The different models yield very different conclusions, shedding further light on the economic forces at work in the various frameworks.

To illustrate another application of search and matching, we study Kiyotaki and Wright's (1993) search model of money. Agents who differ with respect to their taste for different goods meet pairwise and at random. In this model, fiat money can potentially ameliorate the problem of "double coincidence of wants."

28.2. An island model

The model here is a simplified version of Lucas and Prescott's (1974) "island economy." There is a continuum of agents populating a large number of spatially separated labor markets. Each island is endowed with an aggregate production function $\theta f(n)$, where n is the island's employment level and $\theta > 0$ is an idiosyncratic productivity shock. The production function satisfies

$$f' > 0, \qquad f'' < 0, \qquad \text{and} \qquad \lim_{n \to 0} f'(n) = \infty. \qquad (28.2.1)$$

The productivity shock takes on m possible values, $\theta_1 < \theta_2 < \cdots < \theta_m$, and the shock is governed by strictly positive transition probabilities, $\pi(\theta, \theta') > 0$. That is, an island with a current productivity shock of θ faces a probability $\pi(\theta, \theta')$ that its next period's shock is θ'. The productivity shock is persistent in the sense that the cumulative distribution function, $\text{Prob}\,(\theta' \leq \theta_k | \theta) = \sum_{i=1}^{k} \pi(\theta, \theta_i)$, is a decreasing function of θ.

At the beginning of a period, agents are distributed in some way over the islands. After observing the productivity shock, the agents decide whether or not to move to another island. A mover forgoes his labor earnings in the period of the move, whereas he can choose the destination with complete information about current conditions on all islands. An agent's decision to work or to move is taken so as to maximize the expected present value of his earnings stream. Wages are determined competitively, so that each island's labor market clears with a wage rate equal to the marginal product of labor. We will study stationary equilibria.

28.2.1. A single market (island)

The state of a single market is given by its productivity level θ and its beginning-of-period labor force x. In an equilibrium, there will be functions mapping this state into an employment level, $n(\theta, x)$, and a wage rate, $w(\theta, x)$. These functions must satisfy the market-clearing condition

$$w(\theta, x) = \theta f'\big[n(\theta, x)\big]$$

and the labor supply constraint

$$n(\theta, x) \leq x.$$

Let $v(\theta, x)$ be the value of the optimization problem for an agent finding himself in market (θ, x) at the beginning of a period. Let v_u be the expected value obtained next period by an agent leaving the market, a value to be determined by conditions in the aggregate economy. The value now associated with leaving the market is then βv_u. The Bellman equation can then be written as

$$v(\theta, x) = \max\Big\{\beta v_u, \; w(\theta, x) + \beta E\left[v(\theta', x')|\theta, x\right]\Big\}, \tag{28.2.2}$$

where the conditional expectation refers to the evolution of θ' and x' if the agent remains in the same market.

The value function $v(\theta, x)$ is equal to βv_u whenever there are any agents leaving the market. It is instructive to examine the opposite situation when no one leaves the market. This means that the current employment level is $n(\theta, x) = x$ and the wage rate becomes $w(\theta, x) = \theta f'(x)$. Concerning the continuation value for next period, $\beta E\left[v(\theta', x')|\theta, x\right]$, there are two possibilities:

Case i: All agents remain, and some additional agents arrive next period. The arrival of new agents corresponds to a continuation value of βv_u in the market.

Any value less than βv_u would not attract any new agents, and a value higher than βv_u would be driven down by a larger inflow of new agents. It follows that the current value function in equation $(28.2.2)$ can under these circumstances be written as

$$v(\theta, x) \;=\; \theta\, f'(x) + \beta v_u\,.$$

Case ii: All agents remain, and no additional agents arrive next period. In this case $x' = x$, and the lack of new arrivals implies that the market's continuation value is less than or equal to βv_u. The current value function becomes

$$v(\theta, x) \;=\; \theta\, f'(x) + \beta E\,[v(\theta', x)|\theta] \;\le\; \theta\, f'(x) + \beta v_u\,.$$

After putting both of these cases together, we can rewrite the value function in equation $(28.2.2)$ as follows:

$$v(\theta, x) \;=\; \max\Big\{\beta v_u\,,\; \theta\, f'(x) + \min\{\beta v_u\,,\; \beta E\,[v(\theta', x)|\theta]\}\Big\}\,. \qquad (28.2.3)$$

Given a value for v_u, this is a well-behaved functional equation with a unique solution $v(\theta, x)$. The value function is nondecreasing in θ and nonincreasing in x.

On the basis of agents' optimization behavior, we can study the evolution of the island's labor force. There are three possible cases:

Case 1: Some agents leave the market. An implication is that no additional workers will arrive next period, when the beginning-of-period labor force will be equal to the current employment level, $x' = n$. The current employment level, equal to x', can then be computed from the condition that agents remaining in the market receive the same utility as the movers, given by βv_u,

$$\theta\, f'(x') + \beta E\,[v(\theta', x')|\theta] \;=\; \beta v_u\,. \qquad (28.2.4)$$

This equation implicitly defines $x^+(\theta)$ such that $x' = x^+(\theta)$ if $x \ge x^+(\theta)$.

Case 2: All agents remain in the market, and some additional workers arrive next period. The arriving workers must expect to attain the value v_u, as discussed in case i. That is, next period's labor force x' must be such that

$$E\,[v(\theta', x')|\theta] \;=\; v_u\,. \qquad (28.2.5)$$

This equation implicitly defines $x^-(\theta)$ such that $x' = x^-(\theta)$ if $x \leq x^-(\theta)$. It can be seen that $x^-(\theta) < x^+(\theta)$.

Case 3: All agents remain in the market, and no additional workers arrive next period. This situation was discussed in case *ii*. It follows here that $x' = x$ if $x^-(\theta) < x < x^+(\theta)$.

28.2.2. The aggregate economy

The previous section assumed an exogenous value to search, v_u. This assumption will be maintained in the first part of this section on the aggregate economy. The approach amounts to assuming a perfectly elastic outside labor supply with reservation utility v_u. We end the section by showing how to endogenize the value to search in the face of a given inelastic aggregate labor supply.

Define a set X of possible labor forces in a market as follows:

$$
X \equiv \begin{cases}
\left\{ x \in \left\{ x^-(\theta_i),\, x^+(\theta_i) \right\}_{i=1}^m \,:\, x^+(\theta_1) \leq x \leq x^-(\theta_m) \right\}, \\
\qquad\qquad\qquad\qquad\qquad \text{if } x^+(\theta_1) \leq x^-(\theta_m); \\
\left\{ x \in [x^-(\theta_m),\, x^+(\theta_1)] \right\}, \qquad\quad \text{otherwise.}
\end{cases}
$$

The set X is the ergodic set of labor forces in a stationary equilibrium. This can be seen by considering a single market with an initial labor force x. Suppose that $x > x^+(\theta_1)$; the market will then eventually experience the least advantageous productivity shock with a next period's labor force of $x^+(\theta_1)$. Thereafter, the island can at most attract a labor force $x^-(\theta_m)$ associated with the most advantageous productivity shock. Analogously, if the market's initial labor force is $x < x^-(\theta_m)$, it will eventually have a labor force of $x^-(\theta_m)$ after experiencing the most advantageous productivity shock. Its labor force will thereafter never fall below $x^+(\theta_1)$, which is the next period's labor force of a market experiencing the least advantageous shock (given a current labor force greater than or equal to $x^+(\theta_1)$). Finally, in the case that $x^+(\theta_1) > x^-(\theta_m)$, any initial distribution of workers such that each island's labor force belongs to the closed interval $[x^-(\theta_m),\, x^+(\theta_1)]$ can constitute a stationary equilibrium. This would be a parameterization of the model where agents do not find it worthwhile to relocate in response to productivity shocks.

In a stationary equilibrium, a market's transition probabilities among states (θ, x) are given by

$$
\begin{aligned}
\Gamma(\theta', x'|\theta, x) \; = \; \pi(\theta, \theta') \; \cdot \; I\Big(& \big[x' = x^+(\theta) \text{ and } x \geq x^+(\theta)\big] \text{ or} \\
& \big[x' = x^-(\theta) \text{ and } x \leq x^-(\theta)\big] \text{ or} \\
& \big[x' = x \text{ and } x^-(\theta) < x < x^+(\theta)\big]\Big),
\end{aligned}
$$

$$\text{for } x, x' \in X \text{ and all } \theta, \theta';$$

where $I(\cdot)$ is the indicator function that takes on the value 1 if any of its arguments are true and 0 otherwise. These transition probabilities define an operator P on distribution functions $\Psi_t(\theta, x; v_u)$ as follows: Suppose that at a point in time, the distribution of productivity shocks and labor forces across markets is given by $\Psi_t(\theta, x; v_u)$, then the next period's distribution is

$$
\begin{aligned}
\Psi_{t+1}(\theta', x'; v_u) \; &= \; P\Psi_t(\theta', x'; v_u) \\
&= \; \sum_{x \in X} \sum_{\theta} \Gamma(\theta', x'|\theta, x)\, \Psi_t(\theta, x; v_u).
\end{aligned}
$$

Except for the case when the stationary equilibrium involves no reallocation of labor, the described process has a unique stationary distribution, $\Psi(\theta, x; v_u)$.

Using the stationary distribution $\Psi(\theta, x; v_u)$, we can compute the economy's average labor force per market,

$$
\bar{x}(v_u) \; = \; \sum_{x \in X} \sum_{\theta} x\, \Psi(\theta, x; v_u),
$$

where the argument v_u makes explicit that the construction of a stationary equilibrium rests on the maintained assumption that the value to search is exogenously given by v_u. The economy's equilibrium labor force $\bar{x}$ varies negatively with v_u. In a stationary equilibrium with labor movements, a higher value to search is only consistent with higher wage rates, which in turn require higher marginal products of labor, that is, a smaller labor force on the islands.

From an economy-wide viewpoint, it is the size of the labor force that is fixed, let's say $\hat{x}$, and the value to search that adjusts to clear the markets. To find a stationary equilibrium for a particular $\hat{x}$, we trace out the schedule $\bar{x}(v_u)$ for different values of v_u. The equilibrium pair $(\hat{x}, v_u)$ can then be read off at the intersection $\bar{x}(v_u) = \hat{x}$, as illustrated in Figure 28.2.1.

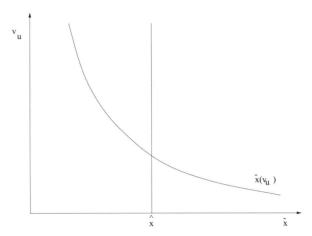

Figure 28.2.1: The curve maps an economy's average labor force per market, $\bar{x}$, into the stationary equilibrium value to search, v_u.

28.3. A matching model

Another model of unemployment is the matching framework, as described by Diamond (1982), Mortensen (1982), and Pissarides (1990). The basic model is as follows: Let there be a continuum of identical workers with measure normalized to 1. The workers are infinitely lived and risk neutral. The objective of each worker is to maximize the expected discounted value of leisure and labor income. The leisure enjoyed by an unemployed worker is denoted z, while the current utility of an employed worker is given by the wage rate w. The workers' discount factor is $\beta = (1+r)^{-1}$.

The production technology is constant returns to scale, with labor as the only input. Each employed worker produces y units of output. Without loss of generality, suppose each firm employs at most one worker. A firm entering the economy incurs a vacancy cost c in each period when looking for a worker, and in a subsequent match the firm's per-period earnings are $y - w$. All matches are exogenously destroyed with per-period probability s. Free entry implies that the expected discounted stream of a new firm's vacancy costs and earnings is equal to zero. The firms have the same discount factor as the workers (who would be the owners in a closed economy).

The measure of successful matches in a period is given by a matching function $M(u, v)$, where u and v are the aggregate measures of unemployed workers and vacancies. The matching function is increasing in both its arguments, concave, and homogeneous of degree 1. By the homogeneity assumption, we can write the probability of filling a vacancy as $q(v/u) \equiv M(u, v)/v$. The ratio between vacancies and unemployed workers, $\theta \equiv v/u$, is commonly labelled the *tightness* of the labor market. The probability that an unemployed worker will be matched in a period is $\theta q(\theta)$. We will assume that the matching function has the Cobb-Douglas form, which implies constant elasticities,

$$M(u, v) = A u^\alpha v^{1-\alpha},$$
$$\frac{\partial M(u, v)}{\partial u} \frac{u}{M(u, v)} = -q'(\theta) \frac{\theta}{q(\theta)} = \alpha, \tag{28.3.1}$$

where $A > 0$, $\alpha \in (0, 1)$, and the last equality will be used repeatedly in our derivations that follow.

Finally, the wage rate is assumed to be determined in a Nash bargain between a matched firm and worker. Let $\phi \in [0, 1)$ denote the worker's bargaining strength, or his weight in the Nash product, as described in the next subsection.

28.3.1. A steady state

In a steady state, the measure of laid-off workers in a period, $s(1 - u)$, must be equal to the measure of unemployed workers gaining employment, $\theta q(\theta)u$. The steady-state unemployment rate can therefore be written as

$$u = \frac{s}{s + \theta q(\theta)}. \tag{28.3.2}$$

To determine the equilibrium value of θ, we now turn to the situations faced by firms and workers, and we impose the no-profit condition for vacancies and the Nash bargaining outcome on firms' and workers' payoffs.

A firm's value of a filled job J and a vacancy V are given by

$$J = y - w + \beta [sV + (1 - s)J], \tag{28.3.3}$$
$$V = -c + \beta \{q(\theta)J + [1 - q(\theta)]V\}. \tag{28.3.4}$$

That is, a filled job turns into a vacancy with probability s, and a vacancy turns into a filled job with probability $q(\theta)$. After invoking the condition that vacancies earn zero profits, $V = 0$, equation $(28.3.4)$ becomes

$$J = \frac{c}{\beta q(\theta)},$$ (28.3.5)

which we substitute into equation $(28.3.3)$ to arrive at

$$w = y - \frac{r+s}{q(\theta)}c.$$ (28.3.6)

The wage rate in equation $(28.3.6)$ ensures that firms with vacancies break even in an expected present-value sense. In other words, a firm's match surplus must be equal to J in equation $(28.3.5)$ in order for the firm to recoup its average discounted costs of filling a vacancy.

The worker's share of the match surplus is the difference between the value of an employed worker E and the value of an unemployed worker U,

$$E = w + \beta\big[sU + (1-s)E\big],$$ (28.3.7)

$$U = z + \beta\{\theta q(\theta)E + [1-\theta q(\theta)]U\},$$ (28.3.8)

where an employed worker becomes unemployed with probability s and an unemployed worker finds a job with probability $\theta q(\theta)$. The worker's share of the match surplus, $E - U$, has to be related to the firm's share of the match surplus, J, in a particular way to be consistent with Nash bargaining. Let the total match surplus be denoted $S = (E - U) + J$, which is shared according to the Nash product

$$\max_{(E-U),J} \ (E - U)^\phi J^{1-\phi}$$ (28.3.9)

$$\text{subject to} \quad S = E - U + J,$$

with solution

$$E - U = \phi S \text{ and } J = (1-\phi)S.$$ (28.3.10)

After solving equations $(28.3.3)$ and $(28.3.7)$ for J and E, respectively, and substituting them into equations $(28.3.10)$, we get

$$w = \frac{r}{1+r}U + \phi\left(y - \frac{r}{1+r}U\right).$$ (28.3.11)

The expression is quite intuitive when seeing $r(1 + r)^{-1}U$ as the annuity value of being unemployed. The wage rate is just equal to this outside option plus the worker's share ϕ of the one-period match surplus. The annuity value of being unemployed can be obtained by solving equation $(28.3.8)$ for $E - U$ and substituting this expression and equation $(28.3.5)$ into equations $(28.3.10)$,

$$\frac{r}{1+r}U \;=\; z + \frac{\phi\,\theta\,c}{1-\phi}\,. \tag{28.3.12}$$

Substituting equation $(28.3.12)$ into equation $(28.3.11)$, we obtain still another expression for the wage rate,

$$w \;=\; z + \phi(y - z + \theta c)\,. \tag{28.3.13}$$

That is, the Nash bargaining results in the worker receiving compensation for lost leisure z and a fraction ϕ of both the firm's output in excess of z and the economy's average vacancy cost per unemployed worker.

The two expressions for the wage rate in equations $(28.3.6)$ and $(28.3.13)$ determine jointly the equilibrium value for θ,

$$y - z \;=\; \frac{r + s + \phi\,\theta\,q(\theta)}{(1-\phi)q(\theta)}\,c\,. \tag{28.3.14}$$

This implicit function for θ ensures that vacancies are associated with zero profits, and that firms' and workers' shares of the match surplus are the outcome of Nash bargaining.

28.3.2. Welfare analysis

A planner would choose an allocation that maximizes the discounted value of output and leisure net of vacancy costs. The social optimization problem does not involve any uncertainty because the aggregate fractions of successful matches and destroyed matches are just equal to the probabilities of these events. The social planner's problem of choosing the measure of vacancies, v_t, and next period's employment level, n_{t+1}, can then be written as

$$\max_{\{v_t, n_{t+1}\}_t} \sum_{t=0}^{\infty} \beta^t \left[y n_t + z(1 - n_t) - c v_t\right], \tag{28.3.15}$$

$$\text{subject to} \quad n_{t+1} = (1 - s)n_t + q\left(\frac{v_t}{1 - n_t}\right)v_t, \tag{28.3.16}$$

$$\text{given} \quad n_0\,.$$

The first-order conditions with respect to v_t and n_{t+1}, respectively, are

$$-\beta^t c + \lambda_t \left[q'(\theta_t) \theta_t + q(\theta_t) \right] = 0, \qquad (28.3.17)$$

$$-\lambda_t + \beta^{t+1}(y-z) + \lambda_{t+1} \left[(1-s) + q'(\theta_{t+1}) \theta_{t+1}^2 \right] = 0, \qquad (28.3.18)$$

where λ_t is the Lagrangian multiplier on equation (28.3.16). Let us solve for λ_t from equation (28.3.17), and substitute into equation (28.3.18) evaluated at a stationary solution,

$$y - z = \frac{r + s + \alpha \theta q(\theta)}{(1-\alpha)q(\theta)} c. \qquad (28.3.19)$$

A comparison of this social optimum to the private outcome in equation (28.3.14) shows that the decentralized equilibrium is only efficient if $\phi = \alpha$. If the workers' bargaining strength ϕ exceeds (falls below) α, the equilibrium job supply is too low (high). Recall that α is both the elasticity of the matching function with respect to the measure of unemployment, and the negative of the elasticity of the probability of filling a vacancy with respect to θ_t. In its latter meaning, a high α means that an additional vacancy has a large negative impact on all firms' probability of filling a vacancy; the social planner would therefore like to curtail the number of vacancies by granting workers a relatively high bargaining power. Hosios (1990) shows how the efficiency condition $\phi = \alpha$ is a general one for the matching framework.

It is instructive to note that the social optimum is equivalent to choosing the worker's bargaining power ϕ such that the value of being unemployed is maximized in a decentralized equilibrium. To see this point, differentiate the value of being unemployed (28.3.12) to find the slope of the indifference in the space of ϕ and θ,

$$\frac{\partial \theta}{\partial \phi} = -\frac{\theta}{\phi(1-\phi)},$$

and use the implicit function rule to find the corresponding slope of the equilibrium relationship (28.3.14),

$$\frac{\partial \theta}{\partial \phi} = -\frac{y - z + \theta c}{\left[\phi - (r+s) q'(\theta) q(\theta)^{-2} \right] c}.$$

We set the two slopes equal to each other because a maximum would be attained at a tangency point between the highest attainable indifference curve and equation (28.3.14) (both curves are negatively sloped and convex to the origin):

$$y - z = \frac{(r+s)\frac{\alpha}{\phi} + \phi\theta\,q(\theta)}{(1-\phi)q(\theta)}\,c. \qquad (28.3.20)$$

When we also require that the point of tangency satisfy the equilibrium condition (28.3.14), it can be seen that $\phi = \alpha$ maximizes the value of being unemployed in a decentralized equilibrium. The solution is the same as the social optimum because the social planner and an unemployed worker both prefer an optimal rate of investment in vacancies, one that takes matching externalities into account.

28.3.3. Size of the match surplus

The size of the match surplus depends naturally on the output y produced by the worker, which is lost if the match breaks up and the firm is left to look for another worker. In principle, this loss includes any returns to production factors used by the worker that cannot be adjusted immediately. It might then seem puzzling that a common assumption in the matching literature is to exclude payments to physical capital when determining the size of the match surplus (see, e.g., Pissarides, 1990). Unless capital can be moved without friction in the economy, this exclusion of payments to physical capital must rest on some implicit assumption of outside financing from a third party that is removed from the wage bargain between the firm and the worker. For example, suppose the firm's capital is financed by a financial intermediary that demands specific rental payments in order not to ask for the firm's bankruptcy. As long as the financial intermediary can credibly distance itself from the firm's and worker's bargaining, it would be rational for the two latter parties to subtract the rental payments from the firm's gross earnings and bargain over the remainder.

In our basic matching model, there is no physical capital, but there is investment in vacancies. Let us consider the possibility that a financial intermediary provides a single firm funding for this investment. The simplest contract would be that the intermediary hand over funds c to a firm with a vacancy in exchange for a promise that the firm pay ϵ in every future period of operation. If the firm cannot find a worker in the next period, it fails and the intermediary writes off the loan, and otherwise the intermediary receives the stipulated interest payment ϵ so long as a successful match stays in business. This agreement with a single firm will have a negligible effect on the economy-wide values of market tightness θ and the value of being unemployed U. Let us examine the consequences for the particular firm involved and the worker it meets.

Under the conjecture that a match will be acceptable to both the firm and the worker, we can compute the interest payment ϵ needed for the financial intermediary to break even in an expected present-value sense,

$$c = q(\theta)\beta \sum_{t=0}^{\infty} \beta^t (1-s)^t \epsilon \quad \Longrightarrow \quad \epsilon = \frac{r+s}{q(\theta)} c. \qquad (28.3.21)$$

A successful match will then generate earnings net of the interest payment equal to $\tilde{y} = y - \epsilon$. To determine how the match surplus is split between the firm and the worker, we replace y, w, J, and E in equations $(28.3.3)$ and $(28.3.7)$, and $(28.3.9)$ by $\tilde{y}$, $\tilde{w}$, $\tilde{J}$, and $\tilde{E}$. That is, $\tilde{J}$ and $\tilde{E}$ are the values to the firm and the worker, respectively, for this particular filled job. We treat θ, V, and U as constants, since they are determined in the rest of the economy. The Nash bargaining can then be seen to yield

$$\tilde{w} = \frac{r}{1+r}U + \phi\left(\tilde{y} - \frac{r}{1+r}U\right) = \frac{r}{1+r}U + \phi\frac{\phi(r+s)}{(1-\phi)q(\theta)}c,$$

where the first equality corresponds to the previous equation $(28.3.11)$. The second equality is obtained after invoking $\tilde{y} = y - \epsilon$ and equations $(28.3.12)$, $(28.3.14)$, and $(28.3.21)$, and the resulting expression confirms the conjecture that the match is acceptable to the worker who receives a wage in excess of the annuity value of being unemployed. The firm will, of course, be satisfied with any positive $\tilde{y} - \tilde{w}$ because it has not incurred any costs whatsoever in order to form the match,

$$\tilde{y} - \tilde{w} = \frac{\phi(r+s)}{q(\theta)}c > 0,$$

where we once again have used $\tilde{y} = y - \epsilon$, equations $(28.3.12)$, $(28.3.14)$, and $(28.3.21)$, and the preceding expression for $\tilde{w}$. Note that $\tilde{y} - \tilde{w} = \phi\epsilon$ with the following interpretation: If the interest payment on the firm's investment, ϵ, was not subtracted from the firm's earnings prior to the Nash bargain, the worker would receive an increase in the wage equal to his share ϕ of the additional "match surplus." The present financial arrangement saves the firm this extra wage payment, and the saving becomes the firm's profit. Thus, a single firm with the proposed contract would have a strictly positive present value when entering the economy of the previous subsection. If there were unlimited entry of new firms having access to intermediaries offering such a contract, those profits would be competed away. We ask the reader to characterize equilibrium outcomes under free entry.

28.4. Matching model with heterogeneous jobs

Acemoglu (1997), Bertola and Caballero (1994), and Davis (1995) explore matching models where heterogeneity on the job supply side must be negotiated through a single matching function, which gives rise to additional externalities. Here, we will study an infinite horizon version of Davis's model, which assumes that heterogeneous jobs are created in the same labor market with only one matching function. We extend our basic matching framework as follows: Let there be I types of jobs. A filled job of type i produces y^i. The cost in each period of creating a measure v^i of vacancies of type i is given by a strictly convex upward-sloping cost schedule, $C^i(v^i)$. In a decentralized equilibrium, we will assume that vacancies are competitively supplied at a price equal to the marginal cost of creating an additional vacancy, $C^{i'}(v^i)$, and we retain the assumption that firms employ at most one worker. Another implicit assumption is that $\{y^i, C^i(\cdot)\}$ are such that all types of jobs are created in both the decentralized steady state and the socially optimal steady state.

28.4.1. A steady state

In a steady state, there will be a time-invariant distribution of employment and vacancies across types of jobs. Let η^i be the fraction of type i jobs among all vacancies. With respect to a job of type i, the value of an employed worker, E^i, and a firm's values of a filled job, J^i, and a vacancy, V^i, are given by

$$J^i = y^i - w^i + \beta\left[sV^i + (1-s)J^i\right], \tag{28.4.1}$$

$$V^i = -C^{i'}(v^i) + \beta\left\{q(\theta)J^i + \left[1 - q(\theta)\right]V^i\right\}, \tag{28.4.2}$$

$$E^i = w^i + \beta\left[sU + (1-s)E^i\right], \tag{28.4.3}$$

$$U = z + \beta\left\{\theta q(\theta)\sum_j \eta^j E^j + \left[1 - \theta q(\theta)\right]U\right\}, \tag{28.4.4}$$

where the value of being unemployed, U, reflects that the probabilities of being matched with different types of jobs are equal to the fractions of these jobs among all vacancies.

After imposing a zero-profit condition on all types of vacancies, we arrive at the analogue to equation (28.3.6),

$$w^i = y^i - \frac{r+s}{q(\theta)}C^{i\prime}(v^i).$$ (28.4.5)

As before, Nash bargaining can be shown to give rise to still another characterization of the wage,

$$w^i = z + \phi\Big[y^i - z + \theta\sum_j \eta^j C^{j\prime}(v^j)\Big],$$ (28.4.6)

which should be compared to equation (28.3.13). After setting the two wage expressions (28.4.5) and (28.4.6) equal to each other, we arrive at a set of equilibrium conditions for the steady-state distribution of vacancies and the labor market tightness,

$$y^i - z = \frac{r + s + \phi\theta\, q(\theta)\dfrac{\sum_j \eta^j C^{j\prime}(v^j)}{C^{i\prime}(v^i)}}{(1-\phi)q(\theta)}\, C^{i\prime}(v^i).$$ (28.4.7)

When we next turn to the efficient allocation in the current setting, it will be useful to manipulate equation (28.4.7) in two ways. First, subtract from this equilibrium expression for job i the corresponding expression for job j,

$$y^i - y^j = \frac{r+s}{(1-\phi)q(\theta)}\left[C^{i\prime}(v^i) - C^{j\prime}(v^j)\right].$$ (28.4.8)

Second, multiply equation (28.4.7) by v^i and sum over all types of jobs,

$$\sum_i v^i(y^i - z) = \frac{r + s + \phi\theta\, q(\theta)}{(1-\phi)q(\theta)}\sum_i v^i C^{i\prime}(v^i).$$ (28.4.9)

(This expression is reached after invoking $\eta^j \equiv v^j / \sum_h v^h$, and an interchange of summation signs.)

28.4.2. Welfare analysis

The social planner's optimization problem becomes

$$\max_{\{v_t^i, n_{t+1}^i\}_{t,i}} \sum_{t=0}^{\infty} \beta^t \left[\sum_j y^j n_t^j + z\left(1 - \sum_j n_t^j\right) - \sum_j C^j(v_t^j) \right], \qquad (28.4.10a)$$

$$\text{subject to} \quad n_{t+1}^i = (1-s)n_t^i + q\left(\frac{\sum_j v_t^j}{1 - \sum_j n_t^j} \right) v_t^i, \quad \forall i,\ t \geq 0, \quad (28.4.10b)$$

$$\text{given} \quad \{n_0^i\}_i. \qquad (28.4.10c)$$

The first-order conditions with respect to v_t^i and n_{t+1}^i, respectively, are

$$-\beta^t C^{i\prime}(v_t^i) + \lambda_t^i q(\theta_t) + \frac{q'(\theta_t)}{1 - \sum_j n_t^j} \sum_j \lambda_t^j v_t^j = 0, \qquad (28.4.11)$$

$$-\lambda_t^i + \beta^{t+1}(y^i - z) + \lambda_{t+1}^i(1-s)$$

$$+ \frac{q'(\theta_{t+1})\,\theta_{t+1}}{1 - \sum_j n_{t+1}^j} \sum_j \lambda_{t+1}^j v_{t+1}^j = 0. \qquad (28.4.12)$$

To explore the efficient relative allocation of different types of jobs, we subtract from equation $(28.4.11)$ the corresponding expression for job j,

$$\lambda_t^i - \lambda_t^j = \frac{\beta^t \left[C^{i\prime}(v_t^i) - C^{j\prime}(v_t^j) \right]}{q(\theta_t)}. \qquad (28.4.13)$$

Next, we do the same computation for equation $(28.4.12)$ and substitute equation $(28.4.13)$ into the resulting expression evaluated at a stationary solution,

$$y^i - y^j = \frac{r+s}{q(\theta)} \left[C^{i\prime}(v^i) - C^{j\prime}(v^j) \right]. \qquad (28.4.14)$$

A comparison of equation $(28.4.14)$ to equation $(28.4.8)$ suggests that there will be an efficient *relative* supply of different types of jobs in a decentralized equilibrium only if $\phi = 0$. For any strictly positive ϕ, the difference in marginal costs of creating vacancies for two different jobs is smaller in the decentralized equilibrium as compared to the social optimum; that is, the decentralized equilibrium displays smaller differences in the distribution of vacancies across types of jobs. In other words, the decentralized equilibrium creates relatively too many "bad jobs" with low y's or, equivalently, relatively too few "good jobs" with high y's. The inefficiency in the mix of jobs disappears if the workers have no bargaining power so that the firms reap all the benefits of upgrading jobs.[1] But from before we know that workers' bargaining power is essential to correct an excess supply of the *total* number of vacancies.

To investigate the efficiency with respect to the total number of vacancies, multiply equation (28.4.11) by v^i and sum over all types of jobs,

$$\sum_i \lambda_t^i v_t^i = \frac{\beta^t \sum_i v_t^i C^{i\prime}(v_t^i)}{q(\theta_t) + q'(\theta_t)\theta_t}. \tag{28.4.15}$$

Next, we do the same computation for equation (28.4.12) and substitute equation (28.4.15) into the resulting expression evaluated at a stationary solution,

$$\sum_i v^i(y^i - z) = \frac{r + s + \alpha\,\theta\,q(\theta)}{(1-\alpha)q(\theta)} \sum_i v^i C^{i\prime}(v^i). \tag{28.4.16}$$

A comparison of equations (28.4.16) and (28.4.9) suggests the earlier result from the basic matching model; that is, an efficient *total* supply of jobs in a decentralized equilibrium calls for $\phi = \alpha$.[2] Hence, Davis (1995) concludes that

---

[1] The interpretation that $\phi = 0$, which is needed to attain an efficient relative supply of different types of jobs in a decentralized equilibrium, can be made precise in the following way: Let v and n denote any sustainable stationary values of the economy's measure of total vacancies and employment rate, that is, $sn = q\left(\frac{v}{1-n}\right)v$. Solve the social planner's optimization problem in equation (28.4.10) subject to the additional constraints $\sum_i v_t^i = v$, $\sum_i n_{t+1}^i = n \forall t \geq 0$, given $\{n_0^i : \sum_i n_0^i = n\}$. After applying the steps in the main text to the first-order conditions of this problem, we arrive at the very same expression (28.4.14). Thus, if $\{v, n\}$ is taken to be the steady-state outcome of the decentralized economy, it follows that equilibrium condition (28.4.8) satisfies efficiency condition (28.4.14) when $\phi = 0$.

[2] The suggestion that $\phi = \alpha$, which is needed to attain an efficient total supply of jobs in a decentralized equilibrium, can be made precise in the following way. Suppose that the social planner is forever constrained to some arbitrary relative distribution, $\{\gamma^i\}$, of types of jobs and vacancies, where $\gamma^i \geq 0$ and $\sum_i \gamma^i = 1$. The constrained social planner's problem is then given by equations (28.4.10) subject to the additional restrictions $v_t^i = \gamma^i v_t$, $n_t^i = \gamma^i n_t \forall t \geq 0$. That is, the only choice variables are now total vacancies and employment, $\{v_t, n_{t+1}\}$. After consolidating the two first-order conditions with respect to v_t and n_{t+1}, and evaluating at a stationary solution, we obtain

$$\sum_j y^j \gamma^j - z = \frac{r + s + \alpha\,\theta\,q(\theta)}{(1-\alpha)q(\theta)} \sum_j \gamma^j C^{j\prime}(\gamma^j v).$$

there is a fundamental tension between the condition for an efficient mix of jobs ($\phi = 0$) and the standard condition for an efficient total supply of jobs ($\phi = \alpha$).

28.4.3. *The allocating role of wages I: separate markets*

The last section clearly demonstrates Hosios's (1990) characterization of the matching framework: "Though wages in matching-bargaining models are completely flexible, these wages have nonetheless been denuded of any allocating or signaling function: this is because matching takes place before bargaining and so search effectively precedes wage-setting." In Davis's matching model, the problem of wages having no allocating role is compounded through the existence of heterogeneous jobs. But as discussed by Davis, this latter complication would be overcome if different types of jobs were *ex ante* sorted into separate markets. Equilibrium movements of workers across markets would then remove the tension between the optimal mix and the total supply of jobs. Different wages in different markets would serve an allocating role for the labor supply across markets, even though the equilibrium wage in each market would still be determined through bargaining after matching.

Let us study the outcome when there are such separate markets for different types of jobs and each worker can participate in only one market at a time. The modified model is described by equations (28.4.1), (28.4.2), and (28.4.3) where the market tightness variable is now also indexed by i and θ^i, and the new expression for the value of being unemployed is

$$U = z + \beta\{\theta^i q(\theta^i)E^i + [1 - \theta^i q(\theta^i)]U\}. \qquad (28.4.17)$$

In an equilibrium, an unemployed worker attains the value U regardless of which labor market he participates in. The characterization of a steady state proceeds along the same lines as before. Let us here reproduce only three equations that will be helpful in our reasoning. The wage in market i and the annuity value of an unemployed worker can be written as

$$w^i = \phi y^i + (1 - \phi)\frac{r}{1+r}U, \qquad (28.4.18)$$

By multiplying both sides by v, we arrive at the very same expression (28.4.16). Thus, if the arbitrary distribution $\{\gamma^i\}$ is taken to be the steady-state outcome of the decentralized economy, it follows that equilibrium condition (28.4.9) satisfies efficiency condition (28.4.16) when $\phi = \alpha$.

$$\frac{r}{1+r} U = z + \frac{\phi \theta^i C^{i'}(v^i)}{1 - \phi}, \tag{28.4.19}$$

and the equilibrium condition for market i becomes

$$y^i - z = \frac{r + s + \phi \theta^i q(\theta^i)}{(1 - \phi)q(\theta^i)} C^{i'}(v^i). \tag{28.4.20}$$

The social planner's objective function is the same as expression $(28.4.10a)$, but the earlier constraint $(28.4.10b)$ is now replaced by

$$n_{t+1}^i = (1 - s)n_t^i + q\left(\frac{v_t^i}{u_t^i}\right) v_t^i,$$

$$1 = \sum_j \left(u_t^j + n_t^j\right),$$

where u_t^i is the measure of unemployed workers in market i. At a stationary solution, the first-order conditions with respect to v_t^i, u_t^i, and n_{t+1}^i can be combined to read

$$y^i - z = \frac{r + s + \alpha \theta^i q(\theta^i)}{(1 - \alpha)q(\theta^i)} C^{i'}(v^i). \tag{28.4.21}$$

Equations $(28.4.20)$ and $(28.4.21)$ confirm Davis's finding that the social optimum can be attained with $\phi = \alpha$ as long as different types of jobs are sorted into separate markets.

It is interesting to note that the socially optimal wages, that is, equation $(28.4.18)$ with $\phi = \alpha$, imply wage differences for *ex ante* identical workers. Wage differences here are not a sign of any inefficiency but rather necessary to ensure an optimal supply and composition of jobs. Workers with higher pay are compensated for an unemployment spell in their job market, which is on average longer.

28.4.4. The allocating role of wages II: wage announcements

According to Moen (1997), we can reinterpret the socially optimal steady state in the last section as an economy with competitive wage announcements instead of wage bargaining with $\phi = \alpha$. Firms are assumed to freely choose a wage to announce, and then they join the market offering this wage without any bargaining. The socially optimal equilibrium is attained when workers as wage takers choose between labor markets, so that the value of an unemployed worker is equalized in the economy.

To demonstrate that wage announcements are consistent with the socially optimal steady state, consider a firm with a vacancy of type i which is free to choose any wage $\tilde{w}$ and then join a market with this wage. A labor market with wage $\tilde{w}$ has a market tightness $\tilde{\theta}$ such that the value of unemployment is equal to the economy-wide value U. After replacing w, E, and θ in equations (28.4.3) and (28.4.17) by $\tilde{w}$, $\tilde{E}$, and $\tilde{\theta}$, we can combine these two expressions to arrive at a relationship between $\tilde{w}$ and $\tilde{\theta}$,

$$\tilde{w} = \frac{r}{1+r}U + \frac{r+s}{\tilde{\theta}q(\tilde{\theta})}\left(\frac{r}{1+r}U - z\right). \tag{28.4.22}$$

The expected present value of posting a vacancy of type i for one period in market $(\tilde{w}, \tilde{\theta})$ is

$$-C^{i'}(v^i) + q(\tilde{\theta})\beta\sum_{t=0}^{\infty}\beta^t(1-s)^t(y^i - \tilde{w}) = -C^{i'}(v^i) + q(\tilde{\theta})\frac{y^i - \tilde{w}}{r+s}.$$

After substituting equation (28.4.22) into this expression, we can compute the first-order condition with respect to $\tilde{\theta}$ as

$$q'(\tilde{\theta})\frac{y^i}{r+s} - \frac{z}{\tilde{\theta}^2} + \left[\frac{1}{\tilde{\theta}^2} - \frac{q'(\tilde{\theta})}{r+s}\right]\frac{r}{1+r}U = 0.$$

Since the socially optimal steady state is our conjectured equilibrium, we get the economy-wide value U from equation (28.4.19) with ϕ replaced by α. The substitution of this value for U into the first-order condition yields

$$y^i - z = \frac{r + s + \alpha\tilde{\theta}q(\tilde{\theta})}{(1-\alpha)q(\tilde{\theta})}\ \frac{\theta^i}{\tilde{\theta}}\ C^{i'}(v^i). \tag{28.4.23}$$

The right side is strictly decreasing in $\tilde{\theta}$, so by equation (28.4.21) the equality can only hold with $\tilde{\theta} = \theta^i$. We have therefore confirmed that the wages in an optimal steady state are such that firms would like freely to announce them and to participate in the corresponding markets without any wage bargaining. The equal value of an unemployed worker across markets ensures the participation of workers, who now also act as wage takers.

28.5. Matching model with overlapping generations

To emphasize again how congestion externalities are such a driving force in matching models, it is useful to study a matching model in which workers are heterogeneous along one or more dimensions. One important source of heterogeneity is aging. Chéron, Hairault and Langot (2008), and Menzio, Telyukova and Visschers (2010) study overlapping generations models in which unemployed workers either enter a single matching function or are assigned to type-specific matching functions. Here, we adopt a framework of Chéron, Hairault and Langot who assume a single matching function and an exogenous retirement age $T + 1$. A retiring generation is replaced by a new generation of the same size, normalized to unity. All newborn enter the labor market as unemployed.

For each newly filled job as well as for each ongoing job, a new productivity is drawn at the beginning of every period from a cumulative distribution function $G(\epsilon)$ with $\epsilon \in [0, 1]$. Upon observing the job productivity, the firm decides whether or not to operate the job. If a job is not operated, the match between the firm and worker is broken, and the worker returns to the pool of unemployed. In an equilibrium, there will be age-specific reservation productivities, denoted R_i for a worker of age i, below which jobs are terminated. This gives rise to intergenerational externalities in the labor market like those that arose because of heterogenous jobs in section 28.4.

28.5.1. A steady state

In a steady state, there are time-invariant unemployment rates $\{u_i\}_{i=1}^T$, where the index i denotes the age of workers. Since newborn workers enter as unemployed, $u_1 = 1$. Given equilibrium market tightness θ and reservation productivities $\{R_i\}_{i=2}^T$, unemployment rates across ages evolve as

$$u_i = u_{i-1}\Big[1 - \theta q(\theta)\big(1 - G(R_i)\big)\Big] + (1 - u_{i-1})\,G(R_i), \qquad (28.5.1)$$

for $i = 2, \ldots, T$. Note that the unemployed of age $i-1$ can be matched to jobs in the subsequent period and hence, the reservation productivity R_i determines which of those jobs are operated. Total unemployment is $u = \sum_{i=1}^T u_i$, with an economy-wide unemployment rate of u/T.

For a job with productivity ϵ that is matched and acceptable to a worker of age i (i.e., $\epsilon \geq R_i$), a firm's value, $J_i(\epsilon)$, and an employed worker's value, $E_i(\epsilon)$, are

$$J_i(\epsilon) = \epsilon - w_i(\epsilon) + \beta\left[\int_{R_{i+1}}^1 J_{i+1}(\epsilon')dG(\epsilon') + G(R_{i+1})V\right], \quad (28.5.2)$$

$$V = -c + \beta q(\theta)\sum_{i=2}^T\left[\frac{u_{i-1}}{u}\left(\int_{R_i}^1 J_i(\epsilon)dG(\epsilon) + G(R_i)V\right)\right]$$
$$+ \beta(1 - q(\theta))V, \qquad (28.5.3)$$

$$E_i(\epsilon) = w_i(\epsilon) + \beta\left[\int_{R_{i+1}}^1 E_{i+1}(\epsilon')dG(\epsilon') + G(R_{i+1})U_{i+1}\right], \quad (28.5.4)$$

$$U_i = z + \beta\theta q(\theta)\left[\int_{R_{i+1}}^1 E_{i+1}(\epsilon')dG(\epsilon') + G(R_{i+1})U_{i+1}\right]$$
$$+ \beta(1 - \theta q(\theta))U_{i+1}$$
$$= z + \beta U_{i+1} + \beta\theta q(\theta)\int_{R_{i+1}}^1\big[E_{i+1}(\epsilon') - U_{i+1}\big]dG(\epsilon'), \quad (28.5.5)$$

where the value of a vacancy, V, reflects a firm's probabilities of being matched with workers of different ages. A free entry condition ensures that a vacancy earns zero expected profits, $V = 0$, and so equation (28.5.3) can be rewritten as

$$q(\theta) = \frac{c}{\beta\sum_{i=2}^T\frac{u_{i-1}}{u}\int_{R_i}^1 J_i(\epsilon)dG(\epsilon)}. \qquad (28.5.6)$$

The rearrangement of the value of an unemployed worker of age i in the second equality of equation $(28.5.5)$ shows that a successful match earns the worker a surplus of employment over the value of remaining unemployed, $E_{i+1}(\epsilon') - U_{i+1}$.

The total surplus of a firm-worker match with job productivity ϵ and worker of age i is

$$S_i(\epsilon) = J_i(\epsilon) + E_i(\epsilon) - U_i \geq 0. \qquad (28.5.7)$$

The surplus is zero for $\epsilon < R_i$ when a job is terminated, but positive for $\epsilon \geq R_i$ when the surplus satisfies

$$S_i(\epsilon) + U_i = \epsilon + \beta \left[\int_0^1 S_{i+1}(\epsilon') dG(\epsilon') + U_{i+1} \right],$$

and hence the surplus functions satisfy

$$S_i(\epsilon) = \max \left\{ \epsilon + \beta \int_0^1 S_{i+1}(\epsilon') dG(\epsilon') - (U_i - \beta U_{i+1}), \, 0 \right\}. \qquad (28.5.8)$$

As before, Nash bargaining determines how the total surplus is divided between a worker and a firm according to

$$E_i(\epsilon) - U_i = \phi S_i(\epsilon) = \frac{\phi}{1 - \phi} J_i(\epsilon). \qquad (28.5.9)$$

The value $\bar{\epsilon}_i$ of the productivity ϵ at which the first argument behind the max operator in equation $(28.5.8)$ is zero is

$$\bar{\epsilon}_i = -\beta \int_0^1 S_{i+1}(\epsilon') dG(\epsilon') + U_i - \beta U_{i+1}$$

$$= z - \beta \left[1 - \theta q(\theta) \phi \right] \int_{R_{i+1}}^1 S_{i+1}(\epsilon') dG(\epsilon'), \qquad (28.5.10)$$

where we have used equation $(28.5.5)$ to eliminate U_i and then invoked the Nash bargaining outcome in equation $(28.5.9)$. Since $S_{i+1}(\epsilon') = 0$ for $\epsilon' < R_{i+1}$, we have also legitimately neglected to integrate over the region at which values are zero. If the cutoff value $\bar{\epsilon}_i$ in equation $(28.5.10)$ is positive, it is the optimal reservation productivity, i.e.,

$$R_i = \max\{\bar{\epsilon}_i, 0\}.$$

Since the surplus function is zero after retiring from the labor market, $S_{T+1}(\epsilon') = 0$, so we see from equation $(28.5.10)$ that the reservation productivity in the last period before retirement is $R_T = z$, i.e., acceptable jobs are all those with a productivity above the value of leisure for an unemployed worker. Moreover, in all periods prior to the worker's last period in the labor market, the reservation productivity is strictly less than z because by staying on the job, the worker is ensured a new productivity draw next period, while if becoming unemployed, the worker would face uncertainty about whether or not he or she will be matched with a firm in the next period.

28.5.2. Reservation productivity is increasing in age

Since the surplus function is weakly decreasing in age, $S_i(\epsilon) \geq S_{i+1}(\epsilon)$, we can conclude from equation (28.5.10) that the reservation productivity is weakly increasing in age. To see this more clearly, suppose that the reservation productivity is at an interior solution for age i, $R_i = \bar{\epsilon}_i \in (0,1)$, and therefore also at interior solutions for older ages, $R_j \in (0,1)$ for $j \geq i$. Next, apply integration by parts to the integral over the future surplus in equation (28.5.10),

$$\int_{R_{i+1}}^{1} S_{i+1}(\epsilon')\, dG(\epsilon')$$

$$= S_{i+1}(1)\,G(1) - S_{i+1}(R_{i+1})\,G(R_{i+1}) - \int_{R_{i+1}}^{1} S_{i+1}'(\epsilon')\,G(\epsilon')d\epsilon'$$

$$= S_{i+1}(1) - \int_{R_{i+1}}^{1} G(\epsilon')d\epsilon' = \int_{R_{i+1}}^{1} [1 - G(\epsilon')]d\epsilon', \qquad (28.5.11)$$

where the second equality invokes $G(1) = 1$, $S_{i+1}(R_{i+1}) = 0$, and $S_{i+1}'(\epsilon') = 1$. The third equality uses the latter derivative and the interior solution to the reservation productivity R_{i+1}, which imply that $S_{i+1}(1) = \int_{R_{i+1}}^{1} 1\, d\epsilon'$. After substituting expression (28.5.11) into equation (28.5.10)), the reservation productivities are determined recursively by

$$R_j = z - \beta\big[1 - \theta q(\theta)\,\phi\big] \int_{R_{j+1}}^{1} [1 - G(\epsilon')]d\epsilon', \qquad (28.5.12a)$$

for $j = i, \ldots, T-1$, and

$$R_T = z. \qquad (28.5.12b)$$

28.5.3. *Wage rate is decreasing in age*

Our finding that the reservation productivity increases with age, means that employed older workers have a higher average productivity than younger ones. However, in terms of wage rates conditional on job productivity, older workers earn less than younger workers, as we shall now show. Recall from wage equation (28.3.11) in the standard matching model that the equilibrium wage is a function of both the job productivity and the worker's value of unemployment. The latter decreases with increased in age in our model.

After substituting expression (28.5.6) for $q(\theta)$ in equation (28.5.5) and utilizing Nash bargaining outcome (28.5.9), the value of an unemployed worker of age i becomes

$$U_i = z + \beta U_{i+1} + \frac{\phi}{1-\phi}\,\theta\,c\,\kappa_{i+1}, \qquad (28.5.13)$$

where

$$\kappa_{i+1} \equiv \frac{\int_{R_{i+1}}^{1} S_{i+1}(\epsilon')dG(\epsilon')}{\sum_{j=2}^{T}\frac{u_{j-1}}{u}\int_{R_j}^{1} S_j(\epsilon)dG(\epsilon)}. \qquad (28.5.14)$$

Since the surplus function is weakly deceasing in age, it follows that $\kappa_i \geq \kappa_{i+1}$ and hence, the value of unemployment in equation (28.5.13) can be shown, starting with the terminal value $U_{T+1} = 0$ and then calculated recursively, to be decreasing in age.

To compute the equilibrium wage for a worker of age i, we start with the Nash bargaining outcome in expression (28.5.9),

$$E_i(\epsilon) - U_i = \phi\big[J_i(\epsilon) + E_i(\epsilon) - U_i\big], \qquad (28.5.15)$$

which can be rearranged to read

$$
\begin{aligned}
(1-\phi)U_i &= E_i(\epsilon) - \phi\big[J_i(\epsilon) + E_i(\epsilon)\big] \\
&= w_i(\epsilon) - \phi\epsilon + (1-\phi)\beta G(R_{i+1})U_{i+1} \\
&\quad + \beta \int_{R_{i+1}}^{1}\Big\{E_{i+1}(\epsilon') - \phi\big[J_{i+1}(\epsilon') + E_{i+1}(\epsilon')\big]\Big\}dG(\epsilon') \\
&= w_i(\epsilon) - \phi\epsilon + (1-\phi)\beta G(R_{i+1})U_{i+1} + (1-\phi)\beta[1 - G(R_{i+1})]U_{i+1} \\
&\quad + \beta \int_{R_{i+1}}^{1}\Big\{E_{i+1}(\epsilon') - U_{i+1} - \phi\big[J_{i+1}(\epsilon') + E_{i+1}(\epsilon') - U_{i+1}\big]\Big\}dG(\epsilon') \\
&= w_i(\epsilon) - \phi\epsilon + (1-\phi)\beta U_{i+1},
\end{aligned}
$$

where the second equality is obtained by eliminating $J_i(\epsilon)$ and $E_i(\epsilon)$ by using equations (28.5.2) and (28.5.4), and the third equality follows from adding and subtracting $(1 - \phi)\beta[1 - G(R_{i+1})]U_{i+1}$. Next, the integral on the right side of the third equality is zero according to Nash bargaining outcome (28.5.15) and after further simplification, we arrive at the last fourth equality. Thus, from the outermost left and right hand sides of the above succession of equalities, the equilibrium wage of a worker of age i satisfies

$$w_i(\epsilon) = \phi\epsilon + (1 - \phi)\left[U_i - \beta U_{i+1}\right],$$

and after eliminating U_i by using equation (28.5.13), we arrive at

$$w_i(\epsilon) = z + \phi[\epsilon - z + \theta c \kappa_{i+1}]. \tag{28.5.16}$$

Since $\kappa_{T+1} = 0$, it follows that the wage in the last period before retirement is $w_T = z + \phi(\epsilon - z)$, i.e., the worker receives the outside payoff to an unemployed worker, z, plus a worker's Nash bargaining share of the surplus from a one-period match, $\phi(\epsilon - z)$. Wages prior to that last period $(i < T)$ are higher because of the higher outside value of younger workers as captured by the term $\phi \theta c \kappa_{i+1}$ in expression (28.5.16), which is decreasing in age.

28.5.4. Welfare analysis

Following Chéron, Hairault and Langot (2008), we consider a planning problem when the weights on subsequent generations are equal and individual agents do not discount the future, $\beta = 1$. The optimal allocation is then attained by maximizing steady-state output net of vacancy costs plus the value of leisure enjoyed by the unemployed. Given an unemployment rate u_{i-1} at the end of age $i - 1$, the output of workers of age i is the product of the fraction of age i workers employed, $\left[u_{i-1}\theta q(\theta) + 1 - u_{i-1}\right]\left[1 - G(R_i)\right]$, and their average productivity, $\left[1 - G(R_i)\right]^{-1} \int_{R_i}^1 \epsilon\, dG(\epsilon)$. Omitting the constant zu_1 from the objective function, the planner's optimization problem becomes

$$\max_{\theta, \{R_i, u_i\}_{i=2}^T} \sum_{i=2}^T \left\{ \left[u_{i-1}\theta q(\theta) + 1 - u_{i-1}\right] \int_{R_i}^1 \epsilon\, dG(\epsilon) + zu_i \right\} - c\theta \sum_{j=1}^T u_j,$$

subject to $\quad u_i = u_{i-1}\left\{1 - \theta q(\theta)\left[1 - G(R_i)\right]\right\} + (1 - u_{i-1})G(R_i)$

$$\text{for } i = 2, \dots, T$$

given $\quad u_1 = 1$.

We confine our analysis to settings for the primitives that give rise to interior solutions for reservation productivities for workers of all ages. The first-order conditions at interior solutions are then

$$
\theta: \quad \sum_{i=2}^{T} u_{i-1} \left[q(\theta) + \theta q'(\theta) \right] \int_{R_i}^{1} \epsilon \, dG(\epsilon) - c \sum_{j=1}^{T} u_j
$$

$$
- \sum_{i=2}^{T} \lambda_i \left[q(\theta) + \theta q'(\theta) \right] \left[1 - G(R_i) \right] u_{i-1} = 0,
$$

$$
R_i: \quad \left[u_{i-1} \theta q(\theta) + 1 - u_{i-1} \right] \left(-R_i \, g(R_i) \right)
$$

$$
+ \lambda_i \left[u_{i-1} \theta q(\theta) g(R_i) + (1 - u_{i-1}) g(R_i) \right] = 0,
$$

$$
u_i: \quad z - c\theta - \lambda_i + \left[\theta q(\theta) - 1 \right] \int_{R_{i+1}}^{1} \epsilon \, dG(\epsilon)
$$

$$
+ \lambda_{i+1} \left\{ 1 - \theta q(\theta) \left[1 - G(R_{i+1}) \right] - G(R_{i+1}) \right\} = 0,
$$

where the last two terms on the left side of the last equation should be understood to be zero when taking the first-order condition with respect to unemployment in the last period before retirement, u_T, i.e., the first-order condition then becomes $z - c\theta - \lambda_T = 0$. After rearranging and simplifying, the first-order conditions can be written as

$$
\theta: \quad c \, u = q(\theta)[1 - \alpha] \sum_{i=2}^{T} u_{i-1} \left\{ \int_{R_i}^{1} \epsilon \, dG(\epsilon) - \lambda_i \left[1 - G(R_i) \right] \right\}, \quad (28.5.17)
$$

$$
R_i: \quad \lambda_i = R_i, \quad (28.5.18)
$$

$$
u_i: \quad \lambda_i = z - c\theta - \left[1 - \theta q(\theta) \right]
$$

$$
\cdot \left\{ \int_{R_{i+1}}^{1} \epsilon \, dG(\epsilon) - \lambda_{i+1} \left[1 - G(R_{i+1}) \right] \right\}, \quad (28.5.19)
$$

where u is total unemployment, $u = \sum_{j=1}^{T} u_j$, and α is the elasticity of matching with respect to unemployment, $\alpha = -q'(\theta) \, \theta / q(\theta)$, as detailed in equation

(28.3.1). By substituting (28.5.18) into (28.5.17) and (28.5.19), and by applying integration by parts,[3] the following equations charactize the social optimum,

$$q(\theta) = \frac{c}{[1-\alpha]\sum_{i=2}^{T}\frac{u_{i-1}}{u}\int_{R_i}^{1}[1-G(\epsilon)]\,dG(\epsilon)}, \qquad (28.5.20)$$

$$R_i = z - c\theta - [1 - \theta q(\theta)]\int_{R_{i+1}}^{1}[1-G(\epsilon)]\,dG(\epsilon), \qquad (28.5.21a)$$

for $i = 2, \ldots, T-1$, and

$$R_T = z - c\theta. \qquad (28.5.21b)$$

Given a market tightness θ, it follows immediately from equations (28.5.21) that the socially optimal reservation productivity is increasing in age, as it is in the decentralized or market economy analyzed above. However, note that the reservation productivity (28.5.21b) in the last period before retirement is lower than the corresponding reservation productivity (28.5.12b) in the market economy. Therefore, an optimal labor market policy would seem to call for a subsidy to employment of older workers in order to lower their reservation productivity. We will confirm this conjecture and also show that the employment of younger workers should be taxed.

To study an optimal labor market policy, it will be useful to substitute expression (28.5.20) for $q(\theta)$ in equation (28.5.21a),

$$R_i + \int_{R_{i+1}}^{1}[1-G(\epsilon)]\,dG(\epsilon) = z - c\theta + \frac{\theta c}{1-\alpha}\hat{\kappa}_{i+1}, \qquad (28.5.22)$$

where

$$\hat{\kappa}_{i+1} \equiv \frac{\int_{R_{i+1}}^{1}[1-G(\epsilon)]\,dG(\epsilon)}{\sum_{j=2}^{T}\frac{u_{j-1}}{u}\int_{R_j}^{1}[1-G(\epsilon)]\,dG(\epsilon)}, \qquad (28.5.23)$$

for $i = 2, \ldots, T-1$. Finally, we add and subtract $\theta c \hat{\kappa}_{i+1}$ to the right side of equation (28.5.22),

$$R_i + \int_{R_{i+1}}^{1}[1-G(\epsilon)]\,dG(\epsilon) = z - c\theta[1 - \hat{\kappa}_{i+1}] + \frac{\alpha}{1-\alpha}\theta c \hat{\kappa}_{i+1}. \qquad (28.5.24)$$

---

[3] Integration by parts yields

$$\int_{R_i}^{1}\epsilon\,dG(\epsilon) = G(1) - R_i G(R_i) - \int_{R_i}^{1}G(\epsilon)\,d\epsilon = R_i[1 - G(R_i)] + \int_{R_i}^{1}[1-G(\epsilon)]\,d\epsilon,$$

where the last equality is obtained by adding and subtracting R_i.

28.5.5. The optimal policy

When the social optimum entails interior solutions for reservation productivities of all ages, the optimal allocation can be supported with an age-specific subsidy to employment, δ_i, (a tax if negative)[4] so long as workers' bargaining strength ϕ satisfies the Hosios condition, $\phi = \alpha$. We will assume that the Hosios condition holds (as well as maintaining our assumption of the previous section that $\beta = 1$).

The introduction of subsidies to employment changes the earlier equation (28.5.8) for the surplus, which now becomes

$$S_i(\epsilon) = \max\left\{\epsilon + \delta_i + \beta \int_0^1 S_{i+1}(\epsilon')dG(\epsilon') - (U_i - \beta U_{i+1}), \, 0\right\}. \quad (28.5.25)$$

Likewise, the earlier equation (28.5.2) for a firm's value of a filled job changes to

$$J_i(\epsilon) = \epsilon + \delta_i - w_i(\epsilon) + \beta \int_{R_{i+1}}^1 J_{i+1}(\epsilon')dG(\epsilon'). \quad (28.5.26)$$

Following the steps of deriving wages in section 28.5.3, we arrive at the expression

$$w_i(\epsilon) = z + \phi[\epsilon + \delta_i - z + \theta c \kappa_{i+1}]. \quad (28.5.27)$$

At an interior reservation productivity R_i, we know that $J_i(R_i) = 0$ and hence from equation (28.5.26) we have

$$\begin{aligned}
0 &= R_i + \delta_i - w_i(R_i) + \beta(1 - \phi) \int_{R_{i+1}}^1 S_{i+1}(\epsilon')dG(\epsilon') \\
&= R_i + \delta_i - w_i(R_i) + \beta(1 - \phi) \int_{R_{i+1}}^1 \big[1 - G(\epsilon')\big]dG(\epsilon'), \quad (28.5.28)
\end{aligned}$$

where the first equality invokes the Nash bargaining outcome (28.5.9), $J_{i+1}(\epsilon') = (1-\phi)S_{i+1}(\epsilon')$, and the second equality uses expression (28.5.11) that holds for the surplus at an interior solution for the reservation productivity R_{i+1}. Similar invocations of relationships (28.5.9) and (28.5.11) in the no-profit condition (28.5.6), and in equation (28.5.14) for κ_{i+1}, establish that

$$q(\theta) = \frac{c}{\beta(1 - \phi) \sum_{i=2}^T \frac{u_{i-1}}{u} \int_{R_i}^1 \big[1 - G(\epsilon)\big]dG(\epsilon)}, \quad (28.5.29)$$

---

[4] We assume that any deficit or surplus from the proposed scheme of employment subsidies and taxes are offset with lump-sum transfers imposed on all agents.

$$\kappa_{i+1} = \hat{\kappa}_{i+1}, \tag{28.5.30}$$

respectively, where $\hat{\kappa}_{i+1}$ is given by equation (28.5.23).

After substituting expression (28.5.27) for $w_i(R_i)$ in equation (28.5.28), and using equation (28.5.30), we find that an equilibrium is characterized by

$$R_i + \beta \int_{R_{i+1}}^{1} \left[1 - G(\epsilon)\right] dG(\epsilon) = z - \delta_i + \frac{\phi}{1 - \phi} \theta \, c \, \hat{\kappa}_{i+1}, \tag{28.5.31}$$

By comparing expressions (28.5.24) and (28.5.31), and recalling our assumptions that $\phi = \alpha$ and $\beta = 1$, it follows that an age-specific employment subsidy of $\delta_i = c\,\theta\left[1 - \hat{\kappa}_{i+1}\right]$ would attain the socially optimal reservation productivity whenever the market tightness is the same. By inspecting equations (28.5.20) and (28.5.29), we can also confirm (via a circular argument) that market tightness θ is indeed the same whenever the reservation productivites are the same.

Thus, we have shown that employment in the last period before retirement, should be subsidized by $\delta_T = c\,\theta$. The subsidy to employment in earlier ages, $\delta_i = c\,\theta\left[1 - \hat{\kappa}_{i+1}\right]$, tapers off with the distance to retirement since $\hat{\kappa}_{i+1}$ is decreasing in age and eventually, at a sufficiently young age, the subsidy becomes negative and turns into a tax on employment of young workers (when $\hat{\kappa}_{i+1} > 1$). Note that, except for one caveat, $\hat{\kappa}_{i+1}$ as defined in (28.5.23) is the expected next-period surplus for an employed worker of age i relative to a weighted average across employed workers of all ages, where the weights are age-specific unemployment, u_i, as a fraction of total unemployment, u. The caveat is that these weights sum to less than one because unemployment of the youngest generation, $u_1 = 1$, is included in u while there are no employed workers in that generation. However, this caveat just reinforces that there is some critical cutoff age i, where $\hat{\kappa}_{j+1} > 1$ for all $j \leq i$, because the expected next-period surplus of such a young employed worker, which tends to be greater than any weighted average in the economy, is compared to something less than a weighted average of expected next-period surpluses of all employed workers.

The rationale for the subsidy $\delta_T = c\,\theta$ to employed workers in the last period before retirement is that if one of them joins the ranks of the unemployed, the economy incurs a vacancy cost per unemployed equal to $c\,\theta$ with no potential gain in terms of future matches. So long as this cost exceeds the worker's value of leisure when unemployed net of the output in the present job, $c\,\theta \geq z - \epsilon$, it is socially optimal for this worker to remain employed, and this is what the subsidy accomplishes by lowering the reservation productivity to $R_T = z -$

$c\theta$. Similarly, employed workers farther away from the retirement age are also subsidized, but to a smaller extent, in order to avoid inefficient congestion in the matching function. However, interestingly, the argument is reversed for sufficiently young workers whose employment should instead be taxed, because they would otherwise fail to internalize the positive externality that they exert in the matching function.

28.6. Model of employment lotteries

Consider a labor market without search and matching frictions but where labor is indivisible. An individual can supply either one unit of labor or no labor at all, as assumed by Hansen (1985) and Rogerson (1988). In such a setting, employment lotteries can be welfare enhancing. The argument is best understood in Rogerson's static model, but with physical capital (and its implication of diminishing marginal product of labor) removed from the analysis. We assume that a single good can be produced with labor, n, as the sole input in a constant returns to scale technology,

$$f(n) = \gamma n, \qquad \text{where} \quad \gamma > 0. \qquad (28.6.1)$$

In a competitive equilibrium, the equilibrium wage is then equal to γ. Following Hansen and Rogerson, the preferences of an individual are assumed to be additively separable in consumption, c, and labor,

$$u(c) - v(n).$$

The standard assumptions are that both u and v are twice continuously differentiable and increasing, but while u is strictly concave, v is convex. However, as pointed out by Rogerson, the precise properties of the function v are not essential because of the indivisibility of labor. The only values of $v(n)$ that matter are $v(0)$ and $v(1)$. Let $v(0) = 0$ and $v(1) = A > 0$. An individual who can supply one unit of labor in exchange for γ units of goods would then choose to do so if

$$u(\gamma) - A \geq u(0),$$

and otherwise the individual would choose not to work.

The proposed allocation might be improved upon by introducing employment lotteries. That is, each individual chooses a probability of working, $\psi \in [0,1]$, and he trades his stochastic labor earnings in contingency markets. We assume a continuum of agents so that the idiosyncratic risks associated with employment lotteries do not pose any aggregate risk, and the contingency prices are then determined by the probabilities of events occurring. (See chapters 8, 13, and 14.) Let c_1 and c_2 be the individual's choice of consumption when working and not working, respectively. The optimization problem becomes

$$\max_{c_1,c_2,\psi} \ \psi\,[u(c_1) - A] \ + \ (1-\psi)\,u(c_2)\,,$$

$$\text{subject to} \quad \psi c_1 + (1-\psi)c_2 \ \leq \ \psi\gamma\,,$$

$$c_1, c_2 \ \geq \ 0\,, \quad \psi \in [0,1]\,.$$

At an interior solution for ψ, the first-order conditions for consumption imply that $c_1 = c_2$,

$$\psi\,u'(c_1) \ = \ \psi\,\lambda\,,$$

$$(1-\psi)\,u'(c_2) \ = \ (1-\psi)\,\lambda\,,$$

where λ is the multiplier on the budget constraint. Since there is no harm in also setting $c_1 = c_2$ when $\psi = 0$ or $\psi = 1$, the individual's maximization problem can be simplified to read

$$\max_{c,\psi} \ u(c) - \psi\,A\,,$$

$$\text{subject to} \quad c \ \leq \ \psi\gamma\,, \quad c \geq 0\,, \quad \psi \in [0,1]\,. \tag{28.6.2}$$

The welfare-enhancing potential of employment lotteries is implicit in the relaxation of the earlier constraint that ψ could only take on two values, 0 or 1. With employment lotteries, the marginal rate of transformation between leisure and consumption is equal to γ.

The solution to expression $(28.6.2)$ can be characterized by considering three possible cases:

Case 1. $A/u'(0) \geq \gamma$.
Case 2. $A/u'(0) < \gamma < A/u'(\gamma)$.
Case 3. $A/u'(\gamma) \leq \gamma$.

The introduction of employment lotteries will only affect individuals' behavior in the second case. In the first case, if $A/u'(0) \geq \gamma$, it will under all circumstances be optimal not to work ($\psi = 0$), since the marginal value of leisure in terms of consumption exceeds the marginal rate of transformation even at a zero consumption level. In the third case, if $A/u'(\gamma) \leq \gamma$, it will always be optimal to work ($\psi = 1$), since the marginal value of leisure falls short of the marginal rate of transformation when evaluated at the highest feasible consumption per worker. The second case implies that expression (28.6.2) has an interior solution with respect to ψ and that employment lotteries are welfare enhancing. The optimal value, ψ^*, is then given by the first-order condition

$$\frac{A}{u'(\gamma\psi^*)} = \gamma.$$

An example of the second case is shown in Figure 28.6.1. The situation here is such that the individual would choose to work in the absence of employment lotteries, because the curve $u(\gamma n) - u(0)$ is above the curve $v(n)$ when evaluated at $n = 1$. After the introduction of employment lotteries, the individual chooses the probability ψ^* of working, and his welfare increases by $\triangle_\psi - \triangle$.

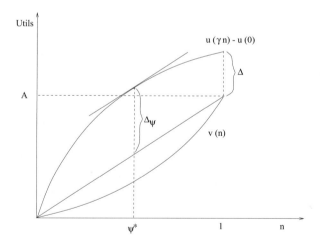

Figure 28.6.1: The optimal employment lottery is given by probability ψ^* of working, which increases expected welfare by $\triangle_\psi - \triangle$ as compared to working full-time, $n = 1$.

28.7. Lotteries for households versus lotteries for firms

Prescott (2005b) focuses on the role of nonconvexities at the level of individual households and production units in the study of business cycles. On the household side, he envisions indivisibilities in labor supply like those in the previous section, while on the firm side, he uses capacity constraints as an example. In spite of these nonconvexities at the micro level, where all units are assumed to be infinitesimal, Prescott points out that the aggregate economy is convex when there are lotteries for households and lotteries for firms that serve to smooth the nonconvexities and that thereby deliver both a stand-in household and a stand-in firm.

Prescott thus recommends an aggregation theory to rationalize a stand-in household that is analogous to better-known aggregation results that underlie the stand-in firm and the aggregate production function. He emphasizes the formal similarities associated with smoothing out nonconvexities by aggregating over firms, on the one hand, and aggregating over households, on the other. Here we shall argue that the economic interpretations that attach to these two types of aggregation make the two aggregation theories very different.[5] Perhaps this explains why this aggregation method has been applied more to firms than to households.[6]

Before turning to a critical comparison of the two aggregation theories, we first describe a simple technology that will capture the essence of Prescott's example of nonconvexities on the firm side, while leaving intact most of our analysis in section 28.6.

---

[5] Our argument is based on Ljungqvist and Sargent's (2005) comment on Prescott (2005b).

[6] Sherwin Rosen often used a lottery model for the household. Instead of analyzing why a particular individual chose higher education, Rosen modeled a family with a continuum of members that allocates fractions of its members to distinct educational choices that involve different numbers of years of schooling. See Ryoo and Rosen (2003).

28.7.1. An aggregate production function

We replace our earlier linear technology (28.6.1) with a production technology based on a strictly concave function $g(\cdot)$,

$$g(0) = 0, \quad g(1) = \gamma, \quad g' > 0, \quad g'' < 0 \implies g'(1) < \gamma.$$

The point of normalization, $g(1) = \gamma$, will be a focal point in our analysis, and the strict concavity of g ensures that $g'(1) < \gamma$.

We assume that there is a continuum of firms and each firm has a production technology given by

$$f(n) = \begin{cases} g(n) & \text{if } n \geq 1, \\ 0 & \text{otherwise,} \end{cases} \tag{28.7.1}$$

where n is the amount of labor employed in the firm. Note that production can take place only if the firm employs at least $n \geq 1$.[7] We normalize the measure of households in section 28.6 to unity and assume that there is a measure of firms equal to $Z < 1$, i.e., there are more households than firms. Each household owns an equal share of all firms.

In a competitive equilibrium where N households are working, there will be $\min\{N, Z\}$ firms in operation. Because of the technological constraint in (28.7.1), firms will choose to operate only if they can employ at least one worker. Moreover, competitive forces will guarantee that the maximum number of firms is operating subject to the constraint that each firm employs at least one worker. This is an implication of our assumption of decreasing returns to scale in each firm. The assumption guarantees that it is profitable to operate many small firms rather than one large firm and that all operating firms will employ the same amount of labor. Thus, in a competitive equilibrium, aggregate output as a function of aggregate employment N is given by

$$F(N) = \begin{cases} N\gamma & \text{if } N < Z, \\ Z\,g\!\left(\frac{N}{Z}\right) & \text{if } N \geq Z. \end{cases} \tag{28.7.2}$$

The aggregate production function in (28.7.2) can be understood as follows. In the first case, only N firms are active and each employs one worker. Hence,

---

[7] As a clarification, note that we do not impose an integer constraint on employment in a firm. It is true that each household in section 28.6 faces the integer constraint of supplying either one unit of labor or no labor at all. However, a household that chooses to work can very well divide its one unit of labor across several firms.

aggregate output is equal to $N g(1) = N \gamma$, and the economy is operating at less than full capacity because there are idle firms ($N < Z$). In the second case, all Z firms are active and each one employs the same amount of labor, $n = N/Z$. Hence, aggregate output is equal to $Z g(N/Z)$, and the economy is operating at full capacity in the sense that there are no idle firms. Note that the aggregate production function in (28.7.2) is convex even though individual firms are subject to a nonconvexity in (28.7.1).

We now turn to an example of time-varying capacity utilization to compare and criticize the aggregation theory underlying the stand-in household in section 28.6 and the aggregation theory underlying the aggregate production function in (28.7.2).[8]

28.7.2. Time-varying capacity utilization

We assume that the stand-in household in section 28.6 is subject to an aggregate preference shock where the disutility of working can take on two different values, $A \in \{0, \bar{A}\}$. The parameters satisfy the following restrictions:

$$\frac{\bar{A}}{u'(0)} < \gamma, \qquad g'\left(\frac{1}{Z}\right) < \frac{\bar{A}}{u'\left[Z g\left(\frac{1}{Z}\right)\right]}. \tag{28.7.3}$$

These parameter restrictions are the analogue to the parameter restriction in case 2 of section 28.6. In particular, restrictions (28.7.3) guarantee an interior solution with respect to the employment lottery when $A = \bar{A}$, i.e., employment, will then satisfy $N \in (0, 1)$ where N is both the measure of households working and the probability of an individual household working since the population of all households is normalized to one.

To see that parameter restrictions (28.7.3) guarantee an interior solution with respect to N when $A = \bar{A}$, we will examine why neither $N = 0$ nor $N = 1$ can constitute an equilibrium. First, we can reject $N = 0$ with the following argument. Whenever $N < Z$, competition among firms drives up

---

[8] There are three differences between Prescott's (2004) example and ours, but none materially effects our illustration of time-varying capacity utilization. First, Prescott postulates an additional production factor, capital, that can also be freely allocated across firms. Second, Prescott assumes a technology for creating new firms. Third, Prescott studies technology shocks while we explore preference shocks.

the equilibrium wage to $w = g(1) = \gamma$. That is, firms are then not a scarce input in production and, therefore, earn no rents. Given the equilibrium wage $w = \gamma$, the first inequality in (28.7.3) states that the stand-in household's first-order conditions would be violated if $N = 0$. Second, we can reject an equilibrium outcome with $N = 1$ as follows. At full employment, all firms are operating and aggregate output is given by $Z\,g(1/Z)$, which is also equal to per capita output since the measure of households is normalized to one. Moreover, according to section 28.6, households will trade in contingent claims prior to the outcome of the employment lottery so that each household's consumption is also given by $c = Z\,g(1/Z)$. The equilibrium wage at full employment is given by $w = g'(1/Z)$, i.e, the marginal product of labor in an individual firm that employs the same amount of labor as all other firms. Given the consumption outcome and wage rate when $N = 1$, we can ask if the stand-in household would indeed choose the probability of working equal to one that would be required in order for this allocation to constitute an equilibrium. According to the second inequality in (28.7.3), the answer is no because the stand-in household would then value a marginal increase in leisure more than the loss of wage income. Thus, we can conclude that parameter restrictions (28.7.3) guarantee an interior solution with respect to the probability of working when $A = \bar{A}$.

In contrast, when the preference shock is $A = 0$, the stand-in household will inelastically supply one unit of labor since there is no disutility of working. The economy will then be operating at full employment with no idle firms. Hence, different realizations of the preference shock $A \in \{0, \bar{A}\}$ will trigger changes in unemployment and potentially changes in capacity utilization, where the latter depends on the size of the given measure of firms. Everything else being equal, a higher Z makes it more likely that the preference shock $\bar{A}$ entails idle firms in an equilibrium. The households and firms that are designated to be unemployed and idle, respectively, are determined by the outcome of lotteries among households and lotteries among firms. Prescott's assertion that the aggregation theory for households is the analogue of the aggregation theory for firms seems to be accurate. So what is the difference between these two aggregation theories?

An important distinction between firms and households is that firms have no independent preferences. They serve only as vehicles for generating rental payments for employed factors and profits for their owners. When a firm becomes inactive, the "firm" itself does not care whether it continues or ceases to exist. Our example of a nonconvex production technology that generates time-varying capacity utilization illustrates this point very well. The firms that do not find any workers stay idle; that is just as well for those idle firms because the firms in operation earn zero rents. In short, whether individual firms operate or remain idle is the end of the story in the aggregation theory behind the aggregate production function in (28.7.2). But in the aggregation theory behind the stand-in household's utility function in (28.6.2), it is really just the beginning. Individual households do have preferences and care about alternative states of the world. So the aggregation theory behind the stand-in household has an additional aspect that is not present in the theory that aggregates over firms, namely, it says how consumption and leisure are smoothed across households with the help of an extensive set of contingent claim markets. This market arrangement and randomization device stands at the center of the employment lottery model. To us, it seems that they make the aggregation theory behind the stand-in household fundamentally different than the well-known aggregation theory for the firm side.

Next, we explore how models with employment lotteries that are used to generate unemployed individuals in a frictionless framework can have very different implications than models embodying frictional unemployment. In particular, models with employment lotteries predict effects from layoff taxes that are opposite to those in search models.

28.8. Employment effects of layoff taxes

The models of employment determination in this chapter can be used to address the question, how do layoff taxes affect an economy's employment? Hopenhayn and Rogerson (1993) apply the model of employment lotteries to this very question and conclude that a layoff tax would reduce the level of employment. Mortensen and Pissarides (1999b) reach the opposite conclusion in a matching model. We will here examine these results by scrutinizing the economic forces at work in different frameworks. The purpose is both to gain further insights into the workings of our theoretical models and to learn about possible effects of layoff taxes.[9]

Common features of many analyses of layoff taxes are as follows: The productivity of a job evolves according to a Markov process, and a sufficiently poor realization triggers a layoff. The government imposes a layoff tax τ on each layoff. The tax revenues are handed back as equal lump-sum transfers to all agents, denoted by T per capita.

Here, we assume the simplest possible Markov process for productivities. A new job has productivity p_0. In all future periods, with probability ξ, the worker keeps the productivity from last period, and with probability $1 - \xi$, the worker draws a new productivity from a distribution $G(p)$.

In our numerical examples, the model period is 2 weeks, and the assumption that $\beta = 0.9985$ then implies an annual real interest rate of 4 percent. The initial productivity of a new job is $p_0 = 0.5$, and $G(p)$ is taken to be a uniform distribution on the unit interval. An employed worker draws a new productivity on average once every two years when we set $\xi = 0.98$.

28.8.1. A model of employment lotteries with layoff taxes

In a model of employment lotteries, a market-clearing wage w equates the demand and supply of labor. The constant-returns-to-scale technology implies that this wage is determined from the supply side, as follows. At the beginning of a period, let $V(p)$ be the firm's value of an employee with productivity p,

$$V(p) \ = \ \max \left\{ p - w + \beta \left[\xi V(p) + (1 - \xi) \int V(p')\ dG(p') \right], \right.$$
$$\left. - \tau \right\}. \tag{28.8.1}$$

Given a value of w, the solution to this Bellman equation is a reservation productivity $\bar{p}$. If there exists an equilibrium with strictly positive employment, the equilibrium wage must be such that new hires exactly break even, so that

$$V(p_0) \ = \ p_0 - w + \beta \left[\xi V(p_0) + (1 - \xi) \int V(p')\, dG(p') \right] \ = \ 0$$

---

[9] The analysis is based on Ljungqvist's (2002) study of layoff taxes in different models of employment determination.

$$\Rightarrow w = p_0 + \beta(1-\xi)\tilde{V}, \tag{28.8.2}$$

where

$$\tilde{V} \equiv \int V(p') \, dG(p').$$

To compute $\tilde{V}$, we first look at the value of $V(p)$ when $p \geq \bar{p}$,

$$
\begin{aligned}
V(p)\Big|_{p \geq \bar{p}} &= p - w + \beta \left[\xi V(p) + (1-\xi)\tilde{V} \right] \\
&= p - w + \beta \xi \left\{ p - w + \beta[\xi V(p) + (1-\xi)\tilde{V}] \right\} + \beta(1-\xi)\tilde{V} \\
&= (1 + \beta\xi) \left[p - w + \beta(1-\xi)\tilde{V} \right] + \beta^2 \xi^2 V(p) \\
&= \frac{p - w + \beta(1-\xi)\tilde{V}}{1 - \beta\xi} = \frac{p - p_0}{1 - \beta\xi}, \tag{28.8.3}
\end{aligned}
$$

where the first equalities are obtained through successive substitutions of $V(p)$, and the last equality incorporates equation $(28.8.2)$. We can then use equation $(28.8.3)$ to find an expression for $\tilde{V}$,

$$
\begin{aligned}
\tilde{V} &= \int_{-\infty}^{\bar{p}} -\tau \, dG(p) + \int_{\bar{p}}^{\infty} V(p) \, dG(p) \\
&= -\tau \, G(\bar{p}) + \int_{\bar{p}}^{\infty} \frac{p - p_0}{1 - \beta\xi} \, dG(p). \tag{28.8.4}
\end{aligned}
$$

From Bellman equation $(28.8.1)$, the reservation productivity satisfies

$$\bar{p} - w + \beta \left[\xi V(\bar{p}) + (1-\xi)\tilde{V} \right] = -\tau.$$

After imposing equation $(28.8.2)$ and $V(\bar{p}) = -\tau$, we find

$$\bar{p} = p_0 - (1 - \beta\xi)\tau \equiv \bar{p}(\tau). \tag{28.8.5}$$

Equations $(28.8.5)$, $(28.8.4)$, and $(28.8.2)$ can be used to solve for the equilibrium wage $w = w(\tau)$.

In a stationary equilibrium, let μ be the mass of new jobs created in every period. The mass of jobs with productivity p_0 that have not yet experienced a new productivity draw can then be expressed as

$$\mu \sum_{i=0}^{\infty} \xi^i = \frac{\mu}{1 - \xi}, \tag{28.8.6}$$

and the mass of jobs that have experienced a new productivity draw and are still operating is given by

$$\sum_{i=0}^{\infty} \xi^i \mu (1-\xi) \left[1 - G(\bar{p})\right] \sum_{j=0}^{\infty} \left\{\xi + (1-\xi)\left[1 - G(\bar{p})\right]\right\}^j$$

$$= \frac{\mu}{1-\xi} \frac{1 - G(\bar{p})}{G(\bar{p})}. \qquad (28.8.7)$$

After equating the sum of these two kinds of jobs to N (which we use to denote the total mass of all jobs), we get the following steady-state relationship:

$$\mu = N G(\bar{p})(1-\xi). \qquad (28.8.8)$$

The firms generate aggregate profits Π. These profits are here computed gross of aggregate layoff taxes, i.e., $\Pi + T$. (Recall that the government hands back layoff tax revenues to the representative agent as a lump-sum transfer T.) Using the masses of jobs in expressions $(28.8.6)$ and $(28.8.7)$, we have

$$\Pi + T = \frac{\mu}{1-\xi}\left(p_0 - w\right) + \frac{\mu}{1-\xi}\frac{1 - G(\bar{p})}{G(\bar{p})} \int_{\bar{p}}^{\infty} \frac{p - w}{1 - G(\bar{p})} \, dG(p)$$

$$= N \left[G(\bar{p})\left(p_0 - w\right) + \int_{\bar{p}}^{\infty} (p - w)\, dG(p) \right], \qquad (28.8.9)$$

where the last inequality invokes relationship $(28.8.8)$.

In a stationary equilibrium with wage w and a gross interest rate $1/\beta$, the representative agent's optimization problem reduces to the static problem:

$$\max_{c,\psi} \; u(c) - \psi A\,,$$

$$\text{subject to} \quad c \leq \psi w + \Pi + T\,, \quad c \geq 0\,, \quad \psi \in [0,1]\,, \qquad (28.8.10)$$

where the profits Π and the lump-sum transfer T are taken as given by the agents.[10] We let ψ^* denote the optimally chosen probability of working. It is equal to N in an equilibrium and the corresponding optimal consumption level is

$$c^* = Nw + \Pi + T = N \left[G(\bar{p})p_0 + \int_{\bar{p}}^{\infty} p \, dG(p) \right], \qquad (28.8.11)$$

---

[10] As above, we are normalizing the measure of agents to unity so that aggregate variables also represent per capita outcomes.

where we have invoked expression (28.8.9). Hence, the steady-state expected
lifetime utility of an agent before seeing the outcome of any employment lottery
is equal to

$$\sum_{t=0}^{\infty} \beta^t \left[u\left(c^*\right) - \psi^* A \right].$$

Following Hopenhayn and Rogerson (1993), the preference specification is
$u(c) = \log(c)$ and the disutility of work is calibrated to match an employment
to population ratio equal to 0.6, which leads us to choose $A = 1.6$. Figures
28.8.1–28.8.5 show how equilibrium outcomes vary with the layoff tax. The
curves labelled L pertain to the model of employment lotteries. As derived
in equation (28.8.5), the reservation productivity in Figure 28.8.1 falls when it
becomes more costly to lay off workers. Figure 28.8.2 shows how the decrease in
number of layoffs is outweighed by the higher tax per layoff, so total layoff taxes
as a fraction of GNP increase over almost the whole range. Figure 28.8.3 reveals
changing job prospects, where the probability of working falls with a higher
layoff tax (which is equivalent to falling employment in a model of employment
lotteries). The welfare loss associated with a layoff tax is depicted in Figure
28.8.4 as the amount of consumption that an agent would be willing to give up
in exchange for a steady state with no layoff tax, and the "willingness to pay"
is expressed as a fraction of per capita consumption at a zero layoff tax.

Figure 28.8.5 reproduces Hopenhayn and Rogerson's (1993) result that employment falls with a higher layoff tax (except at the highest layoff taxes). Intuitively, from a private perspective, a higher layoff tax is like a deterioration
in the production technology; the optimal change in the agents' employment
lotteries will therefore depend on the strength of the substitution effect versus
the income effect. The income effect is largely mitigated by the government's
lump-sum transfer of the tax revenues back to the private economy. Thus, layoff taxes in models of employment lotteries have strong negative employment
implications that are caused by substituting leisure for work. Formally, the logarithmic preference specification gives rise to an optimal choice of the probability
of working, which is equal to the employment outcome, as given by

$$\psi^* = \frac{1}{A} - \frac{T + \Pi}{w}. \tag{28.8.12}$$

The precise employment effect here is driven by profit flows from firms gross
of layoff taxes expressed in terms of the wage rate. Since these profits are to

a large extent generated in order to pay for firms' future layoff taxes, a higher layoff tax tends to increase the accumulation of such funds with a corresponding negative effect on the optimal choice of employment.

Negative employment effect of layoff taxes, when evaluated at $\tau = 0$

Under the assumption that $p_0 = 1$, i.e., the initial productivity of a new job is equal to the upper support of the uniform distribution $G(p)$ on the unit interval $[0, 1]$, we will show that the derivative of equilibrium employment is strictly negative with respect to the layoff tax when evaluated at $\tau = 0$.

Expressions $(28.8.4)$ and $(28.8.9)$ can then be evaluated as follows:

$$\tilde{V} = -\tau \bar{p} + \left[\frac{1 + \bar{p}}{2} - 1\right] \frac{1 - \bar{p}}{1 - \beta \xi}, \qquad (28.8.13)$$

and

$$\Pi + T = N \left[\bar{p} + (1 - \bar{p})\frac{1 + \bar{p}}{2} - w\right]. \qquad (28.8.14)$$

From equations $(28.8.2)$ and $(28.8.13)$,

$$w = 1 + \beta(1 - \xi)\left[-\tau\bar{p} - \frac{(1 - \bar{p})^2}{2(1 - \beta\xi)}\right],$$

and after substituting for $\bar{p}$ from $(28.8.5)$

$$w = 1 - \beta(1 - \xi)\tau\left[1 - \frac{(1 - \beta\xi)\tau}{2}\right] \equiv w(\tau). \qquad (28.8.15)$$

By substituting $(28.8.14)$ into $(28.8.12)$ and using expressions $(28.8.5)$ and $(28.8.15)$, we arrive at an equilibrium expression for N,

$$N(\tau) = \frac{2w(\tau)}{\gamma\left[2\bar{p}(\tau) + 1 - \bar{p}(\tau)^2\right]}$$

with its derivative

$$\frac{dN(\tau)}{d\tau} = \frac{-2\beta(1 - \xi)\bar{p}(\tau)\left[2\bar{p}(\tau) + 1 - \bar{p}(\tau)^2\right] + 4(1 - \beta\xi)\left[1 - \bar{p}(\tau)\right]w(\tau)}{\gamma\left[2\bar{p}(\tau) + 1 - \bar{p}(\tau)^2\right]^2}.$$

Evaluating the derivative at $\tau = 0$, where $\bar{p}(0) = p_0 = 1$, we have

$$\left.\frac{dN(\tau)}{d\tau}\right|_{\tau=0} = \frac{-\beta(1 - \xi)}{\gamma} < 0.$$

This states that in general equilibrium, employment falls in response to the introduction of a layoff tax in our employment lottery model.

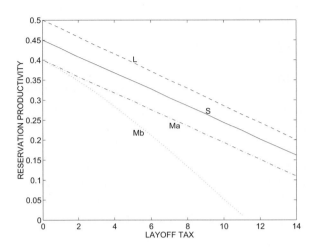

Figure 28.8.1: Reservation productivity for different values of the layoff tax.

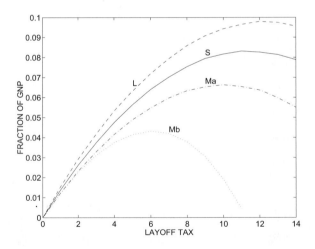

Figure 28.8.2: Total layoff taxes as a fraction of GNP for different values of the layoff tax.

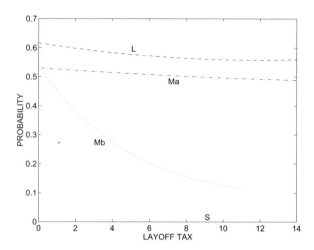

Figure 28.8.3: Probability of working in the model with employment lotteries and probability of finding a job within 10 weeks in the other models, for different values of the layoff tax.

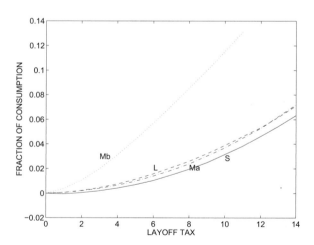

Figure 28.8.4: A job finder's welfare loss due to the presence of a layoff tax, computed as a fraction of per capita consumption at a zero layoff tax.

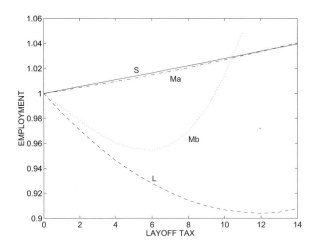

Figure 28.8.5: Employment index for different values of the layoff tax. The index is equal to 1 at a zero layoff tax.

28.8.2. *An island model with layoff taxes*

To stay with the described technology in an island framework, let each job represent a separate island, and an agent moving to a new island experiences productivity p_0. We retain the feature that every agent bears the direct consequences of his decisions. He receives his marginal product p when working and incurs the layoff tax τ if leaving his island. The Bellman equation can then be written as

$$V(p) \;=\; \max \left\{ p \,-\, z \,+\, \beta \left[\xi V(p) \,+\, (1 - \xi) \int V(p')\, dG(p') \right], \right.$$
$$\left. -\, \tau \,+\, \beta^T V(p_0) \right\}, \qquad (28.8.16)$$

where z is the forgone utility of leisure when working and T is the number of periods it takes to move to another island.[11] The solution to this equation is a reservation productivity $\bar{p}$.

---

[11] Note that we have left out the lump-sum transfer from the government because it does not affect the optimization problem.

If there exists an equilibrium with agents working, we must have

$$V(p_0) = p_0 - z + \beta \left[\xi V(p_0) + (1 - \xi) \int V(p') \, dG(p') \right]$$

$$\implies \quad \beta(1 - \xi)\tilde{V} = (1 - \beta\xi)V(p_0) + z - p_0 \,, \qquad (28.8.17)$$

where

$$\tilde{V} \equiv \int V(p') \, dG(p') \,.$$

If the equilibrium entails agents moving between islands, the reservation productivity, by equation (28.8.16), satisfies

$$\bar{p} - z + \beta \left[\xi V(\bar{p}) + (1 - \xi)\tilde{V} \right] = -\tau + \beta^T V(p_0) \,,$$

and, after imposing equation (28.8.17) and $V(\bar{p}) = -\tau + \beta^T V(p_0)$,

$$\bar{p} = p_0 - (1 - \beta\xi) \left[\tau + (1 - \beta^T) V(p_0) \right] \,. \qquad (28.8.18)$$

Note that if agents could move instantaneously between islands, $T = 0$, the reservation productivity would be the same as in the model of employment lotteries, given by equation (28.8.5).

A higher layoff tax also reduces the reservation productivity in the island model; that is, an increase in τ outweighs the drop in the second term in square brackets in equation (28.8.18). For a formal proof, let us make explicit that the value function and the reservation productivity are functions of the layoff tax, $V(p; \tau)$ and $\bar{p}(\tau)$. Consider two layoff taxes, τ and τ', such that $\tau' > \tau \geq 0$, and denote the difference $\triangle\tau = \tau' - \tau$. We can then construct a lower bound for $V(p; \tau')$ in terms of $V(p; \tau)$. In response to the higher layoff tax τ', the agent can always keep his decision rule associated with $V(p; \tau)$ and an upper bound for his extra layoff tax payments would be that he paid $\triangle\tau$ in the current period and every Tth period from there on,

$$V(p; \tau') > V(p, \tau) - \sum_{i=0}^{\infty} \beta^{iT} \triangle\tau \,, \qquad (28.8.19)$$

where the strict inequality follows from the fact that it cannot be optimal to constantly move. In addition, the agent might be able to select a better decision rule than the one associated with τ. In fact, the reservation productivity must fall in response to a higher layoff tax whenever there is an interior solution with respect to $\bar{p}$, as given by equation (28.8.18). By using equations (28.8.18) and (28.8.19), we have

$$\bar{p}(\tau') - \bar{p}(\tau) = -(1 - \beta\xi)\{\triangle\tau + (1 - \beta^T)[V(p_0; \tau') - V(p_0; \tau)]\}$$

$$< -(1 - \beta\xi)\left[\triangle\tau - (1 - \beta^T)\sum_{i=0}^{\infty}\beta^{iT}\triangle\tau\right] = 0.$$

The numerical illustration in Figures 28.8.1 through 28.8.5 is based on a value of leisure $z = 0.25$ and a length of transition between jobs $T = 7$; that is, unemployment spells last 14 weeks. The curves that pertain to the island model are labeled S. The effects of layoff taxes on the reservation productivity, the economy's total layoff taxes, and the welfare of a recent job finder are all similar to the outcomes in the model of employment lotteries. The sharp difference appears in Figure 28.8.5 depicting the effect on the economy's employment. In the island model where agents are left to fend for themselves, a lower reservation productivity is synonymous with both less labor reallocation and lower unemployment. Lower unemployment is thus attained at the cost of a less efficient labor allocation.

Mobility costs also cause employment to rise in the general version of the island model, as mentioned by Lucas and Prescott (1974, p. 205). For a given expected value of arriving on a new island v_u, the value function in equation (28.2.3) is replaced by

$$v(\theta, x) = \max\Big\{\beta v_u - \tau,$$

$$\theta f'(x) + \min\{\beta v_u, \beta E[v(\theta', x)|\theta]\}\Big\}, \qquad (28.8.20)$$

which lies below equation (28.2.3), but with a drop of at most τ. Similarly, equation (28.2.4) changes to

$$\theta f'(n) + \beta E\left[v(\theta', n)|\theta\right] = \beta v_u - \tau. \qquad (28.8.21)$$

An implication here is that $x^+(\theta)$ rises in response to a higher layoff tax. The unchanged expression (28.2.5) means that $x^-(\theta)$ falls as a result of the preceding drop in the value function. In other words, the range of an island's employment levels characterized by no labor movements is enlarged. This effect will shift the curve in Figure 28.2.1 downward and decrease the equilibrium value of v_u. Less labor reallocation maps directly into a lower unemployment rate.

28.8.3. A matching model with layoff taxes

We now modify the matching model to incorporate a layoff tax, and the exogenous destruction of jobs is replaced by the described Markov process for a job's productivity. A job is now endogenously destroyed when the outside option, taking the layoff tax into account, is higher than the value of maintaining the match. The match surplus, $S_i(p)$, is a function of the job's current productivity p and can be expressed as

$$S_i(p) + U_i = \max\left\{ p + \beta \left[\xi S_i(p) + (1-\xi) \int S_i(p')\, dG(p') + U_i \right], \right.$$

$$\left. U_i - \tau \right\}, \qquad (28.8.22)$$

where U_i is once again the agent's outside option, that is, the value of being unemployed. Both $S_i(p)$ and U_i are indexed by i, since we will explore the implications of two alternative specifications of the Nash product, $i \in \{a, b\}$,

$$\left[E_a(p) - U_a \right]^\phi J_a(p)^{1-\phi}, \qquad (28.8.23)$$

$$\left[E_b(p) - U_b \right]^\phi \left[J_b(p) + \tau \right]^{1-\phi}. \qquad (28.8.24)$$

Specification (28.8.23) leads to the usual result that the worker receives a fraction ϕ of the match surplus, while the firm gets the remaining fraction $(1-\phi)$,

$$E_a(p) - U_a = \phi S_a(p) \quad \text{and} \quad J_a(p) = (1-\phi)S_a(p). \qquad (28.8.25)$$

The alternative specification (28.8.24) adopts the assumption of Saint-Paul (1995) that the layoff cost changes the firm's threat point from 0 to $-\tau$, and thereby increases the worker's relative share of the match surplus. Solving for the corresponding surplus-sharing rules, we get

$$E_b(p) - U_b = \phi\big(S_b(p) + \tau\big),$$
$$J_b(p) = (1-\phi)S_b(p) - \phi\tau. \qquad (28.8.26)$$

The worker's continuation value outside of the match associated with Nash product (28.8.23) or (28.8.24), respectively, is

$$U_a = z + \beta\big[\theta q(\theta)\phi S_a(p_0) + U_a\big], \tag{28.8.27}$$

$$U_b = z + \beta\big\{\theta q(\theta)\phi\big[S_b(p_0) + \tau\big] + U_b\big\}. \tag{28.8.28}$$

The equilibrium conditions that firms post vacancies until the expected profits are driven down to zero become

$$(1 - \phi)S_a(p_0) = \frac{c}{\beta q(\theta)}, \tag{28.8.29}$$

$$(1 - \phi)S_b(p_0) - \phi\tau = \frac{c}{\beta q(\theta)}, \tag{28.8.30}$$

for Nash product (28.8.23) or (28.8.24), respectively.

In the calibration, we choose a matching function $M(u, v) = 0.01u^{0.5}v^{0.5}$, a worker's bargaining strength $\phi = 0.5$, and the same value of leisure as in the island model, $z = 0.25$. Qualitatively, the results in Figures 28.8.1 through 28.8.4 are the same across all the models considered here. The curve labeled Ma pertains to the matching model in which the workers' relative share of the match surplus is constant, while the curve Mb refers to the model in which the share is positively related to the layoff tax. However, matching model Mb does stand out. Its reservation productivity plummets in response to the layoff tax in Figure 28.8.1, and is close to zero at $\tau = 11$. A zero reservation productivity means that labor reallocation comes to a halt, and the economy's tax revenues fall to zero in Figure 28.8.2. The more dramatic outcomes under Mb have to do with layoff taxes increasing workers' relative share of the match surplus. The equilibrium condition (28.8.30) requiring that firms finance incurred vacancy costs with retained earnings from the matches becomes exceedingly difficult to satisfy when a higher layoff tax erodes the fraction of match surpluses going to firms. Firms can break even only if the expected time to fill a vacancy is cut dramatically; that is, there has to be a large number of unemployed workers for each posted vacancy. This equilibrium outcome is reflected in the sharply falling probability of a worker finding a job within 10 weeks in Figure 28.8.3. As a result, there are larger welfare costs in model Mb, as shown by the welfare loss of a job finder in Figure 28.8.4. The welfare loss of an unemployed agent is even larger in model Mb, whereas the differences between employed and unemployed agents in the three other model specifications are negligible (not shown in any figure).

In Figure 28.8.5, matching model Ma looks very much like the island model with increasing employment, and matching model Mb displays initially falling employment, similar to the model of employment lotteries. The later sharp reversal of the employment effect in the Mb model is driven by our choice of a Markov process with rather little persistence. (For a comparison, see Ljungqvist, 2002, who explores Markov formulations with more persistence.)

Mortensen and Pissarides (1999a) propose still another bargaining specification where expression (28.8.23) is the Nash product when a worker and a firm meet for the first time, while the Nash product in expression (28.8.24) characterizes all their consecutive negotiations. The idea is that the firm will not incur any layoff tax if the firm and worker do not agree on a wage in the first encounter; that is, there is never an employment relationship. In contrast, the firm's threat point is weakened in future negotiations with an already employed worker because the firm would then have to pay a layoff tax if the match were broken up. We will here show that, except for the wage profile, this alternative specification is equivalent to just assuming Nash product (28.8.23) for all periods. The intuition is that the modified wage profile under the Mortensen and Pissarides assumption is equivalent to a new hire posting a bond equal to his share of the future layoff tax.

First, we compute the wage associated with expression (28.8.23), $w_a(p)$, from the expression for a firm's match surplus,

$$J_a(p) = p - w_a(p) + \beta\left[\xi J_a(p) + (1 - \xi)\int J_a(p')\,dG(p')\right], \qquad (28.8.31)$$

which together with equation (28.8.25) implies

$$w_a(p) = p - (1 - \phi)\,S_a(p) + \beta\Big[\xi(1 - \phi)\,S_a(p)$$
$$+ (1 - \xi)\int(1 - \phi)\,S_a(p')\,dG(p')\Big]. \qquad (28.8.32)$$

Second, we verify that the present value of these wages is exactly equal to that of Mortensen and Pissarides' bargaining scheme for any completed job, under the maintained hypothesis that the two formulations have the same match surplus $S_a(p)$. Let $J_1(p)$ and $J_+(p)$ denote the firm's match surplus with Mortensen and

Pissarides' specification in the first period and all future periods, respectively. The solutions to the maximization of their Nash products are

$$
\begin{aligned}
J_1(p) &= (1 - \phi)S_a(p)\,, \\
J_+(p) &= (1 - \phi)S_a(p) - \phi\tau\,.
\end{aligned}
\tag{28.8.33}
$$

The associated wage functions can be written as

$$
\begin{aligned}
w_1(p) &= p - J_1(p) + \beta\Big[\xi J_+(p) + (1 - \xi)\int J_+(p')\,dG(p')\Big] \\
&= w_a(p) - \beta\,\phi\tau\,, \\
w_+(p) &= p - J_+(p) + \beta\Big[\xi J_+(p) + (1 - \xi)\int J_+(p')\,dG(p')\Big] \\
&= w_a(p) + r\,\beta\,\phi\tau\,,
\end{aligned}
$$

where the second equalities follow from equations (28.8.32) and (28.8.33), and $r \equiv \beta^{-1} - 1$. It can be seen that the wage under the Mortensen and Pissarides' specification is reduced in the first period by the worker's share of any future layoff tax, and future wages are increased by an amount equal to the net interest on this posted "bond." In other words, the present value of a worker's total compensation for any completed job is identical for the two specifications. It follows that the present value of a firm's match surplus is also identical across specifications. We have thereby confirmed that the same equilibrium allocation is supported by Nash product (28.8.23) and Mortensen and Pissarides' alternative bargaining formulation.

28.9. Kiyotaki-Wright search model of money

We now explore a discrete-time version of Kiyotaki and Wright's (1993) search model of money.[12] Let us first study their environment without money. The economy is populated by a continuum of infinitely lived agents, with total population normalized to unity. There is also a number of differentiated commodities, which are indivisible and come in units of size one. Agents have idiosyncratic tastes over these consumption goods as captured by a parameter $x \in (0, 1)$. In

---

[12] Our main simplification is that the time to produce is deterministic rather than stochastic. We also alter the way money is introduced into the model.

particular, x equals the proportion of commodities that can be consumed by any given agent, and x also equals the proportion of agents that can consume any given commodity. If a commodity can be consumed by an agent, then we say that it is one of his consumption goods. An agent derives utility $U > 0$ from consuming one of his consumption goods, while the goods that he cannot consume yield zero utility.

Initially, let each agent be endowed with one good, and let these goods be randomly drawn from the set of all commodities. Goods are costlessly storable, but each agent can store at most one good at a time. The only input in the production of goods is the agents' own prior consumption. After consuming one of his consumption goods, an agent produces next period a new good drawn randomly from the set of all commodities. We assume that agents can consume neither their own output nor their initial endowment, so for consumption and production to take place there must be exchange among agents.

Agents meet pairwise and at random. In each period, an agent meets another agent with probability $\theta \in (0, 1]$ and he has no encounter with probability $1 - \theta$. Two agents who meet will trade if there is a mutually agreeable transaction. Any transaction must be quid pro quo because private credit arrangements are ruled out by the assumptions of a random matching technology and a continuum of agents. We also assume that there is a transaction cost $\epsilon \in (0, U)$ in terms of disutility, which is incurred whenever accepting a commodity in trade. Thus, a trader who is indifferent between holding two goods will never trade one for the other.

Agents choose trading strategies in order to maximize their expected discounted utility from consumption net of transaction costs, taking as given the strategies of other traders. Following Kiyotaki and Wright (1993), we restrict our attention to symmetric Nash equilibria, where all agents follow the same strategies and all goods are treated the same, and to steady states, where strategies and aggregate variables are constant over time.

In a symmetric equilibrium, an agent will trade only if he is offered a commodity that belongs to his set of consumption goods, and then consumes it immediately. Accepting a commodity that is not one's consumption good would only give rise to a transaction cost ϵ without affecting expected future trading opportunities. This statement is true because no commodities are treated as special in a symmetric equilibrium, and therefore the probability of a commodity being accepted by the next agent one meets is independent of the type of commodity one has.[13] It follows that x is the probability that a trader located at random is willing to accept any given commodity, and x^2 becomes the probability that two traders consummate a barter in a situation of "double coincidence of wants."

At the beginning of a period before the realization of the matching process, the value of an agent's optimization problem becomes

$$V_c^n = \theta\, x^2\, (U - \epsilon) + \beta V_c^n,$$

where $\beta \in (0, 1)$ is the discount factor. The superscript and subscript of V_c^n denote a nonmonetary equilibrium and a commodity trader, respectively, to set the stage for our next exploration of the role for money in this economy. How will fiat money affect welfare? Keep the benchmark of a barter economy in mind,

$$V_c^n = \frac{\theta\, x^2\, (U - \epsilon)}{1 - \beta}. \qquad (28.9.1)$$

28.9.1. Monetary equilibria

At the beginning of time, suppose a fraction $\bar{M} \in [0, 1)$ of all agents are each offered one unit of fiat money. The money is indivisible, and an agent can store at most one unit of money or one commodity at a time. That is, fiat money will enter into circulation only if some agents accept money and discard their endowment of goods. These decisions must be based solely on agents' beliefs about other traders' willingness to accept money in future transactions, because fiat money is by definition unbacked and intrinsically worthless. To determine whether or not fiat money will initially be accepted, we will therefore first have to characterize monetary equilibria.[14]

Fiat money adds two state variables in a symmetric steady state: the probability that a commodity trader accepts money, $\Pi \in [0, 1]$, and the amount of

---

[13] Kiyotaki and Wright (1989) analyze commodity money in a related model with nonsymmetric equilibria, where some goods become media of exchange.

[14] If money is valued in an equilibrium, the relative price of goods and money is trivially equal to 1, since both objects are indivisible and each agent can carry at most one unit of the objects. Shi (1995) and Trejos and Wright (1995) endogenize the price level by relaxing the assumption that goods are indivisible.

money circulating, $M \in [0, \bar{M}]$, which is also the fraction of all agents carrying money. An equilibrium pair (Π, M) must be such that an individual's choice of probability of accepting money when being a commodity trader, π, coincides with the economy-wide Π, and the amount of money M is consistent with the decisions of those agents who are initially free to replace their commodity endowment with fiat money.

In a monetary equilibrium, agents can be divided into two types of traders. An agent brings either a commodity or a unit of fiat money to the trading process; that is, he is either a commodity trader or a money trader. At the beginning of a period, the values associated with being a commodity trader and a money trader are denoted V_c and V_m, respectively. The Bellman equations can be written

$$V_c = \theta(1 - M)x^2\big(U - \epsilon + \beta V_c\big) + \theta M x \max_{\pi}\big[\pi \beta V_m + (1 - \pi)\beta V_c\big]$$

$$+ \big[1 - \theta(1 - M)x^2 - \theta M x\big]\beta V_c, \tag{28.9.2}$$

$$V_m = \theta(1 - M)x\Pi\big(U - \epsilon + \beta V_c\big) + \big[1 - \theta(1 - M)x\Pi\big]\beta V_m. \tag{28.9.3}$$

The value of being a commodity trader in equation (28.9.2) equals the sum of three terms. The first term is the probability of the agent meeting other commodity traders, $\theta(1 - M)$, times the probability that both want to trade, x^2, times the value of trading, consuming, and returning as a commodity trader next period, $U - \epsilon + \beta V_c$. The second term is the probability of the agent meeting money traders, θM, times the probability that a money trader wants to trade, x, times the value of accepting money with probability π, $\pi \beta V_m + (1 - \pi)\beta V_c$, where π is chosen optimally. The third term captures the complement to the two previous events when the agent stores his commodity to the next period with a continuation value of βV_c. According to equation (28.9.3), the value of being a money trader equals the sum of two terms. The first term is the probability of the agent meeting a commodity trader, $\theta(1 - M)$, times the probability of both wanting to trade, $x\Pi$, times the value of trading, consuming, and becoming a commodity trader next period, $U - \epsilon + \beta V_c$. The second term is the probability of the described event not occurring times the value of keeping the unit of fiat money to the next period, βV_m.

The optimal choice of π depends solely on Π. First, note that if $\Pi < x$ then equations (28.9.2) and (28.9.3) imply that $V_m < V_c$, so the individual's best response is $\pi = 0$. That is, if money is being accepted with a lower probability than a barter offer, then it is harder to trade using money than barter, so agents would never like to exchange a commodity for money. Second, if $\Pi > x$, then equations (28.9.2) and (28.9.3) imply that $V_m > V_c$, so the individual's best response is $\pi = 1$. If money is being accepted with a greater probability than a barter offer, then it is easier to trade using money than barter, and agents would always like to exchange a commodity for money whenever possible. Finally, if $\Pi = x$, then equations (28.9.2) and (28.9.3) imply that $V_m = V_c$, so π can be anything in $[0, 1]$. If monetary exchange and barter are equally easy, then traders are indifferent between carrying commodities and fiat money, and they could accept money with any probability. Based on these results, the individual's best-response correspondence is as shown in Figure 28.9.1, and there are exactly three values consistent with $\Pi = \pi$: $\Pi = 0$, $\Pi = 1$, and $\Pi = x$.

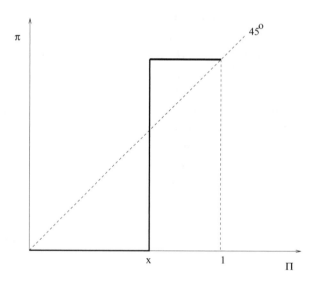

Figure 28.9.1: The best-response correspondence.

We can now answer our first question, namely, how many of the agents who are initially free to exchange their commodity endowment for fiat money will choose to do so? The answer is implicit in our discussion of the best-response correspondence. Thus, we have the following three types of symmetric equilibria:

1. A nonmonetary equilibrium with $\Pi = 0$ and $M = 0$, which is identical to the barter outcome in the previous section: Agents expect that money will be valueless, so they never accept it, and this expectation is self-fulfilling. All agents become commodity traders associated with a value of V_c^n, as given by equation (28.9.1).

2. A pure monetary equilibrium with $\Pi = 1$ and $M = \bar{M}$: Agents expect that money will be universally acceptable. From our previous discussion we know that agents will then prefer to bring money rather than commodities to the trading process. It is therefore a dominant strategy to accept money whenever possible; that is, expectation is self-fulfilling. Another implication is that the fraction $\bar{M}$ of agents who are initially free to exchange their commodity endowment for fiat money will also do so. Let V_c^p and V_m^p denote the values associated with being a commodity trader and a money trader, respectively, in a pure monetary equilibrium.

3. A mixed monetary equilibrium with $\Pi = x$ and $M \in [0, \bar{M}]$: Traders are indifferent between accepting and rejecting money as long as future trading partners take it with probability $\Pi = x$, so partial acceptability with agents setting $\pi = x$ can also be self-fulfilling. However, a mixed monetary equilibrium has no longer a unique mapping to the amount of circulating money M. Suppose the initial choices between commodity endowment and fiat money are separate from agents' decisions on trading strategies. It follows that any amount of money between $[0, \bar{M}]$ can constitute a mixed monetary equilibrium because of the indifference between a commodity endowment and a unit of fiat money. Of course, the allocation in a mixed monetary equilibrium with $M = 0$ is identical to the one in a nonmonetary equilibrium. Let $V_c^i(M)$ and $V_m^i(M)$ denote the values associated with being a commodity trader and a money trader, respectively, in a mixed monetary equilibrium with an amount of money equal to $M \in [0, \bar{M}]$.

28.9.2. Welfare

To compare welfare across different equilibria, we set $\pi = \Pi$ in equations (28.9.2) and (28.9.3) and solve for the reduced-form expressions

$$V_c = \frac{\psi}{1-\beta}\big\{(1-\beta)x + \beta\theta x \Pi\big[M\Pi + (1-M)x\big]\big\}, \qquad (28.9.4)$$

$$V_m = \frac{\psi}{1-\beta}\big\{(1-\beta)\Pi + \beta\theta x \Pi\big[M\Pi + (1-M)x\big]\big\}, \qquad (28.9.5)$$

where $\psi = [\theta(1-M)x(U-\epsilon)]/[1 - \beta(1 - \theta x \Pi)] > 0$. The value V_m is greater than or equal to V_c in a monetary equilibrium, since a necessary condition is that monetary exchange is at least as easy as barter ($\Pi \geq x$),

$$V_m = V_c + \psi(\Pi - x).$$

After setting $\Pi = x$ in equations (28.9.4) and (28.9.5), we see that a mixed monetary equilibrium with $M > 0$ gives rise to a strictly lower welfare as compared to the barter outcome in equation (28.9.1),

$$V_c^i(M) = V_m^i(M) = (1-M)V_c^n.$$

Even though some agents are initially willing to switch their commodity endowment for fiat money, it is detrimental for the economy as a whole. Since money is accepted with the same probability as commodities, money does not ameliorate the problem of "double coincidence of wants" but only diverts real resources from the economy.[15] In fact, as noted by Kiyotaki and Wright (1990), the mixed monetary equilibrium is isomorphic to the nonmonetary equilibrium of another economy where the probability of meeting an agent is reduced from θ to $\theta(1-M)$.

---

[15] This welfare result differs from that of Kiyotaki and Wright (1993), who assume that a fraction $\bar{M}$ of all agents are initially endowed with fiat money without any choice. It follows that those agents endowed with money are certainly better off in a mixed monetary equilibrium as compared to the barter outcome, while the other agents are indifferent. The latter agents are indifferent because the existence of the former agents has the same crowding-out effect on their consumption arrival rate in both types of equilibria. Our welfare results reported here are instead in line with Kiyotaki and Wright's (1990) original working paper based on a slightly different environment where agents can at any time dispose of their fiat money and engage in production.

In a pure monetary equilibrium ($\Pi = 1$), the value of being a money trader is strictly greater than the value of being a commodity trader. A natural welfare criterion is the *ex ante* expected utility before the quantity $\bar{M}$ of fiat money is randomly distributed,

$$W = \bar{M}V_m^p + (1 - \bar{M})V_c^p$$

$$= \frac{\theta(1 - \bar{M})x(U - \epsilon)}{1 - \beta}\left[\bar{M} + (1 - \bar{M})x\right]. \tag{28.9.6}$$

The first and second derivatives of equation (28.9.6) are

$$\frac{\partial W}{\partial \bar{M}} = \frac{\theta x(U - \epsilon)}{1 - \beta}\left\{1 - 2\left[\bar{M} + (1 - \bar{M})x\right]\right\}, \tag{28.9.7}$$

$$\frac{\partial^2 W}{\partial \bar{M}^2} = -2\frac{\theta x(U - \epsilon)}{1 - \beta}(1 - x) < 0. \tag{28.9.8}$$

Since the second derivative is negative, fiat money can only have a welfare-enhancing role if the first derivative is positive when evaluated at $\bar{M} = 0$. Thus, according to equation (28.9.7), money can (cannot) increase welfare if $x < .5$ ($x \geq .5$). Intuitively speaking, when $x \geq .5$, each agent is willing to consume (and therefore accept) at least half of all commodities, so barter is not very difficult. The introduction of money would here only reduce welfare by diverting real resources from the economy. When $x < .5$, barter is sufficiently difficult so that the introduction of some fiat money improves welfare. The optimum quantity of money is then found by setting equation (28.9.7) equal to zero, $\bar{M}^\star = (1 - 2x)/(2 - 2x)$. That is, $\bar{M}^\star$ varies negatively with x, and the optimum quantity of money increases when x shrinks and the problem of "double coincidence of wants" becomes more difficult. In particular, $\bar{M}^\star$ converges to .5 when x goes to zero.

28.10.   Concluding remarks

The frameworks of search and matching present various ways of departing from the frictionless Arrow-Debreu economy where all agents meet in a complete set of markets. This chapter has mainly focused on labor markets as a central application of these theories. The presented models have the concept of frictions in common, but there are also differences. The island economy has frictional unemployment without any externalities. An unemployed worker does not inflict any injury on other job seekers other than what a seller of a good imposes on his competitors. The equilibrium value to search, v_u, serves the function of any other equilibrium price of signaling to suppliers the correct social return from an additional unit supplied. In contrast, the matching model with its matching function is associated with externalities. Workers and firms impose congestion effects when they enter as unemployed in the matching function or add another vacancy in the matching function. To arrive at an efficient allocation in the economy, it is necessary that the bilaterally bargained wage be exactly right. In a labor market with homogeneous firms and workers, efficiency prevails only if the workers' bargaining strength, ϕ, is exactly equal to the elasticity of the matching function with respect to the measure of unemployment, α. In the case of heterogeneous jobs in the same labor market with a single matching function, we established the impossibility of efficiency without government intervention.

The matching model unarguably offers a richer analysis through its extra interaction effects, but it comes at the cost of the model's microeconomic structure. In an explicit economic environment, feasible actions can be clearly envisioned for any population size, even if there is only one Robinson Crusoe. The island economy is an example of such a model with its microeconomic assumptions, such as the time it takes to move from one island to another. In contrast, the matching model with its matching function imposes relationships between aggregate outcomes. It is therefore not obvious how the matching function arises when gradually increasing the population from one Robinson Crusoe to an economy with more agents. Similarly, it is an open question what determines when heterogeneous firms and labor have to be matched through a common matching function and when they have access to separate matching functions.

Peters (1991) and Montgomery (1991) suggest some microeconomic underpinnings to labor market frictions, which are further pursued by Burdett, Shi, and Wright (2001). Firms post vacancies with announced wages, and unemployed workers can apply to only one firm at a time. If the values of filled jobs differ across firms, firms with more valued jobs will have an incentive to post higher wages to attract job applicants. In an equilibrium, workers will be indifferent between applying to different jobs, and they are assumed to use identical mixed strategies in making their applications. In this way, vacancies may remain unfilled because some firms do not receive any applicants, and some workers may find themselves "second in line" for a job and therefore remain unemployed. When assuming a large number of firms that take market tightness as given for each posted wage, Montgomery finds that the decentralized equilibrium does maximize welfare for reasons similar to Moen's (1997) identical finding that was discussed earlier in this chapter.

Lagos (2000) derives a matching function from a model without any exogenous frictions at all. He studies a dynamic market for taxicab rides in which taxicabs seek potential passengers on a spatial grid and the fares are regulated exogenously. In each location, the shorter side determines the number of matches. It is shown that a matching function exists for this model, but this matching function is an equilibrium object that changes with policy experiments. Lagos sounds a warning that assuming an exogenous matching function when doing policy analysis might be misleading.

Throughout our discussion of search and matching models, we have assumed risk-neutral agents. Acemoglu and Shimer (1999), and Gomes, Greenwood, and Rebelo (2001) analyze a matching model and a search model, respectively, where agents are risk averse and hold precautionary savings because of imperfect insurance against unemployment.

Exercises

Exercise 28.1   **An island economy** (Lucas and Prescott, 1974)

Let the island economy in this chapter have a productivity shock that takes on two possible values, $\{\theta_L, \theta_H\}$ with $0 < \theta_L < \theta_H$. An island's productivity remains constant from one period to another with probability $\pi \in (.5, 1)$, and its productivity changes to the other possible value with probability $1 - \pi$. These symmetric transition probabilities imply a stationary distribution where half of the islands experience a given θ at any point in time. Let $\hat{x}$ be the economy's labor supply (as an average per market).

a. If there exists a stationary equilibrium with labor movements, argue that an island's labor force has two possible values, $\{x_1, x_2\}$ with $0 < x_1 < x_2$.

b. In a stationary equilibrium with labor movements, construct a matrix Γ with the transition probabilities between states (θ, x), and explain what the employment level is in different states.

c. In a stationary equilibrium with labor movements, we observe only four values of the value function $v(\theta, x)$ where $\theta \in \{\theta_L, \theta_H\}$ and $x \in \{x_1, x_2\}$. Argue that the value function takes on the same value for two of these four states.

d. Show that the condition for the existence of a stationary equilibrium with labor movements is

$$\beta(2\pi - 1)\theta_H \; > \; \theta_L \,, \tag{1}$$

and, if this condition is satisfied, an implicit expression for the equilibrium value of x_2 is

$$[\theta_L + \beta(1 - \pi)\theta_H] \, f'(2\hat{x} - x_2) \; = \; \beta\pi\theta_H f'(x_2) \,. \tag{2}$$

e. Verify that the allocation of labor in part d coincides with a social planner's solution when maximizing the present value of the economy's aggregate production. Starting from an initial equal distribution of workers across islands, condition (1) indicates when it is optimal for the social planner to increase the number of workers on high-productivity islands. The first-order condition for the social planner's choice of x_2 is then given by equation (2).

Exercise 28.2 **Business cycles and search** (Gomes, Greenwood, and Rebelo, 2001)

Part I *The worker's problem*

Think about an economy in which workers all confront the following common environment: Time is discrete. Let $t = 0, 1, 2, \ldots$ index time. At the beginning of each period, a previously employed worker can choose to work at her last period's wage or draw a new wage. If she draws a new wage, the old wage is lost and she will be unemployed in the current period. She can start work at the new wage in the next period. New wages are independent and identically distributed from the cumulative distribution function F, where $F(0) = 0$, and $F(M) = 1$ for $M < \infty$. Unemployed workers face a similar problem. At the beginning of each period, a previously unemployed worker can choose to work at last period's wage offer or to draw a new wage from F. If she draws a new wage, the old wage offer is lost and she can start working at the new wage in the following period. Someone offered a wage is free to work at that wage for as long as she chooses (she cannot be fired). The income of an unemployed worker is b, which includes unemployment insurance and the value of home production. Each worker seeks to maximize $E_0 \sum_{t=0}^{\infty} (1-\mu)^t \beta^t I_t$, where μ is the probability that a worker dies at the end of a period, β is the subjective discount factor, and I_t is the worker's income in period t; that is, I_t is equal to the wage w_t when employed and the income b when unemployed. Here, E_0 is the mathematical expectation operator, conditioned on information known at time 0. Assume that $\beta \in (0, 1)$ and $\mu \in (0, 1)$.

a. Describe the worker's optimal decision rule. In particular, what should an employed worker do? What should an unemployed worker do?

b. How would an unemployed worker's behavior be affected by an increase in μ?

Part II *Equilibrium unemployment rate*

The economy is populated with a continuum of the workers just described. There is an exogenous rate of new workers entering the labor market equal to μ, which equals the death rate. New entrants are unemployed and must draw a new wage.

c. Find an expression for the economy's unemployment rate in terms of exogenous parameters and the endogenous reservation wage. Discuss the determinants of the unemployment rate.

We now change the technology so that the economy fluctuates between booms (B) and recessions (R). In a boom, all employed workers are paid an extra $z > 0$. That is, the income of a worker with wage w is $I_t = w + z$ in a boom and $I_t = w$ in a recession. Let whether the economy is in a boom or a recession define the *state* of the economy. Assume that the state of the economy is i.i.d. and that booms and recessions have the same probabilities of .5. The state of the economy is publicly known at the beginning of a period before any decisions are made.

d. Describe the optimal behavior of employed and unemployed workers. When, if ever, might workers choose to quit?

e. Let w_B and w_R be the reservation wages in booms and recessions, respectively. Assume that $w_B < w_R$. Let G_t be the fraction of workers employed at wages $w \in [w_B, w_R]$ in period t. Let U_t be the fraction of workers unemployed in period t. Derive difference equations for G_t and U_t in terms of the parameters of the model and the reservation wages, $\{F, \mu, w_B, w_R\}$.

f. Figure 28.1 contains a simulated time series from the solution of the model with booms and recessions. Interpret the time series in terms of the model.

Exercise 28.3 **Business cycles and search again**

The economy is either in a boom (B) or recession (R) with probability .5. The state of the economy $(R$ or $B)$ is i.i.d. through time. At the beginning of each period, workers know the state of the economy for that period. At the beginning of each period, a previously employed worker can choose to work at her last period's wage or draw a new wage. If she draws a new wage, the old wage is lost, b is received this period, and she can start working at the new wage in the following period. During recessions, new wages (for jobs to start next period) are i.i.d. draws from the c.d.f. F, where $F(0) = 0$ and $F(M) = 1$ for $M < \infty$. During booms, the worker can choose to quit and take *two* i.i.d. draws of a possible new wage (with the option of working at the higher wage, again for a job to start the next period) from the *same* c.d.f. F that prevails during recessions. (This ability to choose is what "jobs are more plentiful during booms" means to workers.) Workers who are unemployed at the beginning of

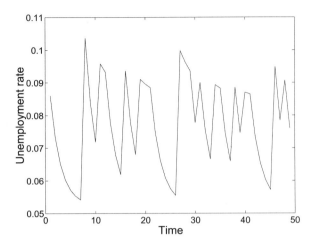

Figure 28.1: Unemployment during business cycles.

a period receive b this period and draw either one (in recessions) or two (in booms) wage offers from the c.d.f. F to start work next period.

A worker seeks to maximize $E_0 \sum_{t=0}^{\infty} (1 - \mu)^t \beta^t I_t$, where μ is the probability that a worker dies at the end of a period, β is the subjective discount factor, and I_t is the worker's income in period t; that is, I_t is equal to the wage w_t when employed and the income b when unemployed.

a. Write the Bellman equation(s) for a previously employed worker.

b. Characterize the worker's quitting policy. If possible, compare reservation wages in booms and recessions. Will employed workers ever quit? If so, who will quit and when?

Exercises 28.4–28.6 **European unemployment**

The following three exercises are based on work by Ljungqvist and Sargent (1998), Marimon and Zilibotti (1999), and Mortensen and Pissarides (1999b), who calibrate versions of search and matching models to explain high European unemployment. Even though the specific mechanisms differ, they all attribute the rise in unemployment to generous benefits in times of more dispersed labor market outcomes for job seekers.

Exercise 28.4   **Skill-biased technological change**   (Mortensen and Pissarides, 1999b)

Consider a matching model in discrete time with infinitely lived and risk-neutral workers who are endowed with different skill levels. A worker of skill type i produces h_i goods in each period that she is matched to a firm, where $i \in \{1, 2, \ldots, N\}$ and $h_{i+1} > h_i$. Each skill type has its own but identical matching function $M(u_i, v_i) = Au_i^\alpha v_i^{1-\alpha}$, where u_i and v_i are the measures of unemployed workers and vacancies in skill market i. Firms incur a vacancy cost $c h_i$ in every period that a vacancy is posted in skill market i; that is, the vacancy cost is proportional to the worker's productivity. All matches are exogenously destroyed with probability $s \in (0, 1)$ at the beginning of a period. An unemployed worker receives unemployment compensation b. Wages are determined in Nash bargaining between matched firms and workers. Let $\phi \in [0, 1)$ denote the worker's bargaining weight in the Nash product, and we adopt the standard assumption that $\phi = \alpha$.

a. Show analytically how the unemployment rate in a skill market varies with the skill level h_i.

b. Assume an even distribution of workers across skill levels. For different benefit levels b, study numerically how the aggregate steady-state unemployment rate is affected by mean-preserving spreads in the distribution of skill levels.

c. Explain how the results would change if unemployment benefits are proportional to a worker's productivity.

Exercise 28.5   **Dispersion of match values**   (Marimon and Zilibotti, 1999)

We retain the matching framework of exercise *28.4* but assume that all workers have the same innate ability $h = \bar{h}$ and any earnings differentials are purely match specific. In particular, we assume that the meeting of a firm and a worker is associated with a random draw of a match-specific productivity p from an exogenous distribution $G(p)$. If the worker and firm agree to stay together, the output of the match is then $p \cdot h$ in every period as long as the match is not exogenously destroyed as in exercise *28.4*. We also keep the assumptions of a constant unemployment compensation b and Nash bargaining over wages.

a. Characterize the equilibrium of the model.

b. For different benefit levels b, study numerically how the steady-state unemployment rate is affected by mean-preserving spreads in the exogenous distribution $G(p)$.

Exercise 28.6 **Idiosyncratic shocks to human capital** (Ljungqvist and Sargent, 1998)

We retain the assumption of exercise *28.5* that a worker's output is the product of his human capital h and a job-specific component which we now denote w, but we replace the matching framework with a search model. In each period of unemployment, a worker draws a value w from an exogenous wage offer distribution $G(w)$ and, if the worker accepts the wage w, he starts working in the following period. The wage w remains constant throughout the employment spell that ends either because the worker quits or the job is exogenously destroyed with probability s at the beginning of each period. Thus, in a given job with wage w, a worker's earnings wh can only vary over time because of changes in human capital h. For simplicity, we assume that there are only two levels of human capital, h_1 and h_2 where $0 < h_1 < h_2 < \infty$. At the beginning of each period of employment, a worker's human capital is unchanged from last period with probability π_e and is equal to h_2 with probability $1 - \pi_e$. Losses of human capital are only triggered by exogenous job destruction. In the period of an exogenous job loss, the laid off worker's human capital is unchanged from last period with probability π_u and is equal to h_1 with probability $1 - \pi_u$. All unemployed workers receive unemployment compensation, and the benefits are equal to a replacement ratio $\gamma \in [0, 1)$ times a worker's last job earnings.

a. Characterize the equilibrium of the model.

b. For different replacement ratios γ, study numerically how the steady-state unemployment rate is affected by changes in h_1.

Comparison of models

c. Explain how the different models in exercises *28.4* through *28.6* address the observations that European welfare states have experienced less of an increase in earnings differentials as compared to the United States, but suffer more from long-term unemployment where the probability of gaining employment drops off sharply with the length of the unemployment spell.

d. Explain why the assumption of infinitely lived agents is innocuous for the models in exercises *28.4* and *28.5*, but the alternative assumption of finitely lived agents can make a large difference for the model in exercise *28.6*.

Exercise 28.7   **Temporary jobs and layoff costs**

Consider a search model with temporary jobs. At the beginning of each period, a previously employed worker loses her job with probability μ, and she can keep her job and wage rate from last period with probability $1 - \mu$. If she loses her job (or chooses to quit), she draws a new wage and can start working at the new wage in the following period with probability 1. After a first period on the new job, she will again in each period face probability μ of losing her job. New wages are independent and identically distributed from the cumulative distribution function F, where $F(0) = 0$, and $F(M) = 1$ for $M < \infty$. The situation during unemployment is as follows. At the beginning of each period, a previously unemployed worker can choose to start working at last period's wage offer or to draw a new wage from F. If she draws a new wage, the old wage offer is lost and she can start working at the new wage in the following period. The income of an unemployed worker is b, which includes unemployment insurance and the value of home production. Each worker seeks to maximize $E_0 \sum_{t=0}^{\infty} \beta^t I_t$, where β is the subjective discount factor, and I_t is the worker's income in period t; that is, I_t is equal to the wage w_t when employed and the income b when unemployed. Here E_0 is the mathematical expectation operator, conditioned on information known at time 0. Assume that $\beta \in (0, 1)$ and $\mu \in (0, 1]$.

a. Describe the worker's optimal decision rule.

Suppose that there are two types of temporary jobs: short-lasting jobs with μ_s and long-lasting jobs with μ_l, where $\mu_s > \mu_l$. When the worker draws a new wage from the distribution F, the job is now randomly designated as either short-lasting with probability π_s or long-lasting with probability π_l, where $\pi_s + \pi_l = 1$. The worker observes the characteristics of a job offer, (w, μ).

b. Does the worker's reservation wage depend on whether a job is short-lasting or long-lasting? Provide intuition for your answer.

We now consider the effects of layoff costs. It is assumed that the government imposes a cost $\tau > 0$ on each worker that loses a job (or quits).

c. Conceptually, consider the following two reservation wages, for a given value of μ: (i) a previously unemployed worker sets a reservation wage for accepting last period's wage offer; (ii) a previously employed worker sets a reservation wage for continuing working at last period's wage. For a given value of μ, compare these two reservation wages.

d. Show that an unemployed worker's reservation wage for a short-lasting job exceeds her reservation wage for a long-lasting job.

e. Let $\bar{w}_s$ and $\bar{w}_l$ be an unemployed worker's reservation wages for short-lasting jobs and long-lasting jobs, respectively. In period t, let N_{st} and N_{lt} be the fractions of workers employed in short-lasting jobs and long-lasting jobs, respectively. Let U_t be the fraction of workers unemployed in period t. Derive difference equations for N_{st}, N_{lt} and U_t in terms of the parameters of the model and the reservation wages, $\{F, \mu_s, \mu_l, \pi_s, \pi_l, \bar{w}_s, \bar{w}_l\}$.

Exercise 28.8 **Productivity shocks, job creation, and job destruction**, donated by Rodolfo Manuelli

Consider an economy populated by a large number of identical individuals. The utility function of each individual is

$$\sum_{t=0}^{\infty} \beta^t x_t,$$

where $0 < \beta < 1$, $\beta = 1/(1+r)$, and x_t is income at time t. All individuals are endowed with one unit of labor that is supplied inelastically: If the individual is working in the market, its productivity is y_t, while if he or she works at home, productivity is z. Assume that $z < y_t$. Individuals who are producing at home can also, at no cost, search for a market job. Individuals who are searching and jobs that are vacant get randomly matched. Assume that the number of matches per period is given by

$$M(u_t, x_t),$$

where M is concave, increasing in each argument, and homogeneous of degree 1. In this setting, u_t is interpreted as the total number of unemployed workers, and v_t is the total number of vacancies. Let $\theta \equiv v/u$, and let $q(\theta) = M(u, v)/v$ be the probability that a vacant job (or firm) will meet a worker. Similarly, let $\theta q(\theta) = M(u, v)/u$ be the probability that an unemployed worker is matched with a vacant job. Jobs are exogenously destroyed with probability s. In order to create a vacancy, a firm must pay a cost $c > 0$ per period in which the vacancy is "posted" (i.e., unfilled). There is a large number of potential firms (or jobs), and this guarantees that the expected value of a vacant job, V, is zero. Finally, assume that when a worker and a vacant job meet, they bargain according to the Nash bargaining solution, with the worker's share equal to φ. Assume that $y_t = y$ for all t.

a. Show that the zero-profit condition implies that

$$w = y - (r + s)c/q(\theta).$$

b. Show that if workers and firms negotiate wages according to the Nash bargaining solution (with worker's share equal to φ), wages must also satisfy

$$w = z + \varphi(y - z + \theta c).$$

c. Describe the determination of the equilibrium level of market tightness, θ.

d. Suppose that at $t = 0$, the economy is at its steady state. At this point, there is a once-and-for-all increase in productivity. The new value of y is $y' > y$. Show how the new steady-state value of θ, θ', compares with the previous value. Argue that the economy "jumps" to the new value right away. Explain why there are no "transitional dynamics" for the level of market tightness, θ.

e. Let u_t be the unemployment rate at time t. Assume that at time 0 the economy is at the steady-state unemployment rate corresponding to θ, the "old" market tightness, and display this rate. Denote this rate as u_0. Let $\theta_0 = \theta'$. Note that change in unemployment rate is equal to the difference between job destruction at t, JD_t and job creation at t, JC_t. It follows that

$$JD_t = (1 - u_t)s,$$
$$JC_t = \theta_t q(\theta_t)u_t,$$
$$u_{t+1} - u_t = JD_t - JC_t.$$

Go as far as you can characterizing job creation and job destruction at $t = 0$ (after the shock). In addition, go as far as you can describing the behavior of both JC_t and JD_t during the transition to the new steady state (the one corresponding to θ').

Exercise 28.9 **Workweek restrictions, unemployment, and welfare**, donated by Rodolfo Manuelli

Recently, France has moved to a shorter workweek of about 35 hours per week. In this exercise you are asked to evaluate the consequences of such a move. To this end, consider an economy populated by risk-neutral, income-maximizing workers with preferences given by

$$U = E_t \sum_{j=0}^{\infty} \beta^j y_{t+j}, \quad 0 < \beta < 1, \quad 1 + r = \beta^{-1}.$$

Assume that workers produce z at home if they are unemployed, and that they are endowed with one unit of labor. If a worker is employed, he or she can spend x units of time at the job, and $(1-x)$ at home, with $0 \le x \le 1$. Productivity on the job is yx, and x is perfectly observed by both workers and firms.

Assume that if a worker works x hours, his or her wage is wx.

Assume that all jobs have productivity $y > z$, and that to create a vacancy firms have to pay a cost of $c > 0$ units of output per period. Jobs are destroyed with probability s. Let the number of matches per period be given by

$$M(u, v),$$

where M is concave, increasing in each argument, and homogeneous of degree one. In this setting, u is interpreted as the total number of unemployed workers, and v is the total number of vacancies. Let $\theta \equiv v/u$, and let $q(\theta) = M(u, v)/v$.

Assume that workers and firms bargain over wages, and that the outcome is described by a Nash bargaining outcome with the workers' bargaining power equal to φ.

a. Go as far as you can describing the unconstrained (no restrictions on x other than it be a number between 0 and 1) market equilibrium.

b. Assume that $q(\theta) = A\theta^{-\alpha}$, for some $0 < \alpha < 1$. Does the solution of the planner's problem coincide with the market equilibrium?

c. Assume now that the workweek is restricted to be less than or equal to $x^* < 1$. Describe the equilibrium.

d. For the economy in part c, go as far as you can (if necessary, make additional assumptions) describing the impact of this workweek restriction on wages, unemployment rates, and the total number of jobs. Is the equilibrium optimal?

Exercise 28.10   **Costs of creating a vacancy and optimality**, donated by Rodolfo Manuelli

Consider an economy populated by risk-neutral, income-maximizing workers with preferences given by

$$U = E_t \sum_{j=0}^{\infty} \beta^j y_{t+j}, \quad 0 < \beta < 1, \quad 1 + r = \beta^{-1}.$$

Assume that workers produce z at home if they are unemployed. Assume that all jobs have productivity $y > z$, and that to create a vacancy firms have to pay p_A, with $p_A = C'(v)$, per period when they have an open vacancy, with v being the total number of vacancies. Assume that the function $C(v)$ is strictly convex, twice differentiable and increasing. Jobs are destroyed with probability s.

Let the number of matches per period be given by

$$M(u, v),$$

where M is concave, increasing in each argument, and homogeneous of degree 1. In this setting, u is interpreted as the total number of unemployed workers, and v is the total number of vacancies. Let $\theta \equiv v/u$, and let $q(\theta) = M(u, v)/v$.

Assume that workers and firms bargain over wages and that the outcome is described by a Nash bargaining outcome with the worker's bargaining power equal to φ.

a. Go as far as you can describing the market equilibrium. In particular, discuss how changes in the exogenous variables, z, y, and the function $C(v)$, affect the equilibrium outcomes.

b. Assume that $q(\theta) = A\theta^{-\alpha}$ for some $0 < \alpha < 1$. Does the solution of the planner's problem coincide with the market equilibrium? Describe instances, if any, in which this is the case.

Exercise 28.11 **Financial wealth, heterogeneity, and unemployment**, donated by Rodolfo Manuelli

Consider the behavior of a risk-neutral worker who seeks to maximize the expected present discounted value of wage income. Assume that the discount factor is fixed and equal to β, with $0 < \beta < 1$. The interest rate is also constant and satisfies $1 + r = \beta^{-1}$. In this economy, jobs last forever. Once the worker has accepted a job, he or she never quits and the job is never destroyed. Even though preferences are linear, a worker needs to consume a minimum of a units of consumption per period. Wages are drawn from a distribution with support on $[a, b]$. Thus, any employed individual can have a feasible consumption level. There is no unemployment compensation.

Individuals of type i are born with wealth a^i, $i = 0, 1, 2$, where $a^0 = 0$, $a^1 = a$, $a^2 = a(1 + \beta)$. Moreover, in the period that they are born, all individuals are unemployed. Population, N_t, grows at the constant rate $1 + n$. Thus, $N_{t+1} = (1+n)N_t$. It follows that, at the beginning of period t, at least nN_{t-1} individuals — those born in that period — will be unemployed. Of the nN_{t-1} individuals born at time t, φ^0 are of type 0, φ^1 of type 1, and the rest, $1 - \varphi^0 - \varphi^1$, are of type 2. Assume that the mean of the offer distribution (the mean offered, not necessarily accepted, wage) is greater than a/β.

a. Consider the situation of an unemployed worker who has $a^0 = 0$. Argue that this worker will have a reservation wage $w^*(0) = a$. Explain.

b. Let $w^*(i)$ be the reservation wage of an individual with wealth i. Argue that $w^*(2) > w^*(1) > w^*(0)$. What does this say about the cross-sectional relationship between financial wealth and employment probability? Discuss the economic reasons underlying this result.

c. Let the unemployment rate be the number of unemployed individuals at t, U_t, relative to the population at t, N_t. Thus, $u_t = U_t/N_t$. Argue that in this economy, the unemployment rate is constant.

d. Consider a policy that redistributes wealth in the form of changes in the fraction of the population that is born with wealth a^i. Describe as completely as you can the effect upon the unemployment rate of changes in φ^i. Explain your results.

Extra credit: Go as far as you can describing the distribution of the random variable "number of periods unemployed" for an individual of type 2.

Chapter 29
Foundations of Aggregate Labor Supply

29.1. Introduction

The section 28.6 employment lotteries model for years served as the foundation of the high aggregate labor supply elasticity that generates big employment fluctuations in real business cycle models. In the original version of his Nobel prize lecture, Prescott (2005a) highlighted the central role of employment lotteries for real business cycle models when he asserted that "Rogerson's aggregation result is every bit as important as the one giving rise to the aggregate production function." But Prescott's enthusiasm for employment lotteries has not been shared universally, especially by researchers who have studied labor market experiences of individual workers. For example, Browning, Hansen, and Heckman (1999) expressed doubts about the employment lotteries model when they asserted that "the employment allocation mechanism strains credibility and is at odds with the micro evidence on individual employment histories." This chapter takes such criticisms of the employment lotteries to heart by investigating how the aggregate labor supply elasticity would be affected were we to replace employment lotteries and complete markets for consumption insurance with the incomplete markets arrangements that seem more natural to labor economists. This change reorients attention away from the fraction of its members that a representative family chooses to send to work at any moment, to *career lengths* chosen by individual workers who self-insure by saving and dissaving. We find that abandoning the employment lotteries coupled with complete consumption insurance claims trading assumed within many real business cycle models and replacing them with individual workers who self-insure by trading a risk-free bond does not by itself imperil that high aggregate labor supply elasticity championed by Prescott. The labor supply elasticity depends on whether shocks and government financed social security retirement schemes leave most workers on or off corners with respect to their retirement decisions, in a model of indivisible labor.

During the last half decade, macroeconomists have mostly abandoned employment lotteries in favor of 'time-averaging' and incomplete markets as an 'aggregation' theory for aggregate labor supply. This is undoubtedly a positive development because now researchers who may differ about the size of the aggregate labor supply elasticity can at least talk in terms of a common framework and can focus on their disagreements about the proper quantitative settings for a commonly agreed on set of parameters and constraints.

To convey these ideas, we build on an analysis of Ljungqvist and Sargent (2007), who in a particular continuous time model showed that the very same aggregate allocation and individual (expected) utilities that emerge from a Rogerson-style complete-market economy with employment lotteries are also attained in an incomplete-market economy without lotteries. In the Ljungqvist-Sargent setting, instead of trading probabilities of working at any point in time, agents choose fractions of their lifetimes to devote to work and use a credit market to smooth consumption across episodes of work and times of retirement.[1]

This chapter studies how two camps of researchers, namely, those who champion high and low labor supply elasticities, respectively, both came to adopt the same theoretical framework.[2] The first part of the chapter revisits equivalence results between an employment lotteries model and a time-averaging model,

---

[1] Larry Jones and Casey Mulligan anticipated aspects of this equivalence result. In the context of indivisible consumption goods, in the original 1988 version of his paper, Jones (2008) showed how timing could replace lotteries when there is no discounting. In the 2008 published version of his paper, he extended the analysis to cover the case of discounting. In comparing an indivisible-labor complete-market model and a representative-agent model with divisible labor, Mulligan (2001) suggested that the elimination of employment lotteries and complete markets for consumption claims from the former model might not make much of a quantitative difference; "The smallest labor supply decision has an infinitesimal effect on lifetime consumption and the marginal utility of wealth in the [divisible-labor] model, and a small-but-larger-than-infinitesimal effect on the marginal utility of wealth in the [indivisible-labor] model – as long as the effect on lifetime consumption is a small fraction of lifetime income *or* the marginal utility of wealth does not diminish too rapidly." However, as we shall learn later in this chapter, these qualifications vanish when time is continuous, as well as for infinitely-lived agents in discrete time. As a discussant of Ljungqvist and Sargent (2007), Prescott (2007) endorsed their incomplete markets, career length model as a model of aggregate labor supply. In addition, he reduced his previous stress on the employment lotteries model by adding a new section, "The life cycle and labor indivisibility," to the final version of his Nobel lecture published in America (Prescott 2006).

[2] This is the theme of Ljungqvist and Sargent (2011).

then pursues various extensions to the time-averaging setup as a model of career length determination. The second half of the chapter retraces the steps that led Chang and Kim's (2007) to discover a high labor supply elasticity in simulations of a Bewley with incomplete markets and indivisible labor. Chang and Kim's agents optimally alternate between periods of work and leisure (they 'time average') to allocate consumption and leisure over their infinite lifespans. The chapter concludes by studying how Ljungqvist and Sargent's (2007) equivalence result in continuous time with finitely-lived agents extends to a deterministic version of Chang and Kim's (2007) discrete-time growth model inhabited by infinitely-lived agents.

29.2. Equivalent allocations

Following Ljungqvist and Sargent (2007), consider an agent who lives in continuous time with a deterministic lifespan of unit length, and lifetime preferences given by

$$\int_0^1 e^{-\rho t} \left[u(c_t) - v(n_t) \right] dt \,, \tag{29.2.1}$$

where $c_t \geq 0$ and $n_t \in \{0, 1\}$ are consumption and labor supply at time t, respectively, and ρ is his subjective discount rate. That $n_t \in \{0, 1\}$ asserts that labor supply is indivisible. The instantaneous utility function over consumption, $u(c)$, is strictly increasing, strictly concave, and twice continuously differentiable. Since labor is indivisible, we need to specify only two points for the disutility of work $v(n)$, so we normalize $v(0) = 0$ and let $v(1) = B > 0$.

Until section 29.8, we assume a given wage rate w and a given interest rate $r = \rho$.

29.2.1. Choosing career length

At each point in time, an agent can work at a wage rate w and can save or dissave at an interest rate r. An agent's asset holdings at time t are denoted by a_t and its time derivative by $\dot{a}_t$. Initial assets are assumed to be zero, $a_0 = 0$, and the budget constraint at time t is

$$\dot{a}_t = ra_t + wn_t - c_t, \tag{29.2.2}$$

with a terminal condition $a_1 \geq 0$. This is a no-Ponzi scheme condition.

To solve the agent's optimization problem, we formulate the current-value Hamiltonian

$$H_t = u(c_t) - Bn_t + \lambda_t \left[ra_t + wn_t - c_t \right], \tag{29.2.3}$$

where λ_t is the multiplier on constraint (29.2.2). It is called the costate variable associated with the state variable a_t. First-order conditions with respect to c_t and n_t, respectively, are:

$$u'(c_t) - \lambda_t = 0, \tag{29.2.4a}$$

$$-B + \lambda_t w \begin{cases} < 0 & \text{if } n_t = 0; \\ = 0 & \text{if indifferent to } n_t \in \{0,1\}; \\ > 0 & \text{if } n_t = 1. \end{cases} \tag{29.2.4b}$$

Furthermore, the costate variable obeys the differential equation

$$\dot{\lambda}_t = \lambda_t \rho - \frac{\partial H_t}{\partial a_t} = \lambda_t [\rho - r]. \tag{29.2.5}$$

When $r = \rho$, Ljungqvist and Sargent (2007) show that the solution to this optimization problem yields the same lifetime utility as if the agent had access to employment lotteries and complete insurance markets (including consumption claims that are contingent on lottery outcomes). First, we note from equation (29.2.5) that when $r = \rho$ the costate variable is constant over time and hence, by equation (29.2.4a), the optimal consumption stream is constant over time, $c_t = \bar{c}$. Then after invoking optimality condition (29.2.4b), there are three possible cases with respect to the agent's lifetime labor supply,

$$-B + u'(\bar{c})w \begin{cases} < 0 & \text{\textit{Case 1}: } n_t = 0 \text{ for all } t; \\ = 0 & \text{\textit{Case 2}: indifference to } n_t \in \{0,1\} \text{ at any} \\ & \quad \text{particular instance in time;} \\ > 0 & \text{\textit{Case 3}: } n_t = 1 \text{ for all } t. \end{cases} \tag{29.2.6}$$

These three cases stand as analogues to the three cases in the section 28.6 static model with employment lotteries. The agent finds it optimal never to work and always to work in the first and third case, respectively. The interesting case is the intermediate one in which the agent is indifferent between work and leisure at any particular instance in time. At such an interior solution for lifetime labor supply, optimality condition (29.2.6) at equality determines the optimal constant consumption stream,

$$u'(\bar{c}) = \frac{B}{w} \, . \tag{29.2.7}$$

Evidently, this is the counterpart to the consumption outcome in the employment lottery model. When utility is logarithmic in consumption, the optimal consumption level in expression (29.2.7) becomes

$$\bar{c} = \frac{w}{B} \, , \qquad \text{if } u(c) = \log(c) \, . \tag{29.2.8}$$

While the agent is indifferent between work and leisure at any particular instance in time, he cares about the integral of his work over his lifetime. His lifetime labor supply is determined by the agent's present-value budget constraint at equality when financing the optimal constant consumption stream in expression (29.2.7). The present-value budget constraint is obtained from budget constraint (29.2.2), and the initial and terminal conditions for asset holdings, $a_0 = a_1 = 0$:

$$w \int_0^1 e^{-rt} n_t \, dt = \bar{c} \int_0^1 e^{-rt} \, dt. \tag{29.2.9}$$

Thus, the optimal plan has the agent working a fraction of his lifetime, where the associated present value of labor income is given by expression (29.2.9). Many streams of lifetime labor supply yield the same present value of labor income in expression (29.2.9). The agent is indifferent among such alternative lifetime labor profiles because constancy of the associated present value of labor income implies constancy of the associated lifetime disutility of work in preference specification (29.2.1) when $\rho = r$. Hence, the agent is indeed indifferent about when he supplies his labor, as we also inferred from the second case of (29.2.6).

In subsequent sections 29.3–29.7, we will assume that $r = \rho = 0$, i.e., no discounting. Under that assumption, the optimal fraction of a lifetime devoted to work, as given by present-value budget constraint (29.2.9), is the same regardless of when the agent supplies his labor,

$$T \equiv \int_0^1 n_t \, dt = \frac{\bar{c}}{w}, \tag{29.2.10}$$

where T denotes an agent's choice of career length. When utility is logarithmic in consumption, equations (29.2.8) and (29.2.10) determine the optimal career length at an interior solution,

$$T = \frac{1}{B}, \qquad \text{if } u(c) = \log(c); \tag{29.2.11}$$

where for an interior solution we require that $B \geq 1$.

Next, we confirm that a corresponding employment lottery model yields the same (expected) lifetime utility to an agent and support the same set of aggregate allocations, i.e., the introduction of lotteries and complete consumption insurance does not matter in this economy.

29.2.2. *Employment lotteries*

Consider a continuum $j \in [0, 1]$ of ex ante identical agents like those in section 29.2.1. When markets are complete and there are employment lotteries to overcome the nonconvexity in labor supply, a decentralized market equilibrium is the solution to a planner problem, in which the planner weights are equal across all the ex ante identical agents. The planner chooses a consumption and employment allocation $c_{jt} \geq 0$, $n_{jt} \in \{0, 1\}$ to maximize

$$\int_0^1 \int_0^1 e^{-\rho t} \left[u(c_{jt}) - B n_{jt} \right] dt \, dj \tag{29.2.12}$$

subject to

$$\int_0^1 \int_0^1 e^{-rt} \left[w n_{jt} - c_{jt} \right] dt \, dj \geq 0. \tag{29.2.13}$$

Here the planner can borrow and lend at the rate r and send agents to work to earn the wage w.

The strict concavity of the utility function $u(\cdot)$ and our assumption that $r = \rho$ imply that the planner sets a constant consumption level across agents and across time, $c_{jt} = \bar{c}$ for all j and t. The planner exposes each agent at time t to a lottery that sends him to work with probability $\psi_t \in [0,1]$. The planner chooses $\bar{c}$ and ψ_t to maximize

$$\int_0^1 e^{-\rho t} \left[u(\bar{c}) - B\psi_t \right] dt \qquad (29.2.14)$$

subject to

$$\int_0^1 e^{-rt} \left[w\psi_t - \bar{c} \right] dt \geq 0. \qquad (29.2.15)$$

This problem resembles the 'time averaging' problem of a single agent in section 29.2.1. At an interior solution, the optimal constant consumption stream is once again given by equation (29.2.7), $u'(\bar{c}) = B/w$. A multitude of employment lotteries can satisfy present-value budget constraint (29.2.15) to finance the optimal consumption choice. Agents would be indifferent among all of those alternative lottery designs. As before, identical present values of labor income for any two labor supply schemes imply identical (expected) lifetime disutilities of work for those two schemes since $\rho = r$.[3]

This argument suffices to establish the equivalence of aggregate allocations and expected utilities between the incomplete-market economy in section 29.2.1 and the employment-lotteries, complete-market economy of the present section. An agent's optimal consumption is uniquely determined and identical across the two economies. For a given present-value of aggregate consumption, the same aggregate present-value of labor income can be attained with a multitude of intertemporal allocations for the aggregate measure of employed agents. Each of those alternative aggregate allocations is associated with either an incomplete-market economy where individual agents engage in time averaging or a complete-market economy with one of a variety of appropriate lottery designs. Since an agent's expected disutility of work is the same under the alternative implementations, it follows that an agent's expected utility is the same in the two economies.

---

[3] For example, at the beginning of time, the planner can randomize over a constant fraction of agents $\bar{\psi}$ who are assigned to work for every $t \in [0,1]$, and a fraction $1-\bar{\psi}$ who are asked to specialize in leisure, where $\bar{\psi}$ is chosen to satisfy the planner's intertemporal budget constraint (29.2.15). An alternative arrangement would be, at each time $t \in [0,1]$, the planner runs a lottery that sends a time invariant fraction $\bar{\psi}$ to work and a fraction $1-\bar{\psi}$ to leisure. Agents are indifferent between these alternative lottery designs since they yield the same expected lifetime disutility of work.

29.3. Taxation and social security

We study taxation and social security in a continuous-time overlapping generations model. At each instance in time, there is a constant measure of newborn ex ante identical agents like those in section 29.2.1 entering the economy. Thus, the economy's population and age structure stay constant over time. Our focus is not on the determination of intertemporal prices in this overlapping generations environment with its possible dynamic inefficiencies (see chapter 9), so we retain our small open economy assumption of an exogenously given interest rate, which also implies a given wage rate if the economy's production technology is constant returns to scale in labor and capital. [4]

We assume that utility is logarithmic in consumption, $u(c) = \log(c)$, and that there is no discounting, $r = \rho = 0$. The assumption of no discounting is inessential for most of our results, and where it matters we will take note. The analytical convenience is that the optimal career length is uniquely determined and does not depend on the timing of an agent's lifetime labor supply, as shown in expressions (29.2.10) and (29.2.11).

As emphasized by Prescott (2005), if labor income is taxed and tax revenues are handed back lump sum to agents, a model with indivisible labor and employment lotteries exhibits a large labor supply elasticity. Under the equivalence result in section 29.2, we follow Ljungqvist and Sargent (2007) and demonstrate that the same high labor supply elasticity arises in the incomplete-market model where career lengths rather than the odds of working in employment lotteries are shortened in response to such a tax system.

In the spirit of Ljungqvist and Sargent (2012), we offer a qualification to the high labor supply elasticity in a model of lifetime labor supply. When a government program such as social security is associated with a large implicit tax on working beyond an official retirement age, there might not be much of an effect of taxation on career length for those agents who could be at a corner solution, strictly preferring to retire at the official retirement age.

---

[4] In the case of a constant-returns-to-scale Cobb-Douglas production function, equation (29.8.5b) shows how the interest rate in international capital markets determines the capital-labor ratio in a small open economy, which in turn determines the wage rate in (29.8.5a).

29.3.1. Taxation

If labor income is taxed at rate $\tau \in [0, 1)$ and tax revenues are not returned to agents as tranfers in any form, there would be no effect on labor supply, for the same reason that equilibrium career length (29.2.11) does not depend on the level of the wage w. The reason is that income and substitution effects cancel with variations in the net-of-tax wage rate under the assumption that preferences are consistent with balanced growth. But if instead all tax receipts are rebated lump sum to agents, the labor supply elasticity will be large.

Let x be the present value of lump-sum transfers that each agent receives over his lifetime, as determined by the government budget constraint

$$\tau w T^{\star} = x, \tag{29.3.1}$$

where $T^{\star}$ is the equilibrium career length. Note that given a zero interest rate and a lifetime of unit length, x is the instant-by-instant per capita lump-sum transfer that satisfies the government's static budget constraint (29.3.1) as well as the present value of total lump-sum transfers paid to an agent over his lifetime.

As in section 29.2.1, an agent again chooses a unique constant consumption $\bar{c}$, and is indifferent among alternative labor supply paths that yield the particular present value of income that is required to finance his consumption choice. Under the present assumption of no discounting, all of those alternative labor supply paths have the same career length, i.e, the same fraction of an agent's lifetime devoted to work, $T = \int_0^1 n_t \, dt$. Hence, an agent's optimization problem becomes

$$\max_{\bar{c}, T} \left\{ \log(\bar{c}) - BT \right\} \tag{29.3.2}$$

subject to

$$\bar{c} \le (1 - \tau)wT + x, \tag{29.3.3}$$
$$\bar{c} \ge 0, \;\; T \in [0, 1].$$

Substitute budget constraint (29.3.3) into the objective function of (29.3.2), then compute a first-order condition with respect to career length at an interior solution,

$$\frac{(1 - \tau)w}{(1 - \tau)wT + x} - B = 0. \tag{29.3.4}$$

Substituting (29.3.1) into first-order condition (29.3.4) shows that equilibrium career length is

$$T^{\star}(\tau) \equiv \frac{1 - \tau}{B}. \tag{29.3.5}$$

We conclude that lifetime labor supply is highly elastic when labor is indivisible. According to expression (29.3.5), the elasticity of lifetime labor supply with respect to the net-of-tax rate $(1 - \tau)$ is equal to one.

The reader can verify that a model with employment lotteries yield the same equilibrium consumption and the same (expected) lifetime utility of an agent. For example, we can adopt the first example of a lottery design in footnote 3, where the planner for each cohort of newborn agents, administers a lifetime employment lottery once and for all at the beginning of life that assigns a fraction $\psi \in [0, 1]$ of agents to work always and a fraction $1 - \psi$ always to enjoy leisure. This planner problem is identical to the time averaging planning problem above, provided that we replace the choice variable T by ψ.

29.3.2. Social security

Instead of returning tax receipts lump sum to agents as in section 29.3.1, we now assume that all revenues are used to finance a social security system in which agents are eligible to retire and collect benefits after an official retirement age R. All labor earnings are subject to a flat rate social security tax $\tau \in (0, 1)$. Benefits *after* the agent's chosen retirement date T, which may or may not equal R, equal a replacement rate ρ times a worker's average earnings, i.e., ρ times the wage rate w. Agents who choose to retire after R collect no benefits until they actually retire.

To construct an equilibrium, we set the two parameters R and τ of the social security system, and then solve residually for a replacement rate ρ that is consistent with a balanced government budget. At an equilibrium career length $\tilde{T}$, the government budget constraint is

$$\tau w \tilde{T} = \left(1 - \max\{R, \tilde{T}\}\right) \rho w, \tag{29.3.6}$$

where the left side is tax revenues and the right side is social security benefits. The first (second) argument of the max operator presumes an equilibrium outcome in which workers retire before (after) the official retirement age. Note that the unit length of a lifetime implies that an age interval corresponds both to a fraction of an agent's lifetime and also to a fraction of the population within that age interval at any point in time. From budget constraint $(29.3.6)$ we can solve for the replacement rate,

$$\rho = \frac{\tau \tilde{T}}{1 - \max\{R, \tilde{T}\}} . \tag{29.3.7}$$

An agent's optimal career length solves

$$\max_{T \in [0,1]} \left\{ \log\left[(1 - \tau)wT + \rho w \min\{1 - R, 1 - T\}\right] - BT \right\}, \tag{29.3.8}$$

where we have substituted the agent's budget constraint into the utility function, and the arguments of the min operator appear in the same order as in the max operator of $(29.3.6)$, i.e., the first (second) argument refers to the case when the agent chooses to work shorter (longer) than the official retirement age.

Case with $\tilde{T} \leq R$

In the case of an optimal career length $T \leq R$, the first-order condition of $(29.3.8)$ at an interior solution (with respect to $T \leq R$) becomes

$$\frac{(1 - \tau)w}{(1 - \tau)wT + \rho w(1 - R)} - B = 0. \tag{29.3.9}$$

By government budget balance in $(29.3.7)$, $\rho = \tau \tilde{T}/(1 - R)$, which can be substituted into $(29.3.9)$ to yield an expression for equilibrium career length,

$$\tilde{T} = \frac{1 - \tau}{B} \equiv T^+(\tau). \tag{29.3.10}$$

Case with $\tilde{T} \geq R$

In the case of an optimal career length $T \geq R$, the first-order condition of $(29.3.8)$ at an interior solution (with respect to $T \geq R$) becomes

$$\frac{(1 - \tau)w - \rho w}{(1 - \tau)wT + \rho w(1 - T)} - B \geq 0, \tag{29.3.11}$$

which holds with equality except under a binding corner solution with $T = 1$. However, such a corner solution can be ruled out as an equilibrium outcome because government budget balance in $(29.3.7)$ would imply that the replacement rate goes to infinity; hence, it must be optimal for a worker to retire prior to the end of his lifetime. After substituting $\rho = \tau \tilde{T}/(1 - \tilde{T})$ into the denominator of $(29.3.11)$ at equality, we obtain an expression for equilibrium career length

$$\tilde{T} = \frac{1 - \tau - \rho}{B} = \frac{1 - \dfrac{\tau}{1 - \tilde{T}}}{B}, \tag{29.3.12}$$

where the second equality follows when we also substitute out for the second appearance of ρ.

Expression $(29.3.12)$ can be rearranged to become

$$B\tilde{T}^2 - (1 + B)\tilde{T} + 1 - \tau = 0. \tag{29.3.13}$$

The smaller root of this quadratic equation determines the equilibrium career length:

$$\tilde{T} = \frac{1 + B - \sqrt{(1 + B)^2 - 4B(1 - \tau)}}{2B} \equiv T^-(\tau), \tag{29.3.14}$$

where $\tilde{T}^-(0) = 1/B$, and $T^-(\tau)$ decreases monotonically to zero as τ goes to one.[5]

From equation $(29.3.10)$ that defines $T^+(\tau)$ and from equation $(29.3.12)$ that implicitly defines $T^-(\tau)$, it follows immediately that $T^+(\tau) > T^-(\tau)$ for $\tau \in (0, 1)$. We can now state a proposition that describes how the retirement age $\tilde{T}$ chosen in equilibrium depends on the official social security retirement age.

PROPOSITION: Given an official retirement age $R \in (0, 1)$ and a tax rate $\tau \in (0, 1)$, the equilibrium career length $\tilde{T}(R, \tau)$ is unique and given by

---

[5] After setting $\tau = 0$ in quadratic equation $(29.3.13)$, the two roots are

$$\frac{1 + B \pm \sqrt{1 + 2B + B^2 - 4B}}{2B} = \frac{1 + B \pm \sqrt{(1 - B)^2}}{2B}$$

$$= \frac{1 + B \pm |1 - B|}{2B} = \frac{1 + B \pm (B - 1)}{2B} = \left(1, \frac{1}{B}\right).$$

where we have invoked our parameter restriction $B \geq 1$ to evaluate the absolute value of $|1 - B| = B - 1$. The smaller root constitutes the equilibrium career length since it agrees with the agent's choice in $(29.2.11)$.

i)   if $R \leq T^-(\tau)$, then $\tilde{T}(R,\tau) = T^-(\tau)$ (retire *after* the official retirement age);
ii) if $R \geq T^+(\tau)$, then $\tilde{T}(R,\tau) = T^+(\tau)$ (retire *before* the official retirement age);
iii) otherwise, $\tilde{T}(R,\tau) = R$ (retire *at* the official retirement age).

Given $R = 0.6$, the solid curve in Figure 29.3.1 displays equilibrium career length as a function of τ. Within a range of tax rates between 16–40 percent, equilibrium career length does not respond to changes in the tax rate because agents are at a corner solution and strictly prefer to retire at the official retirement age R. Away from that corner, career length is highly sensitive to the social security tax rate τ in Figure 29.3.1.

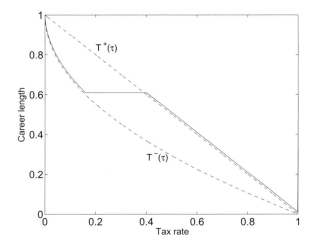

Figure 29.3.1: Social security. Solid curve depicts equilibrium career length as a function of a social security tax rate τ, given an official retirement age $R = 0.6$. At low (high) tax rates, $\tau < 0.16$ ($\tau > 0.40$), an agent retires after (before) the official retirement age, where the actual retirement age lies along the curve $T^-(\tau)$ ($T^+(\tau)$), given a disutility of work $B = 1$.

When an equilibrium has agents retiring *before* the official retirement age, $R > \tilde{T} = T^+(\tau)$, equilibrium career length (29.3.10) is identical to outcome (29.3.5) under the Prescott tax system. The reasons are that (a) under our assumption that average earnings alone determine the replacement rate without regard to career length, agents regard their social security contributions purely as a tax and perceive no extra benefits accruing to them from paying it, while (b) the present value of future social security payments operates like a lump sum transfer when optimal career length falls short of the official retirement age. The sensitivity of career length to social security taxation is even larger in an equilibrium that has agents retiring *after* the official retirement age, $R < \tilde{T} = T^-(\tau)$, because the marginal decision about career length is then also distorted by the loss of benefits incurred from working beyond the official retirement age, as shown by the first equality in expression (29.3.12).

29.4. Earnings-experience profiles

The equivalence of outcomes across models of employment lotteries and time averaging breaks down when human capital can be accumulated. A human capital accumulation technology typically makes career choice in effect induce another indivisibility that will be handled differently by our two types of models. While an agent in a time averaging model will contemplate when to terminate a career during which earnings have increased because of work experience or investments in human capital, the 'invisible hand' in a complete-market economy with employment lotteries will preside over a dual labor market in which some agents specialize in work and others in leisure. Here we adopt a specification of earnings-experience profiles of Ljungqvist and Sargent (2012).[6] An agent with past employment spells totaling $h_t = \int_0^t n_s \, ds$ has the opportunity to earn

$$w_t = W h_t^\phi, \qquad W > 0, \quad \phi \in [0, 1]. \tag{29.4.1}$$

---

[6] We defer an analysis of a Ben-Porath's (1967) human capital technology to section 29.6.

29.4.1. Time averaging

Under the assumption of no discounting, an agent is indifferent about the timing of his labor supply, so we are free to assume that the agent frontloads work at the beginning of life. The present value of labor income for someone who works a fraction T of his lifetime is

$$\int_0^T W t^\phi \, dt = \frac{W\, T^{\phi+1}}{\phi+1} \,.\qquad(29.4.2)$$

As before, since the subjective discount rate equals the market interest rate, an agent chooses a constant consumption stream $\bar{c}$. Hence, an agent's optimization problem becomes

$$\max_{\bar{c},T}\Big\{\log(\bar{c}) - BT\Big\}\qquad(29.4.3)$$

subject to

$$\bar{c} \le \frac{W\, T^{\phi+1}}{\phi+1}\,,\qquad(29.4.4)$$

$$\bar{c} \ge 0,\ \ T \in [0,1].$$

We substitute budget constraint $(29.4.4)$ into the objective function of $(29.4.3)$, and compute a first-order condition with respect to career length at an interior solution,

$$T = \frac{\phi+1}{B}\,;\qquad(29.4.5)$$

where the implicit parameter restriction for an interior solution is that $B \ge \phi+1$.

Because preferences are consistent with balanced growth, the optimal career length $(29.4.5)$ does not depend on the earnings level parameter W. But evidently, career length does increase with the elasticity parameter ϕ. The more elastic the earnings profile is to accumulated working time, the longer is an agent's career.

29.4.2. Employment lotteries

We make three modifications to the planner problem in section 29.2.2. Besides our two specializations of zero discounting and that the instantaneous utility function over consumption is logarithmic, there are now agent-specific wage rates w_{jt} with each agent's earnings increasing in his past experience as given by (29.4.1).

Because an agent's earnings increase with his experience, it follows immediately that an optimal employment allocation has a fraction ψ of agents to work always ($n_{jt} = 1$ for all $t \in [0,1]$ for these unlucky people) and a fraction $1 - \psi$ always to enjoy leisure ($n_{jt} = 0$ for all $t \in [0,1]$ for these lucky ones). Hence, the indeterminacy in lottery designs is now gone. An agent who works throughout his lifetime generates present-value labor income equal to $W/(\phi + 1)$, as defined in (29.4.2).

As before, the planner chooses constant consumption $\bar{c}$ across agents and across time. The planner's problem becomes

$$\max_{\bar{c},\psi}\left\{\log(\bar{c}) - B\psi\right\} \tag{29.4.6}$$

subject to

$$\bar{c} \le \psi\frac{W}{\phi + 1}, \tag{29.4.7}$$

$$\bar{c} \ge 0, \ \psi \in [0,1].$$

We substitute budget constraint (29.4.7) into the objective function of (29.4.6), and compute a first-order condition with respect to the fraction of the population sent to work at an interior solution,

$$\psi = \frac{1}{B}. \tag{29.4.8}$$

We conclude that agents in a complete-market economy with employment lotteries on average work less than agents who are left alone to 'time average' in an incomplete-market economy, as characterized by (29.4.5). The latter agents confront a difficult choice between enjoying leisure and earning additional labor income at the peak of their lifetime earnings potential. This choice is not faced by agents who follow the instructions of the planner who uses lotteries to convexify the indivisibility brought by careers. Of course, in the special ($\phi = 0$)

case when work experience does not affect earnings, the aggregate labor supplies as well as the expected lifetime utilities are exactly the same across the two economies, as asserted in the equivalence result of section 29.2.

29.4.3. Prescott tax and transfer scheme

It is instructive to revisit Prescott's tax analysis in section 29.3.1 for the present environment with earnings-experience profiles. We invite the readers to verify that the equilibrium career length in the time averaging economy is then

$$T^\star = \frac{(1-\tau)(\phi+1)}{B},$$
(29.4.9)

and the employment-population fraction in the employment lotteries economy is

$$\psi^\star = \frac{(1-\tau)}{B}.$$
(29.4.10)

While the labor supplies in (29.4.9) and (29.4.10) differ, we note that the elasticity of the supply with respect to the net-of-tax rate $(1 - \tau)$ is the same and equal to one. This equality is another reflection of broad similarities that typically prevail across incomplete-market and complete-market economies with indivisible labor. We shall encounter another example in section 29.8 when we compare the aggregate labor supply in a Bewley incomplete markets economy with its complete-market counterpart.

29.4.4. No discounting now matters

Recall that under a flat earnings-experience profile ($\phi = 0$) in section 29.2.1, an agent is indifferent about the multitude of labor supply paths that yield the same present-value of labor income in budget constraint (29.2.9). The reason is that two alternative labor supply paths with the same present-value of labor income imply the same lifetime disutility of work when $\rho = r$. Note that for strictly positive discounting, $\rho = r > 0$, a labor supply path that is tilted toward the future means that an agent will have to work for a longer period of time to generate the same present-value of labor income as compared to a labor supply path that is tilted toward the present. But that is acceptable to the agent since future disutilities of work are discounted at the same rate as labor earnings when the subjective discount rate is equal to the market discount rate.

But if there is an upward-sloping earnings-experience profile ($\phi > 0$), an agent is no longer indifferent to the described variation in career length associated with the timing of lifetime labor supply. In particular, when $\rho = r > 0$, an agent strictly prefers to shift his labor supply to the end of life because at a given lifetime disutility of work, working later in life would mean spending more total time working. That would push the worker further up the experience-earnings profile and thereby increase the present value of lifetime earnings.

Features not present in our model would attenuate such a desire to postpone labor supply to the end of life, e.g., borrowing constraints that force an agent to finance consumption with current labor earnings, incomplete insurance markets that compel an agent to resolve career uncertainties earlier, and forecast declines in dexterity with advances in age.

29.5. Intensive margin

Prescott et al. (2009) extend the analysis of Ljungqvist and Sargent (2007) in section 29.2 by introducing an intensive margin in labor supply, i.e., $n_t \in [0, 1]$ is now a continuous rather than a discrete choice variable. However, to retain the central force of indivisible labor, they postulate a nonlinear mapping from n_t to effective labor services, in particular, an increasing mapping that is first convex and then concave. For expositional simplicity, we let the effective labor services associated with n_t be $(n_t - \underline{n})$ where $\underline{n} \in (0, 1)$. As noted by Prescott et al. (2009) such a mapping can reflect costs associated with getting set up in a job, learning about coworkers, and so on.

The preferences are the same as those of Ljungqvist and Sargent (2007) in (29.2.1) but now with no discounting, $\rho = r = 0$. Under the present assumption that n_t is a continuous choice variable, we need to make additional assumptions about the function $v(\cdot)$. The instantaneous disutility function over work, $v(n)$, is strictly increasing, strictly convex, and twice continuously differentiable.

29.5.1. Employment lotteries

We begin by solving a complete-market economy with employment lotteries in a static model. To compute an equilibrium allocation, we posit that a planner chooses consumption and employment $c_j \geq 0$, $n_j \in [0,1]$ for a continuum of agents $j \in [0,1]$ to maximize

$$\int_0^1 [u(c_j) - v(n_j)] \, dj \tag{29.5.1}$$

subject to

$$\int_0^1 c_j \, dj \leq w \int_0^1 [n_j - \underline{n}] \, dj. \tag{29.5.2}$$

Strict concavity of $u(c)$ makes it optimal to assign the same consumption to each agent, $\bar{c}$. Likewise, because of strict convexity of $v(n)$, the planner asks for the same labor supply from each agent who is sent to work, $\bar{n}$. Conditional on working, the labor supply $\bar{n} > \underline{n}$ because it cannot be optimal to have agents incurring disutility of work without earning any income. For an agent j who is not working, $n_j = 0$.

Given this characterization of an optimal allocation, the planner's optimization problem becomes

$$\max_{\bar{c}, \bar{n}, \psi} \left\{ u(\bar{c}) - \psi v(\bar{n}) \right\} \tag{29.5.3}$$

subject to

$$\bar{c} \leq w \, (\bar{n} - \underline{n}) \, \psi, \tag{29.5.4}$$
$$\bar{c} \geq 0, \quad \bar{n} \in [0,1], \quad \psi \in [0,1],$$

where ψ is the fraction of the population that the planner sends to work, the same fraction ψ is also the probability of working in the employment lottery of the decentralized market economy.

As emphasized by Prescott et al. (2009), the interesting case is the one where the solutions for ψ and $\bar{n}$ are both interior. In this case, after substituting budget constraint (29.5.4) at equality into the objective function of (29.5.3), we obtain the following first-order conditions with respect to ψ and $\bar{n}$, respectively,

$$u'\Big(\psi \, w \, (\bar{n} - \underline{n}) \Big) \, w \, (\bar{n} - \underline{n}) = v(\bar{n}), \tag{29.5.5a}$$

$$u'\Big(\psi \, w \, (\bar{n} - \underline{n}) \Big) \, w \, \psi = \psi \, v'(\bar{n}). \tag{29.5.5b}$$

Dividing these equations gives

$$v'(\bar{n}) = \frac{v(\bar{n})}{\bar{n} - \underline{n}}. \tag{29.5.6}$$

This condition for optimality states that the marginal cost to the planner to supply additional effective labor services should be equalized across intensive and extensive margins. The marginal disutility at the intensive margin is $v'(\bar{n})$ when employed agents are asked to increase their hours worked, while the marginal cost at the extensive margin is $v(\bar{n})/(\bar{n} - \underline{n})$, i.e., the *average* disutility per effective hour of an agent who is asked to switch from not working to working.

Note that an employed agent's optimal labor supply $\bar{n}$ can be computed from (29.5.6) and depends on neither $\bar{c}$ nor ψ, except for the supposition of an interior solution for ψ. Given a solution for $\bar{n}$, we can then use either (29.5.5a) or (29.5.5b) to solve for ψ.

29.5.2. *Time averaging*

We now turn to a time averaging economy. An agent's problem is similar to that in section 29.2.1 but with the added intensive margin of Prescott et al. (2009) (and no discounting). An agent chooses lifetime consumption and employment $c_t \geq 0$, $n_t \in [0, 1]$ for $t \in [0, 1]$ to maximize

$$\int_0^1 [u(c_t) - v(n_t)] \, dt \tag{29.5.7}$$

subject to

$$\int_0^1 c_t \, dt \leq w \int_0^1 [n_t - \underline{n}] \, dt. \tag{29.5.8}$$

It is immediate that this problem is identical to the planner's problem in the static model of section 29.5.1, the only difference being that we now integrate across time rather than across agents. Hence, we can reformulate the agent's optimization problem to become

$$\max_{\bar{c}, \bar{n}, T} \left\{ u(\bar{c}) - T v(\bar{n}) \right\} \tag{29.5.9}$$

subject to

$$\bar{c} \leq w (\bar{n} - \underline{n}) T, \tag{29.5.10}$$

$$\bar{c} \geq 0, \quad \bar{n} \in [0, 1], \quad T \in [0, 1],$$

where T is the fraction of an agent's lifetime devoted to work, i.e., his career length.

29.5.3. Prescott taxation

To examine effects of taxation where there are both intensive and extensive margins, we adapt the analysis in section 29.3.1. The government budget constraint becomes

$$\tau\, w\, (\bar{n} - \underline{n})\, T^{\star} = x. \qquad (29.5.11)$$

Under the assumption that utility is logarithmic in consumption, an agent's optimization problem becomes

$$\max_{\bar{c}, \bar{n}, T} \left\{ \log(\bar{c}) - T v(\bar{n}) \right\} \qquad (29.5.12)$$

subject to

$$\bar{c} \le (1 - \tau) w\, (\bar{n} - \underline{n})\, T + x, \qquad (29.5.13)$$

$$\bar{c} \ge 0, \quad \bar{n} \in [0, 1], \quad T \in [0, 1].$$

Substitute budget constraint $(29.5.13)$ into the objective function of $(29.5.12)$, and compute the first-order conditions at interior solutions with respect to T and $\bar{n}$, respectively,

$$\frac{(1 - \tau)\, w\, (\bar{n} - \underline{n})}{(1 - \tau)\, w\, (\bar{n} - \underline{n})\, T + x} - v(\bar{n}) = 0, \qquad (29.5.14a)$$

$$\frac{(1 - \tau)\, w\, T}{(1 - \tau)\, w\, (\bar{n} - \underline{n})\, T + x} - T\, v'(\bar{n}) = 0. \qquad (29.5.14b)$$

Dividing these equations gives

$$v'(\bar{n}) = \frac{v(\bar{n})}{\bar{n} - \underline{n}}. \qquad (29.5.15)$$

This condition is the same as expression $(29.5.6)$ when there is no taxation and hence the intensive margin is not affected by taxation. To compute the equilibrium career length, we substitute $(29.5.11)$ into first-order condition $(29.5.14a)$,

$$T^{\star} = \frac{1 - \tau}{v(\bar{n})}.$$

Along with Prescott et al. (2009), we conclude that the effects of taxation are the same as in Ljungqvist and Sargent (2007), i.e., all the adjustment of labor supply takes place along the extensive margin, and the elasticity of aggregate labor supply with respect to the net-of-tax rate $(1 - \tau)$ is equal to one.

The reason that none of the adjustment takes place along the intensive margin is that any changes in labor when already working occur along an increasing marginal disutility of work, while adjustment along the extensive margin are made at a constant disutility of work by varying the fraction of one's lifetime devoted to work. The constancy of the latter terms of trade between working and not working was the essential ingredient of the famous (or, depending on your viewpoint, infamous) high labor supply elasticity in models of employment lotteries when labor is indivisible.

Rogerson and Wallenius (2009) break the constancy of the terms of trade between working and not working by adding a life cycle earnings profile to the present framework, but in contrast to section 29.4, they take that earnings profile as exogenously given rather than having it be determined as a function of an agent's past work experience. In the Rogerson and Wallenius setup, two results follow immediately: (a) agents choose to work when their life cycle earnings profile is highest, namely, when it exceeds an optimally chosen reservation level; and (b) labor supply n_t at a point in time varies positively with the exogenous earnings level. Taxation in this augmented framework affects labor supply along both the intensive and extensive margins. While an increasing marginal disutility of work continues to frustrate adjustment along the intensive margin, there is now decreasing earnings when extending the career beyond the heights of the exogenous life cycle earnings profile, which then also frustrates adjustment along the extensive margin. The assumed curvatures of the disutility of work at the intensive margin and that of the exogenous lifecycle earnings profile determine how much adjustment occurs along the intensive and extensive margins.

29.6. Ben-Porath human capital

We return to the assumption that labor is strictly indivisible, $n_t \in \{0, 1\}$, and add a Ben-Porath human capital accumulation technology to the framework of section 29.2. We take note of Ben-Porath's (1967, p. 361) observation that if the technology were to exhibit exact constant returns to scale, the marginal cost of additional units of human capital would be constant until all of the agent's current human capital is devoted to the effort of accumulating human capital and hence, the optimal rate of investment at any point in time would be either full specialization or no investment at all. Under our simplifying assumption of no depreciation of human capital, it follows that an agent would specialize and make all of his investment in human capital upfront. Acquiring human capital can be thought of as formal education before starting to work.

To represent the notion of specializing in human capital investments in a simple way, we assume that an agent has access to a technology that can instantaneously determine his human capital through the investment of $m \geq 0$ units of goods in himself, which produces a human capital level

$$h = m^\gamma, \qquad \gamma \in (0, 1), \qquad (29.6.1)$$

and there is no depreciation of human capital. It follows trivially that it will be optimal for an agent to use that technology once and for all before starting to work. Under our assumption of a perfect credit market, an agent chooses investment goods m that maximize his present value labor income, in conjunction with his choice of an optimal career length T.

Papers by Guvenen et al. (2011) and Manuelli et al. (2012) that incorporate Ben-Porath human capital technologies in life cycle models inspire our analysis. Those papers mainly focus on tax dynamics driven, not by the force in the Prescott tax system in section 29.3.1, but instead by wedges that distort an agent's investment in human capital. Guvenen et al. (2011) postulate progressive labor income taxation while Manuelli et al. (2012) assume that investments in human capital are not fully tax-deductible. In both cases, the central force is that the tax rate on returns to human capital is higher than the rate applied to labor earnings foregone while investing in human capital, or the rate at which goods input to human capital can be deducted from an agent's tax liabilities.

Following Manuelli et al. (2012), we assume a flat-rate tax $\tau \in (0, 1)$ on labor income and that only a fraction $\epsilon \in [0, 1]$ of goods input to human capital is tax-deductible. To isolate the key force at work in Manuelli et al. (2012) as well as in Guvenen et al. (2011), we assume no lump sum transfers of tax revenues to agents. However, at the end of the section, we will show how lump sum handovers remain as potent in suppressing the aggregate labor supply.[7]

29.6.1. Time averaging

As mentioned, an agent will find it optimal to invest an amount m in the human capital technology before starting to work. Equality between the subjective discount rate and the market interest rate implies that the agent chooses a constant consumption stream $\bar{c}$, and that he is indifferent to the timing of his labor supply. Moreover, because we assume no discounting so that $\rho = r = 0$, the optimal career length T is unique and does not depend on the timing of the agent's labor supply. Under the postulated human capital technology (29.6.1) and described tax policy, an agent's optimization problem becomes

$$\max_{\bar{c},m,T} \left\{ \log(\bar{c}) - BT \right\} \tag{29.6.2}$$

subject to

$$\bar{c} \leq (1-\tau)w\, m^\gamma\, T - (1-\tau\epsilon)m, \tag{29.6.3}$$
$$\bar{c} \geq 0, \quad m \geq 0, \quad T \in [0,1].$$

We substitute budget constraint (29.6.3) into the objective function (29.6.2), and compute first-order conditions with respect to m and, at an interior solution, T. After some manipulations, these first-order conditions with respect to m and T, respectively, become

$$m^{1-\gamma} = \frac{\gamma(1-\tau)w}{1-\tau\epsilon}T \tag{29.6.4a}$$

$$T = \frac{1}{B} + \frac{1-\tau\epsilon}{1-\tau}\frac{m^{1-\gamma}}{w}. \tag{29.6.4b}$$

---

[7] Given indivisible labor, Manuelli et al. (2012) disarm the potentially large effects of lump sum transfers of tax revenues by modelling social security systems with implicit tax wedges at an official retirement age, which gives rise to corner solutions in agents' career decisions as analyzed in section 29.3.2. In Guvenen et al.'s (2011) analysis of divisible labor as well as in their exploration of indivisible labor in an earlier working paper, the sensitivity of career length to lump sum transfers does not arise because they assume an exogenous retirement age, a common assumption in much the overlapping generations literature.

Substituting $(29.6.4a)$ into $(29.6.4b)$ yields

$$T = \frac{1}{(1-\gamma)B},$$ \hfill $(29.6.5)$

where the implicit parameter restriction for an interior solution is $(1-\gamma)B \geq 1$.

The optimally chosen career length in $(29.6.5)$ is invariant to taxation under our assumption that no tax revenues are handed back lump sum to agents. Any effect of taxation on the stock of human capital and hence the level of labor earnings, do not affect an agent's willingness to work since preferences are consistent with balanced growth, i.e., income and subsitution effects cancel. Given a constant career length in $(29.6.5)$, it follows from expression $(29.6.4a)$ that human capital investments would also be invariant to taxation if all these investments were tax-deductible, i.e., if $\epsilon = 1$. But if $\epsilon < 1$, we see that human capital investments decline in the tax rate τ because human capital returns are taxed at a higher rate than the rate at which rate the goods input to human capital is tax-deductible.

The severity of the tax distortion depends on the curvature parameter γ of the human capital technology. For example, when human capital investments are not tax-deductible, $\epsilon = 0$, we can solve for m from equation $(29.6.4a)$, and compute an agent's human capital stock as given by $(29.6.1)$:

$$h = \left[\gamma(1-\tau)w\,T\right]^{\frac{\gamma}{1-\gamma}}.$$

Thus, the elasticity of human capital, and for that matter, also labor earnings whT, with respect to net-of-tax rate $(1-\tau)$ is equal to $\gamma/(1-\gamma)$, which becomes arbitrarily large as γ approaches one. A high value of γ is associated with strong output effects of taxation because of reasons similar to those in an 'AK model' (output is linear in a single input, capital). The single input in our human capital technology $(29.6.1)$ is reproducible, and exhibits weak diminishing returns when γ is close to one. Likewise, for a standard formulation of the Ben-Porath technology, both the input of purchased goods and the input of an agent's current human capital services are *de facto* reproducible, so similarities with an 'AK model' arise if the human capital technology exhibits close to constant returns to scale (combined with the standard assumption of constant returns to scale in the goods technology that employs human and physical capital).

29.6.2. Employment lotteries

As in the case of an earnings-experience profile in section 29.4, the planner will optimally (and randomly) assign a fraction ψ of agents to work their entire lives while a fraction $1 - \psi$ will specialize in leisure. Needless to say, the planner will invest only in human capital for agents who are sent work. Under the postulated human capital technology (29.6.1) and described tax policy, the planner's problem becomes

$$\max_{\bar{c}, m, \psi} \left\{ \log(\bar{c}) - B\psi \right\} \tag{29.6.6}$$

subject to

$$\bar{c} \leq (1 - \tau) w \, m^\gamma \, \psi - (1 - \tau\epsilon) m \psi, \tag{29.6.7}$$

$$\bar{c} \geq 0, \quad m \geq 0, \quad \psi \in [0, 1].$$

We substitute budget constraint (29.6.7) into the objective function (29.6.6), and compute first-order conditions with respect to m and, at an interior solution, ψ. After some manipulations, these first-order conditions with respect to m and ψ, respectively, become

$$m^{1-\gamma} = \frac{\gamma(1 - \tau)w}{1 - \tau\epsilon}, \tag{29.6.8a}$$

$$\psi = \frac{1}{B}. \tag{29.6.8b}$$

As in the case of an earnings-experience profile, agents in the employment lotteries economy on average work less than do agents in the time-averaging economy: compare expression (29.6.8b) to that of (29.6.5). Agents in the time-averaging economy confront a difficult choice between enjoying leisure and earning additional labor income derived from their past investment in human capital. Once again, this difficult choice is not confronted by agents in the employment lotteries economy where the planner randomly assigns a fraction ψ of the population to work their entire lives, and thereby ensures an efficient use of all human capital. However, the difference in labor supply diminishes as γ approaches zero. In the limit ($\gamma = 0$) when the technology can no longer be used to augment an agent's human capital, labor supplies are the same across the two economies and we are back to our equivalence result in section 29.2.

Given that agents who are sent to work in the employment lotteries economy work their entire lives, it is not surprising that the planner makes a larger human capital investment in each employed worker as compared to an agent's investment decision in the time averaging economy: compare expression $(29.6.8a)$ to that of $(29.6.4a)$. After solving for the planner's choice of m from $(29.6.8a)$ and substituting into human capital technology $(29.6.1)$, the human capital stock per employed worker in the employment lotteries economy is

$$h = \left[\gamma(1-\tau)w\right]^{\frac{\gamma}{1-\gamma}}.$$

Thus, the elasticities of human capital and labor earnings $wh\psi$, with respect to net-of-tax rate $(1-\tau)$ are the same as those in the time averaging economy and equal to $\gamma/(1-\gamma)$.

29.6.3. Prescott taxation

We now add Prescott's assumption that tax revenues, net of any tax deductions on human capital investments, are returned lump sum to agents. Such handouts remain potent in suppressing the aggregate labor supply in the time averaging economy as well as in the employment lotteries economy.

In the time averaging economy, we adapt the analysis of section 29.3.1 as follows. The government budget constraint becomes

$$\tau\, w\, h^{\star}\, T^{\star} - \tau\, \epsilon\, m^{\star} = x, \tag{29.6.9}$$

where $T^{\star}$, $h^{\star}$ and $m^{\star}$ are equilibrium values of career length, human capital stock, and agents' purchase of goods input to the human capital technology, respectively. An agent's budget constraint is augmented to include the lump sum transfer x,

$$\bar{c} \leq (1-\tau)w\, m^{\gamma}\, T - (1-\tau\epsilon)m + x. \tag{29.6.10}$$

After substituting budget constraint $(29.6.10)$ into the objective function $(29.6.2)$, we can compute and verify that our earlier first-order condition $(29.6.4a)$ with respect to m is unchanged, i.e.,

$$m^{1-\gamma} = \frac{\gamma(1-\tau)w}{1-\tau\epsilon}T. \tag{29.6.11}$$

However, the career length T is no longer invariant to taxation. Specifically, after substituting government budget constraint $(29.6.9)$ into the agent's first-order condition with respect to T, at an interior solution, the result can be rearranged to read

$$T = \frac{1-\tau}{B} + \frac{m^{1-\gamma}}{w}, \qquad (29.6.12)$$

into which we substitute expression $(29.6.11)$ to obtain

$$T^\star = \frac{1-\tau}{\left[1 - \gamma \frac{1-\tau}{1-\tau\epsilon}\right] B}. \qquad (29.6.13)$$

Since $\tau \in (0,1)$ and $\epsilon \in [0,1]$, the equilibrium career length is now shorter than it was in expression $(29.6.5)$ when tax revenues, net of any tax deductions on human capital investments, were not handed back lump sum to agents.

For example, when investments in human capital are fully tax-deductible, $\epsilon = 1$, we disarm the key distortionary force that is the focus of Manuelli et al. (2012), i.e., the choice of goods input in expression $(29.6.13)$ is no longer distorted by the differential tax treatment of investments into and returns from human capital, but Prescott's tax distortion, arising from the lump sum handover of tax revenues to agents, is a forceful determinant of equilibrium career length $(29.6.13)$, evaluated at $\epsilon = 1$,

$$T^\star\big|_{\epsilon=1} = \frac{1-\tau}{(1-\gamma)B}. \qquad (29.6.14)$$

Hence, career length is no longer invariant to taxation when the government hands over tax revenues lump sum to agents; in particular, compare expression $(29.6.14)$ to that in $(29.6.5)$. And once again, the elasticity of aggregate labor supply with respect to the net-of-tax rate $(1-\tau)$ is equal to one. But now labor income declines further because of depressed investments in human capital, as determined by the goods input from expression $(29.6.11)$ that varies positively with career length.

We leave it as an exercise to readers to derive the corresponding equilibrium outcomes in the employment lotteries model, by following the same steps as above. It can be verified that the planner's first-order condition with respect to m remains the same as in expression $(29.6.8a)$, while steps analogous to those above, where government budget constraint is substituted into the planner's first-order condition with respect to ψ, yield

$$\psi^\star = \frac{(1-\tau)wm^\gamma - (1-\tau\epsilon)m}{(wm^\gamma - m)\,B} < \frac{1}{B}. \qquad (29.6.15)$$

The strict inequality is implied by $\tau \in (0, 1)$ and $\epsilon \in [0, 1]$, and hence, the equilibrium fraction of population sent to work is smaller as compared to that found in section 29.6.2, where tax revenues were not handed back lump sum to agents, as given by (29.6.8*b*). It is again instructive to consider the case when investments in human capital are fully tax-deductible, $\epsilon = 1$,

$$\psi^{\star}\Big|_{\epsilon=1} = \frac{1-\tau}{B}, \qquad (29.6.16)$$

where the elasticity of aggregate labor supply with respect to the net-of-tax rate $(1 - \tau)$ is again equal to one.

29.7. Earnings shocks

Next we study how earnings shocks affect an agent's choice of career length. Following Ljungqvist and Sargent (2012), we study an unanticipated permanent earnings shock, which will enable us to highlight forces that will also be at work in richer environments. For additional analytical simplicity, we assume a flat earnings profile, no discounting, and that utility is logarithmic in consumption. The parameter restriction $B > 1$ guarantees a strictly interior solution to lifetime labor supply (at least prior to the unanticipated earnings shock).

As in our basic setup in section 29.2.1, an agent can choose to work at the wage rate w, and solves a deterministic lifetime labor supply problem. The optimal solution at an interior solution is the constant consumption level in expression (29.2.8), $\bar{c} = w/B$, and a career length that is given by expression (29.2.11), $1/B$. Since an agent is indifferent to the timing of his labor supply, we are free to assume that he starts to work at time $t_0 \in (0, 1 - 1/B)$, and continues to work his entire optimally chosen career length, i.e., the agent is intent on retiring at age

$$\bar{R} = t_0 + 1/B. \qquad (29.7.1)$$

An employed agent pays off debt and accumulates assets for retirement, with asset holdings at time $t \in [t_0, \bar{R}]$ as given by

$$A(t) = \int_0^{t_0} -\bar{c}\,ds + \int_{t_0}^t (w - \bar{c})\,ds = w(t - t_0) - \frac{w}{B}t, \qquad (29.7.2)$$

where we have invoked the optimal consumption level, $\bar{c} = w/B$. Before starting to work, an agent finances consumption by borrowing and hence, there is some date $\bar{t} \in (t_0, \bar{R})$ at which an employed has just repaid his debt, $A(\bar{t}) = 0$, and starts to accumulate assets for retirement, where

$$A(\bar{t}) = w(\bar{t} - t_0) - \frac{w}{B}\bar{t} = 0 \quad \Longrightarrow \quad \bar{t} = \frac{B}{B-1}t_0. \tag{29.7.3}$$

Consider an unanticipated mid-career earnings shock at time $\hat{t} \in [t_0, \bar{R}]$, when the wage rate unexpectedly jumps from w to $\hat{w}$ for $t \in [\hat{t}, 1]$. Subject to the asset holdings $A(\hat{t})$ that were accumulated under the old optimal plan, the shock prompts the agent to maximize the remainder of his lifetime utility,

$$\int_{\hat{t}}^{1} \Big[\log(\hat{c}_t) - B\hat{n}_t \Big] dt$$

by choosing new values $\hat{c}_t \geq 0$ and $\hat{n}_t \in \{0, 1\}$ of consumption and labor supply, respectively, for $t \in [\hat{t}, 1]$. The agent's revised optimal plan prescribes a constant consumption path over the interval $[\hat{t}, 1]$ and a new retirement age $\hat{R} \in [\hat{t}, 1]$.

For the agent who after the unanticipated wage shock at $\hat{t}$ chooses to work until $R \in [\hat{t}, 1]$, the sum of the financial assets already accumulated at time $\hat{t}$, $A(\hat{t})$, and the present value of future labor income becomes

$$A(\hat{t}) + \int_{\hat{t}}^{R} \hat{w}\, ds = w(\hat{t} - t_0) - \frac{w}{B}\hat{t} + \hat{w}(R - \hat{t}). \tag{29.7.4}$$

This expression divided by $1 - \hat{t}$ is then the constant consumption rate over the remaining lifetime $1 - \hat{t}$, since the time $\hat{t}$ present value of financial plus nonfinancial wealth must equal the present value of consumption over the period $[\hat{t}, 1]$.

The agent's optimal retirement age thus solves

$$\max_{R \in [\hat{t}, 1]} \left\{ (1 - \hat{t}) \log \left[\frac{w(\hat{t} - t_0) - \frac{w}{B}\hat{t} + \hat{w}(R - \hat{t})}{1 - \hat{t}} \right] - B(R - \hat{t}) \right\}.$$

The first-order condition for R is

$$\frac{(1 - \hat{t})\hat{w}}{w(\hat{t} - t_0) - \frac{w}{B}\hat{t} + \hat{w}(R - \hat{t})} - B \begin{cases} < 0, & \text{corner solution } \hat{R} = \hat{t}; \\ = 0, & \text{interior solution } \hat{R} \in [\hat{t}, 1]; \\ > 0, & \text{corner solution } \hat{R} = 1; \end{cases} \tag{29.7.5}$$

where $\hat{R}$ is the optimal retirement age after the wage shock at time $\hat{t}$. At an interior solution to $\hat{R}$, first-order condition (29.7.5) holds with equality,

$$\frac{(1-\hat{t})}{B} = \frac{w}{\hat{w}}(\hat{t} - t_0) - \frac{w}{\hat{w}}\frac{1}{B}\hat{t} + \hat{R} - \hat{t}.$$

After adding t_0 to both sides of this equation, and using expression (29.7.1) for the original retirement age, $\bar{R} = t_0 + 1/B$, the post-shock retirement age $\hat{R}$ at an interior solution relates to the original retirement age $\bar{R}$ in the following way:

$$\hat{R} = \bar{R} + \frac{\hat{w} - w}{\hat{w}}\left[\frac{B-1}{B}\hat{t} - t_0\right] = \bar{R} + \frac{\hat{w}-w}{\hat{w}}(\hat{t}-\bar{t})\frac{B-1}{B}, \qquad (29.7.6)$$

where the second equality is obtained by using expression (29.7.3) to substitute out for t_0.

Evidently, the sign of the revision $\hat{R} - \bar{R}$ to an unanticipated wage shock depends (i) on whether $\hat{w} > w$ or $\hat{w} < w$, and (ii) on whether $\hat{t}$ is greater than or smaller than $\bar{t}$, where $\bar{t}$ defined in (29.7.3) is the point in time when the asset holdings of an employed agent turns from being negative to being positive. In response to a *negative* earnings shock, $\hat{w} < w$, the agent reduces (increases) his lifetime labor supply if his time $\hat{t}$ asset holdings are positive (negative), i.e., if $A(\hat{t}) > 0$ ($A(\hat{t}) < 0$), which means that the shock occurs at a time $\hat{t} > \bar{t}$ ($\hat{t} < \bar{t}$). In contrast, in response to a *positive* wage shock, $\hat{w} > w$, the agent increases (decreases) her lifetime labor supply if her current asset holdings are positive (negative).

These strong predictions based merely on the signs of the earnings shock and an agent's asset holdings, follow from the assumption that preferences are consistent with balanced growth. Ljungqvist and Sargent (2012) generalize the result to a larger class of such preferences and allow for the earnings-experience profile in section 29.4.

29.7.1. *Interpretation of wealth and substitution effects*

For an agent with positive asset holdings at $\hat{t}$, a *negative* wage shock means that returns to working fall relative to the marginal value of his wealth. That induces the agent to enjoy more leisure because doing that has now become relatively less expensive. But with negative asset holdings at $\hat{t}$, a negative wage shock compels the agent to supply more labor both to pay off time $\hat{t}$ debt and to moderate the adverse effect of the shock on his future consumption.

With a *positive* wage shock, leisure becomes more expensive, causing the agent to substitute away from leisure and toward consumption. This force makes lifetime labor supply increase for an agent with positive wealth. But why does a positive wage shock lead to a *reduction* in life-time labor supply when time $\hat{t}$ assets are negative?

In the case of a positive wage shock and negative time $\hat{t}$ assets, consider a hypothetical asset path that would have prevailed if the agent had enjoyed the higher wage rate $\hat{w}$ from the beginning starting at $t = 0$. Along that hypothetical path, the agent would have been even further in debt at $\hat{t}$ (since the optimal constant consumption level would have been equal to $\hat{w}/B$, as given by (29.2.8)). So at $\hat{t}$, the agent actually finds himself *richer* at $\hat{t}$ than he would have in our hypothetical scenario. Because there is less debt to be repaid at $\hat{t}$, the agent chooses to supply less labor than he would have in the hypothetical scenario. In other words, it is not optimal to make up for what would have been past underconsumption relative to our hypothetical path, so the agent chooses instead to enjoy more leisure because he has relatively less debt at $\hat{t}$ than he would along the hypothetical path.

29.8. Time averaging in a Bewley model

In a version of a Bewley model with incomplete markets (see chapter 18), Chang and Kim (2007) demonstrate how indivisible labor is associated with a high labor supply elasticity when the infinitely-lived agents engage in 'time averaging,' i.e., alternating between work and leisure. In such an incomplete-markets model, agents accumulate assets not only because of the standard precautionary motive to self-insure against productivity shocks, but also to finance planned spells of leisure.

We abstract from aggregate productivity shocks, but otherwise postulate the same neoclassical growth model (and its calibration) as that of Chang and Kim (2007). The economy is populated by a continuum (measure one) of agents who have identical preferences but experience different idiosyncratic productivity shocks. An agent's preference specification is similar to that of section 29.2 except that now time is discrete, agents live forever, and there is uncertainty;

$$E_0 \sum_{t=0}^{\infty} \beta^t \left[\log(c_t) - Bn_t \right], \tag{29.8.1}$$

where E_t is the expectation operator conditional on information at time t, and $\beta \in (0,1)$ is the agent's subjective discount factor. An agent who works in period t supplies z_t efficiency units of labor where the idiosyncratic productivity level z_t varies exogenously according to a stochastic process with a transition probability distribution function $\pi(z'|z) = \text{Prob}(z_{t+1} \leq z'|z_t = z)$, which has a unique unconditional stationary cumulative distribution function $G(z)$.[8]

The aggregate production function is Cobb-Douglas and exhibits constant returns to scale,

$$F(L_t, K_t) = L_t^{\alpha} K_t^{1-\alpha},$$

where L_t and K_t are the aggregates of efficiency units of labor and of physical capital. Capital depreciates at rate δ each period.

---

[8] As compared to Chang and Kim (2007), we let our parameter B in preference specification (29.8.1) replace their composite of three parameters, $B\bar{h}^{1+1/\gamma}/(1+1/\gamma)$, since the separate identification of e.g. a curvature parameter γ has no significance under the assumption of indivisible labor. Likewise, our normalization of time supplied when working, $n = 1$, as compared to their separate parameter $\bar{h}$ also lacks significance as long as our disutility parameter B has the same value as their composite of three parameters, and so long as we properly scale the productivity shocks so that the implied processes for an agent's efficiency units of labor are the same.

29.8.1. Incomplete markets

A stationary equilibrium has a constant interest rate r and a constant wage rate w per efficiency unit of labor. The state variables for an agent's problem are then his beginning-of-period assets a, before receiving interest earnings, and his productivity z. The agent's value function is

$$V(a, z) = \max_{n,c,a'} \left\{ \log(c) - Bn + \beta \int V(a', z') \, d\pi(z'|z) \right\}, \qquad (29.8.2)$$

subject to

$$a' = (1 + r)a + wzn - c, \qquad (29.8.3)$$
$$n \in \{0, 1\}, \quad c \geq 0, \quad a' \geq \hat{a},$$

where savings must satisfy a borrowing constraint, $a' \geq \hat{a}$. The solution to this problem includes a decision rule for labor supply, $n(a, z)$, consumption, $c(a, z)$, and asset holdings, $a'(a, z)$.

After substituting budget constraint (29.8.3) into the utility function in (29.8.2), we take a first-order condition with respect to a' and obtain an Euler equation,

$$\frac{1}{c(a, z)} = \beta \int V_1(a', z') \, d\pi(z'|z) = \beta(1 + r) \int \frac{1}{c(a', z')} d\pi(z'|z), \qquad (29.8.4)$$

into which we have substituted the decision rule for consumption, $c(a, z)$, and applied the Benveniste-Scheinkman formula $V_1(a', z') = (1 + r) \, u'(c(a', z'))$.

Firms' profit maximization ensures that

$$w = F_1(L, K) = (1 - \alpha) \left(\frac{K}{L} \right)^{\alpha}, \qquad (29.8.5a)$$

$$r = F_2(L, K) - \delta = \alpha \left(\frac{K}{L} \right)^{\alpha - 1} - \delta. \qquad (29.8.5b)$$

Associated with a stationary equilibrium is a time-invariant distribution of agents across asset holdings and productivities, $J(a, z)$. The invariant distribution satisfies

$$J(a^o, z^o) = \int_{a' \leq a^o, z' \leq z^o} \left\{ \int \mathbf{I}(a' = a'(a, z)) \, d\pi(z'|z) \, dJ(a, z) \right\} da' \, dz',$$

where $\mathbf{I}(\cdot)$ is an indicator function that equals 1 if its argument is true and 0 otherwise.

Markets for labor, capital, and goods clear:

$$L = \int zn(a, z)\, dJ(a, z),$$

$$K = \int a\, dJ(a, z),$$

$$F(L, K) + (1 - \delta)K = \int \left\{ a'(a, z) + c(a, z) \right\} dJ(a, z).$$

29.8.2. Complete markets

An allocation for an economy with complete markets solves an assignment problem that confronts a representative family with a continuum of family members. The family tells each member what to consume and when to work. When preferences are additively separable in consumption and leisure, optimal consumption is the same for everyone, regardless of work status. The family sends the most productive members to work. In particular, the representative family sets a reservation productivity $z^\star$ such that members with productivities greater than or equal to $z^\star$ work while the others do not. The value function of the representative family satisfies

$$V(a) = \max_{z^\star, c, a'} \left\{ \log(c) - B[1 - G(z^\star)] + \beta V(a') \right\}, \qquad (29.8.6)$$

where the maximization is subject to

$$a' = (1 + r)a + w \int_{z^\star}^{\infty} z\, dG(z) - c, \qquad (29.8.7)$$

$$c \geq 0, \quad a' \geq \hat{a}.$$

The representative family solves a deterministic problem because it has a continuum of members. The *ex ante* probability that a single member draws from a particular interval of productivities equals the *post* fraction of the family's members drawing from that interval.

First-order conditions with respect to c, a' and $z^\star$ at interior solutions are:

$$\frac{1}{c} - \lambda = 0, \tag{29.8.8a}$$

$$-\lambda + \beta V'(a') = 0, \tag{29.8.8b}$$

$$BG'(z^\star) + \lambda w[-z^\star G'(z^\star)] = 0. \tag{29.8.8c}$$

In a steady state, consumption is constant over time. Application of the Benveniste-Scheinkman formula gives $V'(a) = (1 + r)u'(c)$, so it follows from first-order conditions $(29.8.8a)$ and $(29.8.8b)$ that

$$1 + r = \frac{1}{\beta} = 1 - \delta + \alpha \left(\frac{K}{L}\right)^{\alpha-1}, \tag{29.8.9}$$

where the second equality invokes profit-maximization condition $(29.8.5b)$. The optimal consumption level is obtained from conditions $(29.8.8a)$ and $(29.8.8c)$:

$$c = \frac{wz^\star}{B}. \tag{29.8.10}$$

The equilibrium capital stock held by the representative family is $K = a = a'$, which together with expressions $(29.8.9)$ and $(29.8.10)$ can be substituted into budget constraint $(29.8.7)$:

$$K = \frac{1}{\beta}K + w \int_{z^\star}^{\infty} z\, dG(z) - \frac{wz^\star}{B}. \tag{29.8.11}$$

After dividing expression $(29.8.11)$ by the integral in that expression, i.e., by the family's supply of efficiency units of labor, which in equilibrium equals L, we obtain

$$\frac{K}{L} = \frac{1}{\beta}\frac{K}{L} + w - \frac{wz^\star}{B\int_{z^\star}^{\infty} z\, dG(z)}. \tag{29.8.12}$$

We can now solve for a stationary equilibrium in three steps. First, we use the second equality in expression $(29.8.9)$ to determine the equilibrium capital-labor ratio, K/L, in terms of parameters. Next, given the capital-labor ratio, we can compute the wage rate from profit-maximization condition $(29.8.5a)$. Finally, with the capital-labor ratio and the wage rate in hand, expression $(29.8.12)$ becomes one equation to be solved for the equilibrium value of the reservation productivity $z^\star$.

29.8.3. Simulations of Prescott taxation

We adopt the calibration of Chang and Kim (2007) except that we shut down aggregate productivity shocks. To highlight differences and similarities across our incomplete- and complete-market versions of the economy, we compute equilibrium outcomes under Prescott's tax and transfer scheme in section 29.3.1.

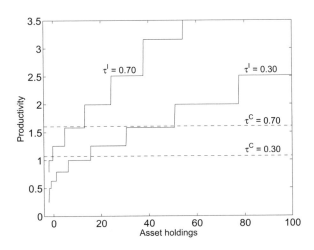

Figure 29.8.1: Reservation productivity as a function of asset holdings in the economy with incomplete markets (solid curves) and complete markets (dashed curves), respectively, where the lower (upper) curve refers to tax rate 0.30 (0.70).

For labor tax rates of 0.30 and 0.70, respectively, reservation productivities as functions of asset holdings are displayed in Figure 29.8.1. In the incomplete-market economy (solid curves), an agent's reservation productivity increases in his asset holdings. A high asset level means that, everything else equal, an agent is poised to enjoy one of his intermittent spells of leisure, which will result in asset decumulation and his ultimate return to work. For an agent with high assets to postpone such a desired spell of leisure, the agent must experience a relatively high productivity to be willing to continue to work for a while. As one would expect, the reservation productivities for the higher tax rate 0.70 lie well above those for the lower tax rate 0.30, since Prescott's tax and transfer scheme is very potent in suppressing agents' labor supply and causing them to choose more leisure. In the complete-market economy, the single productivity cutoff (dashed curve) is indicative of a privately efficient allocation. It is the most productive agents who work at any point in time.

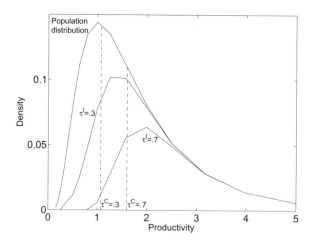

Figure 29.8.2: Productivity distribution. The upper solid curve is the population productivity distribution, while the other two in descending order show the agents thereof who are employed in the incomplete-market economy given tax rate 0.30 and 0.70, respectively. The corresponding masses of employed agents in the complete-market economy are the halves of the population distribution to the right of a vertical dashed line, where the left (right) dashed line refers to tax rate 0.30 (0.70).

The top solid curve in Figure 29.8.2 depicts the stationary distribution of productivities in the population. A dashed vertical line is the productivity cutoff in the complete-market economy, where the left (right) one refers to tax rate 0.30 (0.70), i.e., the same reservation productivity as the corresponding dashed line in Figure 29.8.1. All agents with productivities to the right of the dashed line work in the complete-market economy, and hence, the area under that portion of the population distribution equals the employment-population ratio. In the incomplete-market economy, the endogenous stationary distribution of agents across both productivities and asset holdings, $J(a, z)$, together with the decision rule for whether or not to work, $n(a, z)$, determine how many agents are at work at different productivity levels. Those employed workers in the incomplete-market economy are depicted by the solid curves that lie weakly below the top population curve, which in descending order refer to tax rates 0.30 and 0.70, respectively. As in the complete-market economy, virtually all agents with high productivities are working in the incomplete-market economy. But over a mid-range of productivities, there are significant differences between the two economies. On the one hand, some agents in the incomplete-market do not work but would have been working in the complete-market economy. The reason is that because their asset holdings are relatively high, their shadow value of additional wealth falls below the utility of leisure. On the other hand, other agents in the incomplete-market economy work but would not have worked in the complete-market economy. These agents have low asset holdings and so feel compelled to work despite their low productivities.

The work and asset decisions of individual agents in the incomplete-market economy determine the distribution of asset holdings, and the capital stock. For labor tax rates 0.30 and 0.70, respectively, the solid curves in Figure 29.8.3 depict the cumulative distribution function for asset holdings in the incomplete-market economy. At the high tax rate 0.70 (upper solid curve), asset holdings become concentrated at lower levels. As in the case of the elevated reservation productivities in Figure 29.8.1, taxation suppresses market activity in favor of leisure. In the complete-market economy, tax rate 0.70 is associated with a similar large decline in per capita asset holdings, as depicted by the vertical dashed lines in Figure 29.8.3 where the left (right) one refers to tax rate 0.70 (0.30).

From a production perspective, what matters is the capital stock relative to the aggregate supply of efficiency units of labor. In the complete-market economy, that capital-labor ratio is determined by steady-state relationship (29.8.9) which does not depend on the labor tax rate (but would have depended on any intertemporal tax wedge such as a tax on capital income). Since the wage rate is a function of the capital-labor ratio in (29.8.5a), it follows in Figure 29.8.4 that the wage rate in the complete-market economy (dashed curve) is invariant to the labor tax rate. In contrast, the wage rate in the incomplete-market economy (solid curve) falls with the labor tax rate and lies above the wage rate of the complete-market economy. To understand the latter outcome, we recall that in a Bewley model like ours with infinitely-lived agents, the interest rate must fall below the subjective rate of discounting β^{-1}, which is the steady-state

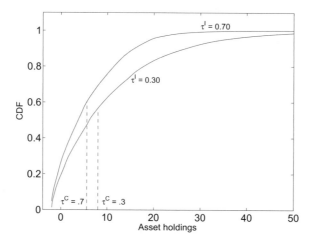

Figure 29.8.3: Asset distribution. The lower (upper) solid curve is the cumulative distribution function for asset holdings in the incomplete-market economy when the tax rate is 0.30 (0.70). The right (left) vertical dashed line is the per capita asset holdings in the complete-market economy when the tax rate is 0.30 (0.70).

interest rate in the complete-market economy. Since the equilibrium interest rate is inversely related to the capital-labor ratio in expression ($29.8.5b$), it follows immediately that the capital-labor ratio is higher in the incomplete-market economy, and therefore by expression ($29.8.5a$), so is the wage rate.

Figure 29.8.5 shows that the fraction of the population employed is higher in the incomplete-market economy than in the complete-market economy. As seen in Figure 29.8.2, there are those agents who work and those who do not work in the incomplete-market economy, but who would have done the opposite if they instead had lived in the complete-market economy. Evidently, the group of agents who work in the incomplete-market economy but would not have worked in the complete-market economy is larger. With no insurance markets, agents on average work more in order to accumulate precautionary savings in the event of low productivity in the future.[9]

---

[9] Marcet et al. (2007) conduct an analysis similar to that of Chang and Kim (2007) but where labor is divisible, $n_t \in [0, 1]$, and the idiosyncratic productivity shock takes on only two values, $z_t \in \{0, 1\}$. In addition to the precautionary savings effect that tends to increase

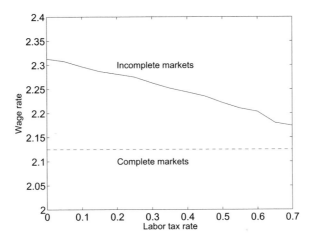

Figure 29.8.4: Wage rate per efficiency unit of labor in the economy with incomplete markets (solid curve) and complete markets (dashed curve), as a function of the labor tax rate.

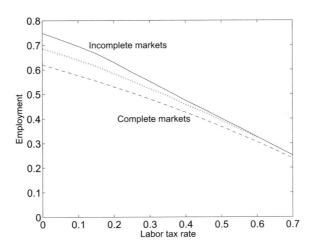

Figure 29.8.5: Employment-population ratio in the economy with incomplete markets (solid line) and complete markets (dashed line), as a function of the labor tax rate. The dotted line represents the former economy with a less persistent productivity process.

the capital stock under uncertainty, they identify an ex post wealth effect on labor supply that

What makes the employment-population ratio to converge across the two economies at higher tax rates in Figure 29.8.5? A key reason is that Prescott's tax and transfer scheme effectively insures the agents by collecting tax revenues and then returning them lump sum as equal amounts to all agents. To explore how precautionary savings drive the employment wedge between the incomplete- and complete-market economies at low tax rates, consider the following perturbation of the idiosyncratic productivity process. Specifically, suppose that agents face a transition probability distribution function

$$\tilde{\pi}(z'|z; \lambda) = (1 - \lambda)\pi(z'|z) + \lambda G(z'), \qquad (29.8.13)$$

where $\lambda \in [0, 1]$. For $\lambda = 0$, the productivity process is the same as that of Chang and Kim (2007), while for $\lambda = 1$, productivities are independent and identically distributed across agents and time, with realizations governed by the stationary unconditional distribution of Chang and Kim's process. Such perturbations do not affect equilibrium outcomes in the complete-market economy because they do not affect the constraints of the representative family. But agents in the incomplete-market are now ex ante relieved when they do not have to bear as much of the risk associated with the persistence of Chang and Kim's productivity process. The dotted line in Figure 29.8.5 shows equilibrium outcomes in the incomplete-market economy for $\lambda = 0.1$, where employment is now closer to that of the complete-market economy.

A striking feature of Figure 29.8.5 is the high elasticity of aggregate labor supply to taxation in the complete-market as well as in the incomplete-market economy. This message is shared with the first part of this chapter when agents were finitely lived and at interior solutions with respect to their choices of career length.

---

can depress the aggregate hours of work as well as the capital stock in an incomplete-market economy. (See section 17.8.)

29.9. L and S equivalence meets C and K's agents

Krusell et al. (2008) conjecture that the equivalence result of Ljungqvist and Sargent (2007) in continuous time described in section 29.2 extends to a deterministic version of Chang and Kim's (2007) discrete-time framework in section 29.8. Remove the productivity shocks and normalize an agent's efficiency units of labor to one in the latter framework ($z_t = 1$ for all t). Then Krusell et al. (2008) show how agents choose the constant consumption stream in expression (29.2.8), $\bar{c} = w/B$, when the steady-state interest rate is equal to the rate of subjective discounting, $1 + r = 1/\beta$. Krusell et al. (2008) indicate the existence of stationary equilibrium in which agents support that constant consumption stream by alternating between spells of working and enjoying leisure ('time averaging'). They argue that the aggregate allocation is the same as if markets had been complete and there had been employment lotteries.

29.9.1. Guess the value function

Since we are removing the productivity shocks from the model of section 29.8, an agent's single state variable in a stationary equilibrium is his beginning-of-period asset level a, before receiving interest earnings. We guess, and will then verify, that an agent's value funcion $V(a)$ takes the form,

$$
V(a) = \begin{cases}
\dfrac{\log\left[\dfrac{1-\beta}{\beta} a + w\right] - B}{1-\beta} & \text{if } a \leq \underline{a}; \\[4ex]
\dfrac{1}{1-\beta}\left(\log\left[\dfrac{w}{B}\right] - 1\right) + \dfrac{B}{\beta w} a & \text{if } a \in (\underline{a}, \bar{a}); \\[4ex]
\dfrac{\log\left[\dfrac{1-\beta}{\beta} a\right] - B}{1-\beta} & \text{if } a \geq \bar{a}.
\end{cases}
\tag{29.9.1}
$$

To appreciate what motivates our guess, we begin with some observations. In a stationary equilibrium with $1 + r = 1/\beta$ and no uncertainty, an agent's intertemporal Euler equation (29.8.4) implies constant consumption over time.[10]

---

[10] Because of the equality between the market interest rate $1 + r$ and the rate of subjective discounting $1/\beta$, any consumption path that varies over time can be improved upon by

Thus, any asset accumulation or decumulation by an agent can only be motivated by that agent's desire to engage in time averaging with respect to his labor supply. For an agent with assets in some range $(\underline{a}, \bar{a})$, we shall show that time averaging is indeed optimal because it enables him to finance an optimal constant consumption level $\bar{c} = w/B$. But first we discuss our guess of the value function outside of this asset range.

If an agent has too little (too much) assets, he will choose to work forever (to never work) and to consume the highest affordable constant consumption level associated with that labor supply plan. Consider an agent whose beginning-of-period assets $a \leq \underline{a}$ are so low that if he works forever and consumes the highest affordable constant consumption, $w + ra$, that consumption level will be less than or equal to $\bar{c} = w/B$. We can verify later that such a poor agent will indeed choose to work forever and to consume $w + ra$ in each period. After invoking $r = (1/\beta) - 1$, the critical asset limit $\underline{a}$ is

$$w + \frac{1 - \beta}{\beta} \underline{a} = \frac{w}{B},$$

$$\underline{a} = \frac{\beta w}{1 - \beta}[B^{-1} - 1]. \tag{29.9.2}$$

If $n_t = 1$ and $c_t = w + ra$ for all t, preference specification (29.8.1) yields lifetime utility given by the conjectured value function (29.9.1) when $a \leq \underline{a}$.[11]

Next, consider an agent whose beginning-of-period assets $a \geq \bar{a}$ are so high that if he never works and consumes the highest affordable constant consumption, ra, that consumption level will be greater than or equal to $\bar{c} = w/B$. We can later verify that such a rich agent will indeed choose never to work and to consume ra in each period. After invoking $r = (1/\beta) - 1$, the critical asset limit $\bar{a}$ is

$$\frac{1 - \beta}{\beta} \bar{a} = \frac{w}{B},$$

$$\bar{a} = \frac{\beta w}{1 - \beta}B^{-1}. \tag{29.9.3}$$

---

shifting consumption from periods of high to periods of low consumption. An agent's employment status does not matter since preference specification (29.8.1) is additively separable in consumption and leisure.

[11] Under the implicit but necessary parameter restriction for an equilibrium with time averaging, $B > 1$, note that asset limit $\underline{a}$ in (29.9.2) is negative, i.e., only agents who are initially indebted, $a < 0$, could conceivably want to choose to work forever with constant consumption equal to $w + ra$.

If $n_t = 0$ and $c_t = ra$ for all t, preference specification $(29.8.1)$ does yield lifetime utility given by the conjectured value function $(29.9.1)$ when $a \geq \bar{a}$.

To complete our guess about the value function, draw a straight line between the end point $V(\underline{a})$ and the starting point $V(\bar{a})$ for the segments with $a \leq \underline{a}$ and $a \geq \bar{a}$, respectively. This guess is motivated by the insight from section 29.2 that time averaging can under some conditions replace employment lotteries and attain linear combinations in the space of utilities when employment is a discrete choice variable. The linear segment for the value function is expressed in $(29.9.1)$ when $a \in (\underline{a}, \bar{a})$. Note that the slope with respect to assets, $B/(\beta w)$, is the same as the derivative at the end point of the preceding segment for the value function $(a \leq \underline{a})$ as well as the derivative at the starting point of the succeeding segment for the value function $(a \geq \bar{a})$:

$$V'(a)\Big|_{a \uparrow \underline{a}} = \frac{\beta^{-1}}{\frac{1-\beta}{\beta}\underline{a} + w} = \frac{B}{\beta w}$$

$$V'(a)\Big|_{a \downarrow \bar{a}} = \frac{1}{(1-\beta)\bar{a}} = \frac{B}{\beta w},$$

where we have invoked expression $(29.9.2)$ and $(29.9.3)$ for $\underline{a}$ and $\bar{a}$, respectively.

29.9.2. Verify optimality of time averaging

Without productivity shocks, Bellman equation $(29.8.2)$ can be simplified to

$$V(a) = \max\left\{ \max_{a'}\left\{ \log((1+r)a + w - a') - B + \beta V(a') \right\}, \right.$$
$$\left. \max_{a'}\left\{ \log((1+r)a - a') + \beta V(a') \right\} \right\}, \tag{29.9.4}$$

where the first max operator selects whether to work, and the budget constraints are substituted into the utility functions. Given our conjectured value function $(29.9.1)$, we solve the optimization problem on the right side of $(29.9.4)$ to verify that our conjectured value function $V(a)$ does indeed emerge on left side of $(29.9.4)$.

Note that the conjectured value function $(29.9.1)$ is (weakly) concave so that the two inner optimization problems (one for working, another for not working) on the right side of $(29.9.4)$ are both concave programming problems. Moreover, since the conjectured value function is continuous and differentiable everywhere, we can solve each optimization problem (for working and for not working) one by one, using first-order conditions, and compare the values. Let $W(a, 1)$ and $W(a, 0)$ denote the value of working and not working, respectively, and hence, $V(a) = \max\{W(a, 1), W(a, 0)\}$.

We start by verifying the conjectured value function for $a \in (\underline{a}, \bar{a})$ when time averaging should be an optimal policy. First, conditional on working, take a first-order condition with respect to a' in the first inner optimization problem on the right side of $(29.9.4)$

$$\frac{1}{(1 + r)a + w - a'} = \beta V'(a'),$$

$$a' = \frac{1}{\beta}a + w - \frac{w}{B}. \tag{29.9.5}$$

Here we have invoked the conjectured steady-state interest rate, $1 + r = \beta^{-1}$, and proceeded as if a' also falls in the range $(\underline{a}, \bar{a})$ where the conjectured value function $(29.9.1)$ has derivative $V'(a') = B/(\beta w)$. Since a' exceeds a, it follows that a must fall below some upper bound $a^\star < \bar{a}$ in order for $a' \in (\underline{a}, \bar{a})$, where that upper bound $a^\star$ is given by[12]

$$a^\star = \bar{a} - \beta w < \bar{a}. \tag{29.9.6}$$

Given the optimal choice of a' in expression $(29.9.5)$, we can compute from the budget constraint that the implied consumption level is $c = w/B$. With

---

[12] Using expression $(29.9.5)$ for a', the upper bound $a^\star$ on asset level a that ensures $a' \leq \bar{a}$, can be solved from

$$\frac{1}{\beta}a^\star + w - \frac{w}{B} = \bar{a}.$$

Multiplying both sides by β, and subtracting and adding $\bar{a}$ on the right side, yield

$$a^\star + \beta w - \frac{\beta w}{B} = \beta \bar{a} - \bar{a} + \bar{a}.$$

After invoking expression $(29.9.3)$ for $\bar{a}$, we find that the last term on the left side is equal to the first two terms on the right side, and hence, we have arrived at the equality in $(29.9.6)$.

choices of both a' and c in hand, we can compute the value of working,

$$W(a, 1) = \log(c) - B + \beta V(a')$$

$$= \log\left(\frac{w}{B}\right) - B + \beta\left\{\frac{1}{1-\beta}\left[\log\left(\frac{w}{B}\right) - 1\right] + \frac{B}{\beta w}\left[\frac{1}{\beta}a + w - \frac{w}{B}\right]\right\}$$

$$= \frac{1}{1-\beta}\left[\log\left(\frac{w}{B}\right) - 1\right] + \frac{B}{\beta w}a = V(a),$$

and hence, we have confirmed that working yields a value equal to our conjectured value function when $a \in (\underline{a}, a^\star)$, where the upper bound $a^\star$ ensures that $a' \in (\underline{a}, \bar{a})$.

Next, conditional on not working, the first-order condition with respect to a' in the second inner optimization problem on the right side of $(29.9.4)$ is

$$\frac{1}{(1+r)a - a'} = \beta V'(a'),$$

$$a' = \frac{1}{\beta}a - \frac{w}{B}, \tag{29.9.7}$$

where we have invoked the conjectured steady-state interest rate, $1 + r = \beta^{-1}$, and proceeded as if a' also falls in the range $(\underline{a}, \bar{a})$ where the conjectured value function $(29.9.1)$ has the derivative $V'(a') = B/(\beta w)$. Since a' falls below $\underline{a}$, it follows that a must exceed some lower bound $a_\star > \underline{a}$ in order for $a' \in (\underline{a}, \bar{a})$, where that lower bound $a_\star$ is given by [13]

$$a_\star = \underline{a} + \beta w > \underline{a}. \tag{29.9.8}$$

Given the optimal choice of a' in expression $(29.9.7)$, we can compute from the budget constraint that the implied consumption level is $c = w/B$. With

---

[13] Using expression $(29.9.7)$ for a', the lower bound $a_\star$ on asset level a that ensures $a' \geq \underline{a}$, can be solved from

$$\frac{1}{\beta}a_\star - \frac{w}{B} = \underline{a}.$$

Multiplying both sides by β, and subtracting and adding $\underline{a}$ on the right side, yield

$$a_\star - \frac{\beta w}{B} = \beta\underline{a} - \underline{a} + \underline{a}.$$

After invoking expression $(29.9.2)$ for $\underline{a}$, we find that the first two terms on the right side is greater than the last term on the left side by an amount βw, and hence, we have arrived at the equality in $(29.9.8)$.

choices of both a' and c in hand, we can compute the value of not working to be

$$W(a,0) = \log(c) + \beta V(a')$$

$$= \log\left(\frac{w}{B}\right) + \beta\left\{\frac{1}{1-\beta}\left[\log\left(\frac{w}{B}\right) - 1\right] + \frac{B}{\beta w}\left[\frac{1}{\beta}a - \frac{w}{B}\right]\right\}$$

$$= \frac{1}{1-\beta}\left[\log\left(\frac{w}{B}\right) - 1\right] + \frac{B}{\beta w}a = V(a),$$

and hence, we have confirmed that not working yields a value equal to our conjectured value function when $a \in (a_\star, \bar{a})$, where the lower bound $a_\star$ ensures that $a' \in (\underline{a}, \bar{a})$.

Following steps similar to above, we leave as an exercise for the reader to complete the verification of conjectured value function (29.9.1) (or consult the appendix of Krusell et al. (2008)). In particular, we can show that an agent with assets $a \in (\underline{a}, a_\star)$ ($a \in (a^\star, \bar{a})$) strictly prefers to work (not to work) so that his next period's assets a' do not fall outside of the asset range $(\underline{a}, \bar{a})$. Thus, we conclude that agents with assets $a \in (\underline{a}, \bar{a})$ find it optimal to engage in time averaging, i.e., to alternate between work and leisure, to finance an optimal consumption level $\bar{c} = w/B$ with asset holdings fluctuating within the range $a \in (\underline{a}, \bar{a})$. Also, we can verify that an agent with assets $a \le \underline{a}$ ($a \ge \bar{a}$) strictly prefers to work forever (never to work) and choose the highest affordable constant consumption of $c = w + ra$ ($c = ra$).

Krusell et al. (2008) assume that $\beta > 0.5$, which is required for $a_\star < a^\star$.[14] Together with expressions (29.9.6) and (29.9.8), it then follows that $\underline{a} < a_\star < a^\star < \bar{a}$. Another implicit assumption of Krusell et al. (2008) is that the preference parameter for the disutility of work is high enough so that there exist interior solutions to an agent's lifetime labor supply problem. In our formulation, that parameter restriction is $B > 1$.

---

[14] To derive a parameter restriction that ensures $a_\star < a^\star$, we substitute expression (29.9.2) and (29.9.3) for $\underline{a}$ and $\bar{a}$, respectively, into expression (29.9.8) and (29.9.6) for $a_\star$ and $a^\star$, respectively,

$$a_\star = \underline{a} + \beta w = \frac{\beta w}{1-\beta}[B^{-1} - 1] + \beta w < \frac{\beta w}{1-\beta}B^{-1} - \beta w = \bar{a} - \beta w = a^\star,$$

which simplifies to $\beta > 0.5$.

29.9.3. Equivalence of time averaging and lotteries

Krusell et al. (2008) argue that there exists a stationary equilibrium for the incomplete-market economy where all agents engage in time averaging with assets in the range $(\underline{a}, \bar{a})$, and the aggregate values of K and L are the same as in a corresponding complete-market economy with employment lotteries.

We have already studied equilibrium outcomes in a more general version of the complete-market economy in section 29.8.2. Under our present assumption that all agents have a constant productivity level that is normalized to one, equation (29.8.10) shows that the optimal consumption level is $c = w/B$, and the aggregate labor supply is given by the appropriate version of equation (29.8.12):

$$\frac{K}{L} = \frac{1}{\beta}\frac{K}{L} + w - \frac{w}{BL}, \tag{29.9.9}$$

where once again the capital-labor ratio K/L and the wage w are determined by equations (29.8.9) and (29.8.5a). Hence, we can solve for the aggregate labor supply L from equation (29.9.9).

In the stationary equilibrium of the incomplete-market economy with time averaging, agents are indifferent to alternative lifetime labor supply paths that yield equal present values of labor income. In a competitive equilibrium, an 'invisible hand' arranges agents' labor and savings decisions so that at every point in time, the aggregate labor supply and aggregate asset holdings equal the same constant aggregates L and K as those in the complete-markets economy. An equilibrium interest equal to $1 + r = 1/\beta$ makes a constant consumption $\bar{c} = w/B$ be the optimal choice for the worker-consumer.

29.10. Concluding remarks

A high aggregate labor supply elasticity hinges on a substantial fraction of agents being at an interior solution with respect to their lifetime labor supplies. This finding emerges from models with finitely-lived agents who choose career length and also in Chang and Kim's (2007) model of infinitely-lived agents who engage in time averaging across periods of work and leisure.

When agents are finitely lived, two forces can lower the labor supply elasticity: (1) government financed social security retirement schemes that leave agents at a corner solution with respect to their choices of career lengths, and (2) large adverse labor market shocks towards the end of working lives that prematurely terminate careers by pushing the shadow value of additional labor earnings below the utility of leisure in early retirement.

It is an occasion to celebrate that two camps of researchers, namely, those who have championed high and low labor supply elasticities, have come together in adopting the same theoretical framework. Nevertheless, the serious division between the two camps about quantitative magnitudes of labor supply elasticities persists. But we see the emergence of agreement over a basic theoretical framework as genuine progress relative to the earlier stalemate when proponents of employment lotteries used macroeconomic observations to build support for their aggregation theory, while opponents brought a different set of microeconomic observations to refute the employment lotteries allocation mechanism.[15]

To illustrate how far we have come, we revisit our own section 28.7 reasoning, where we are concerned about an asymmetry between idle firms and idle workers in a particular model. While idle firms are truly indifferent about their operating status because operating firms are just breaking even without making any profits, "the aggregation theory behind the stand-in household has an additional aspect that is not present in the theory that aggregates over firms, namely, it says how consumption and leisure are smoothed across households with the help of an extensive set of contingent claim markets. This market

---

[15] It would be a mistake to regard the abandonment of a stand-in household with its employment lotteries as unconditional surrender to the other tradition in macroeconomics of overlapping generations models that has commonly postulated incomplete markets. The reason is that earlier work in the overlapping generations tradition has often postulated an exogenous retirement age, shutting down the key choice focused on in time-averaging models of career choice. It is the possibility of interior solutions to lifetime labor supply in combination with indivisible labor that have led real business cycle researchers like Prescott (2006) to embrace lifecycle models of labor supply.

arrangement and randomization device stands at the center of the employment lottery model. To us, it seems that they make the aggregation theory behind the stand-in household fundamentally different than the well-known aggregation theory for the firm side." Well, we now also can assert that this difference is not important for those households who, being at an interior solution for lifetime labor supply, are about to choose whether to supply more of their indivisible labor by extending their careers before retiring.

Having a diverse group of researchers focus on a common set of observations on lifetime labor supply within a common theoretical framework bodes well for the eventual arrival of what we hope will be the "labor supply elasticity accord" foretold by Ljungqvist and Sargent (2011).

Part VII
Technical appendices

Appendix A.
Functional Analysis

This appendix provides an introduction to the analysis of functional equations (functional analysis). It describes the contraction mapping theorem, a workhorse for studying dynamic programs.

A.1. Metric spaces and operators

We begin with the definition of a metric space, which is a pair of objects, a set X, and a function d.[1]

Definition A.1.1. *A metric space is a set X and a function d called a metric, $d\colon X \times X \to R$. The metric $d(x, y)$ satisfies the following four properties:*

M1. *Positivity: $d(x, y) \geq 0$ for all $x, y \in X$.*

M2. *Strict positivity: $d(x, y) = 0$ if and only if $x = y$.*

M3. *Symmetry: $d(x, y) = d(y, x)$ for all $x, y \in X$.*

M4. *Triangle inequality: $d(x, y) \leq d(x, z) + d(z, y)$ for all x, y, and $z \in X$.*

We give some examples of the metric spaces with which we will be working:

Example A.1. $l_p[0, \infty)$. We say that $X = l_p[0, \infty)$ is the set of all sequences of complex numbers $\{x_t\}_{t=0}^{\infty}$ for which $\sum_{t=0}^{\infty} |x_t|^p$ converges, where $1 \leq p < \infty$. The function $d_p(x, y) = (\sum_{t=0}^{\infty} |x_t - y_t|^p)^{1/p}$ is a metric. Often, we will say that $p = 2$ and will work in $l_2[0, \infty)$.

Example A.2. $l_\infty[0, \infty)$. The set $X = l_\infty[0, \infty)$ is the set of bounded sequences $\{x_t\}_{t=0}^{\infty}$ of real or complex numbers. The metric is $d_\infty(x, y) = \sup_t |x_t - y_t|$.

Example A.3. $l_p(-\infty, \infty)$ is the set of "two-sided" sequences $\{x_t\}_{t=-\infty}^{\infty}$ such that $\sum_{t=-\infty}^{\infty} |x_t|^p < +\infty$, where $1 \leq p < \infty$. The associated metric is $d_p(x, y) = (\sum_{t=-\infty}^{\infty} |x_t - y_t|^p)^{1/p}$.

---

[1] General references on the mathematics described in this appendix are Luenberger (1969) and Naylor and Sell (1982).

Example A.4. $l_\infty(-\infty, \infty)$ is the set of bounded sequences $\{x_t\}_{t=-\infty}^\infty$ with metric $d_\infty(x, y) = \sup|x_t - y_t|$.

Example A.5. Let $X = C[0, T]$ be the set of all continuous functions mapping the interval $[0, T]$ into R. We consider the metric

$$d_p(x, y) = \left[\int_0^T |x(t) - y(t)|^p dt \right]^{1/p},$$

where the integration is in the Riemann sense.

Example A.6. Let $X = C[0, T]$ be the set of all continuous functions mapping the interval $[0, T]$ into R. We consider the metric

$$d_\infty(x, y) = \sup_{0 \le t \le T} |x(t) - y(t)|.$$

We now have the following important definition:

Definition A.1.2. *A sequence $\{x_n\}$ in a metric space (X, d) is said to be a Cauchy sequence if for each $\epsilon > 0$ there exists an $N(\epsilon)$ such that $d(x_n, x_m) < \epsilon$ for any $n, m \ge N(\epsilon)$. Thus a sequence $\{x_n\}$ is said to be Cauchy if $\lim_{n,m \to \infty} d(x_n, x_m) = 0$.*

We also have the following definition of convergence:

Definition A.1.3. *A sequence $\{x_n\}$ in a metric space (X, d) is said to converge to a limit $x_0 \in X$ if for every $\epsilon > 0$ there exists an $N(\epsilon)$ such that $d(x_n, x_0) < \epsilon$ for $n \ge N(\epsilon)$.*

The following lemma asserts that every convergent sequence in (X, d) is a Cauchy sequence:

Lemma A.1.1. *Let $\{x_n\}$ be a convergent sequence in a metric space (X, d). Then $\{x_n\}$ is a Cauchy sequence.*

Proof. Fix any $\epsilon > 0$. Let $x_0 \in X$ be the limit of $\{x_n\}$. Then for all m, n one has

$$d(x_n, x_m) \le d(x_n, x_0) + d(x_m, x_0)$$

by virtue of the triangle inequality. Because x_0 is the limit of $\{x_n\}$, there exists an N such that $d(x_n, x_0) < \epsilon/2$ for $n \ge N$. Together with the preceding inequality, this statement implies that $d(x_n, x_m) < \epsilon$ for $n, m \geq N$. Therefore, $\{x_n\}$ is a Cauchy sequence. ∎

We now consider two examples of sequences in metric spaces. The examples are designed to illustrate aspects of the concept of a Cauchy sequence. We first consider the metric space $\{C[0,1], d_2(x,y)\}$. We let $\{x_n\}$ be the sequence of continuous functions $x_n(t) = t^n$. Evidently this sequence converges pointwise to the function

$$x_0(t) = \begin{cases} 0, & 0 \leq t < 1 \\ 1, & t = 1. \end{cases}$$

Now, in $\{C[0,1], d_2(x,y)\}$, the sequence $x_n(t)$ is a Cauchy sequence. To verify this claim, calculate

$$d_2(t^m, t^n)^2 = \int_0^1 (t^n - t^m)^2 dt = \frac{1}{2n+1} + \frac{1}{2m+1} - \frac{2}{m+n+1}.$$

Clearly, for any $\epsilon > 0$, it is possible to choose an $N(\epsilon)$ that makes the square root of the right side less than ϵ whenever m and n both exceed N. Thus $x_n(t)$ is a Cauchy sequence. Notice, however, that the limit point $x_0(t)$ does *not* belong to $\{C[0,T], d_2(x,y)\}$ because it is not a continuous function.

As our second example, we consider the space $\{C[0,T], d_\infty(x,y)\}$. We consider the sequence $x_n(t) = t^n$. In $(C[0,1], d_\infty)$, the sequence $x_n(t)$ is *not* a Cauchy sequence. To verify this claim, it is sufficient to establish that, for any fixed $m > 0$, there is a $\delta > 0$ such that

$$\sup_{n>0} \sup_{0 \leq t \leq 1} |t^n - t^m| > \delta.$$

Direct calculations show that, for fixed m,

$$\sup_n \sup_{0 \leq t \leq 1} |t^n - t^m| = 1.$$

Parenthetically we may note that

$$\sup_{n>0} \sup_{0 \leq t \leq 1} |t^n - t^m| = \sup_{0 \leq t \leq 1} \sup_{n>0} |t^n - t^m| = \sup_{0 \leq t \leq 1} \lim_{n \to \infty} |t^n - t^m|$$

$$= \sup_{0 \leq t \leq 1} \lim_{n \to \infty} t^m |t^{n-m} - 1| = \sup_{0 \leq t \leq 1} t^m = 1.$$

Therefore, $\{t^n\}$ is not a Cauchy sequence in $(C[0,1], d_\infty)$.

These examples illustrate the fact that whether a given sequence is Cauchy depends on the metric space within which it is embedded, in particular on the metric that is being used. The sequence $\{t^n\}$ is Cauchy in $(C[0,1], d_2)$, and more generally in $(C[0,1], d_p)$ for $1 \leq p < \infty$. The sequence $\{t^n\}$, however, is *not* Cauchy in the metric space $(C[0,1], d_\infty)$. The first example also illustrates the fact that a Cauchy sequence in (X, d) need *not* converge to a limit point x_0 belonging to the metric space. The property that Cauchy sequences converge to points lying in the metric space is desirable in many applications. We give this property a name.

Definition A.1.4. *A metric space* (X, d) *is said to be complete if each Cauchy sequence in* (X, d) *is a convergent sequence in* (X, d). *That is, in a complete metric space, each Cauchy sequence converges to a point belonging to the metric space.*

The following metric spaces are complete:

$$(l_p[0, \infty), d_p), \qquad 1 \leq p < \infty$$
$$(l_\infty[0, \infty), d_\infty)$$
$$(C[0, T], d_\infty).$$

The following metric spaces are not complete:

$$(C[0, T], d_p), \qquad 1 \leq p < \infty.$$

Proofs that $(l_p[0, \infty), d_p)$ for $1 \leq p \leq \infty$ and $(C[0, T], d_\infty)$ are complete are contained in Naylor and Sell (1982, chap. 3). In effect, we have already shown by counterexample that the space $(C[0,1], d_2)$ is not complete, because we displayed a Cauchy sequence that did not converge to a point in the metric space. A definition may now be stated:

Definition A.1.5. *A function f mapping a metric space (X, d) into itself is called an operator.*

We need a notion of continuity of an operator.

Definition A.1.6. *Let $f : X \to X$ be an operator on a metric space (X, d). The operator f is said to be continuous at a point $x_0 \in X$ if for every $\epsilon > 0$*

there exists a $\delta > 0$ such that $d[f(x), f(x_0)] < \epsilon$ whenever $d(x, x_0) < \delta$. The operator f is said to be continuous if it is continuous at each point $x \in X$.

We shall be studying an operator with a particular property, the application of which to any two distinct points $x, y \in X$ brings them closer together.

Definition A.1.7. *Let (X, d) be a metric space and let $f : X \to X$. We say that f is a contraction or contraction mapping if there is a real number $k, 0 \le k < 1$, such that*

$$d[f(x), f(y)] \le kd(x, y) \qquad \text{for all} \quad x, y \in X.$$

It follows directly from the definition that a contraction mapping is a continuous operator.

We now state the following theorem:

Theorem A.1.1. *Contraction Mapping*

Let (X, d) be a complete metric space and let $f : X \to X$ be a contraction. Then there is a unique point $x_0 \in X$ such that $f(x_0) = x_0$. Furthermore, if x is any point in X and $\{x_n\}$ is defined inductively according to $x_1 = f(x), x_2 = f(x_1), \ldots, x_{n+1} = f(x_n)$, then $\{x_n\}$ converges to x_0.

Proof. Let x be any point in X. Define $x_1 = f(x), x_2 = f(x_1), \ldots$. Express this as $x_n = f^n(x)$. To show that the sequence x_n is Cauchy, first assume that $n > m$. Then

$$d(x_n, x_m) = d[f^n(x), f^m(x)] = d[f^m(x_{n-m}), f^m(x)]$$
$$\le kd[f^{m-1}(x_{n-m}), f^{m-1}(x)]$$

By induction, we get

$$(*) \qquad\qquad d(x_n, x_m) \le k^m d(x_{n-m}, x).$$

When we repeatedly use the triangle inequality, the preceding inequality implies that

$$d(x_n, x_m) \le k^m[d(x_{n-m}, x_{n-m-1}) + \ldots + d(x_2, x_1) + d(x_1, x)].$$

Applying $(*)$ gives

$$d(x_n, x_m) \le k^m(k^{n-m-1} + \ldots + k + 1)d(x_1, x).$$

Because $0 \le k < 1$, we have

(†) $$d(x_n, x_m) \le k^m \sum_{i=0}^{\infty} k^i d(x_t, x) = \frac{k^m}{1-k} d(x_1, x).$$

The right side of (†) can be made arbitrarily small by choosing m sufficiently large. Therefore, $d(x_n, x_m) \to 0$ as $n, m \to \infty$. Thus $\{x_n\}$ is a Cauchy sequence. Because (X, d) is complete, $\{x_n\}$ converges to an element of (X, d).

The limit point x_0 of $\{x_n\} = \{f^n(x)\}$ is a fixed point of f. Because f is continuous, $\lim_{n\to\infty} f(x_n) = f(\lim_{n\to\infty} x_n)$. Now $f(\lim_{n\to\infty} x_n) = f(x_0)$ and $\lim_{n\to\infty} f(x_n) = \lim_{n\to\infty} x_{n+1} = x_0$. Therefore $x_0 = f(x_0)$.

To show that the fixed point x_0 is unique, assume the contrary. Assume that x_0 and y_0, $x_0 \ne y_0$, are two fixed points of f. But then

$$0 < d(x_0, y_0) = d[f(x_0), f(y_0)] \le k d(x_0, y_0) < d(x_0, y_0),$$

which is a contradiction. Therefore f has a unique fixed point. ∎

We now restrict ourselves to sets X whose elements are functions. The spaces $C[0, T]$ and $l_p[0, \infty)$ for $1 \le p \le \infty$ are examples of spaces of functions. Let us define the notion of inequality of two functions.

Definition A.1.8. *Let X be a space of functions, and let $x, y \in X$. Then $x \ge y$ if and only if $x(t) \ge y(t)$ for every t in the domain of the functions.*

Let X be a space of functions. We use the d_∞ metric, defined as $d_\infty(x, y) = \sup_t |x(t) - y(t)|$, where the supremum is over the domain of definition of the function.

A pair of conditions that are sufficient for an operator $T : (X, d_\infty) \to (X, d_\infty)$ to be a contraction appear in the following theorem:[2]

Theorem A.1.2. *Blackwell's Sufficient Conditions for T to be a Contraction*

Let T be an operator on a metric space (X, d_∞), where X is a space of functions. Assume that T has the following two properties:
(a) Monotonicity: For any $x, y \in X$, $x \ge y$ implies $T(x) \ge T(y)$.
(b) Discounting: Let c denote a function that is constant at the real value c for

---

[2] See Blackwell's (1965) Theorem 5. This theorem is used extensively by Stokey and Lucas with Prescott (1989).

all points in the domain of definition of the functions in X. For any positive real c and every $x \in X$, $T(x+c) \le T(x) + \beta c$ for some β satisfying $0 \le \beta < 1$. Then T is a contraction mapping with modulus β.

Proof. For all $x, y \in X$, $x \le y + d(x, y)$. Applying properties (a) and (b) to this inequality gives

$$T(x) \le T(y + d(x, y)) \le T(y) + \beta d(x, y).$$

Exchanging the roles of x and y and using the same logic implies

$$T(y) \le T(x) + \beta d(x, y).$$

Combining these two inequalities gives $|T(x) - T(y)| \le \beta d(x, y)$ or

$$d(T(x), T(y)) \le \beta d(x, y).$$

∎

A.2.   Discounted dynamic programming

We study the functional equation associated with a discounted dynamic programming problem:

$$v(x) = \max_{u \in R^k} \{r(x, u) + \beta v(x')\}, \qquad x' \le g(x, u), \qquad 0 < \beta < 1. \qquad (A.2.1)$$

We assume that $r(x, u)$ is real valued, continuous, concave, and bounded and that the set $[x', x, u : x' \le g(x, u), u \in R^k]$ is convex and compact.

We define the operator

$$Tv = \max_{u \in R^k} \{r(x, u) + \beta v(x')\}, \qquad x' \le g(x, u), \quad x \in X.$$

We work with the space of continuous bounded functions mapping X into the real line. We use the metric $d_\infty(v, w) = \sup_{x \in X} |v(x) - w(x)|$. This metric space is complete.

The operator T maps a continuous bounded function v into a continuous bounded function Tv. (For a proof, see Stokey and Lucas with Prescott, 1989.)[3]

We now establish that T is a contraction by verifying Blackwell's pair of sufficient conditions. First, suppose that $v(x) \geq w(x)$ for all $x \in X$. Then

$$
\begin{aligned}
Tv &= \max_{u \in R^k} \{r(x, u) + \beta v(x')\}, & x' \leq g(x, u) \\
&\geq \max_{u \in R^k} \{r(x, u) + \beta w(x')\}, & x' \leq g(x, u) \\
&= Tw.
\end{aligned}
$$

Thus, T is monotone. Next, notice that for any positive constant c,

$$
\begin{aligned}
T(v + c) &= \max_{u \in R^k} \{r(x, u) + \beta[v(x') + c]\}, & x' \leq g(x, u) \\
&= \max_{u \in R^k} \{r(x, u) + \beta v(x') + \beta c\}, & x' \leq g(x, u) \\
&= Tv + \beta c.
\end{aligned}
$$

Thus, T discounts. Therefore, T satisfies both of Blackwell's conditions. It follows that T is a contraction on a complete metric space. Therefore the functional equation $(A.2.1)$, which can be expressed as $v = Tv$, has a unique fixed point in the space of bounded continuous functions. This fixed point is approached in the limit in the d_∞ metric by iterations $v^n = T^n(v^0)$ starting from any bounded and continuous v^0. Convergence in the d_∞ metric implies uniform convergence of the functions v^n.

Stokey and Lucas with Prescott (1989) show that T maps concave functions into concave functions. It follows that the solution of $v = Tv$ is a concave function.

---

[3] The assertions in the preceding two paragraphs are the most difficult pieces of the argument to prove.

A.2.1. Policy improvement algorithm

For ease of exposition, in this section we shall assume that the constraint $x' \leq g(x, u)$ holds with equality. For the purposes of describing an alternative way to solve dynamic programming problems, we introduce a new operator. We use one step of iterating on the Bellman equation to define the new operator T_μ as follows:

$$T_\mu(v) = T(v)$$

or

$$T_\mu(v) = r[x, \, \mu(x)] + \beta v\{g[x, \, \mu(x)]\} \ ,$$

where $\mu(x)$ is the policy function that attains $T(v)(x)$. For a fixed $\mu(x)$, T_μ is an operator that maps bounded continuous functions into bounded continuous functions. Denote by C the space of bounded continuous functions mapping X into X.

For any admissible policy function $\mu(x)$, the operator T_μ is a contraction mapping. This fact can be established by verifying Blackwell's pair of sufficient conditions:

1. T_μ is monotone. Suppose that $v(x) \geq w(x)$. Then
$$T_\mu v = r[x, \, \mu(x)] + \beta v\{g[x, \, \mu(x)]\}$$
$$\geq r[x, \, \mu(x)] + \beta w\{g[x, \, \mu(x)]\} = T_\mu w \ .$$

2. T_μ discounts. For any positive constant c
$$T_\mu(v + c) = r(x, \mu) + \beta \left(v\{g[x, \mu(x)] + c\} \right)$$
$$= T_\mu v + \beta c \ .$$

Because T_μ is a contraction operator, the functional equation

$$v_\mu(x) = T_\mu[v_\mu(x)]$$

has a unique solution in the space of bounded continuous functions. This solution can be computed as a limit of iterations on T_μ starting from any bounded continuous function $v_0(x) \in C$,

$$v_\mu(x) = \lim_{k \to \infty} T_\mu^k(v_0)(x) \ .$$

The function $v_\mu(x)$ is the value of the objective function that would be attained by using the stationary policy $\mu(x)$ each period.

The following proposition describes the *policy iteration* or *Howard improvement* algorithm.

Theorem A.2.1.   Let $v_\mu(x) = T_\mu[v_\mu(x)]$. *Define a new policy* $\bar\mu$ *and an associated operator* $T_{\bar\mu}$ *by*

$$T_{\bar\mu}[v_\mu(x)] = T[v_\mu(x)] \; ;$$

that is, $\bar\mu$ *is the policy that solves a one-period problem with* $v_\mu(x)$ *as the terminal value function. Compute the fixed point*

$$v_{\bar\mu}(x) = T_{\bar\mu}[v_{\bar\mu}(x)] \;.$$

Then $v_{\bar\mu}(x) \geq v_\mu(x)$. *If* $\mu(x)$ *is not optimal, then* $v_{\bar\mu}(x) > v_\mu(x)$ *for at least one* $x \in X$.

Proof. From the definition of $\bar\mu$ and $T_{\bar\mu}$, we have

$$T_{\bar\mu}[v_\mu(x)] = r[x, \bar\mu(x)] + \beta v_\mu\{g[x, \bar\mu(x)]\} =$$
$$T(v_\mu)(x) \geq r[x, \mu(x)] + \beta v_\mu\{g[x, \mu(x)]\}$$
$$= T_\mu[v_\mu(x)] = v_\mu(x)$$

or

$$T_{\bar\mu}[v_\mu(x)] \geq v_\mu(x) \;.$$

Apply $T_{\bar\mu}$ repeatedly to this inequality and use the monotonicity of $T_{\bar\mu}$ to conclude

$$v_{\bar\mu}(x) = \lim_{n\to\infty} T_{\bar\mu}^n[v_\mu(x)] \geq v_\mu(x) \;.$$

This establishes the asserted inequality $v_{\bar\mu}(x) \geq v_\mu(x)$. If $v_{\bar\mu}(x) = v_\mu(x)$ for all $x \in X$, then

$$v_\mu(x) = T_{\bar\mu}[v_\mu(x)]$$
$$= T[v_\mu(x)] \;,$$

where the first equality follows because $T_{\bar\mu}[v_{\bar\mu}(x)] = v_{\bar\mu}(x)$, and the second equality follows from the definitions of $T_{\bar\mu}$ and $\bar\mu$. Because $v_\mu(x) = T[v_\mu(x)]$, the Bellman equation is satisfied by $v_\mu(x)$.   ∎

The *policy improvement* algorithm starts from an arbitrary feasible policy and iterates to convergence on the two following steps:[4]

---

[4] A policy $\mu(x)$ is said to be *unimprovable* if it is optimal to follow it for the first period, given a terminal value function $v(x)$. In effect, the policy improvement algorithm starts with an arbitrary value function, then by solving a one-period problem, it generates an improved policy and an improved value function. The proposition states that optimality is characterized by the features, first, that there is no incentive to deviate from the policy during the first period, and second, that the terminal value function is the one associated with continuing the policy.

Step 1. For a feasible policy $\mu(x)$, compute $v_\mu = T_\mu(v_\mu)$.

Step 2. Find $\bar\mu$ by computing $T(v_\mu)$. Use $\bar\mu$ as the policy in step 1.

In many applications, this algorithm proves to be much faster than iterating on the Bellman equation.

A.2.2. A search problem

We now study the functional equation associated with a search problem of chapter 6. The functional equation is

$$v(w) = \max\left\{ \frac{w}{1-\beta}, \beta \int v(w')dF(w') \right\}, \qquad 0 < \beta < 1. \tag{A.2.2}$$

Here, the wage offer drawn at t is w_t. Successive offers w_t are independently and identically distributed random variables. We assume that w_t has cumulative distribution function $\text{prob}\{w_t \le w\} = F(w)$, where $F(0) = 0$ and $F(\bar w) = 1$ for some $\bar w < \infty$. In equation $(A.2.2)$, $v(w)$ is the optimal value function for a currently unemployed worker who has offer w in hand. We seek a solution of the functional equation $(A.2.2)$.

We work in the space of bounded continuous functions $C[0, \bar w]$ and use the d_∞ metric

$$d_\infty(x, y) = \sup_{0 \le w \le \bar w} |x(w) - y(w)|.$$

The metric space $(C[0, \bar w], d_\infty)$ is complete.

We consider the operator

$$T(z) = \max\left\{ \frac{w}{1-\beta}, \beta \int z(w')dF(w') \right\}. \tag{A.2.3}$$

Evidently the operator T maps functions z in $C[0, \bar w]$ into functions $T(z)$ in $C[0, \bar w]$. We now assert that the operator T defined by equation $(A.2.3)$ is a contraction. To prove this assertion, we verify Blackwell's sufficient conditions. First, assume that $f(w) \ge g(w)$ for all $w \in [0, \bar w]$. Then note that

$$Tg = \max\left\{ \frac{w}{1-\beta}, \beta \int g(w')dF(w') \right\}$$

$$\le \max\left\{ \frac{w}{1-\beta}, \beta \int f(w')dF(w') \right\}$$

$$= Tf.$$

Thus, T is monotone. Next, note that for any positive constant c,

$$T(f + c) = \max \left\{ \frac{w}{1 - \beta}, \beta \int [f(w') + c] dF(w') \right\}$$

$$= \max \left\{ \frac{w}{1 - \beta}, \beta \int f(w') dF(w') + \beta c \right\}$$

$$\leq \max \left\{ \frac{w}{1 - \beta}, \beta \int f(w') dF(w') \right\} + \beta c$$

$$= Tf + \beta c.$$

Thus, T satisfies the discounting property and is therefore a contraction.

Application of the contraction mapping theorem, then, establishes that the functional equation $Tv = v$ has a unique solution in $C[0, \bar{w}]$, which is approached in the limit as $n \to \infty$ by $T^n(v^0) = v^n$, where v^0 is any point in $C[0, \bar{w}]$. Because the convergence in the space $C[0, \bar{w}]$ is in terms of the metric d_∞, the convergence is uniform.

Appendix B.
Linear projections and hidden Markov models

B.1. Linear projections

For reference we state the following theorems about linear least-squares projections. We let Y be an $(n \times 1)$ vector of random variables and X be an $(h \times 1)$ vector of random variables. We assume that the following first and second moments exist:

$$EY = \mu_Y, \ EX = \mu_X,$$
$$EXX' = S_{XX}, \ EYY' = S_{YY}, \ EYX' = S_{YX}.$$

Letting $x = X - EX$ and $y = Y - EY$, we define the following covariance matrices

$$Exx' = \Sigma_{xx}, \ E'_{yy} = \Sigma_{yy}, \ Eyx' = \Sigma_{yx}.$$

We are concerned with estimating Y as a linear function of X. The estimator of Y that is a linear function of X and that minimizes the mean squared error between each component Y and its estimate is called the *linear projection of Y on X*.

Definition B.1.1.   The *linear projection* of Y on X is the affine function $\hat{Y} = AX + a_0$ that minimizes $E \operatorname{trace} \{(Y - \hat{Y})(Y - \hat{Y})'\}$ over all affine functions $a_0 + AX$ of X. We denote this linear projection as $\widehat{E}[Y \mid X]$, or sometimes as $\widehat{E}[Y \mid x, 1]$ to emphasize that a constant is included in the "information set."

The linear projection of Y on X, $\widehat{E}[Y \mid X]$ is also sometimes called the *wide sense expectation of Y conditional on X*. We have the following theorems:

Theorem B.1.1.

$$\widehat{E}[Y \mid X] = \mu_y + \Sigma_{yx} \Sigma_{xx}^{-1}(X - \mu_x). \tag{B.1.1}$$

Proof. The theorem follows immediately by writing out E trace $(Y - \hat{Y})(Y - \hat{Y})'$ and completing the square, or else by writing out E trace$(Y - \hat{Y})(Y - \hat{Y})'$ and obtaining first-order necessary conditions ("normal equations") and solving them. ∎

Theorem B.1.2.

$$\hat{E}\left[\left(Y - \hat{E}[Y \mid x]\right) \mid X'\right] = 0.$$

This equation states that the errors from the projection are orthogonal to each variable included in X.

Proof. Immediate from the normal equations. ∎

Theorem B.1.3.   *(Orthogonality principle)*

$$E\left[\left[Y - \hat{E}\left(Y \mid x\right)\right] x'\right] = 0.$$

Proof. Follows from Theorem 21.3. ∎

Theorem B.1.4.   *(Orthogonal regressors)*

Suppose that
$X' = (X_1, X_2, \ldots, X_h)', EX' = \mu' = (\mu_{x1}, \ldots, \mu_{xh})'$, *and* $E(X_i - \mu_{xi})(X_j - \mu_{xj}) = 0$ *for* $i \neq j$. *Then*

$$\hat{E}\left[Y \mid x_1, \ldots, x_n, 1\right] = \hat{E}\left[Y \mid x_1\right] + \hat{E}\left[Y \mid x_2\right] + \ldots + \hat{E}\left[Y \mid x_n\right] - (n - 1)\mu_y.$$
$$(B.1.2)$$

Proof. Note that from the hypothesis of orthogonal regressors, the matrix Σ_{xx} is diagonal. Applying equation $(B.1.1)$ then gives equation $(B.1.2)$. ∎

B.2. Hidden Markov models

This section gives a brief introduction to hidden Markov models, a tool that is useful to study a variety of nonlinear filtering problems in finance and economics. We display a solution to a nonlinear filtering problem that a reader might want to compare to the linear filtering problem described earlier.

Consider an N-state Markov chain. We can represent the state space in terms of the unit vectors $S_x = \{e_1, \ldots, e_N\}$, where e_i is the ith N-dimensional unit vector. Let the $N \times N$ transition matrix be P, with (i, j) element

$$P_{ij} = \text{Prob}(x_{t+1} = e_j \mid x_t = e_i).$$

With these definitions, we have

$$E x_{t+1} \mid x_t = P' x_t.$$

Define the "residual"

$$v_{t+1} = x_{t+1} - P' x_t,$$

which implies the linear "state-space" representation

$$x_{t+1} = P' x_t + v_{t+1}.$$

Notice how it follows that $E\, v_{t+1} \mid x_t = 0$, which qualifies v_{t+1} as a "martingale difference process adapted to x_t."

We want to append a "measurement equation." Suppose that x_t is not observed, but that y_t, a noisy function of x_t, is observed. Assume that y_t lives in the M-dimensional space S_y, which we represent in terms of M unit vectors: $S_y = \{f_1, \ldots, f_M\}$, where f_i is the ith M-dimensional unit vector. To specify a linear measurement equation $y_t = C(x_t, u_t)$, where u_t is a measurement noise, we begin by defining the $N \times M$ matrix Q with

$$\text{Prob}\,(y_t = f_j \mid x_t = e_i) = Q_{ij}.$$

It follows that

$$E\,(y_t \mid x_t) = Q' x_t.$$

Define the residual

$$u_t \equiv y_t - E\, y_t \mid x_t,$$

which suggests the "observer equation"

$$y_t = Q'x_t + u_t.$$

It follows from the definition of u_t that $E\ u_t \mid x_t = 0$. Thus, we have the linear state-space system

$$x_{t+1} = P'x_t + v_{t+1}$$
$$y_t = Q'x_t + u_t.$$

Using the definitions, it is straightforward to calculate the conditional second moments of the error processes v_{t+1}, u_t.[1]

B.3. Nonlinear filtering

We seek a recursive formula for computing the conditional distribution of the hidden state:

$$\rho_i(t) = \text{Prob}\{x_t = i \mid y_1 = \eta_1, \ldots, y_t = \eta_t\}.$$

Denote the history of observed y_t's up to t as $\eta^t = \text{col}\ (\eta_1, \ldots, \eta_t)$. Define the conditional probabilities

$$p(\xi_t, \eta_1, \ldots, \eta_t) = \text{Prob}\ (x_t = \xi_t, y_1 = \eta_1, \ldots, y_t = \eta_t),$$

---

[1] Notice that

$$x_{t+1}x'_{t+1} = P'x_t(P'x_t)' + P'x_tv'_{t+1}$$
$$+ v_{t+1}(P'x_t)' + v_{t+1}v'_{t+1}.$$

Substituting into this equation the facts that $x_{t+1}x'_{t+1} = \text{diag}\ x_{t+1} = \text{diag}\ (P'x_t) + \text{diag}\ v_{t+1}$ gives

$$v_{t+1}v'_{t+1} = \text{diag}\ (P'x_t) + \text{diag}\ (v_{t+1}) - P'\text{diag}\ x_tP$$
$$- P'x_tv'_{t+1}(P'x_t)'.$$

It follows that

$$E\ [v_{t+1}v'_{t+1} \mid x_t] = \text{diag}\ (P'x_t) - P'\text{diag}\ x_tP.$$

Similarly,

$$E\ [u_t\ u'_t \mid x_t] = \text{diag}\ (Q'x_t) - Q'\text{diag}\ x_tQ.$$

and assume $p(\eta_1, \ldots, \eta_t) \neq 0$. Then apply the calculus of conditional expectations to compute[2]

$$
\begin{aligned}
p(\xi_t \mid \eta^t) &= \frac{p(\xi_t, \eta_t \mid \eta^{t-1})}{p(\eta_t \mid \eta^{t-1})} \\
&= \frac{\sum_{\xi_{t-1}} p(\eta_t \mid \xi_t) \, p(\xi_t \mid \xi_{t-1}) p(\xi_{t-1} \mid \eta^{t-1})}{\sum_{\xi_t} \sum_{\xi_{t-1}} p(\eta_t \mid \xi_t) p(\xi_t \mid \xi_{t-1}) p(\xi_{t-1} \mid \eta^{t-1})}.
\end{aligned}
$$

This result can be written

$$
\rho_i(t+1) = \frac{\sum_s Q_{ij} P_{si} \rho_s(t)}{\sum_s \sum_i Q_{ij} P_{si} \rho_s(t)},
$$

where $\eta_{t+1} = j$ is the value of y at $t+1$ We can represent this recursively as

$$
\begin{aligned}
\tilde{\rho}(t+1) &= \ \mathrm{diag}\,(Q_j) P' \rho(t) \\
\rho(t+1) &= \frac{\tilde{\rho}(t+1)}{< \tilde{\rho}(t+1), \underline{1} >}.
\end{aligned}
$$

where Q_j is the jth column of Q, and $\mathrm{diag}\,(Q_j)$ is a diagonal matrix with Q_{ij} as the ith diagonal element; here $< \cdot, \cdot >$ denotes the inner product of two vectors, and $\underline{1}$ is the unit vector.

---

[2] Notice that

$$
p(\xi_t, \eta_t \mid \eta^{t-1}) = \sum_{\xi_{t-1}} p(\xi_t, \eta_t, \xi_{t-1} \mid \eta^{t-1})
$$

$$
= \sum_{\xi_{t-1}} p(\xi_t, \eta_t \mid \xi_{t-1}, \eta^{t-1}) p(\xi_{t-1} \mid \eta^{t-1})
$$

$$
p(\xi_t, \eta_t \mid \xi_{t-1}, \eta^{t-1}) = p(\xi_t \mid \xi_{t-1}, \eta^{t-1}) p(\eta_t \mid \xi_t, \xi_{t-1}, \eta^{t-1})
$$

$$
= p(\xi_t \mid \xi_{t-1}) p(\eta_t \mid \xi_t).
$$

Combining these results gives the formula in the text.

References

Abel, Andrew B., N. Gregory Mankiw, Lawrence H. Summers, and Richard J. Zeckhauser. 1989. "Assessing Dynamic Efficiency: Theory and Evidence." *Review of Economic Studies*, Vol. 56, pp. 1–20.

Abreu, Dilip. 1988. "On the Theory of Infinitely Repeated Games with Discounting." *Econometrica*, Vol. 56, pp. 383–396.

Abreu, Dilip, David Pearce, and Ennio Stacchetti. 1986. "Optimal Cartel Equilibria with Imperfect Monitoring." *Journal of Economic Theory*, Vol. 39, pp. 251–269.

Abreu, Dilip, David Pearce, and Ennio Stacchetti. 1990. "Toward a Theory of Discounted Repeated Games with Imperfect Monitoring." *Econometrica*, Vol. 58(5), pp. 1041–1063.

Acemoglu, Daron. 1997. "Good Jobs versus Bad Jobs: Theory and Some Evidence." Mimeo. CEPR Discussion Paper No. 1588.

Acemoglu, Daron, and Robert Shimer. 1999. "Efficient Unemployment Insurance." *Journal of Political Economy*, Vol. 107, pp. 893–928.

Acemoglu, Daron, and Robert Shimer. 2000. "Productivity gains from unemployment insurance." *European Economic Review*, Vol. 44, pp. 1195–1224.

Adda, Jerome, and Russell W. Cooper. 2003. *Dynamic Economics: Quantitative Methods and Applications.* Cambridge, Mass.: MIT Press.

Aghion, Philippe, and Peter Howitt. 1992. "A Model of Growth through Creative Destruction." *Econometrica*, Vol. 60, pp. 323–351.

Aghion, Philippe, and Peter Howitt. 1998. *Endogenous Growth Theory.* Cambridge, Mass.: MIT Press.

Aiyagari, S. Rao. 1985. "Observational Equivalence of the Overlapping Generations and the Discounted Dynamic Programming Frameworks for One-Sector Growth." *Journal of Economic Theory*, Vol. 35(2), pp. 201–221.

Aiyagari, S. Rao. 1987. "Optimality and Monetary Equilibria in Stationary Overlapping Generations Models with Long Lived Agents." *Journal of Economic Theory*, Vol. 43, pp. 292–313.

Aiyagari, S. Rao. 1993. "Explaining Financial Market Facts: The Importance of Incomplete Markets and Transaction Costs." *Quarterly Review*, Federal Reserve Bank of Minneapolis, Vol. 17(1), pp. 17–31.

Aiyagari, S. Rao. 1994. "Uninsured Idiosyncratic Risk and Aggregate Saving." *Quarterly Journal of Economics*, Vol. 109(3), pp. 659–684.

Aiyagari, S. Rao. 1995. "Optimal Capital Income Taxation with Incomplete Markets and Borrowing Constraints." *Journal of Political Economy*, Vol. 103(6), pp. 1158–1175.

Aiyagari, S. Rao, and Mark Gertler. 1991. "Asset Returns with Transactions Costs and Uninsured Individual Risk." *Journal of Monetary Economics*, Vol. 27, pp. 311–331.

Aiyagari, S. Rao, and Ellen R. McGrattan. 1998. "The Optimum Quantity of Debt." *Journal of Monetary Economics*, Vol. 42(3), pp. 447–469.

Aiyagari, S. Rao, and Neil Wallace. 1991. "Existence of Steady States with Positive Consumption in the Kiyotaki-Wright Model." *Review of Economic Studies*, Vol. 58(5), pp. 901–916.

Aiyagari, Marcet, Albert, Thomas J. Sargent, and Juha Seppälä. 2002. "Optimal Taxation without State-Contingent Debt." *Journal of Political Economy*, Vol. 110(6), pp. 1220–1254.

Albarran, Pedro, and Orazio Attanasio. 2003. "Limited Commitment and Crowding out of Private Transfers: Evidence from a Randomized Experiment" *Economic Journal*, Vol. 113(486), pp. C77–85.

Albrecht, James, and Bo Axell. 1984. "An Equilibrium Model of Search Unemployment." *Journal of Political Economy*, Vol. 92(5), pp. 824–840.

Allen, Franklin. 1985. "Repeated principal-agent relationships with lending and borrowing." *Economics Letters*, Vol. 17(1-2), pp. 27-31.

Altug, Sumru. 1989. "Time-to-Build and Aggregate Fluctuations: Some New Evidence." *International Economic Review*, Vol. 30(4), pp. 889–920.

Altug, Sumru, and Pamela Labadie. 1994. *Dynamic Choice and Asset Markets*. San Diego: Academic Press.

Alvarez, Fernando, and Urban J. Jermann. 2000. "Efficiency, Equilibrium, and Asset Pricing with Risk of Default" *Econometrica*, Vol. 68(4), pp. 775-797.

Alvarez, Fernando, and Urban J. Jermann. 2001. "Quantitative Asset Pricing Implications of Endogenous Solvency Constraints." *Review of Financial Studies*, Vol. 14(4), pp. 1117-1151.

Alvarez, Fernando, and Urban J. Jermann. 2004. "Using Asset Prices to Measure the Cost of Business Cycles." *Journal of Political Economy*, Vol. 112, No. 6, pp. 1223–1256.

Alvarez, Fernando, Patrick J. Kehoe, and Pablo Andrés Neumeyer. 2004. "The Time Consistency of Optimal Monetary and Fiscal Policies." *Econometrica*, Vol. 72, pp. 541–567.

Anderson, Evan W., Lars P. Hansen, Ellen R. McGrattan, and Thomas J. Sargent. 1996. "Mechanics of Forming and Estimating Dynamic Linear Economies." In Hans M. Amman, David A. Kendrick, and John Rust (eds.), *Handbook of Computational Economics, Vol. 1, Handbooks in Economics, Vol. 13*. Amsterdam: Elsevier Science, North-Holland, pp. 171–252.

Anderson, E., L. P. Hansen, and T. J. Sargent. 2003. "A Quartet of Semigroups for Model Specification, Robustness, Prices of Risk, and Model Detection." *Journal of the European Economic Association*, Vol. 1(1), pp. 68–123.

Ang, Andrew, and Monika Piazzesi. 2003. "A No-Arbitrage Vector Autoregression of Term Structure Dynamics with Macroeonomic and Latent Variables." *Journal of Monetary Economics*, Vol. 50(4), pp. 745–288.

Angeletos, George-Marios. 2002. "Fiscal Policy With Noncontingent Debt And The Optimal Maturity Structure" *Quarterly Journal of Economics*, Vol. 117(3), pp. 1105-1131.

Apostol, Tom M. 1975. *Mathematical Analysis*. 2nd ed. Reading, Mass.: Addison-Wesley.

Arrow, Kenneth J. 1962. "The Economic Implications of Learning by Doing." *Review of Economic Studies*, Vol. 29, pp. 155–173.

Arrow, Kenneth J. 1964. "The Role of Securities in the Optimal Allocation of Risk-Bearing." *Review of Economic Studies*, Vol. 31, pp. 91–96.

Åström, K. J. 1965. "Optimal Control of Markov Processes with Incomplete State Information." *Journal of Mathematical Analysis and Applications*, Vol. 10, pp. 174–205.

Atkeson, Andrew G. 1988. "Essays in Dynamic International Economics." Ph.D. dissertation, Stanford University.

Atkeson, Andrew G. 1991. "International Lending with Moral Hazard and Risk of Repudiation." *Econometrica*, Vol. 59(4), pp. 1069–1089.

Atkeson, Andrew, V. V. Chari, and Patrick J. Kehoe. 2010. "Sophisticated Monetary Policies." *Quarterly Journal of Economics*, Vol. 125, pp. 47–89.

Atkeson, Andrew, and Robert E. Lucas, Jr. 1992. "On Efficient Distribution with Private Information." *Review of Economic Studies*, Vol. 59(3), pp. 427–453.

Atkeson, Andrew, and Robert E. Lucas, Jr. 1995. "Efficiency and Equality in a Simple Model of Efficient Unemployment Insurance." *Journal of Economic Theory*, Vol. 66(1), pp. 64–88.

Atkeson, Andrew, and Christopher Phelan. 1994. "Reconsidering the Costs of Business Cycles with Incomplete Markets." In Stanley Fischer and Julio J. Rotemberg (eds.), *NBER Macroeconomics Annual*. Cambridge, Mass.: MIT Press, pp. 187–207.

Attanasio, Orazio P. 2000. "Consumption." In John Taylor and Michael Woodford (eds.), *Handbook of Macroeconomics*. Amsterdam: North-Holland.

Attanasio, Orazio P., and Steven J. Davis. 1996. "Relative Wage Movements and the Distribution of Consumption." *Journal of Political Economy*, Vol. 104(6), pp. 1227–1262.

Attanasio, Orazio P., and Guglielmo Weber. 1993. "Consumption Growth, the Interest Rate and Aggregation." *Review of Economic Studies*, Vol. 60(3), pp. 631–649.

Auerbach, Alan J., and Laurence J. Kotlikoff. 1987. *Dynamic Fiscal Policy*. New York: Cambridge University Press.

Auernheimer, Leonardo. 1974. "The Honest Government's Guide to the Revenue from the Creation of Money." *Journal of Political Economy*, Vol. 82, pp. 598–606.

Azariadis, Costas. 1993. *Intertemporal Macroeconomics*. Cambridge, Mass.: Blackwell Press.

Backus, David K, Patrick J. Kehoe, and , Finn E. Kydland. 1992. "International Real Business Cycles." *Journal of Political Economy*, Vol. 100(4), pp. 745-75.

Backus, David K. and Stanley E. Zin. 1994. "Reverse Engineering the Yield Curve." Mimeo. NBER Working Paper No. 4676.

Balasko, Y., and Karl Shell. 1980. "The Overlapping-Generations Model I: The Case of Pure Exchange without Money." *Journal of Economic Theory*, Vol. 23(3), pp. 281–306.

Barillas, F., A. Bhandari, S. Bigio, R. Colacito, M. Juillard, S. Kitao, C. Matthes, T. J. Sargent, and Y. Shin. 2012. "Practicing Dynare" Mimeo. New York University.

Barillas, Francisco, Lars Peter Hansen, and Thomas J. Sargent. 2009. "Doubts or variability?" *Journal of Economic Theory*, Vol. 144(6), pp. 2388-2418.

Barro, Robert J. 1974. "Are Government Bonds Net Wealth?" *Journal of Political Economy*, Vol. 82(6), pp. 1095–1117.

Barro, Robert J. 1979. "On the Determination of Public Debt." *Journal of Political Economy*, Vol. 87, pp. 940–971.

Barro, Robert J., and David B. Gordon. 1983a. "A Positive Theory of Monetary Policy in a Natural Rate Model." *Journal of Political Economy*, Vol. 91, pp. 589–610.

Barro, Robert J., and David B. Gordon. 1983b. "Rules, Discretion, and Reputation in a Model of Monetary Policy." *Journal of Monetary Economics*, Vol. 12, pp. 101–121.

Barro, Robert J., and Xavier Sala-i-Martin. 1995. *Economic Growth*. New York: McGraw-Hill.

Barseghyan, Levon, Francesca Molinari, Ted O'Donoghue, and Joshua C. Teitelbaum. 2011. "The Nature of Risk Preferences: Evidence from Insurance Choices" Mimeo. Working Paper 11-03, Cornell University, Center for Analytic Economics.

Barsky, Robert B., Gregory N. Mankiw, and Stephen P. Zeldes. 1986. "Ricardian Consumers with Keynesian Propensities." *American Economic Review*, Vol. 76(4), pp. 676–691.

Basar, Tamer, and Geert Jan Olsder. 1982. *Dynamic Noncooperative Game Theory*. New York: Academic Press.

Bassetto, Marco. 2002. "A Game-Theoretic View of the Fiscal Theory of the Price Level." *Econometrica*, Vol. 70(6), pp. 2167–2195.

Bassetto, Marco. 2005. "Equilibrium and Government Commitment." *Journal of Economic Theory*, Vol. 124, No. 1, pp. 79–105.

Bassetto, Marco, and Narayana Kocherlakota. 2004. "On the Irrelevance of Government Debt When Taxes Are Distortionary." *Journal of Monetary Economics*, Vol. 51(2), pp. 299–304.

Baumol, William J. 1952. "The Transactions Demand for Cash: An Inventory Theoretic Approach." *Quarterly Journal of Economics*, Vol. 66, pp. 545–556.

Bellman, Richard. 1957. *Dynamic Programming*. Princeton, N.J.: Princeton University Press.

Bellman, Richard, and Stuart E. Dreyfus. 1962. *Applied Dynamic Programming*. Princeton, N.J.: Princeton University Press.

Benassy, Jean-Pascal. 1998. "Is There Always Too Little Research in Endogenous Growth with Expanding Product Variety?" *European Economic Review*, Vol. 42, pp. 61–69.

Benassy, Jean-Pascal. 2011. *Macroeconomic Theory*. Oxford: Oxford University Press.

Benoit, Jean-Pierre, and Vijay Krishna. 1985. "Finitely Repeated Games." *Econometrica*, Vol. 53, pp. 905–922.

Ben-Porath, Yoram. 1967. "The Production of Human Capital and the Life Cycle of Earnings." *Journal of Political Economy*, Vol. 75, pp. 352–365.

Benveniste, Lawrence, and Jose Scheinkman. 1979. "On the Differentiability of the Value Function in Dynamic Models of Economics." *Econometrica*, Vol. 47(3), pp. 727–732.

Benveniste, Lawrence, and Jose Scheinkman. 1982. "Duality Theory for Dynamic Optimization Models of Economics: The Continuous Time Case." *Journal of Economic Theory*, Vol. 27, pp. 1–19.

Bernheim, B. Douglas, and Kyle Bagwell. 1988. "Is Everything Neutral?" *Journal of Political Economy*, Vol. 96(2), pp. 308–338.

Bertola, Giuseppe, and Ricardo J. Caballero. 1994. "Cross-Sectional Efficiency and Labour Hoarding in a Matching Model of Unemployment." *Review of Economic Studies*, Vol. 61, pp. 435–456.

Bertsekas, Dimitri P. 1976. *Dynamic Programming and Stochastic Control*. New York: Academic Press.

Bertsekas, Dimitri P. 1987. *Dynamic Programming: Deterministic and Stochastic Models*. Englewood Cliffs, N.J.: Prentice-Hall.

Bertsekas, Dimitri P., and Steven E. Shreve. 1978. *Stochastic Optimal Control: The Discrete Time Case.* New York: Academic Press.

Bewley, Truman F. 1977. "The Permanent Income Hypothesis: A Theoretical Formulation." *Journal of Economic Theory,* Vol. 16(2), pp. 252–292.

Bewley, Truman F. 1980. "The Optimum Quantity of Money." In J. H. Kareken and N. Wallace (eds.), *Models of Monetary Economies.* Minneapolis: Federal Reserve Bank of Minneapolis, pp. 169–210.

Bewley, Truman F. 1983. "A Difficulty with the Optimum Quantity of Money." *Econometrica,* Vol. 51, pp. 1485–1504.

Bewley, Truman F. 1986. "Stationary Monetary Equilibrium with a Continuum of Independently Fluctuating Consumers." In Werner Hildenbrand and Andreu Mas-Colell (eds.), *Contributions to Mathematical Economics in Honor of Gerard Debreu.* Amsterdam: North-Holland, pp. 79–102.

Black, Fisher, and Myron Scholes. 1973. "The Pricing of Options and Corporate Liabilities." *Journal of Political Economy,* Vol. 81, pp. 637–654.

Blackwell, David. 1965. "Discounted Dynamic Programming." *Annals of Mathematical Statistics,* Vol. 36(1), pp. 226–235.

Blanchard, Olivier Jean. 1985. "Debt, Deficits, and Finite Horizons." *Journal of Political Economy,* Vol. 93(2), pp. 223–247.

Blanchard, Olivier Jean, and Stanley Fischer. 1989. *Lectures on Macroeconomics.* Cambridge, Mass.: MIT Press.

Blanchard, Olivier Jean, and Charles M. Kahn. 1980. "The Solution of Linear Difference Models under Rational Expectations." *Econometrica,* Vol. 48(5), pp. 1305–1311.

Blume, Lawrence and David Easley. 2006. "If You're so Smart, why Aren't You Rich? Belief Selection in Complete and Incomplete Markets." *Econometrica,* Vol. 74(4), pp. 929-966.

Blundell, Richard and Ian Preston. 1998. "Consumption Inequality and Income Uncertainty." *Quarterly Journal of Economics,* Vol. 113(2), pp. 603-640.

Bohn, Henning. 1995. "The Sustainability of Budget Deficits in a Stochastic Economy." *Journal of Money, Credit, and Banking,* Vol. 27(1), pp. 257–271.

Bond, Eric W., and Jee-Hyeong Park. 2002. "Gradualism in Trade Agreements with Asymmetric Countries." *Review of Economic Studies,* Vol. 69, pp. 379–406.

Braun, R. Anton. 1994. "Tax Disturbances and Real Economic Activity in the Postwar United States." *Journal of Monetary Economics,* Vol. 33(3), pp. 441–462.

Breeden, Douglas T. 1979. "An Intertemporal Asset Pricing Model with Stochastic Consumption and Investment Opportunities." *Journal of Financial Economics,* Vol. 7(3), pp. 265–296.

Brock, William A. 1972. "On Models of Expectations Generated by Maximizing Behavior of Economic Agents Over Time." *Journal of Economic Theory,* Vol. 5, pp. 479–513.

Brock, William A. 1974. "Money and Growth: The Case of Long Run Perfect Foresight." *International Economic Review,* Vol. 15, pp. 750–777.

Brock, William A. 1982. "Asset Prices in a Production Economy." In J. J. McCall (ed.), *The Economics of Information and Uncertainty.* Chicago: University of Chicago Press, pp. 1–43.

Brock, William A. 1990. "Overlapping Generations Models with Money and Transactions Costs." In B. M. Friedman and F. H. Hahn (eds.), *Handbook of Monetary Economics, Vol. 1*. Amsterdam: North-Holland, pp. 263-295.

Brock, William A., and Leonard Mirman. 1972. "Optimal Economic Growth and Uncertainty: The Discounted Case." *Journal of Economic Theory*, Vol. 4(3), pp. 479–513.

Browning, Martin, Lars P. Hansen, and James J. Heckman. 1999. "Micro Data and General Equilibrium Models." In John Taylor and Michael Woodford (eds.), *Handbook of Macroeconomics*. Volume 1A. Amsterdam: North-Holland, pp. 543–633.

Bruno, Michael, and Stanley Fischer. 1990. "Seigniorage, Operating Rules, and the High Inflation Trap." *Quarterly Journal of Economics*, Vol. 105, pp. 353–374.

Bryant, John, and Neil Wallace. 1984. "A Price Discrimination Analysis of Monetary Policy." *Review of Economic Studies*, Vol. 51(2), pp. 279–288.

Buera, Francisco, and Juan Pablo Nicolini. 2004. "Optimal maturity of government debt without state contingent bonds" *Journal of Monetary Economics*, Vol. 51(3), pp. 531-554.

Buiter, Willem H. 2002. "The Fiscal Theory of the Price Level: A Critique." *Economic Journal*, Vol. 112, pp. 459–480.

Bulow, Jeremy, and Kenneth Rogoff. 1989. "Sovereign Debt: Is to Forgive to Forget?" *American Economic Review*, Vol. 79, pp. 43–50.

Burdett, Kenneth, and Kenneth L. Judd. 1983. "Equilibrium Price Dispersion" *Econometrica*, Vol. 51(4), pp. 955-69.

Burdett, Kenneth, and Dale T. Mortensen. 1998. "Wage Differentials, Employer Size, and Unemployment" *International Economic Review*, Vol. 39(2), pp. 257–73.

Burdett, Kenneth, Shouyong Shi, and Randall Wright. 2001. "Pricing and Matching with Frictions." *Journal of Political Economy*, Vol. 109(5), pp. 1060–1085.

Burnside, C., M. Eichenbaum, and S. Rebelo. 1993. "Labor Hoarding and the Business Cycle." *Journal of Political Economy*, Vol. 101(2), pp. 245–273.

Burnside, C., and M. Eichenbaum. 1996a. "Factor Hoarding and the Propagation of Business Cycle Shocks." *American Economic Review*, Vol. 86(5), pp. 1154–74.

Burnside, C., and M. Eichenbaum. 1996b. "Small Sample Properties of GMM Based Wald Tests." *Journal of Business and Economic Statistics*, Vol. 14(3), pp. 294–308.

Caballero, Ricardo J. 1990. "Consumption Puzzles and Precautionary Saving." *Journal of Monetary Economics*, Vol. 25, No. 1, pp. 113-136.

Cagan, Phillip. 1956. "The Monetary Dynamics of Hyperinflation." In Milton Friedman (ed.), *Studies in the Quantity Theory of Money*. Chicago: University of Chicago Press, pp. 25–117.

Calvo, Guillermo A. 1978. "On the Time Consistency of Optimal Policy in a Monetary Economy." *Econometrica*, Vol. 46(6), pp. 1411–1428.

Calvo, Guillermo A. 1983. "Staggered prices in a utility-maximizing framework." *Journal of Monetary Economics*, Vol. 12, No.3, pp. 383–398.

Campbell, John Y., Andrew W. Lo, and A. Craig MacKinlay. 1997. *The Econometrics of Financial Markets*. Princeton, N.J.: Princeton University Press.

Campbell, John Y., and John H. Cochrane. 1999. "By Force of Habit: A Consumption-Based Explanation of Aggregate Stock Market Behavior." *Journal of Political Economy*, Vol. 107(2), pp. 205–251.

Campbell, John Y., and Robert Shiller. 1988. "The Dividend-Price Ratio And Expectations of Future Dividends and Discount Factors." *Review of Financial Studies*, Vol. 1, pp. 195–227.

Campbell, John Y., and Robert Shiller. 1991. "Yield Spreads and Interest Rate Movements: A Bird's Eye View." *Review of Economic Studies*, Vol. 58, pp. 495–514.

Canova, Fabio. 2007. *Method for Applied Macroeconomic Research*. Princeton, New Jersey: Princeton University Press.

Carroll, Christopher D., and Miles S. Kimball. 1996. "On the Concavity of the Consumption Function." *Econometrica*, Vol. 64(4), pp. 981–992.

Casella, Alessandra, and Jonathan S. Feinstein. 1990. "Economic Exchange during Hyperinflation." *Journal of Political Economy*, Vol. 98(1), pp. 1–27.

Cass, David. 1965. "Optimum Growth in an Aggregative Model of Capital Accumulation." *Review of Economic Studies*, Vol. 32(3), pp. 233–240.

Cass, David. 1972. "On Capital Overaccumulation in the Aggregative, Neoclassical Model of Economic Growth: A Complete Characterization." *Journal of Economic Theory*, Vol. 4, pp. 200–233.

Cass, David, and M. E. Yaari. 1966. "A Re-examination of the Pure Consumption Loans Model." *Journal of Political Economy*, Vol. 74, pp. 353–367.

Cass, David, and Karl Shell. 1983. "Do Sunspots Matter?" *Journal of Political Economy*, Vol. 91(2), pp. 193–227.

Chamberlain, Gary, and Charles Wilson. 2000. "Optimal Intertemporal Consumption Under Uncertainty." *Review of Economic Dynamics*, Vol. 3(3), pp. 365–395.

Chamley, Christophe. 1986. "Optimal Taxation of Capital Income in General Equilibrium with Infinite Lives." *Econometrica*, Vol. 54(3), pp. 607–622.

Chamley, Christophe, and Heraklis Polemarchakis. 1984. "Assets, General Equilibrium, and the Neutrality of Money." *Review of Economic Studies*, Vol. 51, pp. 129–138.

Champ, Bruce, and Scott Freeman. 1994. *Modeling Monetary Economies*. New York: Wiley.

Chang, Roberto. 1998. "Credible Monetary Policy in an Infinite Horizon Model: Recursive Approaches." *Journal of Economic Theory*, Vol. 81(2), pp. 431–461.

Chang, Yongsung, and Sun-Bin Kim. 2007. "Heterogeneity and Aggregation: Implications for Labor-Market Fluctuations." *American Economic Review*, Vol. 97, pp. 1939–1956.

Chari, V. V., Lawrence J. Christiano, and Martin Eichenbaum. 1998. "Expectations Traps." *Journal of Economic Theory*, Vol. 81(2), pp. 462–492.

Chari, V. V., Lawrence J. Christiano, and Patrick J. Kehoe. 1994. "Optimal Fiscal Policy in a Business Cycle Model." *Journal of Political Economy*, Vol. 102(4), pp. 617–652.

Chari, V. V., Lawrence J. Christiano, and Patrick J. Kehoe. 1996. "Optimality of the Friedman Rule in Economies with Distorting Taxes." *Journal of Monetary Economics*, Vol. 37(2), pp. 203–223.

Chari, V. V., and Patrick J. Kehoe. 1990. "Sustainable Plans." *Journal of Political Economy*, Vol. 98, pp. 783–802.

Chari, V. V., and Patrick J. Kehoe. 1993a. "Sustainable Plans and Mutual Default." *Review of Economic Studies*, Vol. 60, pp. 175–195.

Chari, V. V., and Patrick J. Kehoe. 1993b. "Sustainable Plans and Debt." *Journal of Economic Theory*, Vol. 61, pp. 230–261.

Chari, V. V., Patrick J. Kehoe, and Edward C. Prescott. 1989. "Time Consistency and Policy." In Robert Barro (ed.), *Modern Business Cycle Theory*. Cambridge, Mass.: Harvard University Press, pp. 265–305.

Chatterjee, Satyajit, and Dean Corbae. 1996. "Money and Finance with Costly Commitment." *Journal of Monetary Economics*, Vol. 37(2), pp. 225–248.

Chen, Ren-Raw, and Louis Scott. 1993. "Maximum Likelihood Estimation for a Multifactor Equilibrium Model of the Term Structure of Interest Rates." *Journal of Fixed Income*, pp. 95–09.

Chéron, Arnaud, Jean-Olivier Hairault, and François Langot. 2008. "Job Creation and Job Destruction over the Life Cycle." Mimeo. University of Maine and Paris School of Economics.

Cho, In-Koo, and Akihiko Matsui. 1995. "Induction and the Ramsey Policy." *Journal of Economic Dynamics and Control*, Vol. 19(5–7), pp. 1113–1140.

Chow, Gregory. 1981. *Econometric Analysis by Control Methods*. New York: Wiley.

Chow, Gregory. 1997. *Dynamic Economics: Optimization by the Lagrange Method*. New York: Oxford University Press.

Christiano, Lawrence J. 1990. "Linear-Quadratic Approximation and Value-Function Iteration: A Comparison." *Journal of Business and Economic Statistics*, Vol. 8(1), pp. 99–113.

Christiano, Lawrence J., and M. Eichenbaum. 1992. "Current Real Business Cycle Theories and Aggregate Labor Market Fluctuations." *American Economic Review*, Vol. 82(3).

Christiano, Lawrence J., M. Eichenbaum, and C. Evans. 2003. "Nominal Rigidities and the Dynamic Effects of a Shock to Monetary Policy." *Journal of Political Economy*, forthcoming.

Christiano, Lawrence J., R. Motto, and M. Rostagno. 2003. "The Great Depression and the Friedman-Schwartz Hypothesis." *Journal of Money, Credit, and Banking*, forthcoming.

Christensen, Bent Jasper and Nicholas M. Kiefer. 2009. *Economic Modeling and Inference*. Princeton, New Jersey: Princeton University Press.

Clower, Robert. 1967. "A Reconsideration of the Microfoundations of Monetary Theory." *Western Economic Journal*, Vol. 6, pp. 1–9.

Cochrane, John H. 1991. "A Simple Test of Consumption Insurance." *Journal of Political Economy*, Vol. 99(5), pp. 957–976.

Cochrane, John H. 1997. "Where Is the Market Going? Uncertain Facts and Novel Theories." *Economic Perspectives*, Vol. 21(6), pp. 3–37.

Cochrane, John H., and Lars Peter Hansen. 1992. "Asset Pricing Explorations for Macroeconomics." In Olivier Jean Blanchard and Stanley Fischer (eds.), *NBER Macroeconomics Annual*. Cambridge, Mass.: MIT Press, pp. 115–165.

Cochrane, John H. 2005. "Money as stock" *Journal of Monetary Economics*, Vol. 52(3), pp. 501–528.

Cogley, Timothy. 1999. "Idiosyncratic Risk and the Equity Premium: Evidence from the Consumer Expenditure Survey." *Journal of Monetary Economics*, Vol. 49(2), pp. 309–334.

Cole, Harold L., and Narayana Kocherlakota. 1998. "Dynamic Games with Hidden Actions and Hidden States." Mimeo. Federal Reserve Bank of Minneapolis: Staff Report 254.

Cole, Harold L., and Narayana Kocherlakota. 2001. "Efficient Allocations with Hidden Income and Hidden Storage." *Review of Economic-Studies*, Vol. 68(3), pp. 523–42.

Correia, Isabel, Emmanuel Farhi, Juan Pablo Nicolini, and Pedro Teles. 2010. "Policy at the Zero Bound." Mimeo. Federal Reserve Bank of Minneapolis.

Constantinides, George M., and Darrell Duffie. 1996. "Asset Pricing with Heterogeneous Consumers." *Journal of Political Economy*, Vol. 104(2), pp. 219–240.

Cooley, Thomas F. 1995. *Frontiers of Business Cycle Research*. Princeton, N.J.: Princeton University Press.

Cooper, Russell W. 1999. *Coordination Games: Complementarities and Macroeconomics*. New York: Cambridge University Press.

Correia, Isabel H. 1996. "Should Capital Income Be Taxed in the Steady State?" *Journal of Public Economics*, Vol. 60(1), pp. 147–151.

Correia, Isabel, and Pedro Teles. 1996. "Is the Friedman Rule Optimal When Money Is an Intermediate Good?" *Journal of Monetary Economics*, Vol. 38, pp. 223–244.

Cox, John C., Jonathan E. Ingersoll, Jr., and Stephen A. Ross. 1985a. "An Intertemporal General Equilibrium Model of Asset Prices." *Econometrica*, Vol. 53(2), pp. 363–384.

Cox, John C., Jonathan E. Ingersoll, Jr., and Stephen A. Ross. 1985b. "A Theory of the Term Structure of Interest Rates." *Econometrica*, Vol. 53(2), pp. 385–408.

Dai, Qiang, and Kenneth J. Singleton. 2000. "Specification Analysis of Affine Term Structure Models." *Journal of Finance*, Vol. LV, pp. 1943-1978.

Davies, James B. and Anthony Shorrocks. 2000. "The Distribution of Wealth." In A.B. Atkinson, F. Bourguignon (eds.), *Handbook of Income Distribution, Vol. 1.*. Amsterdam: North-Holland, pp. 605–675.

Davis, Steven J. 1995. "The Quality Distribution of Jobs and the Structure of Wages in Search Equilibrium." Mimeo. Chicago: University of Chicago Graduate School of Business.

Deaton, Angus. 1992. *Understanding Consumption*. New York: Oxford University Press.

Deaton, Angus and Christina Paxson. 1994. "Intertemporal Choice and Inequality." *Journal of Political Economy*, Vol. 102(3), pp. 437-67.

Debreu, Gerard. 1954. "Valuation Equilibrium and Pareto Optimum." *Proceedings of the National Academy of Sciences*, Vol. 40, pp. 588–592.

Debreu, Gerard. 1959. *Theory of Value*. New York: Wiley.

DeJong, David and Chetan Dave. 2011. *Structural Macroeconometrics, second edition*. Princeton and Oxford: Princeton University Press.

Den Haan, Wouter J., and Albert Marcet. 1990. "Solving the Stochastic Growth Model by Parametering Expectations." *Journal of Business and Economic Statistics*, Vol. 8(1), pp. 31–34.

Den Haan, Wouter J and Albert Marcet. 1994. "Accuracy in Simulations." *Review of Economic Studies*, Vol. 61(1), pp. 3–17.

Diamond, Peter A. 1965. "National Debt in a Neoclassical Growth Model." *American Economic Review*, Vol. 55, pp. 1126–1150.

Diamond, Peter A. 1981. "Mobility Costs, Frictional Unemployment, and Efficiency." *Journal of Political Economy*, Vol. 89(4), pp. 798–812.

Diamond, Peter A. 1982. "Wage Determination and Efficiency in Search Equilibrium." *Review of Economic Studies*, Vol. 49, pp. 217–227.

Diamond, Peter A. 1984. "Money in Search Equilibrium." *Econometrica*, Vol. 52, pp. 1–20.

Diamond, Peter A., and Joseph Stiglitz. 1974. "Increases in Risk and in Risk Aversion." *Journal of Economic Theory*, Vol. 8(3), pp. 337–360.

Diaz-Giménez, J., Edward C. Prescott, T. Fitzgerald, and Fernando Alvarez. 1992. "Banking in Computable General Equilibrium Economies." *Journal of Economic Dynamics and Control*, Vol. 16, pp. 533–560.

Diaz-Giménez, J., V. Quadrini, and J.V. Ríos-Rull. 1997. "Dimensions of Inequality: Facts on the U.S. Distributions of Earnings, Income, and Wealth." *Quarterly Review*, Federal Reserve Bank of Minneapolis, Vol. 21, pp. 3–21.

Dixit, Avinash K., Gene Grossman, and Faruk Gul. 2000. "The Dynamics of Political Compromise." *Journal of Political Economy*, Vol. 108(3), pp. 531–568.

Dixit, Avinash K., and Joseph E. Stiglitz. 1977. "Monopolistic Competition and Optimum Product Diversity." *American Economic Review*, Vol. 67, pp. 297–308.

De Santis, Massimiliano. 2007. "Individual Consumption Risk and the Welfare Cost of Business Cycles" *American Economic Review*, Vol. 97, No. 4, pp. 1488–1506.

Dolmas, J. 1998. "Risk Preferences and the Welfare Cost of Business Cycles." *Review of Economic Dynamics*, Vol. 1(3), pp. 646–676.

Domeij, David, and Jonathan Heathcote. 2000. "Capital versus Labor Income Taxation with Heterogeneous Agents." Mimeo. Stockholm School of Economics.

Doob, Joseph L. 1953. *Stochastic Processes*. New York: Wiley.

Dornbusch, Rudiger. 1976. "Expectations and Exchange Rate Dynamics." *Journal of Political Economy*, Vol. 84, pp. 1161–1176.

Dow, James R., Jr., and Lars J. Olson. 1992. "Irreversibility and the Behavior of Aggregate Stochastic Growth Models." *Journal of Economic Dynamics and Control*, Vol. 16, pp. 207–233.

Duffie, D. and L. G. Epstein. 1992. "Stochastic Differential Utility." *Econometrica*, Vol. 60(2), pp. 353–394.

Duffie, Darrell. 1996. *Dynamic Asset Pricing Theory*. Princeton, N.J.: Princeton University Press, pp. xvii, 395.

Duffie, Darrell, and Rui Kan. 1996. "A Yield-Factor Model of Interest Rates." *Mathematical Finance*, Vol. 6(4), pp. 379–406.

Duffie, Darrell, J. Geanakoplos, A. Mas-Colell, and A. McLennan. 1994. "Stationary Markov Equilibria." *Econometrica*, Vol. 62(4), pp. 745–781.

Eichenbaum, Martin. 1991. "Real Business-Cycle Theory: Wisdom or Whimsy?" *Journal of Economic Dynamics and Control*, Vol. 15(4), pp. 607–626.

Eichenbaum, Martin, and Lars P. Hansen. 1990. "Estimating Models with Intertemporal Substitution Using Aggregate Time Series Data." *Journal of Business and Economic Statistics*, Vol. 8, pp. 53–69.

Eichenbaum, Martin, Lars P. Hansen, and S.F. Richard. 1984. "The Dynamic Equilibrium Pricing of Durable Consumption Goods." Mimeo. Pittsburgh: Carnegie-Mellon University.

Elliott, Robert J., Lakhdar Aggoun, and John B. Moore. 1995. *Hidden Markov Models: Estimation and Control.* New York: Springer-Verlag.

Ellsberg, Daniel. 1961. "Risk, Ambiguity and the Savage Axioms" *Quarterly Journal of Economics*, Vol. 75(4), pp. 643–669.

Engle, Robert F., and C. W. J. Granger. 1987. "Co-Integration and Error Correction: Representation, Estimation, and Testing." *Econometrica*, Vol. 55(2), pp. 251–276.

Epstein, Larry G., and Stanley E. Zin. 1989. "Substitution, Risk Aversion, and the Temporal Behavior of Consumption and Asset Returns: A Theoretical Framework." *Econometrica*, Vol. 57(4), pp. 937–969.

Epstein, Larry G., and Stanley E. Zin. 1991. "Substitution, Risk Aversion, and the Temporal Behavior of Consumption and Asset Returns: An Empirical Analysis." *Journal of Political Economy*, Vol. 99(2), pp. 263–286.

Ethier, Wilfred J. 1982. "National and International Returns to Scale in the Modern Theory of International Trade." *American Economic Review*, Vol. 72, pp. 389–405.

Evans, George W., and Seppo Honkapohja. 2001. *Learning and Expectations in Macroeconomics.* Princeton University Press, Princeton, New Jersey.

Evans, George W., and Seppo Honkapohja. 2003. "Expectations and the Stability Problem for Optimal Monetary Policies." *Review of Economic Studies*, Vol. 70(4), pp. 807–24.

Faig, Miquel. 1988. "Characterization of the Optimal Tax on Money When It Functions as a Medium of Exchange." *Journal of Monetary Economics*, Vol. 22(1), pp. 137–148.

Fama, Eugene F. 1976a. *Foundations of Finance: Portfolio Decisions and Securities Prices.* New York: Basic Books.

Fama, Eugene F. 1976b. "Inflation Uncertainty and Expected Returns on Treasury Bills." *Journal of Political Economy*, Vol. 84(3), pp. 427–448.

Farmer, Roger E. A. 1993. *The Macroeconomics of Self-fulfilling Prophecies.* Cambridge, Mass.: MIT Press.

Fernandes, Ana and Christopher Phelan. 2000. "A Recursive Formulation for Repeated Agency with History Dependence" *Journal of Economic Theory*, Vol. 91(2), pp. 223-247.

Fischer, Stanley. 1983. "A Framework for Monetary and Banking Analysis." *Economic Journal*, Vol. 93, Supplement, pp. 1–16.

Fisher, Irving. 1913. *The Purchasing Power of Money: Its Determination and Relation to Credit, Interest and Crises.* New York: Macmillan.

Fisher, Irving. [1907] 1930. *The Theory of Interest.* London: Macmillan.

Fisher, Jonas. 2006. "The Dynamic Effects of Neutral and Investment-Specific Technology Shocks." *Journal of Political Economy*, Vol. 114, No. 3, pp. 413–451.

2003Technology Shocks Matter.Federal Reserve Bank of Chicago, Working Paper No. 14

Foley, Duncan K. and Martin F. Hellwig. 1975. "Asset Management with Trading Uncertainty." *Review of Economic Studies*, Vol. 42, no. 3, pp. 327-346.

Frankel, Marvin. 1962. "The Production Function in Allocation and Growth: A Synthesis." *American Economic Review*, Vol. 52, pp. 995–1022.

Friedman, Milton. 1956. *A Theory of the Consumption Function.* Princeton, N.J.: Princeton University Press.

Friedman, Milton. 1967. "The Role of Monetary Policy." *American Economic Review*, Vol. 58, 1968, pp. 1–15. Presidential Address delivered at the 80th Annual Meeting of the American Economic Association, Washington, DC, December 29.

Friedman, Milton. 1969. "The Optimum Quantity of Money." In Milton Friedman (ed.), *The Optimum Quantity of Money and Other Essays*. Chicago: Aldine, pp. 1–50.

Friedman, Milton, and Anna J. Schwartz. 1963. *A Monetary History of the United States, 1867–1960*. Princeton, N.J.: Princeton University Press and NBER.

Fuchs, William Martin, and Francesco Lippi. 2006. "Monetary Union with Voluntary Participation" *Review of Economic Studies*, Vol. 73, pp. 437-457.

Fudenberg, Drew, Bengt Holmström, and Paul Milgrom. 1990. "Short-Term Contracts and Long-Term Agency Relationships." *Journal of Economic Theory*, Vol. 51(1).

Fudenberg Drew, and David M. Kreps. 1993. "Learning Mixed Equilibria" *Games and Economic Behavior*, Vol. 5(3), pp. 320–367.

Fudenberg, Drew and David K. Levine. 1998. *The Theory of Learning in Games*. MIT Press, Cambridge, MA..

Gabel, R. A., and R. A. Roberts. 1973. *Signals and Linear Systems*. New York: Wiley.

Gale, David. 1973. "Pure Exchange Equilibrium of Dynamic Economic Models." *Journal of Economic Theory*, Vol. 6, pp. 12-36.

Gali, Jordi. 1991. "Budget Constraints and Time-Series Evidence on Consumption." *American Economic Review*, Vol. 81(5), pp. 1238–1253.

Gali, Jordi. 2008. *Monetary Policy, Inflation, and the Business Cycle: An Introduction to the New Keynesian Framework*. Princeton and Oxford: Princeton University Press.

Gallant, R., L. P. Hansen, and G. Tauchen. 1990. "Using Conditional Moments of Asset Payoffs to Infer the Volatility of Intertemporal Marginal Rates of Substitution." *Journal of Econometrics*, Vol. 45, pp. 145–179.

Genicot, Garance and Debraj Ray. 2003. "Group Formation in Risk-Sharing Arrangements." *Review of Economic-Studies*, Vol. 70(1), pp. 87–113.

Gittins, J.C. 1989. *Multi-armed Bandit and Allocation Indices*. New York: Wiley.

Golosov M., N. Kocherlakota, and A. Tsyvinski. 2003. "Optimal Indirect and Capital Taxation." *Review of Economic Studies*, Vol. 70, no. 3, July, pp. 569–587.

Gomes, Joao, Jeremy Greenwood, and Sergio Rebelo. 2001. "Equilibrium Unemployment." *Journal of Monetary Economics*, Vol. 48(1), pp. 109–152.

Gong, Frank F., and Eli M. Remolona. 1997. "A Three Factor Econometric Model of the U.S. Term Structure." Mimeo. Federal Reserve Bank of New York, Staff Report 19.

Gourinchas, Pierre-Olivier, and Jonathan A. Parker. 1999. "Consumption over the Life Cycle." Mimeo. NBER Working Paper No.7271.

Granger, C. W. J. 1966. "The Typical Spectral Shape of an Economic Variable." *Econometrica*, Vol. 34(1), pp. 150–161.

Granger, C. W. J. 1969. "Investigating Causal Relations by Econometric Models and Cross-Spectral Methods." *Econometrica*, Vol. 37(3), pp. 424–438.

Green, Edward J. 1987. "Lending and the Smoothing of Uninsurable Income." In Edward C. Prescott and Neil Wallace (eds.), *Contractual Arrangements for Intertemporal Trade, Minnesota Studies in Macroeconomics series, Vol. 1*. Minneapolis: University of Minnesota Press, pp. 3–25.

Green, Edward J., and Robert H. Porter. 1984. "Non-Cooperative Collusion under Imperfect Price Information." *Econometrica*, Vol. 52, pp. 975–993.

Greenwood, Jeremy, Zvi Hercowitz, and Gregory W. Huffman. 1988. "Investment, Capacity Utilization, and the Real Business Cycle." *American Economic Review*, Vol. 78, no. 3, pp. 402-17.

Greenwood, J., Z. Hercowitz, and P. Krusell. 1997. "Long Run Implications of Investment-Specific Technological Change." *American Economic Review*, Vol. 78, pp. 342–362.

Grossman, Gene M., and Elhanan Helpman. 1991. "Quality Ladders in the Theory of Growth." *Review of Economic Studies*, Vol. 58, pp. 43–61.

Grossman, Sanford J., and Robert J. Shiller. 1981. "The Determinants of the Variability of Stock Market Prices." *American Economic Review*, Vol. 71(2), pp. 222–227.

Gul, Faruk, and Wolfgang Pesendorfer. 2000. "Self-Control and the Theory of Consumption." Mimeo. Princeton, New Jersey: Princeton University.

Guvenen, Fatih, Burhanettin Kuruscu, and Serdar Ozkan. 2011. "Taxation of Human Capital and Wage Inequality: A Cross-Country Analysis" Mimeo. University of Minnesota, University of Toronto and Federal Reserve Board.

Guidotti, Pablo E., and Carlos A. Vegh. 1993. "The Optimal Inflation Tax When Money Reduces Transactions Costs: A Reconsideration." *Journal of Monetary Economics*, Vol. 31(3), pp. 189–205.

Hall, Robert E. 1971. "The Dynamic Effects of Fiscal Policy in an Economy with Foresight." *Review of Economic Studies*, Vol. 38, pp. 229–244.

Hall, Robert E. 1978. "Stochastic Implications of the Life Cycle-Permanent Income Hypothesis: Theory and Evidence." *Journal of Political Economy*, Vol. 86(6), pp. 971–988. (Reprinted in *Rational Expectations and Econometric Practice*, ed. Robert E. Lucas, Jr., and Thomas J. Sargent, Minneapolis: University of Minnesota Press, 1981, pp. 501–520.).

Hall, Robert E. 1997. "Macroeconomic Fluctuations and the Allocation of Time." *Journal of Labor Economics*, Vol. 15(1), pp. 223-250.

Hall, Robert E., and Dale Jorgenson. 1967. "Tax Policy and Investment Behavior." *The American Economic Review*, Vol. 57(3), pp. 391–414.

Hamilton, James D. 1994. *Time Series Analysis*. Princeton, N.J.: Princeton University Press.

Hamilton, James D., and Marjorie A. Flavin. 1986. "On the Limitations of Government Borrowing: A Framework for Empirical Testing." *American Economic Review*, Vol. 76(4), pp. 808–819.

Hansen, Gary D. 1985. "Indivisible Labor and the Business Cycle." *Journal of Monetary Economics*, Vol. 16, pp. 309–327.

Hansen, Gary D., and Ayşe İmrohoroğlu. 1992. "The Role of Unemployment Insurance in an Economy with Liquidity Constraints and Moral Hazard." *Journal of Political Economy*, Vol. 100 (1), pp. 118–142.

Hansen, Lars P. 1982a. "Consumption, Asset Markets, and Macroeconomic Fluctuations: A Comment." *Carnegie-Rochester Conference Series on Public Policy*, Vol. 17, pp. 239–250.

Hansen, Lars P. 1982b. "Large Sample Properties of Generalized Method of Moments Estimators." *Econometrica*, Vol. 50, pp. 1029–1060.

Hansen, Lars P., Dennis Epple, and Will Roberds. 1985. "Linear-Quadratic Duopoly Models of Resource Depletion." In Thomas J. Sargent (ed.), *Energy, Foresight, and Strategy*. Washington, DC: Resources for the Future, pp. 101–142.

Hansen, Lars P. and Thomas J. Sargent. 1991. "Exact Linear Rational Expectations Models." In Lars P. Hansen and Thomas J. Sargent (eds.), *Rational Expectations Econometrics*. Boulder, Colo.: Westview Press, pp. 45–76.

Hansen, L. P., T. J. Sargent, and N. E. Wang. 2002. "Robust Permanent Income and Pricing with Filtering." *Macroeconomic Dynamics*, Vol. 6(1), pp. 40–84.

Hansen L. P. and T. J. Sargent. 2001. "Robust Control and Model Uncertainty." *American Economic Review*, Vol. 91(2), pp. 60–66.

Hansen, Lars P. and Thomas J. Sargent. 2008. *Robustness*. Princeton University Press, Princeton, New Jersey.

Hansen, Lars P., and Ravi Jagannathan. 1991. "Implications of Security Market Data for Models of Dynamic Economies." *Journal of Political Economy*, Vol. 99, pp. 225–262.

Hansen, Lars P., and Ravi Jagannathan. 1997. "Assessing Specification Errors in Stochastic Discount Factor Models." *Journal of Finance*, Vol. 52(2), pp. 557–590.

Hansen, Lars P., William T. Roberds, and Thomas J. Sargent. 1991. "Time Series Implications of Present Value Budget Balance and of Martingale Models of Consumption and Taxes." In L. P. Hansen and T. J. Sargent (eds.), *Rational Expectations and Econometric Practice*. Boulder, Colo.: Westview Press, pp. 121–161.

Hansen, Lars P., and Thomas J. Sargent. 1980. "Formulating and Estimating Dynamic Linear Rational Expectations Models." *Journal of Economic Dynamics and Control*, Vol. 2(1), pp. 7–46.

Hansen, Lars P., and Thomas J. Sargent. 1981. "Linear Rational Expectations Models for Dynamically Interrelated Variables." In R. E. Lucas, Jr., and T. J. Sargent (eds.), *Rational Expectations and Econometric Practice*. Minneapolis: University of Minnesota Press, pp. 127–156.

Hansen, Lars P., and Thomas J. Sargent. 1982. "Instrumental Variables Procedures for Estimating Linear Rational Expectations Models." *Journal of Monetary Economics*, Vol. 9(3), pp. 263–296.

Hansen, Lars P., and Thomas J. Sargent. 1995. "Discounted Linear Exponential Quadratic Gaussian Control." *IEEE Transactions on Automatic Control*, Vol. 40, pp. 968–971.

Hansen, Lars P., Thomas J. Sargent, and Thomas D. Tallarini, Jr. 1999. "Robust Permanent Income and Pricing." *Review of Economic Studies*, Vol. 66(4), pp. 873–907.

Hansen, Lars P. and Thomas J. Sargent. 2008. *Robustness*. Princeton University Press, Princeton, New Jersey.

Hansen, Lars P. and Thomas J. Sargent. 2011. "Wanting Robustness in Macroeconomics." In Benjamin Friedman and Michael Woodford (eds.), *Handbook of Monetary Economics, Volume 3B*. Elsevier, North-Holland, pp. 1097-1157.

Hansen, Lars P., and Kenneth J. Singleton. 1982. "Generalized Instrumental Variables Estimation of Nonlinear Rational Expectations Models." *Econometrica*, Vol. 50(5), pp. 1269–1286.

Hansen, Lars P., and Kenneth J. Singleton. 1983. "Stochastic Consumption, Risk Aversion, and the Temporal Behavior of Asset Returns." *Journal of Political Economy*, Vol. 91(2), pp. 249–265.

Hansen, Lars P., and Thomas J. Sargent. 2000. "Recursive Models of Dynamic Linear Economies" Mimeo. University of Chicago and Stanford University.

Harrison, Michael, and David Kreps. 1979. "Martingales and Arbitrage in Multiperiod Security Markets." *Journal of Economic Theory*, Vol. 20, pp. 381–408.

Heathcote, Jonathan, Kjetil Storesletten, and Giovanni L. Violante. 2008. "Insurance and opportunities: A welfare analysis of labor market risk." *Journal of Monetary Economics*, Vol. 55(3), pp. 501–525.

Heathcote, Jonathan, Fabrizio Perri, and Giovanni L. Violante. 2010. "Unequal We Stand: An Empirical Analysis of Economic Inequality in the United States: 1967-2006." *Review of Economic Dynamics*, Vol. 13, No. 1, pp. 15–51.

Heathcote, Jonathan, Kjetil Storesletten, and Giovanni L. Violante. 2012. "Consumption and Labor Supply with Partial Insurance: An Analytical Framework." Mimeo. Working paper, Federal Reserve Bank of Minneapolis and NYU.

Heaton, John, and Deborah J. Lucas. 1996. "Evaluating the Effects of Incomplete Markets on Risk Sharing and Asset Pricing." *Journal of Political Economy*, Vol. 104(3), pp. 443–487.

Helpman, Elhanan. 1981. "An Exploration in the Theory of Exchange-Rate Regimes." *Journal of Political Economy*, Vol. 89(5), pp. 865–890.

Hirshleifer, Jack. 1966. "Investment Decision under Uncertainty: Applications of the State Preference Approach." *Quarterly Journal of Economics*, Vol. 80(2), pp. 252–277.

Holmström, Bengt. 1983. "Equilibrium Long Term Labour Contracts." *Quarterly Journal of Economics, Supplement*, Vol. 98(1), pp. 23–54.

Hopenhayn, Hugo A., and Juan Pablo Nicolini. 1997. "Optimal Unemployment Insurance." *Journal of Political Economy*, Vol. 105(2), pp. 412–438.

Hopenhayn, Hugo A., and Edward C. Prescott. 1992. "Stochastic Monotonicity and Stationary Distributions for Dynamic Economies." *Econometrica*, Vol. 60(6), pp. 1387–1406.

Hopenhayn, Hugo, and Richard Rogerson. 1993. "Job Turnover and Policy Evaluation: A General Equilibrium Analysis." *Journal of Political Economy*, Vol. 101, pp. 915–938.

Hosios, Arthur, J. 1990. "On the Efficiency of Matching and Related Models of Search and Unemployment." *Review of Economic Studies*, Vol. 57, pp. 279–298.

Hubbard, R. Glenn, Jonathan Skinner, and Stephen P. Zeldes. 1995. "Precautionary Saving and Social Insurance." *Journal of Political Economy*, Vol. 103(2), pp. 360–399.

Huffman, Gregory. 1986. "The Representative Agent, Overlapping Generations, and Asset Pricing." *Canadian Journal of Economics*, Vol. 19(3), pp. 511–521.

Huggett, Mark. 1993. "The Risk Free Rate in Heterogeneous-Agent, Incomplete-Insurance Economies." *Journal of Economic Dynamics and Control*, Vol. 17(5-6), pp. 953–969.

Huggett, Mark, and Sandra Ospina. 2000. "Aggregate Precautionary Savings: When Is the Third Derivative Irrelevant?" Mimeo. Washington, D.C.: Georgetown University.

İmrohoroğlu, Ayşe. 1992. "The Welfare Cost of Inflation Under Imperfect Insurance." *Journal of Economic Dynamics and Control*, Vol. 16(1), pp. 79–92.

İmrohoroğlu, Ayşe, Selahattin İmrohoroğlu, and Douglas Joines. 1995. "A Life Cycle Analysis of Social Security." *Economic Theory*, Vol. 6(1), pp. 83–114.

Ingram, Beth Fisher, Narayana Kocherlakota, and N. E. Savin. 1994. "Explaining Business Cycles: A Multiple-Shock Approach." *Journal of Monetary Economics*, Vol. 34(3), pp, 415–428.

References

Ireland, Peter N. 1994. "Inflationary Policy and Welfare with Limited Credit Markets." *Journal of Financial Intermediation*, Vol. 3(3), pp. 245–271.

Ireland, Peter N. 1997. "Sustainable Monetary Policies." *Journal of Economic Dynamics and Control*, Vol. 22, pp. 87–108.

Jacobson, David H. 1973. "Optimal Stochastic Linear Systems with Exponential Performance Criteria and Their Relation to Deterministic Differential Games." *IEEE Transactions on Automatic Control*, Vol. 18(2), pp. 124–131.

Johnson, Norman, and Samuel Kotz. 1971. *Continuous Univariate Distributions*. New York: Wiley.

Jones, Charles I. 1995. "R&D-Based Models of Economic Growth." *Journal of Political Economy*, Vol. 103, pp. 759–784.

Jones, Larry E.. 2008. "A Note on the Joint Occurrence of Insurance and Gambling" *Macroeconomic Dynamics*, Vol. 12, pp. 97-111.

Jones, Larry E., and Rodolfo Manuelli. 1990. "A Convex Model of Equilibrium Growth: Theory and Policy Implications." *Journal of Political Economy*, Vol. 98, pp. 1008–1038.

Jones, Larry E., and Rodolfo E. Manuelli. 1992. "Finite Lifetimes and Growth." *Journal of Economic Theory*, Vol. 58, pp. 171–197.

Jones, Larry E., Rodolfo E. Manuelli, and Peter E. Rossi. 1993. "Optimal Taxation in Models of Endogenous Growth." *Journal of Political Economy*, Vol. 101, pp. 485–517.

Jones, Larry E., Rodolfo E. Manuelli, and Peter E. Rossi. 1997. "On the Optimal Taxation of Capital Income." *Journal of Economic Theory*, Vol. 73(1), pp. 93–117.

Jovanovic, Boyan. 1979a. "Job Matching and the Theory of Turnover." *Journal of Political Economy*, Vol. 87(5), pp. 972–990.

Jovanovic, Boyan. 1979b. "Firm-Specific Capital and Turnover." *Journal of Political Economy*, Vol. 87(6), pp. 1246–1260.

Jovanovic, Boyan, and Yaw Nyarko. 1996. "Learning by Doing and the Choice of Technology." *Econometrica*, Vol. 64(6), pp. 1299–1310.

Judd, Kenneth L. 1985a. "On the Performance of Patents." *Econometrica*, Vol. 53, pp. 567–585.

Judd, Kenneth L. 1985b. "Redistributive Taxation in a Simple Perfect Foresight Model." *Journal of Public Economics*, Vol. 28, pp. 59–83.

Judd, Kenneth L. 1990. "Cournot versus Bertrand: A Dynamic Resolution." Mimeo. Stanford, Calif.: Hoover Institution, Stanford University..

Judd, Kenneth L. 1996. "Approximation, Perturbation, and Projection Methods in Economic Analysis." In Hans Amman, David Kendrick, and John Rust (eds.), *Handbook of Computational Economics, Vol. 1*. Amsterdam: North-Holland.

Judd, Kenneth L. 1998. *Numerical Methods in Economics*. Cambridge, Mass.: MIT Press.

Judd, Kenneth L., Sevin Yeltekin, and James Conklin. 2003. "Computing Supergame Equilibria." *Econometrica*, Vol. 71, no. 4, pp. 1239–1254.

Judd, Kenneth L., and Andrew Solnick. 1994. "Numerical dynamic programming with shape preserving splines." Mimeo. Stanford, Calif.: Hoover Institution.

Kahn, Charles, and William Roberds. 1998. "Real-Time Gross Settlement and the Costs of Immediacy." Mimeo. Federal Reserve Bank of Atlanta, Working Paper No.98–21, December.

Kalman, R. E. 1960. "Contributions to the Theory of Optimal Control." *Bol. Soc. Mat. Mexicana*, Vol. 5, pp. 102–119.

Kalman, R. E., and R. S. Bucy. 1961. "New Results in Linear Filtering and Prediction Theory." *J. Basic Eng., Trans. ASME, Ser. D*, Vol. 83, pp. 95–108.

Kandori, Michihiro. 1992. "Repeated Games Played by Overlapping Generations of Players." *Review of Economic Studies*, Vol. 59(1), pp. 81–92.

Kareken, John, T. Muench, and N. Wallace. 1973. "Optimal Open Market Strategy: The Use of Information Variables." *American Economic Review*, Vol. 63(1), pp. 156–172.

Kareken, John, and Neil Wallace. 1980. *Models of Monetary Economies*. Minneapolis: Federal Reserve Bank of Minneapolis, pp. 169–210.

Kareken, John, and Neil Wallace. 1981. "On the Indeterminacy of Equilibrium Exchange Rates." *Quarterly Journal of Economics*, Vol. 96, pp. 207–222.

Kaplan, Greg and Giovanni L. Violante. 2010. "How Much Consumption Insurance beyond Self-Insurance" *American Economic Journal: Macroeconomics*, Vol. 2(4), pp. 53–87.

Kehoe, Patrick, and Fabrizio Perri. 2002. "International Business Cycles with Endogenous Incomplete Markets." *Econometrica*, Vol. 70(3), pp. 907–928.

Kehoe, Timothy J., and David K. Levine. 1984. "Intertemporal Separability in Overlapping-Generations Models." *Journal of Economic Theory*, Vol. 34, pp. 216–226.

Kehoe, Timothy J., and David K. Levine. 1985. "Comparative Statics and Perfect Foresight in Infinite Horizon Economies." *Econometrica*, Vol. 53, pp. 433–453.

Kehoe, Timothy J., and David K. Levine. 1993. "Debt-Constrained Asset Markets." *Review of Economic Studies*, Vol. 60(4), pp. 865–888.

Keynes, John Maynard. 1940. *How to Pay for the War: A Radical Plan for the Chancellor of the Exchequer*. London: Macmillan.

Kihlstrom, Richard E., and Leonard J. Mirman. 1974. "Risk Aversion with Many Commodities." *Journal of Economic Theory*, Vol. 8, pp. 361–388.

Kim, Chang-Jin, and Charles R. Nelson. 1999. *State Space Models with Regime Switching*. Cambridge, Mass.: MIT Press.

Kim, J., and S. Kim. 2003. "Spurious Welfare Reversals in International Business Cycle Models." *Journal of International Economics*, Vol. 60, pp. 471–500.

Kimball, Miles S., and Mankiw, Gregory. 1989. "Precautionary Saving and the Timing of Taxes." *Journal of Poltical Economy*, Vol. 97(4), pp. 863–879.

Kimball, M. S. 1990. "Precautionary Saving in the Small and in the Large." *Econometrica*, Vol. 58, pp. 53–73.

Kimball, M. S. 1993. "Standard Risk Aversion." *Econometrica*, Vol. 63(3), pp. 589–611.

Kimbrough, Kent P. 1986. "The Optimum Quantity of Money Rule in the Theory of Public Finance." *Journal of Monetary Economics*, Vol. 18, pp. 277–284.

King, Robert G., and Charles I. Plosser. 1988. "Real Business Cycles: Introduction." *Journal of Monetary Economics*, Vol. 21, pp. 191–193.

King, Robert G., Charles I. Plosser, and Sergio T. Rebelo. 1988. "Production, Growth and Business Cycles: I. The Basic Neoclassical Model." *Journal of Monetary Economics*, Vol. 21, pp. 195–232.

King, Robert G., and Alexander L. Wolman. 1999. "What Should the Monetary Authority Do When Prices are Sticky." In John B. Taylor (ed.), *Monetary Policy Rules*. Chicago: University of Chicago Press, pp. 349-398.

Kiyotaki, Nobuhiro, and Randall Wright. 1989. "On Money as a Medium of Exchange." *Journal of Political Economy*, Vol. 97(4), pp. 927–954.

Kiyotaki, Nobuhiro, and Randall Wright. 1990. "Search for a Theory of Money." Mimeo. Cambridge, Mass.: National Bureau of Economic Research, Working Paper No. 3482.

Kiyotaki, Nobuhiro, and Randall Wright. 1993. "A Search-Theoretic Approach to Monetary Economics." *American Economic Review*, Vol. 83(1), pp. 63–77.

Kletzer, Kenneth M. and Brian D. Wright. 2000. "Sovereign Debt as Intertemporal Barter." *American Economic Review*, Vol. 90(3), pp. 621–39.

Kocherlakota, Narayana R. 1996a. "The Equity Premium: It's Still a Puzzle." *Journal of Economic Literature*, Vol. 34(1), pp. 42–71.

Kocherlakota, Narayana R. 1996b. "Implications of Efficient Risk Sharing without Commitment." *Review of Economic Studies*, Vol. 63(4), pp. 595–609.

Kocherlakota, Narayana R. 1998. "Money Is Memory." *Journal of Economic Theory*, Vol. 81 (2), pp. 232–251.

Kocherlakota, Narayana R., and Christopher Phelan. 1999. "Explaining the Fiscal Theory of the Price Level." *Quarterly Review*, Federal Reserve Bank of Minneapolis, Vol. 23(4), pp. 14–23.

Kocherlakota, Narayana R., and Neil Wallace. 1998. "Incomplete Record-Keeping and Optimal Payment Arrangements." *Journal of Economic Theory*, Vol. 81(2), pp. 272–289.

Kocherlakota, Narayana R. 2004. "Figuring Out the Impact of Hidden Savings on Optimal Unemployment Insurance." *Review of Economic Dynamics.*, forthcoming.

Koeppl, Thorsten V. 2003. "Differentiability of the Efficient Frontier when Commitment to Risk Sharing Is Limited" Mimeo. Frankfurt, Germany: European Central Bank.

Koopmans, Tjalling C. 1965. *On the Concept of Optimal Growth*. The Econometric Approach to Development Planning.Chicago: Rand McNally

Kreps, D. M. and E. L. Porteus. 1978. "Temporal Resolution of Uncertainty and Dynamic Choice." *Econometrica*, Vol. 46(1), pp. 185–200.

Kreps, David M. 1979. "Three Essays on Capital Markets." Mimeo. Technical Report 298. Stanford, Calif.: Institute for Mathematical Studies in the Social Sciences, Stanford University.

Kreps, David M. 1988. *Notes on the Theory of Choice*. Boulder, Colo.: Westview Press.

Kreps, David M. 1990. *Game Theory and Economic Analysis*. New York: Oxford University Press.

Krueger, Dirk. 1999. "Risk Sharing in Economies with Incomplete Markets." Mimeo. Stanford, California: Stanford University.

Krueger, Dirk, and Fabrizio Perri. 2004. "On the Welfare Consequences of the Increase in Inequality in the United States." *NBER Macroeconomics Annual 2003*, Volume 18, pp. 83–138.

Krueger, Dirk, and Fabrizio Perri. 2006. "Does Income Inequality Lead to Consumption Inequality? Evidence and Theory." *Review of Economic Studies*, Vol. 73(1), pp. 163–193.

Krueger, Dirk, Fabrizio Perri, Luigi Pistaferri, and Giovanni L. Violante. 2010. "Cross Sectional Facts for Macroeconomists." *Review of Economic Dynamics*, Vol. 13(1), pp. 1–14.

Krusell, Per, Toshihiko Mukoyama, Richard Rogerson, Ayşgül Şahin. 2008. "Aggregate Implications of Indivisible Labor, Incomplete Markets, and Labor Market Frictions." *Journal of Monetary Economics*, Vol. 55, pp. 961–979.

Krusell, Per, and Anthony Smith. 1998. "Income and Wealth Heterogeneity in the Macroeconomy." *Journal of Political Economy*, Vol. 106(5), pp. 867–896.

Kwakernaak, Huibert, and Raphael Sivan. 1972. *Linear Optimal Control Systems*. New York: Wiley.

Kydland, Finn E., and Edward C. Prescott. 1977. "Rules Rather than Discretion: The Inconsistency of Optimal Plans." *Journal of Political Economy*, Vol. 85(3), pp. 473–491.

Kydland, Finn E., and Edward C. Prescott. 1980. "Dynamic Optimal Taxation, Rational Expectations and Optimal Control." *Journal of Economic Dynamics and Control*, Vol. 2(1), pp. 79–91.

Kydland, Finn E., and Edward C. Prescott. 1982. "Time to Build and Aggregate Fluctuations." *Econometrica*, Vol. 50(6), pp. 1345–1371.

Labadie, Pamela. 1986. "Comparative Dynamics and Risk Premia in an Overlapping Generations Model." *Review of Economic Studies*, Vol. 53(1), pp. 139–152.

Lagos, Ricardo. 2000. "An Alternative Approach to Search Frictions." *Journal of Political Economy*, 108(5), pp. 851–873.

Laibson, David I. 1994. "Hyperbolic Discounting and Consumption." Mimeo. Cambridge, Mass.: Massachusetts Institute of Technology.

Davig, Troy, Eric M. Leeper, and Todd B. Walker. 2010. ""Unfunded liabilities" and uncertain fiscal financing." *Journal of Monetary Economics*, Vol. 57, No. 5, pp. 600–619.

Leland, Hayne E. 1968. "Saving and Uncertainty: The Precautionary Demand for Saving." *Quarterly Journal of Economics*, Vol. 82(3), pp. 465–473.

LeRoy, Stephen F. 1971. "The Determination of Stock Prices." Ph.D. dissertation, Philadelphia: University of Pennsylvania.

LeRoy, Stephen F. 1973. "Risk Aversion and the Martingale Property of Stock Prices." *International Economic Review*, Vol. 14(2), pp. 436–446.

LeRoy, Stephen F. 1982. "Risk Aversion and the Term Structure of Interest Rates." *Economics Letters*, Vol. 10(3–4), pp. 355–361. (Correction in *Economics Letters*, 1983, Vol. 12(3–4); 339–340.).

LeRoy, Stephen F. 1984a. "Nominal Prices and Interest Rates in General Equilibrium: Money Shocks." *Journal of Business*, Vol. 57(2), pp. 177–195.

LeRoy, Stephen F. 1984b. "Nominal Prices and Interest Rates in General Equilibrium: Endowment Shocks." *Journal of Business*, Vol. 57(2), pp. 197–213.

LeRoy, Stephen F., and Richard D. Porter. 1981. "The Present-Value Relation: Tests Based on Implied Variance Bounds." *Econometrica*, Vol. 49(3), pp. 555–574.

Lettau, Martin and Sydney Ludvigson. 2001. "Consumption, Aggregate Wealth, and Expected Stock Returns." *Journal of Finance*, Vol. LVI(3), pp. 815–849.

Lettau, Martin and Sydney Ludvigson. 2004. "Understanding Trend and Cycle in Asset Values: Reevaluating the Wealth Effect on Consumption." *American Economic Review*, forthcoming.

Levhari, David, and Leonard J. Mirman. 1980. "The Great Fish War: An Example Using a Dynamic Cournot-Nash Solution." *Bell Journal of Economics*, Vol. 11(1).

Levhari, David, and T. N. Srinivasan. 1969. "Optimal Savings under Uncertainty." *Review of Economic Studies*, Vol. 36(2), pp. 153–163.

Levine, David K., and Drew Fudenberg. 1998. *The Theory of Learning in Games.* Cambridge, Mass.: MIT Press.

Levine, David K., and William R. Zame. 2002. "Does Market Incompleteness Matter?" *Econometrica*, Vol. 70(5), pp. 1805-1839.

Ligon, Ethan. 1998. "Risk Sharing and Information in Village Economies." *Review of Economic Studies*, Vol. 65(4), pp. 847–864.

Ligon, Ethan, Jonathan P. Thomas, and Tim Worrall. 2002. "Informal Insurance Arrangements with Limited Commitment: Theory and Evidence from Village Economies." *Review of Economic Studies*, Vol. 69(1), pp. 209–244.

Lippman, Steven A., and John J. McCall. 1976. "The Economics of Job Search: A Survey." *Economic Inquiry*, Vol. 14(3), pp. 347–368.

Ljungqvist, Lars. 2002. "How Do Layoff Costs Affect Employment?" *Economic Journal*, Vol. 112, pp. 829–853.

Ljungqvist, Lars, and Thomas J. Sargent. 1998. "The European Unemployment Dilemma." *Journal of Political Economy*, Vol. 106, pp. 514–550.

Ljungqvist, Lars, and Thomas J. Sargent. 2005. "Lotteries for Consumers versus Lotteries for Firms." In Timothy Kehoe, T.N. Srinivasan and John Whalley (eds.), *Frontiers in Applied General Equilibrium Modeling: In Honor of Herbert Scarf.* Cambridge University Press, pp 119–126.

Ljungqvist, Lars, and Thomas J. Sargent. 2007. "Do Taxes Explain European Employment?: Indivisible Labor, Human Capital, Lotteries, and Savings." In Daron Acemoglu, Kenneth Rogoff, and Michael Woodford (eds.), *Macroeconomics Annual 2006.* Cambridge MA., MIT Press pp 181–224.

Ljungqvist, Lars, and Thomas J. Sargent. 2011. "A Labor Supply Elasticity Accord?" *American Economic Review*, Vol. 101, pp. 487–491.

Ljungqvist, Lars, and Thomas J. Sargent. 2012. "Career Length: Effects of Curvature of Earnings Profiles, Earnings Shocks, Taxes, and Social Security" Mimeo. Stockholm School of Economics and New York University.

Lucas, Robert E., Jr. 1972. "Expectations and the Neutrality of Money." *Journal of Economic Theory*, Vol. 4, pp. 103–124.

Lucas, Robert E., Jr. 1973. "Some International Evidence on Output-Inflation Trade-Offs." *American Economic Review*, Vol. 63, pp. 326–334.

Lucas, Robert E., Jr. 1976. "Econometric Policy Evaluation: A Critique." In K. Brunner and A. H. Meltzer (eds.), *The Phillips Curve and Labor Markets.* Amsterdam: North-Holland, pp. 19–46.

Lucas, Robert E., Jr. 1978. "Asset Prices in an Exchange Economy." *Econometrica*, Vol. 46(6), pp. 1429–1445.

Lucas, Robert E., Jr. 1980a. "Equilibrium in a Pure Currency Economy." In J. H. Kareken and N. Wallace (eds.), *Economic Inquiry.* Vol. 18(2), pp. 203-220. (Reprinted in *Models of Monetary Economies*, Federal Reserve Bank of Minneapolis, 1980, pp. 131–145.)

Lucas, Robert E., Jr. 1980b. "Two Illustrations of the Quantity Theory of Money." *American Economic Review*, Vol. 70, pp. 1005–1014.

Lucas, Robert E., Jr. 1981. "Econometric Testing of the Natural Rate Hypothesis." In Robert E. Lucas, Jr. (ed.), *Studies of Business-Cycle Theory*. Cambridge, Mass.: MIT Press, pp. 90–103. (Reprinted from *The Econometrics of Price Determination Conference*, ed. Otto Eckstein. Washington, D.C.: Board of Governors of the Federal Reserve System, 1972, pp. 50–59).

Lucas, Robert E., Jr. 1982. "Interest Rates and Currency Prices in a Two-Country World." *Journal of Monetary Economics*, Vol. 10(3), pp. 335–360.

Lucas, Robert E. Jr.. 1986. "Principles of Fiscal and Monetary Policy." *Journal of Monetary Economics*, Vol. 17(1), pp. 117–134.

Lucas, Robert E., Jr. 1987. *Models of Business Cycles*. Yrjo Jahnsson Lectures Series. London: Blackwell.

Lucas, Robert E., Jr. 1988. "On the Mechanics of Economic Development." *Journal of Monetary Economics*, Vol. 22, pp. 3–42.

Lucas, Robert E., Jr.. 1990. "Supply-Side Economics: An Analytical Review." *Oxford Economic Papers*, Vol. 42(2), pp. 293-316.

Lucas, Robert E., Jr. 1992. "On Efficiency and Distribution." *Economic Journal*, Vol. 102(4), pp. 233–247.

Lucas, Robert E., Jr.. 2003. "Macroeconomic Priorities." *American Economic Review*, 93(1), pp 1–14.

Lucas, Robert E., Jr., and Edward C. Prescott. 1971. "Investment under Uncertainty." *Econometrica*, Vol. 39(5), pp. 659–681.

Lucas, Robert E., Jr., and Edward C. Prescott. 1974. "Equilibrium Search and Unemployment." *Journal of Economic Theory*, Vol. 7(2), pp. 188–209.

Lucas, Robert E., Jr., and Nancy Stokey. 1983. "Optimal Monetary and Fiscal Policy in an Economy without Capital." *Journal of Monetary Economics*, Vol. 12(1), pp. 55–94.

Luenberger, David G. 1969. *Optimization by Vector Space Methods*. New York: Wiley.

Lustig, Hanno. 2000. "Secured Lending and Asset Prices." Mimeo. Stanford, Calif.: Stanford University, Department of Economics.

Lustig, Hanno. 2003. "The Market Price of Aggregate Risk and the Wealth Distribution" Mimeo. Chicago: University of Chicago, Department of Economics.

Mace, Barbara. 1991. "Full Insurance in the Presence of Aggregate Uncertainty." *Journal of Political Economy*, Vol. 99(5), pp. 928–956.

Mankiw, Gregory N. 1986. "The Equity Premium and the Concentration of Aggregate Shocks." *Journal of Financial Economics*, Vol. 17(1), pp. 211–219.

Mankiw, N. Gregory, and Stephen P. Zeldes. 1991. "The Consumption of Stockholders and Nonstockholders." *Journal of Financial Economics*, Vol. 29(1), pp. 97–112.

Manuelli, Rodolfo, and Thomas J. Sargent. 1988. "Models of Business Cycles: A Review Essay." *Journal of Monetary Economics*, Vol. 22 (3), pp. 523–542.

Manuelli, Rodolfo and Thomas J. Sargent. 2010. "Alternative Monetary Policies in a Turnpike Economy." *Macroeconomic Dynamics*, Vol. 14, No. 5, pp. 727-762.

Manuelli, Rodolfo, Ananth Seshadri, and Yongseok Shin. 2012. "Lifetime Labor Supply and Human Capital Investment." Mimeo. Washington University in St. Louis and University of Wisconsin.

Marcet, Albert, Francesc Obiols-Homs, and Philippe Weil. 2007. "Incomplete markets, labor supply and capital accumulation." *Journal of Monetary Economics*, Vol. 54, no. 8, pp. 2621–2635.

Marcet, Albert, and Ramon Marimon. 1992. "Communication, Commitment, and Growth." *Journal of Economic Theory*, Vol. 58(2), pp. 219–249.

Marcet, Albert, and Ramon Marimon. 1999. "Recursive Contracts." Mimeo. Universitat Pompeu Fabra, Barcelona.

Marcet, Albert, and Juan Pablo Nicolini. 1999. "Recurrent Hyperinflations and Learning." *American Economic Review*, Vol. 93, No. 3, pp. 1476–1498.

Marcet, Albert, and Thomas J. Sargent. 1989. "Least Squares Learning and the Dynamics of Hyperinflation." In William Barnett, John Geweke, and Karl Shell (eds.), *Economic Complexity: Chaos, Sunspots, and Nonlinearity*. New York: Cambridge University Press.

Marcet, Albert, and Thomas J. Sargent. 1989. "Convergence of least squares learning mechanisms in self-referential linear stochastic models." *Journal of Economic Theory*, Vol. 48, number 2, pp. 337-368.

Marcet, Albert, and Kenneth J. Singleton. 1999. "Equilibrium Asset Prices and Savings of Heterogeneous Agents in the Presence of Incomplete Markets and Portfolio Constraints." *Macroeconomic Dynamics*, Vol. 3(2), pp. 243–277.

Marimon, Ramon, and Shyam Sunder. 1993. "Indeterminacy of Equilibria in a Hyperinflationary World: Experimental Evidence." *Econometrica*, Vol. 61(5), pp. 1073–1107.

Marimon, Ramon, and Fabrizio Zilibotti. 1999. "Unemployment vs. Mismatch of Talents: Reconsidering Unemployment Benefits." *Economic Journal*, Vol. 109, pp. 266–291.

Majumdar, Mukul. 2009. *Equilibrium, Welfare, and Uncertainty: Beyond Arrow-Debreu*. London: Routledge.

Mas-Colell, Andrew, Michael D. Whinston, and Jerry R. Green. 1995. *Microeconomic Theory*. New York: Oxford University Press.

Matsuyama, Kiminori, Nobuhiro Kiyotaki, and Akihiko Matsui. 1993. "Toward a Theory of International Currency." *Review of Economic Studies*, Vol. 60(2), pp. 283–307.

McCall, John J. 1970. "Economics of Information and Job Search." *Quarterly Journal of Economics*, Vol. 84(1), pp. 113–126.

McCall, B. P. 1991. "A Dynamic Model of Occupational Choice." *Journal of Economic Dynamics and Control*, Vol. 15(2), pp. 387–408.

McCallum, Bennett T. 1983. "The Role of Overlapping-Generations Models in Monetary Economics." *Carnegie-Rochester Conference Series on Public Policy*, Vol. 18, pp. 9–44.

McCallum, Bennett T. 2001. "Indeterminacy, Bubbles, and the Fiscal Theory of Price Level Determination." *Journal of Monetary Economics*, Vol. 47(1), pp. 19–30.

McCandless, George T., and Neil Wallace. 1992. *Introduction to Dynamic Macroeconomic Theory: An Overlapping Generations Approach*. Cambridge, Mass.: Harvard University Press.

McGrattan, Ellen R. 1994a. "A Note on Computing Competitive Equilibria in Linear Models." *Journal of Economic Dynamics and Control*, Vol. 18(1), pp. 149–160.

McGrattan, Ellen R. 1994b. "The Macroeconomic Effects of Distortionary Taxation." *Journal of Monetary Economics*, Vol. 33, pp. 573–601.

McGrattan, Ellen R. 1996. "Solving the Stochastic Growth Model with a Finite Element Method." *Journal of Economic Dynamics and Control*, Vol. 20(1–3), pp. 19–42.

Mehra, Rajnish, and Edward C. Prescott. 1985. "The Equity Premium: A Puzzle." *Journal of Monetary Economics*, Vol. 15(2), pp. 145–162.

Mendoza, Enrique G, and Linda L. Tesar. 1998. "The International Ramifications of Tax Reforms: Supply-Side Economics in a Global Economy." *American Economic Review*, Vol. 88(1), pp. 226-45.

Menzio, Guido, Irina A. Telyukova, and Ludo Visschers. 2010. "Directed Search over the Life Cycle" Mimeo. University of Pennsylvania, University of California at San Diego and Simon Fraser University.

Mertens, Karel, and Morten O. Ravn. 2011. "Understanding the Aggregate Effects of Anticipated and Unanticipated Tax Shocks." *Review of Economic Dynamics*, Vol. 14(1), pp. 27–54.

Miao, Jianjun. 2003. "Competitive Equilibria of Economies with a Continuum of Consumers and Aggregate Shocks" Mimeo. Boston: Boston University, Department of Economics.

Miller, Bruce L. 1974. "Optimal Consumption with a Stochastic Income Stream." *Econometrica*, Vol. 42(2), pp. 253–266.

Miller, Marcus and Mark Salmon. 1985. "Dynamic Games and the Time Inconsistency of Optimal Policy in Open Economies." *Economic Journal*, Vol. 95, pp. 124-37.

Miller, Robert A. 1984. "Job Matching and Occupational Choice." *Journal of Political Economy*, Vol. 92(6), pp. 1086–1120.

Miranda, Mario J., and Paul L. Fackler. 2002. *Applied Computational Economics and Finance*. Cambridge Mass.: MIT Press.

Modigliani, Franco, and Richard Brumberg. 1954. "Utility Analysis and the Consumption Function: An Interpretation of Cross-Section Data." In K. K. Kurihara (ed.), *Post-Keynesian Economics*. New Brunswick, N.J.: Rutgers University Press.

Modigliani, F., and M. H. Miller. 1958. "The Cost of Capital, Corporation Finance, and the Theory of Investment." *American Economic Review*, Vol. 48(3), pp. 261–297.

Moen, Espen R. 1997. "Competitive Search Equilibrium." *Journal of Political Economy*, Vol. 105(2), pp. 385–411.

Montgomery, James D. 1991. "Equilibrium Wage Dispersion and Involuntary Unemployment." *Quarterly Journal of Economics*, Vol. 106, pp. 163–179.

Mortensen, Dale T. 1982. "The Matching Process as a Noncooperative Bargaining Game." In John J. McCall (ed.), *The Economics of Information and Uncertainty*. Chicago: University of Chicago Press, for the National Bureau of Economic Research, pp. 233–258.

Mortensen, Dale T. 1994. "The Cyclical Behavior of Job and Worker Flows." *Journal of Economic Dynamics and Control*, Vol. 18, pp. 1121–1142.

Mortensen, Dale T., and Christopher A. Pissarides. 1994. "Job Creation and Job Destruction in the Theory of Unemployment." *Review of Economic Studies*, Vol. 61, pp. 397–415.

Mortensen, Dale T., and Christopher A. Pissarides. 1999a. "New Developments in Models of Search in the Labor Market." In Orley Ashenfelter and David Card (eds.), *Handbook of Labor Economics, Vol. 3B*. Amsterdam: Elsevier/North-Holland.

Mortensen, Dale T., and Christopher A. Pissarides. 1999b. "Unemployment Responses to 'Skill-Biased' Technology Shocks: The Role of Labour Market Policy." *Economic Journal*, Vol. 109, pp. 242–265.

Mulligan, Casey B. 2001. "Aggregate Implications of Indivisible Labor" *Advances in Macroeconomics*, Vol. 1(1): Article 4.

Muth, J. F. [1960] 1981. "Estimation of Economic Relationships Containing Latent Expectations Variables." In R. E. Lucas, Jr., and T. J. Sargent (eds.), *Rational Expectations and Econometric Practice*. Minneapolis: University of Minnesota Press, pp. 321–328.

Muth, John F. 1960. "Optimal Properties of Exponentially Weighted Forecasts." *Journal of the American Statistical Association*, Vol. 55, pp. 299-306.

Muth, John F. 1961. "Rational Expectations and the Theory of Price Movements." *Econometrica*, Vol. 29, pp. 315–335.

Naylor, Arch, and George Sell. 1982. *Linear Operator Theory in Engineering and Science*. New York: Springer.

Neal, Derek. 1999. "The Complexity of Job Mobility among Young Men." *Journal of Labor Economics*, Vol. 17(2), pp. 237–261.

Negishi, T.. 1960. "Welfare Economics and Existence of an Equilibrium for a Competitive Economy." *Metroeconomica*, Vol. 12, pp. 92–97.

Nerlove, Marc. 1967. "Distributed Lags and Unobserved Components in Economic Time Series." In William Fellner et al. (eds.), *Ten Economic Studies in the Tradition of Irving Fisher*. New York: Wiley.

Niepelt, Dirk . 2004. "The Fiscal Myth of the Price Level" *Quarterly Journal of Economics*, Vol. 119(1), pp. 276–299.

Noble, Ben, and James W. Daniel. 1977. *Applied Linear Algebra*. Englewood Cliffs, N.J.: Prenctice Hall.

O'Connell, Stephen A., and Stephen P. Zeldes. 1988. "Rational Ponzi Games." *International Economic Review*, Vol. 29(3), pp. 431–450.

Obstfeld, M. 1994. "Evaluating Risky Consumption Paths: the Role of Intertemporal Substitutability." *European Economic Review*, Vol. 38(7), pp. 1471–1486.

Otrok, Christopher. 2001. "On Measuring the Welfare Cost of Business Cycles." *Journal of Monetary Economics*, Vol. 47(1), pp. 61–92.

Paal, Beatrix. 2000. "Destabilizing Effects of a Successful Stabilization: A Forward-Looking Explanation of the Second Hungarian Hyperinflation." *Journal of Economic Theory*, Vol. 15(3), pp. 599–630.

Pavoni, Nicola and Giovanni L. Violante. 2007. "Optimal Welfare-to-Work Programs." *Review of Economic Studies*, Vol. 74, No. 1, pp. 283–318.

Pearlman, J. G.. 1992. "Reputational and Nonreputational Policies Under Partial Information." *Journal of Economic Dynamics and Control*, Vol. 16(2), pp. 339–358.

Pearlman, J. G., D. A. Currie, and P. L. Levine. 1986. "Rational expectations with partial information." *Economic Modeling*, Vol. 3, pp. 90-1-5.

Peled, Dan. 1984. "Stationary Pareto Optimality of Stochastic Asset Equilibria with Overlapping Generations." *Journal of Economic Theory*, Vol. 34, pp. 396–403.

Peters, Michael. 1991. "Ex Ante Price Offers in Matching Games Non-Steady States." *Econometrica*, Vol. 59(5), pp. 1425–1454.

Phelan, Christopher. 1994. "Incentives and Aggregate Shocks." *Review of Economic Studies*, Vol. 61(4), pp. 681–700.

Phelan, Christopher, and Robert M. Townsend. 1991. "Computing Multi-period, Information-Constrained Optima." *Review of Economic Studies*, Vol. 58(5), pp. 853–881.

Phelan, Christopher, and Ennio Stacchetti. 2001. "Sequential Equilibria in a Ramsey Tax Model." *Econometrica*, Vol. 69(6), pp. 1491–1518.

Phelps, Edmund S. 1970. *Introduction to Microeconomic Foundations of Employment and Inflation Theory*. New York: Norton.

Phelps, Edmund S., and Robert A. Pollak. 1968. "On Second-Best National Saving and Game-Equilibrium Growth." *Review of Economic Studies*, Vol. 35(2), pp. 185–199.

Piazzesi, Monika. 2005. "Bond Yields and the Federal Reserve" *Journal of Political Economy*, Vol. 113(2), pp. 311–344.

Piazzesi, Monika, and Martin Schneider. 2006. "Equilibrium Yield Curves." In Daron Acemoglu, Kenneth Rogoff, and Michael Woodford (eds.), *Macroeconomics Annual 2006*. Cambridge MA., MIT Press pp 389-442.

Pissarides, Christopher A. 1983. "Efficiency Aspects of the Financing of Unemployment Insurance and Other Government Expenditures." *Review of Economic Studies*, Vol. 50(1), pp. 57–69.

Pissarides, Christopher A. 1990. *Equilibrium Unemployment Theory*. Cambridge, U.K.: Basil Blackwell.

Pratt, John W. 1964. "Risk Aversion in the Small and in the Large." *Econometrica*, Vol. 32(1–2), pp. 122–136.

Prescott, Edward C. 1977. "Should Control Theory Be Used for Economic Stabilization?" *1977*, Journal-of-Monetary-Economics. Supplementary Series.Vol. 7, pp. 13–38

Prescott, Edward C. 2002. "Richard T. Ely Lecture: Prosperity and Depression." *American Economic Review*, Vol. 92(2), pp. 1–15.

Prescott, Edward C. 2005a. "The Transformation of Macroeconomic Policy and Research." In *Les Prix Nobel 2004*, Stockholm: Almqvist & Wiksell International, pp. 370–395. http://nobelprize.org/economics/laureates/2004/prescott-lecture.pdf.

Prescott, Edward C.. 2005b. "Nonconvexities in Quantitative General Equilibrium Studies of Business Cycles." In Timothy Kehoe, T.N. Srinivasan and John Whalley (eds.), *Frontiers in Applied General Equilibrium Modeling: In Honor of Herbert Scarf*. Cambridge University Press, pp 95–118.

Prescott, Edward C. 2006. "Nobel Lecture: The Transformation of Macroeconomic Policy and Research." *Journal of Political Economy*, Vol. 114 pp. 203–235.

Prescott, Edward C. 2007. "Comment." In Daron Acemoglu, Kenneth Rogoff, and Michael Woodford (eds.), *Macroeconomics Annual 2006*. Cambridge MA., MIT Press pp 233–242.

Prescott, Edward C., and Rajnish Mehra. 1980. "Recursive Competitive Equilibrium: The Case of Homogeneous Households." *Econometrica*, Vol. 48(6), pp. 1365–1379.

Prescott, Edward C., Richard Rogerson, and Johanna Wallenius. 2009. "Lifetime Aggregate Labor Supply with Endogenous Workweek Length." *Review of Economic Dynamics*, Vol. 12 pp. 23–36.

Prescott, Edward C., and Robert M. Townsend. 1980. "Equilibrium Under Uncertainty: Multiagent Statistical Decision Theory." In Arnold Zellner (ed.), *Bayesian Analysis in Econometrics and Statistics.* Amsterdam: North-Holland, pp. 169–194.

Prescott, Edward C., and Robert M. Townsend. 1984a. "General Competitive Analysis in an Economy with Private Information." *International Economic Review*, Vol. 25, pp. 1–20.

Prescott, Edward C., and Robert M. Townsend. 1984b. "Pareto Optima and Competitive Equilibria with Adverse Selection and Moral Hazard." *Econometrica*, Vol. 52, pp. 21–45.

Putterman, Martin L., and Shelby Brumelle. 1979. "On the Convergence of Policy Iteration on Stationary Dynamic Programming." *Mathematics of Operations Research*, Vol. 4(1), pp. 60–67.

Putterman, Martin L., and M. C. Shin. 1978. "Modified Policy Iteration Algorithms for Discounted Markov Decision Problems." *Management Science*, Vol. 24(11), pp. 1127–1137.

Quah, Danny. 1990. "Permanent and Transitory Movements in Labor Income: An Explanation for 'Excess Smoothness' in Consumption." *Journal of Political Economy*, Vol. 98(3), pp. 449–475.

Razin, Assaf, and Efraim Sadka. 1995. "The Status of Capital Income Taxation in the Open Economy." *FinanzArchiv*, Vol. 52(1), pp. 21–32.

Rebelo, Sergio. 1991. "Long-Run Policy Analysis and Long-Run Growth." *Journal of Political Economy*, Vol. 99, pp. 500–521.

Reinganum, Jennifer F. 1979. "A Simple Equilibrium Model of Price Dispersion." *Journal of Political Economy*, Vol. 87(4), pp. 851–858.

Ríos-Rull, Víctor José. 1994a. "Life-Cycle Economies and Aggregate Fluctuations." Mimeo. University of Pennsylvania.

Ríos-Rull, Víctor José. 1994b. "Population Changes and Capital Accumulation: The Aging of the Baby Boom." Mimeo. Philadelphia: University of Pennsylvania, Department of Economics.

Ríos-Rull, Víctor José. 1994c. "On the Quantitative Importance of Market Completeness." *Journal of Monetary Economics*, Vol. 34(3), pp. 463–496.

Ríos-Rull, Víctor José. 1995. "Models with Heterogeneous Agents." In Thomas F. Cooley (ed.), *Frontiers of Business Cycle Research.* Princeton, N.J.: Princeton University Press, pp. 98–125.

Ríos-Rull, Víctor José. 1996. "Life-Cycle Economies and Aggregate Fluctuations." *Review of Economic Studies*, Vol. 63(3), pp. 465–489.

Rodriguez, Santiago Budria, Javier Díaz-Giménez, Vincenzo Quadrini, and Jose-Victor Ríos-Rull. 2002. "Updated Facts on the U.S. Distributions of Earnings, Income, and Wealth." *Quarterly Review*, Federal Reserve Bank of Minneapolis, Summer.

Roberds, William T. 1996. "Budget Constraints and Time-Series Evidence on Consumption: Comment." *American Economic Review*, Vol. 86(1), pp. 296–297.

Robert, Christian P. and George Casella. 2004. *Monte Carlo Statistical Methods, Second Edition.* Springer, New York.

Rogerson, William P. 1985a. "Repeated Moral Hazard." *Econometrica*, Vol. 53(1), pp. 69–76.

Rogerson, William P. 1985b. "The First Order Approach to Principal-Agent Problems." *Econometrica*, Vol. 53(6), pp. 1357–1367.

Rogerson, Richard. 1988. "Indivisible Labor, Lotteries, and Equilibrium." *Journal of Monetary Economics*, Vol. 21, pp. 3–16.

Rogerson, Richard, and Johanna Wallenius. 2009. "Micro and Macro Elasticities in a Life Cycle Model with Taxes." *Journal of Economic Theory*, Vol. 144, pp. 2277–2292.

Rogoff, Kenneth. 1989. "Reputation, Coordination, and Monetary Policy." In Robert J. Barro (ed.), *Modern Business Cycle Theory*. Cambridge, Mass.: Harvard University Press, pp. 236–264.

Roll, Richard. 1970. *The Behavior of Interest Rates: An Application of the Efficient Market Model to U.S. Treasury Bills*. New York: Basic Books.

Romer, David. 1996. *Advanced Macroeconomics*. New York: McGraw-Hill.

Romer, Paul M. 1986. "Increasing Returns and Long-Run Growth." *Journal of Political Economy*, Vol. 94, pp. 1002–1037.

Romer, Paul M. 1987. "Growth Based on Increasing Returns Due to Specialization." *American Economic Review Paper and Proceedings*, Vol. 77, pp. 56–62.

Romer, Paul M. 1990. "Endogenous Technological Change." *Journal of Political Economy*, Vol. 98, pp. S71–S102.

Rosen, Sherwin, and Robert H. Topel. 1988. "Housing Investment in the United States." *Journal of Political Economy*, Vol. 96(4), pp. 718–740.

Rosen, Sherwin, Kevin M. Murphy, and Jose A. Scheinkman. 1994. "Cattle Cycles." *Journal of Political Economy*, Vol. 102(3), pp. 468–492.

Ross, Stephen A. 1976. "The Arbitrage Theory of Capital Asset Pricing." *Journal of Economic Theory*, Vol. 13(3), pp. 341–360.

Rotemberg, Julio J. 1987. "The New Keynesian Microfoundations." In Stanley Fischer (ed.), *NBER Macroeconomics Annual*. Cambridge, Mass.: MIT Press, pp. 69–104.

Rotemberg, Julio J. and Michael Woodford. 1997. "An Optimization-Based Econometric Framework for the Evaluation of Monetary Policy." In Olivier Blanchard and Stanley Fischer (eds.), *NBER Macroeconomic Annual*. Cambridge, Mass.: MIT Press, pp. 297-345.

Rothschild, Michael, and Joseph Stiglitz. 1970. "Increasing Risk I: A Definition." *Journal of Economic Theory*, Vol. 2(3), pp. 225–243.

Rothschild, Michael, and Joseph Stiglitz. 1971. "Increasing Risk II: Its Economic Consequences." *Journal of Economic Theory*, Vol. 3(1), pp. 66–84.

Rothschild, Michael. 1973. "Models of Market Organization with Imperfect Information: A Survey" *Journal of Political Economy*, Vol. 81(6), pp. 1283–1308.

Rubinstein, Mark. 1974. "An Aggregation Theorem for Security Markets." *Journal of Financial Economics*, Vol. 1(3), pp. 225–244.

Runkle, David E. 1991. "Liquidity Constraints and the Permanent-Income Hypothesis: Evidence from Panel Data." *Journal of Monetary Economics*, Vol. 27(1), pp. 73–98.

Ryoo, J. and S. Rosen. 2004. "The Engineering Labor Market." *Journal of Political Economy*, Vol. 112(1), pp. S110–S140.

Saint-Paul, Gilles. 1995. "The High Unemployment Trap." *Quarterly Journal of Economics*, Vol. 110, pp. 527–550.

Samuelson, Paul A. 1958. "An Exact Consumption-Loan Model of Interest with or without the Social Contrivance of Money." *Journal of Political Economy*, Vol. 66, pp. 467–482.

Samuelson, Paul A. 1965. "Proof That Properly Anticipated Prices Fluctuate Randomly." *Industrial Management Review*, Vol. 6(1), pp. 41–49.

Sandmo, Agnar. 1970. "The Effect of Uncertainty on Saving Decisions." *Review of Economic Studies*, Vol. 37, pp. 353–360.

Santos, Manuel S. 1991. "Smoothness of the Policy Function in Discrete Time Economic Models." *Econometrica*, Vol. 59(5), pp. 1365-82.

Santos, Manuel S. 1993. "On High-Order Differentiability of the Policy Function" *Economic Theory*, Vol. 3(3), pp. 565-70.

Sargent, Thomas J. 1979a. *Macroeconomic Theory*. New York: Academic Press.

Sargent, Thomas J. 1980. "Tobin's *q* and the Rate of Investment in General Equilibrium." In K. Brunner and A. Meltzer (eds.), *On the State of Macroeconomics*. Carnegie-Rochester Conference Series 12, Amsterdam: North-Holland, pp. 107–154.

Sargent, Thomas J. 1980. "Lecture notes on Filtering, Control, and Rational Expectations." Mimeo. Minneapolis: University of Minnesota, Department of Economics.

Sargent, Thomas J.. 1979b. "A Note On Maximum Likelihood Estimation of The Rational Expectations Model of The Term Structure" *Journal Of Monetary Economics*, Vol. 35, pp. 245–274.

Sargent, Thomas J. 1987a. *Macroeconomic Theory*,. 2nd ed. New York: Academic Press.

Sargent, Thomas J. 1987b. *Dynamic Macroeconomic Theory*. Cambridge, Mass.: Harvard University Press.

Sargent, Thomas J. 1991. "Equilibrium with Signal Extraction from Endogenous Variables." *Journal of Economic Dynamics and Control*, Vol. 15, pp. 245–273.

Sargent, Thomas J. 1992. *Rational Expectations and Inflation, 2nd ed.*. New York: Harper and Row.

Sargent, Thomas J. 1999. *The Conquest of American Inflation*. Princeton University Press, Princeton, New Jersey.

Sargent, Thomas J. 2008. "Evolution and Intelligent Design" *American Economic Review*, Vol. 98(1), pp. 5–37.

Sargent, Thomas J., and Bruce Smith. 1997. "Coinage, Debasements, and Gresham's Laws." *Economic Theory*, Vol. 10, pp. 197–226.

Sargent, Thomas J., and François R. Velde. 1990. "The Analytics of German Monetary Reform." *Quarterly Review*, Federal Reserve Bank of San Francisco, No. 4, pp. 33–50.

Sargent, Thomas J., and François R. Velde. 1995. "Macroeconomic Features of the French Revolution." *Journal of Political Economy*, Vol. 103(3), pp. 474–518.

Sargent, Thomas J., and François R. Velde. 1999. "The Big Problem of Small Change." *Journal of Money, Credit, and Banking*, Vol. 31(2), pp. 137–161.

Sargent, Thomas J., and Neil Wallace. 1973. "Rational Expectations and the Dynamics of Hyperinflation." *International Economic Review*, Vol. 14, pp. 328–350.

Sargent, Thomas J., and Neil Wallace. 1975. "'Rational' Expectations, the Optimal Monetary Instrument, and the Optimal Money Supply Rule" *Journal of Political Economy*, Vol. 83(2), pp. 241–254.

Sargent, Thomas J., and Neil Wallace. 1981. "Some Unpleasant Monetarist Arithmetic." *Quarterly Review*, Federal Reserve Bank of Minneapolis, Vol. 5(3), pp. 1–17.

Sargent, Thomas J., and Neil Wallace. 1982. "The Real Bills Doctrine vs. the Quantity Theory: A Reconsideration." *Journal of Political Economy*, Vol. 90(6), pp. 1212–1236.

Sargent, Thomas J., and Neil Wallace. 1983. "A Model of Commodity Money." *Journal of Monetary Economics*, Vol. 12(1), pp. 163–187.

Sargent, Thomas J., Noah Williams, and Tao Zha. 2009. "The Conquest of South American Inflation." *Journal of Political Economy*, Vol. 117, No. 2, pp. 211–256.

Schmitt-Grohe, Stephanie, and Martin Uribe. 2004a. "Optimal Fiscal and Monetary Policy Under Sticky Prices." *Journal of Economic Theory*, Vol. 114(2), pp. 198–230.

Schmitt-Grohe, Stephanie and Martin Uribe. 2004b. "Solving dynamic general equilibrium models using second-order approximation to the policy function." *Journal of Economic Dynamics and Control*, Vol. 28, pp. 755-775.

Seater, John J. 1993. "Ricardian Equivalence." *Journal of Economic Literature*, Vol. 31(1), pp. 142–190.

Segerstrom, Paul S. 1998. "Endogenous Growth without Scale Effects." *American Economic Review*, Vol. 88, pp. 1290–1310.

Segerstrom, Paul S., T. C. A. Anant, and Elias Dinopoulos. 1990. "A Schumpeterian Model of the Product Life Cycle." *American Economic Review*, Vol. 80, pp. 1077–1091.

Shavell, Stephen, and Laurence Weiss. 1979. "The Optimal Payment of Unemployment Insurance Benefits Over Time." *Journal of Political Economy*, Vol. 87, pp. 1347–1362.

Shell, Karl. 1971. "Notes on the Economics of Infinity." *Journal of Political Economy*, Vol. 79(5), pp. 1002–1011.

Shi, Shouyong. 1995. "Money and Prices: A Model of Search and Bargaining." *Journal of Economic Theory*, Vol. 67, pp. 467–496.

Shiller, Robert J. 1972. "Rational Expectations and the Structure of Interest Rates." Ph.D. dissertation, Massachusetts Institute of Technology.

Shiller, Robert J. 1981. "Do Stock Prices Move Too Much to Be Justified by Subsequent Changes in Dividends?" *American Economic Review*, Vol. 71(3), pp. 421–436.

Shimer, Robert. 2010. *Labor Markets and Business Cycles*. Princeton, New Jersey: Princeton University Press.

Shin, Yongseok. 2007. "Managing the maturity structure of government debt." *Journal of Monetary Economics*, Vol. 54, pp. 1565-1571.

Sibley, David S.. 1975. "Permanent and Transitory Income Effects in a Model of Optimal Consumption with Wage Income Uncertainty." *Journal of Economic Theory*, Vol. 11, pp. 68–82.

Sidrauski, Miguel. 1967. "Rational Choice and Patterns of Growth in a Monetary Economy." *American Economic Review*, Vol. 57(2), pp. 534–544.

Sims, Christopher A. 1972. "Money, Income, and Causality." *American Economic Review*, Vol. 62(4), pp. 540–552.

Sims, Christopher A. 1989. "Solving Nonlinear Stochastic Optimization and Equilibrium Problems Backwards." Mimeo. Minneapolis: Institute for Empirical Macroeconomics, Federal Reserve Bank of Minneapolis, No. 15.

Sims, Christopher A. 1994. "A Simple Model for the Determination of the Price Level and the Interaction of Monetary and Fiscal Policy." *Economic Theory*, Vol. 4, pp. 381–399.

Siow, A. 1984. "Occupational Choice under Uncertainty." *Econometrica*, Vol. 52(3), pp. 631–645.

Siu, Henry E. 2004. "Optimal Fiscal and Monetary Policy With Sticky Prices." *Journal of Monetary Economics*, Vol. 51, no. 3, pp. 575–607.

Smith, Bruce. 1988. "Legal Restrictions, 'Sunspots,' and Peel's Bank Act: The Real Bills Doctrine versus the Quantity Theory Reconsidered." *Journal of Political Economy*, Vol. 96(1), pp. 3-19.

Smith, Lones. 1992. "Folk Theorems in Overlapping Generations Games." *Games and Economic Behavior*, Vol. 4(3), pp. 426–449.

Solow, Robert M. 1956. "A Contribution to the Theory of Economic Growth." *Quarterly Journal of Economics*, Vol. 70, pp. 65–94.

Sotomayor, Marlida A. de Oliveira. 1984. "On Income Fluctuations and Capital Gains." *Journal of Economic Theory*, Vol. 32(1), pp. 14–35.

Spear, Stephen E., and Sanjay Srivastava. 1987. "On Repeated Moral Hazard with Discounting." *Review of Economic Studies*, Vol. 54(4), pp. 599–617.

Stacchetti, Ennio. 1991. "Notes on Reputational Models in Macroeconomics." Mimeo. Stanford University, September.

Stachurski, John. 2009. *Economic Dynamics: Theory and Computation*. Cambridge, Mass: MIT Press.

Stigler, George. 1961. "The Economics of Information." *Journal of Political Economy*, Vol. 69(3), pp. 213–225.

Stiglitz, Joseph E. 1969. "A Reexamination of the Modigliani-Miller Theorem." *American Economic Review*, Vol. 59(5), pp. 784–793.

Stiglitz, Joseph E. 1987. "Pareto Efficient and Optimal Taxation and the New New Welfare Economics." In Alan J. Auerbach, and Martin Feldstein (eds.), *Handbook of Public Economics, Vol. 2*. Amsterdam: Elsevier/North-Holland.

Stokey, Nancy L. 1989. "Reputation and Time Consistency." *American Economic Review*, Vol. 79, pp. 134–139.

Stokey, Nancy L. 1991. "Credible Public Policy." *Journal of Economic Dynamics and Control*, Vol. 15(4), pp. 627–656.

Stokey, Nancy, and Robert E. Lucas, Jr. (with Edward C. Prescott). 1989. *Recursive Methods in Economic Dynamics*. Cambridge, Mass.: Harvard University Press.

Storesletten, Kjetil, Christopher I. Telmer, and Amir Yaron. 1998. "Persistent Idiosyncratic Shocks and Incomplete Markets." Mimeo. Carnegie Mellon University and Wharton School, University of Pennsylvania.

Storesletten, Kjetil, Christopher I. Telmer, and Amir Yaron. 2004. "Consumption and risk sharing over the life cycle." *Journal of Monetary Economics*, Vol. 51(3), pp. 609-633.

Svensson, Lars E. O. 1986. "Sticky Goods Prices, Flexible Asset Prices, Monopolistic Competition, and Monetary Policy." *Review of Economic Studies*, Vol. 53, pp. 385–405.

Tallarini, Thomas D., Jr. 1996. "Risk-Sensitive Real Business Cycles." Ph.D. dissertation, University of Chicago.

Tallarini, Thomas D., Jr. 2000. "Risk-Sensitive Real Business Cycles." *Journal of Monetary Economics*, Vol. 45(3), pp. 507–532.

Tauchen, George. 1986. "Finite State Markov Chain Approximations to Univariate and Vector Autoregressions." *Economic Letters*, Vol. 20, pp. 177–181.

Taylor, John B. 1977. "Conditions for Unique Solutions in Stochastic Macroeconomic Models with Rational Expectations." *Econometrica*, Vol. 45, pp. 1377–1185.

Taylor, John B. 1980. "Output and Price Stability: An International Comparison." *Journal of Economic Dynamics and Control*, Vol. 2, pp. 109–132.

Thomas, Jonathan, and Tim Worrall. 1988. "Self-Enforcing Wage Contracts." *Review of Economic Studies*, Vol. 55, pp. 541–554.

Thomas, Jonathan, and Tim Worrall. 1990. "Income Fluctuation and Asymmetric Information: An Example of a Repeated Principal-Agent Problem." *Journal of Economic Theory*, Vol. 51(2), pp. 367–390.

Thomas, Jonathan, and Tim Worrall. 1994. "Informal Insurance Arrangements in Village Economies." Mimeo. Liverpool Research Papers in Economics and Finance, 9402.

Tirole, Jean. 1982. "On the Possibility of Speculation under Rational Expectations." *Econometrica*, Vol. 50, pp. 1163–1181.

Tirole, Jean. 1985. "Asset Bubbles and Overlapping Generations." *Econometrica*, Vol. 53, pp. 1499–1528.

Tobin, James. 1956. "The Interest Elasticity of the Transactions Demand for Cash." *Review of Economics and Statistics*, Vol. 38, pp. 241–247.

Tobin, James. 1961. "Money, Capital, and Other Stores of Value" *American Economic Review*, Vol. 51(2), pp. 26–37.

Tobin, James. 1963. "An Essay on the Principles of Debt Management." In William Fellner et al. (eds.), *Fiscal and Debt Management Policies.*. Englewood Cliffs, N.J.: Prentice-Hall, pp. 141–215. (Reprinted in James Tobin, *Essays in Economics*, Vol. 1. Amsterdam: North-Holland, 1971, pp. 378–455.)

Topel, Robert H., and Sherwin Rosen. 1988. "Housing Investment in the United States." *Journal of Political Economy*, Vol. 96(4), pp. 718–740.

Townsend, Robert M. 1980. "Models of Money with Spatially Separated Agents." In J. H. Kareken and N. Wallace (eds.), *Models of Monetary Economies*. Minneapolis: Federal Reserve Bank of Minneapolis, pp. 265–303.

Townsend, Robert M. 1983. "Forecasting the Forecasts of Others." *Journal of Political Economy*, Vol. 91, pp. 546–588.

Townsend, Robert M. 1994. "Risk and Insurance in Village India" *Econometrica*, Vol. 62.539–592.

Trejos, Alberto, and Randall Wright. 1995. "Search, Bargaining, Money and Prices." *Journal of Political Economy*, Vol. 103, pp. 118–139.

Turnovsky, Stephen J., and William A. Brock. 1980. "Time Consistency and Optimal Government Policies in Perfect Foresight Equilibrium." *Journal of Public Economics*, Vol. 13, pp. 183–212.

Uzawa, Hirofumi. 1965. "Optimum Technical Change in an Aggregative Model of Economic Growth." *International Economic Review*, Vol. 6, pp. 18–31.

Villamil, Anne P. 1988. "Price Discriminating Monetary Policy: A Nonuniform Pricing Approach." *Journal of Public Economics*, Vol. 35(3), pp. 385–392.

Wallace, Neil. 1980. "The Overlapping Generations Model of Fiat Money." In J. H. Kareken and N. Wallace (eds.), *Models of Monetary Economies*. Minneapolis: Federal Reserve Bank of Minneapolis, pp. 49–82.

Wallace, Neil. 1981. "A Modigliani-Miller Theorem for Open-Market Operations." *American Economic Review*, Vol. 71, pp. 267–274.

Wallace, Neil. 1983. "A Legal Restrictions Theory of the Demand for 'Money' and the Role of Monetary Policy." *Quarterly Review*, Federal Reserve Bank of Minneapolis, Vol. 7(1), pp. 1–7.

Wallace, Neil. 1989. "Some Alternative Monetary Models and Their Implications for the Role of Open-Market Policy." In Robert J. Barro (ed.), *Modern Business Cycle Theory*. Cambridge, Mass.: Harvard University Press, pp. 306–328.

Walsh, Carl E. 1998. *Monetary Theory and Policy*. Cambridge, Mass.: MIT Press.

Wang, Cheng, and Stephen D. Williamson. 1996. "Unemployment Insurance with Moral Hazard in a Dynamic Economy." *Carnegie-Rochester Conference Series on Public Policy*, Vol. 44, pp. 1–41.

Wang, Neng. 2003. "Caballero meets Bewley: The permanent-income hypothesis in general equilbrium" *American Economic Review*, Vol. 93(3), pp. 927–936.

Watanabe, Shinichi. 1984. *Search Unemployment, the Business Cycle, and Stochastic Growth*. Ph.D. dissertation, University of Minnesota.

Weil, Philippe. 1989. "The Equity Premium Puzzle and the Risk-Free Rate Puzzle." *Journal of Monetary Economics*, Vol. 24(2), pp. 401–421.

Weil, Philippe. 1990. "Nonexpected Utility in Macroeconomics." *Quarterly Journal of Economics*, Vol. 105, pp. 29–42.

Weil, Philippe. 1993. "Precautionary Savings and the Permanent Income Hypothesis." *Review of Economic Studies*, Vol. 60(2), pp. 367–383.

Werning, Ivan. 2002. "Optimal unemployment insurance with hidden savings." Mimeo. University of Chicago Working Paper.

Werning, Ivan. 2007. "Optimal Fiscal Policy with Redistribution." *Quarterly Journal of Economics*, Vol. 122(3), pp. 925–67.

Wen, Y. 1998. "Can a Real Business Cycle Pass the Watson Test?" *Journal of Monetary Economics*, Vol. 42, pp. 185–202.

Whiteman, Charles H. 1983. *Linear Rational Expectations Models: A Users Guide*. Minneapolis: University of Minnesota Press.

Whittle, Peter. 1963. *Prediction and Regulation by Linear Least-Square Methods*. Princeton, N.J.: Van Nostrand-Reinhold.

Whittle, Peter. 1990. *Risk-Sensitive Optimal Control*. New York: Wiley.

Wilcox, David W. 1989. "The Sustainability of Government Deficits: Implications of the Present-Value Borrowing Constraint." *Journal of Money, Credit, and Banking*, Vol. 21(3), pp. 291–306.

Woodford, Michael. 1994. "Monetary Policy and Price Level Determinacy in a Cash-in-Advance Economy." *Economic Theory*, Vol. 4, pp. 345–380.

Woodford, Michael. 1995. "Price-Level Determinacy without Control of a Monetary Aggregate." *Carnegie-Rochester Conference Series on Public Policy*, Vol. 43, pp. 1–46.

Woodford, Michael. 2003. "Optimal Interest-Rate Smoothing." *Review of Economic Studies*, Vol. 70(4), pp. 861–886.

Woodford, Michael. 2003. *Interest and Prices*. Princeton, N.J.: Princeton University Press.

Wright, Randall. 1986. "Job Search and Cyclical Unemployment." *Journal of Political Economy*, Vol. 94(1), pp. 38–55.

Young, Alwyn. 1998. "Growth without Scale Effects." *Journal of Political Economy*, Vol. 106, pp. 41–63.

Zeira, Joseph. 1999. "Informational Overshooting, Booms, and Crashes." *Journal of Monetary Economics*, Vol. 43(1), pp. 237–257.

Zeldes, Stephen P. 1989. "Optimal Consumption with Stochastic Income: Deviations from Certainty Equivalence." *Quarterly Journal of Economics*, Vol. 104(2), pp. 275–298.

Zhang, Harold. 1997. "Endogenous Borrowing Constraints with Incomplete Markets." *Journal of Finance*, Vol. 52(5), pp. 2187–2209.

Zhao, Rui. 2001. "The Optimal Unemployment Insurance Contract: Why a Replacement Ratio?" Mimeo. University of Illinois, Champagne-Urbana, Department of Economics.

Zhu, Shenghao. 2009. "Existence of Equilibrium in an Incomplete Market Model with Endogenous Labor Supply" Mimeo. New York University.

Zhu, Xiaodong. 1992. "Optimal Fiscal Policy in a Stochastic Growth Model." *Journal of Economic Theory*, Vol. 58, pp. 250–289.

Zin, Stanley E. 2002. "Are behavioral asset-pricing models structural?" *Journal of Monetary Economics*, Vol. 49(1), pp. 215-228.

Subject Index

Author Index

Abel, Andrew, 349, 491

Abreu, Dilip, xxxv, xxxvi, 797, 938, 949, 959, 968, 969, 971

Acemoglu, Daron, 1142, 1189

Adda, Jerome, xxii

Aghion, Philippe, xxii, 605

Aiyagari, Rao, 364, 614, 650, 653, 687, 725, 726, 732

Albarran, Pedro, 900

Allen, Franklin, 798, 835

Altug, Sumru, xxii, 489

Alvarez, Fernando, 16, 122, 548, 643, 756, 903

Anant, T.C.A., 605

Anderson, Evan W., 130, 134, 140, 541, 544, 546

Ang, Andrew, 15, 559

Angeletos, George-Marios, 382

Apostol, Tom, 821

Arrow, Kenneth J., 251, 268, 583, 589

Atkeson, Andrew, xxxvi, 833, 1005, 1039, 1069

Attanasio, Orazio, 350, 551, 900

Azariadis, Costas, xxii

Backus, David K., 15, 420, 559, 561, 565

Bagwell, Kyle, 372

Balasko, Yves, 341

Barillas, Francisco, 84, 390, 545

Barro, Robert J., 372, 592, 937, 969, 1075

Barseghyan, Levon, 539

Barsky, Robert, 373

Bassetto, Marco, xxiv, 373, 679, 952, 980, 986, 1068

Baumol, William J., 1047

Bellman, Richard, xx, 103

Ben-Porath, Yoram, 1216, 1225

Benassy, Jean-Pascal, xxii, 605

Benoit, Jean-Pierre, 948

Benveniste, Lawrence, 104, 880

Bernheim, B. Douglas, 372

Bertola, Giuseppe, 1142

Bertsekas, Dimitri, 103, 104, 129

Bewley, Truman, 725, 1093

Bhandari, Anmol, 84, 390

Bigio, Saki, 84, 390

Black, Fischer, 488

Blackwell, David, 197

Blanchard, Olivier J., xxii, 136, 140, 372

Blume, Lawrence, 252

Blundell, Richard, 5, 80

Bohn, Henning, 509

Bond, Eric W., 1005, 1016, 1022

Breeden, Douglas T., 13, 488

Brock, William A., 143, 319, 322, 350, 431, 456, 489, 1087

Browning, Martin, xxviii, 1203

Brumelle, Shelby, 116

Bruno, Michael, 332, 1054

Bryant, John, 1058, 1111

Buera, Francisco, 382

Buiter, Willem H., 1061

Burdett, Kenneth, xxiv, 183, 184, 1189

Burnside, Craig, 60

Caballero, Ricardo J., 761, 1142

Cagan, Phillip, 63

Calvo, Guillermo A., 987

Campbell, John Y., 15, 76, 481

Canova, Fabio, xxii, 56, 60

Carroll, Christopher D., 761

Cass, David, 341, 1065

Chéron, Arnaud, 1149

Chamberlain, Gary, 699, 701, 717

Chamley, Christophe, 12, 613, 620, 622, 623

Champ, Bruce, 350

Chang, Roberto, 938, 980, 986

Chang, Yongsung, 1205, 1235, 1239, 1245

Chari, V.V., 614, 632, 640, 664, 674, 686, 938, 979, 1069, 1070

Chatterjee, Satyajit, 1093

Chen, Ren-Raw, 559

Chow, Gregory, 103, 131, 135

Christensen, Bent Jasper, xxii, 60

Christiano, Lawrence J., 60, 146, 431, 614, 632, 640, 664, 674, 686, 979, 1070

Clower, Robert, 1077

Matlab Index